The Geology of North America
Volume P-2

Economic Geology, U.S.

Edited by

H. J. Gluskoter
U.S. Geological Survey
MS 956
National Center
Reston, Virginia 22092

D. D. Rice
U.S. Geological Survey
MS 971
Denver Federal Center
Box 25046
Denver, Colorado 80225

R. B. Taylor
U.S. Geological Survey
MS 905
Denver Federal Center
Box 25046
Denver, Colorado 80225

1991

Acknowledgment

Publication of this volume, one of the synthesis volumes of *The Decade of North American Geology Project* series, has been made possible by members and friends of The Geological Society of America, corporations, and government agencies through contributions to the Decade of North American Geology fund of the Geological Society of America Foundation.

Following is a list of individuals, corporations, and government agencies giving and/or pledging more than $50,000 in support of the DNAG Project:

Amoco Production Company
ARCO Exploration Company
Chevron Corporation
Cities Service Oil and Gas Company
Diamond Shamrock Exploration Corporation
Exxon Production Research Company
Getty Oil Company
Gulf Oil Exploration and Production Company
Paul V. Hoovler
Kennecott Minerals Company
Kerr McGee Corporation
Marathon Oil Company
Maxus Energy Corporation
McMoRan Oil and Gas Company
Mobil Oil Corporation
Occidental Petroleum Corporation

Pennzoil Exploration and Production Company
Phillips Petroleum Company
Shell Oil Company
Caswell Silver
Standard Oil Production Company
Sun Exploration and Production Company
Superior Oil Company
Tenneco Oil Company
Texaco, Inc.
Union Oil Company of California
Union Pacific Corporation and its operating companies:
    Union Pacific Resources Company
    Union Pacific Railroad Company
    Upland Industries Corporation
U.S. Department of Energy

© 1991 by The Geological Society of America, Inc.
All rights reserved.

All materials subject to this copyright and included in this volume may be photocopied for the noncommercial purpose of scientific or educational advancement.

Copyright is not claimed on any material prepared by government employees within the scope of their employment.

Published by The Geological Society of America, Inc.
3300 Penrose Place, P.O. Box 9140, Boulder, Colorado 80301

Printed in U.S.A.

Front Cover: View to west of maintenance/administrative building, coal processing facilities, reclaimed land, and active pits of the Belle Ayr Mine, AMAX Coal Company, 18 mi south of Gillette, Wyoming. The subbituminous coal seam is 22 m thick. Photo courtesy of AMAX Coal and Horizons, Inc.

Library of Congress Cataloging-in-Publication Data
Economic geology—U.S. / edited by H.J. Gluskoter, D.D. Rice, R.B. Taylor.
    p.   cm. — (The Geology of North America ; v. P-2)
   Includes bibliographical references and index.
   ISBN 0-8137-5214-0
   1. Geology, Economic—United States. I. Gluskoter, Harold J. (Harold Jay), 1935- . II. Rice, Dudley, D. III. Taylor, Richard B. (Richard Bartlett) IV. Series.
QE71.G48   1986   vol. P-2
[TN23]
557 s—dc20
[553'.0973]   91-12933
                                                          CIP

10 9 8 7 6 5 4 3 2 1

Contents

Preface vii

I. MINERAL DEPOSITS

1. Introduction 1
   Richard B. Taylor

Metals

2. Gold and silver deposits of the United States 3
   Roger P. Ashley

3. Copper and molybdenum deposits in the United States 23
   Edwin W. Tooker

4. Lead and zinc deposits 43
   Ernest L. Ohle

5. Iron and manganese 63
   Gary B. Sidder

*6. Deposits containing nickel, cobalt, chromium, and platinum-group
   elements in the United States* 87
   Michael P. Foose

7. Uranium and vanadium deposits 103
   Daniel R. Shawe, J. Thomas Nash, and William L. Chenoweth

8. The other metals 125
   Ronald G. Worl

Industrial metals

*9. Phosphate deposits of the United States—discovery, development;
   Economic geology and outlook for the future* 153
   James B. Cathcart

10. *Evaporites and brines* 165
    Omer B. Raup and Marc W. Bodine, Jr.

11. *Oil shale* 183
    John R. Donnell

12. *Other selected industrial minerals* 189
    Donald A. Brobst

II. OIL AND GAS

13. *Introduction* 213
    Dudley D. Rice

General characteristics

14. *Origin and migration of oil and gas* 225
    Fred F. Meissner

15. *Pore system aspects of hydrocarbon trapping* 241
    William R. Almon and John B. Thomas

16. *Exploration techniques* 255
    Edward A. Beaumont, Norman H. Foster, Richard R. Vincelette, and
    Graham R. Curtis

Synthesis of selected provinces

17. *Petroleum geology of the Appalachian basin* 273
    Wallace de Witt, Jr., and Robert C. Milici

18. *The Michigan basin* 287
    C. R. Reszka, Jr.

19. *The northern Gulf of Mexico basin* 301
    D. M. Curtis

20. *Anadarko basin* 325
    Herbert G. Davis and Robert A. Northcutt

21. *The Permian basin* 339
    Bernold M. Hanson, Brian K. Powers, Chester M. Garrett, Jr.,
    Donald E. McGookey, Ed H. McGlasson, Ralph L. Horak,
    S. J. Mazzullo, Alastair M. Reid, Gerald G. Calhoun,
    John Clendening, and Brenda Claxton

22. *Oil and gas resources of the San Juan basin, New Mexico*
    and Colorado 357
    James E. Fassett

23. *Powder River basin* 373
    James E. Fox, Gordon L. Dolton, and Jerry L. Clayton

24. *Geologic controls on hydrocarbon occurrence, Fossil basin area,*
    Cordilleran thrust belt 391
    Maurice A. Warner

25. *Petroleum potential of the Great Basin* 403
    Norman H. Foster and Richard R. Vincelette

26. *San Joaquin basin, California* 417
    David C. Callaway and Ernest W. Rennie, Jr.

27. *Geologic controls on hydrocarbon occurrence within*
    the Santa Maria basin of western California 431
    John B. Dunham, B. W. Bromely, and Victor J. Rosato

28. *North Slope of Alaska* 447
    Kenneth J. Bird

III. COAL

29. *Coal; A brief overview* 463
    Hal J. Gluskoter

Geology of coal

30. *Geology of coal; Environments of deposition* 469
    Peter J. McCabe

31. *Paleobotany and paleoecology of coal* 483
    Tom L. Phillips and Aureal T. Cross

32. *Coalification in North American coal fields* 503
    Heinz H. Damberger

33. *Pennsylvanian coals of central and eastern United States* 523
    Alan C. Donaldson and Cortland Eble

34. *Cretaceous and Tertiary coals of the Rocky Mountains*
    and Great Plains regions 547
    Romeo M. Flores and Timothy A. Cross

Major coal deposits of the United States

35. *Tertiary coals of the Gulf Coast* 573
    John A. Breyer

36. *Coals of far-western United States* 583
    Aureal T. Cross

37. *Economic Alaskan coal deposits* 591
    Gary D. Stricker

Index 603

Plates

(in accompanying slipcase)

Plate 1. Mineral deposits of the United States I:
  A. Gold and silver
  B. Copper and molybdenum
  C. Lead
  D. Zinc

Plate 2. Mineral deposits of the United States II:
  A. Iron and manganese
  B. Nickel, cobalt, chromium, and platinum-group minerals
  C. Uranium and vanadium
  D. Miscellaneous metals

Plate 3. Mineral deposits of the United States III:
  A. Phosphate
  B. Evaporites
  C. Clays and zeolites
  D. Asbestos, barite, feldspar, fluorspar, kyanite and related minerals, talc, wollastonite

Plate 4. Reconstruction of a Pennyslvanian coal swamp

Plate 5. Oil and gas map of the United States

Plate 6. Cross-sections of major oil and gas basins: I

Plate 7. Cross-sections of major oil and gas basins: II

Plate 8. Coal map of North America (north)

Plate 9. Coal map of North America (south)

Preface

The Geology of North America series has been prepared to mark the Centennial of The Geological Society of America. It represents the cooperative efforts of more than 1,800 individuals from academia, state and federal agencies of many countries, and industry to prepare syntheses that are as current and authoritative as possible about the geology of the North American continent and adjacent oceanic regions.

This series is part of the Decade of North American Geology (DNAG) Project, which also includes seven wall maps at a scale of 1:5,000,000 that summarize the geology, tectonics, magnetic and gravity anomaly patterns, regional stress fields, thermal aspects, and seismicity of North America and its surroundings. Together, the synthesis volumes and maps are the first coordinated effort to integrate all available knowledge about the geology and geophysics of a crustal plate on a regional scale.

The products of the DNAG Project present the state of knowledge of the geology and geophysics of North America through the 1980s, and they point the way toward work to be done in the decades ahead.

In addition to the contributions from organizations and individuals acknowledged at the front of this book, major support has been provided to the editors of this volume by the U.S. Geological Survey.

A. R. Palmer
General Editor for the volumes
published by the Geological Society
of America

J. O. Wheeler
General Editor for the volumes
published by the Geological
Survey of Canada

Chapter 1

Introduction

R. B. Taylor
U.S. Geological Survey, MS 905, Denver Federal Center, Denver, Colorado 80225

This section provides an introduction to the geology of metallic and industrial mineral deposits in the conterminous United States. The section is organized into chapters discussing specific commodities or groups of commodities. Authors of these chapters explore some of the important kinds of mineral resources in the United States, including brief descriptions of uses of particular materials, the mining history, and future prospects. The primary focus is on the geology of important deposits, commonly presented in a historical context. Maps included as plates provide information on locations and distributions of deposits and mines, as selected by the authors for emphasis and explanation. Bibliographic citations provide an introduction to additional literature for the interested reader. Discussions of the geology of particular deposits such as presented here provide needed background for writings that synthesize groups of similar deposits into deposit models. The collection of models for metallic deposits edited by Cox and Singer (1986) provides one-page geological summaries for 85 descriptive models, and graphical summaries of grade and tonnage for 60 of these. Although some grouping by deposit type is the logical outcome of the organization of the present section, refer to the Cox and Singer volume for this type of treatment. The publication *United States Mineral Resources,* (Brobst and Pratt, 1973) can provide an important supplement to this section.

Geologists, economists, and historians have documented the dependence of modern civilization on mineral and energy resources and the changes in needs through time as technology has developed to its present state. Further changes will follow as needs for commodities, commodity prices, and developments in mining and extractive technology change the economic outlook. Our civilization depends on a steady supply of mineral and mineral fuel commodities, which unlike many other kinds of resources, are nonrenewable. Although the geologic processes of ore formation continue to be active, the rate of accumulation of most resources requires geologic time, not the brief span of civilization.

Economic geology is a special domain directly utilizing the techniques of modern science toward meeting the needs of society. It is concerned with the geology of mineral resources, those concentrations of elements in the Earth's crust in such a form that there is promise that at some time a useful mineral commodity might be extractable at a profit. Our use of the term "resource" includes as yet undiscovered deposits that can be predicted to exist on the basis of geological reasoning. Resource classification includes deposits also classified as reserves, but the converse is not necessarily true; moreover, the two terms are frequently confused. Reserves include only well-known deposits from which a mineral commodity can be extracted at a profit under present economic conditions. This distinction is essential to understanding writings about mineral supplies. A useful classification distinguishing different kinds of resources and their relations to reserves was developed by McKelvey (1973), and should be consulted in its original form.

Research in economic geology blends the efforts of scientists from government, academia, and industry. Knowledge of the geology of deposits of metals and industrial minerals is greatly dependent on information in the third dimension gained from drilling and mining. Society's need for materials, and hence its willingness to pay costs of exploration and development, makes this knowledge possible. In turn, we rely on such research to assess the adequacy of resources to meet national needs.

Public policy in the United States to the present time has been to promote mineral development, and also to balance resource development with other land uses. A review of land classification by G. O. Smith and others (1913) points out that mining laws were largely formulated between 1865 and 1875 and that they were based on local mining customs that had attained the force of law in mining camps on the public domain, intended to facilitate orderly development. Discussions of issues related to the "best use" of the nation's land in this report provide important background for today's debates on the best use of public land, balancing development and preservation. The Smith report also points out the need for the best possible information before decisions on land use are made, and in the main has a contemporary tone.

The General Mining Law of 1872, as extended by judicial decisions and modified by legislation, still forms the basis for exploration and development of all minerals on Federal lands exclusive of oil, gas, coal, phosphate, sodium, oil shale, potash, certain other hydrocarbons, and nonmetallics provided for under the Mineral Leasing Act of 1920. The first requirement is the finding of minerals within the limits of a claim; the second is that "the evidence be of such character that a person of ordinary

Taylor, R. B., 1991, Introduction, *in* Gluskoter, H. J., Rice, D. D., and Taylor, R. B., eds., Economic Geology, U.S.: Boulder, Colorado, Geological Society of America, The Geology of North America, v. P-2.

prudence would be justified in the further expenditure of his labor and means, with a reasonable prospect of success, in developing a valuable mine." Details of the laws relating to development of public lands for mineral purposes are not appropriate here, but some understanding of purpose and practices related to the Mining Law of 1872 is important in understanding incentives prompting exploration for mineral deposits, their discovery, and development; the last a theme developed in this section. This law has made it possible for the individual prospector and small exploration company to find and develop ore deposits. Although exploration increasingly requires geological, geophysical and geochemical capabilities, the location system provided by the 1872 law makes participation in exploration possible for those of limited means; commonly they discover significant deposits that subsequently are developed by groups with large capital resources. It provides powerful incentive for individual effort. Debates continue on proper laws for transferring title of resources on public lands to the private sector, on limitation of access to some lands, and on royalties, lease systems, and taxes. Domestic production, as contrasted with foreign sources of raw materials, reduces the need for imports, thus favorably influencing the international balance of payments and providing needed materials at times of national emergency.

Exploration and development of resources have a time and a place shaped by economic factors, mineral endowment, understanding of the geology of ore deposits, and land availability.

The cumulative effect of the withdrawal of lands in the United States from mineral entry, location, or lease, is an increasing concern because of its negative effect on domestic mineral resource development. Article IV of the U.S. Constitution states that the Congress has the power to dispose and make rules respecting the territory belonging to the United States. Advice from the past seems worth quoting: ". . . a wise nation, like a prudent man, learns to husband its resources. Land values are now recognized, the purpose in both legislation and administration has changed, and the highest development alone is sought. With the most and best of the Nation's land already alienated, the national duty is to put to its best use what remains" (Smith, G. O. and others, 1913, p. 7).

I wish to acknowledge the efforts of the authors of the chapters in this section and the editorial assistance of Paul K. Sims.

REFERENCES CITED

Cox, D. P., and Singer, D. A., eds., 1986, Mineral deposit models: U.S. Geological Survey Bulletin 1693, 379 p.

Brobst, D. A., and Pratt, W. P., eds., 1973, United States mineral resources: U.S. Geological Survey Professional Paper 820, 722 p.

McKelvey, V. E., 1973, Mineral resource estimates and public policy, *in* D. A. Brobst, and Pratt, W. P., eds., United States Mineral resources: U.S. Geological Survey Professional Paper 820, p. 9–20.

Smith, G. O., and others, 1913, The classification of the public lands: U.S. Geological Survey Bulletin 537, 197 p.

MANUSCRIPT ACCEPTED BY THE SOCIETY JANUARY 23, 1991

Chapter 2

Gold and silver deposits of the United States

Roger P. Ashley
U.S. Geological Survey, 345 Middlefield Road, Menlo Park, California 94025

INTRODUCTION

Gold and silver share a long history as monetary metals, serving as media of exchange and repositories of wealth. In developed countries, however, industrial uses of both metals have become increasingly important in the twentieth century at the expense of monetary uses. For the past two decades the largest single use of silver in the United States has been photographic products (Reese, 1985). Applications in electrical and electronic equipment now represent the second largest use of silver, and the second largest use of gold as well (Lucas, 1985). Historically, both gold and silver have been used extensively for jewelry and artistic and decorative items, and these uses are still important, although much more so for gold than for silver. Jewelry represents the largest single use of gold, whereas jewelry and sterlingware together constitute the third largest use of silver. Dental supplies represent the third largest use of gold, but a relatively minor use of silver.

Gold and silver have been mined in the United States since early in the nineteenth century. The major period of production for gold began with discoveries in California in 1848 and continued to World War I, with annual output reaching 90 to 120 t (3 to 4 million troy oz) in the last two decades of this period (Simons and Prinz, 1973). Another brief surge of production began in 1935, after the price of gold was increased to $35 per troy ounce, but was cut off in 1942, when gold mines were shut down by War Production Board Limitation Order L-208. The major period of silver production began in 1873 with full development of the Comstock Lode district and continued until 1929, with annual output of 1,500 to 2,400 t (50 to 75 million troy oz) per year during much of this period. Silver production briefly reached these levels again from 1935 to 1942, along with increased gold production (Heyl and others, 1973).

After World War II, gold production resumed at levels of 1.5 to 2 million troy oz per year and declined to about 1 million troy oz per year by the late 1970s. By-product gold, mostly from porphyry copper deposits, was an important component of production from the 1950s through the 1970s. Gold demand greatly exceeded domestic mine production during this period; before 1968 the difference was made up largely by U.S. Treasury sales, and after 1968 mainly by imports. The establishment of a free market in 1968, coupled with political instability, fuels shortages, and high inflation rates in market-economy countries through the 1970s, have led to relatively high gold prices in recent years. These prices have in turn resulted in significant increases in mine production in the United States every year since 1980 (Lucas, 1988a), with scores of new mines being opened during this period. This increase in production, combined with some decrease in gold consumption, has significantly reduced U.S. reliance on imports.

Although silver prices have also increased since 1968, U.S. production has remained close to 1,200 t (40 million troy oz) per year, with some losses of by-product sources balanced by new production from disseminated silver deposits and by-product production from new gold mines (Reese, 1988). For the last three decades, industrial demand for silver has exceeded new mine production worldwide, as well as in the United States. This has led to greater reclaiming of silver and conversion of official stocks and hoarded silver to industrial uses. Reliance on imports has increased somewhat in recent years, and currently about 60 percent of silver consumed in the United States is imported. At the present time the price of silver is relatively low compared to the price of gold.

Cumulative gold production of the United States is about 10,900 t (350 million troy oz), which is about 11 percent of the 103,000 t (3.3 billion troy oz) estimated total world production (Lucas, 1985, 1988b). Archean gold deposits, exploited directly as lodes and indirectly as placers and paleoplacers derived from Archean terranes, account for 70 percent of total world production, yet Archean deposits account for only a small part of U.S. production. Significant gold production in the United States has come from an extraordinary variety of geologic environments and deposit types, including alluvial and beach placers, gold-quartz veins, polymetallic veins and replacement deposits, various epithermal deposits including veins, stockworks, and disseminated types, and iron-rich sedimentary rocks. Substantial by-product gold has also come from copper- and copper molybdenum-porphyry deposits.

Cumulative U.S. silver production is about 172,000 t (5.5 billion troy oz). Epithermal vein deposits such as the Comstock

---

Ashley, R. P., 1991, Gold and silver deposits of the United States, *in* Gluskoter, H. J., Rice, D. D., and Taylor, R. B., eds., Economic Geology, U.S.: Boulder, Colorado, Geological Society of America, The Geology of North America, v. P-2.

Lode and Tonopah, Nevada, dominated domestic silver production before World War I, with important contributions from lead-zinc replacement deposits containing co-product silver, and from all types of gold deposits, which contain more or less by-product silver. Since then, silver production has been mainly from the polymetallic vein deposits of the Coeur d'Alene district of Idaho, and as a by-product of copper- and copper molybdenum-bearing porphyry deposits, a by-product of Mississippi Valley lead-zinc deposits, and a by-product of porphyry, vein, and replacement deposits at Butte, Montana. About two-thirds of identified unmined silver resources in the United States reside in various base- and precious-metal deposits in which silver is a by-product, and only one-third in deposits in which silver is the main product (Heyl and others, 1973). World silver resources are divided between by-product and primary-product deposits in about this same 2:1 proportion (Reese, 1988).

In this chapter, only deposits with silver as a major constituent or co-product are discussed in detail. These include various types of epithermal vein and disseminated deposits, and polymetallic vein and replacement deposits. Other deposit types that are sources of by-product silver are discussed fully in the chapters on copper-molybdenum and lead-zinc.

The purpose of this chapter is to describe ore deposits that account for a significant part of the precious metals endowment of the United States. Although the purpose is not to propose or evaluate genetic ore deposit models, I have used models in the process of selecting deposits that are important and representative, and in organizing the discussion. Sources of deposit models that I have used and sometimes modified are given with a definition of deposit type at the beginning of each section.

This treatment is not comprehensive with respect to geographic distribution of gold and silver deposits, or even distribution of various deposit types. Because more than 1,000 mining districts in the United States have significant past production or reserves of gold or silver, or both, there is not space here for an exhaustive catalog. The occurrence map (Plate 1A) shows geographic ranges of deposit types based on distribution of gold deposits larger than 0.31 t (10,000 troy oz) combined production and reserves, and silver deposits larger than 3.1 t (100,000 troy oz). More detailed maps include those of Tooker and Vercoutere (1986), McKnight and others (1962), and Guild (1981). The mining districts located on Plate 1A are those mentioned in the text; many but not all of the important districts are included. Koschmann and Bergendahl (1968) is an excellent source of summary information and references for mining districts in the United States that produced gold before 1966. It is out of date with respect to many important bulk-minable gold deposits discovered in the last two decades. The perspectives on deposit types, reserves, and resources in the articles by Simons and Prinz (1973) on gold, and Heyl and others (1973) on silver, are still largely valid. Summary geologic descriptions of many U.S. deposits are included in Boyle's treatises on gold (1979) and silver (1968).

MAJOR GOLD AND SILVER DEPOSITS OF THE UNITED STATES

Placer deposits

Placer deposits are gold-bearing sediments formed by disintegration and erosion of rocks containing gold deposits, and subsequent deposition of this detritus, including grains of native gold. The resulting deposits include fluvial sands and gravels and beach sands (alluvial deposits), residual soils or weathering debris found near gold-bearing outcrops (eluvial deposits), and rarely wind-blown sands (aeolian deposits). Older placer deposits in clastic sedimentary rocks are referred to as fossil placers; as these deposits are eroded, gold is released and incorporated in more recent placers. Recycled gold from fossil placers is probably an important component of many placer deposits found in modern river gravels and beach sands.

Placer deposits, particularly those in Europe and the Middle East, have been known and worked since the beginning of recorded history. Alluvial-type deposits, in particular, are so common to human experience, and techniques for mining them have been known so long, that most reports describing occurrences or mining techniques do not include definitions of terms or discussion of geologic characteristics of placer deposits. A recent discussion of the origin and application of the term "placer," and the range of deposits included is given by Boyle (1979).

Placer gold was discovered on both the East and West Coasts in the late eighteenth century. Placer mining began in southeastern California in the late 1770s, while the territory was still under Spanish rule (Clark, 1970), but owing to the isolation and small population of the region, no rush developed. Early discoveries in the southern Piedmont region of Virginia, North and South Carolina, and Georgia attracted only local attention, but a boom finally developed from the late 1820s to the mid-1830s (Pardee and Park, 1948).

Interest abruptly shifted to northern California in 1848 with the discovery of placer gold on the American River by James Marshall. Alluvial gold from placers in all the major river systems of the central and northern Sierra Nevada dominated California production for 15 years, at which time placers of Tertiary age, exploited by hydraulic mining methods, became the major source and remained so until 1884 (Clark, 1970). Placer mining in the modern river systems regained importance in 1898 when dredging began on the lower Feather River. Large-scale mechanized mining of the gravel fields at the western front of the Sierra Nevada continued until the 1960s, and recently resumed on the lower Yuba River.

The California gold rush stimulated exploration of the American West, with subsequent discovery of major placer fields in southeastern Oregon (1851), northeastern Oregon (1861), west-central Idaho (1862), western Montana (1862), and Alaska, including the Yukon River region (1881), Nome (1898), and Fairbanks (1902) (see Koschmann and Bergendahl, 1968). Although the exploitation of the placer gold deposits and subsequent development of lode deposits was of critical importance in economic development of the western conterminous United States, relatively little effort was expended in systematic description and scientific investigation of these placers because the early federal and state surveys devoted most of their efforts to reconnaissance and basic descriptive geology. The major period of discovery, development, and mining was over before the U.S. Geological Survey was created in 1879. In Alaska, where development came later, the placer districts are described relatively well in a number of U.S. Geological Survey reports (see Koschmann and Bergendahl, 1968, for references).

Although placer deposits have produced about one-third of United States gold (Lucas, 1985), and the majority of this gold came from Quaternary alluvial deposits, the only deposits to be described here are the placers found in Tertiary channels of the Sierra Nevada. These deposits were exploited by the hydraulic mining method, in which gravel beds are eroded with high-pressure streams of water conducted from canals through large iron pipes and nozzles. The deposits are of interest for two reasons. First, a large volume of unmined gold-bearing gravel remains in the channels (Yeend, 1974). Second, because the remaining deposits are accessible in natural and artificial exposures and through drilling, they have been relatively well studied since hydraulic mining ceased in 1884 as a result of the "Sawyer decision," an injunction issued by Judge Lorenzo Sawyer in the case of Woodruff vs. North Bloomfield Gravel Mining Company, prohibiting dumping of debris into the Sacramento and San Joaquin River systems. Large amounts of sand and gravel released by hydraulic mining were deposited at the edges of the Central Valley, destroying farmland by silting and flooding, and resulting in the extensive litigation between agricultural and mining interests that led to the Sawyer decision.

The Tertiary gravels occupy river channels in an erosion surface cut on the complex north-to-north-northwest–striking slates, phyllites, greenstones, and marbles of Paleozoic and Mesozoic age that form the metamorphic belts of the Sierra Nevada, and the extensive Mesozoic plutonic rocks of the composite Sierra Nevada batholith. The channels converge westward, revealing a series of west-flowing river systems similar in general features to the modern river systems of the Sierra Nevada, but different in detail. They are overlain by a variety of volcanic and sedimentary rocks of Tertiary age, including volcanic-rich sands and gravels, rhyolitic tuffs and fine-grained tuffaceous sedimentary rocks, and andesite breccias that are probably mainly mudflows. The oldest radiometric and paleontologic dates from the sequence overlying the channel gravels are early Oligocene. The character of bedrock units, the distribution of channels in the northern Sierras, and the character of overlying units were delineated in early work of the U.S. Geological Survey (Lindgren and Turner, 1895; Lindgren, 1900). Lindgren went on to produce a classic comprehensive report on the Tertiary gravels (Lindgren, 1911).

Occurrence of significant gold in the gravels correlates closely with occurrence of gold-bearing veins in the bedrock (Lindgren, 1911). The upper, eastern parts of the channels, which cross granitic rocks of the composite Sierra Nevada batholith, are nearly barren. To the west, gold content increases in each channel as soon as it crosses vein systems in the metamorphic belts. The channel deposits of the ancestral Yuba River consist of a lower, relatively coarse gravel that contains most of the gold, and an upper, finer gravel with more abundant quartz and quartzite clasts (Yeend, 1974). The "pay streak" or "pay lead" occurred near bedrock in all the channels, but did not always occupy the deepest part of the channel, wandering from one side of the channel to the other. Lindgren offered no detailed explanation for the variable location of the pay streak, but indicated that the high grades of the pay streak were critical mainly to the drift miner driving tunnels in the gravel; much gravel outside the pay streak was of grade acceptable for hydraulic mining. Yeend (1974) showed that low but significant gold values persist for considerable distances (tens of meters) above bedrock in the North Columbia diggings of the ancestral Yuba River gravels. The lower part of the lower gravels is bluish gray (the "blue lead" of local mining terminology), and water-saturated. The color results from fine-grained secondary pyrite disseminated through the gravels. Yeend (1974) proposed that anaerobic bacterial decay of vegetation trapped behind temporary dams along the channel was the source of H_2S that produced the pyrite. Carbonized wood is partly to completely replaced by opal or chalcedony, and wood fragments are sites of increased amounts of pyrite. Lindgren (1911) concluded that precipitation of gold in the gravels from solution was insignificant, but conceded that dissolution and redeposition may have taken place on a small scale to explain gold possibly associated with secondary pyrite in the blue gravels. Yeend (1974) discussed the unresolved problem that vein-quartz detritus is concentrated in the upper gravels whereas gold is concentrated in the lower gravels. Clearly, even paleoplacers with a geologic history as simple as that of the Tertiary gravels of the Sierra Nevada have involved hydrologic and geochemical processes that defy easy explanation.

Gold-quartz veins

This deposit type includes quartz veins in which gold is the only important commodity. Although many different sulfide minerals, as well as tellurides and sulfosalts may occur, veins of this type are characterized by small total amounts of these minerals, and thus are also referred to as low-sulfide gold-quartz veins (see Cox and Singer, 1986). Gangue minerals other than quartz include carbonates (ferroan carbonates are particularly characteristic), albite, and sericite, but quartz is by far the most abundant. Wall rocks show carbonate alteration, commonly quite extensive. They are found in terranes consisting of regionally metamorphosed volcanic rocks, graywackes, and shales. Such terranes include Archean greenstone belts and Phanerozoic mobile belts and ophiolites. Mafic and ultramafic dikes and sills, regional fault systems, and syndeformation and postdeformation granitic plutons are common features of these terranes. Gold-quartz veins may be associated with any of them. The deposit class, as defined here, includes veins that Lindgren (1933) interpreted as magmatic-hydrothermal in origin and included in his mesothermal and hypothermal deposit categories. It also includes veins interpreted as having been deposited from fluids of metamorphic origin.

Because placer deposits in northern California, Oregon, western Idaho, and Alaska were formed by erosion of metamorphic-plutonic terranes, gold-quartz veins in these terranes were discovered during the early periods of placer mining, and many deposits of small to moderate size are known in the Paleozoic and Mesozoic metamorphic belts of the western continental margin. Similarly, small gold-quartz veins, together with other types of lode deposits, were also found in the southern Piedmont region in the early 1800s. Among the many small deposits known, a few relatively large deposits stand out, including the Treadwell and Alaska-Juneau deposits in southeastern Alaska and the French Gulch district of northern California. The greatest concentration of deposits and the largest districts, however, are located in the Sierra Nevada of California. The Sierran deposits are divided into five main groups based on geographic proximity and common geologic features. These include the Mother Lode, the West Gold Belt, the East Gold Belt (Clark, 1970), the Grass Valley and Nevada City districts, and the northern Sierra Nevada (Alleghany and Downieville districts and vicinity). The Grass Valley district and Mother Lode belt will be described here as the most important examples. These districts have many characteristics in common with vein and stockwork gold deposits of Archean greenstone belts (Hutchinson and Burlington, 1984; Boyle, 1976).

The Grass Valley district is the most productive mining district in California and one of the most productive in the United States, having yielded more than 390 t (12,600,000 troy oz) of gold in 106 years of mining, ending in 1956. Greenstones and slates of late Paleozoic(?) age are cut by an irregular granodiorite stock (Fig. 1) and metamorphosed to amphibolites and schists around the stock (Lindgren, 1896). A major north-trending regional fault immediately west of the district forms the eastern boundary of the Smartville ophiolite (Schweickert, 1978). A serpentine mass in the northern part of the district is one of several located near or along this regional fault. North- to west-trending, gently dipping veins cut the granodiorite and adjacent amphibolite and schist. Two of the three largest mines, the Empire and North Star, are in this group of gently dipping veins. Another group of west-trending, steeply dipping veins cuts serpentine and amphibolite in the northern part of the district. The third major mine, the Idaho-Maryland, is located here. Throughout the district, northeast-striking, steeply dipping fractures commonly mark changes in width and grade of the veins and ore shoots. The veins consist of multiple generations of quartz with minor sericite, chlorite, and epidote, and late fracture and vug fillings of ankerite and calcite (Johnston, 1940). Sulfide minerals were deposited mainly with the quartz; pyrite and arsenopyrite tend to be earlier, and sphalerite, chalcopyrite, and galena, later. Gold is associated with galena. Carbonates with accompanying sericite, pyrite, and arsenopyrite replace the wallrocks. Individual veins were mined for distances as great as 3,400 m downdip, and the entire vein system was mined over a vertical range of more than 1,200 m. Over this vertical distance the character of veins and ore shoots did not change appreciably; mining ceased because of increasing costs in the post–World War II period of constant gold price.

The Mother Lode belt is a system of en echelon quartz veins 1 to 5 km wide that follows the Melones fault zone for almost 200 km from Georgetown, California, south-southeast to Mariposa, California. Rocks along and within the Melones zone are mainly greenstone, phyllite, and serpentine. The Melones is a major regional structure that separates island-arc sequences of Jurassic age on the west from metasedimentary rocks of Paleozoic age (Calaveras Formation) on the east. Ore bodies include quartz veins and bodies of mineralized wall rock known as "gray ore." The quartz ore bodies are veins and stringer zones, locally ribboned with bands of wall rock, containing free gold with pyrite and minor arsenopyrite, sphalerite, galena, chalcopyrite, tetrahedrite, and locally tellurides (Knopf, 1929). Total sulfides are usually only 1 to 2 percent of the ore. The quartz veins fill fissures formed by high-angle reverse faults of modest displacement that cut the cleavage of the metamorphic wall rocks at acute angles. Ore occurs in the veins in shoots that typically pitch steeply. Wall rocks are extensively altered to ankerite plus sericite, albite, pyrite, and arsenopyrite. Near or adjacent to quartz veins the altered wall rocks may be gold bearing, forming gray ore. Sulfide minerals in gray ore are also mainly pyrite and arsenopyrite, but sulfides are more abundant than they are in either veins or unmineralized altered wall rocks, reaching 10 percent.

Mining began in the southern Mother Lode belt in 1849; many major mines opened within the next few years, and several other mines were expanded into major operations in the 1880s and 1890s. Some mines shut down before World War I owing to rising operating costs, but the increase in the price of gold in the 1930s brought a resurgence of output (Clark, 1970). Few mines reopened after the forced closure of World War II, and production ceased in 1953. In 1986, open-pit operations began at the sites of the Carson Hill and Harvard mines in the central part of the belt. Cumulative production of the entire belt is about 400 t (13,000,000 troy oz). Over half of this came from the Jackson-Plymouth district, a 15-km-long segment near the north end of the belt. Although by World War II mines in the Jackson-Plymouth district were the deepest in the United States, reaching 1,800 m, no consistent or significant changes in mineralization were seen through this vertical interval, given that the veins and ore shoots were both notoriously variable in character laterally and vertically (Knopf, 1929). Considerable potential exists for additional ore, particularly gray ore, which may remain in significant quantities, even near the surface. Mining companies have been active in exploration and development at a number of Mother Lode properties in recent years.

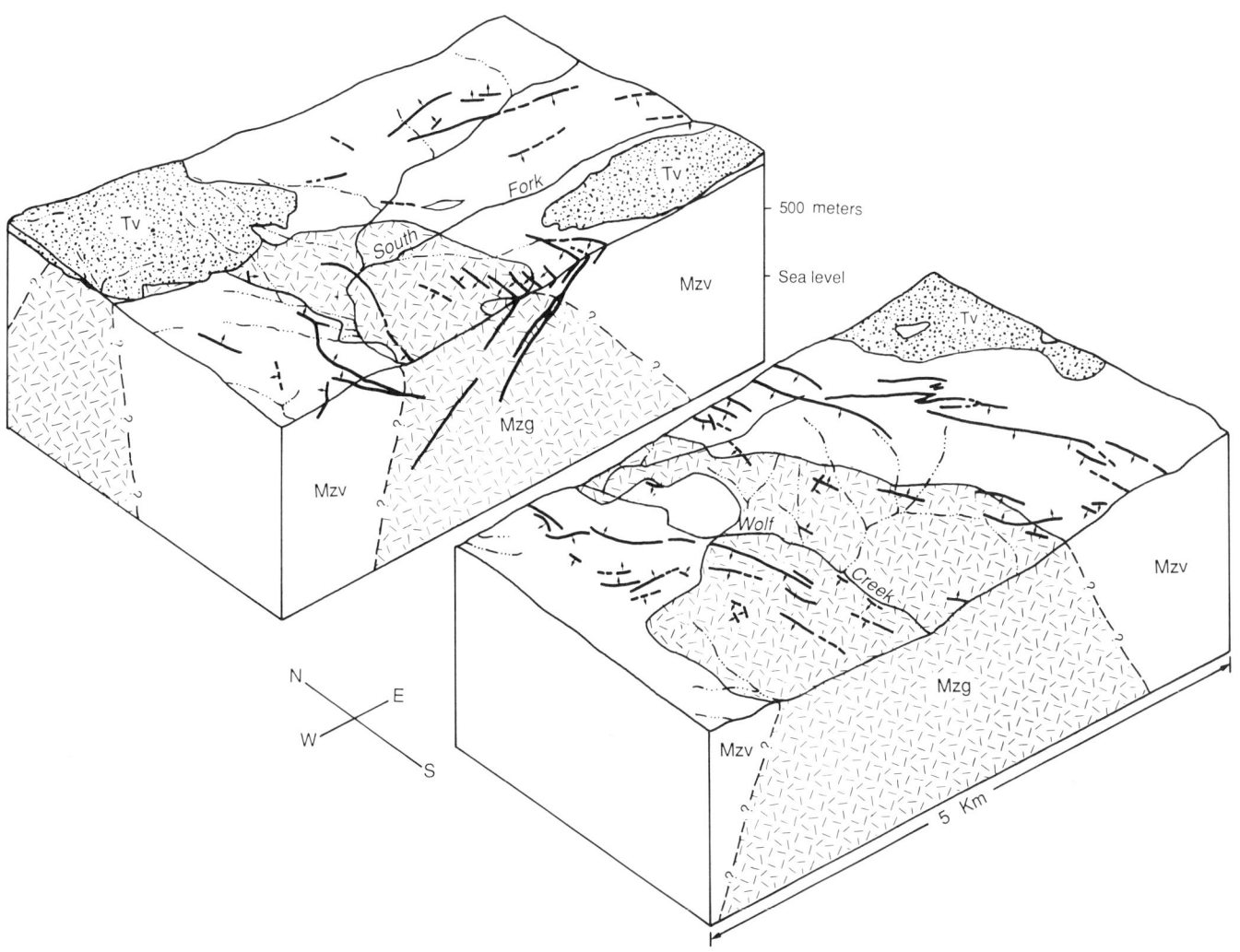

Figure 1. Isometric projection of the Grass Valley Quadrangle, California, cut through the vein system of the Empire mine. Mzv, Mesozoic metavolcanic and metaintrusive rocks of mafic and intermediate composition, including greenstone (mainly porphyritic andesite and diabase), amphibole schist, and serpentine; Mzg, Mesozoic granodiorite; Tv, Tertiary auriferous gravels overlain by andesitic tuff and tuff breccia. Heavy lines are veins (arrows show dip directions); light lines are contacts. Modified from Johnston, 1940.

Recent isotopic studies of gold-quartz veins and altered wall rocks in the Sierra Nevada are concentrated in the Alleghany district, where underground access is relatively good. K-Ar and Rb-Sr age data show that the quartz veins formed 20 m.y. or more after prograde metamorphism and penetrative deformation (Böhlke and Kistler, 1986). The veins are not demonstrably related to local igneous activity, but may be related to upward-moving fluids driven by deep-seated metamorphism or magmatism. Quartz-mica $\delta^{18}O$ fractionations and fluid inclusion data indicate that the veins were deposited at temperatures of 300° to 350°C (Böhlke and Kistler, 1986; Coveney, 1981). The fluids responsible for metasomatic alteration and vein mineral deposition were CO_2-bearing, had relatively high δD and $\delta^{18}O$ values, and may have equilibrated with metamorphic rocks (Marshall and Taylor, 1981; Böhlke and Kistler, 1986).

Gold deposits associated with iron formation

Because very few deposits of this type are known in the United States, most published descriptions are for deposits in Precambrian, mainly Archean terranes of Canada, Australia, and Africa, and characteristics of this deposit type are defined from these foreign deposits. Ore bodies are stratabound or stratiform in iron-rich sedimentary rocks that occur in volcanic-sedimentary sequences metamorphosed to lower greenschist to middle amphibolite facies. Within individual ore bodies, gold may be concentrated in quartz stringers and fracture fillings. Relatively sulfide-rich banded iron formation may also be ore, particularly sulfide- or carbonate-facies iron formation. Silicic and alkalic intrusions, such as quartz porphyry or syenite stocks, occur in many districts, especially in Canada. Districts may also be associated with major shear zones that mark depositional facies boundaries and localize porphyritic stocks, and thus have a long history. Wall-rock alteration is mainly sulfidation of iron formation, involving addition of pyrite, pyrrhotite, or arsenopyrite, and carbonatization of other rocks, involving addition of siderite or ankerite. Carbonatization is most notable in mafic and ultramafic volcanic rocks. Native gold, whether found in iron formation or associated quartz veins, is associated with pyrite, arsenopyrite, pyrrhotite, galena, sphalerite, or chalcopyrite. In iron formation it is also associated with magnetite. Minor minerals vary from deposit to deposit, and include scheelite, wolframite, molybdenite, tetrahedrite, stibnite, and realgar.

The wide range of host rocks and forms of gold ore bodies found in Archean greenstone belts, even within districts, has produced continued debate over the relative importance of gold deposition from sea-floor hydrothermal systems during accumulation of the volcanic-sedimentary sequence versus later epigenetic introduction of gold. Confusion also arises in classifying deposits, because discordant quartz veins in Archean greenstone belts often are similar to low-sulfide gold-quartz veins with respect to ore, gangue, and associated alteration, regardless of whether they are associated with rocks such as komatiites and iron formations that are most common in Archean sequences. These arguments of genesis and classification have not been active in the United States because there are few known deposits in the limited exposure areas of early Precambrian terranes, and only one deposit of major importance, the Homestake mine near Lead, South Dakota. Owing to the prominence of the Homestake deposit, some U.S. authors apply the name "Homestake gold" to Precambrian gold deposits associated with iron formation (Cox and Singer, 1986). The Homestake deposit is clearly stratabound in an iron-rich sedimentary unit, but ideas about its genesis have changed with time.

The Homestake mine is the largest gold mine in the United States, and one of the largest in the world, with combined production and reserves of 1,240 t (40,000,000 troy oz) of gold. It has produced almost continuously since 1876. The first comprehensive report on the geology and ore deposits of the Homestake mine and the surrounding northern Black Hills is that of Irving and others (1904). More recently, a series of articles was published by J. A. Noble and others between 1948 and 1950 (see Noble, 1950). Ore is restricted to the Homestake Formation, a 70- to 100-m-thick unit of quartz-sideroplesite/cummingtonite schist (see Fig. 2); mineralogy varies with metamorphic grade. The Homestake Formation is underlain by the Poorman Formation, 600 m of ankeritic phyllite, and overlain by the Ellison Formation, 800 m of quartzite and quartz-mica phyllite. Other formations above the Ellison include metamorphosed basalt, chert, graywacke, and pelitic sedimentary rocks. All the Precambrian formations in the Lead area are of early Proterozoic age. Parts of the Homestake Formation that constitute ore are chloritized and contain veins and irregular masses of quartz and disseminated to locally massive arsenopyrite, pyrite, and pyrrhotite. Gold is particularly closely associated with arsenopyrite. A

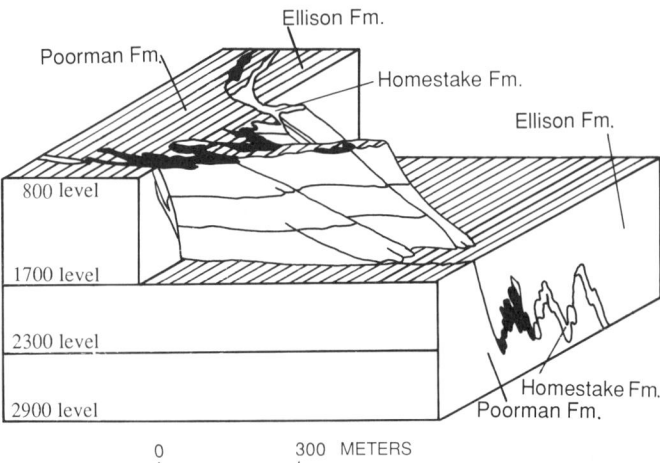

Figure 2. Isometric projection of the Homestake ore bodies (solid black), Homestake mine, South Dakota. Overlying Ellison Formation partially removed over plunging folds defined by the Homestake Formation. Modified from Lindgren, 1933.

series of isoclinal folds in the Homestake and surrounding formations define a south-plunging synclinorium. Ore shoots are rod-shaped bodies because they are restricted to Homestake Formation beds located in the cores of isoclinal folds, where cross folds intersect the isoclinal folds. Steeply dipping dikes of Tertiary age cut the Precambrian metamorphic rocks throughout the mine. Quartz-carbonate-chlorite-pyrite alteration with sphalerite, galena, chalcopyrite, realgar, and minor gold is associated with these dikes, contributing to Noble's (1950) interpretation that the ores are hydrothermal replacement deposits of Tertiary age.

Stable isotope studies by Rye and Rye (1974), including interpretations of sulfur, oxygen, hydrogen, and carbon isotopic data, indicate that the Homestake deposit is of Precambrian age, and probably of syngenetic, exhalative origin. Sulfur from sulfides in the Homestake and adjacent Poorman and Ellison Formations shows a characteristic range of isotopic values for each formation, and is probably syngenetic. Values for sulfides associated with Tertiary ore bodies, in contrast, show no correlation with host-rock lithology. Values for $\delta^{18}O$ and $\delta^{13}C$ obtained by Rye and Rye (1974) for quartz and ankerite of veins in the Homestake Formation correlate with wall-rock lithologies, indicating that the veins were deposited from fluids produced by dehydration reactions in response to conditions transitional from low-grade to medium-grade metamorphsim. Fluid inclusion studies show that the fluids contained abundant CO_2 and some CH_4. A lead-isotope study of galena samples from gold-bearing quartz veins shows that the Homestake ores formed 1.6 b.y. ago from source lead about 2.5 b.y. old (Rye and others, 1974). The conclusions of Rye and Rye (1974) and Rye and others (1974) differ from Noble's interpretation that all the Homestake mineralization is of Tertiary age and hydrothermal origin. Interestingly, however, their ideas are similar in part to earlier interpretations featuring a Precambrian age and hydrothermal origin (see, e.g., Paige, 1923; Irving and others, 1904).

Ore bodies in the Homestake mine have been developed to depths of more than 3,000 m. Clearly, an extraordinarily large ore-forming system produced the mineralization at the Homestake. The limits of the deposit may ultimately be defined by technological and economic factors rather than geologic factors.

Epithermal veins and stockworks

Epithermal precious-metal deposits occur in volcanic arc and back-arc terranes throughout the world. Because these deposits are formed at shallow depths, most with significant production are found in the Cenozoic volcanic belts of the Circum-Pacific, Caribbean, and Mediterranean regions. They show a wide range of gold-silver ratios, including both gold-rich and silver-rich types, and also show significant variations in base-metal content (Heald-Wetlaufer and others, 1983). Because many deposits contain spectacular showings of high-grade ore ("bonanza"-type deposits), they have been an attractive exploration target in the Cordillera since the sixteenth century. Several major epithermal districts in Mexico were discovered during the Spanish colonial period in the early 1500s, but the first major discovery in the United States was the Comstock Lode, in 1859. Epithermal deposits were the major source of domestic silver production until World War I, and were also an important source of gold during the same period.

Volcanic-hosted epithermal precious-metal deposits are associated with volcanic centers of all types, including stratovolcanoes, dome fields, calderas, and maars (Sillitoe and Bonham, 1984). Ore bodies occur in permeable zones, including regional and local fault systems (such as caldera ring fractures), fragmental volcanic and volcaniclastic deposits, volcanic explosion breccias associated with both domes and maars, and hydrothermal explosion breccias. Volcanic host rocks range in composition from intermediate (andesite or rhyodacite) to silicic (rhyolite), and are rarely basaltic. Ore deposition occurs during waning volcanism or shortly after volcanism ceases in the host volcanic center. The vertical extent of the productive ore horizon, whether it is found in veins, stockworks, or breccia bodies, is most commonly between 200 and 600 m. Most districts show an abrupt bottoming of ore shoots at a more or less constant elevation. Geologic considerations indicate paleodepths to top of ore were less than 500 to 600 m for most deposits. So-called hot-springs gold deposits described in recent literature (e.g., Giles and Nelson, 1982) are epithermal deposits that include hot-springs sinter in the host volcanic-sedimentary section, suggesting that they formed at very shallow depths. Because the gold in these deposits is generally located in stockworks or breccias that cut the sinter-bearing section, rather than in sinter itself, they are here considered a variety of epithermal stockwork deposits. Epithermal precious-metal deposits are grouped into two main types: quartz-adularia type and quartz-alunite type.

In the quartz-adularia deposit type, potassium feldspar (adularia) is present with quartz as a gangue mineral in the veins. Other gangue minerals include carbonate (calcite, ankerite, siderite, rhodochrosite), barite, fluorite, or manganese oxides. Altered rocks immediately adjacent to the veins often have the assemblage quartz-adularia. The quartz-adularia zone grades outward into a quartz-sericite (illite) zone, which grades in turn into an argillic zone that includes various combinations of kaolinite, illite, montmorillonite, and mixed layer illite-montmorillonite. The argillic alteration is superimposed on widespread propylitic alteration. Silver-rich deposits (Ag:Au > 30:1) usually contain acanthite as a major silver mineral, and may contain silver sulfosalts or electrum. The accompanying base metal sulfides are typically sphalerite, galena, chalcopyrite, sometimes tetrahedrite, and pyrite. Deposits that are relatively base-metal rich contain these same base-metal sulfides in greater abundance. In gold-rich deposits, native gold is important, and acanthite and electrum may also be present. The suite of base-metal sulfides remains the same, but these minerals are usually unimportant economically. The quartz-adularia type is divided into hot-spring, Creede, Comstock, Sado, and gold-silver-tellurium deposit types by Cox and Singer (1986).

In the quartz-alunite deposit type, alteration adjacent to ore bodies is advanced argillic rather than potassic. The most intensely altered rocks, which always host the ore, contain quartz and any combination of the following: alunite, pyrophyllite, diaspore, kaolinite (dickite), or zunyite. The advanced argillic assemblages give way outward to argillic (quartz-kaolinite) or phyllic-argillic (quartz-sericite-kaolinite) assemblages. These kaolinite-bearing rocks grade into argillic rocks bearing montmorillonite or mixed-layer illite-montmorillonite, or both. The argillic rocks give way to propylitized rocks that predate the main episode of alteration. Unlike quartz-adularia–type deposits, ore-mineral assemblages of quartz-alunite type deposits always include copper sulfosalts, the most prominent being enargite or luzonite-group minerals. Tetrahedrite and tennantite are also important, as are native gold, silver sulfosalts, covellite, bismuthinite, and occasionally tellurides. Pyrite is always abundant. Although this deposit type includes distinct silver-rich (Ag:Au >3:1) and gold-rich subgroups, copper is almost always a by-product. Gangue minerals other than quartz include alunite, kaolinite, and sometimes barite. Carbonates are absent.

Quartz is always an important member of the gangue assemblage in both epithermal deposit types. Quartz and its contained fluid inclusions have yielded much of the fluid composition, oxygen isotope, and hydrogen isotope data presently available for these deposits. Although this data base is limited (see Buchanan, 1981; Hayba, 1983), it shows that the fluids that deposited both gangue and ore minerals had low salinities (typically 1 to 3 weight percent, maximum 14 weight percent NaCl equivalent), and were dominated by meteoric water (less than 10 percent magmatic water component). Sulfur isotope data for various sulfides (Hayba, 1983) show that sulfur is usually of magmatic origin. Fluid inclusion temperatures during main-stage ore deposition show a range of several tens of degrees Centigrade, between 220°C and 290°C, with medium temperature most commonly around 250°C. Evidence of boiling has been observed in at least a dozen deposits.

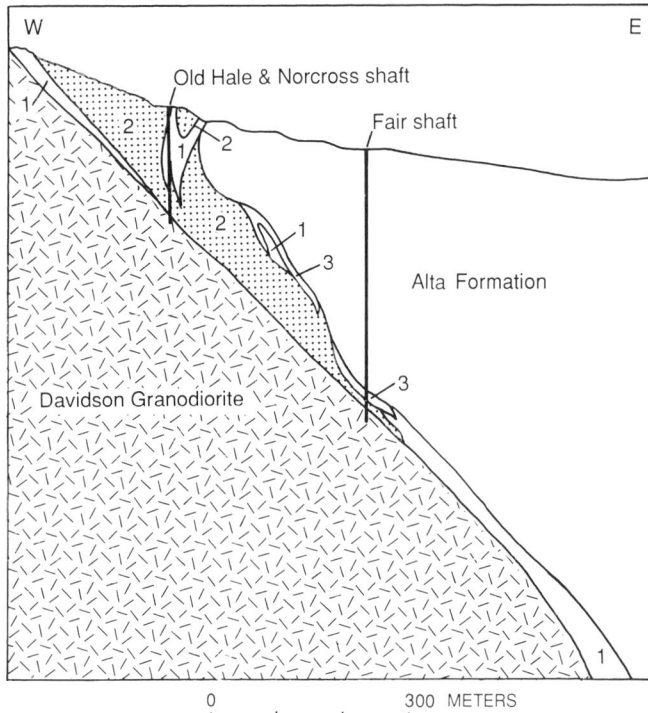

Figure 3. Cross section of the Comstock lode, Nevada, at the Hale and Norcross shaft. Features of the Comstock fault zone and associated hanging-wall splays include: (1) quartz vein and stockwork ore, (2) crushed and altered wall rock, (3) high-grade (bonanza) ore. Modified from Becker, 1882.

Epithermal precious-metal deposits occur in all of the western states, with the possible exception of Wyoming. They are numerous throughout the Basin and Range Province of Nevada, southern Idaho, western Utah, eastern California, western and southern Arizona, and southwestern New Mexico. A few districts are found in the Cascade volcanic arc in Oregon and Washington, and a few, including the new McLaughlin deposit, are associated with very young volcanic rocks and hydrothermal systems in the California Coast Ranges. Only one relatively small district is known in the Aleutian arc. Deposits are also associated with Tertiary volcanic fields and volcanic centers in cratonal terrane of Montana, South Dakota, Colorado, and New Mexico. Several well-known and important deposits are in the latter group, including Creede, in the San Juan volcanic field of southwestern Colorado, and Cripple Creek, which is a mineralized diatreme complex in the Front Range of Colorado (Thompson and others, 1985). The Comstock Lode is described here as a prominent example of the epithermal deposit type, having many features typical of the quartz-adularia deposit subtype.

The Comstock Lode district is located in calc-alkalic volcanic rocks of Miocene age in the Virginia Range, immediately southeast of Reno, Nevada. The section includes porphyritic flows, breccias, and tuffs, mainly of rhyodacitic and trachyandesitic composition (Thompson, 1956; Thompson and White, 1964). Large ore bodies were located along the plane and in hanging-wall splays of the Comstock fault, a north-trending, east-dipping normal fault (Becker, 1882; see Fig. 3). Other normal faults in the vicinity, including the Silver City and Occidental faults, host some ore, but the Comstock fault has the largest strike length (11 km), the largest offset (760 m), and the largest ore bodies. The Davidson Granodiorite, a hypabyssal intrusion, cuts the volcanic pile immediately west of the Comstock Lode, and locally forms the footwall of the Comstock fault. Wall-rock alteration is not only extensive around the Comstock Lode, but is widespread in the Tertiary volcanic rocks of the Virginia Range. Propylitized rocks are particularly common, and several assemblages that include chlorite, montmorillonite, albite, calcite, quartz, pyrite, epidote, and zeolites have been described (Whitebread, 1976). Argillized rocks bearing montmorillonite or kaolinite and, near the surface, silicified rocks consisting of quartz, alunite, pyrite, and jarosite are zonally arranged around permeable features, which include faults, fractures, and volcanic breccias. The bonanza ore bodies were quartz containing argentite, gold, stephanite, and pyrite. Polybasite, electrum, sphalerite, galena, and chalcopyrite were also important in the ores. The quartz in Comstock ores typically appears crushed; the ores have a granular texture, and banding is rare. This ore texture is unusual for epithermal deposits, which more commonly show banded, colloform, and comb textures. Adularia is not reported in the bonanza ore bodies, but does occur in quartz veins along the Comstock fault and in the Occidental vein. Maximum depth of production is about 900 m. The Comstock Lode produced 257 t (8,260,000 troy oz) of gold and 5,970 t (192,000,000 troy oz) of silver from 1859 to 1957. Efforts in the past decade to develop near-surface low-grade ore bodies have not resulted in much additional production.

Direct observations of the bonanza ore bodies recorded in the literature on the Comstock lode are limited because the most productive period (1863 to 1878) had passed by the time Becker began the first systematic geological study of the district, in 1880. The long period of subsequent minor production, however, has provided access and produced sample collections adequate for illuminating, if not exhaustive, modern studies. Determinations of oxygen and hydrogen isotopic composition of veins and wall rocks by Taylor (1973) and O'Neil and Silberman (1974) indicate that the mineralizing hydrothermal system was dominated by meteoric water, although one deep vein sample showed a significant component of magmatic water. Limited fluid inclusion data indicate that the temperature range of ore fluids was about 240° to 300°C, and the salinity around 3 weight percent equivalent NaCl. Modern studies have proved possible in many other epithermal deposits, and although individual studies are often limited in scope, the aggregate of results indicates that epithermal ore deposits are fossil equivalents of modern high-temperature geothermal systems (White, 1981; Henley and Ellis, 1983; Hayba and others, 1985).

Disseminated epithermal gold deposits

The term "disseminated" is now commonly applied to gold deposits in which very fine-grained gold is dispersed through a relatively large volume of rock. Deposits are amenable to bulk-mining methods, allowing profitable extraction from relatively low-grade ores. Although it should be possible to apply the term "disseminated" to a variety of gold deposits without regard to their genesis, the term has come into widespread use in the past 20 years for a group of ore deposits with common characteristics being actively sought and extensively developed and mined in the western United States. Most of these deposits occur in sedimentary host rocks, the most prominent being the Carlin deposit in northern Eureka County, Nevada; hence the term "Carlin-type deposit" is commonly applied to sediment-hosted examples. The name "carbonate-hosted gold-silver deposit" is also used because the sedimentary host rocks are usually calcareous (Cox and Singer, 1986). Because these deposits have many features in common with epithermal gold deposits, I prefer to call them "disseminated epithermal deposits." Some bulk-minable precious-metal deposits in volcanic host rocks also consist of fine-grained disseminated gold that is not accompanied by easily recognized alteration features or introduced minerals; such deposits are included here. If, however, a deposit includes precious metal–bearing veins, stockworks, or breccia fillings visible in the field, I include it with epithermal veins and stockworks.

Sedimentary host rocks are mostly silty limestones and dolomites, sometimes carbonaceous (see summary of features by Tooker, 1985, and Bagby and Berger, 1985). Volcanic host rocks are generally fragmental, including volcaniclastic sediments, ash-flow tuffs, flow breccias, and breccia pipes. The rocks must be permeable, having ample pore space for gangue and ore minerals. Either the rocks must have original permeability, as is the case with fragmental volcanic rocks, or they must have original components soluble in hydrothermal fluids, such as calcite and dolomite in calcareous sedimentary rocks. Owing to the critical importance of permeability, many deposits are stratiform. Faults and fractures may also be important in localizing ore because they provide channelways for hydrothermal solutions. Thus mineralization often spreads out in permeable beds around crosscutting fractures.

The most prominent manifestation of hydrothermal activity is silicification, in which carbonates are replaced by quartz and pores of the rock are filled with quartz, forming the bulk of the gangue. In sedimentary host rocks, quartz is generally accompanied by pyrite, and less commonly by barite, calcite, or fluorite. In volcanic host rocks, quartz is again generally accompanied by pyrite, also commonly by adularia, and at times by barite. Other alteration products include K-mica and clays, which may occur zonally with respect to silicification. Massive jasperoid bodies appear within some deposits and near, but not in, others. These are intensely silicified zones showing complete replacement of the original rock by hydrothermal quartz and pyrite (hematite in the oxidized zone), cut by multiple generations of quartz or chalcedony veinlets. In some sediment-hosted deposits, organic material was mobilized and redeposited in permeable zones, probably during regional heating events that predated ore deposition. In oxidized rocks the main ore mineral is native gold, generally present in grains less than 1 μm in diameter. Gold in unoxidized rocks is accompanied by pyrite, stibnite, realgar, orpiment, cinnabar, and native silver, in various proportions. Argentite appears in some volcanic-hosted deposits. Base-metal sulfides are scarce, but more common and abundant in sediment-hosted deposits. The precious-metal and sulfide mineral suite reflects the typical geochemical association found in these deposits: Au, Ag, Hg, Sb, As, and Tl.

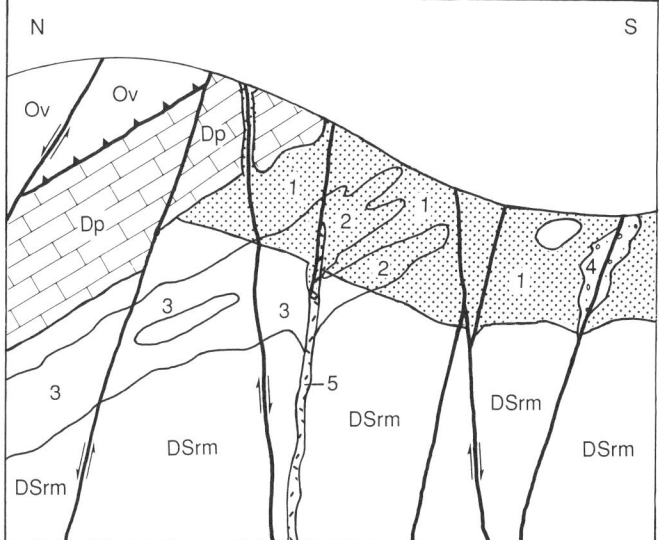

Figure 4. Simplified cross section through the main ore zone, Carlin mine, Nevada. (1) zone of leaching alteration; (2) upper part of main ore zone, oxidized ore; (3) lower part of main ore zone, unoxidized ore; (4) jasperoid body; (5) igneous dike along fault. Ov, Ordovician Vinini Formation; DSrm, Silurian-Devonian Roberts Mountains Formation; Dp, Devonian Popovich Formation. Not to scale. Modified from Radtke, 1985.

Because most disseminated epithermal deposits are relatively new and many are not yet well exposed by mining, few comprehensive studies have been completed. The Carlin will be described here because more published information presently exists than is yet available for any other deposit. Investigations are currently in progress at many deposits.

The Carlin deposit is located in silty dolomitic limestones of the Roberts Mountains Formation of Silurian-Devonian age (Radtke, 1985). The deposit lies immediately beneath the Roberts Mountains thrust, and was probably capped by chert and shale of the upper plate at the time of mineralization. Crudely stratiform ore bodies formed in favorable beds around high-angle faults (Fig. 4). Calcite and some dolomite were leached from these beds by hydrothermal fluids and replaced by quartz, pyrite, kaolinite, and K-mica. Gold was introduced with these hydrothermal minerals; most of the gold is associated with pyrite. Realgar, stibnite, cinnabar, and an uncommon suite of thallium minerals accompanied and followed gold. Prominent also at the Carlin are discordant jasperoid bodies that occur along faults; those in or near the ore bodies are gold bearing. After gold and associated sulfide minerals were deposited, carbonate was again leached from the central part of the ore zone, and pyrite and hydrocarbons were oxidized. Calcite-barite veins formed early, before silicification, and calcite veins with or without barite formed late, after gold deposition but probably before oxidation.

Mineralization at Carlin has not been directly dated, but is considered Tertiary because of the presence of mid-Tertiary igneous rocks in the vicinity, and other indirect evidence (Radtke, 1985).

Fluid inclusion studies (Radtke, 1985) indicate that low-temperature (175° to 200°C), low-salinity fluids accomplished the main-stage carbonate replacement and gold deposition. Deuterium and oxygen isotope studies by Rye (1985) indicate that the fluids were dominated by meteoric water. Barite shows much higher $\delta^{34}S$ values than sulfides. The sulfate in barite is thought to have come essentially unchanged from a sedimentary sulfate source, and the sulfide sulfur is thought similarly to have come essentially unchanged from a sedimentary sulfide source; sulfide and sulfate did not reach isotopic equilibrium in the hydrothermal fluids. The late leaching and oxidation at the Carlin are interpreted by Radtke (1985) as mainly the result of hypogene rather than supergene processes. Whether hypogene leaching and oxidation are a common feature of sediment-hosted disseminated epithermal deposits will be determined by studies in progress.

The Mercur district, in the Oquirrh Mountains of north-central Utah, was the first disseminated epithermal deposit discovered in the United States (Butler, 1920). Silver was discovered there about 1870, and gold in 1883, but a successful ore treatment was not achieved until the cyanidation process was applied in 1890, shortly after its commercial introduction in South Africa. Several other disseminated epithermal deposits were discovered before World War I, one deposit was discovered in the 1930s, and the Carlin deposit was discovered in 1962. Since the Carlin mine went into production in 1965, disseminated precious-metal deposits have been high-priority targets for exploration companies. Deposits have been found throughout the Basin and Range Province, and two have been found in Montana. Most are gold dominated, but a few in volcanic host rocks are silver dominated. Most of the approximately 75 deposits currently in production or with announced reserves are in Nevada. Present reserves total at least 2,500 t (80,000,000 troy oz) gold and 10,000 t (320,000,000 troy oz) silver. It is reasonable to assume, based on recent annual discovery rates, that more deposits are yet to be discovered, and that their geographic range may ultimately include all regions in which epithermal vein deposits are found.

Precious-metal–bearing skarns

Gold and silver are recovered commonly from skarns as by-products of base-metal production. Exploitation of skarns primarily for precious metals is much less common, but such deposits are known in Alaska, California, Colorado, Montana, Nevada, and Utah, associated with granitic stocks of Mesozoic or Tertiary age. These deposits are generally more valuable for gold than silver, and except for the deposits at Battle Mountain, Nevada, and Carr Fork, Utah (Tooker and Vercoutere, 1986), are relatively small.

Both gold and silver are most common in copper and lead-zinc skarns (Boyle, 1979), although silver is much more abundant than gold in some iron skarns. Precious metals are closely associated with sulfides, mainly pyrite, pyrrhotite, arsenopyrite, chalcopyrite, sphalerite, and galena, and less closely associated with molybdenite, magnetite, hematite, and scheelite. Precious metals generally appear late in the ore mineral paragenesis. Gangue minerals are the typical suite of calc-silicates produced by high-temperature contact metasomatism of carbonate-bearing rocks, plus products of retrograde hydrothermal alteration. Ore bodies are irregular but stratabound within calcareous beds; if beds are relatively thin, tabular ore bodies may result. Associated intrusive rocks are intermediate to silicic in composition.

The only major gold-bearing skarn deposits presently in production in the United States are those of the Battle Mountain district, Nevada. In the Copper Canyon area of the Battle Mountain district, copper skarn ore bodies with significant by-product gold and silver occur in calcareous conglomerate and limestone immediately adjacent to the Copper Canyon stock, a granodiorite intrusion of mid-Tertiary age (Blake and others, 1978). In addition to these copper-gold-silver deposits, gold-silver skarn deposits occur to the north and south of the Copper Canyon stock, as much as 1 km away from the intrusive contact, at sites where fractures cut favorable beds (Fig. 5). Sulfide minerals and gold are found in limestone, calcareous siltstone, and the limy matrix of calcareous conglomerate, all of which have been contact metamorphosed to andradite-diopside-epidote or quartz-diopside-epidote assemblages and subsequently altered successively to actinolite/tremolite-quartz and chlorite-quartz-epidote-clay assemblages (Blake and others, 1978; Theodore and others, 1986; Wotruba and others, 1987). The ore bodies are mainly mantos located near the transition zone between the calc-silicate–altered host rocks and unaltered calcareous sedimentary rocks. Pyrrhotite and pyrite are abundant in gold-bearing rocks. Other less abundant sulfides include chalcopyrite, arsenopyrite, sphalerite, galena, and marcasite. Sulfides were introduced during both the actinolitic and chloritic alteration stages. Native gold is associated with pyrrhotite and pyrite. Silver is mostly associated with galena but also forms electrum. Ore deposits of various types show a zonal distribution around the stock (Roberts and Arnold, 1965), with the gold-silver deposits located in a belt outside the copper-gold-silver deposits, and inside a peripheral zone of lead-zinc-silver deposits. Although both base and precious metals in all ore bodies were introduced after skarn formation, mineralization was a continuous process that evolved from thermal metamorphism to hydrothermal alteration and metallization (Blake and others, 1978; Theodore and others, 1986). The entire process was related to emplacement and cooling of the Copper Canyon stock.

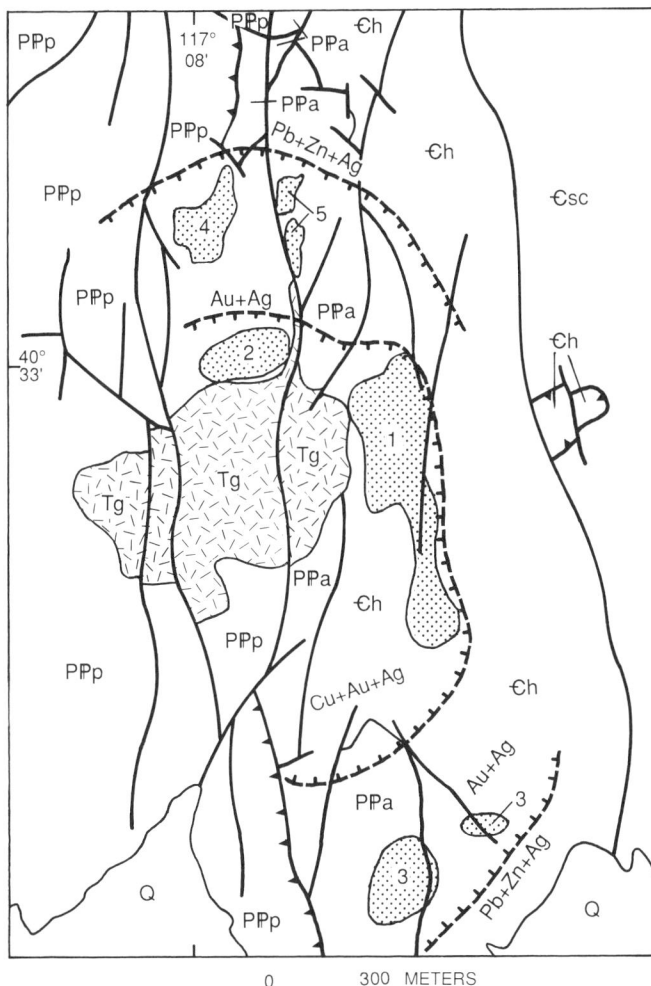

Figure 5. Simplified geologic map of the Copper Canyon area, Nevada, showing district zoning of metals. (1) East ore body (Cu-Au-Ag); (2) West ore body (Cu-Au-Ag); (3) Tomboy-Minnie deposits (Au-Ag); (4) Lower Fortitude deposit (Au-Ag); (5) Upper Fortitude deposits (Au-Ag). ∈sc, Cambrian Scott Canyon Formation; ∈h, Cambrian Harmony Formation; PℙPa, Pennsylvanian-Permian Antler sequence (includes Pennsylvanian Battle Formation, Pennsylvanian-Permian Antler Peak Limestone, and Permian Edna Mountain Formation); PℙPp, Pennsylvanian-Permian Pumpernickel Formation (upper plate of Golconda thrust); Tg, altered granodiorite of Copper Canyon; Q, Quaternary basalt flows and alluvial deposits. Heavy lines are faults (thrust faults shown with sawteeth on upper plate); light lines are contacts; dashed lines with hachures show boundaries between metal zones. Modified from Wotruba and others, 1987.

Many porphyry copper and copper-molybdenum deposits in the western United States, especially those in Arizona and New Mexico, have skarns associated with mineralized plutons, and most of these deposits have yielded more or less by-product gold and silver. Although gold and silver are relatively abundant in the Battle Mountain system, studies to date indicate that neither the geologic environment nor the processes involved in forming the gold-silver deposits there are unusual for porphyry-type systems, suggesting that other porphyry districts have potential for undiscovered skarn precious-metal deposits.

Polymetallic vein and replacement deposits

Polymetallic veins have many features in common with both gold-quartz veins and epithermal veins. They consist of quartz, one or more carbonates, chlorite, and less commonly adularia, sericite, barite, or fluorite. Relative to gold-quartz veins they are sulfide-rich, and have a relatively complex suite of sulfide and sulfosalt minerals, including pyrite, arsenopyrite, chalcopyrite, galena, sphalerite, tetrahedrite-tennantite, silver sulfosalts, and other copper, lead, and bismuth sulfosalts. Wall-rock alteration varies among deposits, but is somewhat predictable knowing the dominant rock type. Deposits in volcanic or plutonic rocks show phyllic, argillic, and propylitic alteration. Deposits in metamorphic rocks may show carbonate alteration, like that common around gold-quartz veins. Vein or replacement ore bodies in carbonate rocks are accompanied by silicification or dolomitization. These deposits may be most valuable either for gold or for silver, and may be mined also for base metals. Indeed, the primary product, co-product, and by-product relationships between gold, silver, copper, lead, and zinc may vary from ore body to ore body within a district, and may vary with changing economic conditions over time.

Polymetallic deposits appear in a variety of geologic settings. Some of the most productive are associated with known or inferred copper-, copper-molybdenum-, or molybdenum-porphyry systems, commonly zonally arranged around a porphyry stock (Tintic). Where wall rocks around such porphyry systems include sedimentary carbonates, these beds often host important replacement deposits. Some districts consist of extensive carbonate-replacement ore bodies, but igneous intrusions are small and scattered or localized (Leadville, Colorado). In the western United States and Alaska, polymetallic veins are associated with many batholiths and stocks that do not have associated porphyry mineralization. These deposits are usually found within the intrusive bodies near their margins, or in adjacent contact-metamorphosed wall rocks, and most are relatively small. They may also be found in volcanic centers along volcano-tectonic structures such as caldera ring-faults or radial faults, probably above subvolcanic intrusions (Sunnyside Mine, Colorado). Finally, sedimentary or metamorphic terranes may host clusters of deposits associated with regional fault zones and subsidiary fractures; accompanying igneous rocks are minor or absent (Coeur d'Alene, Idaho). Lindgren (1933) classified most of these deposits as magmatic-hydrothermal, in the mesothermal temperature range. Of the many diverse examples of polymetallic deposits, Tintic, Utah, and Coeur d'Alene, Idaho, are described here to illustrate the broad range of deposits that have contributed significant amounts of precious metals along with base metals.

The Tintic district of Utah (including the East Tintic subdistrict) has produced 106 t (3,400,000 troy oz) of gold and 8,400 t (269,000,000 troy oz) of silver. It has been the subject of scientific studies for 90 years, from which have emerged the comprehensive reports of Lindgren and Loughlin (1919), Morris (1968), and Morris and Lovering (1979). A summary description of the district is given by Morris and Mogensen (1978).

The Tintic district is located in a thick upper Precambrian–Paleozoic shelf sequence consisting mainly of limestone and dolomite, with lesser amounts of quartzite, shale, and argillite. Structure of these rocks is dominated by north- to northwest-trending folds. The folded sequence forms part of the upper plate of the Midas thrust, which is exposed to the north of the district. Several sets of high-angle faults are associated with the folds. These structures formed during the Sevier orogeny, in Late Cretaceous time. Large parts of the district are covered by latitic and quartz latitic flows and tuffs that represent remnants of caldera fill and a composite volcano of Oligocene age. Numerous small intrusions, including quartz monzonite and monzonite porphyry stocks, were emplaced during the several Tertiary volcanic episodes. Hydrothermal alteration and mineralization accompanied the last of these stocks, which is exposed in the southwestern part of the district (Fig. 6). A set of north-northeast–trending faults formed during and after Tertiary volcanism and intrusion.

Most of the ore bodies are pipes or irregular masses that replace carbonate rocks; some are localized along faults and some are not. In addition to these replacement ore bodies, some ore forms replacement veins and fissure veins along fault zones in various rock types. Ore minerals include galena, sphalerite, argentite, a wide variety of both common and unusual copper and silver sulfosalts, some unusual lead and bismuth sulfosalts, native gold, and tellurides. Gangue minerals include quartz, barite, calcite, dolomite, and rhodochrosite. District zoning is a prominent feature of the ores, especially in the Main Tintic subdistrict (Fig. 6). Proceeding northward from the stock, one crosses successive zones of copper-gold-lead-silver, lead-silver, and lead-zinc ores. Wall-rock alteration includes early hydrothermal dolomitization of limestones and chloritization of volcanic rocks, later argillization of igneous rocks and leaching of carbonate rocks, then deposition of jasperoid along solution conduits, and finally potassic alteration. Although all alteration is related to the same hydrothermal episode, the potassic alteration represents the only alteration event closely associated in time with deposition of ore and gangue minerals; it is seen only in previously argillized rocks, where clays and chlorite are replaced by K-mica and K-feldspar.

The Coeur d'Alene district is located in northern Idaho, about 100 km east of Spokane, Washington. Production to 1984 includes 31,100 t (1 billion troy oz) silver, 15.9 t (510,000 troy oz) gold, 7,600,000 t lead, 2,900,000 t zinc, and 154,000 t copper, plus antimony and cadmium (A. V. Heyl, written communication, 1986). The district, about 40 km east-west by 10 to 15 km north-south, is roughly bisected by the Osburn fault, a major regional structure that extends west-northwest for more than 160 km from western Montana nearly to the Idaho-Washington border (Hobbs and others, 1965). The host rocks are sedimentary formations of the Belt Supergroup of Precambrian age, mainly argillite and quartzite with minor limestone and do-

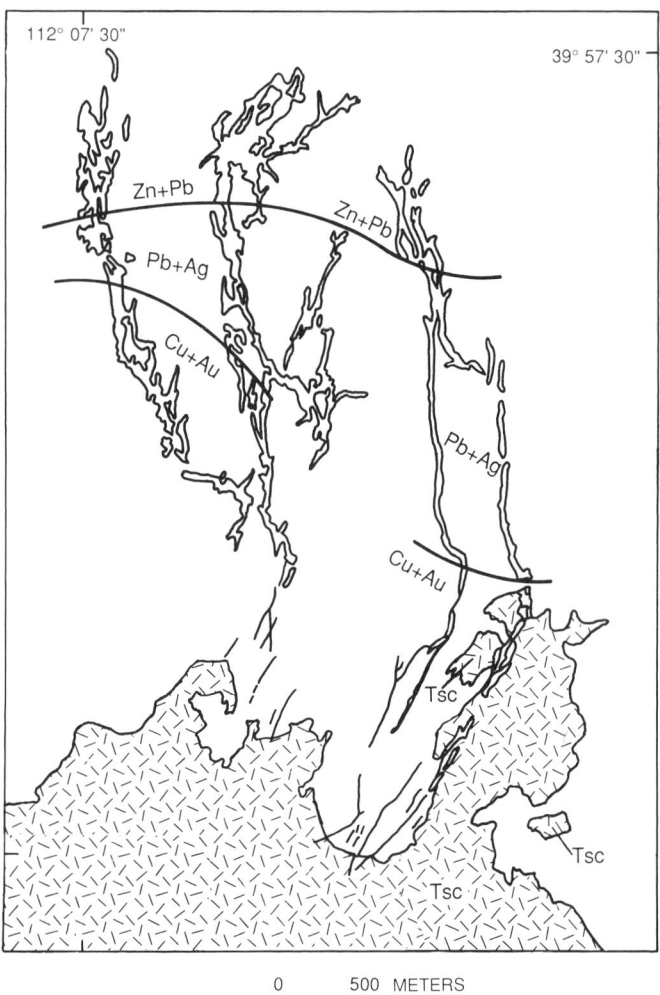

Figure 6. Map showing vertical projection of replacement ore bodies in Paleozoic sedimentary rocks (mostly carbonates), and zoning of metals, Main Tintic subdistrict, Tintic mining district, Utah. Tsc, monzonite of the Silver City stock. Heavy lines delimit zones in which the elements shown predominate in the ores. Modified from Morris, 1968; and Morris and Mogensen, 1978.

lomite. Shallow-water sedimentary structures are found throughout the Belt section, which is at least 6,200 m thick in the district. The rocks are generally fine-grained and quartz-rich; the only other important constituent in addition to quartz is sericite. Chlorite, calcite, and dolomite are each abundant locally but generally are minor. Small amounts of sulfide minerals, mostly pyrite and pyrrhotite, are common. The mineral assemblages reflect low-grade regional metamorphism.

The Belt rocks have been subjected to complex folding and faulting. South of the Osburn fault, most folds and faults trend west-northwest to northwest, parallel or subparallel to the fault. North of the Osburn fault, they trend northwest to north. Folds and faults show 26 km of right-lateral offset across the Osburn fault; most of this movement occurred after the veins formed. Hobbs and others (1965) explained the difference in trends across the Osburn fault as the result of dextral bending of northwest-trending structures preceding breakage and offset on the Osburn fault and other major right-lateral strike-slip faults to the north and south of the Osburn fault.

Igneous rocks include two groups of small stocks, mainly of monzonitic composition, located in the north-central part of the district. The two largest bodies are known as the Gem stocks, and the remaining smaller bodies form another group known as the Dago Peak stocks. They are of Cretaceous age. Lamprophyre and diabase dikes of probable Tertiary age cut the stocks, and the lamprophyre dikes locally cut the veins.

The ore bodies are shoots, generally of greater vertical than lateral dimension, in steeply dipping replacement veins consisting of siderite, quartz, biotite, chlorite, and at times ankerite, calcite, magnetite, specular hematite, or minor barite. Most of the quartz in the veins represents recrystallized detrital quartz in the wall rocks; relatively late vein-filling quartz is minor. Sulfide minerals include pyrite, galena, sphalerite, tetrahedrite, and chalcopyrite (Hobbs and Fryklund, 1968). Other main-stage sulfide minerals found in small amounts or restricted to parts of the district include pyrrhotite, arsenopyrite, boulangerite, bournonite, gersdorffite, and silver sulfosalts. The silver is carried mainly by tetrahedrite, but also by galena. Around the Gem stocks, amphiboles (grunerite and hornblende) are important in the veins, magnetite is particularly abundant, and garnet is a minor mineral. The veins generally follow north-northwest- to north-trending faults subsidiary to the major faults. Mineralization is complex in detail: the veins have been grouped geographically into a dozen belts, grouped chronologically into as many as six periods of mineralization, and divided into many paragenetic stages (Fryklund, 1964). Over the past 80 years, interpretations have varied regarding whether the main-stage lead-zinc-silver mineralization predates, postdates, or is directly associated with the Gem stocks (compare, for example, Ransome and Calkins, 1908; Fryklund, 1964; Landis and others, 1984), and on the most important features of district zoning and their significance.

The only widespread indication of hydrothermal alteration away from the veins is bleaching, which involves removal of hematite, magnetite, chlorite, and possibly carbonaceous material in the Belt rocks, and conversion of some magnetite, hematite, and locally pyrite to goethite (Paul L. Weis, in Fryklund, 1964). Bleached areas occur mostly in the mineral belts, including parts or all of single veins or vein systems. Thus, this weak alteration is geographically associated with mineralization, but shows no consistent geometric relation to the veins, and is probably earlier than the main-stage mineralization. Hydrothermal chlorite occurs in wall rocks beyond the edges of the veins as defined by the limits of quartz recrystallization, but the outer limits of hydrothermal chlorite halos are apparently not easy to delineate in the field. Near the Gem stocks, biotite and garnet are disseminated and form veinlets in zones as much as a few tens of meters wide around the veins. Hydrothermal alteration that is zonally and genetically related to the main-stage sulfide-bearing veins is not a conspicuous feature of the Coeur d'Alene district.

Recent studies by Landis and others (1984) of fluid-inclusion phase relations and compositions, strontium-isotope composition of fluid-inclusion waters, and sulfur-isotope composition of host rocks and vein sulfides indicate that the ores of the Coeur d'Alene district were deposited from metamorphic fluids generated during regional greenschist-facies metamorphism. The fluid-inclusion studies indicate fluid temperature of 350°C, pressure more than 1 to 2 kilobars, and low salinity. Landis and others (1984) inferred that mineralization occurred at or just past the peak of regional metamorphism as the response of the rocks to stress changed from plastic to brittle behavior. Meager geochronologic data from this and earlier studies indicates that the metamorphism and vein formation took place in late Precambrian time. Most of the lead in the ore deposits is of Precambrian age (Zartman and Stacey, 1971). Geochemical dispersion patterns and contact metamorphic minerals in veins around the stocks indicate some remobilization of ore-related elements during the Cretaceous thermal event associated with the monzonitic stocks (Gott and Cathrall, 1980; Landis and others, 1984).

It is unlikely that any major polymetallic district remains to be discovered in the United States. The major districts, however, represent huge accumulations of metals in numerous ore bodies. Because physical and chemical ore controls are complex, exploration for new ore bodies is tedious and costly, but clearly facilitated by detailed geologic information and understanding of ore-forming processes. As long as one or more of the many commodities found in these districts are of commercial interest, they can yield new ore bodies for many years to come.

Gold associated with gneissic metamorphic rocks

Recently, several bulk-minable gold deposits with minor silver have been discovered in medium-grade gneisses in southeastern California, including the Mesquite and Picacho mines in the Chocolate Mountains, and several mines and prospects in the Cargo Muchacho Mountains (see summary by Wilkinson and Wendt, 1986). Many other prospects in southeastern California and western Arizona are currently being explored and evaluated. The host gneisses include both metamorphosed plutonic and supracrustal rocks of volcanic origin. Protolith ages are Jurassic.

Several types of ore occur in these deposits in varying proportions. Most common is intensely brecciated gneiss cemented with limonite, carbonate (generally calcite), and clay. Other ore types include gold disseminated in sulfide-bearing gneiss and gold-bearing quartz veins. Ore currently being developed and mined is mostly oxidized, so detailed comparisons of paragenetic relationships between ore types are not yet possible. Pyrite is common to unoxidized ores of all types, along with lesser amounts of chalcopyrite, sphalerite, and galena. Gold occurs free and associated with sulfides, especially pyrite and chalcopyrite. Several other copper-bearing sulfides and native silver have been reported. Sericitization and chloritization are associated with all ore types. Biotitization and feldspathization have been reported in some disseminated ore bodies. In the Cargo Muchacho district, kyanite-bearing quartz-muscovite schist and quartzite, locally containing tourmaline, are spatially associated with gold mineralization. Rock textures indicate that these aluminous rocks are the result of post-kinematic metasomatism, but the relation of this alteration to gold mineralization is uncertain. Fluorite, calcite, or scheelite occur locally in some veins, and hematite and barite occur in some breccia ore bodies.

Ore bodies of all types are generally tabular, with low to moderate dips; breccia ore bodies follow both high-angle and low-angle fault zones, and veins and disseminated ore bodies follow low-angle faults or foliation in the host gneiss. The low-angle structures include detachment faults of mid-Tertiary age and thrust faults of Mesozoic age. The high-angle faults are generally the latest structures and therefore must be of Tertiary age. At Mesquite they may be related to the San Andreas fault.

Several ideas for genesis of these deposits are summarized by Wilkinson and Wendt (1986) and by Tosdal and Smith (1987). These include: (1) syngenetic gold concentration with some redistribution during Mesozoic and Tertiary igneous and hydrothermal events, (2) hydrothermal introduction of gold during Jurassic plutonism and regional metamorphism, (3) hydrothermal introduction associated with peraluminous granitic intrusive rocks of early Tertiary age (Wilt and Keith, 1984), and (4) hydrothermal introduction associated with mid-Tertiary detachment faulting. Regarding the latter hypothesis, it is important to note that elsewhere in southeastern California and western Arizona, several types of hydrothermal mineral deposits, including copper, iron, and manganese, as well as some gold-silver deposits, are associated with detachment faults, both in brecciated metamorphic host rocks and upper-plate Tertiary volcanic host rocks. Wilkins and others (1986) suggested that metal-bearing brines expelled from synorogenic basins during extreme crustal extension, with associated igneous activity and high heat flow, produced these deposits.

At this time, insufficient information is available to determine which of the above processes are most important or whether all possible processes have been identified. The larger deposits, such as Mesquite, show equivocal relationships to detachment faults, and it seems likely that they are the result of complex and perhaps multiple processes. The deposits are not well enough understood to delineate their geographic range, so none is shown on Plate 1a. If gold must be preconcentrated by syngenetic, metamorphic, or magmatic processes and further concentrated by processes associated with detachment faulting, the favorable terrane could prove to be limited. If, on the other hand, the deposits include several genetic types, each might have sizeable prospective areas within the Basin and Range Province and perhaps elsewhere.

Sandstone silver deposits

Silver Reef, Utah, is the only known sandstone silver deposit, but several other occurrences of this type are known in southern Utah (Heyl and others, 1973). The Silver Reef ore bodies are tabular and stratabound in sandstone beds and lenticular channels. The main silver mineral is cerargyrite. The ores also contain copper in malachite and azurite, and uranium and vanadium in carnotite. Particularly high silver grades are associated with carbonaceous material and with pyritic concretions now oxidized to limonite. Heyl and others (1973) suggested that in addition to oxidized ores such as those mined at Silver Reef, unoxidized sandstone-hosted silver ore bodies should exist. They also suggested that sandstone silver deposits are a variant of sandstone-hosted uranium and copper-uranium deposits of the Colorado-Plateau type, and the potential for discoveries in the Colorado Plateau and elsewhere in the western United States may be significant.

Deposits with by-product gold or silver

Many ore deposits are exploited for more than one metal, and some are exploited for both metals and nonmetallic mineral commodities. As demand for various commodities changes with time, and price differentials between metals vary, the relative economic importance of individual metals in polymetallic deposits varies. Deposits are usually sought, developed, and exploited, however, on the basis of current and projected demand, price, and supply of only one or two commodities. Here gold and silver are considered by-products if the deposit in question was developed on the basis of economic projections made for another commodity. It is implied that the deposit would never have been and probably would not now be developed to recover gold or silver alone, regardless of increased present value of precious metals in the ore. In most cases, continued production of a deposit depends on demand and price for the primary product, but in some deposits, owing to large increases in prices of precious metals relative to base metals in recent years, base and precious metals have become co-products of approximately equal value.

Because gold and silver show important similarities in geochemical behavior, they have a consistent mutual by-product or co-product relationship in many types of deposits. Silver is a by-product of all lode gold deposits, regardless of type, and is also a by-product of all placer deposits, even though placers usually have a higher abundance ratio of gold to silver than associated source lode deposits. Epithermal vein and stockwork deposits span a complete spectrum from gold-rich varieties with by-product silver to silver-rich varieties with by-product gold, although relatively large numbers of deposits cluster at silver-gold ratios less than 1:1 and greater than 90:1 (Graybeal, 1981), suggesting that at least two different sets of chemical controls apply in epithermal deposits. Disseminated epithermal deposits in sedimentary host rocks are mostly gold rich, whereas disseminated epithermal deposits in volcanic host rocks may be gold rich or silver rich. In polymetallic deposits with silver or gold or both as by-products, any of the base metals copper, lead, and zinc may be the primary product.

Whereas all gold deposits have some silver, several types of primary silver deposits have virtually no gold, including the sandstone silver deposits described above, cobalt-silver veins, and cobalt-uraninite silver veins. There are no significant cobalt-bearing silver veins in the United States, and little potential for future discoveries.

The remainder of this section is devoted to deposits containing by-product gold or silver, or both, in which neither metal is usually a primary product.

Porphyry copper and copper-molybdenum deposits. For almost 40 years beginning after World War II and continuing until very recently, porphyry-type deposits have been a major source of both gold and silver in the United States. From the late 1940s to the mid-1970s, the heyday of U.S. copper production, porphyry deposits supplied approximately 40 percent of domestic gold production and 20 percent of domestic silver production. The open-pit mine on the copper-molybdenum porphyry deposit at Bingham, Utah, accounted for about half the by-product gold and 10 percent of the by-product silver produced during this period. Since 1981, the percentage of domestic gold and silver production supplied by porphyry deposits has declined.

Significant gold and silver resources are found in copper and copper-molybdenum porphyry systems; molybdenum-rich porphyry systems contain little of these metals. Large porphyry systems may contain considerable amounts of gold and silver, even if precious-metal grades are not unusually high. Bingham has yielded about 545 t (17,500,000 troy oz) of gold and more than 7,800 t (250,000,000 troy oz) of silver from porphyry and associated vein, replacement, and skarn ores (Tooker and Vercoutere, 1986; A. V. Heyl, written communication, 1986). Features of porphyry systems are summarized by Tooker elsewhere in this volume.

Detailed information on gold and silver grades is not available for the majority of porphyry-type deposits, and information on the distribution of gold and silver within single deposits is scarce. In most deposits, precious metal values are too low relative to copper and molybdenum values to have any significant implications for mining strategy or mill operation; therefore, assays are not done routinely during mining. Gold and silver follow copper through concentration and smelting, are separated from copper during electrolytic refining, and must be recovered from the electrolytic anode slime by a precious-metals refinery. Thus the precious metals are ultimately separated from large volumes of copper ore, and grade variations in the original material cannot be traced. Sparse precious-metals data for porphyry copper deposits compiled by Gilmour (1982) show that gold grades range from 0.007 to 0.2 g/t (0.0002 to 0.006 troy oz/short ton), and silver grades range from about 0.3 to 3 g/t (0.01 to 0.1 troy oz/short ton). The Copper Basin porphyry copper ore body at Battle Mountain, Nevada, is unusually rich in precious metals, with gold and silver grades about double the maximums of the above ranges.

Langton and Williams (1982) show that concentrations of gold and silver in the Dos Pobres ore body, Arizona, correlate closely with copper content. They found blebs of sylvanite and hessite in bornite, and concluded that quartz-orthoclase-apatite-bornite-chalcopyrite veinlets that formed during the main pulse of mineralization carry most of the precious-metal content of the deposit.

Sillitoe (1979) reviewed data for relatively gold-rich copper porphyry deposits from various parts of the world, and concluded that gold, occurring at least in part as native metal, is associated with chalcopyrite and bornite in feldspar-stable potassic alteration zones, so that gold grade correlates with copper grade. Gold-rich deposits are characterized by unusually large amounts of magnetite (as much as 10 percent) in the potassic zone, suggesting that where mineralizing plutons crystallize and generate hydrothermal fluids under relatively high f_{O_2}/f_{S_2} conditions, circumstances are particularly favorable for gold deposition.

Polymetallic vein and replacement deposits. These deposits have been described above in the course of discussion of primary-product gold and silver deposits. The purpose of additional comment here is to point out several districts that have been particularly important producers of by-product precious metals.

The phenomenally large concentration of metals at Butte, Montana, forming a strongly zoned district containing porphyry copper-molybdenum, copper-zinc vein, and copper-zinc-lead-silver vein and replacement deposits (see Tooker, this volume), is the second largest producer of silver in the United States, after Coeur d'Alene, with cumulative production of more than 21,200 t (680,000,000 troy oz). About half this silver came from the central copper-molybdenum porphyry, and the remainder from the peripheral zone of polymetallic vein and replacement ores (Heyl and others, 1973). Butte has also produced 91 t (2,900,000 troy oz) of gold.

Polymetallic vein and replacement deposits have made important contributions to both gold and silver production in the Leadville and Gilman districts of Colorado; Park City, Utah; Eureka and Pioche, Nevada; Superior and Tombstone, Arizona; and Silver City, New Mexico (Heyl and others, 1973; Tooker and Vercoutere, 1986). They have been important for silver but not gold at Kennecott, Alaska, and Darwin, California. In the larger districts, such as Leadville and Gilman, silver was the primary product of lode mining until the 1890s, whereas base metals were the primary products beginning in the early 1900s.

Massive sulfide deposits. Massive sulfide deposits occur in the mobile belts or accreted terranes of both the Cordillera and Appalachians, and also in Precambrian terranes in both the north-central and western United States (see Tooker, this volume). These deposits have been exploited mainly for copper and zinc, but many contain recoverable quantities of precious metals. The largest producer is Jerome, Arizona, with 49 t (1,570,000 troy oz) of gold and about 1870 t (60,000,000 troy oz) of silver (Tooker and Vercoutere, 1986; Heyl and others, 1973). Another significant producer in the Precambrian metasedimentary and metavolcanic rocks of central Arizona is the Big Bug (Iron King) district. The second largest producer is the West Shasta district of northern California, with 16 t (520,000 troy oz) of gold and about 930 t (30,000,000 troy oz) of silver. More than half of this production came from a gossan with relatively high gold and silver grades at the Iron Mountain mine. Gossans, representing weathered and leached parts of pyritic massive sulfide bodies, have accounted for much of the precious-metals production from other smaller districts.

The Appalachian region in Canada includes some large massive sulfide deposits. Most Appalachian deposits known in the United States are relatively small and have not produced much in the way of precious metals, with the exception of Ducktown, Tennessee, which has produced about 0.5 t (17,000 troy oz) of gold and about 125 t (4,000,000 oz) of silver (Tooker and Vercoutere, 1986; A. V. Heyl, written communication, 1986).

Several massive sulfide ore bodies, including the large ore body at Crandon and some smaller deposits, have been discovered recently in the Precambrian of northern Wisconsin. These ore bodies, if and when they are put into production, are expected to yield by-product silver and gold. Grade and reserve figures have not been released.

In spite of the small size of most massive sulfide deposits (Cox and Singer, 1986), mining companies have recently explored both sulfide ore bodies and gossans in many districts in both the eastern and western United States, hoping to exploit them primarily for gold. Few deposits, however, have been brought into production.

Mississippi Valley–type lead-zinc-fluorite deposits. Several of the stratiform galena-sphalerite-fluorite deposits found in Paleozoic carbonate rocks of the central United States are sources of by-product silver, including the Southeast Missouri, Upper Mississippi Valley, and Illinois-Kentucky districts (Heyl and others, 1973). The lead belt of southeastern Missouri has produced around 50 to 65 t (1,500,000 to 2,000,000 troy oz) of silver per year in recent years. Similar ore bodies in the Metaline district of northeastern Washington have also produced significant by-product silver. In these deposits, more silver is often associated with sphalerite than with galena or any other sulfide mineral. For a description of the geology of Mississippi Valley–type deposits, see Ohle (this volume).

Sandstone- and shale-hosted copper deposits. Deposits of this type are important sources of base metals in other parts of the world, particularly Europe. They consist of stratiform bodies of disseminated copper minerals in clastic sedimentary rocks, with no associated igneous rocks. Many are important sources of by-product silver, but none are important for gold. Two prominent examples occur in the United States: the stratiform deposits in Precambrian shale near White Pine, Michigan, and the recently opened Troy deposit in sandstone of the Precambrian Belt Supergroup in northwestern Montana.

At White Pine, Michigan, chalcocite and some native copper are finely disseminated in shale and siltstone with some interbedded sandstone (see description in Tooker, this volume). The copper-bearing section is less than 10 m thick. The only known silver mineral is native silver, which tends to be concentrated in copper-rich beds, and is most conspicuous in the vicinity of native copper. Much of the silver produced has been incorporated in the White Pine Lake–brand copper, which contains 12 troy oz or more of silver per short ton (410 g/t; Heyl and others, 1973).

The Troy mine has been one of the top silver producers in the United States since it opened late in 1981, yielding about 125 t (4,000,000 troy oz) of silver with 18,000 t of co-product copper per year. The ore consists of quartzite with intergranular bornite, digenite, chalcocite, and native silver (Hayes, 1984). The quartzite represents metamorphosed fluvial or shallow-marine sands; ore minerals were deposited in the sands during diagenesis.

Modest deposits and occurrences of sandstone-hosted copper are widespread in the United States, found in Mesozoic sandstones in and around the Colorado Plateau and in sandstones of the eastern Triassic basins, as well as in the Belt Supergroup south of Troy. Silver is unusually abundant in the Belt deposits relative to the deposits of the Southwest and Northeast; therefore the best potential for new deposits of this type appears to be in the Belt basin of Montana and Idaho.

Shale-bearing sedimentary sequences, many of which are metalliferous, are widespread in many parts of the United States. Outside of northern Michigan, however, shale-hosted copper-silver deposits are a rarity. In contrast, shale-hosted lead-zinc deposits are of great current interest, with the recent discovery of large sedimentary-exhalative lead-zinc deposits at Lik and Red Dog Creek in the western Brooks Range of Alaska. Present estimates of silver reserves in these deposits total more than 7,800 t (250,000,000 troy oz; Nokleberg, this series).

Native copper deposits. The native copper deposits in Precambrian basaltic flows and conglomerates of northern Michigan (see Tooker, this volume) contain significant amounts of native silver, and the amount of by-product silver is substantial, owing to the large volume of ore treated (about 330,000,000 t). According to Heyl and others (1973) and A. V. Heyl (written communication, 1986), silver yield averaged about 15 troy oz for each short ton (510 g/t) of copper metal, and recorded production is probably much less than actual production because as much as 8 troy oz of silver was left in each short ton (275 g/t) of copper metal to make the Lake-brand copper produced. Heyl estimates silver production at approximately 1,490 t (48,000,000 troy oz). With average copper grade of 1.48 percent (White, 1968), the average silver grade would be about 7 g/t (0.2 troy oz/short ton). Amygdaloidal basalt ore bodies were relatively rich in silver, and conglomerate ore bodies relatively lean.

Layered mafic-ultramafic intrusions. Small amounts of gold and silver are found in the Stillwater Complex of Montana and the Duluth Complex of Minnesota, associated mainly with low-grade nickel-copper mineralization in both cases. Owing to the large sizes of these presently uneconomic deposits, amounts of by-product precious metals are potentially large. Specific grade and tonnage information, however, are not presently available.

THE FUTURE

The future of gold and silver mining and associated exploration depends mainly on economic factors that are beyond the scope of this geological summary. The assumption made here is that strong demand for precious metals will continue, and that prices will be established by the free market. It follows that the price of gold should at least keep pace with inflation, and the price of silver may improve somewhat.

The United States is well endowed with gold and silver, having about 9 percent of world gold reserves, 12 percent of total world gold resources, 17 percent of world silver reserves, and perhaps as much as 25 percent of world silver resources (Lucas, 1988b; Reese, 1985, 1988; Heyl and others, 1973). The figures for U.S. gold are more impressive when one considers that more than half of world gold reserves and about half of total world gold resources reside in the Republic of South Africa.

With most metals, production is dominated by a few relatively large deposits or districts. Much of the present surge of gold production in the United States is due to numerous relatively small deposits, mostly disseminated epithermal deposits containing less than 25 t (800,000 troy oz) gold, discovered in the last decade in response to higher gold prices. Although the present production surge should last well into the 1990s, and new surges of discovery and production could develop in the future, the level of U.S. production in the long term will be determined by the number of large mines and districts with gold endowment in the range of hundreds of metric tons (more than 3,000,000 troy oz), that can operate for decades. The Carlin trend in Nevada has developed into a district of this magnitude. How many others will emerge from the scores of new discoveries in the western United States remains to be seen.

The fate of silver production from primary-product mines in the United States will be tied mainly to economic conditions in the Coeur d'Alene district for years to come. Although other primary silver districts, mainly with disseminated epithermal ores, are gaining importance, many new major discoveries would be required to sustain the 350 to 600 t/year (11,000,000 to 20,000,000 troy oz/year) level of production that has been typical of the Coeur d'Alene district for many years. The fate of major components of the large by-product silver resources of the United States is presently in doubt because the host base-metal deposits are uneconomic to mine. However, the ratio of silver to base metals produced from base-metal deposits with by-product precious metals may rise in the future; both old and new base-metal deposits may not be mined unless they contain relatively large amounts of precious metals.

REFERENCES CITED

Bagby, W. C., and Berger, B. R., 1985, Geologic characteristics of sediment-hosted, disseminated precious-metal deposits in the western United States, *in* Berger, B. R., and Bethke, P. M., eds. Geology and geochemistry of epithermal systems: Reviews in Economic Geology, v. 2, p. 169–202.

Becker, G. F., 1882, Geology of the Comstock lode and the Washoe district: U.S. Geological Survey Monograph 3, 422 p.

Blake, D. W., Theodore, T. G., and Kretschmer, E. L., 1978, Alteration and distribution of sulfide mineralization at Copper Canyon, Lander County, Nevada: Arizona Geological Society Digest, v. 11, p. 67–78.

Böhlke, J. K., and Kistler, R. W., 1986, Rb-Sr, K-Ar, and stable isotope evidence for the ages and sources of fluid components of gold-bearing quartz veins in the northern Sierra Nevada foothills metamorphic belt, California: Economic Geology, v. 81, p. 296–322.

Boyle, R. W., 1968, The geochemistry of silver and its deposits: Geological Survey of Canada Bulletin 160, 264 p.

——, 1976, Mineralization processes in Archean greenstone and sedimentary belts: Geological Survey of Canada Paper 75-15, 45 p.

——, 1979, The geochemistry of gold and its deposits: Geological Survey of Canada Bulletin 280, 584 p.

Buchanan, L. J., 1981, Precious metal deposits associated with volcanic environments in the Southwest: Arizona Geological Society Digest, v. 14, p. 237–262.

Butler, B. S., 1920, Camp Floyd or Mercur district, *in* Butler, B. S., Loughlin, G. F., Heikes, V. C., and others, eds., The ore deposits of Utah: U.S. Geological Survey Professional Paper 111, p. 382–395.

Clark, W. B., 1970, Gold districts of California: California Division of Mines and Geology Bulletin 193, 186 p.

Coveney, R. M., Jr., 1981, Gold quartz veins and auriferous granite at the Oriental mine, Alleghany district, California: Economic Geology, v. 76, p. 2176–2199.

Cox, D. P., and Singer, D. A., 1986, Mineral deposit models: U.S. Geological Survey Bulletin 1693, 379 p.

Fryklund, V. C., Jr., 1964, Ore deposits of the Coeur d'Alene district, Shoshone County, Idaho: U.S. Geological Survey Professional Paper 445, 103 p.

Giles, D. L., and Nelson, C. E., 1982, Epithermal lode gold deposits of the circum-Pacific rim, *in* Transactions, 3rd Circum-Pacific Energy and Mineral Resources Conference, Honolulu, Hawaii, August 22–28, 1982: American Association of Petroleum Geologists, p. 273–278.

Gilmour, P., 1982, Grades and tonnages of porphyry copper deposits, *in* Titley, S. R., ed., Advances in geology of the porphyry copper deposits, southwestern North America: Tucson, The University of Arizona Press, p. 7–35.

Gott, G. B., and Cathrall, J. B., 1980, Geochemical-exploration studies in the Coeur d'Alene district, Idaho and Montana: U.S. Geological Survey Professional Paper 1116, 63 p.

Graybeal, F. T., 1981, Characteristics of disseminated silver deposits in the western United States: Arizona Geological Society Digest, v. 14, p. 271–281.

Guild, P. W., compiler, 1981, Preliminary metallogenic map of North America: U.S. Geological Survey, scale 1:5,000,000, 4 sheets.

Hayba, D. O., 1983, A compilation of fluid inclusion and stable isotope data on selected precious- and base-metal epithermal deposits: U.S. Geological Survey Open-File Report 83-450, 24 p.

Hayba, D. O., Bethke, P. M., Heald, P., and Foley, N. K., 1985, Geologic, mineralogic, and geochemical characteristics of volcanic-hosted epithermal precious-metal deposits, *in* Berger, B. R., and Bethke, P. M., eds., Geology and geochemistry of epithermal systems: Reviews in Economic Geology, v. 2, p. 129–167.

Hayes, T. S., 1984, The relation between stratabound copper-silver ore and Revett Formation sedimentary facies at Spar Lake, Montana, *in* Hobbs, S. W., ed., The Belt, abstracts with summaries, Belt Symposium II, 1983: Montana Bureau of Mines and Geology Special Publication 90, p. 63–64.

Heald-Wetlaufer, P., Hayba, D. O., Foley, N. K., and Goss, J. A., 1983, Comparative anatomy of epithermal precious- and base-metal districts hosted by volcanic rocks; A talk presented at the GAC/MAC/GGU Joint Annual Meeting, May 11–13, 1983, Victoria, British Columbia: U.S. Geological Survey Open-File Report 83-710, 16 p.

Henley, R. W., and Ellis, A. J., 1983, Geothermal systems ancient and modern—a geochemical review: Earth-Science Reviews, v. 19, p. 1–50.

Heyl, A. V., Hall, W. E., Weissenborn, A. E., Stager, H. K., Puffett, W. P., and

Reed, B. L., 1973, Silver, *in* Brobst, D. A., and Pratt, W. P., eds., United States Mineral Resources: U.S. Geological Survey Professional Paper 820, p. 581–603.

Hobbs, S. W., and Fryklund, V. C., Jr., 1968, The Coeur d'Alene district, Idaho, *in* Ridge, J. D., ed., Ore deposits of the United States, 1933–1967—the Graton–Sales Volume: New York, American Institute of Mining, Metallurgical, and Petroleum Engineers, Inc., p. 1417–1435.

Hobbs, S. W., Griggs, A. B., Wallace, R. E., and Campbell, A. B., 1965, Geology of the Coeur d'Alene district, Shoshone County, Idaho: U.S. Geological Survey Profesional Paper 478, 139 p.

Hutchinson, R. W., and Burlington, J. L., 1984, Some broad characteristics of greenstone belt gold lodes, *in* Foster, R. P., ed., Gold '82—the geology, geochemistry and genesis of gold deposits: Rotterdam, A. A., Balkema, p. 339–371.

Irving, J. D., Emmons, S. F., and Jaggar, T. A., Jr., 1904, Economic resources of the northern Black Hills: U.S. Geological Survey Professional Paper 26, 222 p.

Johnston, W. D., Jr., 1940, The gold quartz veins of Grass Valley, California: U.S. Geological Survey Professional Paper 194, 101 p.

Knopf, A., 1929, The Mother Lode system of California: U.S. Geological Survey Professional Paper 157, 88 p.

Koschmann, A. H., and Bergendahl, M. H., 1968, Principal gold-producing districts of the United States: U.S. Geological Survey Professional Paper 610, 283 p.

Langton, J. M., and Williams, S. A., 1982, Structural, petrological, and mineralogical controls for the Dos Pobres orebody, Lone Star mining district, Graham County, Arizona, *in* Titley, S. R., ed., Advances in the geology of the porphyry copper deposits, southwestern North America: Tucson, The University of Arizona Press, p. 335–352.

Landis, G. P., Leach, D. L., and Hofstra, A. H., 1984, Silver-base metal mineralization as a product of metamorphism—Coeur d'Alene district, Shoshone County, Idaho—concepts of genesis, *in* Hobbs, S. W., ed., The Belt, abstracts with summaries, Belt Symposium II, 1983: Montana Bureau of Mines and Geology Special Publication 90, p. 68.

Lindgren, W., 1896, The gold quartz veins of Nevada City and Grass Valley districts, California: U.S. Geological Survey 17th Annual Report, pt. 2, p. 1–266.

——, 1900, Description of the Colfax quadrangle, California: U.S. Geological Survey Geologic Atlas of the United States, Folio 66, scale 1:125,000, 10 p.

——, 1911, The Tertiary gravels of the Sierra Nevada of California: U.S. Geological Survey Professional Paper 73, 226 p.

——, 1933, Mineral deposits: New York, McGraw-Hill Book Company, Inc., 930 p.

Lindgren, W., and Loughlin, G. F., 1919, Geology and ore deposits of the Tintic mining district, Utah: U.S. Geological Survey Professional Paper 107, 282 p.

Lindgren, W., and Turner, H. W., 1895, Description of the Smartville Quadrangle, California: U.S. Geological Survey Geologic Atlas, Folio 18, 6 p.

Lucas, J. M., 1985, Gold, *in* Mineral Facts and Problems, 1985 edition: U.S. Bureau of Mines Bulltin 675, p. 323–338.

——, 1988a, Gold: U.S. Bureau of Mines Minerals Yearbook 1986, v. 1, Metals and Minerals, p. 421–458.

——, 1988b, Gold: U.S. Bureau of Mines Mineral Commodity Summaries, 1988, p. 62–63.

Marshall, B., and Taylor, B. E., 1981, Origin of hydrothermal fluids responsible for gold deposition, Alleghany district, Sierra County, California: U.S. Geological Survey Open-File Report 81-355, p. 280–293.

McKnight, E. T., Newman, W. L., Klemic, H., and Heyl, A. V., Jr., 1962, Silver in the United States: Mineral Investigations Resource Map MR-34, scale 1:3,168,000.

Morris, H. T., 1968, The Main Tintic mining district, Utah, *in* Ridge, J. D., ed., Ore deposits of the United States, 1933–1967—the Graton–Sales Volume: New York, American Institute of Mining, Metallurgical, and Petroleum Engineers, Inc., p. 1043–1073.

Morris, H. T., and Lovering, T. S., 1979, General geology and mines of the East Tintic mining distrct, Utah and Juab Counties, Utah: U.S. Geological Survey Professional Paper 1024, 203 p.

Morris, H. T., and Mogensen, A. P., 1978, Tintic mining district, Juab and Utah Counties, Utah, *in* Shawe, D. R., and Rowley, P. D., eds., Field excursion C-2, guidebook to mineral deposits of southwestern Utah: Utah Geological Association Publication 7, p. 41–47.

Noble, J. A., 1950, Ore mineralization in the Homestake gold mine, Lead, South Dakota: Geological Society of America Bulletin, v. 61, p. 221–252.

O'Neil, J. R., and Silberman, M. L., 1974, Stable isotope relations in epithermal Au-Ag deposits: Economic Geology, v. 69, p. 902–909.

Paige, S., 1923, The geology of the Homestake mine: Economic Geology, v. 18, p. 205–237.

Pardee, J. T., and Park, C. F., Jr., 1948, Gold deposits of the southern Piedmont: U.S. Geological Survey Professional Paper 213, 156 p.

Radtke, A. S., 1985, Geology of the Carlin gold deposit, Nevada: U.S. Geological Survey Professional Paper 1267, 124 p.

Ransome, F. L., and Calkins, F. C., 1908, Geology and ore deposits of the Coeur d'Alene district, Idaho: U.S. Geological Survey Professional Paper 62, 203 p.

Reese, R. G., Jr., 1985, Silver, *in* Mineral Facts and Problems, 1985 edition: U.S. Bureau of Mines Bulletin 675, p. 729–739.

——, 1988, Silver: U.S. Bureau of Mines Mineral Commodity Summaries, 1988, p. 144–145.

Roberts, R. J., and Arnold, D. C., 1965, Ore deposits of the Antler Peak Quadrangle, Humboldt and Lander Counties, Nevada: U.S. Geological Survey Professional Paper 459-B, p. B1–B94.

Rye, D. M., and Rye, R. O., 1974, Homestake gold mine, South Dakota—I. Stable isotope studies: Economic Geology, v. 69, p. 293–317.

Rye, D. M., Doe, B. R., and Delevaux, M. H., 1974, Homestake gold mine, South Dakota—II. Lead isotopes, mineralization ages, and source of lead in ores of the northern Black Hills: Economic Geology, v. 69, p. 814–822.

Rye, R. O., 1985, A model for the formation of carbonate-hosted disseminated gold deposits based on geologic, fluid-inclusion, geochemical, and stable-isotope studies of the Carlin and Cortez deposits, Nevada, *in* Tooker, E. W., ed., Geologic characteristics of sediment- and volcanic-hosted disseminated gold deposits—search for an occurrence model: U.S. Geological Survey Bulletin 1646, p. 35–42.

Schweickert, R. A., 1978, Triassic and Jurassic paleogeography of the Sierra Nevada and adjacent regions, California and western Nevada, *in* Howell, D. G., and McDougall, K. A., eds., Mesozoic paleogeography of the western United States: Society of Economic Paleontologists and Mineralogists, Pacific Section, Pacific Coast Paleogeography Symposium 2, Sacramento, California, 1976, p. 361–384.

Sillitoe, R. H., 1979, Some thoughts on gold-rich porphyry copper deposits: Mineralium Deposita, v. 14, p. 161–174.

Sillitoe, R. H., and Bonham, H. F., Jr., 1984, Volcanic landforms and ore deposits: Economic Geology, v. 79, p. 1286–1298.

Simons, F. S., and Prinz, W. C., 1973, Gold, *in* Brobst, D. A., and Pratt, W. P., eds., United States Mineral Resources: U.S. Geological Survey Professional Paper 820, p. 263–275.

Taylor, H. P., Jr., 1973, O^{18}/O^{16} evidence for meteoric-hydrothermal alteration and ore deposition in the Tonopah, Comstock lode, and Goldfield mining districts, Nevada: Economic Geology, v. 68, p. 747–764.

Theodore, T. G., Howe, S. S., Blake, D. W., and Wotruba, P. R., 1986, Geochemical and fluid zonation in the skarn environment at the Tomboy–Minnie gold deposits, Lander County, Nevada: Journal of Geochemical Exploration, v. 25, p. 99–128.

Thompson, G. A., 1956, Geology of the Virginia City quadrangle, Nevada: U.S. Geological Survey Bulletin 1042-C, p. 45–77.

Thompson, G. A., and White, D. E., 1964, Regional geology of the Steamboat Springs area, Washoe County, Nevada: U.S. Geological Survey Professional Paper 458-A, p. A1–A52.

Thompson, T. B., Trippel, A. D., and Dwelley, P. C., 1985, Mineralized veins and breccias of the Cripple Creek district, Colorado: Economic Geology, v. 80, p. 1669–1688.

Tosdal, R. M., and Smith, D. B., 1987, Some characteristics of gneiss-hosted gold deposits of southeastern California [abs.], in USGS research on mineral resources—1987, program and abstracts: U.S. Geological Survey Circular 995, p. 71.

Tooker, E. W., 1985, Discussion of the disseminated-gold-ore-occurrence model, in Tooker, E. W., ed., Geologic characteristics of sediment- and volcanic-hosted disseminated gold deposits—search for an occurrence model; U.S. Geological Survey Bulletin 1646, p. 107–148.

Tooker, E. W., and Vercoutere, T. L., 1986, Gold in the conterminous United States, perspective of 1986—preliminary map of selected geographic, economic, and geologic attributes of productive (> 10,000 oz) gold districts: U.S. Geological Survey Open-File Report 86-209, 32 p., and accompanying map, scale 1:5,000,000.

White, D. E., 1981, Active geothermal systems and hydrothermal ore deposits: Economic Geology, 75th Anniversary Volume, p. 392–423.

White, W. S., 1968, The native-copper deposits of northern Michigan, in Ridge, J. D., ed., Ore deposits of the United States, 1933–1967—The Graton-Sales volume: New York, American Institute of Mining, Metallurgical, and Petroleum Engineers, Inc., p. 303–325.

Whitebread, D. H., 1976, Alteration and geochemistry of Tertiary volcanic rocks in parts of the Virginia City quadrangle, Nevada: U.S. Geological Survey Professional Paper 936, 43 p.

Wilkins, J., Jr., Beane, R. E., and Heidrick, T. L., 1986, Mineralization related to detachment faults—a model: Arizona Geological Society Digest, v. 16, p. 108–117.

Wilkinson, W. H., and Wendt, C. J., 1986, Precious metal mineralization, stratigraphy, and tectonics in southeastern California: Arizona Geological Society Digest, v. 16, p. 267–281.

Wilt, J. C., and Keith, S. B., 1984, Metallogeny of Arizona lode gold production: Geological Society of America Abstracts with Programs, v. 16, p. 697.

Wotruba, P. R., Benson, R. G., and Schmidt, K. W., 1987, The Fortitude gold-silver deposit, Copper Canyon, Lander County, Nevada, in Johnson, J. L., and Abbott, E., eds., Bulk mineable precious metal deposits of the western United States, guidebook for field trips: Reno, Nevada, Geological Society of Nevada Symposium, April 6–8, 1987, p. 343–347.

Yeend, W. E., 1974, Gold-bearing gravel of the ancestral Yuba River, Sierra Nevada, California: U.S. Geological Survey Professional Paper 772, 44 p.

Zartman, R. E., and Stacey, J. S., 1971, Lead isotopes and mineralization ages in Belt Supergroup rocks, northwestern Montana and northern Idaho: Economic Geology, v. 66, p. 849–860.

MANUSCRIPT ACCEPTED BY THE SOCIETY OCTOBER 26, 1988

NOTE ADDED IN PROOF: GOLD AND SILVER DEPOSITS OF THE UNITED STATES—1991 UPDATE

Introduction

Considerable literature on gold deposits has appeared since this chapter was written in 1986. The main purpose of this update is to add selected recent references, but I have also included comments on exploration outlook, probable future importance, and classification schemes for various deposit types.

Most statistics presented in this chapter were revised in May, 1989, so extensive updating is not necessary. The total cumulative gold production for the United States through 1989 is about 363 million troy ounces (11,300 t), which represents 12 percent of cumulative world gold production. Total cumulative silver production through 1989 is about 5.6 billion troy ounces (175,000 t). The United States now ranks third in the world in both annual gold and silver production, and U.S. gold and silver reserves both represent 11 percent of world reserves (Lucas, 1990; Reese, 1990). Gold prices have remained relatively high since 1986. U.S. gold production more than doubled between 1986 and 1989 as many new mines opened and existing mines expanded. Silver prices, on the other hand, have been relatively low during this period. The approximately 50-percent increase in silver production during this period is attributed mainly to byproduct silver from gold mines and from base-metal mines opened or re-opened in response to improved base-metal prices.

Gold-quartz veins

In recent studies, most investigators interpret low-sulfide gold quartz veins as products of metamorphic fluids (Groves and Phillips, 1987; Goldfarb and others, 1988), but some propose that they form from deeply-circulating meteoric fluids (Nesbitt, 1988).

Exploration for these deposits continues, notably in the Sierra Nevada province and the Juneau gold belt, and many properties are in development or production in these regions.

Epithermal veins and stockworks

Berger and Henley (1989) summarize characteristics and genetic models for epithermal deposits. Bonham (1989) divides epithermal deposits into low-sulfur, high-sulfur, and alkalic types. His high-sulfur type corresponds to the quartz-alunite type described here, and his low-sulfur and alkalic types are included here with quartz-adularia deposits. Nelson (1988) presents a new summary of hot spring gold deposits.

In the present gold-mining boom the epithermal districts have been re-examined, and they are now second in importance only to sedimentary-rock-hosted deposits (Carlin type). A wide variety of volcano-tectonic environments host epithermal deposits, and many known districts contain environments that are not yet thoroughly explored. On a regional scale, the vast areas of favorable volcanic rocks that occur throughout the western United States can yield new districts, as recent discoveries in southeastern Oregon have shown.

Disseminated epithermal gold deposits

This section of the chapter is concerned mainly with sedimentary-rock-hosted (Carlin-type) deposits. Percival and others (1988) provide an updated summary for these deposits, and Berger and Henley (1989) include them in their summary of epithermal deposits. I included some volcanic-hosted disseminated deposits with this deposit type, but most such deposits can be assigned to the quartz-adularia type described in the section on epithermal veins and stockworks, based on recently published descriptions.

Early work on the origin of the sedimentary-rock-hosted deposits, some of which is cited above, indicated that they are epithermal deposits whose distinguishing characteristics result from interaction between meteoric-hydrothermal fluids and the host sedimentary rocks. Recent work suggests that variations on the epithermal (geothermal analogue) model should be considered. Hofstra and others (1990) propose that Jerritt Canyon and other deposits formed by mixing of metal-bearing isotopically evolved meteoric waters with unexchanged oxidizing meteoric waters. Sillitoe and Bonham (1990) propose that the deposits are distal products of magmatic-hydrothermal systems. Cox and Singer (1990) recognize a group of sedimentary-rock-hosted deposits that are distinctly enriched in silver, base metals, and manganese, which they also interpret as distal products of magmatic-hydrothermal systems.

The Carlin-type deposits make up the majority of new gold discoveries since 1965, and because some are very large, they now account for about 70 percent of U.S. gold reserves. Reserves in the Carlin trend alone presently exceed 40,000,000 oz (1245 t) and continue to increase. Additional major discoveries are likely, especially in other trends. The sedimentary-rock-hosted deposits will dominate the U.S. gold industry for the 1990s, and possibly much longer.

Precious-metal-bearing skarns. Although there is a large body of literature on skarn deposits, summary articles on gold-bearing skarns have appeared only recently (Meinert, 1989; Theodore and others, 1991).

Exploration for gold-bearing skarns has yielded some important new deposits, such as McCoy, Nevada. Exploration of known porphyry and polymetallic vein and replacement districts should reveal more skarn deposits, and there is potential for discovery of new districts in most of the western states.

Polymetallic vein and replacement deposits. Recent articles by Tooker (1990a, b) clarify the relative importance of polymetallic vein and replacement deposits and porphyry copper-molybdenum deposits as sources of precious metals at Bingham and Butte.

Leach and others (1988) propose a metamorphic origin for veins of the Coeur d'Alene district, which are described here in the section on polymetallic vein and replacement deposits.

Porphyry copper and copper-molybdenum deposits. Cox and Singer (1988) compiled data on gold grades in porphyry copper deposits. They found that although molybdenum is dispersed under the ore-forming conditions that tend to retain gold, in some systems neither element was efficiently dispersed, resulting in deposits with modest enrichments of both gold and molybdenum. Several of these porphyry copper-gold-molybdenum deposits occur in the western U.S., including Bingham, Utah, and Ajo, Arizona. The porphyry deposit at Bingham Canyon accounts for about 7 percent of U.S. reserves of both gold and silver (Tooker, 1990a; Reese, 1990). Copper-gold porphyry deposits, which have the highest gold grades, are scarce in the western U.S.

Massive sulfide deposits. The genetic model generally accepted for volcanogenic massive sulfide deposits involves circulation of seawater through volcanic rocks, driven by submarine volcanic systems. Although recent research has not resulted in major changes in this model, the occurrence of gold in massive sulfide deposits has received more attention in recent years. This topic is discussed in detail by Huston and Large (1989).

REFERENCES

Berger, B. R., and Henley, R. W., 1989, Advances in the understanding of epithermal gold-silver deposits, with special reference to the western United States, *in* Keays, R. R., Ramsay, W.R.H., and Groves, D. I., eds., The geology of gold deposits—the perspective in 1988: Economic Geology Monograph 6, p. 405–423.

Bonham, H. F., Jr., 1989, Bulk mineable gold deposits of the western United States, *in* Keays, R. R., Ramsay, W.R.H., and Groves, D. I., eds., The geology of gold deposits—the perspective in 1988: Economic Geology Monograph 6, p. 193–207.

Cox, D. P., and Singer, D. A., 1988, Distribution of gold in porphyry copper deposits: U.S. Geological Survey Open-File Report 88-46, 22 p.

——, 1990, Descriptive and grade-tonnage models for distal disseminated Ag-Au deposits: A supplement to U.S. Geological Survey Bulletin 1693: U.S. Geological Open-File Report 90-282, 7 p.

Goldfarb, R. J., Leach, D. L., Pickthorn, W. J., and Paterson, C. J., 1988, Origin of lode-gold deposits of the Juneau gold belt, southeastern Alaska: Geology, v. 16, p. 440–443.

Groves, D. I., and Phillips, G. N., 1987, The genesis and tectonic control on Archean gold deposits of the western Australian shield—a metamorphic replacement model: Ore Geology Reviews, v. 2, p. 287–322.

Hofstra, A. H., Landis, G. P., Leventhal, J. S., Northrop, H. R., Rye, R. O., Doe, T. C., and Dahl, A. R., 1990, Genesis of sediment-hosted disseminated gold deposits by fluid mixing and sulfidation of iron in the host rocks: Chemical reaction path modeling of ore depositional processes at Jerritt Canyon, Nevada [abs.], *in* Geology and ore deposits of the Great Basin, Program with Abstracts: Geological Society of Nevada and U.S. Geological Survey Great Basin Symposium, April 1–5, 1990, p. 55.

Huston, D. L., and Large, R. R., 1989, A chemical model for the concentration of gold in volcanogenic massive sulphide deposits: Ore Geology Reviews, v. 4, p. 171–200.

Leach, D. L., Landis, G. P., and Hofstra, A. H., 1988, Metamorphic origin of the Coeur d'Alene base- and precious-metal veins in the Belt basin, Idaho and Montana: Geology, v. 16, p. 122–125.

Lucas, J. M., 1990, Gold: U.S. Bureau of Mines Mineral Commodity Summaries, 1990, p. 70–71.

Meinert, L. D., 1989, Gold skarn deposits—geology and exploration criteria, *in* Keays, R. R., Ramsay, W.R.H., and Groves, D. I., eds., The geology of gold deposits—the perspective in 1988: Economic Geology Monograph 6, p. 537–552.

Nesbitt, B. E., 1988, Gold deposit continuum: A genetic model for lode Au mineralization in the continental crust: Geology, v. 16, p. 1044–1048.

Nelson, C. E., 1988, Gold deposits in the hot spring environment, *in* Schafer, R. W., Cooper, J. J., and Vikre, P. G., eds., Bulk mineable precious metal deposits of the western United States: Geological Society of Nevada, Symposium Proceedings, April 6–8, 1987, p. 417–431.

Percival, T. J., Bagby, W. C., and Radtke, A. S., 1988, Physical and chemical features of precious metal deposits hosted by sedimentary rocks in the western United States, *in* Schafer, R. W., Cooper, J. J., and Vikre, P. G., eds., Bulk mineable precious metal deposits of the western United States: Geological Society of Nevada, Symposium Proceedings, April 6–8, 1987, p. 11–34.

Reese, R. G., 1990, Silver: U.S. Bureau of Mines Mineral Commodity Summaries, 1990, p. 154–155.

Sillitoe, R. H., and Bonham, H. F., Jr., 1990, Sediment-hosted gold deposits: Distal products of magmatic-hydrothermal systems: Geology, v. 18, p. 157–161.

Theodore, T. G., Oris, G. J., Hammarstrom, J. M., and Bliss, J. D., 1991, Gold-bearing skarns: U.S. Geological Survey Bulletin 1930, 61 p.

Tooker, E. W., 1990a, Gold in the Bingham district, Utah, *in* Shawe, D. R., Ashley, R. P., and Carter, L.M.H., eds., Geology and resources of gold in the United States, gold in porphyry copper systems: U.S. Geological Survey Bulletin, 1857-E, p. E1–E16.

Tooker, E. W., 1990b, Gold in the Butte district, Montana, *in* Shawe, D. R., Ashley, R. P., and Carter, L.M.H., eds., Geology and resources of gold in the United States, gold in porphyry copper systems: U.S. Geological Survey Bulletin 1857-E, p. E17–E27.

Chapter 3

Copper and molybdenum deposits in the United States

Edwin W. Tooker
U.S. Geological Survey, MS 901, 345 Middlefield Road, Menlo Park, California 94025

INTRODUCTION

Copper and molybdenum resources have been abundantly available, but not completely recognized at any one time, in the United States from the colonial period to the present; they occur in a variety of types of mineral deposits. Increased use of these metals from deposit types that were exploited successively over the years may be correlated with emerging national needs that resulted from the political and economic growth of a new nation and the growth of scientific and technical knowledge. While in a broad sense, in recent years, the porphyry (or dispersed[1]) type deposits have provided the major source of ores of both commodities, they are also available in a number of non-porphyry types of deposits. Composite geologic characteristics for the several types of U.S. porphyry and non-porphyry deposits of copper and molybdenum are sketched briefly here, and reference is made to more detailed descriptions of individual deposits composing the types. However, defined deposit models for the occurrence of these commonly closely associated metals become diffuse in those cases where there is a transition from a single commodity to a coproduct or byproduct relationship or a gradation from one type of deposit into another type within or between mining districts.

Past domestic production of copper and molybdenum has contributed a substantial part of world supplies, and significant domestic reserves remain, but current (1986) U.S. production has been reduced owing to a number of economic factors. The availability of economic and subeconomic deposits, their location, characteristic deposit features, and known reserves demonstrate that the nation has adequate future resources of these metals. However, the outlook for a resumption of U.S. production at past levels seems uncertain in the near future because of world economic and political conditions. Increasing dependency on copper imports is anticipated in spite of seeming domestic abundance. Molybdenum exports will continue to fill foreign demand.

I am especially indebted to D. P. Cox, T. G. Theodore, and D. A. Singer, specialists for these commodities and the analysis of their resources, and have adopted the classification of copper and molybdenum deposit types and the descriptive format introduced in the volume edited by Cox and Singer (1986). I wish to acknowledge the following authors who contributed to this compilation (numbers of the ore deposit models that they described are in parentheses): D. P. Cox (17, 18a, 18b, 20c, 21a, 29b, 30b, 32c), G. M. Jones (18b), S. D. Ludington (16), W. D. Menzie (7a, 18b, 21b), H. T. Morris (19a), D. L. Mosier (16, 17, 19a, 21a, 24a, 28a, 30a), N. J Page (5a, 7a), D. A. Singer (7a, 16, 17, 18a, 19a, 20c, 21a, 24a, 28c, 30b), and T. G. Theodore (16, 18b, 21b). I am also appreciative of the many other detailed data and discussions, particularly of individual authors included in the volumes edited by Ridge (1968), and Titley (1982), who provide descriptions of deposit types and a historical perspective of the geologic and economic significance of copper and molybdenum resources.

GROWTH OF U.S. COPPER AND MOLYBDENUM PRODUCTION

The discovery and use of copper preceded that of molybdenum. Primitive man used copper tools as early as 4000 B.C. and later (about 2500 B.C.) combined copper with tin to make bronze; brass, an alloy of copper and zinc, has been used for about 2,000 years. The Greeks and Romans recognized the soft (leadlike) molybdenum minerals, but the sulfide mineral molybdenite was not identified until 1778. By 1898 small amounts of molybdenum were used in compounding chemicals and dyes and in hardening steel.

The emergence and growth of mineral resources exploration and development in the United States has been chronicled by Rabbitt (1979, 1980) and graphed in Figure 1. The French explorers of North America in the late 1600s knew of masses of native copper in the Lake Superior region, and the local Indians apparently had made use of the metal in adornment and implements. Mining of copper in the United States began about 1709 in Simsbury, Connecticut, and following the discovery of copper in New Jersey in 1719, a small amount was shipped to England.

---

[1]In this chapter a distinction is made between a *dispersed* deposit, in which ore minerals are scattered outward from a known subvolcanic source, and a *disseminated* deposit, in which minerals are distributed irregularly from an as yet undetermined source. Elsewhere these terms have been used interchangeably.

Tooker, E. W., 1991, Copper and molybdenum deposits in the United States, *in* Gluskoter, H. J., Rice, D. D., and Taylor, R. B., eds., Economic Geology, U.S.: Boulder, Colorado, Geological Society of America, The Geology of North America, v. P-2.

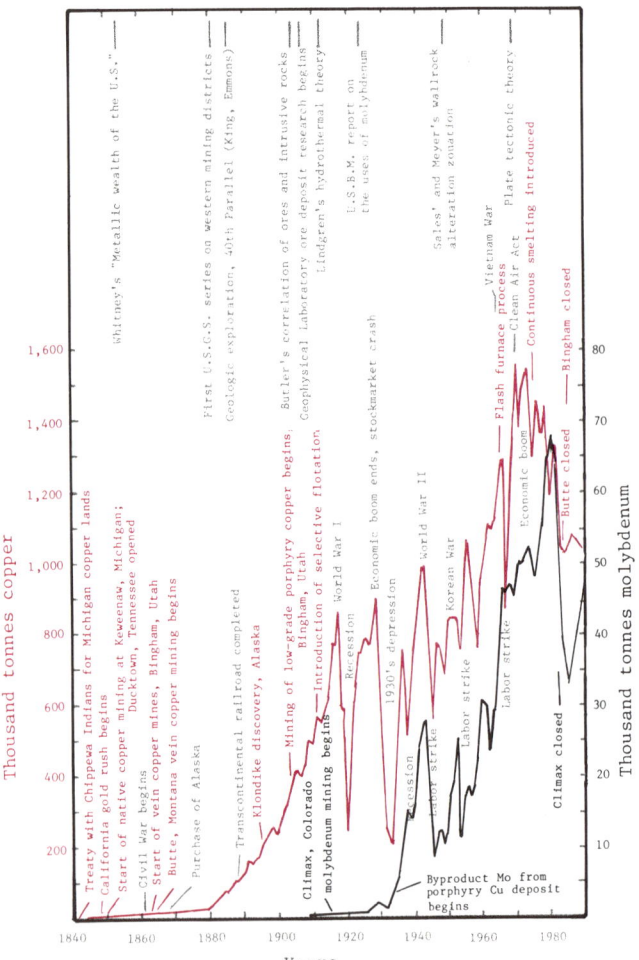

Figure 1. Production data for copper (red) and molybdenum (bold black) from 1840 to the present (U.S. Geological Survey volumes, Mineral Resources of the United States, 1840–1915; Minerals Yearbooks, 1915 to present), and annotations (screened black) of significant associated historical, economic, and political events.

In 1800, mainly because of Navy needs, Congress authorized investigation of the copper resources of the Lake Superior region, and a treaty with the Chippewa Indians in 1843 opened this region for development. By the mid 1800s, the demand for copper, although small, still could not be met from domestic sources, and raw copper ores and copper and brass manufactures were imported.

Discovery of gold placers in California in 1849 shifted the focus of resource exploration from the eastern regions of the country to the west. However, copper was discovered in Tennessee in 1850 by prospectors looking for gold. The search for the lode sources of placer gold in subsequent years spread out from the California gold fields and led to the discovery of many of the Cordilleran base and precious metal vein and replacement deposits. Decreased mining activity in England during this period also provided for a flow of emigrating trained miners, and the completion of a transcontinental railroad facilitated access to the new mining districts.

The Civil War spurred production of copper for railroad construction and for use in rifle and pistol cartridges, bronze fittings, and canned goods; by 1879 the United States had become independent of world copper and was a leading exporter. During this period more than 90 percent of the copper came from east of the 100th meridian, mainly from the Lake Superior area. By 1890, the Lake Superior area yielded first place to Montana, and Arizona had become third in the production of copper. In the case of molybdenum, it was not until about 1900 that limited amounts began to be mined intermittently in the southwest United States; lack of demand halted activity from 1906 until World War I.

Advances in geologic science and mining and metallurgical technology also provided impetus for finding and extracting metal resources and made significant contributions to the success of U.S. industrialization between 1880 and 1916 (see Joralemon, 1973). Jackling's new method for mining low-grade copper dispersed in intrusive porphyritic igneous rock at Bingham, Utah, Butler's and Lingren's hydrothermal ore deposit exploration models, and the careful descriptions of U.S. mining districts in U.S. Geological Survey maps and reports and in publications of universities and scientific and technical societies devoted to economic geology research have all contributed to the stunning success of resource development in the United States and elsewhere.

Production of copper rose rapidly, reaching a peak of nearly 900,000 t (tonnes) per year during World War I, and molybdenum production increased markedly in the early 1930s, in part as a byproduct of porphyry copper production and increased use in steel. Thus the copper and molybdenum production trends in Figure 1 are nearly parallel. A peak production of about 1,600,000 t of copper per year was reached in the 1970s, and a production high of about 68,000 t of molybdenum was achieved in the early 1980s. At the present time, both commodities are experiencing production lows, and a number of major domestic copper and molybdenum mines have been closed, although they are expected to reopen as economic conditions improve.

CURRENT SOURCES, USERS, AND RESERVES

A summary of present U.S. sources of these metals, the principal domestic users, exports, and reserves is derived from the U.S. Bureau of Mines.

Sources of Copper (Jolly, 1985, 1986)
Domestic mines, 53%, as primary, coproduct, and byproduct ores, mainly in Arizona, New Mexico, and Utah.
Imports, 23%, mostly refined or fabricated, mainly from Chile, Canada, Peru, and Mexico.
Recycled scrap, 22%, mainly from smelters and refinerys, brass mills, foundries, and chemical plants.

Users of Copper
Domestic, in building construction (39%), electrical and electronic products (25%), industrial machinery and equipment (10%), transportation (11%), and consumer products (10%).
Exports, an amount nearly equal to imports.

Sources of Molybdenum (Blossom, 1985, 1986)
Domestic mines, 92%, as primary, coproduct, or byproduct minerals, mainly in California, Colorado, Idaho, New Mexico, and Utah.
Imports, 7%, mainly from Chile, Canada, China, and Mexico.
Recycled scrap, small amounts.

Users of Molybdenum
Domestic, 28% of mine production—75% of which is used in iron and steel industries. Major end-use applications occur in machinery (35%), oil and gas (20%), transportation (15%), chemicals (15%), electrical (10%), and others (5%).
Exports, about 72% of mine production.

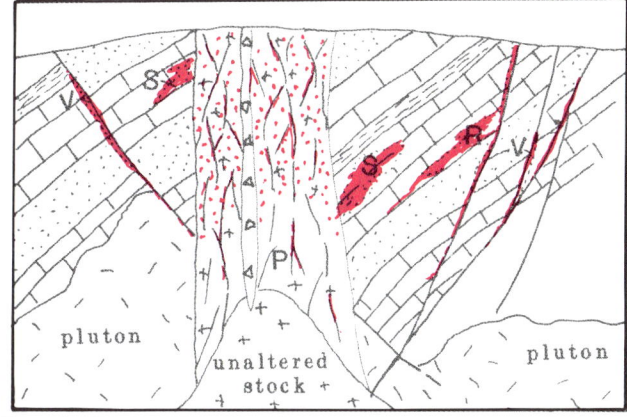

Figure 2. Sketch cross section of the geologic setting for typical porphyry copper ore (P) and commonly associated skarn (S), replacement manto (R), and vein (V) ores; modified from a sketch in Cox and Singer (1986).

U.S. copper reserves constitute about 17 percent of world reserves; more than 90 percent of U.S. reserves contain copper as the principal commodity recovered. Other world reserves occur in Canada and Mexico (10%), Chile and Peru (28%), Zaire and Zambia (19%), Europe (15%), Asia (7%), and Australia and New Guinea (4%). In addition, there are sizable subeconomic potential copper resources in Alaska, Montana, Minnesota, and elsewhere in the world (Cox and others, 1981; Jolly, 1985). Jolly also reports estimates of land-based and deep sea nodule copper resources to be 1.6 and 0.7 billion tonnes, respectively.

The U.S. reserves of about 3 million tonnes of molybdenum metal constitute about one-half of the estimated world reserves. The foreign reserves are located mainly in Canada, Chile, the USSR, and China (Blossom, 1985).

LOCATION OF U.S. DISTRICTS AND DEPOSITS

Copper and molybdenum deposits are concentrated mainly in the western part of the United States (Plate 1B), although search in the eastern part of the country has begun to find previously overlooked or bypassed deposits. On Plate 1B, red lines enclose areas in which deposits of copper are clustered as unique or overlapping concentrations; black lines indicate such areas for molybdenum. Boundaries of the areas of copper and molybdenum deposit concentrations are approximated from more detailed location data of Cobb (1960a, 1960b), Kinkle and Peterson (1962), King (1970), Tooker (1980), and Guild (1981). Districts or deposits mentioned in the text are located individually as numbered spots within these concentration areas. The sizes of spots reflect their past production; the shapes of the spots for large deposits indicate their porphyry or non-porphyry type. Unnumbered spots locate deposits that have produced these commodities in indicated ranges.

TYPES OF U.S. COPPER AND MOLYBDENUM DEPOSITS

Copper and molybdenum deposits in the United States are classified as porphyry and non-porphyry systems in Table 1.

Generalized resource characteristics of the deposit types composing these two systems are based on ore deposit models derived from a large but not exhaustive sampling of type deposits by Cox and Singer (1986). An example deposit for each type in Table 1 becomes an identifying name for that type in the descriptions that follow. However, the model deposit characteristics that follow have been generalized from those of the broad group of that type and may not precisely describe in detail the features of the individual identifying deposit. More detailed comprehensive descriptions of example deposits are found in collections of papers edited by Ridge (1968), Titley and Hicks (1966), and Titley (1982).

Porphyry, stockwork, and breccia pipe systems

Nearly 66 percent of the world's copper and 95 percent of the world's molybdenum have been derived from porphyry ore systems. These deposits generally comprise low-grade (<1.0 percent) copper and/or molybdenum dispersed mainly in silica-rich (some mafic), highly fractured and brecciated, hydrothermally altered, pluglike, subvolcanic porphyritic intrusives.

Copper deposits

Copper porphyry (San Manuel, Arizona). The porphyry copper deposit (Fig. 2) is commonly characterized by chalcopyrite in stockwork veinlets in hydrothermally altered porphyritic intrusives and adjacent country rocks (Titley, 1982). The porphyritic texture consists of closely spaced phenocrysts in a microaplitic quartz-feldspar groundmass. The composition of the intrusive may range from a tonalite to monzogranite or syenite and may intrude granitic, volcanic, calcareous sedimentary, and/or metamorphic host rocks. These porphyritic rocks range in age from the Mesozoic into the Cenozoic. Emplacement usually was along rift zones contemporaneous with island-arc volcanism that occurred along convergent plate boundaries or along shear

TABLE 1. CLASSIFICATION OF DEPOSIT TYPES*

Copper	Molybdenum
A. Porphyry, stockwork, breccia pipe (dispersed) systems	
1. Copper	1. Granite-high F porphyry
2. Copper-gold	2. Calc-alkaline-low F porphyry
3. Copper-molybdenum	3. Coproduct/byproduct molybdenum-copper
4. Complex copper-gold-molybdenum-base metals-silver	
B. Non-porphyry systems	
1. Volcanic red bed-native and sulfide	1. Vein
2. Skarn	2. Skarn
3. Vein	3. Others (minor), includes pegmatite, and sediment-hosted
4. Replacement	
5. Massive sulfide (kuroko, Cyprus, Besshi types)	
6. Sediment-hosted (shale, carbonate, sandstone/quartzite, red bed)	
7. Magmatic segregations or dissemination in mafic rocks	

*Based on Cox and Singer, 1986.

zones that separate or cut across cratonal terrane margins. Uplift and erosion usually have exposed the subvolcanic intrusive.

The copper porphyry ore and gangue occur in stockwork veinlets and scattered grains in the host intrusive and fractured wallrocks (see Model 17 in Cox and Singer, 1986). Individual or superposed ore and gangue mineral assemblages include chalcopyrite, pyrite, and possibly molybdenite; chalcopyrite, magnetite, possibly bornite, and gold. The gangue mineral assemblages may include quartz, K-feldspar, biotite, and sometimes anhydrite; or quartz, sericite, and perhaps clay minerals. Late veins of enargite, tetrahedrite, galena, sphalerite, and barite may occur in or peripheral to some deposits. Wallrock alteration (from the most intense zone outward) is sodic-calcic, potassic, phyllic, argillic, and propylitic. High-alumina minerals may occur in the upper parts of some deposits; propylitic and phyllic alteration may overprint an early potassic assemblage. Green and blue copper carbonates and silicates occur in weathered outcrops; where intensely leached, barren outcrops commonly overlie enriched zones containing chalcocite and other sulfides. These cappings often were important guides for early mining. Iron-stained fractures in leached outcrops commonly contain hematite. Anomalous rutile may occur in hydrothermally altered rock as well as in the overlying soils. Base metal skarn, epithermal vein, polymetallic replacement, and volcanic-hosted massive replacements may be associated laterally with these dispersed deposits.

Thirty-one U.S. porphyry copper districts included in Model 17 (in Cox and Singer, 1986) are located in Arizona, Washington, Nevada, New Mexico, Alaska, Montana, California, Utah, and Wyoming. Example districts include those at San Manuel, Arizona, and Yerrington, Nevada. A median tonnage for the 208 copper porphyry deposits that constitute this model, worldwide, is 140 million t, and the median grade is 0.54 percent copper. A distribution of these porphyry copper deposits by grade and tonnage (Fig. 3a, b) shows that 80 percent of the deposits range in size from 19 to more than 1,100 million t and in grade from less than 0.31 percent to more than 0.94 percent. Because tonnage and grade are independent variables and a median-size deposit may or may not have a median grade, the product of median grade and tonnage may not provide an accurate amount of copper in such a deposit. However, according to D. A. Singer (oral communication, 1986), these values statistically are reasonable for comparison purposes. These diagrams are useful in characterizing deposit types, and where available, I have used the median values of Cox and Singer (1986) to indicate average comparative sizes of the copper and molybdenum type deposits.

Copper-gold porphyry (Dos Pobres, Arizona). The type deposit is composed of stockwork veinlets of chalcopyrite, bornite, and magnetite in extensively brecciated porphyritic intrusions and coeval volcanic rocks (Titley, 1982; Model 20C, *in* Cox and Singer, 1986). The deposits are associated with a variety of

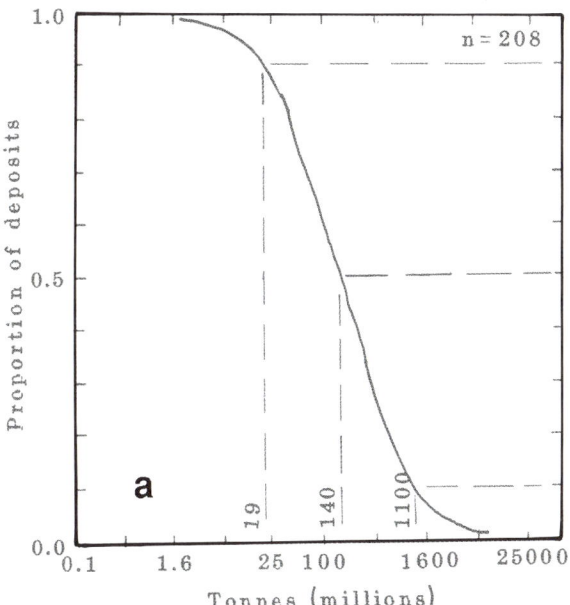
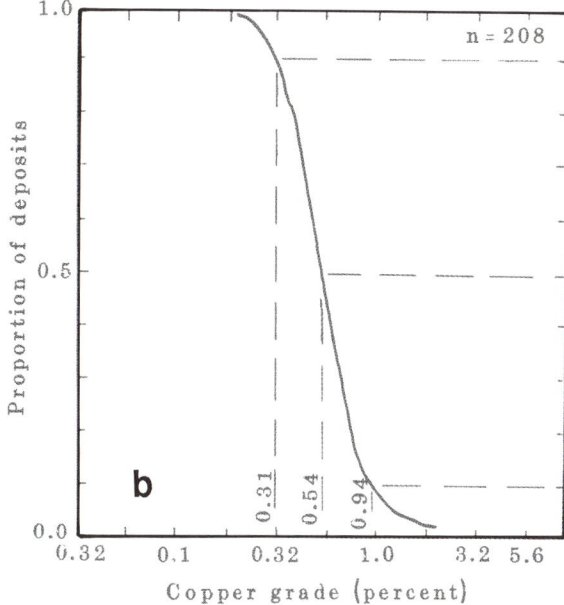

Figure 3. Example curves derived from plots of tonnage (a) and grade (b) data from 208 dispersed copper deposits; modified from Model 17 in Cox and Singer (1986).

igneous rocks, including intrusives ranging in composition from a tonalite to monzogranite, coeval dacites, andesite flows, and tuffs, as well as syenites or monzonites and coeval high K, low Ti volcanic (shoshonite) rocks. Ages of the deposits range from Triassic in British Columbia to Quaternary in the South Pacific. A subvolcanic porphyry (emplaced at 1–2 km depth) may intrude a volcanic center in an island-arc structural setting during the waning stages of a volcanic cycle and also form in a continental margin rift zone.

A bell-shaped ore zone is localized on the top of a volcanic-intrusive center (Fig. 4); highest ore grades commonly occur where the stock branches upward. Ore minerals typically include a network of veinlets, scattered grains of chalcopyrite and bornite, and traces of native gold, electrum, sylvite, and hessite; gangue minerals include quartz, K-feldspar, biotite, magnetite, chlorite, possible actinolite, and anhydrite. Pyrite, sericite, clay minerals, and calcite may occur in late stage veins. Wallrock alteration assemblages include an inner zone of quartz, with or without magnetite, biotite (chlorite), K-feldspar, actinolite, and anhydrite, and an outer propylitic zone. Late quartz, pyrite, sericite, and clay minerals may overprint an early feldspar-stable alteration. Surface iron stains may be weak; residual soils often contain anomalous amounts of rutile, and gold commonly is enriched over the ore bodies.

Dos Pobres, Arizona, is the only one of 40 copper-gold deposit porphyries identified worldwide that occurs in the United States (Cox and Singer, 1986). The median size of these deposits is 100 million t, and median grades are 0.5 percent Cu, 0.38 g/t Au, 1.0 g/t Ag, and minor molybdenite. The deposits may be associated with porphyry Cu-Mo deposits and gold placers. The

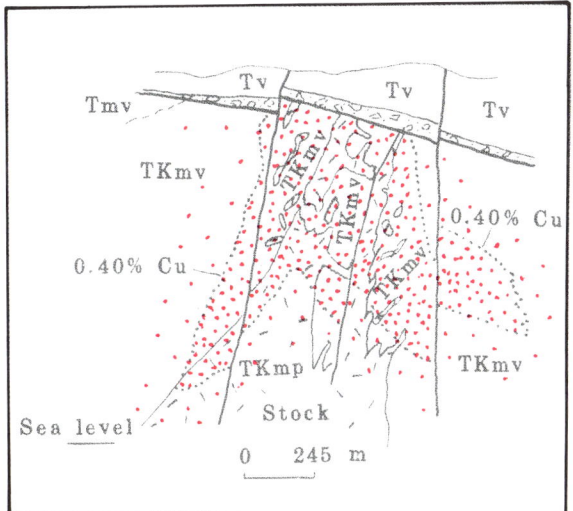

Figure 4. Sketch cross section of the geologic setting for the Cu-Au-type porphyry at Dos Pobres, Arizona, modified from a diagram in Titley (1982). Tv—basalt flows, tuffs, rhyolite, basal conglomerate; Tmv—metavolcanics; TKmp—tonalite-quartz monzonite porphyry; TKmv—metavolcanics, metavolcaniclastics, andesites, and agglomerate, intensely brecciated near intrusive porphyry.

ratio of Au (in g/t) to Mo (in percent) is greater than 30 in deposits of this type.

Copper-molybdenum porphyry (Sierrita, Arizona). The deposits consist of stockwork veinlets and dispersed grains of chalcopyrite and molybdenite in country rocks in or near a porphyritic intrusion (Titley, 1982). The ratio of Au (in ppm) to Mo (in percent) is less than 3 (Model 21a in Cox and Singer, 1986). The porphyry may range in composition from a tonalite to monzogranite and may occur as stock, dike, and associated breccia pipes intrusive into batholithic, volcanic, or sedimentary rocks. The porphyries contain sparse phenocrysts set in a fine- to medium-grained aplitic groundmass; however, porphyritic texture may be restricted to small dikes. Ages of these rocks vary, but they are mainly Mesozoic to Tertiary. The porphyry is emplaced in highly faulted cupolas of batholiths and is contemporaneous with the formation of dikes and breccia pipes. The faults are subduction-induced above volcanic-plutonic arcs, mainly along continental margins but also along convergent plate boundaries.

The ore minerals include chalcopyrite, pyrite, and molybdenite; peripheral vein and replacement deposits contain chalcopyrite, sphalerite, galena, and sometimes gold. The outermost zone may have veins of Cu-Ag-Sb-sulfides, barite, and gold. The wallrock alteration assemblages include quartz, K-feldspar, biotite (chlorite), and possibly anhydrite (potassic alteration), grading outward to propylitic alteration. Late white mica and clay minerals (phyllic alteration) may form a capping or an outer zone or may constitute the whole assemblage. High alumina alteration minerals may be present in the upper levels of the system. Ore grade is controlled by close spacing of veinlets and mineralized fractures. Favorable mineralized country rocks include calcareous sedimentary rocks, diabase, tonalite, or diorite. Surface rocks are intensely leached, resulting in wide areas of iron oxide staining and fractures coated with hematitic limonite. A blanket of supergene copper (chalcocite) occurs below the leached zone. Anomalous amounts of rutile occur in residual soils over the ore body. The ore zone usually is the site of a magnetic low, owing to the alteration and replacement of magnetite.

Six deposits in Arizona and Washington out of 16 worldwide make up this type in Cox and Singer (1986). The median size is 500 million t, and the median grade is 0.42 percent Cu, 0.016 percent Mo, 0.02 ppm Au, and 1.2 ppm Ag. Associated deposit types may include Cu, Zn, or Fe skarns, which may contain high gold concentrations, gold and base metal sulfosalts in veins, gold placers, volcanic-hosted massive replacements, and polymetallic manto replacements. Examples of this type in the United States include the Sierrita, Inspiration, Morenci, Ray, and Twin Buttes deposits in Arizona, Bond Creek, Alaska, and Tyrone, New Mexico, as well as Glacier Peak in Washington.

Copper-molybdenum-gold–base metals–silver complex porphyry/non-porphyry (Bingham, Utah). A few large complex districts, such as Bingham, Utah, are individually distinctive and contain characteristic geologic features of several deposit types. Bingham consists of stockwork veinlets and scattered ore minerals in an altered composite equigranular and porphyritic, hydrothermally altered igneous stock and dike system (in Titley, 1982). Ore host rocks at Bingham consist of equigranular monzonite, intruded by porphyritic quartz monzonite and cut by a latite porphyry dike swarm, and fragmented sedimentary rocks. Coeval skarn, vein, manto replacements of sedimentary rocks, and placers adjoin the dispersed deposits. The age of the subvolcanic Bingham intrusives and hydrothermal system is early Tertiary. At Bingham, the intrusive stocks, dikes, and breccia pipes occur in highly faulted, folded, miogeoclinal predominantly carbonate, and hydrothermally altered craton shelf strata. Deposits of this type commonly are sited near continental margins and in Phanerozoic orogenic belts that overlie presumed basement cratonal rifts.

The zoned sulfide deposit consists of closely spaced stockwork fractures filled with quartz, pyrite, chalcopyrite, molybdenite, and native gold and scattered grains of these same minerals replacing mafic minerals in fractured igneous and sedimentary rocks (Fig. 5). Peripheral copper-gold–bearing skarns occur in metamorphosed carbonates along contacts with the intrusion (Atkinson and Einaudi, 1978). Outer fissure veins and massive manto replacement deposits occur in both the intrusive and sedimentary rocks (described in Ridge, 1968). The ores include sphalerite, galena, silver, manganese, pyritic copper, and native gold. Zoned hydrothermal alteration of wallrocks occurs around the central stock and consists of an inner potassic (quartz-orthoclase-phlogopite) and local sericite-quartz cap that is surrounded by an outer propylitic (actinolite-chlorite-epidote) zone. In quartz-rich sediments, an inner quartz, biotite, orthoclase, and superposed outer sericite-argillic zone occurs. Thick limestones have an inner garnet zone followed by a diopside, talc, and tremolite zone.

The central intrusive controlled the zonal distribution of ores formed from highly saline metal-bearing fluids at high tempera-

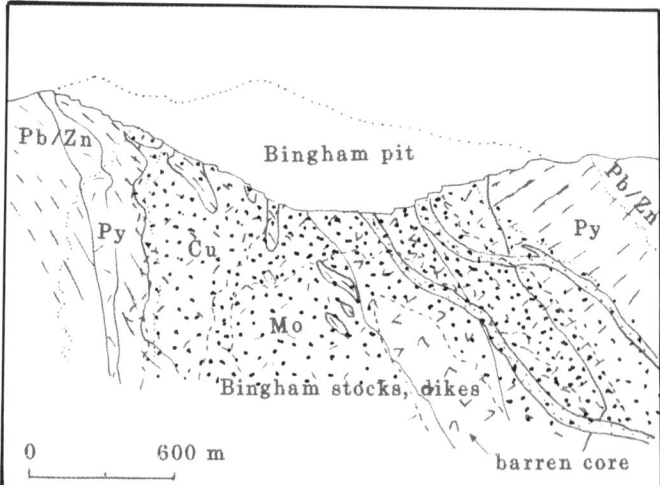

Figure 5. Sketch cross section of the geologic setting for the ores at Bingham, Utah, modified from John (1978). Dash lines indicate bedding in host sediments. Overlapping ore zones: Cu, in red; Mo, in black; pyrite (py) enclosed by dot-dash lines; and a Pb-Zn-Ag outer zone (Pb/Zn).

tures (400°–600°C) in a region of intense fracturing and brecciation. The formation temperature of sulfides in the outer zone decreases to 300°C. A hydrostatic load of about 3 km indicates a subvolcanic environment, and mixing of hydrothermal and meteoric fluids and boiling led to venting along breccia pipes. Isotopes suggest a mantle source for these hydrothermal solutions, which were introduced in pulses over a span of about 1 m.y. following intrusion of the stock 39.8 m.y. ago. Weathering and leaching produced copper and iron-oxide–stained fracture surfaces; some sulfides occur at the surface, but secondary sulfide minerals formed concentrations at depth along the main fractures. The intrusive body composing the deposit is characterized geophysically by an aeromagnetic low.

The porphyry and associated (coeval) non-porphyry ore systems such as this are generally large. Median tonnage and grade data for them are not available, but the production averages at Bingham provide an order of magnitude. Production through 1972 (James, 1978, p. 1219) includes 11,856,000 t Cu, 504,700 kg Au, 2,473,000 t Pb, 1,038,000 t Zn, and 8,421,000 kg Ag. Between 1938 and 1955, 22,000 t Mo were produced. The average grade of copper was 0.92 percent; of molybdenum, 0.036 percent; and of gold, 0.34 g/t. Copper reserves in 1976 were estimated at 9.5 million t at an average grade of 0.70 percent (Cox and others, 1981). Other examples of this type in the United States include deposits at Ely, Nevada, and Bisbee, Arizona.

Molybdenum deposits

Granite-high F porphyry (Climax, Colorado). The typical deposit, such as the one at Climax, Colorado, consists of an umbrella-shaped stockwork of quartz and molybdenite (Fig. 6)

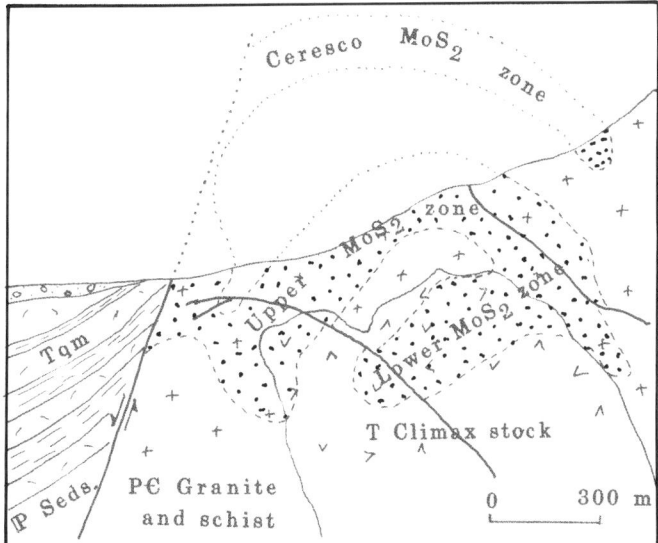

Figure 6. Sketch cross section of the geologic setting of ore at Climax, Colorado, modified from a diagram in Ridge (1968). Climax stock and numerous dikes at its top (not shown) are intrusive into Precambrian granite. Molybdenite ore zones (black stipple) occur in the fractured and faulted hood zones.

associated with fluorite in granite porphyry (in Ridge, 1968). The host porphyritic granite-rhyolite stock generally is a multistage hypabyssal intrusive that contains more than 75 percent SiO_2 and is cut by radial dikes and small breccias. Ages of the intrusions are generally middle Tertiary, and for the most part they are located in extensional fracture zones in the craton, far from continental margins.

The ores comprise molybdenite, quartz, fluorite, and possibly K-feldspar, pyrite, wolframite, cassiterite, and topaz (pyrite and topaz are late phases). These minerals occur mostly in veinlet fractures and as minor scattered grains in host rocks. The ore zone has been subjected to intense pre-ore quartz and quartz and K-feldspar vein filling and phyllic and propylitic alteration. Rhodochrosite, rhodonite, and spessartine garnet halos surround the ore zone. Greissen veins occur below the ore zone. Weathered outcrops are stained yellow to shallow depth by ferrimolybdite. Erosion and glaciation have removed much of the sedimentary or metamorphic rock cover, as well as ore, at Climax.

A grade and tonnage model based on nine world deposits of this type (Model 16 in Cox and Singer, 1986) indicates a median size of 200 million t and a grade of 0.19 percent Mo. All but one of the districts surveyed are in the United States—in Colorado, Utah, Nevada, and Montana. Climax, one of the largest districts, had an estimated production through 1983 of more than 430 million t of ore, from which 832 thousand t of Mo have been produced (Sutulov, 1983). This was about 38 percent of all molybdenum extracted in the world to that time. All deposits are considered to have substantial reserves; those at Climax are estimated to be 500 million t of ore containing more than 1 million t of molybdenum (King and others, 1973). Examples of this type in the United States include Henderson and Mount Emmons, Colorado; Questa, New Mexico (lower part); and Pine Grove, Utah. The deposits are associated with silver-base metal ores and fluorspar. Rhyolite-hosted tin deposits may overlie them, and porphyry tungsten deposits may also be related to this system.

Calc-alkaline-low F porphyry (Buckingham, Nevada). The deposits are composed of stockwork quartz-molybdenite veinlets and disseminations in felsic porphyritic intrusives and the adjacent host rocks (Westra and Keith, 1981). The intrusive may be a porphyritic tonalite, granodiorite, or monzogranite with fine aplitic groundmass, intruded within the age range of Mesozoic through Tertiary (Fig. 7). The deposits occur in very faulted orogenic belts associated with calc-alkaline intrusions (Model 21b in Cox and Singer, 1986).

Ore minerals include molybdenite, pyrite, and possibly scheelite, chalcopyrite, and argentian tetrahedrite. The gangue minerals include quartz, biotite, and sometimes K-feldspar, calcite, white mica, and clay minerals. Wallrock alteration minerals include inner potassic and outer propylitic zones, often with a phyllic and argillic overprint. Weathering produces yellow ferrimolybdite and secondary copper minerals from associated copper ores. The main ore controls are close-spaced fractures in the porphyry and adjoining country rock.

The grade and tonnage model in Cox and Singer (1986) is

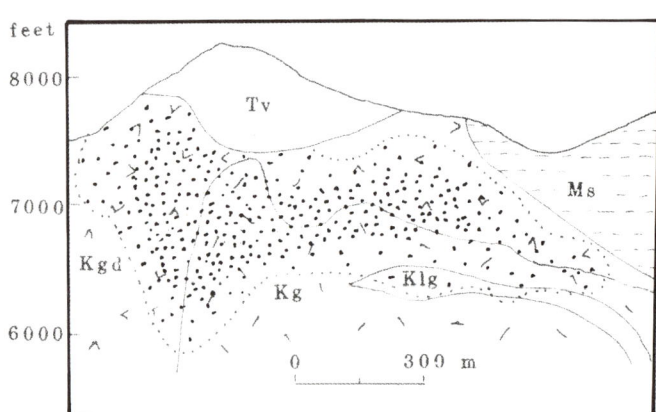

Figure 7. Sketch cross section of the geologic setting of a calc-alkaline-low F–type molybdenum porphyry, Thompson Creek, Idaho, modified from Hall and others, 1984. Tv, volcanic rocks; Klg, leucogranite; Kg, biotite granite (in part porphyritic); Kgd, biotite granodiorite; Ms, metasedimentary rocks; Mo ore in black.

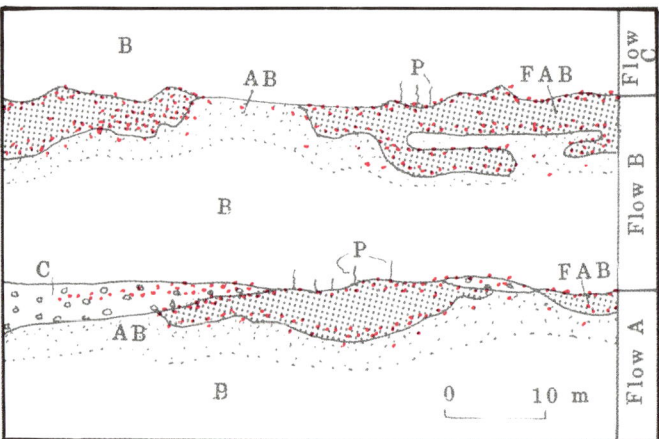

Figure 8. Sketch cross section of the geologic setting at Keweenaw, Michigan deposits, modified from a diagram in Ridge (1968). B, massive basalt; AB, nonfragmented amygdular basalt; FAB, fragmental amygdaloid; P, pipe amygdules; C, conglomerate. Copper mineralization in red.

based on 33 world deposits, eight of which are in the United States, in Nevada, Montana, Idaho, Alaska, and Washington. A median size for this model is 94 million t, and a median grade is 0.085 percent Mo. These deposits are associated with Cu-Mo porphyries, Cu skarns, and volcanic-hosted Cu-As-Sb deposits. In addition to Buckingham, Nevada, examples of these deposits in the United States include Mount Tolman, Washington; Pine Nut, Nevada; Thompson Creek, Idaho; and Quartz Hill, Alaska.

Copper-molybdenum byproduct/coproduct porphyry. Copper-molybdenum porphyries contain nearly 40 percent of the molybdenum reserves. Sutulov (1978) lists 34 deposits in the United States, 21 of which are producers of molybdenum as a byproduct. These districts occur in Arizona, Colorado, New Mexico, Utah, Montana, California, Alaska, Washington, Idaho, and Nevada. The average grade of molybdenum in these deposits is 0.018 percent, compared with the average grade of molybdenum porphyries at 0.178 percent. Description of the Cu-Mo type appears in the preceding section on copper deposits.

Non-porphyry systems

About one-third of the world's copper supplies occur in non-porphyry types of deposits (Table 1), mostly as stratabound ores in sedimentary rocks (25%), as volcanogenic massive sulfides (5%), and as Ni-Cu ores in mafic intrusives, native copper ores, and in other types such as sea floor nodules, metal-rich deep ocean basin sediments, and brines (4%) (Cox and others, 1973). Many of these types of deposits are large but currently are subeconomic in the United States. Sediment-hosted and massive sulfides make up 7 and 2 percent of U.S. reserves respectively.

Copper deposits

Volcanogenic-sedimentary red bed (Keweenaw, Michigan). Stratabound disseminated native copper and copper sulfides occur in the upper parts of thick sequences of basalts, and copper sulfides may occur in the overlying clastic sediments. The host rock terrane includes subaerial to shallow marine interlayered basalt flows, breccias, and tuffs; interbedded red bed sandstones; tuffaceous sandstone; and conglomerate overlain by tidal limestones and black shales (Fig. 8). The favored ore horizons are in fragmentary and porous amygdular layers, flow-top breccias, and faults in the basalts and overlying laminated algal carbonates and clastic sediments. The deposit ages range from the Proterozoic, to Triassic and Jurassic, to Tertiary. The ore-forming environment was along a continental margin rift zone near the marine-continental interface in an equatorial region (from Model 23 in Cox and Singer, 1986).

These deposits include native copper and minor silver in the flows and coarse calcitic beds and chalcopyrite and other Cu_2S minerals in faults and fractures. Locally, chalcocite, bornite, and chalcopyrite occur in the overlying shales and carbonate rocks. Fine-grained pyrite is common but is less abundant with copper sulfides. The copper distribution was controlled by the permeability of host rocks and tectonic fracturing in the weak basalt flow tops and sedimentary beds. Ore textures vary from flow-top breccia and amygdule fillings in basalt, to finely disseminated grains and coatings along interstices and partings of clastic and shale sediments, to massive replacement of carbonates. Fine hematite provides the red color; alteration minerals include quartz, calcite, zeolite (laumontite), chlorite, epidote, pumpellyite, prehnite, and K-feldspar; a zonal pattern suggests a regional metamorphism. Copper nuggets in stream beds result from the weathering of the deposits.

Four U.S. deposits of this type occur at Keweenaw and Calumet, Michigan, and Kennecott and Denali, Alaska. The Michigan deposits were the main domestic sources of copper for many years. It was estimated (in Ridge, 1968) that the Michigan native copper districts produced more than 5.95 mil. t of copper

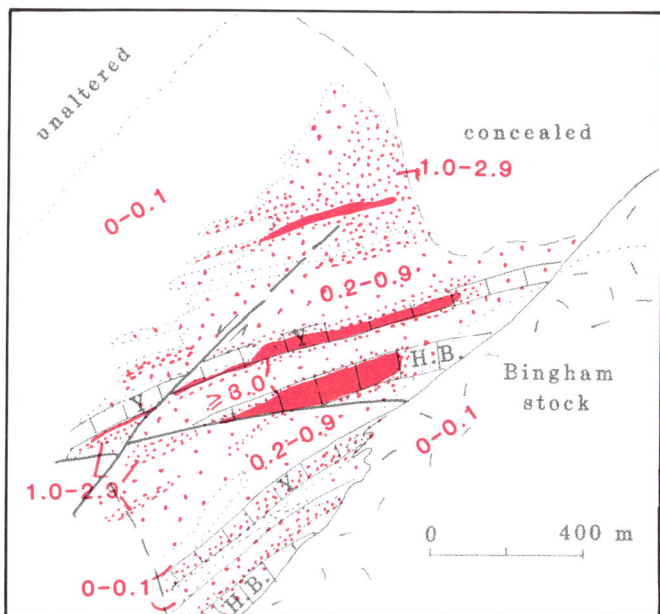

Figure 9. Sketch cross section of the geologic setting of skarn ores, Carr Fork, Utah; modified from Atkinson and Einaudi (1978). Percent total copper sulfides (in red) in the Bingham stock and faulted host Paleozoic carbonate rocks (HB, Highland Boy limestone; Y, Yampa limestone are marker beds within the Oquirrh Formation).

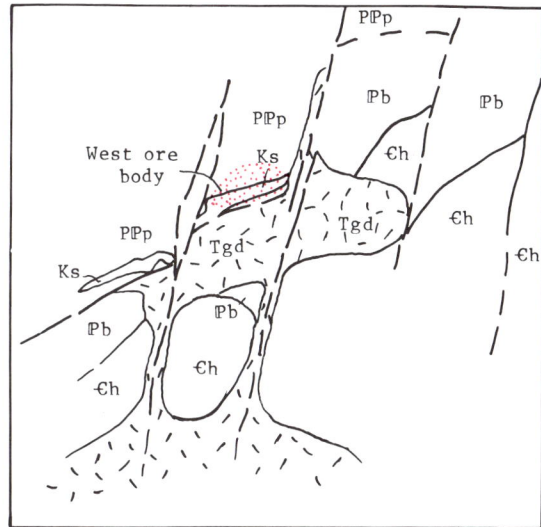

Figure 10. Sketch cross section of the geologic setting of copper skarn deposit at Copper Canyon, Nevada, modified from Theodore and Blake (1978). No scale. PIPp, Pumpernickel Formation; IPp; Battle Formation; €h, Harmony Formation; Tgd, granodiorite of Copper Canyon; Ks, andesite-diopside skarn in the Pumpernickel Formation.

having an average grade of 1.48 percent. Production at Kennecott was about 618,000 t of copper (Berg and others, 1964).

Skarn (Carr Fork, Utah; Copper Canyon, Nevada). At least two distinctive types of skarn deposits containing copper and/or molybdenum ores occur in the United States. The more common type is associated with porphyry copper deposits (e.g., Carr Fork, adjoins Bingham, Utah); the second type includes skarns associated with barren or weakly mineralized stocks (Copper Canyon, Nevada).

Carr Fork, Utah. Skarns adjoining porphyry copper deposits, such as that at Carr Fork, Utah (Fig. 9), are related to intrusives ranging in composition from tonalite to monzogranite and ranging in age from the Mesozoic into the Tertiary (Atkinson and Einaudi, 1978). The porphyries intrude fractured metamorphosed carbonate or clastic calcareous sedimentary rocks along the margin of rifts in the cratonal continental shelf. The ores include chalcopyrite, pyrite, and magnetite in an inner garnet pyroxene zone; bornite, chalcopyrite, and possibly sphalerite and tennantite in an outer wollastonite zone. Scheelite and traces of molybdenite and galena may be present. The skarn zone is composed of fine granular calc-silicate and quartz and sulfide veinlets. Potassic alteration is associated with andradite in the intrusive rocks and, with diopside in the calcareous rocks. Farther from the contact are zones of wollastonite or tremolite and minor garnet, idocrase, and clinopyroxene. These in turn grade outward into marble. The distribution of phyllic alteration occurs where retrograde actinolite, chlorite, and clay minerals occur in the skarn copper ores, along close-spaced stockwork veins in the skarn and igneous rock and along favorable strata in the sediments.

Model 18a in Cox and Singer (1986) is based on 18 world deposits. It indicates a median size of 80 million t and a median grade of 0.98 percent Cu. Type deposits commonly are associated with porphyry copper and replacement base and precious metal deposits. Some of the other examples in the United States include Christmas, Arizona; Ely, Nevada; Santa Rita, New Mexico; and Twin Buttes, Arizona.

Copper Canyon, Nevada. Ores in skarns adjoining barren or weakly mineralized granitic and porphyritic stocks and breccia pipes intrusive into carbonate strata, in part comparable with the deposit at Copper Canyon, Nevada (Fig. 10), are composed characteristically of chalcopyrite, pyrite, and possibly hematite, magnetite, bornite, pyrrhotite, minor molybdenite, and bismuthinite, sphalerite, galena, cosalite, arsenopyrite, enargite, tennantite, loellingite, colbaltite, and tetrahedrite (Theodore and Blake, 1978; Model 18b in Cox and Singer, 1986). Gold and silver may be important byproducts. Wall rock alteration of the carbonates mainly consists of diopside and andradite in the central part, possibly wollastonite and tremolite in an outer zone, and marble in a peripheral zone. Igneous rocks may or may not be altered to epidote, pyroxene, and garnet. Retrograde alteration to actinolite, chlorite, and clay minerals is also sometimes observed. Ores occur in irregular or tabular bodies in the carbonate and calcareous clastic rocks near the contact with the intrusion, in xenoliths in the igneous stocks, and in breccia pipes that cut skarn. The associated igneous rocks are commonly barren. Weathering produces copper carbonates, silicates, and an iron-rich gossan. On the basis of 64 world deposits characterizing Model 18b (in Cox and Singer, 1986), including 10 in the U.S., this type has a median size of 0.56 million t and a median grade 1.7 percent Cu. In addition to Copper Canyon, examples in the U.S. include Victoria, Nevada, and Oracle Ridge, Arizona.

Vein (Butte, Montana, upper part). Copper occurs as a primary mineral or as a byproduct in many large to small polymetallic vein deposits. More than 8 million t of copper as well as substantial byproduct silver, gold, zinc, manganese, and lead have been produced from the Main Stage veins at Butte, Montana (in Ridge, 1968). These veins are superimposed on porphyry copper-molybdenum mineralization that predates the veins by about 5 m.y. (Miller, 1973). Elsewhere, vein deposits may be associated with replacement deposits and with vein and manto deposits that occur peripheral to a number of porphyry copper deposits, such as Bingham, Utah, from which about 370,361 t of byproduct copper were produced through 1964 (in Ridge, 1968).

The Main Stage polymetallic vein deposits at Butte occur in a quartz monzonite stock, associated aplite and pegmatite dikes and plugs, and quartz porphyry dikes (Fig. 11). These rocks intrude folded and metamorphosed cratonal platform miogeoclinal sedimentary and volcanic rocks of Proterozoic ages. The stock was intruded along a continental rift system during the Late Cretaceous Laramide orogeny. Main Stage veins are composed mainly of chalcocite, enargite, bornite, chalcopyrite, and pyrite in the central and intermediate zones, and sphalerite, galena, rhodochrosite, and barite with silver and gold in the intermediate and peripheral zones. Extensive argillization of wall rocks occurs in the central zone, decreasing outward. Supergene enrichment produced a chalcocite blanket. Beneath the Main Stage veins, in deeper workings in the stock, is a low grade pre–Main Stage dispersed quartz, pyrite, molybdenite, chalcopyrite, and gold assemblage, which is accompanied by biotitic wall rock alteration. Main-Stage ore controls are along intersecting systems of veins that have an average width of 6 to 9 m but that locally are as much as 30 m wide in the central zone. These veins have been mineralized over long horizontal and vertical distances and persist structurally into the molybdenite mineralization below. Extensive systems of deep-reaching veins provided access for the circulation and mixing of meteoric and magmatic hydrothermal waters. K-Ar ages on alteration minerals of the pre–Main Stage and Main Stage are 52.8 and 57.5 Ma, respectively. Hydrothermal fluids of the Main Stage were about 300°C, whereas those of the pre–Main Stage were 600°C. Erosion exposed the vein system and formed the adjacent placer gold deposits.

Production of 452 million t of ore from the underground veins and from open pit disseminated secondary ores between 1880 and 1972 was distributed as follows (Miller, 1973):

Copper (average grade 2%)	9,000,000 t
Zn	245,000 t
Mn	1,900,000 t
Pb	43,000 t
Ag	23,255,000 kg
Au	102,000 kg

Mining of the deep dispersed molybdenum and copper ores has been minimal thus far. While Butte is a unique deposit, from which primary copper ores were produced, a number of vein deposits have produced significant but smaller amounts of byproduct copper. Examples include San Francisco (Oatman), Arizona, as a byproduct of gold; Central City–Idaho Springs, Colorado, as a byproduct of gold and silver; and Philipsburg, Montana, as a byproduct of silver-gold-manganese ores (see Siders, this volume).

Replacement (Superior, Arizona; Tintic, Utah.) Copper occurs as the primary metal sought in a number of polymetallic replacement deposits, such as those at Superior, Arizona, and also as a byproduct of replacement base and precious metal ores, as at Tintic, Utah.

Superior, Arizona. The replacement copper at Superior is accompanied by gold, silver, sphalerite, and galena as byproducts of the ores (Short and others, 1943). The ores occur in a series of disconnected shoots in replaced shattered carbonate, quartzite, and diabase host rocks lying between two east-trending shear zones. The diabase was more susceptible to replacement than the sedimentary rocks (Fig. 12). Alteration of wall rocks consists of silicification and sericitization, which is more intense in the diabase. A quartz monzonite porphyry dike may be the source of altering and ore fluids.

The principal ore minerals at Superior are pyrite, bornite, chalcopyrite, and enargite; subordinate ore minerals are tennantite and hypogene chalcocite. Locally, sphalerite predominates and is accompanied by galena. Supergene enrichment was important. Gold is associated with malachite and chyrscolla in an iron and manganese oxide gangue. Silver accompanies copper and gold. About 311,700 t of copper from ore having an average grade of 6.3 percent was produced between 1911 and 1943.

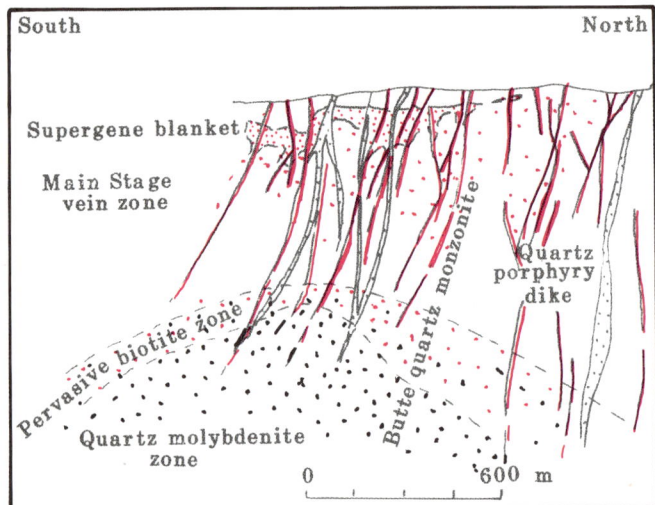

Figure 11. Sketch cross section of the geologic setting of vein and porphyry deposits at Butte, Montana; modified from a diagram in Ridge (1968).

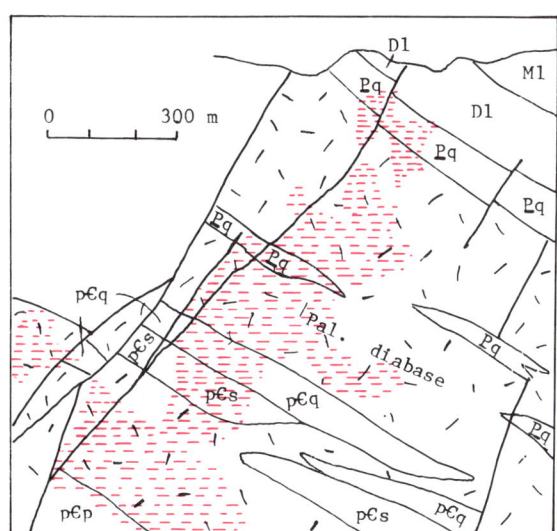

Figure 12. Sketch cross section of the geologic setting of replacement deposits (in red) at Superior, AZ, modified from Short and others (1943); p€p, Pinal schist; p€q, Precambrian quartzite; Pq, Proterozoic quartzite (Troy Quartzite); Dl, Devonian limestone (Martin Limestone); Ml, Mississippian limestone (Escabrosa Limestone).

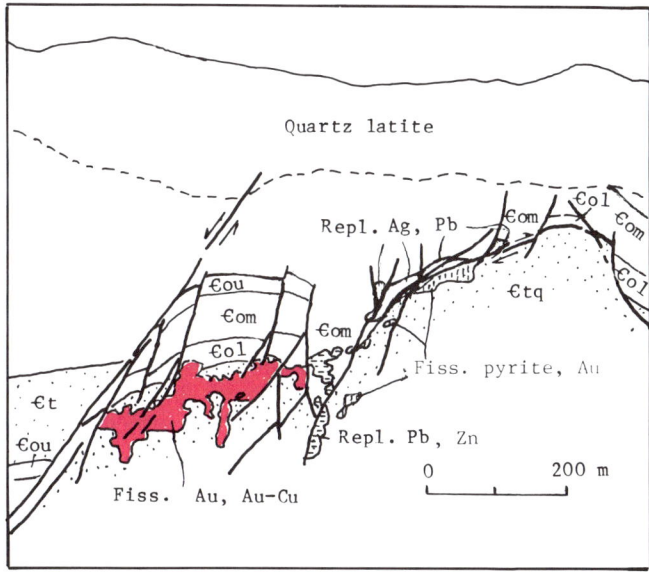

Figure 13. Sketch cross section of the geologic setting of zoned vein and replacement ore deposits, East Tintic, Utah, modified from a diagram in Ridge (1968). Cambrian sedimentary host rocks include: €t, limestone; €ol, €om, €ow, shale; and €tq, quartzite. Copper-bearing veins (in red) are typically restricted to the quartzite units.

Tintic, Utah. Byproduct copper also has been obtained from a number of replacement deposits, such as at Tintic, Utah, which was primarily a base and precious metal district (*in* Ridge, 1968). Here the deposits are of hydrothermal epigenetic origin and consist of massive lenses, pipes, and veins in limestone, dolomite, and permeable rock adjoining an igneous center. The sedimentary rocks were formed on the Phanerozoic miogeoclinal platform shelf; they are overlain by volcanics and intruded by porphyritic calc-alkaline plutons. Deformation of strata by thrust and normal faults and the formation of volcanic and intrusive centers prepared the region, provided a source of heat, and facilitated movement of metals in hydrothermal fluids. The age range of ore deposition varies but is concentrated in the late Mesozoic into the early Cenozoic.

These deposits are usually zoned: The central zone ore minerals contain enargite, sphalerite, argentite, tetrahedrite, digenite, and chalcopyrite; an intermediate zone includes galena, sphalerite, argentite, tetrahedrite, possible proustite, and pyrargyrite; and an outer zone contains sphalerite and rhodochrosite. Quartz, pyrite, marcasite, and barite compose the gangue. Local rare gold, sylvanite, and calaverite also occur. The replacement ores range from massive to vuggy, the limestone is dolomitized and silicified (including jasperoid), and shales and igneous rocks are chloritized, argillized, and pyritized. Ore bodies are localized along faults and bedding, bedding-plane faults, and karst solution channelways (Fig. 13). Where weathering has exposed ores, they commonly are oxidized.

On the basis of 52 large (>100,000 t) world deposits sampled, 22 of which are in the United States (in Nevada, Utah, California, and Arizona), the median tonnage is 1.8 million t. The median grade of ore is 5.2 percent Pb, 3.9 percent Zn, 0.094 percent Cu, 150 g/t Ag, and 0.19 g/t Au (Model 19a in Cox and Singer, 1986). Among other U.S. examples of this type are those at Darwin, California; Eureka, Nevada; Ophir and Park City, Utah; and Tombstone, Arizona. Commonly the deposits are associated with base metal skarns and porphyry copper deposits, as in the Lark Mine at Bingham.

Massive sulfide (kuroko-type, West Shasta, California; Cyprus-type; and Besshi-type, Ducktown, Tennessee). At least three distinctive types of copper-bearing massive (>60 percent) sulfide deposits have been recognized in the United States by Cox and Singer (1986). The kuroko-type is the most common and occurs in accreted as well as in cratonal platform terranes. The Cyprus type occurs more rarely in ophiolite terranes in areas of plate convergence. Known Besshi-type massive sulfides also are of limited occurrence in the U.S., and may have formed along a spreading ridge beneath a sedimentary sequence at the continental slope in older areas of plate convergence.

Kuroko-type. The deposits of West Shasta, California, and Crandon, Wisconsin, are examples of the kuroko-type that occurs in marine volcanogenic rocks of felsic to intermediate composition and are copper- and zinc-bearing massive sulfides (Franklin and others, 1981). Host rocks may consist of marine rhyolite, dacite, and lesser basalt associated with organic-rich mudstones or pyritic siliceous shale. The ages of these deposits may range from the Archean through the Cenozoic. Mineralization is related to marine volcanic-derived hot springs in an anoxic marine environment along island-arc belts; Archean greenstones in the craton are characterized by local tensional faults and fractures and may represent ancient accreted marine terranes. The massive deposits are characterized by an upper stratified (black ore) zone, a lower stratiform (yellow ore) zone, and an underlying dispersed stockwork feeder zone (Fig. 14). Black ore contains pyrite, sphalerite, chalcopyrite, pyrrhotite, and sometimes galena, barite, tetrahedrite, and bornite. Yellow ore contains pyrite and chalcopyrite, with occasional sphalerite, pyrrhotite, and magnetite. The stockwork contains veinlets and disseminations of pyrite, chalcopyrite, gold, and silver in altered wall rocks. Alteration assemblages include zeolite, smectite, and chlorite (?) adjacent to or in the blanket deposits; silica, chlorite, and sericite are associated with the stockwork zones. In metamorphosed kuroko deposits, cordierite and anthophyllite occur in the footwall with graphitic schist in the hanging wall of metamorphosed deposits. Massive and stockwork ore occurs in a center of felsic volcanics, where local fracturing and brecciation are associated with hot springs, organic-rich mudstone, pyritic siliceous shale, sulfide clasts, and volcanic breccia fragments. Some deposits are gravity transported to depressions in the sea floor, where weathered yellow, red, and brown gossans form.

A median deposit size is 1.5 million t, and the median grade is 1.3 percent Cu, 2.0 percent Zn, 0.16 g/t Au, and 13 g/t Ag, based on 432 world deposits of this type (Model 28a in Cox and Singer, 1986). Among the U.S. deposits, most occur in accreted oceanic-derived terranes in California, Oregon, Alaska, New England, and Georgia. A few cratonal deposits in Proterozoic accreted(?) terrane deposits occur in Arizona, New Mexico, and Wisconsin. Typical deposits in the United States include those at West Shasta, California; Bald Mountain, Maine; Orange Point, Alaska; Crandon, Wisconsin (Ohle, this volume); and Jerome, Arizona.

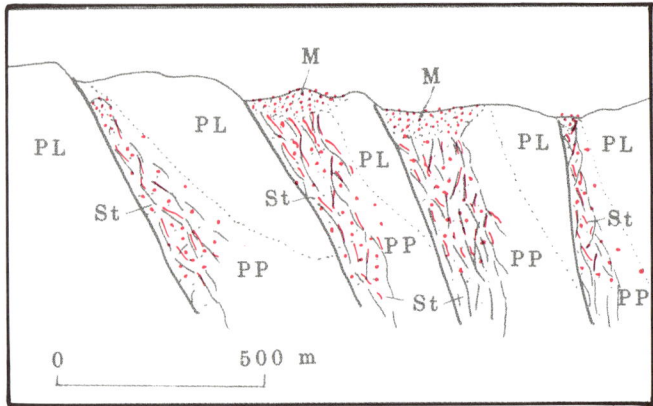

Figure 15. Sketch cross section of the geologic setting for a Cyprus-type massive sulfide, modified from Adamides (1980). PL, pillow lava; PP, propylitized and pyritized lava; M, massive pyrite; St, stockwork zone.

Cyprus type. The Turner-Albright, Oregon, deposit is a domestic example of the Cyprus type. These deposits contain massive pyrite, chalcopyrite, and sphalerite in pillow volcanics and sometimes in brecciated basalts (Franklin and others, 1981). Host rocks represent the typical ophiolite (mafic) assemblage and range in age from the Archean (?) to the Tertiary; most are of Ordovician to Cretaceous age. Deposits form in pillow basalts or mafic volcanic breccias adjacent to submarine hot springs, along faults in oceanic axial grabens, or in back-arc spreading ridges (Fig. 15). Only rarely do ores occur in the overlying sediments. The deposits are often associated with manganese and iron-rich cherts. Massive sulfide ores at the top are composed of pyrite, chalcopyrite, and sphalerite; marcasite and pyrrhotite are also often present. Underlying stockwork ores are composed of pyrite, pyrrhotite, minor chalcopyrite, sphalerite, and Co, Au, and Ag minerals. Sulfides are brecciated and recemented. Stockwork zone alteration includes feldspar destruction and formation of abundant quartz, chalcedony, chlorite, illite, and calcite. Some deposits may be overlain by Mn-poor and Fe-rich ochre zones. Weathering has produced massive limonite gossans and placer gold.

Model 24a in Cox and Singer (1986) is based on 49 world deposits and includes three in the United States. Deposits have a median tonnage of 1.6 million t and a median grade of 1.7 percent Cu. Silver, gold, lead, and zinc are recovered from a few of the larger deposits. U.S. deposits include the Big Mike, Nevada; Rua Cove, Alaska; and Turner-Albright, Oregon.

Besshi-type. Domestic examples of the Besshi-type massive sulfide deposits occur at Ducktown, Tennessee, and Mountain City, Nevada (in Ridge, 1968). These deposits (Fig. 16) typically

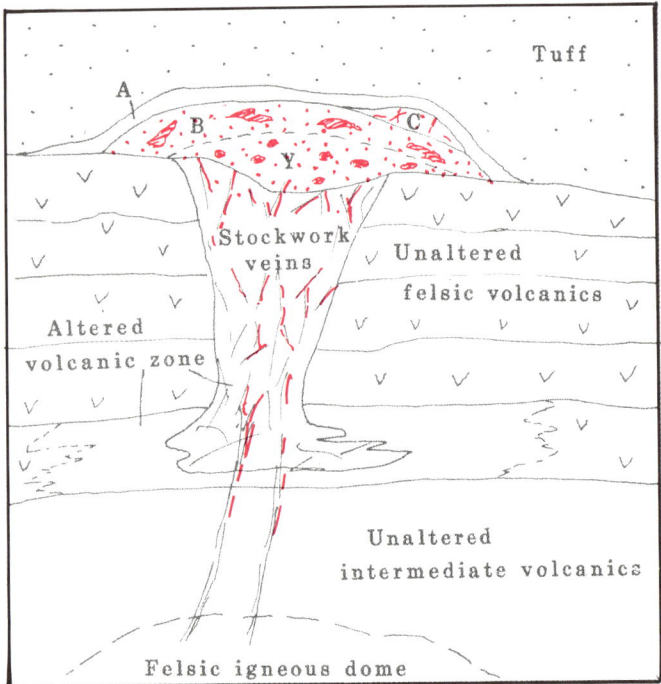

Figure 14. Sketch cross section of the geologic setting for a kuroko-type massive sulfide deposit, modified from Franklin and others (1981). B, black ore, includes massive pyrite, sphalerite, galena, barite, and pyrrhotite; Y, yellow ore, includes massive pyrite, chalcopyrite, and minor sphalerite and magnetite; A, chert, felsite, zeolite, clays, carbonates, and hematite; C, barite; stockwork contains veins with quartz, pyrite, chalcopyrite, and possibly sphalerite.

consist of thin, sheetlike bodies of massive to well-laminated pyrite, pyrrhotite, and chalcopyrite in host rock that is composed of thin laminated clastic, sometimes deformed and metamorphosed sediments and mafic tuffs (see Model 24b in Cox and Singer, 1986). Locally the host is black shale, iron formation, or chert. Basaltic volcanism may provide the source of submarine hot springs; ores are localized in permeable or fractured sediments or volcaniclastics. The deposits may have formed in rifted basins, in island-arc or back-arc settings, or along a spreading ridge at the continental slope.

Ores are composed of pyrite, pyrrhotite, chalcopyrite, sphalerite, and possibly magnetite, valleriite, galena, bornite, tetrahedrite, cobaltite, cubanite, stannite, and molybdenite. Quartz, carbonate, albite, white mica, chlorite, amphibole, and tourmaline compose the gangue. Pyrite may be framboidal to colliform in brecciated or stringer ore. A weathered gossan often remains.

A median deposit size is 220,000 t, and a median grade is 1.5 percent Cu, based on 44 deposits in Model 24b in Cox and Singer (1986). Silver, gold, and zinc are potential byproducts.

Sediment-hosted disseminations in shale, carbonate, sandstone/quartzite, and red beds, White Pine, Michigan; replacements in carbonate, Ruby Creek, Alaska. A number of stratabound or stratiform disseminated and massive replacement copper sulfide deposits occur in the reduced parts of permeable clastic red beds and in carbonate breccias.

White Pine, Michigan. Deposits of disseminated copper-bearing green or gray shale, siltstone, sandstone, thinly laminated carbonate, evaporate beds, local channel conglomerate, and laminated silty dolomite, exemplified by the deposits at White Pine, Michigan (in Ridge, 1968), form in and along the margins of shallow marine basins (Fig. 17). The deposits range in age from the Middle Proterozoic to early Mesozoic and occur along intracontinental rifts or along spreading areas as well as along passive continental margins. Algal mats, mud cracks, cross bedding, and scour-and-fill channel structures are prominent and provide permeable host sediments. Ore minerals include chalcocite and other

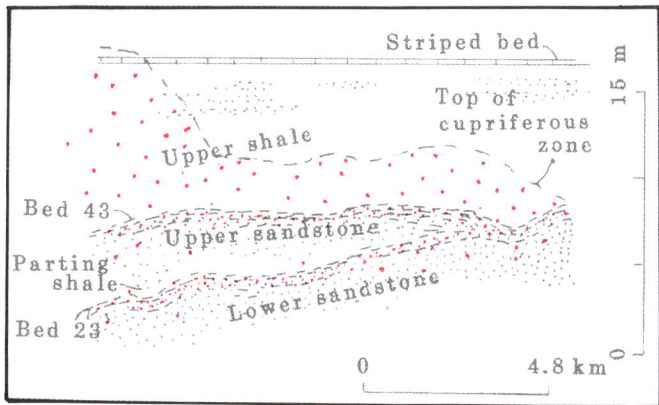

Figure 17. Schematic cross section of the form and regionally transgressive top of the copper-bearing zone (dashed line) in the White Pine district, Michigan, modified from a diagram in Ridge (1968). Note vertical scale exaggeration.

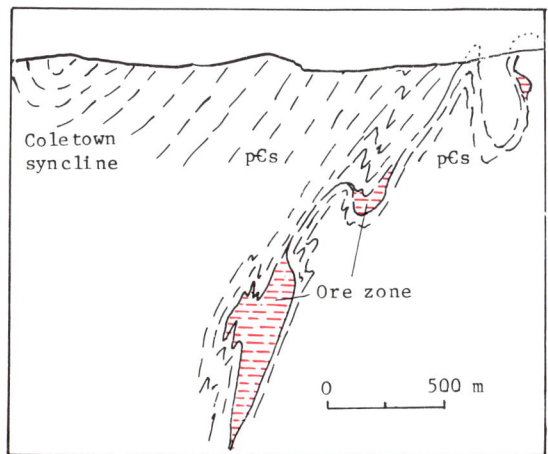

Figure 16. Sketch cross section of the geologic setting for a Besshi-type massive sulfide deposit at Ducktown, Tennessee, modified from a diagram in Ridge (1968). pЄs—Precambrian biotite-sericite schist.

Cu_2S minerals replacing early pyrite (framboidal or colloform) and possibly bornite and native silver. Zoned ores may consist of a central chalcocite and bornite zone, rims of chalcopyrite, and peripheral galena and sphalerite. Cobalt and germanium minerals also may be present. Altered red bed wall rocks near ore deposits are white or gray; regionally they may be altered to purple. Surface exposures may be completely leached, and secondary chalcocite deposited down dip. Ore controls include a reducing (low pH) environment as evidenced by fossil wood and algal mats, abundance of biogenic sulfur, pyritic sediments, and petroleum in paleoaquifers. The footwalls were highly permeable and not faulted. Production plus reserves and resources at White Pine are estimated to aggregate more than 8 million t of copper in ores of average grade of about 1.2 percent (from data in Ridge, 1968; and Cox and others, 1981). Ohle (written communication, 1986) reports that by the end of 1986 the total production of copper at White Pine is expected to exceed that of the Keweenan deposits.

Model 30b in Cox and Singer (1986), based on 57 deposits (9 in the U.S., in Oklahoma, Texas, New Mexico, Maine, Montana, and Michigan), indicates a median size of 22 million t and median grade of 2.1 percent Cu. Examples of these deposits in the U.S. include Creta, Oklahoma, and Nacimiento, New Mexico (red beds); Spar Lake, Montana (quartzite); and White Pine, Michigan (shale). Associated deposits may include layers of halite, sylvite, gypsum, and anhydrite as well as sandstone uranium, volcanogenic red bed and replacement copper.

Ruby Creek, Alaska. Carbonate-hosted copper contains massive copper and base metal sulfide mineral replacements, fillings, and disseminations in limestone and dolomite breccia. Metamorphosed host rocks are fine-grained massive carbonate, carbonaceous laminated stromatolite, dolomite, and shale that may range in age from the Proterozoic to the Pennsylvanian. Deposits form as tabular or pipelike zones along faults, as fillings in karst breccia zones, and as massive replacements of carbonate. The minerals include pyrite, bornite, chalcocite, chalcopyrite, carrollite, sphalerite, and tennantite. Germanium and gallium minerals are commonly present. Dolomitization, siderization, and silicification of the wall rocks may accompany or predate mineralization. Early pyrite or arsenopyrite occurs as fracture fillings and disseminations. The presence of bituminous segregations suggests a reducing environment. Weathering of the ores produced copper carbonates and silicates, iron and cobalt oxides, and cobalt arsenates. The copper reserves at Ruby Creek are estimated at 100 million t, containing 1.2–1.6 percent copper (1.4 million t Cu) (Berg and others, 1964; Nokleberg and others, 1987). Apex Mine in Utah probably is an oxidized example of this type of deposit (Bernstein, 1986).

Magmatic segregations or disseminations in mafic rocks (Duluth, Minnesota; Stillwater, Montana; Gap, Pennsylvania). Large, low-grade occurrences in which copper is generally a byproduct of nickel are mainly subeconomic future resources in the U.S. In addition to the subtypes listed, there undoubtedly are undiscovered additional types.

Duluth, Minnesota. The Duluth Cu-Ni-PGE type includes sporadically distributed massive to disseminated sulfides associated with the basal portions of large layered intrusions found in a cratonal rift zone (Foose, this volume). Potential ore minerals include pyrrhotite, pentlandite, chalcopyrite, cubanite, possible platinum group minerals (PGE), and graphite; locally the sulfides may show evidence of hydrothermal remobilization. Ages of these deposits may range from the Precambrian to the Tertiary (Model 5a in Cox and Singer, 1986).

Stillwater, Montana. The Stillwater Ni-Cu type is a large layered mafic to ultramafic intrusive containing nickel and copper sulfides at the base (Foose, this volume). The deposits may range in age from the Precambrian to the Tertiary. They are found in cratonal shield terranes intruding granitic gneiss or volcanogenic-sedimentary rocks. Potential ore minerals include pyrrhotite, chalcopyrite, pentlandite, cobalt sulfide, and byproduct PGE. Locally the ores are massive, but they may be interstitial to silicates or disseminations. The Stillwater type is associated with the Merensky Reef–type PGE deposits and is commonly underlain by pyritic shales, argillites, graywacke, anhydrite, or other sources of sulfur (Model 1 in Cox and Singer, 1986).

Gap, Pennsylvania. The Gap Ni-Cu type is a synorogenic-synvolcanic Ni-Cu deposit that consists of deformed massive lense, matrix, and disseminated sulfides in small- to medium-size gabbroic intrusives in greenstone belts (Foose, this volume). The age of deposits is mostly Precambrian, but they may have formed as late as the Tertiary. The gabbro was intruded synvolcanically during an orogenic period in a mobile belt or in metamorphosed terranes containing volcanic and sedimentary rocks. Potential ore minerals include pyrrhotite, pentlandite, chalcopyrite, and possibly pyrite, titaniferous magnetite, chrome magnetite, graphite, and byproduct cobalt and PGE minerals.

A grade and tonnage model (Model 7a in Cox and Singer, 1986) based on 32 deposits (4 of which are in the U.S.) indicates a median size of 2.1 million t and a grade of 0.87 percent Ni and 0.47 percent Cu. Other U.S. examples occur in Alaska.

Molybdenum deposits

Less than 5 percent of U.S. molybdenum production comes from non-porphyry types of deposits. These mainly include the vein (upper part of the Questa, New Mexico) and skarn (Pine Creek, California) types. Minor production has come from pegmatites and sediment-hosted types.

Vein (Questa, New Mexico—upper part). Small, high-grade molybdenum-quartz veins in the upper part of the Questa, New Mexico deposit (Fig. 18) were formed along fractures in and along contacts of porphyritic aplite dikes and andesite (in Ridge, 1968). The vein deposits are underlain at depth by a biotite granite pluton, a shallower aplite intrusive, and near-surface porphyry dikes, which intrude a middle Tertiary volcanic field; the larger low-grade dispersed molybdenum deposit occurs mainly in the aplite intrusive. The veins consist mainly of molybdenite and quartz, but some zoned veins contain quartz, K-feldspar, and biotite along the walls and a banded quartz, molybdenite, and pyrite in an inner zone. Where the central part of the vein is reopened locally, fluorite, rhodochrosite, quartz, and calcite occur with minor galena, sphalerite, chalcopyrite, and molybdenite. The andesite was regionally propylitized. In the vicinity of the intrusive center, feldspars in aplite dikes were replaced by silica, sericite, and kaolinite; the andesite was strongly biotized, grading outward into chlorite and epidote. Mineralization was preceded by metamorphism and the introduction of K-spar followed by fracturing and pre-ore hydrothermal alteration, which included decalcification, silicification, sericitization, pyritization, and kaolinization in the vicinity of vent faults. Ore was controlled mainly by three major intersecting shear systems formed and reopened during surges of the intrusive system. Near the surface, molybdenite is oxidized to ferrimolybdite and molybdenum-bearing limo-

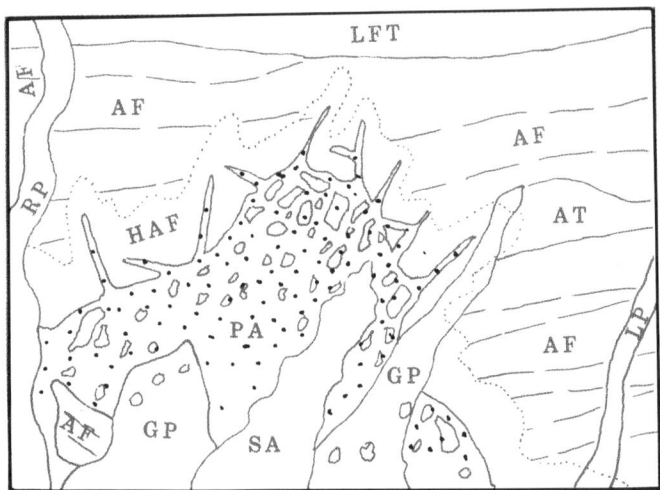

Figure 18. Sketch cross section of the geologic setting at Questa, New Mexico, modified from a diagram in Ridge (1968); no scale given. LFT, latite flows and tuffs; AF, andesitic flows and tuffs; AT, andesitic tuffs; RP, rhyolite porphyry dike; LP, latite porphyry dike; GP, granite porphyry; PA, porphyritic aplite (includes dispersed MoS_2); SA, highly siliceous aplite; and HAF, halo of aplitic flooding (vein zone).

nite and is accompanied by manganese oxides. The ore grade of the vein deposits was about 5 percent, and 9,072 t of molybdenum were produced. Current mining (1986) is in the underlying low-grade (0.25–0.30 percent MoS_2) porphyry ores. Other examples of vein deposits in the U.S. that have either been mined out or are yet a resource include those at Mammoth, Arizona; Cave Peak, Texas; Red Mountain, Colorado (near Urad); and several small deposits in Maine, Pennsylvania, and North Carolina. The Shakan Bay, Alaska, deposit contains reserves of 150–300 t MoS_2.

Skarn (Pine Creek, California). Byproduct molybdenum and copper were produced at Pine Creek, California, from scheelite-bearing skarn (Fig. 19), which was formed by pyrometasomatic replacement of calcareous sedimentary rocks to marble and skarn assemblages along the contact with an acidic granitic intrusive (in Ridge, 1968; Newberry, 1982). The deposits occur intermittently over long horizontal and vertical distances; their form is controlled by the irregularities of quartz-monzonite contacts and the bedding in the marble. The molybdenum grade is higher in the upper parts of the ore zone where a grade of 0.60 to 1.0 percent occurs locally; the average grade is 0.50 percent. Copper increases in the lower parts, grading to as much as 1 percent locally. The Pine Creek Mine contains the main deposit in a roof pendant composed of folded meta-sedimentary and meta-volcanic rocks. Ore is controlled by the geometry of the contact between intrusive rocks and marble layers. Alaskite dikes postdate the skarn alteration. Zones outward from the quartz monzonite include silicified quartz monzonite, epidote skarn, amphibolite skarn, normal and light skarn, a thin rind of calc-silicate, and marble. In marble the ore is composed of pyrite, pyrrhotite, possible chalcopyrite, and molybdenite, but no scheelite, which occurs in the calc-silicate-skarn zones. Amphibolite skarn may contain chalcopyrite replacing magnetite and bornite rims on chalcopyrite. Epidote skarn contains molybdenite and chalcopyrite but no scheelite. Cross-cutting quartz veins are unmineralized. Scheelite was formed early during the contact metamorphism phase; the sulfides were produced during subsequent hydrothermal alteration. Production from this deposit (in Ridge, 1968) was 8,530 t of Mo, 9,799 t of Cu, and 3.4 million units of WO_3 from a total of 6.6 million t of ore. Other examples of this type include Strawberry, California; Lost Creek, Montana; and Osgood Mountain, Nevada (Newberry, 1982).

Pegmatite and sediment-hosted. Molybdenum occurring in pegmatites and stratabound sedimentary rocks is mostly of subeconomic quality, although potentially of large quantity. Pegmatite bodies and aplite dikes may contain erratically distributed coarsely crystalline rosettes, thick books, or flakes of molybdenum but are seldom of ore grade (King and others, 1973). On the other hand, lignitic and arkosic sandstones in the Dakotas and eastern Montana contain minable thicknesses and grades (0.1 to 0.2 percent Mo). King and others (1973) also report that bedded sandstone uranium deposits in Arizona, New Mexico, Utah, Wyoming, South Dakota, and the Gulf Coast of Texas (Shawe, this volume) contain irregular concentrations in

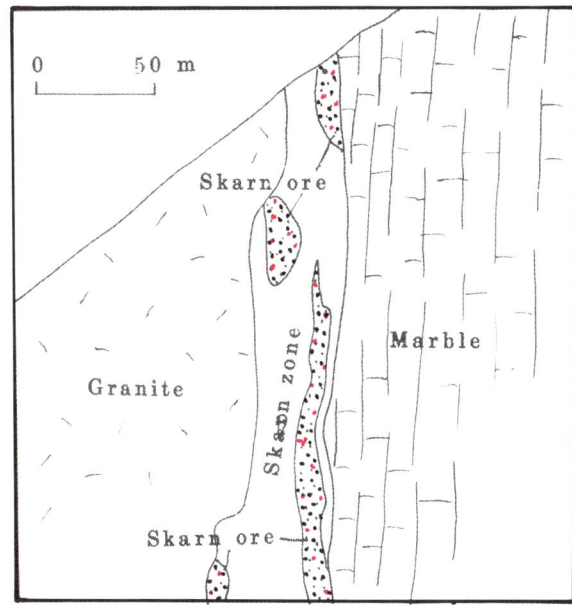

Figure 19. Sketch cross section of the geologic setting for skarn ores at Pine Creek, California, modified from Bateman (1982).

significant amounts, which may be considered as potential sources of byproduct molybdenum. Erickson and others (1981, 1985) noted extensive molybdenum-rich organic layers in Paleozoic limestone units in Missouri, and a similar occurrence was described in Georgia by Foss and others (1983). However, none of these currently are of mineable thickness or grade.

COPPER AND MOLYBDENUM RESOURCES IN GEOLOGIC TERRANES

A promising area of future resource research suggested by this review of copper and molybdenum is the correlation of resource occurrence (terranes) with geologic terranes that were produced sequentially by plate tectonic processes. Formation of the continental crust, the generation of oceanic crust, and distribution of terranes along the margin of the continent by accretionary (subduction, obduction, and/or lateral displacement) processes (Jones and others, 1983) also were accompanied by formation of distinctive types of ores in each. The types and occurrences of deposits associated in generally geologically simpler accreted crustal terranes of Phanerozoic ages were described by Albers (1981). More complex cratonal terranes, which contain Precambrian basement and Phanerozoic cover rock and may have formed originally by accretionary processes, were subjected to one or more orogenic cycles, and their constituent ore deposits were modified or remobilized. Recognition of them will be more difficult. However, the characteristic lithologic and structural attributes of individual geologic terranes and associated resources within the craton shields, platforms, and shelves are now beginning to be detected (e.g., Bennett, 1986; Sims and Peterman, 1986).

Exploration for and discovery of conventional and new types of copper and molybdenum deposits and the assessment of their resources in the United States undoubtedly will benefit from research efforts to discern and correlate the presence of model deposit types that occur in "model" cratonal and accreted terranes. Guild (1972, 1978) and Albers (1983) were among those who indicated ways to begin such studies in the western United States. The distribution of copper and molybdenum districts in Plate 1 and the model deposit types described in this chapter should provide an additional base for extending these studies nationwide.

WORLD SUPPLIES AND U.S. DEMAND FOR COPPER AND MOLYBDENUM

Adequate world supplies of copper and molybdenum are contingent primarily on their geologic and economic availability and substitutability, factors that are important in assuring expected demands for these commodities in the U.S.

Geologic Availability

The average amount of copper in the earth's crust is about 58 ppm, most of which is not readily available. Molybdenum is a much rarer element, ranging from 1 to 1.5 ppm. Skinner and others (1975) point out that most of the metals in the crust are locked in solid solution within silicate minerals. Enormous amounts of energy are required to disrupt the host silicate structures and to concentrate and recover these metals from crustal rocks. Fortunately, local metal concentrations (around 1 percent) also occur as sulfide and oxide minerals, from which copper and molybdenum can be separated by currently available economic methods. The challenge is to locate minable concentrations in limited parts of the earth's crust from which most of the near-surface high-grade ores of conventional types have already been found.

Nearly half of the world's identified resources of molybdenum occur in the U.S., and 58 percent of the world's copper resources occur in the Western Hemisphere (Table 2). More than 70 percent of U.S. copper and most molybdenum resources are located in the western U.S. Recent exploration also has revealed porphyry and massive sulfide type deposits in the Appalachian regions of the U.S. Large subeconomic resources of copper or molybdenum also occur in Alaska, South Dakota, and Minnesota. Extensive subeconomic resources of copper (Cox and others, 1973) and molybdenum (King and others, 1973) are already known or predicted on lands around the Pacific Basin. Skinner and others (1975) conclude that the magnitude of recoverable copper-bearing manganese nodules on the deep sea floor of the Pacific Ocean near Hawaii is as large as those identified on land (Foose, this volume).

Thus, a reasonable conclusion is that adequate supplies of copper and molybdenum are geologically available in the U.S. and elsewhere in the world and that the prospect for additional discoveries is excellent. The problem then becomes one of assuring their deliverability.

Economic availability

A number of economic factors have had an impact on the production of copper and molybdenum as well as on where it has occurred: available and accessible deposits of minable size and grade, favorable metal prices and assured user demand, adequate financial backing for production of new-mined metal in competition with substitute or recycled materials, manageable production and environmental costs, continued investment in improving discovery and production technology, and a viable continuing exploration program to assure adequate reserves. During the last three decades, domestic copper production has been marked by periods of boom and bust caused by fluctuations in world economic conditions, financial manipulations, periodic labor unrest, the needs of wartimes, and more recently, governmental requirements for environmental protection (Fig. 1), which are discussed more fully by Jolly (1985) and Blossom (1985). Several factors are directly or indirectly related to geologic considerations.

The size, grade, and minability of primary and by-product copper and molybdenum ores have become serious domestic mining considerations. Jolly (1985) points out that the shift toward the mining of lower-grade ores by less costly open-pit and block-caving methods has increased the tonnages mined and permitted lowering the average grade of copper from 2 percent in the 1900s to 0.6 percent in 1983. One result has been that some undepleted higher grade ores in underground mines have become subeconomic. Byproduct metals such as molybdenum, gold, silver, cobalt, and uranium from copper mines and tin, tungsten, pyrite, and rhenium from molybdenum mining of porphyry deposits may have helped to keep some domestic mines open. A more serious fact is that many of the foreign porphyry copper deposits are of higher grade than those remaining in the U.S.; they are larger, and have correspondingly much lower mining costs.

TABLE 2. WORLD RESOURCES OF COPPER AND MOLYBDENUM

	Copper* (mil. t)	Molybdenum† (mil. Kg)
North America	150	14,500
South America	140	6,100
Europe	70	2,120
Africa	70	50
Asia	35	3,000
Oceania	35	180
Total	500	25,000

*Jolly, 1985.
†Blossom, 1985

Pollution control and mined-land reclamation are significant domestic economic factors. Open pit mining requires large land areas for the extraction and disposal of waste material, the reclamation of mined areas, and the disposition of effluents. Protection of the quality of underground water supplies as well as maintenance of pollution-free clean air is now required by law, adding domestic production costs not encountered by many foreign producers. Loss of domestic smelting capacity in the U.S. is a major problem that has resulted, in part, from inability to economically resolve pollution controls.

Substitutes

Substitute materials for both copper and molybdenum are available; the impact of substitution on copper demand is greater than that for molybdenum. Copper and its alloys may be replaced by aluminum, titanium, carbon steel, stainless steel, zinc, glass, and plastics; many of these are more costly than copper. Molybdenum substitutes, possible for numerous metallurgical and chemical uses, include boron, columbium, chromium, manganese, nickel, graphite, tantalum, tungsten, plastics, and ceramics. However, as with copper, these may not be acceptable substitutes for some uses because molybdenum provides special properties that have superior performance characteristics. In addition, in the case of molybdenum, the small amount used may make any cost difference negligible.

Demand

Forecast estimates of U.S. and foreign demands vary from an anticipated modest increased use of copper (primarily in the electrical, electronic, and construction industries) and of molybdenum (for alloys, pigments, and lubricants) to continuing stagnation of metal consumption (Jolly, 1985, 1986; Blossom, 1985, 1986; Mining Journal, 1986). Miniaturization, substitute uses of other materials, and recovery of scrap undoubtedly will tend to moderate future demands for mine production. Increased conservation in use of metals has also been an important factor. The key to expanding future needs for these metals is an increase in the rate of economic growth in the world, barring the extraordinary requirements of wartime uses.

OUTLOOK FOR FUTURE U.S. ACCESS TO COPPER AND MOLYBDENUM RESOURCES

The preceding data indicate no anticipated near-term shortage of geologically available copper and molybdenum in the United States. But a present oversupply of these metals and resulting low prices for them (below domestic production costs) have resulted in closure of many domestic mines and smelters and an increased dependence on imports. Is the current downturn a temporary condition or a continuing trend? Among potential threats to long-term resource availability, according to Tilton and Landsberg (1983), are the failure to replace depleted resources as well as insufficient investments in mine and processing facilities. Will reduced industry activity in the U.S. have an impact on future access to these resources and on sustained exploration and discovery of new copper and molybdenum domestic reserves?

The trend toward decreasing domestic production and reliance on imports of copper to meet anticipated demand may not be reversible in the near future. Hewett (1929) observed that early development of base and precious metal resources in several northern European nations proceeded through a sequence of discrete stages of resource activity. Plots of mining activity showed successive overlapping peaks: (1) early mine production, mostly for export, to be refined and fabricated elsewhere; (2) increased mining activity—a peak in the number of mines brought into production that stimulated the next peak; (3) a peak development of indigenous smelting and refining facilities. This resulted in (4) a maximum in the production of metals from ores. A final stage (5) occurred when mining declined, owing to the depletion of reserves or the availability of lower-priced/better quality resources elsewhere. Waters (1986) pointed out that an expanding copper resource base combined with new, more efficient mining technology in the late nineteenth century to lower the price of metal and displace the more costly European production to areas such as the United States. After nearly 100 years of operation, the U.S. copper mining industry finds it necessary to compete with substantially richer foreign ore grades and is transferring its exploration and production efforts overseas. It is ironic that the present oversupply of these metals is, in part at least, the result of oversuccess in discovering new deposits. The shift to imports of copper also seems to be occurring at the same time that the nation is moving away from its historical industrial production base toward a more service-oriented economy. Domestic production of molybdenum ores, on the other hand, is expected to resume once current surplus supplies are exhausted. Substantial amounts will continue to be exported. The high U.S. endowment of molybdenum resources compared with those elsewhere, molybdenum's essential (75 percent) use in iron and steel products, and the low substitutability and recyclability of molybdenum are among the factors expected to reverse the downward trend in demand for molybdenum and cause a resumption of domestic production.

If history is repeated then, except for discovery of new bonanza copper deposits or higher metal prices which would permit U.S. deposits to compete once more, users of copper will adjust to importation. Unmined domestic copper ore reserves will become a resource "bank account" until a time when domestic mining may once again be able to compete with the then depleted or inaccessible foreign ores. Fortunately, assured supplies of copper for import are available from politically stable and secure nations such as Canada and Australia, as well as from stage 1 developing nations that are anxious to encourage trade with the U.S.

Porphyry-type deposits, and particularly those that closely adjoin higher grade skarn, vein, and replacement deposits (e.g., Bingham, Utah), remain the best sources of copper and molybdenum in terms of size, grade, and minability, as shown in

TABLE 3. COMPARATIVE MEDIAN GRADES AND MEDIAN TONNAGES OF COPPER
AND MOLYBDENUM ORES IN MODEL DEPOSIT TYPES*

No.	Types of deposits	Median grade (percent)	Median tonnage (millions)	Median tonnes metal (millions)
1.	Cu porphyry	0.54	140	0.76
2.	Cu-Au porphyry	0.5	100	0.5
3.	Cu-Mo porphyry	0.42	500	2.1
4.	Granite-high F Mo porphyry	0.19	200	0.38
5.	Cal-Alk-low F Mo porphyry	0.085	94	0.08
6.	Cu (mineraliz. porph) skarn	0.98	80	0.78
7.	Cu (non-mineraliz. porph) skarn	1.7	0.56	0.01
8.	Polymetal vein Cu	0.094	1.8	0.02
	Massive sulfide:			
9.	kuroko	1.3	1.5	0.02
10.	Cyprus	1.7	1.6	0.03
11.	Besshi	1.5	0.22	0.003
12.	Sediment-hosted Cu	2.1	22	0.46
13.	Magmatic segg. (Gap) Cu	0.87	2.1	0.02

*Based on data from the compilation edited by Cox and Singer (1986).

Tables 3 and 4, and Figure 20. The most promising areas for porphyry copper deposit exploration seem to be along the rifted margins of the cratonal platform and associated volcanic belts in accreted oceanic crustal terranes around the Pacific Ocean basin. The accreted terranes also are favorable sites for the accumulation of copper in smaller but abundant volcanogenic massive sulfide deposits. The largest dispersed molybdenum deposits, however, occur within the craton platform terrane.

New types of copper and molybdenum deposits are also being sought. The Ruby Creek, Alaska, carbonate-hosted ores represent an important new type of large non-porphyry copper deposit that may be expected to occur elsewhere in Alaska and the southwestern Great Basin. Potentially large carbonate-hosted stratabound deposits of molybdenum have been identified but not fully studied and evaluated in Phanerozoic rocks in Missouri (Erickson and others, 1981) and in Georgia (Foss and others, 1983). Sims (oral communication, 1986) reports geologic and geophysical evidence emerging from research in Precambrian platform terranes that strongly suggests the possible existence of an Olympic Dam–type (Australian) deposit (see Model 29b in Cox and Singer, 1986) in the midcontinent craton. The bottom line is that we still have much to learn about the geologic occurrences and distribution of metals such as copper and molybdenum, even in a nation as well-explored as the United States.

Increasingly we live on an interdependent earth. Mineral deposits are both where you find them or can produce them—here or there. If domestic production of copper continues to be economically unfeasible, we must continue to depend on imports.

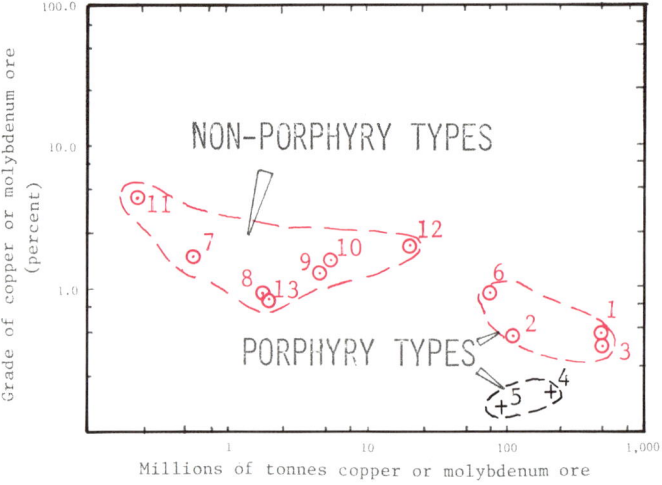

Figure 20. Plot showing comparative ranges in median ore grades and median tonnages for model porphyry and non-porphyry copper (in red) and molybdenum (in black) ore types described in this report. Total metal for a hypothetical median-size deposit is shown.

Future access to supplies from abroad undoubtedly will require coordinated planning between the domestic industry and governmental and private resource research organizations. Most competing industrialized nations have already adopted that strategy.

TABLE 4. COMPARATIVE TONNAGES OF COPPER AND/OR MOLYBENUM PRODUCTION AND RESERVES PLUS RESOURCES IN U.S. DEPOSITS*

Deposit/type	Average grade (percent)	Production	Reserves and resources	Total metal
		(in millions of tonnes)		
Bingham, porphyry Cu	0.70	10.7	9.5	20.2
Mo	0.034	0.07		
Climax porphyry Mo	0.4	0.832	1.6	2.43
Kennecott volc. red bed- native Cu	1.3	0.54	none(?)	0.54
Carr Fork, skarn Cu	1.8	minor	0.99	0.99
Pine Creek, skarn Mo	0.12	0.008	none(?)	0.008
Butte, vein Cu	0.65	9.0	14.6	23.6
Questa, vein Mo	5	0.010	none(?)	0.010
Superior, replacement Cu	4.5	0.31	0.45	0.76
Crandon, kuroko massive sulfide Cu	1.0	none	0.60	0.60
Ducktown, Besshi massive sulfide Cu	1.0	0.55	0.4	0.95
White Pine, shale-hosted Cu	1.2	0.46	7.6	8.06
Ruby Creek, carbonate-hosted Cu	1.4	none	1.4	1.4
Duluth, Mag. segg Cu	0.9	none	1.8	1.8

*This report.

REFERENCES

Adamides, N. C., 1980, The form and environment of formation of the Kalaoos ore deposits, Cyprus, *in* Panayiotou, A., ed., Ophiolites, Proceedings International Ophiolite Symposium, Nicosia, Cyprus, 1979: Cyprus Geological Survey Department, p. 663–674.

Albers, J. P., 1981, A lithologic tectonic framework for the metallogenic provinces of California: Economic Geology Bulletin, v. 76, no. 4, p. 765–790.

——, 1983, Distribution of mineral deposits in accreted terranes and cratonal rocks of western United States: Canadian Journal of Earth Sciences, v. 20, no. 6, p. 1019–1029.

Atkinson, W. W., Jr., and Einaudi, M. T., 1978, Skarn formation and mineralization in the contact aureol at Carr Fork, Bingham, Utah: Economic Geology v. 73, no. 7, p. 1326–1365.

Bateman, P. C., 1982, Scheelite-bearing skarns of east-central Sierra Nevada, California, U.S.A., *in* Tungsten geology symposium, Beijing, China: Jiangxi, China Geological Publishing House, p. 23–31.

Bennett, E. H., 1986, Relationship of the trans-Challis fault system in central Idaho to Eocene and Basin and Range extensions: Geology, v. 14, no. 6, p. 481–484.

Berg, H. C., Eberlein, G. D., and MacKevett, E. M., Jr., 1964, Metallic mineral resources, *in* Mineral and water resources of Alaska: Committee Print, Committee on Interior and Insular Affairs, U.S. Senate, 88th Congress, 2d Session, p. 95–124.

Bernstein, L. R., 1986, Geology and mineralogy of the Apex mine, Washington County, Utah: U.S. Geological Survey Bulletin 1577, 20 p.

Blossom, J. W., 1985, Molybdenum, *in* Mineral facts and problems, 1985 ed.: U.S. Bureau of Mines Bulletin 675, 14 p.

——, 1986, Molybdenum, *in* Mineral commodity summaries, 1986: Washington, D.C., U.S. Bureau of Mines, p. 106–107.

Cobb, E. H., 1960a, Copper, lead, and zinc occurrences in Alaska: U.S. Geological Survey Mineral Investigations Resource Map MR-9, scale 1:2,500,000.

——, 1960b, Molybdenum, tin, and tungsten occurrences in Alaska: U.S. Geological Survey Mineral Investigations Resources Map MR-10, scale 1:2,500,000.

Cox, D. P., and Singer, D. A., eds., 1986, Mineral deposit models: U.S. Geological Survey Bulletin 1793, 379 p.

Cox, D. P., Schmidt, R. G., Vine, J. D., Kirkemo, H., Tourtelot, E. B., and Fleischer, M., 1973, Copper, *in* Brobst, D. A., and Pratt, W. P., eds., United States mineral resources: U.S. Geological Survey Professional Paper 820, p. 163–190.

Cox, D. P., Wright, N. A., and Coakley, G. J., 1981, The nature and use of copper resource and resource data: U.S. Geological Survey Professional Paper 907-F, p. F1–F20.

Erickson, R. L., Mosier, E. L., Odland, S. K., and Erickson, M. S., 1981, A favorable belt for possible mineral discovery in subsurface Cambrian rocks in southern Missouri: Economic Geology, v. 76, no. 4, p. 921–973.

Erickson, R. L., Erickson, M. S., Mosier, E. L., and Chazin, B., 1985, Summary geochemical and generalized geologic maps of the Springfield 1° × 2° quadrangle and adjacent area, Missouri: U.S. Geological Survey Miscellaneous Field Studies Map, MF-1830A, Pamphlet, 51 p.

Foss, D. W., Gatten, O. J., and Young, R. S., 1983, Shiloh Church molybdenum deposit, Polk County, Georgia: Society of Mining Engineers of AIME, Preprint No. 83-86, 7 p.

Franklin, J. M., Sangster, D. M., and Lydon, J. W., 1981, Volcanic-associated massive sulfide deposits: Economic Geology 75th Anniversary Volume, p. 485–627.

Guild, P. W., 1972, Metallogeny and the new global tectonics: International

Geologic Congress 24th, Montreal 1972, Proceedings, p. 17–24.

——, 1978, Metallogenesis in the western United States: London, Journal Geological Society, v. 35, part 4, p. 355–376.

——, 1981, Preliminary metallogenic map of North American; A numerical listing of deposits: U.S. Geological Survey Circular 858-A, 93 p., scale 1:5,000,000.

Hall, W. E., Schmidt, E. A., Howe, S. S., and Broch, M. J., 1984, The Thompson Creek, Idaho, porphyry molybdenum deposit; An example of a fluorine deficient molybdenum granodiorite system: Proceedings of the Quadrennial IAGOD Symposium, Stuttgart, Germany, p. 349–357.

Hewett, D. F., 1929, Cycles in metal production: American Institute of Mining and Metallurgical Engineers Technical Publication no. 183, p. 3–31.

James, L. P., 1978, The Bingham copper deposits, Utah, as an exploration target; History and pre-excavation geology: Economic Geology, v. 73, no. 7, p. 1218–1227.

John, E. C., 1978, Mineral zones in the Utah copper orebody: Economic Geology, v. 73, no. 7, p. 1250–1259.

Jolly, J.L.W., 1985, Copper, *in* Mineral facts and problems, 1985 ed.: U.S. Bureau of Mines Bulletin 675, p. 197–221.

——, 1986, Copper, *in* Mineral commodity summaries, 1986: Washington, D.C., U.S. Bureau of Mines, p. 42–43.

Jones, D. L., Howell, D. G., Coney, P. J., and Morgen, H.W.H., 1983, Recognition character and analysis of tectonic stratigraphic terranes in western North American: Journal of Geological Education, v. 31, p. 295–303.

Joralemon, I. B., 1973, Copper, the encompassing story of mankind's first metal: Berkeley, California, Howell-North Books, 407 p.

King, R. V., 1970, Molybdenum in the United States exclusive of Alaska and Hawaii: U.S. Geological Survey Mineral Investigations Resource Map MR-55, scale 1:3,168,000.

King, R. V., Shawe, D. R., and Mackevett, E. M., Jr., 1973, Molybdenum, *in* Brobst, D. A., and Pratt, W. P., eds., United States mineral resources: U.S. Geological Survey Professional Paper 820, p. 425–435.

Kinkle, A. R., Jr., and Peterson, N. P., 1962, Copper in the United States exclusive of Alaska and Hawaii: U.S. Geological Survey Mineral Investigations Resource Map MR-13, scale 1:3,168,000.

Miller, R. N., 1973, Production history of the Butte district and geological function, past and present, *in* Miller, R. N., ed., Guidebook of the Butte field meeting of S.E.G., Butte, Montana, August, 1973: Butte, Montana, Anaconda Co., p. F1–F10.

Mining Journal (London), 1986, Sticking with copper: Mining Journal, v. 306, no. 7849, p. 49–51.

Newberry, R. J., 1982, Tungsten-bearing skarns of the Sierra Nevada; I. The Pine Creek mine, California: Economic Geology, v. 77, no. 4, p. 823–844.

Nokleberg, W. J., and 7 others, 1987, Significant metalliferous lode deposits and placer districts of Alaska: U.S. Geological Survey Bulletin 1786, 104 p.

Rabbitt, M. C., 1979, Minerals, lands, and geology for the common defense and general welfare; Vol. 1, Before 1879: Washington, D.C., U.S. Government Printing Office, 331 p.

——, 1980, Minerals, lands, and geology for the common defense and general welfare; Vol. 2, 1879–1904: Washington, D.C., U.S. Government Printing Office, 405 p.

Ridge, J. D., ed., 1968, Ore deposits of the United States, 1933–1967; The Graton-Sales volume: New York, American Institute of Mining, Metallurgical and Petroleum Engineers, 1880 p.

Short, M. N., Galbraith, F. W., Harshman, E. N., Kuhn, T. H., and Wilson, E. D., 1943, Geology and ore deposits of the Superior Mining area, Arizona: Arizona Bureau of Mines Bulletin 151, Geology Series 16, 159 p.

Sims, P. K., and Peterman, Z. E., 1986, Early Proterozoic central plains orogen; A major buried structure in the north-central United States: Geology, v. 14, no. 6, p. 488–491.

Skinner, B. J., and others, 1975, Resources of copper, *in* Skinner, B. J., Chairman, Committee on mineral resources and the environment report, Mineral resources and the environment: Washington, D.C., National Academy of Sciences, p. 127–186.

Sutulov, A., ed., 1978, International molybdenum encyclopedia, 1778–1978; Vol. I, Resources and production: Santiago, Chile, International Publishing Co., 402 p.

——, 1983, International molybdenum yearbook 1983: Santiago, Chile, International Publications, p. 64–66.

Theodore, T. G., and Blake, D. W., 1978, Geology and geochemistry of the West ore body and associated skarns, Copper Canyon porphyry copper deposits, Lander County, Nevada: U.S. Geological Survey Professional Paper 798-C, p. C1–C85.

Tilton, J. E., and Landsberg, H. H., 1983, Non-fuel minerals; The fear of shortages and the search for policies, *in* Castle, E. N., and Price, K. A., eds., U.S. interests and global natural resources, energy, mineral, food: Washington, D.C., Resources for the Future, p. 48–80.

Titley, S. R., ed., 1982, Advances in geology of the porphyry copper deposits: Tucson, University of Arizona Press, 559 p.

Titley, S. R., and Hicks, C. L., eds., 1966, Geology of the porphyry copper deposits, southwestern North America: Tucson, University of Arizona Press, (Wilson Volume), 287 p.

Tooker, E. W., 1980, Preliminary map of copper provinces in the conterminous United States: U.S. Geological Survey Open-File Report OF-79-576D, scale 1:5,000,000.

Waters, P. D., 1986, International mineral development; A bankers view: Bulletin of Canadian Institute of Mining, v. 79, no. 887, p. 100–103.

Westra, G., and Keith, S. B., 1981, Classification and genesis of stockwork molybdenum deposits: Economic Geology, v. 76, no. 4, p. 844–873.

Manuscript Accepted by the Society May 16, 1987

NOTE ADDED IN PROOF:

A somewhat pessimistic outlook for future domestic copper production, from the mid 1980s perspective of the original assessment above, may have been somewhat premature. By 1990 the picture had improved. According to U.S. Bureau of Mines (1990) information, several favorable economic factors occurred between 1986 and 1989 to cause this improvement. These included sustained higher prices for the metal, lower energy costs, more favorable interest rates, increased copper consumption, the depletion of inventories, improved productivity, and the development of environmental-compliance technology. These prompted the reopening of important copper mines, such as Bingham, Utah (Tooker, 1990a), and Butte, Montana (Tooker, 1990b), and resulted in both increased domestic copper production and increased recycling of copper scrap. The results of advanced materials research, like the recently discovered and potentially important ceramic superconducting material (containing copper and other metals) that can operate at very high temperatures, will continue to spur development of new uses for copper.

There was a similar increase in domestic molybdenum production, owing primarily to expanding export markets during this period. This industry also continued research to develop new alloys from molybdenum.

World resource estimates of copper and molybdenum have been revised only slightly since 1985 (Table 2, above, and USBM, 1990). The amounts of copper reported for North America and Africa have decreased, while those for South America and Oceania increased. The world resource estimates of molybdenum have not changed appreciably in this time interval.

Whether the apparent economic revival, which began in the late 1980s, is only a minor aberration in an ongoing, long-term, downward trend in the domestic production of these metals remains to be determined. It is an encouraging break in the long-term pattern. However, without substantial improvement in world political and economic conditions (including replacement of copper and molybdenum supplies depleted by a prolonged extension of the international war effort in the Persian Gulf region), the original long-term prognosis probably is still inescapable.

Tooker, E. W., 1990a, Gold in the Bingham district, Utah, *in* Shawe, D. R., Ashley, R. P. and Carter, L.M.H., eds., Gold in porphyry copper systems: U.S. Geological Survey Bulletin 1857-E, p. E1–E16.

——, 1990b, Gold in the Butte district, Montana, *in* Shawe, D. R., Ashley, R. P., and Carter, L.M.H., eds., Gold in porphyry copper systems: U.S. Geological Survey Bulletin 1857-E, p. E17–E27.

U.S. Bureau of Mines, 1990, Mineral commodity summaries, 1990: Washington, D.C., U.S. Bureau of Mines, p. 52–53, 114–115.

Printed in U.S.A.

Chapter 4

Lead and zinc deposits

Ernest L. Ohle
8989 Escalante, No. 120, Tucson, Arizona 85730

THE LEAD INDUSTRY

Lead has been a much-sought-after commodity in North America since early Colonial times. Although more sophisticated uses, such as in pipes and vessels in water systems, coins, pottery glaze, and roofing, were common in Babylonian, Egyptian, and Roman cultures, the principal need for lead on the American frontier was to make bullets. Military success in the frontier wars and in the Civil War was affected by the ready availability, or lack, of ammunition. Hence, as the population moved westward from the Atlantic seaboard and spread along the Ohio and Mississippi River systems from New Orleans, localities where lead mineralization cropped out soon attracted attention and became population centers. The Wisconsin-Illinois District (discovered in 1682), southeast Missouri (1720), Austinville, Virginia (1750), and Rosiclaire, Illinois (1830), are notable examples. There were also numerous showings that never developed into large producers.

Early discovery of lead deposits was facilitated by the fact that galena, the most common lead mineral, is resistant to weathering and persists in the clayey residuum. In southeast Missouri, the greatest North American lead-producing region, all of the production for the first 150 years was from such residuum rather than from solid rock.

Plate 1C shows the locations of the deposits that have produced over 95 percent of all the lead mined to date in the United States. The productive longevity of some of the early-found eastern deposits is truly remarkable.

Ammunition is still a significant use for lead, but since the beginning of the Industrial Revolution in the 19th century, the applications of lead have become more diverse and consumption has grown. Its valuable characteristics are: low melting point, high density, softness, ease of forming, and corrosion resistance. The relative importance of various uses has changed over time in response to new inventions such as telephone and electric systems, which use lead for cable sheathing; the automobile, which uses lead for batteries and as a gasoline additive; and in a negative way, in response to environmental legislation. Laws passed since 1974 have sharply reduced the permissible lead content in paint and gasoline. In 1973, the lead content in gasoline averaged 0.5 g/l (1.9 g/gal). On January 1, 1986, the legal limit became 0.026 g/l (0.1 g/gal). This reduction in use between 1973 and 1986 resulted in an annual market loss to the lead industry of 172,000 metric tonnes (190,000 short tons), or 42 percent of the current annual mine production. Lead demand also has been much affected in recent years by competition from substitute materials such as plastics, which have taken much of the cable and pipe markets, and titanium oxide, which now is the dominant white pigment in paints. Satellite television and radio transmission have sharply reduced the need for cables that previously might have required lead.

Indeed, the lead industry is much in need of large new uses to replace these lost markets. Batteries now consume 70 percent of all the lead used, and much of this is recycled metal from a large secondary industry. It just recirculates. Forty-five percent of the U.S. lead supply is reclaimed metal. Loss of the lead-acid battery business to one of the possible substitutes using cadmium, mercury, nickel, iron, silver, or zinc would be a serious blow to the industry. However, for the moment, the lead-acid battery market is secure and growing at nearly 20 percent per year. In 1985, nearly 75 million lead-acid batteries were sold. An interesting recent development is the commissioning by Duke Power Company of an experimental lead-acid battery installation that will be used as a standby unit to meet peak power demands.

The price of lead, and also of zinc, have moved in close tandem with industrial activity. Times of war brought great demand, but prices were fixed by the government. Both metals are world commodities, so U.S. producers compete with overseas sources. Tariff protection is minimal for concentrates but is significant for metal. For lead, Canada, Mexico, Australia, and Peru are the chief foreign shippers to the United States, which produces about 14 percent of the world's newly mined lead and consumes 21 percent. An interesting discussion of international patterns of lead trade may be found in Easton (1986).

Development of lead-bearing deposits is much influenced by the not-infrequent presence of silver in the lead minerals. Most of the lead-bearing ore in the eastern United States has low silver content. But, as westward expansion spread to the Rockies and beyond, much of the lead ore that was found had more silver, and indeed, the exploration that found it was looking for precious metals. Silver occurs with, and often in, the lead mineral galena,

Ohle, E. L., 1991, Lead and zinc deposits, *in* Gluskoter, H. J., Rice, D. D., and Taylor, R. B., eds., Economic Geology, U.S.: Boulder, Colorado, Geological Society of America, The Geology of North America, v. P-2.

TABLE 1. U.S. MINE PRODUCTION OF LEAD BY STATES*

State	1906	1920	1940	1960	1983
Colorado	45	21	10	16	withheld
Idaho	106	108	95	39	24
Montana	2	14	21	5	1
Utah	51	61	69	35	......
Washington	......	3	2	7	......
Missouri	101	147	156	102	371
Oklahoma	......	58	19	1	......
Others	12	39	42	18	11
Total	317	451	414	223	407

*Thousands of metric tons.
Data from U.S. Bureau of Mines *Mineral Trade Notes*.

and this added value has always been a major economic difference between eastern and western mines. The economic success of western lead properties has to a large degree been determined by the amount of silver present because it offsets the lower ore production costs in the huge, shallow, eastern mines. Table 1 shows the lead production by states at five intervals over the past 80 years. It is evident that the preponderance of production has come from east of the Rocky Mountains, with significant amounts of western ore coming only from Utah, Idaho, Colorado, and Montana. Each of these four states produces silver-bearing ore. Even so, Missouri, with virtually no silver, has been the largest lead-producing state every year since 1904, with one exception (a strike year). In 1985, Missouri produced 90 percent of all of the domestic primary lead.

THE ZINC INDUSTRY

The demand for zinc was much slower to develop than that for lead, and it was not until late in the 19th century that its use rose to significant levels; it now stands third among the nonferrous metals in annual consumption. Today zinc is a vital and versatile component of modern living. The automobile industry uses about one-third of U.S. consumption, as die-cast hardware and ornaments and in protective undercoating. Other important uses are as a protective coating on steel in other manufactured products (galvanizing), and as a chemical component (usually zinc oxide) in rubber and paints. In 1984, the United States produced 5 percent of the world's zinc and consumed 17 percent. The United States was the world's largest producer from 1901 to 1971.

Zinc mineralization occurs in many states (Plate 1D), but most of the production has come from Tennessee, New York, New Jersey, Missouri, Oklahoma, Kansas, Wisconsin, Colorado, Idaho, Utah, and Montana. In recent years, Tennessee, New York, and Missouri have yielded by far the greatest tonnages (Table 2), and this should continue for several decades. Table 2 also records the rise and fall of production from Colorado; the Coeur d'Alene district in Idaho; Franklin Furnace, New Jersey; and the Tri-State District of Missouri-Kansas-Oklahoma; as well as the growth in output at Balmat, New York; Viburnum, Missouri; and the East and Central Tennessee Districts.

The most common zinc-bearing mineral, sphalerite, commonly occurs with lead, but the important deposits in Tennessee and at Franklin Furnace, New Jersey, are essentially lead free. The zinc/lead ratio in the now-closed Tri-State District was 5 to 1, and in the Upper Mississippi Valley District (Wisconsin and Illinois) it was 1.5 to 1. In the undeveloped volcanogenic deposit at Crandon, Wisconsin, zinc and copper occur together in a ratio of 5 to 1. Thus, zinc more often than not is a co-product or by-product. In the western states it generally is of secondary importance to the lead-silver values. A major technological advance in the early part of this century was the development of the froth-flotation process, which made it possible to separate different minerals in ore, and to make cleaner, higher-grade concentrates. Prior to that time, zinc was quite frequently regarded as a lead-ore contaminant rather than a valuable co-product.

In contrast to galena, sphalerite does not tend to persist in weathered residuum. Rather it is oxidized to smithsonite and hemimorphite, zinc carbonate and silicate, respectively. For many years in the midcontinent region, important production of oxidized ore took place in north Arkansas, east Tennessee, and the Tri-State District. Some early smelters were designed to treat only oxidized zinc concentrates; today zinc plants are designed to treat only zinc sulfide.

In recent decades, the zinc industry, as well as the lead industry, has become concentrated in fewer and fewer big corporations. A half-dozen companies have dominated the scene. In the western United States, small independent miners could be successful as long as there were custom mills and smelters to ship to, but closure of these plants in the last 20 years has made it difficult for these small mines to operate. In the Midwest and East, virtually all producers are now affiliated with giant corporations. This contrasts with the pre-1950 situation when dozens of small companies were active in the Tri-State, Upper Mississippi Valley, and

TABLE 2. U.S. MINE PRODUCTION OF ZINC BY STATES*

State	1906	1920	1940	1960	1983
Arizona	......	......	14	33	......
Colorado	29	22	5	28	withheld
Idaho	......	13	64	34	withheld
Kansas	4	55	52	2	......
Missouri	118	23	12	1	52
Montana	......	83	48	12	......
New Jersey	10	70	82	......	15
New York	......	65	33	60	52
Oklahoma	......	199	148	......	......
Tennessee	......	17	32	83	100
Wisconsin	......	24	6	16	......
Other	17	16	67	91	29
Total	180	531	603	392	248

*Thousands of metric tons.
Data from U.S. Bureau of Mines *Mineral Trade Notes*.

Illinois-Kentucky Districts. As production in these areas tapered off or ceased, custom mills and smelters closed, making it difficult for a small new operator to get started. Virtually the only recourse today is to find a big company to take over his prospect.

Smelting of zinc concentrates is the major source of cadmium, germanium, gallium, and indium, which occur trapped in the sphalerite lattice. Cadmium is the most abundant of these and varies from 0.69 percent in concentrates from the Illinois-Kentucky fluorspar district (Hall and Heyl, 1968) to 0.12 percent in concentrates from Balmat-Edwards (Lenker, 1962). These by-products have been a source of revenue to zinc smelters for 50 years, but credit has rarely been given to the mine owners.

The change from smelting of near-surface oxidized ores to sulfide ores was mentioned above. Over the years, technological advances and environmental legislation also have affected smelting procedures. Several physical and chemical properties of zinc make its extraction from concentrates somewhat more difficult than the relatively simple smelting of copper or lead. For years, pyrometallurgical processes in so-called "fire smelters" dominated the industry. Later, electrothermic and finally electrolytic processes have been favored. The newer smelters are all electrolytic. These are more efficient and cleaner and allow better recovery of precious metals from the concentrates.

The zinc industry, like the lead industry, has experienced problems with rapidly rising production costs and continually increasing proportions of imported metal. Canada, Spain, Mexico, and West Germany have been important exporters in recent years. As more and more U.S. smelters are closed, often due to environmental reasons, imports are mostly in metal form rather than in concentrates.

At mid-1980s production costs, and selling prices for zinc and lead of 35 cents and 20 cents per pound, respectively, none of the domestic producers are making a satisfactory profit. In constant dollars, both metals are selling today for less than they did in 1910.

As indicated by Easton (1986), a major change in zinc trade patterns will occur in 1990, when opening of the Red Dog Mine in Alaska will make the United States a net exporter of concentrates for the first time. The United States will remain the major metal importer. Consumption of lead worldwide has risen at the rate of 2 percent a year since 1900, and for zinc about 3 percent. However, at the present time, Hissock (1986) says there is no indication that the peak consumption of 4.2 million tonnes for lead in 1979, and 4.8 million tonnes for zinc in 1973, will be exceeded in the near future.

Additional statistical information on production, consumption, prices, costs, and reserves can be found in publications of the U.S. Bureau of Mines and the Lead and Zinc Industries Association (292 Madison Avenue, New York, NY 10017). Reference should also be made to Bush (1986).

TYPES OF DEPOSITS

Lead and zinc deposits occur in many diverse geologic environments. Representatives of most of the types have been found in the United States, but the so-called Mississippi Valley type (MVT) has yielded by far the largest domestic production. The locations of MVT deposits are indicated on Plates 1C and 1D and Figure 1. In 1985 they produced about 20 percent of the world's newly mined lead and 10 percent of the zinc. Outside the United States, other types of geologic settings yield larger percentages of the metals than do those of the MVT.

In the western United States, MVT deposits are surprisingly scarce. Important quantities of lead and zinc have been or will be produced, however, from other ore types: manto-like deposits such as those in the Leadville Limestone in Colorado, which have some characteristics of MVT ores but were formed at higher temperature; silver-bearing veins, such as those at Tintic, Utah, and Coeur d'Alene, Idaho; contact metasomatic deposits in altered limestone adjacent to mineralized igneous stocks, such as at Bingham, Utah, and Santa Rita, New Mexico; and in a few important but relatively scarce massive sulfide occurrences, such as Crandon, Wisconsin, and Jerome, Arizona; and the so-far-uncommon stratiform deposits in sedimentary rocks, such as the huge deposit near Red Dog, Alaska.

In the northeastern United States, a small number of significant deposits occur in Precambrian marble at Franklin Furnace and Sterling Hill, New Jersey, and at Balmat, Edwards, and Pierrepont, New York (Plate 1D). These are metamorphosed deposits, perhaps originally of the Mississippi Valley or sedimentary exhalative types.

Mississippi Valley type

This ore deposit type is so-named because some of the greatest examples are located in the U.S. midcontinent region. Four of the world-class mineral districts are clustered around the Ozark Mountains, as are some of the literally hundreds of mineralized areas of lesser importance. The relative proportions of the valuable minerals are different in each area, but lead, and sometimes zinc, was recovered in each of them.

MVT deposits occur in many places around the world where there are intracratonic basins covered with relatively thin successions of sedimentary rocks. Northern Canada, Poland, England, France, Italy, and North Africa are notable examples. In the United States, the host rocks are all of Paleozoic age, but in Europe and North Africa, Mesozoic formations also have ore. Generally the rocks are undisturbed and flat-lying, but in the Appalachians and Alps the rocks were involved in post-ore thrusting and folding. Carbonate rocks are the principal ore hosts, but in Sweden, France, and parts of the Southeast Missouri District, galena occurs in sandstone. There are no major MVT deposits in the western United States, for unknown reasons, and significant production has come only from Metaline Falls, Washington.

Although all MVT deposits do not show all of the characteristics listed below, they all have enough of the distinctive features to make assignment to the classification relatively easy:

1. Most of the deposits occur in passive structural regions inland from present-day cratonic margins.
2. Absence of associated igneous rocks that are potential sources of the ore solutions.
3. Simple mineralogy and coarse grain size of ore and gangue minerals.
4. Low precious metal content.
5. Occur mostly in carbonate rocks, especially dolostone, but also occur in sandstone or conglomerate.
6. Occur at relatively shallow depths.
7. Many bear a relationship to positive structures.
8. Have evidence of host-rock dissolution both before and during ore-solution circulation.
9. Each district is largely limited to a certain stratigraphic interval. (An excellent graphic description of this "favored bed" relationship is given by Moore [in Bastin, 1939]).
10. Fluid inclusion temperatures are in the range of 60 to 200°C, but most are between 75 and 150°C.
11. Extremely saline fluid inclusions: 10 to 50 percent NaCl equivalent.
12. The ore minerals were deposited mainly as open-space filling, but in some districts, replacement predominates.
13. In some districts, such as the Southeast Missouri, Tri-State, and Upper Mississippi Valley Districts, the galena is "too radiogenic" for its age, the so-called "J-lead."
14. Each district has a characteristic mineral deposition sequence.
15. Districts occupy huge areas measured in tens, even hundreds of square kilometers.
16. Alteration types include dolomitization, silicification, and recrystallization of limestone to a coarse grain size.

Important examples. *Southeast Missouri District.* This district has one of the world's greatest lead concentrations. It also has produced much zinc and some copper, cobalt, nickel, silver, and cadmium. By the time ore is exhausted, sometime in the 21st century, it will have produced over 27 million tonnes (30 million tons) of metallic lead.

The district is also noteworthy as one of the finest examples of an ore deposit closely related to a lithofacies pattern in carbonate rocks (Fig. 1; after Kisvarsanyi, 1982). When Paleozoic sedimentation began in the Upper Cambrian, the St. Francois Mountains at the eastern end of the Ozarks stood above a widespread peneplain surface as a cluster of monadnocks as much as 600 m (2,000 ft) high. The first Paleozoic formation deposited on this surface consisted of as much as 130 m (400 ft) of Lamotte Sandstone. This sandstone was not thick enough to cover the tops of some of the higher Precambrian knobs, which therefore were ringed by what are locally known as "pinchouts." The principal ore host, the Bonneterre Formation, overlies the Lamotte, followed by a succession of largely limestone and dolostone with some shale and sandstone. Immediately above the Bonneterre in all the mining areas is a facies of the Davis Formation that consists of much green shale and shaly limestone. As each of the successively younger formations was deposited, more and more of the ancient hills were covered until finally the youngest Cambrian formation, the Eminence, concealed the highest knob. Each formation that did not spread over the tops of the knobs pinched out against it; those beds that did pass over the tops of the knobs reflected the underlying Precambrian highs as domes, which were accentuated by compaction and draping of the sediments.

Knowledge of the topography on the Precambrian surface is important because ore tends to occur in the Bonneterre above Lamotte pinchouts (Fig. 2). In fact, even in places where a local ore trend may be controlled by some secondary feature, as described below, almost without exception, the ore trend extends back to a knob. This observation has been of great importance in exploration.

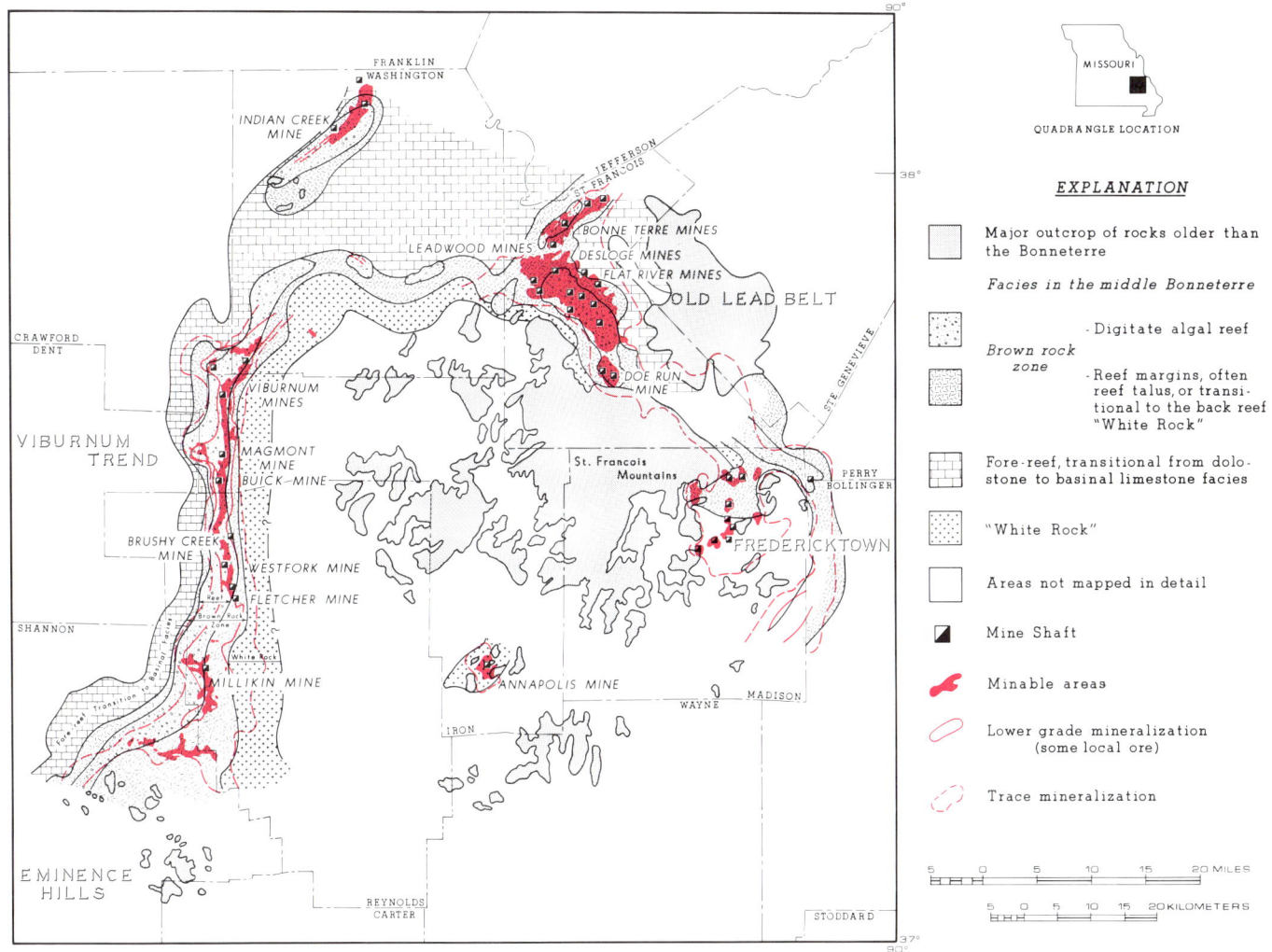

Figure 1. Location of mineralization in the southeast Missouri lead–zinc district superimposed on a geologic map showing the major facies of the middle Bonneterre Formation (Upper Cambrian). The district is in three principal parts: the Fredericktown area, the Old Lead belt, and the Viburnum trend (modified from Kisvarsanyi, 1977, 1982).

For descriptive purposes, the Southeast Missouri District can be divided into three principal parts (Fig. 1), each with overall similarity but interesting local differences. These parts are: the Fredericktown area, the Old Lead belt, and the Viburnum trend. Near Fredericktown, there is a moderately productive area for lead (it is well known because it contains the richest cobalt-nickel concentrations in the region), in which virtually all of the ore closely overlies Lamotte pinchouts. Inasmuch as several of the knobs are rudely conical, many of the orebodies are doughnut shaped, or at least arcuate (Fig. 2). Residual lead was first found in the district about 1720. There was sporadic production in the area until 1961; some exploration interest still exists, although the individual orebodies are relatively small.

The part of the district most productive to date is called the Old Lead belt (Fig. 1). Ore was first discovered here in 1864, at the town of Bonne Terre (again, originally as galena in clay). Over the next 50 years, discoveries were made in a triangular area encompassing Flat River, Desloge, and Leadwood, Missouri. It is an area of about 260 km^2 (100 mi^2), not all underlain by ore but with huge orebodies that in total yielded about 235 million tonnes (285 million tons) before the district was abandoned in 1972.

The Old Lead belt, as elsewhere, contains knobs. The Bonne Terre Mine produced some 30 million tonnes (34 million tons) from the largest strictly knob-related orebody ever discovered. In addition to the knobs, huge carbonate sand bars (Fig. 3; and Figs. 6 and 7 in Ohle and Brown, 1951) were developed during Bonneterre time. These bars consist of the 19, 15, and 10 beds. All trend northeast, a reflection of strong currents from the northwest that prevailed during the lower half of Bonneterre deposition. When deposition of the Bonneterre began, the Lamotte surface was nearly flat except near knobs; accordingly, the carbonate

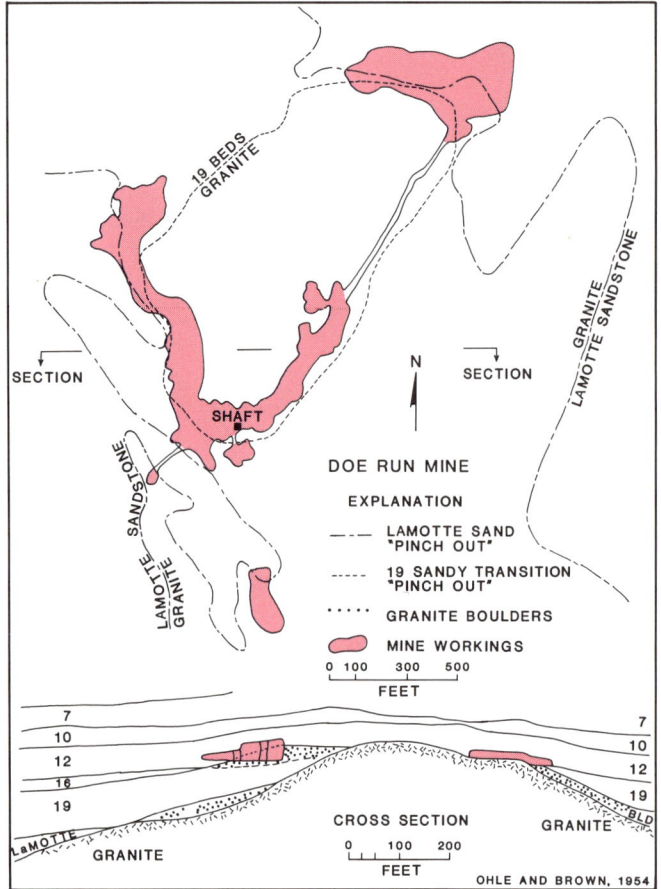

Figure 2. Relationship of a Precambrian knob and the orebodies at Doe Run Mine, southeast Missouri. Cross section is generalized (after Ohle and Brown, 1951).

sand bars are strictly features of lower to middle Bonneterre sedimentation. Bars are lacking in the upper part of the formation.

Nearly a dozen large carbonate sand bars, or sand ridges, were formed near Flat River, Desloge, and Leadwood. A few are several miles long, and those near Flat River have an amplitude of as much as 30 m (100 ft; Fig. 3). They are rudely parallel and are known in great detail because the orebodies followed them. Ore trends were locally developed on both flanks of the bars; accordingly, two long parallel stopes related to the same structure commonly exist (Ohle and Brown, 1951).

The sand bars are high-energy features, and the central cores always contain coarser carbonate than the laterally equivalent material on the flanks. The rock in the bars today is dolomite, but the long basinal depressions between bars are composed of shaly, fine-grained, limey material. Commonly, the ore-bearing dolomite beds pass into limestone a few tens of feet outside the orebodies.

During periods of quiet water in the building of the ridges, algal growth was dominant over clastic deposition and formed darker-colored layers a meter or so thick. These layers are generally brown and are characterized by oval or digitate shapes that led to their being called "marble." Such algal layers commonly passed laterally into black shale bands a few centimeters thick, which were especially rich in galena. This passage at stope margins from a few centimeters of shale to a meter of "marble" was called a "fan structure" (Ohle and Brown, 1951, Plate 2C).

After about one-fourth (30 m, 100 ft) of the Bonneterre had been deposited, the sea floor had a corrugated appearance resulting from the sand ridges. At that time, there was a great influx of algae that lodged on knobs and the tops of sand ridges—places where the water depth was favorable for their growth (Fig. 3). Organic growth continued until as much as 30 m (100 ft) of digitate algal reef had been formed. The reef mounds reflected the shapes of the underlying topographic highs and, hence, were either doughnut shaped if the knob was high, or linear if implanted atop a long sand bar. Well-developed northeast-trending reefs invariably had a wave-resistant northwest side with reef debris piled up at the base. Off-reef sediment with dips as great as 40° lapped onto the rigid structures (Ohle, 1985).

For some reason, sand ridges at Desloge and Leadwood ceased building earlier in Bonneterre time and were not colonized by algae as were those at Flat River. Since ore favored both reef and ridge structures, many Flat River stopes were nearly 60 m (200 ft) high, as compared to the 6 to 10-m-high (20 to 30 ft) stopes that were typical of Desloge and Leadwood.

The upper half of the Bonneterre Formation in the Old Lead belt contained ore only at the Bonne Terre Mine. The northwest currents that were so prominent in earlier time had ceased, and ridges did not form, although there was an ample supply of oolite that could have been piled up. The beds in the upper Bonneterre are flat-lying, and the ore is stratiform, the galena selectively replacing certain dolostone beds.

More than 75 percent of the ore in the Old Lead belt was mined before the first geological study of the area, which followed the establishment in 1947 by the St. Joseph Lead Company of a geology department at the instigation of Francis Cameron and under the direction of John S. Brown. Although earlier workers had recognized favorable rocks and lithofacies patterns, none of these features were understood. Fortunately, almost all of the old workings (15 km^2; 6 mi^2) of open stopes and 650 km (400 mi) of drifts made during 83 years of mining were still accessible to mapping, and it was possible to gain an understanding of the ore controls of the entire district. Data from some 75,000 drill holes were also studied. The information that was assembled by a dozen geologists in a 10-year period was instrumental in guiding the drilling that resulted in the discovery of the Viburnum trend in 1956.

The general impression in the industry is that the Old Lead belt was huge but low grade. It certainly was huge, as stopes a few kilometers long, a hundred meters or so wide, and 3 to 60 m (10 to 200 ft) high will testify. But it was not all really low grade. It did have a large low-grade halo that produced 2 to 4 percent lead ore for the last 30 years of mining, but as recently as 1912, ore in

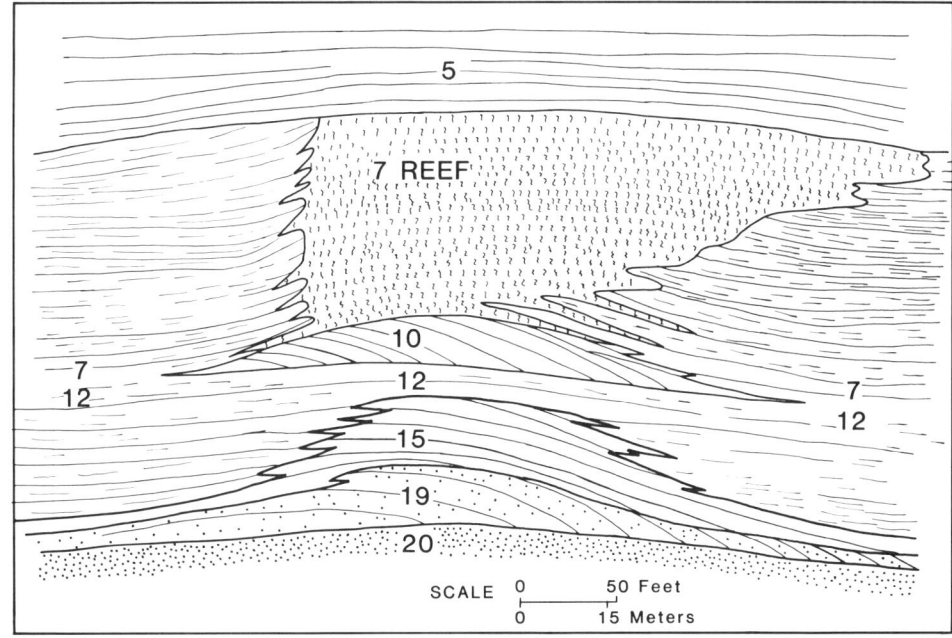

Figure 3. Cross section looking northeast of a typical carbonate sand bar in the Old Lead belt of southeast Missouri, surmounted by a digitate algal reef (after Snyder and Emery, 1956). The numbers refer to sub-units in the Bonneterre Formation.

the Bonne Terre mine averaged 9 percent lead, a figure that is still respectable even after the rich discoveries at Viburnum. Viburnum probably will have a much shorter period of slowly declining ore grade because it has a much smaller halo.

The Viburnum trend (Fig. 1) was found by an exploration program in which several thousand holes were drilled during a 12-year period in an area of about 2,500 km² (1,000 mi²). The relatively small Indian Creek Mine (Fig. 1) was found during this period and provided encouragement to continue exploration. The trend was truly a magnificent prize. By the time of its ultimate exhaustion, a north-south belt about 65 km (40 mi) long will have produced more than 225 million metric tons (250 million short tons) of 5 to 19 percent lead plus zinc. Large areas assay 20 to 30 percent combined. Interconnection of the mines of the five operating companies may produce a continuous opening for the entire 65-km length.

In the same way as the Fredericktown and Old Lead belt areas in the Southeast Missouri District, the Viburnum trend has its distinctive features. Unlike the other two, however, most of the better ore is in the upper part of the Bonneterre Formation where the layers are thick and often oolitic. Ore extends upward to the contact with the overlying Davis Formation, which here is largely green shale. Another distinctive feature is the occurrence of much of the better ore in long linear collapse breccias, which were not known in the older mining areas. Rogers and Davis (1977) described three nearly parallel but partially intertwining breccia bodies at the Buick Mine. Each body has a distinctive mineralogy, which suggests that the ore-solution composition changed with time, and that its circulation perhaps was in pulses rather than as steady-state flow. About 50 percent of the ore for the trend as a whole is in such collapse breccias.

Knobs were also important at Viburnum. They served as the starting points for growth of a great north-south barrier reef along the west side of the main St. Francois Mountains Precambrian igneous area (Fig. 1). Some ore actually is in the reef and some is associated with Lamotte pinchouts, but the majority of higher grade mineralized rock is in the upper oolites located just above the reef (Fig. 4). In one 16-km (10-mi) segment of the trend in the Buick and Magmont Mines, knobs are lacking and the Lamotte is quite thick. In this area, the reef grew across the mouth of a large bay from knob attachments at the north and south ends.

The location of the Viburnum ore trend clearly is related to the reef, for it is closely superimposed over it. The reason for this association is not known; the reef could have been the main ore-solution circulation channel that fed upward into the top beds of the Bonneterre, or more likely, the rigid barrier reef may have acted as a fulcrum, so that when the more compactable beds around it subsided, fracture channelways were created in the layers that now contain ore.

Drilling of the Viburnum area has resulted in recognition of a facies pattern that correlates almost perfectly with ore distribution (Larsen, 1977). Three distinct concentric rings of facies zones exist. In the ring closest to the main Precambrian highland (Figs. 1 and 2; Larsen, 1977, Fig. 1), the central part of the formation, and in places all of it, consists of white, coarse-grained, vuggy dolomite locally called "White Rock." Some regard this as a back-reef deposit because, seaward, the "White Rock" is surrounded by a brown rock ring that is in considerable part a

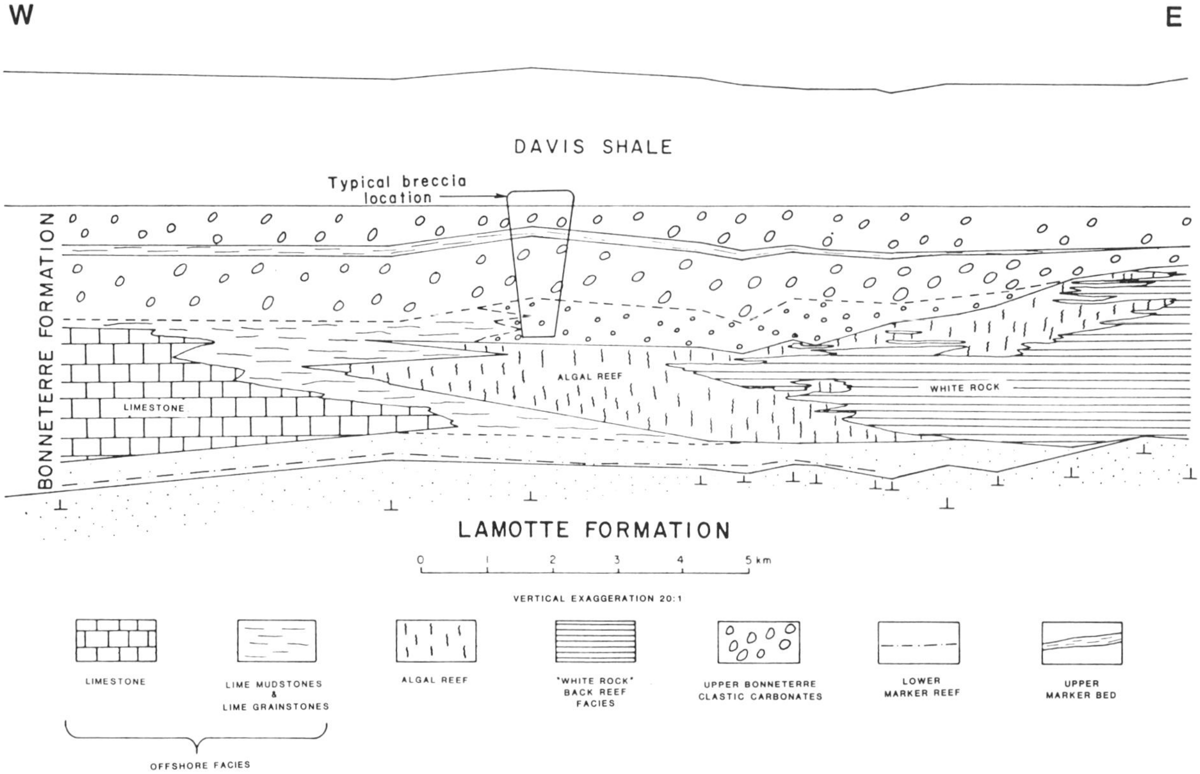

Figure 4. Generalized cross section, looking north, of the Viburnum trend, southeast Missouri, showing the facies relationships of the "White Rock," the reef (brown rock), and the basinal limestone with the characteristic slump breccia location over the reef. The best ore lies above the reef (modified from Gerdemann and Myers, 1972).

digitate algal reef. In the Old Lead belt, the fringing reef contained a large tonnage of low-grade ore, but as noted above, the ore that has made the Viburnum trend famous is predominantly just above the reef.

Outside the ring of brown rock, the middle Bonneterre beds grade into shaly limestone. As shown in the cross section (Fig. 4), ore is closely confined to the lithofacies deposited between the white rock and the limestone. Recognition of this facies pattern is obviously of prime importance in exploration. Review of the older mining areas shows that this same pattern is present all the way around the St. Francois Mountains area; it is simply more tightly drawn on the Viburnum trend.

Discovery of large amounts of high-grade ore outside the Old Lead belt coincided with great advances in the study of carbonate petrology. As a result, many investigations have been made by academic and company geologists. From the profusion of excellent literature on this important and interesting district, the following are especially recommended: Volume 72, Number 3, of *Economic Geology* (The Viburnum trend issue; Skinner, 1977), and the numerous maps and publications of the U.S. Geological Survey in the Rolla 2° study (Erickson and others, 1978; Pratt, 1981). For descriptions of the Old Lead belt, see Ohle and Brown (1951) and Snyder and Gerdemann (1968). An interesting account of the discoveries of the various parts of the Southeast Missouri District will be published by the Missouri Geological Survey (Ohle and Gerdemann, 1989).

Tri-State District. The Tri-State District is an area in Missouri, Kansas, and Oklahoma about 120 km (75 mi) long and as much as 30 km (20 mi) wide. Less than 1 percent of the district's area contains ore, but all of it has a distinctive appearance and geologic setting.

The district is now abandoned, but between 1848 and 1955 it was a major producer of both lead and zinc. The metals occurred in a zinc/lead ratio of 5 to 1. For many years the district dominated the U.S. zinc mining industry. Mining began in the shallower eastern end of the district in Missouri and, about 1890, spread westward into Kansas and Oklahoma. The Picher Field in Oklahoma was opened between 1910 and 1915 and proved to be the largest single subdistrict. It produced more than 60 percent of the 500+ million tons of ore. Although some orebodies averaged over 6 percent zinc plus lead, the overall district average was less than 3 percent.

Although the Tri-State District, like the Southeast Missouri District, is on the flanks of the Ozark dome, the two have very different geologic settings. Tri-State ore occurs in the cherty limestone of the Mississippian Boone Formation which is much younger than the Cambro-Ordovician rocks that host almost all of the other deposits of MVT in the central and eastern United

States. Dips are 2 to 3 m/km (10 to 15 ft/mi) to the northwest. Much of the ore, especially at Picher, was in spectacular collapse breccias. The principal alteration is silicification in the form of black jasperoid, which together with galena, sphalerite, pink dolomite, and calcite, composes the matrix for most of the breccia. Excellent photos of typical breccia and ore are given in McKnight and Fischer (1970) and Ohle (1985).

The host Boone Formation was subdivided by Fowler and Lyden (1932), who identified the beds from top to bottom with letters from B to R. Each bed, where mineralized, had distinctive characteristics in the nature of the chert, appearance of the breccia, and shape of the orebodies. Near the base of the Boone, in O, P, and Q beds, the ore was not brecciated; instead, sphalerite occurred in a thin stratigraphic interval 3 to 5 m (8 to 15 ft) thick in layers a few centimeters thick and parallel to the bedding. This type of ore covered many hectares and was called "Sheet Ground." All of it was low grade, about 2 percent combined lead and zinc, but it was very consistent and could be mined cheaply. The principal areas of Sheet Ground ore were in Missouri between Dueneweg and Oronogo, and in southeastern Kansas.

In the overlying G, H, and K beds in the Missouri part of the district, the orebodies were limited in size to a few hectares but had thicknesses of up to 50 m (150 ft). These bodies had the shape of chimneys, and they had a close spatial relationship to karst developed on the pre-Pennsylvanian surface. Some older publications, such as Smith and Siebenthal (1907), describe in detail these very interesting structures, some of which contain doughnut-shaped orebodies around the periphery of collapsed sink holes.

The great Picher Field in Oklahoma and Kansas was by far the most valuable part of the district. In a 23-km² (9-mi²) area, the M-bed was crisscrossed by a lacy pattern of mineralized collapse breccia from which more than 225 million tonnes (250 million tons) were extracted. Beds G, H, and K, overlying M, were also mineralized locally and, in strong structural centers, stopes 30 m (100 ft) high removed all the beds from G to M. McKnight and Fischer's (1970) maps and sections should be consulted for an understanding of the ore distribution in the various stratigraphic intervals.

The "holes" in the lacy pattern are either coarse crystalline white dolomite or "remnant islands" of unaltered, fine-grained, fossiliferous Mississippian limestone (Fig. 5). The brecciated chert-jasperoid ore trends tend to encircle the dolomite "islands," and it is common for a stope to have a dolomite wall and a limestone wall on opposite sides of the ore. Excellent descriptions of this most interesting pattern are given by McKnight and others (1944), McKnight and Fischer (1970), and Brockie and others (1968). The origin of the lacy pattern is not totally understood, but the most likely explanation is that the dolomite "islands" are relicts of patch reefs (Hagni, 1976, 1982).

A well-developed graben structure called the Miami trough (McKnight and Fischer, 1970) crosses the Picher Field in a northeast direction; ore has a minor tendency to spread along the trough, and lead/zinc ratios are higher close to it. These relations suggest that the structure preceded the mineralization.

The specific controls that localized many of the Tri-State orebodies are not clear. It has been suggested that some bodies are related to the pre-Pennsylvanian karst surface and that the Picher Field pattern may reflect reef distribution. In a broad way, the location of the Picher Field is related to the intersection of the northeast-trending Miami trough and the northwest-trending Bendelari monocline (McKnight and Fischer, 1970). Many early-mined deposits around Joplin, as well as the trend of the largest Sheet Ground orebody, may be related to the Joplin anticline (Siebenthal, 1915). However, there is much unmineralized area in seemingly similar structural positions, so that other factors obviously are involved. Efforts to explain the concentration of ore-solution flow and ore deposition have emphasized the dip off the Ozark Mountains (Siebenthal, 1915), the karstic and reef structures mentioned above, and the distribution of shale units in the stratigraphic column. All of the important Tri-State deposits are in or near a triangular area located north of the pinchout of the Devonian Chattanooga Shale, southwest of the pinchout of the Mississippian Northview Shale, and southeast of the eroded edge of the Pennsylvanian Cherokee Sahle (Brockie and others, 1968, Fig. 1). The presumption is that rising ore solutions would have had easier passage through the porous and permeable rocks in this area.

Excellent descriptions and discussions of the Tri-State District are given by Siebenthal (1915), Brockie and others (1968), McKnight and Fischer (1970), Lyden (1950), and Hagni (1976, 1982).

Appalachian deposits. Zinc and lesser lead deposits occur at various places along the eastern and southern margins of the North American craton from Newfoundland to Oklahoma (Plates I-C and I-D). Important producing areas are in Newfoundland; Nova Scotia; Austinville-Ivanhoe, Virginia, Friedensville, Pennsylvania; Mascot–Jefferson City, Tennessee; Copper Ridge, Tennessee; and northern Arkansas (Hoagland, 1976). Slightly inland from the craton edge is the Central Tennessee District. All of these districts, except Austinville-Ivanhoe, are in carbonate rocks of Ordovician (Beekmantown) age. The Austinville-Ivanhoe District is in carbonate rocks of the Cambrian Shady Formation. At the present time, in the U.S. part of this long belt, only the Tennessee districts are producing, and that state is currently the most productive in the nation. Grades are not high, but the orebodies are large and permit extraction by economical high-capacity equipment.

An outstanding feature of all these deposits is the impressive development of spectacular collapse breccias (Ohle, 1985). Much of the breccia forms elongate trends many meters wide, 3 to 13 m (10 to 40 ft) high, and in many cases, hundreds of meters long. These breccia trends faithfully follow certain stratigraphic zones and connect areas where brecciation has broken upward across the bedding to form "breakthroughs" (Fig. 6). The clasts in the breccias are largely unmineralized "original" dolomite, which has

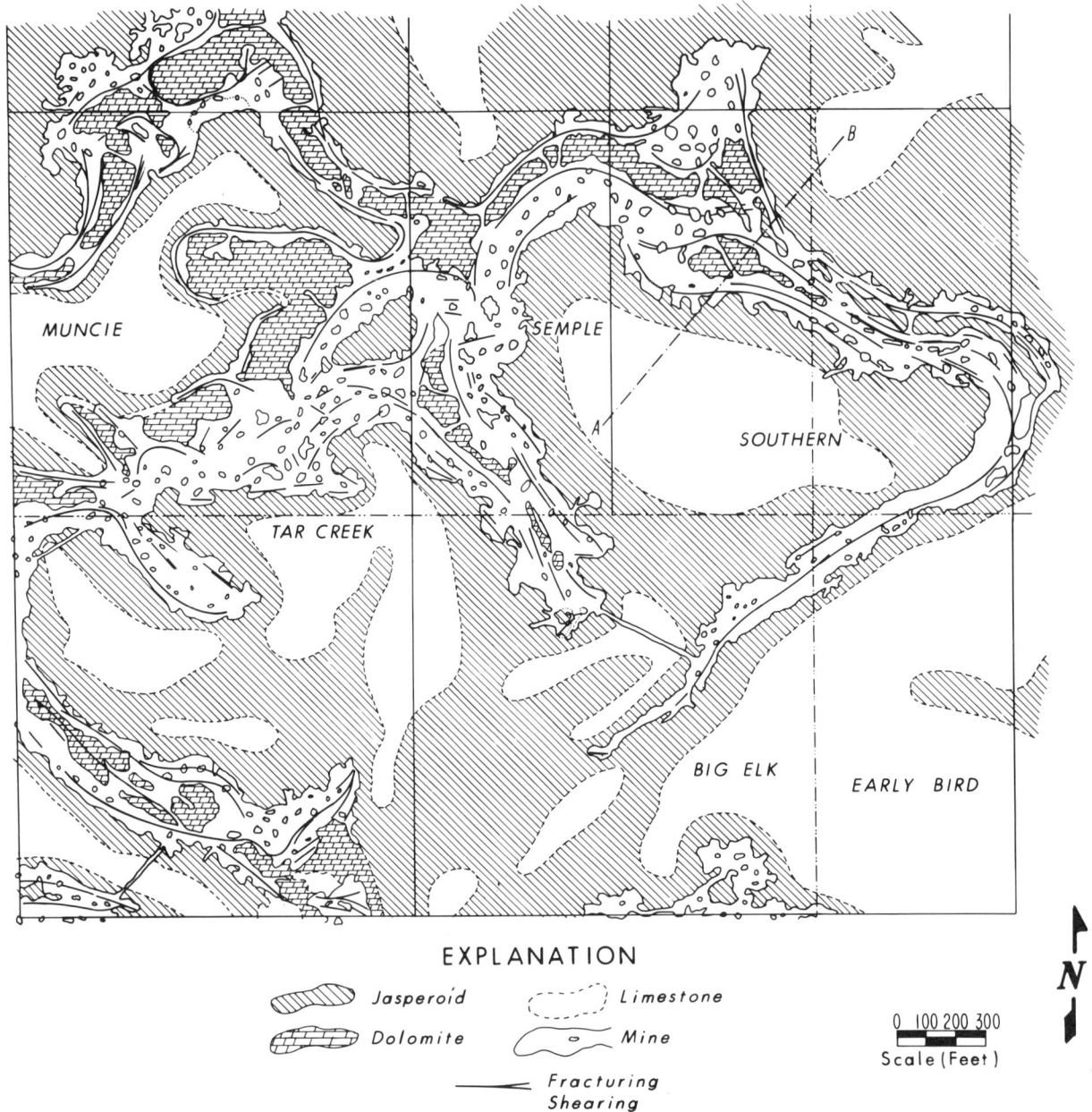

Figure 5. Detail of part of the Picher Field in Oklahoma showing the relationship of "dolomite islands" and limestone remnants to the ore trends (modified after Brockie and others, 1968).

been undermined by solution-removal of interbedded limestone layers. Collapse also occurs where limestone layers are greatly thinned during dolomitization and recrystallization. Limestone blocks are virtually absent in the breccias, except close to the extreme outer edges. The matrix filling is surprisingly complete; it is primarily white dolomite and sphalerite and, less commonly, galena, barite, and calcite. The orebodies are surrounded by halos of varying widths in which the limestone layers are altered to the coarse dolomite called "recrystalline." Outward from the main breccia bodies, fracturing of the "original" dolomite drops off rapidly, as does the ore grade.

The pattern of the mineralized breccia trends may be a rectangular trellis, reflecting a pre-existing fracture system, or it may be highly irregular and vermicular (Ohle, 1985). Not uncommonly, stope patterns look much like unmineralized caves. And, indeed, it seems quite likely that many of the mineralized breccia trends were at one time parts of a cave system, later utilized and perhaps enlarged by ore-solution flow. This belief is supported by the presence in some of the pipe-like breccia domes of a central mass of early unmineralized karst breccia (Fig. 6). In places, this "rock matrix breccia," as it is called locally, has been fractured and locally mineralized. Volume 66, Number 5, of *Economic Geology* (Bateman, 1971) was devoted to a description of this paleo-aquifer concept.

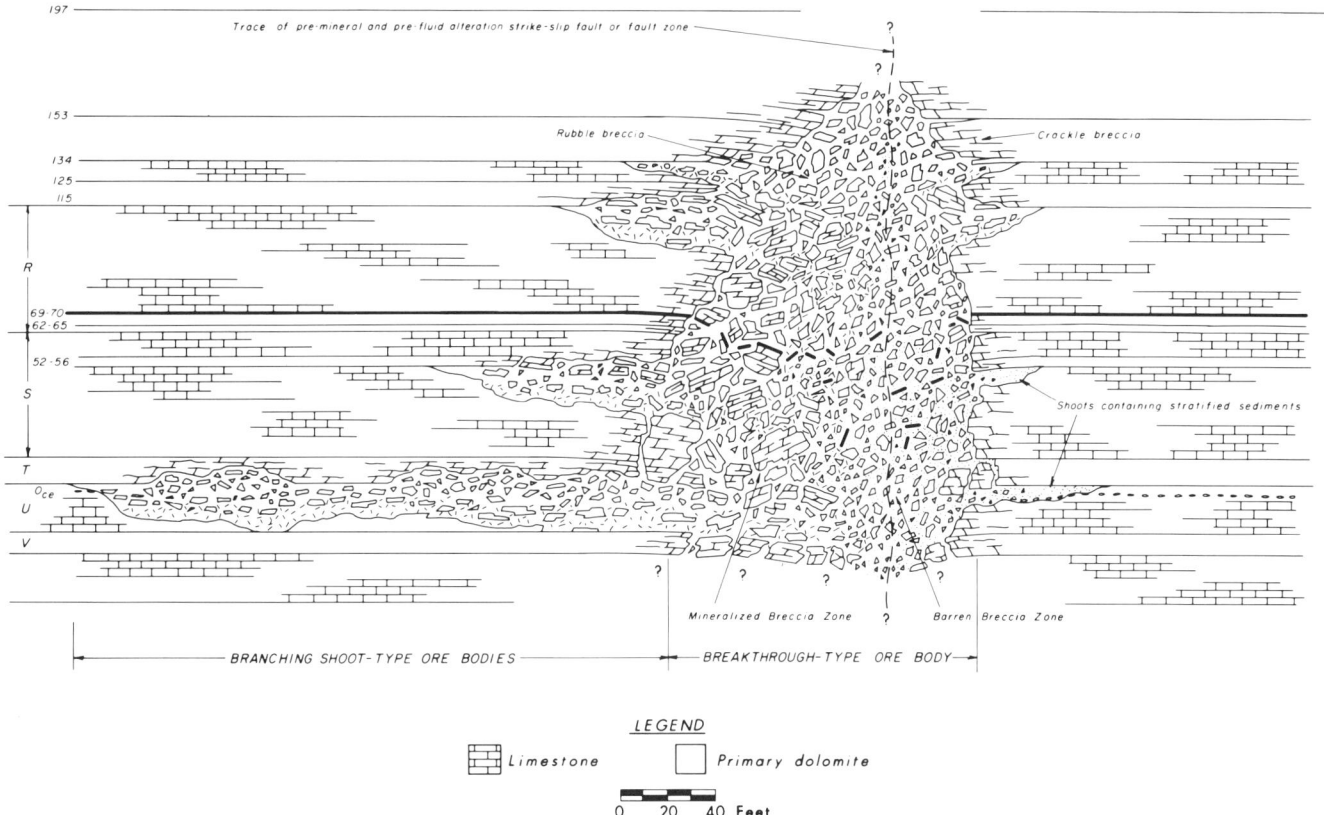

Figure 6. Cross section of a "breakthrough" structure in the East Tennessee Zinc District (after Crawford and others, 1969).

Many ideas have been advanced over the years as to the origin of the breccias in the Appalachian deposits (Ohle, 1985). Significant facts are: (1) all the fragments are now lower stratigraphically than they were originally (Fig. 6); (2) the fine-grained "original" dolomite clasts are highly angular; (3) some of the breccia-filled bodies are only a few feet across; and (4) pockets of breccia occur in which all the fragments are only a few millimeters (less than one-half inch) across. Collectively, these led me to conclude that although limestone solution and collapse was the prime cause of the fragmentation, other agencies must have been active. I concluded that extensional forces caused by compression of cavity walls and chemical brecciation may have been important.

These orebodies are believed by most geologists to have formed prior to the Appalachian Revolution orogeny; (Kendall, 1960). In Virginia and Tennessee, they were transported some distance by the tremendous overthrusting that produced the Valley and Ridge Province. Shortening in a northwest-southeast direction may have been as much as tens of miles. The beds now dip steeply southeast for the most part. This makes it difficult to reconstruct the original sedimentary facies pattern and the structures that influenced the location of the ore. Excellent descriptions of these deposits may be found in Hoagland (1976), Crawford and others (1969), Crawford and Hoagland (1968), and McCormick and others (1969).

Genesis of MVT deposits. The question of how MVT deposits formed has been studied and argued by geologists for 150 years (Moore, 1939; Ohle, 1959, 1980; Crook, 1933). The problem is still unresolved, and there are modern advocates of each of the following hypotheses as to the source of the ore solutions and the metals:

1. Deposition from magmatic solutions emanating from igneous intrusions.

2. Syn-sedimentary or early diagenetic deposition by sedimentary processes, probably with a localized contribution from hot springs.

3. Weathering of overlying rocks to release the metals into downward-moving ground waters (the Hydatogene process).

4. Leaching, transportation, and deposition by ground water moving in artesian flow.

5. Deposition from basinal brines expelled from compacting sedimentary piles; the metals are extracted from the sediments by the fluids during maturation and are carried by them to the points of deposition.

Each of these hypotheses has problems that have prevented universal acceptance. For example, the lack of obviously related igneous intrusions that could be the solution source is a difficulty for the magmatic idea; and the high fluid inclusion temperatures are a problem for ground-water proposals or the syn-sedimentary hypothesis: an abnormally high temperature gradient would be necessary. The high salt content of the inclusions needs explaining by each school as does the question of how the sulfur and the metals get together to form the ore minerals. Did they travel together or did the metals move alone to points where reduced sulfur was available? Furthermore, how can solutions that may have traveled tens of miles through carbonate rocks have the capability of dissolving limestone or dolostone to cause collapse breccias?

Unfortunately, thus far no genetic proposal fits all the evidence. Over the past 150 years, the consensus of geological opinion about the origin of MVT deposits has swung back and forth from one hypothesis to another. During the first half of the 20th century the magmatic concept held sway, but in recent years it has lost support. Today most U.S. geologists (but not Europeans) call on basin brines as the key. This has been much influenced in recent decades by the recognition of: (1) metals in Salton Sea geothermal brines (White and others, 1963; White, 1968; McKebbon and Elders, 1985), (2) the Red Sea metal-bearing brines (Swallow and Crease, 1963; Brewer and others, 1965; Degens and Ross, 1969), and (3) galena and metallic lead deposited on oil well casings in Russia (Lebedev, 1967) and at several places in the southern United States (Carpenter and others, 1974; Carpenter, 1979). Great progress has been made in recent years in accumulating chemical and physical facts about MVT ore solutions and the orebodies, but much more is needed before general agreement will be reached as to how these great deposits were formed.

Stratiform zinc deposits in Precambrian carbonate rocks in the northeastern United States

Five deposits in Precambrian Grenville limestone in the northeastern United States have contributed significantly to domestic zinc output over a long period of time. Two are in New Jersey, at Franklin and Sterling Hill, and three are in upper New York State, near Gouverneur. The largest and richest was at Franklin, a deposit that was the keystone of the New Jersey Zinc Company for nearly 100 years. It was depleted in 1954, but the nearby Sterling Hill deposit still has a small production. The New York District, consisting of the Balmat, Edwards, and Pierrepont deposits, is still a major factor in U.S. zinc production, although the Edwards mine is now closed. Pierrepont is a relatively recent find, some 35 km (30 mi) from Balmat, and its discovery much enlarged the area known to be favorable. The deposits in these two states are dissimilar, but they occur in rocks of approximately the same age and they all have been subjected to intense folding, recrystallization, and remobilization of the metallic minerals.

Franklin and Sterling Hill. The Franklin deposit was truly remarkable and, except for the smaller Sterling Hill orebody nearby, is unique. The mineralogy of the ore is especially unusual—50 percent franklinite (Fe, Mn, Zn_3O_4), 20 to 30 percent willemite (Zn_2SiO_4), 2 to 6 percent zincite (ZnO), and 3 to 11 percent calcite, with about 100 other mineral species, some of them quite rare (Lindgren, 1933). The willemite, zincite, and calcite are brilliantly fluorescent, and specimens from the mine are found in many museums. A paucity of sulfide is especially notable. Because of the franklinite, manganese was a valuable co-product. Total ore production from Franklin was about 22 million tonnes (25 million tons) grading 20 percent zinc, making it one of the richest, most compact zinc concentrations in the Earth's crust.

The ore at both Franklin and Sterling Hill occurs in the form of gently plunging hook-shaped folds. The ore bed at Franklin varied from 3 to 30 m (12 to 100 ft) in thickness, and the plunge length was about 1,000 m (3,200 ft). Ore graded into coarse crystalline limestone and was bounded laterally by coarse orthogneiss. Pegmatite dikes cut ore, limestone, and gneisses and contributed to the unusual mineral assemblage.

The origin of these two unique orebodies has been much debated, and the near-absence of sulfides has been especially difficult to explain. Metamorphism and remobilization of the ore minerals into the axes of structural troughs obscure the original relationships. Despite extensive exploration, similar occurrences have not been found elsewhere.

Balmat-Edwards-Pierrepont. This area in upper New York State began significant production near Edwards in 1915 and has been an important part of the U.S. zinc industry since 1930 when mining began at Balmat. Today the area accounts for about 20 percent of U.S. zinc output. Over the life of the district, more than 25 million tonnes (22 million tons) of ore have been produced containing slightly over 10 percent zinc. Excellent descriptions of the ore and its structural relations are given in Lea and Dill (1968), from which most of this account is taken.

The Balmat orebodies are localized at various stratigraphic intervals throughout some 550 m (1,800 ft) of a carbonate succession on the northwest side of a layer of gneiss that extends 15 km (10 mi) between the two deposits. The Edwards mine is on the southeast side of the gneiss. A recognizable stratigraphic sequence has been noted in each area, but it has not been possible to correlate details of the Balmat section with those at Edwards.

The orebodies are found within certain stratigraphic zones consisting of complexly folded alternating competent (strongly silicated limestone) and incompetent (dolomite) beds. The orebodies are localized by structural crossings of primary and secondary folds. Individual orebodies at both properties are highly irregular in shape, reflecting the influences of the stratigraphic and structural controls. However, each orebody (there are a half-dozen or more in each of the two areas) has a generally distinct shape in plan that may persist for hundreds or even thousands of meters down the northward plunge. Lea and Dill (1968, Figs. 5 through 9) show the three-dimensional shapes and the relationships to regional and local structures.

The ore mineralogy is simple—abundant pyrite and sphalerite and minor galena, pyrrhotite, and chalcopyrite. At Balmat #2 Mine, supergene hematite and magnetite persist downplunge for hundreds of meters. Gangue minerals are quartz, diopside, tremolite, serpentine, talc, carbonate, barite, and anhydrite. Some of the ore is very coarse grained and has cleavage faces as much as several centimeters in dimension. Wall-rock alteration is almost nonexistent.

Because structural deformation and metamorphism have been so intense, most original features of the orebodies have been destroyed. All workers agree that the ore is pre-metamorphism in age, but there are disagreements as to its original character. Solomon (1963) favored syngenetic deposition of the ore metals as low-grade disseminations that were remobilized and concentrated into orebodies during metamorphism and folding. Lea and Dill (1968) seem to agree. Doe (1960), on the other hand, suggests, on the basis of trace element and isotope studies, that the sulfides were not derived from the surrounding rocks. Still others (Briskey, personal communication, 1985) believe the ore was deposited as a sedimentary exhalative (SEDEX) concentration in a carbonate environment, similar to the Irish deposits, and was then reorganized by metamorphism.

Lead-zinc-silver mantos in and near the Leadville Limestone

An important group of deposits that were once very productive occur in dolostones and limestones in central Colorado. They share some characteristics with MVT deposits and, indeed, have been grouped with them by some geologists (Callahan, 1964; DeVoto, 1983). However, their mineral composition is more diverse, and their temperatures of deposition were hotter; consequently, a separate classification seems more appropriate. These differences led to early designation of the Colorado deposits as mesothermal.

Lead ore was discovered at Leadville in 1874 in the form of argentiferous cerussite. Mining of this near-surface oxidized lead ore occupied the next decade, after which sulfide ore predominated. At first, zinc was discarded, but after 1899 it began to be recovered. An interesting discovery in 1899 was that the walls of previously mined sulfide stopes were zinc carbonate. From 1910 to 1925, mining of this carbonate ore, at an 18 percent zinc cut-off grade, made zinc the principal product of the district. Since that time, ore production has been sporadic in response to variable economic conditions. Some mining goes on today. Of the total district output, 37 percent of the value has come from silver, 22 percent from zinc, 21 percent from lead, and 13 percent from gold (Tweto, 1968b).

The important examples of this ore group in central Colorado are shown in Figure 7. The orebodies are distributed along the Paleozoic outcrop belt wrapped around the Sawatch anticline. Other, quite similar ore districts, although not in the Leadville Limestone, are found in Nevada (Eureka, Cortez, Pioche) and Utah (Park City, Tintic). There are many similarities also to the mantos in northern Mexico (Prescott, 1916; Hewitt, 1968). The Colorado deposits lie within the northeast-trending Colorado mineral belt. Leadville is the biggest of the deposits. It has yielded about 30 million tonnes (33 million tons) of 12 to 15 percent combined lead and zinc with 2 to 4 oz of silver per tonne (Tweto, 1968b). Gilman, Aspen, and Tincup are other good-sized productive areas. All are predominantly in Paleozoic carbonate units.

These lead, zinc, and silver deposits are distinguishable from yet another important carbonate-hosted ore type, contact metasomatic deposits, by their lack of obviously related igneous intrusives. The Leadville rocks, while much altered, are not silicated. The lead-zinc-silver mineral suite commonly is arranged in a zonal pattern around a centralized "source" or "hot spot." The Breece Hill area at Leadville is such a center, and was classified as contact metamorphic or hypothermal by Emmons and others (1927) and Loughlin and Behre (1934). These investigators recognized, however, that intrusive sills in this area were not the source of the mineralizing solutions, although the sills are much sericitized. In the carbonate host rocks, there is strong dolomitization, recrystallization, and disaggregation of the carbonate grains called "sanding." Development of an interesting texture called "zebra rock" is widespread. There is a close association of the deposits with a plumbing system resulting from karsting similar to that developed in the Tennessee zinc districts. It seems probable that, in the Leadville mantos as in Tennessee, the ore solution simply utilized (and perhaps enlarged) a pre-existing system of subterranean water courses. Rock within and adjacent to the flow channels was intensely attacked by the hot metal-bearing brines.

A striking feature of each of these deposits is their limitation to narrow stratigraphic zones. The ore-bearing beds are selectively attacked by the ore solutions for reasons not completely understood, although considerable research has been applied to the question (Rove, 1947). The Leadville formation is a superb example of a "favored formation;" it contains most of the larger orebodies and, together with specific dolomite stratigraphic zones in the Dyer Member of the Devonian Chaffee Formation and the Ordovician Manitou Formation, accounts for the preponderance of the mine production.

Because of the strong stratigraphic influence, the orebodies are tabular mantos. The thickness of the ore beds at Leadville is as much as 40 m (125 ft), and some orebodies have roots extending downward along what seem to be feeder veins. Horizontal dimensions are greater—up to 1,000 m (3,300 ft) in length and 130 m (500 ft) in width—than vertical dimensions. At Gilman (Figs. 8 and 9; after Radabaugh and others, 1968), 60 percent of the 9 million tonnes (10 million tons) of ore has come from four mantos arranged in a fork-like pattern as much as 2,800 m (8,500 ft) long. Cross sections and mine plans presented by Emmons and others (1927, Fig. 18), Radabaugh and others (1968), and DeVoto (1983) are helpful in visualizing the three-dimensional shape of the orebodies.

Localization of ore in the Leadville Formation is not simply a result of the greater chemical reactivity of this unit. Physical factors are also involved. For example, the mantos at Leadville and Gilman spread out as multiply stacked orebodies beneath igneous sills, the Belden Shale, or other impervious strata. Tweto (1968b) reported that in well-mineralized parts of the Leadville District, where there are multiple igneous sills, the miners encountered as many as 11 mineralized "contacts," as they called them, in sinking the shafts.

An important unconformity between the Mississippian

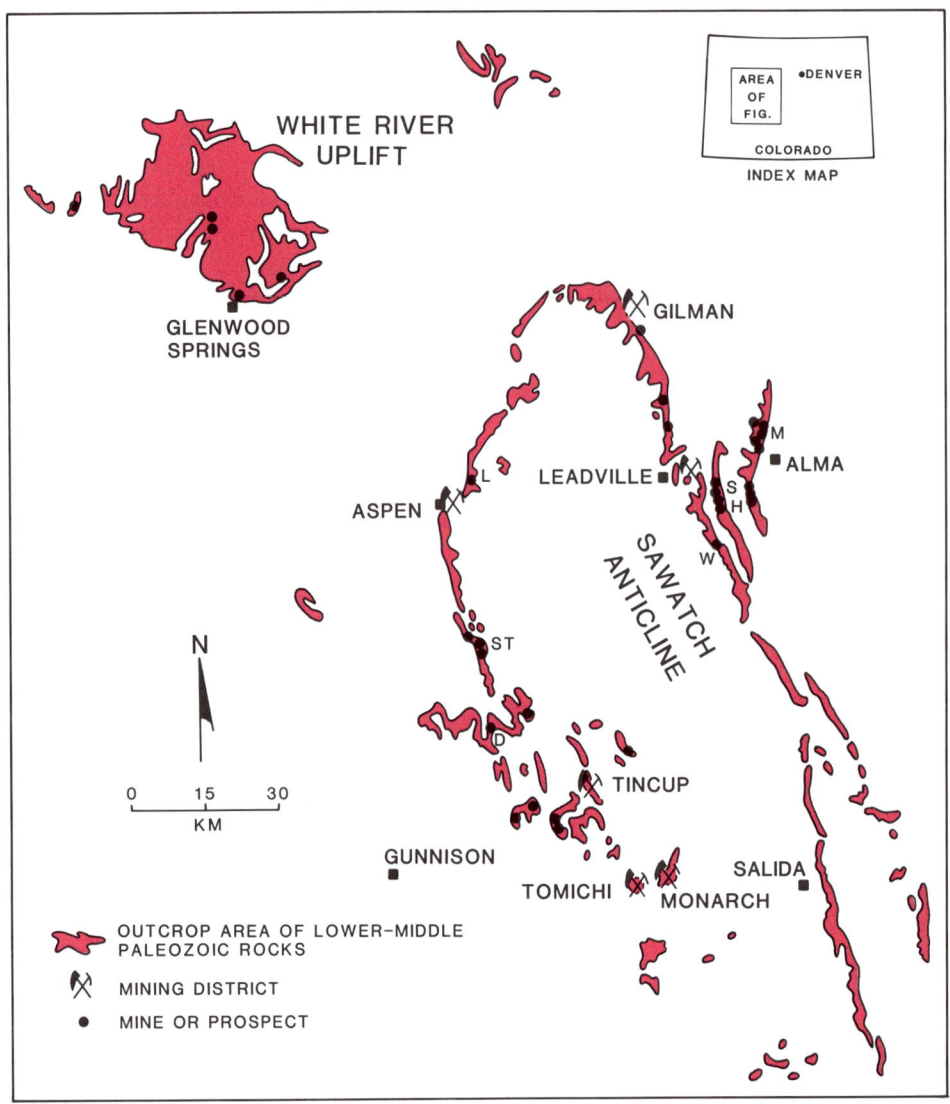

Figure 7. Map of central Colorado lead–zinc districts (modified from DeVoto, 1983; Heyl, 1964).

Leadville Limestone and the Pennsylvanian Belden Shale is the site of the deep karsting that thinned the Leadville erratically. There are deep, trash-filled valleys with erosion locally cutting downward as far as the Dyer. DeVoto (1983, p. 461) has emphasized the importance of this erosion surface and pointed out that all of the significant ore is within 200 ft of the unconformity. He believed the lead-zinc-silver ore predates the Laramide intrusives, contrary to the conclusions of most others who have attributed ore deposition to Laramide magmatic solutions. Whichever school is correct, it is a fact that the post-Leadville unconformity and its associated karsting exerted a great influence on ground preparation by creating open spaces that were later mineralized. Observation of the amount of solution thinning that took place in the Leadville District is a prime exploration tool.

Leadville and the other lead-zinc-silver districts of the Colorado mineral belt have been important parts of one of the most prominent mining areas in the western United States for more than 100 years. Consequently, they have been the subject of extensive investigations to which the reader is referred. Excellent summaries are found in Tweto (1968a and b), Radabaugh and others (1968), and Vanderwilt (1947). Despite the vast amount of study, there are still significant disagreements about important aspects of the mineralizing process.

SEDEX lead-zinc-barite deposits in the Brooks Range, Alaska

Between 1953 and 1955, I. L. Tailleur of the U.S. Geological Survey observed evidence of a sedimentary exhalative (SEDEX) lead-zinc-barite deposit along Drenchwater Creek on the south flank of the Brooks Range in Alaska (Lange and others, 1985; Mayfield and others, 1979). The locality is some 800 km (500 mi) northwest of Fairbanks. Subsequently, he found similar

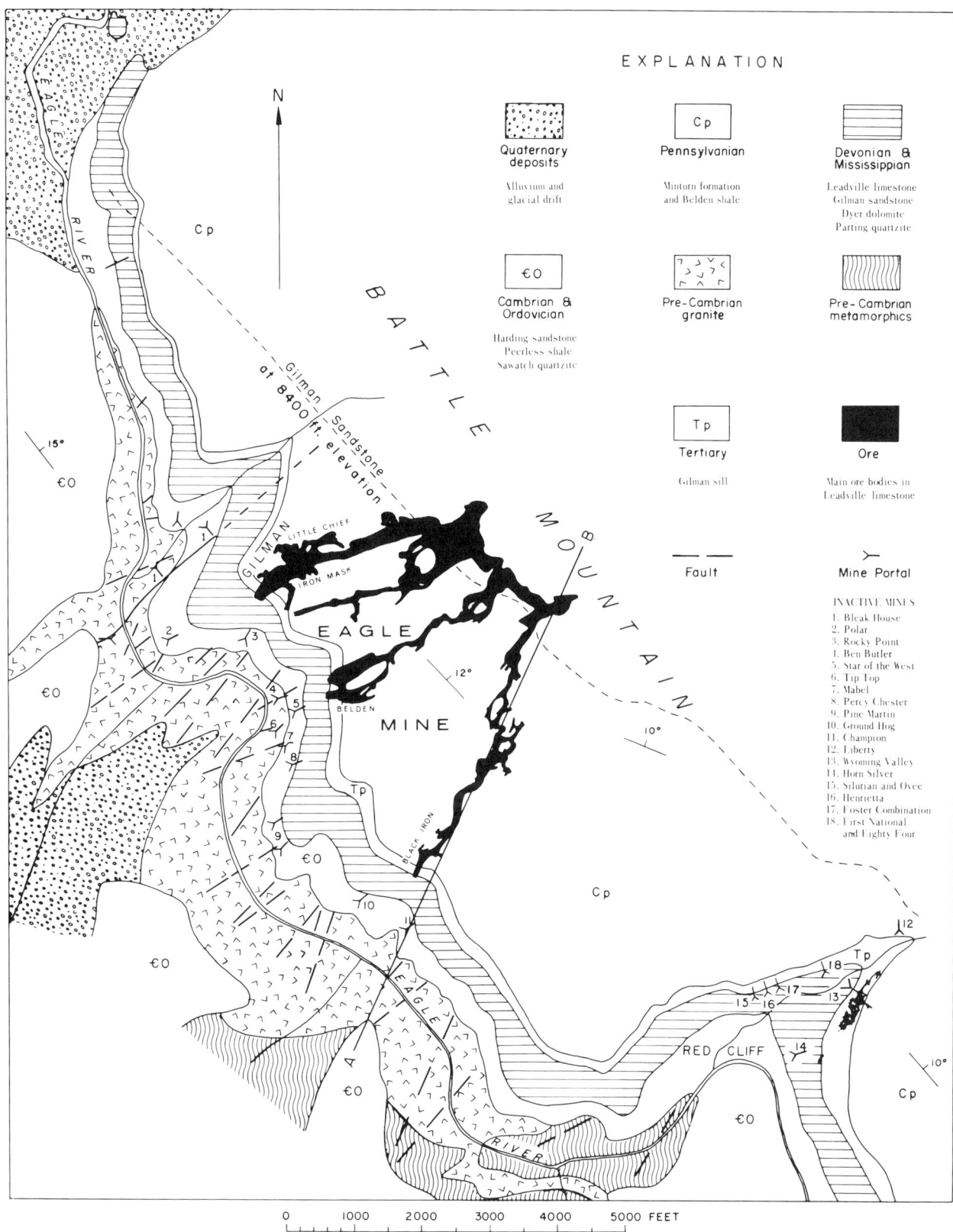

Figure 8. Geologic map of the Gilman District, Eagle County, Colorado, showing the plan of the Gilman mine workings (after Radabaugh and others, 1968).

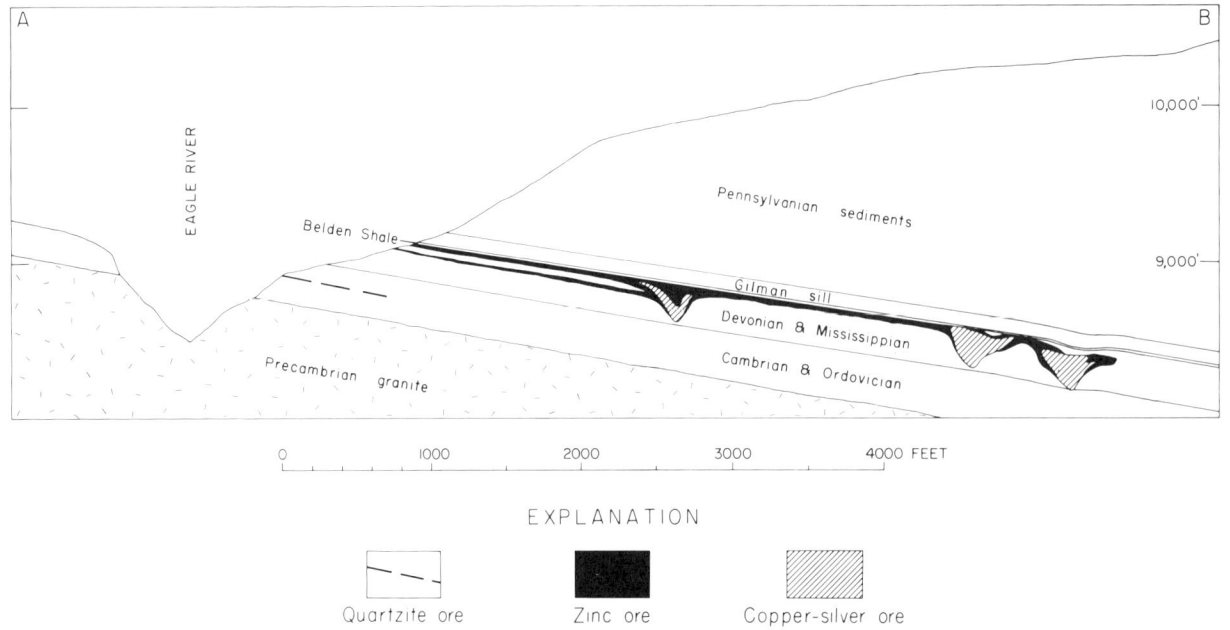

Figure 9. Generalized cross section of the Gilman orebody showing the chimney, limestone manto, and quartzite manto along line A-B of Figure 8 (after Radabaugh and others, 1968, Fig. 13).

deposits along Red Dog Creek 185 km (115 mi) west of Drenchwater. These were the first major lead-zinc occurrences of this geologic type to be found in the United States although western Canada has several major examples in the Selwyn Basin, the Kechika trench, and the Purcell Supergroup. Good examples also occur in Europe at Meggen and Rammelsberg, in Ireland, and in Australia at Mt. Isa and McArthur River. SEDEX deposits in Arkansas and Nevada have supplied most of the U.S. barite in recent decades, but these deposits do not contain significant amounts of sulfides. The closest counterparts to Red Dog in the lower 48 states are in the small Wood River District of Idaho (Briskey, personal communication, 1986).

The Red Dog deposit has an announced ore reserve of 76 million tonnes (85 million tons), grading 17.1 percent zinc, 5 percent lead, and 82 g/tonne of silver, and potentially commercial amounts of barite. The deposit is now being readied for production by Cominco American, Inc.

As indicated by the term "sedimentary exhalative," these deposits are now accepted by most geologists as being truly syngenetic; they are contemporaneous with most of the clastic sedimentary sequences in which they occur. According to the most widely accepted model, metal-rich solutions flowed into small third-order basins gradually being filled with clastic rocks. The basins are thought to be associated with rifting. Violent interruptions of the normally quiet sedimentary in-filling are indicated by zones of locally derived breccia believed to be the result of penecontemporaneous faulting. The Irish deposits, especially Tynagh and Navan, are exceptional in that the surrounding rocks are limestone with only a minor clastic component.

In some SEDEX deposits, copper is the main metal, but in many areas, as at Red Dog, lead and zinc are the valuable elements. The orebodies are pancake- or lense-shaped with two dimensions measured in hundreds of meters, whereas the thickness is not more than a few tens of meters at most. Some deposits have several mineralized beds, reflecting pulsations in ore fluid flow. Delicate banding with laminae of the different ore minerals, broad mineral zoning laterally across the deposits, and a very fine opaline chert are common features. Characteristically, the sulfide grains are extremely small and present a metallurgical problem.

A few descriptions of Red Dog have appeared since 1978. These are listed in Lange and others (1985). Accounts of Canadian and Australian occurrences should be consulted to learn more of the details of the SEDEX ore type (Large, 1983; Carne and CAthro, 1982; Gustafson and Williams, 1981; Williams, 1978). An excellent general presentation of the SEDEX geological environment and mineralization processes is in the short course text of the Mineralogical Association of Canada (Sangster, 1983).

Although SEDEX deposits have not yet contributed significantly to U.S. zinc and lead production, it is obvious that they will soon. By 1990, Red Dog is expected to make the United States a net exporter of concentrates, although still an importer of zinc metal. Further exploration of the Red Dog–Drenchwater belt may reveal other, thinly covered deposits in addition to those already known. In the lower 48 states, active exploration has tried to find minable deposits in seemingly similar geologic settings but, so far, has not been successful.

Volcanogenic massive sulfides

Orebodies now recognized as volcanic-associated massive sulfide deposits have been mined for hundreds, even thousands, of years, but they were not recognized as a genetic type until the mid-1950s. Prior to that time, the concensus of geological opinion was that they were of hydrothermal replacement origin. An excellent summary of the modern interpretation is given in Franklin and others (1981).

A large percentage of massive sulfide deposits contain less than five million tons of ore, but some are very large and valuable, such as the Horne (50+ million tonnes), Kidd Creek (300 million tonnes), Flin Flon (60 million tonnes), Brunswick (109 million tons), and Rio Tinto (750 million tonnes).

Volcanogenic massive sulfide deposits are stratabound and stratiform, commonly layered accumulations that normally contain at least 60 percent sulfide minerals. They were formed in a submarine volcanic environment. The surrounding rocks may be almost entirely volcanic or largely sedimentary, a factor that seems to be determined by the proximity of deposition to the point of solution discharge from the sea floor and the location of the volcanic vent. Where the vent is far removed from the site of metallic deposition, the proportion of sediments rises, and the volcanogenic ore type grades into the sediment-hosted category.

Massive sulfide deposits range in age from Archean to the present-day "black smokers" recently discovered at sea-floor–spreading centers. Zinc is present in nearly all of them but, in many, is too minor to be recovered. Copper and lead also are very important, and either one may be dominant in local areas. A lead-zinc combination is common in the related sediment-hosted massive sulfide ores.

For more than 60 years, volcanogenic massive sulfide deposits have been very productive of zinc, copper, lead, and precious metals in the greenstone belts of the Precambrian Shield north of the U.S.–Canadian border. But, despite intense exploration efforts in the U.S. to find extensions of the productive belts, there was no success until 1968 when Kennecott found a small deposit called Flambeau at Ladysmith, Wisconsin (May, 1977; Schwenk, 1977). This discovery stimulated even more intense efforts, which bore fruit in three more finds: Thornapple by Kennecott, Pelican River by Noranda, and Crandon by Exxon. Whereas Flambeau is copper-rich, the other three are zinc-rich deposits with subordinate copper and lead. Crandon is by far the largest. Its announced tonnage is 69 million tonnes (78 million tons) averaging 5.2 percent zinc, 1.18 percent copper, 0.46 percent lead, 38.1 g/tonne of silver, and 1.05 g/tonne of gold.

It is now known that the world-class copper-zinc deposits long since mined out at Jerome, Arizona, are of volcanogenic origin, although they were once classified as magmatic hydrothermal replacements. Elsewhere in the United States, small deposits in California, Washington, Idaho, Nevada, and Oregon, and in the Appalachian Mountains also are volcanogenic and have produced some zinc as well as copper. The deposit at Crandon, however, is by far the largest zinc-rich representative ever found in the United States. It is not yet in production because of environmental and political delays, so its geology is known only from 218 drill holes. The orebody is covered by 30 to 60 m (100 to 200 ft) of glacial cover; it was discovered by geophysical prospecting. However, the geological interpretation has been much aided by the similarity to other well-known volcanogenic deposits worldwide. Preliminary descriptions of Crandon are by May and Schmidt (1982) and by Lambe and Rowe (1987).

Essential facts known about Crandon as summarized from Lambe and Rowe are:

The orebody occurs in an east-west–trending greenstone belt that extends for 240 km (160 mi) across central Wisconsin.

The belt is composed of Precambrian felsic to mafic volcanic rocks, sedimentary rocks, and younger granite intrusives—all metamorphosed to greenschist to amphibolite facies. The indicated age of the ore deposit is 1,800 Ma.

The massive ore consists of 54 percent pyrite, 15 percent sphalerite, and lesser amounts of chalcopyrite and galena. Gangue minerals are quartz, chlorite, and minor sericite and dolomite.

The sulfide body is a massive, tabular lens 1,300 m (4,000 ft) long and as much as 60 m (190 ft) wide. It was originally horizontal but now dips 70 to 90° to the north. The massive sulfide consists of laminae of pyrite and sphalerite with minor chalcopyrite, galena, quartz, chlorite, sericite, and dolomite. Tuff, chert, and argillite occur in thin interbeds.

Beneath the massive sulfide lense is a series of volcanic breccias indicating a complex history of explosive volcanism. Within these breccias there is a concentration of veins and veinlets consisting of chalcopyrite and quartz or pyrite, sphalerite, and chalcopyrite in what is called "stringer ore."

The Crandon deposit is one of the most significant discoveries of recent years. Not only is it economically valuable in itself, but its existence is an indication that similar deposits may await discovery in the midcontinent region. Crandon has provided great stimulation to exploration in the covered Precambrian rocks of this area.

Other domestic sources of zinc and lead

In addition to the deposit types discussed above, which include the largest (MVT), the most unique geologically (Franklin), and those yet to be developed (Red Dog and Crandon), there are a number of other deposits in still different geologic environments that have contributed or are contributing to the U.S. zinc and lead supply. These are described in other chapters in this volume because their primary economic interest is a metal or mineral other than zinc or lead:

1. The fluorspar-zinc-lead veins and mantos in southern Illinois and western Kentucky. For years these deposits dominated the fluorspar business; in the early decades the veins yielded considerable lead, and after 1940, much zinc was produced from the mantos (Brobst, this volume).

2. Silver veins with sphalerite and galena (Coeur d'Alene, Idaho; Tintic, Utah; Ashley, this volume).

3. Silicated limestone replacement deposits adjacent to porphyry copper deposits (Bingham-Lark, Utah; Chino-Hanover, New Mexico; Bisbee, Arizona; Tooker, this volume).

LOCATIONS OF POSSIBLE FUTURE DISCOVERIES

Future lead and zinc discoveries in the United States, with the probable exception of Alaska, will occur mainly in places where the deposits do not crop out. Mention has already been made of the relatively recent finds at Viburnum, Missouri, and in central Tennessee where exploration was based on projections of known favorable geologic environments. More successes of these kinds are possible; they will result from a shrewd analysis of sedimentary environments, alert follow-up of chance intersections in oil or water wells, intelligent application of the latest geophysical and geochemical techniques, and good luck.

A number of orebodies, some not yet developed, were found by the New Jersey Zinc Company in a superbly applied soil geochemical program in the deeply weathered Appalachians. The area lent itself beautifully to this approach, and those who realized it were rewarded.

As has been mentioned, Crandon was discovered by an airborne geophysical survey in an area covered with glacial till. Three smaller deposits are known in the same greenstone belt. Are there others? And what of the rest of the vast midcontinent region where thick sedimentary rocks as well as till cover the basement formations? New techniques that will define exploration targets more precisely are needed to guide this expensive exploration. But they surely will come, as will better understanding of the Precambrian history in this large area. Geologists with vision are ready, but society and government must give more support to those who are willing to take the risks. Greater public appreciation is needed of the contribution that great mineral discoveries give to our national well-being. Something is wrong with the system when a magnificent deposit like Crandon cannot be developed because of public restraints.

ACKNOWLEDGMENTS

Many friends and colleagues over the years have contributed to the facts and ideas expressed in this chapter, and their help in building an understanding of lead and zinc deposits is gratefully acknowledged. Special thanks are due R. B. Taylor, J. Briskey, and C. H. Maxwell for critical reviews of the manuscript, which contributed to its accuracy and readability. G. Kisvarsanyi, P. Gerdemann, H. Myers, K. Larsen, and P. Sweeney provided the maps and section of Viburnum. R. Radabaugh is due thanks for permission to reprint the map of Gilman, and R. DeVoto for the map of the Leadville Limestone ore districts. W. Nokleberg supplied information on Red Dog in advance of publication as did R. Lambe on Crandon. W. Woodbury and J. Jolly provided statistical information from U.S. Bureau of Mines files. Appreciation is extended to these and to all others on whom I have drawn for assistance.

REFERENCES CITED

Bastin, E. S., ed., 1939, Contributions to a knowledge of the lead and zinc deposits of the Mississippi Valley region: Geological Society of America Special Paper 24, 149 p.

Bateman, A. M., ed., 1971, A paleoaquifer and its relation to economic mineral deposits; The Lower Ordovician Kingsport Formation and Mascot Dolomite: Economic Geology, v. 66, p. 695–810.

Brewer, P. G., Riley, J. P., and Culkin, F., 1965, The chemical composition of the hot salty water from the bottom of the Red Sea: Deep-Sea Research, v. 12, p. 497–503.

Brockie, D. C., Hare, E. K., Jr., and Dingess, P. R., 1968, The geology and ore deposits of the Tri-State District of Missouri, Kansas, and Oklahoma, in Ridge, J. D., ed., Ore deposits of the United States, 1933–1967, v. 1: New York, American Institute of Mining, Metallurgical, and Petroleum Engineers, p. 400–430.

Bush, W. R., ed., 1986, Economics of internationally traded minerals: New Orleans, Louisiana, Society of Mining Engineers Annual Meeting, 303 p.

Callahaan, W. H., 1964, Paleophysiographic premises for prospecting for stratabound base metal mineral deposits in carbonate rocks: Ankara, CENTO Symposium on mining, geology, and base metals, p. 230–235.

Carne, R. C., and Cathro, R. J., 1982, Sedimentary exhalative (sedex) zinc-lead-silver deposits, northern Canada cordilleran: Canadian Institute of Mining and Metallurgy Bulletin, v. 75, p. 66–78.

Carpenter, A. B., 1979, Interim report on lead and zinc in oil-field brines in the central Gulf Coast and in southern Michigan: Society of Mining Engineers of the American Institute of Mining, Metallurgical, and Petroleum Engineers, New Orleans meeting, (preprint) 5 p.

Carpenter, A. N., Trost, M. L., and Pickett, E. E., 1974, Preliminary report on the origin and chemical evolution of lead and zinc-rich oil field brines in central Mississippi: Economic Geology, v. 69, p. 1191–1206.

Crawford, J., and Hoagland, A. D., 1968, The Mascot-Jefferson City zinc district, Tennessee, in Ridge, J. D., ed., Ore deposits of the United States, 1933–1967, v. 1: New York, American Institute of Mining, Metallurgical, and Petroleum Engineers, p. 242–256.

Crawford, J., Fulweiler, R. E., and Miller, H. W., 1969, Mine geology of the New Jersey Zinc Company's Jefferson City Mine: Tennessee Division of Geology Report of Investigations 23, p. 64–75.

Crook, T., 1933, History of the theory of ore deposits: London, Thomas Murby and Company, 163 p.

Degens, E. T., and Ross, D. A., eds., 1969, Hot brines and recent heavy metal deposits in the Red Sea: New York, Springer-Verlag, 600 p.

DeVoto, R. A., 1983, Central Colorado karst-controlled lead-zinc-silver deposits (Leadville, Gilman, Aspen, and others), a late Paleozoic Mississippi Valley-type district, in Kisvarsanyi, G., Grant, S. K., Pratt, W. P., and Koenig, J. W., eds., Proceedings, International Conference on Mississippi Valley-type lead-zinc deposits: Rolla, Missouri, p. 459–485.

Doe, B., 1960, The distribution and composition of sulfide minerals at Balmat, New York [Ph.D. thesis]: Pasadena, California Institute of Technology, 151 p.

Easton, A. W., 1986, Changes in trade patterns for lead and zinc; A regional analysis, in Bush, W. R., ed., Economics of internationally traded minerals: New Orleans, Louisiana, Society of Mining Engineers Annual Meeting, p. 215–250.

Emmons, S. F., Irving, J. D., and Loughlin, G. F., 1927, Geology and ore deposits of the Leadville mining district, Colorado: U.S. Geological Survey Professional Paper 148, 368 p.

Erickson, R. L., Mosier, E. L., and Viets, J. G., 1978, Generalized geologic and summary geochemical maps of the Rolla 1° × 2° Quadrangle, Missouri: U.S. Geological Survey Miscellaneous Field Studies Map MF-1004-A, scale 1:250,000.

Fowler, G. M., and Lyden, J. P., 1932, The ore deposits of the Tri-State district (Missouri-Kansas-Oklahoma): American Institute of Mining, Metallurgical, and Petroleum Engineers Transactions, v. 102, p. 206–225.

Franklin, J. M., Sangster, D. M., and Lydon, J. W., 1981, Volcanic associated massive sulfide deposits: Economic Geology 75th Anniversary Volume, p. 485–627.

Gerdemann, P. E., and Myers, H. E., 1972, Relationship of carbonate facies pattern to ore distribution and ore genesis in the southeast Missouri lead district: Economic Geology, v. 67, p. 426–433.

Gustafson, L. B., and Williams, N., 1981, Sediment-hosted stratiform deposits of copper, lead, and zinc: Economic Geology 75th Anniversary Volume, p. 139–178.

Hagni, R. D., 1976, Tri-State ore deposits; The character of their host rocks and genesis, in Wolf, K. H., ed., Handbook of stratabound and stratiform ore deposits: New York, Elsevier, v. 6, p. 457–494.

——, 1982, The influence of original host rock character upon alteration and mineralization in the Tri-State district of Missouri, Kansas, and Oklahoma, U.S.A., in Amstutz, G. C., El Goresy, A., Franzel, G., Kluth, C., Moh, G., Wauschkuhn, A., and Zimmerman, R. A., eds., Ore genesis; The state of the art: Heidelberg, Springer-Verlag, p. 97–107.

Hall, W. E., and Heyl, A. V., 1968, Distribution of minor elements in ore and host rock, Illinois-Kentucky fluorspar district and Upper Mississippi Valley zinc-lead district: Ecnomic Geology, v. 63, p. 655–670.

Hewitt, W. P., 1968, Geology and mineralization of the Santa Eulalia District, Chihuahua, Mexico: Society of Mining Engineers Transactions, v. 241, p. 228–260.

Heyl, A. V., 1964, Oxidized zinc deposits of the United States; Part 3, Colorado: U.S. Geological Survey Bulletin 1135-C, 98 p.

Hissock, S. A., 1986, Demand pattern for lead and zinc in the mature economies, in Bush, W. R., ed., Economics of internationally traded minerals: New Orleans, Louisiana, Society of Mining Engineers Annual Meeting, p. 220–233.

Hoagland, A. D., 1976, Appalachian zinc-lead deposits, in Wolf, K. H., ed., Handbook of stratabound and stratiform ore deposits: New York, Elsevier, v. 6, p. 495–534.

Kendall, D. L., 1960 (1961), Ore deposits and sedimentary features; Jefferson City Mine, Tennessee: Economic Geology, v. 55, p. 985–1003; v. 56, p. 1137–1138; Discussion, v. 56, p. 444–446.

Kisvarsanyi, G., 1977, The role of the Precambrian igneous basement in the formation of the stratabound lead-zinc-copper deposits in southeast Missouri: Economic Geology, v. 72, p. 435–442.

——, 1982, Regional depositional facies of the Cambrian Bonneterre Formation, Rolla 1° × 2° Quadrangle, Missouri: U.S. Geological Survey Miscellaneous Field Studies Map MF–1002–I, scale 1:250,000.

Lambe, R. N., and Rowe, R. G., 1987, Volcanic history, mineralization, and alteration of the Crandon massive sulfide deposit, Wisconsin: Economic Geology, v. 82, p. 1204–1238.

Lange, I. M., Nokleberg, W. J., Plahuta, J. T., Krouse, H. R., and Doe, B. R., 1985, Geologic setting, petrology, and geochemistry of stratiform sphalerite-galena-barite deposits, Red Dog Creek and Drenchwater Creek areas, northwestern Brooks Range, Alaska: Economic Geology, v. 80, p. 1896–1926.

Large, D. E., 1983, Sediment-hosted massive sulfide lead-zinc deposits, in Sangster, D.F., ed., Short course in sediment-hosted stratiform lead-zinc deposits: Victoria, British Columbia, Mineralogical Association of Canada, p. 1–30.

Larsen, K. G., 1977, Sedimentology of the Bonneterre Formation, southeast Missouri: Economic Geology, v. 72, p. 408–419.

Lea, E. R., and Dill, D. B., Jr., 1968, Zinc deposits of the Balmat-Edwards District, New York, in Ridge, J. D., ed., Ore deposits of the United States, 1933–1967, v. 1: New York, American Institute of Mining, Metallurgical, and Petroleum Engineers, p. 20–48.

Lebedev, L. M., 1967, On contemporaneous deposits of native lead from thermal brines of Cheleken: Akad Nauk USSR, Doklady, v. 174, p. 197–200.

Lenker, E. S., 1962, A trace element study of selected sulfide minerals from the eastern United States [Ph.D. thesis]: University Park, Pennsylvania State University, 151 p.

Lindgren, W., 1933, Mineral deposits: London and New York, McGraw-Hill Book Company, 930 p.

Loughlin, G. F., and Behre, C. H., Jr., 1934, Zoning of ore deposits in and adjoining the Leadville District, Colorado: Economic Geology, v. 29, p. 215–234.

Lyden, J. P., 1950, Aspects of structure and mineralization used as guides in the development of the Picher Field: American Institute of Mining, Metallurgical, and Petroleum Engineers Transactions, v. 187, p. 1251–1260.

May, E. R., 1977, Flambeau; A Precambrian supergene enriched massive sulfide deposit: Geoscience Wisconsin, v. 1, p. 1–26.

May, E. R., and Schmidt, P. G., 1982, The discovery, geology, and mineralogy of the Crandon Precambrian massive sulfide deposit, Wisconsin, in Hutchinson, R. W., Spence, C. D., and Franklin, J. M., eds., Precambrian sulfide deposits: Geological Association of Canada Special Paper 25, p. 447–480.

Mayfield, C. F., Curtis, E. M., Ellerstuk, I. R., and Tailleur, J. L., 1979, Reconnaissance geology of the Ginny Creek zinc-lead-silver and Miniuktuk barite deposits, northwestern Brooks Range, Alaska: U.S. Geological Survey Open-File Report 79–1091, scale 1:63,360, 2 sheets, 20 p.

McCormick, J. E., Evanss, L. L., Palmer, R. A., Rasnick, F. D., Quarles, K. G., Mellen, W. V., and Riner, B. G., 1969, Mine geology of American Zinc Company's Young Mine: Tennessee Division of Geology Report of Investigations 23, p. 45–75.

McKebben, M. A., and Elders, W. A., 1985, Re-Zn-Cu-Pb mineralization in the Salton Sea geothermal system, Imperial Valley, California: Economic Geology, v. 80, p. 539–559.

McKnight, E. T., and Fischer, R. P., 1970, Geology and ore deposits of the Picher Field, Oklahoma and Kansas: U.S. Geological Survey Professional Paper 588, 165 p.

McKnight, E. T., Fischer, R. P., Addison, C. C., Bowie, K. R., Thiel, J. M., Owens, M. F., and Wells, F. G., 1944, Maps showing structural geology and dolomitized areas in part of the Picher zinc-lead field, Oklahoma and Kansas: U.S. Geological Survey Tri-State zinc-lead investigations, preliminary maps 1-6.

Moore, R. C., 1939, Significance of the stratigraphic distribution of Mississippi Valley ore deposits, in Bastin, E. S., ed., Contributions to a knowledge of the lead and zinc deposits of the Mississippi Valley region: Geological Society of America Special Paper 24, p. 29–38.

Ohle, E. L., 1959, Some considerations in determining the origin of ore deposits of the Mississippi Valley type: Economic Geology, v. 54, p. 769–789.

——, 1980, Some considerations in determining the origin of ore deposits of the Mississippi Valley type, Part 2: Economic Geology, v. 75, p. 161–172.

——, 1985, Breccias in Mississippi Valley-type deposits: Economic Geology, v. 80, p. 1736–1752.

Ohle, E. L., and Brown, J. S., eds., 1951, Geologic problems in the southeast Missouri lead district: Geological Society of America Bulletin, v. 65, p. 201–222, 935–936.

Ohle, E. L., and Gerdemann, P. E., 1989, Recent exploration history in southeast Missouri: Missouri Geological Survey Bulletin (in press).

Pratt, W. P., ed., 1981, Metallic mineral-resource potential of the Rolla 1° × 2° Quadrangle, Missouri, as appraised in September 1980: U.S. Geological Survey Open-File Report 81–518, 77 p.

Prescott, B., 1916, The main mineral zone of the Santa Eulalia District: American Institute of Mining, Metallurgical, and Petroleum Engineers Transactions, v. 51, p. 57–99.

Radabaugh, R. E., Merchant, J. S., and Brown, J. M., 1968, Geology and ore deposits of the Gilman (Red Cliff, Battle Mountain) district, Eagle County, Colorado, in Ridge, J. D., ed., Ore Deposits of the United States, 1933–1967, v. 1: American Institute of Mining, Metallurgical, and Petroleum Engineers, v. 642–644.

Rogers, R. K., and Davis, J. H., 1977, Geology of the Buick Mine, Viburnum trend, southeast Missouri: Economic Geology, v. 72, p. 372–380.

Rove, O., 1947, Some physical characteristics of certain favorable and unfavorable ore horizons: Economic Geology, v. 42, p. 57–77, 161–193.

Sangster, D. F., ed., 1983, Short course in sediment-hosted stratiform lead-zinc deposits: Victoria, British Columbia, Mineralogical Association of Canada, 309 p.

Schwenk, C. G., 1977, Discovery of the Flambeau Deposit, Rusk County, Wisconsin; A geophysical case study: Geoscience Wisconsin, v. 1, p. 27–42.

Siebenthal, C. E., 1915, Origin of the zinc and lead deposits of the Ozark region: U.S. Geological Survey Bulletin 606, 283 p.

Skinner, B. J., ed., 1977, An issue devoted to the Viburnum trend: Economic Geology, v. 72, no. 3, p. 337–525.

Smith, W.S.T., and Siebenthal, C. E., 1907, U.S. Geological Survey Geologic Atlas: Joplin District Folio, no. 148.

Snyder, F. G., and Emery, J. A., 1956, Geology in development and mining in the southeast Missouri lead belt: Mining Engineering, v. 8, p. 1216–1224.

Snyder, F. G., and Gerdemann, P. E., 1968, Geology of the southeast Missouri lead district, *in* Ridge, J. D., ed., Ore deposits of the United States, 1933–1967, v. 1: American Institute of Mining, Metallurgical, and Petroleum Engineers, p. 326–358.

Solomon, P. J., 1963, Sulfur isotopes and textural studies of the ores at Balmat, N.Y., and Mt. Isa, Queensland [Ph.D. thesis]: Cambridge, Massachusetts, Harvard University, 162 p.

Swallow, J. C., and Crease, J., 1963, Hot salty water at the bottom of the Red Sea: Nature, v. 205, p. 165–166.

Tweto, O., 1968a, Geologic setting and interrelationships of mineral deposits in the mountain provinces of Colorado and south-central Wyoming, *in* Ridge, J. D., ed., Ore deposits of the United States, 1933–1967, v. 1: American Institute of Mining, Metallurgical, and Petroleum Engineers, p. 552–588.

——, 1968b, Leadville District, Colorado, *in* Ridge, J. D., ed., Ore deposits of the United States, 1933–1967, v. 1: American Institute of Mining, Metallurgical, and Petroleum Engineers, p. 681–705.

Vanderwilt, J. W., ed., 1947, Mineral resources of Colorado: State of Colorado Mineral Resources Board, 547 p.

White, D. E., 1968, Environments of generation of some base metal ore deposits: Ecnomic Geology, v. 65, p. 301–335.

White, D. E., Anderson, E. T., and Grubbs, D. K., 1963, Geothermal brine well; Mile-deep drill hole may tap ore-bearing magmatic water and rocks undergoing metamorphism: Science, v. 139, p. 919–922.

Williams, N., 1978, Studies of the base metal sulfide deposits at McArthur River, Northern Territory, Australia; 1, The Cooley and Ridge deposits: Economic Geology, v. 73, p. 1036–1056.

MANUSCRIPT ACCEPTED BY THE SOCIETY FEBRUARY 10, 1989

Chapter 5

Iron and manganese

Gary B. Sidder
Central Mineral Resources, U.S. Geological Survey, MS 905, Denver Federal Center, Denver, Colorado 80225

INTRODUCTION

Iron is the least expensive and most widely used metal, and iron ore is its primary source for the manufacture of steel and pig iron. Manganese is the least expensive metal to alloy with iron and, therefore, is also an essential raw material for the production of pig iron and steel. Although vital domestic industries and ultimately our economy are dependent on these two strategic and critical metals, their availability within the United States is entirely divergent. Iron ore has been produced from 45 of the 50 states, and in 1988 the U.S. was the fifth largest producer in the world. The domestic reserve base is estimated to be about 25.2 billion metric tons (mt) of ore containing approximately 5.4 billion mt of iron. This is sufficient to supply the nation's needs for about 380 years, based on the average apparent consumption for the last five years. In sharp contrast, the U.S. has not produced any manganese ore since 1970, nor are there significant reserves (Klemic and others, 1973; Jones, 1989a; Kuck, 1989a). This chapter will explore the distribution and types of deposits as well as the past histories and future potential of these metals in the United States.

Abundance, occurrence, and use

Iron is the fourth most abundant element by weight in the Earth's crust and first in the whole Earth. Manganese is the twelfth and eleventh, respectively. The average concentration of iron in crustal rocks is about 5 wt %, and that of manganese is about 0.1 wt %. The clarke of concentration for iron ranges from about 4 to 14 (ore with 20 to 70 wt % Fe) and for manganese ore is more than 350. The overall abundance of iron and its ability to exist in more than one oxidation state account for its distribution in about 300 minerals. However, only seven minerals are mined for ore: hematite (Fe_2O_3), magnetite (Fe_3O_4), goethite ($Fe_2O_3 \cdot H_2O$) or limonite (variable compositions of hydrated iron oxide), and less commonly, siderite ($FeCO_3$), chamosite [(Fe^{2+}, Fe^{3+}, Mg, Al)$_6$ (Si, Al)$_4O_{10}(OH)_8$], pyrite (FeS_2), and pyrrhotite ($Fe_{1-x}S$) (Klemic and others, 1973). Manganese has geochemical characteristics similar to iron, and it substitutes for some iron ions in many minerals. Manganese oxides, carbonates, and hydroxides such as pyrolusite (MnO_2), psilomelane [$BaMn^{2+}Mn^{4+}_8O_{16}(OH)_4$ and variable compositions for hard hydrous manganese oxides], rhodochrosite ($MnCO_3$), and wad (soft hydrous manganese oxides) form the predominant ore minerals (Dorr and others, 1973).

Iron ore has been mined in the U.S. from several types of deposits that formed in vastly different geologic environments over a wide range of conditions. For example, bedded sedimentary deposits such as banded iron formations, ironstones, and other iron-rich sediments were formed at low temperatures as chemical precipitates, whereas massive deposits derived from magmatic and/or hydrothermal processes such as magmatic segregations, injections, and replacements were emplaced at relatively high temperatures and pressures, and secondary enrichments such as laterites and enriched iron formation were produced at surface or near-surface conditions (Gross, 1970; Klemic and others, 1973; Guilbert and Park, 1986). Plate 2A shows the location and types of iron mining districts in the United States (from Carr and others, 1967; Gross, 1970; Klemic and Tooker, 1979; and other references cited in text).

The manufacture of iron and steel consumes more than 98 percent of the iron ore produced in the United States. Other commodities such as cement, heavy-medium materials, pigments, high-density concrete, and others account for the remaining percentage of consumption (Kuck, 1989b).

The largest deposits of manganese ore in the world are marine chemical sediments, or sedimentary manganese, and secondary enrichment types. Other manganese deposits such as hydrothermal and volcanogenic types and deep-sea nodules contribute less than 5 percent of the world's present manganese production and resources. The distribution of former and potential producing districts in the U.S. is shown on Plate 2A (from Crittenden and Pavlides, 1962; Tooker and Cannon, 1980; and other references cited in text). The three major grades of manganese ore (more than 35 wt % Mn) are: metallurgical, battery, and chemical. Metallurgical-grade ore is used in the manufacture of ferromanganese and special manganese alloys for iron- and steel-making, which accounts for more than 90 percent of domestic manganese consumption. Battery-grade ore is used to produce dry-cell batteries, and chemical-grade ore and concentrates act as oxidizing agents in chemical processes such as the production of hydroquinone for photographic developers and the coloring of bricks, ceramics, and glass (DeYoung and others, 1984; Jones, 1985a).

Production of iron ore in the U.S. during 1988 totalled

Sidder, G. B., 1991, Iron and manganese, *in* Gluskoter, H. J., Rice, D. D., and Taylor, R. B., eds., Economic Geology, U.S.: Boulder, Colorado, Geological Society of America, The Geology of North America, v. P-2.

about 59 million mt, averaging about 64 wt % Fe, from 21 open pits and 1 underground mine. The value of iron ore shipped was about $1.9 billion. Low-grade (as low as 25 wt % Fe) taconite-type ores that are hosted by banded iron formation of Precambrian age in the Lake Superior region of Minnesota and Michigan (Plate 2A) accounted for about 99 percent of the total crude ore production in 1988, and they form more than 95 percent of the U.S. reserve base. Figure 1 depicts the contribution of these ores to total U.S. production since 1875. Imports of iron ore amounted to more than 22 million mt in 1988, or 22 percent of domestic consumption (Kuck, 1989a, b). The U.S. did not produce any manganese ore in 1988, and it imported about 485,000 mt. Ferruginous manganese ore (10 to 35 wt % Mn) that is a co-product with iron ore in the Cuyuna Range, Minnesota, and manganiferous iron ore (2 to 10 wt % Mn) from the Battleground Schist in Cherokee County, South Carolina (Plate 2A), are the only manganese-bearing ores that have been shipped in the 1980s. Together, these districts have produced a total of about 18,000 mt that has averaged about 12 wt % Mn. These domestic manganiferous ores have supplied less than 1 percent of the annual primary demand for manganese (Jones, 1985b, 1989a).

Historical perspective

Iron is one of the seven biblical metals along with gold, silver, copper, tin, lead, and mercury. The Iron Age dawned in the 14th century B.C., and the earliest use of iron was for agricultural tools. Manganese, too, had historical uses for the ancient Egyptians and Romans, who used oxide ore to control the color of glass (pink, purple, or black, depending on the amount added). By the 1600s, England had become the leader in the production and manufacture of iron. The first ore from the U.S. was mined in 1608 from Virginia, where a total of about 32 mt of ore yielded about 15.4 mt of iron when processed in England. The first regular production of iron ore in the colonies was established in 1643 near Boston at Lyon, Massachusetts. The Hammersmith ironworks built on the Saugus River utilized bog and pond ore dug from swamps and ponds. The main products manufactured included the first fire engine in Boston, pots, nails, and farm tools. Other ironworks were established in Connecticut, Rhode Island, and New Jersey by 1674. The primary source of iron in these areas was limonite from bog ores (Rabbitt, 1979; Morral, 1984; Jones, 1985a; Klinger, 1985a).

New England was the hub of the iron industry during the early 1700s, at which time ores other than soft limonitic bog ores were developed. Magnetite ore was exploited about 1710 by New Jersey ironmakers, and by 1750, hematite, hard limonite, and siderite ores were also being produced. All 13 colonies except Georgia had an iron industry at mid-century, and they manufactured muskets and cannons in addition to kitchen utensils and farm tools. By the beginning of the Revolutionary War, the Colonial iron industry was greater than that of England; however, English ironmakers began to use bituminous coal instead of charcoal in blast furnaces during the war and regained the lead in the

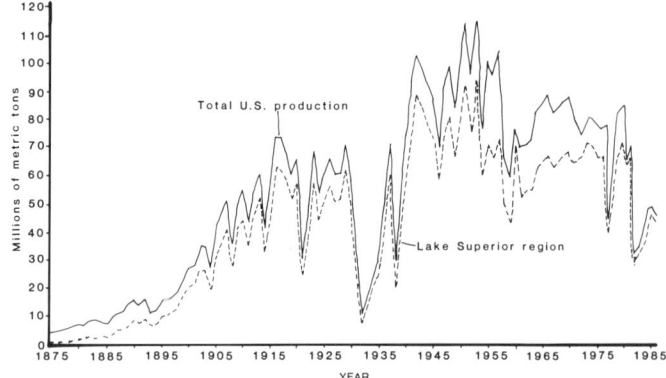

Figure 1. Production of iron ore in the U.S. and in the Lake Superior region, 1875–1985. Data from Klemic and others (1973, p. 295) and U.S. Bureau of Mines Minerals Yearbooks. Data for Lake Superior region in 1982 are incomplete because Michigan withheld information for that year.

manufacture of iron. Anthracite coal and hot blast were finally adopted for the domestic smelting of iron ore during the 1830s, and together they allowed manufacturers to produce higher-grade materials such as iron rails. The first domestic mining of manganese, which was used to color earthenware, took place at Paddy Run, Frederick City, Virginia, in 1832 and in Tennessee during 1837. It was also discovered in 1839 that the addition of manganese made steel more malleable and forgeable (Rabbitt, 1979; Morral, 1984; Jones, 1985a).

Deposits of magnetite, hematite, and brown ores of soft limonite became the dominant source of iron ore in the 1840s as supplies of bog ore were exhausted. Pennsylvania was the leading mining state because of its rich iron and coal resources. However, iron ore in the Lake Superior region was discovered in 1844 on the Marquette Range of the Upper Peninsula of Michigan, and production became significant in 1856 with the completion of a shipping canal at Sault Ste. Marie. The adaptation of the Bessemer steel-making process, discovered in 1856 by Robert Mushet, brought new impetus to the iron industry and allowed the U.S. to become self-sufficient in metal for weapons and other high-quality products. Prior to 1856, the majority of manganese ore was consumed in the production of chlorine. However, after the implementation of the Bessemer process, which requires the addition of manganese to avoid brittleness, its dominant use has been in steel-making. Also, Leclanche filed a patent application in 1866 for his manganese dioxide–ammonium chloride–zinc dry cell battery. Pennsylvania, Michigan, New York, New Jersey, Ohio, and Missouri were the leading producers of iron ore in the 1870s, with the four former states accounting for about 75 percent of all U.S. production (Klemic and others, 1973; Rabbitt, 1979; Morey, 1983; Morral, 1984; Jones, 1985a; Klinger, 1985a).

Through the last quarter of the 19th century and into the 20th, the production of iron ore doubled every 15 years. The Birmingham, Alabama, district was established as a major iron-making center in the 1880s, during which its iron production increased about tenfold. By 1890, the U.S. was the world's leading iron- and steel-making country. It produced more than 28 percent of the iron ore and more than 34 percent of both the pig iron and steel in the world. Michigan had captured 45 percent of the iron ore industry and Alabama was second, ahead of Pennsylvania, New York, Wisconsin, and Minnesota. Ore was discovered in the Menominee (1874), Gogebic (1880), Mesabi (1890), and Cuyuna (1904) Ranges of Michigan, Wisconsin, and Minnesota. By 1907, the Lake Superior region accounted for 80 percent of the national production (Fig. 1). This shift of production to the Great Lakes area and the southeast caused most of the eastern mines to be closed by the turn of the century. Minnesota equalled and quickly surpassed Michigan in production of iron ore after the Mesabi ores came onstream. The Mesabi Range has continued to the present as the largest producer of usable iron ore in the Lake Superior region (Klemic and others, 1973; Rabbitt, 1980; Morey, 1983; Klinger, 1985b).

The nature of iron-ore mining changed dramatically after World War II. High-grade (60 to 70 wt % Fe) direct-shipping ores and ores that were easily beneficiated by washing were the dominant source mined until the 1950s. However, the high rate of production depleted these resources, and new technological innovations for processing low-grade ore into high-grade agglomerates as pellets made it feasible to mine low-grade taconite orebodies. The impact of this process has been remarkable. Several hundred iron mines closed between 1955 and 1980, and in 1988, 10 mines accounted for 99 percent of the iron ore output. Moreover, other countries have benefited from this technology as much as or more than the United States. The USSR surpassed the U.S. in total production of iron ore to become the world leader in the 1950s, and presently the U.S. ranks fifth behind the USSR, Brazil, Australia, and China and accounts for only about 6 percent of the world's production (Klemic and others, 1973; Kuck, 1989a).

The production of manganese from domestic deposits has not amounted to more than a few percent of annual consumption, except for periods when government subsidies artificially stimulated production. The period of highest production lasted from World War II through the Korean War, and total production then supplied only about 20 percent of domestic consumption. The major proportion of ore mined has come from the manganese-silver-zinc deposits at Butte and Philipsburg, Montana, as well as from manganese associated with banded iron formation such as in the Cuyuna Range (Plate 2A; Klemic and others, 1973; Jones, 1985a).

Rising costs, overcapacity, the increase of less expensive imports, and a reduction in demand led to a drastic reorganization of the iron and steel industries in the U.S. during the mid-1980s. The domestic iron-ore industry is almost totally dependent on the U.S. steel industry because domestic steel companies directly control about 79 percent of domestic iron-ore production; exports account for only about 10 percent of production, with more than 95 percent shipped to Canadian steel manufacturers.

Thus, U.S. producers have been more competitive in the Great Lakes market than in international trade. As a consequence, several companies such as Reserve Mining Company and LTV Corporation, owner of the nation's second largest steel company, as well as Sharon Steel Corporation and Wheeling-Pittsburgh Steel Corporation, filed for bankruptcy and reorganized in the last five years. Nonetheless, major restructuring of the iron-ore industry, with modernization and conversion of pellet plants and drastic reductions in employment and pelletizing capacity, has lowered operating costs and improved productivity. In addition, a weakened U.S. dollar in Europe and Japan and restrictions on imports of finished steel have combined to increase demand for domestic iron ore and steel in the last three years. The need for iron ore is expected to continue into 1990, and production is predicted to remain steady for several years (Kuck, 1989a, 1989b; Skillings, 1989).

CLASSIFICATION OF IRON DEPOSITS

Iron deposits have been classified in several ways by previous authors. Some classifications are descriptive, based on the rock associations, and others are genetic, based on geologic processes. Unfortunately, many deposits of iron are not as easily categorized as those of some other metals such as chromium. Iron deposits are hosted by many different types of rocks, sometimes within the same deposit, and they are formed by processes that seem to span genetic associations. Moreover, particular aspects of ore formation in many iron deposits are still unknown or are at least equivocal.

Recent summaries of mineral deposit types by the U.S. Geological Survey and the Geological Survey of Canada (Cox and Singer, 1986, and Eckstrand, 1984, respectively) class iron deposits by their rock associations and other common geologic characteristics, regardless of the state of knowledge on ore genesis for a particular deposit. These synopses, as well as the descriptive classification used by Gross (1970), are the basis for the classification used in this chapter.

BEDDED, IRON-RICH SEDIMENTARY STRATA

Iron ore hosted by sedimentary rocks has been mined from many deposits. Among these are some of the world's largest ore deposits—the banded iron formations—as well as ironstones and other iron-rich sediments such as bog ore, black band ore, and others.

Banded iron formation

Iron formation has been defined in many ways. Moreover, many terms such as "itabarite," "jaspilite," and "ironstone" have been used in different countries as local names for rocks otherwise known as "iron formation." Therefore, in this chapter, iron formation is defined as: a chemical sediment, typically thin-bedded or laminated, that contains 15 percent or more iron (modified from James, 1954, 1966). Banded iron formation (BIF), as

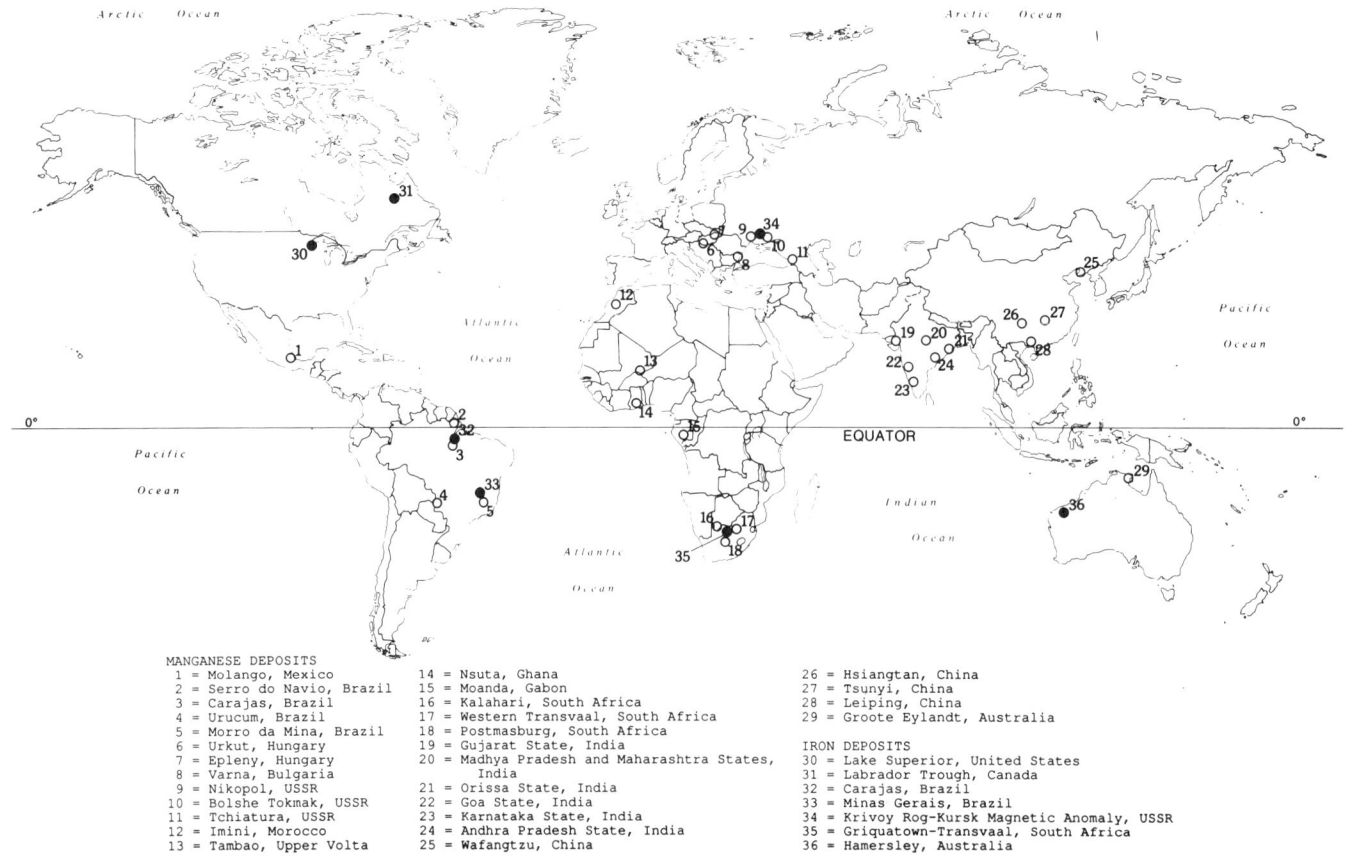

Figure 2. Location of major manganese and iron deposits in the world (from DeYoung and others, 1984, p. 6).

used here, means a thinly layered or laminated sequence of rocks that consists predominantly of alternate bands of chert (or silica-rich material, regardless of crystallinity) and iron-rich minerals. The presence of chert thus distinguishes these rocks from other iron-rich sedimentary rocks such as ironstones. BIF is further characterized by banding on several scales: beds that are generally a meter or more thick, layers that are typically 0.5 to 3 cm thick, and laminations that are on a scale of a millimeter or fractions of a millimeter. A single bed may extend continuously for hundreds of kilometers and cover tens of thousands of square kilometers. The iron content (as Fe) typically ranges from 20 to 35 percent, and silica (as SiO_2) is in the range from 40 to 50 wt % (James and Sims, 1973; James and Trendall, 1982; James, 1983).

Deposits of banded iron formation contain the largest concentrations of iron in the world. These deposits presently supply most of the world's iron ore and account for the bulk of its resources. It is estimated that the amount of iron ore initially contained in these deposits exceeded 1,000,000,000,000,000 mt that graded about 30 percent Fe. Present world resources exceed 800 billion mt of ore that contains more than 230 billion mt of iron. Moreover, although the number of known occurrences of BIF is in the thousands, seven districts contain more than 90 percent of all BIF-hosted ore. These seven districts, the Lake Superior region in the U.S., the Labrador trough in Canada, the Hamersley Range in Western Australia, Minas Gerais and Carajas in Brazil, Transvaal-Griquatown in South Africa, and Krivoy Rog–Kursk magnetic anomaly (KMA) in the USSR, are shown in Figure 2. Deposition of banded iron formation took place over a vast time span from Early Archean to the Phanerozoic. However, the majority of BIF deposits in the world cluster into four age brackets: middle Archean (3,500 to 3,000 Ma); Late Archean (2,900 to 2,600 Ma); early Proterozoic (2,500 to 1,900 Ma); and Late Proterozoic to early Phanerozoic (750 to 450 Ma). Most remarkably, the ages of the seven largest districts fall between 2,500 and 1,900 Ma (James and Sims, 1973; James and Trendall, 1982; James, 1983; Kuck, 1989a).

Gross (1965) subdivided deposits of iron formation into six classes and separated BIF into two: the Lake Superior type and the Algoma type. A third type of BIF (known as the Rapitan type), more recently identified by James and Trendall (1982) and James (1983), accounts for most BIF deposits of late Precambrian to early Paleozoic age. Rapitan-type BIF is spatially and temporally related to glaciogenic sediments; the best example of this type is in the Late Proterozoic Rapitan Group of the Snake River area, northwestern Canada. However, deposits of the Rapitan-type BIF have not been identified in the U.S., and they will not be discussed further in this chapter.

Lake Superior–type BIF is characterized by interbedded

chert (or quartz) and iron-rich minerals with a granular or oolitic texture, as summarized in Table 1. Clastic material is virtually absent within the BIF. The alternate or rhythmic bands range in thickness from fractions of a millimeter to as much as a meter. In the Hamersley Range, continuous bands are believed to extend for hundreds of kilometers and cover tens or hundreds of thousands of square kilometers. An individual bed may pinch and swell, and nodules or lenses of chert and jasper, rare crossbeds, or algal-like growth colonies may disrupt the continuity of a layer. These deposits are further characterized by fairly constant concentrations of the major constituents. For example, SiO_2 averages about 45 wt %, Fe is about 30 wt %, and the trace-element content is low. Lake Superior–type BIF is commonly associated with a succession of rocks that includes dolomite, quartzite, black carbonaceous shale, argillite, chert, chert breccia, and minor volcanic rocks such as tuffs and flows. More than one horizon of BIF may be present within this package of rocks, and the major deposits generally occur in the basal and middle parts of the sequence. All deposits of Lake Superior–type BIF are older than about 1700 Ma, and they dominate the Early Proterozoic period of mineralization. Indeed, all seven of the largest BIF mentioned previously are of the Lake Superior type. These deposits apparently formed by chemical precipitation of iron and silica along anorogenic continental shelves and the margins of miogeoclinal epicontinental basins. These characteristics suggest that the genesis of Lake Superior–type banded iron formation is perhaps unique and may be intimately linked with a major phase or transition in the evolution of the Earth's atmosphere, hydrosphere, and biosphere (Gross, 1965; Eichler, 1976; Button and others, 1982; James and Trendall, 1982).

Algoma-type BIF is chemically and mineralogically similar to the Lake Superior–type BIF; however, significant differences exist (Table 1). For example, Algoma-type BIF is more limited in thickness and areal extent, oolitic and granular textures are absent or inconspicuous, and perhaps most importantly, submarine volcanic rocks such as pillowed andesite-basalt, tuffs, other pyroclastic rocks, or rhyolitic flows, and graywacke, gray-green slate, and black carbonaceous slate are intimately associated. Most stratigraphically distinct occurrences of BIF are present in Late Archean rocks, and virtually all of these deposits are Algoma-type associated with greenstone belts. However, Algoma-type deposits are also found in submarine volcanogenic sequences of almost all post-Archean to Phanerozoic ages. Thus, Algoma-type BIF was deposited in volcanically active environments, and unlike Lake Superior–type BIF, its genesis is directly related to submarine volcanic exhalative activity (Gross, 1973; Button and others, 1982; James, 1983). Some greenstone belts host large stratabound and stratiform gold deposits that are associated with sulfide- and carbonate-facies banded iron formation (Algoma-type) within regionally more extensive oxide ± silicate facies. The Homestake, South Dakota, gold mine, which accounts for about 25 percent of the gold produced in the U.S., is hosted by arsenic-bearing sulfide-facies BIF (Ashley, this volume; Thorpe and Franklin, 1984).

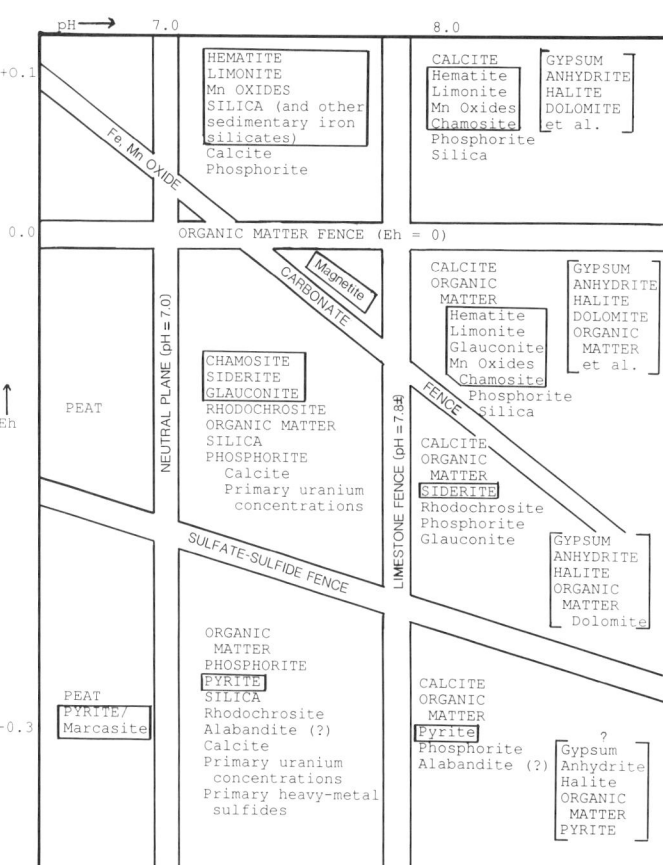

Figure 3. Eh–pH diagram that shows fields of occurrence of chemical end members in nonclastic sediments. Associations in brackets refer to hypersaline solutions (salinity >200‰). Pertinent iron minerals have been circled. The sulfide facies lies below the sulfate–sulfide fence and at pH lower than the limestone fence. The oxide facies is above the oxide–carbonate fence. Magnetite has been added immediately above the oxide–carbonate fence. (From Guilbert and Park, 1986, p. 609.)

James (1954) recognized that banded iron formation in the Lake Superior region contains mineral assemblages that apparently formed under restrictive environmental conditions. These physicochemical conditions correspond to fields of stability that are defined by Eh and pH factors, as shown in Figure 3. James (1954) defined four ideal facies of iron formation: oxide, silicate, carbonate, and sulfide. The oxide facies represents deposition in shallow water and an environment of relatively high Eh and high fugacity of oxygen, as illustrated in Figure 4. The other facies, in order, indicate deposition in progressively deeper water and more negative Eh conditions. These facies patterns are developed in both major types of banded iron formation as well as in Clinton-Minette-type ironstone, although minor differences do exist in their distribution (Table 1). It should be noted that these facies commonly overlap, and distinction of the ideal end-member facies is not always possible (James, 1954; Gross, 1965; Eichler, 1976).

TABLE 1. COMPARISON OF BANDED IRON FORMATIONS AND CLINTON-MINETTE–TYPE IRONSTONES*

	Lake Superior-type BIF	Algoma-type BIF	Clinton-Minnette–type Ironstones
Distribution	All continents	All continents	All continents
Age	Pre–1700 Ma	Dominantly pre–2600 Ma; also Proterozoic and Phanerozoic	Post–2000 Ma; predominantly Phanerozoic, especially lower Paleozoic and Jurassic
Geologic Association	Chert, chert breccia, dolomite, quartzite, graphitic black shale, argillite; clastic rocks below or above BIF, not within; volcanic rocks may or may not be present	Chert, graywacke, carbonaceous shale and slate, mafic to felsic pyroclastic and flows, commonly pillowed andesite-basalt and other submarine volcanogenic rocks; greenstone belts	Shale, sandstone, graywacke, and carbonate rocks; volcanic rocks commonly absent; lack chert
Sedimentary Environment	Miogeosynclinal; stable continental shelves, restricted intra-cratonic basins; shallow water	Eugeosynclinal; water >100 m in depth	Epicontinental, shallow marine, near-shore
Areal Extent	BIF persistent over 100s to 1000s of km	BIF persistent for few km	Beds may be persistent for 10 km
Thickness of Beds	Major units 10s to several 100s of m	Major units less than 1 to 10s of m	Major beds about 1 to 10s m
Sedimentary Structures	Layered, banded, commonly oolitic and granular	Layered, banded; granular and oolitic textures inconspicuous or absent	Massive with oolitic textures; banding poor to absent
Sedimentary Facies	Oxide-facies most abundant; silicate- and carbonate-facies commonly intergradational; sulfide-facies insignificant to absent	Oxide-facies predominant; carbonate- and sulfide-facies thin and discontinuous; silicate-facies commonly masked by metamorphism	Oxide-facies dominant; silicate- (chamosite) and carbonate-facies commonly present; sulfide-facies rare
Mineralogy	Magnetite, hematite, chert, siderite greenalite, dolomite, and pyrite; goethite, chamosite, glauconite, pelletal collophane, and calcite rare or absent	Magnetite, hematite, chert, siderite, greenalite, minnesotaite, stilpnomelane, chlorite, dolomite, and pyrite; goethite, chamosite, glauconite, pelletal phosphate, and calcite rare to absent	Goethite, hematite, chamosite, glauconite, siderite, calcite, dolomite, pelletal collophane, and pyrite; quartz (chert) and magnetite rare, greenalite absent
Chemistry	Low P_2O_5 (<0.45%), low Al_2O_3 (0.1 to 1.5%), low alkalies (<2%)	Low P_2O_5 (<0.45%), low Al_2O_3 (0.1 to 2.5%), low alkalies (<2%), Mg higher vs. Lake Superior-type BIF); MgO/CaO >1	P_2O_5 commonly 0.25 to >1.5% [Al_2O_3 about 2 to 5%], and to 17%, alkalies to >10%, SiO_2/Fe << BIF, MgO/CaO <1
Source of Iron	Ocean; transported from terrigenous and submarine volcanic sources	Submarine volcanic exhalative activity	Continental weathering

*Adapted from James, 1966; Stanton, 1972, Eichler, 1976; Holland, 1984; and others (see references in sections on banded iron formations and Ironstones).

The oxide facies is the most recognizable in the field. Not only does it form distinct ridges of banded rocks, but the abundance of magnetite gives a strong magnetic high. This facies predominantly contains hematite and magnetite. Both minerals occur as primary constituents, and they are present in separate bands or in various proportions within a single band. Thus, they define two subfacies of the oxide facies: hematite-banded and magnetite-banded. The hematite-banded subfacies consists of finely crystalline hematite interlayered with gray chert or reddish jasper. Ooliths and granules are common in most layers. The hematite-banded subfacies was deposited in shallow, strongly oxidized, weakly acidic to alkaline waters in a near-shore environment. The initial precipitate was probably a hydrated ferric oxide such as $Fe_2O_3 \cdot nH_2O$ or $Fe(OH)_3$. The magnetite-banded subfacies contains layers of magnetite and alternate layers of dark chert, green iron silicates, and carbonate in varying proportions. Hematite is less commonly associated with magnetite in rocks that contain silicate or carbonate minerals. Magnetite may have formed as hydromagnetite ($Fe_3O_4 \cdot nH_2O$) or a mixture of $Fe(OH)_2$ and $Fe(OH)_3$ in upper well-oxygenated waters and settled into a weakly oxidized to moderately reduced bottom environment in which hematite was not stable. The formation of both magnetite and hematite requires a low fugacity of CO_2; otherwise siderite would form. It might be noted that magnetite is not predicted to be a primary mineral from the Eh-pH diagram (Fig. 3). However, textural relations and its abundance in unmetamorphosed and diagenetically unaltered rocks indicate that at least some magnetite is primary in origin (James, 1954; Gross, 1965; Goodwin, 1973; Guilbert and Park, 1986).

The silicate facies is recognized by its distinct, olive green to yellowish brown color. Although metamorphism commonly masks the origin of the iron silicates, Gross (1965) agreed with James (1954) that greenalite [$(Fe^{2+}, Fe^{3+})_{5-6}Si_4O_{10}(OH)_8$] is primary in origin and suggested that some minnesotaite [$(Fe^{2+}, Mg)_3Si_4O_{10}(OH)_2$], stilpnomelane [$K(Fe^{2+}, Fe^{3+}, Mg,Al)_3Si_4O_{10}(OH)_2 \cdot xH_2O$], chamosite and other iron chlorites, and perhaps glauconite [$K(Fe^{2+}, Fe^{3+}, Mg, Al)_2Si_4O_{10}(OH)_2$] may also be primary. James (1954) distinguished between granular and nongranular subfacies on the basis of dense, massive, cryptocrystalline to microcrystalline, oval-shaped granules in the former and laminated or thin streaky layers of iron silicate minerals in the latter. Both may contain abundant magnetite and siderite. These characteristics indicate that the silicate facies was deposited as an amorphous ferrous silicate [probably $Fe_3Si_2O_5(OH)_4$] in a moderately oxidized to moderately reduced environment. The presence of hematite and pyrite in the silicate facies further emphasizes the wide range of physicochemical conditions in which this facies is stable (Fig. 3). Moreover, the number and complexity of iron silicate minerals and the effect of varying intensities of metamorphism in different BIF deposits make it difficult to define a specific depositional environment. Rather, the silicate facies apparently formed in an environment common to parts of both the oxide and carbonate facies (James, 1954; Gross, 1965; Eichler, 1976).

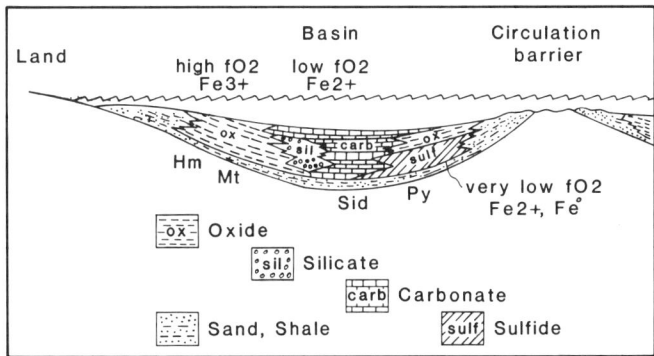

Figure 4. Schematic section of depositional zones of iron formation facies in a barred basin. The relative fugacity of oxygen, valence state of iron, and the dominant iron mineral are also shown. Hm, hematite; Mt, magnetite; Sid, siderite; Py, pyrite. (From Eichler, 1976, p. 175; and Guilbert and Park, 1986, p. 612.)

In the carbonate facies, siderite is the most prominent mineral. It is commonly interlayered with gray chert in roughly equal proportions. Ankerite, calcite, dolomite, magnetite, hematite, iron silicates, and pyrite may also be present within this facies. Ooliths and granules are generally absent, and siderite appears to have accumulated as a fine-grained mud. This facies probably formed in water that was undisturbed by waves or currents and was deeper, cooler, and more reducing than that of the oxide facies. The transition zone between the carbonate and oxide facies may contain granules or ooliths of siderite with iron silicate, magnetite, and chert. Stylolites are a distinctive feature of the carbonate facies and indicate a diagenetic origin for some of the carbonate minerals. Moreover, carbonate replacement of primary oxide and associated silicate minerals are typical diagenetic effects in BIF. However, it is unlikely that all of the carbonate is secondary in origin (James, 1954; Gross, 1965; Dimroth, 1979; Walker, 1984).

The sulfide facies occurs as disseminated pyrite in black carbonaceous shale, pyrite and pyrrhotite intermixed with siderite and other carbonates in banded chert, and as distinct beds of pyrite and pyrrhotite with minor siderite. It is the least abundant of the facies of iron formation and is commonly not recognized in the field, in part because it is easily eroded. Granular textures are not evident. The sulfide facies formed in a strongly reducing environment with abundant H_2S or HS^- and weakly acidic to alkaline waters. This environment is most characteristic of the deeper parts of sedimentary basins where the water is not agitated, organic carbon is preserved, and H_2S is generated by bacterial action on the organic material. The zone of deposition of the sulfide facies may overlap or alternate with that of the carbonate facies in a transitional environment marked by carbonate-rich sulfide facies and sulfide-rich carbonate facies (James, 1954; Gross, 1965).

Although the chemistry of BIF as a whole is relatively uniform from deposit to deposit, variations do exist between the different facies. For example, CO_2 is distinctly higher in the carbonate facies, Fe^{2+} is greater, and SiO_2 is slightly less than in the oxide and silicate facies. However, MgO consistently ranges between 2 and 4 percent in all three facies, and perhaps surprisingly, the total iron contents of these facies are subequal. These patterns are present because deposits of banded iron formation do not conform to an ideal distribution of end-member facies as described above, but rather the facies are intergradational and form a mixture with local dominance of one.

James (1966), Gross and McLeod (1980), James and Trendall (1982), and Davy (1983) have compiled the most detailed major-element analyses of iron formation and their facies. All have noted that systematic geochemical investigations have not yet been adequately conducted to define and interpret chemical variations in BIF. Nonetheless, the major deposits generally show a small range in the overall concentration of the dominant constituents. For example, silica varies between 40.7 and 48.5 percent, and iron ranges from 24.5 to 37.9 percent. Unlike many ore deposits, the only other elements concentrated above average crustal values are Mn, C, Mg, and perhaps Ca. Others, such as Al, K, Na, Ti, Sr, Ba, V, Cr, Co, Ni, and Cu, are depleted with respect to their crustal average. Fryer (1983) noted that the rare earth element (REE) abundances in BIF are extremely low, with absolute REE contents less than normal crustal abundances. The REE may ultimately be used to identify primary precipitates and diagenetic changes as well as the genesis of BIF. However, data are scant, mostly of reconnaissance nature, and more detailed studies and documented sequences are required to interpret the histories of these rocks (James, 1966; Lepp, 1975; Eichler, 1976; Davy, 1983; Fryer, 1983; Kimberley, 1983).

Banded iron formation is present in two principal regions of the United States: the Lake Superior region of Minnesota, Michigan, and Wisconsin; and the northern Rocky Mountains in Montana, Wyoming, and South Dakota. Small deposits also occur in Arizona, New Mexico, and Colorado (Plate 2A). Both Lake Superior–type and Algoma-type BIF are represented by these deposits (Bayley and James, 1973).

Lake Superior–type banded iron formation. Lake Superior–type banded iron formation dominates the past and present iron-ore production in the U.S., and the Lake Superior region is preeminent among all the districts. BIF occurs in six districts, or ranges: the Mesabi, Cuyuna, and Gunflint Ranges in Minnesota; the Marquette and Menominee Ranges of Michigan; and the Gogebic Range in Michigan and Wisconsin (Plate 2A). The Menominee Range is subdivided into the Menominee and Iron River–Crystal Falls districts. The Mesabi has produced about 72 percent of the 4.9 billion mt of iron ore shipped from this region, and the three ranges in Michigan account for about 26 percent. The Gunflint Range does not contain any deposits of commercial grade and size (Morey, 1983; Klinger, 1985a).

A brief review of the geology in the Lake Superior region is given in this chapter. More detailed descriptions are contained in Medaris (1983), Reed and others (1990), and in references cited here. Rocks of Early Proterozoic age (2500 to 1600 Ma) were deposited unconformably on two contrasting Archean crustal segments: the Superior province greenstone–granite terrane (2750 to 2600 Ma) to the north, and the migmatitic gneiss and amphibolite terrane (in part about 3600 Ma) to the south. Between 2200 and 1900 Ma, clastic and chemical sedimentary rocks, including banded iron formation, and mafic to felsic volcanic rocks were deposited in an intracontinental rift basin (termed the Animikie Basin) on and parallel to the Great Lakes tectonic zone, which juxtaposes the two Archean terranes. These rocks form stratigraphic sequences up to 10 km thick over an area of about 700 km east to west and 400 km north to south, and they record a complete transition from a stable miogeoclinal shelf environment to a deep-water environment. Rocks of these sequences were deformed, metamorphosed, and intruded by calc-alkalic, intermediate to felsic, plutonic rocks during the Penokean orogeny at about 1860 ± 50 Ma (Sims and others, 1981; Morey, 1983; Stille and Clauer, 1986).

Stratigraphic successions for each of the six ranges in the Lake Superior region have been correlated throughout much of the Animikie Basin, even though physical continuity between rocks of the several ranges has not been firmly established. The strata in the northwestern part of the region make up the Mille Lacs and Animikie Groups, whereas those in the southeastern segment form the Marquette Range Supergroup, which is divided into the Chocolay, Menominee, Baraga, and Paint River Groups (Morey, 1983). Correlations among these rocks are presented in Figure 5.

Early Proterozoic rocks of the Mille Lacs and Animikie Groups and the Marquette Range Supergroup represent three grossly fining-upward depositional cycles. Irregularities on the basement surface and erosion have resulted in variable thicknesses of the deposits throughout the region. The basal units of each cycle, where present, consist predominantly of quartz-rich conglomerate and arenitic sandstone (now quartzite). Deposits of platform sediments, stromatolitic dolomite, and locally shale accumulated on the coarse sands to form the Chocolay Group. The Mille Lacs Group contains rocks of similar lithologies that were also deposited under shallow-water conditions along the fringes of the basin. In addition, mafic and intermediate subaqueous volcanogenic rocks, black carbonaceous shale, and minor amounts of cherty banded iron formation accumulated toward the axial part of the basin (Bayley and James, 1973; Sims and others, 1981; Morey, 1983).

Strata of the Animikie and Menominee Groups, which contain the main BIF, were deposited unconformably on Archean basement rocks or on eroded remnants of the Mille Lacs and Chocolay Groups. Banded iron formations are the principal rock types in the two former groups. The major iron-bearing units are the Gunflint, Biwabik, Trommald, Ironwood, Negaunee, and Vulcan Formations (Fig. 5). The Biwabik and Ironwood Iron Formations represent nearly contemporaneous, transgressive shelf sedimentation near strandlines on opposite sides of the Animikie Basin, whereas the other units, because of differences in thickness, stratigraphic continuity, and sedimentary associations and struc-

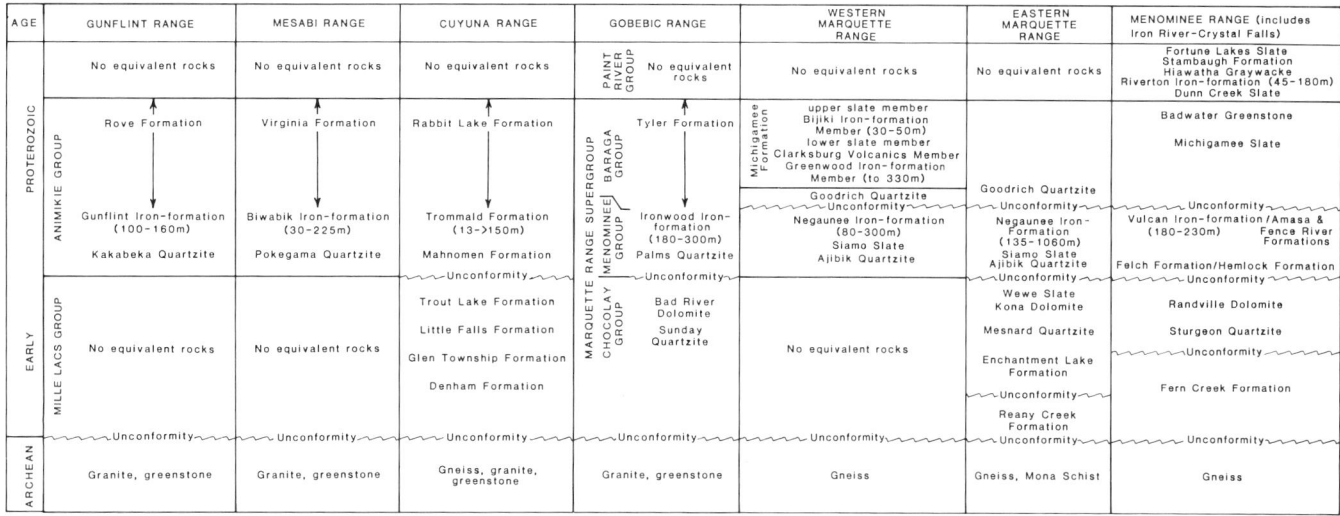

Figure 5. Correlation chart of early Proterozoic bedded rocks in the Lake Superior region (from Sims and others, 1981, p. 382; Morey, 1983, p. 26; Morey and Van Schmus, 1988, Fig. 7).

tures, appear to have been deposited simultaneously either in separate basins or in isolated second-order troughs within the same larger basin. The accumulation of deeper-water sediments such as intercalated carbonaceous mudstone, graywacke, and siltstone, and volcanogenic rocks accompanied minor deformation and uplift and effectively terminated the deposition of iron formation throughout the region (Bayley and James, 1973; Schmidt, 1980; Sims and others, 1981; Morey, 1983).

The Baraga Group of the Marquette Range Supergroup and the upper part of the Animikie Group (Fig. 5) are characterized by submarine mafic to felsic volcanic rocks, graywacke turbidites, and shale. Several minor horizons of banded iron formation are present within this sequence; however, their deposition appears to have taken place during lulls or cessations in clastic sedimentation and volcanism. Deposition in a deep-water environment continued with the accumulation of turbiditic graywacke (flysch), shale, and small lenses of iron formation (most notably the Riverton Iron Formation) to form the Paint River Group of the Marquette Range Supergroup. However, this group is preserved only in the Iron River–Crystal Falls district of the Menominee Range. These rocks may have originally had much greater thickness and extent, or they may have been deposited in a small (about 50 km) depression within the main basin. Deformation, metamorphism, and magmatism associated with the Penokean orogeny terminated the cycles of sedimentation (Bayley and James, 1973; Morey, 1983).

Differences in the intensity of deformation and metamorphism coupled with a cover of glacial drift and vegetation impede correlations of stratigraphic members of BIF between ranges in the Lake Superior region. For example, strata in the Mesabi and Gunflint Ranges dip gently southward and are weakly metamorphosed from the zeolite facies to the lower greenschist facies, whereas BIF in the other ranges dips steeply, is intricately folded and faulted, and is locally metamorphosed to the lower amphibolite facies. In places, metamorphic grade increases parallel to strike of BIF, which at least allows progressive changes within a horizon or unit to be traced. One of the more interesting aspects of these banded iron formations is that they do not consistently follow or precede any other rock type, nor are they restricted to any one sedimentological regime within the Animikie Basin. A large part of the banded iron formation in Michigan and Minnesota represents the oxide facies and facies intermediate to the carbonate and silicate facies, as defined by James (1954). A representative stratigraphic succession through the Biwabik Iron Formation is presented in Figure 6. Cherty taconite, or the low-grade BIF protore, is generally composed of granular and oolitic chert with magnetite, siderite, and iron silicates in varying proportions representative of the oxide facies. Slaty taconite typically consists of chert, siderite, ankerite, magnetite, and iron silicate minerals that are finer grained, nongranular, and thin-bedded or laminated, and indicative of the carbonate-silicate facies (James, 1954; Gair, 1973; Bayley and James, 1973; Morey, 1973, 1983).

Most of the iron ore mined in the Lake Superior region prior to 1955 was soft hematite- or goethite-rich deposits known as natural orebodies. These secondary-enriched deposits of BIF contained 60 percent Fe or greater, and their vast tonnage formed the basis for the rise of the U.S. to world economic leadership. At the present time, however, taconite that grades about 30 percent Fe composes more than 95 percent of the U.S. production and reserves (Morey, 1983; Klinger, 1985a; Morris, 1985).

The genesis of Lake Superior–type banded iron formation has been the subject of many discussions and theories. In addition to characteristics discussed above, such as variations in facies and sedimentary associations, other characteristics of BIF units that differ regionally and worldwide make it difficult to formulate a uniform genetic model for the origin of BIF. Moreover, complexities imposed by secondary processes further promulgate the many difficulties of defining BIF genesis. Among the significant questions that have been asked are: Is the source of iron and silica volcanic or continental? What factors and conditions (e.g., tectonics, Eh, pH, f_{O_2}) controlled the transport and accumulation of such enormous quantities of iron and silica? Why was iron virtually co-precipitated with only silica over a relatively short interval of geologic time during the Precambrian? What role did microbiota play in the precipitation of BIF? What is the reason that these deposits were primarily confined to the coasts and shelves of basins; i.e., what was the sedimentary environment? Was the composition of the waters fresh, brackish, or seawater? What is the cause of the banding and layering? How important were diagenetic processes in the formation of BIF? It is beyond the scope of this chapter to discuss each theory and defend or criticize specific elements. Rather, an attempt will be made to summarize the most recent ideas of BIF genesis, and references will be given to direct the reader to more exhaustive discussions.

The source of iron and silica in BIF was both terrigenous and volcanic. Weathering of continental cratons and submarine volcanic emanations provided iron and silica to the oceans over a long period of geologic time in the Archean. Ultimately, then, the oceans themselves were the direct sources of iron and silica to the Early Proterozoic Lake Superior–type BIF. It is inferred that silica accumulated in seawater to levels near saturation because of the absence of organisms that form siliceous skeletons and would deplete silica from solution. Iron remained soluble as ferrous (Fe^{2+}) iron because of the essentially anoxic reducing conditions in the hydrosphere and atmosphere. During the Early Protero-

zoic, previously stable cratons were subjected to extension, resulting in development of shallow-marine marginal basins and intracontinental troughs. Also, an upper, more oxygenated layer of seawater and a more oxidized atmosphere (to about 1 percent of the present atmospheric level of O_2) developed between 2300 and 2000 Ma with the proliferation of photosynthesizing organisms and/or abiotic photodissociation of $FeOH^+$ complexes in the upper mixed layer of the ocean, in addition to photolytic dissociation of water vapor in the atmosphere. Precipitation of iron in BIF was initiated by increased oxidation and perhaps by seawater evaporation, which further enhanced the concentration of iron and silica. Direct oxidation and precipitation of iron by stromatolites and algal-like microorganisms may also have been important. The composition of the iron precipitates depended on local conditions. For example, cherty oxide-facies BIF formed in relatively unrestricted shallow basins on a subtidal shelf, whereas slaty carbonate, silicate, and sulfide facies were favored in deeper water and in barred basins and troughs with lower oxygen fugacities and oxidation potentials. The mechanism that transferred reduced, ferrous-rich bottom waters to more oxidized upper layers appears to involve upwelling. Such upwelling may have been in part climatically induced. As a consequence, seasonal upwelling and subsequent growth of cyanobacteria may have produced more oxygen and resulted in the oxidation and hydrolysis of ferrous iron to form ferric hydroxide in upper surface waters that settled to the sea floor. Precipitation of iron ceased when oxygen production was reduced during the cold season. Fine-scale banding may have resulted from this seasonal deposition of iron colloids coupled with nearly continuous inorganic precipitation of amorphous silica. Layers of BIF and overlap of facies developed as the environment of deposition shifted with time. Subsequent diagenetic changes, such as bacterial reduction of ferric to ferrous iron and unmixing of metastable amorphous assemblages into crystalline phases, as well as metamorphic effects such as recrystallization and increased crystal size, caused the varied nature of BIF. Lake Superior–type BIF deposition terminated with an influx of clastic sediments or volcanic material, or with slight deformation and uplift. The relatively sudden disappearance of this type of BIF by about 1700 Ma appears to be related to the establishment of a stable aerobic ocean and atmosphere (Eichler, 1976; Button and others, 1982; Ojakangas, 1983; Towe, 1983; Trendall and Morris, 1983; Holland, 1984; Walker, 1984; Baur and others, 1985; Birnbaum and Wireman, 1985; Pelymskiy and Shishova, 1985; Walker and Brimblecombe, 1985; Drozdovskaya, 1986; Francois, 1986).

The natural orebodies formed from the oxidation and leaching of primary BIF by circulating waters during periods of intense chemical weathering. Enrichment took place in porous and permeable zones of BIF, or along structures such as fractures, faults, and joints or dikes associated with upward-facing structural traps (e.g., synclines). Initial oxidation and hydration of primary, diagenetic, or low-grade metamorphic minerals in carbonate- and oxide-facies BIF to several iron oxides and hydroxides occurred under conditions of extreme aridity and deep ground-water circu-

Virginia Formation		ferruginous slate
Biwabik Iron Formation	Upper Slaty Member	even bedded siliceous, cherty carbonate and chert, conglomerate, granular
	Upper Cherty Member	wavy bedded, granular cherty, siliceous, carbonate
		algal, chert, jasper, conglomerate
		wavy bedded, granular, cherty, siliceous, carbonate, conglomerate
	Lower Slaty Member	even bedded siliceous, carbonate and wavy bedded granular chert, siliceous
		black slate
	Lower Cherty Member	lean cherty, siliceous, carbonate
		wavy bedded, granular chert, siliceous oxide
		even bedded cherty magnetite-hematite
		algal, jasper, conglomerate
	Pokegama Quartzite	

Figure 6. Stratigraphic succession of the Biwabik Iron formation (from Morey, 1983, p. 51).

lation (i.e., deep water table) within the zone of aeration and vadose water. This process resulted in oxidized BIF with increased secondary porosity and permeability. Water was then entrapped in upward-opening structures during wet cycles, and further leaching and dissolution of chert and iron minerals formed solutions saturated with silica and perhaps ferrous iron and carbon dioxide that were flushed from the system. Some chert was replaced by siderite due to gradual loss of carbon dioxide through evaporation. Repeated stagnation of water flow and periodic expulsion of these silica-saturated waters in an artesian-like system caused the elimination of silica and the concentration of iron ore both by formation and oxidation of secondary siderite and by infiltration of colloidal iron oxides. Thus, the orebodies grew upward from depth toward the surface. This theory, proposed by James and others (1968), differs from one that invokes more typical meteoric deep weathering by accounting for a mechanism to remove silica on a large scale and to allow circulation of meteoric waters to depths of 1,000 m or more. Morey (1983) noted that neither these theories nor ones that involve hydrothermal fluids have been totally proved or disproved. However, Morey (1983) accepted the hypothesis of James and others (1968) as the one that accounts for most of the geologic attributes of the soft ores and suggested that, rather than there being just one period of ore formation in the Precambrian, as proposed by James and others (1968), cyclic ground-water activity took place over at least three episodes between the Early Proterozoic, soon after BIF deposition, and the Late Cretaceous. It should be noted that secondary magnetite hard ore in the Marquette Range may have had a different origin from that proposed for the soft ores (see Cannon, 1976).

Algoma-type banded iron formation. Algoma-type banded iron formation is named for rocks in the Michipicoten area, Ontario, Canada, of the Algoma mining district (Plate 2A). In the U.S., the most significant iron ore production from Algoma-type BIF has been from the Vermilion district of Minnesota (Plate 2A), which produced about 104 million mt of ore before 1977 when the mines closed. The Atlantic City district in the Wind River Range, Wyoming (Plate 2A), has proven reserves of more than 100 million mt. Other areas of Algoma-type BIF include the Late Archean Dickinson Group north of the Felch trough of Michigan and the Early Proterozoic Yavapai Series of central Arizona (Plate 2A). The Vermilion Range (2750 to 2700 Ma), Wind River Range (>2600 Ma), and Dickinson Group (>2700 Ma) banded iron formations as well as the Yavapai Series BIF (1820 to 1775 Ma) are present within sequences of volcanic and sedimentary rocks. The volcanic rocks include dominantly mafic to intermediate flows, tuffs, and other pyroclastics, and the sediments are in part volcaniclastic with characteristic graywacke and slate in addition to arkose and conglomerate. BIF occurs at many horizons within this volcano-sedimentary pile. The overall chemistry of Algoma-type BIF is similar to that of the Lake Superior type. However, increased concentrations of Al_2O_3, and Co, Ni, Cr, Cu, and Mg reflect the closer association with clastic sediments and volcanic rocks, respectively (Bayley and James, 1973; James, 1983; Klinger, 1985b).

The Archean Soudan Iron Formation Member of the Ely Greenstone is the most extensive unit that was mined in the Vermilion district. It is interbedded with basaltic to andesitic flows and tuffs and is overlain by dacitic tuffs, agglomerates, and other volcaniclastic rocks of the upper member of the Ely Greenstone and the Archean Lake Vermilion Formation. The Soudan BIF consists of several types of fine-grained ferruginous chert that are interbedded with finely crystalline tuffaceous rocks and, less commonly, metabasalt and epiclastic rocks. These units are at most about 100 m thick with strike lengths less than about 2 km. Ferruginous chert is composed dominantly of assemblages of the oxide facies. For example, jaspilite is a thin-bedded or laminated rock that contains alternate layers of chert or jasper and magnetite (or martite) and specular hematite, and massive chert or jasper contains disseminated magnetite and/or hematite. The carbonate (chert-siderite), silicate (chert-chlorite), and sulfide (chert-pyrite) facies are present locally (Sims, 1972; Bayley and James, 1973; Sims and James, 1984).

The genesis of these deposits is less equivocal than that of the Lake Superior–type BIF. The iron was apparently derived predominantly from submarine volcanic exhalations. Gross (1983b) and others have suggested that banded silica and iron-rich muds in the Red Sea, and iron and manganese oxide deposits on mid-oceanic ridges are the prototypes for Algoma-type BIF. The deposits probably accumulated in small, relatively shallow basins during periods of quiescence in volcanism. The lack of stromatolites and other microorganisms indicate that these deposits accumulated in water more than 100 m deep below the photic zone and that precipitation did not involve direct microbial oxidation of iron. The interbedded fine-grained volcaniclastic rocks may have been deposited as ash falls or reworked pyroclastic debris. The environment of deposition strongly favored formation of the oxide facies with lesser carbonate, silicate, and sulfide facies. In general, the oxide facies is present farther from the source of volcanism than the other facies. A general zonation from sulfide through carbonate to oxide facies is common in many Algoma-type BIFs of Canada, although a similar zonation has not been delineated in the Vermilion or other districts in the United States. The sulfide facies occurs as black carbon-rich mudstone with intermixed volcanic ash and tuff and disseminated pyrite, pyrrhotite, and possibly lead, zinc, and copper sulfides. These mudstones are commonly associated closely with stratiform base-metal sulfide deposits. Figure 7 depicts the possible setting of the Algoma-type BIF and its relation to Lake Superior–type BIF (Sims, 1972; Goodwin, 1973; Button and others, 1982; Gross, 1983a; James, 1983; Walter and Hofmann, 1983). Gross (1983b) noted that separation of BIF from volcanogenic massive sulfide deposits may be related to the degree of seawater dilution in the hydrothermal fluids that emanate from the submarine vents. Secondary hematite hard ore in the Vermilion district may have had an origin similar to that in the Marquette Range (Cannon, 1976).

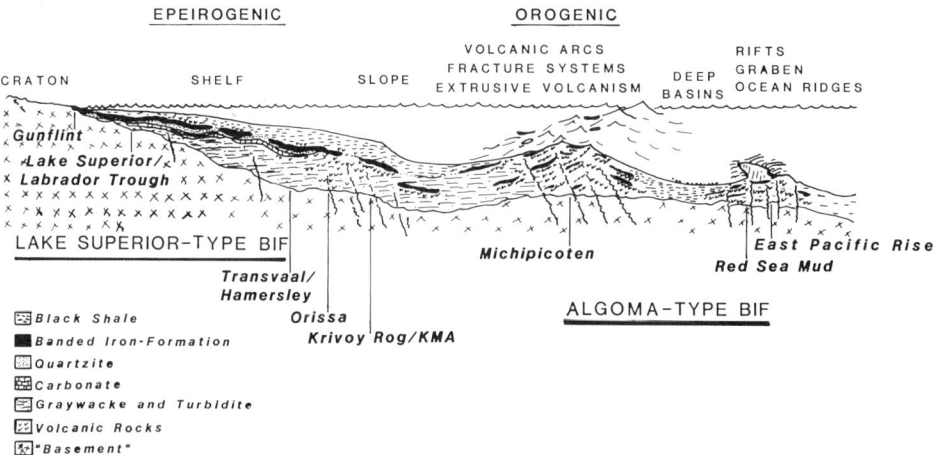

Figure 7. Tectonic environments for deposition of banded iron formation (from Gross, 1983b, p. 174).

Ironstones

Deposits of ironstone represent the second largest source of iron ore from past production and for future reserves in the U.S. and around the world. In contrast to the banded iron formations discussed above, ironstones are predominantly Phanerozoic in age, banding is indistinct or absent, and perhaps most importantly the deposits are devoid of associated chert. Silica is present mainly within iron silicate minerals such as chamosite and less commonly as clastic grains of quartz. Iron occurs dominantly in hematite and goethite; magnetite is rare. In the U.S., oolitic iron ore of the Clinton Formation, Middle Silurian in age, is typical of these ironstones. Elsewhere, the minette ores (Jurassic in age) of Alsace-Lorraine, France, and the Northampton Sand, Marlstone, Cleveland, and Frodingham iron-bearing formations (of Jurassic age) and the Claxby Iron Formation (Cretaceous) of England are representative of these oolitic ironstones (Stanton, 1972; Kimberley, 1981).

Clinton-Minette-type ironstone. The Clinton Formation crops out intermittently along the Appalachian Mountain chain from upstate New York to Alabama. Ore is best developed in the Birmingham district of Alabama (Plate 2A), where more than 300 million mt of ore have been produced and potential resources are estimated to be more than 2 billion mt. There, the Clinton Formation (known locally as the Red Mountain Formation) consists of oolitic red hematite interbedded with thin beds of iron-stained sandstone, carbonaceous shale, and impure limestone. These rocks are typically deep red to purplish in color because of their high concentration of hematite, whereas the minette ores are more commonly brownish to greenish brown due to more abundant siderite, chamosite, and goethite. The ore occurs as seams less than about 7 m thick in lenticular beds that are generally less than 15 m thick. Three seams have been mined in the Birmingham district. Primary ore is present as: (1) oolitic ore, with hematite oolites (1 to 2 mm in diameter) in a calcite matrix, or hematite-coated pebbles or grains cemented with calcite and hematite; (2) fossil ore, in which hematite has coated and replaced fragments of bryozoans, mollusks, and trilobites, and with hematite cement; and (3) flaxseed ore, which occurs as small, flattened concretions of hematite with hematite cement (Simpson and Gray, 1968). These ores are known commonly as hard or soft, depending on the degree of supergene leaching of matrix material. Soft ore is generally of higher grade, friable, unconsolidated material, whereas hard ore is dense and compact. Other iron minerals in ore include siderite and chamosite. Mineral assemblages of the oxide, silicate, and carbonate facies are commonly developed in ironstones (Table 1), although the facies are not well segregated, and they may be intermixed on a microscopic scale. The latter two facies are present as chamosite ooliths with hematite and beds of siderite with chamosite and hematite, respectively. The sulfide facies is uncommonly associated with deposits of ironstone, although pyrite may be present in the carbonate or silicate facies, and pyritic black shale may be a minor component of the stratigraphic section. The ores average about 25 to 35 percent Fe and rarely exceed about 45 percent Fe. Ironstone can be distinguished chemically from Lake Superior-type BIF by its high concentration of phosphorous (0.25 to more than 1.5 percent versus generally less than 0.45 percent) and alumina (typically 2 to 5 percent versus less than 1.5 percent) (James, 1966; Simpson and Gray, 1968; Gross, 1970; Stanton, 1972; Maynard, 1986).

It is generally well accepted that ironstones are chemically precipitated sediments, and that volcanism and volcanic exhalations were not involved in their formation or at most contributed only a minor proportion of iron. Moreover, most ironstones formed along the coastlines of shallow inland seas and on the shelves of cratons in sedimentary basins rarely more than 160 km in maximum dimension. However, unlike the Precambrian banded iron formations, ironstones formed within an oxidized hydrosphere and under an oxidized atmosphere, which means that a vast reservoir of reduced iron did not reside in the oceans awaiting oxidation. Also, the composition of the atmosphere may have changed during the late Precambrian and the Phanerozoic with an increase of the partial pressure of oxygen (P_{O_2}) and a decrease of P_{CO_2}, whereas the overall composition of seawater did not vary greatly, although fluctuations did occur. Thus, the transport of iron from a presumed weathered continental source via rivers to the sea throughout the Phanerozoic requires a unique explanation. It is most likely that iron was transported as colloids or detrital grains in the suspended load of streams rather than as dissolved ferrous iron. Molecules such as $FeCl_2 \cdot nH_2O$ or $Fe(OH)_3 \cdot nH_2O$ may have been the dominant iron compounds during transport. Iron may then have been deposited as a flocculate, biochemical precipitate, or direct chemical precipitate from seawater. Goethite or hematite and a chamositic clay were apparently the earliest primary minerals formed in ironstones (Simpson and Gray, 1968; Kimberley, 1981; Holland, 1984; Maynard, 1986).

On the contrary, Kimberley (1981) proposed that the principal concentrating process for oolitic ironstones (which he termed sandy, clayey, and oolitic, shallow-inland-sea iron formation, or SCOS-IF) was electrolytic precipitation due to dissolution of aragonitic mud. Therefore, the Clinton–Minette ores would be early diagenetic replacements. Many other theories have been proposed for the origin of ironstones (see Kimberley, 1981, p. 46); however, a combination of direct chemical precipitation of iron transported by streams and rivers and diagenetically replaced calcareous ooze appears to account for most factors involved in the genesis of ironstones. Van Houten (1986) and Boucot and Gray (1986) concluded that correlations consistent with presently considered geological and biological factors do not exist for the origin of ironstones.

Other iron-rich sediments

Other types of iron-rich sedimentary deposits contribute a small percentage to the world's iron resources, although some have had historical importance. These include black-band ore, Lahn-Dill–type ore, and bog ore. Sedimentary beds that contain siderite as nodules, discontinuous lenses, or thin layers (<1 m) compose the black-band ores; oolites are not present. Siderite may also form cement for quartz and other clastic particles and organic matter. Black-band ore in the eastern U.S. is characteristically present in association with coal measures. Thus, deposits are most extensive in rocks of Carboniferous and Permian age, and small amounts were mined during the colonial period in Pennsylvania, Massachusetts, and Ohio. Sideritic ores in eastern Texas (Plate 2A) have been mined recently. The ore is associated with the greensand facies of the Weches Formation in the Eocene Claiborne Group. The Weches Formation ranges from a few cm to about 30 m thick, and it consists of siderite, glauconitic sand, quartz sand, and clay. Sideritic iron ore forms nodules and irregular lenticular beds in a zone 1.5 to 9 m thick, and weathered "brown ore" forms beds up to 1 m thick. In general, these carbonate ores contain about 20 to 50 percent Fe, depending on the degree of secondary enrichment, and phosphate is subequal to that in Clinton-Minette–type ironstones. The origin of black-band ore is not well understood. The siderite is in part primary, diagenetic, and secondary. In eastern Texas, the iron is interpreted to be derived diagenetically from the alteration of glauconite. Siderite formed in areas of methane production, and goethite formed from oxidation of siderite and from reprecipitation of leached iron above the water table. The environment of deposition appears to be in brackish waters and marine swamps (Carr and others, 1967; Gross, 1970; Stanton, 1972; Foos, 1984).

Lahn-Dill–type iron ore deposits are syngenetic accumulations of predominantly hematite within submarine volcanic sequences of spilitized tuffs and lavas. They are located primarily in rocks of the central European Variscan (or Hercynian) system, which is Devonian and Early Carboniferous in age. They are not known, however, in the United States. The reader is referred to Quade (1976) and references contained therein for further discussion.

Deposits of bog ore have been mined longer than any other source of iron ore, and prior to the mid-19th century they were the most important source. Although bog ore is currently uneconomic to mine, some is accumulating at present; its genesis is relatively unequivocal. Bog ores develop in swamps and lakes of recently glaciated, commonly crystalline-shield areas and in volcanic lakes such as those in Japan and the Kurile Islands. Ores mined in the U.S. were formed in swamps and lakes, and they will be discussed exclusively here.

The mineralized lakes and bogs in glaciated regions of North America are younger than about 11,000 years old, and they contain abundant sand and gravel deposits in shallow-water areas. The deposits of iron ore are generally hosted by these coarse sediments. Ore varies in form from nodules, oolites, and pisolites that are less than 1 mm to more than 2 cm in diameter, to coatings and crusts, to beds of fine-grained lacustrine sediments around the margin of the lake. Goethite and earthy limonite are the major iron-bearing minerals, and manganese-bearing minerals such as birnessite [$(Na,Ca)Mn_7O_{14} \cdot 3H_2O$], todorokite [$(Mn,Ba,Ca,Mg,Zn)Mn_3O_7 \cdot H_2O$], and psilomelane are also present. Nodules may have alternating iron- and manganese-rich bands around a nucleus of quartz, feldspar, glacial clay, limestone, or wood. Concentrations of iron and manganese range from about 20 to 35 percent and 5 to about 25 percent, respectively. The iron:manganese ratio in these fresh-water nodules is generally greater than 1 and ranges to about 10, as opposed to deep-sea nodules with ratios less than 1. However, some bog ore forms manganese deposits, and a gradation from bog iron to bog manganese ores does exist (Stanton, 1972; Callender and Bowser, 1976). The fresh-water ferromanganese nodules contain much lower contents of Co, Ni, and Cu (<1 percent combined) than the oceanic nodules (1 to 3 percent combined Cu + Co + Ni). More detailed comparisons between marine and fresh-water nodules may be found in Callender and Bowser (1976).

Tundra environments, which maintain a cold humid climate, high water table, and closed drainage systems, are the most favorable for the accumulation of bog iron deposits. The metals are derived from glacial debris, which consists predominantly of Precambrian crystalline rocks, and are transported in acidic ground and surface waters (pH < 6). These waters are able to transport substantial iron and manganese (probably as dissolved complexes such as bicarbonate [HCO_3^-] and organic compounds and as suspended particles) because of abundant humic acid in the water from decayed vegetation. These metals are oxidized and precipitated once the surface water empties into the oxygenated more alkaline lake waters, or the ground water percolates through the coarse, near-shore bottom sediments of the lake and is oxidized. The content of minor elements is proportional to the accumulation or accretion rate of the deposit because the ferromanganese oxides and hydroxides scavenge ions from solution. It is not clear at present what the role of bacteria is to the oxidation and precipitation of iron and manganese. However, comparison with marine nodules suggests that biological activity may also be an important mechanism in the deposition and growth of iron and manganese crusts and nodules (Callender and Bowser, 1976; Heath, 1981; Dean, 1983).

MASSIVE, MAGMATIC, AND HYDROTHERMAL DEPOSITS

Deposits of iron that are related to igneous processes constitute about 1 to 2 percent of the iron ore mined in the U.S. and the world. Their classification and genesis remain problematic because of the overlap between processes (magmatic versus hydrothermal) and environments (intrusive versus extrusive). The discussion in this chapter will start with those deposits most clearly related to magmatic processes (mafic intrusion-hosted, titaniferous magnetite deposits) and end with those formed predominantly from hydrothermal fluids (replacement deposits).

Mafic intrusion-hosted, titaniferous magnetite

Deposits of titaniferous magnetite are present in two distinct environments: layered mafic intrusions and plutonic anorthosites. However, magnetite-rich layers in layered mafic intrusions within the U.S.—such as the Archean Stillwater Complex, Montana, and zoned ultramafic complexes in Alaska—do not form economic concentrations (Foose, this volume). Perhaps the best-known, best-documented occurrence of discrete, vanadium-bearing, titaniferous magnetite layers is in the "Upper Zone" of the Bushveld Complex, South Africa. The distribution and geologic relations of the oxide-rich layers indicate that they are intimately related to processes of fractional crystallization that formed the sequence of layered mafic rocks, especially anorthosites, and iron-enriched, late-stage, residual magma in the uppermost part of the complex. Readers are referred to a special issue of *Economic Geology* (1985, v. 80, no. 4) for further discussion of the nature and origin of the magnetite-rich layers in the Bushveld Complex.

Anorthosite massifs are major sources of titanium and minor sources of iron. Significant U.S. deposits of titaniferous magnetite and ilmenite that are hosted by anorthosite include Sanford Lake, New York; the Laramie Range, Wyoming; and the Roseland district, Virginia (Plate 2A). Anorthosite refers to rocks composed of 90 percent or more plagioclase that has a composition of andesine to labradorite (about An 35 to An 65). It is commonly associated with granulite-facies metamorphic rocks in addition to gabbro and charnockite-series rocks. Most massif-type anorthosites are between 1,700 and 1,100 m.y. old. These rocks host concordant layers or sills of Fe-Ti oxide minerals, vein-like massive bodies, veins or dikes, and disseminations. Ilmenite is the dominant oxide in andesine-type anorthosites, whereas titaniferous magnetite is more prevalent in the labradorite-type. The concentration of iron ranges from about 20 to 45 percent and that of titanium between about 10 and 50 percent in both types; vanadium, chromium, and phosphorous can also be present in anomalously high concentrations. The Fe:Ti ratio in ore varies from about 2:1 in the former to about 5:1 or more in the latter. Accessory (or gangue) minerals include apatite, plagioclase, pyroxene, olivine, biotite, hornblende, garnet, pyrrhotite, pyrite, and chalcopyrite. Alteration of wall rocks adjacent to orebodies is not evident (Gross, 1968; Gross and Rose, 1984; Guilbert and Park, 1986).

The origin of massif-type anorthosites and their related plutonic rocks and ore deposits is one of the major unsolved and controversial problems of petrology, according to Philpotts (1981) and Ashwal (1982). Hypotheses for the origin of the Fe-Ti oxide deposits include hydrothermal replacement of anorthosite, metamorphic migration, residual liquid segregation, and liquid immiscibility. The best explanation at present invokes a magmatic origin for both anorthosite and the ores. In this model, plagioclase was the first phase to crystallize from a primary magma of possible basaltic or high-Al gabbroic composition, and it accumulated gravitationally to form anorthosite. Minor pyroxene (augite and pigeonite) and olivine joined plagioclase at slightly lower temperatures. Thus, the residual magma became enriched in iron, titanium, phosphorous, alkalies, and silica. Some of this liquid may have been interstitial to or trapped within a meshwork of plagioclase crystals. The residual magma then split into immiscible fractions, with alkalies and alumina entering the silica-rich liquid, and iron, titanium, phosphorous, and other metals and volatiles forming a second liquid. Some of this iron-rich liquid remained in interstitial sites to crystallize as disseminated oxide and apatite; some may have intruded or sunk downward, because of its high density, to produce veins and a sill or a layer on top of earlier accumulated crystals; and some was injected tectonically into surrounding anorthositic rocks to form dikes or veins of oxide-apatite ore. Hence, the Fe-Ti oxide-apatite ores are considered to be comagmatic with anorthosite, and their formation involves residual liquid segregation and immiscibility (Philpotts, 1981; Ashwal, 1982; Goldberg, 1984; Herz and Force, 1984).

Magnetite intrusions and extrusions

Bodies of massive magnetite and hematite within intermediate to felsic plutonic and volcanic rocks have been mined in the U.S. since the 1840s. Production has been principally from Precambrian rocks in Missouri. Deposits at Iron Mountain, Pilot Knob, and Pea Ridge (Plate 2A) have produced more than 35 million mt from open-pit and underground mines. Pea Ridge is the only underground iron mine that currently operates in the U.S., and it accounts for all of Missouri's production: about 800,000 mt of iron ore at an average grade of about 66 percent Fe in 1989 (Skillings, 1989). These deposits resemble those at Kiruna, Sweden, and Olympic Dam, Australia, and their genesis has been attributed to magmatic intrusion, extrusion, and hydrothermal replacement processes. As noted by Wright (1986) and Panno and Hood (1983), deposits of all three origins may be present within a district or province.

The orebody at Pea Ridge occurs as a vertical, dike-like, tabular mass of magnetite with varying amounts of specular hematite, quartz, apatite, pyrite, chlorite, barite, fluorite and monazite intimately intermixed. Overall, the orebody is about 1,000 m long, as much as 270 m wide, and at least 700 m deep. The ore has relatively sharp contacts at low angles (to 30°) with altered host rocks, which consist of porphyritic rhyolite tuffs. The hanging wall, or south, side of the deposit is composed of several zones of altered porphyry breccia fragments that are cemented by magnetite. Rb-Sr age dates from the rhyolite porphyries (1310 ± 70 Ma) and aplitic dikes (1290 ± 110 Ma) that cut the orebody and porphyries indicate that ore probably formed nearly contemporaneously with the host rocks (Emery, 1968; Snyder, 1969).*

Hematite-magnetite orebodies at Iron Mountain are hosted by Precambrian porphyritic andesitic flow rocks that are at least 350 m thick. The Main Orebody resembles a dome-shaped shell or an inverted cone, and the Northwest Orebody is shaped like a sickle with a vertical part at depth that flattens to horizontal and curls under at the surface. Both orebodies extend to at least 200 m deep, and their maximum dimensions are about 400 m long and 50 m wide or thick. Hematite makes up 80 and 95 percent of the ore in the Main and Northwest Orebodies, respectively, with magnetite the other ore mineral. Gangue minerals associated with the ore are andradite garnet, quartz, calcite, actinolite after salitic pyroxene, apatite, dolomite, fluorite, pyrite, chalcopyrite, and barite. This ore assemblage is present in veins, as disseminations in andesite, and as cement in brecciated andesite. However, most of the wall rocks and breccia blocks are relatively unaltered, but where present, alteration consists of epidote and chlorite replacement of andesite. Contacts between ore and wall rocks are sharp in most areas, with minor areas of disseminated (or replacement) ore between barren porphyritic andesite and ore. Thus, the ore formed predominantly as open-space filling of fractures. Post-ore andesitic dikes that are Precambrian in age and a basal Paleozoic conglomerate with hematite pebbles indicate that the ore is of Precambrian (1350 to 1100 Ma) age (Murphy and Ohle, 1968; Snyder, 1969).

The Pilot Knob area contains a surface deposit of hematite and a subsurface deposit of magnetite. The hematite orebody is within a sequence of Precambrian felsite flow rocks, laminated tuffs, and volcanic agglomerates. The total thickness of these rocks is more than 150 m. A lower laminated tuff unit hosts the hematite orebody, which is about 5 to 9 m thick. Ore is present as a bed of finely laminated specular hematite that preserves some sedimentary features such as ripple marks and mudcracks. It is about 240 m stratigraphically above the top of the subsurface magnetite deposit. The magnetite orebody occurs within a sequence of Precambrian rhyolitic to andesitic ash-flow tuffs. The footwall consists of a rhyolitic ash-flow tuff, the uppermost part of which contains some magnetite as groundmass to lapilli-sized rock fragments and as veins and veinlets. The hanging wall is an andesitic ash-flow tuff about 125 m thick that is not altered or mineralized. The deposit is roughly tabular and concordant, with a crescentic shape in plan view due to subsequent folding. Three types of ore are present: (1) basal matrix ore, or massive, coarse-grained magnetite, with minor amounts of quartz, feldspar, and angular fragments of wall rocks; (2) uppermost disseminated ore, a zone about 37 m thick composed of fine-grained magnetite with quartz, feldspar, chlorite, fluorite, flattened pumice fragments, and relict glass shards; and (3) intermediate mixture ore, a framework breccia of disseminated ore fragments in a stockwork of matrix ore. A zone of relatively unmineralized rock (known as internal rock) about 11 m thick is present near the stratigraphic center of the disseminated ore. Both the hematite and magnetite orebodies are interpreted to be predominantly of replacement origin. The internal rock represents a more densely welded zone between porous unwelded parts of a simple cooling unit of an ash-flow tuff, which was not as intensely mineralized because of its low porosity and permeability. However, the matrix ore does not fit a replacement model. Rather, its dissimilarity in texture from other volcanic rocks in the area, its uniform appearance (angular fragments of disseminated ore as xenoliths), its partly cross-cutting relation to disseminated ore, variable thickness, and distinct trace-element concentrations suggest that it is of magmatic-injection origin (Snyder, 1969; Panno and Hood, 1983).

The deposits described above are good examples of the complexity of this class of iron ore. Different types of ore are located within the same deposit, and the source of the iron is not readily apparent. The source for the Missouri deposits is interpreted to be the rhyolitic and andesitic magmas that produced the volcanic host rocks. Iron concentration possibly took place by differentiation and segregation. This process was perhaps aided by high sodium and phosphorous contents and a low fugacity of oxygen during the early stages of crystallization. Faults, in part produced by magmatic pressures as well as by tectonism, caused the escape of late magmatic iron-enriched fluids that filled fracture and breccia zones. Where open spaces were not present, the fluids flowed along bedding and formational contacts within the sequence of volcanic rocks and replaced porous horizons or were extruded onto the surface. Such extrusive magnetite flows include El Laco, Chile, and Cerro de Mercado, Mexico (Snyder, 1969; Guilbert and Park, 1986; Lyons, 1988). The world-famous iron deposits at Kiruna, Sweden, had for years been considered as

---

*See note added in proof.

the type example of intrusive magmatic segregation iron ore. However, recently it has been suggested that the orebodies in the Kiruna district may have had multiple origins that include magmatic intrusive, volcano-sedimentary exhalative, and subaerial flow processes (Parak, 1985, Wright, 1986; and references contained therein). As summarized by Wright (1986), the quandry involves apparently contradictory evidence for both intrusive and volcanic origins of the ore.

Replacement deposits

Numerous replacement deposits of iron are present throughout the United States. Most are hydrothermal magnetite deposits that are related to nearby intrusions. Such deposits are commonly referred to as skarn, contact metasomatic, or pyrometasomatic deposits. Examples include the large Cornwall and Grace (or Morgantown) mines as well as about 45 smaller deposits in Pennsylvania, the St. Lawrence County magnetite district, New York, and numerous deposits in the Cordilleran region such as Iron Springs, Utah, the Central district, New Mexico, and the Eagle Mountains, California (Plate 2A). Some deposits, such as Iron Springs and some of those in St. Lawrence County, lack significant skarn formation. However, they are included in the following discussion because of their hydrothermal metasomatic character and their similarities to other iron skarn deposits, except as noted.

Skarn refers to rocks that contain Ca-Fe-Mg-Mn silicate minerals that formed by replacement during contact or regional metamorphism and metasomatism. Iron skarns exhibit the most widely diversified geologic settings of all the skarn-type metal deposits. They are found in Mesozoic and Tertiary oceanic island-arc, continental arc, and rifted continental margin terrains. Ore is present as magnetite in disseminations, irregular massive replacement bodies, or veins and breccias within skarn, limestone, or intrusive rocks. Pyrite and chalcopyrite are the most abundant sulfide minerals (generally less than 3 to 5 percent total), and cobaltite, cubanite, pyrrhotite, arsenopyrite, molybdenite, and sphalerite may also be present. Gangue minerals commonly include apatite, sodic scapolite, diopside, phlogopite, garnet, actinolite, and quartz. Large iron skarn deposits such as Cornwall contain 40 to 300 million mt, whereas small deposits range in size from less than 2 to 10 million mt. Average grades are generally >40 percent iron; cobalt (to 0.1 percent), nickel (to 0.02 percent), copper (0.05 to 2 percent), gold (to 1.4 g/mt), and zinc (to 0.4 percent) may be present in anomalous recoverable concentrations. Cornwall was the principal domestic source of cobalt until it ceased production in 1971. Host rocks for iron skarns include carbonate, calcareous clastic, and intermediate to mafic volcanic rocks. Some iron ore in St. Lawrence County is hosted by granite gneiss without any skarn alteration. Replacement deposits may form in any of these reactive rocks immediately adjacent to intrusive contacts, as conformable lenses at a distance from any pluton, or within the intrusive. Compositions of plutonic rocks associated with iron-replacement deposits range from diorite, gabbro, (Buena Vista Hills, Nevada) and diabase (as at Cornwall) to quartz monzonite (Iron Springs), granite (St. Lawrence County), and syenite, with their volcanic equivalents commonly present as tuffs and flows. The intrusions are generally medium-grained equicrystalline to slightly porphyritic in texture, have a wide range of silica contents, and have total alkali concentrations similar to base-metal skarn deposits, but they are generally more mafic and have higher Na_2O/K_2O ratios. Also, the iron content of the intrusions commonly is inversely proportional to that of associated skarn minerals. Endoskarn, or skarn formed in the intrusive rocks, may be extensive, and it is characterized by epidote-pyroxene ± garnet and sodium metasomatism represented by albite and marialitic scapolite. Zones of calc-silicate minerals in both endoskarn and exoskarn are poorly developed. In general, epidote, diopsidic-salitic pyroxene, sphene, and apatite are more typical in altered igneous rocks, whereas grandite garnet associated with magnetite is most common in replaced limestone, and forsterite is present in skarn developed after dolomite. Retrograde alteration minerals include actinolite, chlorite, calcite, quartz, ilvaite, and less commonly phlogopite, tourmaline, serpentine, potassium feldspar, ludwigite, sericite, and kaolinite (Einaudi and others, 1981; Meinert, 1984; Guilbert and Park, 1986; Foose, this volume).

As with other iron deposits already discussed, the source of iron in these replacement deposits is a controversial subject. Most investigators have invoked a magmatic hydrothermal model, with the nearby intrusion the source of the metals and fluids. Kwak and others (1986) and Meinert (1984) suggested that the inverse relation between iron concentrations in skarn and intrusive rocks; the association of cobalt, nickel, gold, and copper; high temperatures (between about 700° and 350°C) of skarn formation and mineralization; and high salinities in the fluids (>30 wt % equivalent NaCl) are evidence of an orthomagmatic source. Whitney and others (1985) confirmed experimentally that high concentrations of iron may evolve in chloride solutions that equilibrate with natural magmatic (quartz monzonitic) systems at submagmatic temperatures (700° to 400°C). However, Mackin (1968) believed that the iron at Iron Springs was deuterically leached from mafic phenocrysts in the quartz monzonite porphyry (see Guilbert and Park, 1986, p. 460–465), and others have suggested that volcanic wall rocks adjacent to deposits in other areas might be the source of metals. Rose and others (1985) analyzed oxygen, sulfur, and carbon isotopes from magnetite, silicate, and carbonate minerals in skarn, ore, Paleozoic limestone host rocks, and the Triassic diabase (gabbro) sheets associated with the ores at Cornwall and other magnetite deposits in Pennsylvania. High values of $\delta^{18}O$ calculated for the fluids (13 to 16 per mil) suggested that the ore-bearing solutions were formed by heating and circulation of meteoric, connate, or magmatic waters near the contact of the diabase. They concluded that a large nonmagmatic component was most dominant; however, a magmatic component either via a magmatic hydrothermal phase or leached diabase must have been involved in order to explain the copper and cobalt content in the ore at Cornwall.

SECONDARY ENRICHMENTS

Enrichment of iron oxide protore by supergene weathering processes has enhanced the grade and tonnage of many deposits. For example, residual brown ores developed from ironstone deposits were mined extensively in the southeastern United States. Other types of secondary enrichments include laterites and gossans. Laterites are iron-rich blanket deposits that result from deep weathering of rocks in tropical and subtropical climates. They represent the residual accumulation of oxidized and relatively insoluble constituents of the host rocks, which are commonly mafic to ultramafic in original composition. Nodules of hematite and goethite are the predominant iron oxides that formed. Commonly, other metals such as nickel and chromium are also enriched. The only recently active nickel mine in the U.S. is hosted by laterites near Riddle, Oregon (Gross, 1970; Foose, this volume). Other lateritic iron deposits are present in Washington, Oregon, and Hawaii (Plate 2A).

Gossans (or leached capping) are residual deposits of iron oxides and hydroxides that form during decomposition and oxidation of sulfide deposits by chemical weathering processes. Gossans are readily identified by their bright red, yellow, and brownish colors. Porphyry copper deposits and massive sulfide deposits are renowned for their leached capping. Common iron minerals in gossan include goethite, hematite, jarosite ($KFe_3(SO_4)_2(OH)_6$), and limonite. Some gossans are mined for their use as paint pigment, and they may be used as indicators of the nature of sulfide mineralization and the grade of copper ore that lie beneath the surface (Gross, 1970; Anderson, 1982).

MANGANESE DEPOSITS

The classification of manganese deposits is made difficult, as with that of iron, by the overlap of processes that form them and the gradations between environments that host them. Moreover, many deposits are modified by supergene processes, and unlike deposits of iron, different types of manganese deposits do not appear to be restricted to specific periods of time, even though the Cenozoic Era hosts the largest number. Also, similar to iron, one type of deposit dominates the worldwide production and reserves of manganese. Marine chemical sediment, or sedimentary-nonvolcanogenic manganese, deposits account for more than 85 percent of the identified, economic manganese resources of the world and about 60 percent of cumulative production. Although deposits of this type have not been mined in the U.S., they are present on all the continents. Among the largest sedimentary manganese deposits are Groote Eylandt in Australia; Molango in Mexico; Nikopol, Tchiatura, and Bolshe Tokmak in the Soviet Union; the Kalahari field in South Africa; the Urkut district in Hungary; and several districts in India, China, and Brazil (Fig. 2). Secondary enrichment–type deposits constitute the bulk of the remainder of manganese production and resource totals. The larger high-grade enriched deposits typically occur in tropical regions as residual concentrations. Examples include Moanda in Gabon; Nsuta in Ghana, and others in Africa; Serro do Navio and Carajas in Brazil; and Orissa State and others in India (Fig. 2). Altogether, about 29 deposits of these two types contain the total of 3,600 million mt of manganese metal in the worldwide reserve base, and South Africa and the Soviet Union account for more than 80 percent of this total. Again, the U.S. currently has neither domestic production nor significant reserves of manganese ore. Thus, the classification of manganese deposits used here will be kept as simple as possible. Besides the two classes already identified, other types to be discussed include hydrothermal, sedimentary-volcanogenic, those associated with banded iron formation, and deep-sea nodules. Past production of manganese in the U.S. has been predominantly from hydrothermal deposits and from those associated with BIF (National Materials Advisory Board, 1981; Roy, 1981; DeYoung and others, 1984; Jones, 1985b, 1989b).

Hydrothermal deposits

Hydrothermal deposits of manganese, as discussed here, include only those that would be considered subaerial (continental) and hypogene in origin. This class then encompasses hypothermal to epithermal (including hot spring) vein, replacement, and strata-meteoric, bound deposits. The fluids may be magmatic, meteoric or connate in origin, or a combination of these, and the source of the manganese is predominantly leached country rocks, but may involve a magmatic component as well. Butte and Philipsburg, Montana (Plate 2A), have produced the most manganese, although many deposits of this type in the southwestern U.S. were mined primarily for other metals such as silver, lead, and zinc. Butte is a classic example of a zoned porphyry copper deposit. Manganese is present in veins with silver and zinc in the outermost peripheral zone, which rings the intermediate zone (Cu + Zn) and the central zone (Cu + Mo) (Tooker, this volume). A total of more than 1.7 million mt of manganese was produced from the east-west–striking Emma-Travona vein and other veins of the Anaconda system. The veins contained nearly pure, coarsely crystalline rhodochrosite with lesser rhodonite in widths as great as 30 m or more. The Anaconda system is the oldest series of Main Stage veins that cut the Upper Cretaceous Butte Quartz Monzonite, which hosts porphyry copper-molybdenum ore. Alteration envelopes around veins in the peripheral zone consist of an adjacent sericitic zone, which passes outward to the kaolinitic subzone of the argillic zone. The montmorillonitic subzone and a fringe propylitic subzone form the outermost parts of the argillic zone and border fresh quartz monzonite. The alteration and mineralization associated with the Main Stage veins is interpreted to be the result of meteoric water circulation through earlier-formed, pre–Main Stage, porphyry-hosted protore with remobilization and redeposition of some metals (Meyer and others, 1968; Brimhall, 1980; Tooker, this volume).

The Philipsburg district produced more than 645,000 mt of manganese ore from replacement deposits in lower Paleozoic carbonate rocks. The orebodies are tabular in thin-bedded host rocks and form irregular vertical pipes in more massive, thick-bedded units. Rhodochrosite and manganoan dolomite are the main hypogene minerals. However, most of the manganese that was mined in the district was from oxide deposits that formed by supergene oxidation of the manganoan carbonate deposits. Most of the carbonate-hosted ore is along quartz veins that produced zinc, lead, silver, gold, and copper, and rhodochrosite is present as gangue in the quartz veins. Thus, the manganese deposits are believed to be genetically associated with the base- and precious-metal sulfide deposits, and the nearby Philipsburg batholith (Laramide in age) was the probable source of hydrothermal fluids and possibly metals (Prinz, 1967).

Numerous small hydrothermal deposits of manganese are present in the western United States. Most are vein occurrences of psilomelane, with lesser rhodonite and rhodochrosite, in Tertiary volcanic rocks, and they are commonly spatially and genetically associated with deposits of barite, fluorite, and epithermal base-metal sulfide-silver-gold (Hewett, 1964). For example, Eggleston and others (1983) and Norman and others (1983) reported that ores in the Luis Lopez district, New Mexico (Plate 2A), which produced more than 100,000 mt of manganese ore, were deposited in veins and fault breccia zones from heated meteoric waters. Precipitation took place by boiling at temperatures between 375° and 150°C and depths of 1,400 to 400 m below the surface. The source of the manganese was rhyolitic ash-flow tuff in the Oligocene Hells Mesa Formation of the Datil Group, and it was depleted by an order of magnitude due to hydrothermal leaching (Eggleston and others, 1983; Norman and others, 1983). The San Juan Mountains, Colorado (Plate 2A), contain deposits of rhodonite and rhodochrosite that are associated with gold-silver-lead-zinc-copper veins. These deposits host potential resources of about 44 million mt that grade about 8 percent Mn (National Materials Advisory Board, 1981). One of the largest known manganese orebodies in the U.S. is in the Artillery district, Arizona (Plate 2A). Manganese oxide minerals are strata-bound within, and they cement Miocene siltstone, sandstone, conglomerate, and felsic tuff and form veins along faults. A total of about 221,000 mt of ore that graded 20 percent Mn was mined prior to the cessation of production in 1960, and an estimated resource of about 160 million mt that averages 3.9 percent Mn remains. Hein and Koski (1983), Yeh and others (1985), and Spencer and Welty (1986) suggested that volcanic or heated meteoric or saline connate (seawater) hydrothermal fluids leached manganese from upper crustal country rocks and deposited the oxide minerals at temperatures of about 120°C or less.

Manganese in banded iron formation

Some banded iron formations host deposits of manganese. The Cuyuna Range, Minnesota, for example, contains beds of BIF that are enriched in manganese. This ferruginous manganese ore (average grade about 15 percent Mn) has been mined as a co-product with iron ore, and an estimated 247 million mt of ore with 8 percent Mn and 32 percent Fe remain as a future resource.

These manganiferous rocks are typically within the carbonate or, less commonly, oxide facies of BIF, and they are more common in Phanerozoic Algoma-type BIF sequences than in Proterozoic Lake Superior–type rocks. The manganese protore is generally of low grade (<20 percent Mn), although metamorphism and supergene enrichment may mask the nature of its primary occurrence. Manganese minerals such as braunite ($3Mn_2O_3 \cdot MnSiO_3$) and ferroan rhodochrosite are present both as individual laminae, pellets, and pods within BIF and as disseminated grains throughout the iron-rich beds. Even though most of the braunite and rhodochrosite may be secondary in origin, manganese was originally incorporated in primary manganoan siderite and manganese oxide minerals before release and reprecipitation during diagenesis and metamorphism. Thus, the manganese:iron ratio ranges from less than 0.5 to more than 5 in manganiferous facies of BIF. Manganiferous banded iron formation in Aroostock County, Maine (Plate 2A), is another example of this type of deposit. Total resources of about 341 million mt of ore that contain an average of about 9 percent Mn and 20 percent Fe have been calculated for three districts in a discontinuous belt of Paleozoic metasedimentary and metavolcanic rocks about 106 km long. However, these deposits have not been mined because of the extremely fine grain size of the manganese minerals, their intimate admixture with hematite and other gangue minerals, the lack of significant secondary enrichment, the thinness (an average of about 17 m) and lens-like shape of mineralized beds, and the overall cost involved in mining, milling, and extraction of these low-grade manganiferous ores (Pavlides, 1962; National Materials Advisory Board, 1981; Gross, 1983b). In contrast, supergene-enriched and protore deposits of BIF in the Carajas district, Brazil (Fig. 2), contain about 65 million mt of ore that grades about 44 percent Mn and 3,220 million mt of iron ore with 66 percent Fe. Most of these ores do not require any treatment beyond simple crushing and washing (Dayton and Sassos, 1985).

The genesis of manganese-rich BIF is presumably the same as that for iron deposits in BIF discussed earlier. However, the presence of both separate manganese-rich laminae and disseminated manganese minerals in iron-rich rocks within some deposits is possibly caused by variations in the rates of metal and fluid supply to the environment of chemical sedimentation, the distance from the centers of volcanic exhalative activity, and perhaps local or seasonal variations in Eh and pH, which would affect the solubilities of the two metals (Gross, 1983b; Guilbert and Park, 1986).

Sedimentary-volcanogenic deposits

Sedimentary-volcanogenic manganese refers to those deposits that are generally hosted by submarine volcano-sedimentary sequences of rocks and whose source of manganese is volcanic exhalative activity or volcanic rocks that have been depleted by hydrothermal fluids. Host rocks include Phanerozoic chert, jasper, argillite, shale, graywacke, conglomerate, limestone, and intermediate to mafic volcanic tuffs and flows. Some have been metamorphosed to the greenschist facies and are typical greenstones, whereas others have been spilitized or have undergone higher grades of metamorphism. Manganese minerals such as hausmannite (Mn_3O_4), rhodochrosite, rhodonite, braunite, pyrolusite, and others are present as thin (<2 m), small (about 15 m in diameter) lenses and discontinuous beds that are most commonly within sections of bedded chert and argillite. Deposits of sedimentary-volcanogenic manganese are generally low tonnage (<10,000 mt) and low to high grade (20 to 55 percent Mn, with an average grade of about 30 percent Mn). Hundreds of occurrences in the Mesozoic and Cenozoic Franciscan Complex of California (Plate 2A) have been identified, but only about 123,000 mt of manganese ore have been produced. Other occurrences in the U.S. include those in the Eocene Crescent Formation on the Olympic Peninsula, Washington; in the upper Paleozoic Calaveras Formation of the Sierra Nevada, California; and in the upper Paleozoic Pumpernickel and Havallah Formations, Nevada (Plate 2A). More notable examples around the world include the Troodos ophiolite in Cyprus and the Nicoya Complex in Costa Rica. All these deposits appear to be derived from hydrothermal fluids related to volcanic centers. Manganese oxide minerals associated with the ophiolitic part of the Franciscan Complex may have formed in a mid-ocean ridge environment similar to that of modern hydrothermal systems at sea-floor spreading centers, whereas deposits on the Olympic Peninsula and in the Calaveras Formation represent a seamount and an island-arc environment, respectively (Hein and Koski, 1983; Rona, 1984; Fehn, 1986; Mosier and Page, 1988). Boctor (1985) showed experimentally that the f_{H_2} in chloride-bearing hydrothermal fluids may be the significant factor in the separation of manganese from iron in such deposits. Moreover, rocks in these deposits have undergone variable amounts of diagenesis and metamorphism that mask the primary nature of deposition and mineralization and complicate their genetic interpretation. For example, Hein and others (1987) suggested that the Franciscan deposits formed by diagenetic and replacement processes at depths of 500 ± 300 m and temperatures of 70° ± 20°C within the chert-argillite section. These host sediments are interpreted to have been deposited in a deep-water basin at or near a continental margin (Hein and others, 1987).

Sedimentary manganese deposits

Sedimentary-nonvolcanogenic, or marine chemical sediment, deposits of manganese account for more than 85 percent of the world's manganese reserves. These high-grade (to 55 percent Mn), large-tonnage (greater than 200 million mt) deposits occur in marine-transgressive sedimentary rock sequences that were deposited in shallow water on stable platforms. Marine rocks such as carbonate, clay, marl, and fine-grained glauconite-bearing clastic sediments host manganese ore within a few hundred meters above widespread unconformities or transitions from continental to marine rocks. Interestingly, many of the deposits are associated with anoxic conditions in the oceans that formed during and just after high sea-level stands. These anoxic events are best documented for the early Paleozoic, Late Jurassic, and mid-Cretaceous. Thus, the major deposits of this type, such as Groote Eylandt, Australia; Nikopol, U.S.S.R.; and Molango, Mexico (Fig. 2) are predominantly Phanerozoic in age (Cannon and Force, 1983; DeYoung and others, 1984; Frakes and Bolton, 1984; Force and Cannon, 1988).

Sedimentary manganese deposits contain a wide variety of ore minerals and textures and are interbedded with a wide range of rock types. Manganese oxide (cryptomelane [$K(Mn^{2+}, Mn^{4+})_8O_{16}$], pyrolusite, manganite [$MnO(OH)$], and psilomelane) and carbonate (rhodochrosite, kutnahorite [$CaMn(CO_3)_2$], and manganocalcite) minerals are the dominant ore minerals. The carbonate minerals generally formed in more reduced offshore areas, whereas manganese oxides precipitated in a more aerated nearshore environment. Oolites and pisolites in a matrix of sandy glauconitic clay or a cement of manganese oxide or carbonate are the most common textural varieties of ore. Beds of ore may be 10 m or more thick and several kilometers or more long, and graded bedding (both inverse and normal) within the ore zone may also be developed. Commonly, the ore horizon is underlain and overlain by sub-ore manganiferous rocks. Diagenetic changes, such as dolomitization, solution-collapse brecciation, and secondary-ore formation, and supergene enrichment may obscure the primary depositional features (Cannon and Force, 1983; DeYoung and others, 1984; Bolton and Frakes, 1985; Force and others, 1986).

The genesis of these sedimentary manganese deposits is directly related to anoxic conditions that developed during transgression. A warm climate and lower thermal gradients from the poles to the equator that are associated with marine transgressions diminish deep oceanic circulation and the rate of oxygen replenishment to deep water. Thus, widespread stratified oceans that are anoxic below the zone of surface turbulence (to about 200 m deep) form. The solubility of manganese in the anoxic seawater is about 500 times greater than that of oxygenated seawater. Manganese is precipitated on the more-oxidized and more alkaline side of the Eh-pH interface during transgression across a shallow shelf. The greatest concentration of ore takes place at the feather edge of peak transgression. Such a depositional setting may exist as an embayed shoreline of the open sea (as at Groote Eylandt), an arm of the sea (Imini, Morocco), and bays in restricted basins (Nikopol). The Black and Baltic Seas may be modern analogs to this model. It might be noted that the source of manganese may include weathered continents, leached submarine volcanic rocks, and/or submarine volcanic exhalative activity. In effect, the anoxic waters of the ocean, as in Lake Superior–type BIF, are the immediate source of manganese (Cannon and Force, 1983; Frakes and Bolton, 1984; Bolton and Frakes, 1985; Force and others, 1986; Force and Cannon, 1988).

Although deposits of this type have not been identified in the U.S., Cannon and Force (1983) identified two regions that might contain such sedimentary deposits. The Chamberlain deposit,

South Dakota (Plate 2A), is within the Oacoma zone in the DeGrey Member of the Upper Cretaceous Pierre Shale. Carbonate concretions with an average grade of about 15 percent Mn are irregularly distributed throughout a zone about 15 m thick. The host shale averages about 0.7 percent Mn. Although this stratigraphic member is exposed in outcrop for hundreds of kilometers, its low grade and the uneven distribution of the nodules prevent it from being classed as a resource at this time. However, Cannon and Force (1983) and Force and Cannon (1988) suggested that the Chamberlain deposit may be the low-grade distal part of higher-grade deposits in the Pierre Shale to the east. They also noted that transgressive sediments of Cretaceous to Tertiary age in the Atlantic and Gulf Coastal Plains are broadly favorable for this type of deposit.

Deep-sea nodules and crusts on the ocean floors represent potential resources for manganese, cobalt, copper, and nickel. They are discussed by Foose (this volume).

Secondary enrichment deposits

Large deposits that result from supergene weathering processes of manganese protore typically occur in tropical regions. In the U.S., minor secondary enrichment has formed numerous deposits of low-grade ore and added small tonnages to other deposits, as noted above. Many enriched deposits occur in residual clay derived from weathering of limestone, dolomite, and shale. For example, residual clay deposits at Batesville, Arkansas (Plate 2A), formed from protore of braunite and hausmannite that is disseminated in calcite gangue and forms replacement pods or lenses in the Upper Ordovician Fernvale Limestone. These residual deposits produced about 298,000 mt of manganese ore before production ceased in 1959. Estimated resources of about 178 million mt that grade 4.0 percent Mn remain as protore. More than 100 mines and prospects of manganese are present in the Appalachian belt from Alabama to Virginia (Plate 2A). Manganese oxide minerals (pyrolusite, psilomelane, and wad) occur as supergene fracture fillings, replacements, and botryoidal incrustations in Paleozoic quartzite, chert, dolomite, and shale, as well as pellets and nodules in residual clay deposits. The orebodies are irregular pockets, lenticular masses, and stringers that produced more than 100,000 mt of ore that graded 7 to more than 45 percent Mn. Some of these deposits contain anomalously high concentrations of cobalt that range from 0.2 to 1.5 percent, and goethite or brown iron ore commonly is closely associated with the manganese ore (Knechtel, 1944; Pierce, 1944; Crittenden and Pavlides, 1962; National Materials Advisory Board, 1981).

SUMMARY

Iron and manganese are two metals that are strategic and critical to the United States. Both are major ingredients in steel manufacture, and satisfactory economic substitutes do not exist. The U.S. has sufficient iron ore in reserve; however, the nation is totally reliant on imports for manganese ore. Development of domestic sources is made difficult by problems encountered in exploration for this metal. Manganese deposits do not give a uniquely identifiable geophysical or geochemical signature. Moreover, many manganiferous rocks are not readily distinguished in the field. Thus, a better understanding of the genesis of manganese deposits must be attained in order to apply mineral-deposit models in exploration.

Both iron and manganese are present in a wide variety of geologic environments, including some that are devoid of ore. The largest deposits of both are in chemically precipitated sedimentary rocks. Although the metals commonly occur together, ore deposits of both typically form separately from each other. The control of their distribution in environments that range from magmatic to marine and terrestrial sedimentary systems is related to conditions of temperature, Eh, pH, and f_{O_2}, as well as the concentrations of trace elements and complexes such as phosphorous, chlorine, and bicarbonate that are important for the transport of iron and manganese. However, the genesis of some iron and manganese deposits remains enigmatic.

REFERENCES CITED

Anderson, J. A., 1982, Characteristics of leached capping and techniques of appraisal, *in* Titley, S. R., ed., Advances in geology of the porphyry copper deposits, southwestern North America: Tucson, The University of Arizona Press, p. 275–295.

Ashwal, L. D., 1982, Mineralogy of mafic and Fe-Ti oxide-rich differentiates of the Marcy anorthosite massif, Adirondacks, New York: American Mineralogist, v. 67, p. 14–27.

Baur, M. E., Hayes, J. M., Studley, S. A., and Walter, M. R., 1985, Millimeter-scale variations of stable isotope abundances in carbonates from banded iron-formations in the Hamersley Group of western Australia: Economic Geology, v. 80, p. 270–282.

Bayley, R. W., and James, H. L., 1973, Precambrian iron-formations of the United States: Economic Geology, v. 68, p. 934–959.

Birnbaum, S. J., and Wireman, J. W., 1985, Sulfate-reducing bacteria and silica solubility; A possible mechanism for evaporite diagenesis and silica precipitation in banded iron formations: Canadian Journal of Earth Sciences, v. 22, p. 1904–1909.

Boctor, N. Z., 1985, Rhodonite solubility and thermodynamic properties of aqueous $MnCl_2$ in the system $MnO-SiO_2-HCl-H_2O$: Geochimica et Cosmochimica Acta, v. 49, p. 565–575.

Bolton, B. R., and Frakes, L. A., 1985, Geology and genesis of manganese oolite, Chiatura, Georgia, U.S.S.R.: Geological Society of America Bulletin, v. 96, p. 1398–1406.

Boucot, A. J., and Gray, J., 1986, Comment on "Oolitic ironstones and contrasting Ordovician and Jurassic paleogeography": Geology, v. 14, p. 634–635.

Brimhall, G. H., Jr., 1980, Deep hypogene oxidation of porphyry copper potassium-silicate protore at Butte, Montana; A theoretical evaluation of the copper remobilization hypothesis: Economic Geology, v. 75, p. 384–409.

Button, A., Brock, T. D., Cook, P. J., Eugster, H. P., Goodwin, A. M., James, H. L., Margulis, L., Nealson, K. H., Nriagu, J. O., Trendall, A. F., and Walters, M. R., 1982, Sedimentary iron deposits, evaporites, and phosphorites; State-of-the-art report, *in* Holland, H. D., and Schidlowski, M., eds., Mineral deposits and the evolution of the biosphere: New York Springer-Verlag, p. 259–273.

Callender, E., and Bowser, C. J., 1976, Freshwater ferromanganese nodules, *in* Wolf, K. H., ed., Handbook of strata-bound and stratiform ore deposits; 2, Regional studies and specific deposits: New York, Elsevier Scientific Publishing Company, v. 7, Au, U, Fe, Mn, Hg, Sb, W, and P deposits, p. 341–394.

Cannon, W. F., 1976, Hard iron ore of the Marquette Range, Michigan: Economic Geology, v. 71, p. 1012–1028.

Cannon, W. F., and Force, E. R., 1983, Potential for high-grade shallow-marine manganese deposits in North America, *in* Shanks, W. C., III, ed., Cameron volume on unconventional mineral deposits: Society of Mining Engineers of the America Institute of Mining, Metallurgical, and Petroleum Engineers, Inc., p. 175–189.

Carr, M. S., Guild, P. W., and Wright, W. B., 1967, Iron in the United States, exclusive of Alaska and Hawaii: U.S. Geological Survey Map MR-51, 20 p., scale 1:3,168,000.

Cox, D. P., and Singer, D. A., eds., 1986, Mineral deposit models: U.S. Geological Survey Bulletin 1693, 379 p.

Crittenden, M. D., and Pavlides, L., 1962, Manganese in the United States, exclusive of Alaska and Hawaii: U.S. Geological Survey Map MR-23, 8 p., scale 1:3,168,000.

Davy, R., 1983, Part A. A contribution on the chemical composition of Precambrian iron-formations, *in* Trendall, A. F., and Morris, R. C., eds., Iron-formation; Facts and problems: Amsterdam, Elsevier, p. 325–343.

Dayton, S. H., and Sassos, M. P., 1985, Carajas; New district, new mines, and a big future: Engineering and Mining Journal, v. 186, no. 11, p. 24–82.

Dean, W. E., 1983, Geochemistry of deep-sea manganese nodules; Organic involvement, *in* Shanks, W. C., III, ed., Cameron volume on unconventional mineral deposits: Society of Mining Engineers of the American Institute of Mining, Metallurgical, and Petroleum Engineers, Inc., p. 123–132.

DeYoung, J. H., Jr., Sutphin, D. M., and Cannon, W. F., 1984, International strategic minerals inventory summary report; Manganese: U.S. Geological Survey Circular 930–A, 22 p.

Dimroth, E., 1979, Facies models 16; Diagenetic facies of iron formation, *in* Walker, R. G., ed., Facies models: Geoscience Canada Reprint Series 1, p. 183–189.

Dorr, J.V.N., II, Crittenden, M. D., Jr., and Worl, R. G., 1973, Manganese, *in* Brobst, D. A., and Pratt, W. P., eds., United States mineral resources: U.S. Geological Survey Professional Paper 820, p. 385–399.

Drozdovskaya, A. A., 1986, Physiochemical conditions of global iron accumulation in the early Proterozoic: Doklady-Earth Sciences Section, v. 278, nos. 1-6, p. 191–194.

Eckstrand, O. R., ed., 1984, Canadian mineral deposit types; A geological synopsis: Geological Survey of Canada Economic Geology Report 36, 86 p.

Eggleston, T. L., Norman, D. I., Chapin, C. E., and Savin, S., 1983, Geology, alteration and genesis of the Luis Lopez manganese district, New Mexico, *in* Chapin, C. E., ed., Socorro region II: New Mexico Geological Society 34th Annual Field Conference Guidebook, p. 241–246.

Eichler, J., 1976, Origin of the Precambrian banded iron-formations, *in* Wolf, K. H., ed., Handbook of strata-bound and stratiform ore deposits; 2, Regional studies and specific deposits: New York, Elsevier, Scientific Publishing Co., v. 7, p. 157–201.

Einaudi, M. T., Meinert, L. D., and Newberry, R. T., 1981, Skarn deposits: Economic Geology, 75th Anniversary Volume, p. 317–391.

Emery, J. A., 1968, Geology of the Pea Ridge iron ore body, *in* Ridge, J. D., ed., Ore deposits of the United States, 1933–1967: American Institute of Mining, Metallurgical, and Petroleum Engineers, Inc., v. 1, p. 359–369.

Fehn, U., 1986, Evolution of low-temperature convection cells near spreading centers; A mechanism for the formation of the Galapagos mounds and similar manganese deposits: Economic Geology, v. 81, p. 1396–1407.

Foos, A. M., 1984, The mineralogy, petrography, and geochemistry of the Eocene Lone Star iron ores, east Texas, and the Ordovician Hooker Ironstone, northwest Georgia [Ph.D. thesis]: University of Texas at Dallas, 286 p.

Force, E. R., Back, W., Spiker, E. C., and Knauth, L. P., 1986, A ground-water mixing model for the origin of the Imini manganese deposit (Cretaceous) of Morocco: Economic Geology, v. 81, p. 65–79.

Force, E. R., and Cannon, W. F., 1988, Depositional model for shallow-marine manganese deposits around black shale basins: Economic Geology, v. 83, p. 93–117.

Frakes, L. A., and Bolton, B. R., 1984, Origin of manganese giants; Sea-level change and anoxic-oxic history: Geology, v. 12, p. 83–86.

Francois, L. M., 1986, Extensive deposition of banded iron formations was possible without photosynthesis: Nature, v. 320, p. 352–354.

Fryer, B. J., 1983, Part B, Rare earth elements in iron-formation, *in* Trendall, A. F., and Morris, R. C., eds., Iron-formation; Facts and problems: Amsterdam, Elsevier, p. 345–358.

Gair, J. E., 1973, Iron deposits of Michigan (United States of America) *in* Genesis of Precambrian iron and manganese deposits; Proceedings of the Kiev Symposium, August, 1970: Paris, UNESCO, Earth Sciences 9, p. 365–375.

Goldberg, S. A., 1984, Geochemical relationships between anorthosite and associated iron-rich rocks, Laramie Range, Wyoming: Contributions to Mineralogy and Petrology, v. 87, p. 376–387.

Goodwin, A. M., 1973, Archean iron-formations and tectonic basins of the Canadian Shield, *in* James, H. L., and Sims, P. K., eds., Precambrian iron-formations of the world: Economic Geology, v. 68, p. 915–933.

Gross, G. A., 1965, Geology of iron deposits in Canada; V. 1, General geology and evaluation of iron deposits: Geological Survey of Canada Economic Geology Report 22, 181 p.

—— , 1970, Nature and occurrence of iron ore deposits, *in* United Nations, Department of Economic and Social Affairs, 1970, Survey of world iron ore resources: New York, United Nations, p. 13–31.

—— , 1973, The depositional environment of principal types of Precambrian iron-formations; Proceedings of the Kiev Symposium, August 1970: Paris, UNESCO, Earth Sciences 9, p. 15–21.

—— , 1983a, Tectonic systems and the deposition of iron-formation: Precambrian Research, v. 20, p. 171–187.

—— , 1983b, Low grade manganese deposits; A facies approach, *in* Shanks, W. C., III, ed., Cameron volume on unconventional mineral deposits: Society of Mining Engineers of the American Institute of Mining, Metallurgical, and Petroleum Engineers, Inc., p. 35–46.

Gross, G. A., and McLeod, C. R., 1980, A preliminary assessment of the chemical composition of iron formations in Canada: Canadian Mineralogist, v. 18, p. 223–229.

Gross, G. A., and Rose, E. R., 1984, Mafic intrusion-hosted titanium-iron, *in* Eckstrand, O. R., ed., Canadian mineral deposit types; A geological synopsis: Geological Survey of Canada Economic Geology Report 36, p. 46.

Gross, S. O., 1968, Titaniferous ores of the Sanford Lake district, New York, *in* Ridge, J. D., ed., Ore deposits of the United States, 1933–1967: American Institute of Mining, Metallurgical, and Petroleum Engineers, Inc., v. 1, p. 140–153.

Guilbert, J. M., and Park, C. F., Jr., 1986, The geology of ore deposits: New York, W. H. Freeman and Company, 985 p.

Heath, G. R., 1981, Ferromanganese nodules of the deep sea: Economic Geology 75th Anniversary Volume, p. 736–765.

Hein, J. R. and Koski, R. A., 1983, Volcanic manganese deposits in the western Cordillera; Lithologic associations and paleoceanographic settings and economic deposits associated with biogenic siliceous rocks, *in* Cronin, T. M. Cannon, W. F., and Poore, R. Z., eds., Paleoclimate and mineral deposits: U.S. Geological Survey Circular 822, p. 32–33.

Hein, J. R., Koski, R. A., and Yeh, H.-W., 1987, Chert-hosted manganese deposits in sedimentary sequences of the Franciscan Complex, Diablo Range, California, *in* Hein, J. R., ed., Siliceous sedimentary rock-hosted ores and petroleum: New York, Van Nostrand Reinhold Company, p. 206–230.

Herz, N., and Force, E. R., 1984, Rock suites in Grenvillian terrane of the Roseland district, Virginia, *in* Bartholomew, M. J., ed., The Grenville event in the Appalachians and related topics: Geological Society of America Special Paper 194, p. 187–214.

Hewett, D. H., 1964, Veins of hypogene manganese oxide minerals in the southwestern United States: Economic Geology, v. 59, p. 1429–1472.

Holland, H. D., 1984, The chemical evolution of the atmosphere and oceans: Princeton, New Jersey, Princeton University Press, 582 p.

James, H. L., 1954, Sedimentary facies of iron-formation: Economic Geology, v. 49, p. 235–293.

—— , 1966, Chemistry of the iron-rich sedimentary rocks: U.S. Geological Survey Professional Paper 440-W, 61 p.

—— , 1983, Distribution of banded iron-formation in space and time, *in* Trendall, A. F., and Morris, R. C., eds., Iron-Formation; Facts and problems: Amsterdam, Elsevier, p. 471–490.

James, H. L., and Sims, P. K., 1973, Introduction, *in* James, H. L., and Sims, P. K., eds., Precambrian iron-formations of the world: Economic Geology, v. 68, p. 913–914.

James, H. L., and Trendall, A. F., 1982, Banded iron formation; Distribution in time and paleoenvironmental significance, *in* Holland, H. D., and Schidlowski, M., eds., Mineral deposits and the evolution of the biosphere: New York, Springer-Verlag, p. 199–217.

James, H. L., Dutton, C. E., Pettijohn, F. J., and Wier, K. L., 1968, Geology and ore deposits of the Iron River–Crystal Falls district, Iron County, Michigan: U.S. Geological Survey Professional Paper 570, 134 p.

Jones, T. S., 1985a, Manganese, *in* Mineral facts and problems, 1985 edition: U.S. Bureau of Mines Bulletin 675, p. 483–498.

—— , 1985b, Manganese, *in* U.S. Bureau of Mines Minerals Yearbook 1984, volume I, Metals and minerals: U.S. Bureau of Mines, p. 627–640.

—— , 1989a, Manganese, in U.S. Bureau of Mines Minerals Yearbook 1987, volume I, Metals and minerals: U.S. Bureau of Mines, p. 601–615.

—— , 1989b, Manganese; Mineral Commodity Summaries 1989: U.S. Bureau of Mines, p. 100–101.

Kimberley, M. M., 1981, Oolitic iron formations, *in* Wolf, K. H., ed., Regional studies and specific deposits: New York, Elsevier, Handbook of strata-bound and stratiform ore deposits, part 3, v. 9, p. 25–76.

—— , 1983, Constraints on genetic modeling of Proterozoic iron formation, *in* Medaris, L. G., Jr., Byers, C. W., Mickelson, D. M., and Shanks, W. C., eds., Proterozoic geology; Selected papers from an international Proterozoic symposium: Geologic Society of America Memoir 161, p. 227–235.

Klemic, H., and Tooker, E. W., 1979, Preliminary map of iron provinces in the conterminous United States: U.S. Geological Survey Open-File Report 79-576H.

Klemic, H., James, H. L., and Eberlein, G. D., 1973, Iron, *in* Brobst, D. A., and Pratt, W. P., eds., United States Mineral Resources: U.S. Geological Survey Professional Paper 820, p. 291–306.

Klinger, F. L., 1985a, Iron ore, *in* Mineral facts and problems, 1985 edition: U.S. Bureau of Mines Bulletin 675, p. 385–403.

—— , 1985b, Iron ore, *in* U.S. Bureau of Mines Minerals Yearbook 1984, volume I, Metals and minerals: U.S. Bureau of Mines, p. 475–496.

Knechtel, M. M., 1944, Manganese deposits of the Lyndhurst–Vesuvius district, Augusta and Rockbridge Counties, Virginia: U.S. Geological Survey Bulletin 940-F, p. 163–198.

Kuck, P. H., 1989a, Iron ore; Mineral Commodity Summaries 1989: U.S. Bureau of Mines, p. 80–81.

—— , 1989b, Iron ore, in U.S. Bureau of Mines Minerals Yearbook 1987, volume I, Metals and minerals: U.S. Bureau of Mines, p. 471–494.

Kwak, T.A.P., Brown, W. M., Abeysinghe, P. B., and Tan, T. H., 1986, Fe solubilities in very saline hydrothermal fluids; Their relation to zoning in some ore deposits: Economic Geology, v. 81, p. 447–465.

Lepp, H., ed., 1975, Geochemistry of iron, *in* Benchmark papers in geology: Stroudsburg, Pennsylvania, Dowden, Hutchinson, and Ross, v. 18, 464 p.

Lyons, J. I., 1988, Volcanogenic iron oxide deposits, Cerro de Mercado and vicinity, Durango, Mexico: Economic Geology, v. 83, p. 1886–1906.

Mackin, J. H., 1968, Iron ore deposits of the Iron Springs district, southwestern Utah, *in* Ridge, J. D., ed., Ore deposits of the United States, 1933–1967: American Institute of Mining, Metallurgical, and Petroleum Engineers, Inc.,

v. 2, p. 992–1019.

Maynard, J. B., 1986, Geochemistry of oolitic iron ores: An electron microprobe study: Economic Geology, v. 81, p. 1473–1483.

Medaris, L. G., Jr., ed., 1983, Early Proterozoic geology of the Great Lakes region: Geological Society of America Memoir 160, 141 p.

Meinert, L. D., 1984, Mineralogy and petrology of iron skarns in western British Columbia, Canada: Economic Geology, v. 79, p. 869–882.

Meyer, C., and 10 others, 1968, Ore deposits at Butte, Montana, *in* Ridge, J. D., ed., Ore deposits of the United States, 1933–1967: American Institute of Mining, Metallurgical, and Petroleum Engineers, Inc., v. 2, p. 1371–1416.

Morey, G. B., 1973, Mesabi, Gunflint, and Cuyuna ranges, Minnesota, *in* Genesis of Precambrian iron and manganese deposits; Proceedings of the Kiev Symposium, August, 1970: Paris, UNESCO, Earth Sciences 9, p. 193–208.

—— , 1983, Animikie Basin, Lake Superior region, U.S.A., *in* Trendall, A. F., and Morris, R. C., eds., Iron-formation; Facts and problems: Amsterdam, Elsevier, p. 13–67.

Morey, G. B., and Van Schmus, W. R., 1988, Correlation of Precambrian rocks of the Lake Superior region, United States: U.S. Geological Survey Professional Paper 1241-F, 36 p.

Morral, F. R., 1984, Manganese; A chronology and bibliography: CIM Bulletin, v. 77, no. 862, p. 72–75.

Morris, R. C., 1985, Genesis of iron ore in banded iron-formation by supergene and supergene-metamorphic processes; A conceptual model, *in* Wolf, K. H., ed., Regional studies and specific deposits: New York, Elsevier, Handbook of strata-bound and stratiform ore deposits, part 4, v. 13, p. 73–235.

Mosier, D. L., and Page, N. J., 1988, Descriptive and grade-tonnage models of volcanogenic manganese deposits in oceanic environments—a modification: U.S. Geological Survey Bulletin 1811, 28 p.

Murphy, J. E., and Ohle, E. L., 1968, The Iron Mountain Mine, Iron Mountain, Missouri, *in* Ridge, J. D., ed., Ore deposits of the United States, 1933–1967: American Institute of Mining, Metallurgical, and Petroleum Engineers, Inc., v. 1, p. 287–302.

National Materials Advisory Board, 1981, Manganese reserves and resources of the world and their industrial implications; Report of the panel on manganese supply and its industrial implications of the committee on technical aspects of critical and strategic materials: National Materials Advisory Board Publications NMAB-374, 334 p.

Norman, D. I., Bazrafshan, K., and Eggleston, T. L., 1983, Mineralization of the Luis Lopez epithermal manganese deposits in light of fluid inclusion and geologic studies, *in* Chapin, C. E., ed., Socorro region II: New Mexico Geological Society 34th Annual Field Conference Guidebook, p. 247–251.

Ojakangas, R. W., 1983, Tidal deposits in the early Proterozoic basin of the Lake Superior region; The Palms and the Pokegama Formations; Evidence for subtidal-shelf deposition of Superior-type banded iron-formation, *in* Medaris, L. G., Jr., ed., Early Proterozoic geology of the Great Lakes region: Geological Society of America Memoir 160, p. 49–66.

Panno, S. V., and Hood, W. C., 1983, Volcanic stratigraphy of the Pilot Knob iron deposits, Iron County, Missouri: Economic Geology, v. 78, p. 972–982.

Parak, T., 1985, Phosphorous in different types of ore, sulfides in the iron deposits, and the type and origin of ores at Kiruna: Economic Geology, v. 80, p. 646–665.

Pavlides, L., 1962, Geology and manganese deposits of the Maple and Hovey Mountains area, Aroostock County, Maine: U.S. Geological Survey Professional Paper 362, 116 p.

Pelymskiy, G. A., and Shishova, S. F., 1985, Evolution of iron mineralization in the Precambrian: International Geology Review, v. 27, p. 505–513.

Philpotts, A. R., 1981, A model for the generation of massif-type anorthosites: Canadian Mineralogist, v. 19, p. 233–253.

Pierce, W. G., 1944, Cobalt-bearing manganese deposits of Alabama, Georgia, and Tennessee: U.S. Geological Survey Bulletin 940-J, p. 265–285.

Prinz, W. C., 1967, Geology and ore deposits of the Philipsburg district, Granite County, Montana: U.S. Geological Survey Bulletin 1237, 66 p.

Quade, H., 1976, Genetic problems and environmental features of volcano-sedimentary iron-ore deposits of the Lahn-Dill type, *in* Wolf, K. H., ed.,

Regional studies and specific deposits: New York, Elsevier, Handbook of strata-bound and stratiform ore deposits, 2, v. 7, Au, U, Fe, Mn, Hg, Sb, W, and P deposits, p. 255–294.

Rabbitt, M. C., 1979, Minerals, lands, and geology for the common defence and general welfare; Volume 1, Before 1879: Washington, D.C., U.S. Government Printing Office, 331 p.

—— , 1980, Minerals, lands, and geology for the common defence and general welfare; Volume 2, 1879–1904: Washington, D.C., U.S. Government Printing Office, 407 p.

Reed, J. C., Jr., Bickford, M. E., Houston, R. S., Link, P. K., Rankin, D. W., Sims, P. K., and Van Schmus, R., eds., 1990, Precambrian; Conterminous U.S.: Boulder, Colorado, Geological Society of America, The Geology of North America, v. C-2 (in press).

Rona, P. A., 1984, Hydrothermal mineralization at seafloor spreading centers: Earth-Science Reviews, v. 20, p. 1–104.

Rose, A. W., Herrick, D. C., and Deines, P., 1985, An oxygen and sulfur isotope study of skarn-type magnetite deposits of the Cornwall type, southeastern Pennsylvania: Economic Geology, v. 80, p. 418–443.

Roy, S., 1981, Manganese deposits: New York, Academic Press, 458 p.

Schmidt, R. G., 1980, The Marquette Range Supergroup in the Gogebic mining district, Michigan and Wisconsin: U.S. Geological Survey Bulletin 1460, 96 p.

Simpson, T. A., and Gray, T. R., 1968, The Birmingham red-ore district, Alabama, *in* Ridge, J. D., ed., Ore deposits of the United States, 1933–1967: American Institute of Mining, Metallurgical, and Petroleum Engineers, Inc., v. 1, p. 187–206.

Sims, P. K., 1972, Banded iron-formations in Vermilion district, *in* Sims, P. K., and Morey, G. B., eds., Geology of Minnesota; A centennial volume: St. Paul, Minnesota Geological Survey, p. 79–81.

Sims, P. K., and James, H. L., 1984, Banded iron-formations of late Proterozoic age in the central Eastern Desert, Egypt; Geology and tectonic setting: Economic Geology, v. 79, p. 1777–1784.

Sims, P. K., Card, K. D., and Lumbers, S. B., 1981, Evolution of early Proterozoic basins of the Great Lakes region, *in* Campbell, F.H.A., ed., Proterozoic basins of Canada: Geological Survey of Canada Paper 81-10, p. 379–397.

Skillings, D. N., Jr., 1989, North American iron ore industry to approach 100 million gross ton production mark in 1989: Skillings' Mining Review, v. 78, no. 30, p. 14–28.

Snyder, F. G., 1969, Precambrian iron deposits in Missouri, *in* Wilson, H.D.B.,

ed., Magmatic ore deposits: Lancaster, Pennsylvania, Economic Geology Publishing Company, Economic Geology Monograph 4, p. 231–238.

Spencer, J. E., and Welty, J. W., 1986, Possible controls of base- and precious-metal mineralization associated with Tertiary detachment faults in the lower Colorado River trough, Arizona and California: Geology, v. 14, p. 195–198.

Stanton, R. L., 1972, Ore petrology: New York, McGraw-Hill Book Company, 713 p.

Stille, P., and Clauer, N., 1986, Sm-Nd isochron-age and provenance of the argillites of the Gunflint Iron Formation in Ontario, Canada: Geochimica et Cosmochimica Acta, v. 50, p. 1141–1146.

Thorpe, R. I., and Franklin, J. M., 1984, Chemical-sediment-hosted gold, *in* Eckstrand, O. R., ed., Canadian mineral deposit types; A geological synopsis: Geological Survey of Canada Economic Geology Report 36, p. 29.

Tooker, E. W., and Cannon, W. F., 1980, Preliminary map of manganese provinces in the conterminous United States: U.S. Geological Survey Open-File Report 79–0576-0, scale 1:5,000,000.

Towe, K. M., 1983, Precambrian atmospheric oxygen and banded iron formations; A delayed ocean model: Precambrian Research, v. 20, p. 161–170.

Trendall, A. F., and Morris, R. C., eds., 1983, Iron-formation; Facts and problems: Amsterdam, Elsevier, 558 p.

Van Houten, F. B., 1986, Reply to Comment on "Oolitic ironstones and contrasting Ordovician and Jurassic paleogeography": Geology, v. 14, p. 635.

Walker, J.C.G., 1984, Suboxic diagenesis in banded iron formations: Nature, v. 309, p. 340–342.

Walker, J.C.G., and Brimblecombe, P., 1985, Iron and sulfur in the pre-biologic ocean: Precambrian Research, v. 28, p. 205–222.

Walter, M. R., and Hofmann, H. J., 1983, The paleontology and paleoecology of Precambrian iron-formations, *in* Trendall, A. F., and Morris, R. C., eds., Iron-formation; Facts and problems: Amsterdam, Elsevier, p. 373–400.

Whitney, J. A., Hemley, J. J., and Simon, F. O., 1985, The concentration of iron in chloride solutions equilibrated with synthetic granitic compositions; The sulfur-free system: Economic Geology, v. 80, p. 444–460.

Wright, S. F., 1986, On the magmatic origin of iron ores of the Kiruna type; An additional discussion: Economic Geology, v. 81, p. 192–194.

Yeh, H.-W., Hein, J. R., and Koski, R. A., 1985, Stable-isotope study of volcanogenic- and sedimentary-manganese deposits: U.S. Geological Survey Open-File Report 85–662, 16 p.

MANUSCRIPT ACCEPTED BY THE SOCIETY FEBRUARY 10, 1989

NOTE ADDED IN PROOF

Interest in and exploration for Olympic Dam–type Cu-U-Au-REE-Fe deposits in the midcontinent area of the U.S. have increased significantly in the last two years. Sims and others (1987) and Kisvarsanyi and Kisvarsanyi (1989) identified the St. Francois and the Spavinaw granite-rhyolite terranes in southern Missouri as being favorable for an Olympic Dam–type deposit. All of the iron deposits in the Missouri iron province are hosted by anorogenic volcanic rocks of the St. Francois terrane, which has been dated as 1.38 to 1.48 Ga by U-Pb ages on zircons (Sims and others, 1987). The mineralization at Pea Ridge has a minimum age of 1.465 Ga, based on a U-Pb date from xenotime in a late quartz vein (data of R. Van Schmus, as reported in Marikos and others [1989]), and major, minor, and trace-element concentrations in trachyte, host rhyolite, and aplite dikes that temporally bracket the mineralization confirm that the deposit was emplaced during the same regional igneous event as the host rocks. Indeed, the ore may be genetically related to the host anorogenic rhyolite (Day and others, 1989). Recent underground sampling and drilling at Pea Ridge have identified anomalous gold concentrations in barite-rich REE-bearing breccias on the periphery of the magnetite orebody, in hematite-rich portions on the margins of the orebody, and in silicified zones in the host rocks (Husman, 1989; Nuelle and others, 1989).

Day, W. C., Kisvarsanyi, E. B., Nuelle, L. M., Marikos, M. A., and Seeger, C. M., 1989, New data on the origin of the Pea Ridge iron-apatite deposit, southeast Missouri—implications for Olympic Dam–type deposits [abs.]: Geological Society of America, Abstracts with Programs, v. 21, no. 6, p. A132.

Husman, J. R., 1989, Gold, rare earth element, and other potential by-products of the Pea Ridge iron ore mine, Washington County, Missouri: Missouri Department of Natural Resources Open-File Report 89-78-MR, Contribution to Precambrian Geology No. 21, 18 p.

Kisvarsanyi, G., and Kisvarsanyi, E. B., 1989, Precambrian geology and ore deposits of the southeast Missouri iron metallogenic province, *in* Brown, V. M., Kisvarsanyi, E. B., and Hagni, R. D., eds., "Olympic Dam–type" deposits and geology of Middle Proterozoic rocks in the St. Francois Mountains terrane, Missouri: Guidebook Series of the Society of Economic Geologists, v. 4, p.1–40.

Marikos, M. A., Nuelle, L. M., and Seeger, C. M., 1989, Geology of the Pea Ridge Mine (Field trip stop no. 1), *in* Brown, V. M., Kisvarsanyi, E. B., and Hagni, R. D., ed., "Olympic Dam–type" deposits and geology of Middle Proterozoic rocks in the St. Francois Mountains terrane, Missouri: Guidebook Series of the Society of Economic Geologists, v. 4, p. 41–54.

Nuelle, L. M., Seeger, C. M., Marikos, M. A., and Day, W. C., 1989, Mineral assemblages of the Pea Ridge Fe-REE deposit: Implications of an Olympic Dam-type variant in southeast Missouri: Geological Society of America 1989 Annual Meeting, Abstracts with Programs, v. 21, no. 6, p. A34.

Sims, P. K., Kisvarsanyi, E. B., and Morey, G. B., 1987, Geology and metallogeny of Archean and Proterozoic basement terranes in the northern midcontinent, U.S.A.—An overview: U.S. Geological Survey Bulletin 1815, 51 p.

Printed in U.S.A.

Chapter 6

Deposits containing nickel, cobalt, chromium, and platinum-group elements in the United States

Michael P. Foose
U.S. Geological Survey, 913 National Center, Reston, Virginia 22092

INTRODUCTION

As the United States evolved from an agricultural to an industrial society, some metals became indispensable or critical. Because of national security, some were also deemed to be strategically important. Mostly these strategic and critical metals are essential elements for which the United States is import dependent. Included among them are nickel, chromium, cobalt, and the platinum-group elements (PGEs), the subjects of this chapter.

The justification for treating this diverse group of elements together is that they exhibit a joint association within some mafic and ultramafic rocks. For this reason, discussion of deposits begins with those in mafic/ultramafic rocks and is followed by descriptions of deposits in other types of rocks. Deposits in the United States discussed or cited in the text are located on Plate 2b.

Nickel. Nickel is a critical component in the manufacture of stainless steels and a variety of ferrous and non-ferrous alloys. It is obtained principally from magmatic sulfides within mafic and ultramafic rocks and from laterites developed by weathering of ultramafic rocks. In 1987, the principal primary nickel producers were Canada (25 percent), the U.S.S.R. (23 percent), Australia (9 percent), New Caledonia (8 percent), and Indonesia (6 percent). In the recent past, United States production has come from a laterite deposit located near Riddle, Oregon. This mine, however, closed in 1987.

Chromium. The most important uses of chromium are in stainless steels and non-ferrous alloys. The mineral chromite is also important as a refractory. Chromite is the only commercial source of chromium and is obtained from two different types of deposits. Most (98 percent) of the world's resources are in stratiform deposits in layered intrusions. Examples are the Bushveld Complex (South Africa), the Great Dyke (Zimbabwe), and the Stillwater Complex (U.S.A.). The generally small (<100,000 tons of ore) podiform deposits that occur in alpine-type ultramafic complexes make up the second type. Significant deposits of this group are in Turkey, the Philippines, New Caledonia, and the U.S.S.R. World production of chromite is dominated by South Africa and the U.S.S.R., which in 1987, respectively accounted for 33 and 29 percent of the world's primary mine production. The United States has no primary chromite production and, in 1987, was supplied principally by South Africa (61 percent), Turkey (10 percent), and Zimbabwe (10 percent).

The platinum-group elements. The platinum-group elements (PGEs) consist of platinum, palladium, iridium, osmium, rhodium, and ruthenium. Of these, platinum and palladium are the most important. They are used mostly as catalysts to (a) control automobile exhaust, (b) refine gasoline, and (c) manufacture chemicals, and for electrical applications. In 1987, virtually all primary mine production of PGEs came from the U.S.S.R. (48 percent), South Africa (45 percent), and Canada (4 percent); the principal producing deposits were, respectively, magmatic segregations in the mafic and ultramafic rocks of the Norilsk district (U.S.S.R.), the Bushveld Complex (South Africa), and the Sudbury Complex (Canada). The United States has commonly produced small amounts of PGEs as by-products from copper ores and occasionally from placer deposits. This production pattern changed in 1987 with the opening of a new platinum deposit in the Stillwater Complex, Montana. The initially planned production capacity of this mine represents about 3 and 8 percent of the United States' demand for platinum and palladium, respectively.

Cobalt. Cobalt is an essential component in superalloys used in jet engines as well as an important part of many magnets, cutting tools, and electrical products. It differs, however, from most other strategic and critical commodities in that it is almost always obtained as a by-product, chiefly from deposits mined primarily for either nickel or copper. Major world sources of cobalt in 1987 were the stratabound copper deposits of Zaire and Zambia (respectively 57 and 11 percent), nickel-bearing sulfides and laterite deposits in the U.S.S.R. (9 percent), and nickel-bearing sulfide deposits in Canada (9 percent). The United States has no primary cobalt production.

DEPOSITS IN MAFIC AND ULTRAMAFIC ROCKS

Mafic and ultramafic rocks that host deposits of Ni, Cr, PGEs, and Co may be divided into five groups. These are: (a) stratiform intrusions, (b) rift-related intrusions or flows, (c) syn-

Foose, M. P., 1991, Deposits containing nickel, cobalt, chromium, and platinum-group elements in the United States, *in* Gluskoter, H. J., Rice, D. D., and Taylor, R. B., eds., Economic Geology, U.S.: Boulder, Colorado, Geological Society of America, The Geology of North America, v. P-2.

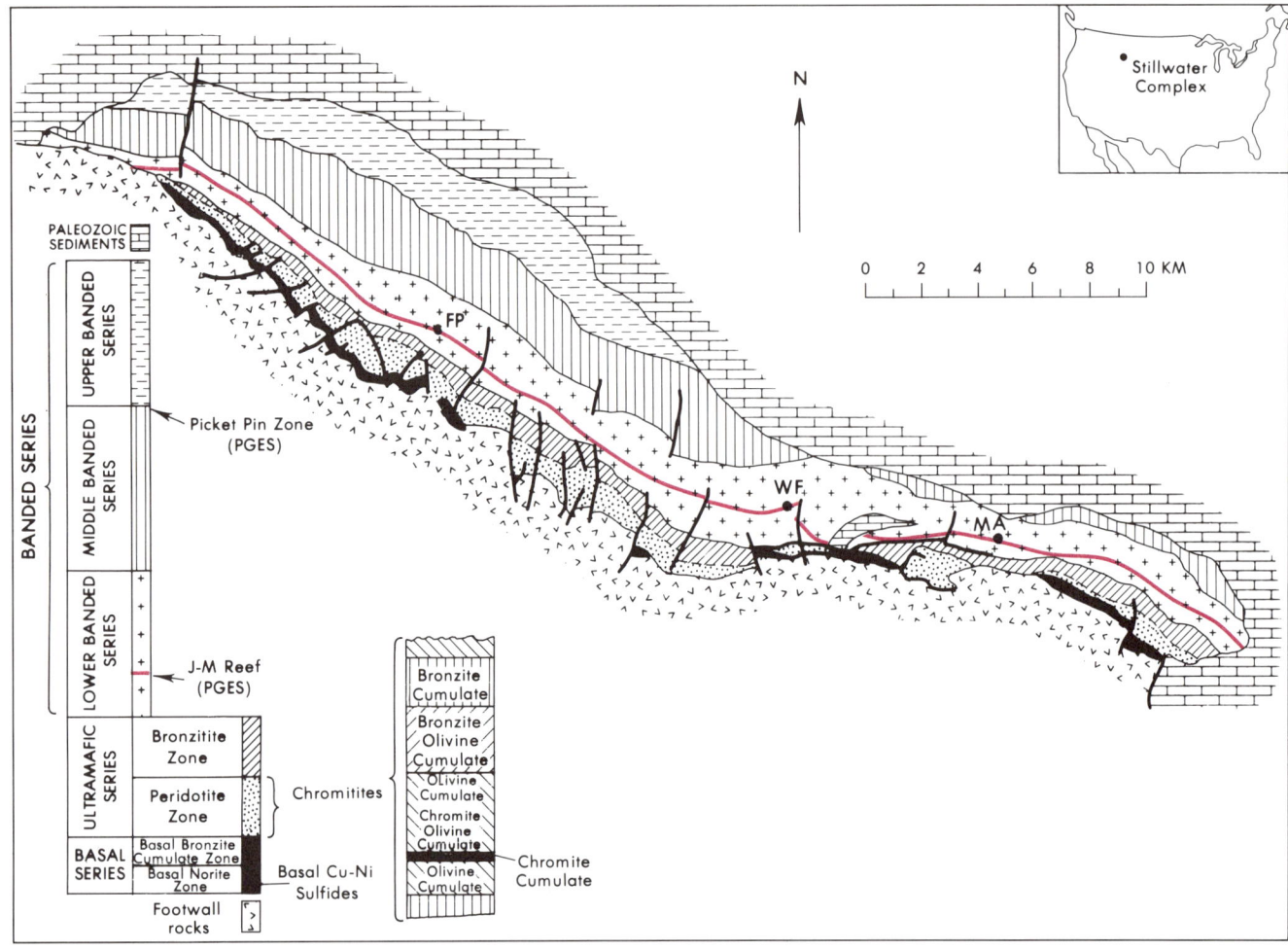

Figure 1. Generalized geologic map of the Stillwater Complex, Montana, and stratigraphic section shows the position of the Cu-Ni, chromite, and PGE deposits. The main PGE-bearing zone is shown in red. Also shown are locations of the Minneapolis (MA), West Fork (WF), and Frog Pond (FP) adits. An idealized cyclic unit in the Ultramafic series illustrates the setting of chromitites. Modified from Jackson (1968) and Page (1979).

orogenic intrusions, (d) ophiolites, and (e) komatiites. All but deposits hosted in komatiites have been identified within the United States. Of the other four deposit types, the Stillwater layered intrusive complex (group a) is by far the most important.

The Stillwater Complex

The Stillwater Complex is an Archean (approximately 2.7 Ga old) layered intrusion located in southwestern Montana (No. 1, Plate 2B; Fig. 1). The ultramafic and mafic rocks forming this complex trend approximately east-west and dip moderately (40° to 60°) to the north. The exposed portion of the complex consists of cumulate rocks up to 5,500 m thick that extend over a strike-length of about 48 km. These rocks are bounded to the east and west by Laramide-age faults, and are overlain unconformably by Paleozoic sediments. The original aerial extent of the complex is not known, but gravity data indicates that the exposed rocks represent only a small part of the original layered intrusion.

The Stillwater Complex is divided into three series (Fig. 1), each of which has distinctly different types of mineral deposits. From the base upward, these are: (1) the Basal series, with low-grade copper-nickel sulfides; (2) the Ultramafic series, which contains chromite; and (3) the Banded series, which has PGE–bearing sulfides.

The Basal series. The Basal series is between 0 and 160 m thick. It overlies hornfels, which is intruded by dikes and sills of diabase and sulfide-bearing norites. Its upper contact is marked by the first major occurrence of laterally continuous olivine cumulate.

The Basal series is subdivided into two zones (Fig. 1). The lower, Basal norite zone, consists of multiphase cumulates composed primarily of bronzite (Opx), olivine (Ol), and plagioclase (Pl), with minor amounts of chromite (Chr) and inverted pigeonite. Locally, noncumulate textured rocks containing plagioclase phenocrysts are present. This zone grades upward into the Basal bronzite zone, which is almost entirely Opx cumulates. Sulfides are present throughout both zones, but are most abundant toward the base of the series.

Discovered in 1883, the sulfides of the Basal series were the first potentially economic mineral deposits to be recognized in the Stillwater Complex. However, metal recovery has not been sufficient to sustain mining operations, even though relatively recent work has defined several hundred million tons of mineralized rock that might be feasible for mining. This material is relatively low grade, showing typical values of 0.25 percent Cu and 0.25 percent Ni (Page and others, 1985); Co is not reported, but unpublished data indicate a Co/Ni ratio of about 0.01.

The sulfides are predominantly pyrrhotite (90.4 percent) and lesser amounts of chalcopyrite (5.6 percent) and pentlandite (4.0 percent; Page, 1979). Typically the sulfides either fill the interstices between or are globular inclusions within high-temperature silicate grains.

Abundance of sulfides varies greatly. The top of the Basal bronzite cumulate zone typically contains sparse sulfides, whereas lower parts of the Basal norite zone have more abundant disseminated and net-textured grains of sulfide minerals. Massive sulfides are relatively rare. Page (1979) has shown that the vertical size distribution of some of the sulfide grains approximates distributions found in rain drops, a relation that suggests some of this vertical variation may be due to the settling of sulfides through a silicate liquid. Sulfide abundance also varies laterally, with the greatest concentrations in areas that were initially depressions at the base of the intrusion.

The sulfide-silicate textures are typical of magmatic sulfides and indicate that the sulfides initially separated as a high-temperature immiscible liquid. The separation of such a liquid could be in response to internal changes in the magmatic system, or could be induced by reactions between the magma and the country rock. In the Stillwater Complex, values for sulfur isotopes obtained from the sulfide minerals do not correspond to values for sulfur from the country rocks, indicating that the country rock was not an important source of sulfur used in forming these sulfides (Zientek, 1983). In addition, sulfide droplets in sills and dikes in the country rock, which were emplaced before the Basal series, show that immiscible sulfide liquids were present before the onset of the emplacement of the Stillwater Complex.

The Ultramafic series. The Ultramafic series consists chiefly of cumulates of olivine, bronzite, and chromite. Its base is defined by the first appearance of laterally continuous cumulus olivine, and generally overlies Basal series rocks. Locally, the Basal series is absent, and the Ultramafic series rests directly on footwall rocks. It is overlain by the Banded series (Fig. 1).

The lower part of the Ultramafic series is the Peridotite zone. It is a layered succession of rocks consisting of cycles that are ideally, from bottom up, Ol cumulates some of which contain chromite, Ol-Opx cumulates, and Opx cumulates (Fig. 1). These cyclic units are interpreted to be the result of fractional crystallization of a basaltic liquid. New cycles are thought to mark the influx of fresh magma. In some cases, new magma was injected before completion of the previous cycle, resulting in a "be-

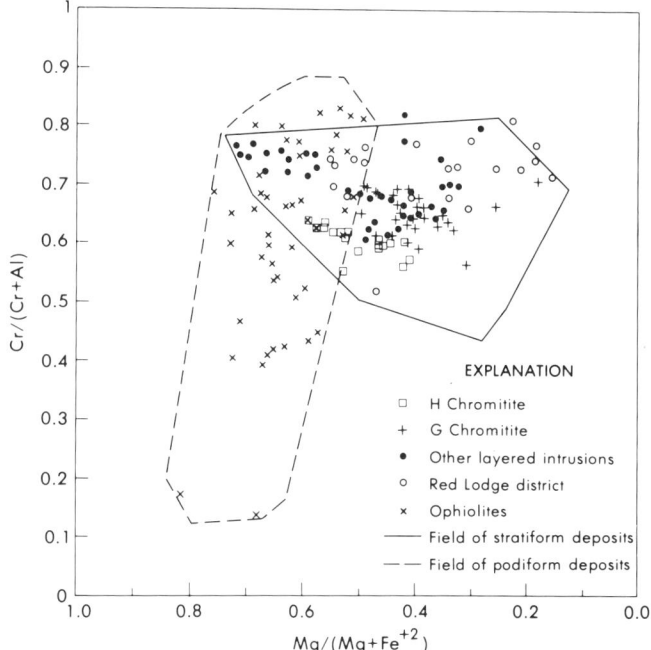

Figure 2. Composition of chromites from the G and H plotted with data from the Red Lodge district, some other layered intrusions, and some ophiolite-hosted deposits. Also shown are the compositional fields for both stratiform (solid line) and podiform chromite (dashed line) deposits. Data from Irvine (1967), Jackson (1968), Loferski (1980), and Duke (1983).

headed" cycle in which the uppermost (Opx) layer and, in places, the middle (Ol+Opx) layer is missing. The Peridotite zone has between 8 and 21 of these cyclic units, which range in thickness from 3 to 381 m (Page, 1977).

The cyclically layered Peridotite zone is overlain by a mostly monotonous sequence of Opx cumulates that make up the Bronzitite zone. The combined thickness of the two zones ranges from 850 to 2,759 m.

Chromite is ubiquitous in each cyclic unit of the Peridotite zone, but is usually less than 2 modal percent. Some olivine cumulates, however, contain economically significant layers of chromite. The thickest chromitites typically have sharp basal contacts, massive lower parts, and tops that grade through disseminated chromite into barren Ol cumulates.

The chromitites have been labeled according to their stratigraphic position from A (bottom) to K (top). With the exception of the lowest, they form laterally persistent layers. The thickest seams are the G (1 to 8 m), H (about 1.3 m), B (3 layers, each 20 cm to 1 m), A (0.3 m), and K (2 layers, each 2 to 4 cm); all the other chromitites are less than 4 cm thick (Jackson, 1968). Only the G and the H have been mined.

The compositions of the G and H chromitites are plotted in Figure 2 along with compositions of chromites from some other stratiform intrusions and from ophiolite-hosted (podiform) chromite deposits. Chromites from the stratiform and podiform deposits display a considerable range of compositions, which plot in two somewhat overlapping fields. Chromites from stratiform deposits have relatively constant Cr/Al, but widely varying Mg/Fe. Those from podiform deposits show variations in Cr/Al that are slightly negatively correlated with changes in Mg/Fe. Other differences, not shown in Figure 2, are a generally lower proportion of Fe^{+++} and lower TiO_2 (<0.3 weight percent) in podiform chromites. Compositions of chromite from the G and H horizons fall chiefly in the central part of the field for stratiform deposits. Chromites from the small deposits of the Red Lodge district (No. 2, Plate 2b), 40 km southeast of the Stillwater Complex, also have compositions typical of stratiform intrusions and are thought to be part of a disrupted and metamorphosed stratiform complex.

The economic potential of chromite deposits is determined largely by thickness, continuity, and grade of ore. The G and H chromitites of the Stillwater Complex are laterally continuous and are comparable in thickness with seams currently being mined in the Bushveld Complex. Furthermore, the Cr/Fe ratio is about 1.6:1, near the minimum ratio of 1.5:1 acceptable for metallurgical uses, and is similar to values for Bushveld chromites. The Cr_2O_3 content of the Stillwater chromite concentrates, however, is only 22.5 percent, much lower than the 46.0 to 47.6 percent Cr_2O_3 typical of chromite shipped from the Bushveld.

As a result of their low grade, significant production from the Stillwater chromites occurred on only two occasions. The first was during World War II when the United States government was forced to turn to domestic sources of chromite. The second was between 1952 and 1962 when the Federal government purchased Stillwater chromite for the Korean War and subsequent stockpiling program.

A wide variety of mechanisms have been called upon to explain the formation of chromitites in layered intrusions. Initially, analogies with sedimentary processes were made because of the well-developed layering shown by some deposits. For example, Jackson (1961) explained the layering in the Ultramafic series to be principally due to settling of grains nucleated near the magma chamber floor. Subsequent studies have pointed out problems with crystal settling in many situations. As a result, processes which form layered cumulates through in situ nucleation have recently been suggested.

There appear to be three principal ways in which chromitites may be formed. The first is precipitation of chromite as a result of decreases in magma temperature. Cooling of basaltic liquids quickly causes supersaturation with chromite, thereby inducing chromite precipitation. However, olivine will usually precipitate along with chromite, and thus cooling alone would not form massive chromitites. The second possibility involves changes in either oxygen fugacity or total pressure. An increase in f_{O_2} within a basaltic liquid decreases chromite solubility; therefore, variations in f_{O_2} may cause chromite to form (Ulmer, 1969). Cameron (1980) has also suggested that changes in total pressure may produce similar results. Such variations would act virtually instantaneously throughout a magma and thus could have formed of laterally extensive chromite layers. It is difficult, however, to envision a process whereby changes in either f_{O_2} or total pressure would occur with sufficient repetitiveness to form the numerous chromite seams found in the Stillwater Complex.

A third way in which chromitites may form is through magma mixing (Irvine, 1977). Because the olivine-chromite cotectic in basaltic systems is concave toward the chromite field, the mixing of two liquids that lie on different parts of the cotectic would produce a hybrid liquid that is within the chromite field. The result would be to precipitate a monomineralic chromitite. In a recent study, Murck (1985) has concluded that magma mixing best explains formation of the Stillwater chromitites.

The Banded series. The approximately 4,000-m-thick Banded series rests on the Ultramafic series (Fig. 1). Its basal contact is sharp and is defined by the first appearance of cumulus plagioclase. With the exception of a few thin horizons, plagioclase is a cumulus phase throughout the remainder of the complex, occurring either alone or in combinations with olivine, clinopyroxene (Cpx), or orthopyroxene. The resulting succession is divided into: (1) the Lower Banded series—a sequence made up predominantly of Pl-Opx and Pl-Opx-Cpx cumulates, but in which there is a thin zone of PGE–bearing olivine cumulates; (2) the Middle Banded series—a sequence in which two Pl cumulate layers 300 to 500 m thick are separated by a heterogeneous and complexly layered group of rocks containing Pl-Cpx, Pl-Ol, Pl-Cpx-Opx, Pl-Cpx-Ol, and Pl-Cpx-Ol-Opx cumulates, (3) the Upper Banded series—a sequence that is mostly Pl-Opx-Cpx cumulates but in which pigeonite replaces orthopyroxene in the upper part.

Sulfides and PGE minerals were reported from the Banded series in 1936 (Howland and others, 1936). The sulfides, however, are very disseminated and were generally not considered to be of economic interest. Later, speculations that the Stillwater Complex might hold an analog to the Bushveld's Merensky Reef resulted in exploration for PGEs in 1967. By 1971, several targets had been identified, and by 1974, ore-grade material had been found by drilling. Additional drilling has proven a laterally continuous zone of PGE-bearing sulfides within the lower part of the Banded series (Bow and others, 1982; Todd and others, 1982), which is now known as the J-M Reef. Discovery of the J-M Reef stimulated close examination of the other sulfide-bearing zones within the Banded series. Of these other horizons, the most promising and most thoroughly explored occurs at the very top of the Middle banded series, and is informally known as the Picket Pin zone (Fig. 1).

The J-M Reef. Approximately 400 to 500 m above the top of the Ultramafic series, the predominantly gabbroic rocks of the Lower Banded series are interrupted by a thin sequence of rocks that contain olivine cumulates. This sequence, informally known as the Reef Package, contains a sulfide-bearing zone 1 to 3 m thick that is the J-M Reef. The Reef Package is 4 to 130 m thick, shows internally complicated stratigraphy, and has been explored in detail at three localities. From east to west, these are the Minneapolis, West Fork, and Frog Pond adits (Fig. 1).

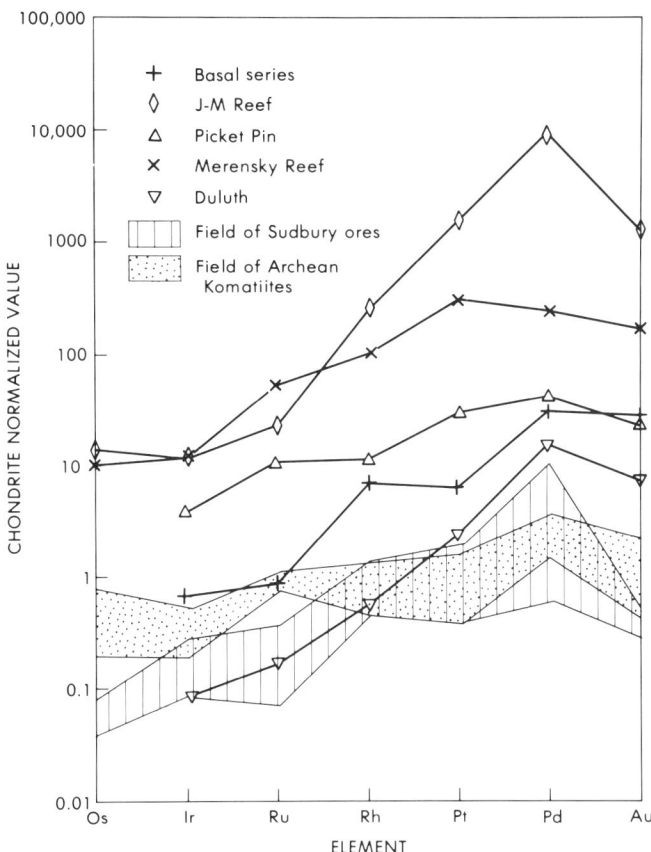

Figure 3. Chondrite normalized platinum-group element values for selected magmatic sulfide deposits. PGE values are amounts expected in 100 percent sulfide that have been normalized to abundances found in chondrites. Data from the Stillwater Complex (Basal series - crosses, J-M Reef - diamonds, and Picket Pin zone - triangles) are compared with data from the Duluth Complex (inverted triangles), Merensky Reef (x), the field of ores from the Sudbury Complex (vertical lines), and data from Archean komatiites (stippled). Data from Barnes and Naldrett (1985), Naldrett (1981), Page and others (1985), and Foose (unpublished data); values in chondrites are those given by Naldrett (1981).

To some extent, the lateral stratigraphic variations within the Reef Package can be related to the configuration of the footwall of the Stillwater Complex (Turner and others, 1985). In the western part of the complex, where the magma chamber floor was apparently stable, the Reef Package has a fairly uniform internal stratigraphy. Here, it is approximately 100 m thick, and consists of ten olivine-bearing zones that are typically a few meters thick. These zones are interlayered with Pl, Pl-Opx, and Pl-Opx-Cpx cumulates (Todd and others, 1982).

Farther east, disruptions of the footwall occurred during crystal accumulation, with the result that the Reef Package thins and thickens over basement highs and lows and is complicated by slump structures. At the West Fork adit, this sequence contains only four olivine-bearing zones (Mann and Chong-Pin, 1985). Farther east, at the Minneapolis adit, it generally consists of only one olivine-bearing zone and varies in thickness from 4 to 25 m (Turner and others, 1985).

The location of the J-M Reef within the Reef Package varies laterally and vertically. At the Frog Pond adit, sulfides are concentrated in a 0.3- to 2-m-thick olivine cumulate that makes up the base of the fifth and most laterally continuous of the olivine-rich layers (LeRoy, 1985). In the West Fork adit the PGEs are in Pl-Ol cumulates slightly higher in the section, but locally they extend into underlying Ol cumulates and overlying Pl cumulates. At the Minneapolis adit, the PGEs are concentrated in either the single olivine-bearing layer or are associated with a "mixed rock" that consists of "boulders" of Ol cumulate, Pl-Ol cumulate, and Pl cumulate (Turner and others, 1985). In detail, the PGE-bearing sulfides occur either in footwall depressions, as disseminations that extend outward from these depressions, as laterally continuous concentrations associated with the "mixed" rock, and as crosscutting veins of sulfide.

The most abundant sulfide minerals are chalcopyrite, pyrrhotite, pentlandite, and pyrite. Pentlandite contains most of the palladium, whereas platinum is generally in discrete minerals such as moncheite, copperite, braggite, and some arsenides. Sulfide minerals generally make 0.2 to 0.5 volume percent, and rarely exceed 2 volume percent of the rock (Irvine and others, 1983; Turner and others, 1985). At places, the sulfides exhibit a vertical gradation from finely disseminated grains at the top through net-textured masses, to coarse-grained blebs at the base. Sulfide grains occur both interstitial to, or included in, the silicate phases, and may be associated with concentrations of graphite.

Although sulfides are sparse, they contain large amounts of PGEs. Published grades for the J-M Reef are about 14.7 ppm (equals approximately 14.7 g/t or 0.44 troy oz per ton) combined PGEs with a Pt:Pd ratio of about 3.5 (Todd and others, 1982; LeRoy, 1985). Another way of expressing the abundance of PGEs is to calculate their expected concentration in pure sulfide and normalize this value to concentrations found in chondrites (Naldrett, 1981). This approach facilitates comparison among deposits by eliminating variations due to differing abundances of sulfides. The results (Fig. 3) show extreme concentrations of PGEs in the J-M Reef sulfides, values that are two to three orders of magnitude greater than found in typical magmatic sulfides. The mechanism by which such high concentrations were obtained is central to understanding the origin of the J-M reef.

Although it appears possible that hydrothermal fluids have been involved in concentrating some of these PGEs (Boudreau and others, 1986; Boudreau, 1988), most attention has focused on magmatic processes to explain the formation of this deposit. Models that rely on mixing of two, or possibly even three, different magmas explain many of the features observed in the reef. The involvement of compositionally distinct liquids was first suggested in order to explain differences in crystallization sequences (Todd and others, 1982) and has subsequently been elaborated upon by Barnes and Naldrett (1986). The portion of the complex below the J-M reef is thought to have crystallized from a liquid with ultramafic affinities (crystallization order of Ol: Opx: Pl: Cpx). The sudden appearance of olivine with plagioclase in the J-M Reef departs from this crystallization order and marks the influx of a new liquid, one that had anorthositic characteristics (crystallization order of Pl: Ol: Cpx: Opx). Additional evidence for the mixing of compositionally different liquids is provided by the embayed textures of olivine in much of the reef, textures which indicate that after crystallization these olivines were immersed in a liquid with which they were not in equilibrium. Other studies (Lambert and Simmons, 1988) have shown abrupt changes in abundances of REE between adjacent rocks in the Reef Package changes that indicate the existence of compositionally different liquids.

The manner in which these liquids mixed is less certain and directly relates to how the high concentrations of PGEs are produced. Irvine and others (1983) propose that differences in density caused the two liquids to form separate layers in the magma chamber, which then mixed through double diffusive convection. Double diffusive convection is a process that has only relatively recently been applied to layered intrusions. Detailed explanations are given by McBirney and Noyes (1979) and Irvine and others (1983). Simply summarized, a stratified liquid may result from the initial juxtaposition of liquids having two or more components that diffuse at markedly different rates. Most typically these components are heat and chemical concentrations. Each layer of this stratified liquid establishes its own convective pattern, which causes heat and liquid to be exchanged with adjacent layers. These layers are oriented horizontally and intersect the subhorizontal magma chamber floor at shallow angles. Different phases precipitate simultaneously from each compositionally distinct liquid layer, where they intersect the chamber floor. The result is a succession of cumulates that are somewhat analogous to rings in a bathtub. Continued crystallization simultaneously adds to the inside part of each layer so that the layering grows inward from the margin of the intrusion. Irvine and others (1983) and Todd and others (1982) suggest that double diffusive convection mixed an ultramafic liquid, in which fractional crystallization had produced relatively high concentrations of PGEs, with a relatively sulfur-rich anorthositic liquid. The result was the J-M Reef.

Although elegant in theory, the ability of double diffusive convection to operate effectively in mafic intrusions is still uncertain (McBirney, 1985). An alternative mechanism for forming the J-M Reef depends on the very large (1,000 to 10,000) partition coefficients that PGEs have for sulfide liquids. As a result of these large coefficients, immiscible sulfides are extremely effective collectors of PGEs in a silicate melt. However, for them to obtain very high PGE concentrations from melts with ordinary PGE contents, they must equilibrate with, and therefore scavenge PGEs from, a very large volume of silicate liquid. It has been suggested (Campbell and others, 1983; Barnes and Naldrett, 1985, 1986) that the influx of a new and compositionally different liquid entered as a buoyant plume, which rose through and mixed turbulently with the preexisting liquid. In doing so, sulfides precipitated and equilibrated with large volumes of melt, thereby obtaining high concentrations of PGEs. The settling of these sulfides to the magma chamber floor formed the J-M Reef. Crystallization in this turbulent environment also may have formed the "boulders" found in the "mixed" rock that occurs along parts of the reef.

The Picket Pin zone. Within the Stillwater Complex, the only other identified zone of sufficient lateral extent and abundance of PGEs to have received sustained exploration is the Picket Pin zone (Fig. 1). This zone is near the top of the Middle Banded series, about 3,000 m above the J-M Reef. The sulfide minerals are concentrated just below the top of a thick layer of Pl cumulate, which is overlain by Pl-Ol cumulate. In detail, the sulfides are most abundant in that part of the zone where the Pl cumulates change from coarse- to medium-grained (Boudreau and McCallum, 1985).

Sulfides are mostly disseminated and occur in pod-like or lenticular concentrations that are up to 20 to 30 m in length. Some concentrations are pipe-like, crosscutting bodies that may merge upward with more stratigraphically conformable lenses. The principal sulfide minerals are pyrrhotite, chalcopyrite, and pentlandite. Published analyses for the Picket Pin zone give 0.27 percent Cu, 0.14 percent Ni, 270 ppb Pt, and 340 ppb Pd (Todd and others, 1982).

The presence of discordant sulfide pipes merging upward with more concordant lenses is consistent with concentration through a filter-pressing process. PGEs tend to be excluded from the silicate crystal structure and thus are concentrated in the residual, intercumulus, sulfide-rich liquid. It is probable that PGE-rich interstitial liquids were expelled upward within a compacting pile of cumulus plagioclase. These liquids became concentrated and precipitated sulfides at a relatively impermeable boundary resulting from the reduction of grain size at the top of the plagioclase cumulate layer.

The Duluth Complex

The United States' second major concentration of metals in mafic and ultramafic rocks occurs as sulfides in the 1.1-Ga Duluth Complex of northeastern Minnesota (No. 3, Plate 2B; Fig. 4). Unlike the Stillwater Complex, this complex has been of interest primarily for its Ni and Cu potential. This complex also differs from the Stillwater in its relatively young, post-Archean age and its setting within a rift environment (group B deposit).

Geologic setting. The Duluth Complex, which is over 200 km long and up to 60 km wide, is one of the world's larger intrusive complexes. It is associated with the Midcontinent gravity high, which is the geophysical expression of the 1.1-Ga rifting event that formed this complex. This gravity high extends more than 1,000 km to the southwest, suggesting that the Duluth Complex is only part of a much larger magmatic system.

The complex is a composite intrusion consisting chiefly of anorthosite and troctolite (Fig. 4B). These rocks have been slightly tilted (10° to 20°), so that footwall rocks are now exposed to the west and the overlying North Shore Volcanic Group to the east. The overlying volcanic rocks are the same age as the complex and are presumed to be the extrusive derivatives of the underlying plutonic rocks.

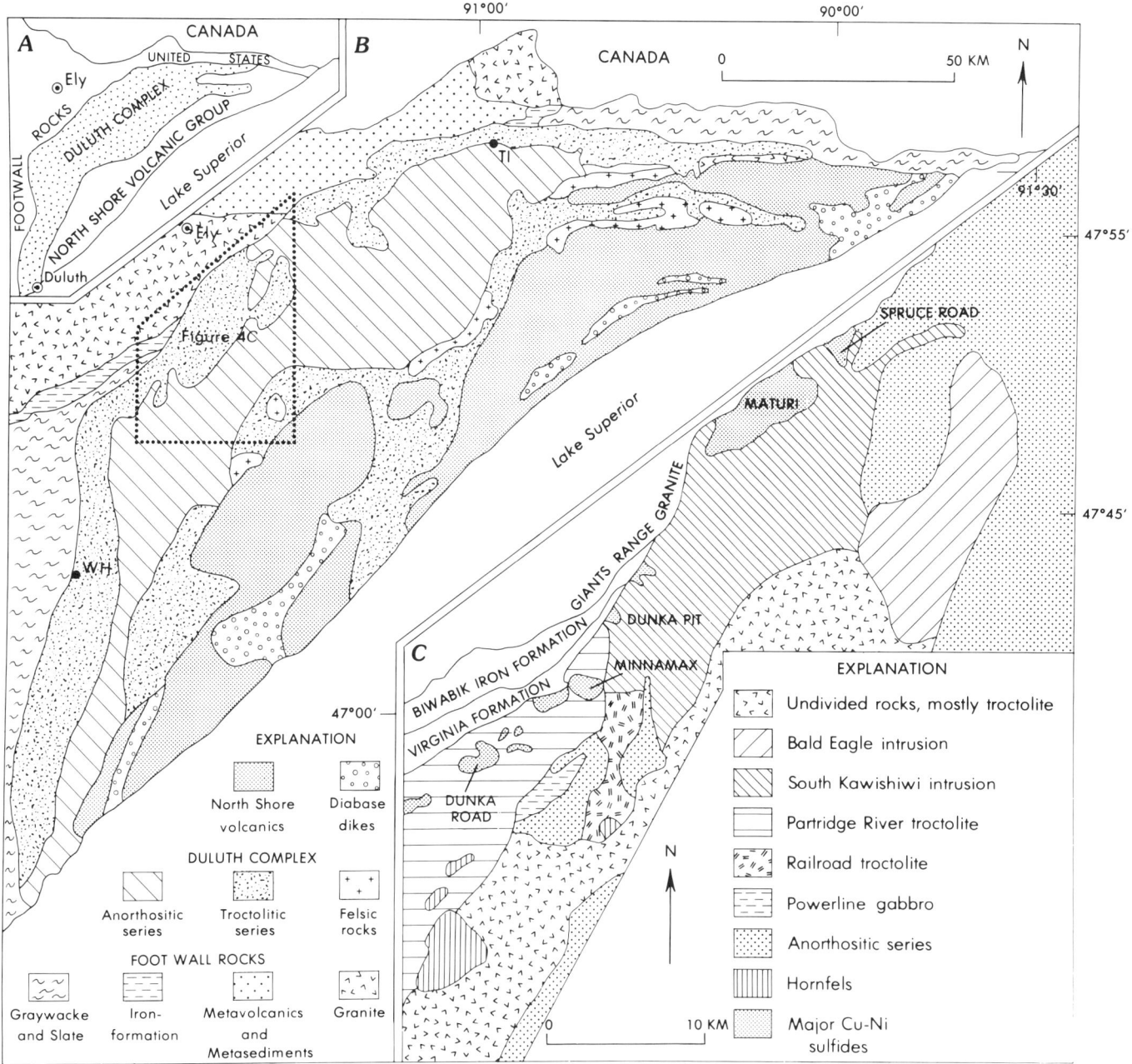

Figure 4. Generalized geologic map of the Duluth Complex. (A) Index map; (B) General geology of the Duluth Complex showing location of deposits in the Water Hen (WH) and Tuscarora (TI) intrusions; (C) Geologic map of the area southeast of Ely, Minnesota (outlined in 4 B), containing most of the sulfide deposits. Modified from Weiblen and Morey (1980) and Foose and Weiblen (1986).

Rocks of the Duluth Complex belong to two separate series, neither of which show the lithologic diversity typical of stratiform-layered complexes. The oldest (anorthositic series) is composed almost entirely of Pl cumulates and is intruded by rocks of the younger troctolitic series that are predominantly Pl-Ol cumulates with some interlayered Pl cumulates. Rocks of both series were injected into voids formed by rift-related faulting (Weiblen and Morey, 1980). These chambers received repeated injections of fresh magmas, with the result that individual intrusions of either anorthositic- or troctolitic-series rocks generally show little evidence for much overall chemical evolution.

Aside from their obvious mineralogic differences, there are two other important distinctions between these two series. First, the anorthositic series is more chemically evolved than the troctolitic series; second, these anorthositic rocks were emplaced mostly as crystal mushes, while the troctolitic-series liquids evidently contained relatively few intratelluric crystals (Miller, 1986). Miller suggests that both series were derived from a similar high-

Al olivine tholeiitic liquid and that, during the early stages of rifting, plagioclase separated from this liquid, floated upward, and concentrated at the base of the crust. The anorthositic series formed from injections of this crystal mush along early faults. Later, as further rifting opened the magmatic system, primitive liquids from which plagioclase had not separated were intruded to form the troctolitic series.

Sulfide mineralization. There are two general controls on the occurrence of sulfides in the Duluth Complex. First, significant amounts of sulfides have been found only in rocks of the troctolitic series. Second, all major occurrences are along the northwest margin of the complex where troctolites are in contact with footwall rocks. Along this contact, at least six major concentrations of sulfides have been identified (Fig. 4B, C). These are the Water Hen intrusion, the Dunka Road deposit, the Minnamax deposit, the Dunka Pit, the Spruce Road deposit, and the Tuscarora intrusion. None has yet gone into production.

Although they occur over a distance in excess of 100 km, these deposits appear to be generally similar in composition, mineralogy, and setting. In part this appearance of similarity may be deceiving because relatively few studies have been published for the area outside that southeast of Ely, Minnesota (Fig. 4C). This area, which contains most of the identified mineral occurrences, has more than 4.4 billion tons of mineralized rock averaging 0.66 percent copper and 0.2 percent nickel. The Co/Ni ratio is approximately 0.10 to 0.15. Values of about 0.67 ppm Pt and 0.20 ppm Pd have been reported in some deposits.

The sulfide minerals in this area occur within two intrusions that are informally known as the Partridge River Troctolite (PRT) and the South Kawishiwi Intrusion (SKI; Fig. 4). These bodies cut across the country rocks so that the footwall changes from sulfide-bearing slates of the Virginia formation in the south to Biwabik Iron-Formation near the PRT-SKI contact, to Giants Range Granite in the north (Fig. 4C). The basal parts of both intrusions, in which the sulfides are concentrated, are predominantly troctolite with minor amounts of norite, picrite, dunite, and oxide-rich cumulates. Biotite is a minor but significant constituent of many of the sulfide-bearing rocks, and trace amounts of graphite are often present. Inclusions of country rock are widespread and are commonly associated with the sulfide-rich zones.

The basal zones of these two intrusions differ somewhat in detail. The basal part of the PRT consists of at least five laterally traceable sequences (Tyson and Chang, 1984). Sulfides are concentrated toward the base, but also occur as "cloud zones" higher in the section. In contrast, the basal part of the SKI is extremely heteroegeneous with no discernible stratification. Sulfides occur throughout the basal part, which is abruptly overlain by a sequence of sulfide-free troctolitic rocks that has a well-defined stratigraphy (Bonnichsen and others, 1980; Foose and Weiblen, 1986).

Two of the six major sulfide occurrences are outside the area described above. These are in the Water Hen and Tuscarora intrusions (Fig. 4B). The Water Hen is known only from drill holes, which show it to differ from the PRT and SKI in having large amounts of olivine cumulates. Sulfides are concentrated in a basal olivine cumulate unit as well as in olivine cumulates that are interlayered with the overlying troctolites and anorthosites (Mainwaring and Naldrett, 1977). The Tuscarora intrusion is even less well known, but mineralized exposures show it to have rock types and stratigraphic settings similar to those of the PRT and SKI.

Despite some differences in host rocks, the mode of occurrence of all the sulfides appears to be similar. The sulfide minerals are predominantly pyrrhotite, chalcopyrite, cubanite, and pentlandite, which occur as disseminated grains. Proportions of these minerals vary both vertically and laterally. One of the most striking changes occurs southeast of Ely, where a decrease in the abundance of chalcopyrite and cubanite relative to pentlandite results in a decrease in the Cu/Ni ratio from 5:1 at the Dunka Road deposit to 2:1 at the Spruce Road deposit (Fig. 4C).

The sulfide minerals occur in four ways. Most common are interstitial grains in the roughly triangular voids between plagioclase laths. More rarely, the sulfide grains are in silicate grains. These included grains are generally in rims of plagioclase, but in the Water Hen intrusion some are in cumulus olivine. Thirdly, sulfides occur as intergrowths with hydrous silicates, typically biotite. Finally, sulfides form thin crosscutting veins (Foose and Weiblen, 1986).

The distribution of the sulfide minerals is inconsistent with simple magmatic accumulation. Specifically, the occurrence of most sulfides as interstitial masses or inclusions in silicate rims show that they formed after the crystallization of most of the silicate minerals. The intergrowths of sulfide minerals with hydrous minerals suggests that water-rich volatiles were involved in sulfide nucleation. Finally, a close spatial association of sulfides with footwall rocks or country-rock inclusions indicates that these rocks influenced accumulation of the sulfides.

Isotopic data clearly show that assimilation of country rock provided volatiles to the surrounding magma and that sulfur in the copper-nickel sulfides was derived from the country rock. The Water Hen (Mainwaring and Naldrett, 1977), Dunka Road (Ripley, 1981), and Minnamax (Ripley and Al-Jassar, 1987) deposits show relatively heavy $\delta^{34}S$ values that are around +7 to +15 per mil. These are much heavier than the approximately 0 per mil values typical of sulfides that have precipitated from uncontaminated magmas. They are, however, similar to values found in underlying sulfide-bearing slates of the Virginia Formation. Similarly, whole-rock and plagioclase $\delta^{18}O$ of up to 9.6 per mil occur in troctolites near inclusions. Again, these are higher than typical magmatic values of 6 to 7 per mil, but are in the range observed from hornfels inclusions (Rao and Ripley, 1983).

The above features indicate that sulfides in the Duluth Complex generally formed late, after about 75 percent of the melt had crystallized. At this time, sulfur-rich volatiles produced by magma interactions with footwall rocks or inclusions caused sulfides to precipitate in the interstices between silicate grains. The result was sporadically disseminated and compositionally diverse sulfides that concentrated near the footwall or around inclusions.

Faults were active during this stage of magma consolidation, and the presence of sulfide concentrations near structural breaks in the footwall suggests they may have provided pathways along which some of the sulfur-rich volatiles were introduced (Weiblen and Morey, 1980).

Synorogenic intrusions

The Stillwater and Duluth Complexes contain the most important magmatic sulfide deposits in the United States, and together hold the vast majority of the nation's resources of Ni, Cr, and PGEs, as well as having a significant potential for by-product Co. However, many other smaller intrusions contain concentrations of sulfide minerals that may be mineable. Some of these magmatic deposits, like the Friday, Gap, Keywest, Great Eastern, and Salt Chuck deposits (Nos. 4, 5, 6, 7; Plate 2B) are poorly known, and it is not clear how they should be classified. There are, however, deposits in two separate areas that deserve special mention. These are the La Perouse intrusion in Alaska and the Moxie and Katahdin plutons in Maine (Nos. 8, 9, and 10; Plate 2b). All of these are synorogenically emplaced intrusions. Several other synorogenic bodies are shown on Plates 2B (Nos. 11, 12, and 13).

The 27-km-long by 12-km-wide La Perouse intrusion is in the Fairweather Range of southeast Alaska. It was emplaced during the middle Tertiary at the edge of a continental plate that was undergoing compressional strike-slip deformation. As a result of synintrusion deformation, the approximately 6,000 m of exposed cumulates of this complex were folded into a syncline.

The base of the intrusion consists of 680 m of sulfide-bearing Ol-Chr, Ol-Pl, and Ol-Pl-Cpx cumulates. They are overlain by interlayered Pl-Cpx-Ol-Opx, Pl-Cpx-Ol, and Pl-Opx-Cpx cumulates. Layers range in thickness from a few centimeters to tens of meters, but generally lack lateral continuity. Chemical patterns through these rocks are complicated with many reversals, and like the Duluth Complex, there is no evidence of a consistent evolution in magma composition upward through the complex. These observations are consistent with repeated influxes of liquids having slightly different compositions (Himmelberg and Loney, 1981). In addition, the ubiquitous presence of graphite through the complex, in places making seams up to 30 cm thick, attests to relatively low f_{O_2} in these liquids.

The sulfides in the La Perouse are chiefly pyrrhotite, pentlandite, and chalcopyrite, which occur in the basal cumulate unit. They are exposed in only two or, occasionally under light snow conditions, three nunatacks. Exploration has shown more than 100 million tons of mineralized rock averaging 0.5 percent Ni and 0.3 percent Cu; typical ranges for PGEs are 0.1 to 0.8 ppm Pd, 0.02 to 0.2 ppm Pt, and <0.005 to 0.006 ppm Rh (Czamanske and others, 1981). Because the area is within an active glacier, moving ice makes drilling impossible in many areas so that the full extent of this resource remains unknown.

The sulfides occur mostly as disseminated interstitial grains and as trapped globules within euhedral olivines. The latter provide unequivocal evidence for an early magmatic origin for these sulfides. Furthermore, the presence of such inclusions throughout the entire complex suggests that the intrusion was saturated in sulfide throughout its entire evolution.

Syntectonic gabbroic intrusions also were emplaced in northern New England during the Devonian Acadian orogeny. Two of these bodies, the Moxie and Katahdin plutons, contain sulfide deposits that are hosted predominantly in olivine gabbros and some peridotite.

Thompson and Naldrett (1984) have modeled the process of sulfide formation in both of these intrusions, utilizing the partition coefficients between silicate liquid and sulfide melt to predict patterns of chalcophile elements. Since chalcophile elements greatly prefer the sulfide melt to the silicate liquid, their abundance in rocks from which an early immiscible sulfide liquid has separated should be depleted as compared to those in which sulfide separation did not occur. In practice, depletion is best detected by measuring the nickel content of olivine.

In the Moxie pluton, olivines are strongly depleted in nickel (160 to 700 ppm versus expected values of 300 to 2,000 ppm), and provide evidence of a relatively early separation of sulfide liquid. The sulfur that saturated the silicate liquid was derived both from magmatic and sedimentary sources, as shown by variations in $\delta^{34}S$ values from 0 per mil (typical of magmatic values) to values between −8 to −25 per mil (values similar to those of sulfides in the country rock). Additional confirmation that sulfur was added to this magma is provided by estimates—derived from the nickel contents of the olivines—of the proportion of silicate liquid to immiscible sulfide liquid. For the Moxie body, this ratio is about 100:1. The large amounts of sulfur indicated by this ratio could not have been carried in a silicate liquid, and therefore, some external sulfur must have been added.

The Katahdin pluton contains large amounts (more than 200 million tons) of nearly massive pyrrhotite. Grades are about 0.2 percent Ni, 0.1 percent Cu, and 0.17 percent Co. As in the Moxie pluton, low-sulfur isotopes (−11 to −22 per mil) indicate addition of large amounts of country rock sulfur. Olivines from this body are extremely depleted in nickel (70 to 200 ppm versus expected values of 700 to 1,000 ppm). These low nickel contents indicate a very low ratio of silicate liquid to immiscible sulfide melt, consistent with the early separation of a relatively large amount of sulfide liquid. The separation of large amounts of sulfide quickly depleted the silicate liquid in nickel and copper. Olivines crystallized from this depleted liquid were nickel-poor, and the large amounts of sulfides that continued to separate were iron-rich.

Thompson and Naldrett (1984) further showed that the nickel content of olivines could be used to estimate the grade of sulfide deposits. They calculated that the deposit in the Moxie pluton should contain between 0.5 and 2 percent nickel. In contrast, the sulfides associated with the more severely nickel depleted olivines from the Katahdin gabbro were estimated to contain less than 0.5 percent nickel.

The ability to generate models that reasonably predict observed compositional patterns in both mineralized and unmineralized rocks represents a major advance in understanding magmatic sulfide deposits. It also provides a useful exploration tool, making possible the detection of depletion patterns due to the separation of immiscible sulfides in rocks away from areas where the sulfides are actually concentrated.

Ophiolites

Ophiolites constitute the fourth major group of mafic and ultramafic rocks. For reasons that are not completely known, these fragments of oceanic crust seldom contain magmatic sulfide deposits. Instead, they are principally important as sources of chromite. Chromite generally occurs in small (less than 1 million tons) discontinuous bodies. These are termed podiform deposits to distinguish them from their more laterally continuous counterparts in layered intrusions.

Because most ophiolites occur as dismembered and altered fragments, features associated with individual chromite deposits are often obscured. A combination of information obtained from numerous deposits, however, reveals the basic features of podiform deposits. An ideal ophiolite consists of a base of tectonized harzburgite, which is generally thought to be depleted mantle from which basaltic melts have been removed. It is overlain by a succession of ultramafic cumulates, mafic cumulates, noncumulate gabbros, sheeted dikes, and pillow lavas. The contact between the ultramafic cumulates and the harzburgite represents the petrologic Moho and is also the zone along which podiform chromite deposits are concentrated (Stowe, 1987).

Unlike the chromites in stratiform deposits, these chromites have relatively constant Mg/Fe, but Cr/Al varies widely (Fig. 2). Formation of high-Al chromites is favored by increasing pressure, so that at least some of this variation may be depth related. In the Oman ophiolite, for example, chromites with the highest Al contents occur in the underlying harzburgite.

Most podiform deposits are enclosed within dunite. The dunite that occurs in the cumulate section may be interlayered with other ultramafic cumulates that show cyclic layering and size grading similar to that of stratiform complexes. More commonly, the chromite deposits are enclosed in a highly deformed dunite envelope that may be from a few centimeters to several meters thick and is surrounded by tectonized harzburgite.

Chromites in both settings are commonly highly deformed. Grains often show pull-apart textures, and deposits may contain a penetrative secondary foliation. Some deposits in both environments, however, show typical cumulate textures (Thayer, 1969), a feature that leaves little doubt of a primary magmatic origin.

The generally similar appearance of chromites occurring in the cumulate part of ophiolites and those in stratiform intrusions suggests that both formed by similar magmatic processes. The lack of lateral continuity of the deposits in ophiolites is most likely due to the tectonically active environment of deposition. On the other hand, the genesis of podiform deposits found in the underlying tectonized harzburgite is less easily explained. Although the cumulate textures in some deposits indicate a magmatic origin, the surrounding harzburgites are generally interpreted to be the residuum left by extensive partial melting. Explanations for the origin of these deposits include the following: (1) the harzburgites are not depleted mantle, but rather are deformed cumulates that contained primary chromite segregations (Thayer, oral communication, 1980); (2) the magmatic chromite deposits and enclosing dunite envelopes either sank (Dickey, 1975) or were folded (Greenbaum, 1977) into the underlying depleted tectonites; or (3) the chromites were deposited in subsequently deformed conduits through which melts that formed the overlying cumulate section and their chromite deposits had passed (Lago and others, 1982). This latter hypothesis appears to be mechanically the most sound.

Podiform deposits in the United States occur along the Pacific and Atlantic margins (Nos. 14, 15, 16, 17, 18, 19, 20; Plate 2b). In the west, the principal deposits occur in four belts. These are (1) the Klamath Mountain and (2) Sierra Nevada belts in which ophiolites form west-facing arcs that range in age from Devonian in the east to Middle Jurassic in the west, the (3) Upper Jurassic Coast Range, and (4) the Lower Permian ophiolites that are exposed in the Blue Mountains of Oregon. Of the few deposits in Alaska, the podiform deposit at Red Mountain is the most important. In the east, some podiform deposits occur in ophiolites that are Late Cambrian to Early Ordovician in age.

These deposits are almost all small. The more than 650 mines in the Klamath, Coast Range, Sierra Nevada, and Blue Mountains provinces produced a total of less than 650,000 tons of chromite and have a median size of only about 100 tons. In contrast, the median size of economic podiform deposits in Turkey, the Philippines, New Caledonia, and Cuba is about 25,000 tons. The largest single mine in the United States was the Wood mine in Pennsylvania, which yielded between 100,000 and 200,000 tons of ore. There has been no significant production from any domestic podiform deposit since 1978.

Unrepresented deposits

Two important types of deposits have not been found in the United States. The deposits of the Norilsk district of the U.S.S.R. represent one type that is somewhat similar to the deposits in the Duluth Complex. Both of these deposits occur in rift environment, and in both the assimilation of country-rock sulfur was important in precipitating sulfides. The Norilsk deposits differ, however, by occurring in shallow intrusions and in having large amounts of PGEs. The other type are deposits occurring in komatiites, a suite of rocks that was essentially unrecognized before 1969. Those komatiites that host nickel deposits have several characteristic features. Most important, they can usually be linked to ultramafic magmas that were either extruded as flows or emplaced as shallow intrusions. The magmas are MgO-rich (generally MgO >20 percent) and commonly display spinifex, a texture made by platy and skeletal olivine and clinopyroxene formed as a result of magma supersaturation during cooling. Important depos-

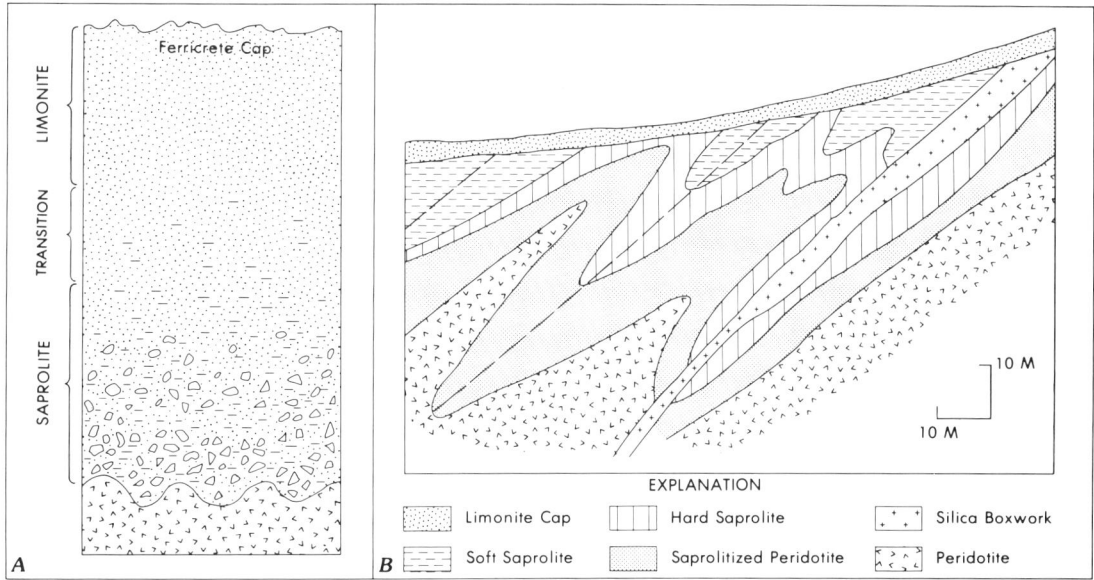

Figure 5. Laterites: (A) A typical laterite profile; (B) cross section of the Riddle nickel deposit (after Cumberlidge and Chace, 1968).

its of this type occur in Australia, Canada, and Zimbabwe. Although some komatiites have been found in the United States, none are the high-MgO types associated with nickel deposits elsewhere, and no significant nickel sulfide concentrations have been identified.

SECONDARY DEPOSITS

Weathering

Lateritic soils developed by the tropical weathering of ultramafic rocks are commonly enriched in nickel, cobalt, and chromium. The processes by which this enrichment occurs vary among deposits and are largely controlled by features such as the mineralogy of the bedrock, amount of rainfall, and nature of the drainage. In a general way, ground water moving through the soil dissolves and reprecipitates minerals to establish a vertical soil profile zoned from the least soluble minerals at the top to more soluble phases at the base. An idealized profile consists of four zones (Fig. 5):

1. Limonite. Predominantly a zone composed of goethite and amorphous ferric hydroxides with sparse gibbsite and manganese oxide. Some insoluble minerals like spinel, magnetite, and maghemite may be present. Ferricrete containing goethite and hematite veinlets, and pisolites may form a cap. Nickel occurs mainly in iron-hydroxide phases and in manganese oxide, which may also be rich in cobalt. Nickel content generally increases downward, but seldom exceeds 2 percent.

2. Transition. Limonite grades downward into saprolite. Under certain climatic and drainage conditions, this transition zone also contains concentrations of nontronite and boxworks of quartz.

3. Saprolite. A zone composed of isovolumetrically altered and partly decomposed bedrock in which primary textures are preserved. Decomposed olivine and pyroxene are typically concentrated along joints and fractures. A wide variety of secondary minerals are present in which nickel substitutes for Mg. These are commonly referred to as garnierite or garnierite group minerals, and include pimelite (nickel-montmorillonite), nickel-serpentines, schuchardite (nickel-chlorite), kerolite (nickel-talc), and nickel sepiolites. Nickel content commonly exceeds 2 percent.

4. Bedrock. Predominantly serpentinized peridotite or dunite.

Data on mineral stability (Golightly, 1981) reveal a general order of solubilities consistent with the observed weathering profile. At a pH of between 5 and 6, typical of near-surface waters, goethite is by far the least soluble mineral. Deeper in the laterite profile, pH increases to values of between 7 and 9 near the bedrock. Accompanying this increase are large decreases in the solubility of the secondary Mg-bearing minerals that form in the saprolite zone. Ni in ground water is much less soluble than magnesium, so that it replaces Mg in the secondary minerals. In order to make a high-grade deposit, ground water must transport Ni from the low pH, near-surface zone, to deeper and higher-pH zones where it substitutes for Mg. The necessary ground-water movement generally can occur only in well-drained areas. Thus, poorly drained or incompletely developed laterites commonly lack high-grade Ni ores.

The process used to recover metals from a laterite significantly affects the economics of a deposit. Two basic recovery processes are generally employed. Pyrometallurgical processing involves the smelting of the laterite to form either ferronickel or nickel matte; cobalt is not separated and recovered. In contrast, hydrometallurgical processing utilizes either ammonium carbonate (Caron process) or sulfuric acid leaching to remove nickel and possibly cobalt. The choice between these two processes is primarily dictated by the ore composition. The high Mg content of saprolitic ores generally makes them unsuitable for chemical leaching, so that they must be refined by the more energy-intensive pyrometallurgical process. On the other hand, poor slag formation results from melting of the Fe-rich limonitic ores so that they are best processed hydrometallurgically.

Early Tertiary tropical weathering and peneplanation formed laterites on some of the ophiolite belts of the northwestern United States (Nos. 14, 15, 16, 17, 18; Plate 2b). Of these, the most important is the deposit near Riddle, Oregon (No. 21; Plate 2b), which formed on a Jurassic ophiolite fragment (Cumberlidge and Chace, 1968). Some laterites also developed over ophiolites in the Appalachian belt, the largest being the deposit at Webster, North Carolina (No. 22, Plate 2B).

Initially, the Riddle deposit probably was a laterally continuous blanket with a profile similar to that shown in Figure 5A. Subsequent late Tertiary uplift, erosion, and gravity faulting broke up the deposit so that it now consists of two main ore bodies from which most of the original limonitic ore has been eroded. Further extensive faulting during uplift formed permeable zones along which extensive leaching and redeposition occurred. The resulting deposit (Fig. 5B) consists of a parent rock that varies from fresh to serpentinized peridotite in which is a nearly orthogonal set of fault-related veins and fractures having extensive silica boxworks. Saprolite makes up about 70 percent of the deposit, and is subdivided by MgO content into saprolitized peridotite (34 to 40 percent MgO), hard saprolite (24 to 34 percent MgO), and soft saprolite (<24 percent MgO). The saprolite is capped by a relatively thin limonite zone that probably formed during recent weathering. Until its closing in 1987, the silica boxwork zone and overlying saprolite were the principal materials mined at Riddle.

The Riddle deposit was first discovered in 1864, but it was not until 1954 that significant production of nickel began. In large part, production was made possible by government price and purchase guarantees. Because of the high Mg content, the ores were treated by pyrometallurgy, and ferronickel containing about 45 percent nickel and 55 percent iron was produced. No cobalt was recovered.

Many other nickel laterites occur in the northwest United States (Plate 2B), mostly in areas that are deeply dissected. These deposits, therefore, tend not to be laterally extensive. Typical profiles are similar to that shown in Figure 5A, having a cap of reddish soil and iron-oxide pellets, a limonitic zone of yellow-orange soil without rock fragments, a yellow-brown saprolite zone that grades down into weathered peridotite, and then into fresh peridotite. The lower saprolitic zone, however, is commonly thin or absent; few of these deposits contain signficant enrichments of Ni in the saprolite zone such as is found at the Riddle deposit.

The lack of Mg-rich laterites in these deposits has two important economic implications. First, the grades of Ni are generally lower than at Riddle, and second, the ores are amenable to less energy-intensive hydrometallurgical refining. Recent metallurgical research has focused on ways in which such hydrometallurgical processes could efficiently recover both nickel and cobalt from these deposits, as well as produce by-product chromite.

Chemical weathering also has formed the over 100 Cobalt-bearing manganese deposits that lie in a belt extending from Alabama to Virginia (No. 23; Plate 2B; Pierce, 1944). These occurrences (also discussed by Sidder, this volume) are composed dominantly of manganese oxides that are either supergene fracture fillings or replacements, botryoidal incrustations, or residual nodules and pellets. Cobalt contents typically are between 0.3 and 1.5 percent. These deposits are all small, and although carefully examined during World War II, have produced only a few tons of ore.

Placer deposits

Physical erosion of ultramafic rocks and their weathering products may result in placer concentration of some residual minerals. The phases most susceptible to secondary concentration are chromite and PGEs.

Some small chromite deposits in stream sands were once mined in North Carolina and Maryland, but the largest concentration of placer chromite occurs in beach sands along a 200-km strip of coastline near the California-Oregon border (No. 24; Plate 2B). These deposits occur as lenses and layers that range from very small to bodies more than 1,000 m long, 300 m wide, and 3 m thick. They are generally of low grade, only about 5 percent Cr_2O_3, and separation of chromite from other heavy minerals is difficult. After the Stillwater Complex, these deposits are the United States' second largest resource of chromium.

Minor amounts of PGEs have been recovered from many placer deposits, chiefly as a byproduct of gold mining. The Goodnews Bay deposit (No. 25; Plate 2B), in contrast, has been mined principally for PGEs. The PGEs occur chiefly in two alloys, the most abundant alloy consists predominantly of Pt, Fe, and Ir; the other is composed mostly of Ir and Os. In addition, small amounts of cooperite, laurite, sperrylite, and mertieite are present. The weighted mean average of metals obtained from this deposit are Pt (82.5%), Ir (11.32%), Os (2.15%), Ru (.17%), Rh (1.3%), Pd (.38%), and Au (2.43%) (Mertie, 1976). The predominance of Pt-Fe alloys in this deposit is characteristic of deposits derived from zoned ultramafic complexes and contrasts with the predominantly Os-Ir-Ru alloys in placers derived from ophiolites.

The Goodnews Bay placers were derived from the weathering of the adjacent Red Mountain zoned ultramafic complex. This body, composed mostly of serpentinized dunite, is less than 2 km from some of the claims. The PGE-bearing alloys and minerals that were eroded from this complex were subsequently concentrated in stream gravels. Grain sizes vary from less than .06 mm to greater than 3.96 mm, with over 80 percent falling between 0.2 and 1.0 mm. Many show signs of abrasion due to transport.

STRATABOUND DEPOSITS

Most of the deposits discussed so far are important primarily for Ni, Cr, and PGEs, and have also been clearly associated with ultramafic rocks, either as primary deposits or as secondary derivatives. The Blackbird deposit near Salmon, Idaho (No. 26; Plate 2B), is strikingly different because it is principally a Co deposit and because its relation to ultramafic/mafic rocks is much less clear. This deposit consists of copper and cobalt which occur in the Early and Middle Proterozoic (approximately 1.7 Ga) Yellowjacket Formation. This formation consists of a sequence of thinly laminated micaceous quartzites, garnet schists, phyllite, argillite, and quartz-biotite schist. It also contains abundant interlayered chloritic schists that are interpreted to have been volcanic tuff layers. Although these rocks have undergone local intense folding, faulting, and metamorphism, primary sedimentary features are still preserved.

Copper and cobalt occur mostly as chalcopyrite and cobaltite (CoAsS). Other minerals are pyrrhotite, pyrite, cobaltian pyrite, safflorite, and some gold (Bennett, 1977). Gangue is principally quartz and biotite. Tourmaline is present locally.

The sulfide minerals occur in large quartz veins that locally crosscut layering, and as disseminated grains in layers that are conformable with the enclosing rock. Some of the sulfides are structurally controlled, as shown by veins in axial zones of folds and along fractures. Because of these features, many of the early studies concluded that the deposit was formed by high-temperature hydrothermal fluids. However, more recent study shows that, despite local discordances, the mineralization is stratabound and that several other similar deposits occur along strike in the same stratigraphic sequence. Subsequent deformation and metamorphism must therefore have locally remobilized some of the minerals to form the vein mineralization.

The recognition of the importance of stratigraphic controls prompted careful restudy of the deposit. It is now considered to have formed in a rift zone (Hughes, 1983). Syndepositional faults are thought to have disrupted the unconsolidated sediments, and provided conduits along which heated solutions associated with the volcanic activity moved. Cobalt and copper were deposited from these solutions either at or near the sea floor; exhalites rich in tourmaline and chert also were precipitated. The origin of the cobalt and copper has not been clearly established. The finely disseminated minerals, which show no evidence of having been disturbed by later deformation or metamorphism, show a close spatial association with the altered mafic volcanics (Nash and Hahn, 1986). The metals, therefore, appear to have been derived from the volcanic rocks.

The Blackbird deposit was found in 1893. Although some mining occurred at that time, the main period of mining was from 1949 to 1959 when as much as 1,000 tons of cobalt were produced annually. The loss of government contracts in 1960 caused operations to be discontinued. In 1977 and 1978, military disruptions temporarily shut down the cobalt mines of Zaire and thus disrupted the United States' principal cobalt source. The resulting increases in Co price and national concern over establishing a domestic cobalt source led to extensive work on the Blackbird deposit, but the failure to obtain government price support for domestically produced cobalt has caused most work on the deposit to cease. Although currently inactive, the deposit remains the nation's major cobalt resource.

The Cu-Co deposits of central Africa, which are the world's principal source of cobalt, are also stratabound mineralization in Proterozoic sediments. However, they differ significantly from the Blackbird deposit. Particularly, the Blackbird has large amounts of As, which does not occur in the African deposits, and an exhalative origin is indicated for part of the Blackbird, whereas replacement texturese in the African deposits show them to be early diagenetic.

Perhaps closer analogs to the Blackbird deposit are the several small deposits in the Sykesville district, Maryland (No. 27; Plate 2B). These deposits are in schists, mafic volcanics, and serpentine, which probably represent deep-water sediments and mixed ophiolitic debris. Chalcopyrite and lesser amounts of carrollite ($Cu(Co,Ni)_2S_4$) are the principal sulfide minerals present. They occur in both conformable layers and discordant veins. The conformable layers are closely associated with ultramafic rocks (talc and chlorite schists) and may occur within quartzites that contain thin layers of magnetite. In contrast, veins tend to be in fractures and generally appear related to metamorphism and deformation. The small amounts of ore produced typically contained 10 to 20 percent Cu and 0.1 to 0.2 percent Co. Locally, the Co content was as much as 5 percent.

As at Blackbird, the Sykesville deposits were initially considered to be of hydrothermal origin. Also, as at Blackbird, the stratigraphic concordance of some of the sulfide minerals and their close association with layers of ultramafic rocks and magnetite-bearing quartzites prompted a reinterpretation of the deposit as syngenetic and stratabound. The ultramafic rocks are now considered to be part of an olistostrome, and the banded magnetite-quartz rock is interpreted to be an exhalite, formed in part from hydrothermal alteration of the ultramafic rocks (Candela and Wylie, 1989). Hot-spring activity at the ocean floor is thought to have leached Cu and Co from the ultramafic rocks, which were subsequently reprecipitated as sulfides. The deposit is thus similar to the Blackbird in its association with mafic/ultramafic rocks, and association with exhalites. It differs notably in having no arsenic.

HYDROTHERMAL AND RELATED DEPOSITS

Hydrothermal and related deposits occasionally contain some Ni, Co, and PGEs, but the potential recovery of metals from most of these deposits is small. Some deposits of this type are shown on Plate 2B, and include the Co-bearing silver-zinc veins of the Coeur d'Alene district, Idaho (28); PGEs in sheared ultramafics at the New Rambler Mine, Wyoming (29); PGEs in limestone replacement bodies and associated Ni- and Co-bearing supergene deposits of the Goodsprings district, Nevada (30); Ag-

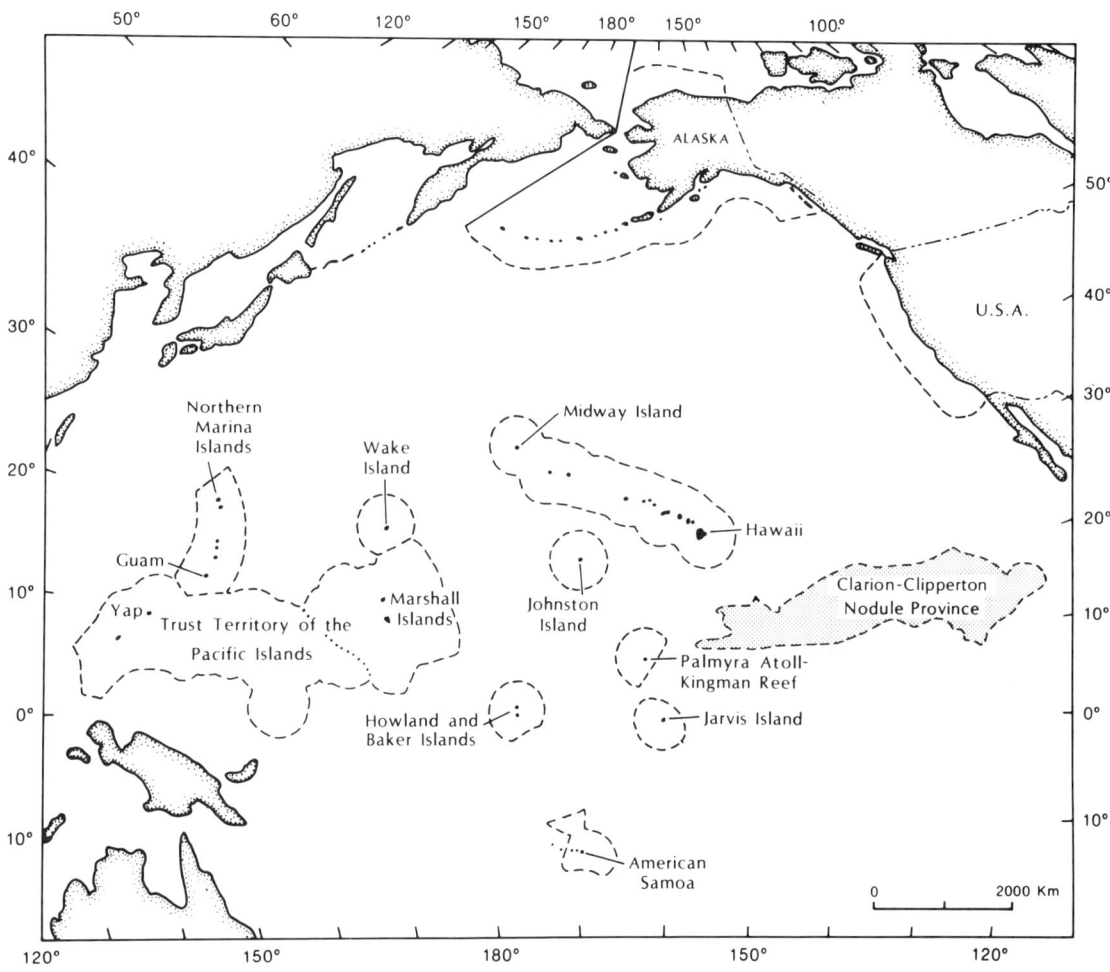

Figure 6. Location of some marine cobalt deposits. The Clarion-Clipperton manganese nodule area is shaded. Dashed lines mark parts of the United States' Exclusive Economic Zone that have cobalt crusts (modified from Hein and others, 1987).

Ni-Co-U–bearing veins of the Blackhawk district, NM (31); Co in altered mafic and ultramafic rocks at the Mackinaw mine, Washington (32) and the Gold Hill deposit, Colorado (33); and Co in the skarn deposits near Cornwall, Pennsylvania (34). The Cornwall deposits are iron ores formed in Paleozoic limestones near contacts with Jurassic diabase (see also Sidder, this volume). Cobalt was recovered from pyrite, and at the time production ceased in 1971, these deposits were the United States' principal source of domestic cobalt. Some porphyry copper deposits contain small amounts of PGEs, and by-product recovery of PGEs has occurred at Bingham, Utah (35) (described by Tooker, this volume) and Ely, Nevada (36).

Potentially important concentrations of cobalt and nickel occur in some of the Mississippi Valley–type lead-zinc deposits of southeast Missouri (No. 37; Plate 2B). The genesis of the Mississippi Valley type of deposit is discussed by Ohle (this volume). The deposits in southeast Missouri contain on the order of 6 percent Pb, 1 percent Zn, 0.01 to 0.04 percent Co, and 0.01 to 0.06 percent Ni. Deposits near Fredericktown, Missouri, however, are much richer in cobalt (0.1 to 0.6 percent) and nickel (0.2 to 0.9 percent). The nickel and cobalt are mostly in siegenite ($(Ni,Co)_3S_4$). The reason for these locally higher concentrations of Ni and Co is not clear, but they may be due to compositional differences in source areas from which the metal-transporting fluids were derived.

Because the cobalt and nickel are harmful to the electrowinning recovery of zinc, the Co- and Ni-rich ores have generally not been mined or have been discarded in the tailings. Efforts to reprocess these tailings and to open some of the old mines were undertaken in the early 1980s when Co prices were high. After the Blackbird deposit in Idaho, the deposits in southeast Missouri are the nation's principal domestic resource of Co.

DEEP-OCEAN DEPOSITS

Although they are outside the geographic confines of North America, two types of deposits occurring in the ocean may prove to be important sources of nickel and cobalt. Deep-sea nodules rich in cobalt and manganese have been known since the HMS Challenger expedition (1872 to 1876). These nodules, 1 to 3 cm in diameter, cover large areas of the abyssal plane (4 to 6 km of water) in areas with low sedimentation rates. Nodules are precipitated from seawater and grow at rates of only several mm per million years. The most economically attractive nodules occur in the east Pacific Region between the Clarion and Clipperton fracture zones (Fig. 6). Nodules in this area could average about 25 percent Mn, 1.3 percent Ni, 1.0 percent Cu, and 0.22 percent Co. Efforts to recover these metals have been hampered both by the technological problems of mining in deep water and by the lack of international legal agreements.

A second type of Co-rich deposit is represented by manganese crusts on the flanks of seamounts, ridges, and plateaus. These crusts are 2 to 4 cm thick, and in contrast to nodules, occur in shallower water (generally 1,000 to 1,500 m), and have 3 to 6 times as much Co. Typical crusts contain 20 to 30 percent Mn, 0.6 to 1.1 percent Co, 0.3 to 0.7 percent Ni, and 0.05 to 0.15 percent Cu. As much as 1.3 ppm Pt has also been reported (Hein and others, 1987). The location of these crusts is one of their most economically attractive features. Not only are the crusts in much shallower water than the nodules, many occur within the United States' Exclusive Economic Zone. Access to these deposits would, therefore, not be complicated by the legal uncertainties that have hampered development of deep-sea nodules.

SUMMARY AND CONCLUSIONS

Deposits of Ni, Co, Cr, and PGEs are predominantly associated with mafic and ultramafic rocks. The largest U.S. deposits are those in the Stillwater and Duluth Complexes. Other magmatic deposits occur in the La Perouse Complex, several gabbroic bodies in Maine, and in some ophiolites. These elements also occur in secondary deposits derived from mafic and ultramafic rocks. Most notable are Ni-Co laterites and PGE and chromite placer deposits. The stratabound Cu-Co deposits at Blackbird, Idaho, and Sykesville, Maryland, have a less clearly established genetic link with mafic/ultramafic rocks, but are spatially associated with either mafic volcanics or altered ultramafic rocks. Finally, the Co- and Ni-bearing hydrothermal deposits in southeast Missouri and the supergene Co deposits in the southeast United States appear to have formed without any direct involvement of mafic or ultramafic rocks.

In reviewing the deposits of Ni, Co, Cr, and PGEs in the United States, two important points are evident. First, at least two important deposit types are not known in the United States. These are deposits in komatiites and those of the Norilsk type in the U.S.S.R. Undiscovered deposits of these types may exist, and these known deposits should provide useful models for their exploration. Second, understanding and locating ore deposits is an evolutionary process. The Stillwater Complex provides a good example of such an evolution. The earliest phase of exploration in that complex was concerned with the reasonably well-exposed and easily found nickel sulfides of the Basal series. As knowledge increased, the chromite deposits of the Ultramafic series were located and developed. However, only in recent years has exploration located the important disseminated PGEs in Banded series rocks. The increased ability to discover deposits as a result of new knowledge indicates that continued work should generate new prospects for domestic deposits of Ni, Co, Cr, and PGEs.

REFERENCES CITED

Barnes, S. J., and Naldrett, A. J., 1985, Geochemistry of the J-M (Howland) Reef of the Stillwater Complex, Minneapolis Adit area; 1. Sulfide chemistry and sulfide-olivine equilibrium: Economic Geology, v. 80, p. 627–645.
—— , 1986, Geochemistry of the J-M Reef of the Stillwater Complex, Minneapolis Adit area; 2. Silicate mineral chemistry and petrogenesis: Journal of Petrology, v. 27, p. 791–825.
Bennett, E., 1977, Reconnaissance geology and geochemistry of the Blackbird Mountain-Panther Creek region, Lemhi County, Idaho: Idaho Bureau of Mines and Geology Pamphlet 167, 108 p.
Bonnichsen, B., Fukui, L. M., and Chang, L.L.Y., 1980, Geologic setting, mineralogy, and geochemistry of magmatic sulfides, South Kawishiwi intrusion, Duluth Complex, Minnesota, *in* Ridge, J. D., Proceedings 5th Quadrennial IAGOD Symposium, v. 1: Stuttgart, West Germany, E. Schweizerbart'sche Verlagsbuchhandlung, p. 545–565.
Boudreau, A. E., 1988, Investigations of the Stillwater Complex; 4. The role of volatiles in the petrogenesis of the J-M Reef, Minneapolis Adit section: Canadian Mineralogist, v. 26, p. 193–208.
Boudreau. A. E., and McCallum, I. S., 1985, Features of the Picket Pin Pt-Pd deposit, *in* Czomanske, G. K., and Zientek, M. L., eds., The Stillwater Complex, Montana—Geology and guide: Montana Bureau of Mines and Geology Special Publication 92, p. 346–357.
Boudreau, A. E., Mathez, E. A., and McCallum, I. S., 1986, Halogen geochemistry of the Stillwater and Bushveld Complexes—Evidence for transport of platinum-group elements by Cl-rich fluids: Journal of Petrology, v. 27, p. 967–986.
Bow, C., and 6 others, 1982, Investigations of the Howland Reef of the Stillwater Complex, Minneapolis adit area—Stratigraphy, structure, and mineralization: Economic Geology, v. 77, p. 1481–1492.
Cameron, E. N., 1980, Evolution of the lower Critical Zone, central sector, eastern Bushvelt Complex, and its chromite deposits: Economic Geology, v. 75, p. 845–871.
Campbell, I. H., Naldrett, A. J., and Barnes, S. J., 1983, A model for the origin of the platinum-rich sulfide horizons in the Bushveld and Stillwater Complexes: Journal of Petrology, v. 24, p. 133–165.
Candela, P. A., and Wylie, A. G., 1989, The ultramafite-associated Fe-Cu-Co-Ni-Zn deposits of the Sykesville district Maryland piedmont: Washington, D.C., American Geophysical Union, 18th International Geological Congress Field Trip Guidebook T241, 10 p.
Cumberlidge, J. T., and Chace, F. M., 1968, Geology of the Nickel Mountain mine, Riddle, Oregon, *in* Ore Deposits in the United States, 1933–1967, Graton-Sales Volume: New York, American Institute of Mining, Metallurgical, and Petroleum Engineers, v. 2, p. 1650–1672.
Czamanske, G. K., Haffty, J., and Nabbs, S. W., 1981, Pt, Pd, and Rh analyses and beneficiation of mineralized mafic rocks from the La Perouse layered gabbro, Alaska: Economic Geology, v. 76, p. 2001–2011.
Dickey, J. S., Jr., 1975, A hypothesis of origin for podiform chromite deposits: Geochimica et Cosmochimica Acta, v. 39, p. 1061–1074.
Duke, J. M., 1983, Ore deposit models 7. Magmatic segregation deposits of chromite: Geoscience Canada, v. 10, p. 15–24.
Foose, M. P., and Weiblen, P. W., 1986, The physical and petrologic setting and textural and compositional characteristics of sulfides from the South Kawishiwi Intrusion, Duluth Complex, Minnesota, U.S.A., *in* Friedrich, G. H.,

and others, eds., Geology and metallogeny of copper deposits: Society for Geology Applied to Mineral Deposits Special Publication 4, Heidelberg, Springer-Verlag, p. 7–24.

Golightly, J. P., 1981, Nickeliferous laterite deposits: Economic Geology 75th Anniversary Volume, p. 710–735.

Greenbaum, D., 1977, The chromitiferous rocks of the Troodos ophiolite complex, Cyprus: Economic Geology, v. 72, p. 1175–1194.

Hein, J. R., Morgenson, L. A., Clague, D. A., and Koski, R. A., 1987, Cobalt-rich ferromanganese crusts from the exclusive economic zone of the United States and nodules from the oceanic Pacific, *in* Scholl, D. W., Grantz, A., and Vedder, J. G., eds., Geology and resource potential of the continental margin of western North America and adjacent ocean basins—Beaufort Sea to Baja California: Circum-Pacific Council for Energy and Mineral Resources Earth Science Series, v. 6, p. 753–771.

Himmelberg, G. R., and Loney, R. A., 1981, Petrology of the ultramafic and gabbroic rocks of the Brady Glacier nickel-copper deposit, Fairweather Range, southeastern Alaska: U.S. Geological Survey Professional Paper 1195, 26 p.

Howland, A. L., Peoples, J. W., and Sampson, E., 1936, The Stillwater igneous complex and associated occurrence of nickel and planinum metals: Montana Bureau of Mines and Geology Miscellaneous Contribution 7, 15 p.

Hughes, G. J., 1983, Basinal setting of the Idaho cobalt belt Blackbird Mining district, Lemhi County, Idaho—an interim report: U.S. Geological Survey Open-File Report 86-430, 29 p.

Irvine, T. N., 1967, Chromian spinel as a petrogenetic indicator—part 2. Petrologic applications: Canadian Journal of Earth Sciences, v. 4, p. 71–103.

—— , 1977, Origin of chromite layers in the Muskox intrusion and other stratiform intrusions—a new interpretation: Geology, v. 5, p. 273–277.

Irvine, T. N., Keith, D. W., and Todd, S. G., 1983, The J-M platinum-palladium reef of the Stillwater Complex, Montana; 2, Origin by double-diffusive convective magma mixing and implications for the Bushveld Complex: Economic Geology, v. 78, p. 1287–1334.

Jackson, E. D., 1961, Primary textures and mineral associations in the Ultramafic zone of the Stillwater Complex, Montana: U.S. Geological Survey Professional Paper 358, 106 p.

—— , 1968, The chromite deposits of the Stillwater Complex, Montana, *in* Ore deposits in the United States, 1933–1967, Graton-Sales Volume: New York, American Institute of Mining, Metallurgical, and Petroleum Engineers, v. 2, p. 1495–1510.

Lago, B. L., Rabinowicz, M., and Nicholas, A., 1982, Podiform chromite ore bodies—a genetic model: Journal of Petrology, v. 23, p. 103–125.

Lambert, D. D., and Simmons, E. C., 1988, Magma evolution in the Stillwater Complex, Montana; 2. Rare earth element evidence for the formation of the J-M Reef: Economic Geology, v. 83, no. 6, p. 1109–1126.

LeRoy, L. W., 1985, Troctolite-anorthosite zone 1 and the J-M Reef—Frog pond adit to the Graham Creek area, *in* Czamanske, G. K., and Zientek, M. L., eds., The Stillwater Complex, Montana—Geology and Guide: Montana Bureau of Mines and Geology Special Publication 92, p. 325–333.

Loferski, P. J., 1980, Petrology of chromite-bearing metamorphosed ultramafic rocks from the Beartooth Mountains, Montana [M.S. thesis]: Blacksburg, Virginia Polytechnic Institute and State University, 136 p.

Mann, E. L., and Lin, Chong-Pin, 1985, Geology of the West Fork adit, *in* Czamanske, G. K., and Zientek, M. L., eds., The Stillwater Complex, Montana—Geology and guide: Montana Bureau of Mines and Geology Special Publication 92, p. 210–230.

Mainwaring, P. R., and Naldrett, A. J., 1977, Country-rock assimilation and the genesis of Cu-Ni sulfides in the Water Hen intrusion, Duluth Complex, Minnesota: Economic Geology, v. 72, p. 1269–1284.

McBirney, A. R., 1985, Further considerations of double-diffusive stratification and layering in the Skaergaard Intrusion: Journal of Petrology, v. 26, p. 993–1001.

McBirney, A. R., and Noyes, R. M., 1979, Crystallization and layering in the Skaergaard intrusion: Journal of Petrology, v. 20, p. 487–554.

Mertie, J. B., 1976, Platinum deposits of the Goodnews Bay district, Alaska: U.S. Geological Survey Professional Paper 938, 42 p.

Miller, J. D., Jr., 1986, The geology and petrology of anorthositic rocks in the Duluth Complex, Snowbank Lake Quadrangle, northeastern Minnesota [Ph.D. thesis]: Minneapolis, University of Minnesota, 525 p.

Murck, B. W., 1985, Factors influencing the formation of chromite seams [Ph.D. thesis]: University of Toronto, 137 p.

Naldrett, A. J., 1981, Nickel sulfide deposits—classification, composition, and genesis: Economic Geology 75th Anniversary Volume, p. 628–685.

Nash, J. T., and Hahn, G. A., 1986, Volcanogenic character of sediment-hosted Co-Cu deposits in the Blackbird mining district, Lemhi County, Idaho: Geological Association of Canada, Mineralogical Association of Canada, Canadian Geophysical Union Annual Meeting Program with Abstracts, v. 11, p. 107.

Page, N. J, 1977, Stillwater Complex, Montana—Rock Succession, Metamorphism, and Structure of the Complex and Adjacent Rocks: U.S. Geological Survey Professional Paper 999, 79 p.

—— , 1979, Stillwater Complex, Montana—Structure, mineralogy, and petrology of the Basal zone with emphasis on the occurrence of sulfides: U.S. Geological Survey Professional Paper 1038, 69 p.

Page, N. J, Zientek, M. L., Czamanske, G. K., and Foose, M. P., 1985, Sulfide mineralization in the Stillwater Complex and underlying rocks, *in* Czamanske, G. K., and Zientek, M. L., eds., The Stillwater Complex, Montana—Geology and guide: Montana Bureau of Mines and Geology Special Publication 92, p. 93–96.

Pierce, W. G., 1944, Cobalt-bearing manganese deposits of Alabama, Georgia, and Tennessee, U.S. Geological Survey Bulletin 940-J, p. 265–285.

Rao, B. V., and Ripley, E. M., 1983, Petrochemical studies of the Dunka Road Cu-Ni deposit, Duluth Complex, Minnesota: Economic Geology, v. 78, p. 1222–1238.

Ripley, E. M., 1981, Sulfur isotopic studies of the Dunka Road Cu-Ni deposit, Duluth Complex, Minnesota: Economic Geology, v. 76, p. 610–620.

Ripley, E. M., and Al-Jassar, T. J., 1987, Sulfur and oxygen isotope studies of melt-country rock interaction, Babbitt Cu-Ni deposit, Duluth Complex, Minnesota: Economic Geology, v. 82, p. 87–107.

Stowe, C. W., ed., 1987, Evolution of chromium ore fields: New York, Van Nostrand Reinhold, 340 p.

Thayer, T. P., 1969, Gravity differentiation and magmatic re-emplacement of podiform chromite deposits: Economic Geology Monograph 4, p. 132–146.

Thompson, J.F.H., and Naldrett, A. J., 1984, Sulfide silicate reactions as a guide to Ni-Cu-Co mineralization in central Maine, U.S.A., *in* Buchanan, D. L., and Jones, M. J., eds., Sulfide deposits in mafic and ultramafic rocks: Institute of Mining and Metallurgy Special Publication, p. 103–113.

Todd, S. G., Keith, D. W., LeRoy, L. W., Schissel, D. J., Mann, E. L., and Irvine, T. N., 1982, The J-M platinum-palladium Reef of the Stillwater Complex, Montana; 1. Stratigraphy and petrology: Economic Geology, v. 77, p. 1454–1480.

Turner, A. R., Wolfgram, D., and Barnes, S. J., 1985, Geology of the Stillwater County sector of the J-M Reef, including the Minneapolis adit, *in* Czamanske, G. K., and Zientek, M. L., eds., The Stillwater Complex, Montana—Geology and Guide: Montana Bureau of Mines and Geology Special Publication 92, p. 210–230.

Tyson, R. M., and Chang, L.L.Y., 1984, The petrology and sulfide mineralization of the Partridge River troctolite, Duluth Complex, Minnesota: Canadian Mineralogist, v. 22, p. 23–38.

Ulmer, G. C., 1969, Experimental investigations of chromite spinels: Economic Geology Monograph 4, p. 114–131.

Weiblen, P. W., and Morey, G. B., 1980, A summary of the stratigraphy, petrology, and structure of the Duluth Complex: American Journal of Science, v. 280–A, p. 88–133.

Zientek, M. L., 1983, Petrogenesis of the basal zone of the Stillwater Complex, Montana [Ph.D. thesis]: Stanford, California, Stanford University, 246 p.

MANUSCRIPT ACCEPTED BY THE SOCIETY DECEMBER 9, 1988

Printed in U.S.A.

Chapter 7

Uranium and vanadium deposits

D. R. Shawe and J. T. Nash
U.S. Geological Survey, MS 905 and 912, Box 25046, Denver Federal Center, Denver, Colorado 80225
W. L. Chenoweth
707 Brassie Drive, Grand Junction, Colorado 81506

INTRODUCTION

Uranium and vanadium deposits have been mined in the United States since the latter part of the 19th century. Early mining of uranium deposits provided uranium used mainly for coloring glass and ceramic glazes, and associated radium needed in the field of medicine. Development of the atomic bomb by the United States during World War II, and burgeoning energy demands following the war, led to greatly increased production and use of uranium. The high rates of domestic production and use in the late 1950s and early 1960s were due to the need for uranium in nuclear weapons. Production in the United States again increased in the late 1970s owing to the increased development of nuclear power, but it declined sharply after 1980 because of environmental concerns over further use of nuclear power and the availability of cheaper foreign sources (Fig. 1).

Vanadium also had early use for coloring glass and glazes. Today, the principal use of vanadium is for toughening and strengthening various steel alloys. Domestic use of vanadium has grown steadily, with surges in war years, to recent time, when economic reversals and the advent of cheaper foreign sources have substantially reduced domestic production of steel. Minor uses of vanadium include alloying with titanium, and as a catalyst.

Discussion of the geology of deposits of uranium and/or vanadium in the United States is presented here in the context of the historical discovery and development of the deposits. We first describe deposits, mainly of vein and sandstone types, that produced uranium and vanadium initially and through World War II. For two decades following the war, uranium and vanadium production has come from sedimentary rocks, metamorphic rocks, and surficial deposits. In recent years, with the decline in production from uranium-vanadium deposits in sandstone, significant vanadium production has come from phosphorite, residues of imported petroleum, and imported vanadiferous slags. Uranium and vanadium production data for significant districts and areas in the United States are given in Table 1. We describe districts selected to demonstrate the principal geologic variations among uranium-vanadium deposits, or that have particular historical significance. Several important districts are not described in the text because of space limitations. Locations of deposits discussed in the text or listed in Table 1 are shown on Plate 2C. Finally, this chapter considers possible future resources of uranium and vanadium in the United States.

EARLY URANIUM AND VANADIUM PRODUCTION, THROUGH WORLD WAR II

Central City District, Colorado (uranium)

The Central City District was the earliest producer of uranium ore in the United States. Pitchblende was discovered in 1871, and most production occurred from 1872 up to World War I (Sims and others, 1955). Renewed interest in uranium following WWII led to recognition of other uranium-bearing veins, but little production resulted.

Host rocks of the pitchblende veins are interlayered Precambrian microcline and biotite gneisses and pegmatite (Sims and others, 1963a). The Precambrian rocks are intruded by abundant dikes and irregular plutons of early Tertiary (Laramide) porphyritic igneous rocks, including uranium-rich bostonitic rocks (Sims and others, 1963b; Sims and Sheridan, 1964). Faults that formed during Laramide deformation are the sites of vein fillings. Rocks that border the veins were altered to quartz-sericite with locally abundant disseminated pyrite grading outward to argillized rocks.

The productive uranium-bearing veins in the Central City District are small; one of the larger veins is about 100 m long, a maximum of 3 m wide, and it was mined to a depth of about 100 m. Another vein was mined for a length of about 330 m, but to lesser depth. Pitchblende-bearing uranium-rich (grade range 0.3 to 12 percent U_3O_8) veins in the Central City District are the earliest of a sequence that included later deposition of ores of lead, zinc, copper, silver, and gold. The pitchblende veins exhibit no relationship to district-wide zoning of the gold-silver veins. Uranium-lead isotopic data on pitchblende indicate that the veins

---

Shawe, D. R., Nash, J. T., and Chenoweth, W. L., 1991, Uranium and vanadium deposits, *in* Gluskoter, H. J., Rice, D. D., and Taylor, R. B., eds., Economic Geology, U.S.: Boulder, Colorado, Geological Society of America, The Geology of North America, v. P-2.

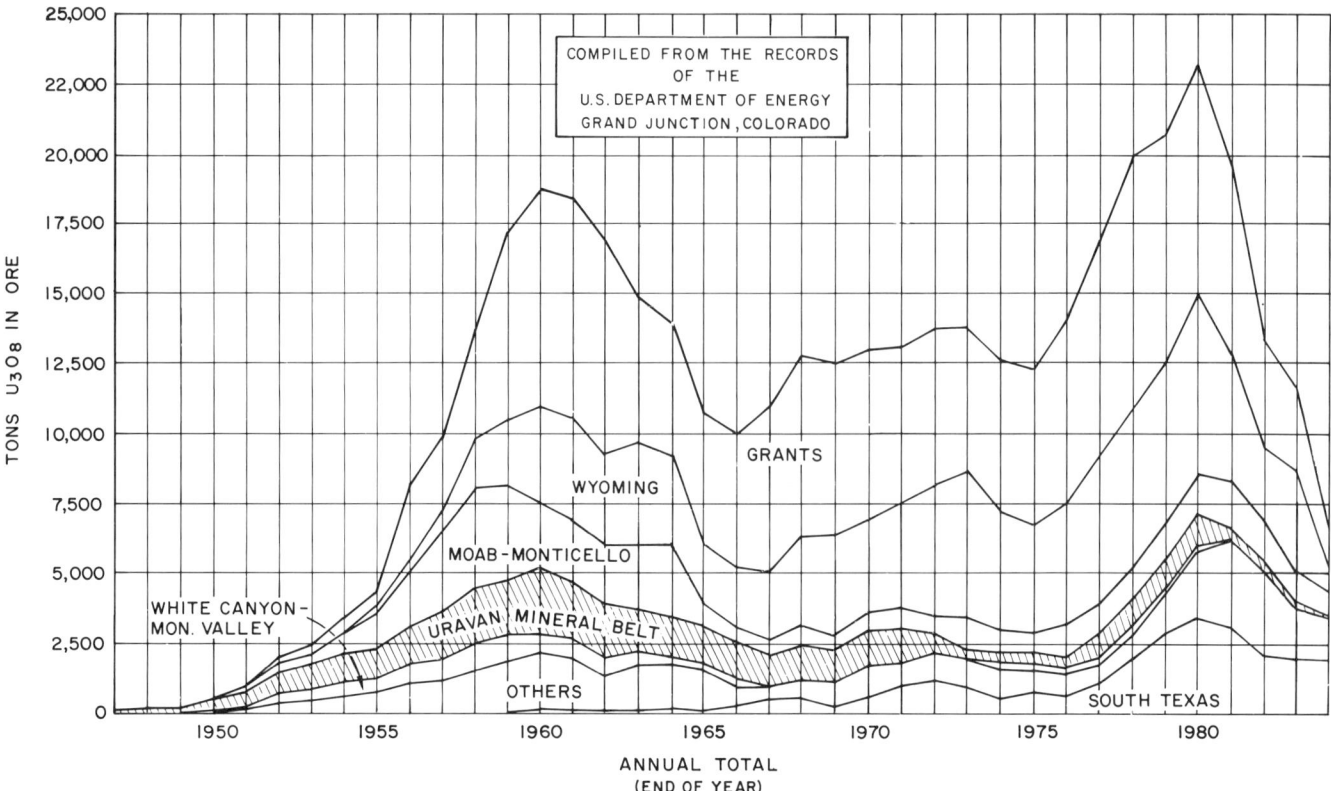

Figure 1. Uranium ore production by area, 1947 through 1983. "Others" indicates all other uranium deposits not included in the indicated groups.

were deposited about 60 Ma. Their proximity to the uranium-rich Laramide bostonitic porphyries indicates a genetic association; possibly the uranium was derived directly from the bostonite magmas.

Baringer Hill, Texas (uranium)

A rare-earth– and uranium-bearing pegmatite at Baringer Hill was discovered in 1887 (Hess, 1908). It produced a small amount of rare-earth minerals, and minor uranium minerals from which radium was recovered. The pegmatite was emplaced in coarse, porphyritic granite of Precambrian age as an asymmetrical body or pipe (Lindgren, 1933). The pegmatite was characterized by large crystals of smoky quartz, microcline and albite intergrowths, fluorite, biotite, and ilmenite. Other minerals, constituting a small fraction of 1 percent of the mass, include silicates, niobates, titanates, and uranates of cerium, yttrium, and other rare metals, as well as chalcopyrite, pyrite, sphalerite, molybdenite, and powellite.

Uravan mineral belt, Colorado and Utah (uranium and vanadium)

Mining of the uranium-vanadium mineral carnotite commenced on the Colorado Plateau where the existence of this yellow mineral in the area was known before 1880 (Coffin, 1921). In 1898, 10 t of ore containing more than 20 percent U_3O_8 and 15 percent V_2O_5 was shipped from the Rock Creek area near Paradox Valley (Kimball, 1904), but not until 1910 did substantial mining of carnotite deposits begin in southwestern Colorado and adjacent parts of Utah, in the region that became known as the Uravan mineral belt.

During the first decade (early 1900s), mining in the Uravan mineral belt was mainly for uranium for use as a coloring agent in glass and glazes. Thereafter, until about 1923, the ores were worked primarily for their radium content, and vanadium was recovered as a by-product. Following a period of relative inactivity the carnotite ores were mined for their vanadium content, from the early 1930s to about 1944. During the later years of WWII, uranium was recovered mainly from processing of tailings at vanadium mills. The Atomic Energy Commission (AEC) established a market for uranium in 1947, and thereafter, mining again focused on the recovery of uranium, with vanadium as a by-product.

The carnotite ores of the Uravan mineral belt were studied in the 1930s, primarily by the U.S. Geological Survey and U.S. Bureau of Mines, because of their value as the principal domestic source of vanadium. These early vanadium investigations provided the background of geological data on the carnotite deposits that in WWII and following years made possible the production of the significant amounts of uranium needed in the war effort (Manhattan Project; D. Foster Hewett, oral communication, ca.

1955) and in the subsequent development of nuclear energy. Between 1943 and 1947 about 1,350 t of U_3O_8 was produced for the Manhattan Project on the Colorado Plateau (U.S. Department of Energy, 1982).

A detailed description of the ore deposits of the Uravan mineral belt is deferred to a later section of this chapter, where other sandstone-type uranium deposits are discussed, because most production came after WWII.

Vanadium deposits of the Colorado Plateau (vanadium and minor uranium)

Substantial vanadium has been produced from numerous sandstone-hosted deposits on the Colorado Plateau outside of the Uravan mineral belt. The deposit near Rifle, Colorado, and deposits near Placerville, Colorado, were mined primarily for vanadium; they have low uranium content. Vanadium ore was discovered near Rifle in about 1909 (Burwell, 1932). The Rifle Mine was worked sporadically until 1953 and again in 1965 through 1977. The nearby smaller Garfield Mine was located in 1929 and it was worked on and off until about 1945 (Fischer, 1960), and again in 1966 through 1977. Vanadium production commenced in the Placerville District in 1909, and most production was from 1910 to 1920 and 1940 to 1944 (Fischer, 1942; Bush and others, 1959).

Vanadium ore in the Rifle District (Plate 2C, locality 29) occurs in the Triassic-Jurassic Glen Canyon Sandstone and in the lower part of the overlying Jurassic Entrada Sandstone (Fischer, 1960; Chenoweth, 1982). These rocks are fine-grained, "clean" eolian sandstones that aggregate about 30 to 60 m in thickness. The district is on the flank of the south-dipping Grand Hogback monocline and about 1 to 2 km north of a major east-trending, nearly vertical fault with maximum displacement of about 150 m, downthrown to the south. The sandstone formations dip about 20 to 30° south at the deposit, and they are flexed into gentle folds near the major fault south of the district. The normally light-reddish-brown sandstone of the Glen Canyon and Entrada is bleached to light gray in the vicinity of the vanadium deposit, and similar alteration in shales of the Chinle Formation that underlies the Glen Canyon is centered on the deposit and forms a halo considerably broader than the deposit itself. Barren sandstone surounding the vanadium deposit shows evidence of dissolution of sand grains as well as local silica overgrowths on quartz grains; carbonate minerals locally cement the sandstone.

The ore deposit forms an east-trending elongate body more than 3 km long and less than 200 m wide. It is transected near its west end by a canyon. The Rifle Mine exploited the larger east segment of the body, and the Garfield Mine worked the smaller west end. The body has a much-flattened and greatly attenuated S-shaped form whose principal limbs have been referred to as the No. 1, No. 2, and No. 3 "veins." The veins (tabular layers) are connected by curved ore layers—the flexed segments of the flattened S—that are called "rolls." Other roll-like forms that are transitional from the tabular layers appear to be controlled by diastems in the sweeping eolian crossbeds of the host sandstone. Although the ore layers at Rifle are generally massive and uniform in grade, ore is locally streaked, where darker colored and higher grade zones follow the bedding of the sandstone.

Vanadian illite-smectite and chlorite are the principal ore minerals. They form mats that fill pore spaces between sand grains, and in high-grade ore they replace corroded margins of detrital sand grains. Bladed montroseite is a minor interstitial component, as radiating clusters perched on quartz grains. Minor amounts of tuyuyamunite or carnotite and bayleyite as secondary minerals are the only uranium minerals recognized in the deposit. Vanadium content is highest, commonly more than 2 percent V_2O_5, near one side of ore, and lowest, 1 percent or less V_2O_5, on the opposite, diffuse side. The reverse is true for uranium, whose maximum content rarely exceeds 0.1 percent U_3O_8. The ore layers are bounded on the sharply defined side by a thin layer a few millimeters thick of an interstitial solid-solution mixture of galena and clausthalite. This layer is in turn bordered by a thicker layer that averages about 0.5 m thick of greenish-gray sandstone impregnated with chromium-bearing illite-smectite. The bounding layers form a continuous rind on the downdip side of the flattened S-shaped ore deposit of tabular layers and connecting rolls.

Vanadium deposits in the Placerville District (Plate 2C, locality 27) have some similarities to the deposit at Rifle. Ore occurs in the Entrada Sandstone where the normally light-reddish-brown rock was bleached to light gray (buff at the surface where weathered). Red-bed sections overlie and underlie the Entrada in the Placerville District. Several north-trending grabens form a zone that coincides with the general area of ore deposits (Bush and others, 1959, 1960). Vanadium-mineralized rock forms a wavy, virtually continuous, north-trending layer about 15 km long and 2 km wide in the Entrada. Individual ore bodies are widely separated and they occur where the generally thin layer thickens to more than about 0.3 m to as much as 6 m. Ore bodies commonly are circular to elliptical in plan, and their upper, lower, and lateral margins locally are transitional into rolls similar to those seen at Rifle. Ore consists of sandstone impregnated with vanadian illite-smectite, as at Rifle. A chromium-bearing illite-smectite–impregnated layer underlies the vanadium-bearing layer in the Placerville District. Both the vanadium layer and chromium layer underlie the depositional west edge of the Pony Express Limestone Member of the Wanakah Formation, which immediately overlies the Entrada.

The Rifle and Placerville deposits probably precipitated at an interface between a pore solution already present in host sandstone and an introduced solution that penetrated the sandstone (Fischer, 1960). At Placerville the pore solution may have been trapped beneath the west edge of the Pony Express Limestone Member. Zonation of chromium on one side of a deposit probably resulted from fractionation of metals out of the introduced solution at the solution interface. We suggest that at Rifle the introduced solution may have been conducted into the porous sandstone layer from the fault zone that lies downdip and to the

TABLE 1. UNITED STATES PRODUCTION OF URANIUM AND VANADIUM*

District or Area	County/State	Period†	Tons Ore	Tons U_3O_8	Grade (%)	Tons V_2O_5	Grade (%)	Notes	Source
Central City District	Gilpin, Colo.	1872-1914	300	55	High	......	......		Sims and others, 1955
		1945-1955	......	7	......	......	......		Sims and Sheridan, 1964
		1954-1959	49	0.2	0.40	0.08	0.21		AEC files
Uravan mineral belt	Mesa, Montrose,	1898-1944	648,000	......	......	12,300	1.90	Extends into San Juan and Grand Counties, Utah. Includes production from Gateway, Uravan, Bull Canyon, and Slick Rock districts, Colo., and Thompson's, La Sal, and Dry Valley Districts, Utah	Chenoweth, 1980a, b
	San Miguel, Colo.	1947-1968	13,988,000	35,000	0.25	179,700	1.29		Chenoweth, 1980a, b
Slick Rock District	San Miguel, Colo.	to 1978	......	~9,000	......	~50,000	......	Tons V_2O_5 estimated from average grade for Uravan mineral belt	Chenoweth notes from DOE files
Uravan District	Montrose, Colo.	to 1978	......	~8,000	......	~45,000	......	Same as above	Same as above
Bull Canyon District	Montrose, Colo.	to 1978	......	~7,700	......	~43,000	......	Same as above	Same as above
Rifle District	Garfield, Colo.	1909-1953	750,000	......	......	~12,500	1-3		Fischer, 1960
		1951-1970	394,000	280	0.07	5,950	1.51		Chenoweth, 1980a, b
Placerville District	San Miguel, Colo.	1910-1944	250,000	......	......	6,250	2.5		Fischer, 1942; Bush and others, 1959
Front Range area	Jefferson, Larimer, Boulder, Gilpin, Clear Creek, Park, Colo.	1948-1966	26,000	13	0.05	520	2.0		Chenoweth, 1980a, b
		1953-1982	1,364,700	6,660	0.49	......	......	Mostly Schwartzwalder Mine	DOE records compiled by Chenoweth
Schwartzwalder Mine	Jefferson, Colo.	1949-1985	......	8,800	0.13	......	......		Wallace and Karlson, 1985
Maybell District	Moffat, Colo.	1954-1981	1,628,600	2,600	......	......	......	Tons U_3O_8 includes 500 tons from heap leaching, 1976-1981	Chenoweth, 1986
Marshall Pass District	Saguache, Colo.	1955-1982	277,800	1,130	0.39	......	......	Tons U_3O_8 includes 60 tons from leaching, 1970-1973	DOE records compiled by Chenoweth
Tallahassee Creek District	Fremont, Colo.	1957-1972	95,400	240	0.25	......	......	Large undeveloped resources in area	Chenoweth, 1980a, b
Temple Mountain	Emery, Utah	1914-1964	......	650	......	1,900	......		Hawley and others, 1965
Delta Mine	Emery, Utah	1948-1970	687,100	1,730	0.25	1,450	0.28	Includes Delta Mine	AEC files
White Canyon area	Emery, Utah	1953-1967	145,500	410	0.28	90	0.10		AEC files
	San Juan, Utah	1948-1974	1,924,000	4,980	0.26	......	......	Includes Happy Jack Mine, Red Canyon, Elk Ridge, Deer Flat	Chenoweth, 1975
Green River District	Emery, Utah	to 1979	670,000	1,320	0.20	1,270	0.19		DOE records compiled by Chenoweth
Lisbon Valley District	San Juan, Utah	1948-1979	10,703,000	34,250	0.32	......	......		Huber, 1981
Henry Mountains District	Garfield, Utah	1914-1944	485	......	......	16	3.20		Chenoweth, 1980c

TABLE 1. UNITED STATES PRODUCTION OF URANIUM AND VANADIUM* (continued)

District or Area	County/State	Period†	Tons Ore	Tons U_3O_8	Grade (%)	Tons V_2O_5	Grade (%)	Notes	Source
Henry Mountains District	Garfield, Utah	1948-1978	79,500	240	0.30	850	1.35	Large undeveloped resources in area	Chenoweth, 1980c
Marysvale District	Piute, Utah	1949-1969	307,300	670	0.22	......	......		AEC files
Spor Mountain District	Juab, Utah	1955-1968	104,100	210	0.20	......	......	Mostly Yellow Chief Mine	AEC files
St. Anthony Mine	Pinal, Arizona	1934-1944	......	......	......	1,270	......		Creasey, 1950
Monument Valley area	Navajo, Apache, Arizona	1942-1946	4,100	2	4.03	78	1.92	Extends into San Juan County, Utah. Only 52 tons assayed for U_3O_8	Chenoweth, 1985a
		1947-1969	1,362,000	4,370	0.32	12,400	0.94	Includes Moonlight and Monument No. 1, Mitten No. 2 Mines	Chenoweth and Malan, 1973
Moonlight Mine	Navajo, Arizona	1956; 1959-1966	223,200	590	0.26	470	0.60		AEC files; Scarborough, 1981
Cane Valley area	Apache, Arizona	1947-1969	777,400	2,660	0.34	11,600	1.42	Includes Monument No. 2 Mine	AEC files
Monument No. 2 Mine	Apache, Arizona	1943-1946	490	......	......	7	1.40		Chenoweth, 1985a
Lakachukai Mountains	Apache, Arizona	1950-1968	724,800	1,740	0.24	7,390	1.02		Chenoweth and Malan, 1973
Grand Canyon breccia pipes	Coconino, Mohave, Arizona	1956-1969	496,500	2,130	0.43	0.9	1.02	Includes Orphan, Hack, Ridenour, and Chapel Mines; V_2O_5 data for Hack and Ridenour only	AEC files
	Coconino, Mohave, Arizona	1980-1985	380,000	2,500	0.65	......	......		Energy Fuels Nuclear, 1986, communication to Chenoweth
Orphan Mine	Coconino, Arizona	1956-1969	495,100	2,130	0.43	......	......	Also produced 107,000 oz silver and 3,340 tons copper	Chenoweth report to USGS, 1986
Date Creek Basin	Yavpai, Arizona	1955-1959	10,800	17	0.15	5	0.05	Mostly Anderson Mine; large undeveloped resources in areal	AEC files
Grants mineral belt (includes Gallup, Smith Lake, Ambrosia Lake, and Laguna Districts)	McKinley, Cibola, New Mexico	1950-1970	39,133,200	85,180	0.22	3,560	0.14	V_2O_5 grade calculated only from ore analyzed for vanadium; 80% of vanadium from Jackpile ores paid for by AEC but not recovered; production mostly from the Westwater Canyon Member, and the Jackpile sandstone of the Morrison Formation	AEC files
	Cibola, McKinley, New Mexico	1950-1981	......	3,340	0.20	......	......	Production from the Todilto Limestone Member	Chenoweth, 1985b
Wyoming Tertiary basins	Campbell, Converse, Wyoming	1952-1970	16,151,300	36,440	0.23	240	0.03	Includes Sand Wash Basin, Moffat County, Colorado	AEC files
	Carbon, Fremont, Johnson, Natrona, Sweetwater, Wyoming	1952-1982	60,737,400	98,560	0.16	......	......		DOE files compiled by Chenoweth
Powder River Basin, Pumpkin Buttes area	Campbell, Johnson, Campbell, Wyoming	1952-1970	42,100	110	0.27	60	0.20		AEC files
Monument Hill District	Converse, Campbell, Wyoming	1954-1970	483,600	890	0.18	100	0.05		AEC files

TABLE 1. UNITED STATES PRODUCTION OF URANIUM AND VANADIUM* (continued)

District or Area	County/State	Period†	Tons Ore	Tons U_3O_8	Grade (%)	Tons V_2O_5	Grade (%)	Notes	Source
Gas Hills District	Fremont, Natrona, Wyoming	1954-1962	4,703,000	11,160	0.24	......	......		King and others, 1965
Great Divide Basin, Crooks Gap area	Fremont, Wyoming	1954-1970	1,742,100	4,070	0.23	......	......		AEC files
Copper Mountain	Fremont, Wyoming	1954-1971	118,000	190	0.16	......	......		AEC files
Shirley Basin	Carbon, Wyoming	1960-1970	2,122,100	5,190	0.22	......	......	Tons U_3O_8 includes 470 tons from solution mining	AEC files
Northern Black Hills	Crook, Wyoming	1953-1968	707,900	1,550	0.22	2,150	0.37	Extends into Butte County, South Dakota	AEC files
Southern Black Hills, Edgemont District	Custer, Fall River, South Dakota	1952-1972	914,100	1,220	0.13	940	0.20	Extends into Weston County, Wyoming	AEC files
Cave Hills-Slim Buttes area	Harding, South Dakota	1954-1966	89,700	370	0.41	......	......	Production from lignite	AEC files
South Texas Coastal Plain	Karnes, Live Oak, Duval, Bee, DeWitt, Atascosa, Jim Hogg, Wells, Starr, Gonzales Texas	1954-1982	32,992,500	23,770	0.12	......	......	U_3O_8 grade estimated only for ore mined; since 1979 the bulk of production has been by solution mining	DOE files compiled by Chenoweth
	Karnes, Live Oak, Texas	1958-1970	1,166,200	2,060	0.18	......	......		AEC files
Austin District	Lander, Nevada	1954-1966	22,100	55	0.25	......	......	Mostly Apex Mine	AEC files
Midnite Mine	Stevens, Washington	1955-1970	1,207,400	2,860	0.24	......	......		AEC files
		1955-1982	......	7,250	0.16-0.18	......	......		Chenoweth notes
North and South Belfield area	Stark, Billings, North Dakota	1956-1968	80,200	290	0.36	32	0.19	Production from lignite	AEC files
Bokan Mountain	Alaska	1957-1973	70,000	420	0.60	......	......		Finch and others, 1973
Western United States		1947-1982	210,700,000	......	......	401,300	0.69	Tons ore is total mined; V_2O_5 grade calculated only for ore analyzed for vanadium; all V_2O_5 was purchased by AEC, but not all was recovered at mills	AEC files
United States		1947-1982	230,647,000	410,300	0.18	......	......	Tons U_3O_8 includes miscellaneous non-ore sources; ore production from New Mexico, Wyoming, Alaska, Arizona, California, Colorado, Florida, Idaho, Montana, Nevada, New Jersey, North Dakota, Oklahoma, Oregon, South Dakota, Texas, Utah, and Washington	U.S. Department of Energy, 1983

TABLE 1. UNITED STATES PRODUCTION OF URANIUM AND VANADIUM* (continued)

District or Area	County/State	Period†	Tons Ore	Tons U_3O_8	Grade (%)	Tons V_2O_5	Grade (%)	Notes	Source
Wilson Springs	Garland, Arkansas	1967–1985	......	......	......	~62,000	1.0		G. N. Breit, written communication, 1986
Swanson deposit	Pittsylvania, Virginia	......	......	24,500	0.113	......	......	Reserves; deposit has not been mined	Halladay, 1989
Petroleum residues		1977–1985	......	......	......	23,000	......	Mostly imported crude oils	Kuck, 1985

*...... = not applicable, or no data.
†Some mines continued to produce beyond the period for which production is given.

south of the deposit. At Placerville, introduced fluids may have entered the Entrada along faults that bound the grabens that coincide with the zone of alteration and mineral deposition in the Entrada.

Other vanadium deposits

Other sources that contributed significant vanadium production during this early period were base-metal vanadate deposits (Fischer, 1975a), and the Phosphoria Formation (Busch, 1961). The St. Anthony Mine in Arizona (Plate 1, locality 39) produced more vanadium than any other base-metal deposit in the United States (Table 1). There, vanadate minerals and wulfenite coat and partly replace supergene base-metal minerals in the veins, and are thought to have been introduced by hypogene solutions following oxidation of primary base-metal sulfides (Creasey, 1950). The Phosphoria Formation, a marine deposit of Permian age that contains rich phosphorite beds and is widely exposed in the northwestern United States (Plate 2C), yielded some vanadium during WWII.

TWO DECADES FOLLOWING WORLD WAR II

Uravan mineral belt (uranium and vanandium)

Most uranium and vanadium production in the mineral belt has come from the uppermost sandstone unit of the Salt Wash Sandstone Member of the Upper Jurassic Morrison Formation, commonly referred to as the "ore-bearing sandstone" or "upper rim." The ores of the Uravan mineral belt, referred to initially as carnotite ores, were recognized near the beginning of this period to have been composed initially of low-valent (unoxidized) minerals of uranium and vanadium. The high-valent (oxidized) carnotite was recognized to have formed as a near-surface weathering product. The Salt Wash Member was deposited as a broad, fan-shaped alluvial apron by aggrading streams (Craig and others, 1955). Structural warping of the region during deposition resulted in development of a lesser alluvial fan, the toe of which coincides with the position of the mineral belt (Shawe, 1962). The ore-bearing sandstone was deposited across the toe of the fan as a broad, more or less continuous unit of porous fine-grained sand, which contained abundant detrital plant debris that became carbonized or silicified following burial. Folding and faulting of the region, in part related to development of the salt anticlines in the region of the Paradox Basin, took place during and following deposition of the Salt Wash, resulting in the development of zones of permeability that facilitated subsequent flow of ground water into and through the sedimentary rocks.

Shortly after deposition the ore-bearing sandstone underwent local oxidizing alteration where oxygenated meteoric waters destroyed detrital iron-bearing silicate minerals and leached iron from detrital iron-titanium oxide minerals, and deposited hematite. Local reducing environments in the vicinity of carbonized plant debris resulted in destruction of iron-bearing silicate miner-

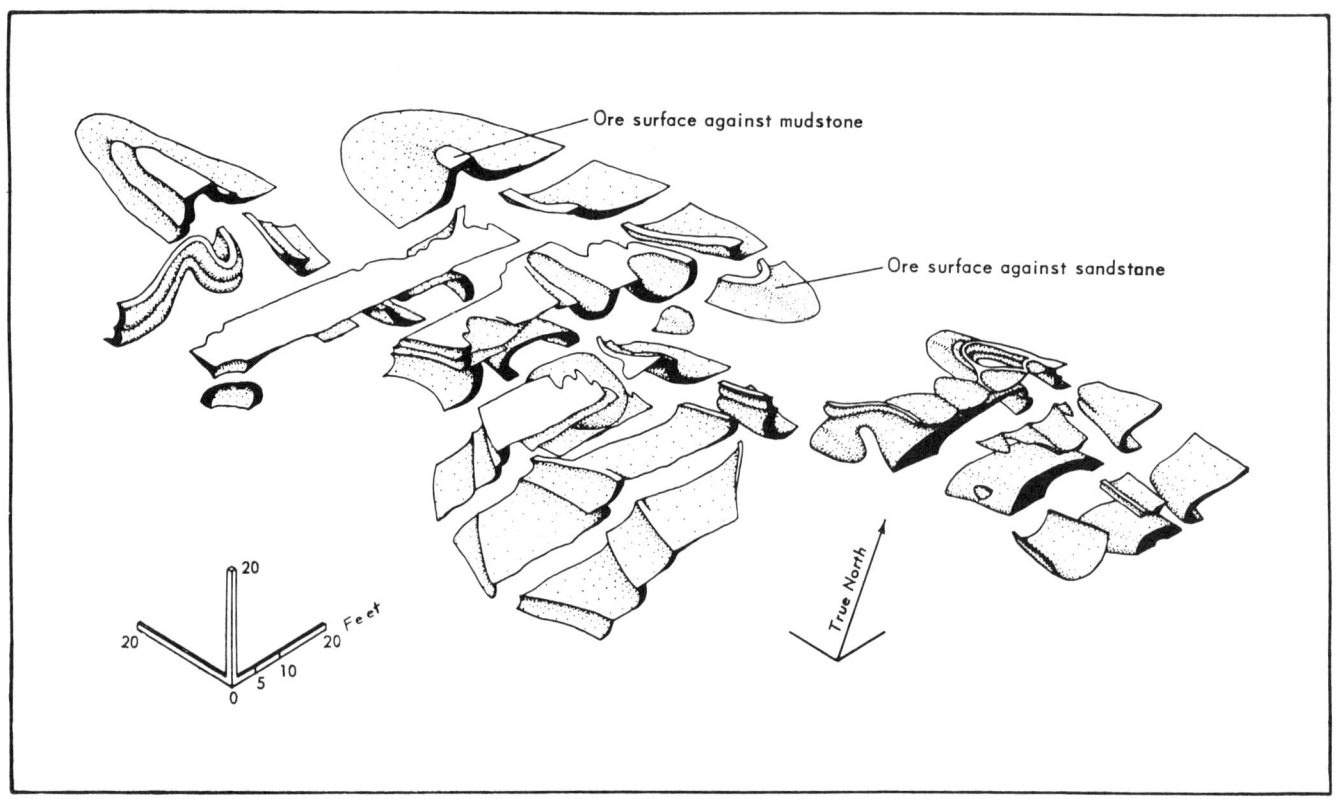

Figure 2. Cutaway-perspective diagram of roll orebodies in the Slick Rock District, Uravan mineral belt, Colorado. From Shawe and others (1959).

als, partial destruction of iron-titanium oxide minerals, and deposition of a small amount of pyrite. Later, a more intense reducing alteration that resulted from introduction of an extraneous fluid into the sandstone occurred close to fractures and in permeable parts of the sandstone. Uranium-vanadium deposits formed in zones of epigenetic alteration where the zones contained previously reduced rocks in the vicinity of carbonized plant debris (Shawe, 1976).

Most of the ore mined has come from tabular deposits that range in size to many tens of thousands of tons. Widely distributed roll-shaped orebodies tend to be smaller, ranging in size from a few tens of tons to several thousand tons of ore. The highly convoluted rolls (Fig. 2) in many places are transitional from tabular ore layers where fluvial-crossbedded host sandstone is segmented into complex zones of permeability separated by thin layers of claystone and mudstone. Ore minerals are not uniformly distributed in much of the ore, except locally where ore occurs in generally massive sandstone. More commonly, ore is enriched along bedding layers, adjacent to carbonized plant material, against the sharply bounded curving "inside" edge of rolls, and against either upper or lower sharply bounded surfaces of tabular orebodies. In unweathered primary deposits, the principal uranium minerals are uraninite and coffinite, and the principal vanadium minerals are montroseite and vanadian clay minerals. Where the ores are weathered, a great variety of secondary (higher valent) minerals occur, notably carnotite (Weeks and others, 1959).

The Uravan mineral belt ores are also enriched in copper, lead, zinc, molybdenum, silver, cadmium, chromium, arsenic, selenium, and other elements relative to barren sandstones. As in the vanadium deposits at Rifle and Placerville, these elements show zoned distributions in the orebodies. Lead and selenium commonly occur in a thin layer adjacent to the "inside" edge of roll bodies and against the upper or lower surfaces of tabular ore layers (Shawe, 1966). In the Slick Rock District, some elements, particularly copper, occur in greatest abundance in deposits that are close to a zone of faults, which is centered on the principal zone of ore deposits within the district (Shawe and others, 1959).

The age of epigenetic alteration and, by inference, the associated uranium–vanadium ores in the Uravan mineral belt, based on geologic relationships of altered zones in the sedimentary rocks, is most likely early Tertiary. Lead isotope ages determined on minerals from the ores, however, range from about 70 to 115 Ma (Stieff and others, 1953; K. R. Ludwig, cited by Granger and others, 1985). Filling temperatures of about 120 to 130°C of fluid inclusions in diagenetic calcite and barite from organic-rich Cretaceous Mancos Shale, which overlay the Morrison Formation during the inferred time of mineralization in the mineral belt, indicate that the temperature of ore formation was not above about 150°C (Shawe, 1976).

We infer that pore fluids driven from compacting Mancos Shale (Shawe, 1976), or circulating ground waters that encountered the underlying organic-rich Pennsylvanian Paradox evaporite section (G. N. Breit, written communication, 1986), penetrated the sedimentary rocks in the mineral belt along zones of faults (Shawe, 1969). The fluids caused epigenetic alteration of the rocks and deposition of uranium-vanadium ores at a solution interface with pore water in altered zones where detrital organic material provided a suitable reducing environment. Source of the uranium is debatable; the source of the vanadium was probably detrital iron-titanium oxide minerals.

Deposits in Triassic rocks (uranium and vanadium)

A group of widely scattered uranium deposits is known in the Shinarump and Moss Back Members and related strata of the Upper Triassic Chinle Formation of the Colorado Plateau. Several of these have had large production, and numerous deposits have produced small amounts of ore.

Shinarump Conglomerate Member.
The Monument No. 2 Mine in the Monument Valley District, Arizona (Plate 2C, locality 35) (Witkind and Thaden, 1963), the Moonlight Mine in Monument Valley, Arizona (Plate 2, locality 34), and the Happy Jack Mine in the White Canyon area, Utah (Plate 2C, locality 19; Trites and Chew, 1955), have been large uranium producers in the Shinarump Member. The Monument No. 2 operated from 1942 until about 1969; the Moonlight Mine was mined from 1956 to 1966. The Happy Jack Mine was located as a copper prospect in 1900 and it was mined briefly in 1906 and 1914. From 1949 until about 1965, it was mined for uranium, and copper was recovered (Isachsen and Evensen, 1956). The three deposits in the Shinarump occur where host strata are gently dipping, and all are in channels cut several meters deep into underlying strata where the member normally is about 10 to 15 m thick. Channel fill consists of interlayered lenses of fluvial sandstone, conglomerate, siltstone, and claystone. Uranium deposits were localized around accumulations of carbonized wood "trash" that is more abundant within channels than it is in the Shinarump beyond the channels. Strata in the channels are bleached gray in contrast to the normal reddish brown of the member, and of underlying strata. Faults are minor or absent at the deposits.

Ore occurs near the base of the channels as tabular layers that are irregular in shape but generally elongated parallel to channel trends. The mineralized zone at the Monument No. 2 is more than 1 km long. Ore may also occur as smooth rolls similar to those in the Uravan mineral belt, and at the Monument No. 2 as "rods," which are high-grade rodlike forms that commonly enclose silicified logs and tend to be aligned with the channel. Individual tabular orebodies may be as much as 200 m long, 25 m wide, and 6 m thick. Much of the ore in the Monument Valley area has grades of a few tenths of a percent U_3O_8 and a few percent of V_2O_5. Locally, as in some rods, grade may be as high as about 17 percent U_3O_8. Vanadium is quite low, and copper grades are a few percent at the Happy Jack Mine (Trites and Chew, 1955). Uranium-vanadium ratios vary considerably throughout all the mineralized zones. Uraninite and montroseite are the only primary uranium and vanadium minerals at the Monument No. 2 and Moonlight Mines, and they are accompanied by pyrite, galena, sphalerite, and bornite. At the Happy Jack Mine, chalcopyrite and covellite also are present; montroseite is absent. There, the sulfide minerals form an extensive zone wider than that of the uranium minerals. Numerous secondary minerals of uranium and vanadium or copper are present where the Shinarump deposits have been weathered.

Lead-uranium isotope analyses of uraninite suggest that the age of the deposits is in the range 70 to 210 Ma (Stieff and others, 1953; Miller and Kulp, 1963; Granger and others, 1985), indicating that some uranium may have been introduced into or deposited in host strata shortly after sedimentation, and some long after.

Moss Back Member and related strata.
Significant uranium and/or vanadium production has also come from the Moss Back Member of the Chinle in the Lisbon Valley District, Utah (Plate 2C, locality 21), and the Temple Mountain District, Utah (Plate 2C, locality 16), and from a 10-m-thick sandstone layer that underlies the Moss Back Member at the Delta Mine in Utah (Plate 2C, locality 17). Uranium ore was first recognized on the Lisbon Valley anticline in 1913 in a basal sandstone of the Chinle Formation (Wood, 1968). In 1948, low-grade uranium deposits were discovered in the upper part of the Permian Cutler Formation below the Chinle, and in 1952 a large high-grade uranium deposit was discovered at the Mi Vida Mine at the base of the Chinle and top of the Cutler. By mid-1965, several other large deposits had been recognized at this stratigraphic horizon on the anticline throughout a belt 15 km long (partly removed by erosion), and production from the district has continued to the present. Ore averaged about 0.4 percent U_3O_8; much of the ore in the south part of the belt averaged at least 0.4 percent V_2O_5, but vanadium is negligible in ore deposits in the north part of the district. Uranium minerals were recognized in the Temple Mountain District as early as 1898, but no significant ore was produced until 1914 (Hawley and others, 1965). By 1920, ore that averaged about 1.75 percent U_3O_8 and 4.0 percent V_2O_5 had been mined; it provided the only early production of radium from Triassic rocks on the Colorado Plateau. Extensive production in the district commenced in 1948, and it had practically ceased by about 1965. The uranium-vanadium deposit at the Delta (Hidden Splendor) Mine was discovered in 1952 (Isachsen and Evenson, 1956); the mine had been virtually exhausted by about 1967.

The Moss Back Member—0 to 40 m thick—and related strata consist of mostly fluvial, fine- to coarse-grained, grayish green to light gray arkosic sandstone interbedded with thinner layers of mudstone and calcarenite conglomerate. Carbonized and silicified plant material is locally abundant in all of the rock types. Mineralized sandstone at the top of the Cutler and directly below the Moss Back in the Lisbon Valley District is in a unit as much as 15 m thick of fine- to medium-grained arkose.

The structural settings of the three districts are varied. At

Lisbon Valley, ore deposits occur at a constant level on the southwest flank of the Lisbon Valley anticline, in beds that dip 10 to 20° southwest. The ore trend lies within about 3 km of the Lisbon Valley fault and it is subparallel to it. At Temple Mountain the ores are in gently dipping strata on the southeastern flank of the San Rafael swell anticlinal uplift near collapse structures that are crudely oval pipes about 400 m across. Brecciated continental beds, including the Moss Back, within the pipes have collapsed as much as 100 m as a result of solution of underlying limestone strata. Several faults transect the district. The Delta Mine is in gently dipping sedimentary rocks that are slightly displaced by a swarm of high-angle faults at the south-plunging nose of the San Rafael swell.

The Moss Back Member and related strata are bleached gray or gray-green in the vicinity of the ore deposits, in part as a result of a reducing environment related to contained carbonized plant material that localized ore, and also to mineralizing solutions. At Temple Mountain the Moss Back also contains substantial petroleum residue that has been removed in zones through which mineralizing solutions passed.

Ore deposits are mostly crudely oval to irregular tabular bodies that range in thickness to as much as 15 m and in length to several hundred meters; the largest orebodies in the Lisbon Valley district are about 1 km long. Elongation of the bodies is subparallel with sediment transport directions. Local variations in lithology have resulted in small irregularities in shape, including "splits" and roll-like forms that locally are transitional from the generally massive tabular layers of ore. Roll orebodies similar to those in the Uravan mineral belt are particularly well developed in the Temple Mountain District, where they formed at the contact between altered sandstone and petroliferous sandstone. Textural variations in the ores also reflect local variations in lithology of the host rocks.

Uraninite is the principal uranium mineral in the deposits; it is accompanied variously by coffinite and by the vanadium minerals montroseite, paramontroseite, doloresite, and vanadium-bearing illite-smectite. Also present in places are minor amounts of several sulfides and sulfates, ferroselite, calcite, and sericite. Numerous secondary minerals are present where the deposits have been weathered. The ore minerals and associated gangue minerals impregnate sandstone, filling pore spaces and replacing detrital grains on their margins and along fractures in them. They also have replaced carbonized wood; rarely, coalified wood that has been replaced by uraninite contains as much as 80 percent U_3O_8. At Temple Mountain, uraniferous asphaltite derived from petroleum permeates mineralized rock and imparts a dark gray to black color to the ore. The orebodies in the Moss Back Member, particularly at Temple Mountain, commonly are zoned so that elements such as lead, selenium, and chromium are enriched in concentric or parallel layers at or near the edges of orebodies, much like in the Rifle District and the Uravan mineral belt. District-wide zonation of elements, reflecting the same sequence seen in individual zoned orebodies, also occurs at Temple Mountain, centered on the collapse structures.

Hawley and others (1965) proposed that at Temple Mountain, mineralizing solutions entered the Moss Back Member and spread laterally from the Temple Mountain collapse structures, removing petroleum and altering the rocks. Ore minerals were precipitated with asphaltite at the margins of zones of alteration.

Uranium-lead isotopic data on uraninite from ores in the Moss Back Member indicate discordant ages that range from about 85 to 210 Ma (Miller and Kulp, 1963). Sulfur isotopic data indicate low-temperature biogenic activity. Uranium may have started accumulating in the Chinle not long after the sediments were deposited, and where temperatures were low enough for sulfate-reducing bacteria to produce H_2S that precipitated uranium. However, in the Lisbon Valley District, the apparent structural control of the ore deposits that is related to the present configuration of the anticline suggests that the ores formed much later than the immediate post-depositional period when Chinle beds were nearly horizontal. The San Rafael swell and the Temple Mountain collapse structures formed in early Tertiary time (Hawley and others, 1965), and we infer, therefore, that the uranium deposits could not have formed earlier.

Grants mineral belt, New Mexico (uranium and minor vanadium)

The Grants mineral belt in northwestern New Mexico is the region of greatest uranium production and reserves in the United States (Plate 2, localities 40, 41, and 43). Uranium minerals were known in the Grants region since 1920, but a major deposit was not discovered until 1950 (Kelley and others, 1968). Subsequently, discoveries were made throughout the mineral belt. Most production has come from the Westwater Canyon Member of the Morrison Formation, and significant production has come from the so-called Jackpile sandstone at the top of the Brushy Basin Member of the Morrison. Most of the remainder of the district's production has been from the Jurassic Todilto Limestone Member of the Wanakah Formation, a limestone-gypsum unit that overlies the Entrada Sandstone, about 50 to 180 m below the Morrison Formation. A little production has come from the Entrada Sandstone and the Dakota Sandstone.

Gentle east-trending folds formed along the south margin of the San Juan Basin during deposition of the ore-bearing formations. Following erosion, and after deposition of the overlying Upper Cretaceous Dakota Sandstone, the region was widely but not severely faulted and folded, and in late Tertiary time the Mount Taylor volcanic pile was emplaced (Granger and others, 1961).

The Westwater Canyon Member and the Jackpile sandstone are units of fine- to coarse-grained fluvial arkosic sandstone interstratified with thin layers of mudstone. The sandstone units were deposited on an alluvial plain by a braided-stream system at the south margin of the San Juan Basin; mudstone units reflect periodic lacustrine conditions. The Westwater Canyon sandstone is mostly reddish, but near uranium deposits it is light gray. The overlying Brushy Basin mudstones are dominantly light greenish

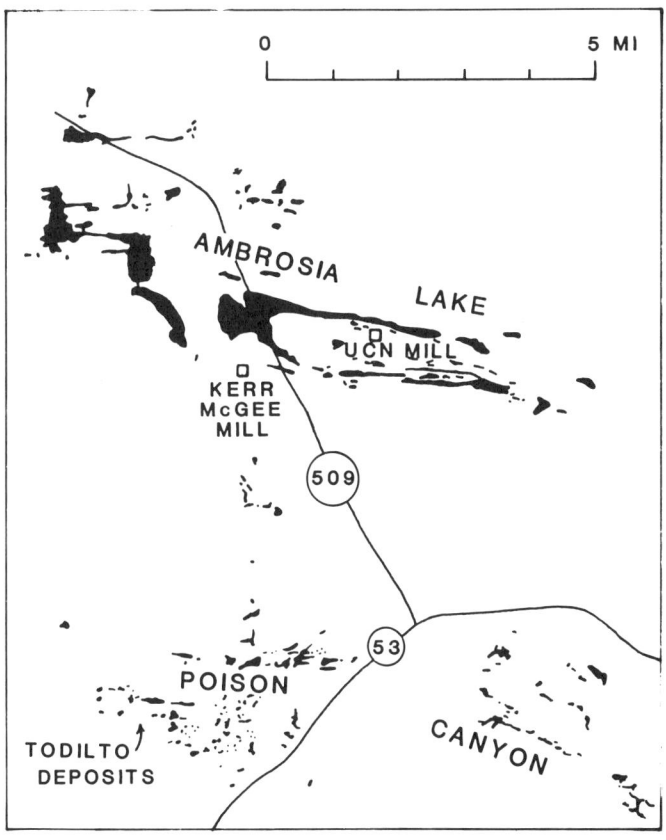

Figure 3. Uranium deposits of the Ambrosia Lake–Poison Canyon area, Grants mineral belt, New Mexico. Modified from Kelley and others (1968).

gray to grayish green, and the Jackpile sandstone is light gray. The light gray ore-bearing parts of the sandstone units contain locally abundant fossilized plant remains, some of which are carbonized and some of which are silicified (Moench and Schlee, 1967; Granger and others, 1961). Alteration effects associated with the ores in the Jackpile sandstone are spatially related to the Jackpile–Brushy Basin contact (Adams and others, 1978). Other alteration effects that are related to the pre–Dakota Sandstone erosion surface disrupt alteration patterns related to the ores, suggesting that ore deposition and its associated alteration took place before the pre-Dakota erosion occurred. Other geologic evidence indicates that ore deposition took place before Late Cretaceous–early Tertiary tilting at the south margin of the San Juan Basin (Moench and Schlee, 1967). Uranium-lead isotopic data suggest a minimum age of 130 Ma (Ludwig and others, 1984).

The so-called prefault or "trend" orebodies in the West-water Canyon, in the Poison Canyon Tongue, and in the Jackpile in the Grants belt are flat-lying mantolike bodies, elongated parallel with sediment depositional trends (Fig. 3). They occur singly or in clusters, in places in stacked fashion (Fig. 4). Individual bodies are as much as 1 km long, more than 100 m wide, and generally no thicker than about 5 m. They commonly have transitions into rolls shaped similar to those elsewhere on the Colorado Plateau, and they display "splits" where local variations in lithology, and hence in permeability, modified the flow of depositing solutions. Ore is dark gray, generally uniformly colored in featureless sandstone, and streaked and mottled where deposition was controlled in detail by bedding and by minor variations in lithology and distribution of detrital carbonized plant material. Distribution of uranium in the ores is locally irregular, the grade ranging from less than 0.20 percent to more than 1.0 percent U_3O_8, and very low V_2O_5.

Coffinite and uraninite are the primary uranium minerals. Uraniferous carbonaceous material, ubiquitous in ore, imparts the dark gray color. It was deposited as humates with the ore metals, and its presence suggests the role of humic acids in the transport of uranium into the orebodies (Fishman and others, 1985), or trapping of uranium in the orebodies. Pyrite, marcasite, calcite, jordisite, ilsemannite, and ferroselite are associated with ore minerals. Secondary minerals formed where the deposits are weathered. Some elements are zoned in the Grants deposits, as for example, concentrations of molybdenum and selenium at or near the upper or lower surfaces of tabular ore layers. Sulfur isotope data suggest that the sulfur in the deposits had a complex origin, including possible fractionation by bacterial reduction of sulfate (Jensen and others, 1960).

In addition to the trend orebodies in the Grants belt, several so-called redistributed orebodies formed as a result of the encroachment of oxygenated ground water into the primary ores, and by mobilization of uranium along the encroaching oxidation front. The orebodies that resulted have been referred to as roll orebodies, but their forms are quite irregular and the orebodies generally appear as an uneven series of coalescing stacked lenses. These orebodies formed following the post–Dakota Sandstone faulting that took place in the region, where the faults allowed access of meteoric water into the host sandstones. Uranium-lead isotopic data indicate that the redistributed ores are as young as 1 Ma (Ludwig and others, 1982, 1984).

Ore deposits along the southwestern edge of the Grants belt have been classed as remnant ore deposits, having been left as the unoxidized remnants of primary-trend ore deposits following the encroachment of oxygenated ground water (Kirk and Condon, 1986).

Primary orebodies that formed in the Todilto Limestone were controlled by the positions of gentle to tight and irregular folds, locally brecciated. Bodies as much as 450 m long parallel to fold axes, and containing more than 20 t of U_3O_8 at grades of about 0.20 percent U_3O_8 have been mined, although most mined deposits in the Todilto were much smaller. Unoxidized deposits in the Todilto contain uraninite, pyrite, fluorite (locally abundant in massive replacements of limestone), and barite. Small, tabular orebodies, elliptical in shape, formed in the bleached upper part of the Entrada Sandstone just below the Todilto Limestone.

The Woodrow Mine in the north Laguna area (Plate 2C, locality 43) is an unusual uranium deposit formed in a nearly vertical breccia pipe about 10 m in diameter, of Jackpile sandstone that has collapsed about 15 m stratigraphically. The pipe was mined to a depth of 60 m by 1956, producing ore that averaged 1.26 percent U_3O_8 and 0.05 percent V_2O_5. Ore with a grade as high as 20 percent U_3O_8 was mined from carbon-rich clay gouge in the ring fracture around the pipe. In addition to uraninite and coffinite, primary minerals identified from the pipe are chalcopyrite, galena, pyrite, marcasite, and barite.

The prevailing view of genesis of the uranium deposits in the Grants mineral belt is that formation waters in mudstones of the Brushy Basin, charged with uranium derived from leached volcanic materials and with humic acid, on compaction moved out of the mudstones into adjacent sandstone units (sandstone of the underlying Westwater Canyon Member and the overlying Jackpile sandstone). The sandstones became altered, and uranium and associated minerals were deposited in an appropriate reducing environment (Adams and others, 1978; Turner-Peterson and others, 1980).

Wyoming Tertiary basins (uranium and minor vanadium)

Several significant uranium districts have been developed in Eocene continental sedimentary rocks that were deposited in basins formed during Laramide deformation. Some vanadium has been produced from the Wyoming Tertiary ores, but in general, uranium exceeds vanadium in amount by a factor of 20 to 25.

Uranium accumulations were discovered at Lost Creek in the Great Divide Basin–Crooks Gap area (Plate 2C, locality 11) in 1936 (Bailey, 1969), and uraniferous coal and shale beds were recognized in the basin in 1945. Uranium was recognized in faulted Cambrian shale at Crooks Gap in 1953, and in the Eocene Battle Spring Formation in 1954. Uranium deposits were discovered at Pumpkin Buttes in the north part of the Powder River Basin (Plate 2C, locality 7) in 1951 (Mrak, 1968), in the Gas Hills at the south margin of the Wind River Basin (Plate 2C, locality 10) in 1953 (Anderson, 1969), and in the Shirley Basin area (Plate 2, locality 12) in 1955 (Harshman, 1972).

Uranium deposits in the Great Divide Basin–Crooks Gap area are mostly in the Eocene Battle Spring Formation; in the Powder River Basin they are in the Eocene Wasatch Formation; and in the Gas Hills and the Shirley Basin they are in the Eocene Wind River Formation. The mineralized Eocene strata are widespread, gently dipping to nearly flat sandstone units as much as 60 m thick. They consist of poorly consolidated, coarse-grained, arkosic, fluvial sandstone interbedded with finer-grained clastic rocks, and in places with carbonaceous shales and coal layers. At Crooks Gap, carbonaceous mudstone and boulder conglomerate are also mineralized. The mineralized interval in the Eocene formations is about 100 to 200 m thick except at Crooks Gap where it is about 450 m thick. Mineralized strata are devoid of faults except in the Crooks Gap area.

The Wyoming Tertiary basin orebodies are at the contact between so-called altered and unaltered sandstone. Light gray, unaltered sandstone, typically on the downdip side of orebodies, contains pyrite, calcite, iron-titanium oxide minerals, and carbonaceous "trash" derived from detrital woody material. Yellowish

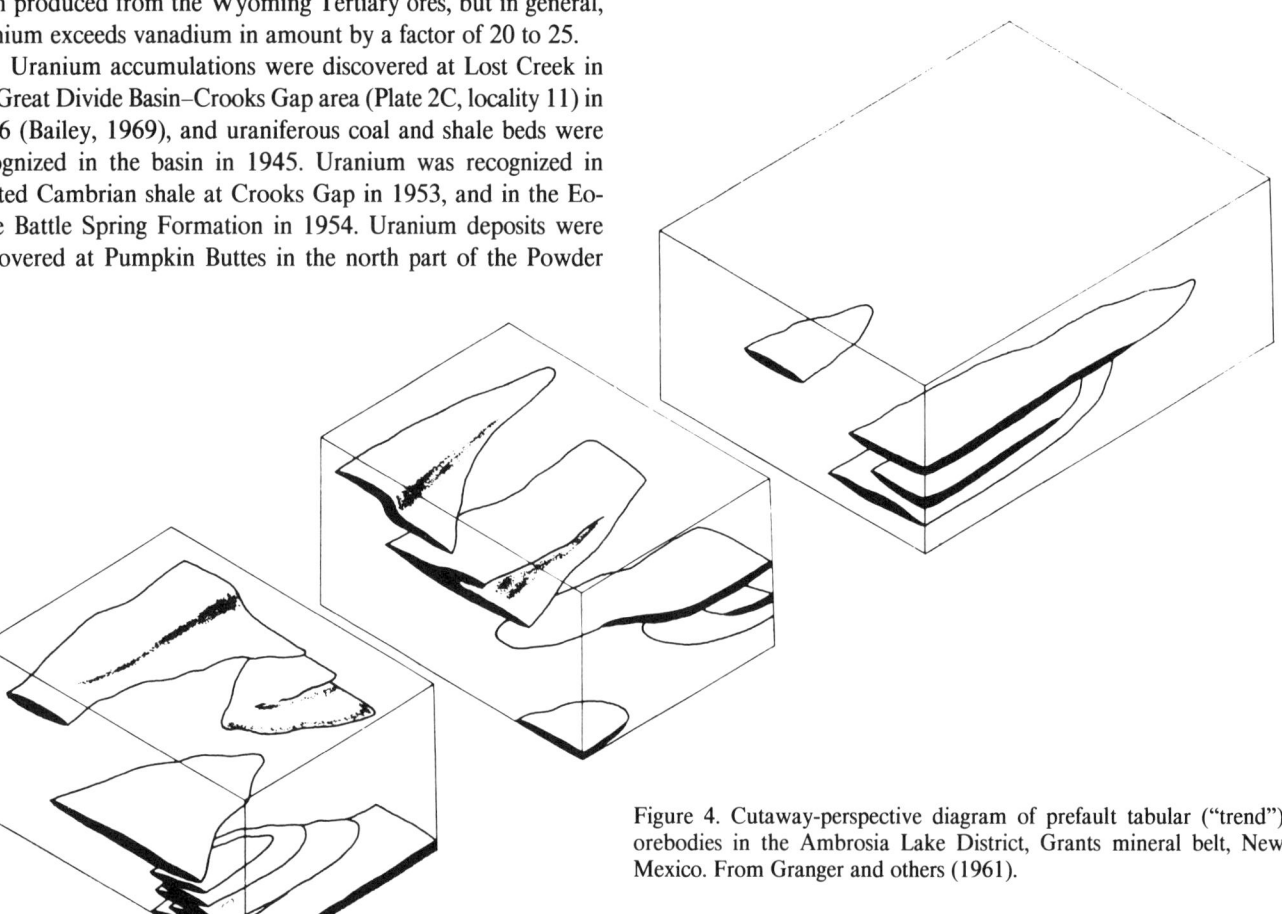

Figure 4. Cutaway-perspective diagram of prefault tabular ("trend") orebodies in the Ambrosia Lake District, Grants mineral belt, New Mexico. From Granger and others (1961).

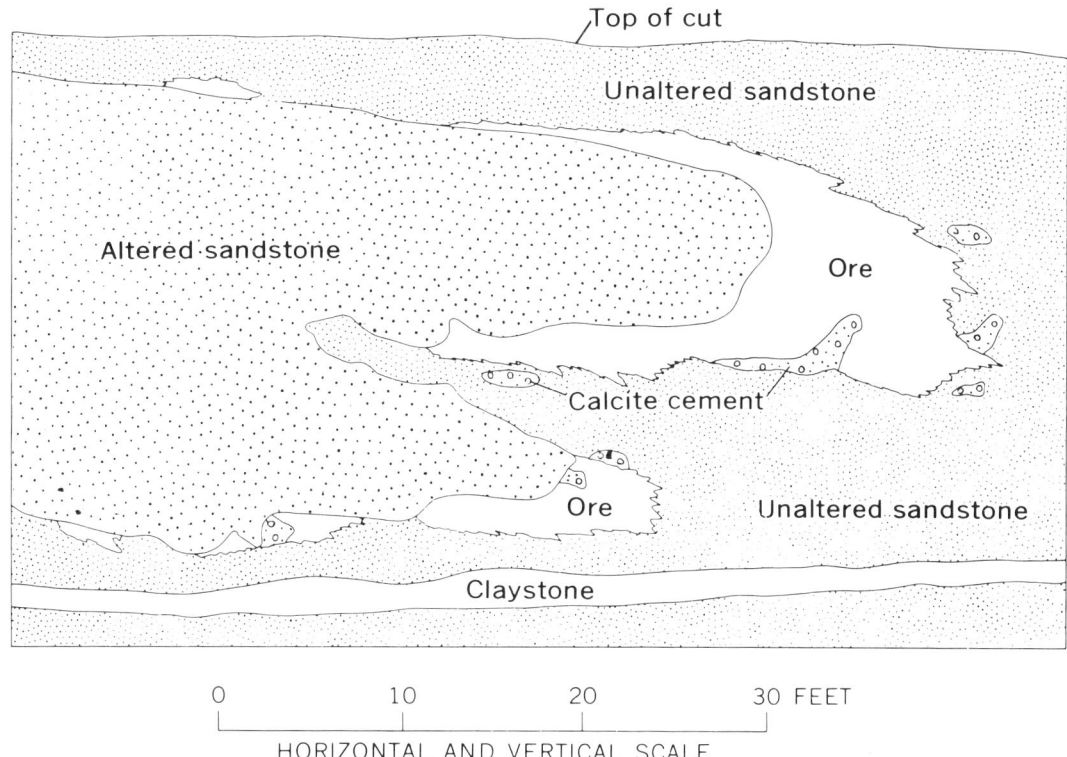

Figure 5. Section of a complex roll orebody in the Shirley Basin, Wyoming. From Harshman (1972).

to reddish altered sandstone contains detrital iron-titanium oxide minerals but it is devoid of pyrite and carbonaceous trash, and it contains only minor calcite. Altered transmissive sandstone forms large tongues several kilometers long that penetrate unaltered sandstone downdip. In the Powder River Basin, altered sandstone contains about 6 ppm uranium, compared to unaltered sandstone in which uranium averages only about 2 ppm (Davis, 1969).

The orebodies are characteristically large roll-like forms that connect upper and lower limbs that are nearly parallel to bedding. Although often stated to have C-shaped cross sections, the Wyoming rolls may be quite irregular (Fig. 5), and mining-grade zones may not be continuous throughout a roll. The irregularities are related to local variations in host lithology, particularly permeability. The largest orebodies are as much as 60 m wide from an edge against altered sandstone, 10 m thick, and a kilometer or more long, parallel to the edge of altered sandstone. Average grades range from 0.1 to 0.7 percent U_3O_8, and high-grade zones may contain as much as 0.5 percent organic carbon. Highest V_2O_5 grades are about 0.27 percent, but they are generally much lower. Slightly mineralized low-grade sandstone (0.01 to 0.10 percent U_3O_8) extends into unaltered sandstone beyond the orebodies by as much as 100 m or more, and in places it constitutes significant segments of the roll fronts between accumulations sufficient to be considered ore deposits. As with roll orebodies on the Colorado Plateau, elements are zoned in the Wyoming roll deposits. For example, selenium is most abundant in a zone of enrichment that overlaps the contact of ore and altered sandstone (Harshman, 1972). In places, "remnant" deposits occur in altered barren sandstone behind the solution-front roll orebodies, where chemically reducing conditions near abundant lignite and carbonaceous shale precipitated uranium.

Principal primary (epigenetic) minerals are pyrite, marcasite, uraninite, ferroselite, native(?) selenium, hematite, and calcite. In the Shirley Basin, detrital hornblende and epidote are significantly depleted in ore, and detrital minerals such as feldspars show extensive replacement by calcite and ore minerals, but they are not replaced in altered or unaltered sandstone on either side of ore (Harshman, 1972).

The age of the Wyoming ores is about 15 to 30 Ma, based on uranium-lead isotope studies (Harshman, 1972; Davis, 1969; Ludwig, 1979; Santos and Ludwig, 1983). The age is appropriate to Miocene uplift of ranges bounding the basins, an event believed to have been favorable for influx of oxygenated ground waters capable of carrying uranium from a source area.

It is generally thought that the uranium in the Wyoming deposits was leached from granite in mountains that bound the Tertiary basins or from volcanic materials in sedimentary rocks that overlie the ores. The uranium was transported by oxygenated ground water moving down the hydraulic gradient as tongues conducted through transmissive parts of the sandstone layers. The oxygenated ground water altered previously reduced rocks, and the rolls formed along redox fronts migrating downdip (Melin, 1964; Harshman, 1972). Uranium and associated elements accumulated incrementally, owing to reduction and precipitation by both biogenic and nonbiogenic sulfide (Reynolds and Goldhaber, 1983). However, the altered condition of detrital feldspars, hornblende, and epidote in ore but not in adjacent barren rocks in the Shirley Basin proves that the orebodies there formed at static interfaces, analogous to the formation of roll orebodies on the Colorado Plateau. Favorable zones for ore deposition at the interfaces were probably where indigenous carbonized plant material was most abundant.

South Texas Coastal Plain (uranium)

Uranium was discovered in western Karnes County, Texas (Plate 2C, locality 45), in 1954. Shortly thereafter, additional deposits were found southwest of Karnes County. Production began in 1958. The host rocks for the south Texas uranium ores are mostly tuffaceous sands and sandstones of the Miocene-Pliocene Goliad Sand, Miocene Oakville Sandstone, Oligocene-Miocene Catahoula Tuff, and Whitsett Formation of the Eocene Jackson Group. All units were deposited in fluvial-lacustrine coastal plain and beach environments (Eargle and others, 1971; Galloway, 1978; Adams and Smith, 1980; Reynolds and others, 1982). The strata dip gently southeastward. The sandstone units in the Catahoula and the Oakville do not contain significant carbonized plant debris or lignite, whereas those in the Whitsett Formation of the Jackson Group do. Northeast-trending faults typically lie close and parallel to elongate ore bodies. Oil and gas fields in the region also are elongate and oriented northeasterly, and many uranium deposits are localized above them. The normal color of the ore-bearing sandstones is gray, reflecting their chemically reduced character. Authigenic pyrite and marcasite are abundant downdip from ore deposits. Altered rocks, updip from ore deposits and against which the ores were deposited, have light reddish to yellowish colors, reflecting their chemically oxidized state.

Many of the Texas Coastal Plain ore deposits are small, but some individual orebodies, such as the compound Felder-Lamprecht deposit in Live Oak County (Plate 2C, locality 46), are large, as much as 5 km long, in somewhat sinuous configurations. As seen in cross sections, the orebodies are crudely C-shaped rolls with concave side against altered rock that lies updip of the orebodies. Nevertheless, some of the so-called Texas roll deposits consist of a set of irregularly shaped orebodies distributed randomly along an alteration boundary. Some orebodies occur along the margins of fluvial channels entirely within reduced pyritic clayey sand to sandy mud, and they do not have C-shaped roll forms (Galloway, 1978). Others with C-shaped forms are entirely within reduced rocks. Uranium in unweathered primary deposits occurs as coffinite and uraninite(?). Where the ores are oxidized from weathering, minerals such as meta-autunite and meta-tyuyamunite are present. Other elements besides uranium, including selenium and molybdenum, occur in the deposits, and they may be zoned as in roll orebodies in sandstone elsewhere, but the zoned distribution commonly is not well evidenced. The Felder deposit in the Miocene Oakville Sandstone has an age of about 5 Ma, possibly a final period of mineralization not precluding earlier phases (Ludwig and others, 1982a).

Uranium deposits in the sandstone units of Oligocene and Miocene age that lack indigenous carbonized plant material probably formed as a result of reduction of uranium carried in oxygenated ground waters, which flowed downdip and encountered sulfide-bearing fluids that leaked up fault planes from underlying oil and gas pools (Reynolds and Goldhaber, 1978; Goldhaber and others, 1983). Uranium deposits in sandstone of Eocene age, however, in which carbonized plant material is locally abundant, probably were precipitated as a result of reduction related to the accumulations of plant material (Reynolds and others, 1982). Much of the uranium ore in the south Texas Coastal Plain, in which pre-ore, ore-stage, and post-ore alteration processes affected the host sandstones, appears to have had a complex history (Goldhaber and others, 1978).

Schwartzwalder Mine, Colorado (uranium)

Discovered in 1949, the Schwartzwalder uranium mine (Plate 2C, locality 30) has been in production most of the time since; it is the largest hardrock uranium producer in the United States (Wallace and Karlson, 1985). It is one of a cluster of similar vein-type deposits in Proterozoic rocks not far from an unconformity beneath overlying Phanerozoic red beds. The Proterozoic rocks are submarine sedimentary and volcanic rocks and interbedded iron formation, metamorphosed to amphibolite grade and deformed several times. A key wall rock unit at the Schwartzwalder Mine, garnet-biotite gneiss, is thickened at the nose of a nearly vertical isoclinal synform and cut by subsequent faults that were active in the Proterozoic, Paleozoic, and Laramide to provide optimum structural permeability for a deeply circulating hydrothermal system. The Schwartzwalder vein system is principally steeply dipping and it has been mined through a vertical extent of 700 m. Other important veins are in tension fractures with flat dips in the hanging wall of the high-angle fault-vein system. A pre-uranium stage of deposition at the Schwartzwalder was rich in base-metal sulfides, adularia, and hematite. Fine-grained uraninite, along with ankerite, adularia, and sulfides, was deposited in a main ore stage that was interrupted by several stages of brecciation. Molybdenum is notably enriched in the uranium stage. The hydrothermal character of the ores is supported by a mineral assemblage that does not change significantly over the large vertical extent of the veins, and by fluid inclusions that indicate temperatures near 200°C (Wallace and Whelan, 1986). Lead and stable isotopic data indicate that the vein-forming constituents were derived from the Proterozoic metasedimentary rocks, with no geologic or isotopic indication of a contribution from Laramide igneous rocks such as at Central City, 30 km to the west (Ludwig and others, 1985; Wallace and Whelan, 1986).

Apex Mine, Austin District, Nevada (uranium)

The largest uranium producer in Nevada, the Apex or Rundberg Mine in the Austin District (Plate 2C, locality 2) was found in 1953 (Garside, 1973). Structurally controlled deposits occur along the faulted contact of Jurassic quartz monzonite with Cambrian quartzite and shale. The ore zone is in several fracture and breccia zones in quartzite, quartz monzonite, and also strongly sericitized aplite dikes that were intruded near the contact. Most of the mining was above the water table in intensely argillized rocks that contain locally rich pockets of autunite and torbernite. Some coffinite and uraninite have been reported, but they are not typical of the ore as mined. Tertiary dikes intrude the mineralized fault zone, and high-angle faults of Tertiary age offset mineralized zones and cause problems for mining and exploration. The ore zone has undergone multiple periods of faulting and alteration, and probably one or more periods of supergene modification. Granitic rocks in the area are not notably radioactive and contain normal amounts of uranium and thorium, whereas Tertiary volcanic rocks have high radioactivity and uranium contents in excess of 10 ppm (Nash, unpublished data, 1976).

The Apex deposit may have formed by hydrothermal processes along the intrusive granite contact, or it could have formed under the structural control of the contact zone during alteration of Tertiary volcanic rocks. Structurally controlled deposits of oxidized uranium minerals are common in the Great Basin, but none are known to contain large tonnages of uranium. Leaching of uranium from widespread silicic volcanic rocks and deposition in structurally prepared traps is an attractive hypothesis of ore formation. However, relatively recent Basin-range faulting and range tilting has affected the Apex deposit and may have tended to destroy other uranium deposits elsewhere in the region.

Marysvale District, Utah (uranium)

The Marysvale mining district (Plate 2C, locality 14) has yielded base and precious metals since the late 1860s, but it was not until 1949 that uranium was discovered in the district (Kerr, 1968). Uranium occurs in a set of veins that was emplaced chiefly in quartz monzonite, granite, and rhyolite. The quartz monzonite is a porphyry dated at about 23 Ma, with which is associated fine-grained granite that is dated at about 21 Ma; both are overlain by rhyolite tuff that is dated at about 19 Ma and intruded by rhyolite dikes that are dated at 19 to 18 Ma (Cunningham and others, 1982). Nearly vertical east-northeast–trending faults host the veins. Argillic alteration of wallrocks that formed chlorite, montmorillonite, kaolinite, calcite, sericite, fluorite, and quartz, accompanied precipitation of uranium in the veins.

Ore-bearing parts of the veins range in horizontal length from less than 30 to more than 450 m, and in width between 1 and 2 m. Where cross veins intersect the northeasterly veins, width may exceed 6 m (Kerr, 1968). The veins commonly are lenticular and discontinuous, with intervals of clay gouge, and in places they consist of wall-rock breccia cemented by vein minerals (Walker and Osterwald, 1956). The veins contain pitchblende and coffinite associated with fine-grained silica, fluorite, pyrite, and adularia. Other minerals in or near the pitchblende veins are calcite, molybdenite, jordisite, umohoite, hematite, magnetite, and marcasite. A large number of secondary uranium minerals are present near the surface. The veins increase in molybdenum content with depth. U-Pb isotope age determinations indicate an age of mineralization of 19 Ma (Cunningham and others, 1982). The uranium veins may be related to a possible molybdenum porphyry deposit deep beneath the Marysvale District (Cunningham and Steven, 1979).

Bokan Mountain, Alaska (uranium)

Uranium was discovered at Bokan Mountain (Plate 2C, locality 51) in 1955. At the Ross Adams Mine, the principal producer, uranothorite and uranothorianite occur in veinlets in peralkaline granite (MacKevett, 1963). Average uranium grade is about 0.60 percent U_3O_8, and the uranium-thorium ratio is about 1 to 1.

Grand Canyon breccia pipes, Arizona (uranium)

Breccia pipes in Phanerozoic sedimentary rocks of the Grand Canyon area (Plate 2C, locality 37 and vicinity) produced minor copper in the 19th century; they were found to contain uranium in the 1950s. In recent years they have emerged as among the most successful U.S. producers of uranium because of their higher grades and significant by-product copper, and about 5 breccia pipes are currently (1986) in production. The Orphan Lode Mine (Plate 2C, locality 37) on the south side of the Grand Canyon has been the principal producer; its ores average 0.43 percent U_3O_8 and it produces many times as much copper (Gornitz and Kerr, 1970; Wenrich, 1985). Rocks above the Mississippian Redwall Limestone contain breccia pipes that are about 30 to 150 m in diameter and have been explored for more than 450 m vertically. The breccias were formed by downward collapse into solution caverns developed in the Redwall. Uraninite and base-metal sulfide minerals, particularly chalcopyrite, fill the matrix of breccias and ring fractures. Bleaching of red beds was the chief alteration associated with deposition of ore; carbonate minerals cement the breccias and the undisturbed rocks outside of the pipes. Base metals, rare earth elements, and thorium are enriched in the ores. Fluid inclusions yield filling temperatures of 80° to 173°C and have salinities of 10 to 17 weight percent NaCl equivalent. Uranium-lead isotopic studies suggest uranium deposition occurred at about 200 Ma (Wenrich, 1985).

Midnite Mine, Washington (uranium)

The Midnite Mine (Plate 2C, locality 1) is the second-largest hardrock uranium mine in the U.S.; it produced between 1955 and 1982. Reduced and oxidized uranium minerals are disseminated and in thin veinlets in Proterozoic phyllites and calc-silicate hornfels adjacent to a uraniferous Late Cretaceous porphyritic quartz monzonite (Nash and Lehrman, 1975). The chief structural controls on ores are depressions in the roof of the pluton. Faults and stockwork fractures have localized ore in some localities. Three generally tabular orebodies have been mined.

Pyrite and marcasite are intergrown with uraninite and coffinite, commonly with colloform texture; wallrocks were mildly altered to chlorite and sericite in reactions that reflect retrograde metamorphism rather than uranium mineralization.

Geochronologic studies (Ludwig and others, 1981) demonstrate that the ores formed at 51 Ma, about 24 m.y. after intrusion of the adjacent pluton, and at the same time as emplacement of volcanic rocks near the mine. The high uranium content of the pluton (about 17 ppm), as well as the very high uranium content of zircons from the pluton, suggests that it was the original source of the uranium. The deposits seem to have formed from a warm, weakly oxidizing, near-surface solution (less than 100° C) by leaching of metasedimentary rocks, possibly beneath a capping of volcanic rocks. Deposition of uranium together with marcasite probably was similar to that postulated for deposition of uranium and marcasite in sandstone-type deposits in the presence of metastable sulfur oxyanions (Granger and Warren, 1969).

Pitch Mine, Marshall Pass District, Colorado (uranium)

Unusual deposits of uraninite plus coffinite in brecciated dolomite and other rock types at the Pitch Mine (Plate 2C, locality 33) are along a major Laramide reverse fault on the west side of the Sawatch Range (Nash, 1985). Most ore is in Mississippian dolomite, but rocks ranging from Precambrian granite to Pennsylvanian black shale and sandstone also contain ore. Exploration in the mid 1970s defined a zone of reserves about 100 m wide and at least 1,400 m long. Alteration consisted chiefly of deposition of fine-grained pyrite, marcasite, and silica as matrix cements. Prominent jasperoid zones along the fault possibly were formed at the time of uranium mineralization. Uniformly shattered dolomite in the fault zone has grades in excess of 0.5 percent U_3O_8, whereas less intensely fractured and sheared granite and black shale contain less than 0.1 percent U_3O_8 in large zones. Part of the ore is oxidized, but most is reduced and associated with pyrite-marcasite. Many elements are enriched in the ore zone; the most important associations of uranium are with sulfur, iron, and molybdenum. The most likely time of formation of the deposits was in the Oligocene(?) when volcanic rocks covered the fault zone and were hyddrothermally altered.

The Pitch Mine ore deposits may have been formed because of reduction by pre-ore sulfide minerals in dolomites and black shales, or by biogenic reduction of sulfate. The source of uranium may have been the overlying Tertiary volcanic rocks that are somewhat enriched in uranium.

Spor Mountain, Utah (uranium)

Uranium was discovered in surface exposures in The Dell near Spor Mountain, Thomas Range (Plate 2C, locality 13) in 1953; substantial mining at the Yellow Chief Mine commenced in 1959, and most of the ore was mined by 1962 (Bowyer, 1963; Lindsey, 1978). The host rock for the deposit is water-laid tuff of the 21-Ma Topaz Mountain tuff, part of an extensive pile of Tertiary volcanic rocks exposed in the Thomas Range. The Yellow Chief deposit is in the lower part of the Topaz Mountain tuff below bentonite layers and within local lenses of tuffaceous sandstone and conglomerate that were derived principally from the erosion of volcanic rocks and lesser amounts of Paleozoic carbonate and quartzite rocks.

Uranium ore lies along the east side of a normal fault, a segment of the ring-fracture system of a caldera (Shawe, 1972), where the host rocks have been downthrown against the fault and preserved from erosion. In the vicinity of the fault the tuff has been extensively zeolitized, leached of alkali metals, and probably also of uranium, perhaps by ground waters that concentrated uranium in the Yellow Chief deposit (Lindsey, 1978).

The only ore mineral in the uranium deposit is beta-uranophane. It is more or less uniformly disseminated in lenses as much as 6 m thick and 90 m long that are nearly concordant with bedding in the tuffaceous sedimentary rocks. Weeksite is found in minor amounts in limestone conglomerate that overlies the ore zone. The uranium minerals fill interstices and fractures, and coat sand grains and pebbles.

The uranium deposit probably formed by precipitation of secondary uranium minerals from meteoric water (Lindsey, 1978, 1982). The uranium may have been derived from ground-water leaching of nearby hydrothermal deposits or by leaching of uranium-rich volcanic rocks.

RECENT TWO DECADES

Date Creek Basin, Arizona (uranium)

The uranium deposit at the Anderson Mine in Date Creek Basin near Artillery Peak (Plate 2C, locality 38) was discovered in 1955 (Sherborne and others, 1979; Mueller and Halbach, 1983). Small amounts of ore were mined, but exploration that demonstrated significant resources did not take place until two decades later.

Host rocks for the deposit are gently dipping lacustrine facies of basin-fill sediments of the Miocene Chapin Wash Formation, consisting of tuffaceous mudstone, fine-grained sandstone, and silicified marlstone. The lacustrine formation in the mine area is about 150 m thick, including two basal carbonaceous units each about 35 m thick, separated by a layer of conglomeratic sandstone about 30 m thick. The carbonaceous lacustrine units contain abundant carbonized plant material and they are extensively mineralized, with low-grade (100 to 200 ppm) uranium in an area of about 5 km^2. In the upper of the two units are irregularly shaped masses of uranium ore ranging between 0.1 and 0.3 percent U_3O_8 in intervals as much as 2 m thick. Ore crosscuts bedding in places. Locally where ore layers are stacked, aggregate thickness reaches 11 m. Uranium occurs in coffinite, poorly crystallized uraninite, and colloform humates. Pyrite is common in the mineralized zones. Uranium mineralized rocks are locally enriched in vanadium (to 0.9 percent), molybdenum, and arsenic. Some stratigraphic horizons within the host formation that are low in uranium may contain as much as 0.27 percent Li and 3.5 percent F (Otton, 1985). Abundant rhyolite ash in the lacustrine rocks was altered to smectite clay minerals and heulandite-clinoptilolite zeolites, and at the same time silicified and calcified. A set of northwest-trending normal faults of Basin-range type, mostly of post–Chapin Wash age, characterizes the area surrounding the mine. During weathering, uranium was redistributed into carnotite and uraniferous silica along fractures. Age of the deposits is considered to be Miocene, slightly younger than the enclosing sedimentary rocks; Basin-range faults of late Miocene and early Pliocene age displace the deposits.

Uranium probably was preconcentrated during early diagenesis by sorption onto colloidal humic substance, silica gel, and zeolites (Mueller and Halbach, 1983). Subsequently it was precipitated principally as coffinite by bacteriogenic H_2S.

Copper Mountain, Wyoming (uranium)

Small uranium deposits were recognized and mined in Tertiary and Precambrian rocks at Copper Mountain in the Owl Creek Mountains (Plate 2C, locality 9) between 1954 and 1971 (Yellich and others, 1978; AEC records). Only recently has it been shown that several of the deposits have potential for possible bulk tonnage production based on grades in the range 0.2 to 0.01 percent U_3O_8. The deposits of high-valent (oxidized) and low-valent (reduced) uranium minerals occur chiefly in highly fractured granite gneiss and schist, but also in overlying Eocene clastic rocks. High-angle tension fractures adjacent to reverse faults controlled the distribution of uranium. Host rocks were altered to sericite, epidote, and chlorite; pyrite and hematite occur on fractures with uranium minerals of many types. Fluid inclusions thought to reflect uranium mineralization homogenize in the range 100° to 169°C (Shrier and Perry, 1982). The deposits are thought to have formed initially by hydrothermal leaching of uranium from uraniferous Archean granitic rocks, and transport of uranium into structurally prepared sites along reverse fault zones, with deposition caused by mixing of hydrothermal solutions with basinal brines containing hydrocarbons. We suspect that supergene processes also were important and possibly produced some secondary enrichment. Although a large reserve of U_3O_8 was outlined in the mid-1970s (Rocky Mountain Energy, oral communication, 1986), grades were too low to permit mining after prices fell.

McDermitt caldera, Nevada (uranium)

The Moonlight Mine at the west edge of the McDermitt caldera complex (Plate 2C, locality 2) produced a small amount of uranium in the 1970s. Uranium ore occurs along a brecciated contact between intrusive rhyolite and wall rocks of granodiorite and andesite (Rytuba and Glanzman, 1979). Pitchblende is associated with pyrite, quartz, and fluorite. Anomalous amounts of molybdenum, beryllium, and mercury are found in and near ore. Mineralization probably was related to the episode of rhyolite emplacement at about 14 Ma. Other mineralized zones are known at the northeastern margin of the McDermitt caldera complex in Oregon (Wallace and Roper, 1981). At the Bretz deposit, uranium minerals accompany earlier-deposited mercury minerals in fractured silicic lavas and ash-flow tuffs in the footwall of the ring-fracture zone of the caldera. Just south, at the Aurora deposit, uraninite and coffinite occur with pyrite and an argillic-alteration mineral assemblage in fractured rhyolite lavas.

Swanson deposit, Virginia (uranium)

One of the most significant uranium deposit discoveries of the past decade was the Swanson deposit, near Chatham, in the Piedmont belt of Virginia (Plate 2C, locality 99). The deposit was discovered about 1978 during reconnaissance radiometric traverses of highways; outcropping mineralized rock in a roadside ditch contained ore-grade material. The deposit occurs in highly sheared, cataclastic Precambrian-Paleozoic gneiss in the footwall of the Chatham fault zone at the west border of the Danville Triassic basin (Halladay, 1989). Host rocks are deformed biotite-feldspar augen gneisses and amphibolites that contain disseminated coffinite and uraninite. Several stages of fracture-filling mineralization followed the formation of the disseminated type. Host rocks were altered to hematite, chlorite, and sericite, and extensive sodium metasomatism has formed secondary albite in the rock. Uranium, chiefly in the form of coffinite, is associated with albite, chlorite, apatite, quartz, and calcite. The high rock content of albite (about 8 percent Na_2O) and apatite (about 1.8 percent P_2O_5) are unusual features of the ore zones. Metals other than uranium are not enriched in ore. The deposit formed probably by several stages of metamorphic-hydrothermal upgrading of uranium in clastic sedimentary rocks. The Swanson deposit is estimated to contain 49 million lbs of U_3O_8 at an average grade of 0.113 percent (Halladay, 1989).

Wilson Springs, Arkansas (vanadium)

Production of vanadium from the Wilson Springs deposit at the Potash Sulfur Springs intrusive complex near Hot Springs (Plate 2, locality 48) commenced in about 1967, following recognition in the early 1960s of a substantial vanadiferous contact metamorphic deposit (Hollingsworth, 1967). The deposit consists of 4 separate major orebodies (G. N. Breit, oral communication, 1986) within a contact zone of fenite, feldspathic breccias, and metamorphosed sedimentary rocks adjacent to niobium-bearing undersaturated alkalic igneous rocks of a carbonatite-cored complex emplaced into Paleozoic sedimentary rocks. A somewhat similar deposit is associated with the alkalic complex at nearby Magnet Cove, Arkansas. Vanadium at Wilson Springs occurs in clays, goethite, montroseite, and secondary vanadium minerals in a gossanlike aggregate of quartz, limonite, and clay, in the near-surface weathered part of the contact zone. Limonite gives way to pyrite in depth. Grade of the mined ore is about 1.0 percent V_2O_5, and vanadium is the only product recovered.

Phosphate rock (uranium and vanadium)

Uranium and vanadium were recovered in small amounts during the 1940s and 1950s from marine phosphorite deposits, and in larger amounts in recent years, particularly uranium from the Florida land-pebble phosphate deposits of Miocene and Pliocene ages, and vanadium from the northwestern United States Phosphoria Formation of Permian age (Plate 2C). Marine phosphorite contains an average of about 0.01 percnt U_3O_8 (range 0.005 to 0.05). The average V_2O_5 content of phosphate rock in the Phosphoria Formation is about 0.05 percent, and in Florida phosphate it is about 0.02 percent (Cathcart and Gulbrandsen, 1973). In the early 1980s, before the current depression in the uranium market, a phosphoric acid processing plant in Florida marketed about 125 t annually of U_3O_8 recovered from phosphate rock. The Phosphoria Formation contains the largest domestic reserves of vanadium, and it presently is the chief domestic source of vanadium (Kuck, 1985). The metal is recovered from vanadium-rich ferrophosphorus, a by-product of elemental phosphorous production. In recent years, on the order of about 1,800 t V_2O_5 has been produced annually from phosphate rock.

Other vanadium sources

In the past decade, about 1,800 to 3,500 t of V_2O_5 has been recovered annually in the United States from petroleum residues (Kuck, 1985). Most of the vanadium is recovered from foreign petroleum, inasmuch as U.S. petroleum generally has a low vanadium content (generally less than 15 ppm V in crude oils from the eastern and midcontinent U.S., and less than 150 ppm V in most crude oils from western states; some California and Alaska heavy crude oils contain as much as 400 ppm V). Venezuelan crude oils average about 200 ppm V, and commonly they contain as much as 350 ppm V.

Vanadium also has been recovered in the United States in substantial amounts recently from vanadiferous slag imported from South Africa (Kuck, 1985), the western world's leading producer of vanadium. The slag is a product of processing titaniferous magnetite mined from the Bushveld layered igneous complex. For example, in 1983 about 200 t of V_2O_5 was obtained from imported South African slags. For the decade prior to 1983, substantially more vanadium per year was obtained from South African vanadiferous slags, and since 1983, volume has again increased.

RESOURCES

Resources of uranium and vanadium in the United States include known reserves in many districts and mining areas, as well as undiscovered resources. Potential resources of uranium and vanadium are considered here, the use of which will depend on economic, technological, and political factors not discussed. Although other workers have attempted to quantify potential (unknown) resources, we make qualitative assessments intended to indicate only a relative likelihood that particular mineralized geologic environments could supply the commodities in the future.

We speculate that the classic environments of vein- and sandstone-type deposits in known districts could continue to produce uranium and vanadium from as yet undiscovered ores. Also, similar districts remain to be discovered elsewhere, where the classic environments have not been adequately examined. At Central City, Colorado, additional uranium resources likely will not be found in large amounts at and near the surface, but resources might exist at depth along some of the more productive veins. Deposits of the Central City type may occur in other regional mineral belts similar to the Colorado mineral belt in which the Central City District lies, and where other types of uranium deposits are known, as for example in New Mexico, Utah, Nevada, Idaho, and Montana.

Despite the intense exploration for sandstone-type uranium-vanadium deposits on the Colorado Plateau, in the Wyoming Tertiary basins, and in the Texas Coastal Plain up until the recent drastic drop in the price of uranium, some of the districts may yet contain undiscovered deposits, e.g., in untested stratigraphic horizons and in proven horizons at greater depths. Certain regional elements, such as major fault zones that have provided increased permeability to sedimentary strata and appear to have influenced localization of uranium-vanadium ores in parts of the Colorado Plateau, probably have not been employed sufficiently to guide exploration. New districts may be discovered when this and other underused parameters are applied in the search for sandstone-type deposits. Despite many similarities among sandstone-type deposits throughout the Colorado Plateau, significant differences from district to district suggest that unique local conditions prevailed in their formation. From the standpoint that somewhat varied geologic environments have resulted in formation of economic deposits of uranium and vanadium in sandstone, we are encouraged that there will be newly recognized environments that will yet yield additional substantial resources of these metals.

Mineralized breccia pipes in sedimentary rocks in the Grand Canyon area of the Colorado Plateau, although generally small, are numerous and of relatively high grade, which makes them competitive in a depressed market. The pipes that crop out are relatively easy to recognize; mineralized pipes are well expressed by their chemical and geophysical characteristics, and pipes can be tested efficiently with a few drill holes.

At the Midnite contact-metamorphic deposit, Washington, only higher grade zones were mined, but substantial zones of lower grade (about 0.1 percent U_3O_8) also are present, and they provide a significant possible resource for the future. Deposits of the Midnite type contain on the order of 15,000 t U_3O_8 of mixed high and low grades, offering a tempting exploration target where an appropriate geologic environment can be identified. Also, fault-controlled breccia deposits similar in size and grade to those at the Pitch Mine, Colorado, can be anticipated where similar favorable geologic environments of appropriate structure, host lithologies, and uranium source exist. On the other hand, sedimentary deposits like that at the Yellow Chief Mine, Utah, may occur in an appropriate geologic environment elsewhere, but none are likely to be large. Because of their secondary nature, they would be of young age and near the surface, but owing to their tectonically active environment, older deposits may have been destroyed by erosion, or buried to depths that would be uneconomical to mine.

Sediment-hosted uranium deposits such as those recently discovered in northwestern Nebraska may be widespread in the northern Great Plains. Probably numerous geologically young, low-grade, lacustrine, sediment-hosted uranium deposits of the Date Creek Basin, Arizona, type are present in Quaternary basins of the western United States, but their discovery and production await substantial improvements in economic and technological factors. The large low-grade (below 0.1 percnt U_3O_8) deposits in fractured gneiss, schist, and clastic rocks at Copper Mountain, Wyoming, likewise await improved economic conditions before they become viable. Deposits of this type probably are present elsewhere in the Rocky Mountain and Great Basin regions where favorable uranium sources in Precambrian granitic or Tertiary volcanic terranes are extensive. However, formation and preservation of these, as well as deposits of the Pitch Mine type, like the Yellow Chief Mine, are dependent on appropriate timing of structural events and source-rock availability. Calderas in which alkali rhyolitic magmatism was active, as at the McDermitt caldera complex and near the Yellow Chief Mine, might contain uranium deposits of a variety of types. The Swanson deposit near Chatham, Virginia, somewhat similar to those at Copper Mountain, suggests a possibly large unevaluated uranium resource in the Southeast. Metasedimentary rocks elsewhere in the southeastern U.S. Piedmont are geologically similar to those in the Chatham area. Some are characterized by geochemical anomalies but they have not been tested for uranium deposits. Vein deposits also are possible future sources of uranium in the Southeast.

Numerous occurrences of uranium in the northern Appalachian Mountains, particularly in Pennsylvania and in the New Jersey highlands (not described in this chapter), such as sandstone-type deposits at Mauch Chunk, Pennsylvania, as well as lode-type deposits, indicate a province of potentially large uranium resources. Because this region is extensively urbanized, however, there are severe constraints on developing resources there.

Potential resources of vanadium in contact-metamorphic deposits associated with niobium-bearing, undersaturated, alkalic-rock carbonatite complexes may exist elsewhere in the United States besides the Wilson Springs–Hot Springs area of Arkansas.

Very large resources of uranium and vanadium exist in marine phosphorites and in black shales, but large production of vanadium, from phosphorite, has begun only recently. Successful extraction of the metals from phosphate rock has been possible only because the metals are by-products of the processing of phosphate for elemental phosphorous. Economic extraction of vanadium from titaniferous magnetite in layered mafic complexes has been demonstrated in the Bushveld complex of South Africa, but economic and technological factors must change before deposits in the United States such as those in the Duluth Gabbro, Minnesota; Iron Mountain, Wyoming (Fischer, 1975b); and smaller titaniferous magnetite deposits such as at MacIntyre, New York, can be exploited. The world's largest resources of vanadium occur in titaniferous magnetite deposits.

Surficial deposits of uranium (and in part, vanadium) have not been exploited widely in the United States, but they offer the potential of large low-grade resources. Valleys and playas in desert regions may contain calcrete or related deposits characterized by minerals such as carnotite that precipitate from migrating oxygenated ground waters. Uraniferous peat bogs in temperate valleys and lake bottoms are known in New England, the Pacific Northwest, and some other mountainous regions of the country (Otton, 1986).

Quite speculative are the possibilities for occurrence in the United States of large, uranium-rich, so-called unconformity-type deposits such as those in the Athabasca Basin, Canada, and in the Alligator Rivers area of northern Australia. Perhaps the Belt Basin in northern Idaho and northwestern Montana offers certain aspects of the environments considered favorable for such deposits. The immense, uranium-bearing Olympic Dam polymetallic deposit in southern Australia might have analogs in Precambrian rocks concealed beneath the Phanerozoic cover of the U.S. midcontinent.

The United States has adequate resources of uranium and vanadium in a wide variety of geologic environments for probably long into the future. Whether or not these resources will be exploited in lieu of acquiring the commodities from foreign sources will depend on a combination of economic, technological, and even political factors. The fate of the uranium market hangs on decisions to be made principally regarding uses of nuclear energy to generate electricity. Major improvement in the vanadium market will depend on increasing economic prosperity and the ability of the U.S. steel industry to compete with the rest of the world.

REFERENCES CITED

Adams, S. S., and Smith, R. B., 1980, Geology and recognition criteria for sandstone uranium deposits in mixed fluvial-shallow marine sedimentary sequences, south Texas: U.S. Department of Energy Open-File Report GJBX–4(81), 146 p.

Adams, S. S., Curtis, H. S., Hafen, P. L., and Salek-Nejad, H., 1978, Interpretation of post-depositional processes related to formation and destruction of the Jackpile-Paguate uranium deposit, northwest New Mexico: Economic Geology, v. 73, p. 1635–1654.

Anderson, D. C., 1969, Uranium deposits of the Gas Hills, in Parker, R. B., and Harshman, E. N., eds., Wyoming uranium issue: University of Wyoming Contributions to Geology, v. 8, no. 2, pt. 1, p. 93–103.

Bailey, R. V., 1969, Uranium deposits in the Great Divide Basin–Crooks Gap area, Fremont and Sweetwater Counties, Wyoming: University of Wyoming Contributions to Geology, v. 8, no. 2, pt. 1, p. 105–120.

Bowyer, B., 1963, Yellow Chief uranium mine, Juab County, Utah, in Sharp, B. J., and Williams, N. C., eds., Beryllium and uranium mineralization in western Juab County, Utah: Utah Geological Society Guidebook to the Geology of Utah, no. 17, p. 15–22.

Burwell, B., 1932, Mining methods and costs at the vanadium mine of the U.S. Vanadium Corp., Rifle, Colorado: U.S. Bureau of Mines Information Circular 6662, 10 p.

Busch, P. M., 1961, Vanadium; A materials survey: U.S. Bureau of Mines Information Circular 8060, 95 p.

Bush, A. L., Bromfield, C. S., and Pierson, C. T., 1959, Areal geology of the Placerville Quadrangle, San Miguel County, Colorado: U.S. Geological Survey Bulletin 1072–E, p. 299–384.

Bush, A. L., Marsh, O. T., and Taylor, R. B., 1960, Areal geology of the Little Cone Quadrangle, Colorado: U.S. Geological Survey Bulletin 1082–G, p. 423–492.

Butler, A. P., Jr., Finch, W. I., and Twenhofel, W. S., 1962, Epigenetic uranium in the United States: U.S. Geological Survey Mineral Investigations Resource Map MR–21, scale 1:3,168,000.

Cathcart, J. B., and Gulbrandsen, R. A., 1973, Phosphate deposits, in Brobst, D. A., and Pratt, W. P., eds., United States mineral resources: U.S. Geological Survey Professional Paper 820, p. 515–525.

Chenoweth, W. L., 1975, Uranium deposits of the Canyonlands area, in Fassett, J. E., ed., Canyonland Country: Four Corners Geological Society Guidebook of 8th Field Conference, p. 253–260.

——, compiler, 1980a, Guidebook Colorado uranium field trip: Denver, Colorado, American Association of Petroleum Geologists Energy Minerals Division, 93 p.

——, 1980b, Uranium in Colorado, in Kent, H. C., and Porter, K. W., eds., Colorado geology: Rocky Mountain Association of Geologists, p. 217–224.

——, 1980c, Uranium-vanadium deposits of the Henry Mountains, in Picard, M. D., ed., Henry Mountains symposium: Utah Geological Association, p. 299–304.

——, 1982, The vanadium-uranium deposits of the East Rifle Creek area, Garfield County, Colorado, in Averett, W. R., ed., Southeastern Piceance Basin, western Colorado: Grand Junction Geological Society Field Trip Guidebook, p. 79–81.

——, 1985a, Early vanadium-uranium mining in Monument Valley, Apache and Navajo Counties, Arizona, and San Juan County, Utah: Arizona Bureau of Geologic and Mineral Technology Open-File Report 85–15, 13 p.

——, 1985b, Historical review of uranium production from the Todilto Limestone, Cibola and McKinley Counties, New Mexico: New Mexico Geology, v. 7, p. 80–83.

——, 1986, Geology and production history of the uranium deposits in the Maybell, Colorado, area, in Stone, D. S., ed., New interpretation of northwest Colorado geology: Rocky Mountain Association of Geologists, p. 289–292.

Chenoweth, W. L., and Malan, R. C., 1973, The uranium deposits of northeastern Arizona, in James, H. L., ed., Guidebook of Monument Valley and vicinity, Arizona and Utah: New Mexico Geological Society 24th Field Conference, p. 139–149.

Coffin, R. C., 1921, Radium, uranium, and vanadium deposits of southwestern Colorado: Colorado Geological Survey Bulletin 16, 231 p.

Craig, L. C., and others, 1955, Stratigraphy of the Morrison and related formations, Colorado Plateau region; A preliminary report: U.S. Geological Survey Bulletin 1009–E, p. 125–168.

Creasey, S. C., 1950, Geology of the St. Anthony (Mammoth) area, Pinal County, Arizona, in Arizona zinc and lead deposits, Part 1: Arizona Bureau of Mines Bulletin 156, Geologic series 18, p. 63–84.

Cunningham, C. G., and Steven, T. A., 1979, Uranium in the central mining area, Marysvale District, west-central Utah: U.S. Geological Survey Miscellaneous Investigations Series Map I–1177, scale 1:24,000.

Cunningham, C. G., Ludwig, K. R., Naeser, C. W., Weiland, E. K., Mehnert, H. H., Steven, T. A., and Rasmussen, J. D., 1982, Geochronology of hydrothermal uranium deposits and associate igneous rocks in the eastern source area of the Mount Belknap Volcanics, Marysvale, Utah: Economic Geology, v. 77, p. 453–463.

Davis, J. F., 1969, Uranium deposits of the Powder River Basin, in Parker, R. B., and Harshman, E. N., eds., Wyoming uranium issue: University of Wyoming Contributions to Geology, v. 8, no. 2, pt. 1, p. 131–141.

Eargle, D. H., Hinds, G. W., and Weeks, A.M.D., 1971, Uranium geology and mines, south Texas; American Association of Petroleum Geologists Convention, Houston, Texas, April 1–2, 1971: Houston Geological Society Field Trip Guidebook, 59 p.

Finch, W. I., Butler, A. P., Jr., Armstrong, F. C., and Weissenborn, A. E., 1973, Uranium, in Brobst, D. A., and Pratt, W. P., eds., United States mineral resources: U.S. Geological Survey Professional Paper 820, p. 456–468.

Fischer, R. P., 1942, Vanadium deposits of Colorado and Utah; A preliminary report: U.S. Geological Survey Bulletin 936–P, p. 363–394.

——, 1960, Vanadium-uranium deposits of the Rifle Creek area, Garfield County, Colorado, with a section on Mineralogy, by Theodore Bottinelly: U.S. Geological Survey Bulletin 1101, 52 p.

——, 1975a, Geology and resources of base-metal vanadate deposits: U.S. Geological Survey Professional Paper 926–A, 14 p.

——, 1975b, Vanadium resources in titaniferous magnetite deposits: U.S. Geological Survey Professional Paper 926–B, 10 p.

Fishman, N. S., Reynolds, R. L., and Robertson, J. F., 1985, Uranium mineralization in the Smith Lake District of the Grants Uranium region, New Mexico: Economic Geology, v. 80, p. 1348–1364.

Galloway, W. E., 1978, Uranium mineralization in a coastal-plain fluvial aquifer system; Catahoula Formation, Texas: Economic Geology, v. 73, p. 1655–1676.

Garside, L. J., 1973, Radioactive mineral occurrences in Nevada: Nevada Bureau of Mines and Geology Bulletin 81, 121 p.

Goldhaber, M. B., Reynolds, R. L., and Rye, R. O., 1978, Origin of a south Texas roll-type uranium deposit; 2, Petrology and sulfur isotope studies: Economic Geology, v. 73, p. 1690–1705.

——, 1983, Role of fluid mixing and fault-related sulfide in origin of the Ray Point uranium district, South Texas: Economic Geology, v. 78, p. 1043–1063.

Gornitz, V., and Kerr, P. F., 1970, Uranium mineralization and alteration, Orphan Mine, Grand Canyon, Arizona: Economic Geology, v. 65, p. 751–768.

Granger, H. C., and Warren, C. G., 1969, Unstable sulfur compounds and the origin of roll-type uranium deposits: Economic Geology, v. 64, p. 160–171.

Granger, H. C., Santos, E. S., Dean, B. G., and Moore, F. B., 1961, Sandstone-type uranium deposits at Ambrosia Lake, New Mexico; An interim report: Economic Geology, v. 56, p. 1179–1210.

Granger, H. C., Finch, W. I., and others, 1985, The Colorado Plateau uranium province: International Atomic Energy Agency Technical Committee on Recognition of Uranium Provinces, Geological Museum, London, 38 p.

Halladay, C. R., 1989, The Swanson uranium deposit, Virginia: a structurally controlled U-P albitite deposit [abs.], in Uranium resources and geology of North America: Saskatoon, Canada, Proceedings of a technical committee meeting organized by the International Atomic Energy Agency, September 1987, p. 519.

Harshman, E. N., 1972, Geology and uranium deposits, Shirley Basin Area, Wyoming: U.S. Geological Survey Professional Paper 745, 82 p.

Hawley, C. C., Wyant, D. G., and Brooks, D. B., 1965, Geology and uranium deposits of the Temple Mountain District, Emery County, Utah: U.S. Geological Survey Bulletin 1192, 154 p.

Hess, F. L., 1908, Minerals of the rare-earth metals at Baringer Hill, Llano County, Texas, in Contributions to economic geology, 1907, Part 1, Metals and nonmetals, except fuels: U.S. Geological Survey Bulletin 340, p. 286–294.

Hollingsworth, J. S., 1967, Geology of the Wilson Springs vanadium deposits, Garland County, Arkansas: Geological Society of America Guidebook, Central Arkansas Economic Geology and Petrology Field Conference, November 18–19, 1967, p. 22–28.

Huber, G. C., 1981, Geology of the Lisbon Valley uranium district, southeastern Utah, in Epis, R. C., and Callender, J. F., eds., Western slope Colorado, western Colorado and eastern Utah: New Mexico Geological Society,

Thirty-second field conference, p. 177–182.

Isachsen, Y. W., and Evensen, C. G., 1956, Geology of uranium deposits of the Shinarump and Chinle Formations on the Colorado Plateau, *in* Page, L. R., and others, eds., Contributions to the geology of uranium and thorium: U.S. Geological Survey Professional Paper 300, p. 263–280.

Jensen, M. L., Field, C. W., and Nakai, N., 1960, Sulfur isotopes and the origin of sandstone-type uranium deposits; U.S. Atomic Energy Commission Biennial Progress Report for 1959–1960, Contract AT(30–1)–2261: New Haven, Connecticut, Yale University, Department of Geology, 281 p.

Kelley, V. C., Kittel, D. F., and Melancon, P. E., 1968, Uranium deposits of the Grants region, *in* Ridge, J. D., ed., Ore deposits of the United States, 1933–1967, Graton-Sales Volume, v. 1: New York, American Institute of Mining, Metallurgical, and Petroleum Engineers, p. 747–769.

Kerr, P. F., 1968, The Marysvale, Utah, uranium deposits, *in* Ridge, J. D., ed., Ore deposits of the United States, 1933–1967, The Graton-Sales Volume: New York, American Institute of Mining, Metallurgical, and Petroleum Engineers, v. II, p. 1020–1042.

Kimball, G., 1904, Discovery of carnotite: Engineering and Mining Journal, v. 77, p. 956.

King, J. W., Noble, E. A., Russell, R. T., and Austin, S. R., 1965, Preliminary report on the geology and uranium deposits of the Gas Hills area, Fremont and Natrona Counties, Wyoming: U.S. Atomic Energy Commission Open-File Report RME–200, 60 p.

Kirk, A. R., and Condon, S. M., 1986, Structural control of sedimentation patterns and the distribution of uranium deposits in the Westwater Canyon Member of the Morrison Formation, northwestern New Mexico; A subsurface study, *in* Turner-Peterson, C. E., Santos, E. S., and Fishman, N. S., eds., A basin analysis case study; The Morrison Formation, Grants Uranium region, New Mexico: American Association of Petroleum Geologists Studies in Geology 22, p. 105–143.

Kuck, P. H., 1985, Vanadium, *in* Mineral facts and problems, 1985: U.S. Bureau of Mines Bulletin 675 preprint, 21 p.

Lindgren, W., 1933, Mineral deposits: New York and London, McGraw-Hill, 930 p.

Lindsey, D. A., 1978, Geology of the Yellow Chief Mine, Thomas Range, Juab County, Utah, *in* Shawe, D. R., ed., Guidebook to mineral deposits of the central Great Basin: Nevada Bureau of Mines and Geology Report 32, p. 65–68.

—— , 1982, Tertiary volcanic rocks and uranium in the Thomas Range and northern Drum Mountains, Juab County, Utah: U.S. Geological Survey Professional Paper 1221, 71 p.

Ludwig, K. R., 1979, Age of uranium mineralization in the Gas Hills and Crooks Gap districts, Wyoming, as indicated by U-Pb isotope apparent ages: Economic Geology, v. 74, p. 1654–1668.

Ludwig, K. R., Nash, J. T., and Naeser, C. W., 1981, U-Pb isotope systematics and age of uranium mineralization, Midnite Mine, Washington: Economic Geology, v. 76, p. 89–110.

Ludwig, K. R., Goldhaber, M. B., Reynolds, R. L., and Simmons, K. R., 1982a, Uranium-lead isochron age and preliminary sulfur isotope systematics of the Felder uranium deposit, south Texas: Economic Geology, v. 77, p. 557–563.

Ludwig, K. R., Rubin, B., Fishman, N. S., and Reynolds, R. L., 1982b, U-Pb ages of uranium ores in the Church Rock uranium district, New Mexico: Economic Geology, v. 77, p. 1942–1945.

Ludwig, K. R., Simmons, K. R., and Webster, J. D., 1984, U-Pb isotope systematics and apparent ages of uranium ores, Ambrosia Lake and Smith Lake Districts, Grants mineral belt, New Mexico: Economic Geology, v. 79, p. 322–337.

Ludwig, K. R., Wallace, A. R., and Simmons, K. R., 1985, The Schwartzwalder uranium deposit; 2, Age of uranium mineralization and lead isotope constraints on genesis: Economic Geology, v. 80, p. 1858–1871.

MacKevett, E. M., Jr., 1963, Geology and ore deposits of the Bokan Mountain uranium-thorium area, southeastern Alaska: U.S. Geological Survey Bulletin 1154, 125 p.

Melin, R. E., 1964, Description and origin of uranium deposits in Shirley Basin, Wyoming: Economic Geology, v. 59, p. 835–849.

Miller, D. S., and Kulp, J. L., 1963, Isotopic evidence on the origin of the Colorado Plateau uranium ores: Geological Society of America Bulletin, v. 74, p. 609–630.

Moench, R. H., and Schlee, J. S., 1967, Geology and uranium deposits of the Laguna district, New Mexico: U.S. Geological Survey Professional Paper 519, 117 p.

Mrak, V. A., 1968, Uranium deposits in the Eocene sandstones of the Powder River Basin, Wyoming, *in* Ridge, J. D., ed., Ore deposits of the United States, 1933–1967, Graton-Sales Volume: New York, American Institute of Mining, Metallurgical, and Petroleum Engineers, p. 838–848.

Mueller, A., and Halbach, P., 1983, The Anderson Mine (Arizona); An early diagenetic uranium deposit in Miocene lake sediments: Economic Geology, v. 78, p. 275–292.

Nash, J. T., 1985, Geology and genesis of uranium deposits at the Pitch Mine, Saguache County, Colorado, *in* Fuchs, H., ed., Vein type uranium deposits: Vienna, International Atomic Energy Agency, IAEA–TECDOC–361, p. 169–180.

Nash, J. T., and Lehrman, N. J., 1975, Geology of the Midnite Mine area, Stevens County, Washington; A preliminary report: U.S. Geological Survey Open-File Report 75–402, 36 p.

Otton, J. K., 1985, Geologic environment of uranium in lacustrine host rocks in the western United States, *in* Report of the working group on uranium geology organized by the International Atomic Energy Agency, Geologic environments of sandstone-type uranium deposits: International Atomic Energy Agency Technical document 328, p. 229–241.

—— , 1986, Uranium: Geotimes, v. 31, no. 2, p. 56–57.

Reynolds, R. L., and Goldhaber, M. B., 1978, Origin of a south Texas roll-type uranium deposit; 1, Alteration of iron-titanium oxide minerals: Economic Geology, v. 73, p. 1677–1689.

—— , 1983, Iron disulfide minerals and the genesis of roll-type uranium deposits: Economic Geology, v. 78, p. 105–120.

Reynolds, R. L., Goldhaber, M. B., and Carpenter, D. J., 1982, Biogenic and nonbiogenic ore-forming processes in the south Texas uranium district; Evidence from the Panna Maria deposit: Economic Geology, v. 77, p. 541–556.

Rytuba, J. J., and Glanzman, R. K., 1979, Relation of mercury, uranium, and lithium deposits to the McDermitt caldera complex, Nevada-Oregon, *in* Ridge, J. D., ed., Papers on mineral deposits of western North America: Nevada Bureau of Mines and Geology Report 33, p. 109–117.

Santos, E. S., and Ludwig, K. R., 1983, Age of mineralization at the Highland Mine, Powder River Basin, Wyoming, as indicated by U-Pb isotope analyses: Economic Geology, v. 78, p. 498–501.

Scarborough, R. B., 1981, Radioactive occurrences and uranium production in Arizona: Arizona Bureau of Geology and Mineral Technology Open-File Report 81–1, 297 p.

Shawe, D. R., 1962, Localization of the Uravan mineral belt by sedimentation, Geological Survey Research 1962: U.S. Geological Survey Professional Paper 450–C, p. C6–C8.

—— , 1966, Zonal distribution of elements in some uranium-vanadium roll and tabular deposits on the Colorado Plateau, *in* Geological Survey Research 1966: U.S. Geological Survey Professional Paper 550–B, p. B169–B175.

—— , 1969, Possible exploration targets for uranium deposits, south end of the Uravan mineral belt, Colorado-Utah, Geological Survey Research 1969: U.S. Geological Survey Professional Paper 650-B, p. B73–B76.

—— , 1972, Reconnaissance geology and mineral potential of the Thomas, Keg, and Desert calderas, central Juab County, Utah: U.S. Geological Survey Professional Paper 800–B, p. B67–B77.

—— , 1976, Sedimentary rock alteration in the Slick Rock District, San Miguel and Dolores Counties, Colorado: U.S. Geological Survey Professional Paper 576–D, 51 p.

Shawe, D. R., Archbold, N. L., and Simmons, G. C., 1959, Geology and uranium-vanadium deposits of the Slick Rock District, San Miguel and

Dolores Counties, Colorado: Economic Geology, v. 54, p. 395–415.

Sherborne, J. E., Buckovic, W. A., DeWitt, D. B., Hellinger, T. S., and Parlak, S. J., 1979, Major uranium discovery in volcaniclastic sediments, Basin and Range province, Yavapai County, Arizona: American Association of Petroleum Geologists Bulletin, v. 63, p. 621–646.

Shrier, T., and Perry, W. T., 1982, A hydrothermal model for the North Canning uranium deposit, Owl Creek Mountains, Wyoming: Economic Geology, v. 77, p. 632–645.

Sims, P. K., and Sheridan, D. M., 1964, Geology of uranium deposits in the Front Range, Colorado: U.S. Geological Survey Bulletin 1159, 116 p.

Sims, P. K., Osterwald, F. W., and Tooker, E. W., 1955, Uranium deosits in the Eureka Gulch area, Central City District, Gilpin County, Colorado: U.S. Geological Survey Bulletin 1032–A, 29 p.

Sims, P. K., and others, 1963a, Geology of uranium and associated ore deposits, central part of the Front Range mineral belt, Colorado: U.S. Geological Survey Professional Paper 371, 119 p.

Sims, P. K., Drake, A. A., Jr., and Tooker, E. W., 1963b, Economic geology of the Central City District, Gilpin County, Colorado: U.S. Geological Survey Professional Paper 359, 231 p.

Stieff, L. R., Stern, T. W., and Milkey, R. G., 1953, A preliminary determination of the age of some Colorado Plateau uranium ores by the lead-uranium method: U.S. Geological Survey Circular 271, 19 p.

Trites, A. F., Jr., and Chew, R. T., III, 1955, Geology of the Happy Jack Mine, White Canyon area, San Juan County, Utah: U.S. Geological Survey Bulletin 1009–H, p. 235–248.

Turner-Peterson, C. E., Gundersen, L. C., Francis, D. S., and Aubrey, W. A., 1980, Fluvio-lacustrine sequences in the Upper Jurassic Morrison Formation and the relationship of facies to tabular uranium ore deposits in the Poison Canyon area, Grants mineral belt, New Mexico, *in* Turner-Peterson, C. E., ed., Uranium in sedimentary rocks; Application of the facies concept to exploration: Society of Economic Paleontologists and Mineralogists, Rocky Mountain Section, p. 177–211.

U.S. Department of Energy, 1982, American sources of uranium acquired by the Manhattan Project: U.S. Department of Energy Open-File Report TM–350, 4 p.

—— , 1983, Statistical data of the uranium industry: U.S. Department of Energy Report No. GJO–100(83), 77 p.

Walker, G. W., and Osterwald, F. W., 1956, Relation of secondary uranium minerals to pitchblende-bearing veins at Marysvale, Piute County, Utah, *in* Page, L. R., and others, eds., Contributions to the geology of uranium and thorium: U.S. Geological Survey Professional Paper 300, p. 123–129.

Wallace, A. B. and Roper, M. W., 1981, Geology and uranium deposits along the northeastern margin, McDermitt caldera complex, Oregon, *in* Goodell, P. C., and Waters, A. C., eds., Uranium in volcanic and volcaniclastic rocks: American Association of Petroleum Geologists Studies in Geology 13, p. 73–79.

Wallace, A. R., and Karlson, R. C., 1985, The Schwartzwalder uranium deposit; 1, Geology and structural controls on mineralization: Economic Geology, v. 80, p. 1842–1857.

Wallace, A. R., and Whelan, J. F., 1986, The Schwartzwalder uranium deposit; 3, Alteration, vein mineralization, light-stable isotopes, and genesis of the deposit: Economic Geology, v. 81, p. 872–888.

Weeks, A. D., Coleman, R. G., and Thompson, M. E., 1959, Summary of the ore mineralogy, *in* Garrels, R. M., and Larsen, E. S., eds., Geochemistry and mineralogy of the Colorado Plateau uranium ores: U.S. Geological Survey Professional Paper 320, p. 65–79.

Wenrich, K. J., 1985, Mineralized breccia pipes in northern Arizona: Economic Geology, v. 80, p. 1722–1735.

Witkind, I. J., and Thaden, R. E., 1963, Geology and uranium-vanadium deposits of the Monument Valley area, Apache and Navajo Counties, Arizona: U.S. Geological Survey Bulletin 1103, 171 p.

Wood, H. B., 1968, Geology and exploitation of uranium deposits in the Lisbon Valley area, Utah, *in* Ridge, J. D., ed., Ore deposits of the United States, 1933–1967, Graton-Sales Volume, v. 1: New York, American Institute of Mining, Metallurgical, and Petroleum Engineers, p. 770–789.

Yellich, J. A., Cramer, R. T., and Kendall, R. G., 1978, Copper Mountain, Wyoming, Uranium deposit—Rediscovered: Wyoming Geological Association, Thirtieth Annual Field Conference Guidebook, p. 311–327.

Manuscript Accepted by the Society December 30, 1988

ACKNOWLEDGMENT

This manuscript was improved by the reviews of G. N. Breit and W. I. Finch.

NOTE ADDED IN PROOF

Recent studies (Finch and others, 1990) of uraniferous breccia pipes in the Grand Canyon region of the Colorado Plateau in northern Arizona and adjacent Utah indicate that the pipes contain significant resources of uranium. The newly recognized potential in the breccia pipe deposits adds substantially to U.S. uranium resources.

REFERENCES CITED

Finch, W. I., Sutphin, H. B., Pierson, C. T., McCammon, R. B., and Wenrich, K. J., 1990, The 1987 estimate of undiscovered uranium endowment in solution-collapse breccia pipes in the Grand Canyon region of northern Arizona and adjacent Utah: U.S. Geological Survey Circular 1051, 19 p.

Chapter 8

The other metals

Ronald G. Worl
U.S. Geological Survey, Box 25046, Denver Federal Center, Denver, Colorado 80225

INTRODUCTION

There are many metals of strategic or critical importance to the United States other than those covered in the preceding chapters. The purpose of this chapter is to provide a brief description of the other metals: tungsten, tin, niobium, tantalum, titanium, rare-earth elements, thorium, beryllium, mercury, and aluminum. Each metal will be discussed separately, following the same general outline.

The introductory section for each metal gives background information, with emphasis on the metal's importance to society and the United States. Availability, ease of workability, and unique physical and chemical properties are the main original reasons for each metal being integrated into modern society. Now that these elements are established as essential, however, other factors such as politics, physical distribution, relationship to co-products, environmental and health concerns, market stability, future use, industry infrastructure, and military applications are important in determining national concern. Highlights and milestones in the development of the metal into an essential commodity are given, both those that increase demand and those that lessen it.

The United States is totally reliant on imports of raw-material sources for four of the metals (Sn, Ta, Nb, and Al), and to a large degree, for sources of five others (W, Ti, REE, Th, and Hg). We are self sufficient in only one of the "other" metals (Be). The reasons for a general lack of domestic sources are varied. For some of the metals, domestic reserves are lacking or of such a low grade that the expense to mine them would be prohibitive. Widely fluctuating prices and supply-and-demand situations with some of the metals keep domestic exploration and development to a minimum. Enormous reserves and market control by other countries also have a curtailing effect on the domestic mining industry, even if there are viable domestic reserves. Domestic production of some metals (W, Sn, Th, and REE) is wholly or in part as a by-product; the major commodity controls production, which often does not provide for a stable supply of the by-product metal.

A section on geology for each metal emphasizes the distribution and setting and description of the ores of the metal. The metals occur in a wide variety of geologic environments, but two in particular contain economic deposits of several of the metals discussed: alkalic rock-carbonatite complexes, and residual and placer deposits. Felsic plutonic rock environments also contain important economic deposits. Some of the metals are of low crustal abundance, but still form exploitable deposits in several environments, whereas other metals of greater crustal abundance form limited deposits in very few environments. Some metals are restricted in deposit types and associations, whereas others form a great variety of types with many associations. Many examples of foreign deposits are given to illustrate deposit model types that may be present, but unexploited, in the United States.

A section on resources covers current production, known reserves and resources, problems, and controlling factors. The distribution of current production varies, with the free world and national sources of niobium and some of the rare-earth elements coming essentially from one country, or one mine, whereas production of others comes from many countries and several types of deposits. Environmental and health hazards are becoming a major concern, especially in the United States, and have curtailed some usages and will probably have a pronounced impact in the future. Reserves and resources must always be considered in the context of other factors. A country holding a large reserve may not be a current major producer or have the potential of becoming one because of the overriding influence of these other factors. Control of the exploitation of a metal, in addition to demand, comes in many different forms. Some are political, such as economic control by countries with overwhelming resources, imposed international agreements, and the lack of stable markets and supplies. Many are physical and related to production of co-products and by-products that affect the metal in question.

The future of the group of "other" metals varies from a probable great increase in demand to a decline in demand. Plate 2D, which accompanies this volume, shows the location and metal content of some of the important mines and prospects within the conterminous United States. For more detailed information, the reader is referred to a series of maps published by the U.S. Geological Survey. These show metal provinces and individual deposits and prospects based on size and production. An accompanying table gives the genetic setting in each deposit or district. References are given at the end of the chapter. Measurements in this chapter are metric, except where convention dictates otherwise.

Worl, R. G., 1991, The other metals, *in* Gluskoter, H. J., Rice, D. D., and Taylor, R. B., eds., Economic Geology, U.S.: Boulder, Colorado, Geological Society of America, The Geology of North America, v. P-2.

TUNGSTEN

Introduction

Tungsten is a critical and vital element in the aerospace and other advanced industries, and Stafford (1985a, p. 894) estimates that as much as 50 percent of this usage is essential and nonsubstitutable. The United States has imported 40 to 60 percent of its consumption needs since 1960. The domestic tungsten industry has suffered critical supply-and-demand problems in the past; these have been particularly severe since 1982. In 1983, domestic production was at its lowest level since 1934, but U.S. imports were the highest since 1956.

The major end-product uses of tungsten in the United States are cutting and wear-resistant materials (63 percent), mill products (20 percent), specialty steels (6 percent), and as miscellaneous (11 percent; Stafford, 1985b, p. 929). Tungsten carbide, the most common tungsten product, is one of the hardest cutting agents. Some uses of tungsten are in cutting edges and surfaces subject to intense wear and abrasion, filaments in electric lamps, cathodes for electronic tubes, contact points in distributors, specialty steels, as additives to paints, dyes, toners and glass, and numerous other applications. Tungsten metal is an important element in the aerospace industry because it retains its strength at elevated temperatures under most conditions.

Some uses of tungsten, such as filaments in electric lamps, have no viable substitutes, and in most other uses, substitutions mean a loss in performance, especially at higher temperatures. Titanium carbide compounds are gaining acceptance in cutting tools, and aluminum oxide and industrial diamonds can replace tungsten carbides in some applications.

Primary sources of tungsten are ores of minerals of the wolframite group, termed "black ores," and scheelite. A considerable amount of tungsten is recycled, commonly in alloy or carbide form, but tungsten recovery data are incomplete. Tungsten concentrates are processed chemically to produce ammonium paratungstate (APT) from which tungsten metal powder is made. Other intermediate products are tungistic acid, sodium tungstate, ferrotungsten, and tungsten carbide powder. Tungsten concentrate is sold in units of tungsten trioxide (WO_3), and has a standard grade of about 60 percent WO_3. Scheelite concentrate, natural or synthetic, can be added directly to steel melts, but wolframite concentrates cannot because of their tin and manganese contents.

Tungsten is designated a strategic and critical mineral and is stockpiled by the U.S. government. The stockpile was built to large proportions after the Korean War when supplies from China were not available. Since 1965 a large overstock, based on changing stockpile criteria, has been released. Tungsten was first used commercially in the production of tungsten-manganese steel in the middle of the nineteenth century, which accounted for most tungsten consumption prior to 1900. Tungsten carbide tools became commercial in 1929. Domestic production has fluctuated greatly since it began in 1898. The fluctuations relate to times of need and to acquisitions and releases of the U.S. government stockpile. The United States produced 7,182 tons of concentrate in 1955 and 980 tons in 1983. In 1955 there were more than 750 active mines; in 1983, production essentially came from 2 mines in California (Stafford, 1985a, p. 881).

Geology

Tungsten has an estimated average crustal abundance of 1 to 1.3 ppm and is one of the rarer elements in the crust (Hobbs and Elliott, 1973, p. 669). In igneous rock series, tungsten is more abundant in felsic end members, mainly granite. In the tin-tungsten-fluorite belts of the circum-Pacific, tin and tungsten values increase; the typical differentiation sequence is quartz diorite, granodiorite, hornblende-biotite granite, biotite granite, and muscovite-rich granite, concomitant with increases in Li, Rb, Be, Mo, and F. During magmatic crystallization, tungsten concentrates in the residual fluids, with little going into the common rock-forming minerals. Some of the accessory iron-oxide, titanium, and niobium-tantalum minerals may contain tungsten, and the hypogene manganese oxides locally contain significant amounts. Molybdenum has geochemical characteristics similar to those of tungsten, but the two elements are partitioned in nature mineralogically; molybdenum tends to form sulfides, and tungsten tends to form tungstates. In the surface environment the major tungsten minerals remain as resistates, and the small amount of tungsten released into solution is adsorbed onto iron and manganese oxides. The major tungsten minerals have high specific gravities, but because they are brittle they generally break down and are lost as fine sediment in transport. Therefore placer deposits are not common. The commercially important minerals are scheelite ($CaWO_4$) and minerals of the wolframite group: huebnerite ($MnWO_4$), wolframite [(Fe, Mn)WO_4], and ferberite ($FeWO_4$). There are about 20 other tungsten-bearing minerals, most of which, except for powellite [Ca(Mo,W)O_4] and tungstite (H_2WO_4), are uncommon or rare.

Primary tungsten deposits occur in a wide range of felsic plutonic-hydrothermal environments ranging from high-temperature deep-seated to low-temperature near-surface. Most production has come from contact metasomatic (skarn) deposits, stockwork and related deposits, and tungsten-bearing quartz veins (Hobbs and Elliott, 1973, p. 673). Tungsten skarns occur throughout the world, generally associated with calc-alkaline plutonism of mid-Paleozoic to Late Cretaceous age. Einaudi and others (1981, p. 333) distinguished tungsten skarns from iron, copper, and zinc-lead skarns associated with calc-alkaline intrusives. The important differences are that in tungsten skarns there is a reduced calc-silicate and opaque mineralogy, and the associated intrusive rocks lack evidence of forceful injection. Tungsten skarns are generally stratiform, with ore-grade scheelite limited to metasomatized marble (Fig. 1). There is evidence for an early anhydrous skarn formation at 600° to 500°C, followed by a hydrous retrograde alteration at 450° to 300°C (Einaudi and others, 1981, p. 339). Highest tungsten values are associated with

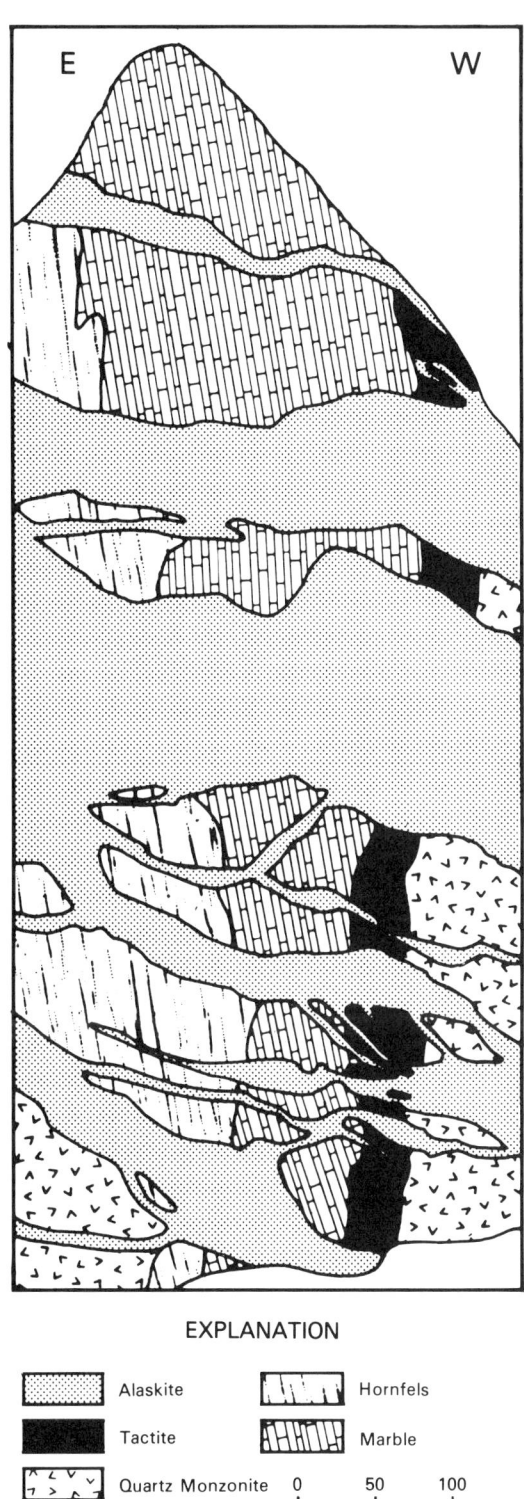

Figure 1. Vertical cross section through the Brownstone Mine, California. (After Gray and others, 1968.)

the retrograde phase, as are hydrothermal sphene and apatite. Scheelite, including the molybdenum-bearing type, is essentially the only tungsten mineral in skarns. Magnetite and fluorite are generally present, and chalcopyrite, pyrite, pyrrhotite, sphalerite, molybdenite, tetrahedrite, stibnite, and bornite are common. The presence of these minerals is generally indicative of an introduction of mineralizing material from the magma. Tungsten skarn deposits are common in the western United States: examples include Mill City and Tempiute, Nevada, and Pine Creek, California (Plate 1). At Mill City, the ore bodies are vein like, generally 1 to 2 m thick and 150 to 300 m in length, contain about 1 percent WO_3, and are composed mainly of garnet, epidote, and quartz. The tungsten is in scattered small grains of white to light brown scheelite.

Stockwork and related bodies that contain some tungsten, generally in scheelite, are numerous throughout the world. These are usually small deposits and are often associated with fissure veins. The important deposits of this type are porphyry molybdenite deposits with associated tungsten, such as at Climax, Colorado. Tungsten at Climax is in huebnerite, and although the tungsten content of the ore is low, only a few hundredths of one percent, the large amount of material mined for molybdenum allows tungsten to be recovered as a by-product. This has been a major source of tungsten in the United States in recent years. Yan and others (1980) report a porphyry-type tungsten deposit in China.

Tungsten-bearing quartz veins contain the bulk of tungsten reserves in the world, and most of the major world-class mines are of this type. The largest concentration of tungsten-bearing quartz veins is in southeast China. Tungsten vein deposits, like the skarn deposits, are generally associated spatially and genetically with granitic rocks. These are mainly S-type granites of the ilmenite series (Hutchison, 1983, p. 144). The vein deposits are mineralogically simple; they consist mainly of quartz with wolframite and/or scheelite. Gangue can include feldspar, muscovite, tourmaline, topaz, beryl, fluorite, barite, and calcite. Ore commonly includes cassiterite in China, and may also include molybdenite, bismuthinite, pyrite, chalcopyrite, sphalerite, galena, arsenopyrite, monazite, and pyrrhotite, and locally stibnite and native gold. The tungsten vein deposits commonly show multistage mineralization related to multistage intrusions of granite (Hutchison, 1983, p. 145). Tungsten vein deposits in China show four major stages; early silicate, oxide, sulfide, and carbonate minerals, each with a distinct accompanying alteration type (Yan and others, 1980). There are also distinct zoning patterns. Individual fissure veins may be as much as 3 m wide and 1,000 m long. Examples of vein deposits in the United States are at Atolia, California; Boulder District, Colorado; Ima mine, Idaho; and Hamme, North Carolina (Plate 2D). Major vein deposits outside China are in Bolivia, Malaysia, Thailand, Peru, and Portugal. The Panasqueira deposit in Portugal is composed of many horizontal ferberite-quartz veins cutting greenstone, and radiating out from the cap of a granite cupola. This deposit represents a mixing of magmatic and meteoric water, with the earlier stages predominantly magmatic and deposited at 360° to 230°C. Later stages were predominantly meteoric and were deposited at 120°C or lower (Hutchison, 1983, p. 152). Annual output is about 1,250 t of wolframite and 62 t of cassiterite concentrates.

Tungsten also occurs in several other types of deposits that may have resource potential in the future. Pipe-like breccia deposits are locally important in southeast Asia, and one of the larger deposits in China is of this type. They are not well described in the literature. Irregular-shaped pipe-like bodies of breccia ore, termed "carbonas," constitute a minor amount of the ore in the Cornwall District, southwest England. A controversial type of deposit is strata-bound tungsten, which may be synsedimentary and formed by submarine exhalations, or may be hydrothermal selective replacement. Examples occur in southwest Africa, Austria, Norway, South Korea, and Australia. Pegmatites have provided some tungsten from many parts of the world. Tungsten is a common constituent of hot spring and epithermal manganese deposits in the western United States, and was mined in at least one locality near Golconda, Nevada. Anomalous amounts of WO_3 occur in saline brines of Searles Lake and Owens Lake, California (Hobbs and Elliott, 1973, p. 675), and although the concentration is low, the large amount of brines constitutes a large tungsten resource. Placer deposits on or next to the source materials are exploited, mainly in China, Burma, and Atiola, California, but are not as significant a source as lode deposits.

There are several potential co-products and by-products associated with tungsten, including tin, molybdenum, copper, beryllium, silver, fluorspar, columbite-tantalite, bismuth, antimony, and gold. In the circum-Pacific belts, tin, tungsten, and bismuth are common co-products, and China has some districts that co-produce tin, tungsten, and fluorspar. Columbite-tantalite are by-products of tin and tungsten production in Africa. Tungsten, antimony, and gold are co-products in some Chinese deposits and at the Yellow Pine District, Idaho. The most important by-product production of tungsten in this country is from porphyry molybdenite deposits at Climax, Colorado. By-products produced from tungsten mining include copper, gold, silver, and molybdenum.

Exploration should be oriented toward the several geologic environments known to contain tungsten concentrations (Hobbs and Tooker, 1980). The most obvious are terranes of S-type granites in areas where late solutions could have migrated and depositional controls existed. Other environments should also be considered for potential large, low-grade deposits such as bedded syngenetic deposits, hot springs and near-surface hydrothermal deposits, and saline-lake brines. Geophysics can be a useful tool to indicate regional structures, outlines, and contacts of individual intrusive bodies, and some skarn deposits. Geochemistry of plutonic bodies may be indicative of associated mineralization, and is also useful in detailed examinations where associated elements are used as pathfinders. Mineral lights indicate the presence of scheelite in outcrop, and panned concentrates would indicate a nearby source of the tungsten minerals.

Resources

World production in 1984 of an estimated 44,939 tons of concentrate came from 31 countries, of which China produced 30 percent and the USSR produced 20 percent. Other important producers are Canada, Korea, Bolivia, Australia, Portugal, Austria, the United States, and Burma (Stafford, 1985b, p. 940). The United States accounted for about 3 percent of the total world production, which is about 14 percent of U.S. consumption. The United States is second only to the USSR in amount of consumption. The world has more than adequate reserves for the near future, approximately 50 percent of which are in China. Known U.S. reserves are not adequate to meet cumulative demands to the year 2000 (Stafford, 1985a, p. 893), and imports are expected to supply the bulk of U.S. needs. Significant tungsten deposits currently under evaluation could add substantially to the reserve base. A large tungsten-molybdenum-fluorspar deposit in China is reported to hold reserves of 190 million tons, averaging 0.35 percent WO_3; in Canada, a large scheelite deposit is estimated to contain 63 million tons, averaging 0.96 percent WO_3 (Ho, 1984, p. 71).

The U.S. tungsten mining industry has the capability of satisfying much more of the U.S. tungsten demand, but unstable world price and market conditions had U.S. producers operating at 10 percent of capacity in 1984. This impact, plus the dominance of the market by China, have brought a halt to most tungsten development and exploration in the United States. This could have an effect on the future domestic supply. There are no significant environmental or other problems connected with the mining and use of tungsten.

Tungsten is an element that has unique uses in modern society and is readily available at reasonable economic and environmental cost. The threat of a supply cutoff is minimal because of the amount of resources in the United States and the abundance of supply outside the United States. Demand is expected to grow at an annual world growth rate of about 2 percent to the year 2000 (Stafford, 1985a, p. 891). Most of the demand growth is keyed to uses that involve cutting, wear-resistant, and hard facing materials, the current major uses of tungsten. New technology could develop uses that would increase the demand, a likely possibility because of the unique properties of tungsten.

TIN

Introduction

Tin has been an essential element to industrial society since the Bronze Age, and today there are no satisfactory substitutes in some of its uses. Major uses are in the container, transportation, machinery, electrical, construction, and chemical industries. Tin is of critical and strategic importance to the United States because of small domestic reserves and reliance on imports for primary tin.

The major use of tin in 1984 was in solder (29 percent), followed by tin plate (27 percent), chemical (13 percent), and babbitt (7 percent; Carlin, 1985a, p. 850). This reflects a major change in usage: for many years, tinplate was dominant. Tin is pervasive throughout the machinery of modern society: in the modern automobile, most industrial machinery, computers, electrical machinery, electronic equipment, appliances, bronze and brass, and plumbing and heating applications. The use of tin in the chemical market has grown recently, and includes the use of organotin compounds in a variety of products and as fungicides and biocides. There are many other uses for tin, and as some established markets are lost, such as the tin can, others are developed.

Substitutions for tin that have had the most impact are aluminum in cans and foil, plastics, paper, and glass for packaging. The tin can is so commonplace that it seems a natural part of society, but it is being replaced. Nonmetallic materials and products coated with copper, aluminum, and zinc have replaced terneplate and tin plate in construction. Aluminum and other alloys and plastics can be substituted for bronze. There is no satisfactory substitute for tin in solder, although the increased use of integrated circuits and miniaturization in the electronics industry has resulted in the use of less tin per unit.

The primary source of tin is the mineral cassiterite, with minor amounts coming from stannite, cylinderite, franckeite, canfieldite, and teallite. An unknown, but significant, amount of tin is supplied from secondary recycling. Commercially pure tin, designated "straits" or "grade A" has a 99.85 percent minimum tin content and 0.030 percent maximum bismuth content (Carlin, 1985a, p. 847–898). Common tin has a 99 percent minimum tin content, hard tin 99.6 percent, and electrolytic tin 99.95 percent. Tin babbitt is a thin-copper-antimony alloy of varying composition. White metals such as pewter and jeweler's metal are tin alloys usually hardened with antimony. High-grade pewter contains 90 to 95 percent tin. Terneplate is an alloy of 3 to 4 parts lead to 1 part tin.

Tin is designated a strategic and critical material and is stockpiled by the U.S. government. The stockpile was greatly in excess of the goal in 1984, and the General Services Administration has been selling tin from the stockpile for several years. Tin, used as early as 3500 B.C. in bronze tools, has played a major role in the development of civilization. Early production was from old alluvial deposits such as those at Cornwall, England, and the locations of tin sources greatly influenced trade cultures. Tin has always been available, and any disruptions have been political, rather than technical. There have been major discoveries through time, and all districts, including Cornwall, still have significant reserves.

The International Tin Council (ITC) regulates tin in world trade by manipulations of a buffer stock of tin and by imposed export quotas. Several nations, including the United States, are not members of the current agreement because they feel it resembles a cartel. Consumption of tin has declined steadily since 1979, and smelter production of tin in 1984 was about 208,000 tons, the lowest level since the mid-1960s. Most of the decline in production has been from member nations, while some of the nonmember nations have increased production and the U.S. government has sold large quantities from the stockpile (International Tin Council, 1984, p. 41).

Geology

Tin is an element of several isotopes and relatively low crustal abundance, in the range of 2 to 3 ppm (Sainsbury and Reed, 1973, p. 641). Similarities in ions allow tin to substitute for iron, titanium, or calcium in some of the common minerals such as biotite, rutile, and andradite garnet. In the magmatic cycle, tin concentrates in the felsic rocks, especially in or associated with certain granites and their extrusive equivalents. In sedimentary rocks the highest concentrations are in shales and clays and phosphorite. There are numerous tin minerals (Sainsbury and Reed, 1973, p. 642), but few have commercial value, and cassiterite (SnO_2) has historically provided almost all of the tin for industry. Others with minor commercial production include the tin sulfides stannite (Cu_2FeSnS_4), cylinderite ($Pb_3Sn_4FeSb_2S_{14}$), teallite ($PbSnS_2$), canfieldite (Ag_8SnS_6), and franckeite ($Pb_5Sn_3Sb_2S_{14}$). Cassiterite is a heavy mineral that is chemically inactive, and in the weathering and erosion process it concentrates in placer deposits, which are the world's main source of tin.

Tin occurrences of the world are concentrated in elongate zones (tin belts). The most obvious, and geologically most recent, tin belts are the circum-Pacific tin-tungsten-fluorine provinces (Mitchell and Garson, 1972). These provinces are in belts of granite intrusives that parallel continental margins or island arcs, and the granites and related mineralization are thought to have formed during subduction of a lithospheric slab. Taylor (1979) prefers to define important stanniferous areas based on production, and shows that some belts are actually a series of adjacent provinces.

Primary lode tin deposits are of several types, all in association with silicic igneous rocks. The deposits take various forms that developed in plutonic, subvolcanic, and volcanic environments. Taylor (1979) and Sainsbury and Reed (1973) present several contrasting classification schemes. Deposits in the plutonic environment are associated with granites and take the form of pegmatites and aplites, disseminations, contact metamorphic deposits (pyrometasomatic), and hydrothermal fissure veins and replacements (Fig. 2). The associated granites are biotite or two-mica varieties, in which a considerable amount of tin may be in the biotite. Tin-pegmatites contain cassiterite commonly in association with beryl, columbite-tantalite, and spodumene, and are commonly large. The Manono pegmatite in Zaire is 10 km long and 100 to 400 m wide. Dissemination of small amounts of cassiterite in altered granite are not commercial deposits, but are one of the main sources of large, rich tin placers in many parts of the world. Contact metamorphic deposits generally consist of cassiterite or malayaite ($CaSnSiO_5$) in tactites with magnetite, garnet, pyroxene, fluorite, tourmaline, sulfide minerals, and in the area of Lost River, Alaska, beryllium minerals (Sainsbury and Reed, 1973, p. 644; Plate 2D). Many of the major lode tin deposits

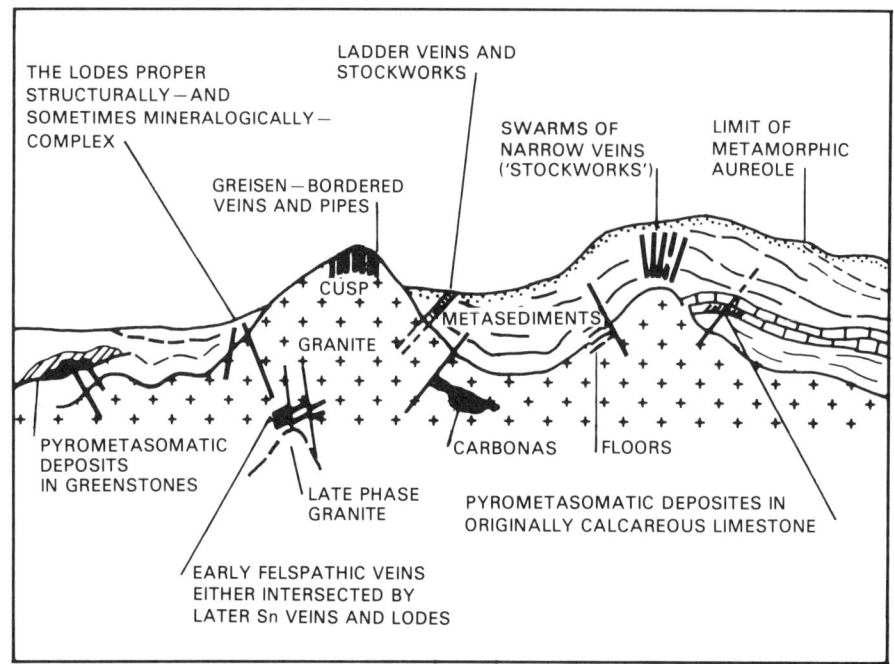

Figure 2. Diagrammatic representation of primary tin deposits in the southwest of England (Cornwall). (After Hosking, 1969.)

are fissure veins or replacement deposits in or associated with granite, termed pneumatolytic-hydrothermal by Sainsbury and Reed (1973, p. 643). Tin deposits at Cornwall, England, are of this type (Fig. 3). They are high-temperature veins and part of a distinct zoning pattern of silver-lead veins at the surface grading downward to copper veins and then to tin veins at depth. Cornwall has produced more than 3 million tons of tin beginning in historic time, with one of the larger ore bodies yielding 80,000 tons of tin (Sainsbury and Reed, 1973, p. 643).

In the volcanic-subvolcanic environment, tin deposits take the form of hydrothermal stockworks, pipes, and fissure-filling veins with pronounced accompanying hydrothermal alteration, and are the main primary deposits in the tin-tungsten-fluorine provinces of the circum-Pacific belt (Mitchell and Garson, 1972, p. B18). The deposits are mainly cassiterite-wolframite in a gangue of quartz, carbonates, fluorite, and tourmaline. Sulfide minerals of copper, lead, zinc, arsenic, bismuth, and molybdenum are common, as is alteration of the tourmalization and greisenization types. Tin-tungsten ore generally occupies the central zones of local and regional zoning patterns, grading outward through lead-zinc-silver veins to quartz-stibnite veins. Tin deposits formed in the near-surface volcanic environment are mineralogically complex, and there is apparent telescoping of tin-tungsten and porphyry-type sulfide deposits. The large, rich deposits of Bolivia are of this type. The ores from several centers in Bolivia average 2 to 4 percent tin with silver, and some of the earlier ores averaged 8 to 12 percent tin. In addition to cassiterite, much tin is in stannite, teallite, cylindrite, franckeite, and canfieldite. Quartz, pyrite, and marcasite are the main gangue minerals with pyrrhotite, bismuthinite, wolframite, sphalerite, arsenopyrite, siderite, chalcopyrite, wavellite, and tourmaline. Fluid-inclusion studies of Bolivian tin deposits estimate that temperatures declined from about 530°C in the early vein stages to about 70°C in the late stages and that salinities decreased from about 46 percent to a few percent (Kelly and Turneaure, 1970). Another type of deposit formed in the volcanic environment is small fracture fillings in Tertiary rhyolites termed "fumarole deposits" (Sainsbury and Reed, 1973, p. 644). Cassiterite occurs intergrown with specular hematite and as disseminations in altered wall rocks. Examples occur in the Black Range, New Mexico (Plate 2D).

The important placer deposits occur in the tin belts of the world close to the primary source areas. Cassiterite concentrations can be found in all types of placers; residual, eluvial, alluvial, and marine, although alluvial placers and combination alluvial-marine placers are the largest and richest. Taylor (1979, p. 290) listed 30 varieties of tin placers. The world's major tin placers are in the tin belt, some 1,600 km long and 200 km wide, along the Malaysian peninsula. Humid tropical weathering has encouraged the release of cassiterite and other heavy minerals from the numerous tin granites, and there is a continuous occurrence of the heavy minerals in residual to eluvial to alluvial to marine placers. The greatest concentration of cassiterite is in the transition zone from eluvial to alluvial, in alluvial, and in marine reworked alluvial placers. Because of sea-level fluctuations during the Pleistocene, many of the landward placers are buried beneath sediments; and alluvial placers that formed seaward of the present shore lines are now beneath sea water. Offshore placers that are the main source of cassiterite in Indonesia are sheet-like deposits formed by a transgressive sea across alluvial placers and primary sources (Sainsbury and Reed, 1973, p. 645).

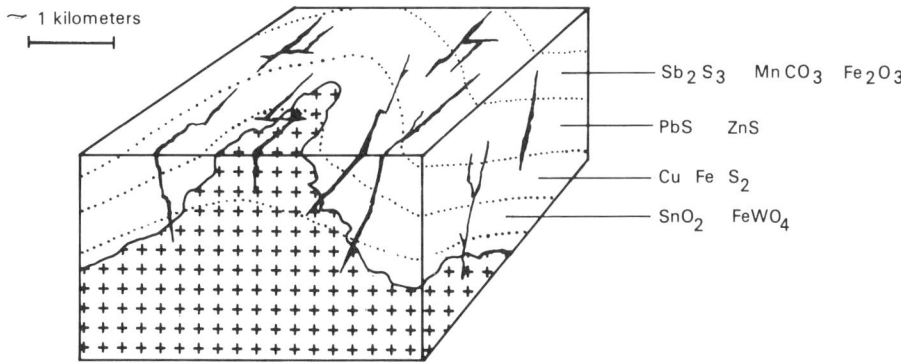

Figure 3. Block diagram showing relation of ore zones to granite-stock contacts, Cornwall, England. (From Hosking, 1969.)

There are numerous co-products and by-products associated with tin. The better known by-products are the large amounts of silver produced from lode tin production in Bolivia, Mexico, and China, and the heavy minerals monazite and columbite-tantalite from tin placers. Tungsten and tin are associates in the circum-Pacific tin-tungsten-fluorine provinces. By-product tin in North America has come from lead-zinc mines in Canada and molybdenum mines and gold placers in the United States.

In exploration for tin deposits, one must first consider the concentration of tin occurrences in linear belts and their association with specialized granites (Reed and Tooker, 1980). Specific geologic environments for ore deposition from late-igneous hydrothermal solutions, such as "cusps" in granite bodies and post-granite fractures, must be present. Analysis of mineralogic zonal patterns must be done only within a province and with care (Taylor, 1979, p. 133). The most important exploration tool is panned concentrates, and in the case of sulfide-cassiterite ores, stream-sediment geochemical samples. Sainsbury and Reed (1973, p. 646), pointing out the fact that new tin deposits are invariably similar to those mined for thousands of years, do not speculate on possible new types of tin deposits, such as those recently found for beryllium and niobium.

Resources

In 1984, 88 percent of the world's tin production of 208,000 tons came from eight countries: Malaysia, the USSR, Thailand, Indonesia, Bolivia, Brazil, China, and Australia. Twenty-five countries made up the other 12 percent of production. The world reserve figure of about 3 million tons is very close to the probable cumulative demand for tin in the world for the years 1983 to 2000: 3.2 million tons (Carlin, 1985a, p. 857). The potential for major discoveries to increase the reserve base is very good. There have been several major discoveries since similar predictions were made by Sainsbury and Reed (1973, p. 647). The possibility of developing large, low-grade deposits has improved in recent years because of advances in metallurgical procedures. Carlin (1985a, p. 858) has suggested that a deposit of about 40 million tons of ore containing 0.25 percent tin as cassiterite could become economic. Major reserves are in Malaysia, Indonesia, Thailand, Bolivia, United Kingdom, the USSR, China, and Brazil. In the United States, the picture is much bleaker. Carlin (1985a) listed 20,000 tons reserve and 700,000 tons as the probable primary tin need to the year 2000. The bulk of this tin supply will have to come from imports. Resources in the United States are mainly as a by-product of the mining and extraction of molybdenum and the base metals in porphyry deposits. Small primary deposits occur in Arizona, New Mexico, Alaska, Idaho, California, and Montana. Large multicommodity deposits like the fluorite-tin-tungsten-beryllium deposits at Lost River, Alaska, could someday be a source.

The United States is almost entirely reliant on imports of refined metal, as there is only one smelter within the United States. In times of crisis the government stockpile will be the source of tin concentrates, but there will still be the need to develop a refining capability. Even the small amount of by-product tin obtained in the United States is not a stable source. Temporary closure of the molybdenum mines in Colorado recently illustrate the point. Most tin in the recycling process is in alloys and is not extracted. In some parts of the world, placer mining of tin is in direct conflict with agriculture; considerable reserves may not be available because of this. The high price of tin, approximately ten times that of aluminum, copper, lead, or zinc, is causing consumers to seek cheaper substitutes.

Tin has been a stable industrial-use metal since metals were first used, and probably will remain so far into the future. At present there is a world surplus, but the predictions to the year 2000 show a minor shortfall between reserves and probable use. Research on increasing recovery during beneficiation, more efficient mining methods, and increased competition will probably help bring the price down. As more attention is applied to primary deposits, the very large-tonnage, low-grade deposits may be found and exploited.

NIOBIUM

Introduction

Niobium (columbium) is an important element in the production of high-strength low-alloy steels (HSLA) and superalloys, and 80 to 85 percent of the western world's niobium consumption is by the steel industry (Tron, A. R., 1984, p. 72). Niobium is of special importance to the United States because of total dependence on imports. Domestic resources are low-grade, and there has been little domestic production since 1959. Niobium and tantalum are generally discussed together because of similarities in source and availability. However, their sources and worldwide availabilities are currently quite different; therefore, they are discussed separately. In general, niobium is the term used in discussing geology, mineralogy, and chemistry, while columbium is used in discussing metallurgy, mineral economics, and mining. In this report, niobium will be used exclusively.

Known since 1801, niobium had no commercial value until the 1930s when it was used as an alloying agent in steel. It is currently used in iron-, cobalt-, and especially nickel-based superalloys in applications such as gas turbine engines, rocket subassemblies, and nuclear-reactor hardware and as ferrocolumbium in stainless steel. Research and development programs into future energy generation and transmission are using superconducting single-phase niobium-titanium alloys and the compound niobium-tin.

Titanium can replace niobium in stainless steel to improve corrosion resistance, and tantalum is a costly substitute in superalloys. Substitution of niobium by vanadium, titanium, and molybdenum in steels and superalloys is a possibility, but may result in lower performance and/or cost effectiveness.

Current primary sources of niobium are mineralized rock containing columbite or pyrochlore-group minerals and Nb-bearing Sn slags. Niobium raw materials are ferrocolumbium, columbium oxide (columbium pentoxide), and columbium metal (Cunningham, 1985b, p. 187). Ferrocolumbium comes in three grades, designated low-alloy steel, alloy and stainless steel, and high purity. Columbium pentoxide is a stable, white to buff-colored compound produced in metallurgical, ceramic, and optical grades.

Niobium is classified as a strategic and critical material with national defense consideration because of its uses. The General Services Administration maintains a stockpile that constitutes a limited emergency supply. There was limited demand for niobium until World War II, at which time a critical shortage developed because of the increased use in steel alloys. Niobium-bearing super alloys are used in jet-airplane engines, and this use after World War II and especially during the Korean conflict caused another critical shortage and spurred intensive exploration programs, both domestic and foreign. This exploration activity led to discovery of several large, low-grade carbonatite deposits and large placer deposits (Parker and Adams, 1972, p. 445). One of the placers, at Bear Valley, Idaho, was the site of the only significant niobium and tantalum production in the United States. Production ceased in 1959.

Prior to 1960, the source for niobium was columbite, mainly from pegmatites and alluvial and eluvial deposits in association with tantalite and other heavy minerals. During the 1960s and 1970s, exploration and beneficiation technology increased; at present, the major free-world supply of niobium is from large deposits of pyrochlore in Brazil and Canada. Niobium is also obtained as a by-product of tantalum processing.

Geology

Niobium and tantalum have a strong chemical coherence because of similar ionic states and ionic radii, and generally occur in the same minerals, commonly those that contain titanium, zirconium, tin, and tungsten. The minerals are chiefly oxides, multiple oxides, and hydroxides, some silicates, and one borate. The more important niobium minerals are columbite $(Fe,Mn)(Nb,Ta)_2O_5$, euxenite $(Y,Ca,Ce,U,Th)(Nb,Ti,Ta)_2O_6$, and pyrochlore $(Na,Ca,Ce)_2(Nb,Ti,Ta)_2(O,OH,F)_7$. Crustal abundance of niobium is about 20 ppm (Parker and Adams, 1972, p. 445). In igneous rock series, niobium concentrates in alkalic granites, nepheline syenites, pegmatites, and carbonatites. Nepheline syenites are commonly considerably enriched in niobium: as much as 900 ppm in some complexes in the USSR (Parker and Adams, 1972, p. 446). In agpaitic nepheline syenites (characterized by an excess of alkalies over alumina), niobium is largely fixed in titanium and zirconium minerals. In the "common" or miaskitic nepheline syenites, pyrochlore is the common niobium-tantalum mineral. Economic deposits are mainly associated with massive carbonatite bodies and carbonatite dikes related to miaskitic complexes. The niobium minerals are heavy resistates in the weathering environment and tend to concentrate in place or be worked into nearby placer deposits.

The most extensive known resources of niobium are now in carbonatite complexes. The main niobium minerals are columbite and pyrochlore, of which the latter is economically the most important. Grades range from below 0.2 percent to more than several percent Nb_2O_5 (Parker and Adams, 1972, p. 447). Carbonatites are found mainly in stable crustal areas and in association with alkalic igneous complexes. For years, pegmatites were the only source of niobium along with tantalum from the mineral series columbite-tantalite. The higher grade pegmatites are complexly zoned and characterized by abundant albite and lithium and beryllium minerals. Few deposits have been worked solely for their niobium or tantalum contents; generally the columbite-tantalite was obtained as a by-product or co-product of mica, beryl, or cassiterite operations.

Another important group of deposits is the erosional derivatives of alkalic igneous complexes. Resistate heavy minerals, such as columbite-tantalite and euxenite in placers and residual deposits, have been significant sources of niobium and tantalum. These minerals are not common in sea-beach placers. The major placer in the United States, at Bear Valley, Idaho (Tooker and Parker, 1980; Plate 2D), contains euxenite and subordinate columbite together with ilmenite, magnetite, zircon, garnet, and monazite. This placer was mined from 1955 through 1959; operations ceased when stockpiling by the General Services Administration was completed. Deeply weathered lateritic and bauxitic zones over alkalic complexes may have undergone niobium enrichment several times. At Araxa, Minas Gerias, Brazil, the residual material overlying carbonatite constitutes the ore presently being mined for pyrochlore and phosphate (Nation Materials Advisory Board, 1982, p. 18).

Tantalum is a common co-product with niobium in pegmatitic and placer deposits, but not the pyrochlore deposits in carbonatites. Niobium, along with tantalum, is a common by-product of tin mining, and also from certain tin slags. Associated elements in both the primary and secondary environments include uranium, thorium, titanium, and the rare-earth elements, all of which have been recovered along with niobium in the past and may be co-extracted in the future. Apatite, a fertilizer phosphate source, is a possible by-product or co-product in the pyrochlore deposits in Brazil and Canada.

Exploration has to be viewed in the context of world supply, and of the U.S. supply in time of emergency. There is an overabundance available in the world, but the United States has few viable reserves. Recognition of alkalic rock complexes and associated carbonatites is the first step. Once the target rocks have been identified, there are no special techniques that will lead to recognition of niobium minerals. Conventional geochemistry and mineralogy are needed to determine the presence of niobium minerals.

Resources

At Araxa, Brazil, the deposit contains 460 million tons of ore averaging 2.5 percent Nb_2O_5, or a reserve of 11.5 million tons of Nb_2O_5 in the weathered zone; reserves in unweathered carbonatite are not given. This is enough to satisfy world demand for many years. Brazil has not exported pyrochlore ore since 1981 when it developed its own capabilities for developing ferrocolumbium and niobium oxide. With a large, high-grade deposit in operation, combined with the capabilities of producing upgraded niobium products, Brazil has a strong control of the market. Canada is the only current source of pyrochlore concentrates, but the deposits are smaller, of lower grade, and more difficult to mine (Cunningham, 1985b, p. 193). The only other country, except for China and the USSR, presently mining significant amounts of niobium raw materials is Nigeria. This is in the form of columbite as a by-product of tin mining. Known reserves are mainly in Brazil, with significant amounts in Canada and several African countries. Known resources in the United States are small and of low grade. Several domestic deposits are pyrochlore in carbonatites; the largest, at Powderhorn, Gunnison County, Colorado, contains over 700 million tons of ore at a grade of 0.057 percent Nb_2O_5 (Armbrustmacher, 1980, p. B10). Other domestic resources are in placers in Idaho and Oklahoma and bauxite in Arkansas (Parker and Adams, 1972, p. 451).

Within the United States there are two strategic concerns about niobium. First, the United States imports all niobium raw materials. Second, because of the increasing difficulty in obtaining raw materials, the United States has to import upgraded niobium products from Brazil, which will lead to the loss of domestic processing capabilities. Peacetime niobium supplies for the United States will continue to come mainly from imports. The low-grade domestic sources are available for emergencies, but at a higher cost.

Brazil now accounts for over 80 percent of the world's niobium supply, and ultimately may account for something like 90 percent (National Materials Advisory Board, 1982, p. 104). This single low-cost source, dedicated to integrated production of upgraded material, is a strong controlling factor. If this source was suddenly disrupted, it would take the United States a considerable period of time to adjust.

The future of niobium is closely connected to that of the steel industry. This connection is reflected in the niobium consumption pattern of the last few years, when the lowest level in many years was reached in 1982, paralleling the low ebb in steel production. The fact of unlimited supply combined with several unique properties should make niobium much used in research applications.

TANTALUM

Introduction

Tantalum is an important element in the electronics, military, and communications industries and also has applications in many other fields. Approximately 60 percent of the tantalum consumed in the United States is used in the production of electronic components, mainly capacitors (Cunningham, 1985a, p. 814). Tantalum is of special importance to the United States because of total dependence on imports and shortage of domestic deposits.

Tantalum has been known since 1802, but had little commercial use until 1940 when tantalum powder capacitors were developed. These are compact, highly reliable, and very efficient. Other applications include use in computers and communications systems; instrumentation in ships, aircraft, missiles, and weapons systems; in superalloys that are used in aerospace structures, jet engines, and gas-turbine parts; and as tantalum carbide in wear-resistant parts. There are numerous other applications for tantalum where corrosion resistance and strength at high temperatures is important.

Aluminum electrolytics and ceramics are competitors for use in capacitors, but neither has the overall capabilities of metallic tantalum powder. An alloy of tantalum and niobium is an acceptable replacement for tantalum in many corrosive environments. Molybdenum and tungsten are possible substitutes for tantalum used in chemical environments, and if halogens and halides are involved, they are superior.

Primary sources of tantalum are concentrates of the minerals tantalite and wodginite, and Ta-bearing tin slags. Tantalum in the tin slags is in strueverite, a variation of rutile, that contains about 12 percent each of Ta_2O_5 and Nb_2O_5. Capacitor-grade powders are minus 325 mesh and 99.9 percent pure. Tantalum carbides are 93.3 to 93.7 percent tantalum and 6.2 to 6.3 percent carbon (Cunningham, 1985a, p. 813).

Tantalum is classified as a strategic and critical mineral because of its applications in aerospace, communications, and military industries. The General Services Administration maintains a stockpile of tantalum and tantalum materials. This stockpile was considerably short of its goal of 7.2 million tons of tantalum on March 31, 1984 (Cunningham, 1985a, p. 817). A critical shortage of tantalum developed in World War II, and again during the Korean conflict. Increased exploration efforts at that time for tantalum and niobium led to the discovery of some large placer deposits. One of these placers, at Bear Valley, Idaho, was the only major tantalum producer in the United States; production ceased in 1959.

World production of tantalum raw materials has averaged about 2.1 million pounds per year for several years, with roughly 1.1 million pounds of the production coming from tin slags (Cunningham, 1985a, p. 815). In 1983, tantalum production was suspended in Canada for the entire year because of low prices. Tantalum consumption in some recent years has exceeded the production of raw materials. The Tantalum Producers International Study Center (TIC), established in 1974 with the general purpose of disseminating information on tantalum, is headquartered in Brussels, Belgium.

Geology

Tantalum and niobium have a strong chemical coherence and generally are found together in nature in many of the same minerals. These minerals are chiefly oxides, multiple oxides, and hydroxides, and commonly contain titanium, zirconium, tin, and tungsten. The more important tantalum minerals are tantalite $(Fe,Mn)(Ta,Nb)_2O_6$, microlite $(Ca,Na)_2(Ta,Nb)_2O_6(O,OH,F)$, euxenite $(Y,Ca,Ce,U,Th)(Nb,Ti,Ta)_2O_6$, wodginite $(Ta,Sn,Mn,Nb,Fe)_{16}O_{32}$, and strueverite $(Ti,Ta,Fe)_3O_6$. Tantalite is a solid-solution series with columbite the niobium-rich end member, and microlite forms a solid-solution series with pyrochlore, the niobium-rich end member; however, only the two end members are common. Crustal abundance of tantalum is considered to be about 2 ppm (Parker and Adams, 1972, p. 445). Highest values are found in the late differentiates of granites, namely albitized granite and pegmatites.

Primary deposits of tantalum are almost exclusively in complex pegmatites characterized by abundant albite and presence of lithium and beryllium minerals. A good example is the Tanco Mine at Bernic Lake, Manitoba, Canada, the single largest primary tantalum source in the world. The major ore mineral is wodginite with lessor amounts of tantalum in tantalite, microlite, and tapiolite. The pegmatite, 1,440 m long by 100 m wide, also contains beryllium, cesium, lithium, rubidium, and tin (National Materials Advisory Board, 1982, p. 23). The Harding pegmatite in the Petaca District, New Mexico, provided microlite as a tantalum source during World War II. The district covers an area about 24 km long and 6 km wide that contains numerous pegmatites sought mainly for mica, but also containing significant concentrations of beryllium, tantalum, niobium, bismuth, uranium, thorium, and rare-earth elements. Tantalum is not concentrated in carbonatites like niobium, and no large relatively low-grade deposits are known.

Another important group of deposits is that produced by the erosion derivatives of granite and pegmatite terrains. Tantalum-bearing resistate minerals, mainly columbite-tantalite and euxenite are common in placers and residual deposits, but not in sea-beach placers. Columbite-tantalite-bearing alluvial deposits and saprolites in central Nigeria are derived from the weathering of altered biotite granites. The euxenite placers at Bear Valley, Idaho (Plate 2D) were derived from an area underlain by quartz diorite and pegmatite related to the Idaho batholith. The Greenbushes mine in Australia was originally worked as a tin mine with tantalum produced as a by-product. The original ore body was alluvial materials derived from a large complex pegmatite. The pegmatite ore body, approximately 3,000 m long and 50 to 80 m wide, is now considered to be a long-term tantalum source. Reserves are estimated at 9.7 million tons grading 0.06 percent Ta_2O_5 (National Materials Advisory Board, 1982, p. 24).

Niobium and tantalum are common co-products in the pegmatitic and alluvial and eluvial deposits, but not in the large, carbonatite-hosted pyrochlore deposits that are sources of niobium. Tantalum is generally a co-product or by-product of tin, and is also associated with deposits of titanium, zirconium, lithium, beryllium, uranium, tungsten, and rare-earth minerals. Commercial minerals that are associated with tantalum in pegmatites include mica, feldspar, beryl, lepidolite, and spodumene.

The best exploration approach is to define the geologic environments where tantalum concentrations occur, and examine them on the basis of associated metals and minerals. Most past exploration has viewed tantalum as a by-product of tin, the major commodity being sought. Panning is a useful technique since tantalum minerals all have high specific gravity. Future exploration should consider some of the lesser known, high-tantalum-content minerals in geologic environments more conductive to large, low-grade deposits.

Resources

Known tantalum reserves in the United States are negligible, and they probably will not be a major source unless there is a national emergency. Natural resources in the world are largely in Australia, Canada, Brazil, and Egypt, much in currently subeconomic deposits. Resources in tin slags from Malaysia and Thailand will continue to be a source for many years. Canada, Brazil, and Australia are the main suppliers of concentrates, mainly as a by-product of tin mining. Thailand and Malaysia are the main sources of tin slags.

Tantalum is of special concern to the United States because of reliance on imports and the lack of a capability for upgrading low-grade tin slags into synthetic concentrate. There is also the lack of price stability. Price fluctuations since 1965 have alternatively spurred exploration, then closed mines. The price fluctuations probably reflected the lack of dependability in long-term supplies and the inability of suppliers to react to sudden shifts in demand (National Materials Advisory Board, 1982, p. 25). Until recently the production of tantalum was dependent on the production of niobium or tin. There are now large supplies of niobium that do not contain tantalum, and the niobium-tantalum deposits cannot compete in the present market for either Nb or Ta. Tin production is controlled by the International Tin Council, and imposed quotas have a strong effect on tantalum production.

The expected growth rate for tantalum consumption is predicted at a much greater rate than that of tin, which portends future supply problems for tantalum. The future will depend on improved technology that will help make some of the larger lower grade deposits and tin slags viable sources, or on finding of new deposits in which tantalum is the primary product.

TITANIUM

Introduction

Titanium is an important strategic and critical material because of its use in the aerospace industry, especially in airframes and the modern turbine engine. About 5 percent of the titanium produced goes to this use. The bulk of the production, 92 percent, is to make titanium dioxide white pigment (Titanium white).

Titanium, because of the high strength-to-weight ratios and high corrosion resistance it imparts to alloys, was of major importance in development of the turbine engine and modern aircraft and spacecraft. There are numerous nonaerospace applications in the chemical, petroleum, power generation, and other industries where superior resistance to corrosion and heat are required (Lynd and Lefond, 1983, p. 1311; Gadden, 1984, p. 54). Titanium dioxide (TiO_2), because of its high refractive index, imparts whiteness, opacity, and brightness to paints, paper, and plastics. There are numerous other applications of TiO_2 pigments and other compounds (Lynd, 1985, p. 867).

Titanium alloys can be replaced in some applications by high-strength low-alloy steels, aluminum, stainless steels, other alloys, and nonmetals. These substitutions, however, may result in lower performance at a higher cost. Disadvantages in use of titanium metal alloys are high cost and difficulty in fabrication. There seems to be no viable substitute for titanium dioxide pigments at this time, although several other materials—calcium, strontium and barium titanates, silcon carbide, and others—also have high refractive indices (Lynd and Lefond, 1983, p. 1351).

The raw material sources for titanium are rutile (including anatase and brookite), ilmenite, leucoxene, and titaniferous slag developed from smelting ilmenite-iron ore. Ilmenite theoretically contains about 53 percent TiO_2, but commercial material contains as much as 70 percent, depending on the degree of alteration of the ilmenite; beyond 70 percent the material is termed leucoxene. Commercial concentrates of rutile contain about 95 percent TiO_2. Rutile concentrates are the preferred material source for chloride-process pigment production, but with the depletion of readily available rutile sources, substitutes are being sought. Synthetic rutile is made from ilmenite by beneficiation to remove the iron and raise the TiO_2 content to that of natural rutile. Ilmenite from sand deposits works best because it is of the correct particle size and contains less contaminants. Commercially available slags contain 70 to 85 percent TiO_2 with a low iron content. Titanium sponge is the primary metal form obtained by reduction of titanium tetrachloride; it is named for its appearance. Impurities are an important consideration in developing titanium dioxide pigment; chromium, vanadium, columbium, manganese, and phosphorus can seriously affect the properties of the pigment. Particle size is critical in the pigments, and must be about 0.2 to 0.4 micrometer. Two major types of pigments are produced: rutile for outdoor paints, and anatase for indoor paints and paper.

Rutile and titanium sponge are included in the list of strategic and critical materials for stockpiling, with the objectives set by the Federal Emergency Management Agency (FEMA). Inventories on March 31, 1984, were 37 percent of the goal for rutile and 17 percent of the goal for sponge (Lynd, 1985). Titanium was used as an alloy additive to iron and steel domestically in 1906, the first titanium dioxide was produced in 1918, and use of titanium compounds in welding-rod coating began in the mid-1930s. Titanium metal has been commercially produced since 1948. Titanium metal consumption has been erratic, with major declines in consumption in 1958 when production of military aircraft was cut back, in 1971 when the supersonic transport program was cancelled, and in 1977 when the B-1 bomber program was cancelled.

Major growth in titanium production took place from 1950 to 1980. In 1950, world production amounted to 788,000 tons of ilmenite and 25,000 tons of rutile. In 1980, production was 4,107,000 tons ilmenite, 480,000 tons rutile, and 1,343,000 tons titanium slag. Production has declined slightly in the early 1980s. The United States now imports most rutile and much of the ilmenite for consumption, mostly from Australia and Canada.

Geology

Titanium is the ninth most abundant element in the Earth's crust and has an average crustal abundance of about 1.07 percent TiO_2, being slightly higher in oceanic than continental crust (Force and others, 1976, p. A1). Few sedimentary, metamorphic, or igneous rocks have titanium contents much different than the average, and mineralogy is probably more important for mineral deposit consideration than geochemical distribution (Force and others, 1976, p. A1). Economic concentrations of titanium occur in the igneous rocks anorthosite, norite, and nepheline syenite, all of which have lower than average TiO_2 content. Much of the titanium in most rocks is apparently in the silicate minerals, mainly sphene, biotite, hornblende, and titanaugite, because of their much greater abundance than the titanium-rich accessory minerals. Geologic factors that partition titanium mainly into the oxide form determine the formation of the ore deposits. An example given by Force and others (1976, p. A8) is the rutile sands of Australia that have TiO_2 values less than the crustal average.

They are ores because of their mineralogy and ease of extraction and beneficiation.

Deposits of titanium are classified on the basis of mineralogy—rutile, ilmenite, or altered ilmenite; and on mode of occurrence, primary or secondary. Rutile is the most common of the three TiO_2 polymorphs, including anatase and brookite. There are numerous known primary deposits of rutile, but few have been of commercial significance. Rutile occurs primarily in alkalic igneous rocks, in noritic-anorthositic complexes, and in granite and syenitic dikes and pegmatites. It is also formed in aluminum-rich high-grade metamorphic rocks and in high- and low-grade metamorphosed mafic and ultramafic rocks (Klemic and others, 1973, p. 658). Production of rutile from primary deposits has been largely as a by-product and co-product of ilmenite mining. Rutile was mined intermittently from 1900 to 1949 at Roseland, Virginia, where the ore was disseminated rutile in anorthosite and in dike-like bodies of rutile, ilmenite, and apatite (nelsonite). Some rutile was also taken from deposits near Magnet Cove, Arkansas, where rutile, anatase, and brookite are associated with agerine phonolite porphyry and novaculite. Considerable studies have been done on the Magnet Cove deposits, but the larger deposits are relatively low grade (3 to 6 percent recoverable TiO_2), and the intimate mixture of titanium minerals and gangue causes major beneficiation problems (Lynd and Lefond, 1983, p. 1320). Rutile in amounts as much as 10 percent has been reported in dikes and disseminations in and associated with gabbros and anorthosites elsewhere in the world (Klemic and others, 1973, p. 658).

Deposits associated with carbonatite-alkalic igneous complexes are potential resources. Brazil recently started mining anatase and ilmenite from weathered carbonatite bodies in Minas Gerais where anatase rock has been stockpiled for years in conjunction with phosphate (apatite) mining. Magnetite-ilmenite-perovskite ($CatiO_3$) segregations in pyroxenite at Powderhorn, Colorado, have been evaluated as a titanium and rare-earth metals resource (Force and Lynd, 1984, p. B5).

There are several other geologic environments that contain potential rutile resources, including the following (Force and others, 1976, p. F1). Rutile in the alteration zones of porphyry copper and other types of metal deposits has by-product potential. An eclogite in northwest Italy contains about 5 percent disseminated rutile. A sillimanite-topaz gneiss that contains a potential resource of titanium (2 to 4 percent rutile) occurs in the Front Range, west of Denver, Colorado.

Ilmenite is the most abundant titanium mineral and is the major raw material source for titanium. The three main types of primary ilmenite deposits—ilmenite-magnetite (titaniferous magnetite), ilmenite-hematite, and ilmenite-rutile—are mainly associated with anorthositic or gabbroic rocks. The deposits are disseminations, pods, lenses, and beds and are mostly thought to be late magmatic in origin. Gangue minerals include plagioclase, pyroxene, biotite, chlorite, apatite, spinel, and chlorite. The ilmenite-magnetite ores may be composed of granular intergrowths, in which case separate homogeneous concentrates can be obtained, or the two minerals may be intimately intergrown. In the Sanford Lake District, New York, and at the Otanmaki deposit, Finland, ilmenite and magnetite occur as independent grains in granular aggregates and disseminations. Magnetite in the Sanford Lake District contains as much as 35 percent ilmenite as exsolution intergrowths, and the ilmenite at Otanmake has minor amounts of intergrown hematite. In the Sanford Lake District, at one time the leading producer of ilmenite in the world, ore grade ranged from 9.5 to 30.0 percent TiO_2. The United States has numerous deposits of titaniferous magnetite in the western states, but further metallurgical research would be needed to develop beneficiation processes for the ores of intimately intergrown ilmenite and magnetite.

The Lac Tio deposit near Allard Lake, Canada, is of the ilmenite-hematite type and of magmatic origin. Ore is composed of exsolution intergrowths of ilmenite and hematite with a gangue of plagioclase, pyroxene, and minor amounts of pyrite, pyrrhotite, and chalcopyrite, in anorthosite and anorthositic gabbro. The ore is in a tabular, flat-lying body more than 1,000 m long and 1,000 m wide. Estimated tonnage is 125,000,000, averaging 32 percent TiO_2 and 36 percent Fe (Harpen and Bates, 1984, p. 127). An example of the ilmenite-rutile type of ores is the nelsonite deposits in Virginia that were mined intermittently from 1900 to 1971. Ilmenite and rutile occur as disseminations in anorthosite and bordering gneiss, and with apatite in dike-like bodies. Most of the commercial mining was in the soft saprolite upper layers of the ore bodies (Lynd and Lefond, 1983, p. 1319).

Secondary titanium deposits provide the bulk of world production, and hold most of the known reserves. Rutile and ilmenite are resistant minerals to weathering and because of their relatively high specific gravity tend to concentrate in placers with other heavy minerals. The deposits are of several types, but by far the most important are beach sands. The coastal areas of all continents have significant ilmenite placer deposits, and local rutile placer deposits. Other heavy minerals present in varying amounts include zircon, monazite, staurolite, tourmaline, spinel, kyanite, sillimanite, corundum, garnet, epidote, and topaz. Composition of the sands depends on the source materials. Ilmenite in placer deposits commonly exhibits some degree of weathering due to the oxidation and loss of iron. The end product is leucoxene, whitish, opaque, fine-grained rutile crystallites.

Australia has extensive titanium deposits along both west and east coasts. They occur in modern beach sands, fossil beaches, buried strand lines, and coastal dunes, and in recently explored offshore occurrences. The heavy mineral suites vary with location as does the chemical composition of ilmenite. At some localities, ilmenite constitutes 90 percent of the concentrate (Lynd and Lefond, 1983, p. 1330). When mining began in 1934, some of the sands contained as much as 50 percent heavy minerals, mainly zircon and rutile. The average heavy-mineral content mined in 1980 was about 3 percent on the east coast and 5 to 10 percent on the west coast. The sands are mainly quartz and heavy minerals, with few light minerals such as mica and feldspars. This suggests a source that was a considerable distance inland. The source has been identified as Mesozoic and Tertiary sandstones that were in turn derived from granulite-facies metamorphic rocks farther inland (Harpen and Bates, 1984, p. 135).

Similar titanium deposits occur along the Atlantic and Gulf coasts of the United States (Tooker and Force, 1980). Deposits in Florida and Georgia are in Recent to Pleistocene sands, in New Jersey they are in Miocene, Pliocene, and Pleistocene sands, and in Tennessee they are in the Cretaceous McNairy Sand, a poorly sorted sandstone. The deposit at Green Cove Springs, Florida, one of the current producers, is described below under Rare-Earth Elements. The Trail Ridge deposit, Florida, is lacking in monazite, garnet, and epidote, and is mined primarily for altered ilmenite and leucoxene, with by-products of zircon and staurolite. The deposit is 25 km long, 2 to 3 km wide, and 12 m thick. Heavy minerals make up about 4 percent, and high-titanium minerals 45 percent of the heavies.

Other secondary titanium occurrences are in residual deposits formed by lateritic weathering of titaniferous rocks. Although not of commercial value at this time, they are widespread, occurring in Hawaii and Oregon in the United States.

Potential by-products/co-products of titanium mining are many. In primary ores, magnetite and hematite are recovered, and in some deposits, also vanadium and a sulfide concentrate as well. In the beach placer deposits, any of the heavy minerals present are potential by-products. Zircon and monazite are the most common; in the United States this is the only commercial source of zirconium and thorium. In tin dredging, ilmenite is a by-product, and titanium minerals are potential by-products in the mining of several other commodities (Force and others, 1976, p. F1). Very large potential resources exist in the Athabasca tar sands, Canada, and in some of the sandstones of the western United States (Harpen and Bates, 1984, p. 141).

Exploration for titanium deposits should be oriented toward the geologic model of the type of deposit sought. In primary ilmenite deposits, magnetic anomalies associated with known host rocks, mainly gabbros and anorthosites, could be the first indication. In beach placer deposits, the source area must contain the granitic rocks and high-grade metamorphic rocks likely to contain ilmenite and rutile. The deposits are most likely to be in well-sorted sands associated with recent and ancient shore lines. The magnetic and radiometric properties of the heavy minerals can be used for rapid identification of panned concentrates.

Resources

The United States has resources of titanium, but production has been declining recently because of competition from imports. The known resources in the United States are about equal in beach placers and in hard-rock deposits associated with igneous rocks (Force and Lynd, 1984, p. B1). A major unevaluated titanium resource may be in extensive sand bodies on the Atlantic Continental Shelf (Grosz and Escowitz, 1983). Significant heavy-mineral concentrations occur in areas offshore of land areas that have no similar economic concentrations. Vertical extent of these deposits is not known. Known resources in the world appear to be adequate to meet demands beyond this century. Countries with significant known resources are Brazil, Canada, South Africa, Australia, Norway, China, India, United States, USSR,, Sri Lanka, and Sierra Leone. Current major producers are Australia, Norway, United States, USSR, Malaysia, Finland, India, China, Sri Lanka, and Brazil for ilmenite and leuocoxene; Australia, Sierra Leone, South Africa, USSR, Sri Lanka, and India for rutile; and Canada and South Africa for titaniferous slag (Lynd, 1985, p. 864).

The higher grade rutile placer deposits are being depleted, and more lower grade ilmenite ores are being used for raw materials. This means upgrading and cleaning the ilmenite ores to produce synthetic rutile, or modifying the $TiCl_4$ production process to handle lower grade ilmenite ores with contaminants. Geographic and environmental problems may have a negative effect on utilization of some of the resources in the United States. Much of the beach placer areas are prime land for development and recreational areas, and will probably be dedicated to those purposes. Other environmental problems are the reclamation of land mined for beach placers, containment and control of particulate matter in air and water, and disposal of sulfates and other chemical accumulates from beneficiation and processing.

Titanium has been called the metal of the future because of its use in the aerospace industry, and this demand will probably increase in the future. The rate of increase for metal will depend on new developments and the economic health of the aerospace industry. Titanium dioxide pigment will probably remain the major titanium use for many years, although the growth rate of this use has declined in the last few years. The probable annual growth rate for titanium metal demand in the United States to the year 2000 has been projected at 5.5 percent from 1983; and the rate of nonmetal forms at 1.8 percent (Lynd, 1985, p. 874).

RARE-EARTH ELEMENTS

Introduction

The rare-earth elements (REE) and yttrium have diverse applications that utilize their unique properties, and demand is expected to increase in the future. Yttrium is technically not one of the rare-earth elements, but is included with them because of similar properties and modes of occurrence. In this chapter the rare-earth elements and yttrium are discussed together and termed the rare earths.

Hedrick (1985a, p. 651) listed approximately 80 uses for the rare earths under the general headings of metallurgy, glass, ceramics, illumination, electronics, chemicals, magnets, nuclear, and other. Consumption in 1984 was 59 percent for petroleum catalysts, 23 percent for metallurgical uses, 15 percent in the ceramics and glass industries, and 3 percent miscellaneous uses (Hedrick, 1985b, p. 752). Lanthanum, neodymium, and praseodymium chlorites are the main rare-earth compounds used in catalysts. The rare earths are used in a variety of ways in metallurgy, but mainly as mischmetal (98 percent rare-earth metals) and silicide (33 percent rare-earth metals) additives in steel. An expanding use for REE, mainly samarium alloyed with cobalt, is in the production of permanent magnets. In the glass industry, cerium oxide is used in fine polishing compounds. Neodymium oxide, praseodymium oxide, erbium oxide, lanthanum oxide, and cerium are special additives to glass production. In ceramics, material composed of 90 percent yttrium oxide and 10 percent thorium oxide is used in high-temperature applications, and yttrium-oxide-stabilized zirconia forms one of the best refractory materials, stable under conditions of oxidation and reduction. Other REE applications include yttrium-iron garnets, gadolinium-iron garnets, yttrium-aluminum garnets, neodymium, erbium, and holmium in laser and microwave applications, rare-earth fluoride in carbon-arc lights, europium oxide in color television picture tubes, and gadolinium-gallium garnets in computer and communication applications (Robjahns, 1984).

Substitutes are available for many rare-earth uses, but they are generally significantly less effective (Hedrick, 1985a, p. 658). Palladium and ultrastable zeolites can be used in petroleum-cracking catalysts, and magnesium and calcium alloys and compounds may be used in place of mischmetal and rare-earth silicide. Spinel-type ferrite compounds are a less expensive substitute for yttrium-iron garnet, and thorium oxide is a less costly refractory than yttrium oxide.

Raw material sources for the rare earths are mainly the minerals bastnaesite, monazite, and xenotime. Other sources include by-product from uranium mining, oxides in residual clay deposits, and from the minerals apatite and loparite in phosphate deposits. Bastnaesite theoretically contains about 75 percent combined rare-earth oxides (REO) and very minor amounts of yttrium. Concentrates contain about 60 percent REO from flotation, 70 percent REO from acid leaching, and 85 percent REO from a combination of acid leaching and calcining. Monazite can contain as much as 70 percent REO and about 2 percent yttrium oxide, but most concentrates contain 55 to 65 percent REO. Xenotime concentrates average about 25 percent Y_2O_3, which can be upgraded to an yttrium concentrate of 60 percent Y_2O_3 and 40 percent REO.

Some of the rare earths have been known and used for more than one hundred years, whereas one, promethium, was not discovered in the natural state until 1965. The incandescent gas mantle was developed in the 1880s, which gave rise to the mining of placer sands for monazite to obtain thorium and cerium oxide. Monazite was mined primarily to obtain thorium until after World War II, with India and Brazil dominating production. Rare earths were of little value then, but uses were gradually discovered. In the 1950s there was an oversupply of the rare earths, but now thorium has become the oversupply commodity because of the extraction of rare earths from monazite. A major bastnaesite deposit was discovered at Mountain Pass, California, in 1949 (Plate 2D). Large-scale mining began in the mid-1960s, and bastnaesite became the principal source of the rare earths. This deposit constituted about half of the world's rare-earth production in 1984. Hedrick (1985a, p. 648) listed 51 processors of rare earths in the world, 20 of which are in the United States. The United States consumes about half of the world production of rare earths, which in 1984 amounted to about 24,800 tons REO in various forms (Hedrick, 1985b, p. 752).

Geology

The rare-earth elements (REE) are those with atomic numbers 57 to 71: lanthanum, cerium, praseodymium, neodymium, promethium, samarium, europium, gadolinium, terbium, dysprosium, holmium, erbium, thulium, ytterbium, and lutetium. Yttrium has atomic number 39. The REE and yttrium are referred to as the rare earths. The cerium subgroup refers to the rare-earth elements derived from bastnaesite and monazite—cerium, lanthanum, neodynium, and praseodymium. The light rare-earth elements (LREE) are those from lanthanum (57) through europium (63), and the heavy rare-earth elements (HREE) are those from gadolinium (64) through lutetium (71). Yttrium is included with the HREE. The most abundant rare earths are cerium, yttrium, neodymium, and lanthanum, with the LREE abundance greatly exceeding that of the HREE (Henderson, 1984).

The rare-earth elements all have similar chemical and physical properties, but still are partially fractionated from one another by several petrologic and mineralogic processes. The degree of fractionation can be used as an indicator of genesis.

The rare earths are lithophile, concentrating in silicate rather than metal or sulphide phases, and are dispersed, occurring in trace amounts in most of the common rock-forming minerals. There are about 200 minerals with minor, but not essential amounts of REE and over 70 with major and usually essential contents of REE (Clark, 1984, p. 34). The economically important minerals are monazite $(Ce,La,Th,Y)PO_4$, bastnaesite $(Ce,La)(CO_3)F$, gadolinite $(YCe)_2Fe^{2+}Be_2Si_2O_{10}$, and xenotime YPO_4. By-product sources of rare earths have also been apatite, bannerite, euxenite, loparite, pyrochlore, and uraninite.

Primary concentrations of rare earths occur in a variety of geologic environments, including veins, gneisses and migmatites, skarns, pegmatites, and alkaline rock complexes and related carbonatites (Adams and Staatz, 1973, p. 551). Current production from primary deposits is almost entirely from alkaline rocks and carbonatites. Alkaline rocks in general, and particularly associated carbonatites, are enriched in REE, both in minerals in which rare earths are a major component and in those in which they are a minor component. Carbonatites contain the bulk of the reserves held in primary deposits.

The best-known rare-earth-bearing carbonatite, and the western world's major source of rare earths, is at Mountain Pass, California (Staatz and Armbrustmacher, 1981). The major ore body, the Sulphide Queen, is 730 m long and as much as 210 m wide. The ore is 60 percent carbonate, mainly calcite, with barite, quartz, and silicate minerals, and an average 12 percent bastnaesite. Grade ranges from 5 to 15 percent and averages 7 percent

REO (Neary and Highley, 1984, p. 427). This deposit, mined primarily for its rare-earth content, contains about 3.6 million tons REO (Hedrick, 1985a, p. 651). The largest identified reserves in the world are associated with the Bayan Obo iron ore deposits, Nei Monggol Autonomous Region, China. The two largest ore bodies contain 15 and 20 million tons of REO (Neary and Highley, 1984, p. 429). The deposits are tabular bodies and lenses in steeply dipping dolomites, and are composed mainly of magnetite, specularite, and hematite with intercalated rare-earth minerals. Bastnasite and monazite and a complex series of niobium-thorium-titanium minerals contain the rare earths. The deposits are thought to have a carbonatite association and genetic relationship to nearby alkaline granite, syenite, and gabbroic rocks. Other carbonatite-hosted deposits include Mrima Hill, Kenya; Karonge, Burundi; Kangankunde, Malawi; Khibiny, USSR; and Araxa, Brazil.

Vein deposits have been important sources of rare earths in the past. The monazite vein at Steenkampskraal, South Africa, was the major source of monazite from 1952 to 1959 and 1962 to 1963, and was worked primarily for its thorium content (Adams and Staatz, 1973, p. 551). Fluorspar vein deposits at the Buffalo Mine, South Africa, contain as much as 2 percent monazite and bastnaesite, and consideration is being given to recovery of by-product monazite concentrates (Neary and Highley, 1984, p. 430). Adams and Staatz (1973, p. 553) list several potential by-product sources of rare earths in veins sought for thorium, tin, and fluorspar.

There are many other primary sources of the rare earths. Apatite in metasomatic iron deposits near Mineville, New York, contains about 11 percent total rare earths, and the dumps from previous mining contain large quantities of apatite (Adams and Staatz, 1973, p. 552). Deposits of igneous apatites and associated loparite in the Kola Peninsula, USSR, mined for their phosphate content, are by-product sources of the rare earths. Gadolinite occurrences in pegmatitic bodies in Canada are potential beryllium-yttrium deposits. Uranium deposits at Elliott Lake, Canada, have produced by-product rare earths from uraninite and bannerite.

In the weathering environment the rare earths concentrate in the detrital fraction, with only minor amounts going into solution. That small portion released is trapped by adsorption on clays. Secondary concentrations of the rare earths occur in residual and placer deposits. Residual deposits include pyrochlore, apatite, and the rare-earth minerals monazite and goyazite in soils overlying the Araxa carbonatite complex, Brazil, the world's largest source of niobium. The deposit has large resources of the rare earths, but there has been no production because of difficulties in beneficiation of the ore (Neary and Highley, 1984, p. 428). Hedrick (1985b, p. 759) reported a residual clay deposit containing samarium and europium in the Jiangxi Province, China.

The most important placers are monazite-bearing beach sands, although some fluviatile, lacustrine, and deltaic deposits contain rare-earth resources. Beach placer sands mined primarily for titanium and zirconium, and locally for gold or cassiterite, are the main source of monazite. Minor amounts of xenotime are also recovered. Monazite commonly constitutes only a very small percentage of the heavy mineral fraction, which is 5 to 15 percent of the sands. The Eneabba deposits, Western Australia, are currently the world's major source of monazite. The deposits are mined mainly for ilmenite and leucoxene with rutile, zircon, and monazite by-products. Monazite generally composes less than 1 percent of the heavy mineral concentrate produced but varies from 0 to 35 percent of the heavy mineral fraction throughout the deposit (Neary and Highley, 1984, p. 432). The Eneabba heavy mineral sands are along a late Tertiary or early Pleistocene shoreline as much as 50 km inland and more than 75 m above sea level. The deposits are in an area 12 km long and as much as 1,200 m wide. In the United States, recent monazite production has been from beach placer sands as a by-product of ilmenite, leucoxene, and zircon recovery at Green Cove Springs, Florida. The deposit is 16 to 21 km long, as much as 1,200 m wide and approximately 6 m thick. Heavy mineral content is 3 to 4 percent, and the deposits are in elevated Pleistocene beach sands. Monazite, and more importantly xenotime, are recovered from "amang," the heavy-mineral residue of alluvial tin mining.

Rare-earths have been or potentially can be produced as by-products from a wide variety of deposits. Placer and alluvial mining for titanium, zirconium, gold, tin, and most of the heavy minerals both obtain monazite and other heavy rare-earth minerals as part of the concentrate. Residual deposits sought for niobium and other elements contain significant amounts of rare earths. Bastnaesite deposits in the United States contain few accessory minerals, and there is no by-product or co-product production. In other primary deposits, rare earths have been produced as by-products of iron-ore mining, beneficiation of apatite, uranium mining, gold mining, beneficiation of tungsten minerals, and have co-product or by-product potential in newly discovered copper–uranium–gold–rare earth, and beryllium–rare earth deposits.

The Mountain Pass, California, bastnaesite deposit, and many monzonite deposits, were originally located because of local radioactivity anomalies. Monazite is radioactive because of its thorium content, and bastnaesite and other rare-earth minerals are generally accompanied by thorium or uranium minerals; thus, radioactivity is a useful exploration tool for these deposits. However, many occurrences of rare-earth minerals are not accompanied by radioactive elements. In many occurrences, and probably in most of the undiscovered deposits, rare-earth materials may not be recognized by mineralogical or radioactive methods, especially where the rare earths are carried in one of the common minerals such as apatite. Analytical chemistry is of great importance in exploration for these deposits. Measurement of individual REE amounts to very low concentrations is now routine because of technical advances in analytical chemistry in recent years. Neutron activation analysis and mass spectrometric isotope dilution are the two most common techniques, although inductively coupled plasma (ICP) emission spectrometry is becoming more popular.

Resources

The United States (1985) is the major producer of rare-earth concentrates and the major source of bastnaesite ore. Australia, China, India, the USSR, and Brazil are the other major producers. The United States also consumes about half of the world production of rare earths (Hedrick, 1985a, p. 656). Reserves in the United States and North America are significant. Bastnaesite deposits at Mountain Pass, California, and monazite-bearing beach sands in Florida will continue to be major rare-earth sources for many years. A large potential resource is the monazite in extensive sand bodies on the Atlantic Continental Shelf (Grosz and Escowitz, 1983). China has large reserves in several types of deposits; Australia, India, and Sri Lanka have considerable reserves in beach placer sands; Brazil in beach placers and carbonatite deposits; and Thailand and Malaysia in tin placers. Many other countries have unevaluated resources (Adams and Staatz, 1973, p. 553).

The United States has more than adequate supplies of rare-earth ores for the production of the cerium subgroup, but currently does not produce an yttrium concentrate. The main source of yttrium presently is xenotime, a by-product of alluvial tin mining in Malaysia and Thailand. Potential sources in the United States are in deposits termed the "Olympic Dam" type in the Midcontinent region, certain iron ore tailings, and in sedimentary phosphate rock. In Canada the newly discovered gadolinite deposits could be a major source, and yttrium is a potential by-product of uranium mining at Elliot Lake.

There are two problems involved with production of the rare earths: the association of radioactive thorium, and the difference in natural abundance and demand for the various rare-earth elements. The radioactive thorium residue from processing of monazite for the rare earths causes increasing storage and disposal problems, thus increasing the beneficiation cost. The natural distribution abundance of the rare earths obtained from the ores is not necessarily the same as the demand distribution. Relatively higher demands for high-purity separated rare earths lead to an oversupply of others that occur in much greater abundance, or are in less demand. The example given by Adams and Staatz (1973, p. 555) illustrates the point. One ton of bastnaesite ore will produce 700 pounds of cerium oxide, but only 1.5 pounds of europium oxide, although separated europium is in relatively greater demand. The 1984 price for cerium oxide was $8.00 per pound; europium oxide was $650 per pound (Hedrick, 1985b, p. 754).

The amount of research and development involving the rare earths in so many different applications indicates a solid future with increased demand. The amount of growth for each individual element is hard to predict, and some of the less utilized and less abundant elements may become of great importance in the future.

THORIUM

Introduction

Thorium is a metal of many special uses and, perhaps more important, is a potential energy source for the future. There has been considerable recent research into extraction and application involving thorium because of its properties, availability, and association with the rare-earth elements.

Thorium is a soft, heavy, ductile, silver-gray metal that is mildly radioactive. The radioactive decay of thorium is the basis for its use, like uranium, for production of energy. Most current uses, however, are not for energy production, and therefore radioactivity is a detriment. The current main uses of thorium are in refractories (57 percent), lamp mantles (17 percent), and aerospace alloys (10 percent) (Hedrick, 1985c, p. 839). Thorium oxide (thoria) is used in refractories in high-temperature applications because it has the highest melting point (3,300°C) of any oxide. Production of mantles for incandescent lamps and oil and gas lamps includes coating with thorium nitrite. Thorium used in aerospace alloys is mainly alloyed with magnesium to impart high strength at elevated temperatures. There are numerous other nonenergy uses of thorium. The only domestic use of thorium for energy production presently is in a nuclear reactor in Colorado (Hedrick, 1984, p. 857).

Substitution of thorium in many nonenergy uses, even by less satisfactory alternative materials, will probably increase because of public concern and increased governmental regulation of radioactive materials. Beryllium, aluminum, yttrium oxides, and yttria-stabilized zirconia and silicon nitride are other high-temperature, lightweight, high-purity alloys that can replace thorium alloys in the aerospace industry.

The main commercial source of thorium is from the mineral monazite. Other potential sources include the minerals thorite, bannerite, brockite, and Tn-bearing pyrochore. The ThO_2 content of monazite varies from 0 to 31.5 percent (Staatz and Olson, 1973, p. 469). Metallurgical-grade thorium oxide is more than 99.8 percent pure, and reactor-grade material, oxide or metal, is more than 99.9 percent pure.

Thorium, known since 1828, was not in commercial demand until the incandescent lamp mantle was invented in 1891. Prior to 1920, thorium was the main commodity obtained from monazite mined with placer deposits. Since then most thorium has been produced as a by-product and co-product of the extraction of rare-earths from monazite. There was little production of monazite in the United States from 1910 to 1948, but in recent years monazite has been recovered in Colorado, Georgia, North and South Carolina, Idaho, and Florida as a by-product of mining heavy mineral sands for titanium and zirconium minerals (Hedrick, 1985c, p. 835). In the 1950s, there was extensive exploration for thorium because of the potential for nuclear energy, and there was a surplus of the rare-earth elements from the processing of monazite. Since the mid-1960s the opposite has been true; there has been a surplus of thorium. World monazite concentrate production in 1984 was 26,555 metric tons.

Geology

Thorium is a lithophile element occurring in nature mainly in oxide minerals. The chief thorium minerals are monazite [(Ce, La,Nd,Th)PO$_4$], thorite (ThSiO$_4$), thorianite (ThO$_2$), and numerous thorium-bearing multiple-oxide minerals with Ti, U, Ca, Fe, Y, Ce, P, Nb, Ta, Sn, and Pb. The abundance of thorium in the crust is estimated to be about 10 ppm; it becomes concentrated by magmatic differentiation, metamorphism, and weathering and erosion. Thorium content increases toward the more silicic alkali-rich end members of a rock series, and is notably high in alkalic rock complexes with associated carbonatites and fenites (Staatz and Olson, 1973, p. 471). In metamorphic rocks, high metamorphic-grade terrains with associated felsic igneous rocks have monazite with the highest thorium content. During weathering and erosion, monazite and several other thorium-bearing heavy minerals remain as detrital grains and form placers. Thorium that is released during weathering is adsorbed on clay minerals and is disseminated in clayey sediments.

In the United States, thorium deposits occur in veins, disseminated deposits, carbonatite bodies, stream placers, beach placers, unconsolidated and consolidated sedimentary rocks, and granite and alkalic igneous rocks (Staatz and others, 1979; Staatz and Armbrustmacher, 1982). Thorium veins consist mainly of quartz, potassic feldspar, and iron oxide. Thorite is the most abundant thorium mineral; monazite, allanite, and brockite are common in some localities. Numerous other thorium-bearing minerals occur, but in minor amounts. The larger and better known vein deposits in the United States are at Lemhi Pass, Montana-Idaho, and the Wet Mountains, Colorado (Plate 2D). These districts contain numerous individual veins ranging from a few centimeters to several meters in thickness and a few meters to 350 m or more in length. In some areas the veins are adjacent to alkalic complexes and carbonatites. Grade within the veins is erratic, usually averaging several tenths, and in some areas several percent, thorium. Rare-earth elements generally accompany thorium and commonly exceed it in abundance.

Disseminated deposits are lower in grade than the veins but contain much larger resources of thorium. Major deposits of this type are in the Bear Lodge Mountains, Wyoming, and Hicks Dome, Illinois (Plate 2D). Disseminated thorium deposits in the Bear Lodge Mountains are in altered trachyte and phonolite host rocks cut by a stockwork of brown to black veinlets consisting of potassic feldspar, fine-grained quartz, cristobalite, various iron and manganese oxides, and rare-earth and thorium-bearing minerals. Monazite, thorite, and brockite are the identified thorium-bearing minerals. Mineralized rock contains as much as 1,200 ppm Th (Staatz and others, 1979, p. 24). Thorium-bearing veins occur in the same area. Thorium at Hicks Dome is in a poorly exposed breccia, which based on radiometric data, is thought to underlie an extensive area at depth (Staatz and others, 1979, p. 22). The breccia consists mostly of limestone and dolomite fragments in a finer grained matrix of similar rock fragments and rounded quartz grains. Hydrothermal minerals, including fluorite, calcite, quartz, barite, pyrite, sphalerite, galena, apatite, florencite, and monazite occur in the matrix of the breccia, as veinlets, and as replacement of fragments. The breccia body is thought to be a diatreme associated with a deep-seated alkalic intrusive rock.

Massive carbonatite bodies are potential large low-grade sources of thorium. Examples are at Iron Hill, Powderhorn district, Colorado, and the Sulfide Queen body in the Mountain Pass district, California. The Iron Hill body is primarily dolomitic carbonatite with thorium-bearing pyrochlore and rare-earth-bearing fluorapatite. Other minerals include barite, goethite, hematite, calcite, quartz, pyrite, magnetite, biotite, rutile, fluorite, bastnaesite, aegrine, anatose, sphalerite, synchisite, and zircon. The average grade of the carbonatite stock is 0.0041 percent ThO$_2$, but the deposit contains more than 2 billion tons of potential resources (Armbrustmacher, 1980, p. B10). Other recoverable commodities include niobium, rare-earth elements, and uranium. The Sulfide Queen carbonatite is one of the world's largest resources of rare earths. The average thorium content is calculated at 0.026 percent ThO$_2$, which would be recovered only as a by-product of extraction of rare-earth elements.

Monazite-bearing placers in North and South Carolina were the world's first source of thorium, and were mined prior to 1917. The placers are at the headwaters of streams that drain monazite-bearing crystalline rocks. There are numerous thorium-bearing stream placers in the central part of Idaho along drainages on rocks of the Idaho batholith (Staatz and others, 1980, p. 9). Monazite was first noted in 1896 during sluicing operations for placer gold. Two of the larger placers were mined during the 1950s—Long Valley primarily for monazite, and Bear Valley primarily for euxenite and columbite to recover niobium and tantalum. Several of the placers are large enough to consider as potential resources for heavy minerals because they could be dredged, but many are too small. The most common heavy mineral is ilmenite, followed by zircon, garnet, and lesser amounts of euxenite, hematite, pyrite, biotite, amphibole, pyroxene, epidote, rutile, anatase, cassiterite, allanite, columbite, smarskite, topaz, xenotime, and gold. Thorium is in monazite, uranothorite, and euxenite. Heavy-mineral content of various placers averages about 0.5 percent, with monazite content ranging from a trace to 16 percent of the heavy minerals.

Monazite-bearing beach placers in Florida and Georgia have been the only domestic source of thorium since about 1960, and this as a by-product of the mining of the placers for titanium and the extraction of rare earths from monazite. Other heavy minerals found in beach placers include zircon, rutile, kyanite, spinel, xenotime, corundum, topaz, epidote, sillimanite, staurolite, and garnet. The exploited deposits are mostly Pliocene and Pleistocene beach placers along relic shorelines 3 to 35 m above sea level and as much as 80 km inland (Staatz and others, 1980, p. 5). The source of the heavy minerals in the beach placers was from the crystalline rocks of the Piedmont province, with perhaps several sedimentary cycles before being deposited in the beach bars. Mineable heavy-mineral deposits range in length from about 1 to 30 km, 150 m to 3 km in width, and 1.5 to 21 m in depth. The heavy-mineral content ranges from 0.5 to 6 percent, with monazite making up trace to 2 percent of the heavy minerals.

Thorium is produced as a by-product of recovery of heavy minerals from placer mining for titanium and zirconium and the extraction of the rare earths from monazite. Potential by-products/co-products include rare earths in all types of deposits, most heavy minerals and metals in placer deposits, and niobium, fluorine, and beryllium in carbonatite-hosted deposits.

Prospecting for thorium deposits would vary depending on the type of deposit sought, with the most useful methods relying on the radioactivity of thorium. Because of the limited use of thorium and the oversupply from by-product production, there has been little recent exploration activity for thorium.

Resources

The U.S. Geological Survey (Staatz and others, 1979–1985; Staatz and others, 1979, 1980) conducted a definitive study of thorium resources during the 1970s. The U.S. Bureau of Mines has investigated potential new uses of thorium, especially in terms of substitution for domestically scarce commodities (Hedrick, 1985c, p. 839).

The major producer of thorium-bearing monazite is Australia, followed by India, Brazil, and China. Numerous other countries have resources and limited production of monazite. Thorium is a domestically available commodity that at present is in oversupply. Thorium supply is determined by the demand for rare-earth elements, and to a lesser degree, for titanium, zirconium, and tin. The relationship between thorium and the rare earths is of interest. Monazite mined in Australia, the largest producer, is sent to the United States or France for processing. In the United States, monazite from Australia and domestic mines is processed only for its rare-earth content. Domestic monazite is exported to France for processing for thorium. All thorium compounds used domestically in 1984 were imported, mainly from France or Great Britain (Hedrick, 1985c, p. 896). Because of governmental restrictions, all thorium produced in India and Brazil is for internal consumption only. Thorium-bearing residue from rare-earth extraction in the United States is stockpiled, and will probably be the first domestic resource used if the demand for thorium increases suddenly. Another source is rare-earth- and thorium-bearing apatite in millions of tons of iron-ore tailings near Mineville, New York (Hedrick, 1984, p. 858).

The major problem influencing the use of thorium in industry is its radioactivity. This is a growing concern to the public and government. The health and environmental concerns and the economic burdens of compliance with associated regulations have prompted a search to find substitutes for nonenergy uses of thorium. The abundant supply and new technology that will lower thorium separation costs may not lead to increased demand until health and disposal problems can be solved.

The future of nonenergy use of thorium in the United States will probably be limited to applications where a viable substitute cannot be found. The use of thorium in energy production will depend on improved technology to deal with nuclear materials and on major policy decisions to pursue this course. There are no supply problems in the United States or the world. By-product production of thorium will be more than adequate to meet nonenergy demands, and there are sufficient known resources if thorium were to be developed as an energy source.

BERYLLIUM

Introduction

Beryllium is an important element in nuclear, aerospace, electronic, electrical, and military applications, with few viable substitutes and some applications that are unique. In 1983, U.S. consumption was estimated at 40 percent for nuclear reactors and aerospace applications, 36 percent electrical equipment, 17 percent electronic components, and 7 percent miscellaneous uses (Farr, 1984, p. 88).

Beryllium is used commercially in metal, oxide, and alloy form. It is one of the lightest metals stable in air, but is brittle, and fabrication use requires expensive powder metallurgy and machining. In the aerospace industry, beryllium is used in specialized structural elements, brake shoes, and heat shields because of its high melting point. Its nuclear reflecting and moderating properties make it useful in the nuclear industry. Beryllium is used in electrical and electronic applications such as insulators, substrata for electronic circuits, and heat sinks because of its very low electrical conductivity. Beryllium-copper alloys have great strength, wear, and corrosion resistance in addition to being excellent conductors of heat and electricity, and are used extensively in the electronics industry.

Phosphor bronze is a possible substitute for beryllium-copper alloys, but with substantial loss of performance because of inferior conductivity, fatigue, and formability characteristics. Beryllium oxide is unique for some applications, but sintered alumina is a possible substitute for some uses in ceramics. Steel, titanium, and graphite composites may be used as substitutes for beryllium metals in the future.

The ores of beryllium are mineralized rock containing bertrandite and beryl. Bertrandite, which recently has become the major source of commercial beryllium, contains 15.1 percent Be (42 percent beryllium oxide). Beryl contains 5 percent Be (14 percent beryllium oxide) when pure and is hand cobbed and sorted to obtain a commercial ore averaging 4 percent Be (11 percent beryllium oxide). Beryllium oxide is extracted with difficulty, depending on the type and mineralogy of the ore. Beryllium-copper master alloy contains 4 percent Be, although the most common commercial alloy contains 2 percent Be. Purchase specifications for the General Services Administration stockpile specify a minimum beryllium metal content of 98 percent, and maximum impurity limits for beryllium oxide of 1.5 percent by weight.

Beryl, beryllium-copper master alloy, and beryllium metal are stockpiled by the General Services Administration under the

Strategic and Critical Material Stock Piling Act of 1946. Beryllium is a critical element for military uses and in the development of new technology in the electronics industry. Beryllium was first produced in the United States in 1932, and had a slow increase in production until a major jump at the start of the space age, culminating in the highest consumption year in 1960 of nearly 9,000 tons of equivalent beryl ore. Since then consumption has varied from 3,600 to 8,000 tons per year. Prior to 1969 nearly all of the U.S. demand was met by imported beryl. Domestic production of the mineral bertrandite began at Spor Mountain, Utah, in 1969, which since has become the major world producer of beryllium ore. Approximately 25 percent of the beryllium consumed by the United States in the years 1979 through 1983 was imported in the form of beryl; only minor amounts of beryllium metal or compounds are imported (Carlin and Petkof, 1984, p. 143). Total world production in 1984 was about 8,700 tons of equivalent beryl ore, of which 5,400 came from the United States (Kramer, 1985, p. 156).

Geology

Beryllium is widely dispersed in the Earth's crust because of its ability to replace silicon, which has a similar ionic radius. Crustal abundance is estimated at 2 to 3.5 ppm (Griffiths, 1973, p. 88). Most Be in the crust is in the common rock-forming minerals, mainly plagioclase feldspar, micas, and clays. Beryllium is an essential element in about 40 minerals and a minor constituent in about 50 others. However, only 6 are considered common or important: beryl, bertrandite, phenakite, chrysoberyl, helvite, and barylite chrysoberyl (Be Al_2 O_4), helvite [(Mn, Fe, Zn)$_4$ Be$_3$ (SiO$_4$)$_3$S], and barylite (Ba Be$_2$ Si$_2$O$_7$) (Griffiths, 1973, p. 90; 1982, p. 63). In igneous rock series, beryllium content generally increases toward the more siliceous, alkali-rich members. There is little dispersion of Be in the weathering environment, and clays and residual minerals contain most of the beryllium of the original rock. This is reflected in the sedimentary rocks: shales contain 2 to 5 ppm Be, sandstones less than 1 ppm, and limestones very little. Metamorphism has little effect on the distribution of Be, and the Be content of metamorphic rocks reflects that of the original rock type.

Griffiths (1973, p. 89) divided beryllium deposits into two broad categories: pegmatitic and nonpegmatitic or hydrothermal. The pegmatitic category includes fine-grained unzoned deposits and coarse-grained zoned deposits. The fine-grained deposits are unzoned aggregates of albite, microcline, spodumene, quartz, beryl, and muscovite. Beryl constitutes about 0.5 percent of these deposits. Although the fine-grained deposits are commercial sources for spodumene, scrap mica, and feldspar, they have not been a commercial source for beryl, mainly because of recovery problems. The coarse-grained zoned deposits are composed of the same minerals; only the proportions and textures vary. The internal structure of the zoned pegmatites is the same everywhere, the beryl crystals are marginal to a center core of quartz (Fig. 4). The zoned deposits have been the source of nearly all the beryl used in industry and also of beryl used as precious gems.

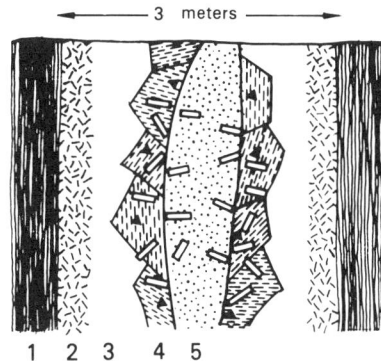

Figure 4. Cross section of pegmatite, Morada Nova beryl mine, Brazil. 1, schist; 2, muscovite zone; 3, pegmatite; 4, microline with beryl crystals; and 5, quartz core. (After Johnston, J. R., 1945.)

The nonpegmatitic or hydrothermal deposits have become the dominant source for beryllium and will probably continue to be so in the future. Most deposits of this type are in structurally tensional areas characterized by high-angle faults such as along the Rio Grande Rift and the Basin and Range Province of the western United States. Beryllium nonpegmatitic deposits are in fluorite mineralogenetic provinces, and fluorite is an accessory to the major component in most deposits (Griffiths, 1982, p. 62). Other associated minerals include quartz; magnetite, maghemite, or hematite; siderite; bismuthinite; wolframite; scheelite; cassiterite; and minor pyrite. Base and precious metals are not common. Potassic-feldspar and phyllosilicates are persistent associates in deposits of this type.

Hydrothermal beryllium deposits are mainly either hypothermal or epithermal. Quartz-rich veins, tactites, and beryl-bearing granites constitute the hypothermal deposits. These are mined mainly for tungsten, but those near Lake George, Colorado, were mined for beryl and bertrandite. Some of the veins are coarse grained and feldspar rich. Pegmatites contain an upper limit of about 0.5 percent beryl, but hypothermal veins have no upper limit (Griffiths, 1982, p. 64). Beryllium in tactite deposits is mainly in idocrase and other low-beryllium content minerals. Mesothermal deposits are not common, and the major base-metal deposits are notably poor in beryllium.

Epithermal deposits are generally associated with silicic volcanic fields, especially those with rhyolite rich in Be, K, Si, and F. The world's major producing mine at Spor Mountain, Utah, is in an epithermal deposit in water-laid tuff (Lindsey, 1977). Fluorite, silica, montmorillonite, adularia, and bertrandite replace limestone and dolomite cobbles in a rhyolite ashbed. The deposit is associated with a topaz-bearing rhyolite. The ore mined at Spor Mountain contains 1 to 2 percent bertrandite (Petkof, 1985, p. 76). Fine-grained adularia and kaolinite or smectite are present in most epithermal deposits together with fluorite and quartz or jasperoid.

Beryl is generally obtained as a co-product of mining of pegmatites for feldspar, mica, lithium minerals, columbite, tantalite, or cassiterite. In nonpegmatitic deposits, potential coproducts and by-products with beryllium, and close associates in nature, are those elements that form soluble complex ions with fluoride; including W, Sn, Nb, U, Th, Bi, and rare-earth metals. There is no recorded by-product production of Be, probably because of the high expense involved in extraction. Uranium oxide is recovered as a by-product of beryllium mining at Spor Mountain, Utah.

Exploration for beryllium deposits must consider many factors, as outlined by Griffiths (1982). None of the beryllium minerals have conspicuous appearances, and the most prominent minerals in outcrops of beryllium deposits are fluorite and quartz. The fluorite is generally purple, and the quartz is a "bull quartz" in hypothermal veins and chalcedonic in epithermal deposits. Regional targets would be based on a combination of fluorite province, siliceous volcanic fields, receptive host rocks (those that would cause deposition of fluoride complexes), and presence of distinctive associated elements. Geochemical exploration, especially careful stream-sediment surveys, would be the best tool in target areas (Griffiths, 1982, p. 66). A neutron activation field instrument is helpful in locating beryllium mineralization. Attention must be given to the less common beryllium minerals. A few years ago, bertrandite was considered a rare mineral, yet it is now the major commercial source of beryllium.

Resources

The world reserve base for beryllium is not well delineated but appears to be more than adequate for predicted future consumption. The major known reserves are in Brazil, the USSR, India, the United States, Canada, Argentina, South Africa, Uganda, Rwanda, and Australia (Petkof, 1985, p. 81). China apparently has large reserves, as suggested by recent exports, although there are no substantiating figures. Reserves in the United States are mainly in bertrandite ore, whereas reserves for the rest of the world are in beryl. A large deposit is currently under development in Canada. The property, at Thor Lake, Northwest Territories, is estimated to contain 1.8 million tons of ore averaging 0.86 percent beryllium oxide. Current major producers are the United States for bertrandite ore, and the USSR and Brazil for beryl.

There are two major factors that will influence the use of beryllium in industry: the high cost of extraction and processing, and potential health and environmental problems. A considerable amount of attention has been given to the causes of berylliosis, a serious chronic lung disorder. Exposure to high levels of beryllium-rich dust is associated with this illness and with the development of lung cancer. There seems to be a strong contrast between the apparent lack of effect by beryllium in natural materials and the highly toxic effects of many beryllium-containing products of industry (Griffiths and others, 1977, p. 7). Proper preventative measures control the problem, but they add significantly to the final cost of the beryllium materials.

The future of beryllium has some uncertainties. There seems to be an adequate supply for current and predicted consumption rates, which have been essentially level since 1960, and the United States will probably continue to be the major producer and consumer. The uncertainties stem from possible development of new uses, improvement of extraction fabrication techniques, and degree of control by enactment of proposed beryllium health standards. Beryllium has unique properties that would make it useful in many future developments, but at this time it is a costly, high-energy-demand commodity, and there are health and environmental concerns.

MERCURY

Introduction

Because of its chemical and physical properties, mercury (quicksilver) has become an essential commodity to modern society, and many of the thousands of uses for mercury have no acceptable alternatives. However, mercury is also a pollutant because it is a highly toxic substance; the amount of mercury in the environment and foodstuffs is carefully regulated.

Mercury is unique among the commercially used metals in that it is a liquid at room temperature and, in addition, it expands and contracts uniformly and conducts heat and electricity efficiently. The battery industry accounted for 54.4 percent of the total consumption in 1984, followed by consumption in chlorine and caustic soda manufacture (13.5 percent), paints (8.5 percent), measuring and control devices (5.2 percent), wiring devices and switches (5.0 percent), electric lighting (3.7 percent), dental equipment and supplies (3.6 percent), and others (6.1 percent; Carrico, 1985b, p. 643). The properties of mercury-cell batteries make it critical to many industrial, military, and aerospace applications. Mercuric oxide is used as the cathode. Prior to World War I, the main use of mercury was in the amalgamation process for the recovery of gold and silver from ores; in 1917, the main use was as an antisyphilitic (Bailey and others, 1973, p. 402).

Substitutions for mercury have been found for several applications that in the past were significant consumers, including amalgamation,, antisyphilitics, munitions, felt manufacture, pattern making, paper manufacture, vermillion, fireworks, and cosmetics. Because of the concern over mercury as a pollutant, efforts continue to find viable substitutes for mercury in other uses. Nickel/cadmium and lithium batteries, organotin compounds in paint, solid-state devices in measuring and control instruments, and the use of diaphragm and/or membrane cells for mercury cells in the chlor-alkali industry are examples of substitutions becoming effective. There are many applications, mainly in electrical and industrial instruments, in which substitution is impractical or results in sacrifice of quality or loss of economy of production.

The main sources of mercury are ores containing the mineral cinnabar and to a lesser extent native mercury and the minerals corderoite, livingstonite, calomel, and metacinnabar. Mercury obtained from mining operations is called prime virgin or virgin metal and is usually 99.9 percent pure; it is acceptable for most end uses. The standard basis for pricing and resource discussions of mercury is by the flask containing 76 pounds. Redistilled mercury is of a higher purity and is sold in smaller containers at a premium price. Secondary mercury is recovered from recycling scrap and sludge from industry. There are no figures as to the amount recycled.

Mercury was known and used by the ancient Egyptians, Greeks, Romans, and Chinese in a variety of ways. Consumption of mercury increased substantially in the sixteenth century when the amalgamation process for recovery of silver was developed. Prior to 1850, three established mines provided most production: the Almaden Mine, Spain, the Idria Mine, Yugoslavia, and the Santa Barbara Mine, Peru. In the second half of the nineteenth century, the Monte Amiata District in Italy and the New Almaden and New Idria mines, California, became major producers. These six districts have accounted for most of the world's accumulative production. Development of the mercury mining industry in California was associated with the discovery and development of gold mining, and the peak year for mercury production was 1877 at 79,917 flasks (Carrico, 1985a, p. 499). In 1984, the United States produced about 19,000 flasks from three mines, two of which produced mercury as a by-product (Carrico, 1985b, p. 642). World production in 1984 was about 175,000 flasks, down from 211,000 in 1983, and from the high of 290,000 flasks in 1969.

Mercury is classified as a strategic and critical commodity, essential to the economy. The U.S. government maintains a stockpile of mercury that is being reduced to a goal of 10,500 flasks, down considerably from a stockpile that was more than 10 times that amount prior to 1976. The government also released 146,923 flasks of secondary mercury during the period 1963 to 1981 (Carrico, 1985a, p. 505).

Geology

Mercury occurs in small amounts in all natural substances. The average content of continental crustal rocks is on the order of 80 ppb (parts per billion), being slightly higher in shales, some limestones and coals, and considerably higher in petroleum (Bailey and others, 1973, p. 407). Cinnabar (HgS), 86.2 percent mercury, is by far the most common mercury mineral. Metacinnabar is a black form of cinnabar. Native mercury occurs in some deposits, and corderoite ($Hg_3S_2Cl_2$), calomel (Hg_2Cl_2), and livingstonite ($HgSb_4S_7$) are common locally. The more than 20 other mercury-bearing minerals are not common.

Mercury deposits have several factors in common: they all formed at shallow depths (less than 1,000 m) from low-temperature hydrothermal solutions (50°C to 200°C) in Tertiary to Recent times, and are located along the continent side of post-Jurassic subduction zones. It appears that continental-margin plutonism and subduction mobilized mercury on a large scale into younger formations. Individual deposits may occur in almost any rock type that provided a physical or chemical depositional site. Physical types of deposits include fissure fillings, breccia fillings, pore-space fillings, stockworks, and replacement bodies. World distribution of mercury deposits is concentrated into two orogenic regions: the circum-Pacific volcano-plutonic system and the southern margins of the Alpine fold belt in the northern borderlands of the Mediterranean. Silica and carbonate minerals are common in introduced gangue, and pyrite and marcasite are common locally. Mercury has a common association with gold and antimony. Associated minerals in deposits in southeast Asia include ores of cinnabar and stibnite with orpiment, realgar, and native arsenic. Mercury minerals will withstand an erosional cycle, and placer deposits of cinnabar are not uncommon; in one major district in Borneo the greatest production was from eluvial placers (Hutchison, 1983, p. 97).

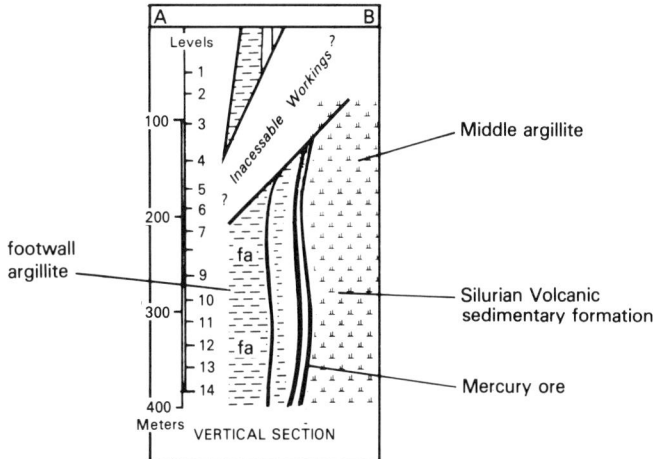

Figure 5. Cross section of the Almaden Mine, Spain. (After Saupé, 1973.)

The Almaden deposit, Spain, has yielded about 7.5 million flasks of mercury, and historically is the world's major producer. The deposit is on the south limb of an anticline in Silurian orthoquartzites and siltstones overlain by a thick sequence of pyroclastic rocks. Mineralization is in lenses 3 to 5 m thick and as much as 400 m in length (Fig. 5). The ore is quartzite with cinnabar filling pore spaces, microfractures, fissures, and other open spaces. Other constituents include minor amounts of pyrite, kaolinite, organic matter, calcite, and dolomite. The mine is worked at an average grade of 2 to 3 percent from ore that ranges from 0.6 to 20 percent mercury. Genesis of the deposit is in question, as there are observable epigenetic, diagenetic, and syngenetic features to the ore bodies. The deposits are variously thought to be replacement bodies, syngenetic sedimentary, submarine hydrothermal, and diagenetic remobilization from within the depositional basin.

The New Almaden district, California, is historically the greatest mercury producer in the United States, although most production was before World War I (Bailey and others, 1973, p. 408). Production has been in excess of 1.1 million flasks. The district is in a tectonically active mélange belt, where Franciscan

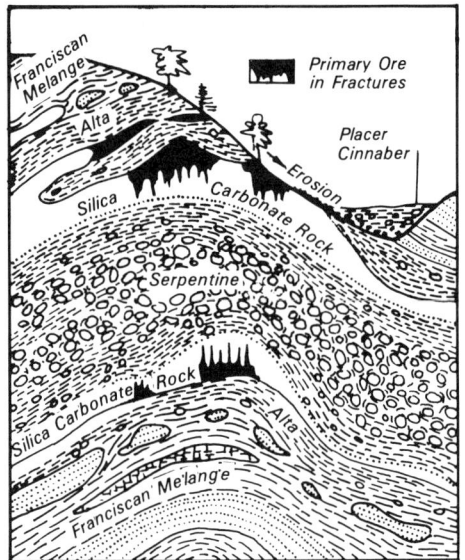

Figure 6. Cross section of the New Almaden Mine, California. (After Bailey and Everhart, 1964.)

mélange is cut by tabular serpentinite bodies (Fig. 6). Parts of the serpentinite have been hydrothermally altered to quartz-carbonate rock, perhaps as an early stage of mercury mineralization. The deposits are cinnabar and native mercury fracture fillings, stockworks, and replacements of the quartz-carbonate rock. Associated minerals are minor amounts of pyrite, stibnite, chalcopyrite, sphalerite, galena, and bornite in a gangue of quartz and dolomite with some hydrocarbons. Some of the ore bodies were exceptionally large and rich; the largest was 60 m wide, 4.8 m thick, and extended downdip 470 m. In the first few years of operation the ore contained more than 20 percent mercury (Hutchison, 1983, p. 100). Grade in recent years is closer to 0.5 percent. The ores and alteration were thought to have formed from low-temperature (50° to 130° C) meteoric water solutions rising along steep faults and fractures along the contact between serpentinite and the Franciscan mélange.

The major producer, and only operating mercury mine, in the United States in 1984 was at McDermitt, Nevada. The ore is in Miocene volcanic ash and lake beds, and is 70 percent cinnabar and 30 percent corderoite. Grade is about 4 kg of mercury per ton of ore, and the mine has about 6 years reserves, based on an annual production rate of 19,000 flasks (Carrico, 1985a, p. 505). The mine produced an average 26,800 flasks a year during the period 1976–1983, which accounted for about 46 percent of the average United States consumption.

The major mercury deposits are mined only for their mercury contents, although in Southeast Asia, mercury, antimony, and gold are common co-products or by-products. Mercury is also a component of other types of ore and is extracted as a by-product of copper mining in Ireland, copper and zinc mining in Czechoslovakia, and zinc mining in New York State, U.S. (Bailey and others, 1973, p. 407). Mercury is becoming an important by-product of gold mining from the large low-grade deposits in the western United States, the same area where mercury is currently being deposited in some hot springs. Mercury in the gold ores was a problem in that it interfered with extraction of gold and presented a potential health hazard. Modification of the processing procedures allowed separation (Carrico, 1985b, p. 646).

Exploration for mercury deposits should be directed toward the restricted geologic environment in which they occur, the continentward side of post-Jurassic subduction zones. Better potential for the future probably lies in finding new resources in existing districts in younger uneroded and covered deposits, or in developing larger low-grade deposits, perhaps with co-products. Panning of stream sediments or slope debris is the best and most widely used prospecting technique. Geochemical techniques to detect mercury have improved greatly over the past few years, in part in response to demands to monitor the environment, and geochemistry is probably a viable approach in exploring for hidden ore bodies.

Resources

The USSR, Spain, China, and the United States accounted for about 82 percent of mercury mine production in 1984. The other producing countries were Mexico, Czechoslavakia, Turkey, Yugoslavia, Federal Republic of Germany, Finland, and the Dominican Republic (Carrico, 1985b, p. 645). The world has a reserve base of about 7.2 million flasks (Carrico, 1985a, p. 500), which amounts to a 35-year supply at the current mine production rate. Spain and Italy account for 65 percent of the reserve base, with an unrealistically small amount assigned to China and the USSR. Bailey and others (1973, p. 409) estimated mercury resources based on a much higher than current price at 17.1 million flasks, about a 90-year supply. The United States has a reserve base of 200,000 flasks, which is not adequate to meet the probable cumulative primary demand to the year 2000 of 700,000 flasks. Therefore, a large part of the U.S. supply of prime virgin mercury will be from imports.

The mercury-mining industry is small in relation to that of many other commodities, but it supports a large end-product industry. Almost every person uses a little mercury each day. About 405 U.S. plants consumed mercury in 1984 (Carrico, 1985b, p. 643). The domestic mercury-mining industry has difficulties competing with foreign producers mining richer ores with cheap labor. Mercury prices also tend to fluctuate greatly in response to erratic demand and overproduction, which also affects the stability of the domestic industry. Most of the world's reserves are concentrated in a few low-cost mines that have the capability of controlling the market. In addition, China, which probably has large reserves, could again enter the free market with large quantities of mercury and undercut most other producers.

The major domestic problem concerning mercury is that it is a pollutant, and its vapors and most of its organic and inorganic compounds are toxic. Precautions must be taken at all levels of mining, manufacture, and end use to ensure that there is no human or environmental contamination. Several past uses of mercury have been replaced by substitutes and some existing uses have become more efficient in controlling mercury, but there are still some uses that will have to control the dissipation of mercury in the environment or be replaced.

The demand for primary virgin mercury in the United States is expected to decline in the future in response to increased output of secondary mercury. The consumption of mercury is also likely to decline because of substitution and continued concern over contamination of the environment and the food chain. The annual consumption growth rate for the rest of the world is expected to be about 1.8 percent (Carrico, 1985a, p. 506). A substantial increase in the use of mercury will depend on developing uses that are not hazardous or refining current uses and fabrication so that the loss of mercury to the environment is controlled.

ALUMINUM

Introduction

Aluminum is experiencing a tremendous growth in demand and is now second only to iron in consumption. It is important in all segments of society and is of critical importance to the United States because of nearly total reliance on imports as a source of primary aluminum. There is also an explosion of technical and scientific information concerning the raw material sources of aluminum (Patterson and others, 1986).

Consumption of aluminum by the United States in 1983 was packaging (37 percent), transportation (17 percent), building (17 percent), electrical (9 percent), consumer goods (8 percent), and other uses (10 percent). The largest user is aluminum beverage cans, which in 1984 accounted for 1.8 million tons of metal (McCawley and Stephenson, 1985, p. 94). Many of the more important uses of aluminum are commonplace. Some of the less known uses include the production of synthetic rubies and sapphires used in the construction of lasers and as jewel bearings, as thin films for mirror surfaces, and as high-capacity electrical transmission lines. A small portion of the bauxite mined and the alumina intermediate material produced are used directly. Refractory-grade calcined bauxite is the major refractory material used in the steel, copper, aluminum, and glass industries.

Possible substitutes include those commodities that aluminum replaced during its rapid growth in usage since World War II—wood, steel, tin, copper, and glass. These commodities would probably be used again only if the aluminum supply were to become critically short. New high-technology materials such as graphite-fiber epoxies, plastics, and titanium and magnesium compounds are replacing aluminum for some uses, especially where they are more cost effective.

The major source of primary aluminum today is bauxite, a general term for mixtures of aluminum hydroxide minerals and impurities. Bauxite ores are discussed on the basis of mineralogy, depending on the type of aluminum hydroxides present or the physical character of the ore. Ores that are primarily composed of the mineral gibbsite are termed trihydrate, those of boehmite and/or diaspore are termed monohydrate, and ores with substantial amounts of each are called mixed ores. Common commercial forms of bauxite are pisolite or oolite, sponge ore, and amorphous or clay ore. Alunite and nepheline are sources of alumina in the USSR (McCawley and Baumgardner, 1985, p. 9). Commercially pure aluminum contains about 99.5 percent aluminum, and super purity aluminum contains a minimum of 99.99 percent aluminum. Bauxite is marketed in five different grades: refractory, abrasive, chemical, cement, and metal, each with different specifications.

Two major milestones paved the way for major commercial use of aluminum. The first was development of a successful electrolytic process for production of aluminum from alumina, the Hall/Heroult process. The second was discovery of a chemical process to produce alumina in large quantities from bauxite, the Bayer process. Both were developed in the late 1800s and, with minor refinements, are the basis for the modern aluminum industry. The aluminum industry started to expand in World War II because of military applications and has been expanding since. Bauxite is a high-volume ore, 76 million tons in 1983 for world production, with some mines accounting for more than 10 million tons. This amounts to about 16 million tons aluminum equivalent.

The U.S. government maintains a stockpile of aluminum and bauxite for civilian and defense use in national emergencies. Aluminum is one of four commodities given defense order priorities under the Defense Protection Act.

Geology

Aluminum is second only to silicon as the most abundant metallic element in the Earth's crust. Average crustal abundance is about 8 percent, and most of the common rock- and soil-forming minerals have aluminum as a major component. It is always in combination with other elements and mostly in silicate minerals. Commercial availability of primary aluminum is dependent more on extractability of alumina from the raw material than it is on abundance and concentration. There are two groups of raw material sources to be considered: bauxitic raw materials, of which the United States has few resources, and nonbauxitic sources, of which the United States has considerable resources. At the present time the production of alumina in the world is almost entirely from bauxitic ores, because the energy cost for breakdown of bauxitic ores is considerably less than that for nonbauxitic ores.

Bauxitic raw materials include bauxitic clay, aluminous laterite, and bauxite, of which the latter two are the major commercial sources (Shaffer, 1983, p. 503). Origin of bauxitic materials is from chemical weathering of aluminum-bearing rocks, during which aluminum is retained in resistant hydrated aluminum

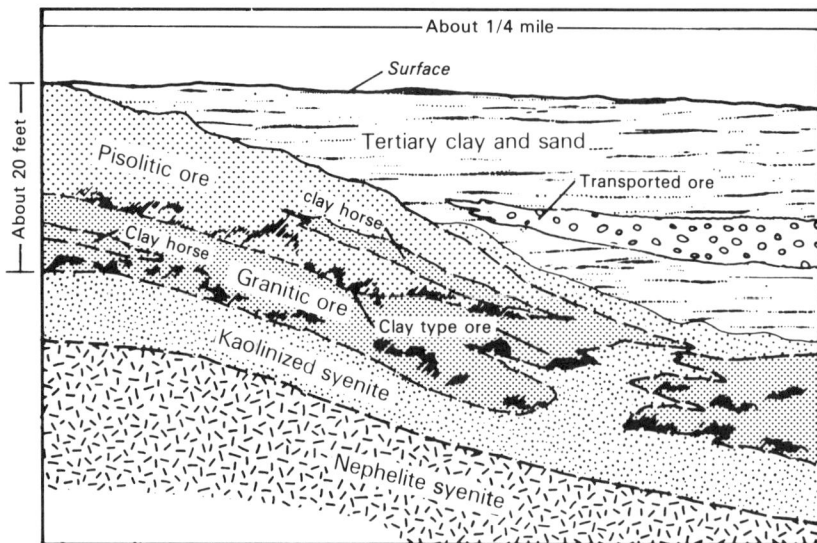

Figure 7. Cross section of the bauxite deposits of Arkansas. (After Branner, 1933).

oxide minerals, and the other elements, mainly silica, are removed. There is no universal agreement on the importance of the original aluminous source, the mechanics and chemistry of aluminum enrichment and desilication, and the duration and intensity of the weathering process. The long-held theory is that bauxite is a residual weathering product of tropical climates; aluminum and iron are released from weathering of the silicate minerals and are reprecipitated as hydroxides and oxides, and the more mobile alkalies, alkaline earths, and silicon are leached from the area in solution. Another theory suggests that aluminum, together with iron, was transported in solution to swampy areas where it was deposited as colloidal gels. The aluminum content of the parent rock probably is not as important as the duration and intensity of the weathering process. Bauxitic raw materials occur on all continents except Antarctica, and are of all ages from Precambrian to Recent (Shaffer, 1983, p. 510).

Bauxite is a variable material of almost any color that can take many different forms, including earthy, clayey, massive, oolitic, pisolitic, nodular, fine-grained, vermicular, brecciated, saccharoidal, vesicular, botryoidal, or platy. Mineralogy consists of the aluminum hydroxide minerals gibbsite, boehmite, and diaspore, with impurities of clay minerals, mica, quartz, hematite, goethite, pyrite, siderite, marcasite, anatase, leucoxene, rutile, ilmenite, and sphene. The varying proportions of clay minerals, aluminum hydroxides, and iron oxides determine if the material is bauxite, bauxitic clay, aluminous laterite, or ferruginous laterite. Classification of deposits can be on the basis of geologic origin, chemical composition, mineral composition, texture, topographic position, underlying rock type, and mode of occurrence. The better known are blanket deposits at or near the surface, pocket deposits in karst depressions, interlayered deposits on unconformities that represent older buried deposits, and detrital deposits formed from debris of eroding bauxitic deposits. Silica content determines grade as much as aluminum content.

High-grade ore should be about 50 percent alumina and less than 5 percent silica; low-grade ore can have the same alumina content and 5 to 10 percent silica.

Bauxite deposits in Arkansas are interlayered deposits along a buried Tertiary weathered surface and detrital deposits in the overlying Tertiary sedimentary rocks (Fig. 7). The bauxite was formed on weathered nepheline syenite intrusive rocks of Cretaceous age. Ore zones average 4 m thick and are overlain by 10 to 50 m of overburden. Grade varies from 40 to 52 percent Al_2O_3, 7 to 17 percent SiO_2, and 6 to 12 percent Fe_2O_3; gibbsite is the dominant mineral. The Cove Peninsula deposits in Australia are blanket deposits on a plateau underlain by Lower Cretaceous shales (Fig. 8). The bauxite averages 3 to 4 m thick and consists mostly of gibbsite with minor boehmite. A reported grade, not necessarily representative, is 48.7 percent Al_2O_3, 3.6 percent SiO_2, and 17 percent Fe_2O_3 (Shaffer, 1983, p. 512).

There are several potential nonbauxitic sources of primary aluminum, some of which have provided minor amounts of raw material. Murry (1983) listed the following for the United States: (1) alunite, common in the western states; (2) high-alumina clays, mainly kaolinite found in sedimentary rocks in Georgia and South Carolina; (3) dawsonite, found in oil shales of the Green River Basin; (4) anorthosites, common in the western states; and (5) nepheline and alkali feldspars in nepheline syenite in Arkansas and the western states. Other sources include aluminum phosphate rock in Florida, aluminous shale and slate from throughout the United States, coal ash, and copper-leach solutions (Patterson and Dyni, 1973, p. 40). The USSR has produced alumina from alunite ores with by-products of potassium sulfate and sulfuric acid since the mid-1960s (Hall and Bauer, 1983, p. 417). The USSR has also produced alumina from nepheline.

By-products from bauxitic ores are not common. A small amount of gallium, iron, and vanadium are produced in some areas. Nonbauxitic ores have more co-product and by-product

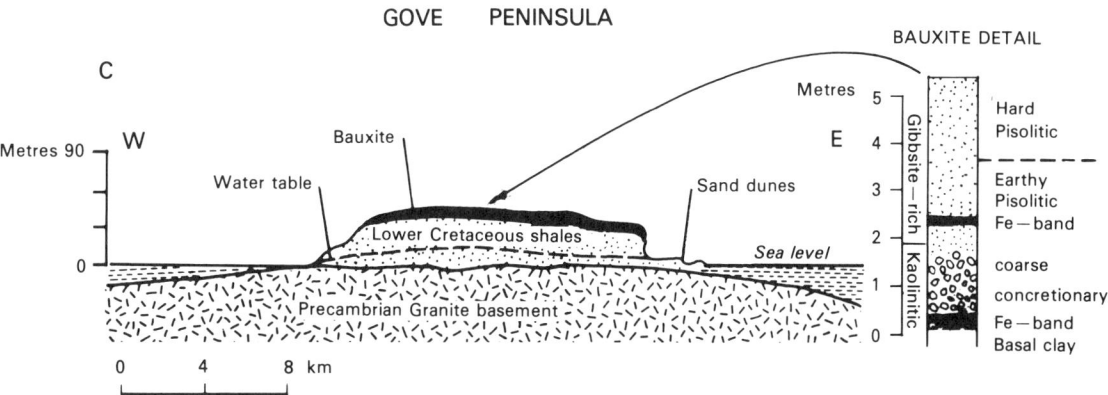

Figure 8. Cross section through the Grove Peninsula, Australia. (Modified after Grubb, 1970.)

potential, and the USSR produces potash, soda ash, and cement materials as co-products of processing nonbauxitic ores.

A thorough understanding of the regional geology, geologic history, and physiography of an area is the basis for exploration for bauxitic deposits. Remote-sensing techniques using enhanced images have been successful by outlining areas of vegetation types associated with aluminous laterites. Bauxite minerals are difficult to identify, and quantitative analysis is needed to determine mineralogy and contents of alumina and silica.

Resources

The United States produced 856,000 tons of bauxite ore in 1984 out of a total world production of 84,664,000 tons (Baumgardner and Hough, 1985, p. 141). Major producers are Australia, Guinea, Jamaica, Brazil, USSR, Surinam, and Yugoslavia. The United States is a major producer of primary aluminum, with a production of about 4 million tons in 1984 out of a capacity of 5 million tons. World production in 1984 was 15.5 million tons (McCawley and Stephenson, 1985, p. 91). Recycled scrap aluminum is expected to amount to about 15 percent of the aluminum in end products for the next decades. Known world bauxite reserves are adequate to meet the predicted cumulative demand for the foreseeable future. In the United States, however, domestic bauxite reserves are estimated to account for only about 8 percent of the cumulative domestic demand to the year 2000 (McCawley and Baumgardner, 1985, p. 29).

Reserves of bauxitic ore in the United States are mainly in Arkansas but include some in Alabama and Georgia. Potential resources include large low-grade ferruginous bauxitic deposits in weathered basalts in Hawaii, ferruginous bauxitic and aluminous lateritic deposits formed from lateritization of basalts in the Columbia River Basalt Group, and bauxitic clay deposits in Arkansas, Georgia, and South Carolina. Nonbauxitic sources for alumina are receiving considerable attention in the United States, and alunite deposits in the western United States are currently being evaluated.

The United States is almost totally dependent on imports for aluminum raw materials and alumina, and has few viable reserves for the future. Energy costs of producing aluminum are also escalating, and disposal of caustic waste materials is causing potential environmental problems. Transportation costs of raw materials are a major concern since the major sources are a considerable distance from the United States. Alumina imports have been increasing, and bauxite raw material imports have been decreasing recently, because the source countries have entered into production of alumina and the United States has been closing refineries. This change reduces the transportation costs, but may eventually leave the United States without the capability of refining bauxitic ores to alumina.

The demand for aluminum in the United States is expected to increase at an annual growth rate of 3.2 percent per year from 1983 to 2000 (McCawley and Baumgardner, 1985, p. 26). Aluminum is an available and useful commodity that probably will have many new applications in the future. New technology and a means of reducing energy costs are needed for the United States to become a major supplier of raw materials, mainly from nonbauxitic ores.

REFERENCES CITED

Adams, J. W., and Staatz, M. H., 1973, Rare-earth elements, in Brobst, D. A., and Pratt, W. P., eds., Mineral resources of the United States: U.S. Geological Survey Professional Paper 820, p. 547–556.

Armbrustmacher, T. J., 1980, Abundance and distribution of thorium in the carbonatite stock at Iron Hill, Powderhorn district, Gunnison County, Colorado: U.S. Geological Survey Professional Paper 1049-B, 11 p.

Bailey, E. H., and Everhart, D. L., 1964, Geology and quicksilver deposits of the New Almaden District, Santa Clara County, California: U.S. Geological Survey Professional Paper 360, 266 p.

Bailey, E. H., Clark, A. L., and Smith, R. M., 1973, Mercury, in Brobst, D. A., and Pratt, W. P., eds., Mineral resources of the United States: U.S. Geological Survey Professional Paper 820, p. 401–414.

Baumgardner, L. H., and Hough, R. A., 1985, Bauxite and Alumina, in Minerals yearbook, 1984: U.S. Bureau of Mines, v. 1, p. 141–152.

Branner, G. C., 1933, The Arkansas Bauxite deposits: Washington, D.C., 16th International Geological Congress Guidebook 2, p. 92–103.

Carlin, J. F., 1985a, Tin, *in* Minerals facts and problems: U.S. Bureau of Mines Bulletin 675, p. 847–858.

——, 1985b, Tin, *in* Minerals yearbook 1984: U.S. Bureau of Mines, v. 1, p. 901–912.

Carlin, J. F., Jr., and Petkof, B., 1984, Beryllium, *in* Minerals yearbook 1983: U.S. Bureau of Mines, v. 1, p. 143–146.

Carrico, L. C., 1985a, Mercury, *in* Mineral facts and problems: U.S. Bureau of Mines Bulletin 675, p. 499–508.

——, 1985b, Mercury, *in* Minerals yearbook 1984: U.S. Bureau of Mines, v. 1, p. 641–646.

Clark, A. M., 1984, Mineralogy of the rare earth elements, *in* Henderson, P., ed., Rare earth element geochemistry: Amsterdam, Elsevier, Developments in Geochemistry, p. 33–61.

Cunningham, L. D., 1985a, Tantalum, *in* Mineral facts and problems: U.S. Bureau of Mines Bulletin 675, p. 811–822.

——, 1985b, Columbian, *in* Minerals facts and problems: U.S. Bureau of Mines Bulletin 675, p. 185–196.

Einaudi, M. T., Meinert, L. D., and Newberry, R. J., 1981, Skarn deposits, *in* Skinner, B. J., ed., Seventy-fifth anniversary volume 1905–1980: Economic Geology, p. 317–391.

Farr, P., 1984, Beryllium, *in* Mining annual review 1984: London, Mining Journal, June 1984, p. 88.

Force, E. R., and Lynd, L. E., 1984, Titanium mineral resources of United States—definitions and documentation: U.S. Geological Survey Bulletin 1558 B, 11 p.

Force, E. R., and others, 1976, Geology and resources of titanium: U.S. Geological Survey Professional Paper 959 A–F, 49 p.

Gadden, P., 1984, Titanium, *in* Mining annual review 1984: London, Mining Journal, June 1984, p. 54–55.

Gray, R. F., Hoffman, V. J., Bagan, R. J., and McKinley, H. L., 1968, Bishop tungsten district, California, *in* Ridge, J. D., ed., Ore deposits of the United States, 1933–1967: American Institute of Mining, Metallurgical and Petroleum Engineers, Inc., p. 1531–1554.

Griffiths, W. R., 1973, Beryllium, *in* Brobst, D. A., and Pratt, W. P., eds., Mineral resources of the United States: U.S. Geological Survey Professional Paper 820, p. 85–93.

——, 1982, Diagnostic features of fluoride-related beryllium deposits, *in* Erickson, R. L., compiler, Characteristics of mineral deposit occurrences: U.S. Geological Survey Open-File Report 82–795, p. 62–66.

Griffiths, W. R., Allaway, W. H., and Groth, D. H., 1977, Beryllium, *in* Geochemistry and the environment; Volume 2, The relation of other selected trace elements to health and disease: Washington, D.C., National Academy of Sciences, p. 7–10.

Grosz, A. E., and Escowitz, E. C., 1983, Economic heavy minerals of the United States Atlantic Continental Shelf, *in* Tanner, W. F., ed., Near-shore sedimentology: Proceedings of the Sixth Symposium on Coastal Sedimentology, p. 1–10.

Grubb, P.L.C., 1970, Mineralogy, geochemistry, and genesis of the bauxite deposits on Grove and Mitchell Plateaux, Northern Australia: Mineral Deposita, v. 5, p. 248–272.

Hall, R. B., and Bauer, C. W., 1983, Alunite, *in* Lefond, S. J., ed., Industrial minerals and rocks, 5th edition: New York, Society of Mining Engineers, p. 417–434.

Harpen, R. W., and Bates, R. L., 1984, Geology of the nonmetallics: New York, Metal Bureau, Inc., 392 p.

Hedrick, J. B., 1984, Thorium, *in* Mineral yearbook 1983: U.S. Bureau of Mines, v. 1, p. 857–862.

——, 1985a, Rare-earth elements and yttrium, *in* Minerals facts and problems: U.S. Bureau of Mines Bulletin 675, p. 647–664.

——, 1985b, Rare-earth minerals and metals, *in* Minerals yearbook 1984: U.S. Bureau of Mines, v. 1, p. 751–762.

——, 1985c, Thorium, *in* Mineral facts and problems: U.S. Bureau of Mines, Bulletin 675, p. 835–846.

Henderson, P., 1984, General geochemical properties and abundances of the rare earth elements, *in* Henderson, P., ed., Rare earth element geochemistry: Amsterdam, Elsevier, Developments in geochemistry, p. 1–32.

Ho, C. E., 1984, Tungsten, *in* Mining annual review 1984: London, Mining Journal, June 1984, p. 41–42.

Hobbs, S. W., and Elliott, J. E., 1973, Tungsten, *in* Brobst, D. A., and Pratt, W. P., eds., Mineral resources of the United States: U.S. Geological Survey Professional Paper 820, p. 667–678.

Hobbs, S. W., and Tooker, E. W., 1980, Preliminary map of tungsten provinces in the conterminous United States: U.S. Geological Survey Open-File Report 79–576.C, scale 1:5,000,000.

Hosking, K.F.G., 1969, The nature of the primary tin ores of the southwest of England, *in* A second technical conference on tin: International Tin Council, v. 1, p. 1157–1243.

Hutchison, C. S., 1983, Economic deposits and their tectonic setting: Singapore, John Wiley and Sons, 365 p.

International Tin Council, 1984, Tin, *in* Mining annual review 1984: London, Mining Journal, June 1984, p. 41–42.

Kelly, W. C., and Turneauve, F. S., 1970, Mineralogy, paragenesis, and geothermometry of tin and tungsten deposits of the eastern Andes: Economic Geology, v. 65, p. 609–680.

Klemic, H., Marsh, S. P., and Cooper, M., 1973, Titanium, *in* Brobst, D. A., and Pratt, W. P., eds., Mineral resources of the United States: U.S. Geological Survey Professional Paper 820, p. 653–665.

Kramer, D. A., 1985, Beryllium, *in* Minerals yearbook, 1984: U.S. Bureau of Mines, v. 1, p. 153–157.

Lindsey, D. A., 1977, Epithermal beryllium deposits in water-laid tuff, western Utah: Economic Geology, v. 72, no. 2, p. 219–232.

Lynd, L. E., 1985, Titanium, *in* Minerals facts and problems: U.S. Bureau of Mines Bulletin 675, p. 859–879.

Lynd, L. E., and Lefond, S. J., 1983, Titanium minerals, *in* Lefond, S. J., ed., Industrial minerals and rocks, 5th edition: New York, Society of Engineers, AIME, p. 1303–1362.

McCawley, F. S., and Baumgardner, L. H., 1985, Aluminum, *in* Minerals facts and problems: U.S. Bureau of Mines, v. 1, p. 9–31.

McCawley, F. S., and Stephenson, P. A., 1985, Aluminum, *in* Minerals yearbook 1984: U.S. Bureau of Mines, v. 1, p. 91–108.

Mitchell, A.H.G., and Garson, M. S., 1972, Relationship of porphyry copper and circum-Pacific tin deposits to palaeo-Benioff zones: Institute of Mining and Metallurgy Transactions, v. 81, p. B10–B25.

Murry, H. H., 1983, Nonbauxitic alumina resources, *in* Shanks, W. C., III, ed., Cameron volume on unconventional mineral deposits: Society of Mining Engineers, p. 111–120.

National Materials Advisory Board, 1982, Tantalum and columbium supply and demand outlook: National Academy of Sciences Publication NMAB–391, 173 p.

Neary, C. R., and Highley, D. E., 1984, The economic importance of the rare earth elements, *in* Henderson, P., ed., Rare earth element geochemistry: Amsterdam, Elsevier, Developments in geochemistry, p. 423–466.

Parker, R. L., and Adams, J. W., 1972, Niobium (columbium) and tantalum, *in* Brobst, D. A., and Pratt, W. P., eds., Mineral resources of the United States: U.S. Geological Survey Professional Paper 820, p. 443–454.

Patterson, S. H., and Dyni, J. R., 1973, Aluminum and bauxite, *in* Brobst, D. S., and Pratt, W. P., eds., Mineral resources of the United States: U.S. Geological Survey Professional Paper 820, p. 35–43.

Patterson, S. H., Kurtz, H. F., Olson, J. C., and Neeley, C. L., 1986, World bauxite resources (Geology and resources of aluminum): U.S. Geological Survey Professional Paper 1076-B, 151 p.

Petkof, B., 1985, Beryllium, *in* Mineral facts and problems: U.S. Bureau of Mines Bulletin 675, p. 75–82.

Reed, B. L., and Tooker, E. W., 1980, Preliminary map of tin occurrence areas in the conterminous United States: U.S. Geological Survey Open-File Report 79–576L, scale 1:7,500,000.

Robjohns, N., 1984, Rare earths, *in* Mining annual review 1984: London, Mining Journal, June 1984, p. 88–89.

Sainsbury, C. L., and Reed, B. L., 1973, Tin, *in* Brobst, D. A., and Pratt, W. P., eds., Mineral resources of the United States: U.S. Geological Survey Professional Paper 820, p. 637–651.

Saupé, F., 1973, La geologie du gisement d'Almaden: Science de la Terre, Memoir 29, Nancy, University of Nancy, 342 p.

Shaffer, J. W., 1983, Bauxitic raw materials, *in* Lefond, S. J., ed., Industrial minerals and rocks, 5th edition: Society of Mining Engineers, p. 503–527.

Staatz, M. H., and Armbrustmacher, T. J., 1981, Preliminary map of rare-earth provinces in the conterminous United States: U.S. Geological Survey Open-File Report 79–576T, scale 1:5,000,000.

—— , 1982, Preliminary map of the thorium provinces in the conterminous United States: U.S. Geological Survey Open-File Report 79–576U, scale 1:5,000,000.

Staatz, M. H., and Olson, J. C., 1973, Thorium, *in* Brobst, D. A., and Pratt, W. P., eds., Mineral resources of the United States: U.S. Geological Survey Professional Paper 820, p. 469–476.

Staatz, M. H., Armbrustmacher, T. J., Olson, J. C., Brownfield, I. K., Brock, M. R., U.S. Geological Survey, Lemons, J. F., Jr., Coppa, L. V., and Clingan, B. V., U.S. Bureau of Mines, 1979, Principal thorium resources in the United States: U.S. Geological Survey Circular 805, 42 p.

Staatz, M. H., Hall, R. B., Macke, D. L., Armbrustmacher, T. J., and Brownfield, I. K., 1980, Thorium resources of selected regions in the United States: U.S. Geological Survey Circular 824, 32 p.

Staatz, M. H., and others, 1979–1985, Geology and resources of thorium in the United States: U.S. Geological Survey Professional Paper 1049 A–E, various paging.

Stafford, P. T., 1985a, Tungsten, *in* Mineral facts and problems: U.S. Bureau of Mines Bulletin 675, p. 881–894.

—— , 1985b, Tungsten, *in* Minerals yearbook 1984: U.S. Bureau of Mines, v. 1, p. 927–940.

Taylor, R. G., 1979, Geology of Tin Deposits, *in the collection* Developments in economic geology: Amsterdam, Elsevier Publishing Co., v. 11, 543 p.

Tooker, E. W., and Force, E. R., 1980, Preliminary map of titanium provinces in the conterminous United States: U.S. Geological Survey Open-File Report 79–576–K, scale 1:5,000,000.

Tooker, E. W., and Parker, R. L., 1980, Preliminary map of columbium (niobium) and tantalum provinces in the conterminous United States: U.S. Geological Survey Open-File Report 79–576–N, scale 1:5,000,000.

Tron, A. R., 1984, Niobium and tantalum, *in* Mining annual review 1984: London, Mining Journal, June 1984, p. 72–73.

Yan, M. Z., Wu, Y. L., and Li, C. Y., 1980, Metallogenetic systems of tungsten in southern China and their mineralization characteristics, *in* Ishihara, S., and Takenouchi, S., eds., Granitic magmatism and related mineralization: Japan, Mining Geology Special Issue no. 8, p. 215–221.

MANUSCRIPT ACCEPTED BY THE SOCIETY FEBRUARY 12, 1989

Printed in U.S.A.

Chapter 9

Phosphate deposits of the United States—discovery, development; Economic geology and outlook for the future

James B. Cathcart
U.S. Geological Survey (retired), 17225 West 16th Place, Golden, Colorado 80401

INTRODUCTION

The world's production of phosphate rock is from three sources: (1) guano or guano derived, (2) apatite deposits of igneous origin, and (3) marine sedimentary phosphorite. Occurrences of the three types are known in the United States, but only the sedimentary phosphorites are economically important (Plate 3A).

Guano or guano-derived deposits account for only a few percent of the world's production. Occurrences of guano in the United States are confined to bat caves in the arid southwest and to bird rookeries in Florida and the Hawaiian Islands. A few thousand tons of guano were produced from bat caves in Texas and New Mexico, but resources are limited and of little economic importance.

Igneous apatite from mines in the USSR, Africa, and Brazil accounts for about one-fifth of the world's production. Deposits in the United States include apatite-ilmenite dikes in Virginia; apatite-magnetite in New York, New Jersey, and Massachusetts; magnetite-apatite bodies in Utah; and apatite-rich iron ores in Missouri. Other igneous apatite occurrences are known in Nevada, Wyoming, Colorado, and Arkansas. Only small tonnages of by-product apatite have been produced from these deposits, and they are of little economic importance.

Sedimentary phosphorite is present in rocks of all ages from Precambrian to Holocene, but economic deposits in the United States are confined to rocks of Ordovician, Permian, and Miocene ages. Large resources of phosphate are present in rocks of Mississippian and Triassic ages in Alaska, but the deposits are so remote geographically that they are not economic at the present time.

About three-fourths of the world's production and virtually all of the United States' production of phosphate rock is from deposits of marine sedimentary phosphorite.

HISTORY OF DISCOVERY AND DEVELOPMENT

The use of natural fertilizer (animal and human excrement, guano, bones, and fish) to increase crop yields is thousands of years old. For example, Wright (1893, p. 12) pointed out that in the days of the Roman Empire the excrement from pigeon houses brought a high price, and as early as the twelfth century the Incas of Peru used guano as a fertilizer. The use of ground bones in England dates at least from 1786 (Gray, 1944), and from the early to mid-1800s there was a brisk trade bringing bone from Europe to fertilize British soils. In 1840, the German chemist Liebig suggested that treating pulverized bones with sulfuric acid would make the contained phosphorus more available to plants, and at about the same time, John Bennet Lawes of Great Britain used ground phosphate rock that he mixed with sulfuric acid to make the phosphate available (Gray, 1944). Lawes used phosphate nodules (called coprolites) from marine rocks of Cretaceous age in the south of England as well as nodules imported from Estremadura in Spain. The first commercial mining of phosphate was in England and Spain in 1847. Production averaged a few thousand tons per year.

Production of phosphate rock in the United States began in 1867 when 6 tons were shipped from deposits in South Carolina. Production in 1868 (the first full year of production) was 12,000 tons. The discovery and production of sedimentary phosphorite in South Carolina led to the discovery of hardrock phosphate (phosphatized limestone) in Florida in 1879 and of river-pebble phosphate in bars along the Peace River in Florida in 1881. The first production of river pebble was in 1887 and of hardrock in 1889. The discovery (in 1888) and production of high-grade and low-cost land pebble phosphate in Florida set off a prospecting boom, and by the middle 1890s there were about 100 mines that produced about 600,000 tons of phosphate rock per year.

Prospecting was extended throughout the southeastern United States, and occurrences were located in Georgia, Alabama, Mississippi, and North Carolina and later in the west. The residual "brown rock" deposits of Tennessee were discovered, and mining began in 1894. The phosphate deposits of the Phosphoria Formation (Permian) of the western United States were discovered in 1889, but the first mining of these deposits was not until 1906.

Deposits or occurrences of marine phosphorite were subse-

Cathcart, J. B., 1991, Phosphate deposits of the United States—discovery, development; Economic geology and outlook for the future, *in* Gluskoter, H. J., Rice, D. D., and Taylor, R. B., eds., Economic Geology, U.S.: Boulder, Colorado, Geological Society of America, The Geology of North America, v. P-2.

quently located in many states; the phosphate deposits of California were discovered in the 1920s, the large phosphorite deposits of Alaska were discovered in the 1940s, and the phosphorite deposits of the Pungo River Formation (Miocene) of North Carolina were not discovered until the 1950s.

South Carolina—Charleston area

Holmes discovered phosphate rock in South Carolina in 1837, and reported (1870) that in searching for marl to be used as a fertilizer he found rounded nodules of a rocky material. The first chemical analysis of the nodules was in 1850 by J. Lawrence Smith and C. U. Shepard, Sr., who reported that a sample contained ". . . 9.2 percent of phosphates of calcium, magnesium, and peroxides of iron . . ." (Willis, 1892, p. 682). During the Civil War, marl beds in the area were mined, and exposures in these quarries showed that the nodules were in pockets at the contact of the surficial sand and the marl beds.

The war made it impossible to exploit the discovery, and it was not until 1867 that the Charleston Mining and Manufacturing Company was organized. It was an immediate success, and according to Wyatt (1892, p. 47), the industry progressed so rapidly that ". . . it has raised the status of South Carolina to that of the most productive phosphate field yet known to industry."

Phosphate (land rock and river rock), produced from 1867 to 1920, totaled about 13 million metric tons. Maximum production was in 1889 when about 540,000 tons were shipped. Production then gradually declined as the higher grade and lower cost phosphate of Florida took over the market.

River rock was mined by dredge or at low tide by hand labor. The rock was taken to shore where it was screened, the coarse phosphate was dried and shipped, and the fines were dumped back into the river.

Land rock was mined by hand until about 1918 when thick overburden required the use of steam shovels. The phosphate was mined by hand throughout the history of the district. The phosphate rock was screened, dried, and shipped, and the fines were discarded.

Florida

The first report of phosphate in Florida was in 1879 (Day, 1886; Wright, 1893). Day (1885, p. 452) states that Dr. C. A. Simmons discovered that the ". . . principal building block of central Florida contained a considerable amount of phosphoric acid. In 1879, he located a quarry and in 1883 began mining the rock and converting it into fertilizer." Day also reported that the rock from the quarry contained about 45 percent B.P.L. (B.P.L. = percent $P_2O_5 \times 2.185$). Definitions of Florida mining terms are given in detail by Cathcart (1963, p. 6–12). There are no records of the amount of rock that was mined, but it probably was used as a fertilizer for at least a few years.

The river-pebble deposits of Florida were discovered by Captain J. Francis LeBaron in 1881. LeBaron (1892, p. 153) stated that he found many phosphatized fossils below little Charley Apopka Creek in ". . . drift gravel overlying limestone." LeBaron left Florida in 1882 and did not return until 1886, at which time he found abundant phosphate gravel along the Peace River and also on Horse Creek. This discovery was reported to the public; the Arcadia Phosphate Company was formed and made the first shipment of phosphate rock from Florida in May 1888. At about the same time the Peruvian Phosphate Company of Tampa, Florida, was mining a deposit on the Alafia River in Hillsborough County.

River-pebble phosphate was mined by dredge from the Peace and Alafia Rivers through 1908, although some material was shipped from stocks until 1914. Peak production was about 120,000 tons per year, and total production was about 1.3 million tons. Production declined because of competition from the higher grade hardrock and land-pebble deposits. Mansfield (1942) estimated that about 50 million tons of river-pebble phosphate remained, although only about 5 million tons was classified as "known." It is unlikely that these deposits will ever be mined because of environmental restrictions, the low P_2O_5 content (about 25 percent), and the small size and irregular habit of the deposits.

The discovery of high-grade phosphatized limestone (hardrock phosphate) was made by Mr. Albertus Vogt who, in digging a well, found a "white subsoil" that was determined to be high in phosphate. The discovery was reported either at the close of 1888 (Day, 1890) or in June of 1889 (Eldridge, 1893). The Dunnellon Phosphate Company and the Baldwin Fertilizer Company began mining late in 1889, and 11,000 tons were shipped in 1890, the first full year of production.

The discovery of high-grade deposits of hardrock phosphate set off a prospecting boom in Florida, and by 1892 there were 105 mines producing phosphate. The hardrock belt extended from just north of Polk County almost to Tallahassee, a distance of more than 200 miles (Eldridge, 1893, map, p. 197).

Hardrock deposits were mined from 1889 to 1966 by hand and with small draglines. Production reached a maximum of about 650,000 tons in 1907 and then declined as production of the lower cost and more uniform land pebble increased. Total production of hardrock was nearly 15 million metric tons.

Prospecting for land-pebble phosphate began late in 1890 (Eldridge, 1893, p. 198), although the term land pebble was first used by Day (1892, p. 452). The first prospecting and mining of land pebble was along the headwaters of the Peace and Alafia Rivers. The mining was by dredge in the flood plains of the rivers, and much of what was mined in the first few years probably should have been classified as river pebble. Four of the seven land-pebble mines shown on a map by Eldridge (1893, p. 197) are at or close to the rivers. As mining moved away from the rivers, the overburden was removed with small draglines or steam shovels, and the ore was mined using high-pressure hydraulic monitors. Only the coarse material was saved; fines were discarded. With the advent of methods to separate sand-sized quartz and phosphate particles, all of the material mined was pumped to the washer and flotation plants in 18- to 24-in-diameter pipelines. Today, mining is done by large draglines with 30- to 54-cubic yard buckets. The mined material is pumped to the washer where the coarse material (pebble) is separated by screening and shipped as a product. The fine material is deslimed, the −150 mesh is discarded as waste, and the sand-sized material is treated by flotation to separate the phosphate pellets from the quartz sand.

The first shipment of land-pebble phosphate rock was in 1891 (Sellards, 1909). Production increased gradually at first, then accelerated rapidly, reaching 1.1 million tons in 1908, 10 million tons in 1956, and 42 million tons in 1979. Due to poor market conditions, production declined to about 30 million tons in 1983.

The phosphate discoveries attracted foreign capital, and until World War I, French, English, Belgian, and domestic companies mined phosphate in Florida. From that time through the 1970s the phosphate companies were all domestic, but in the early 1980s there has again been some foreign capital invested in Florida phosphate.

Total production of land-pebble phosphate rock from 1891 to 1985 is about 1 billion tons of product, averaging about 32 percent P_2O_5.

Tennessee

The phosphate deposits of central Tennessee were named "brown rock," "blue rock," and "white rock," on the basis of their predominant color, in the first published descriptions of the deposits in the 1890s, and the names are still used. Only the brown-rock deposits have any economic importance and have accounted for almost all of the total production of phosphate rock from Tennessee.

The first published information on the Tennessee deposits was by Hayes (1895) who reported that phosphate had been discovered in Lewis and Hickman Counties near the close of 1893. Safford (1894) first described these deposits that came to be called "blue rock." Hayes (1895) noted that the initial production of blue rock was during the last half of 1894, and about 20,000 tons was mined in that year. Production of blue rock reached a maximum of about 90 thousand tons per year and continued until 1961, although only very small tonnages were shipped after 1939. Total production was about 1.9 million tons.

The white-rock deposits were located after the discovery of the blue rock by Florida prospectors (Hayes, 1895). White-rock phosphate was never commercially important, although small tonnages were mined, probably for local use, from 1905 to 1936, and about 20 thousand tons were produced during those years.

Brown-rock phosphate was discovered late in 1895 (Day, 1896), when a Mr. S. Weatherby picked up a piece of laminated brown rock that when analyzed proved to be high-grade (30 percent P_2O_5) phosphate. Mining began in January 1896 near Mt. Pleasant, Tennessee. The discovery was disastrous to the blue-rock industry because the brown rock is higher grade, mining is less costly, and the mines are close to the railroad. In the first full year of mining (1897) about 34,000 tons was produced; by 1900 the total had reached about 400,000 tons, but it was not until 1940 that production reached one million tons. The maximum production of about 4 million tons was in the early 1970s, and since that time, annual production has declined to about 2 million tons. Total production of brown rock is about 100 million tons.

The mining of blue rock and white rock was by open pit, using blasting and pick and shovel. The pits were carried into the outcrop until the depth was about 20 ft, when room and pillar underground methods were used. The rock was shipped without beneficiation, so only the highest grade material was mined. Brown rock has been mined entirely by open-pit methods. Early mining was by pick and shovel, and only the coarse material (+2 in) was saved. By the early 1900s, mining of overburden was done by steam shovel, but the ore was mined by hand. Small draglines were used for overburden removal starting in 1915, and at that time the ore was mined by pick and shovel, hydraulic methods, or with the dragline. Today all mining is with small draglines. Ore is transported to the plants by truck or rail, and fine material (−325 mesh) is removed by washing. Oversize material is nodulized in rotary kilns, mixed with quartz gravel and coke, and treated in electric furnaces to produce elemental phosphorus for use in the chemical industry.

Western states (Idaho, Montana, Wyoming, Utah)

The discovery of phosphate in the western United States was in 1889 in Cache County, Utah (Mansfield, 1917, p. 33), but exploration and development did not begin until 1904. Weeks and Ferrier (1907) published the first of a continuing series of reports by the U.S. Geological Survey on the phosphate deposits of the western states. The first report on specific deposits in Idaho and Wyoming was by Gale and Richards (1910) and on Montana by Gale (1911). Van Horn (1911) was the first to note that the western deposits contained the largest reserves of phosphate in the United States.

The first production of phosphate from the western field was from a mine near Montpelier, Idaho, in 1906 (Service, 1967, p. 175), and some mining was done in Wyoming in the same year (Mansfield, 1917). However, the first marketable production was recorded in 1911 when Van Horn (1911) noted that 0.5 percent of the production of phosphate in the United States was from the western deposits. The first recorded production from Montana was not until 1929. Production started slowly; about 16,000 tons were shipped in 1914, about 65,000 tons in 1925, and production reached 5 million tons in the 1970s.

Resources of phosphate in the western field are extremely large, but tonnages that can be mined by open-pit methods are limited. Thus, the mining of the bulk of the material will have to be by underground methods. Most of the resource is more than 300 m below entry level, too deep to be mined economically in the foreseeable future.

Mining of the phosphate is predominantly by open pit, using

large-earth-moving equipment and blasting to break up the rock, but there are underground mines in Montana.

Alaska

Phosphorite deposits were discovered along the north front of the Brooks Range between 1944 and 1953 during geological investigations of the Naval Petroleum Reserve No. 4 (Patton and Matzko, 1959). Total resources are thought to be very large (Table 1), but the remoteness of the area makes it unlikely that any material will be mined in the near future.

During World War I, when the supply of bone material (then used extensively as a fertilizer) was much diminished, it was reported (Anonymous, 1916) that ". . . the largest known bone deposits in the world . . ." were present in the Pribiloff Islands in the Bering Sea. So far as is known, nothing was done with this bone material.

California

Phosphate pellets in marine sediments in California have been known at least since the 1920s (Galliher, 1931; Hoots and others, 1935). Most of the known occurrences are in rocks of Miocene age (Gower and Madsen, 1965).

The first production of phosphate from California was in 1969, and mining continued intermittently until the first quarter of 1977 (Stowasser, 1978). Because only one company mined in California, production data are combined with the data from the other western states, and total production is not known.

Environmental restrictions and the general low P_2O_5 content of the material make it unlikely that mining will be resumed, at least in the near future.

Georgia—Savannah River

The phosphate deposits of the Savannah River area in northern Georgia and adjacent southern South Carolina were discovered during extensive prospect drilling in the Atlantic Coastal Plain in the late 1950s and early 1960s (Furlow, 1969). The deposit was outlined by drilling, and a few holes drilled under the Atlantic Ocean proved that the deposit extends at least 15 mi offshore. Phosphate companies applied for leases to mine under the Savannah River, but the leases were denied on the basis that mining would cause excessive damage to the river and adjacent tidelands. Although it is not likely that this deposit can be mined by conventional methods, it is certainly a mineable resource for the future.

North Carolina

Prospecting for phosphate in North Carolina began in 1883, the result of interest stimulated by the discovery of high-grade phosphate in Florida (Dabney, 1884). The only phosphate discovered then was a pebbly layer at the base of the Eocene Castle Hayne Formation. Small tonnages of this phosphate were produced intermittently from 1886 to 1899, for use as a soil additive.

The economic deposits of Beaufort County were discovered in the 1950s when samples from water wells drilled in the county were found to contain abundant phosphate pellets. P. M. Brown of the U.S. Geological Survey recognized the potential value of the material and prepared a report showing the distribution of the phosphate as it was then known (Brown, 1958). He suggested that the use of gamma-ray logs of water wells would delimit the deposit. In the late 1950s phosphate companies began prospect drilling and logging water wells, and the economic deposit was outlined early in the 1960s. Mining began in 1964 and has continued to the present. Reserves and resources are large, so mining will continue well into the future.

Mining is by open pit, using large draglines, but the top 12 to 15 m of overburden is removed by dredge, and the draglines are put on the bench created by the dredging operation.

Offshore deposits

Atlantic Ocean. The first report of phosphate in the Atlantic Ocean was in 1883 (Manheim and others, 1980). Today, it is known that phosphate pellets and nodules are commonplace on the floor of the Atlantic Ocean from North Carolina to the southern tip of Florida. The deposits of offshore North Carolina were first described by Pilkey and Luternauer (1967), and the area has been explored in some detail in the years since then (Riggs and others, 1982). Deposits have been described from the Blake Plateau, the Miami Terrace, and the Pourtales Terrace (Manheim and others, 1980).

Pacific Ocean. Phosphate pellets and nodules are widespread off the California Coast from San Francisco to the boundary with Baja California (Dietz and others, 1942).

Offshore resources in the Pacific and Atlantic Oceans are very large, totaling billions of metric tons. No serious efforts to mine these resources have been proposed or undertaken.

Origin

The phosphate of sedimentary phosphorite deposits originates in sea water. McKelvey (1967) pointed out that the ocean is nearly saturated with phosphate, but the phosphate content is not uniform. Deep, cold water contains about 0.3 parts per million (ppm), while warm surface water contains only about 0.01 ppm. Thus, if deep water or cold polar water is brought into contact with warmer surface water by oceanic circulation, phosphate can be precipitated. Phosphate can precipitate inorganically, by organic processes, or by a combination of the two processes. This formed the basis for the upwelling hypothesis of Kazakov (1937), which was elaborated by the work of McKelvey and others on the Phosphoria Formation of the western United States (McKelvey and others, 1959).

The best deposits are formed in warm latitudes in areas of upwelling caused by divergence. These areas are in trade wind belts on the west coasts of continents, as in the deposits of the

Phosphoria Formation. Deposits in the modern ocean (as off the coast of Peru) are characterized by organic matter, siliceous deposits (diatomites), phosphorite, and carbonate rock in coastal water. Onshore, in the arid coast, saline deposits and red beds are formed. This suite of rocks is matched very closely in the deposits of the Phosphoria Formation.

Deposits also form on the west sides of poleward-moving warm currents along the eastern coasts of continents where there is turbulent mixing with cool coastal countercurrents, as in the deposits of the Atlantic Coastal Plain. Positive areas (rising domes or topographic highs) caused turbulent mixing of the cool, southward-moving currents with the warm water of the Gulf Stream, where phosphate was precipitated in basins on the flanks of the highs. These deposits are characterized by light-colored diatomaceous rock, sandstone, shale, and carbonate rock, all containing phosphate grains or pellets. Deposits in Tennessee were formed by northward-moving ocean currents that upwelled and were localized on the rising Nashville dome, where the phosphate precipitated. Deposits on the crest and northeast of the dome do not contain phosphate.

GEOLOGY

Atlantic Coastal Plain

Economic or potentially economic phosphate deposits of the Atlantic Coastal Plain are known only in rocks of Miocene age or in younger rocks that derived much of their phosphate from rocks of Miocene age.

The economic deposits are in part structurally controlled. They are in basins on the flanks of positive areas (either structural or topographic) that were present at the time of deposition. All the deposits are in positions suggesting that phosphorus may have precipitated because of the turbulent mixing of the warm Gulf Stream with cool southward-moving currents (Cathcart, 1968) or that diverting of the Gulf Stream by the positive areas may have promoted upwelling of deeper nutrient-laden water into surficial water, which caused the precipitation of phosphate (Riggs, 1984).

The economic deposits of Florida, Georgia, and North and South Carolina are similar in gross features: they are unconsolidated or partly consolidated sediments, the phosphate is pelletal, the phosphate mineral is a carbonate fluorapatite (francolite), and the principal gangue minerals are quartz, calcite, dolomite, and clay and iron oxide minerals. Differences include variation in the chemical composition of the apatite mineral, the size distribution of the phosphate particles, and the amount and intensity of reworking and leaching by acid ground water.

North Carolina.
The phosphorite deposit of North Carolina in the Pungo River Formation of Miocene age is in a basin on the flank of an unnamed high that is connected with the Cape Fear Arch (Kimrey, 1965). The western or landward edge of the basin was marginal to a low land mass, so only minor amounts of land-derived clastics were deposited in this part of the basin. The ore body consists of thin-bedded quartz and phosphate sand and minor clay and carbonate rock that, according to Riggs and others (1982), is a series of cyclic depositional sequences. To the east, in the deeper, offshore part of the basin, the formation thickens markedly by intercalation of beds of dolomite, clay, and sand.

In the mining area the Pungo River Formation averages about 15 m in thickness, while to the east the formation is as much as 75 m thick. Resources are measured in billions of tons of recoverable phosphate pellets, but most of the reserve is in the eastern part of the basin where the ore deposit and the overburden are so thick that mining by conventional open pit methods is not economic by present-day standards.

The Pungo River Formation is separated by major unconformities from the underlying Castle Hayne Formation of Eocene age and the overlying Yorktown Formation of Pliocene age. Phosphate particles in the Pungo River Formation are black or dark brown, rounded pellets, most of which range in size from 0.1 to 1.0 mm. Coarser phosphate grains that range up to several millimeters in diameter are not common, and the +1 mm fraction is so mixed with shell material and dolomite fragments that it is discarded as a waste product.

The base of the overlying Yorktown Formation contains coarse phosphate pebbles up to several centimeters in diameter, but the Yorktown is discarded as a part of the overburden.

The overburden of the phosphate deposit consists of the upper carbonate-rich part of the Pungo River Formation, the clayey, fossiliferous sand of the Yorktown, and the sand and clay of Pleistocene age. The overburden in the mining area is from 20 to about 35 m thick.

The phosphate pellets are relatively uniform in their P_2O_5 content, averaging about 30 percent; the phosphate mineral is a carbonate fluorapatite (francolite) that contains large amounts of CO_3 substituting for PO_4 in the apatite structure.

The offshore material in Onslow Bay and Frying Pan Shoals is a continuation of the onshore material and is similar in all respects to the onshore material.

South Carolina. Charleston area.
Two types of phosphate deposits, land rock and river rock, occur in the Charleston area. The land-rock deposits are in the Ladson Formation of Pleistocene age (Malde, 1959). The deposits rest on the Cooper Marl of Oligocene age or the Hawthorn Formation of Miocene age, and according to Malde (1959) were reworked from replacement bodies in the Cooper Marl. The time of phosphate replacement was late Pliocene to Pleistocene. The source of the phosphate probably was the Hawthorn Formation. The deposits consist of irregularly shaped phosphate nodules up to several centimeters in diameter that are replaced carbonate rock and sand-size phosphate pellets that probably were reworked from the Hawthorn Formation.

The river-rock deposits of Holocene age are composed of coarse, irregular-shaped but rounded phosphate nodules in bars along and in the flood plains of the modern streams. These deposits formed when phosphate nodules eroded from the Ladson Formation were concentrated in the streams.

Deposits of both types are small and erratic in distribution.

Land-rock deposits range in thickness from 10 cm to about 1.5 m and average about 50 cm. Total resources are estimated by Malde (1959, p. 73) to be 8.8 million tons.

The phosphate of both the river rock and land rock is low in P_2O_5 content (average about 25 percent) but is high in uranium content. Analyzed samples of coarse rock contain from 0.011 to 0.042 percent uranium, and sand-sized phosphate particles range from 0.011 to 0.025 percent uranium. Altschuler and others (1958) reported that secondary apatite hardpan on top of the Cooper Marl has as much as 0.1 percent uranium. In the 1950s the high uranium content prompted several mining companies to obtain mining leases from the state of South Carolina. However, no mining was done, and no applications for leases have been made in recent years. It is unlikely that the phosphate deposits of the Charleston area can be economically mined in the foreseeable future.

Coosawhatchie area. The Coosawhatchie district is in southeastern South Carolina on the Georgia border. The phosphate deposit, a continuation or extension of the Savannah River deposit of Georgia, is in a northeast-trending, troughlike basin on the flank of the Ridgeland high. Data on the geology of the Coosawhatchie deposit are from unpublished company reports and are quite generalized.

The phosphate deposits are in the Hawthorn Formation, which rests unconformably on the Eocene Santee Limestone or the Oligocene Cooper Marl, and is overlain by unconsolidated sand and clay of Pleistocene age and swamp deposits of Holocene age. There are two distinct phosphorite beds—the upper is about 1.6 m thick and consists of phosphate pellets and nodules in a clayey sand matrix. The basal 30 cm of this zone is a phosphate gravel. The lower phosphorite is separated from the upper by about 12 m of gray-green sand, clayey sand, dolomite, and doloclay or dolosilt containing some phosphate pellets. The lower phosphorite averages about 6 m in thickness and consists of olive-green sandy clay or clayey sand containing abundant black and brown phosphate pellets. The basal 30 cm of the lower phosphorite unit is a gravelly phosphate sand.

The upper sandy phosphorite is lithologically similar to the upper part of the phosphorite of the Savannah River. It is likely that this bed is upper Miocene or Pliocene and is correlative with the Duplin Marl of Georgia (Furlow, 1969) and the Bone Valley Formation of Florida (Scott, 1985). Beds below the upper phosphorite and including the lower phosphorite are lithologically identical with the Hawthorn Formation in Georgia and Florida.

The phosphate pellets are dark colored; the phosphate mineral is a carbonate fluorapatite (francolite) with moderate CO_3 substitution. The P_2O_5 content of the separated phosphate particles (concentrate) of the lower phosphorite averages about 30 percent, but the particles in the upper bed are somewhat enriched, averaging about 32 percent P_2O_5. The lower phosphorite is similar to the phosphorite of North Carolina, while the upper bed may be reworked and is similar to phosphorite of the Bone Valley Formation.

Georgia. *Savannah River deposit.* The Savannah River deposit of northeast Georgia is included in rocks of the upper Miocene Duplin Marl, according to Furlow (1969). However, Furlow points out that the distinction between the Duplin Marl and the underlying Hawthorn Formation is only in the increased amount of phosphate in the Duplin Marl. Potentially economic phosphate deposits (Furlow, 1969, Plates 1 and 2) range from the uppermost part of the Miocene to near the bottom of the Miocene, and it is likely that phosphorite occurs in both the Hawthorn Formation and the Duplin Marl. That part of the phosphorite that is in the uppermost part of the Miocene (Duplin) may be correlative with the Bone Valley Formation of Florida, while the lower parts of the phosphorite may be correlative with the Statenville Formation of the Hawthorn Group of North Florida (Scott, 1985).

The phosphorite is gray-green to olive-green dolomitic clayey sand or sandy clay, with thin beds of dolomite, all containing abundant brown and black phosphate pellets. The pellets range from 0.1 to 1.0 mm in diameter, and the potentially economic section ranges from 5 to 15 m in thickness. Overburden ranges in thickness from about 16 to about 45 m. The phosphate mineral is carbonate fluorapatite (francolite) with moderate to large amounts of CO_3 substituting for PO_4 in the apatite structure.

The phosphorite extends to the east as much as 25 km offshore into the Atlantic Ocean, as indicated by a few, scattered drill holes. Logs of drill holes through the deposits do not seem to show the cyclicity of the Pungo River Formation of North Carolina, but this may be due to lumping together of lithologic units in the published drill logs.

The resources of recoverable phosphate pellets are estimated to be almost 2 billion tons (Furlow, 1969), but the total resources are much greater, particularly if the offshore material is considered. It is probable that there is at least as much phosphate offshore as there is onshore.

Florida. The phosphorite deposits of Florida occur in rocks of Miocene age (Hawthorn Formation or Group) or in younger rocks in which much of the phosphate has been derived from the Hawthorn. Economic deposits are known in the north Florida–south Georgia district, in the land-pebble district of central Florida, in east and south Florida, and in northeast Florida. Only the deposits in the land-pebble district and in north Florida are being mined, but deposits in the Osceola National Forest in northeast Florida, in the southern extension of the land-pebble district in south Florida, and in east Florida are probably mineable.

The deposits throughout Florida and in south Georgia are similar. They are in unconsolidated marine sedimentary rocks—sand, clay, and soft carbonate—all containing abundant pelletal phosphate grains. There are two distinct phosphate beds. The lower is in the upper clastic unit of the Hawthorn, the Statenville, and Peace River Formations of Scott (1985). The upper bed is in the Bone Valley Formation of the land-pebble district, and equivalent, unnamed formations of late Miocene and early Pliocene age in the other districts.

The phosphorite of the Hawthorn Formation consists of olive-green clayey sand to sandy clay that contains abundant black and dark-brown phosphate pellets. The pellets contain from 25 to 32 percent P_2O_5, and the apatite mineral is francolite with moderate to abundant CO_3 substituting for PO_4. Thin beds of dolomite, doloclay, or dolosilt are present. The lower part of the Hawthorn Group is the Arcadia Formation of Scott (1985), also called the lower carbonate unit of the Hawthorn. Although this unit also contains abundant phosphate, methods for extracting the phosphate from the hard carbonate rock are not economic today.

In the land-pebble district, the Bone Valley Formation disconformably overlies dolomite of the Hawthorn—the upper clastic unit of the Hawthorn is present as a southward-thickening wedge that pinches out in the northern half of the district. The Bone Valley Formation is late Miocene and early Pliocene in age and consists of a lower phosphorite unit and an upper clayey sand unit that contains only minor phosphate (Altschuler and others, 1964). The Bone Valley Formation pinches out to the north of the land-pebble district on the southern flank of the Ocala uplift. The lower phosphorite unit consists of gray-green, gray-brown, and gray sand and clay that contain abundant phosphate particles ranging from about 0.1 mm to several centimeters in diameter. The coarse fraction (>1 mm, called pebble) is dominant at the base of the formation and on subsurface highs in the central part of the district, while fine sand-size phosphate is dominant in the flatwoods adjacent to the modern streams. The modern streams have cut through the Bone Valley, and finer grained materials have been removed by the streams, thereby concentrating the coarse fraction as bars and in the flood plains of the streams. These coarse phosphate deposits, Pleistocene and Holocene in age, are the so-called river pebbles.

The phosphate in the Bone Valley Formation was derived from the underlying Hawthorn and is enriched in P_2O_5 and concentrated. The phosphate particles contain from 30 to 40 percent P_2O_5, and the phosphate mineral is carbonate fluorapatite, but with only small amounts of CO_3 substituting for PO_4. Weathering after deposition formed aluminum phosphate minerals from the original apatite and altered the clay minerals to kaolinite. The lower phosphorite unit of the Bone Valley Formation ranges in thickness from 0 to as much as 15 m. The formation is present in an area that was limited on the north by the Ocala uplift, on the east by the Lake Wales ridge, and on the west by the Valrico ridge, forming a bay or estuary open to the sea on the south. The Bone Valley Formation, as a conglomeratic phosphorite, is not present to the south where it is replaced by a deeper water, more open ocean facies—the Tamiami Formation (Pliocene).

Upper Miocene rocks, probably equivalent to the Bone Valley, overlie the Hawthorn in east, northeast, and north Florida. The rocks are similar in lithology and contain considerable coarse phosphate, and vertebrate fossils found in the Suwannee River Mine in north Florida are the same as the fossils in the Bone Valley Formation in central Florida. The beds equivalent to the Bone Valley Formation in north and east Florida are thin; they range from a few centimeters to about 1.5 m and average about 0.5 m.

Total resources of phosphate in the Bone Valley Formation are limited, but resources in the upper clastic unit of the Hawthorn Formation are very large, and future mining of phosphate in Florida will be almost entirely from this unit.

Tennessee

Phosphatic limestone. Ordovician limestones in the Central basin of Tennessee are exposed on the Nashville dome. The limestones are phosphatic on the west flank of the dome, with the most phosphatic parts of each younger formation extending successively farther west. The dome was rising during Ordovician time; thus the phosphatic limestones were deposited at roughly the same depth of water. Source of the phosphate was to the south and west in deeper and cooler water. Currents moving north and east brought nutrient-rich water to the shallow water on the flank of the rising dome, and phosphate was precipitated. The dome acted as a barrier, preventing the spread of the nutrient-laden water to the east.

Brown rock. The brown-rock deposits of Tennessee were formed in Pleistocene and Holocene weathering cycles by leaching of the phosphatic limestone by acid ground water. The phosphate is residually concentrated by removal of carbonate. Deposits are described as "cutter," "rim," and "blanket" by Smith and Whitlatch (1940, p. 44–47). Cutter deposits formed where ground water moved downward through closely spaced parallel joints and left the less soluble apatite, quartz, and clay minerals in a series of subparallel, long, narrow deposits. After the carbonate was removed, phosphate was taken into solution, moved downward, and precipitated as secondary apatite in lower beds. Cutter deposits range from 2 to about 5 m in width, from 1 to 5 m thick, and up to 50 m in length.

Rim or collar deposits occur where the phosphatic limestone is exposed only on the face of a cliff or around a hill. Only the edge of the bed is weathered, and the deposit forms a lens along the cliff or extends around the hill to form a "collar."

Blanket deposits are formed where the phosphatic limestone was at the surface over a large area and where weathering can extend deeper and more uniformly into the limestone. Blanket deposits are as much as 15 m thick and may extend over several score acres.

Overburden overlying the brown rock deposits consists of less phosphatic surficial soil; here the economic contact is determined by chemical analyses of drill-hole samples. Gravel and sand deposits of Holocene age overlie the deposit in many places.

White rock. White-rock deposits occur as joint or crack fillings in Devonian or older siliciclastic rocks or as an apatite precipitate or replacement of Silurian limestone. The source of the phosphate is in the overlying Devonian or Mississippian formations (Devonian and Mississippian Chattanooga Shale, Lower

Mississippian Maury Shale, or the Lower Mississippian Fort Payne Chert). Phosphate, dissolved by acid ground water, moves downward and is precipitated at a change in pH. The deposits are related to the present surface, the modern weathering cycle, and are Holocene in age.

Blue rock. Blue-rock deposits are in the Hardin Sandstone Member (Upper Devonian) of the Chattanooga Shale. The Hardin contains abundant phosphate only where it unconformably overlies the highly phosphatic part of the Leipers Limestone (Upper Ordovician). Hayes and Ulrich (1903) suggested that phosphate particles derived from weathering of the Leipers were concentrated in a shallow sea and incorporated into the basal sandstone (Hardin) of the Chattanooga Shale.

Northwestern United States

The phosphorite deposits of the northwestern United States, in the Permian Phosphoria Formation, crop out over an area of 350,000 km^2 in Idaho, Wyoming, Montana, Utah, and Nevada. The type locality of the Phosphoria Formation is in southeast Idaho where it consists of dark chert, phosphatic and carbonaceous mudstone, cherty mudstone, and dark-colored carbonate rock (McKelvey and others, 1959). The best phosphate is in the Meade Peak Member at the base of the formation, but commercial deposits are also known in the Retort Phosphatic Shale Member at the top of the formation. The Phosphoria was deposited on the continental shelf of the North American Continent at the edge of the ancient Pacific Ocean. The thick deposits in Idaho are geosynclinal, while the thinner deposits to the east and north are on the platform. The deposits in the west are structurally complex, but deformation is much less intense in the east. The facies that were deposited were determined, in large part, by water depth. Thus, from west (deepest water) to east (shallowest water) the sequence is black mudstone, black dolomite and phosphorite, chert, carbonate rock, and finally, sandstone. Transgression and regression produced overlap and offlap of these facies. The highly phosphatic Phosphoria Formation in southeast Idaho grades laterally into a more sandy sequence to the north and east in Montana and Wyoming and to a carbonate facies to the south and east in Wyoming and Utah. Farther to the east the carbonate sequence grades laterally into a clastic facies that contains red beds and gypsum (McKelvey and others, 1959).

Total resources of phosphate in the Phosphoria Formation are extremely large, about 25 billion tons of economic and marginally economic material. Most of the resource is in the thickest part of the section in southeast Idaho and is in the Meade Peak Member. To the north, in Montana, the formation is thinner, and much of the phosphate is in the Retort Member. The formation also thins to the east, in Wyoming, and is in an area that is not as structurally complex.

California

Phosphate pellets are present in marine sedimentary rocks in the Coast Ranges of California from San Francisco to Los Angeles. Most of the occurrences are in rocks of Miocene age—the Monterey Formation and the overlying Santa Margarita Formation (Gower and Madsen, 1964). The rocks are highly siliceous and include diatomite, siliceous shale, mudstone, tuffaceous beds, and phosphorite. The phosphorite and associated siliceous shale are highly organic. The phosphate is pelletal, the phosphate mineral is a carbonate fluorapatite, and the gangue minerals include quartz, clay, and tuffaceous material.

The best phosphate is in southern California where the phosphatic zone in the Santa Margarita Formation is about 100 m thick. Only the central 30 m, which contains an average of about 8 percent P_2O_5, is thought to be economic. Selected beds in this section of the formation contain as much as 17 percent P_2O_5.

Total resources of phosphate in California are probably very large but have not been estimated. Potentially economic phosphate beds have been estimated to contain several hundred million tons of phosphate rock that averages about 8 percent P_2O_5 (Rouse, 1979).

Alaska

Deposits of phosphorite in Alaska occur in rocks of the Mississippian and Pennsylvanian Lisburne Group and the Triassic Shublik Formation. Phosphate rock in the Lisburne Group is associated with black chert and shale of the Upper Mississippian Alapah Limestone, near the top of the group. The phosphate zone is up to 10 m thick, but the best material is about 4 m thick and contains an average of about 16 percent P_2O_5. Lithology and phosphate content vary markedly along strike, and the area is structurally complex. The deposits occur over a lateral distance of about 80 km. Resources have been estimated to be about 300 million tons (Patton and Matzko, 1959).

Phosphate deposits of Triassic age are in the lower part of the Shublik Formation. The few measured sections indicate that the black shaly phosphorite is 3 to 7 m thick and may contain an average of about 12 percent P_2O_5. The sampled outcrop belt is about 80 km long and 16 km wide. Total resources, estimated from data in Patton and Matzko (1959), are about 5 billion tons.

The phosphate rock in these deposits consists of phosphate pellets in a carbonate matrix. The phosphate mineral is a carbonate fluorapatite; gangue minerals include calcite, dolomite, quartz, clay, fluorite, and carbonaceous material.

The location of the deposits, in the geographically inaccessible north slope of the Brooks Range, and the fact that they are in areas withdrawn from mineral development, make it unlikely that the material will be mined.

ECONOMIC GEOLOGY

The characteristics of the mineable phosphate deposits in the Atlantic Coastal Plain, the northwestern United States, and Tennessee are considerably different. The economic parameters for the three types of deposits are dependent on geology. Because the deposits of the Atlantic Coastal Plain are flat lying, unconsolidated, and cover large areas, they can be mined with draglines that have bucket capacities of as much as 70 cubic yards, while the much smaller unconsolidated deposits of Tennessee are mined with draglines that have bucket capacities of 2 to 5 cubic yards. The deposits of the northwestern United States are in consolidated rocks in structurally complex areas. Mining of outcropping deposits is done with heavy earth-moving equipment, and blasting may be necessary to break the rocks. Most of the remaining resource is too deep to be mined by open pit. These deep deposits must be mined by underground techniques.

The economic limits for each of the deposits, both maxima and minima, are estimated from current industry standards.

Atlantic Coastal Plain

The economic parameters for deposits of the Coastal Plain are based on primary geologic factors: the phosphate beds are unconsolidated, widespread, and flat lying, and overburden thicknesses are not excessive. Mining can be done cheaply with large draglines, and the disaggregation and separation of the phosphate particles from the gangue minerals is simple and cheap. The economic parameters are discussed in detail by Cathcart and others (1984).

The total depth that can be mined (overburden plus ore) is about 50 m, and because of the large size of the buckets, the ore bed must be greater than 1 m in thickness. The thickness of the overburden that can be mined may be expressed as the ratio of this thickness to the thickness of the ore bed (the maximum is about 5 to 1), but is more commonly computed as the ratio of total cubic yards moved per ton of product recovered. The maximum is about 30 to 1.

The phosphate product (concentrate) should contain more than 29 percent P_2O_5, less than 5 percent $Fe_2O_3 + Al_2O_3$, less than 1.5 percent MgO, and the ratio of percent CaO to percent P_2O_5 should be less than 1.55 (the maximum ratio for the apatite mineral).

The total tonnage of product for a new mine must be a minimum of 40 million tons, but additions to an existing mine can be almost any amount. To be mineable, the phosphate product in the ore zone should be a minimum of about 20 percent by weight, or about 400 tons of product per acre per foot of thickness. Thus a 10-ft ore bed should contain at least 4,000 tons per acre of product.

Western United States

The economic parameters are based on strippable resources; most underground resources are subeconomic at the present time.

The phosphorite bed (total rock) must contain more than 18 percent P_2O_5, less than 3 percent $Fe_2O_3 + Al_2O_3$, less than 1.5 percent MgO, and a CaO to P_2O_5 ratio of less than 1.55. The ratio of cubic yards moved per ton of ore mined must be less than 3.5 to 1, and the ore zone must be more than 5 ft thick. The minimum size for a new deposit should be at least 20 million tons (Cathcart and others, 1984).

Tennessee

The phosphate deposits of Tennessee are economic only when the phosphate is concentrated and enriched by removal of carbonate. Individual deposits are small and erratic in distribution. Because small mining equipment is used, the total overburden thickness cannot exceed 10 m, and the total thickness to the base of the ore should be less than 20 m. Phosphate rock mined in Tennessee is treated in the electric furnace to produce elemental phosphorus, and the following chemical requirements are based on this style of treatment. Phosphate content must be more than 14 percent P_2O_5, although rock that can be upgraded to 30 percent P_2O_5 or more with at least 30 percent recovery can also be mined. Upgrading is done by removing –325 mesh material by disaggregation and washing. Fe_2O_3 content must be less than 10 percent, although it is desirable for the iron content to be as low as possible. Contents of CaO and MgO are not critical, except that if they are excessive, additional silica must be added to the furnace charge to combine with the CaO and MgO, so they can be discarded as slag.

RESOURCE SUMMARY

Resources of phosphate in the United States are extremely large, totaling hundreds of billions of tons (Table 1), but tonnages of economic deposits are much less, about ". . . 17 billion metric tons of identified, recoverable phosphate rock exist in the United States, of which about 7 billion metric tons are thought to be economic or marginally economic" (Cathcart and others, 1984, p. 1). The summary table of resources does not split out the potentially economic reserves; they are included in the total resources. The total resources of phosphate, onshore and offshore, in the eastern United States is estimated to be 33 billion metric tons. Resources in the western United States of material that can be mined by open pit, or is less than 300 m below entry level, total almost 25 billion tons. Phosphate rock that is more than 300 m below entry level in the western states is estimated to be about 500 billion tons, but none of this material can be mined or processed using present technology, so it is a resource only for the distant future.

OUTLOOK FOR THE FUTURE

Economic and marginally economic resources of phosphate rock in the United States are large enough to provide fertilizer for agriculture for the foreseeable future, but the high-grade, cheaply mineable deposits of the Atlantic Coastal Plain and the best of the strippable resources of the western United States are limited.

Subeconomic resources and resources that cannot be mined and processed by conventional methods are extremely large. The mining and processing of deeply buried phosphate rock and rock

TABLE 1. SUMMARY OF PHOSPHATE RESOURCES OF THE UNITED STATES*

Onshore	
Atlantic Coastal Plain	
North Carolina	9×10^9
South Carolina–North Georgia	2×10^9†
North Florida–South Georgia	3×10^9
Central and South Florida	4×10^9
East Florida	4×10^9
Total, Atlantic Coastal Plain	22×10^9
Tennessee	
Brown, blue, white rock	0.05×10^9
Phosphatic limestone (Ordovician)	5×10^9§
Western States	
Idaho, Wyoming, Montana, Utah, Nevada	
Amenable to open pit mining	11.6×10^9
Underground, less than 300 m below entry level	13×10^9
Underground, more than 300 m below entry level	500×10^9§
California	0.3×10^9†
Total, Western States	25×10^9
Alaska	
Brooks Range	5×10^9†
Total, onshore resources	557×10^9
Potentially minable onshore resources	17×10^9

Others**

Precambrian (Michigan, Tennessee, Montana)
Cambrian (Vermont, Nevada)
Ordovician (Kentucky, Iowa, Arkansas, Nevada)
Devonian–Mississippian (Tennessee, Georgia, New York,
    Pennsylvania, Virginia)
Pennsylvanian (Oklahoma, Kansas, Missouri, Illinois)
Cretaceous (Alabama, Georgia, Mississippi, Texas, Colorado, California)
Eocene (North Carolina, Georgia, Texas)
Oligocene (South Carolina)

that contains large amounts of the deleterious elements (Mg, Fe, Al, etc.) is not possible today, and new or improved methods of mining and processing must be devised so that production of phosphate can continue into the future. Methods of mining phosphate rock that is under the Atlantic and Pacific Oceans have not been proven to be economic, and further research is needed in this area.

Open-pit mining and chemical processing of phosphate rock can damage the environment, and the potential for this damage is of increasing concern. Federal and state laws now mandate that land be reclaimed, and laws regulating air and water pollution are stringent and may become more so. The granting of mining permits by county, state, and Federal government agencies is a long procedure, and it is incumbent on mining companies to prove that their activities will not permanently damage the environment.

Thus, although there are abundant resources of phosphate in the Atlantic Coastal Plain, it may be difficult or impossible to mine part, and perhaps a large part, of these resources in the near future. For example, the state of Georgia has refused to allow mining of the Savannah River deposit, and the U.S. Department of the Interior refused to grant mining permits for the Osceola

TABLE 1. SUMMARY OF PHOSPHATE RESOURCES OF THE UNITED STATES*
(continued)

Offshore	
Atlantic Ocean	
Onslow Bay and Frying Pan Shoals	1.5×10^9
Blake Plateau	3.0×10^9
Savannah	1.0×10^9
Jacksonville, Florida	
Miami Terrace	3.5×10^9
Pourtales Terrace	
Total, Atlantic Ocean	9×10^9
Pacific Ocean	
Total, offshore California	2×10^9
Grand total, offshore resources	11×10^9

*All data in metric tons.
†Unmineable because of potential environmental damage.
§No methods exist for mining and processing these resources.
**Tonnages of individual deposits vary widely. Totals are very large, but tons per unit thickness and area are too small to be minable. These are resources for the far distant future.

National Forest. Some of the counties in southern Florida have been reluctant to grant mining permits for lands within their jurisdiction, and if these tendencies persist, it is possible that the tonnage mined from Florida will decrease substantially by the end of this century. Thus, the future for mining of phosphate in Florida is clouded, even though mineable resources are sufficient for many years.

The outlook for mining of offshore material is dependent on the development of new or improved mining techniques. For example, it is possible that slurry mining may be usable, both offshore and onshore (Popper and others, 1985), particularly for the deep onshore deposits that cannot be mined by open pit. There is also the possibility that this method may be acceptable for mining much of the shallower onshore material for which mining permits have not been granted.

BY-PRODUCTS

The apatite mineral of phosphate deposits is amenable to many substitutions: uranium and the rare earths for calcium, and VO_4, SO_4, and CO_3 for PO_4. Fluorine, an integral part of the apatite mineral, is recovered as fluorosilicic acid from gaseous effluents of phosphoric acid plants, and uranium is recovered from phosphoric acid as urano-organic complexes. Vanadium is recovered from "ferro-phos" slag formed in furnace processing of phosphate rock. Rare earths are or have been recovered in the chemical processing of igneous apatite.

By-products may be of much greater importance in the future as the high-grade marine phosphorite deposits are depleted. Costs of mining and processing lower grade material that contains deleterious elements could be balanced by the recovery and sale of by-products.

REFERENCES CITED

Anonymous, 1916, Commerce Reports: February 21, 1916, p. 728.
Altschuler, Z. S., Clarke, R. S., Jr., and Young, E. J., 1958, Geochemistry of uranium in apatite and phosphorite: U.S. Geological Survey Professional Paper 314-D, p. 45-90.
Altschuler, Z. S., Cathcart, J. B., and Young, E. J., 1964, The geology and geochemistry of the Bone Valley Formation and its phosphate deposits, West Central Florida: Geological Society of American Annual Meeting Guidebook, Field Trip no. 6, Nov. 1964, 68 p.
Brown, P. M., 1958, The relation of phosphorites to ground water in Beaufort County, North Carolina: Economic Geology, v. 53, p. 85-101.
Cathcart, J. B., 1963, Economic geology of the Keysville Quadrangle, Florida: U.S. Geological Survey Bulletin 1128, 82 p.
——, 1968, Phosphate in the Atlantic and Gulf Coastal Plains, *in* Brown, L. F., Jr., ed., Fourth Forum on Geology of Industrial Minerals: Austin, University of Texas Press, p. 23-34.
Cathcart, J. B., Sheldon, R. P., and Gulbrandsen, R. A., 1984, Phosphate-rock resources of the United States: U.S. Geological Survey Circular 888, 48 p.
Dabney, C. W., Jr., 1884, North Carolina resources for commercial fertilizers. I. Phosphates: Raleigh, North Carolina, North Carolina Experiment Station Annual Report for 1883, 104 p.
Day, D. T., 1885, Phosphate rock, *in* Mineral Resources of the United States, 1883-1884: U.S. Geological Survey, p. 783-808.

———, 1886, Phosphate rock, *in* Mineral Resources of the United States, 1885: U.S. Geological Survey, p. 445–455.

———, 1890, Phosphate rock, *in* Mineral Resources of the United States, 1888: U.S. Geological Survey, p. 586–596.

———, 1892, Phosphate rock, *in* Mineral Resources of the United States, 1889–1890: U.S. Geological Survey, p. 449–455.

———, 1896, Phosphate, *in* Mineral Resources of the United States, 1895: U.S. Geological Survey, p. 703–712.

Dietz, R. S., Emery, K. O., and Shepard, F. P., 1942, Phosphorite deposits on the sea floor off Southern California: Geological Society of America Bulletin, v. 53, p. 815–848.

Eldridge, G. H., 1893, A preliminary sketch of the phosphates of Florida: American Institute of Mining and Metallurgical Engineers Transactions, v. 21, p. 196–231.

Furlow, J. W., 1969, Stratigraphy and economic geology of the eastern Chatham County phosphate deposit: Georgia Geological Survey Bulletin 82, 40 p.

Gale, H. S., 1911, Rock phosphate near Melrose, Montana: U.S. Geological Survey Bulletin 470-H, p. 440–451.

Gale, H. S., and Richards, R. W., 1910, Preliminary report on the phosphate deposits in southeastern Idaho and adjacent parts of Wyoming and Utah: U.S. Geological Survey Bulletin 430-H, p. 457–535.

Galliher, E. W., 1931, Collophane from Miocene brown shales of California: American Association of Petroleum Geologists Bulletin, v. 15, p. 257–269.

Gower, H. D., and Madsen, B. M., 1965, The occurrence of phosphate rock in California: U.S. Geological Survey Professional Paper 501-D, p. 79–85.

Gray, A. N., 1944, Phosphates and superphosphates: London, E. T. Heron and Co., Ltd., 416 p.

Hayes, C. W., 1895, The Tennessee phosphates: U.S. Geological Survey Annual Report 16, pt. 4, p. 610–630.

Hayes, C. W., and Ulrich, E. O., 1903, Description of the Columbia Quadrangle, Tennessee: U.S. Geological Survey Geologic Atlas, Columbia Folio, no. 95.

Holmes, F. S., 1870, The phosphate rocks of South Carolina: Charleston, South Carolina, Holmes Book House, 87 p.

Hoots, H. W., Blount, A. L., and Jones, P. H., 1935, Marine oil shale in Playa del Rey field, California: American Association of Petroleum Geologists Bulletin, v. 19, no. 2, p. 172–205.

Kazakov, A. V., 1937, The phosphorite facies and the genesis of phosphorites: U.S.S.R. Science Institute Fertilizers and Insectofungicides, Translation no. 142, p. 95–113 [English translation].

Kimrey, J. O., 1965, Description of the Pungo River Formation in Beaufort County, North Carolina: North Carolina Division of Mineral Resources Bulletin 79, 131 p.

LeBaron, J. F., 1892, Discussion of paper by Davidson, Notes on the geological origin of phosphate of lime in the United States and Canada: American Institute of Mining and Metallurgical Engineers Transactions, v. 33, p. 152–157.

Malde, H. E., 1959, Geology of the Charleston phosphate area, South Carolina: U.S. Geological Survey Bulletin 1079, 105 p.

Manheim, F. T., Pratt, R. M., and McFarlin, P. F., 1980, Composition and origin of phosphorite deposits of the Blake Plateau, *in* Bentor, Y. K., ed., Marine phosphorites—geochemistry, occurrence, genesis: Economic Paleontologists and Mineralogists Special Publication 29, p. 117–138.

Mansfield, G. R., 1917, The phosphate resources of the United States: Second Pan American Scientific Congress, Washington, D.C., Dec. 27, 1915–Jan. 8, 1916, 38 p.

———, 1942, Phosphate resources of Florida: U.S. Geological Survey Bulletin 934, 82 p.

McKelvey, V. E., 1967, Phosphate deposits: U.S. Geological Survey Bulletin 1252, 21 p.

McKelvey, V. E., Williams, J. S., Sheldon, R. P., Cressman, E. R., Cheney, T. M., and Swanson, R. W., 1959, The Phosphoria, Park City, and Shedhorn Formations in the Western Phosphate Field: U.S. Geological Survey Professional Paper 313-A, 46 p.

Patton, W. W., Jr., and Matzko, J. J., 1959, Phosphate deposits in northern Alaska: U.S. Geological Survey Professional Paper 302-A, 17 p.

Pilkey, O. H., and Luternauer, J. L., 1967, A North Carolina shelf phosphate deposit of possible commercial interest: Southeastern Geology, v. 8, p. 33–51.

Popper, G. H., Godesky, D. J., and Giambra, J. J., 1985, Phosphate resource potential for borehole mining in the Southeastern Coastal Plain: U.S. Bureau of Mines Information Circular 9043, 20 p.

Riggs, S. R., 1984, Paleoceanographic model of Neogene phosphorite deposition, U.S. Atlantic Continental margin: Science, v. 233, no. 4632, p. 123–131.

Riggs, S. R., Hine, A. C., Snyder, S. W., Lewis, D. W., Ellington, M. D., and Stewart, T. L., 1982, Phosphate exploration and resource potential on the North Carolina Continental Shelf: Proceedings of the 1982 Offshore Technology Conference, May 1982, p. 737–746.

Rouse, G. E., 1979, "Other" uraniferous phosphate resources of the United States, *in* DeVoto, R. H., and Stevens, D. N., eds., Uraniferous phosphate resources, United States and free world: U.S. Department of Energy Publication GJEX–110(79), p. 310–313.

Safford, J. W., 1894, Phosphate-bearing rocks in Middle Tennessee, preliminary notice: American Geologist, v. 13, p. 107–109.

Scott, T. M., 1985, The lithostratigraphy of the Central Florida phosphate district and its southern extension, *in* Cathcart, J. B., and Scott, T. M., eds., Florida land-pebble phosphate district: Geological Society of America Annual Meeting Guidebook, Orlando, Florida, October 1985, p. 28–37.

Sellards, E. H., 1909, Mineral industries: Florida Geological Survey, 2nd Annual Report, p. 232–250.

Service, A. L., 1967, History and development of the phosphate industry in southeast Idaho, *in* Hale, L. A., ed., Anatomy of the western phosphate field: Intermountain Association of Geologists 15th Annual Field Conference Guidebook, September 1967, p. 176–185.

Smith, R. W., and Whitlatch, G. I., 1940, The phosphate resources of Tennessee: Tennessee Division of Geology Bulletin 48, 444 p.

Stowasser, W. F., 1978, Phosphate rock: U.S. Bureau of Mines Minerals Yearbook, 1977, 24 p.

Van Horn, F. B., 1911, Phosphate rock, *in* Mineral Resources of the United States for 1910: U.S. Geological Survey, p. 735–746.

Weeks, F. B., and Ferrier, W. F., 1907, Phosphate deposits in Western United States: U.S. Geological Survey Bulletin 315-P, p. 449–462.

Willis, E., 1892, Phosphate rock, *in* Day, D. T., Report on the mineral industries in the United States at the 11th Census, 1890, p. 681–691.

Wright, C. D., 1893, The phosphate industry of the United States: 6th Special Report of Commissioner of Labor, 145 p.

Wyatt, F., 1892, The phosphates of America, 3d ed.: New York, The Scientific Publishing Co., 187 p.

Manuscript Accepted by the Society October 27, 1988

Chapter 10

Evaporites and brines

Omer B. Raup and Marc W. Bodine, Jr.*
U.S. Geological Survey, MS 939, Box 25046, Denver Federal Center, Denver, Colorado 80225

INTRODUCTION

Mineral commodities produced from evaporite deposits share a common origin in their precipitation from surficial brines. Brines result from the evaporative concentration of solutes in either marine or continental waters in closed or highly restricted basinal environments. When the brine reaches saturation for a given saline mineral, that mineral precipitates and is deposited. If appropriate climatic, hydrologic, geomorphic, and postdepositional geologic conditions prevail, commercially significant quantities of the salt or salts may accumulate and be preserved for exploitation. In addition, some useful trace solutes, particularly those such as lithium, bromine, and iodine, that do not form their own salts with evaporative concentration, are commercially extracted from the natural brines themselves.

Table 1 is an abbreviated list of the most common evaporite minerals and includes those of economic importance mined in the United States. More extensive lists of evaporite minerals, their properties, and occurrence can be found in Stewart (1963), Borchert and Muir (1964), Braitsch (1971), Holser (1979a), Eugster (1980), Harvie and others (1984), and Sonnenfeld (1984). Table 2 contains the chemical analyses of seawater and an array of representative natural brines; these brines are typical source fluids of evaporite minerals or hosts for the commercially important dissolved constituents.

In this chapter the major features of marine and continental evaporite mineral deposition will be discussed along with a brief summary of the sources of natural brines. This is followed with a discussion of each of the evaporite commodities produced domestically. We have omitted the nitrate salts (soda niter and niter), which are extensively mined from continental evaporite deposits in Chile (Erickson, 1981; Garrett, 1985) but are not found in commercial concentrations in this country. We have also excluded any discussion of the zeolite minerals that are mined from a number of domestic deposits. Although many zeolites formed through the devitrification of volcaniclastic sediments deposited in alkaline saline evaporite environments (Surdam and Sheppard, 1978), they are more appropriately described by Brobst (this volume).

MARINE EVAPORITES

Marine evaporites are sedimentary rocks that result from the evaporation of seawater and can be observed forming today at a number of localities. Although these modern evaporites are thin and cover relatively small areas, their origin and the resultant salt succession are the same as for those that formed large thick deposits throughout the Phanerozoic. Because of erosion, diagenetic dissolution, and extensive regional metamorphism, Precambrian evaporites are rare.

Mineral paragenesis

The sequence of minerals precipitated from evaporating seawater was determined by Usiglio (1849). He carefully analyzed Mediterranean Sea water at successive stages of evaporation, and his findings formed the basis for our understanding of marine evaporite deposits. This work was later refined, first by van't Hoff in the early twentieth century and shortly thereafter by D'Ans through an extensive array of detailed experiments. The results of these investigations and those of others have been excellently systematized and summarized in Stewart (1963) and Braitsch (1971). Recently, in a series of reports from Eugster and others (1980) through Harvie and others (1984), the saline mineral paragenesis and coexisting brine compositions have been rigorously reexamined using thermodynamic modeling. These results amplify and, with some exceptions as noted below, generally confirm the earlier experimental results. In addition, trace-element geochemistry, such as the distribution of bromine between halite and the coexisting brine (Kühn, 1968; Holser, 1979b) and stable-isotope investigations (Holser, 1979b; Claypool and others, 1980) have contributed much to our understanding of marine evaporite genesis and subsequent diagenesis.

The typical mineral salt succession that precipitates from evaporating seawater (Fig. 1) initially consists of small amounts of carbonate minerals, usually dolomite with some calcite. With continued evaporation, calcium sulfate saturation is reached and gypsum precipitates. When 90 percent of the water has been lost by evaporation, halite begins to crystallize; in a few cases evaporation proceeds to such an extent that the highly soluble "bitter" salts of magnesium and potassium are deposited. Brine composi-

---
*Deceased.

Raup, O. B., and Bodine, M. W., 1991, Evaporites and brines, *in* Gluskoter, H. J., Rice, D. D., and Taylor, R. B., eds., Economic Geology, U.S.: Boulder, Colorado, Geological Society of America, The Geology of North America, v. P-2.

TABLE 1. SOME COMMON EVAPORITE MINERALS

Mineral Name	Chemical Formula
Halite* (rock salt)	NaCl
Sylvite*	KCl
Langeinite	$K_2Mg_2(SO_4)_3$
Carnallite	$KMgCl_3 \cdot 6H_2O$
Polyhalite	$K_2MgCa_2(SO_4)_4 \cdot 2 \cdot 2H_2O$
Kieserite	$MgSO_4 \cdot H_2O$
Bischofite	$MgCl_2 \cdot 6H_2O$
Gypsum*	$CaSO_4 \cdot 2H_2O$
Anhydrite	$CaSO_4$
Sulfur*	S
Trona*	$Na_3(CO_3)(HCO_3) \cdot 2H_2O$
Nahcolite	$NaHCO_3$
Gaylussite	$Na_2Ca(CO_3)_2 \cdot 5H_2O$
Dawsonite	$NaAl(CO_3)(OH)_2$
Mirabilite*	$Na_2SO_4 \cdot 10H_2O$
Thenardite	Na_2SO_4
Aphthitalite	$(K,Na)_3Na(SO_4)_2$
Glauberite	$Na_2Ca(SO_4)_2$
Burkeite	$Na_6(CO_3)(SO_4)_2$
Borax*	$Na_2B_4O_7 \cdot 10H_2O$
Tincalconite	$Na_2B_4O_7 \cdot 5H_2O$
Kernite	$Na_2B_4O_7 \cdot 4H_2O$
Ulexite	$NaCaB_5O_9 \cdot 8H_2O$
Probertite	$NaCaB_5O_9 \cdot 5H_2O$
Colemanite	$Ca_2B_6O_{11} \cdot 5H_2O$

*A major mineral commodity.

tion throughout the succession varies in response to the composition of the salts precipitated and the extent of evaporation. Thus, until calcium sulfate saturation is reached, there is little change in solute proportions from the seawater composition (no. 1 in Table 2) except for the increase in total concentration. Gypsum precipitation results in a relative loss of calcium (no. 2 in Table 2) in the brine. With the onset of halite crystallization, sodium concentration decreases markedly relative to potassium and magnesium (no. 3 in Table 2) as chloride concentration increases. Bitter salt precipitation begins with epsomite, $MgSO_4 \cdot 7H_2O$, when 1.4 percent of the original water remains. Further evaporation leads to carnallite saturation, and eventually the terminal invariant brine composition is reached when only 0.4 percent of the water remains (Fig. 1). The terminal brine is essentially a 5.7 molal magnesium chloride solution containing only minor quantities of the other solutes; further evaporation deposits the bischofite-carnallite-kieserite-halite-anhydrite salt assemblage. Progressive evaporation causes the density of the brine to increase from 1.03 g/cc (seawater) to >1.3 g/cc as the terminal brine composition is approached.

The bitter salt paragenesis that was determined from the earlier experimental work in the Mg-K-Na-Cl-SO$_4$-H$_2$O system (Stewart, 1963; Braitsch, 1971) begins with the precipitation of bloedite, $Na_2Mg(SO_4) \cdot 4H_2O$, then epsomite, followed by precipitation of coexisting epsomite and kainite, $KCl \cdot MgSO_4 \cdot 3H_2O$. Further concentration leads to carnallite-kieserite saturation, and this pair of bitter salts precipitates until the terminal brine composition is reached. Thermodyanmic modeling, on the other hand, for the Ca-Mg-K-Na-Cl-SO$_4$-H$_2$O system (Eugster and others, 1980) indicates that the earlier precipitation of polyhalite and recrystallization of preexisting anhydrite to polyhalite during rock-salt deposition (Fig. 1) modifies the brine composition such that epsomite is the initial bitter salt precipitated and that no potassium salt is formed until carnallite-kieserite saturation is reached. This latter succession more closely parallels observations in natural deposits. Both bloedite and kainite are rare minerals, much rarer than would be expected from the earlier defined experimental succession.

In most deposits, the earlier products, dolomite with some limestone and gypsum or anhydrite (Fig. 1), are common. In many, the extent of evaporation produced substantial quantities of halite, but in most cases the brines never reached the concentrations necessary to deposit the bitter salts. In some instances, however, these very soluble bitter salts were deposited but later were removed by leaching (Stewart, 1963). Furthermore, a number of different depositional and diagenetic processes can occur, and commonly—particularly in evaporite deposits with bitter-salt assemblages—produce minor to substantial deviation from the ideal marine evaporite mineral paragenesis (Fig. 1; see Stewart, 1963; Borchert and Muir, 1964; Raup, 1970; Braitsch, 1971; Holser, 1979a; Sonnenfeld, 1984). These processes include, among others, magnesium depletion through widespread dolomitization or diagenetic incorporation into clays; sulfate deficiency through reduction, with or without the crystallization of authigenic pyrite and other sulfide minerals; precipitation produced by brine mixing; incongruent mineral dissolution and other reactions through the introduction of foreign basinal ground waters and their interaction with the evaporite assemblage; and a variety of recrystallization phenomena associated with burial and rising temperatures. It is through such processes, for example, that relatively rich beds of sylvite or langbeinite (minerals not observed in the static isothermal succession in Fig. 1) occur in commercial quantities; such beds are the principal sources of potash.

Geology

Grabau (1920) suggested the following geologic settings for the formation of marine evaporites: marginal salt pans, marine salinas, lagoons, and relict seas.

TABLE 2. CHEMICAL COMPOSITION OF SEA WATER AND SOME REPRESENTATIVE MARINE EVAPORITE,
NONMARINE EVAPORITE, AND SUBSURFACE BRINES*

	Sea Water	Marine		Nonmarine Evaporite			Subsurface Brines				
	1 (mg/kg)	2 (mg/kg)	3 (mg/kg)	4 (mg/kg)	5 (mg/kg)	6 (mg/kg)	7 (mg/kg)	8 (mg/L)	9 (mg/L)	10 (mg/kg)	11 (mg/kg)
Mg	1,293	8,340	39,612	11,800	32	15	9,960	4,040	5,770	n.d.	100
Ca	411	1,040	300	148	11	18	74,800	36,300	33.200	n.d.	40,000
Na	10,760	58,550	67,500	79,900	21,400	95,500	22,500	63,300	68,900	119,000	54,000
K	399	1,912	9.962	7.040	1,120	3,760	9,120	1,370	640	15,600	23,800
Cl	19,350	107,920	190,200	147,000	15,100	31,700	208,000	197,600	184,000	105,000	184,000
SO_4	2,709	13,489	46,417	23,100	7.530	152,000	40	350	tr.	49,400	16
HCO_3	143	245	948	484	26,430	5,330	n.d.	200	0	n.d.	n.d.
CO_3	n.d.	n.d.	n.d.	n.d.	n.d.	6,640	n.d.	n.d.	n.d.	38,400	n.d.
Sr	8	n.d.	n.d.	n.d.	96	n.d.	2,650	n.d.	n.d.	n.d.	n.d.
Li	<1	n.d.	n.d.	52	9	n.d.	70	170	17	30	320
Br	67	n.d.	n.d.	n.d.	n.d.	n.d.	2,910	4,800	204	580	700
I	<<1	n.d.	n.d.	n.d.	n.d.	n.d.	40	5	19	25	n.d.
B	4	n.d.	n.d.	37	157	0	380	140	67	4,090	498
TDS	35,150	191,496	354,939	289,000	71,900	293,000	331,000	345,235	293,000	335,000	304,000

*n.d. = not determined, absent, or not reported; tr. = trace.
1. Average sea water (Riley and Chester, 1971).
2. Marine evaporite brine precipitating gypsum, Boccana de Virrila, Peru (Morris and Dickey, 1957)
3. Marine evaporite brine precipitating halite, Boccana de Virrila, Peru (Morris and Dickey, 1957).
4. Nonmarine evaporite brine, Great Salt Lake (north arm), Utah (Hahl and Handy, 1969).
5. Nonmarine evaporite brine, Mono Lake, California (Livingstone, 1963, p. 17).
6. Nonmarine evaporite brine, Deep Springs, California (Jones, 1965).
7. Deep basin formation water, Sylvania Sandstone, Michigan (White and others, 1963, p. 32).
8. Deep basin formation water (evaporite residual brine), Smackover Formation, Arkansas (Collins, 1975).
9. Deep basin formation water (evaporite resolution brine), Frio Formation, Louisiana (Collins, 1975).
10. Pore fluid in "lower salt" body, Searles Lake, California (White and others, 1963, p. 56).
11. Geothermal brine (340°C), Salton Sea, California (Ellis, 1979).

Marginal salt pans are relatively small coastal depressions into which seawater floods by overwash during high tides or storms. Stewart (1963) cites the Rann of Cutch in northwest India as an excellent example of this type of deposit; flood waters cover 18,130 km² during the summer and evaporate during the winter to leave a thin crust of salt.

Kinsman (1969) has demonstrated that evaporating shallow-marine ground waters are far more important than evaporating surface waters in shallow pans in forming extensive evaporites in modern arid, supratidal-flat ("sabkha") environments. Along the Trucial Coast of the Persian Gulf and the northwestern shore of the Gulf of California, storms or tidal surges periodically flood the extensive supratidal flats with seawater that infiltrates into the porous sediments. After the flood waters recede the pore fluids evaporate, and interstitial diagenetic gypsum and halite are precipitated in the sediment. Although the succeeding influx of seawater undoubtedly redissolves most or all of the interstitial halite, much of the gypsum is preserved. Widespread occurrences of sabkha deposition in ancient evaporites have been suggested by a number of workers (Schenk, 1967; Treesh and Friedman, 1974).

Marine salinas, like the salt pans, are small surface depressions on land which seawater enters by percolation through a near-shore barrier. An example is the Lake of Larnaca on Cyprus where water seeps in from the Mediterranean Sea and salt is deposited in the summer (Grabau, 1920). Gypsum and halite are the common minerals found in salinas; the more soluble salts of potassium and magnesium are rare.

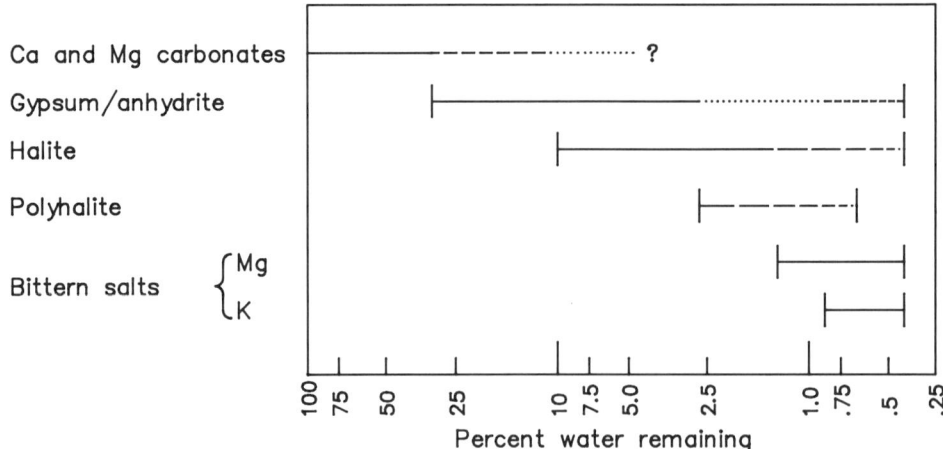

Figure 1. Succession of evaporite minerals coexisting with evaporating sea water at 25°C; solid lines are major minerals; dashed and dotted lines are minor minerals in the evaporite assemblages. The final assemblage at extreme concentration (0.4 percent of water remaining) is the stable assemblage at the terminal invarient brine composition; further evaporation to total dryness precipitates chiefly bischofite and minor quantitites of halite, carnallite, kieserite, and anhydrite (Table 1). Data for all but the Ca and Ca-Mg carbonates from Eugster and others (1980); metastable precipitation and solubility dependence on pH and dissolved carbon dioxide concentration preclude defining unequivocal limits of carbonate mineral precipitation.

Some of the large deposits of marine evaporites are too thick and too areally extensive to have been deposited in either marginal salt pans or marine salinas. Although there are no modern analogs, these large deposits were probably deposited in *lagoons* or *relict seas.* As Stewart (1963) pointed out, the evaporation of a 1,300-m column of seawater of normal composition would produce only a 1-m-thick bed of anhydrite. To achieve the enormous thicknesses of anhydrite and salt found in deposits in many parts of the world, the quantity of seawater required would be much greater than that which could be contained in an isolated inland sea of reasonable size. To explain these vast thicknesses of salt and anhydrite, Ochsenius (1877), developing an earlier idea of Bischof (1864), proposed his "bar theory." This model assumes that a lagoon or sea was connected to the ocean by a narrow opening through a bar or over a shallow sill. Seawater lost through evaporation in the area of the lagoon would be replenished by inflow through the bar or over the sill. In this way, enormous quantities of evaporitic sediments could be concentrated in one area.

Ochsenius (1877) cites the Gulf of Karaboghas as a modern analog. It is on the east side of the Caspian Sea and is separated from it by sand spits. A narrow channel through the spits admits Caspian Sea water into the gulf. According to Urasov and Polyakov (1956), approximately 130×10^6 tons of salt are carried into the gulf each year, and gypsum, halite, and various sulfates of sodium and magnesium are deposited. The analogy to a marine lagoonal system is incomplete; the gulf is not strictly a marine evaporite deposit, because the water is partly derived from meteoric waters in several rivers that discharge into the Caspian Sea.

Most geologists accept some version of the bar theory. Various deposits require some combination of inflow and restricted circulation in a subsiding area of deposition to allow for accumulation of thick marine evaporite deposits. For example, King (1947) suggests that continuous inflow of normal seawater across a partial barrier or sill at the entrance into the restricted basin generates two currents, an upper inflow current of the lighter normal seawater that evaporites, precipitates salts, and at some point sinks to become a lower outflow current of denser, more saline water. In this manner it is possible to maintain a roughly constant salinity in the more concentrated brine that is precipitating evaporites and thereby generating unusual thicknesses of a single evaporite facies such as the extremely thick succession of bedded calcium sulfate characteristic of the Permian Castile Formation in the Delaware basin. Scruton (1953) and, later, Briggs (1958), using variations of King's mechanism, allow more complete and progressive evaporation of the upper inflow layer along its flow path to explain the lateral, sometimes concentric, zoning of evaporite facies in some deposits, such as the Silurian Salina Group evaporites in the Michigan basin (Briggs and Pollack, 1967). Schmalz (1969) has suggested that evaporite accumulation can occur in deep water with appropriate basin geometry and a submerged sill; a deep-water depositional environment eliminates the requirement of basin subsidence accompanying essentially parallel evaporite deposition, which would be necessary to produce thick evaporite sequences in shallow-water environments.

Smith and others (1973) list 70 marine evaporite deposits in the United States. Of these, 61 contain gypsum and 67 contain anhydrite. Halite occurs in 28, only 7 contain bitter salts (sylvite, carnallite, or langbeinite), and 1 contains polyhalite (Fig. 2). The total thickness of the deposits plotted by age is given in Figure 3. Deposits of Permian age are the most extensive and have the greatest total thickness. These are represented by the deposits in the Permian basin in the midcontinent. Other thick deposits are the Louann Salt of probable Jurassic age in the Gulf Coast basin, the Paradox basin evaporites of Pennsylvanian age, and the Salina Group salt deposits of Silurian age in the Michigan and Appalachian basins. Salts of Cretaceous and Tertiary age are also significant. Distribution of these salt deposits is shown on the map on Plate 3B.

NONMARINE EVAPORITES

Nonmarine evaporites are deposited in closed-basin lakes or playas that capture and evaporate local runoff and prevent it from escaping to the sea. The drainage area must be sufficiently large with a terrain that furnishes abundant solutes to the streams. Most nonmarine evaporites are geologically young—Quaternary or Tertiary—reflecting their terrestrial origin and susceptibility to subsequent erosion. Pre-Tertiary deposits, such as intervals in the Triassic and Jurassic Newark Supergroup in New Jersey (Van Houten, 1964; Olsen, 1980), are rare and not commercially important. The Great Basin of the western United States has, since the early Tertiary, hosted a large number of small-to-large closed-basin drainages, many of which have deposited a wide variety of continental evaporite mineral assemblages. A substantial number of these are actively depositing salts today and are invaluable as natural laboratories for studying the hydrology and mineralogy of continental closed-basin systems.

Mineral assemblages in continental evaporites vary markedly from one basin to another (Eugster, 1980). As with marine evaporites, the mineralogy is controlled by the extent of evaporation, but in addition, unlike the marine evaporites, the mineralogy is also strongly dependent on widely differing compositions of the surficial and shallow-subsurface inflow waters. The chemistries of the dilute inflow waters are determined by (1) the lithologies that constitute the terrain throughout the basin, and (2) the extent and character of water-rock interaction that occurs during weathering throughout the basin and during diagenesis within the evaporite environment itself. The origin, evolution, and classification of closed-basin saline waters have been comprehensively developed by Jones (1966), Hardie and Eugster (1970), Eugster and Hardie (1978), and Eugster and Jones (1979).

Two principal types of weathering reactions introduce solutes into meteoric waters in addition to straight-forward mineral dissolution. One is carbonic acid hydrolysis of silicate minerals that results from the atmospheric or soil carbon dioxide dissolved in water. The hydrolysis of dissolved carbon dioxide produces hydrogen ion (reaction 1a), and the resultant reaction with albite, for example, yields kaolinite, dissolved silica (or quartz), and the aqueous sodium and bicarbonate ions (reaction 1b).

$$CO_2 + H_2O \rightarrow H_2CO_3 \rightarrow H^+ + HCO_3^- \tag{1a}$$

$$2NaAlSi_3O_8 + 2CO_2 + 11H_2O \rightarrow Al_2Si_2O_5(OH)_4 + 4H_4SiO_4 + 2Na^+ + 2HCO_3^- \tag{1b}$$

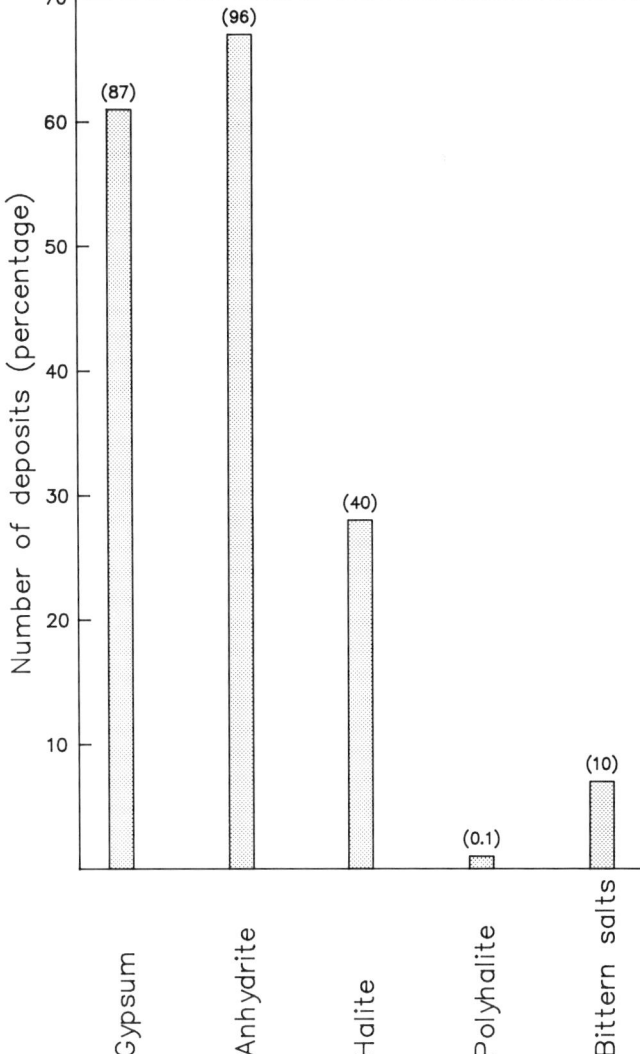

Figure 2. Number of marine evaporite deposits in the United States plotted by lithology of maximum salinity (data from Smith and others, 1973).

The second weathering reaction is a redox reaction between atmospheric oxygen and sulfide minerals that results in sulfuric acid hydrolysis of silicate minerals. Again the kaolinitization of albite illustrates this mechanism. Typically, sulfuric acid is generated by the oxidation of pyrite to produce a ferric oxide mineral and sulfuric acid (reaction 2a), and albite is hydrolyzed to yield kaolinite, dissolved silica, and the aqueous sodium and sulfate ions (reaction 2b).

$$4FeS_2 + 15O_2 + 8H_2O \rightarrow 2Fe_2O_3 + 16H^+ + 8SO_4^{2+} \tag{2a}$$

$$2NaAlSi_3O_8 + H_2SO_4 + 9H_2O \rightarrow Al_2Si_2O_5(OH)_4 + 4H_4SiO_4 + 2Na^+ \; SO_4^{2+} \tag{2b}$$

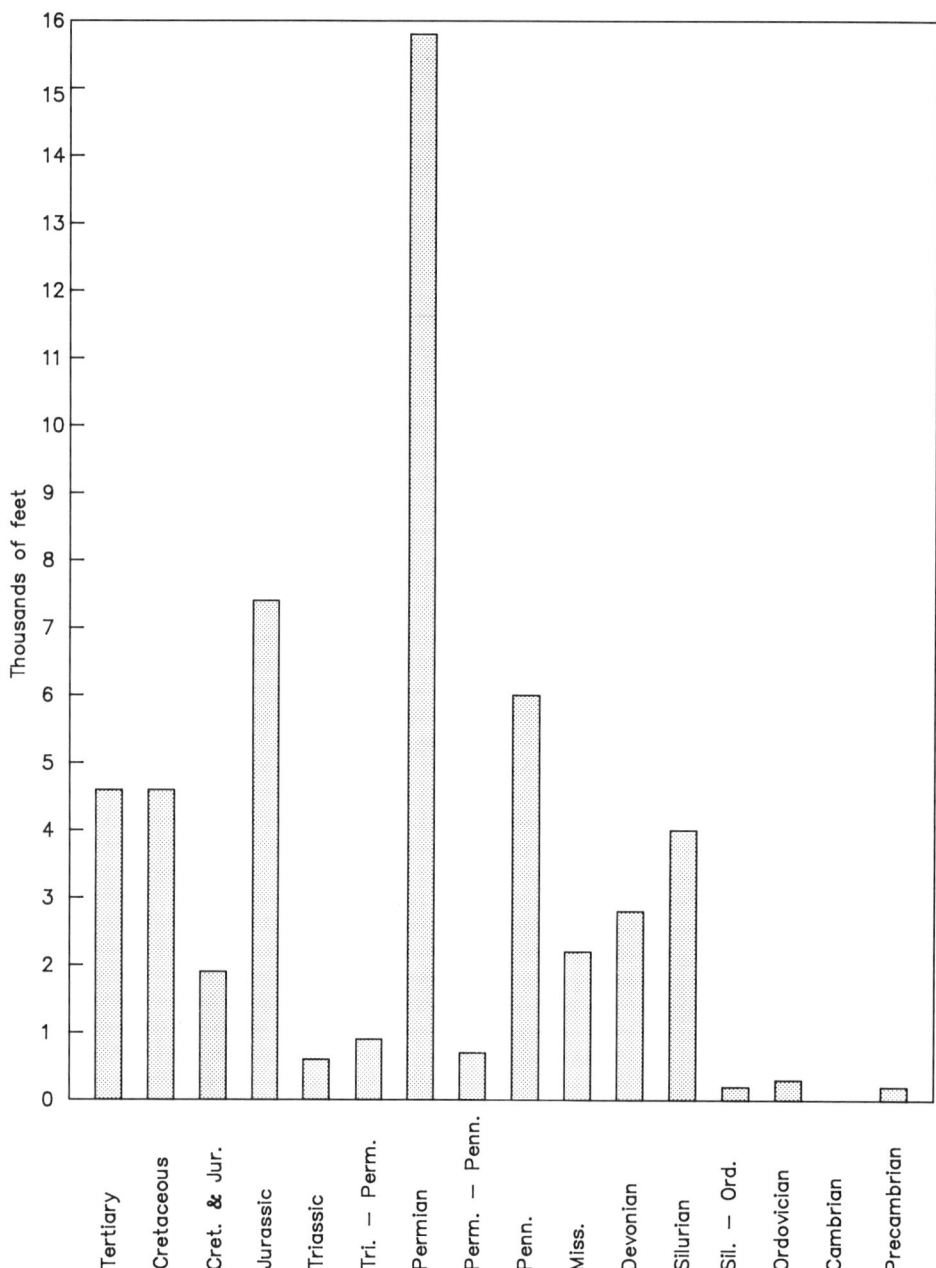

Figure 3. Thickness of marine evaporite deposits in the United States plotted by geologic time (data from Smith and others, 1973).

The wide variation in the compositions of the dilute inflow waters is illustrated in Figure 4a (Eugster, 1980). Anion proportions cluster toward the bicarbonate-carbonate apex, emphasizing the importance of carbonic acid hydrolysis with or without carbonate-rock dissolution. Sulfate proportions are modest, except for the few waters that lie close to the sulfate apex. Sulfate in the waters is principally derived from sulfuric-acid hydrolysis and, particularly in those waters with abundant sulfate, through dissolution of preexisting sulfate-rich evaporite assemblages. Chloride proportions are characteristically low, generally less than 20 percent of the anion assemblage; the chloride is obtained from many sources such as aerosol sea spray and volcanic emanations in the meteoric precipitation, and particularly when chloride is a dominant anion, the dissolution of chloride-bearing evaporites. Cation content in the dilute inflow is highly variable (Fig. 4a). There is a continuous belt of compositions from the alkali (Na + K) apex to the calcium apex with characteristically low to moderate magnesium proportions.

With evaporative concentration in the basin lake or playa environment the least soluble salts precipitate first. The precipitation of these, most commonly the Ca-Mg carbonates (calcite, magnesian calcite, dolomite) with or without gypsum, depletes the evolving brines in alkaline earths, the ionized carbonate species, and sulfate relative to the alkalis and chloride. These changes are reflected in the closed-basin brine compositions (Fig. 4b) as so convincingly documented by Hardie and Eugster (1970). The chloride proportion among the anions is strikingly enhanced when compared with the inflow water compositions (Fig. 4a) and is due to the loss of carbonate-bicarbonate and sulfate to the precipitated minerals. Cation proportions show an equally marked relative loss of calcium from Figure 4a to 4b to the precipitated calcium carbonate and sulfate. Other than a small number of Mg-rich compositions, the brines are characteristically alkali chloride-rich, and many contain substantial, occasionally dominant proportions of carbonate/bicarbonate and sulfate.

As suggested by Jones (1966), in inflow waters with dissolved $(2HCO_3 + CO_3) >> (Ca+Mg)$, that is, waters in which carbonic acid hydrolysis has played a major weathering role (no. 5 in Table 2), evaporation has produced alkaline (pH > 10 not unusual) brines that may lead to the precipitation of gaylussite, trona (Table 1), and other alkali-bearing bicarbonate and carbonate minerals. If the inflow waters contain $SO_4 >> Ca$, that is, waters in which sulfide hydrolysis dominated the weathering regime, evaporation results in the evolution of sulfate-rich brines (no. 6 in Table 2) and the likely precipitation of mirabilite, thenardite, aphthitalite (Table 1), the magnesium sulfate salts, and a variety of mixed alkali-alkaline earth sulfate salts (Eugster, 1980). Brines with an excess of both carbonate/bicarbonate and sulfate may precipitate both alkali sulfate and alkali-bearing carbonate/bicarbonate minerals, oftimes including the double anion salt burkeite (Table 1). Halite frequently coexists with any of the alkali-bearing sulfate or bicarbonate/carbonate salts in these assemblages.

Alternatively, if discharge into the system includes a substantial fraction of connate or evaporite-resolution ground waters that mix with the local meteoric inflow, the resultant brines are likely to be chloride-rich and may, at least superficially, resemble the marine evaporite brines (no. 4 in Table 2). Halite usually dominates in evaporite deposits associated with these brines, and the alkali-bearing carbonate/bicarbonate and sulfate salts are comparatively rare or absent.

If contemporaneous volcanic ash falls and other volcaniclastic sediments are deposited in the evaporite basin, as is characteristic of a number of Great Basin continental evaporite deposits, brine-sediment interaction (1) enriches the brines in lithium, boron, and possibly bromine and other trace, soluble devitrification constituents; and (2) commonly zeolitizes or argillizes the vitric particulates. Further evaporation of the brine and diagenesis may lead to significant accumulations of borate salts (Table 1) in the evaporite mineral assemblage.

The most important economic deposits of nonmarine evaporites are in basins caused by faulting or downwarping. To persist, they require a tectonic setting that will either lower the

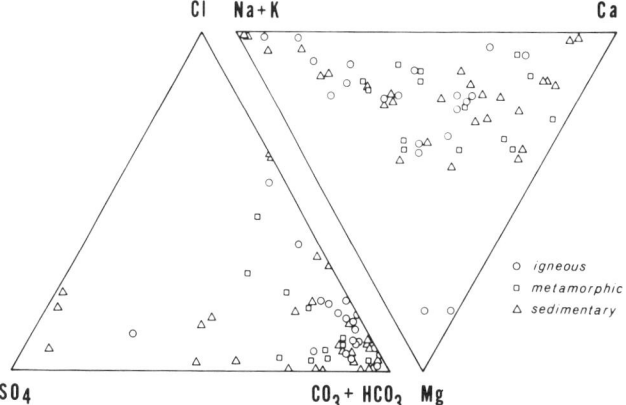

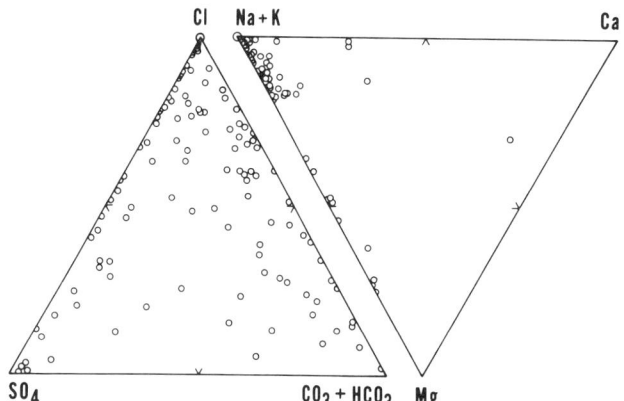

Figure 4. Trilinear diagrams comparing major cation and anion compositions of dilute inflow waters grouped by drainage basin lithology (a) and those of continental closed-basin brines (b) after Eugster and Jones (1979) and Eugster (1980) with data from Hardie and Eugster (1970). Reproduced, with permission from the Annual Review of Earth and Planetary Sciences, vol. 8; © 1980 by Annual Reviews, Inc.

floor of the basin as rapidly or more rapidly than sedimentation fills the basin, or elevate possible outlets faster than erosion can destroy them. Most, but not all, closed basins today are in arid or semiarid areas where sedimentation rates are low. Comparable basins in the geologic past probably had a similar geologic setting (Smith and others, 1973). Furthermore, enough water must have flowed into the basin to introduce large quantities of solutes. Such geologic environments are rare because arid basins generally receive large volumes of water only where they adjoin high mountain ranges or where they are at the end of an integrated drainage system from adjacent less arid regions (Smith and others, 1973).

The nonmarine evaporite deposits are economic sources of sodium and sodium-calcium carbonate and bicarbonate minerals, sodium sulfate minerals, halite, and the borate minerals. At some deposits the concentrations of aqueous potassium, boron, lithium, bromine, iodine, and magnesium allow their economic recovery from the brines. Some of the more important Quaternary deposits are Searles Lake and Death Valley in California and Great Salt Lake in Utah. The Green River Formation in Colorado and Wyoming contains important Tertiary nonmarine evaporites.

NATURAL BRINES

Natural brines occur in several geologic settings and are attributed to a variety of origins (Bodine and Jones, 1986). The most common types of brines are: (1) deep-basin pore fluids in sedimentary rocks, frequently referred to as "oil field" brines, that are often associated with marine evaporite rocks; (2) springs and shallow subsurface brines, frequently at elevated temperatures, which may or may not be associated with recent magmatic activity; and (3) surface and subsurface brines that are associated with Pleistocene or Holocene saline lakes and playas.

Some deep-basin pore fluids, typically the highly concentrated chloride brines (nos. 7, 8, and 9 in Table 2) with variable concentrations of sodium, magnesium, calcium, and potassium, frequently contain substantial strontium, bromine, iodine, lithium, rubidium, and cesium (Collins, 1975, 1976). The origin of these "oil field" brines has long been debated (Rittenhouse, 1967; Lerman, 1970; Collins, 1975; Carpenter, 1978, 1979; Graf, 1982; Stoessel and Moore, 1983) and is still not thoroughly understood today. Many appear to be connate residual brines from marine evaporite deposition (nos. 7 and 8 in Table 2) or, alternatively, evaporite resolution brines (no. 9 in Table 2). It has also been suggested by Bredehoeft and others (1963) that shales may act as semipermeable membranes or "ion filters," and with reverse osmotic pressures, may fractionate connate ground waters, leaving a highly saline residuum strongly enriched in chloride and divalent cations (Hanshaw and Coplen, 1973; Kharaka and Smalley, 1976; Graf, 1982). Deep-basin brines, regardless of their origin, have generally been modified by a variety of diagenetic interactions with the surrounding rocks. For example, some are extremely rich in calcium (nos. 7, 8, and 9 in Table 2), presumably through dolomitization of limestone (Carpenter, 1978) or through albitization of plagioclase (Carpenter, 1978; Stoessel and Moore, 1983). Also, deep-basin brines may be depleted in potassium (no. 8 in Table 2) through illitization or crystallization of authigenic potassium feldspar (Carpenter, 1978).

Brine springs or shallow subsurface brines not associated with recent igneous activity or evaporite deposition are usually upwellings of evaporite resolution brines. Through favorable geometry of the hydrologic flow system they approach or reach the surface. Some of the brines may be at elevated temperatures, reflecting the geothermal gradient to the greatest depth of their flow path.

Thermal fluids, both in the subsurface and as springs, however, are most common in areas of recent magmatic and tectonic activity. Brines, such as the Salton Sea brines of southeastern California (no. 11 in Table 2), are chiefly meteoric waters with, at most, only minor fractions of juvenile water (White, 1974). Their compositions, because of water-rock interaction at elevated temperatures imposed by the igneous activity (Muffler and White, 1969; Honda and Muffler, 1970), are varied and often distinctive. They frequently have relatively high concentrations of boron, lithium, and other unusual components (Ellis, 1979). These waters are economically important as geothermal energy reservoirs (Truesdell and White, 1973), and are frequently cited as examples of active hydrothermal ore deposition systems (Weissberg and others, 1979; White, 1981).

Brine springs and subsurface brines are also found associated with Tertiary and Quaternary saline lake and playa systems (no. 10 in Table 2). They, like their continental evaporite-brine progenitors, are frequently rich in alkali, carbonate/bicarbonate, and sulfate, and may, if occurring within a volcanic terrain, contain unusually high concentrations of boron, lithium, and other devitrification solutes (Smith and others, 1973).

MINERAL PRODUCTS FROM EVAPORITES AND BRINES

Salt

History of production. From the time of the discovery of the New World, salt has been in reasonably plentiful supply (Multhauf, 1978). Much of the Atlantic coast was suited to the production of solar sea salt. Trails from the coastal plain through the "trackless forests" showed that the native Americans and wild animals had found salt springs in the foothills of the Appalachian Mountains. In the early years of Colonial America, salt was produced in numerous salt works. In the early 1600s, salt was produced at Jamestown, at Plymouth Colony, and at Coney Island in New Amsterdam. Toward the late 1600s, as the colonies imported increasing amounts of salt from England, the American salt works withered away. The American Revolution "found the rebellious colonies bereft of salt" (Multhauf, 1978, p. 36), which led to a frantic reestablishment of domestic salt production, and salt works reappeared along the east coast. Salt was produced from seawater by both boiling and solar evaporation. After the Revolution, the United States returned to dependence on imported salt, most of it from Great Britain.

Salt was produced from salt springs by French settlers near what is now Shawneetown, Illinois, in the early 1600s. Salt springs were discovered by other French settlers in the Onondaga Indian territory, near modern Syracuse, New York, in the middle 1600s. This became an important salt-producing area and by 1800 was yielding nearly 2 million kilograms per year. Salt was produced from brine that contained about 10 percent NaCl. It was boiled in 300-liter cast iron pots that were set in furnaces large enough to hold as many as 40 pots. This was the standard practice for salt production in the United States for many years. Casting of these pots, which were also used for the manufacture of potash and soap, was a significant output of the early American iron industry.

The most important post-Revolution site for salt production was at Kanawha, Virginia (now West Virginia), where salt boiling at a buffalo lick began in 1797. Salt production reached nearly 14 million kilograms in 1810, making Kanawha the most important source of salt in the country. Drilling to increase brine flow in 1806 made this the first site for "earth boring" in the United States (Multhauf, 1978).

The production of salt at Onondaga, New York, surpassed that at Kanawha in the 1820s. In 1829 the most important salt works were those at Onondaga, Kanawha, Barnstable County on Cape Cod, Massachusetts, and Shawneetown, Illinois. These four produced 70 percent of the national production.

Another major salt-producing area in the east was at appropriately named Saltville, Virginia. Salt marshes were known to late Pleistocene mammals (Cooper, 1966) but these natural brines were first noted by Thomas Jefferson in 1781. Brine production started in the early 1800s, and salt from this area was a strategic commodity during the Civil War. Saltville is still a major center for production of chemicals and industrial minerals based on salt brine, limestone, and gypsum.

Production from salt springs decreased with the discovery of rock salt in the United States. The initial discovery at Avery Island, Iberia Parish, Louisiana, in 1862 (Lefond, 1969), was at a depth of only 5 m, and was the first of many salt domes found in the Gulf Coast Basin. These domes are still among the nation's leading sources of salt. Bedded salt deposits were later found in Texas, Ohio, Michigan, Kansas, and New York.

Major marine deposits. The seven major marine evaporite basins in the United States (Plate 3B) are the (1) Gulf Coast basin of Alabama, Arkansas, Louisiana, Mississippi, and southeastern Texas; (2) Paradox basin of western Colorado and Utah; (3) Permian basin of southeastern Colorado, Kansas, Oklahoma, western Texas, and New Mexico; (4) Michigan basin of Michigan; (5) Appalachian basin of Maryland, New York, Ohio, Pennsylvania, and West Virginia; (6) Supai basin of Arizona and New Mexico; and (7) Williston basin of Montana, North Dakota, South Dakota, and Wyoming. However, only five, the Gulf Coast, Permian, Michigan, Appalachian, and Williston basins, are commercial sources of salt. In 1983, about 29 percent of United States salt was produced from rock salt, 55 percent came from natural brines and by solution mining of marine evaporite deposits (Morse, 1985a), and the remainder from all other sources.

Gulf Coast basin. The Gulf Coast salt basin (Plate 3B), one of the largest in the United States, underlies approximately 300,000 km^2 in the states of Alabama, Arkansas, Louisiana, and Texas (Lefond, 1969), where the deposit is the Louann Salt. The Louann Salt was thought by Imlay (1943) to be Jurassic, whereas Hazzard and others (1947) believed it to be Permian. The prevailing view, as expressed by Kirkland and Gerhard (1971), favors a Jurassic age for the Louann Salt.

The Gulf Coast basin is characterized by at least 300 salt domes (Lefond, 1969). Salt domes are pillars of salt that have "grown" upward from a thick layer of bedded salt because of the relatively low density of the salt as compared to the enclosing sediments. The tops of some of the domes are within a few meters of the ground surface and are marked by topographic features, whereas the tops of other domes are hundreds of meters below the surface.

Mining at Avery Island was begun immediately after discovery in 1862, but was stopped a year later when the facilities were destroyed by Union forces (Lefond, 1969). The mine works were rebuilt after the Civil War, and today Avery Island remains an important source for rock salt.

The discovery of oil at Spindletop Dome, Texas, in 1901, stimulated exploration for salt domes in the Gulf Coast region (Lefond, 1969). As a result, several domes in Louisiana and Texas are being mined for rock salt, and salt is being produced on a large scale from brine wells and by solution mining in Alabama, Arkansas, Louisiana, and Texas.

Permian basin. The Permian basin (Plate 3B) also has some very extensive salt deposits. The salt beds become progressively younger from northeast to southwest; thus the deposits in Kansas and Oklahoma are Lower Permian, whereas the deposits in southeastern New Mexico and southwestern Texas are Upper Permian.

Salt in Kansas occurs in three formations. The uppermost salt is in the Blaine Formation, the middle salt is in the Stone Corral Formation, and the lowermost is the Hutchinson Salt Member of the Wellington Formation. Production of salt in Kansas is confined to the Hutchinson Salt Member where it is mined by both underground and solution methods.

Minor amounts of salt are produced in Oklahoma from the Cimarron Salt, which is equivalent to the Stone Corral, occurring between the Hutchinson Salt Member and the Blaine Formation, and also by solar evaporation of brine from salt springs.

The largest concentration of Permian salt in the basin is in west Texas and eastern New Mexico. This salt is in three formations, which are from top to bottom, Rustler, Salado, and Castile. The Rustler contains only minor amounts of salt. The Salado Formation, which contains large deposits of potash minerals in addition to halite, underlies an area of about 65,000 km^2 in and adjacent to the Delaware basin, which is part of the Permian basin (Lefond, 1969). The Castile Formation is confined to the Delaware basin and contains as much as 200 m of bedded salt that has not yet been commercially exploited. Although there are enormous tonnages of salt in this part of the Permian basin, the small production is merely a by-product of potash mining.

Michigan and Appalachian basins. The Michigan and Appalachian basins (Plate 3B), which cover parts of New York, Pennsylvania, West Virginia, Ohio, and Michigan, and part of Ontario, Canada, share one of greatest accumulations of salt in the world (Lefond, 1969). These deposits of the Salina Group are of Silurian age and underlie an area of approximately 210,000 km^2. Deposition of salt began in the Michigan basin and eventually spread into the much wider area of the adjacent Appalachian basin to the east (Matthews and Egleson, 1974).

The Silurian salt of the Michigan basin has a maximum thickness of 550 m (Matthews and Egleson, 1974). Two mines near Detroit produce salt from a 170-m-thick bed that is 240 m below the surface. Brine wells produce salt mostly from locations around the edge of the basin, but there are a few in the center of the basin. Salt is also produced from mines and brine wells near Cleveland, Ohio, from various localities in western New York state, and from deep solution mining in West Virginia. Although salt deposits underlie much of Pennsylvania, none are produced because they are too deep.

Williston basin. The Williston basin (Plate 3B), which underlies about 390,000 km^2 in North Dakota, South Dakota, and Montana, and in Manitoba and Saskatchewan in Canada, contains salt deposits of several ages and formations. These are the Ordovician Stonewall Formation, the Devonian Prairie Formation of the Elk Point Group, the Mississippian Charles Formation of the Madison Group, the Permian Opeche Formation, and the Permian and Triassic Spearfish Formations. Although there is a large resource of salt in this basin, production in the United States is limited to solution mining operations near Williston, North Dakota, that recover salt from the Charles Formation at a depth of 2,600 m. The Canadian part of the Williston basin, however, contains one of the world's largest deposits of potash, in the Devonian Prairie Formation.

Major nonmarine deposits. Great Salt Lake, Utah. Great Salt Lake (Plate 3B) is the largest source of nonmarine salt in the United States; six companies are producing halite there by solar evaporation of brine. Rising lake levels due to heavy precipitation in the early 1980s have drastically curtailed production (Tripp, 1985).

Solar extraction from seawater. California. Salt production by the solar evaporation of seawater is done on a large scale in California. Two of the largest operations are at the south end of San Francisco Bay and at San Diego Bay (Ver Planck, 1958). Both operations are based on fractional crystallization. Seawater is introduced into a series of large, shallow ponds situated between high- and low-tide marks so that initial filling can be done by gravity flow at high tide. During the first phase of concentration the brine is brought to saturation with respect to sodium chloride and the less soluble salts are precipitated. The saturated brine, called "pickle," is about 10 percent of the initial volume of seawater (Fig. 1). The pickle is then pumped into a series of ponds where continuing evaporation causes salt to precipitate. In order to avoid precipitation of the very soluble magnesium salts, the concentration of the liquor in the ponds is kept at a constant specific gravity by withdrawing "mother liquor" or bitterns and replacing it with fresh pickle (Ver Planck, 1958). It is not possible to prevent the precipitation of some gypsum and some magnesium compounds during salt precipitation. The crude salt, with careful control, is about 99.4 percent sodium chloride, which is further purified to 99.9 percent.

Machines harvest the salt by scraping it from the bottom of the ponds after the bitterns are drained, and convey it to waiting cars. When the salt is 10 cm (4 in) thick, the harvesting is at a rate of 150 tons per hour. Most of this salt is used in the chemical industry.

The late-stage bittern brines are rich in magnesium and bromine, and have at times been sold to chemical plants for recovery of these components.

Uses. While domestic uses and food preparation constituted the major consumption of salt in the early years of this country, these uses now account for only about 6 percent of production. Now the major uses of salt and brine are in the chemical industry, primarily in the production of chlorine, sodium hydroxide, and sodium carbonate.

Chlorine is used to make vinyl chloride, which is the basis for polyvinyl chloride resins, which in turn are used in a variety of plastic products for use in construction, home furnishings, packaging, and other applications (Morse, 1985a).

Sodium hydroxide and hydrogen gas are coproducts with chlorine in the electroysis of brine. The major uses of sodium hydroxide are in chemical manufacturing, in pulp and paper, in soaps and detergents, in the removal of sulfur dioxide from petroleum, and in the manufacture of glass.

Demand for manufactured sodium carbonate had declined since the development of deposits of natural soda ash.

Other uses for salt and sodium chloride brines include the manufacture of sodium metal, hydrochloric acid, and fluxes for production of high-purity aluminum, dyes, and pharmaceuticals; and in de-icing of highways, water conditioning, and sewage treatment (Morse, 1985a).

Large masses of subsurface salt have economic use or proposed use as large-scale storage facilities. Solution cavities in salt domes of the Gulf Coast are in use as storage for fluid hydrocarbons (Jacoby, 1970). Both salt domes and bedded salts have been proposed as suitable radioactive waste-disposal media (Johnson and Gonzales, 1978). As a result, the Waste Isolation Pilot Plant (WIPP) in the bedded Ochoan or Upper Permian salts of southeastern New Mexico is currently under construction for the storage of military nuclear waste (Powers and others, 1978).

Potash

History of production. Potash is the commercial name for a chemical compound that contains potassium. It was produced during Colonial American times by leaching wood ashes in large cast iron pots. Trees, which were plentiful along the east coast, were burned and the ashes soaked in water; the resultant solution was then evaporated by boiling to precipitate the dissolved salts. This product, used in the manufacture of soap and glass, probably was a mixture of potassium and sodium carbonates and hydroxides (Searls, 1985).

Potash was found to be of benefit to growing plants by the Germans in the early 1800s, and by the late 1800s the United States was importing potash from Germany for use as fertilizer. However by 1910, because of trade disagreements, the potash trade contracts were cancelled. That, plus the trend toward political instability in western Europe, prompted the United States to make a concerted effort to find domestic sources for potash (Smith, 1938). In response to this need, potash was recovered from numerous small alkali lakes in western Nebraska in the years around World War I. This industry died as a result of the discovery of potash at Searles Lake, California, in 1917. Searles Lake was the major United States producer of potash until the 1930s.

The knowledge of an immense deposit of salt in the south-central part of the United States, which was thought to be geologically similar in both age and stratigraphy to the famous potash deposits of Germany, led geologists of both the U.S. Geological Survey and the Texas Bureau of Economic Geology and Technology to believe that potash of commercial quality and thickness might be found somewhere in this area (Hoots, 1925). Arrangements were made by the Texas Bureau and the U.S. Geological Survey to encourage oil drillers to collect and send in cuttings for analysis of any salt deposits that were penetrated, and in 1925, Hoots reported encouraging finds of polyhalite and potassium-rich brines in areas of western Texas and eastern New Mexico.

The first commercial discovery of bedded potash in the United States is credited to V. H. McNutt who found sylvite in cuttings from an oil test in 1925 near Carlsbad, New Mexico (Magraw, 1938). After confirmation by analyses of the cuttings, McNutt secured authorization to prospect the area for potash mineral by diamond drilling, which subsequently proved the occurrence of a commercial deposit. The United States Potash Company was formed to take over the development and operation of the property, and mining started in 1931. Two other mines were opened in the area, by the Potash Company of America in 1934 and by the International Minerals and Chemical Corporation in 1940 (Dunlop, 1951). The United States became independent of European supplies by the beginning of World War II.

Major marine deposits. Carlsbad district, New Mexico (Permian basin). The potash deposits in the Carlsbad district in southeastern New Mexico (Plate 3B) are in evaporites of Late Permian age in the Delaware basin, a subbasin of the much larger Permian basin. These evaporites are in the Ochoan Series and are divided into three formations. From youngest to oldest they are the Rustler, Salado, and Castile. The total thickness of these formations is about 1,200 m (Jones and Madsen, 1968). Potash, mostly in the form of polyhalite (Table 1), occurs widely in the southwestern area of the basin; the only deposits of economic importance, however, are beds of sylvite and langbeinite with halite in the Salado Formation of the Carlsbad district.

The potash occurs in several ore zones within the Salado Formation in an area of approximately 30 km^2, 24 km east and northeast of Carlsbad, New Mexico. Twelve beds contain various amounts and proportions of the potash minerals sylvite and langbeinite (Table 1), in a halite host, and these ore zones are interbedded with layers of barren halite. These and other bittern minerals—kieserite, carnallite, polyhalite, kainite, leonite, picromerite, bloedite, vanhoffite, loeweite, and aphthitalite—collectively reflect a complex and not well understood history of postdepositional alteration of a primary marine succession (Fig. 1). Only six of the twelve mineralized zones are rich and thick enough for commercial production; sylvite and langbeinite are the only minerals sought in these ores. Sylvite is the major potash constituent in commercial fertilizers, and langbeinite is used for fertilizing citrus crops, where the magnesium is a necessary beneficial constituent.

The potash ores are recovered by underground mining with machines that excavate and pulverize the relatively soft rock. The rock is hoisted to the surface where it is ground to monomineral size, and the sylvite and langbeinite are then separated from the halite by flotation.

Moab, Utah (Paradox basin). The potash deposit mined near Moab, Utah, is in evaporites of the Paradox Member of the Hermosa Formation of Middle Pennsylvania age. The Paradox evaporites underlie an area of approximately 28,000 km^2, and had a maximum depositional thickness of about 2,100 m (Hite, 1960). Differential loading during and after evaporite deposition, combined with regional stress, produced a series of northwest-southeast–trending salt anticlines and synclines. Flow thickened the salt in the anticlines and thinned it in the synclines.

The Paradox Member, in areas of maximum thickness, consists of 29 evaporite cycles that contain rhythmic alternations of so-called black shale (organic-rich carbonate), dolomite, anhydrite, and halite. Potash, primarily sylvite and some carnallite, occurs in 18 of the 29 salt beds (Hite, 1960).

Mining of the Paradox basin potash deposits began near Moab, Utah, in 1964 near the crest of the Cane Creek anticline. The sylvite-halite ore was mined at a depth of about 975 m by conventional room-and-pillar methods. Difficulties in mining caused the company to convert to solution mining (Phillips, 1975), and flooding operations started on Christmas Day, 1970, using 5 injection wells. It was expected that it would take four and a half months to fill the mine, but on January 13, 1971, while drilling another injection well, water from several aquifers broke into the mine and it was flooded by January 23. While this incident saved considerable pumping time, the new well had to be plugged so that flow in the mine could be controlled.

Brines from the mine are pumped to surface ponds where halite and sylvite are precipitated by solar evaporation. The ponds cover an area of 160 ha and are lined with heavy plastic covered with a 15-cm layer of salt. The halite-sylvite precipitate is harvested by 20-ton scraper loaders, the blades of which are controlled by an automatic laser-beam system that regulates scraping depth and prevents cutting of the plastic liners. The harvested salts are pumped as a slurry approximately 5 km by a brine-conveying system to the processing plant where sylvite and halite are separated by flotation.

Major nonmarine deposits. Great Salt Lake area, Utah. Potash has been produced from near-surface brines at the Bonneville salt flats near Wendover, Utah, since 1933, and from brines of the Great Salt Lake, Utah, since 1970. Both operations have been interrupted by excessive precipitation and rising lake levels since 1983 (Searls, 1985).

Searles Lake, California. Searles Lake, in the Mojave desert of southern California (Plate 3B), has been a source of potash since 1917 (Searls, 1985). Searles Lake is a playa containing a crystal mush with a complex mixture of sodium carbonates, sulfates, and chloride with interstitial brines rich in potassium, boron, phosphorus, bromine, and lithium. Searles Lake, first worked for borax in 1843, was the largest producer of potash in the United States prior to the development of the bedded potash deposits in New Mexico in 1931. Potash production at Searles Lake has fluctuated in response to competition from the mines in New Mexico, Utah, and Canada, but it has continued through the years because of the diversity of other products (Multhauf, 1978).

Uses. About 94 percent of the United States potash consumption is for agricultural fertilization. The remaining 6 percent is used in a variety of industrial applications. Most of the potash used for fertilizer is potassium chloride, the mineral sylvite. Its high solubility makes it easily assimilated by crops. Some crops require potash in the form of sulfate or nitrate; potassium magnesium sulfate is needed for citrus and some fruits. Potassium compounds are also used in the manufacture of ceramics, glass, soaps, detergents, and gypsum wallboard (Searls, 1985).

Gypsum

History of gypsum production. Gypsum has been used throughout recorded history. The Chinese, Greeks, and Assyrians used it for decorative carvings, and the ancient Egyptians used it as a mortar in the construction of the pyramids. The early Greek philosopher and natural scientist Theophrastus, around 300 B.C., described the burning of gypsum in the preparation of plaster, which is still one of the mineral's principle uses (Pressler, 1985). When gypsum is heated to about 350°C, three-quarters of the water is driven off. When water is subsequently added to this material, it recrystallizes as gypsum. This material became known as Plaster of Paris because of the renowned gypsum deposits in the Montmartre district of Paris, France.

Gypsum was first discovered in New York in 1792, Virginia in 1835, Michigan in 1840, Ohio in 1850, Iowa in 1872, and Colorado and California in 1875.

Major sources. Commercial gypsum deposits in the United States occur in five major areas: (1) the Michigan and Appalachian basins of the Great Lakes and Middle Atlantic areas; (2) the Permian basin of New Mexico, northern Texas, Oklahoma, and Kansas; (3) the Gulf Coast Embayment of southern Texas, Louisiana, and Mississippi; (4) several small basins in the Rocky Mountain area including those in Montana, Wyoming, Colorado, Utah, and New Mexico; and (5) southern California and southern Nevada.

Some gypsum deposits are extremely large. For example, one deposit extends 320 km from Sweetwater, Texas, into Oklahoma, is 30 to 80 km wide and as much as 6 m thick. In Culberson County of western Texas, gypsum crops out over 1,550 km^2 in a belt 80 km long, 24 km wide, and as much as 18 m thick (Pressler, 1985).

Geology. Most gypsum and anhydrite were deposited as part of a sequence of marine evaporites (Fig. 1); both minerals occur widely and abundantly in most marine evaporite basins (Withington, 1962). Gypsum is normally more abundant than anhydrite at or near the surface and grades into anhydrite at depths ranging from a few meters to a few tens of meters. Geologic evidence indicates that some near-surface gypsum was formed by hydration of primary or secondary anhydrite where shallow ground water came in contact with anhydrite. The depth of complete hydration is important in exploitation because the presence of more than a few percent anhydrite renders gypsum unfit for many purposes (Smith and others, 1973).

Some gypsum is mined from deposits formed in continental basins by the evaporation of nonmarine waters. Most of these deposits probably originated as gypsum, and their lack of deep burial has prevented the formation of anhydrite. Most nonmarine deposits are smaller than marine deposits and tend to be nearer the surface (Smith and others, 1973).

Uses. In 1885 a commercial method was developed for retarding the setting time of gypsum plaster; this revolutionized the industry by permitting the use of plaster in construction. The development of prefabricated plaster wallboard ("drywall"), first produced in the United States in 1918, again revolutionized the industry. By 1929, wallboard had attained large-scale use in the construction industry, and by the 1950s, had become the standard material for interior walls and ceilings. In 1983, 72 percent of the gypsum marketed was as prefabricated products. In the same year, six major companies produced 80 percent of the crude gypsum in the United States from 32 open-pit or underground mines, while 36 smaller companies produced the remaining 20 percent from 37 mines. Most of the gypsum is calcined and used in the manufacture of wallboard. The various types of wallboard products are manufactured by continuous automatic machines that can be adjusted to any of the standard products (Pressler, 1985).

The remaining crude gypsum is marketed for use in cement, agriculture, and fillers. Gypsum is used to retard the setting of concrete made from Portland cements. In agriculture, gypsum is used to neutralize alkaline and saline soils and to improve the permeability of clay-rich soils. The sulfur in gypsum is beneficial for some crops (Pressler, 1985).

Sulfur

History of production. Sulfur, known to the ancient world as brimstone—the stone that burns—has been used for various purposes for thousands of years. The Egyptians used sulfur in the bleaching of linen in 2,000 B.C., and Homer, in the *Odyssey,* wrote of its use as a fumigant. During the Peloponnesian War in the fifth century B.C., the Greeks used burning sulfur and pitch to produce suffocating gases; later, the Romans mixed sulfur with tar, pitch, and other flammable materials to produce the first incendiary weapons. Sulfur is a necessary ingredient in gunpowder, which was developed in China in the tenth century. When gunpowder was introduced into Europe in the fourteenth century, it made sulfur an important mineral commodity (Morse, 1985b). The birth of the science of chemistry in the 1700s and the growth of the chemical industries in the 1800s increased the demand for sulfur. Sulfuric acid soon became one of the fundamental process chemicals for a wide range of products. Initial sources of native sulfur were from deposits associated with volcanoes. Early production also came from "pyrites," which referred to pyrite, marcasite, or pyrrhotite. Base-metal sulfides were an additional source.

Major sources. The major sources of sulfur are associated with evaporite deposits. One type occurs in the cap rock of salt domes. The native sulfur in these deposits resulted from the bacterial reduction of anhydrite in the presence of either gaseous or liquid hydrocarbons. Significant deposits of this type were discovered during exploration for oil. In the Frasch process, which made recovery of this type of sulfur possible, superheated water is pumped into the deposit to melt the sulfur, which is then pumped to the surface.

One of the largest sulfur deposits in the free world, in Culberson County in the Delaware basin of western Texas, is in the Castile Formation of Permian age. The origin of this native sulfur deposit is the same as that of the salt-dome deposits, and the sulfur is also produced by the Frasch process (Mussey and Tyree, 1984).

Sulfur is being produced in steadily increasing quantities as a by-product in the processing of sour natural gas and sulfur-rich crude oil. Sour gas contains large quantities of hydrogen sulfide, and environmental constraints require removal of sulfur before the gas and oil can be sold. By-product sulfur from stack gas at metal smelters also furnishes additional supplies.

Uses. Eighty percent of the sulfur produced in the United States is made into sulfuric acid, which in turn is used in the manufacture of hundreds of other chemical products. A major use for sulfuric acid is in the manufacture of phosphate fertilizer, where it is used to digest phosphate rock to produce phosphoric acid plus calcium sulfate. Phosphoric acid is an intermediate product used in the production of high-grade fertilizers such as dicalcium phosphate, diammonium phosphate, and triple superphosphate. Other uses of sulfur are in such diverse products of the chemical industry as plastics, textiles, paper, and paints, and in nonferrous metal production and petroleum refining.

Sodium carbonate

Sodium carbonate (Na_2CO_3), known in industry as "soda ash," is derived from natural sources, as a product of chemical synthesis, and as a by-product of several manufacturing processes.

History of production. The ancient Egyptians used impure sodium carbonate in the manufacture of glass and soap as early as 3,500 B.C. They obtained it from mineral incrustations around alkaline lakes in Lower Egypt. Other early peoples obtained soda ash from the burning of seaweed and other marine plants, by leaching the soluble materials from the ashes and then evaporating the solutions to dryness (Kostick, 1985).

A method to manufacture synthetic sodium carbonate using salt, sulfuric acid, coal, and limestone was devised by Nicolas Le Blanc in France in 1791. In the early 1860s, Ernest and Alfred Solvay, two Belgian brothers, invented an improved process for making soda ash using salt, coke, and limestone, with ammonia as a catalyst. This process became widely used in Europe and is used throughout much of the world today. The Solvay process was introduced into the United States at Syracuse, New York, in 1884 (Kostick, 1985); this plant was just recently shut down because of competition from sources of natural sodium carbonate. Commercial production of natural sodium carbonate was begun at Searles Lake, California, in 1887. A huge deposit of trona (Table 1), another natural source, was discovered in the Green River basin of southwestern Wyoming in 1938 while drilling for oil and gas (Robbins, 1986). All the United States production of natural sodium carbonate is from Searles Lake and the Green River deposits.

Major deposits. Green River basin, Wyoming. The trona of the Green River basin (Plate 3B) is part of a thick sequence of lake sediments and is interbedded with oil-shale deposits of the Eocene Green River Formation. Some of the trona beds are as much as 10 m thick. Where the trona is fairly close to the surface, it is mined by conventional room-and-pillar, or long- or shortwall techniques, and where it is deeper, it is mined by solution. The U.S. Bureau of Mines has estimated that the Wyoming soda ash reserves would last about 3,200 years at the 1983 rate of production (Kostick, 1985).

Oil-shale deposits of the Green River Formation in the Piceance Creek basin of northwestern Colorado contain vast deposits of the mineral nahcolite (Table 1), which is a potential source of sodium carbonate. The mineral dawsonite (Table 1) is also present in these rocks and is a potential source for aluminum. Both of these minerals could be recovered as by-products during the mining and processing of the oil shale (Donnell, this volume).

Searles Lake, California. The saturated subsurface brines at Searles Lake are rich in sodium carbonate, sodium sulfate, and sodium chloride, and also contain potassium, bromine, and lithium. Soda ash is the major product from the plants that produce chemicals from these brines.

Uses. Sodium carbonate is used primarily in the manufacture of glass, but also in detergents, pulp and paper, water treatment, ceramics, and chemicals. A newly developing use for soda ash is in the removal of sulfur from stack gases in coal-fired power plants.

Sodium sulfate

Sodium sulfate, Na_2SO_4, known in the industry as "salt cake," is produced in the United States from several natural sources and as a by-product of many manufacturing processes.

Sodium sulfate has been used for its medicinal value since the fifteenth century. The chemical compound was first accurately described in 1658 by J. R. Glauber (Kostick, 1985), and the hydrated crystalline form mirabilite ($Na_2SO_4 \cdot 10H_2O$) is also called Glauber's salt. The mineral glauberite, also named after Glauber but not to be confused with Glauber's salt, is the double sulfate salt of calcium and sodium (Table 1).

Sodium sulfate is a common constituent in the incrustations around countless saline lakes in arid valleys of the western United States, although the only three commercial sources of natural sodium sulfate are in California, Texas, and Utah.

Most of the natural sodium sulfate produced in the United States comes from Searles Lake, California, where subsurface brines are refrigerated to precipitate mirabilite, which is then refined by recrystallization.

Natural sodium sulfate has been produced from subsurface brines at Brownfield in western Texas since 1933 and from a plant near Seagraves, Texas. Production from these plants fluctuates due to competition from manufacturers of synthetic by-product sodium sulfate.

Natural sodium sulfate is also produced at Ogden, Utah, where brines from Great Salt Lake are concentrated by solar evaporation and the sodium sulfate is recovered from harvesting ponds. Rising lake waters in recent years have flooded the harvesting ponds, but supplies of sodium sulfate are still available from stockpiles.

Uses. Most of the sodium sulfate in the United States is used in the kraft paper industry, but it is also used in the manufacture of glass, ceramic glazes, detergents, and in several other processes.

Boron minerals

Boron occurs in a vast array of minerals, but most of the commercially valuable minerals are hydrated borates of calcium or sodium (Table 1).

History of production. Borax, one of the most common boron compounds, was used hundreds of years ago by artisans in Asia Minor and the Far East as a flux for welding and brazing precious metals and for glazing pottery. Marco Polo started western trade by bringing Tibetan borax crystals to Europe in the thirteenth century. Boron minerals were discovered in Chile in 1852, and during the latter half of the nineteenth century, Chile was the principal world producer. Turkish deposits, which had been worked in ancient times, became a major source in 1865 and remain so today (Lyday, 1985a).

Commercial production of borax in the United States began in 1864 from mineral springs and lakes north of San Francisco, California. Significant discoveries of borates in the playa deposits of Nevada and California were made in the 1870s and 1880s, and the extensive deposits of ulexite (Table 1) were found at Death Valley in 1881. Between 1883 and 1889, borax was transported out of Death Valley by the famous 20-mule teams. Several other major discoveries in southern California were made in the early 1900s (Lyday, 1985a).

Major deposits. All the commercial boron deposits in the United States are in three localities in California. One of the largest is the Kramer deposit at Boron, California, that is primarily borax with associated kernite, colemanite, and ulexite (Table 1). The ore occurs in a flat-lying, irregular tabular mass 3 km long, 0.8 km wide, and 25 to 75 m thick. Borax, which alternates with layers of clays and siltstones, was formed in a Miocene-age lake fed by geothermal springs. The ore is mined by open-pit methods. The crushed borax ore is refined by dissolving in hot borax-rich brines, followed by filtration to remove impurities, and finally by recrystallization. Various grades of refined pentahydrate, decahydrate, and anhydrous sodium borate are produced by repeated dissolution, recrystallization, and dehydration processes (Lyday, 1985a).

Searles Lake is a major producer of borates as well as potash, sodium carbonate, and sodium sulfate. This 106-km^2 playa contains two salt bodies: an upper one 23 m thick, and a lower one 11 m thick. The two bodies are separated by 3.6 m of impervious mud. The porous salt bodies are filled with saturated brines that contain 1 to 2 percent borax, 3 to 4 percent each of soda ash and sodium sulfate, and 16 to 17 percent sodium chloride. The different salts are removed from the brine by precipitation and fractional crystallization through carbonation, refrigeration, and evaporation. The principal borate products are pentahydrate, decahydrate, and anhydrous sodium borate (Lyday, 1985a).

The Furnace Creek deposits in Death Valley National Monument contain ulexite, probertite, and colemanite (Table 1) ores of Miocene age. Mining is by open pit, and the ulexite-probertite ores are beneficiated by screening. The colemanite ore is beneficiated by crushing and flotation, and the concentrate is further enriched by calcining to decrepitate the colemanite, which is separated by air-cyclone. This method results in a 43-percent B_2O_3 product.

Uses. Boron compounds have a wide variety of uses. The major use for boron is in a variety of glass products. Borosilicate glass can withstand severe temperature changes without cracking; thus, it is valuable for cookware, laboratory ware, automotive headlights, and so on. Fiberglass made with boron has excellent flame-retardant properties. Boron is a major constituent in frits for protective and decorative coating on sinks, stoves, refrigerators, etc. It is also used in soaps, detergents, herbicides, fertilizers, soil sterilizers, and fluxes for welding, soldering, and brazing.

Boron nitride has some properties remarkably similar to carbon. At normal temperatures it is a soft, white, highly refractory solid with a waxy luster. It crystallizes in thin hexagonal plates resembling graphite and can withstand oxidation to temperatures as high as 650°C. In fibrous form, boron nitride is as strong as fiberglass but is lighter in weight and more resistant to high temperature. When boron nitride is subjected to extremely high pressure and temperature it forms cubic crystals that rival the hardness of diamond. Boron carbide also is a highly refractory material, is one of the hardest substances known, and is widely used as an abrasive (Lyday, 1985a).

Lithium

Lithium is the lightest of the metallic elements. It occurs in several minerals in pegmatites and in brines related to continental evaporites. The pegmatite occurrences are discussed by Brobst (this volume).

Major deposits. Commercial recovery of lithium from brines in the United States began at Searles Lake, California, in 1938 as a by-product of the production of sodium carbonate and sulfate, potassium and magnesium chlorides, and borates. The brine contains 70 ppm lithium, which is extracted as dilithium sodium phosphate and then converted to lithium carbonate (Rykken, 1976).

The major production of lithium from brines in the United States is at Silver Peak in Clayton Valley, Nevada. The brines of this playa contain 230 ppm lithium. According to Kunasz (1980) the three requirements for the occurrence of lithium in a playa of this type are occurrence in (1) Tertiary or Holocene volcanic terrain, (2) closed structural depressions, and (3) desert areas. The first requirement establishes the source of the lithium. The volcanic terrain supplies the lithium either directly through hot springs or geothermal solutions, or indirectly through the leaching of lithium-bearing volcaniclastic sediments or by the recycling of trapped lithium-bearing solutions. The second requirement provides the necessary mechanism for retaining the dilute solutions carried into the basin. The third requirement provides the environment for the evaporative concentration of the brines.

Uses. The uses of lithium are varied. The major use is as an additive to the cryolite bath in the electric potlines used for the production of metallic aluminum. Lithium fluoride decreases the melting point of the bath. This permits a lower operating temperature, as well as increasing the electrical conductivity of the bath, both of which result in energy savings.

Lithium consumption in the glass and ceramic industries is second only to that in the aluminum industry. As an additive in glass and ceramics, lithium reduces melting temperatures and the thermal coefficient of expansion. Cookware and laboratory vessels that contain lithium compounds are resistant to thermal shock.

The third major use of lithium is in the manufacture of multipurpose greases. Lithium-based greases are stable at high temperatures and have good water resistance, and therefore work well under a wide range of temperatures and severe atmospheric conditions. Thus, they are used extensively in military, industrial, automotive, aircraft, and marine applications.

Batteries with lithium as a major component are finding their way into the marketplace, as they have low density, high tolerance to temperature extremes, and a long shelf life. A small but important use for lithium carbonate is in the treatment of manic depression (Ferrell, 1985).

A potential important use for lithium in the future could be in fuel elements of certain types of controlled thermonuclear fusion reactors. If such reactors are developed, large quantities of the rarer isotope lithium-6 would be required (Bogart, 1976).

Bromine

Elemental bromine is a dark, reddish-brown liquid, and is the only nonmetallic element that is liquid at room temperature and pressure. Bromine was discovered in seawater in 1826 by the French chemist Antoine-Jerome Balard. Since then, bromine has been found in deep-basin "oil field" brines and in brines related to evaporite deposits. Bromine is unusual in that it does not occur as a major constituent in any mineral. For more than 40 years, bromine was produced from seawater, but by 1969, all seawater plants in the United States had been shut down in favor of those producing bromine from subsurface brines. Seawater contains 65 ppm bromine, whereas some subsurface brines contain more than 5,000 ppm bromine (Lyday, 1985b).

The two major bromine-producing areas in the United States are in Michigan and Arkansas. In Michigan, bromine is recovered from brines in the Sylvania Sandstone of the Detroit River Group of Devonian age. In Arkansas, bromine is recovered from brines in the limestones of the Smackover Formation of Jurassic age. Some bromine is also produced at the brine operations at Searles Lake, California (Lyday, 1985b).

The major use for bromine is the gasoline additive ethylene dibromide (EDB). EDB removes residues of lead from engines that use leaded gasoline. Other uses for bromine include flame retardants, photographic chemicals, dyes, fragrances, vitamins, and pharmaceuticals (Lyday, 1985b).

Iodine

Iodine is a gray to purplish black crystalline element. It rarely occurs in the native state but is found in several minerals, particularly in association with nitrate compounds in Chile. The most widespread occurrence of iodine, however, is in oil-field–type brines.

Iodine, discovered in 1813 by Joseph Gay-Lussac, the noted French chemist and physicist, was found as an impurity in seaweed-derived soda ash. He named it "iode" from the Greek word for violet. For many years, the only commercial source for iodine was from the ashes of seaweed harvested along the coasts of France, Norway, and Scotland. The seaweed industry declined after 1868, when iodine production began from nitrate deposits in the Atacama Desert of northern Chile. Chile was the world's major iodine producer until 1966, when Japan began to produce iodine from brines associated with natural gas fields in the Chiba Peninsula (Smith and others, 1973; Lyday, 1985c).

Iodine production in the United States is from brine wells in Pennsylvanian-age rocks in the Anadarko basin of Oklahoma and from brines of the Sylvania Sandstone, of Devonian age, in Michigan (Lyday, 1985c).

The major uses for iodine include photography, pharmaceuticals, catalysts, food additives, and sanitary uses. Other uses include herbicides, cloud-seeding chemicals, detectors in radiation counters, airport luggage scanners, and high-intensity quartz-halogen lights for automobiles, sports stadiums, and television studios (Lyday, 1985c).

Magnesium

Magnesium, a light metallic element, is used both as a metal and in numerous chemical compounds. Magnesium is present in a wide variety of minerals, but only magnesite, $MgCO_3$; dolomite, $Ca,Mg(CO_3)_2$; brucite, $Mg(OH_2)$; and olivine, $(Mg,Fe)_2SiO_4$, are of commercial interest. Magnesium occurs in brines related to evaporite deposits, and in seawater that contains enough magnesium to make it commercially extractable. Only the evaporitic and seawater sources of magnesium are discussed here.

Magnesium compounds have been known for centuries. The term "magnesite" was first used in 1795 by J. C. Delanethrie for a series of magnesium salts including carbonate, sulfate, nitrate, and chloride. D.L.G. Karsten in 1808 restricted the term to magnesium carbonate, and this usage was gradually accepted.

In the early 1940s, magnesium chloride began to be recovered from seawater, which became a major source for magnesium metal (Kramer, 1985). Magnesium compounds were produced from seawater in California, Delaware, Florida, and Texas. Magnesium compounds also were recovered from brine wells in Michigan and from brines of the Great Salt Lake in Utah (Kramer, 1985).

Approximately 85 percent of the magnesium consumed in the United States is in the form of magnesium compounds, and the remainder as the metal. The major use for magnesium compounds is as refractory materials in the iron and steel industry where they are used for such things as liners for blast furnaces, converters, and crucibles. Magnesium compounds are also used in cement, rayon, fertilizer, insulation, rubber, fluxes, medicines, paints, glass, ink, and ceramics. Magnesium metal is alloyed with aluminum to increase the hardness and corrosion resistance of the pure magnesium metal. Aluminum-magnesium alloys are used in beverage cans; structural components in automobiles, aircraft, and military vehicles; and bumpers, wheels, and decorative trim in automobiles (Kramer, 1985).

REFERENCES CITED

Bischof, G., 1864, Lehrbuch der chemischen und physikalischen Geologie, 2nd ed.: Bonn, Germany, v. 2.

Bodine, M. W., Jr., and Jones, B. F., 1986, The salt norm—a quantitative chemical-mineralogic characterization of natural waters: U.S. Geological Survey Water Resources Investigations Report 86-4086, 130 p.

Bogart, S. L., 1976, Fusion power and the potential lithium requirement, in Vine, J. D., ed., Lithium resources and requirements by the year 2,000: U.S. Geological Survey Professional Paper 1005, p. 12–21.

Borchert, H., and Muir, R. O., 1964, Salt deposits—the origin, metamorphism and deformation of evaporites: New York, D. Van Nostrand, 338 p.

Braitsch, O., 1971, Salt deposits, their origin and composition: New York, Springer-Verlag, 297 p.

Bredehoeft, J. D., Blyth, C. R., White, W. A., and Maxey, G. B., 1963, A possible mechanism for the concentration of brines in subsurface formations: American Association of Petroleum Geologists Bulletin, v. 47, p. 211–223.

Briggs, L. I., 1958, Evaporite facies: Journal of Sedimentary Petrology: v. 28, p. 46–56.

Briggs, L. I., and Pollack, H. N., 1967, Digital model of evaporite sedimentation: Science, v. 155, p. 453–456.

Carpenter, A. B., 1978, Origin and chemical evolution of brines in sedimentary basins: Oklahoma Geological Survey Circular 79, p. 60–78.

—— , 1979, Interim report on lead and zinc in oil-field brines in the central Gulf Coast and in southern Michigan: Society of Mining Engineers of the American Institute of Mining and Metallurgical Engineers, Preprint 79-95, 15 p.

Claypool, G. E., Holser, W. T., Kaplan, I. R., Sakai, H., and Zak, I., 1980, The age curves of sulfur and oxygen isotopes in marine sulfate and their mutual interpretation: Chemical Geology, v. 28, p. 199–260.

Collins, A. G., 1975, Geochemistry of oilfield waters: New York, Elsevier, 496 p.

—— , 1976, Lithium abundances in oilfield waters, in Vine, J. D., ed., Lithium resources by the year 2,000: U.S. Geological Survey Professional Paper 1005, p. 116–123.

Cooper, B. N., 1966, Geology of the salt and gypsum deposits in the Saltville area, Smyth and Washington Counties, Virginia, in Rau, J. L., ed., Second symposium on salt: Northern Ohio Geological Society, v. 1, p. 11–34.

Dunlap, J. C., 1951, Geologic studies in a New Mexico potash mine: Economic Geology, v. 46, no. 8, p. 909–923.

Ellis, A. J., 1979, Explored geothermal systems, in Barnes, H. L., ed., Geochemistry of hydrothermal ore deposits, 2nd ed.: New York, John Wiley, p. 954–973.

Erickson, G., 1981, Geology and origin of the Chilean nitrate deposits: U.S. Geological Survey Professional Paper 1188, 37 p.

Eugster, H. P., 1980, Geochemistry of evaporitic lacustrine deposits: Annual Review of Earth and Planetary Sciences, v. 8, p. 35–63.

Eugster, H. P., and Hardie, L. A., 1978, Saline lakes, in Lerman, A., ed., Lakes—chemistry, geology, physics: New York, Springer-Verlag, p. 237–293.

Eugster, H. P., and Jones, B. F., 1979, Behavior of major solutes during closed-basin brine evolution: American Journal of Science, v. 279, p. 609–631.

Eugster, H. P., Harvie, C. E., and Weare, J. H., 1980, Mineral equilibria in the six-component sea water system, Na-K-Mg-Ca-SO$_4$-Cl-H$_2$O, at 25°C: Geochimica et Cosmochimica Acta, v. 44, p. 1355–1348.

Ferrell, J. E., 1985, Lithium, in Knoerr, A. W., ed., Mineral facts and problems: U.S. Bureau of Mines Bulletin 675, p. 461–470.

Garrett, D. E., 1985, Chemistry and origin of the Chilean nitrate deposits, in Schreiber, B. C. and Harner, H. L., eds., Sixth international symposium on salt: Alexandria, Virginia, Salt Institute, v. 1, p. 285–302.

Graf, D. L., 1982, Chemical osmosis, reverse chemical osmosis, and the origin of subsurface brines: Geochimica et Cosmochimica Acta, v. 46, p. 1431–1446.

Grabau, A. W., 1920, Geology of the nonmetallic mineral deposits other than silicates: Vol. 1—Principles of salt deposits: New York, McGraw-Hill, 435 p.

Hahl, D. C., and Handy, A. H., 1969, Great Salt Lake, Utah—a chemical and physical variation of the brine 1963-1966: Utah Geologic and Mineral Survey Water Resources Bulletin 12, 33 p.

Hanshaw, B. B., and Coplen, T. B., 1973, Ultrafiltration by a compacted clay membrane: Geochimica et Cosmochimica Acta, v. 37, p. 2311–2327.

Hardie, L. A., and Eugster, H. P., 1970, The evolution of closed-basin brines, in Morgan, B. A., ed., Fiftieth anniversary symposia—mineralogy and petrology of the upper mantle; sulfides; mineralogy and geochemistry of nonmarine evaporites: Mineralogical Society of America Special Paper 3, p. 273–290.

Harvie, C. E., Møller-Weare, N., and Weare, J. H., 1984, The prediction of mineral solubilities in natural waters—The Na-K-Mg-Ca-H-Cl-SO$_4$-OH-HCO$_3$-CO$_3$-CO$_2$-H$_2$O system to high ionic strengths at 25°C: Geochimica et Cosmochimica Acta, v. 48, p. 723–751.

Hazzard, R. T., Spooner, W. C., and Blanpied, B. W., 1947, Notes on the stratigraphy of the formations which underlie the Smackover Limestone in south Arkansas, northeast Texas, and north Louisiana: Shreveport Geological Society 1945 Reference Report, v. 2, p. 483–503.

Hite, R. J., 1960, Stratigraphy of the saline facies of the Paradox Member of the Hermosa Formation of southeastern Utah and southwestern Colorado, in Smith, K. G., ed., Geology of the Paradox Basin fold and fault belt: Four Corners Geological Society Guidebook, 3rd Field Conference, p. 86–89.

Holser, W. T., 1979a, Mineralogy of evaporites, in Burns, R. G., ed., Marine

minerals: Mineralogical Society of America Short Course Notes, v. 6, p. 211–294.
——, 1979b, Trace elements and isotopes in evaporites, *in* Burns, R. G., ed., Marine minerals: Mineralogical Society of America, Short Course Notes, v. 6, p. 295–346.
Honda, S., and Muffler, L.P.T., 1970, Hydrothermal alteration in core from research drill hole Y-1, Upper Geyser Basin, Yellowstone National Park, Wyoming: American Mineralogist, v. 55, p. 1714–1737.
Hoots, H. W., 1925, Geology of a part of western Texas and southeastern New Mexico, with special reference to salt and potash: U.S. Geological Survey Bulletin 780–B, p. 33–126.
Imlay, R. W., 1943, Jurassic formations of the Gulf regions (of North America including the United States, Mexico, and Cuba): American Association of Petroleum Geologists Bulletin, v. 27, p. 1407–1533.
Jacoby, C. H., 1970, Storage of hydrocarbons in cavities in bedded salt deposits formed by hydraulic fracturing, *in* Rau, J. L., and Dellwig, L. F., eds., Third symposium on salt: Northern Ohio Geological Society, v. 1, p. 463–469.
Johnson, K. S., and Gonzolas, S., 1978, Salt deposits in the United States and regional geologic characteristics important for storage of radioactive waste: Athens, Georgia, Earth Research Associates, 188 p.
Jones, B. F., 1965, The hydrology and mineralogy of Deep Springs Lake, Inyo County, California: U.S. Geological Survey Professional Paper 502–A, 56 p.
——, 1966, Geochemical evolution of closed basin waters in the western Great Basin, *in* Rau, J. L., ed., Second symposium on salt: Northern Ohio Geological Society, v. 1, p. 181–200.
Jones, C. L., and Madsen, B. M., 1968, Evaporite geology of the fifth ore zone, Carlsbad district, southeastern New Mexico: U.S. Geological Survey Bulletin 1252–B, 21 p.
Kharaka, Y. K., and Smalley, W. C., 1976, Flow of water and solutes through compacted clays: American Association of Petroleum Geologists Bulletin, v. 60, p. 973–980.
King, R. H., 1947, Sedimentation in Permian Castile Sea: American Association of Petroleum Geologists Bulletin, v. 28, p. 46–56.
Kinsman, D.J.J., 1969, Modes of formation, sedimentary associations, and diagenetic features of shallow-water and supratidal evaporites: American Association of Petroleum Geologists Bulletin, v. 51, p. 830–840.
Kirkland, D. W., and Gerhard, J. E., 1971, Jurassic salt, central Gulf of Mexico, and its temporal relation to circumgulf evaporites: American Association of Petroleum Geologists Bulletin, v. 55, p. 680–686.
Kostick, D. S., 1985, Soda ash and sodium sulfate, *in* Knoerr, A. W., ed., Mineral facts and problems: U.S. Bureau of Mines Bulletin 675, p. 741–755.
Kramer, D. A., 1985, Magnesium, *in* Knoerr, A. W., ed., Mineral facts and problems: U.S. Bureau of Mines Bulletin 675, p. 471–482.
Kühn, R., 1968, Geochemistry of German potash deposits, *in* Mattox, R. B., ed., Saline deposits: Geological Society of America Special Paper 88, p. 427–504.
Kunasz, I. A., 1980, Lithium in brines, *in* Coogan, A. H., and Hauber, L., eds., Fifth symposium on salt: Northern Ohio Geological Society, v. 1, p. 115–117.
Lefond, S. J., 1969, Handbook of world salt resources: New York, Plenum Press, 384 p.
Lerman, A., 1970, Chemical equilibria and evolution of chloride brines, *in* Morgan, B. A., ed., Fiftieth anniversary symposia—mineralogy and petrology of the upper mantle; sulfides; mineralogy and geochemistry of non-marine evaporites: Mineralogical Society of America Special Paper 3, p. 291–306.
Livingstone, D. A., 1963, Chemical composition of rivers and lakes, *in* Fleischer, M., ed., Data of geochemistry, 6th ed.: U.S. Geological Survey Professional Paper 440–G, 64 p.
Lyday, P. A., 1985a, Boron, *in* Knoerr, A. W., ed., Mineral facts and problems: U.S. Bureau of Mines Bulletin 675, p. 91–102.
——, 1985b, Bromine, *in* Knoerr, A. W., ed., Mineral facts and problems: U.S. Bureau of Mines Bulletin 675, p. 103–110.
——, 1985c, Iodine, *in* Knoerr, A. W., ed., Mineral facts and problems: U.S. Bureau of Mines Bulletin 675, p. 377–384.

Magraw, R. M., 1938, New Mexico sylvinite: Industrial and Engineering Chemistry, v. 30, no. 8, p. 861–864.
Matthews, R. D., and Egleson, G. C., 1974, Origin and implications of a mid-basin potash facies in the Salina salt of Michigan, *in* Coogan, A. H., ed., Fourth symposium on salt: Northern Ohio Geological Society, p. 15–34.
Morris, R. C., and Dickey, P. A., 1957, Modern evaporite deposition in Peru: American Association of Petroleum Geologists Bulletin, v. 41, p. 2467–2474.
Morse, D. E., 1985a, Salt, *in* Knoerr, A. W., ed., Mineral facts and problems: U.S. Bureau of Mines Bulletin 675, p. 679–688.
——, 1985b, Sulfur, *in* Knoerr, A. W., ed., Mineral facts and problems: U.S. Bureau of Mines Bulletin 675, p. 783–797.
Muffler, L.P.T., and White, D. E., 1969, Active metamorphism of Upper Cenozoic sediments in the Salton Sea geothermal field and the Salton Trough, southeastern California: Geological Society of America Bulletin, v. 80, p. 157–182.
Multhauf, R. P., 1978, Neptures's gift—a history of common salt: Baltimore, Maryland, Johns Hopkins University Press, 325 p.
Mussey, J. W., and Tyree, P. O., 1984, Geology and production of west Texas–type sulphur deposits: Society of Mining Engineers of the American Institute of Mining and Metallurgical Engineers, Preprint no. 84–380, 6 p.
Ochsensius, C., 1877, Die Bildung der Steinsalzlager und ihrer Mutterlaugensalze: Halle, Germany, C.E.M. Pfeffer, 172 p.
Olsen, P. E., 1980, The latest Triassic and Early Jurassic formations of the Newark Basin: New Jersey Academy of Science Bulletin, v. 25, p. 25–51.
Phillips, M., 1975, Cane Creek mine solution mining project, Moab potash operations, Texasgulf, Incorporated, *in* Fassett, J. E., ed., Canyonlands: Four Corners Geological Society Guidebook, 8th Field Conference, p. 261.
Powers, D. W., Lambert, S. J., Shaffer, S. E., Hill, L. R., and Weart, W. D., eds., 1978, Geological characterization report, Waste Isolation Pilot Plant (WIPP) Site, southeastern New Mexico: Albuquerque, New Mexico, Sandia Laboratories, SAND78–1596, v. I, 469 p.; v. II, 755 p.
Pressler, J. W., 1985, Gypsum, *in* Knoerr, A. W., ed., Mineral facts and problems: U.S. Bureau of Mines Bulletin 675, p. 349–356.
Raup, O. B., 1970, Brine mixing—an additional mechanism for formation of basin evaporites: American Association of Petroleum Geologists Bulletin, v. 54, p. 2246–2259.
Riley, J. P., and Chester, R., 1971, Introduction to marine chemistry: New York, Academic Press, 465 p.
Rittenhouse, G., 1967, Bromine in oil-field waters and its use in determining possibilities of origin of these waters: American Association of Petroleum Geologists Bulletin, v. 51, p. 2430–2440.
Robbins, J., 1986, Soda ash, what solution?: Industrial Minerals, no. 223, p. 39–61.
Rykken, L. E., 1976, Lithium production from Searles Valley, *in* Vine, J. D., ed., Lithium resources and requirements by the year 2,000: U.S. Geological Survey Professional Paper 1005, p. 33–34.
Schenk, P., 1967, The Macumber Formation of the Maritime Provinces of Canada: Journal of Sedimentary Petrology, v. 37, p. 365–376.
Schmalz, R. F., 1969, Deep-water evaporite deposition—a genetic model: American Association of Petroleum Geologists Bulletin, v. 53, p. 798–823.
Scruton, P. C., 1953, Deposition of evaporites: American Association of Petroleum Geologists Bulletin. v. 37, p. 2498–2512.
Searls, J. P., 1985, Potash, *in* Knoerr, A. W., ed., Mineral facts and problems: U.S. Bureau of Mines Bulletin 675, p. 617–633.
Smith, G. I., Jones, C. L., Culbertson, W. C., Erickson, G. E., and Dyni, J. R., 1973, Evaporites and Brines, *in* Brobst, D. A., and Pratt, W. P., eds., United States mineral resources: U.S. Geological Survey Professional Paper 820, p. 197–216.
Smith, H. I., 1938, Potash in the Permian salt basin (Texas and New Mexico): Industrial and Engineering Chemistry, v. 30, no. 8, p. 854–860.
Stewart, F. H., 1963, Marine evaporites, *in* Fleischer, M., ed., Data of geochemistry, 6th ed.: U.S. Geological Survey Professional Paper 440–Y, 53 p.
Stoessel, R. K., and Moore, C. H., 1983, Chemical constraints and origins of four

suggested groups of Gulf Coast reservoir fluids: American Association of Petroleum Geologists Bulletin, v. 67, p. 896–906.

Surdam, R. C., and Sheppard, R. A., 1978, Zeolites in saline, alkaline-lake deposits, *in* Sand, L. B., and Mumpton, F. A., eds., Natural zeolites; Occurrence, properties, uses: New York, Pergamon, p. 145–174.

Treesh, M. I., and Friedman, G. M., 1974, Sabkha deposition of the Salina Group (Upper Silurian) of New York State, *in* Coogan, A. H., ed., Fourth symposium on salt: Northern Ohio Geological Society, v. 1, p. 35–46.

Tripp, B. T., 1985, Industrial commodities in Utah: Utah Geological and Mineral Survey, Survey Notes, v. 19, no. 3, p. 3–8.

Truesdell, A. H., and White, D. E., 1973, Production of superheated steam from vapor-dominated geothermal reservoirs: Geothermics, v. 2, p. 154–173.

Urasov, G. G., and Polyakov, V. D., 1956, Salts of Kara–Boghaz–Gol: Priroda, no. 9, p. 61 [in Russian].

Usiglio, J., 1849, Analyse de l'eau de la Mediterranée sur le côtes de France: Annalen der Chemie, v. 27, p. 92–107, 172–191.

Van Houten, F. B., 1964, Cyclic lacustrine sedimentation, Upper Triassic, Lockatong Formation, central New Jersey and adjacent Pennsylvania: Kansas Geological Survey Bulletin, v. 169, p. 497–531.

Ver Planck, W. E., 1958, Salt in California: California Department of Natural Resources Bulletin 175, 168 p.

Weissberg, B. G., Browne, P.R.L., and Seward, T. M., 1979, Ore metals in active geothermal systems, *in* Barnes, H. L., ed., Geochemistry of hydrothermal ore deposits (2nd ed.): New York, John Wiley, p. 738–780.

White, D. E., 1974, Diverse origins of hydrothermal ore fluids: Economic Geology, v. 69, p. 954–973.

——, 1981, Active geothermal systems and hydrothermal ore deposits, *in* Skinner, B. J., ed., 75th anniversary volume: Economic Geology, v. 392–423.

White, D. E., Hem, J. D., and Waring, G. A., 1963, Chemical composition of subsurface waters, *in* Fleischer, M., ed., Data of geochemistry, 6th ed.: U.S. Geological Survey Professinal Paper 440-F, 67 p.

Withington, C. F., 1962, Gypsum and anhydrite in the United States, exclusive of Alaska and Hawaii: U.S. Geological Survey Mineral Investigations Resource Map MR-33, scale 1:3,168,000.

MANUSCRIPT ACCEPTED BY THE SOCIETY OCTOBER 27, 1988

Chapter 11

Oil shale

John R. Donnell
6035 South Milwaukee Way, Littleton, Colorado 80121

INTRODUCTION

Oil shale is a fine-grained sedimentary rock containing a large concentration of solid organic matter, called kerogen, that is in great part derived from aquatic organisms. Kerogen is not generally soluble in petroleum solvents but will yield considerable quantities of oil when subjected to destructive distillation at an optimum temperature of 500°C. Kerogen contains a high percentage of sapropelic material derived mostly from phytoplankton existing in open bodies of water. This contrasts with coal deposits that consist mostly of humic material derived from woody plants grown mainly in a paludal environment. Phytoplankton, particularly algae, may form tremendous quantities of organic material given an adequate supply of nutrients, including decomposed organic matter and chemical compounds derived from weathering and solution of mineral matter. Algae thrive in the zone of photosynthesis in a tropical or subtropical climate. Anoxic conditions in the bottom waters and the sediment preserve the organic remains of the phytoplankton, and the biochemical alteration by anaerobic bacteria forms insoluble kerogen.

Kerogen content in the Earth's crust is estimated at 6.5×10^{15} tons (United Nations, 1966). This kerogen content is about 1,000 times more than coal in the Earth's crust. Quality of oil shale is dependent on the percent of mineral matter mixed with kerogen in the sediment. Research on recovery of oil from oil shale indicates there is less energy recovered than expended in retorting oil shale, which will yield less than 30 liters per ton. Therefore, the volume of oil-shale resources worldwide that may be considered of possible economic interest, although extremely large, is only a small fraction of the 6.5×10^{15} tons of kerogen.

Oil shale may form in marine, brackish water, or nonmarine environments (Schlatter, 1969). Some of the more widespread low-grade deposits are formed in open-marine areas during regional transgressions over tectonically stable platforms. Other smaller but richer deposits may be formed in silled basins, flood plains, lagoons, and estuaries. However, most of the richer oil shales were formed in a lacustrine environment. The occurrence of oil shale has been reported in more than 50 countries located on all continents but the Antarctic (Figs. 1 and 2; United Nations, 1966).

MAJOR FOREIGN DEPOSITS

The Irati Formation of Permian age crops out in a narrow band 16 to 64 km wide that parallels the east coast of Brazil. It is exposed almost continuously for a distance of about 1,600 km from the northern border of Sao Paulo State to the Uruguayan border. The oil shale occurs in two zones, one about 6 m and the other 3 m thick. Each zone will yield about 76 liters of oil per ton. The upper sequence in many areas is under less than 30 m of overburden and is separated from the lower zone by about 8.5 m of barren rock. It is estimated that this deposit contains more than 800 billion barrels of oil.

An area of 250,000 km² of the Eromanga Basin in Australia is underlain by Cretaceous oil shales in the Toolebuc Formation. The major part of the deposit is located in Queensland, with smaller segments in New South Wales, South Australia, and Northern Territory. In much of the area, the oil-shale zone is more than 6 m thick, yields an average of 78 liters of oil per ton, and is anomalously high in concentrations of vanadium. The deposit has not been completely evaluated; it is estimated, however, that there is an in-place resource of more than a trillion barrels of oil. The oil shale in a large area near Julia Creek in north-central Queensland is under shallow overburden and thus lends itself to extraction by surface mining methods.

OTHER FOREIGN DEPOSITS

Several oil-shale deposits that are much smaller in areal extent and in total contained resource than the Toolebuc or Irati Formations have received considerable attention because of their present commercial importance or their possible future importance.

Middle Ordovician kukersite underlies about 50,000 km² in the Baltic area, including eastern Estonia and westernmost Russia in the vicinity of Leningrad. This marine oil shale has a maximum thickness of 3 m, an average value of about 170 liters per ton, and shale-oil reserves of 11 billion barrels. Oil shale has been produced from this deposit almost continuously since 1916.

At Fushun, Manchuria, an Oligocene oil-shale deposit that is more than 90 m thick has an average value of 57 liters of oil per ton and an in-place resource of 2 billion barrels of oil. This

---

Donnell, J. R., 1991, Oil shale, *in* Gluskoter, H. J., Rice, D. D., and Taylor, R. B., eds., Economic Geology, U.S.: Boulder, Colorado, Geological Society of America, The Geology of North America, v. P-2.

Figure 1. Significant oil-shale deposits of the world, exclusive of the United States. 1, Baltic Kukersite deposit; 2, Irati deposit; 3, Toolebuc deposit; 4, Fushun deposit; 5, Maoming deposit. Modified from Duncan and Swanson (1965).

oil-shale unit overlies a bituminous coal bed that is between 6 and 140 m thick. Since 1931 the oil-shale overburden has been mined to provide access to the underlying coal, and the oil shale processed to obtain oil.

During the Upper Jurassic Lias, a large shallow sea covered several hundred thousand km^2 of western Europe. Organic-rich rock that will yield an average of 38 to 57 liters of oil per ton was deposited in the sea. Most of the oil shale was removed by erosion; however, in northern Germany near Shandelah, extensive core drilling has delineated a mineable reserve of 2 billion tons of oil shale. Similar exploration around the eastern and southern rim of the Paris Basin in France has outlined large reserves of Liassic oil shale of comparable value to those in Germany.

American companies are aiding the Moroccan government in its attempt to make a detailed evaluation of a large deposit of Cretaceous oil shale near Timahdit in the Middle Atlas Mountains. A sequence of oil shale about 150 m thick will yield more than 38 liters of oil per ton. Within this sequence a 20-m-thick zone averages more than 76 liters per ton. It is estimated the deposit contains an in-place resource of 15 billion barrels of oil.

Japanese and American companies are funding investigations of a series of Tertiary oil-shale deposits paralleling the east coast of Queensland, Australia, north of Brisbane. The deposits are in lake beds formed in several north-trending grabens. Inplace resources in each deposit range from 1 to 3 billion barrels from mineable thicknesses of oil shale that average 76 to 95 liters of oil per ton.

Recent interest has also been shown in several other deposits, including those in Yugoslavia, Israel, and Jordan.

MAJOR DOMESTIC DEPOSITS

Oil shale in the Green River Formation (Eocene) was deposited in two large lakes. They each occupied an area of about 52,000 km^2 on the north and south flank of the Uinta Mountains in the western part of the United States during the early to late Eocene. These were initially fresh-water lakes that became increasingly saline. The alkaline waters provided the nutrients necessary for blue-green algae to flourish throughout most of the history of the lakes. The algae and other plant life formed kerogen that interlayered with other bottom sediments to form oil shale. At the depositional center of the lakes, a 600-m-thick sequence of oil shale was formed, containing varying concentrations of organic material. The eroded remnants of the Green River Formation now underlie an area of about 41,000 km^2 in northwestern Colorado, northeastern Utah, and southwestern Wyoming. Oil-shale beds 0.3 m thick in places will yield as much as 378 liters of oil per ton, and in Colorado near the depositional center of the Piceance Creek Basin, a continuous sequence of oil shale 300 m thick will yield an average of 114 liters of oil per ton. A total resource of 2 trillion barrels of oil is contained in the Green River Formation in sequences at least 5 m thick that will yield an average of more than 57 liters of oil per ton. During the highly saline phase of the lakes, sodium salts were precipitated out of the supersaturated brine; thick deposits of halite and trona were deposited in Wyoming, and halite and nahcolite were deposited in Colorado and Utah. Large quantities of shortite in Wyoming and dawsonite in Colorado are finely disseminated in the oil shale. Trona is now being mined and utilized for soda-ash production in Wyoming. The large deposits of nahcolite in Colorado have economic potential as a scrubbing agent in coal-fired powerplants, and alumina from dawsonite may be recovered as a by-product in oil-shale processing.

In the eastern and central United States, oil shale was formed during the latter part of the Devonian and early part of the Mississippian periods, within a large intracratonic sea that covered parts of 22 states (Conant and Swanson, 1965). Coarse

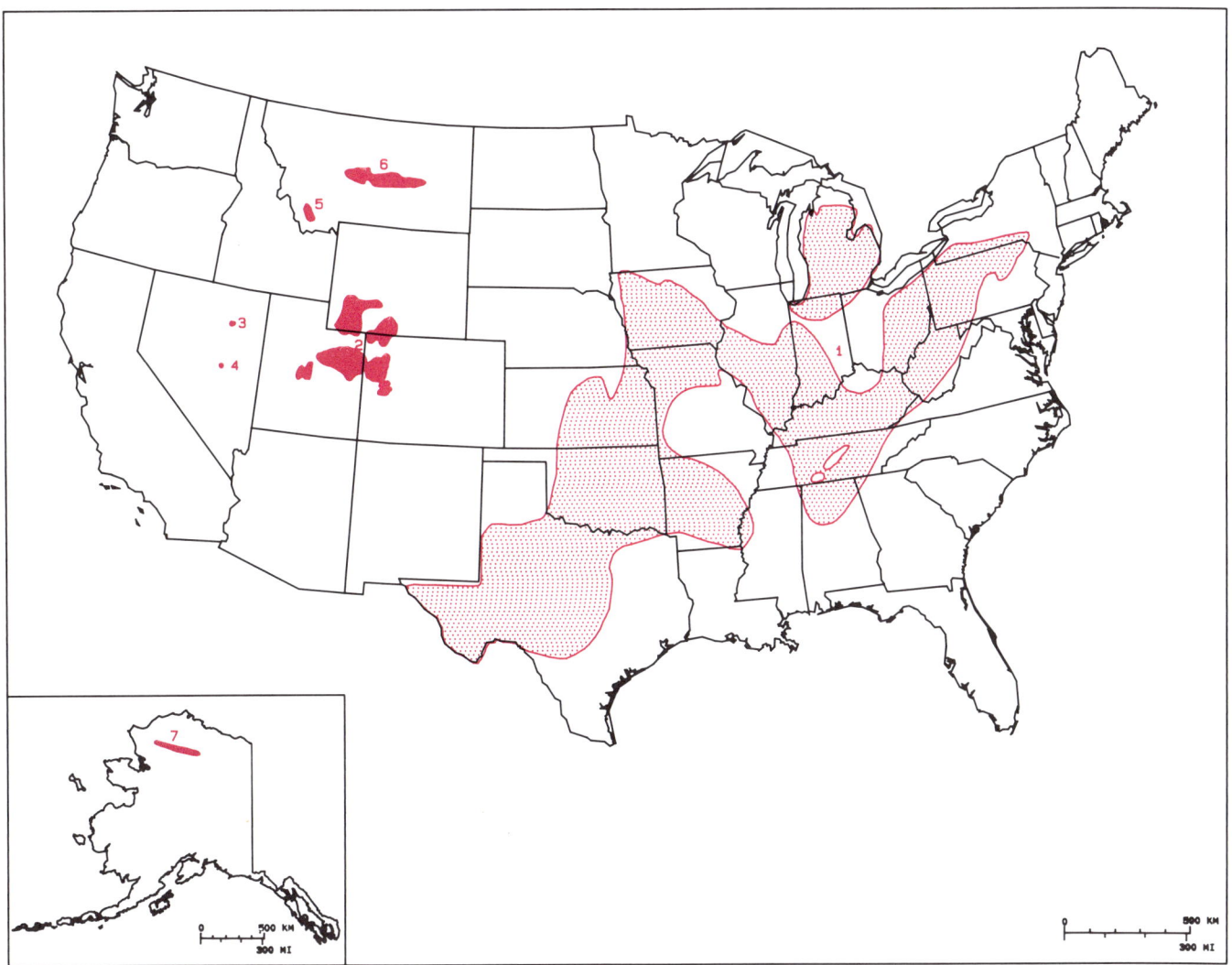

Figure 2. Oil-shale deposits of the United States. 1, Devonian and Mississippian deposits; 2, Green River deposits; 3, Elko deposit; 4, Woodruff deposit; 5, Dillon deposit; 6, Heath deposit; 7, Mesozoic deposits. Modified from Duncan and Swanson (1965).

sediment admixed with a small percentage of organic terrestrial debris was deposited along the eastern margin. This was derived from the large, high land mass that formed the eastern margin of the sea. Other land masses within and adjacent to the sea were more topographically subdued and furnished finer clastic and terrestrial matter to the sedimentary sequence. The humic material derived from land plants combined with sapropelic organic matter from phytoplankton and inorganic clay particles formed organic-rich shale deposits over an area of about 650,000 km^2. In areas where they have been evaluated, oil shales averaging more than 38 liters of oil per ton are estimated to contain a resource of 200 billion barrels (Janka and Dennison, 1979). The greater part of this resource has a 2.5 to 1 overburden to ore ratio in a 15,500 km^2 area of Kentucky, Ohio, Tennessee, and Indiana. The average Devonian shale contains as high a concentration of organic material as the average Green River shale; however, the Devonian shale will yield only 45 percent as much oil upon retorting because it is hydrogen deficient. A retorting method processing the Devonian shale in a hydrogen atmosphere at high pressure with controlled heating rates will produce 2.5 times as much oil as a conventional retorting method. Analyses show that many deposits of humic-rich organic material tend to concentrate trace metals. The Devonian oil shales in some areas contain abnormally high concentrations of uranium. If the demand for either oil or uranium greatly increases in the future with a resultant large product price increase, they could be produced as co-products.

OTHER DOMESTIC DEPOSITS

Several Tertiary lake basins in the western part of the United States contain oil shale. Assay samples from some of these beds show a high yield of oil. A 0.3-m thick bed in the Elko Formation of Eocene age, at Elko, Nevada, assayed 227 liters of oil per ton. A sample of the Oligocene lake beds near the town of Dillon, in southwestern Montana, yielded 136 liters of oil per ton. Other samples from the Kishenehn Formation of Oligocene age near Lake McDonald in Glacier Park, Montana, yielded as much as 140 liters of oil per ton. The rich oil-shale beds in all three of these areas are relatively thin and are generally separated by a sequence of beds lacking oil shale. The total in-place resource in each of these three known deposits is small, and therefore, shale-oil recovery would be difficult and expensive.

A unique, rich oil-shale marine deposit is found in the Jurassic Tigluckpuk Formation in the southern foothills belt north of the Brooks Range in Alaska. Initially deposited as a widespread thin organic-rich bed consisting mainly of the algae tasmanites, the bed was later severely deformed as a result of intrusive activity and repeated periods of thrust and normal faulting. The tasmanite is found in scattered exposures paralleling the Brooks Range for a distance of more than 480 km. Tasmanite at each locality occurs as a lenticular pod usually not exceeding 0.3 m^3 in volume. At many localities, the tasmanite will assay more than 378 liters per ton.

There are several marine black shale localities similar to but smaller in areal extent than the eastern and midwestern Devonian black shale deposit. The best known of these is the phosphate-bearing Phosphoria Formation of Permian age that covers parts of Idaho, Montana, Utah, and Wyoming. At some localities, parts of the Phosphoria that are at least 5 m thick will yield an average of 57 or more liters per ton. Unfortunately, an inverse relationship exists between the phosphate content and the oil yield.

The Devonian Formation near Eureka in central Nevada, the Mississippian Heath Formation near Lewistown in central Montana, and Jurassic black shales north of the Brooks Range in Alaska contain moderate concentrations of a humic-sapropelic mix of organic matter that will yield, in mineable thicknesses, an average of 38 to 76 liters of oil per ton (Desborough and others, 1981). In addition, they also contain abnormally high concentrations of V_2O_5, Zn, Se, Mo, and Ni. Based on 1980 prices, analyses of one sample of the Heath Formation contained trace metals worth $88 per ton. The average trace-metal value of a ton of oil from four samples of the Heath Formation, taken from widely spaced localities, is $33 per ton. The Woodruff Formation in central Nevada, and Triassic black shales in Alaska contain similar metalliferous concentrations. Analyses of a 30-m interval in two cores from the Woodruff showed an average value of $34 and $39 per ton. The Alaskan black shale has a lesser concentration of metals with a value of $24 per ton. The Alaskan black shale and the Woodruff Formation occur in fault blocks of limited areal extent and contain a very small shale-oil resource. The Heath Formation in Montana is more than 50 m thick and underlies a 7,000 km^2 area that is structurally uncomplicated. The Heath is estimated to contain a shale-oil resource of 180 billion barrels in zones that yield an average of more than 38 liters of oil per ton.

PAST OIL-SHALE DEVELOPMENT

Shale oil was produced as early as the 1350s from the Tyrol area in Austria. This oil was processed into a medicinal salve that is still being marketed. In 1838, France mined oil shale and from it produced lamp oil and town gas. As many as 13 countries mined oil shale in the interval between the mid-1800s and the 1960s. Scotland produced shale oil continuously from 1848 to 1964. Some countries ceased shale-oil production following the advent of crude oil production from wells in 1859, and nearly all countries stopped producing shale oil with the availability of low-cost crude oil from the Middle East in the early 1960s.

RECENT FOREIGN OIL-SHALE ACTIVITIES

Much of the present interest in oil-shale development is in countries other than the United States. Russia is producing about 90,000 tons of oil shale per day, mainly from the Tallin-Leningrad area. Eighty percent of the mined shale is direct feed for electrical power generation. About 28 thousand barrels of shale oil a day are being produced in China from installations at Fushun, Manchuria, and at Maoming in Canton Province. Much smaller quantities of oil shale are mined for the generation of electrical power and the spent shale residue used in the making of cement at Dortenhausen, West Germany. About 1,000 tons of oil shale a year are mined in France and transported to a plant near Innsbruck, Austria, where it is converted to a medicinal salve.

Several countries with inadequate domestic supplies of crude oil are proceeding with plans for oil-shale development. Yugoslavia, Israel, and Jordan are actively researching retorting methods to extract oil from their deposits. Japanese companies have an option to develop a large Tertiary lacustrine deposit along the east coast of Queensland, Australia. Brazil has been proceeding in orderly fashion, since the early 1970s, in developing the Irati deposit. Petrobras, the national oil company of Brazil, has produced as much as a thousand barrels a day on a pilot basis in a Petrosix retort, with a capacity of 2,200 tons of shale a day. Petrobras has produced as much as a half-million barrels in a single sustained run. A larger scale retort is now under construction.

About a billion tons of oil shale has been produced worldwide to date. Ninety percent of the production is from Russia and China. Most of the oil shale is converted to oil; however, sizable quantities are used as direct feed in power plants, and smaller quantities are converted to natural gas and petrochemicals.

DOMESTIC OIL-SHALE ACTIVITIES

Prior to the drilling of the Drake oil well in Titusville, Pennsylvania, in 1859, small quantities of lamp oil and illuminating gas were produced from low-grade oil shales mined in the eastern United States. With the advent of adequate supplies of crude oil produced from wells, mining of oil shale in the United States ceased. A renewed period of oil-shale activity began in the early 1920s, stimulated by the high price of petroleum combined with the fear of a shortage of crude oil. The Catlin plant in Elko, Nevada, produced several thousand barrels a day during this period. The discovery of the East Texas oil field in 1926 flooded the country with low-cost oil, making shale oil noncompetitive. Interest in oil shale was renewed during World War II by concern for an adequate supply of petroleum for our military.

During the period 1945 to 1973, the U.S. Bureau of Mines and several oil companies were active in northwestern Colorado developing mining and retorting techniques on a pilot scale. In 1973 the federal government offered several 5,000-acre oil-shale tracts for lease in Colorado, Utah, and Wyoming. The timing of the lease offer coincided with the Middle East oil embargo, and as a consequence of this, the tracts were leased for a bonus bid of about a dollar a square foot. Interest remained high through the 1970s in the Colorado-Utah area with the addition of several more pilot projects. Oil-shale activity diminished during the recession of 1982 to 1983, and essentially ceased when the price of oil fell below $15 a barrel in 1986.

An American company (Unocal) has spent more than $600 million in the construction of a plant built to produce 10,000 barrels of shale oil a day from the Green River Formation in northwestern Colorado. The federal government agreed to purchase upgraded shale oil from Unocal for $42.50 per barrel up to a total of $400 million. Construction of the plant was completed in 1983; however, various unanticipated production problems have thus far prevented commercial development. The Unocal operation is the only oil-shale project presently active in the United States.

FUTURE DOMESTIC SHALE-OIL PRODUCTION POTENTIAL

The United States has not been self-sufficient in petroleum products since 1970. However, we have been consuming increasing amounts of petroleum products each year. This culminated in an average consumption of about 17 million barrels per day in 1977, of which 47 percent was imported. The increasing price of oil in the late 1970s and early 1980s, compounded by the economic recession in the early half of the 1980s, resulted in a concerted worldwide energy conservation effort. This lowered the total demand for crude oil and petroleum products in the United States to 15 million barrels a day. The higher price also stimulated exploration, which increased domestic petroleum production and resulted in a lesser dependence on foreign imports.

The worldwide cartel of oil-producing countries, OPEC, has for years operated under rigid production quotas that in effect determined the world price of petroleum. Recently, in the first half of 1986, the OPEC group greatly exceeded its production, glutting the world market, and causing a precipitous drop in the price of crude oil. Oil that was selling for as much as $35 a barrel prior to 1986 was priced between $10 and $13 a barrel in mid-1986. This abnormally low price has caused many small companies to file for bankruptcy, larger companies to curtail exploration, domestic production to decline, and consumption to increase. Oil imports in the United States that were about 30 percent of total demand in 1985 approached 40 percent in mid-1986 and are predicted to be more than 50 percent by the year 2000.

The oil-shale deposits of the United States, particularly the Green River oil shales, are a known secure resource of oil for many years in the future. Under the present price structure, no synthetic fuels industry can be initiated. However, if prices escalate to $35 a barrel or higher, an industry may be established. The conventional extraction approach is more similar to a mining extraction operation than to that of typical oil production. Front-end costs involving large capital expenditures for mining, crushing, and retorting equipment exceed several billion dollars for an installation producing 50,000 or more barrels per day. Payout for an oil-shale project will be ten or more years, in contrast to a 3- or 5-year payout time for a typical oil field.

True in situ recovery of shale oil is more nearly akin to crude oil recovery. Capital expenditures are far less than those for a conventional oil-shale project; therefore, payout of costs will be in a relatively short period of time. Most true in situ concepts involve conventional fracturing techniques, similar to oil-well fracturing, between the heat input well and the product recovery holes. This technique has not been satisfactory in establishing communication through fracturing. One method used in a thin, rich sequence of oil shale under a thin cover of overburden in the southeastern part of the Uinta Basin, in Utah, has proved to be fairly successful. The process involves drilling and blasting the ore zone with sufficient explosive to rubbleize the oil shale (Lekas, 1981). The rubble is ignited through wells drilled in one end of the horizontal retort and the liquid product collected at the other end. Thus far, shale-oil recovery has averaged about 50 percent of the resource contained in retorts under less than 20 m of overburden. An area containing several billion barrels of oil and under less than 30 m of overburden along the southern and eastern edge of the Uinta Basin is amenable to this extraction method. The ore zone is at least 3 m thick and averages more than 87 liters per ton.

The western margin of the Piceance Creek Basin, Colorado, and the Green River Basin, Wyoming, contain smaller oil-shale deposits with similar characteristics. A small industry in these areas based on the rubbleizing in situ approach could stimulate interest in research and development in other shale-oil recovery processes that may be applied to thicker, richer oil-shale deposits, thus leading to a significant shale-oil supplement to our existing domestic crude oil supply.

PROCESSING METHODS

The conventional approach to shale-oil extraction involves mining, crushing, and heating the shale in a large surface retort. An alternative method, modified in situ, involves underground mining of a room to create void space, rubbleizing the oil shale in the roof to fill the void space, and retorting the rubble in place by heat applied through wells drilled into the top of the rubble pile. The most simple, least expensive, and thus far the most ineffective extraction method is true in situ. This does not involve mining or crushing but does involve detonating explosives in shot holes within the ore zone to create communication between heat input holes for retorting and holes drilled to collect the liquid product.

The shale oil derived from any of the extraction processes is high in sulfur and nitrogen and deficient in hydrogen. Before being accepted as a feedstock at most refineries, the shale oil must be upgraded by the addition of hydrogen and the removal of nitrogen and sulfur, which are recovered as the by-products ammonia and elemental sulfur. The upgraded oil is a premium product greatly desired by any refinery.

REFERENCES

Conant, L. C., and Swanson, V. E., 1961, Chattanooga shale and related rocks of central Tennessee and nearby areas: U.S. Geological Survey Professional Paper 357, 91 p.

Desborough, G. A., Poole, F. G., and Green, G. N., 1981, Metalliferous oil-shales in central Montana and northeastern Nevada: U.S. Geological Survey Open-File Report 81–121, 14 p.

Duncan, D. C., and Swanson, V. E., 1965, Organic rich oil-shale of the United States and the world land areas: U.S. Geological Survey Circular 523, 30 p.

Janka, J. C., and Dennison, J. M., 1979, Devonian oil-shale a major energy resource, *in* Symposium Papers Synthetic Fuels from Oil-shale: Chicago, Illinois, Institute of Gas Technology, p. 21–116.

Lekas, M. A., 1981, The geokinetics horizontal in situ retorting process, *in* 14th Oil-shale Symposium Proceedings: Golden, Colorado School of Mines Press, p. 146–153.

Schlatter, L. E., 1969, Deposition, formation, and classification of oil shale, *in* Selected Papers U.N. Oil-shale Symposium: Cameron and Jones, Inc., v. 2, p. 44–49.

United Nations, 1966, Utilization of oil-shale progress and prospects: United Nations publication 67.11.B.20, 112 p.

MANUSCRIPT ACCEPTED BY THE SOCIETY FEBRUARY 12, 1989

Chapter 12

Other selected industrial minerals

Donald A. Brobst
2268 Wheelwright Court, Reston, Virginia 22091

INTRODUCTION

In the context of this volume, other industrial minerals include some of those not encompassed in the two previous sections on phosphate and evaporites, and brines. The term industrial minerals may be defined in general terms as any rock, mineral, or other naturally occurring substance of economic value, but generally exclusive of metallic ores and mineral fuels. Some anomalies arise in this use of the term. Some materials such as bauxite, chromite, and ilmenite are used both as ores of metals and sources of important nonmetallic products and are commonly included with industrial minerals. Industrial minerals also generally include some manufactured products such as abrasives, cement, lime, and refractory materials. The list of some industrial minerals that have had or probably will have economic value in the United States is given in Table 1.

It is impossible to describe the geology of all industrial minerals in this chapter. Therefore, I will first discuss such commodities as limestone and dolomite, sand and gravel, and clays, because of their widespread production, use, and large total annual value. Selected other commodities, chosen because they are examples of minerals important to society now or possibly in the future, will then be discussed. The discussion of each industrial mineral commodity in this chapter is in the descending order of the value of its domestic production in 1985. Asterisks before the mineral and rock commodities in Table 2 indicate those that are discussed in this chapter. Table 2 also shows the estimated average value in dollars per metric ton at the mine or plant. The total value of the commodities listed is more than $16 billion for 1985, a sum nearly three times the value of domestically produced metallic ores.

The discussion that follows is focused on the geological aspects (mineralogy, physical and chemical properties, occurrence and distribution of deposits) of industrial minerals that make them economically suitable and available for their many uses. The location and distribution of occurrences, deposits, and districts of some selected industrial minerals are shown on Plates 3C and 3D.

Some general references that summarize the geology, uses, technology, and economics of various commodities are useful sources of information on industrial minerals, including those discussed here. The American Institute of Mining, Metallurgical, and Petroleum Engineers volume, *Industrial Minerals and Rocks* (Lefond, 1983), has information on more than 50 commodities and contains a section by R. B. Hoy (p. 393–414) on world sources of information on industrial minerals. The U.S. Geological Survey report, *United States Mineral Resources* (Brobst and Pratt, 1973), emphasizes the geology and resources of many nonmetallic as well as metallic commodities. *Mineral Facts and Problems,* a bulletin of the U.S. Bureau of Mines which is revised every five years, is another source of information on economic and technical aspects of mineral commodities. The latest comprehensive summary of many industrial minerals by Harben and Bates (1984) also contains many citations to major deposits in the U.S. and other nations.

The work of many authors provided the information summarized here. The contributions from all sources are gratefully acknowledged, as are the reviews of this manuscript by C. E. Brown and S. H. Patterson.

In the past, the study of the geology of industrial minerals generally lagged behind that of the metallic commodities. Earlier, many industrial minerals operations were relatively small and supplied local markets, and companies were not pressed to explore: resources* were abundant, and converting them to reserves was a relatively simple task. Thus, only limited interest in research on topics related to industrial minerals was stimulated. More recently, research on the subject has increased.

Gillson's (1960) address to the Society of Economic Geologists highlighted some problems for research on asbestos, barite, clays, fluorspar, lime, cement, pegmatites, and sand and gravel. Problems for further research were specifically discussed by many authors in the volume *United States Mineral Resources* (Brobst

---

*A resource is now defined by the U.S. Geological Survey and the U.S. Bureau of Mines as a concentration of naturally occurring solid, liquid, or gaseous material in or on the Earth's crust in such a form and amount that economic extraction of a commodity from the concentration is currently or potentially feasible. A reserve is that part of an identified resource that meets the minimum physical and chemical criteria related to current mining and production practices that can be extracted or produced at the time of determination (U.S. Bureau of Mines and U.S. Geological Survey, 1980).

Brobst, D. A., 1991, Other selected industrial minerals, *in* Gluskoter, H. J., Rice, D. D., and Taylor, R. B., eds., Economic Geology, U.S.: Boulder, Colorado, Geological Society of America, The Geology of North America, v. P-2.

and Pratt, 1973). From these sources alone it is evident that much research with direct application to the field of industrial minerals remains to be done.

Difficult problems with industrial mineral production lie in the environmental and personal health sectors and in zoning and environmental regulations, which already are restricting mining in many places on both public and private lands, especially near metropolitan areas. The cost of the search for new deposits in more remote areas and the additional transportation costs will surely be reflected in the price of future materials delivered to the market. The rough correlation between the value per ton and the abundance of geologic availability for a given commodity that has been obtained for years probably will not continue indefinitely into the future. Unless the public is aware of the basic scientific and economic problems and the trade-offs necessary to maintain the "good life," some economic pinches ahead may be neither avoided nor blunted.

LIMESTONE AND DOLOMITE

Limestone and dolomite are among the principal industrial minerals in terms of domestic production and value annually. They are the basic raw materials of the construction industry, as crushed stone, aggregate, cement, lime, and building stone. In addition, they and their derivatives are used as fillers, fluxes, refractories, abrasives, soil conditioners, glass raw materials, and ingredients in many chemical processes. Data from the U.S. Bureau of Mines (1986) suggest that more than one billion tons of limestone and dolomite valued at more than $9 billion were produced domestically for all uses in 1985. In recent years, 70 to 75 percent of all stone mined or quarried in the U.S. was limestone or dolomite.

The literature on the geology and commercial production of limestone and dolomite is extensive. The review by Carr and Rooney (1983) contains nearly 400 references to the geology, mining, and uses of limestone and dolomite. Graf and Lamar (1955) described the properties of calcium and magnesium carbonates and their bearing on the uses of these rocks. The annotated bibliographies of Davis (1957) on high-magnesium dolomites and Gazdik and Tagg (1957) on high-calcium limestone deposits also are informative.

Limestone and dolomite are sedimentary carbonate rocks composed principally of the minerals calcite and dolomite. Aragonite has the same composition as calcite, but it has a different crystal structure that reverts to that of calcite in time. Aragonite is thus economically important only in younger deposits of shells and oolites. Minor amounts of siderite and ankerite are associated with limestone and dolomite. The most common impurities in carbonate rocks are clay minerals, chert, and organic matter.

Economically important limestones formed in relatively shallow marine environments by the interaction of seawater and organisms. Limestones that formed in low-energy environments are more likely to be adulterated with clay impurities than those formed in high-energy zones. Post-depositional alterations and modifications are common. Most dolomite is now considered to be an alteration of calcium carbonate sediments by hypersaline brines, although some dolomite might be formed by direct precipitation from seawater. The almost pure dolomite strata in southern Michigan, northern Illinois, Indiana, and Ohio are examples of rocks derived by the brine process.

The explosion of knowledge and understanding of the deposition and diagenesis of carbonate rocks in recent years has led to considerable discussion about the classification of these rocks. Most classifications are based on composition (mineralogy, types of fossils or grains, and chemical constituents) and texture (both depositional and post-depositional features). Compositional criteria are based on variations in content of calcite, dolomite, and noncarbonate materials. Such categorization is insufficient, however, for many industrial requirements. In many uses, limestone or dolomite can be substituted equally well, but certain applications have special needs that are best expressed in terms of chemical compositions that specify the quantity of CaO or $CaCO_3$ or $MgCO_3$, as well as a maximum percentage of allowable impurities. Ultrahigh-calcium limestones contain more than 97 percent $CaCO_3$, high-calcium limestones contain more than 95 percent $CaCO_3$, and high-purity carbonate rock contains more than 95 percent combined $CaCO_3$ and $MgCO_3$. High-magnesium dolomite contains more than 43 percent $MgCO_3$; theoretically pure dolomite contains 45.7 percent $MgCO_3$.

Classification of dolomite leads to problems not found with limestone, especially when the textures of the original rock are obliterated. In such cases, classifications based on crystal size may be required.

TABLE 1. SELECTED OTHER INDUSTRIAL MINERALS*

Asbestos	Meerschaum
Barite	Mica
Cement	Scrap
Clays	Sheet
Bentonite–fuller's earth	Olivine
Kaolin–halloysite	Perlite
Miscellaneous clay	Pumice and pumicite
Refractory or fire clay	Quartz crystals
Corundum	Sand and gravel
Diamonds, industrial	Staurolite
Diatomite	Stone
Feldspar	Crushed
Fluorspar	Dimension
Garnet	Strontium minerals
Gems	Sulfur
Graphite	Talc and pyrophyllite
Helium	Tripoli
Kyanite, sillimanite, and	Vermiculite
andalusite	Wollastonite
Lime	Zeolites
Limestone and dolomite	Zirconium and hafnium minerals

*Exclusive of phosphate rock, the products of evaporites and brines and some metallic ores described elsewhere in this volume that also have uses as industrial minerals, such as bauxite, ilmenite, and chromite.

TABLE 2. ESTIMATED VALUE, IN U.S. DOLLARS, OF SOME INDUSTRIAL MINERALS PRODUCED IN THE UNITED STATES IN 1985

	Average value* (dollars per metric ton)	Total value (in million dollars)
†Limestone and dolomite	–	9,398
Cement	58.92	
Crushed stone	4.56	
Lime	55.55	
Evaporite and brine products	–	2,875
†Sand and gravel	3.30	2,400
Phosphate rock	25.80	1,200
†Clays	–	1,100
Miscellaneous clays	4.67	
Kaolin	91.50	
Ball clay	38.52	
Refractory clay	17.30	
Bentonite	31.13	
Fuller's earth	61.39	
Sulfur	115.14	1,100
Industrial silica sand	14.19	390
Gypsum	9.13	120
Diatomite	229.90	133
Helium	37.50§	65
Lightweight aggregates	–	54
Vermiculite	112.30	
Pumice and pumicite	9.90	
Perlite	37.40	
†Barite	30.80	25
†Talc and pyrophyllite	20.90	24
†Feldspar	37.70	24
Peat, agricultural	26.97	21
†Asbestos	398.20	20
†Fluorspar (1984 value)	–	12
Acid grade	190.30	
Metspar	137.00	
Scrap Mica	–	6
Scrap and flake	48.40	
Garnet	197.00	6
†Kyanite	100.00	–
†Wollastonite	–	–
†Zeolites, natural	–	–
	Total	18,973

Data from U. S. Bureau of Mines (1985, 1986)
*Value calculated from dollars per short ton. Most industrial mineral commodities are sold by the short ton in the United States. For some commodities, special grades bring higher prices.
†Indicates industrial mineral discussed in this chapter.
§Price per 1,000 cubic feet.

Limestone and dolomite ranging in age from Precambrian to Holocene are widespread in the United States, although most of the deposits were formed in the last 500 m.y. (Phanerozoic time). Nearly every state has deposits geologically adequate for construction materials and other uses that do not require chemically pure rocks. Generally, only a small portion of the limestone and dolomite, however, is ultrahigh- or high-calcium limestone or high-magnesium dolomite. Minable carbonate formations occur most extensively in the relatively flat-lying rocks of the midcontinent, in Kentucky, Tennessee, Missouri, and the states adjacent to the Great Lakes. In the folded and faulted parts of the Appalachians and Rocky Mountains, minable limestone and dolomite are more restricted in extent.

The greatest quantities of high-calcium limestone occur in widely distributed formations of Mississippian age. The greatest quantities of high-magnesium dolomite occur in formations of Ordovician and Silurian age. Smaller deposits of high-calcium limestone and high-magnesium dolomite occur in formations of other ages. Some Ordovician formations contain deposits exceptionally high in calcite, such as the Holston Limestone of East Tennessee, the Valentine Member of the Curtin Formation of central Pennsylvania, and the New Market Limestone of western and southwestern Virginia. The Tomstown Formation in the eastern panhandle of West Virginia contains much-exploited deposits of both high-purity limestone and dolomite (Ericksen and Cox, 1968). The carbonate rocks of the Niagara Group in western Ohio have been the chief domestic source of high-magnesium dolomite.

Hubbard and Ericksen (1973) reported that the United States is self-sufficient in limestone and dolomite resources. Reserves, however, of the chemically purest varieties are limited and are decreasing because some deposits are mined out; users continue to establish more stringent specifications, and zoning regulations prevent or restrict mining operations in populous areas. Most mining operations are now open-pit, but in the future, more selective mining will be done and underground operations will increase.

Crushed stone

About 75 percent of the nearly 1 billion tons of material marketed in 1985 as crushed stone (valued at about $3 billion) came from carbonate rocks, according to the U.S. Bureau of Mines (1986). The remaining crushed stone products are granite (14 percent), traprock (9 percent), and sandstone (2 percent). About 69 percent of the crushed stone was used as construction aggregate, mostly in highway construction and maintenance, and the remaining 31 percent went to a broad series of uses, including cement and lime manufacture (12 percent). Carbonate rocks that are to be crushed for construction purposes must be clean, hard, strong, and free of friable material. The stone should break into irregular, mostly equidimensional fragments rather than into thin, platy, elongate fragments. Road metal requires resistance to abrasion and low porosity to prevent breakdown in freezing and thawing. Concrete aggregate must be low in reactive materials such as alkalies, soluble sulfides, and certain types of silica, especially chert (Dunn and Ozol, 1962; Kneller and others, 1968).

Cement

Portland cement accounts for more than 95 percent of all cement manufactured in the U.S. in recent years. Domestically produced portland cement in 1985 was 69 million tons in a sum of 72 million tons of all types of cement having a total value of $4.4 billion. In 1985, the major portland cement–producing states, in descending order, were Texas, California, Pennsylvania, Michigan, Missouri, and Alabama.

In the manufacture of cement, limestone rich in calcium and low in MgO (less than 5 percent) is mixed with silica and aluminum and burned to form a clinker. The ideal raw material is an argillaceous limestone that contains 60 to 80 percent $CaCO_3$ and 15 to 30 percent clay material, which contains the necessary silica, alumina, and iron oxide. The classic American cement rock is that of the Jacksonburg Formation of Ordovician age in eastern Pennsylvania. Also, freshwater marls may be used as sources of calcium carbonate. Beds of oyster shells are dredged in the Gulf of Mexico and San Francisco Bay as sources of carbonate. Silica may be derived from chert in the limestone or may be added as sand from other sources. Generally the silica-alumina balance is achieved by addition of aluminum silicates as clay or shale, although many other materials have been used, such as bauxite (in high-Al cement), staurolite and aluminum dross, volcanic materials, blast-furnace slag, and laterite.

Lime

Lime is a versatile product made by calcining limestone to drive off carbon dioxide and leave a residue of calcium oxide or calcia (CaO), also known as quicklime; if water is added, the product is calcium hydroxide ($Ca[OH]_2$) or slaked lime. Lime is generally made from high-calcium limestone that has less than 5 percent MgO. The lime product made from a high-magnesium limestone is referred to as dolomitic lime.

The domestic chemical and metallurgical process industries now use more than 90 percent of the lime as flux, acid neutralizer, causticizing agent, flocculent aid, hydrolizer, bonding agent, absorbent, and raw material. The value of lime for most purposes is based on the available lime-content (CaO) in the total oxide-content of dolomitic lime. The principal lime-producing states are Ohio, Pennsylvania, Missouri, Kentucky, Alabama, and Texas, each of which produced over 1 million tons in 1985, and together, these states accounted for 56 percent of the domestic yield.

SAND AND GRAVEL

Sand and gravel have two major categories of use: about 97 percent is used in construction, and 3 percent is used in other specialized industrial applications, such as glass-making and foundry sands. In construction, sand and gravel are used in concrete (44 percent), roadbases and coverings (25 percent), asphaltic concrete aggregates and other bituminous mixtures (15 percent), construction fill (11 percent), and other uses (5 percent). In 1985 an estimated 725 million tons were produced domestically that had a value of $2.4 billion. The unit value is low (Table 2), requiring production close to markets that are concentrated around large, growing, urban areas and temporarily around the sites of large construction projects such as dams, bridges, highways, and power plants. Transportation is a major factor in cost. Sand and gravel are produced in every state. In 1985, California, Texas, Michigan, Arizona, Ohio, Alaska, New York, Colorado, Illinois, and Minnesota yielded 52 percent of the total production.

Sand and gravel are composed of different amounts of minerals and rocks and, therefore, have various chemical compositions. Silica is the major constituent of most commercial sands; lesser amounts of feldspar, mica, iron oxides, and various heavy minerals are common. Most users have their own specifications for chemical composition, grain size, and other physical characteristics. Goldman (1961) suggested that the ideal sand and gravel deposit should contain 60 percent gravel and 40 percent sand. This combination provides ample coarse material to crush for roadbase or bituminous aggregate and sand in correct sizes and proportions to use in concrete.

Sandstones and limestones are satisfactory source materials for sand and crushed rock because they are generally dense and hard. Shales are generally poor materials because they commonly are soft, lightweight, and absorptive. Most igneous rocks are satisfactory because they normally are dense, hard, and tough. Marble, quartzite, and gneiss ordinarily are tough and durable.

The major geologic environments that contain sand and gravel deposits of economic importance are: stream deposits in channels or on flood plains and terraces, alluvial fans, glacial and fluvial-glacial deposits on outwash plains, glacial kames and eskers, bench and dune deposits associated with large lakes, and moraines, dredge tailings, marine beach deposits, and rarely, some pre-Quaternary formations (Goldman and Reining, 1983, p. 1153–1155). The sand and gravel deposits of glacial origin in the United States are associated with the continental ice sheets that moved south out of Canada during Pleistocene time. The southern limit of continental glacial deposits is shown on Plate 3D.

Geologic environments favorable as sources of sand and gravel differ in the various regions of the United States: northeastern states, glacial outwash and till (Plate 3D); Atlantic and Gulf Coast states, marine terraces; southeastern and south-central states, river deposits; mountain states, stream and fan deposits; Great Plains, stream deposits; and Pacific Coast states, alluvial fans, river deposits, river terraces, marine beaches, and dunes (Yeend, 1973, p. 562). Resources of sand and gravel within the U.S. are huge, but the amount available, or potentially available, in a specific market at a competitive price is a complex matter that is now and will be more and more controlled by economic rather than geologic factors.

CLAYS

Clays are naturally occurring consolidated or unconsolidated materials composed of very fine particles of clay minerals that are principally layered hydrous aluminum (or magnesium) silicates that may contain various amounts of iron, potassium or sodium, or other ions. The major component of most clay deposits is one of the following clay minerals: kaolinite, halloysite, montmorillonite, palygorskite (attapulgite), and illite. Some deposits contain minor amounts of one or more of these clay minerals. All clay deposits contain some other minerals, which generally include some of the following: quartz, feldspar, mica, carbonates, and iron oxides. The term clay is also applied as a rock term for any earthy, natural, fine-grained argillaceous material, such as clay, shale, argillite, and some argillaceous soils. The term clay also is used as an indication of extremely small particle size, the maximum of which is differently designated in the various scales for particle size. Wentworth (1922) defined clay as material finer than about 4 micrometers, but many soil scientists and mineralogists concede the maximum size for clay particles as 2 micrometers.

Clays rank high in annual tonnage produced and value (Table 2) among the industrial minerals, according to data published by the U.S. Bureau of Mines (1986). In 1985, domestic clay production was more than 40 million tons. As industrial minerals, clays are a complex group that consists of several mineral commodities, each with somewhat different mineralogy, geologic occurrence, technology, and uses. The U.S. Bureau of Mines has been reporting six categories of clay based more on the product than the nature of the raw material. These include, in decreasing order of production in 1985: (1) miscellaneous or common clay, (2) kaolin, (3) bentonite, (4) fuller's earth, (5) fire clay, and (6) ball clay.

Miscellaneous clays

The miscellaneous clays compose the largest annual tonnage of all categories of clay products reported; 27 million tons were produced in 1985. Their value is only a few dollars per ton and generally much less than that of other clays (Table 2). Illite is the dominant mineral in the miscellaneous clays, especially in the shales. In 1985, about 95 percent of these clays produced domestically were fired and used for construction purposes. These products include many molded items, such as structural and fire bricks, drain tile, flue tile, vitrified pipe, conduit, pottery, stoneware, and roofing tiles. Many tons of these clay materials are bloated by firing to form lightweight aggregates used in the manufacture of portland cement. Other uses do not require firing of the finished product. Some clays are used as filler in paint and other products. Shale and clay may be used to pack dynamite in blast holes and for plugging oil and gas wells no longer in use.

The many uses of miscellaneous clay require a broad range of physical properties that relate to plasticity, green strength (strength of the clay when wet or in a plastic state), dry strength (strength after dried), drying and firing shrinkage, vitrification range, and fired color.

The raw materials of the miscellaneous clays occur in many types of rocks ranging in age from Precambrian to Holocene. The rocks include glacial clay, soils, alluvium, loess, shale, fresh and weathered schist, slate, and argillite. The most common mineral in many deposits is a member of the mica group; generally the dominant one is illite, sericite, or one of the normally coarser crystalline micas such as muscovite or biotite, mixed layer varieties, and chlorite. Some raw materials used in this category actually contain more quartz and other detrital materials than clay minerals.

Miscellaneous clays are produced in virtually every state largely because of the wide distribution of useful clay deposits and the competitive costs, which limit the shipment of the finished products to a maximum radius of 200 miles from their point of origin. In recent years, Texas and North Carolina have been leading producing states, along with Alabama, California, Georgia, Illinois, Indiana, Iowa, Louisiana, Maryland, Michigan, Missouri, New York, Ohio, Pennsylvania, South Carolina, and Virginia, all of which produce about 1 million tons annually.

Kaolin, halloysite, ball clay, and refractory clay

The clay products marketed as kaolin, halloysite, ball clay, and refractory or fire clay have the common property of consisting chiefly of minerals of the kaolinite group, generally kaolinite or its close relatives, halloysite or meta-halloysite. The other clay minerals of this group, nacrite and dickite, rarely occur in economic deposits.

Kaolin. Kaolin was first produced in China centuries ago and named "kauling," meaning high ridge, for the hill that was mined near Jauchau Fu. The term kaolin is now used for the group of clay minerals already discussed; as a rock term, which implies more than one mineral present; as the name for the industrial mineral commodity; and as a word synonymous with china clay. The definition for commercial kaolin by Patterson and Murray (1983, p. 612) that follows is commonly used. Kaolin is a clay consisting substantially of pure kaolinite, or related minerals, that is naturally or can be beneficiated to be white or nearly white, will fire white or nearly white, and is amenable to beneficiation by known methods to be suitable for use in whiteware, paper, rubber, paint, and similar uses.

The production of kaolin has a long history in the U.S. The sedimentary kaolins of the coastal plains in Georgia and South Carolina have been known and worked for ceramic uses since Colonial times. Georgia and South Carolina have ranked first and second nationally in kaolin production for more than a century. Although old statistical records are incomplete, Kesler (1956) estimated that 2.7 million tons of kaolin had been produced in Georgia prior to 1932. By 1980, annual production in Georgia had risen to more than 5.5 million tons, and that in South Carolina to more than 540,000 tons. Production from Georgia and South Carolina makes the United States the world's leading producer of kaolin.

The paper industry uses nearly 50 percent of the kaolin production. Much of the paper in magazines with colored pictures contains as much as 30 percent kaolin to impart smoothness, gloss, brightness, opacity, and printability. The rubber industry uses kaolin as filler and extender in natural and synthetic rubber. Kaolin for ceramics now consumes only a small part of the annual production, which goes into whiteware, wall tile, insulators, refractories, and some face brick where white color is desired.

Kaolin and its associated clays are common in tabular lenses and discontinuous beds in sedimentary rocks of Cretaceous age or younger. The beds and lenses vary greatly in shape and extent, but deposits as much as 18 m thick and more than 1.5 km in lateral extent are known. Deposits of this type are abundant in the Georgia–South Carolina belt, extending from Twiggs County northeastward to Lexington County, central South Carolina. These deposits also contain the nation's principal reserves and resources of kaolin. Most of the mining is currently concentrated in Twiggs, Wilkinson, and Washington Counties, Georgia, and Aiken County, South Carolina, where the deposits generally contain 85 to 95 percent kaolinite with associated quartz and small amounts of mica, smectite, ilmenite, anatase, rutile leucoxene, and traces of zircon, tourmaline, kyanite, and graphite. The deposits occur in the sedimentary units of the Coastal Plain a short distance seaward from the crystalline rocks of the Piedmont province. This same belt extends southwest from Macon to Andersonville, Georgia (Zapp, 1965), and westward to Eufaula, Alabama, where extensive kaolin resources are recognized (Warren and Clark, 1965).

The origin of kaolin deposits in sedimentary rocks is uncertain. Geologic evidence shows that some sedimentary deposits of kaolin have either been transported and deposited as kaolin; altered from previously transported rocks by processes related to surficial weathering, submarine or subaqueous alteration, or diagenetic (post-burial) changes; or some combination of these possibilities. Among the deposits least understood are those in the Georgia–South Carolina belt. Virtually all investigators agree that either the kaolin or the material from which it was derived was transported. The disagreements focus on how such large deposits of fine-grained white clay could have formed in environments in which other sediments being deposited consisted of sand containing much more iron than does the clay. Whatever processes explain the origin of these deposits, they must account for these facts: some deposits of kaolin are more fine grained than others; some contain accordion-like booklets of kaolinite and some do not; gibbsite ($Al[OH]_3$) occurs in some deposits but not in others; lignite layers or layers rich in organic matter are in some deposits; and smectite occurs in variable amounts in some deposits. Studies of the sedimentary environment raise questions of deposition of these materials in fresh, brackish, or marine waters. Patterson and Murray (1983, p. 619–622) have summarized the current understanding of this topic.

Other types of kaolin deposits include residual deposits, which generally occur in weathered igneous rocks and, therefore, are irregularly shaped and grade downward through partly weathered zones to the parent rock, and kaolin deposits in hydrothermally altered rocks.

Halloysite. Halloysite, the hydrated clay mineral, is generally a dense, porcelain-like, hard clay that upon dehydration to metahalloysite takes on a cottony texture, is friable, porous, and of light to white color.

The first domestic halloysite was mined near Rising Fawn, Georgia, in the late nineteenth century for use as a filler in food. Other deposits in Georgia and Indiana were mined for pottery clay and alum about the time of World War I. The major source of halloysite in the United States today is the Dragon Mine in the Tintic district, 3.2 km south of Eureka, Juab County, Utah, which since 1949 has supplied over one million tons for processing into petroleum-cracking catalyst. Other deposits in Utah have supplied material for light-colored bricks, firebrick, tile, and paper filler; in Nevada for portland cement; in Idaho for refractories; and in North Carolina for ceramics.

Ball clay. Ball clay is a term derived from an early English mining practice of rolling the highly plastic clay into balls of 14 to 23 kg for transport in wagons to sites for further processing. Ball clays, in addition to high plasticity, generally have high dry strength, long vitrification range, and light color when fired. The difference between the light color required by ball clay standards and that for the white firing of kaolin provides little basis for the distinction of these two types of clay. Ball clay is truly a kaolin composed generally of more than 70 percent kaolinite, but is still commercially classified as a separate product because of the continued use of the name in marketing and in the ceramic industry. Most of the ball clays are consumed in the manufacture of vitreous china sanitaryware, electrical porcelain, dinnerware, wall tile, artware, as well as in some refractory products, ceramic glazes, and porcelain enamel slips. Production of ball clay has been rising for more than 60 years. The annual production in recent years has been over 815,000 tons, of which about 70 percent came from Tennessee; the remainder has been produced from Texas, Kentucky, Mississippi, Maryland, and California. The distribution of the major ball clay deposits in the United States is shown on Plate 3C.

Refractory clay. The name "refractory" or "fire clay" derives from the property that allows it to be fired at high temperatures without warping. Fire clay was first mined domestically at Woodbridge, New Jersey, for shipment to Boston in 1816. Later, the industry expanded to centers in Pennsylvania, the Ohio River Valley, Missouri, and then into the western states. Fire clay was the leading clay produced annually until 1943, although production continued to increase until 1957 when it peaked at 10.7 million tons. The annual decline since then has resulted mostly from the change from open-hearth to basic oxygen furnaces in steel production. By 1970, production had declined to 5.8 million tons, and by 1983 to 900,000 tons. Refractory clays are now used in making firebrick, insulating brick, refractory mortars and cement monolithic and castable materials, ramming- and air-gun mixes, and many other products. The major producing states are Alabama, California, Colorado, Illinois, Kentucky, Missouri, Ohio, Pennsylvania, Texas, and West Virginia. The distribution of some major deposits is shown on Plate 3C.

All deposits of refractory clay occur in sedimentary rocks of Pennsylvanian to Tertiary age. Many commercially valuable deposits are in the coal basins of the eastern half of the nation in rocks of early Pennsylvanian age, and they occur as underclays, the clay beds that lie just under coal beds (Plate 3C). The origin of these underclays is debatable, but many workers consider them to have developed by the alteration of aluminous sediments deposited in a swampy environment (Keller, 1970; Patterson and Hosterman, 1963). Others, however, have postulated that these clays were transported and that some sedimentary winnowing process resulted in the accumulation of rather pure deposits of kaolin.

Major districts and deposits in the Appalachian region (Hosterman and others, 1968, Fig. 59 and Table 54) include the Clearfield, Somerset, Allegheny Valley, and Beaver Valley districts of Pennsylvania; Cordova district, Alabama; East Liverpool district, Ohio and West Virginia; Tuscarowas Valley, Hocking Valley, and Oak Hill districts, Ohio; and Olive Hill district, Kentucky. Fire-clay deposits in Missouri have been described by McQueen (1943) and by Keller (1952) and Keller and others (1954). In the Rocky Mountain region, fire-clay deposits of good quality in lenses 1 to 6 m thick occur in the Purgatoire Formation and the Dakota Sandstone of Cretaceous age that crop out at the western edge of the High Plains in Fremont, Pueblo, Custer, Huerfano, and Las Animas Counties, Colorado (Waagé, 1953). On the West Coast, fire-clay deposits mostly of Cenozoic age occur in the Castle Rock area, Cowlitz County, and in King County, Washington; the Molalla area, Clackamas County, Oregon; and the Alberhill area, Riverside County, California.

Bentonite and fuller's earth

The clay products sold as bentonite and fuller's earth have the common property of containing clay minerals of the smectite group, chiefly montmorillonite of both the sodium and calcium varieties; although some bentonite deposits contain large amounts of saponite, a hydrous magnesium clay, and hectorite, a hydrous lithium-magnesium variety. Fuller's earth deposits of other than bentonite types contain palygorskite (attapulgite), a hydrous magnesium clay. The nature of the clays in these deposits makes some of them particularly well suited for use in drilling muds, bonds for foundry sand and iron ore (taconite) pelletizing, and others as absorbent granules, and in the bleaching of oils. In drilling mud applications, bentonite is preferred where fresh water occurs in the penetrated rocks, and palygorskite-type fuller's earth is preferred where salt water is involved.

The term bentonite, named for the Benton Shale in the Upper Cretaceous of the western plains, was originally proposed by Knight (1898) for a claylike rock material that was later recognized as the product of the devitrification and alteration of the glassy material in tuff and volcanic ash. As the results of x-ray study of the clay minerals were better understood, it became apparent that the clay minerals in various deposits were closely related, although perhaps the origins of the deposits differed. The first commercial shipments of bentonite came from Rock Creek, Wyoming, in 1888. Production rose steadily to a peak of 4.4 million tons in 1981, but by 1983 production had decreased to 2.8 million tons. In 1985, drilling mud consumed 44 percent of the production, foundry sand bond 21 percent, and iron ore pelletizing 14 percent. The distribution of the principal deposits of bentonite and fuller's earth is shown on Plate 3C.

Fuller's earth is a term whose origin is lost in antiquity but undoubtedly refers to the clays used to clean and "full" wool, which involved removing dirt and lanolin from the fibers. The term now has no compositional or mineralogical meaning.

Fuller's earth was used to clean wool and hides in Colonial times, and the Indians probably used it even earlier to clean blankets and other items. The first successful commercial mining in the United States began in 1895 after an Alsatian immigrant farm laborer recognized a clay near Quincy, Florida, as similar to fuller's earth mined in Germany. The clay was used to decolorize and purify animal, vegetable, and mineral oils for many years, but use with mineral oils has declined greatly with advancements in the technology of petroleum refining. Fuller's earth used in petroleum processing is commonly referred to as "naturally active clay" if used in its natural state, or as "activated clay" if first treated with acid to improve its desirable properties, generally those pertaining to its ability to bleach. In the 1930s, fuller's earth was introduced successfully as a dispersant in insecticides and fungicides. After World War II, use increased as an absorbent for grease, oil, water, and other undesirable substances on the floors of factories, gas stations, ship engine rooms, and other installations. Much is now sold annually for litter and bedding for pets, poultry, and other animals, and as soil conditioners in greenhouses and golf courses. In 1985, 1.8 million tons of clay were sold as fuller's earth, 66 percent for filter and absorbent uses, and 11 percent for insecticides.

The clays with sodium ions as the principal exchangeable ion are the high-swelling bentonites that expand 15 to 20 times when wetted with water. These clays also are known as the Wyoming or Western bentonites because of their abundance there. The sodium bentonites have excellent colloidal properties and are much in demand for well drilling mud and other applications that require thixotropic suspensions. The sodium bentonites have a high dry strength, which makes them excellent bonds for foundry sand and pelletizing taconitic iron ore.

The bentonites with calcium as the principal exchangeable ion have little swelling capability, which renders them of little use in drilling mud unless treated with soda ash or other chemicals, but even then they are not as efficient to use as the sodium varieties. "Calcium bentonites" have a high green strength and low to moderate dry strengths, which may be dependent on particle size or other characteristics.

The principal fuller's earth deposits not of the bentonite type consist of palygorskite, but also include sepiolite. Both palygorskite and sepiolite occur in fibrous and elongate, lath-like particles.

Most bentonite occurs in sedimentary rocks and was formed in place by the alteration of tuff or volcanic or other igneous rocks. Some deposits of nearly pure smectite may be accumulations that were transported and deposited in marine or alkaline lake waters. A waxy or soapy texture is common to virtually all bentonites that occur in lenticular bodies or beds along defined stratigraphic zones. The lenticular bodies generally are only a few hundred yards in diameter, but some beds have been traced for 320 km. Contacts generally are sharp on the bottom and gradational on top. Hydrothermal deposits tend to be irregular in shape and have gradational contacts in all directions. The fresh bentonite is generally gray with tints of yellow and green, but some deposits under deeper overburden tend to be blue-gray. Weathered outcrops of high-sodium bentonite characteristically exhibit a frothy surface material. Weathered outcrops of the high-calcium bentonite take on a cracked appearance resembling alligator hide.

The extensive fuller's earth deposits in Georgia and Florida occur as tabular lenses and discontinuous beds in the Hawthorn Formation of Miocene age. The clay is waxy in texture and has the same color as the other bentonites. Nobody has found convincing evidence that these deposits are of volcanic origin. The palygorskite (attapulgite) is considered by Patterson (Patterson and Murray, 1983, p. 549) to have formed by the evaporation of seawater from a tidal flat.

Most of the high-swelling sodium bentonite produced domestically comes from beds in the Mowry Shale of Cretaceous age that crop out extensively over the High Plains of Wyoming, South Dakota, and Montana. The major producing districts are: (1) the northern and western Black Hills districts, Wyoming, Montana, and South Dakota; (2) Kaycee-Midwest, Wyoming; (3) Greybull-Lovell, Wyoming, Montana; (4) Vananda, Montana; and (5) Chinook-Malta-Glasgow, Montana. Most of the bentonite from the Black Hills districts is produced from the Clay Spur Bentonite bed in the uppermost Mowry Shale. The bentonite from the Vananda and Chinook-Malta-Glasgow districts of Montana is produced from the Bearpaw Shale, which is also of Cretaceous age but younger than the Mowry.

The calcium or southern bentonite or fuller's earth is mined in Alabama, Mississippi, Louisiana, Texas, Oklahoma, Missouri, and Tennessee. Alabama has yielded fuller's earth from the Porters Creek Clay of Paleocene age. The unit is a mixture of montmorillonite and other clay minerals that is exposed from west-central Alabama across northeastern Mississippi, through western Tennessee and Kentucky, southern Illinois, and into southeastern Missouri. Bentonite beds in Texas lie in an extensive belt in the Gulf Coastal Plain where useful deposits occur in formations of Late Cretaceous to Tertiary age. Deposits in Vernon Parish, Louisiana, have been worked in the Fleming Formation of Miocene age, which also contains bentonite in Texas. Most of the bentonite in Mississippi has been mined from beds in the Eutaw Formation of Late Cretaceous age near Aberdeen, Monroe County. Extensive beds of Tertiary high-calcium bentonite have been mined in Arizona and California, and smaller deposits occur in all western states.

The leading fuller's earth–producing area is the Meigs-Attapulgus-Quincy district of Georgia and Florida, where the deposits occur in the Hawthorn Formation of Miocene age. The clay in the northern part of the district is mostly diatomaceous montmorillonite. The southern part of the district yields the purest palygorskite (attapulgite) known. The Hawthorn Formation also yields a fuller's earth of montmorillonite in the central part of the Florida peninsula. The Twiggs Clay Member of the Barnwell Formation of Eocene age contains large deposits of fuller's earth rich in montmorillonite and cristobalite in some areas in central Georgia, seaward from the fall line.

BARITE

The mineral barite is the major source of barium and its compounds, whose uses are many. In recent years more than 90 percent of the annual domestic production has been consumed as the weighting agent in the fluids circulated in the drilling of oil and gas wells. The remainder is used in glass manufacture, where it adds brilliance and clarity to the final product; as a weight-adding extender and filler in plastics, rubber, and paper; as a white paint pigment; and in barium chemicals, a small but important market. A compendium by the Food Machinery and Chemical Company in 1961 listed more than 2,000 industrial applications in 17 categories for barium chemicals. One of the better-known uses of pure barium sulfate is as an indicator in medical x-ray photography. Among the more exotic uses of barite are its inclusion in the concrete aggregate shields of nuclear installations, because it is a good gamma-ray absorber, and of that used to bury pipelines in marshy areas. Barium titanate ceramics have found application in the electronics industry. Metallic barium is used as a "getter" (oxidizer) to degasify television and other vacuum tubes.

Domestic barite production began in the vein deposits of the Cheshire area, Connecticut, in about 1840 (Fritts, 1962). The barite was sent to New York for use in paint and as filler. Prior to the passage of the pure food and drug acts early in this century, the use of barite was widespread as a weight-adding diluent in flour, sugar, candy, and other prepared foods. At the end of World War I the domestic barite mining industry was firmly established, mostly in the residual deposits of the southern Appalachian states and Missouri. The oil industry began to use barite in well drilling fluids in 1926. The large, high-grade bedded barite deposit on Chamberlain Creek, near Malvern, Arkansas, came into production at the beginning of World War II and supplied much mud-grade barite for the oil and gas wells that provided fuel to move the Allied military forces. Despite that wartime increase in barite production, it was only the threshold of the great demands placed on the industry in post-war economic expansion at home and abroad. Domestic production rose annually, with a few exceptions, to reach a high in 1981 of 2.5 million tons. The price varies with preparation and use, but the average price per ton was about $30 in 1985, the lowest in many years. In the past two decades the major producing states have been Nevada, Arkansas, Missouri, Georgia, and Tennessee. The distribution of some of the major barite districts and deposits is shown on Plate 3D.

The history of the increase in barite production in the United

TABLE 3. WORLD AND UNITED STATES BARITE PRODUCTION AND UNITED STATES IMPORTS, 1850–1985*

	Production		Imports USA
	World	USA	
1850–1914	5.9	1.1	0.2
1915–1918	1.1	0.6	0.0
1919–1944	15.1	6.4	0.8
1945–1975	78.1	25.0	14.4
1976–1985	60.6	14.6	14.2
Totals	160.8	47.7	29.6

*Millions of metric tons.

States is shown in Table 3. Production from the earliest days of mining to virtually the end of World War II was about 8 million tons of barite, augmented by about one million tons of imported barite. In the next 30 years, nearly 24.5 million tons were produced, and imports reached half that tonnage. In the past 10 years, domestic production and imports have been nearly equal. The United States has produced 83 percent of its total barite since 1945, and 30 percent of its yield since 1975. Despite the close ties of barite production to drilling mud, these same trends in production, imports, and consumption are seen in the statistics for many mineral commodities.

Barite, also called barytes, tiff, cawk, and heavy spar, is the most abundant barium-rich mineral. When pure it contains 58.8 percent Ba (or 65.7 percent BaO and 34.3 percent SO_3) and has a calculated specific gravity of 4.5, although inclusions of other material commonly reduce that value considerably. White to gray and even black are the most common colors in commercial deposits. The hardness is 2.5 to 3.5 on Mohs' scale. Barite is relatively insoluble in water and acid, making it virtually inert.

Barite is commonly associated with quartz, chert, jasperoid, calcite, dolomite, siderite, rhodochrosite, celestite, and fluorite, as well as various sulfide minerals, such as pyrite, chalcopyrite, galena, sphalerite, and their oxidation products. Ferruginous clays make up the bulk of many residual deposits of barite. Barite is a common gangue mineral in many types of ore deposits mined principally for other types of commodities, including lead, zinc, gold, silver, fluorite, and rare earth minerals.

Other barium-rich minerals are rare, but small amounts of witherite (barium carbonate, $BaCO_3$) occur as accessory minerals in many barite deposits and occurrences. The only commercial witherite mine in the United States was worked at El Portal, California, until about 1950. Witherite makes excellent feed for a chemical plant because it dissolves easily in acid. Witherite is but a small part of the world's resources of barium. The barium silicate sanbornite ($BaSi_2O_5$), considered to be rare, has been found in abundance with other barium silicates in contact metamorphosed rocks adjacent to the Sierra Nevada batholith in Fresno County, California. It is one of the rare silicates that is acid soluble, making it of interest as a feed stock for chemical plants.

The commercial deposits of barite occur in rocks of sedimentary, igneous, and metamorphic origin with a wide range in age. Most deposits may be classified as one of three major types: (1) vein and cavity-filling deposits, (2) residual deposits, and (3) bedded deposits (Brobst, 1983). Vein and cavity-filling deposits contain barite and associated minerals along fault grabens, joints, bedding planes, breccia zones, and solution channels, the latter two of which are most common in limestones. The deposits of the western states commonly are associated with igneous rocks of Tertiary age, with some notable exceptions such as the barite and rare-earth deposit associated with an alkalic igneous complex of Precambrian age at Mountain Pass, San Bernardino County, California, as well as some vein deposits in the Wet Mountains, Custer County, Colorado. The deposits commonly have sharp contacts with the wall rocks, and large-scale replacements of the wall rocks beyond the controlling structures are rare. Barite deposits in collapse and sink structures are common in the carbonate rocks in the Appalachian region and in central Missouri, where they are known as circle deposits. These deposits are roughly circular, rubble-filled, collapse structures that are bell- or cone-shaped with the apex up. Most of these deposits are less than 75 m in diameter and 40 m in depth. In the early days of domestic mining, the vein and cavity-filled deposits supplied most of the barite produced, but by the late nineteenth century they were overtaken by the production from the residual deposits of the midcontinent and Appalachian regions.

Residual deposits occur in unconsolidated material and are formed by weathering of preexisting deposits. Many valuable deposits have been mined in the clayey residuum derived from Cambrian and Ordovician carbonate rocks in southeastern Missouri and the Appalachian region from Pennsylvania to Alabama, but especially in the Cartersville district, Georgia, and the Sweetwater district, Tennessee. The barite in residual deposits is mainly white and translucent to opaque, and occurs in dense, fine-grained masses. Most residual masses are 2.5 to 15 cm in diameter; small particles are common, and boulders weighing up to several hundred pounds are rare. Small amounts of pyrite, galena, and sphalerite occur in or on some of the barite and are recovered as a by-product from some deposits. Yellow, brown, and red ferruginous illitic clays and incompletely weathered rock fragments make up the bulk of most deposits. The residual barite-bearing clays of the productive Washington County district, Missouri, are commonly 3 to 4.5 m thick, but may be as much as 15 m thick. The same is true in many districts in the Appalachian region, but in the Cartersville district, Georgia, the clayey residuum is as much as 45 m thick. Some residual deposits may have ribs of bedrock protruding into the clayey residuum containing veins of barite with or without calcite, quartz, and small amounts of fluorite and sulfides of common base metals. In other deposits, the bedrock contains only widely disseminated pods or thin discontinuous veins of barite. The size and shape of the residual deposits varies greatly. Deposits in Georgia, Tennessee, and Missouri can extend over several hundreds of hectares, but many commercially valuable deposits are only a few acres in extent.

The barite content of residual deposits varies greatly not only from one deposit to another but also within deposits. The grade is generally expressed in pounds of barite recovered per cubic yard of residuum processed. Under the best of economic conditions, it is probably not practical in the United States to process residuum that contains less than 60 kg of barite per cubic meter, although for many years after World War II, 120 kg of barite per cubic meter was a common lower limit.

Bedded deposits are those in which barite occurs as the principal mineral or cementing agent in stratiform bodies in layered sequences of sedimentary rock. The major bedded deposits of commercial value are concentrations of generally dark gray, fine-grained, fetid barite. These deposits are now the world's premier producers of drilling mud and chemical-grade barite because they tend to be large and high grade, capable of producing hundreds of thousands to millions of tons of ore that is easily beneficiated. Such deposits are, or have been, mined principally in Nevada, Arkansas, California, Idaho, and Alaska. The total barite production since World War II from the large bedded deposits in Nevada and Arkansas alone has now surpassed that from all other types of deposits since the start of domestic barite mining.

The barite beds are massive to laminated and range in thickness from a few centimeters to 15 m; they extend for hundreds of meters and underlie many hectares. Some barite-rich beds are about 30 m thick, but most are thinner and commonly are intercalated with dark chert and siliceous shale and siltstone. The barite grains generally are less than 0.1 mm in diameter. Barite nodules and rosettes are a substantial part of the associated shaley and silty beds. These nodular beds appear to form envelopes around the groups of beds richest in barite. The nodules are generally of two types, radially or concentrically structured. Conglomeratic beds of nodules and fragments of barite, chert, phosphorite (apatite), and enclosing rocks occur particularly in the bedded deposits of Nevada but have been observed in many other deposits. Many small-scale textural and structural features have been described, including rhythmic alterations of beds.

Many layers of bedded deposits contain 50 to 95 percent barite. Quartz is the major associated mineral, generally in proportions inverse to that of barite. Small amounts of clay and pyrite may be present. Carbonate minerals are uncommon accessory constituents. Spectrographic analyses of many samples of bedded barite indicate less than 1 percent calcium and magnesium. The trace element suite in most bedded barite is small. The barite beds commonly contain several percent organic matter that includes hydrocarbons and fatty acids (Miller and others, 1977). Many beds give off the odor of hydrogen sulfide when struck by a hammer. Sulfur isotope studies (Hanor and Baria, 1977; Rye and others, 1978) indicate that the sulfate in the deposits of Arkansas and Nevada is of marine origin formed in mid-Paleozoic time.

In some places, such as the Selwyn Basin, Yukon Territory, Canada, massive sulfide deposits of commercial significance are associated with bedded barite (Gardner and Hutcheon, 1985). Although similar associations have been reported rarely in the conterminous United States, a few small deposits occur in Nevada (Papke, 1984, p. 28), and the association should not be overlooked by those searching for barite or massive sulfides.

The origin of barite deposits is equivocal. Papke (1984) has summarized the major geologic features bearing on the origin of bedded deposits. The sedimentary character and origin of much of the material in bedded deposits has been observed by many workers. The simplest origin suggested, precipitation directly from seawater, does not, however, explain all the features observed. It appears that chemical precipitation from the fluids of submarine springs reacting with marine water may trigger the local accumulation of deposits of barium sulfate. Subsequent diagenetic processes and, in some cases, hydrothermal alteration may reconstitute the deposits to the condition in which we find them today. Understanding the origin of these deposits is essential to establishing the geologic environment of their deposition, which is a valuable tool in exploration. The discovery and study 40 years ago of the huge Chamberlain Creek bedded deposit in Arkansas has influenced further discoveries of similar deposits in the United States, especially in Nevada, and abroad, notably in the Cuddapah district, Andhra Pradesh, India, and in the Selwyn Basin in the Yukon region, Canada. Many bedded barite deposits probably remain to be found in the siliceous rocks in many sedimentary basins of the world, including the Appalachian region of the United States.

The barite deposits of Nevada have yielded an estimated 17.7 million tons of barite from major bedded and vein deposits. Almost all of the bedded deposits lie in a belt of northeast trend about 95 km wide and extending for nearly 500 km across central Nevada. This belt coincides roughly with that of the Antler orogenic belt that was active in Late Devonian and Early Mississippian time (Stewart, 1980). Almost all of the bedded deposits in Nevada occur in siliceous rocks of the "western assemblage," principally chert, argillite, shale, and minor limestone, quartzite, and mafic volcanic rocks, which compose the Comus, Valmy, Vinini, and Palmetto Formations of Ordovician age and the Slaven Chert of Devonian age. About 70 percent of the bedded deposits occur in Lander and Eureka Counties. The first of the Nevada bedded deposits to be worked was the Pleasant View, Lander County, about 1930. The host rocks for the 61 vein barite deposits of Nevada (Papke, 1984) range in age from Precambrian to Miocene and include: limestone, mostly in Esmeralda County; dolomite, mostly in Elko and Lander Counties; and various igneous rocks—intrusive, volcanic, and metavolcanic—mostly in Mineral County. Most of these veins contain coarse-grained (1 to 10 mm) material of high purity, the directly shippable variety. Quartz is the chief impurity, as in the bedded deposits.

The major deposits of Missouri are clustered in the Washington County area, about 80 km southeast of St. Louis, which has yielded more than 8.5 million tons of barite in more than a century of operations. The barite is recovered from residual clays developed on the Potosi and Eminence Dolomites of Cambrian age. The barite deposits of the Central district (Cole, Miller, Moniteau, and Morgan Counties) are of the circle type, developed mostly in the Gasconade and Jefferson City Dolomites of Ordovician age and the Burlington Limestone of Mississippian age.

In Arkansas, more than 7.6 million tons of barite was mined between 1940 and 1972 from the bedded deposit in the Chamberlain Creek syncline, Hot Spring County (Scull, 1958). The orebody, nearly 1.2 km long and 0.8 km wide, was mined in a zone about 18 meters thick in the bottom of the Stanley Shale (Mississippian age). The richest parts of the body contained more than 70 percent barite. Other bedded deposits in the Stanley Shale and the underlying Arkansas Novaculite also have been mined, but no other deposits like the large one at Chamberlain Creek have been found.

Barite has been mined for many years in the Cartersville district, Bartow County, Georgia, where residual deposits over the Weisner, Shady, and Rome Formations of Cambrian age have been the most productive (Kesler, 1950). Most of the more than 3.5 million tons of barite have been produced in the Cartersville district, but from time to time other deposits in Polk, Floyd, Cherokee, Gordon, Murray, and Whitfield Counties have been mined.

Tennessee has yielded more than 2 million tons of barite, mostly from the Sweetwater district in Loudon, McMinn, and Monroe Counties. The barite occurs in residual deposits overlying the carbonate rocks of the Knox Group of Ordovician age. The district has three belts separated by thrust faults of southeasterly dip. Other sources of barite include the Falls Branch–Greenville district, Sullivan and Washington Counties; the Del Rio district, Cocke County; the Pall Mall mine, Fentress County; and the central Tennessee area where barite occurs in veins with base-metal sulfides in rocks of Ordovician age (Maher, 1970).

California has yielded more than a million tons of barite, mostly from five deposits in the Sierra Nevada region and as a by-product from the Mountain Pass rare-earth deposit of Precambrian age, San Bernardino County. The barite in the latter deposits constitutes 20 to 25 percent of rare earth–bearing carbonate bodies (Olson and others, 1954).

TALC

Industrial talc is a versatile mineral product that ranges in composition from talc of high purity and a theoretical formula of $Mg_6Si_8O_{20}(OH)_4$ to mixtures that have properties like the mineral but contain little talc. Other minerals that occur in talcose mixtures and are sold together commonly include anthophyllite, chlorite, diopside, serpentinite, and tremolite and less commonly calcite, dolomite, magnesite, and quartz. Steatite was the original mineral name for talc, but now that name generally refers to the massive variety of talc especially suitable for electric insulators. Block steatite can be machined to various shapes. Impure varieties of massive or blocky talc are commonly referred to as soapstone. Soft, massive talc such as that used in the manufacture of crayons is called French chalk. Pyrophyllite, $Al_4Si_8O_{20}(OH)_4$, is closely related to talc in structure, properties, and uses, and its statistics of production and consumption generally are combined with those of talc. In 1985, pyrophyllite was produced only in California and North Carolina and constituted only 7 percent of combined production.

Talc has a steadily increasing number of uses. As the uses become more specialized, the quality of the material demanded also has become more restrictive. Most talc is ground before use. About 90 percent is consumed in ceramics, paint, paper, roofing, plastics, cosmetics, rubber, and insecticides. Talc in ceramics assists in the control of thermal expansion of the product. Mixtures of talc and cordierite are especially good as electrical insulators. Talc is a prime filler and extender in paint, but also contributes other desirable qualities such as retention of pigment dispersion and prevention of sagging of the paint film on freshly covered surfaces. Talc is increasingly used in paper as a coating to increase gloss, opacity, and brightness. In roofing materials, talc is a filler that stabilizes the melted asphaltic constituents and increases resistance to weathering. Talc sprinkled on shingles and rolled roofing reduces sticking during manufacture and storage before use. Talc is an important filler in plastics that especially improves resistance to chemicals, heat, and impact. About 5 percent of talc is used annually in cosmetics. The standards for cosmetic talc are stringent and based first on the softness of the mineral (1 on Mohs' hardness scale) and then on color, freedom from grit, slipperiness (because of its sheet structure), and its ability to retain fragrance.

Soapstone objects were made and traded by the natives of North America for centuries. Soapstone bodies in the western foothills of the Sierra Nevada Range, California, were mined by the settlers from the mid-1800s for building stone and furnace foundations and linings. After World War I, production in California expanded significantly. In the eastern United States, talc mining began in New York in 1878 and in Vermont about 1900. In 1985, domestic talc production was about 1 million tons from 30 mines in 10 states, 84 percent of which came from Montana, Texas, Vermont, and New York, in descending order of production.

Commercially important talc deposits occur in metamorphosed dolomite and altered mafic and ultramafic igneous rocks (Roe and Olson, 1983). Deposits of these types are found in metamorphosed terranes the world over. In the United States, talc occurs in the older Appalachian and Piedmont regions from Alabama to New England and in California, Montana, Nevada, and Washington. In the midcontinent region, only Texas has deposits of commercial significance.

More than half the domestic talc production is from regionally metamorphosed deposits of dolomite and associated siliceous rocks. Initially, diopside, forsterite, and tremolite develop from the high-grade metamorphism; subsequently, retrograde metamorphism of these rocks forms talc and serpentine. In the Gouverneur district, St. Lawrence County, New York, talc, tremolite, and serpentine occur in metamorphosed Precambrian sedimentary rocks. The talc zones are stratigraphic units that have been tectonically swelled in the axial parts of folds. Shearing also has developed zones of talc schist. Such deposits occur in the

Murphy marble belt, Cherokee County, North Carolina; the Dillon-Ennis district, Beaverhead and Madison Counties, Montana; and the Panamint and Kingston Ranges, Inyo and San Bernardino Counties, California. The deposits in Montana are in pre-Beltian rocks of the Cherry Creek Group of Precambrian age; they are large but require selective mining and sorting to avoid the interlayered, unaltered wallrocks. The deposits in Montana and California yield some material of steatite or near-steatite grade. The talc deposits of the Chatsworth district, Murray County, Georgia, and the now major Allamore district, Hudspeth County, Texas, are associated with phyllites, schists, and quartzites, which were extensively metasomatized. Residual talc derived from little-metamorphosed Cambrian and Ordovician dolomite occurs near Winterboro, Talladega County, Alabama, where the talc is virtually unfoliated and preserves the crystallinity of the parent rock.

Granite plutons and diabasic dikes intruded into favorable dolomitic beds provide heat and solutions necessary to steatize the sedimentary rocks. At some places, even the granite has been replaced by talc. More than 40 deposits in California formed by this process; they occur along the upper contacts of diabase sills with cherty dolomite of Precambrian age in a belt that extends from the southern Panamint Range, Inyo County, eastward across Death Valley to the Kingston Range, San Bernardino County. The orebodies, which are 3 to 7 m wide and 300 m to 1.5 km long, have been the leading sources of talc in California.

Talc deposits associated with serpentinized ultramafic rocks typically occur in the Appalachian and Piedmont regions of the eastern United States. The deposits form rinds on serpentine-rich bodies, or nearly completely replace them. Some deposits formed where sodic pegmatites intruded a serpentinite body, as in Harford County, Maryland, and the Spruce Pine district, Avery, Mitchell, and Yancey Counties, North Carolina. The largest producing talc mines in altered ultramafic rocks are in Vermont, where as least 145 occurrences and deposits are known in a belt about 250 km long and 8 to 30 km wide on the east side of the Green Mountains. Many similar but smaller deposits have been mined in the Appalachian region and in California, Texas, and other western states. Some talc deposits also have formed by pervasive steatitization of mafic rocks associated with ultramafic rocks, as near Schuyler, Nelson County, Virginia, where the interlocking and felted texture of the minerals makes the material desirable for sawed shapes, such as laboratory tables and sinks.

FELDSPAR

Commercial feldspar includes the three anhydrous alkali aluminum silicates: microcline or orthoclase, albite, and anorthite. The product is a beneficiated mixture of feldspar minerals with ideally less than 5 percent, but sometimes as much as 20 percent, quartz and minor amounts of other minerals. Soda spar is a mixture assaying 7 percent or more Na_2O, and potash spar contains 10 percent or more K_2O. The amount of CaO in feldspar is generally of slight importance to the user. Glass spar is, with few exceptions, a soda spar ground to minus 20-mesh, but as fine as minus 40-mesh, which is primarily used to add alumina to the melt for durability by retarding devitrification of the final product. About 55 kg of feldspar are consumed with other ingredients to produce a ton of container glass, and 50 kg are included per ton of flat-glass. Pottery feldspar is a product generally high in potash and ground to minus 200-mesh, that is, used to stimulate fluxing action in shaped and fired ceramic products. In recent years, about 55 percent of the feldspar production was used in glass and about 40 percent in ceramic products. The remaining 4 percent was used to make fillers for latex paint and some urethane and acrylic materials, abrasive cleaners and polishes, coatings for welding rods, and other miscellaneous uses. About 630,000 tons of feldspathic material were produced domestically in 1985. North Carolina produced 70 percent of the domestic supply. Other producing states in recent years include Connecticut, Georgia, California, Oklahoma, South Dakota, Arizona, Wyoming, Colorado, and Maine. The distribution of some of the major feldspar-producing areas and deposits is shown on Plate 3D.

The commercial terms and specifications are based on the mineralogy and geologic occurrence of the feldspathic materials (Rogers and others, 1983). Orthoclase and microcline have the same composition but different crystal habits, monoclinic and triclinic, respectively. There is isomorphous substitution between potash and sodium, and between sodium and calcium, but much less between potassium and calcium, which results in considerable variation in the chemical composition of natural feldspar. The intergrowths of orthoclase and microcline with albite yield perthite, which is common in pegmatite, especially in graphic granite (corduroy spar), a distinctively textured (cuneiform) quartz and feldspar rock with a high content of potash. Most natural orthoclase and microcline contain 10 to 25 percent $NaAlSi_3O_8$, and most plagioclase contains 5 to 15 percent $KAlSi_3O_8$.

Feldspars are produced commercially from pegmatite, quartz, and associated rocks, as well as from some beach sands and alluvium. Sheet and scrap muscovite mica and many other comparatively rare minerals such as beryl (emerald), lepidolite, spodumene, amblygonite, pollucite, cassiterite, and columbite-tantalite, occur in some pegmatite bodies abundantly enough to be commercially recovered at times. Many pegmatites are zoned and show the tendency to contain different sequences of mineral assemblages (Cameron and others, 1949).

The principal pegmatite districts of the United States are in New England, especially Connecticut, Maine, and New Hampshire; in the Appalachian Piedmont, from Alabama to Virginia; and in the Blue Ridge Mountains of North Carolina and Georgia. Other small districts include the Black Hills, South Dakota; the Front Range of Colorado, notably in Jefferson and Douglas Counties; the Petaca area, northern New Mexico; the Cottonwood district, western Arizona; and the Pala and other districts, southern California. A few even smaller districts are scattered in Idaho, Montana, Nevada, Wyoming, and Washington.

Prior to 1946, it was customary to mine some zones in pegmatites that consisted of large crystals of perthite and oligoclase, quartz, and little else from which the "block spar" was hand cobbed for direct sale to the ceramic industry. Other deposits yielded feldspathic materials that needed beneficiation, which commonly involved grinding to minus 20-mesh followed by a pass through a magnetic separator to get the salable product. By 1935, the manufacturers of container glass were seeking larger supplies of feldspar. The U.S. Bureau of Mines in the meantime began studies of the concentration of feldspar by table agglomeration and flotation (O'Meara and others, 1939), because separation of quartz and feldspar by gravity methods is unsuccessful due to the close similarity of their specific gravities. The first commercial plant to successfully produce feldspar for glass by flotation opened in Kona, Mitchell County, in the Spruce Pine district, North Carolina, in 1946. This technical advance allowed not only production of feldspar for glass, but also by-product quartz sand and flake muscovite mica at low cost from finer grained pegmatite bodies. This new technology opened the large intrusive bodies of Paleozoic age in the Spruce Pine district to use as excellent feed for flotation mills. These rocks, called alaskite (Parker, 1952; Brobst, 1962), have a uniform grain-size of 6.3 to 12.7 mm and a consistency in the mineralogy and composition of the individual minerals.

Flotation also provided a way to produce some other useful minerals from pegmatitic bodies such as at Kings Mountain, North Carolina, which yields spodumene, a useful source of lithium, as well as feldspar and feldspathic sand as secondary products (Broadhurst, 1956).

Some beach and alluvial sands derived principally from granitic bodies and feldspathic metamorphic rocks yield commercial feldspar. Normal decomposition of feldspar results in kaolinization of the soda-spar, leaving behind an enriched potash-spar component, although most felspar concentrated from such sources rarely exceeds 7 percent K_2O. The dune sands of Pacific Grove, Monterey County, California, form a belt about 9.5 km long and 1.5 km wide that consists of about 53 percent quartz and 46 percent feldspar and less than one percent other minerals.

Authigenic potash feldspar has been reported in many types of sedimentary rocks of many ages. The feldspar content may range from less than 10 percent to as much as 90 percent in some tuffaceous beds (Sheppard and Gude, 1965). Studies of Cenozoic tuffs in saline lake beds in California, Arizona, Nevada, and Oregon have revealed a zonal pattern in the distribution of various minerals (see the section on zeolites). Near the center of the former lakes, however, nearly pure beds of potash feldspar occur where the water was more saline. The tuff beds are 0.6 to 2.4 m thick and extend over tens of square kilometers in the Barstow Formation of Miocene age, San Bernardino County, California, and in Lake Tecopa, Inyo County (Sheppard and Gude, 1968). The feldspar, however, seems to be finer grained and contains more iron than is currently specified for use in glass.

ASBESTOS

Asbestos is the name applied to naturally occurring fibrous mineral silicates that are incombustible, can be separated mechanically to fibers of various lengths and cross sections, and have various compositions. Corollary to this definition is that of asbestiform particles: those with a crystal-size ratio of 3:1 (length to diameter). In the late 20th century, this definition has lost its validity as a complete scientific term, although it still has commercial and health significance. Ross and others (1984) have proposed that the term asbestos include the serpentine mineral chrysotile and the amphibole minerals grunerite asbestos (amosite), riebeckite asbestos (crocidolite), anthophyllite asbestos, tremolite asbestos, and actinolite asbestos when the mineral particles have a length-to-diameter ratio of at least 20:1.

Widespread movement of asbestos in world commerce has come only in the last century, beginning with the need for insulation materials in the increasing application of steam technology. As the supply increased, the uses also increased, and after World War II more than 300 manufactured products contained asbestos. Major uses for asbestos products are based on its resistance to heat from flame and friction, abrasion, and corrosion. Asbestos is a good insulator of electricity, has a good capacity to absorb resins, and is excellent for lubricants in packings and gaskets. Paper and textile products with asbestos are resistant to attack by bacteria, vermin, heat and flame, and frequent wetting and drying. Finely divided asbestos makes excellent filters for liquids and gases. In 1973, the peak year of domestic production and consumption, 136,000 tons of asbestos were produced and 862,000 tons were consumed. Since then, production and consumption have declined annually. By 1985 domestic production had declined to 61,600 tons and consumption to 208,600 tons. All of the domestic production came from California and Vermont. The world's production leaders are the USSR, Canada, and the Republic of South Africa. Some of the major domestic asbestos belts, areas, and deposits are shown on Plate 3D.

Chrysotile is the only asbestos mineral in the serpentine group. It composes about 94 percent of the world asbestos production, of which virtually all is mined from deposits in serpentinized peridotite, pyroxenite, and dunite; the remainder is from serpentinized dolomitic limestone. Most of the deposits in the Soviet Union, Canada, and the United States yield chrysotile asbestos.

The amphibole asbestos minerals include grunerite (amosite) asbestos, riebeckite (crocidolite) asbestos, anthophyllite asbestos, tremolite asbestos, and actinolite asbestos (Deer and others, 1963a; Zoltai, 1981; Mann, 1983). These amphibole asbestos minerals commonly have a length to diameter ratio of 20:1 or greater. Grunerite asbestos is a rare variety mined only in the Transvaal Province of South Africa, where it occurs in metamorphosed sedimentary banded iron-formation. The mineral is known commercially as amosite. Crocidolite, the asbestos form of riebeckite, has been mined in only a few places, including the Transvaal and Cape Provinces, South Africa; the Hammersley Range, Western Australia; and the Cochabamba district, Bolivia. In recent years, amosite and crocidolite have accounted for only about 5 percent of world asbestos production. Anthophyllite asbestos has been the product of the amphibolitized and serpentinized ultrabasic rocks of the Paakila and Maljasalmi districts near Outokompu, eastern Finland. Tremolite and actinolite never have been major sources of commercial asbestos.

Asbestos minerals occur as cross-fiber veins in which the fibers are about perpendicular to the vein walls; as slip-fiber veins in which the often matted fibers are about parallel to the walls; or as mass-fiber which consists of an aggregate of variously oriented fibers or stellate, radially arranged fibers. Most cross-fiber veins are split by one or more partings that parallel the vein walls, resulting in fibers considerably shorter than total vein width. Cross-fiber asbestos is the easiest to remove from the enclosing rock. Slip-fibers may be of considerable length and flexibility, but are the most difficult to mine and prepare. Chrysotile, crocidolite, and amosite occur chiefly as the cross-fiber type. Anthophyllite, tremolite, and actinolite asbestos occur mainly as mass- or slip-fiber and are least amenable to separation. In addition, these latter three minerals have brittle, weak fibers of limited usefulness. Some cross-fiber veins are 15 cm thick, but all but a few percent are greater than 1.3 cm thick. Crocidolite veins generally are less than 7.5 cm thick, and most are between 0.5 and 2 cm. Amosite veins, by contrast, are as much as 30 cm thick. The grade of the veins varies. An overall grade of 5 to 7 percent asbestos has been sufficient for mining in many places, although some deposits as low as 1.5 to 2 percent have been mined.

The chrysotile deposits of the eastern United States lie in a major belt of serpentinized ultramafic rocks, stretching through the Appalachians region from Alabama northward through New England and into the highly productive areas of the eastern townships of Quebec, and then into Newfoundland. The asbestos bodies generally are stockworks of cross-fibered veins in serpentinite, which are now recognized as parts of great ophiolite complexes. The deposits of Lamoille and Orleans Counties, Vermont, are a southern extension of the Quebec district. The deposit at Eden, Vermont, has continued to yield much slip-fiber asbestos from a body some 900 m wide at the foot of Mt. Belvidere. The host rock is a highly serpentinized peridotite; the fiber occurs along a contact between gneiss and amphibole schist. The deposits of this eastern belt are assigned an Ordovician to Devonian age and are associated with the early folding of the Appalachian region.

A second belt of serpentinite bodies parallels the Pacific Coast from California to Alaska. Several areas in California have been productive (Rice, 1963). A northwest-trending belt of serpentinized ultrabasic rocks about 32 km long and 6.4 km wide lies in the Sierra foothills about 200 km east of San Francisco. One body of short-fiber asbestos in a shear zone in antigorite schist near Copperopolis, Calaveras County, is about 550 m long and 115 m wide. The deposit is estimated to contain 18 million tons of ore containing 6.5 percent chrysotile similar to that of Quebec. The New Idria ultramafic body near Coalinga, Fresno County, is 24 km long, 5 to 8 km wide, and composed of highly sheared serpentinite recrystallized to chrysotile of extremely platy, slickensided character. This enormous deposit alone probably contains more than 90 million tons of ore. Other altered ultrabasic rocks are known elsewhere in the Coast Ranges, Sierra Nevada and Klamath Mountains of California. The ultrabasic bodies were intruded into Jurassic sedimentary rocks in Late Jurassic or Early Cretaceous time.

Chrysotile deposits in serpentinized dolomitic limestone have yielded small tonnages of high-quality, long-fiber asbestos that is free of magnetite, a most common accessory mineral in the asbestos from rocks of ultrabasic origin. Chrysotile deposits of this type in the Salt River region, Gila County, Arizona, contain asbestos-bearing zones 15 to 20 cm thick and of small extent where the carbonate rocks of the Mescal Limestone (Precambrian age) were metamorphosed adjacent to sills of gabbro and diabase, also of Precambrian age (Shride, 1969).

The decline in demand and use of asbestos since 1973 is tied to the health hazards associated with exposure to asbestos and products made of it. The six minerals called asbestos in commerce are not identical in crystal structure, chemical composition, and geologic occurrence. Ross (1984) has shown that the different asbestos dusts do not have the same effect on human health. Ross further concluded that health problems associated with minerals need to be examined carefully as separate problems with respect to their potential for causing disease and their usefulness to society.

FLUORSPAR

Fluorspar is the commercial name for the mineral fluorite (CaF_2), which is the major source of fluorine, the lightest of the halogen elements. Fluorine is also recovered from the fluorapatite, which is a major component of some phosphate rock (see Cathcart, this volume). A minor source of fluorine has been cryolite, mined only near Ivigtut, western Greenland, until 1962, when the deposits were exhausted.

About 70 percent of the fluorspar consumed in the United States is treated with sulfuric acid to produce hydrogen fluoride (hydrofluoric acid), which is an intermediate product in the manufacture of elemental fluorine and various fluorine compounds widely used in industry. Uranium hexafluoride, for example, is used to separate U^{235} from U^{238} by the diffusion process. Sulfur hexafluoride, a stable high-dielectric gas, is used in coaxial cables, transformers, and radar wave guides. Inorganic fluorides are used as insecticides, preservatives, antiseptics, ceramic additives, and electroplating solutions, antioxidants, and many other applications. Organic fluorides, chemically stable and of low toxicity, are useful as refrigerants and materials for fluorcarbon resins and elastomers. Emulsified perfluorochemicals, organic compounds in which all the hydrogen atoms have been replaced by those of fluorine, are being studied as potential blood substitutes. The manufacture of one ton of virgin aluminum requires 25 to 30 kg of synthetic cryolite or aluminum fluoride, but advances in technology are reducing the amount needed significantly.

Acid-grade fluorspar contains a minimum of 97 percent CaF_2, although material of 96 percent CaF_2 can be used if other impurities are minimal. Users generally set upper limits on the content of silica, calcium carbonate, phosphorous, sulfide, sulfur, arsenic, lead, and some other unwanted materials. Moisture content of dried concentrates for acid (and ceramic) grade generally must not exceed 0.10 percent H_2O.

The steel industry consumed about 26 percent of the fluorspar in 1985; an average of 3 kg of fluorspar is added as flux to the charge of open-hearth, basic oxygen, and electric steel furnaces per ton of steel manufactured. Fluorspar promotes fluidity at lower temperature and enhances the removal of phosphorous and sulfur from the steel into the slag. It serves the same purpose in iron foundries, where 7.5 to 10 kg of fluorspar are added to the cupola charge per ton of metal melted. The metallurgical grade, metspar, must contain a minimum of 60 "effective" percent fluorspar and generally not more than 0.30 percent sulfide sulfur and 0.25 to 0.50 percent lead. Metspar is commonly sold as "lump" or "gravel" spar, which must pass a 2.5- to 3.8-cm screen and contain less than 15 percent of particles that pass a 1.58-mm screen. With the decline in availability of lump and gravel spar of metallurgical grade, the steel industry is turning more and more to square briquettes in sizes up to 5 cm on a side that are commonly bound by molasses and lime, the use of which requires no baking ovens. Pellets bound by sodium silicate also are used. Some materials containing 25 percent CaF_2 and utilizing mill wastes, such as iron ore fines, mill scale, shredded scrap, and flue dust, are currently in use to economize use of fluorspar and dispose of troublesome waste materials.

The remaining 4 percent fluorspar is used in the ceramic industry chiefly to make flint glass (requiring about 3 percent fluorspar by weight), white or colored opal glass (requiring 10 to 20 percent), and the opaque enamels used to cover the steel cases of major appliances and face brick and tile (requiring 3 to 10 percent). Ceramic-grade fluorspar generally is supplied in two grades: No. 1 contains 95 to 96 percent CaF_2, and No. 2 contains 85 to more than 90 percent CaF_2, although some users contract for material of their own standards. Some users specify products containing low limits in amounts of calcite, ferric oxide, and the barest traces of lead and zinc sulfides.

When fluorapatite is processed to phosphoric acid, a fluosilicic acid by-product may be recovered that is equal to recovering 35 percent of the fluorine in the original rock, which is about 10 percent of the P_2O_5 content. The fluosilicic acid may be used directly in water fluoridation or used to prepare chemicals for the recovery of aluminum, and a host of miscellaneous products.

Fluorspar mining in the United States began between 1820 and 1840. Production increased significantly with the development of the basic open-hearth process in steelmaking and its need for fluorspar flux. Between the two World Wars, use increased considerably as the steel, aluminum, chemical, and ceramic industries expanded. Fluorocarbons were introduced to market in 1931, and anhydrous hydrofluoric acid was first used as a catalyst in the manufacture of alkylate for high octane auto and aircraft fuel in 1942. Technical advances in beneficiation just before World War II allowed the separation of fluorspar from base-metal sulfides and common gangue minerals by flotation, and the concentration of fluorspar by heavy media methods permitted the treatment of low-grade areas. Each of these advances increased the domestic supply of fluorspar, although large imports of fluorspar annually have been necessary for many years. More recent advances such as pelletizing and briquetting flotation concentrates for use in steel furnaces and beneficiating ores with abundant dolomite and barite have been helpful in increasing domestic supplies.

In 1985, about 63,000 tons of fluorspar were mined domestically, and the equivalent of another 100,000 tons were added as a by-product of domestically mined phosphate rock. Additional imports indicated an apparent consumption of 652,000 tons. Most of the domestic fluorspar mining districts and the major resources lie in Illinois, Kentucky, Texas, New Mexico, Nevada, Utah, Colorado, Idaho, Montana, and Alaska, although at least 13 states have recorded production (Fulton and Montgomery, 1983). The distribution of some principal fluorspar districts and deposits is shown on Plate 3D. Because of perennial gaps between domestic production and consumption, fluorspar is considered a critical commodity. A national fluorspar stockpile provides the material for defense use in case war or some other national emergency should cut off imports.

Fluorite contains 48.9 percent fluorine and 51.1 percent calcium, although small amounts of cerium and yttrium may substitute for some of the calcium. Fluorite forms cubes or octahedra but also occurs in fine-grained, earthy forms or as crusts and globular aggregates exhibiting radial, fibrous textures. Colorless to shades of purple are most common. Color zoning is common, but exposure to sun may fade the original colors. Some varieties fluoresce blue to violet under ultraviolet light, and some phosphoresce after exposure to sun or ultraviolet light. The fluorspar of commerce is commonly mixed with other minerals such as calcite, quartz, barite, celestite, phosphates, and various sulfides.

Fluorspar deposits form under a broad range of physical and chemical conditions in a wide variety of geologic environments that overlap and intergrade. Fluorspar deposits are associated with igneous rocks and include disseminations, those in pegmatites, carbonatites, and contact aureoles. Deposits associated with sedimentary rocks include those in marine carbonates, evaporite, and phosphorite rocks, and lacustrine and volcaniclastic beds. Some deposits are associated with regionally metamorphosed terranes. The commercially most important deposits, however, are those of hydrothermal origin that occur principally in fissure veins and stratiform deposits (mantos), stockworks and pipes, and many types of zones of alteration.

Fluorspar deposits are widely scattered in the United States.

The major domestic producing district lies in southern Illinois and adjacent Kentucky and has yielded about 10.7 million tons of fluorspar concentrates of all grades from 1880 to 1985. Vein deposits are most numerous and are associated with a complex and extensive system of steeply dipping faults in sedimentary rocks of Mississippian age (Grogan and Bradbury, 1968). The normal faults trend northeast and have displacements ranging from less than 1 m to more than 300 m, but the major deposits are associated with vertical faults of 15 to 150 m of displacement. The veins are as much as 9 to 18 m wide and have been mined to a depth of 275 m along a distance of 3.2 km. Most veins, however, are much smaller and pinch and swell along strike and dip. The grade varies greatly. The major replacement deposits of the Illinois-Kentucky district have now outproduced the vein deposits of the district. The mines around Cave in Rock, Hardin County, Illinois, occupy a zone of minor structural disturbance about 1.5 km wide and 9.5 km long that parallels a major northeast-trending fault having 300 m of displacement. The long, narrow deposits lie along each side of the fault. Typical deposits are 1 to 4.5 m thick and 15 to 45 m wide, but some extreme widths are 105 m and 60 to 450 m long. The deposits occur preferentially in four sets of limestone beds in a zone 55 m thick. The deposits extend to more than 300 m below their outcrops. Fluorite is the principal ore mineral; sphalerite and galena are commonly associated, as are quartz, calcite, barite, pyrite, marcasite, witherite, and strontianite. The crude ores are 15 to 90 percent calcium fluoride and some contain as much as 3.5 percent zinc and 5 percent lead.

In the eastern half of the nation, some fluorspar has been mined from fissure veins southwest of Lexington, Kentucky, and others with barite and sulfide in central Tennessee, where the deposits occur in gently dipping Ordovician limestone on the Nashville dome. Van Alstine and Sweeney (1968) reported 38 occurrences of fluorspar in the Appalachian region from Alabama to Pennsylvania, and they concluded that the region has not yet been adequately prospected for fluorspar. Fluorspar deposits are widely distributed in the western United States (Peters, 1958).

The Lost River deposits on the western Seward Peninsula, Alaska, were worked intermittently for tin between 1904 and 1955. Since then, the deposits have been explored for fluorspar, tin, and beryllium. The valuable minerals occur in felsic dikes, in skarn deposits along contacts of granite and limestone, and in brecciated limestone adjacent to thrust faults. One zone is estimated to contain 24.5 million tons of ore with a grade of 16.2 percent CaF_2, 0.15 percent tin, and 0.03 percent tungsten, but there has been no production (Sainsbury, 1964).

In 1955, small amounts of metspar were produced from several veins of fluorspar in shear and breccia zones in coarse quartz monzonite of Mesozoic age in the Orocopia Mountains, Riverside County, and elsewhere in California. Metspar also was produced from fluorite-sericite veins in the Goodsprings Dolomite in Clark Mountain and the volcanic rocks of the Cave Canyon district near Afton, San Bernardino County, during the 1950s and 1960s. Other vein deposits are known in San Bernardino, Inyo, and Riverside Counties.

Colorado has yielded about 2.3 million tons of fluorspar mostly from five major districts, currently all inactive. In the Jamestown district, Boulder County, fluorspar occurs in veins and pipelike bodies, and breccia zones in granite and granodiorite. The fluorspar in the Northgate district, Jackson County, occurs in open spaces in shear and breccia zones in granite and schist. Metspar was produced there during World War II, but the ore was concentrated to an acid-grade product after a new mill was built in 1952. Vein deposits in the Browns Canyon district, Chaffee County, are as much as 14 m thick and 800 m long in rhyolite and granite.

The Bayhorse deposit, Custer County, Idaho, contains fluorspar as open-space fillings of Eocene or post-Eocene age in collapse breccia in the Bayhorse Dolomite of Ordovician age. Stratabound minerals occur at two principal stratigraphic levels. This deposit was estimated to contain 2.9 million tons of ore between 25 and 36 percent fluorspar (Snyder, 1978). Fluorspar has been produced at Myers Cove from ore in volcanic breccia and fissure fillings.

Fluorspar mining in Montana has been centered in the Crystal Mountain pegmatite deposits near Darby, Ravalli County, where massive white to green and purple fluorspar occurs in three east-trending bodies in coarse-grained biotite granite. The associated minerals are altered feldspar, sericite, quartz, and biotite. The coarse fluorspar was handsorted at the mine for a metspar product from 1952 to 1973. The fines stockpiled at the mine subsequently were shipped for production of briquettes for metallurgical uses. By 1980 the deposits had yielded 513,000 tons of metspar. Fluorspar has many other occurrences in Montana, but none has been worked profitably.

The fluorspar deposits of Nevada (Papke, 1979) are hydrothermal or pyrometasomatic and occur as replacement bodies, veins, replacement and/or fillings in jasperoid bodies, stockworks, and breccia pipes. The Daisy deposit in the Fluorine district near Beatty, Nye County, has been the most productive in the state and has yielded 204,000 tons of the state's total production of 521,000 tons of fluorspar. The deposit occurs in structurally complex, generally steeply dipping, replacement bodies in dolomite. The Baxter deposit, Nye County, yielded 165,000 tons of fluorite for an acid-grade product from 1952 to 1957 for the Kaiser Aluminum and Chemical Corporation from a vein in andesitic rocks developed to a depth of 183 m. The Wells Cargo mine, southeastern Lincoln County, yielded 41,000 tons of fluorspar from four open-pit mines in manto-like replacement bodies generally concordant with the dolomite host rock.

Fluorspar is abundant in southwestern and central parts of New Mexico; seven areas have produced virtually all of the 608,000 tons produced to 1954, the last year of significant production: the Zuni Mountains deposits, Valencia County; the northern Sierra Caballo deposits, Sierra County; the Gila, Burro Mountains, and Cooks Peak districts, Grant County; Fluorite

Ridge district, Luna County; and the Tortugas deposit, Dona Ana County. Fluorspar has been mined from 17 other areas of the state. The deposits of the major areas are related to intrusive and extrusive igneous rocks of Late Cretaceous or Tertiary age emplaced near the edge of uplifts or depressions of the same age. The fluorspar is localized chiefly in Tertiary igneous rocks, Precambrian granites, and Paleozoic sedimentary rocks. Most of the deposits are epithermal veins and fillings in breccias associated with faults. Wall rocks are commonly silicified, sericitized, and fluoritized. The ores range in grade from 35 to 85 percent CaF_2; typical ore had 60 percent CaF_2. Individual bodies were small and commonly contained 18,000 to 32,000 tons of fluorspar (Van Alstine, 1965).

The fluorspar content of some Tertiary lake beds near Rome, Malheur County, Oregon, is as much as 16 percent in some zones, although the grade is generally lower. The fluorspar occurs as fine spherical grains in tuff, tuffaceous mudstone, and mudstone that have proved to be difficult to recover (Sheppard and Gude, 1969b). The deposit is estimated to contain 10 million tons of fluorspar that may eventually become of commercial value.

Fluorspar has been produced in the Christmas Mountains, southern Brewster County, Texas. The ore occurs as bedded replacement mantos adjacent to intrusive bodies of rhyolite. Other replacement bodies of fluorspar in limestone and fissure deposits in rhyolite southwest of Van Horn in the Eagle Mountains have yielded about 13,500 tons of metallurgical and acid-grade material from 1942 to 1950.

The largest fluorspar district in Utah lies on Spor Mountain in the Thomas Range, Juab County, where about a dozen mines yielded 130,000 tons of ore from 1943 to 1962. The deposits occur in a north-trending belt of pipes, veins, and disseminated bodies along faults and in breccias, mostly in dolomites of Ordovician and Silurian age and to a lesser extent in volcanic rocks of Tertiary age. The Spor Mountain area has beryllium deposits on its east and west flanks, but only traces of the metal occur with the high-grade fluorspar deposits in the central part of the mountains (Staatz and Griffitts, 1961; see Worl, this volume).

In the Indian Peak district, Beaver County, Utah, fluorspar has been produced from faults and shear zones in altered Tertiary volcanic rocks and along the contact with a quartz diorite stock. In the Pine Grove district, in the Wah Wah Mountains of western Beaver County, fluorspar, locally coated with uranium minerals, occurs along the fractured contact of a Tertiary intrusive rhyolite porphyry and Cambrian limestone. Fluorspar in the Star district, central Beaver County, is associated with sulfide minerals in fissure fillings in quartzite, intrusive quartz monzonite, and Paleozoic limestones. The Rain Bow mine, Millard County, and the Silver Queen mine, Tooele County, have produced some fluorspar from fissure veins in Paleozoic limestone (Dasch, 1964).

KYANITE AND RELATED MINERALS

Kyanite, andalusite, and sillimanite are anhydrous aluminum silicates with the same chemical composition ($Al_2O_3 \cdot SiO_2$ or Al_2SiO_5) and different properties that put them in commercial demand for high-quality refractories and ceramic products. Upon heating, kyanite begins to break down noticeably at 1,370°C, andalusite at 1,380°C, and sillimanite at 1,547°C, to form a mixture of mullite ($3Al_2O_3 \cdot SiO_2$) and vitreous silica. The mullite generally consists of interlocking acicular crystals, and it remains stable until the temperature attains at least 1,808°C. This property enables mullite to be an advantageous component in refractory shapes and furnace linings for many industrial applications. Above 1,800°C, mullite begins to dissociate into corundum (Al_2O_3) and silica, which recombine as mullite and silica as the temperature drops.

The minerals topaz ($Al_2SiO_4[F,OH]_2$) and dumortierite ($Al_7[BO_3][SiO_4]_3O_3$) lose their fluorine and boron, respectively, in the firing process and also yield a useful mullite material. These minerals are commonly grouped with the other three, although they are much more scarce.

Refractories containing mullite have benefited industry to such a degree that mullite for special uses is now synthesized by fusing very pure alumina and silica at temperatures approaching 1,900°C in an electric furnace. The use of these extra high-purity materials guarantees a minimum of fluxing reactions from impurities. Most of the synthetic mullite, however, is made by sintering less expensive, easily available materials such as kaolin, bauxite, or diaspore clays with the necessary alumina at temperatures of about 1,535 to 1,760°C.

The need for ceramic materials suitable for aircraft spark plug insulators during World War I encouraged research to synthesize mullite. The search for a commercial synthetic sillimanite failed then, as did exploration for minable deposits of sillimanite. The White Mountain andalusite deposit in California and the Oreana dumortierite in Nevada soon became the source of material for spark plug porcelain for the growing automobile industry in Detroit. Later, kyanite was found to meet these refractory and electrical needs, and it soon became the dominant commercial mineral of the group, as it is today. Significant domestic production of kyanite and related minerals in California (1922–1946), Florida (1968–1973), Georgia (1932–1944, 1963–present), Nevada (1925–1948), North Carolina (1934–1943), South Carolina (1948–1969), and Virginia (1922–present) has made the United States a world leader in production, although domestic statistics in recent years are incomplete. The estimated domestic demand for kyanite and synthetic mullite in 1983 was 91,000 tons. In 1985, 90 percent of these materials were consumed in refractory uses: 55 percent in ferrous metal smelting, 20 percent in nonferrous metal smelting, and 15 percent in glass making and ceramic production. The exports in 1983 were estimated at 36,000 tons. Some of the major deposits of these minerals are shown on Plate 3D.

Kyanite, andalusite, and sillimanite are common in many aluminous metamorphic rocks that have been deformed at elevated temperatures. Each of these minerals forms in distinct ranges of temperature and pressure. Kyanite typically forms in schists and gneisses under the high temperatures and pressures of dynamic metamorphism. Andalusite forms at lower pressures and is common in zones of contact or thermal metamorphism surrounding large intrusions of igneous rocks. Sillimanite forms under higher temperature and pressure conditions during both types of metamorphism than do kyanite and andalusite. Natural mullite is rare but can form under conditions that exceed those for sillimanite. Topaz and dumortierite are rare compared to kyanite, andalusite, and sillimanite, but they are locally abundant in some quartzose deposits that appear to be hydrothermally altered volcanic rocks.

The principal deposits of kyanite and sillimanite in schist and gneiss in the eastern United States are in the southern Blue Ridge, extending from northern Georgia to western North Carolina. In the Clarksville district, Habersham and Rabun Counties, Georgia, residual soil and weathered graphitic mica schist containing 6 to 8 percent kyanite were mined between 1932 and 1949. Kyanite was recovered from kyanite-garnet-mica gneiss on the Bowlens Pyramid and Celo Knob at the north end of the Black Range, Yancey County, North Carolina. The deposit lies in the Swannanoa-Burnsville kyanite belt that extends for about 50 km northeastward across Buncombe, Yancey, and Mitchell Counties. Early in the hardrock mining operation, the kyanite content was 10 to 11 percent, but it had declined to about 7 percent by the time mining ceased. East of the Blue Ridge lies an extensive region in the piedmont from eastern Georgia to central North Carolina in which sillimanitic biotite-quartz schists are abundant. Included, however, are many rocks in which sillimanite is scarce or absent. One belt of sillimanitic rocks about 16 km wide trends northeasterly in Hart County, Georgia. Sillimanite nodules of marginal value occur in the Peltzer area, Greenville County, South Carolina.

In the western United States, mica schist and gneiss containing kyanite and related minerals occur in Idaho, Montana, Colorado, the Black Hills region of South Dakota and Wyoming, California, Washington, and Oregon, but only a few deposits are sufficiently large to be of possible commercial importance. A deposit on the south slope of Goat Mountain, Shoshone County, Idaho, contains kyanite and andalusite in schistose rocks of the Belt Series in the contact zone with rocks of the Idaho batholith; and another on Woodrat Mountain, near Kamiah, Idaho, contains kyanite and sillimanite (Van Noy and others, 1970). A great problem with western deposits is not only their low quality but their great distance from major markets.

Less common than the schists and gneisses are the quartz-rich rocks that contain abundant kyanite, or more rarely andalusite, sillimanite, dumortierite, and topaz, in recoverable quantities. These rocks commonly contain more of the aluminosilicate minerals than the mica schists and gneisses, about 20 percent or more, but the individual deposits tend to be smaller. Deposits of this type have been productive for many years and still are today. Kyanitic quartzites are the prime source of domestic kyanite. Kyanite constitutes 15 to 40 percent of the rocks accompanied by quartz and only about 5 percent other minerals, such as pyrite, rutile, mica, and lazulite. The alumina content of these rocks ranges from 10 to 25 percent and averages about 18 percent, which is similar to that of the schistose rocks that enclose the kyanite-bearing layers.

The kyanite-bearing quartzites form a narrow belt in the Piedmont region of the eastern United States, extending from northeastern Georgia to central North Carolina and into southeastern Virginia (Espenshade and Potter, 1960). The belt contains at least 13 deposits and is enclosed within a wider belt of mildly metamorphosed acidic metavolcanic rocks and sediments known in Georgia as the Little River Series and in Virginia as the Volcanic Slate Belt. In 1985, two firms were mining at Graves Mountain, Georgia (Hurst, 1959), and at Willis and Baker Mountains and East Ridge in Virginia. The large deposit at Henry's Knob in the Kings Mountain district, North and South Carolina, has been idle since 1970. A monadnock of andalusite-pyrophyllite-sericite rock has been mined near Hillsboro, Orange County, North Carolina, for use in refractory products. The ore contains 15 to 20 percent disseminated pink andalusite. A belt of metavolcanic rocks of similar composition and lithology to that in North Carolina lies in southeastern California, and adjacent southwestern Arizona also contains kyanitic quartzites. The kyanite, which makes up 25 to 35 percent of these rocks, contains too many fine-grained inclusions of quartz to yield an economically competitive kyanite product. The Ogilby deposit, Imperial County, California, has produced kyanite from some masses in quartzite and quartz-mica schist units exposed discontinuously for about 1.5 km in a zone 120 m in maximum width (Henshaw, 1942). The rock is estimated to contain at least 15 percent kyanite, although the mined rock was said to average 35 percent or more.

The White Mountain deposit, Mono County, California, contains andalusite and diaspore in irregular veins and masses in a zone several miles long and less than 1.5 km wide in quartzite and quartz-sericite schist associated with metaporphyry and quartz monzonite. The mining yielded about 18,000 tons of handcobbed ore that contained 53 percent andalusite.

The only known commercial deposit of dumortierite occurs on the west flank of the Humboldt Range, 9.6 km east of Oreana, Pershing County, Nevada (Kerr and Jenney, 1935). More than 4,500 tons of dumortierite was produced between 1925 and 1945 from the hydrothermally altered rhyolitic rocks of Permian age for shipment to Detroit. Two and one-half miles farther south, at Lincoln Hill on the south side of Rochester Canyon, dumortierite occurs in veinlets with brown tourmaline in a quartz-andalusite-sericite host rock.

An uncommon topaz-sillimanite gneiss of Precambrian age occurs near Evergreen, Jefferson County, Colorado. The gneiss is 3 to 30 m thick and is exposed about 2,100 m along strike. The topaz content in three composite chip samples was 23 to 67 percent, and rutile was 2.2 to 2.4 weight percent. The rock is considered to result from fluorine metasomatism of a sillimanitic quartz gneiss. Sheridan and others (1968) considered the possibility that the rock could be processed to yield a rutile and topaz product.

Kyanite and related minerals also occur in veins, pegmatites, and sedimentary materials, both consolidated and unconsolidated, but generally in small quantities. Topaz has been considered a potential by-product at the molybdenum mine at Climax, Colorado. Stow (1968) estimated that 180,000 tons of heavy minerals containing 16 percent kyanite and sillimanite are discarded annually from phosphate mining in the Bone Valley Formation of Pliocene age in Florida (see Cathcart, this volume). Small amounts of kyanite and sillimanite occur with other heavy minerals, such as rutile, ilmenite, zircon, monazite, and staurolite in coastal sand deposits worked in Florida, Georgia, and New Jersey.

WOLLASTONITE

The calcium silicate, wollastonite ($CaSiO_3$), has special properties as an industrial mineral that make it of increasing importance in the plastics, ceramics, coatings, and other industries (Elevatorski and Roe, 1983). A short-lived mining operation began in 1933 near Randsburg, Kern County, California, to supply wollastonite for the manufacture of mineral wool. A successful mining operation began near Willsboro, Essex County, New York, in 1953. Since 1958, several deposits in California have been worked intermittently. By 1980, the United States was producing about 75 percent of the world's annual yield. The locations of some deposits of wollastonite are shown on Plate 3D.

Wollastonite is an orthosilicate mineral characteristic of contact metamorphism and occurs most commonly within impure limestone near intrusive bodies of granite and other silica-rich igneous rocks. It also may form by crystallization of certain magmas and by the metasomatism of calcareous sedimentary rocks. It occurs in coarse, bladed masses that rarely show good crystal form, although the mineral is acicular or fibrous even in the smallest particles after crushing and grinding. The fragments commonly have a length-to-diameter ratio of about 8:1. These needles have a high strength, which is the key to the desirability of wollastonite for many uses, especially in plastics where high length-to-diameter ratio (20:1) material is used to reinforce thermoplastics and thermoset polymers. The use of wollastonite in ceramic products reduces warping and cracking. Chemical inertness makes wollastonite useful as a filler and coating agent. The naturally high pH (9.9 in a 10 percent water slurry) is a prime property for the coatings industry. Milled grades of wollastonite are good pH stabilizers in PVA and acrylic latex systems, where they replace ammonia and chemical buffers. Most of these paints have about 115 g (4 oz) of wollastonite per 3.78 l (1 gal) to prevent corrosion of the can and lid and to keep the pigments in suspension.

The domestic production of wollastonite is centered in New York and California. Three main deposits with more than 9 million tons of proved reserves containing 55 to 65 percent wollastonite occur in a belt of contact-metamorphosed limestone and metasomatized sedimentary rocks 10 km long and 0.4 km wide near Willsboro, New York. The largest band of rock is 10 to 20 m thick in a host rock of Precambrian limestone. Diopside and garnet rich in iron are associated with the wollastonite. In California, wollastonite has been produced from high-grade pods in a Paleozoic crystalline limestone and from an underlying zone of metamorphic rocks in the Big and Little Maria Mountains, about 32 km northwest of Blythe, Riverside County. Wollastonite also has been mined about 10 km southeast of Ubehebe Peak in the Panamint Range, Inyo County, California, where a large high-grade deposit in calc-silicate rocks was formed in a coarsely crystalline limestone of Devonian or Permian age by contact metamorphism from an intrusive quartz monzonite body. The wollastonite is fine grained, acicular, and associated with diopside, idocrase, tremolite, quartz, and calcite. Other occurrences of wollastonite are known in Arizona, California, Idaho, Nevada, and Utah.

ZEOLITES

The zeolites are a group of more than 40 distinct, naturally occurring hydrated aluminosilicates of alkalis and alkaline earths. The term zeolite stems from the Greek words for boil and stone, which relates to the expansion and loss of water when the minerals are heated. Further study has shown that these silicates have infinitely extended framework structures that enclose interconnected cavities occupied by relatively large exchangeable cations and water molecules. The water and cations have considerable ability to move within the structure, which gives the zeolites their cation exchange and reversible-dehydration properties that are the key to their limited commercial uses today and their promise for increased use in the future.

On the basis of structure and morphology, the zeolites may be subdivided into three main divisions, the natrolite, heulandite, and phillipsite groups. The natrolite group has more numerous linkages in one crystallographic direction than in the plane at right angles, which generally results in a fibrous cleavage and morphology. This group includes natrolite, thomsonite, gismondine, and gonnardite. The heulandite group has linkage more numerous in one plane than in a direction at right angles, which results in platy cleavage. The members of this group include heulandite, clinoptilolite, stilbite, and ferrierite. In the phillipsite group, the binding has the same strength in all directions. This group includes phillipsite, chabazite, harmotome, faujasite, and erionite, although the latter mineral has some chain-like characteristics. Analcime, although not considered a zeolite by some workers, has many of their characteristics and commonly occurs with them in sedimentary rocks. Among other minerals classed as zeolites are mordenite and laumontite-leonhardite. Mordenite has a different structure with 5-membered rings of tetrahedra but approaches the composition of erionite. Laumontite is a calcium aluminosilicate whose structure is not yet completely understood.

All of the zeolite minerals have great similarities and subtle differences in physical and chemical characteristics, and all are commonly found in fine grains in many sedimentary rocks. The zeolite minerals are most easily identified accurately by x-ray diffraction methods (Deer and others, 1963b).

The zeolites have been known since 1756, when the Swedish mineralogist Cronstedt first described stilbite. By the mid-nineteenth century the principle of reversible base-exchange had been discovered. As a result of those studies, any aluminosilicate that demonstrated the property of ion exchange was referred to as a zeolite, and even now the term is applied to clay minerals and synthetic organic ion-exchange resins. Early in the 20th century, zeolites were used in experiments to "soften" water by ion-exchange, but that market was filled by synthetic preparations having good cation selectivities developed for that purpose. An important technical advancement occurred in 1948, when zeolites were synthesized for use as molecular sieves (porous solid materials that exhibit the property of acting as sieves toward gas molecules). Dehydrated zeolite crystals act as sieves by selective absorption or rejection of molecules because of differences in their sizes and other structural factors (Breck, 1974). These synthetic zeolites have found a ready market for use as catalysts, selective sorbents, and desiccants. About 90 percent of the synthetic zeolites are used in the catalytic cracking of petroleum, which has greatly increased the recovery of gasoline. Air components may be separated, and oxygen-enriched streams can be produced with these sieves. Japan has been a pioneer in the development of uses for the large sedimentary deposits of zeolites discovered there in 1949. Clinoptilolite and mordenite from these deposits have been used not only in animal husbandry as an absorber of noxious odors, but also in agriculture as a soil conditioner to increase the effectiveness of chemical fertilizer (Minato and Utada, 1969).

Natural zeolites have found only limited use in the United States, but could be utilized more extensively (Mumpton, 1983). Zeolitic tuff has been used as pozzolan cement, chabazite to desiccate mildly acid natural gas, and clinoptilolite to remove cesium from radioactive wastes. Pollution control may prove to be a stimulant to the demand for natural zeolites because mordenite could be suitable to remove sulfur dioxide from stack gases. Domestic natural clinoptilolite may be useful as an ammonia (NH_4^+) collector from waste water and from organic wastes of barnyards and feedlots (Sheppard and Gude, 1982). The resources of natural zeolites in the United States are large, diverse, and widely distributed, so that domestic and foreign demands could be filled. The location of many deposits and occurrences of clinoptilolite and other zeolites is shown in Plate 3C. The production of natural zeolites in the U.S. has been so small that virtually no data on production are available, although there has been some small production almost every year since the mid-1960s from deposits near Bowie, Arizona; Hector, California; and Jersey Valley, Nevada.

Zeolites occur in many rocks of different types, geologic environments, and ages. In the past, most zeolites were known largely as fillings in fractures and vesicles in igneous rocks, chiefly basalts, from which came most of the museum specimens. In 1891, phillipsite was found extensively in deep-sea sediments, which firmly established its low-temperature origin. In the last 25 years, however, zeolites have been recognized as an important rock-forming component of sedimentary rocks and low-grade metamorphic rocks (Hay, 1966). Zeolites are common authigenic silicate minerals in marine and nonmarine rocks and are especially abundant in altered vitric tuffs. Large volumes of zeolites also have been found in nontuffaceous deposits of saline alkaline lakes. Zeolites grow rapidly in a favorable chemical environment. Zeolites are chemically reactive and are readily transformed to other zeolites or other minerals at low temperatures and pressures. Only nine zeolites—analcime, chabazite, clinoptilolite, erionite, ferrierite, heulandite, laumontite, mordenite, and phillipsite—make up a major part of sedimentary rocks (Table 4). Analcime and clinoptilolite are the most common of all, but of these two, only the latter seems to have significant potential for commercial use. Each of these nine minerals, common in sedimentary rocks, shows a considerable range in Si:Al ratio and cation contents. Except for heulandite and laumontite, they are generally alkalic and more siliceous than their counterparts in mafic igneous rocks.

Most zeolites in sedimentary deposits formed after burial of the enclosing sediments by the reaction of pore water with aluminosilicate materials. Silicic volcanic glass most commonly is the precursor for the zeolites, although clay minerals, feldspars, feldspathoids, and gels also have reacted locally to form zeolites. Hay (1966) found that authigenic zeolites and associated minerals can be correlated with three major factors: (1) composition, grain size, permeability, and age of the rock; (2) composition of pore water (pH, salinity, and proportion of dissolved ions); and (3) depth of burial of the host rock. The common zeolites, except for laumontite and possibly some heulandite, generally occur in tuffaceous sedimentary rocks that have not been buried deeply or exposed to hydrothermal solutions.

Sheppard (1973) suggested a classification of zeolitic sedimentary rocks based on five broad categories of geologic settings: (1) hydrothermal, (2) burial metamorphic, (3) weathering, (4) open system, and (5) closed system. The hydrothermal type includes those zeolites associated with metallic deposits, such as those at East Tintic, Utah, and with hot-spring deposits, such as those at Yellowstone National Park, Wyoming. Zeolites in these areas commonly show a vertical zonation and downward succes-

TABLE 4. ZEOLITES REPORTED IN SEDIMENTARY ROCKS*

Zeolites	Formula
Analcime	$NaAlSi_2O_6 \cdot H_2O$
Chabazite	$(Ca,Na_2)Al_2Si_4O_{12} \cdot 6H_2O$
Clinoptilolite	$(Na_2,K_2,Ca)_3Al_6Si_{30}O_{72} \cdot 24H_2O$
Erionite	$(Na_2,K_2,Ca)_4Al_9Si_{27}O_{72} \cdot 24H_2O$
Ferrierite	$(K,Na)_2(Mg,Ca)_2Al_6Si_{30}O_{72} \cdot 18H_2O$
Heulandite	$(Ca,Na_2)_4Al_8Si_{28}O_{72} \cdot 24H_2O$
Laumontite	$Ca_4Al_8Si_{16}O_{48} \cdot 16H_2O$
Mordenite	$(Na_2,K_2,Ca)Al_2Si_{10}O_{24} \cdot 7H_2O$
Phillipsite	$(K_2,Na_2,Ca)Al_4Si_{12}O_{32} \cdot 12H_2O$

*From Sheppard (1973), Table 148, p. 691.

sion of mineral assemblages that seem to correlate with an increase in temperature. Burial metamorphic type of zeolites was first recognized by Coombs (1954) in Triassic sedimentary rocks of the Southland syncline, New Zealand. Coombs and others (1959) reported a vertical zonation there characterized by a downward succession of clinoptilolite-heulandite-analcime, laumontite-albite, and then prehnite-pumpellynite-albite, which grades downward into rocks typical of the greenschist facies of metamorphosed rocks. Marine volcaniclastic beds are typical host rocks for the zeolites of the burial metamorphic type. These rock sequences are generally more than 3,000 m thick and are locally as much as 12,000 m thick. Laumontite-bearing rocks of this type occur in central Oregon (Dickinson, 1962; Brown and Thayer, 1963); in Mount Rainier National Park, Washington (Fiske and others, 1963); and near Cache Creek, California (Dickinson and others, 1969). Zeolites of the weathering type are volumetrically minor and may have been overlooked in many places. Analcime has been described in alkaline saline soils in the eastern San Joaquin Valley, California (Balder and Whittig, 1968), where analcime occurs in decreasing abundance to a depth of about 1.2 m.

The zeolite deposits of the open- and closed-system varieties are the most voluminous and potentially economically valuable of all types. "Open" and "closed" are used in the hydrologic rather than the thermodynamic sense. Deposits form in an open system by the reaction of volcanic glass with subsurface water of meteoric origin. The original volcanic material commonly was deposited in marine or fluviatile environments or was air-laid on a land surface. Deposits form in a closed system by the reaction of volcanic glass with connate water trapped during sedimentation in a saline alkaline lake.

Deposits of the open system form in thick tuffaceous strata and show a vertical zonation of authigenic silicate minerals. The formation of clinoptilolite in tuff and tuffaceous claystone in the lower part of the Miocene John Day Formation, central Oregon, was attributed by Hay (1963) to hydrolysis and solution of silica glass. The upper part of the formation contains unaltered glass or montmorillonite. A more complex zeolite zonation was found in tuffs of Tertiary age at the Nevada Test Site (Hoover, 1968). An upper zone contains unaltered glass and local concentrations of clay minerals and chabazite. The zeolitic tuffs are as much as 1,200 m thick and show a downward succession of zones rich in clinoptilolite, mordenite, and then analcime. These zones cut across stratigraphic boundaries.

Deposits of the closed-system type form during the diagenesis of sediments of alkaline saline lakes, commonly of the sodium carbonate-bicarbonate variety. Brines of this composition have a pH greater than 9, which may account for the relatively rapid solution of vitric material and the precipitation of zeolites. The authigenic silicate minerals can be correlated with the salinity in deposits of the closed system. Typical of these deposits are those of Pleistocene age at Lake Tecopa, California (Sheppard and Gude, 1968), where the tuff deposited in fresh water near shore or close to inlets contains vitric material unaltered or partly altered to clay minerals. The tuffs deposited in moderately saline water contain zeolites, and those in highly saline and alkaline water in the central part of the basin contain potassium feldspar. Individual tuffs show a lateral zonation toward the center of the basin of unaltered glass to zeolites to potassium feldspar. The zeolites at Lake Tecopa consist chiefly of phillipsite, clinoptilolite, and erionite. Chabazite is a minor constituent of tuffs at Lake Tecopa, but is a locally major constituent of zeolitic tuffs of other saline lacustrine deposits. In other closed-system deposits, such as those in the Barstow Formation of Miocene age in California (Sheppard and Gude, 1969a) and in the Green River Formation of Eocene age in Wyoming (Surdam and Parker, 1972), a zone of analcime separates other zeolites from the zone of potassium feldspar. Other authigenic minerals in deposits of closed systems locally include opal or chalcedony, searlesite ($NaBSi_2O_6 \cdot H_2O$), fluorite (CaF_2), or dawsonite.

Although the genesis of zeolites in sedimentary rocks is reasonably well understood, further geological and industrial research might considerably assist in the commercial development of these deposits. Phillipsite and clinoptilolite are the major zeolites in young deep-sea deposits, but the relative importance of the precursor materials and the interstitial fluids in their formation is unknown. The tuffs of rhyolitic composition in deposits in saline alkaline lakes generally contain a variety of zeolite minerals, but what controls the formation of each mineral is still poorly understood. The relationship of the chemistry of particular varieties of zeolites to the geologic setting of the host rock is certainly worth further investigation. Industrial research worthy of consideration includes development of techniques to separate a zeolite from other zeolites and their gangue minerals and to make chemical and structural modifications of natural zeolites to increase their commercial usefulness at competitive costs.

REFERENCES CITED

Baldar, N. A., and Whittig, L. D., 1968, Occurrence and synthesis of soil zeolites: Soil Science Society of America, Proceedings, v. 32, p. 235–238.

Breck, D. W., 1974, Zeolite molecular sieves; Structural chemistry and use: New York, John Wiley, 771 p.

Broadhurst, S. D., 1956, Lithium resources of North Carolina: North Carolina Division of Mineral Resources Information Circular 15, 37 p.

Brobst, D. A., 1962, Geology of the Spruce Pine district, Avery, Mitchell, and Yancey Counties, North Carolina: U.S. Geological Survey Bulletin 1122–A, p. A1–A26.

——, 1983, Barium minerals, in Lefond, S. J., ed., Industrial minerals and rocks, 5th ed.: New York, American Institute of Mining, Metallurgical, and Petroleum Engineers, p. 485–501.

Brobst, D. A., and Pratt, W. P., eds., 1973, United States mineral resources: U.S. Geological Survey Professional Paper 820, 722 p.

Brown, C. E., and Thayer, T. P., 1963, Low-grade mineral facies in Upper Triassic and Lower Jurassic rocks of the Aldrich Mountains: Journal of Sedimentary Petrology, v. 33, p. 411–425.

Cameron, E. N., and others, 1949, Internal structure of granite pegmatites: Economic Geology Monograph 2, 115 p.

Carr, D. D., and Rooney, L. F., 1983, Limestone and dolomite, in Lefond, S. J., ed., Industrial minerals and rocks, 5th ed.: New York, American Institute of Mining, Metallurgical, and Petroleum Engineers, p. 833–868.

Coombs, D. S., 1954, The nature and alteration of some Triassic sediments from Southland, New Zealand: Royal Society of New Zealand Transactions, v. 82, p. 65–109.

Coombs, D. S., and others, 1959, The zeolite facies, with comments on the interpretation of hydrothermal syntheses: Geochimica et Cosmochimica Acta, v. 17, p. 53–107.

Dasch, M. D., 1964, Fluorine, *in* Mineral and water resources of Utah: Utah Geological and Mineralogical Survey Bulletin 74, p. 162–168.

Davis, R. E., 1957, Magnesium resources of the United States; A geological summary and annotated bibliography to 1953: U.S. Geological Survey Bulletin 1019-E, p. 373–515.

Deer, W. A. and others, 1963a, Rock-forming minerals; Chain silicates: New York, John Wiley and Sons, v. 2, 379 p.

Deer, W. A., and others, 1963b, Rock-forming minerals; Framework silicates: New York, John Wiley and Sons, v. 4, 435 p.

Dickinson, W. R., 1962, Petrology and diagenesis of Jurassic andesitic strata in central Oregon: American Journal of Science, v. 260, p. 481–500.

Dickinson, W. R., and others, 1969, Burial metamorphism of the late Mesozoic Great Valley sequence, Cache Creek, California: Geological Society of America Bulletin, v. 80, p. 519–526.

Dunn, R. J., and Ozol, M. A., 1962, Deleterious properties of chert: New York Department of Public Works Physical Research Report 12, 121 p.

Elevatorski, E. A., and Roe, L. A., 1983, Wollastonite, *in* Lefond, S. J., ed., Industrial minerals and rocks, 5th ed.: New York, American Institute of Mining, Metallurgical, and Petroleum Engineers, p. 1383–1390.

Ericksen, G. E., and Cox, D. P., 1968, Limestone and dolomite, *in* U.S. Geological Survey and U.S. Bureau of Mines, Mineral resources of the Appalachian region: U.S. Geological Survey Professional Paper 580, p. 227–251.

Espenshade, G. H., and Potter, D. B., 1960, Kyanite, sillimanite, and andalusite deposits of the southeastern states: U.S. Geological Survey Professional Paper 336, 121 p.

Fiske, R. S., and others, 1963, Geology of Mount Rainier National Park, Washington: U.S. Geological Survey Professional Paper 444, 93 p.

Fritts, C. E., 1962, The barite mines of Cheshire: Cheshire, Connecticut, The Cheshire Historical Society, 36 p.

Fulton, R. B., and Montgomery, G., 1983, Fluorspar and cryolite, *in* Lefond, S. J., ed., Industrial minerals and rocks, 5th ed.: New York, American Institute of Mining, Metallurgical, and Petroleum Engineers, p. 723–744.

Gardner, H. D., and Hutcheon, I., 1985, Geochemistry, mineralogy, and geology of the Jason Pb-Zn deposits, Macmillan Pass, Yukon, Canada: Economic Geology, v. 80, p. 1257–1276.

Gazdik, G. C., and Tagg, K. M., 1957, Annotated bibliography of high-calcium limestone deposits in the United States: U.S. Geological Survey Bulletin 1019-I, p. 675–713.

Gillson, J. L., 1960, Intriguing examples of geology applied to industrial minerals: Economic Geology, v. 55, p. 629–644.

Goldman, H. B., 1961, Sand and gravel in California; Part A, northern California: California Division of Mines and Geology Bulletin 180–A, 38 p.

Goldman, H. B., and Reining, D., 1983, Sand and gravel, *in* Lefond, S. J., ed., Industrial minerals and rocks, 5th ed.: New York, American Institute of Mining, Metallurgical, and Petroleum Engineers, p. 1151–1168.

Graf, D. L., and Lamar, J. E., 1955, Properties of calcium and magnesium carbonates and their bearing on some uses of carbonate rocks, *in* Bateman, A. M., ed., Fiftieth anniversary volume: Economic Geology, p. 639–713.

Grogan, R. M., and Bradbury, J. C., 1968, Fluorite-lead-zinc deposits of the Illinois-Kentucky mining district, *in* Ridge, J. D., ed., Ore deposits of the United States, 1933–1967, volume 1: New York, American Institute of Mining, Metallurgical, and Petroleum Engineers, p. 370–399.

Hanor, J. S., and Baria, L. R., 1977, Controls on the distribution of barite deposits in Arkansas, *in* Symposium on the geology of the Ouachita Mountains, Little Rock, Arkansas, 1977, volume 2: Arkansas Geological Commission, p. 42–49.

Harben, P. W., and Bates, R. L., 1984, Geology of the nonmetallics: New York, Metals Bulletin, Inc., 392 p.

Hay, R. L., 1963, Stratigraphy and zeolitic diagenesis of the John Day Formation of Oregon: California University Publications, Geological Science, v. 42, no. 5, p. 199–262.

—— , 1966, Zeolites and zeolitic reactions in sedimentary rocks: Geological Society of America Special Paper 85, 130 p.

Henshaw, P. C., 1942, Geology and mineral deposits of Cargo Muchacho Mountains, Imperial County, California: California Journal of Mines and Geology, v. 38, p. 147–196.

Hoover, D. L., 1968, Genesis of zeolites; Nevada Test Site, *in* Eckel, E. B., ed., Nevada Test Site: Geological Society of America Memoir 110, p. 275–284.

Hosterman, J. W., and others, 1968, Clay, *in* U.S. Geological Survey and U.S. Bureau of Mines, Mineral resources of the Appalachian region: U.S. Geological Survey Professional Paper 580, p. 167–188.

Hubbard, H. A., and Erickson, G. E., 1973, Limestone and dolomite, *in* Brobst, D. A., and Pratt, W. P., eds., U.S. mineral resources: U.S. Geological Survey Professional Paper 820, p. 357–364.

Hurst, V. J., 1959, The geology and mineralogy of Graves Mountain, Georgia: Georgia Geological Survey Bulletin 68, 33 p.

Keller, W. D., 1952, Observations on the origin of Missouri high-aluminum clays, *in* Problems of clay and laterite genesis: New York, American Institute of Mining, Metallurgical, and Petroleum Engineers, p. 115–134.

—— , 1970, Environmental aspects of clay minerals: Journal of Sedimentary Petrology, v. 40, p. 788–813.

Keller, W. D., and others, 1954, The origin of Missouri fire clays, *in* Swineford, A., ed., Proceedings, Second National Conference on Clay Minerals, 1953: Washington, National Research Council Publication 327, p. 7–46.

Kerr, P. F., and Jenney, C. P., 1935, The dumortierite-andalusite mineralization at Oreana, Nevada: Economic Geology, v. 30, p. 287–300.

Kesler, T. L., 1950, Geology and mineral deposits of the Cartersville district, Georgia: U.S. Geological Survey Professional Paper 224, 97 p.

—— , 1956, Environment and origin of the Cretaceous kaolin deposits of Georgia and South Carolina: Economic Geology, v. 51, p. 541–554.

Kneller, W. A., and others, 1968, The properties and recognition of deleterious cherts which occur in aggregates used by Ohio concrete producers: University of Toledo Research Foundation Final Report 1014, 201 p.

Knight, W. C., 1898, Bentonite: Engineering and Mining Journal, v. 66, p. 491.

Lefond, S. J., ed., 1983, Industrial minerals and rocks, 5th ed.: New York, American Institute of Mining, Metallurgical, and Petroleum Engineers, 1446 p.

Maher, S. W., 1970, Barite resources of Tennessee: Tennessee Division of Geology Report of Investigations 28, 40 p.

Mann, E. L., 1983, Asbestos, *in* Lefond, S. J., ed., Industrial minerals and rocks, 5th ed.: New York, American Institute of Mining and Metallurgical, and Petroleum Engineers, p. 435–484.

McQueen, H. S., 1943, Geology of the fire clay districts of east-central Missouri: Missouri Geological Survey and Water Resources, v. 28, 2nd ser., 250 p.

Miller, B. L., 1941, Lehigh County, Pennsylvania, geology and geography: Pennsylvania Geological Survey Bulletin C39, 4th ser., 492 p.

Miller, R. E., and others, 1977, The organic geochemistry of black sedimentary barite; Significance and implications of trapped fatty acids: Organic Geochemistry, v. 1, p. 11–26.

Minato, H., and Utado, M., 1969, Zeolite; The clays of Japan; International Clay Conference: Japan Geological Survey, p. 121–134.

Olson, J. C., and others, 1954, Geology of the rare earth deposits of the Mountain Pass district, San Bernardino County, California: U.S. Geological Survey Professional Paper 261, 75 p.

O'Meara, R. G., and others, 1939, Froth flotation and agglomerate tabling of feldspar: American Ceramic Society Bulletin, v. 18, p. 286–292.

Papke, K. G., 1979, Fluorspar in Nevada: Nevada Bureau of Mines and Geology Bulletin 93, 77 p.

—— , 1984, Barite in Nevada: Nevada Bureau of Mines and Geology Bulletin 98, 125 p.

Parker, J. M., 1952, Geology and structure of the Spruce Pine district, North Carolina: North Carolina Division of Mineral Resources Bulletin 65, 26 p.

Patterson, S. H., and Hosterman, J. W., 1963, Geology and refractory clay deposits of the Halderman and Wrigley Quadrangle, Kentucky: U.S. Geological Survey Bulletin 1122-F, 113 p.

Patterson, S. H., and Murray, H. H., 1983, Clays *in* Lefond, S. J., ed., Industrial minerals and rocks, 5th ed.: New York, American Institute of Mining, Metallurgical, and Petroleum Engineers, p. 585–651.

Peters, W. C., 1958, Geologic characteristics of fluorspar deposits in the western United States: Economic Geology, v. 53, p. 663–688.

Rice, S. J., 1963, California asbestos industry: California Division of Mines and Geology Mineral Information Service, v. 16, no. 9, p. 1–7.

Roe, L. A., and Olson, R. H., 1983, Talc, *in* Lefond, S. J., ed., Industrial minerals and rocks, 5th ed.: New York, American Institute of Mining, Metallurgical, and Petroleum Engineers, p. 1275–1301.

Rogers, C. P., Jr., and others, 1983, Feldspar, *in* Lefond, S. J., ed., Industrial minerals and rocks, 5th ed.: New York, American Institute of Mining, Metallurgical, and Petroleum Engineers, p. 709–722.

Ross, M., 1984, A survey of asbestos-related disease in trades and mining occupations and in factory and mining communities as a means of prohibiting health risks of nonoccupational exposure to fibrous materials, *in* Levadie, B., ed., Definitions for asbestos and other health-related silicates: Philadelphia, American Society for Testing and Materials, ASTM-STP 834, p. 51–104.

Ross, M., and others, 1984, A definition of asbestos, *in* Levadie, B., ed., Definitions for asbestos and other health-related silicates: Philadelphia, American Society for Testing Metals, ASTM-STP 834, p. 139–147.

Rye, R. O., and others, 1978, Stable isotope studies of bedded barite at East Northumberland Canyon in Toquima Range, central Nevada: U.S. Geological Survey Journal of Research, v. 6, p. 221–229.

Sainsbury, C. L., 1964, Geology of the Lost River Mine area, Alaska: U.S. Geological Survey Bulletin 1129, 80 p.

Scull, B. J., 1958, Origin and occurrence of barite in Arkansas: Arkansas Geological and Conservation Commission Information Circular 18, 101 p.

Sheppard, R. A., 1973, Zeolites in sedimentary rocks, *in* Brobst, D. A., and Pratt, W. P., eds., United States mineral resources: U.S. Geological Survey Professional Paper 820, p. 689–695.

Sheppard, R. A., and Gude, A. J., III, 1965, Potash feldspar of possible economic value in the Barstow Formation, San Bernadino County, California: U.S. Geological Survey Circular 500, 7 p.

——— , 1968, Distribution and genesis of authigenic silicate minerals in tuffs of Pleistocene Lake Tecopa, Inyo County, California: U.S. Geological Survey Professional Paper 597, 38 p.

——— , 1969a, Diagenesis of tuffs in the Barstow Formation, Mud Hills, San Bernardino, California: U.S. Geological Survey Professional Paper 634, 35 p.

——— , 1969b, Authigenic fluorite in Pliocene lacustrine rocks near Rome, Malheur County, Oregon, *in* Geological Survey Research 1969: U.S. Geological Survey Professional Paper 650-D, p. D69–D74.

——— , 1982, Mineralogy, chemistry, gas absorption, and NH_4^+-exchange capacity for selected zeolitic tuffs from the western United States: U.S. Geological Survey Open-File Report 82-969, 16 p.

Sheridan, D. M., and others, 1968, Rutile and topaz in Precambrian gneiss, Jefferson and Clear Creek Counties, Colorado: U.S. Geological Survey Circular 567, 7 p.

Shride, A. F., 1969, Asbestos, *in* Mineral and water resources of Arizona: U.S. Congress, 2nd session, Senate Interior Insular Affairs, p. 301–311.

Snyder, K. D., 1978, Geology of the Bayhorse fluorite deposit, Custer County, Idaho: Economic Geology, v. 73, p. 207–214.

Staatz, M. H., and Griffitts, W. R., 1961, Beryllium-bearing tuff in the Thomas Range, Juab County, Utah: Economic Geology, v. 56, p. 941–950.

Stewart, J. H., 1980, Geology of Nevada: Nevada Bureau of Mines and Geology Special Publication 4, 136 p.

Stow, S. H., 1968, The heavy minerals of the Bone Valley Formation and their potential value: Economic Geology, v. 63, p. 973–975.

Surdam, R. C., and Parker, R. D., 1972, Authigenic alumino-silicate minerals in tuffaceous rocks of the Green River Formation, Wyoming: Geological Society of America Bulletin, v. 83, p. 689–700.

U.S. Bureau of Mines, 1985, Mineral facts and problems: U.S. Bureau of Mines Bulletin 675, 996 p.

——— , 1986, Mineral commodity summaries: U.S. Bureau of Mines, 187 p.

U.S. Bureau of Mines and U.S. Geological Survey, 1980, Principles of a resource/reserve classification for minerals: U.S. Geological Survey Circular 831, 5 p.

Van Alstine, R. E., 1965, Fluorspar, *in* Mineral and water resources of New Mexico: New Mexico Bureau of Mines and Mineral Resources Bulletin 87, p. 260–267.

Van Alstine, R. E., and Sweeney, J. W., 1968, Fluorspar, *in* U.S. Geological Survey and U.S. Bureau of Mines, Mineral resources of the Appalachian region: U.S. Geological Survey Professional Paper 580, p. 286–288.

Van Noy, R. M., and others, 1970, Kyanite resources in the northwestern United States: U.S. Bureau of Mines Report of Investigations 7426, 81 p.

Waagé, K. M., 1953, Refractory clay deposits of south-central Colorado: U.S. Geological Survey Bulletin 993, 104 p.

Warren, W. C., and Clark, L. D., 1965, Bauxite deposits of the Eufaula district, Alabama: U.S. Geological Survey Bulletin 1099-E, 31 p.

Wentworth, C. K., 1922, A scale of grade and class forms for clastic sediments: Journal of Geology, v. 30, p. 377–392.

Yeend, W., 1973, Sand and gravel, *in* Brobst, D. A., and Pratt, W. P., eds., United States mineral resources: U.S. Geological Survey Professional Paper 820, p. 561–565.

Zapp, A. D., 1965, Bauxite deposits of the Andersonville district, Georgia: U.S. Geological Survey Bulletin 1199-G, 37 p.

Zoltai, T., 1981, Amphibole asbestos mineralogy, *in* Veblen, D. R., ed., Amphiboles and other hydrous pyriboles; Mineralogy: Mineralogical Society of America Reviews in mineralogy, v. 9A, p. 237–278.

MANUSCRIPT ACCEPTED BY THE SOCIETY DECEMBER 12, 1988

Printed in U.S.A.

Chapter 13

Introduction

Dudley D. Rice
U.S. Geological Survey, Box 25046, Denver Federal Center, Denver, Colorado 80225

INTRODUCTION

Energy is an important part of the economic strength of the United States, which has been and continues to be the largest consumer of energy in the world. Oil and gas replaced coal as the main energy sources in the 1940s. At present, about two-thirds of the energy consumed is provided by oil and natural gas (Fig. 1). To be specific, oil supplies about 43 percent of the United States' primary energy; about 97 percent is used for transportation. Gas accounts for about one-quarter of the energy use in the country and is most commonly used for heating and domestic purposes.

Domestic oil and gas production has been declining since the early 1970s; the trend for oil production is shown in Figure 2. The United States now consumes much more oil than it produces, as much as 50 percent of which is imported. Gas consumption has fallen because of the switch from natural gas to coal in the generation of electric power in the late 1970s. About 7 percent of the natural gas used in the United States is imported, mainly from Canada.

The decline in oil and gas production in the United States has been brought about by several factors. Drilling for oil and gas decreased dramatically in 1984 (Fig. 3) as a result of a drop in oil prices. The decrease in drilling resulted in the failure to add reserves of oil and gas to offset consumption. The drop in oil prices also resulted in the shut-in of low-volume "stripper" wells and wells that required expensive recovery technology to provide economic flow rates. Finally, because the United States is generally in a mature stage of exploration, most of the largest fields and wells with best potential have been discovered and are in the later, declining stages of production.

This chapter reviews (1) volumes of discovered and undiscovered hydrocarbons in the United States and a comparison of its hydrocarbon resources with the rest of the world, (2) geographic and geologic distribution of oil and gas accumulations, (3) petroleum geology of 12 selected provinces, and (4) future potential for hydrocarbon resources. The following three chapters in the section deal with (1) origin and migration, (2) entrapment, and (3) exploration techniques of oil and gas. The concepts reviewed in these three chapters are general in nature and are not specific to the United States. The last chapters in this section are

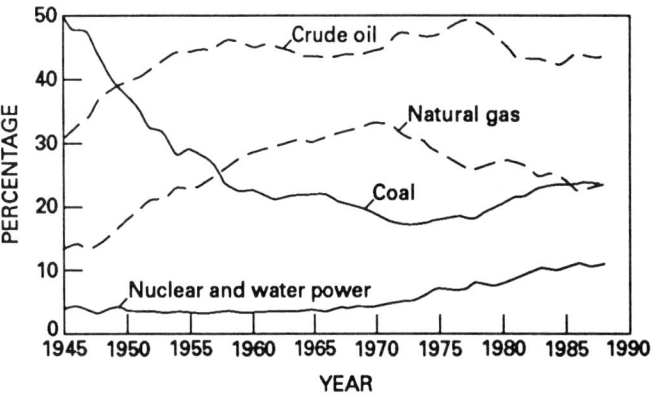

Figure 1. Relative consumption of energy from crude oil, natural gas, coal, and nuclear and water power in the United States from 1945 to 1988. From DeGolyer and MacNaughton (1989).

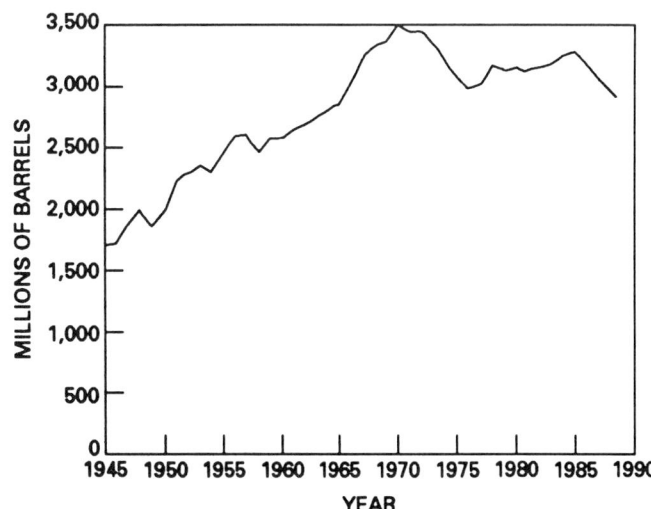

Figure 2. Production of crude oil in the United States from 1945 to 1988. From DeGolyer and MacNaughton (1989).

Rice, D. D., 1991, Introduction *in* Gluskoter, H. J., Rice, D. D., and Taylor, R. B., eds., Economic Geology, U.S.: Boulder, Colorado, Geological Society of America, The Geology of North America, v. P-2.

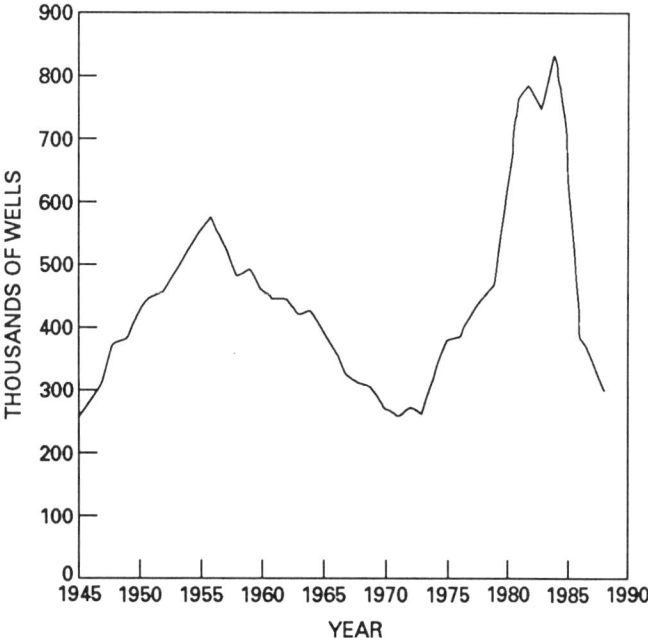

Figure 3. Number of wells drilled for oil and gas in the United States from 1945 to 1988. From DeGolyer and MacNaughton (1989).

summaries of the petroleum geology of 12 selected provinces in the United States.

RESOURCE VOLUMES

The resource classification scheme used in this paper is simplified from the classification by the U.S. Bureau of Mines and U.S. Geological Survey (1980). Discovered hydrocarbon resources include both produced volumes and reserves that have been identified in discovered fields. Undiscovered resources, which are those estimated to exist based on geologic framework, occur outside known accumulations. Ultimate resources include both the discovered and undiscovered volumes.

Hydrocarbon resources are commonly divided into conventional and unconventional. Conventional oil and gas resources are those in discrete accumulations that can be extracted by established technology. Unconventional hydrocarbon resources require different approaches, commonly artificial stimulation, to produce economic flow rates. Unconventional oil includes heavy oil, natural bitumen, and oil shales. Unconventional gas occurs in low-permeability reservoirs, fractured shales, coal beds, or hydrates, or dissolved in geopressured brines.

The United States and the world oil and gas resource bases were assessed recently by Masters and others (1987) and U.S. Geological Survey and Minerals Management Service (1989) and are summarized in Table 1. About 14 percent of the world's ultimate conventional oil resources (243 billion barrels) and 15 percent of the ultimate conventional gas resources (1,402 Tcf) are estimated to occur in the United States. The United States, however, is the largest consumer of energy in the world, and about 75 percent of its discovered conventional oil resources (143 billion barrels) and 70 percent of its discovered conventional gas resources (698 Tcf) have been produced. In comparison, only 34 percent of the discovered oil (381 billion barrels) and 12 percent of the discovered gas (475 Tcf) in the rest of the world have been produced. Further, it is estimated that about 80 percent of the ultimate oil resources (194 billion barrels) and about 72 percent of ultimate gas resources (1,003 Tcf) in the United States have been discovered. In the rest of the world, only 75 percent of the estimated ultimate oil resources (1,125 billion barrels) and 52 percent of the ultimate gas resources (4,078 Tcf) have been discovered. Thus, most of the oil and gas in the United States has been produced and/or discovered, and many companies, particularly the majors, are concentrating their exploration efforts overseas.

The remaining oil and gas resources in the United States, both discovered and undiscovered, will be expensive and difficult to produce and to find. The United States is in a mature stage of exploration; more than 3 million wells have been drilled. As a result, most of the large fields, which contain most of the resources, have been found, and few frontier areas remain. Resources of unconventional oil and gas are estimated to be larger than conventional resources. Although some production has been established from these unconventional sources, higher prices and improved technology are needed to make them reliable, long-term sources of oil and gas.

Geographic and geologic distribution

The distribution of oil and gas accumulations and resources in the United States (Fig. 4 and Plate 5) can be related to the geologic framework of North America. North America consists of a stable interior craton surrounded by unstable margins. The

TABLE 1. COMPARISON OF CONVENTIONAL OIL AND GAS RESOURCES OF THE UNITED STATES AND THE REST OF THE WORLD*

	Discovered		Undiscovered	Ultimate
	Produced	Reserves		
UNITED STATES				
Oil	143 (59%)	51 (21%)	49 (20%)	243
Gas	698 (50%)	305 (22%)	399 (28%)	1,402
WORLD (excluding U.S.)				
Oil	381 (25%)	744 (50%)	376 (25%)	1,501
Gas	475 (46%)	3,603 (46%)	3,800 (48%)	7,878

*Oil is in billion barrels and gas is in trillion cubic feet.
Data from Masters and others (1987) and U.S. Geological Survey and Minerals Management Service (1989).

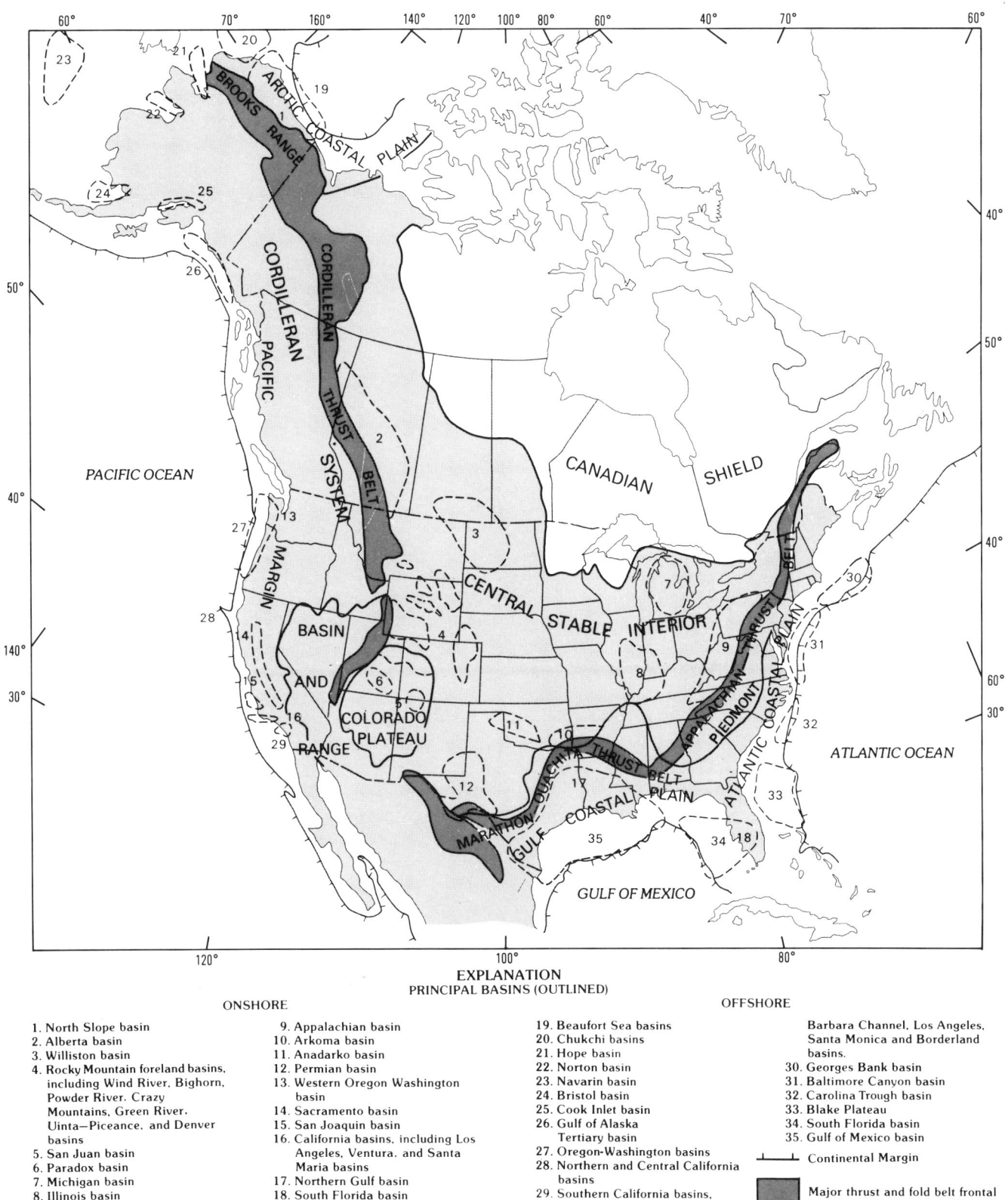

Figure 4. Generalized map of North America showing main sedimentary basins. From U.S. Geological Survey and Minerals Management Service (1989).

Canadian Shield is the core of the continent and is made up of igneous and metamorphic rocks, which are not considered prospective for commercial oil and gas accumulations. The Central Stable Interior is covered by sedimentary rocks that fill numerous basins and contain abundant hydrocarbons. Basins at the margin of the Central Stable Interior, such as the Anadarko and Permian, are generally deeper and structurally more complex than those in the interior and commonly produce more hydrocarbons.

Thrust belts, such as the Cordilleran, Appalachian, Marathon, and Ouachita, flank the Central Stable Interior, and some contain significant hydrocarbon accumulations. Crystalline rocks of the Piedmont, east of the Appalachian thrust belt, are lapped by sedimentary rocks of Mesozoic and Cenozoic age that thicken seaward. In a similar manner, Mesozoic and Cenozoic sedimentary rocks of the Gulf Coastal Plain lap onto the Marathon and Ouachita thrust belts, thicken into the Gulf of Mexico, and are highly productive of hydrocarbons.

The structurally complex Cordilleran System of western North America, which includes the Basin and Range and Pacific margin components, contains several of the basins that produce hydrocarbons. The Brooks Range, part of the Cordilleran thrust belt, forms the central part of Alaska and divides the state into two main structural elements. The Arctic Coastal Plain, the northern element, consists of a thick section of sedimentary rocks that are highly productive of hydrocarbons.

Hydrocarbons are unevenly distributed geographically and geologically in the United States and, for that matter, around the world. Although the United States is a maturely explored country, all of its provinces do not contain hydrocarbons, the yield of oil and gas per volume of rock is highly variable in the productive provinces (Table 2), and most (>98 percent) of the resources occur in fields having ultimate recoveries greater than 1 million barrels of oil equivalent.

About 93 percent of the discovered oil (181 billion barrels) and 87 percent of the discovered gas (872 Tcf) are estimated to be onshore (U.S. Geological Survey and Minerals Management Service, 1989). Further, approximately 67 percent of the estimated undiscovered recoverable oil resources (33 billion barrels) and about 64 percent of the undiscovered recoverable natural gas resources (254 Tcf) are also probably onshore. The most important onshore regions for ultimate oil resources, in decreasing order, are the Gulf Coast, Permian basin, Alaska (in particular the Arctic Coastal Plain), and the Paific Coast (U.S. Geological Survey and Minerals Management Service, 1989). For ultimate gas resources, the most important onshore regions, in decreasing order, are the Gulf Coast, Mid-Continent (in particular the Anadarko basin), and the Permian basin. The only offshore area with significant production and reserves is the Gulf of Mexico. The potential for discoveries exists in all of the offshore areas, but is highest in the Gulf of Mexico followed by Alaska, California, and the Atlantic.

The Significant Oil and Gas Fields data base (NRG Associates, Inc., 1986) was used to investigate the depth, lithology, and age distribution of discovered oil and gas resources in the United States. A significant field, in this case, is defined as one having ultimate recovery greater than 1 million barrels of oil equivalent. The data base has information on 9,344 significant fields that account for more than 98 percent of the discovered resources in the United States. Because the file contains no information on the Appalachian basin and few production data for Oklahoma, the distributions were calculated here with respect to number of significant fields, instead of the percentage of discovered resources.

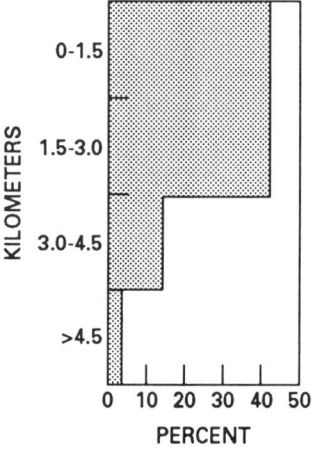

Figure 5. Depth of primary reservoir of significant (>1 million barrels of oil equivalent) oil and gas fields in the United States. Data from NRG Associates, Inc. (1986).

Most significant fields in the United States (84 percent) occur at depths less than 3.0 km (Fig. 5). Furthermore, the percentage of fields that occur in the depth range of 0 to 1.5 km is equal to that of the fields in the depth range of 1.5 to 3.0 km (both 42 percent; Fig. 5). At greater depths, the percentage of the total fields is minor; less than 2 percent of significant fields lie at depths greater than 4.5 km. In addition, these deeper (>3.0 km) fields are predominantly gas, whereas the shallower fields are both oil and gas (Dyman and others, 1990). This depth distribution can be explained in two ways. First, shallower fields are easier and cheaper to find and were developed first. Deeper oil and gas fields are more costly to develop and are commonly associated with production problems, such as abnormal pressures and low permeabilities. Second, the decrease in percentage of deep fields may also be influenced by processes that control hydrocarbon generation; that is, with increasing depth and pressure, oil becomes unstable and only gas is preserved. Exploration for gas has generally lagged behind oil because of economics.

Sandstone is the predominant lithology in the primary reservoirs for about 70 percent of the significant fields, whereas carbonate rocks make up only about 27 percent of the primary reservoirs (NRG Associates, Inc., 1986; Fig. 6). All other lithologic types account for about 3 percent of the primary reservoirs in the significant fields. Limestone composes 61 percent of the carbonate reservoirs, whereas dolomite makes up 39 percent;

TABLE 2. CHARACTERISTICS OF SELECTED PROVINCES*

	Structural Setting†	Area§ (km²)	Volume of Sedimentary Rock§ (km³)	Oil Yield** (barrels/km³)	Gas Yield** (million ft³/km³)	Undiscovered Recoverable Oil‡ (billion barrels)	Undiscovered Recoverable Gas‡ (trillion ft³)
Appalachian basin	Foreland basin	469,000	1,850,000	26,000	400	0.15	6.46
Michigan basin	Craton interior basin	425,000	455,000	37,000	100	1.05	7.78
Northern Gulf of Mexico basin (onshore Western Gulf, East Texas, and Louisiana-Mississippi salt basins)	Passive continental margin basin	615,000	3,263,000	21,000	1,500	4.24	82.47
Anadarko basin	Foreland basin	150,000	456,000	190,000	5,200	0.92	25.12
Permian basin	Foreland basin	222,000	713,000	540,000	1,800	1.89	17.74
San Juan basin	Foreland basin	58,000	126,000	27,000	3,200	0.09	2.00
Powder River basin	Foreland basin	91,000	252,000	15,400	160	2.25	2.78
Cordilleran (Wyoming-Utah-Idaho) thrust belt	Thrust belt	39,000	353,000	52,000	500	0.58	15.81
San Joaquin basin	Forearc/foreland basin	37,000	126,000	143,000	1,700	1.53	3.27
Santa Maria basin (onshore)	Active continental basin	7,100	4,700	335,000	3,200	0.27	0.24
Arctic Coastal Plain (North Slope of Alaska)	Foreland basin	66,000	271,000	603,000	1,700	6.00	22.11
Basin and Range (Eastern)	Extensional basin landward of active continental basin	282,000	§§	...	...	0.29	0.17

*Locations of provinces shown on Figure 4 and Plate 5.
†Classification according to Perry (1989).
§From Varnes and Dolton (1982).
**Generalized yields based on unpublished U.S. Geological Survey cumulative production and proved reserve data as of 1980.
‡From U.S. Geological Survey and Minerals Management Service (1989).
§§Estimate not available because distribution of sedimentary rock poorly known.
***Volume of sedimentary rock not known and only 25 million barrels of oil have been produced.

these percentages are similar to the total volume of these two rock types (Schmoker and others, 1985). Most carbonate reservoir rocks are of Paleozoic age, with the exception of Cretaceous and Tertiary limestones in the Gulf Coast, and were mostly deposited in stable craton settings with low terrigenous input. In contrast, clastic reservoirs represent a wide range of geologic ages and depositional settings.

Based on data from NRG Associates, Inc. (1986), the age distribution of primary reservoirs in significant fields is bimodal, with the major peak having reservoirs of Cretaceous and Tertiary age and a secondary peak having reservoirs of Pennsylvanian age (Fig. 7). This age distribution of reservoirs of significant fields is roughly comparable to the age distribution of ultimate reserves of oil and gas in the world (Bois and others, 1982). However, if the plot had been done on the basis of reserves instead of number of fields, the Permian would also be part of the peak because of the large reserves in the Mid-Continent Panhandle-Hugoton field. Significant fields of Pennsylvanian age are mainly in the Permian basin and Mid-Continent area, whereas those of Cretaceous age are mainly in the Gulf Coast and Rocky Mountain areas and

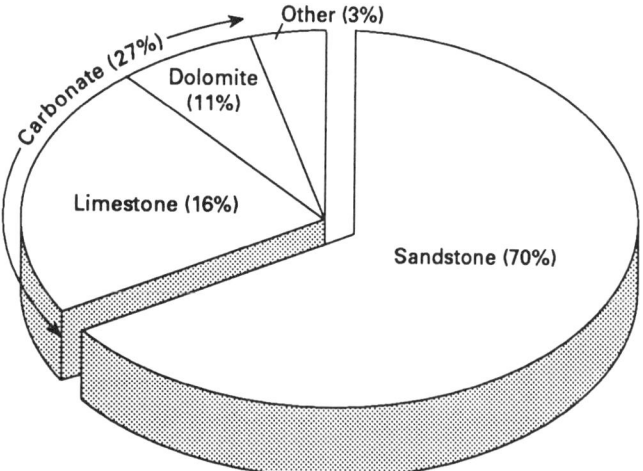

Figure 6. Lithology of primary reservoir of significant (>1 million barrels of oil equivalent) oil and gas fields in the United States. Data from NRG Associates, Inc. (1986).

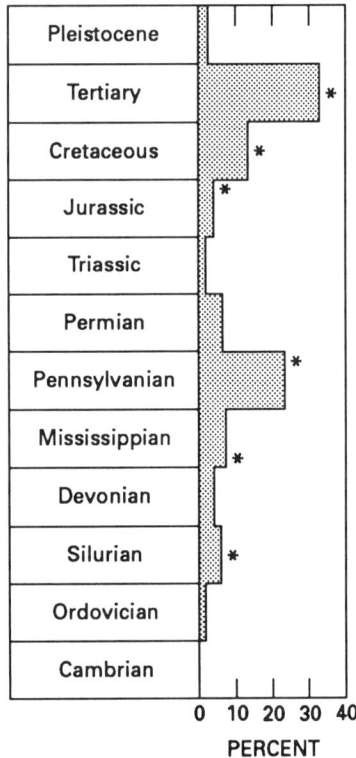

Figure 7. Age of primary reservoir of significant (>1 million barrels of oil equivalent) oil and gas fields in the United States. Data from NRG Associates, Inc. (1986). Six major source-rock intervals as identified worldwide by Ulmishek and Klemme (1990) are marked by asterisks.

those of Tertiary age are primarily in the Pacific Coast and the Gulf Coast areas.

An important factor controlling the age distribution of reservoirs of significant oil and gas fields is the stratigraphic distribution of major source-rock intervals. According to Ulmishek and Klemme (1990), six source-rock intervals (Fig. 7) account for more than 90 percent of the world's discovered original reserves. Source rocks are also well developed in these intervals in the United States. The Silurian, Upper Devonian–Lower Mississippian, Upper Jurassic, and middle Cretaceous intervals contain mainly oil-prone organic matter, whereas the Pennsylvanian–Lower Permian and Oligocene-Miocene intervals contain both oil-prone and gas-prone organic matter. The sporadic deposition of organic-rich sediments was probably related to global sea-level changes (Vail and others, 1977). The source rocks deposited during the Mesozoic (Upper Jurassic and middle Cretaceous) and Tertiary (Oligocene-Miocene) were prolific sources of hydrocarbons and account for about 67 percent of the world's reserves of petroleum (Ulmishek and Klemme, 1990). The largest percentage of reservoirs of significant fields in the United States is of these same ages.

Another controlling factor of age distribution is preservation, which has a tendency to decrease quantities of hydrocarbons in older reservoirs. Younger source-rock intervals are generally more widespread and are at lower levels of thermal maturity. Hydrocarbons are progressively destroyed at higher levels of thermal maturity common in older rocks. In addition, hydrocarbons are commonly redistributed into shallow, younger reservoirs. The large number of fields of Pennsylvanian age (Fig. 7) can be explained by their relatively shallow depths, low levels of thermal maturity, and excellent evaporite seals of Permian age.

OIL AND GAS PROVINCES

The petroleum geology of 12 selected onshore oil and gas provinces (Fig. 4 and Plate 5) is summarized below. A more detailed discussion of the petroleum geology of each of these provinces is provided in later chapters. These 12 provinces were chosen for a variety of reasons, including geographic location, type of basin, and hydrocarbon resources. Most of the provinces have been maturely explored except for the Basin and Range and parts of the Arctic Coastal Plain (Arctic National Wildlife Refuge and the National Petroleum Reserve of Alaska), which are probably the only onshore frontier areas. Cross sections of these basins at both exaggerated (10:1) and true scales are shown in Plates 6 and 7. Other characteristics of the provinces, such as size, hydrocarbon yield, and resource potential, are shown in Table 2.

Appalachian basin

The Appalachian basin is an important hydrocarbon-prducing province located near the major population centers of the eastern United States. The Appalachian basin is an elongate foreland basin located on an aulacogen of early Paleozoic age represented by the Rome trough. An extensive regional homocline dips from west to east, and the eastern margin of the basin is covered by the Blue Ridge and Piedmont thrust sheets. The basin is filled with sedimentary rocks of Cambrian to Permian age; hydrocarbon production is from reservoirs of Cambrian to Pennsylvanian age. In general, siliciclastic rocks are found to the north and east, and marine carbonate rocks are restricted to the south and west. The major source-rock unit in the basin is Devonian shales. The organic-rich shales are also estimated to contain large, essentially undeveloped resources of unconventional gas adsorbed on the organic matter. Although the province has an enormous volume of sedimentary rock, the oil and gas yields are low (Table 2); oil production has decreased steadily since 1964, and gas production has decreased since 1947.

Michigan basin

The Michigan basin is a craton-interior or interior-sag basin, similar to the Illinois and Williston basins (Fig. 4). The basin is probably sited on a Proterozoic rift system and is filled mainly with Paleozoic rocks; all Paleozoic systems are present except for the Permian. Pre-Pennsylvanian strata are dominantly carbonate rocks, and Pennsylvanian strata are mainly nonmarine sandstones. Production is from reservoirs ranging in age from Ordovician to Mississippian; accumulations are both stratigraphically and structurally controlled. Recent discoveries and production have been made in Middle Silurian reefs. Marine source rocks are present throughout the section, and hydrocarbons were probably generated in late Paleozoic time when maximum burial was achieved. Yields of both oil and gas are low (Table 2), which is typical of craton-interior basins.

Northern Gulf of Mexico basin

The northern Gulf of Mexico basin developed as a post-Paleozoic passive margin on the Ouachita fold belt and has been characterized by extensional and gravitational faulting since Triassic time. Unlike other basins developed on passive continental margins, the Gulf basin is characterized by the flow of salt (Jurassic) which, in combination with growth faults, resulted in abundant structural traps and provided conduits for the movement of hydrocarbons. Some production occurs in Mesozoic sandstones and carbonate rocks along the inner rim of the basin; source rocks are interbedded shales and carbonate rocks. About 75 percent of the oil and gas in the province occurs in Cenozoic deltaic sandstones and is trapped by structures related to salt and shale flow and gravitational growth faults. Source rocks for these major resources have not been positively identified; the hydrocarbons may have migrated from depth. The province contains an enormous volume of sedimentary rocks and has the largest resources, both discovered and undiscovered, of any province in the United States. Surprisingly, the yields of hydrocarbons per volume of rock are relatively low (Table 2); only one oil field (East Texas) containing more than one billion barrels of oil has been discovered.

Anadarko basin

The Anadarko basin is the deepest (>12 km) cratonic basin in the United States filled mostly with Paleozoic sedimentary rocks. The present-day foreland basin was sited on an early Paleozoic aulacogen and formed as the result of convergent tectonics in Early Pennsylvanian time. Major quantities of oil and gas have been produced in clastic and carbonate reservoirs ranging in age from Cambrian to Permian, with major production coming from Pennsylvanian and Permian reservoirs. The main oil source rocks are shales of Late Devonian–Early Mississippian and Middle and Late Pennsylvanian age. Thick sequences of gas-prone shales are part of the Upper Mississippian and Lower Pennsylvanian section. Geochemical studies have shown that large volumes of both oil and gas generated in the central part of the basin have migrated long distances both vertically and horizontally. The Panhandle-Hugoton field, located along the shallow western flank of the basin, is the largest gas field in the United States, with estimated ultimate reserves greater than 80 Tcf. The gas occurs in Permian carbonate rocks at depths less than 900 m. Because of this gas field, the basin has the highest gas yield of any of the described provinces (Table 2). In addition, the basin also has major resources of gas at depths greater than 4.5 km.

Permian basin

The greater Permian basin, which includes the Midland and Delaware basins and Central Basin platform, is a foreland basin that attained its present structural configuration during tectonic activity in Pennsylvanian and Early Permian time. It is filled mainly with Paleozoic sedimentary rocks, and production is from reservoirs ranging in age from Ordovician to Permian. The major part of production, however, is from reef-associated sediments of Permian age developed on carbonate shelves and margins around the Midland and Delaware basins and on the Central Basin platform. Source rocks are basinal shales; the hydrocarbons generally migrated from the deeper parts of the basins to uplifts and platforms. Oil generation and migration probably occurred during maximum burial in Permian time. Deep gas generation in the Delaware basin was enhanced by Tertiary igneous activity. The Permian basin has produced the second largest amount of oil in the United States, second only to the northern Gulf of Mexico basin (U.S. Geological Survey and Minerals Management Service, 1989). Compared to the Gulf basin, the volume of sedimentary rock in the Permian basin is much less, the oil yield is much higher, and, due to exploration maturity, the undiscovered potential is much lower (Table 2). The Delaware basin has significant deep gas production and potential from fractured Ordovician dolomites. The Puckett and Gomez fields, with original in-place gas resources of 3.8 and 6 Tcf, respectively, are the two largest, deep gas fields in the United States.

San Juan basin

The Wester Interior region of the Rocky Mountains was the site of an extensive foreland basin located east of the Cordilleran thrust belt during Cretaceous time. This foreland basin, which was periodically occupied by an epicontinental seaway, was broken by vertical movement into several intermontane basins during the Laramide orogeny of Late Cretaceous and early Tertiary time. The circular San Juan basin is one of the more important hydrocarbon-producing intermontane basins in the region. Commercial oil and gas production is from Pennsylvanian, Jurassic, Tertiary, and, most importantly, Cretaceous reservoirs. The Cretaceous reservoirs are mainly sandstones deposited in and along the epicontinental seaway. Oil accumulations are located in structural and stratigraphic traps on the outer part of the basin. Nonassociated gas is generally restricted to a huge "basin center" accumulation in the central part of the basin that is referred to as the Blanco gas field. With an estimated ultimate recovery of 23 Tcf, it is the second largest gas field in the United States. The trapping mechanism for this "basin center" accumulation is probably a combination of hydrodynamics, low permeability, and low pressure. Marine and nonmarine shales and coals of Cretaceous age are the main source rocks for both the oil and gas. Recently, the basin has been the site of intensive exploration and development efforts for methane-rich gas in coal beds of Late Cretaceous age.

Powder River basin

The Powder River basin of Wyoming and Montana is the largest intermontane basin in the northern Rockies, and its present form is the result of the Laramide orogeny. The basin is a north-trending asymmetric basin with its axis along the west side and major thrusting along the western and southern margins. Sedimentary rocks are mainly clastic, of Cretaceous and Tertiary age, and were deposited in association with the Laramide orogeny; carbonate rocks are predominant in the Paleozoic section. Production is mainly oil and is found in stratigraphic and structural (basin-margin anticlines) traps in Cretaceous sandstones deposited in the epicontinental seaway and in Pennsylvanian and Permian rocks associated with an eolian sand sea. Source rocks for Cretaceous accumulations are associated marine shales and for Permian and Pennsylvanian accumulations are shales of the same age.

Cordilleran thrust belt (Wyoming-Utah-Idaho)

The Cordilleran thrust belt extends about 5,000 km (Fig. 4), although only southwestern Wyoming and northern Utah have been found favorable for oil and gas accumulations. Prior to Late Jurassic time, the present-day thrust belt was the site of eastward-thinning shallow marine miogeosynclinal deposition. From Late Jurassic to early Eocene time, episodic thin-skinned compressional tectonics dominated the geologic history and resulted in low-angle thrust faults and elongate concentric folds that were critical to trapping the major hydrocarbon accumulations. Hydrocarbons are trapped in Mississippian carbonate and Triassic sandstone reservoirs in mainly anticlinal accumulations. Shales of Cretaceous age, in juxtaposition with older reservoirs as a result of faulting, are the source rocks. The province is relatively young in terms of exploration maturity because the techniques to explore this structurally complex province were not available until the 1970s. As a result, the oil and gas yields shown on Table 2 are not as reliable as those calculated for more maturely explored provinces.

San Joaquin basin

The San Joaquin basin of California is a north-trending foreland basin with an active fold and thrust belt. The basin developed in late Mesozoic time as a fore-arc basin and was a product of tectonic interactions between the North American Plate and plates of the Pacific Ocean. Economic basement for hydrocarbon accumulation is rocks of Jurassic and Cretaceous age. The reservoirs are mainly of Tertiary age, consisting primarily of marine sandstones, some of which are deep water, and also fractured siliceous shales. Gas accumulations are generally restricted to the northern part of the basin and continue into the Sacramento basin, whereas oil occurs in the southern part of the basin. Important source rocks for gas are Cretaceous and possibly Eocene shales. Siliceous rocks of the Miocene Monterey Formation are the major source for oil. The Monterey is composed of biogenic siliceous rocks deposited during global eustatic high stand and/or anoxic events and is widespread in the coastal basins of California, such as Santa Maria, which will be discussed next. Much of the oil in the basin is low gravity (<20°), and the recent oil production and added reserves in the basin have resulted from improved recovery of this heavy oil.

Santa Maria basin

The Santa Maria basin is one of several oil-rich Tertiary basins along the active continental margin of California. The oil yield of the Santa Maria basin is the third highest of the described provinces (Table 2); the yield in the nearby Los Angeles basin (Fig. 4) is the highest in the United States. The hydrocarbon accumulations in these coastal basins resulted from the favorable combination of tectonics and sedimentation controlled by plate-tectonic interactions along the western edge of the North American Plate. The combination of a deep, detritus-starved marine basin in early Neogene (Miocene) time and oceanographic conditions resulted in deposition of biogenic siliceous sediments of the Monterey Formation that served as both the source and reservoir for the major oil accumulations. With burial diagenesis, the fine-grained biogenic sediments became brittle and fractured to become reservoirs. Compressional tectonics in late Neogene (Pliocene) time resulted in broad folds and tectonic fracturing, thus trapping the hydrocarbons.

Arctic Coastal Plain (North Slope of Alaska)

The North Slope is a composite foreland basin consisting of two wedges. The older wedge is a thin sequence of carbonate and clastic sedimentary rocks of late Paleozoic and early Mesozoic age derived from a northern cratonic source that has been subsequently separated tectonically. The thicker younger wedge of late Mesozoic and Cenozoic rocks is derived from a southern source, the ancestral Brooks Range. The important reservoir rocks are nonmarine to shallow-marine sandstones of Triassic, Cretaceous, and early Tertiary age. Marine shales of Triassic to Cretaceous age are the main source rocks. The province has the highest oil yield of any province described in detail (Table 2) and accounts for about 25 percent of total U.S. oil production. The high oil yield and prolific production are primarily the result of the giant Prudhoe Bay field, with reserves of about 10 billion barrels. The coastal plain is also estimated to have large resources of natural gas; they are presently uneconomic because of their remoteness from markets and the lack of pipelines. The Arctic National Wildlife Refuge, in the eastern part of the province, is considered to be the largest remaining frontier for hydrocarbons in the onshore United States.

Basin and Range (Great Basin)

The Great Basin is that part of the Basin and Range Province with internal surface drainage. The province is characterized by complex geology resulting from late Tertiary extensional faulting and extrusive igneous activity. The province is sparsely explored; only 300 oil and gas wells have been drilled, and only ten oil fields have been discovered in two of the many fault-controlled valleys. Sandstone, carbonate, and volcanic (ignimbrite) reservoirs range in age from Devonian to Oligocene and are highly fractured, with some matrix porosity. Traps are truncation fault blocks within grabens. The grabens contain Tertiary valley-fill sequences and form the seals for the oil fields. Source rocks range in age from Ordovician to Eocene and are both marine and lacustrine in origin. Many aspects of the individual basins are not known, such as their size, volume of sedimentary rocks, and thermal history, and only minor amounts of oil (25 million barrels) have been produced.

FUTURE POTENTIAL

Fewer wells are being drilled, the discovery rate of new oil and gas accumulations is declining, and reserves are decreasing in the United States. These discouraging statistics can be explained by economics and the maturity of hydrocarbon exploration in the United States. Most of the large fields have been discovered, and frontier basins, such as the Arctic National Wildlife Refuge on the North Slope of Alaska and offshore areas, with significant hydrocarbon potential and possibly large fields, are few. Small fields (<1 million barrels of oil equivalent) are estimated to account for 94 percent of undiscovered oil and gas fields in the lower 48

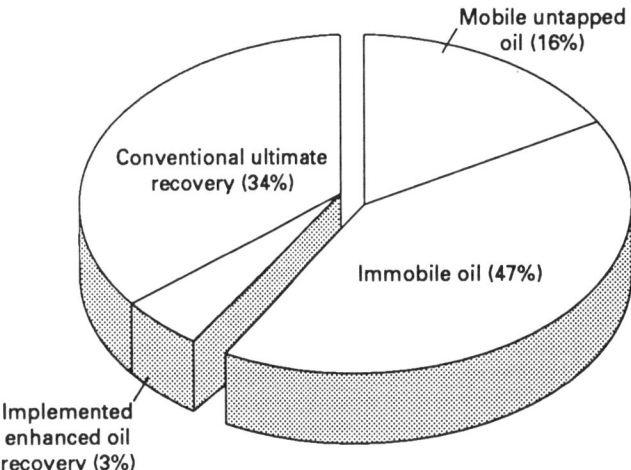

Figure 8. Distribution of oil in existing reservoirs. From Fisher (1987).

states (U.S. Geological Survey and Minerals Management Service, 1989). These small fields will be difficult to find, however, and the total estimated resources will be small (about 25 percent of the total undiscovered oil and 13 percent of the total undiscovered gas.

Possible solutions for problems created by decreasing production in the United States are improved recovery and exploration for unconventional resources. Recoverability of oil based on the study of large fields in Texas has been investigated by Fisher (1987; Fig. 8). He concluded that only about one-third of the original oil in places will be recovered by conventional techniques, such as natural reservoir energy and water flooding. The remaining two-thirds of the oil will remain in developed fields where the infrastructure is already established. Part of this oil is residual and immobile oil, and will require expensive enhanced oil recovery (EOR) methods for extraction. However, Fisher estimated that about 25 percent of the remaining resource is untapped mobile oil that can be recovered by conventional methods, such as infill drilling, aided by studies of geologic reservoir characterization. Tyler (1988) concluded that the amount of unrecovered mobile oil is a function of depositional environment, which has a direct influence on reservoir heterogeneity. One recent innovation to improve primary recovery is horizontal drilling (Montgomery, 1990). So far, the technique has been used for fractured reservoirs, but it will undoubtedly be used in the future to recover untapped mobile oil resulting from reservoir heterogeneity and compartmentalization.

Significant amounts of natural gas in discovered fields are also bypassed because of reservoir heterogeneities. This gas can be recovered by infill drilling designed with the aid of geological studies (Finley and others, 1988).

As domestic supplies of conventional oil and gas become smaller, the United States will have to develop its unconventional sources. Although the resource potential of these unconventional resources is estimated to be large, the amount that can be recovered is limited and will be strongly influenced by technology and economics. Several unconventional gas resources that are closest to being economic are discussed below.

Low-permeability gas reservoirs

A large resource of natural gas is trapped in low-permeability (tight) reservoirs that require artificial stimulation by hydraulic fracturing to produce economic flow rates, as reviewed by Spencer (1989). As defined by the U.S. Federal Energy Regulatory Commission (FERC), tight reservoirs have in situ permeability values less than 0.1 millidarcies. Low-permeability reservoirs composed of sandstone, siltstone, shale, chalk, or limestone result either from deposition of fine-grained sediment (shallow, high porosity) or by diagenesis (deep, low porosity). For geological and engineering purposes, tight reservoirs are subdivided into blanket and lenticular reservoirs. Blanket reservoirs generally respond to artificial fracturing in a predictable manner, whereas lenticular reservoirs respond unpredictably. The National Petroleum Council (NPC) (1980) estimated the technically and economically recoverable gas from tight gas reservoirs in the United States to be in the range of 192 to 574 Tcf, which is in the range of recoverable resources of conventional gas. However, the largest part of the resource is in the Rocky Mountain area where pipelines to handle the developed resource are not in place. Tight gas reservoirs are commonly interbedded with conventional reservoirs, and most of the production is from "sweet spots" with enhanced porosity and permeability because of fracturing and/or development of secondary porosity. The most promising areas evaluated by the NPC are the Green River basin, with deep thermogenic gas, and the northern Great Plains, with shallow bacterial gas.

Coal-bed gas

The presence of methane-rich gas in coal beds has long been recognized because of explosions that occur during coal mining. In addition, coal has been considered as a major source of nonassociated gas that has been expelled and accumulated in adjacent reservoirs. Only recently has coal-bed gas been recognized as a large untapped energy resource, with the coal serving as both the source and the reservoir rock. About 2,000 wells are producing from coal beds, and the recent surge in exploration has been partly driven by a federal tax credit for unconventional fuels. Most of the coal-bed gas resources in the United States are in 13 basins that contain large coal resources as summarized by Rightmire and others (1984). Coals in eastern basins are mainly Pennsylvanian in age, whereas those in the western basins are Cretaceous and early Tertiary in age. In-place coal-bed gas resources are estimated at about 400 Tcf, but a reliable estimate of recoverable resources is not yet available (Kuuskraa and Brandenberg, 1989). The greatest resource potential for coal-bed gas is in the San Juan, Piceance, and northern Appalachian basins. Most of the current activity is in the San Juan and Black Warrior basins, where relatively high-rank (medium- and low-volatile bituminous) coals are at depths less than 1 km and reservoir properties are tectonically enhanced.

Deep gas

Drilling and production data from deep (>4.5 km) wells and reservoirs in the United States indicate that deep gas is an important energy resource and is widely distributed in reservoirs of different lithologies and ages. However, technical problems associated with high pressures and temperatures, low porosity and permeability, and low-quality, corrosive gas make the resource costly to develop. As of 1985, approximately 50 Tcf of gas had been produced from deep reservoirs in the United States, which accounts for about 8 percent of the cumulative production (Dyman and others, 1990). The Potential Gas Committee (1989) estimated that about 33 percent of undiscovered gas resources in the onshore United States occur at depths greater than 4.5 km. This resource, essentially undeveloped, is significant. The most favorable areas for deep gas are the Gulf Coast, Mid-Continent, Permian basin, and Rocky Mountains. The geologic controls and resource potential of deep gas in the United States have been summarized by Rice (1989).

Shale gas

Devonian and Mississippian shales of the eastern United States (Appalachian, Illinois, and Michigan basins) are considered to be a major resource of unconventional gas as reviewed by Roen (1984) and deWitt (1986). The dark gray to black shales contain organic matter ranging from 0.5 to 20 percent and are considered to be the source of the thermogenic gas. The shales have low permeability (0.1 to 10 microdarcies) and low porosity (1 to 3 percent). The gas is adsorbed on the organic matter and also occurs in the pores and fractures. Commercial production of the gas is commonly tied to the occurrence of silty zones or natural fractures in the shale that act as conduits to the wellbore. The resource is attractive because it occurs in heavily populated areas where conventional resources are dwindling. Estimates of in-place shale gas resources range from 206 to 3,900 Tcf, with most of it occurring in the Appalachian basin (Finley and others, 1988). Low recovery rates cause production to be economically sensitive, and only about 3 Tcf of shale gas have been produced.

Large resources of oil and gas have been estimated for other unconventional sources, such as tar sands, oil shales, gas hydrates, and geopressured methane (Cox and Baughman, 1980; Smith, 1980; Finley and others, 1988); however, economic methods of extracting these resources are presently not available.

CONCLUSIONS

Oil and gas are the major sources of energy in the United States. The United States is a hydrocarbon-rich country, but much of its conventional oil and gas has been discovered and produced. The United States rate of consumption remains the highest in the world, but as much as 50 percent of the nation's demand for oil is filled by imports. Natural gas development is about 10 years behind oil development, and natural gas will probably be the most important source of energy by the end of the century. Future supplies of oil and gas within the United States will depend on improved recovery of discovered resources and the costly development of unconventional resources. An added concern to future supplies of domestic oil and gas will be environmental issues related to their development and production.

REFERENCES CITED

Bois, C., Bouche, P., and Pelet, R., 1982, Global geologic history and distribution of hydrocarbon reserves: American Association of Petroleum Geologists Bulletin, v. 66, p. 1248–1270.

Cox, C. H., and Baughman, G. L., 1980, Oil sands; Resource, recovery, and industry: Minerals and Energy Resources, v. 23, no. 4, p. 1–12.

de Witt, W., Jr., 1986, Devonian gas-bearing shales in the Appalachian basin, *in* Spencer, C. W., and Mast, R. F., eds. Geology of tight gas reservoirs: American Association of Petroleum Geologists Studies in Geology 24, p. 1–8.

DeGolyer and McNaughton, 1989, Twentieth century petroleum statistics: Dallas, Texas, 126 p.

Dyman, T. S., Nielson, D. T., Obuch, R. C., Baird, J. K., and Wise, R. A., 1990, Summary of deep oil and gas wells and reservoirs in the U.S.: U.S. Geological Survey Open-File Report 90–305, 35 p.

Finley, R. J., and 12 others, 1988, An assessment of the natural gas resource base of the United States: Texas Bureau of Economic Geology Report of Investigations 179, 69 p.

Fisher, W. L., 1987, Can the U.S. oil and gas resource base support sustained production?: Science, v. 236, p. 1631–1636.

Kuuskraa, V. A., and Brandenberg, C. F., 1989, Coalbed methane sparks a new energy industry: Oil and Gas Journal, v. 87, no. 41, p. 49–56.

Masters, C. D., and 5 others, 1987, World resources of crude oil, natural gas, natural bitumen, and shale oil: Proceedings of 12th World Petroleum Congress, v. 5, p. 3–27.

Montgomery, S. L., 1990, Techniques of horizontal drilling: Petroleum Frontiers, v. 7, no. 1, 48 p.

National Petroleum Council, 1980, Tight gas reservoirs, unconventional gas sources: Washington, D.C., National Petroleum Council, v. 5, part 1, 222 p.

NRG Associates, Inc., 1986, The significant oil and gas fields of the United States (through December 31, 1983): Available from Nehring Associates, Inc., P.O. Box 1655, Colorado Springs, CO 80901.

Perry, W. J., Jr., 1989, Structural settings of deep natural gas accumulations in the conterminous United States, *in* Rice, D. D., ed., Distribution of natural gas and reservoir properties in the continental crust of the United States: Final Report of GRI Contract No. 5087-260-1607, Department of Commerce NTIS Report No. 89/0188, p. 62–72.

Potential Gas Committee, 1989, Potential supply of natural gas in the United States (December 31, 1988): Golden, Colorado, Potential Gas Agency, 160 p.

Rice, D. D., 1989, Distribution of natural gas and reservoir properties in the continental crust of the United States: Final Report for GRI Contract No. 5087-260-1607, Department of Commerce NTIS Report No. 89/0188, 125 p.

Rightmire, C. T., Eddy, G. E., and Kirr, J. N., eds., 1984, Coalbed methane resources of the United States: American Association of Petroleum Geologists Studies in Geology 17, 378 p.

Roen, J. B., 1984, Geology of the Devonian black shales of the Appalachian Basin: Organic Geochemistry, v. 5, p. 241–254.

Schmoker, J. W., Krystinik, K. B., and Halley, R. B., 1985, Selected characteristics of limestone and dolomite reservoirs in the United States: American Association of Petroleum Geologists Bulletin, v. 69, p. 733–741.

Smith, J. W., 1980, Oil shale resources of the United States: Mineral and Energy Resources, v. 23, no. 6, p. 1–20.

Spencer, C. W., 1989, Review of characteristics of low-permeability gas reservoirs in western United States: American Association of Petroleum Geologists Bulletin, v. 73, p. 613–629.

Tyler, N., 1988, New oil from old fields: Geotimes, v. 33, no. 7, p. 8–10.

Ulmishek, G. F., and Klemme, H. D., 1990, Depositional controls, distribution, and effectiveness of world's petroleum source rocks: U.S. Geological Survey Bulletin 1931, 59 p.

U.S. Bureau of Mines and U.S. Geological Survey, 1980, Principles of a resource/reserve classification for minerals: U.S. Geological Survey Circular 831, 5 p.

U.S. Geological Survey and Minerals Management Service, 1989, Estimates of undiscovered conventional oil and gas resources in the United States; A part of the nation's energy endowment: U.S. Department of Interior, 44 p.

Vail, P. R., Mitchum, R. M., Jr., and Thompson, S., III, 1977, Seismic stratigraphy and global changes in sea level; Part 4, Global cycles of relative changes of sea level, *in* Payton, C. E., ed., Seismic stratigraphy; Applications to hydrocarbon exploration: American Association of Petroleum Geologists Memoir 26, p. 83–97.

Varnes, K. L., and Dolton, G. L., 1982, Estimated areas and volumes of sedimentary rock in the United States by province; Statistical background data for U.S. Geological Survey Circular 860: U.S. Geological Survey Open-File Report 82–666c, 11 p.

Manuscript Accepted by the Society October 8, 1990

Printed in U.S.A.

Chapter 14

Origin and migration of oil and gas

Fred F. Meissner
Bird Oil Corporation, 1801 California, Suite 4500, Denver, Colorado 80202

INTRODUCTION

Liquid petroleum ("crude oil" or "oil"), combustible earth gas ("natural gas" or "gas"), and certain related solid bituminous substances (asphaltite, gilsonite, etc.) occur widely and somewhat erratically within subsurface rocks of the Earth and in surface outcrops and seeps. Most occurrences are associated with sedimentary rocks, but the substances are also more rarely found in those of igneous or metamorphic origin. Archaeological evidence and early written documents show that both primitive man and pre-industrial societies utilized bituminous materials for simple and limited purposes. With the rise of modern civilization and the accelerated use of liquid and gaseous fuels, chemicals, and other useful materials derived from oil and gas, man has made a concerted effort to find and exploit commercial accumulations of the substances. Numerous, often conflicting, explanations for their origin, emplacement, and occurrence have also been devised. The greatest body of knowledge on these subjects has been developed in the last two to three decades. The purpose of the following discussion is to synthesize this knowledge into a summary of prevalently accepted theories and concepts concerning the origin and migration of oil and gas. Understanding these specific aspects of the overall science of petroleum geology should enable the prediction of volumes that will move along specific migrational pathways to charge potential sites of accumulation. Such understanding may also aid in making potential resource appraisals, and even guide the development of overall resource policy.

Physical chemical nature of oil and gas

The physical and chemical properties of oil and gas are directly related to their origin and how they migrate. Although the roots of the term "petroleum" (Latin: "rock oil") are specific, the term is somewhat ambiguously used to encompass the science of "petroleum geology," which includes both oil and gas produced as fluids from a well bore or present as such in a surface seep. Liquid oil and natural gas should be distinguished as mobile or producible substances as distinct from immobile nonproducible material of similar composition found in a finely disseminated or microscopic state within some rocks (especially the solvent extractable "bitumen" found in source rocks).

Oil is a complex liquid-phase mixture of "organic" compounds (e.g., those that contain carbon). It consists dominantly of hydrocarbons (e.g., various combinations of carbon and hydrogen representing families of paraffin, naphthene, and aromatic compounds), along with generally lesser, highly variable amounts of nonhydrocarbons (i.e., compounds containing sulfur, nitrogen, oxygen, and other trace elements including some metals, in addition to carbon and/or hydrogen). Several hundred compounds have been identified, and a thousand or more may be present in some oils. Molecular complexity of degassified liquid oils ranges from those that contain only 5 carbon atoms (pentane) to those that contain hundreds. The presence of increasing numbers of compound species accompanied by increasing proportions of large complex molecules is associated with increasing oil density (or decreasing "API gravity"—a common industry measurement inversely proportional to density). Molecular compositions are generally identified through modern methods of chromotography and mass spectrometry. Variations in composition are often useful in correlating oils with common or dissimilar origins through either comparison between oils or comparison of oils to solvent extracts obtained from probable source rocks. Compositional variations may also be related to processes of migration or secondary reservoir alteration. The presence of certain diagnostic compounds ("biomarkers") may be indicative of certain precursor materials found in living organisms or to early depositional and shallow burial environments that altered the precursor material.

Most combustible natural gases consist dominantly of methane (the simplest hydrocarbon), but some may also contain lesser, highly variable amounts of other low-molecular-weight hydrocarbons (ethane, propane, butane, etc.). They may also contain highly variable portions of nonhydrocarbon gases, such as hydrogen sulfide, carbon dioxide, and nitrogen in addition to trace amounts of helium, argon, and rarely hydrogen. Dry gases consist dominantly of methane as the principle hydrocarbon component. Wet gases are generally considered to contain 0.4 or more cubic meters of condensible liquid hydrocarbons per 1,000 m^3 of gas; however, the term is often used in a geochemical sense to include those containing more than about 5 wt % of the heavier hydrocarbon gases ethane, propane and butane, even without the presence of condensible liquids.

Because of temperature- and pressure-dependent behavior

Meissner, F. F., 1991, Origin and migration of oil and gas, *in* Gluskoter, H. J., Rice, D. D., and Taylor, R. B., eds., Economic Geology, U.S.: Boulder, Colorado, Geological Society of America, The Geology of North America, v. P-2.

involving complex mixtures of compounds having greatly different properties, the phase of oil and gas may vary between subsurface and surface conditions. Oil and gas are mutually soluble in each other within certain limits. Gas may exsolve from oil, and oil may condense from gas as they migrate from a deep source or are produced from a deep reservoir toward the Earth's surface. Oils that contain large portions of waxy paraffins may be liquids at high subsurface temperatures, but may solidify at lower temperatures at or near the surface.

ORIGIN OF OIL AND GAS

General theories regarding origin

A large number of principles, hypotheses, and theories have been discovered or proposed to explain the existence and origin of oil and gas. The range of concepts may be considered as representing two major categories: abiogenic (inorganic) and biogenic (organic).

Abiogenic theories. Inorganic theories explaining the origin of oil and gas have been largely based on the occurrence of hydrocarbons in extraterrestrial bodies (planets and meteorites), in volcanic emanations, and in reservoirs within igneous intrusions, and the fact that they are composed of elements common to inorganic compounds that can be converted to hydrocarbons by laboratory processes. Comprehensive discussions of abiogenic concepts can be found in Hedberg (1964), Dott and Reynolds (1969), and Porfir'ev (1974). A number of the proposed mechanisms involve processes that occur deep within the Earth, where they cannot be sampled or measured as a matter of proof. The long-term thermal instability of all hydrocarbons, except possibly methane, at great depth would seem to invalidate many of the hypotheses (Hunt, 1975; Barker, 1977). Many petroleum geologists and geochemists discount the possibility of inorganic hydrocarbon genesis, and most question its significance with respect to that derived from sedimentary organic matter; however, some recent interest has been shown by the work of Gold and Soter (1980), Gold (1986), and Morency and others (1986).

Biogenic theories. A wide variety of both specific and general organic, animal, and vegetable substances have been considered by various researchers (Dott and Reynolds, 1969) as precursor material to the direct or indirect generation of oil and gas. Problems arising from the variation in concepts formulated by individual investigators center around: (1) whether organisms themselves directly generate oil and/or gas or whether they are derived indirectly from their remains; (2) the types of processes that generate oil and/or gas indirectly from organic matter (i.e., diagenetic, thermokinetic, etc.); (3) the time of the process (early or late); (4) the depth of the process (shallow or deep); and (5) the types of products (oil versus gas) generated from specific types of organic matter.

Although by no means unanimous, the general consensus among most geologists and geochemists is:

1. Significant amounts of natural gas (principally methane) are generated in the shallow subsurface by bacterial processes acting on primary organic matter derived from aquatic and terrestrial organisms after their death and shortly after deposition.

2. Oil and large amounts of both wet and dry gas are generated by time- and temperature-dependent metamorphic processes acting on sedimentary organic matter that has been buried to considerable depth. Most oil and gas derived by this process is believed to originate within sediments containing unusual concentrations of organic material. These sediments are commonly referred to as source rocks.

The following discussion emphasizes basic principles and concepts related to this general consensus. More detailed description of the scheme to be considered is contained in standard reference textbooks by Hunt (1979), Tissot and Welte (1984), and Waples (1985).

Modern concepts of oil and gas generation from buried organic matter

The conversion of primary organic matter to oil and/or gas occurs along one of the several possible paths involved in the carbon cycle (Fig. 1). While most organic matter is created by living organisms in the biosphere and is subsequently destroyed and recycled to the atmosphere and hydrosphere in the form of carbon dioxide and water, some of it may escape destruction and be incorporated into sediments that are subsequently buried to increasing depths and temperatures. During burial, this organic matter first becomes chemically altered and restructured, then thermokinetically dissociates and fractionates off volatile fluids (including oil and gas), and finally ends up at great depth as elemental carbon in the form of graphite. The following discussion describes details concerning this evolutionary path.

Creation, preservation, and deposition of organic matter. The creation of organic matter constituting the living biomass of the Earth is triggered by the process of photosynthesis, which utilizes the energy of sunlight to combine carbon dioxide and water in the formation of plant tissues. Much of the plant material is subsequently utilized in the complex life and food chains of animals. When both plants and animals die, the resulting biomass is altered and, in most cases, completely destroyed, by scavenging organisms (saprophytes). Primary organic matter that escapes total destruction, together with matter that may be added secondarily by the scavenger biomass, may be incorporated into sediments. Sedimentary organic material is principally derived from three sources: (1) simple aquatic (marine or lacustrine) animal and plant plankton, algae, etc.; (2) terrestrial vascular plants; and (3) secondary scavengers (principally bacteria).

Sediments that contain appreciable quantities of organic matter (0.5 wt % carbon or more) may eventually become effective sources for thermally generated oil and gas. These organic-rich sediments are relatively rare and are associated with highly specialized environments that are conducive to: (1) the abundant creation of living organic material, (2) the preservation of this material from saprophyte destruction before its incorporation into

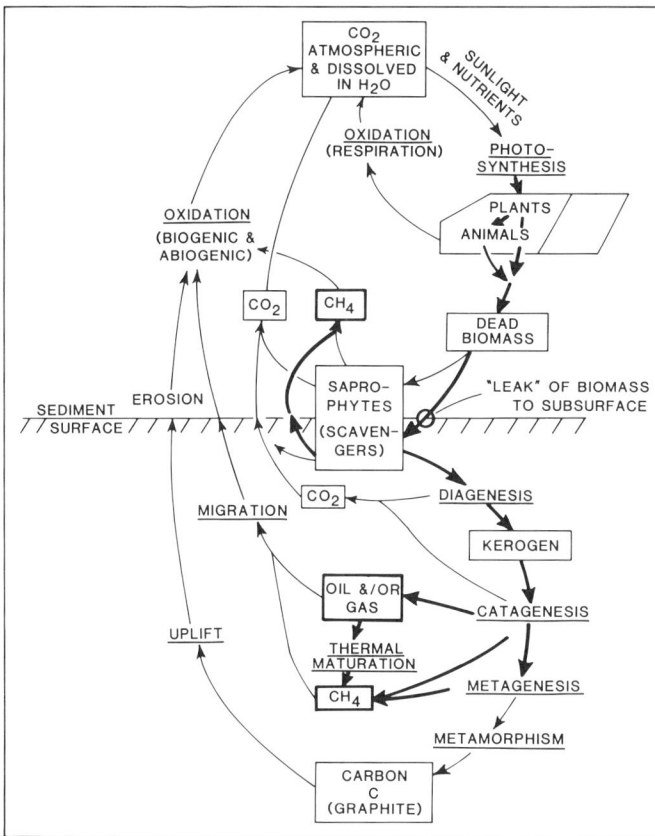

Figure 1. Pathways leading to oil and gas generation within Earth's carbon cycle.

sediments, and (3) a lack of dilution by inorganic mineral sediments (clay, sand, carbonates). High organic productivity is triggered by the presence of sunlight, the availability of nutrient "fertilizers" (potassium, phosphorus, etc.), an abundance of oxygen, and a lack of toxic poisons (hydrogen sulfide, tannic acids, and phenols, etc.). Preservation is accomplished by the absence of oxygen and/or the presence of toxins. Preservation and undiluted sedimentation require low energy levels. Some dimension of lateral and/or vertical transport may be required to link environments of organic productivity with those of sedimentation. Specific environments associated with organic sedimentation include areas of oceanic upwelling, lakes with stagnant bottoms, silled basins, and peat swamps (Trask, 1932; Demaison and Moore, 1980; Parrish, 1982; Ibach, 1982; Lyons and Rice, 1986).

Subsurface alteration of organic matter. The alteration of organic material in the subsurface may be divided into progressive stages (Fig. 2) characterized as diagenesis, catagenesis, and metagenesis (Tissot and Welte, 1984).

Diagenesis, the first stage of subsurface organic transformation, occurs concurrently with sediment consolidation and lithification, and generally extends to depths ranging from 1,800 to 4,500 m, with the range being dependent on burial rate and thermal conditions accompanying burial. In general, the shallower depth limit is related to slow or long-term burial and/or high temperatures. The deeper limit is related to rapid burial and/or low temperatures. Early diagenetic processes take place in the upper few centimeters or meters beneath the sediment surface and are generally characterized by anaerobic microbial saprophyte alteration of original organic material. In this initial phase, biologic building blocks derived from primary material and consisting of biopolymer compounds (e.g., proteins, carbohydrates, lipids, and lignins) are converted to simpler biomonomers (e.g., sugars) and organic acids, and to a more complex residual material called humin. As diagenesis proceeds, biomonomers condense and combine with humin to form large, complicated geo-macromolecules collectively called kerogen. In some cases, large amounts of methane may be generated during early phases of organic diagenesis. This bacterial methane may be an important source for economically significant shallow gas accumulations (Rice and Claypool, 1981). These accumulations are characterized by dry methane gas enriched in light carbon isotope ^{12}C.

Catagenesis, the second stage in organic evolution, occurs during increased burial and exposure to temperatures at which kerogen created during diagenesis becomes thermally unstable. During catagenesis, geo-macromolecules containing kerogen are partially converted to simpler geomonomers, initially including carbon dioxide and water, and phasing into oil and/or gas. During late phases, oil generation from oil-prone kerogen gives way to condensate and wet gas. Specific processes of catagenic transformation include thermal cracking, decarboxylation, and hydrogen disproportionation. This is the principle stage of oil and gas formation, and its presence in subsurface settings with respect to organic-rich sediments (source rocks) is of utmost importance in the creation of accumulations of these substances.

During metagenesis, the third stage of organic evolution, residual kerogen becomes highly condensed and enriched in carbon. Considerable quantities of dry methane may be evolved in early phases, both directly from the residual kerogen and also from the thermal cracking of unmigrated oils generated during the preceding stage of catagenesis. The only remnants of the original sedimentary organic matter are thermally stable carbon in a form resembling graphite and unmigrated methane gas.

Types of kerogen and their generated oil gas products. Several basic types of kerogen may be produced from organic material near the end of its diagenetic stage of evolution. These types are related to the origin and composition of the primary and biogenically altered material from which they were derived. They are also related to the types of products (oil or gas) they produce during catagenesis or within maturity "windows." Several different classification schemes are commonly used for specific kerogen types (Fig. 3). These depend on their genetic origin, geochemical composition, and character of petrographically distinctive organic matter (macerals). Based on their elemental composition, which consists almost entirely of carbon, hydrogen, and oxygen, four "pure" kerogen types may be recognized. These types and their evolutionary maturity changes may be distinguished on a van

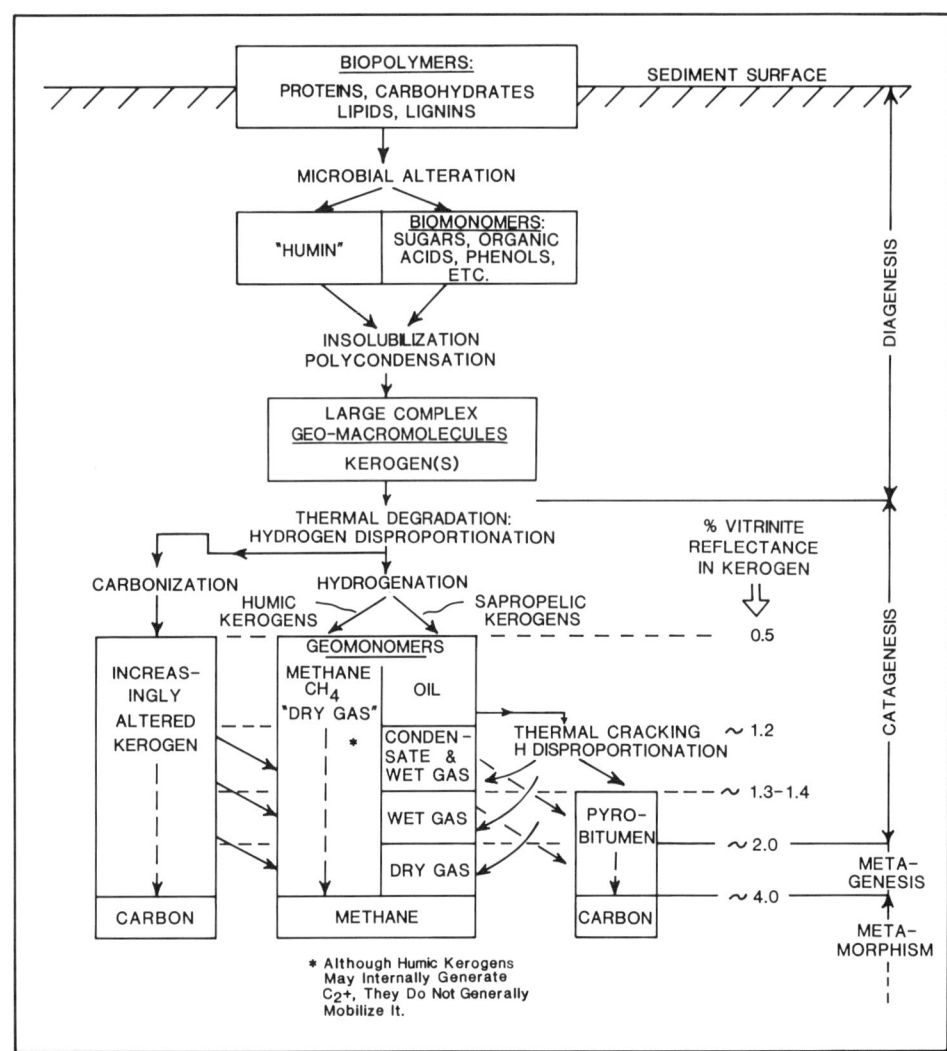

Figure 2. Alteration of organic matter in the subsurface, showing processes and products (modified from Hunt, 1979; Tissot and Welte, 1984).

Krevelen-type diagram (van Krevelen, 1950; Tissot and Welte, 1984), which utilizes the atomic carbon, hydrogen, and oxygen present in kerogen in order to characterize both kerogen-type affinity and compositional changes that occur during its subsurface alteration. Fields for the four "pure" kerogen types and their evolutionary paths (indicated by arrows) through diagenesis, catagenesis, and metagenesis are depicted in Figure 4. Mixtures of any of the types depicted may occupy the spaces between the evolutionary, pure kerogen-type bands, and these mixtures are common in nature. The decrease in atomic hydrogen/carbon and oxygen/carbon ratios along evolutionary paths reflects the preferential loss of hydrogen and oxygen from kerogens in the form of hydrocarbons, carbon dioxide, and water.

Sapropelic type I and II kerogens are derived principally from simple marine and lacustrine aquatic organisms, and contain relatively large amounts of hydrogen; type I contains the most. These kerogens generate mostly oil within the early phases of catagenesis, and condensate and wet gases during later stages. They generate only dry methane gases during metagenesis. Humic type III kerogens, of which coals are representative, are derived principally from terrestrial vascular plants and contain lesser amounts of hydrogen than types I and II. They are generally thought to generate mostly dry methane gases through the catagenic and metagenic stages, although there is some evidence that liquid oils and condensates may also be generated within the catagenic stage under certain conditions of slightly varying composition or burial and heating history (Durand and Paratte, 1983; Tissot, 1984). Type IV kerogen represents organic material that has been severely altered and carbonized or oxidized prior to sedimentary burial. It contains little hydrogen and thus generates only negligible amounts of dry gas during subsurface alteration.

The thermally controlled alteration of kerogens may be simulated by laboratory pyrolysis of source-rock samples, and the types and amounts of generated products as well as their kinetic reaction rates determined (Barker, 1974). Commonly used techniques include those of the Fischer assay (used primarily on oil shales) and the Rock Eval method (Espilalié, 1986). Pyrolysis measurements indicate that at the beginning of catagenesis, source rocks containing type I kerogen are capable of generating the largest amount of hydrocarbons per gram of organic carbon, and those containing type III generate the least, with type II being intermediate. Fischer assay data for type I kerogen found in the Green River oil shales near the start of catagenesis (Cook, 1974) are found to yield aproximately 0,007 m^3 of oil per metric ton of rock per percent organic carbon. Similar data for type II kerogen found in the Devonian shales of Kentucky (Maynard, 1981) indicate yields of approximately 0.004 m^3 per metric ton per percent organic carbon. These laboratory yield values are believed to be somewhat high, in that the reaction kinetics associated with short-term high-temperature pyrolysis may be different than those associated with long-term lower-temperature subsurface generation (Burham and Singleton, 1983). Pyrolysis yields from type III kerogens (considered to generate mostly gas) may not be meaningful in that their higher oxygen content is thought to be released as hydrogen-bearing water, leading to carbon enrichment of residual kerogen in short-term high-temperature laboratory heating rather than as carbon dioxide in long-term lower-temperature earth processes (Tissot, 1984). Data for a more-empirical model of methane generation from humic coals (Jüntgen and Karweil, 1966) indicate a generation capacity of approximately 240 m^3 of gas per metric ton of coal within the stages of catagenesis and metagenesis. This value equates to approximately 3 m^3 per metric ton per percent organic carbon for disseminated type III kerogen based on 86 percent carbon in humic coal at the approximate start of thermally induced methane generation.

Organic maturity indicators and their relation to generation efficiency. During the progressive alteration or "maturation" of organic material with increasing depth and temperature, certain physical and chemical changes take place that allow the material to be placed in a spectrum of diagenetic/catagenetic/metagenetic alterations. Héroux and others (1979)

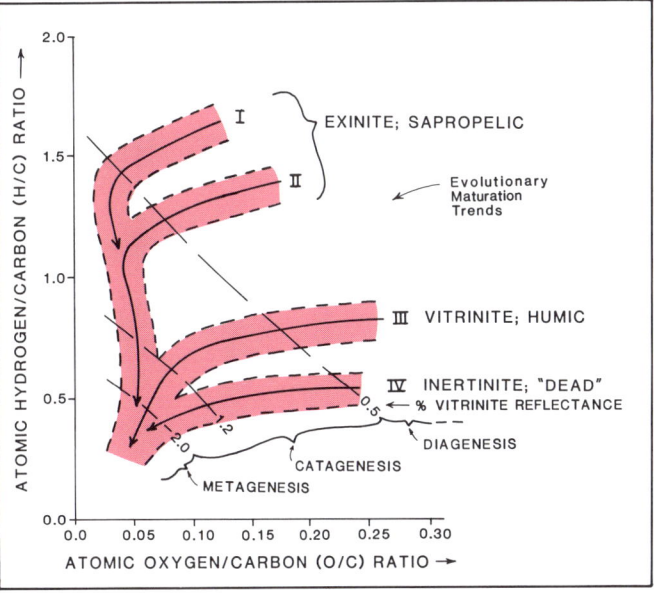

Figure 4. van Krevelen-type diagram showing fields and evolution paths for various types of kerogen (after Tissot and Welte, 1984, and others).

have presented a comprehensive summary of the physical/chemical changes that are used to establish alteration positions within this spectrum. A few of the more commonly used maturity scales (Fig. 5) include those of coal rank, spore and general organic matter coloration, and vitrinite reflectance. Coal rank (as measured by BTU heat and volatile matter content) and vitrinite reflectance (the percentage of light reflected by a certain type of maceral called vitrinite) increase with increasing organic maturity. The colors of spores and other organic matter change from light yellow to brown and then black with increasing maturity. The zone of catagenesis and significant oil/gas generation is shown to occur in: (1) the high-volatile bituminous C through low-volatile bituminous rank range, (2) the darker yellow-to-brown spore and general organic matter color range, and (3) the 0.5 to 2.0 percent vitrinite reflectance range. Source rocks within the catagenic stage that have generated sufficient oil and gas to have achieved expulsion into a reservoir are said to be "mature"; those that have not yet reached this stage are said to be "immature"; and those that have reached advanced stages where a particular product (i.e., oil, condensate, wet gas, dry gas) is no longer being generated are said to be "over-mature" or "over-cooked" for that particular product. Although the onset of incipient oil generation is generally believed to occur at approximately 0.5 percent vitrinite reflectance, the onset of effective expulsion from most average source rocks containing approximately 1 to 10 percent organic carbon generally occurs in the range of 0.6 to 0.7 percent. The floor of liquid oil and the onset of condensate and wet gas generation generally occurs in the approximate range of 1.2 to 1.3 percent reflectance; the floor of condensate generally is about 1.4 percent reflectance; and the top of dry gas generation is

"GENETIC"	GEOCHEMICAL	PETROGRAPHIC MACERALS	THERMAL PRODUCTS
SAPROPELIC	TYPE I	EXINITE (LIPTINITE)	OIL → CONDENSATE → WET GAS → DRY GAS WITH INCREASING THERMAL ALTERATION
	TYPE II		
HUMIC	TYPE III	VITRINITE	MOSTLY DRY GAS –Possibly some oil/wet gas in certain cases.
"DEAD"	TYPE IV	INERTINITE (FUSINITE)	INSIGNIFICANT DRY GAS

Figure 3. Relationship of various terminologies and classifications applied to kerogen, with a summary of oil and gas types mobilized during their thermal alterations.

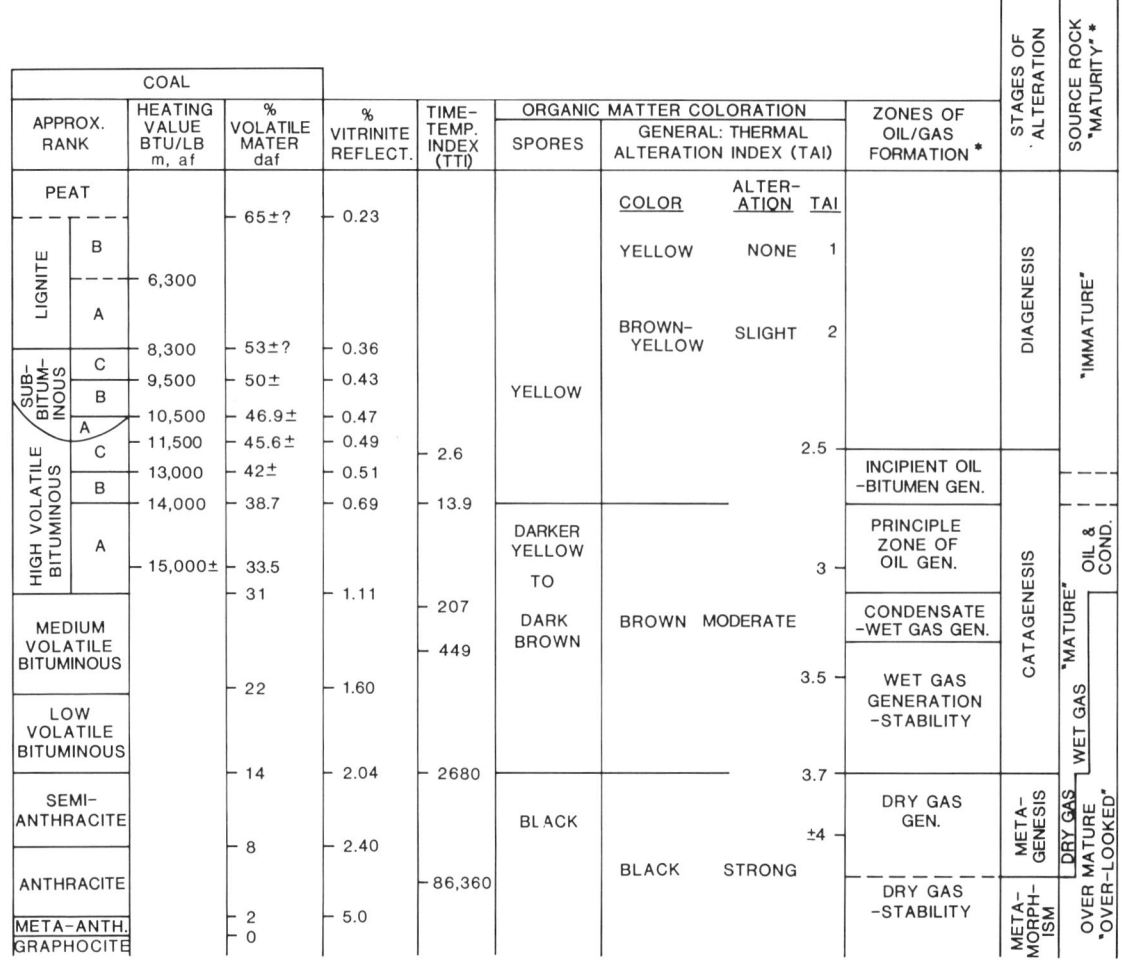

Figure 5. Commonly used indicators of evolutionary change in subsurface organic matter, with related stages of alteration and oil/gas generation. Data compiled from various sources, including Meissner (1984; coal rank data and vitrinite reflectance), Bond (1984; time-temperature index related to vitrinite reflectance), Hood and others (1975, spores and organic matter coloration), Tissot and Welte (1984; general concept of maturity and alteration stages).

about 2.0 percent reflectance. These maturity "thresholds" or generation "windows" based on vitrinite reflectance are approximate guidelines and may vary somewhat, due to the fact that reaction kinetics of type I and II kerogens may be different than those type III kerogen (Tissot and others, 1987).

Another commonly used maturity indicator is the Tmax value obtained during Rock Eval pyrolysis (Espilalié, 1986). This value represents the temperature at which maximum pyrolytic hydrocarbon generation is achieved. It tends to increase with increasing thermal alteration and defines boundary limits for the various generation thresholds of specific kerogen types.

Laboratory pyrolysis also shows that generation capacity from a given kerogen type decreases with increasing levels of catagenic/metagenic alteration. This observation allows an estimation of the amount of oil and/or gas that has been generated at any particular level. Figure 6 shows an example wherein the percentage of ultimate oil and gas generation for general kerogen types are related to given stages of catagenesis through the use of a "maturity" scale based on vitrinite reflectance. The curve depicted for oil is somewhat hypothetical. More specific curves for given kerogen types may be constructed utilizing the "transformation ratio" described by Espilalié (1986), or as determined by kinetic parameters (Tissot and others, 1987).

Influence of temperature and time on oil and gas generation. The generation of oil and gas from kerogen may be described as a collection of independent, first-order, thermokinetic chemical reactions in which both time and temperature are important. Techniques have been devised by which the depths, subsurface temperatures, and times at which certain maturity stages occur may be quantitatively determined (Lopatin, 1971; Waples, 1980; Royden and others, 1980; Tissot and others, 1987). The methods involve a burial reconstruction of the sedimentary section, an estimation of the geothermal gradient through time, and values for the kinetic reaction rates. Figure 7 shows a simple burial/temperature history diagram that indicates the times, depths, and temperatures at which certain critical oil and gas generation thresholds occur. The use of burial history and time/temperature modeling coupled with information on kerogen-specific generation capacities, source-rock thicknesses, and areal extents allows quantitative estimates to be made concerning the amounts of oil and gas that may have been generated in a basin.

MIGRATION OF OIL AND GAS

General considerations

Oil and gas (as well as ground water) are potentially mobile fluids that are intrinsically capable of migrating through a fixed solid rock framework, providing that suitable mechanisms, pathways, and driving energies are present. As discussed in the preceding text, most of the Earth's oil and gas are believed to have formed within mature source rocks that are generally situated within the deeper and hotter parts of sedimentary basins. Although exploitable oil and gas accumulations are occasionally found within fractured reservoirs, which also served as the probable source rock, the following observations provide evidence for their outward migration.

1. Most mature source rocks do not contain the amounts of hydrocarbons and related bituminoids that they would have theoretically generated.
2. Reservoirs for many accumulations are in rocks that are thermally immature and/or do not contain sufficient organic matter to have generated the amount of oil and/or gas contained in them.
3. Oils in a number of accumulations have been geochemically correlated through their compositional similarity to bitumens extracted from distant source rocks from which they were derived.
4. Petroleum fluids move (migrate) through the reservoirs of accumulations toward well bores during conventional production operations.

Because of the evidence that oil and gas do move through certain types of subsurface rocks, most petroleum geologists consider the processes and pathways of migration to be an essential link between their area of origin and site of accumulation. However, it may be shown that some plausible migration processes and pathways may lead to dispersal or destruction rather than concentration into exploitable deposits. Accumulation is commonly thought to occur in areas of "static entrapment," where dynamic migration ceases. It may also occur where bottlenecks to dynamic migration are present, wherein rates of migration into a reservoir area exceed those out of it, or where active leakage out of a potentially static trap is taking place.

The mechanism(s) by which migration is accomplished is currently controversial. Many petroleum geologists believe that the subject of migration is not understood as well as that of either

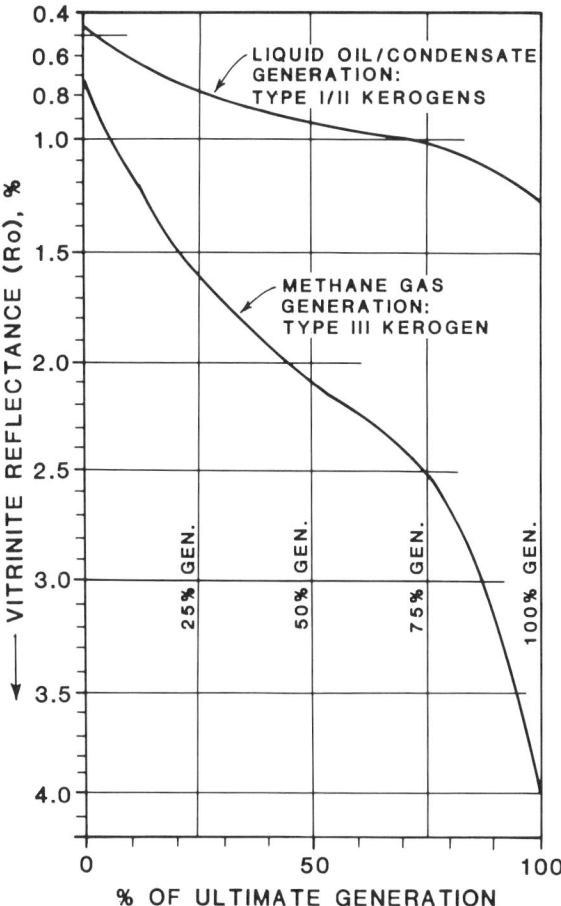

Figure 6. Percentage amounts of ultimate oil or gas generation capacity for different types of kerogen as a function of vitrinite reflectance. Oil-generation curve for an approximate average of type I and type II kerogens adapted from Waples (1979). Gas-generation curve for type III kerogen based on coal data from Jüntgen and Karweil (1966).

its origin or accumulation. A large number of theories and hypotheses within a wide range of plausibility have been offered to explain migration. Multiple mechanisms undoubtedly exist; however, they may operate at different scales and levels of efficiency within the realm of varying geological constraints. Comprehensive discussions concerning principles and problems of migration have been presented by Levorsen (1967), Dott and Reynolds (1969), Hunt (1979), Roberts and Cordell (1980), Perrodon (1983), Tissot and Welte (1984), England and others (1987), and Doligez (1987).

Stages of migration

The terms primary and secondary migration (Illing, 1938) are commonly used to describe stages of oil and gas or protopetroleum movement between a source area and an accumulation. Primary migration is generally used to refer to movement from a fine-grained source rock characterized by low permeability and

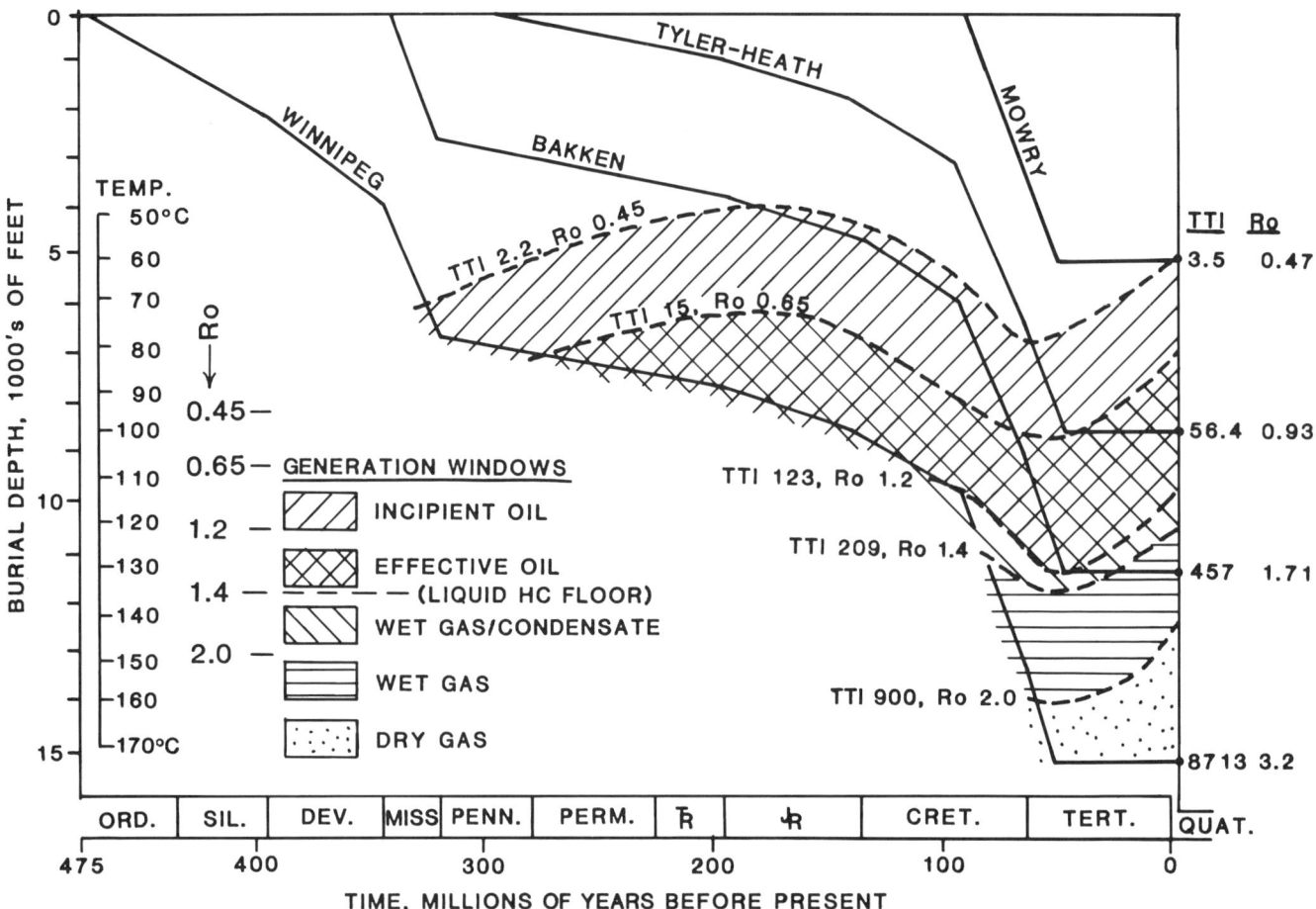

Figure 7. Burial and thermal maturation history for several source-rock units present in the deepest part of the Williston Basin, North Dakota. Time–temperature index (TTI) values calculated after the method of Waples (1980). Relation between TTI and equivalent vitrinite reflectance (R_o) is calculated after the method of Bond (1984).

small pore size toward a carrier/reservoir rock system characterized by higher permeability and larger pore size. The term is often equated to the expulsion of petroleum fluids from a source rock; however, it should also be taken to include movement through any other fine-grained nonsource rocks that may exist between the actual source and a carrier/reservoir system. Secondary migration is involved with the movement of petroleum fluids within a generally porous and permeable carrier-rock system that contains the properties of a potential reservoir in which accumulation may take place. The synonymous terms "remigration," "tertiary migration," or "dismigration" are used to denote oil and gas movement within the carrier/reservoir system resulting from the disturbance of an established accumulation. Remigration may be called "flushing" if it is caused by water movement. The principal differences between the various stages of migration concern grain size and its effects on permeability and pore size.

Probable mechanisms of migration

Only the most plausible, significant, and generally accepted concepts are considered in the following discussion.

Under subsurface conditions, oil and gas may be dissolved in ground water, absorbed in inorganic mineral or kerogen surfaces, or present in a state resembling solid solution within kerogen. They may also be present as a free bulk phase within rock porosity, as is the case where they accumulate in exploitable reservoirs. These states impose constraints on the possible mechanisms that may mobilize them.

In accordance with a scheme proposed by Roof and Rutherford (1958), the following mechanisms, listed in order of increasing maximum rate efficiency, appear to represent the most likely mechanisms of oil and gas migration:

1. Diffusion in a dissolved state within static ground water or kerogen, or along surface films on kerogen surfaces.
2. Transport by moving ground water containing dissolved oil and/or gas.
3. Transport as a free bulk phase through rock porosity originally occupied by water under conditions where ground water is either static or dynamic.

Diffusion migration. This mechanism of migration is governed by Fick's Law:

$$J = D \times \frac{dc}{z},$$

where J = mass rate of transport across an area (mass/time/area); D = total effective diffusion coefficient = d × t, where d = free path diffusion coefficient and t = a "tortuosity" factor; dc/z = concentration gradient: change in concentration (dc) through distance z.

Significant papers concerning diffusion migration include those of Uspenski (1962), Watts (1963), Smith and others (1971), Pandey and others (1974), Bogrodia and Katz (1977), Leythaeuser and others (1982), and Welte and others (1984). Although evidence suggests that diffusion of hydrocarbons dissolved in ground water may be the slowest of plausible migration mechanisms, its effectiveness through magnitudes of geologic time may be significant. Methane has the highest diffusion rate of any hydrocarbon dissolved in ground water, and for normal paraffins, rates decrease logarithmically with increasing number of carbon atoms per molecule. Little is known about the diffusion of oil and gas along surface films formed on kerogen or through a kerogen matrix with the exception of methane movement through humic coals (van Krevelen, 1961). Diffusion is considered to be an important mechanism in production of coal-bed methane (Kissel and Edwards, 1975). Similar diffusion of methane and other petroleum species may occur through or along a continuous kerogen network in more conventional source rocks, and the process may be important for achieving expulsive primary migration from them. The driving force for diffusion migration is the concentration gradient, and migration directions are the same as those for the direction of decreasing concentration. Solution diffusion through ground water must involve an exsolution process to convert it to the free bulk phase stage in which it is found in reservoir accumulations.

Migration of oil and gas dissolved in moving ground water. This mechanism is governed by Darcy's Law:

$$q = (k \times \frac{\rho}{\mu} \times \frac{d\Phi}{z}),$$

where q = mass rate of transport across an area (mass/time/area); k = rock permeability; ρ = water density; μ = water viscosity; $d\Phi/z$ = potential energy gradient: change in potential ($d\Phi$) through distance (z).

Significant papers concerning this migration process include those of Meinshein (1959), McAuliffe (1978), and Price (1976, 1977, 1978). The solubilities of the various molecular hydrocarbon species in water are a major consideration. In any homologous series of paraffins, naphthenes, or aromatics, solubilities tend to decrease substantially with increasing number of carbon atoms in the molecule. For the same number of carbon atoms, aromatics are the most soluble, naphthenes are intermediate, and paraffins are the least soluble. Among the range of individual hydrocarbon species, methane is the most soluble (approximately 7,500 ppm in fresh water at 25°C, 1 atmosphere), followed by ethane (2,500 ppm), and benzene (1,300 ppm). Solubility of the C_{14} n-paraffin tetradecane is only 0.008 ppm, and the C_{14} aromatic phenanthrene is only 1 ppm. Values for even larger molecules are indicated to be extremely low. Solubilities generally increase with increasing temperature and decrease with increasing water salinity. Higher pressures generally increase the solubilities of hydrocarbon gases, but decrease those of liquids. Solubilities of whole crude oils are quite low at laboratory temperatures (less than 15 ppm), but increase to 45 to 450 ppm between 165 and 180°C.

Because methane is the most soluble of hydrocarbons, it would appear to be the most likely species affected by solution migration in moving water (Magara, 1980). Dissolved hydrocarbons have been observed in actual subsurface formation waters (Buckley and others, 1958), and up to 2.5 m^3 of hydrocarbon gas per m^3 of water has been observed to exsolve from some of those found in the Gulf Coast area. Extremely large ground-water volumes appear necessary to transport dissolved oil. Water may be supplied as an initial pore fluid in compacting rocks, as a diagenetic fluid created from clay or gypsum dehydration or from deep circulation systems involving meteoric, metamorphic or igneous processes. The major portion of conventional shale, sandstone, and carbonate rock compaction appears to occur before the onset of catagenic hydrocarbon generation, although significant compaction may be delayed in some instances by the occurrence of abnormally high formation fluid pressures. Some studies have demonstrated insufficient water availability to transport observed volumes of oil associated with well-defined migration systems; other studies have indicated a favorable material balance (Jones, 1980; Bonham, 1980).

The driving force for the transport of oil and gas dissolved in ground water is related to the presence of a potential energy gradient (Hubbert, 1940). Ground-water flow may be initiated by a variety of processes, the most significant of which are compaction, water loss caused by clay diagenesis, and circulation induced by differential topography. Migration directions for oil and gas are the same as those of ground water, and are in directions of decreasing "hydraulic head" or potentiometric surface elevation. As in the case for solution-diffusion in static ground water, the mechanism must involve some process of exsolution to convert it to the bulk-phase pore-saturating state in which it is found in all conventional reservoir accumulations. Such exsolution may be caused by changes in saturation temperature/pressure or groundwater salinity, or perhaps even by molecular-size/pore-throat-size filtration. Exsolution may occur at any point along either primary or secondary migration paths. The fact that the ratios of individual molecular species found in oils contained in bulk-phase reservoir accumulations are not in accord with their water solubilities suggests that this migration mechanism is not compatible with that which has implaced most of Earth's oil.

Transportation of oil and gas as a free bulk phase. This mechanism is governed by Darcy's Law, as is the case for solution transport in moving ground water; however, values for fluid density (ρ) and viscosity (μ), permeability (k), and potential energy gradient ($d\Phi/z$) are those of migrating oil and gas phases rather than ground water.

The entry and partial saturation of oil and gas phases into

and within rock pores originally occupied only by water are associated with capillary behavior that depends on: (1) interfacial or surface tensions between the various fluid species (oil, gas, water); and (2) surface attractions (wettability) the fluid species have for adjacent solid inorganic mineral or kerogen particles (Amyx and others, 1960). Significant papers considering the importance of capillarity in controlling primary and secondary migration of oil or gas, as well as their accumulation and remigration, include those of Hubbert (1953), Aschenbrenner and Auchauer (1960), Arps (1964), Berg (1975), and Schowalter (1979). Capillary phenomena create differential (capillary) pressures between the different fluid phases that occupy porosity. The magnitude of capillary pressure may be related to a number of factors, including the sizes of pores and pore throats, entry or expulsion of nonwetting fluids, and relative amounts (percent saturation) and effective permeability of fluid species. Capillary behavior determines which rocks will permit fluid entry and which will prevent it, depending on available pressures. In general, the matrix mineral framework of most rocks (shales, sandstones, carbonates) are preferentially nonwetting to oil and gas. This means that pressures higher than those in surrounding water are required to cause entry and migration, with larger amounts of pressure being required for fine-grained rocks with small pore throats than for coarser-grained rocks with large pore throats or for fractures. Thus, fine-grained rocks (shales) often form effective barriers to migration and trap seals for accumulations, while coarse-grained rocks form carrier/reservoir systems. Studies indicate that migration is more likely to be accomplished by filaments or slugs of oil and gas with critical vertical dimensions that are in continuous phase continuity between several pores rather than as small discrete bubbles (Berg, 1975). Approximately 10 percent of pore volume must be saturated by oil and gas before a continuous phase network through a rock can be established (Schowalter, 1979). Capillary behavior requires that residual immobile oil/gas saturation should remain in rock pores after active movement through them has ceased. This fact may explain the many occurrences of trace residual saturations along probable migration paths. Residual capillary saturation along a migration path may result in dissipation of migrating fluids before a site of effective accumulation is reached. In many fine-grained rocks, through which migration must take place, pressures sufficient to cause capillary entry and continuous phase saturation are not present; however, pressures may be high enough to cause tensional fractures to form and permit migration (Momper, 1978; Meissner, 1978; Palcilauskis and Domenico, 1980; du Rouchet, 1981). In contrast to most inorganic minerals, kerogen found in source rocks appears to be preferentially oil/gas wet. This factor facilitates the establishment of a continuous phase between pores, and may lead to easy expulsion along a continuous network of kerogen (McAullife, 1978).

Pressure- and temperature-controlled phases of petroleum fluids may change during migration (Silverman, 1965; England and others, 1987), such as in the case where an upwardly migrating phase of liquid gas–saturated oil dissociates into separate oil and gas phases. The dissociated phases may consequently separate and migrate independently. Studies also indicate that significant nonphase changes in composition may occur during both primary and secondary migration due to chromatographic effects (Tissot and Welte, 1984; Leythaeuser and others, 1984) and ground-water solution.

Primary migration resulting in expulsion of oil and gas generated in source rocks is currently thought to be largely due to the development of a continuous bulk phase as generation achieves certain critical levels of saturation within them (Momper, 1978; Jones, 1980). Initial migration of oil and gas out of generative kerogen and into the fine-grained conventional pore system of the source rock is controlled by the sorption capacity of the kerogen. When sorption capacity is exceeded, the generated material forms films on kerogen surfaces and/or droplets within the pores. When films or pore saturation levels reach limits to permit continuous bulk phase movement through the fine-pore/microfracture system of the rock, expulsion to the realm of continued primary or subsequent secondary migration may take place. Both sorption and surface/pore saturation processes are controlled by: (1) kerogen type and distribution within the source rock, (2) amount of kerogen in the rock and its degree of maturity, and (3) the environmental factors of temperature and pressure. The minimum amount of generative saturation required to achieve oil expulsion appears to be in the range of 825 to 850 ppm, based on the amount of extractable bitumen found in source rocks that are known to have achieved outward migration. This factor imposes certain minimum limits on the amounts and maturity of kerogen that must be present in source rocks, in that they will not achieve continuous bulk phase expulsion unless or until they have generated sufficient material to achieve mobilization. If the generative capacity and content of oil and gas in a source rock are known, it should be possible to predict the amounts of oil and gas that have been expelled from it. Figure 8 shows an example of when expulsion from a given source rock has occurred based on an assessment of the amounts of oil and gas that it has generated, retained, and expelled.

Driving forces for the free bulk phase transport of oil and gas are related to the presence of a potential energy gradient for these substances, whereby the forces will move oil and gas between positions of maximum energy toward those of minimum energy within a continuously permeable rock framework. Rigorous mathematical treatment of the physical principles governing this subject (Hubbert, 1953; England and others, 1987) shows that potential energy factors controlling migration may be considerably different between cases in which ground water is (a) not flowing (static) or (b) is flowing (dynamic).

Under static ground-water conditions, oil and gas migration may be initiated by a variety of additive processes, including: (1) capillarity, (2) compaction resulting either from conventional depth-related mechanical processes or from that related to the conversion of solid kerogen to hydrocarbon-filled porosity (Meissner, 1978; Ungerer and others, 1983), (3) possible volume increases associated with the conversion of kerogen to oil and gas

(Momper, 1978), and (4) buoyancy. The first three processes appear to be dominant in the primary phase of migration. The processes of kerogen-to-fluid-oil/gas conversion may cause the generation of abnormally high pressures (Spencer, 1987), facilitating source rock expulsion and migration through fine-grained confining beds through the formation of fractures (Meissner, 1978). Buoyancy is the dominant mechanism of secondary migration under hydrostatic ground-water conditions and forms the basis for the widely accepted theory of anticlinal accumulation. Primary migration directions from an expulsive source rock under static water conditions may be either in an upward (most likely) or downward direction, depending on potential energy gradients and permeability paths connecting to overlying or underlying carrier/reservoir systems. The direction of secondary migration is controlled by buoyancy, in which the density of migrating oil and gas is almost universally less than that of ground water, and results in upward movement within the confines of the carrier/reservoir system to a top-barrier unit, which because of its capillary entry pressure properties is impermeable to oil and gas entry (Fig. 9). Once a top-barrier unit is encountered, migration continues laterally updip until a site of accumulation is encountered, the migrational charge dissipates itself to residual capillary saturation or ground-water solution, or the ground or water surface is reached. Lateral updip migration directions controlled by the structural altitude of a top-barrier may create areas of migrational concentration or bypass. Density differences in a series of oils and gases expelled from a source rock undergoing increasing increments of maturity may also lead to the preferential spillage and continued updip migration of dense oil and gas phases at the spill point of a top-seal trap configuration in the process of gravity segregation, described by Gussow (1954).

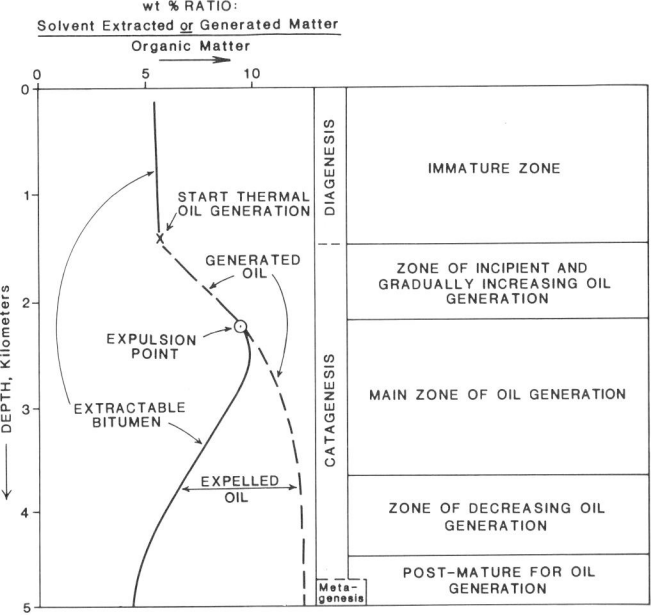

Figure 8. Extractable-bitumen and oil-generation curves illustrating the point at which initial expulsion occurs and the amount of expelled oil (modified from Kontorovich, 1984).

Bulk phase migration of oil and gas under dynamic ground-water conditions may result in migration directions and rates considerably different than those produced under hydrostatic conditions. In addition to the factors of capillarity, buoyancy, etc. that control hydrostatic migration as described above, an additional force that may be conceptually related to water movement must also be considered. As a result of this force, upward and downward migration may depart from the vertical direction dictated by buoyancy (Fig. 10), and lateral migration directions beneath a confining top-barrier may be in any direction controlled by the hydrodynamic potential energy field. The lateral migration direction is generally not in a truly updip direction and may actually be downdip or in some component of a downdip direction. Density contrasts between petroleum fluids and ground water greatly affect the magnitude of hydrodynamic influence on migration. Large contrasts produced by high-density (low API gravity) oil and/or low-density fresh water produce the most profound affects, while those involving low-density gas or high-density salt water produce the least. Dynamic ground-water flow also greatly influences capillary entry pressures of oil and gas into rocks and the efficiencies of migration barriers and trap seals (Schowalter, 1979). Flow directed from a barrier/seal increases the entry pressure and trapping capacity of the rock, while an opposite sense of flow acts conversely.

Summary evaluation of possible migration mechanisms. All of the processes described above are believed to operate at various degrees and places within the subsurface. Most investigators believe that primary and secondary migration are dominated by a continuous bulk phase mechanism when generation rates are high and material must be expelled from source rocks and transported through carrier/reservoir systems at faster rates than those that can be achieved by alternative mechanisms. However, when bulk oil/gas phases cannot be transported through rocks that are impermeable to them because of capillary entry-pressure considerations, they may be transported by solution in moving ground water through rocks that remain permeable to water. If neither bulk phase nor moving water solution transport occurs, oil and gas may move by the slower-rate process of diffusion. Transport through ground-water solution, either by ground-water flow or diffusion, is basically dissipative in nature and may through time destroy accumulations created along a path of bulk phase migration (Bishop and others, 1983). Because of its greater solubility, gas may be particularly sensitive to these mechanisms, and it appears that gas accumulations may not remain through long periods of geologic time.

Distances and directions of migration

The position of mature source rocks relative to those of related oil and gas accumulations indicates a wide range of migrational distances and directions. Fractured reservoirs within mature source rocks represent essentially in situ accumulations where migration distance is simply between the source-rock matrix and the fractures that transect it. Somewhat larger distances

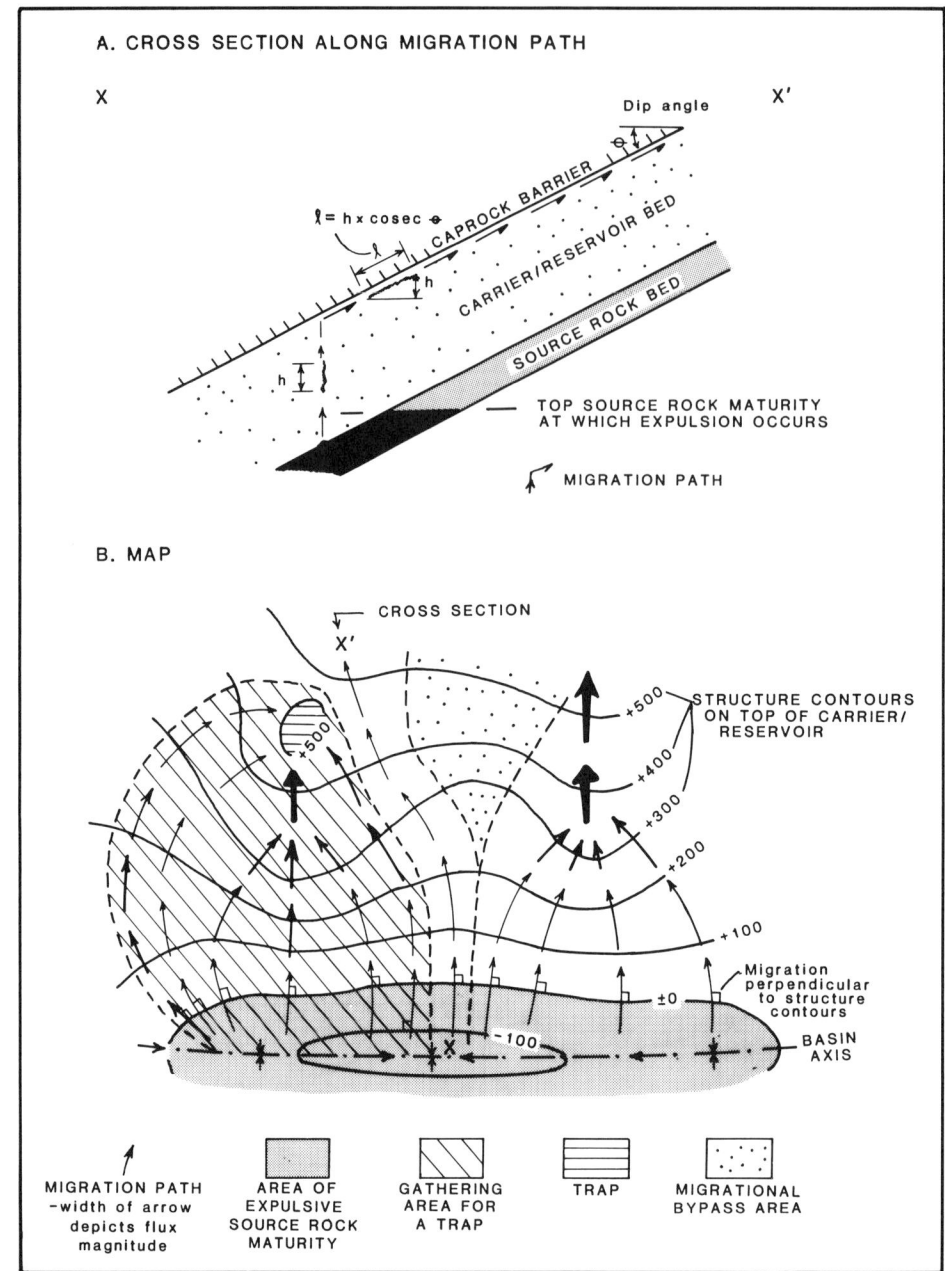

Figure 9. Cross section and map showing directions of bulk, free-phase, oil and gas, secondary migration under hydrostatic ground-water conditions. Critical heights, h, (Berg, 1975) and lengths, l, (Aschenbrenner and Achauer, 1960) of oil and gas filaments shown in the cross section are those required to overcome capillary forces and permit migration.

may be involved where accumulation occurs in a locally developed stratigraphic reservoir (reef, sandstone lens) adjacent to or surrounded by mature source rocks. Migration distances in this case may be equated to the volume of source rock (thickness × distance) required to generate and mobilize the volume of oil and gas found in the accumulation. Distances may be much larger where reservoir accumulations are removed from sites of generation. Lateral distances may be only a few kilometers in migrational systems disrupted by stratigraphic or structural discontinuity (rift basins); they may be very great in nondisrupted, areally extensive carrier/reservoir systems (broad cratonic platforms). Lateral distances of 300 km or more have been indicated by some detailed case studies (Momper, 1978; Meissner, 1978). Upward migration paths from a source rock toward a distantly overlying reservoir accumulation have often invoked the principle of buoyancy to explain an approximately vertical migration system; however, instances have been noted where migration has been stratigraphically downward into an underlying reservoir. Upward

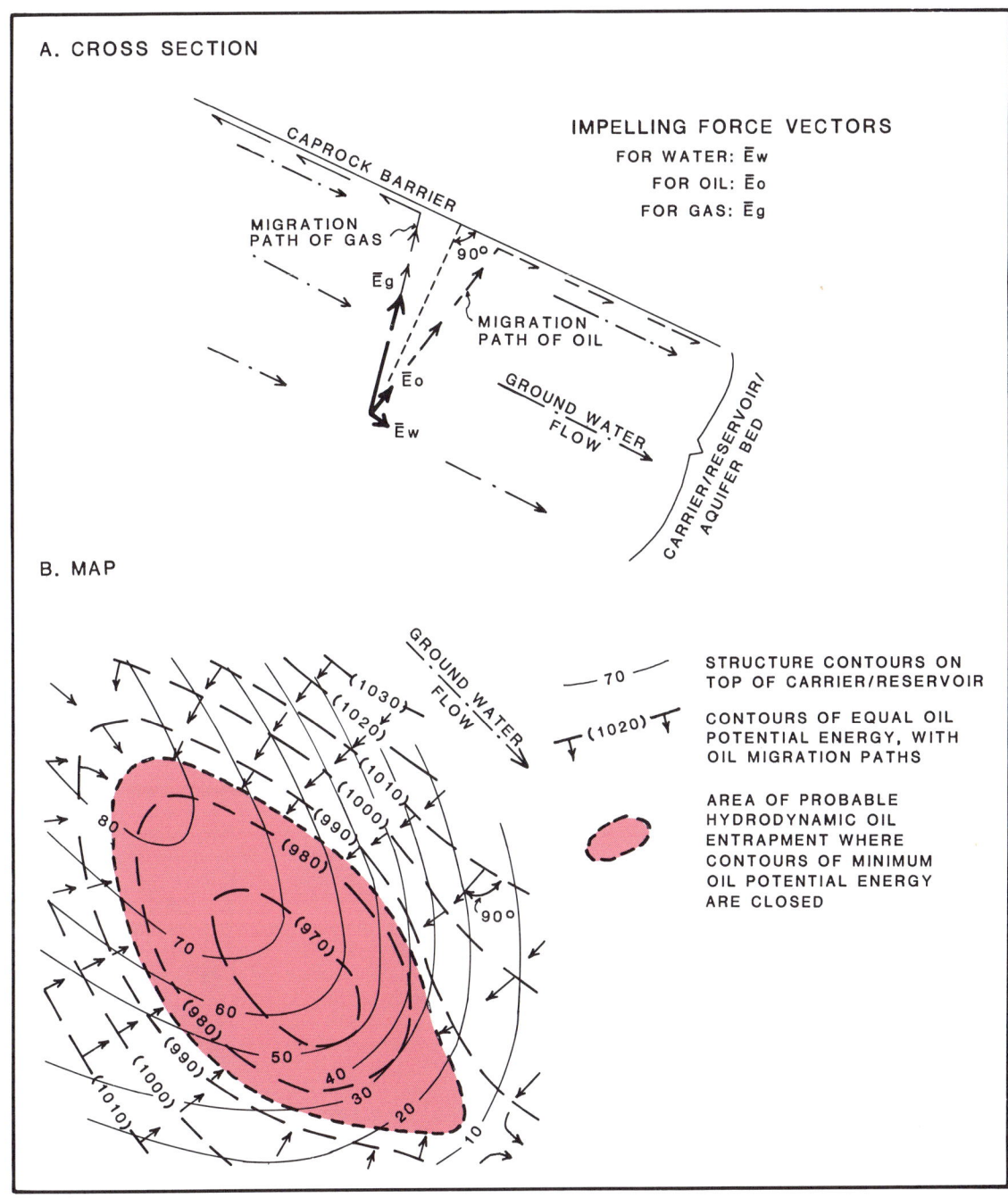

Figure 10. Cross section and map showing secondary migration of oil and gas under hydrodynamic ground-water conditions. Cross section (not related to map) illustrates a special case where downdip ground-water flow allows low-density gas to migrate updip, but causes high-density oil to migrate downdip (after Hubbert, 1953).

vertical distances 300 m or more have been reported in case studies (Momper, 1978; Meissner, 1978; Momper and Williams, 1984), and even larger distances have been postulated in some basins where migration is believed to occur up regional faults and fractures. Since the ultimate migrational goal of oil and gas is to reach minimum potential mechanical or free-energy positions at the Earth surface, possible maximum distances appear to be constrained only by the availability of migration paths through which fluids may move or diffuse.

CONCLUSIONS AND OPPORTUNITIES FOR FUTURE RESEARCH

Although some oil and gas may conceivably be created by abiogenic processes, most of Earth's petroleum fluids appear to have been generated from organic material originally created near its surface by biologic life processes. Significant quantities of dry gas may be generated by shallow bacterial alteration of primary organic matter; however, nearly all oil and large amounts of gas appear to have been generated by time- and temperature-dependent processes acting on diagenetically altered organic matter (kerogen) concentrated into source rocks that have been buried to considerable depth. Several types of kerogen have been identified, and each of these or their apparent mixtures may be shown to generate different types and amounts of oil and/or gas depending on their constitution and degree of thermal alteration or maturation. Although the overall process appears to be reasonably well understood, considerable refinements remain to be made concerning the actual kinetics of the thermal generation process as it applies to specific kerogen types, the relation of certain biomarker compounds in oil and gas that may relate to specific precursor organic matter and/or variable generation rates, and the actual quantitative amounts of oil and gas that have been generated.

Migration of oil and gas occurs between areas of generation in mature source rocks and sites of accumulation. Commonly recognized stages of migration include those of source-rock expulsion and migration through fine-pore rocks (primary migration) and subsequent movement through large-pore rocks (secondary migration) that constitute a carrier/reservoir system in which viable economic accumulations may occur. The most plausible mechanisms for migration, listed in apparent order of increasing time-rate effectiveness or "favorability" are: (1) diffusion, (2) solution transport by moving ground water, and (3) bulk phase transport under conditions where ground water is either static or dynamic. Bulk phase transport is affected by capillary and phase-change phenomena. It appears to be the dominant mechanism creating most accumulations. If this mechanism is blocked, less favored mechanisms may become effective. Mechanisms involving solution in ground water require an exsolution process to create the types of bulk phase accumulations commonly exploited, and unless exsolution occurs, the mechanisms may result in dissipation rather than accumulative concentration. Most investigators believe that migration is not as well understood as the processes of generation and accumulation. Considerable future research opportunity exists in quantifying certain aspects of the various mechanisms, specifically those that relate to diffusion coefficients and related tortuosity factors, particularly in kerogen-rich rocks; geologic realms where exsolution may occur; capillary or solution losses along bulk phase migration paths; and solution/diffusion losses from established accumulations.

An understanding of the origin of oil and gas and how they migrate may have considerable impact in assessing exploration opportunity and potential resource volume at scales ranging from an individual prospect to those of a basin, region, or country.

REFERENCES CITED

Amyx, J. W., Bass, D. M., Jr., and Whiting, R. T., 1960, Petroleum reservoir engineering: New York, McGraw-Hill, 610 p.

Arps, J. J., 1964, Engineering concepts useful in oil finding: American Association of Petroleum Geologists Bulletin, v. 48, p. 157–165.

Aschenbrenner, B. C., and Achauer, C. W., 1960, Minimum conditions for migration of oil in water-wet carbonates: American Association of Petroleum Geologists Bulletin, v. 44, p. 235–243.

Barker, C., 1974, Pyrolysis techniques for source rock evaluation: American Association of Petroleum Geologists Bulletin, v. 58, p. 2349–2361.

—— , 1977, Discussion to Aqueous solubility of petroleum as applied to its origin and primary migration: American Association of Petroleum Geologists Bulletin, v. 61, p. 2146–2149.

Berg, R. R., 1975, Capillary pressure in stratigraphic traps: American Association of Petroleum Geologists Bulletin, v. 59, p. 939–956.

Bishop, R. S., Gehman, H. M., Jr., and Young, A., 1983, Concepts for estimating hydrocarbon accumulation and dispersion: American Association of Petroleum Geologists Bulletin, v. 67, p. 337–348.

Bogrodia, V., and Katz, D. L., 1977, Gas migration by diffusion in aquifer storage: Journal of Petroleum Technology, February, p. 121–122.

Bond, W. A., 1984, Application of Lopatin's method to determine burial history, evolution of the geothermal gradient, and timing of hydrocarbon generation in Cretaceous source rocks in the San Juan Basin, northwestern New Mexico and southwestern Colorado in Woodward, J., Meissner, F. F., and Clayton, J. L., eds., Hydrocarbon source rocks of the greater Rocky Mountain region: Rocky Mountain Association of Geologists, p. 433–447.

Bonham, L. C., 1980, Migration of hydrocarbons in compacting basins in Roberts, W. H., III, and Cordell, R. J., eds., Problems of petroleum migration: American Association of Petroleum Geologists Studies in Geology 10, p. 69–88.

Buckley, S. E., Hocott, C. R., and Taggart, M. S., Jr., 1958, Distribution of dissolved hydrocarbons in subsurface waters in Weeks, L. G., ed., Habitat of oil: American Association of Petroleum Geologists, p. 850–882.

Burnham, A. K., and Singleton, M. F., 1983, High pressure pyrolysis of Green River oil shale, in Miknis, F. P., and McKay, J. F., eds., Geochemistry and chemistry of oil shales: Symposium Series 230, American Chemical Society Series 230, p. 335–351.

Cook, E. W., 1974, Green River oil shale yields: Correlation with elemental analysis: Fuel, v. 53, p. 16–30.

Demaison, G. J., and Moore, G. T., 1980, Anoxic environments and oil source bed genesis: American Association of Petroleum Geologists Bulletin, v. 64, p. 1179–1209.

Doligez, B., ed., 1987, Migration of hydrocarbons in sedimentary basins; Proceedings of the 2nd IPF Exploration Research Conference, Carcans, France, June 15–19, 1987: Paris, Éditions Technip, Collection Colloque et Séminaires 45, 681 p.

Dott, R. H., and Reynolds, M. J., 1969, Sourcebook for petroleum geology: American Association of Petroleum Geologists Memoir 5, 471 p.

Durand, B., and Paratte, M., 1983, Oil potential of coals; A geochemical approach, in Brooks, J., ed., Petroleum geochemistry and exploration of Europe: Oxford, Blackwell, Geological Society of London Special Publication 12, p. 255–265.

du Rochet, J., 1981, Stress fields; A key to oil migration: American Association of Petroleum Geologists Bulletin, v. 65, p. 74–85.

England, W. A., Mackenzie, A. S., Mann, D. M. and Quigley, T. M., 1987, The movement and entrapment of petroleum fluids in the subsurface: Journal of the Geological Society of London, v. 194, p. 327–347.

Espitalié, J., 1986, Use of Tmax as a maturation index for different types of organic matter; Comparison with vitrinite reflectance, in Burrus, J., ed., Thermal modelling in sedimentary basins: Paris, Éditions Technip, Collection Colloques et Séminaires 44, p. 475–496.

Gold, T., 1986, Swedish meteorite crater due 15,000-foot hole: Oil and Gas Journal, v. 84, no. 6 (February 3), p. 72–74.

Gold, T., and Soter, S., 1980, The deep earth-gas hypothesis: Scientific American, January, p. 154–161.

Gussow, W. C., 1954, Differential entrapment of oil and gas; A fundamental principle: American Association of Petroleum Geologists, v. 38, p. 816–853.

Hedberg, H. D., 1964, Geologic aspects of origin of petroleum: American Association of Petroleum Geologists Bulletin, v. 48, p. 1755–1803.

Héroux, Y., Chagnon, A., and Bertrand, R., 1979, Compilation and correlation of major thermal maturation indicators: American Association of Petroleum Geologists Bulletin, v. 63, p. 2128–2144.

Hood, A., Gutjahr, C.C.M., and Heacock, R. L., 1975, Organic metamorphism and the generation of petroleum: American Association of Petroleum Geologists Bulletin, v. 59, p. 986–996.

Hubbert, M. K., 1940, The theory of groundwater motion: Journal of Geology, v. 48, p. 785–944.

—— , 1953, Entrapment of petroleum under hydrodynamic conditions: American Association of Petroleum Geologists Bulletin, v. 37, p. 1954–2026.

Hunt, J. M., 1975, Is there a geochemical depth limit for hydrocarbons: Petroleum Engineer, March, p. 112–124.

—— , 1979, Petroleum geochemistry and geology: San Francisco, California, W. H. Freeman and Company, 617 p.

Ibach, L.E.J., 1982, Relations between sedimentation rate and total organic carbon content in ancient marine sediments: American Association of Petroleum Geologists Bulletin, v. 66, p. 170–188.

Illing, V. C., 1938, The migration of oil in Dunston, A. E., Nash, A. W., Brooks, B. T., and Tizard, H., eds., The science of petroleum: London, Oxford University Press, v. 1, p. 209–215.

Jones, R. W., 1980, Some mass balance and geological constraints on migration mechanisms, in Roberts, W. H., III, and Cordell, R. J., eds., Problems of petroleum migration: American Association of Petroleum Geologists Studies in Geology 10, p. 47–68.

Jüntgen, H., and Karweil, J., 1966, Gasbildung and gaspeicherung in steinkohlenflozen, Part I and II: Erdol and Kohle, Erdgas, Petrochemie, v. 19, p. 251–258 and 339–344.

Kissel, F. N., and Edwards, J. C., 1975, Two phase flow in coal beds: U.S. Bureau of Mines RI 8066, 16 p.

Kontorovich, A. E., 1984, Geochemical methods for the quantitative evaluation of the petroleum potential of sedimentary basins, in Demaison, G., and Murvis, R. J., eds., Petroleum geochemistry and basin evaluation: American Association of Petroleum Geologists Memoir 35, p. 79–109.

Levorsen, A. I., 1967, Geology of petroleum, 2nd ed.: San Francisco, California, W. H. Freeman and Company, 724 p.

Leythaeuser, D., Schaefer, R. G., and Yukler, A., 1982, Role of diffusion in primary migration of hydrocarbons: American Association of Petroleum Geologists Bulletin, v. 66, p. 408–429.

Leythaueser, D., Mackenzie, A., Schaefer, R. G., and Bjory, M., 1984, A novel approach for recognition and quantification of hydrocarbon migration effects in shale-sandstone sequences: American Association of Petroleum Geologists Bulletin, v. 68, p. 196–219.

Lopatin, N. V., 1971, Temperature and geologic time as factors in coalification: Akad, Nauk. SSSR Isv. Ser. Geol., no. 3, p. 95–106 (in Russian); English translation by N. H. Bostik, Illinois State Geological Survey, 1972.

Lyons, P. C., and Rice, C. L., eds., 1986, Paleoenvironmental and tectonic controls of coal-forming basins of the United States: Geological Society of American Special Paper 210, 208 p.

Magara, K., 1980, Agents for primary hydrocarbon migration; A review, in Roberts, W. H., III, and Cordell, R. J., Problems of petroleum migration: American Association of Petroleum Geologists Studies in Geology 10, p. 33–46.

Maynard, J. B., 1981, Some geochemical properties of the Devonian–Mississippian shale sequence, in Roberts, T. G., ed., Geological Society of America, Cincinnati '81 Field Trip Guidebooks, v. II; Economic geology, structure: Falls Church, Virginia, American Geological Institute, p. 336–361.

McAuliffe, C. D., 1978, Chemical and physical constraints on petroleum migration with emphasis on hydrocarbon solubilities in water: American Association of Petroleum Geologists Continuing Education Note Series 8, p. C-1–C-39.

Meinshein, W. G., 1959, Origin of petroleum: American Association of Petroleum Geologists Bulletin, v. 43, p. 925–943.

Meissner, F. F., 1978, Petroleum geology of the Bakken Formation, Williston Basin, North Dakota and Montana, in Williston Basin Symposium: Montana Geological Society, p. 207–227.

—— , 1984, Cretaceous and lower Tertiary coals as sources for gas accumulations in the Rocky Mountain area, in Woodward, J., Meissner, F. F., and Clayton, J. L., eds., Hydrocarbon source rocks of the greater Rocky Montain region: Rocky Mountain Association of Geologists, p. 401–431.

Momper, J. A., 1978, Oil migration limitations suggested by geological and geochemical considerations: American Association of Petroleum Geologists Continuing Education Course Note Series 8, p. B-1 to B-60.

—— , and Williams, J. A., 1984, Geochemical exploration in the Powder River Basin, in Demaison, G., and Murris, R. J., ed., Petroleum geochemistry and basin evaluation: American Association of Petroleum Geologists Memoir 35, p. 181–179.

Morency, M., Mineau, R., Zeller, E., and Dreschoff, 1986, Are fossil fuels really fossils?: Oil and Gas Journal, v. 84, no. 23, June 2, p. 92–95.

Palciauskas, V. V., and Domenico, P. A., 1980, Microfracture development in compacting sediments; Relation to hydrocarbon-maturation kinetics: American Association of Petroleum Geologists Bulletin, v. 64, p. 927–937.

Pandey, G. N., Tek, M. R., and Katz, D. L., 1974, Diffusion of material through porous media with implications in petroleum geology: American Association of Petroleum Geologists Bulletin, v. 58, p. 291–303.

Parrish, J. T., 1982, Upwelling and petroleum source beds, with reference to Paleozoic: American Association of Petroleum Geologists Bulletin, v. 66, no. 6, p. 750–774.

Perrodan, A., 1983, Dynamics of oil and gas accumulations: Bulletin des centres de Recherches Exploration–Production Elf-Aquitaine Memoir 5, 368 p.

Porfir'ev, V. S., 1974, Inorganic origin of petroleum: American Association of Petroleum Geologists Bulletin, v. 58, p. 3–33.

Price, L. C., 1976, Aqueous solubility of petroleum as applied to its origin and primary migration: American Association of Petroleum Geologists Bulletin, v. 60, p. 213–244.

—— , 1977, Aqueous solubility of petroleum as applied to its origin and migration; Reply to discussion: American Association of Petroleum Geologists Bulletin, v. 61, p. 2149–2156.

—— , 1978, New evidence for a hot, deep origin and migration of petroleum: American Association of Petroleum Geologists Continuing Education Note Series 8, p. F-9.

Rice, D. D., and Claypool, G. E., 1981, Generation, accumulation, and resource potential of biogenic gas: American Association of Petroleum Geologists Bulletin, v. 65, p. 5–25.

Roberts, W. H., and Cordell, R. J., eds., 1980, Problems of petroleum migration: American Association of Petroleum Geologists Studies in Geology 10, 273 p.

Roof, V. G., and Rutherford, W. M., 1958, Rate of migration of petroleum by proposed mechanisms: American Association of Petroleum Geologists Bulletin, v. 42, p. 693–980.

Royden, L., Sclater, L. G., and Von Herzen, R. P., 1980, Continental margin subsidence and heat flow; Important parameters in formation of petroleum hydrocarbons: American Association of Petroleum Geologists Bulletin, v. 64, p. 173–187.

Schowalter, T. T., 1979, Mechanics of secondary migration and hydrocarbon entrapment: American Association of Petroleum Geologists Bulletin, v. 63, p. 723–760.

Silverman, S. R., 1965, Migration and segregation of oil and gas, in Young, A., and Galley, J. E., eds., Fluids in subsurface environments: American Association of Petroleum Geologists Memoir 4, p. 53–65.

Spencer, C. W., 1987, Hydrocarbon generation as a mechanism for overpressuring in Rocky Mountain region: American Association of Petroleum Geologists Bulletin, v. 71, p. 368–388.

Smith, J. E., Erdman, J. G., and Morris, D. A., 1971, Migration, accumulation, and retention of petroleum in the Earth: Proceedings of the 8th World Petroleum Conference, v. 2, p. 13–26.

Tissot, B. P., 1984, Recent advances in petroleum geochemistry applied to hydrocarbon exploration: American Association of Petroleum Geologist Bulletin, v. 68, p. 545–563.

Tissot, B. P., and Welte, D. H., 1984, Petroleum formation and occurrence, 2nd ed.: Berlin, Springer-Verlag, 699 p.

Tissot, B. P., Pelet, R., and Ungerer, Ph., 1987, Thermal history of sedimentary basins, maturation indices, and kinetics of oil and gas generation: American Association of Geologists Bulletin, v. 71, p. 1445–1466.

Trask, P. D., 1932, Origin and environment of source beds of petroleum: Houston, Texas, Gulf Publishing Company, 323 p.

Ungerer, P., Behar, E., and Discamps, D., 1983, Tentative calculation of the overall volume expansion of organic matter during hydrocarbon genesis from geochemistry data; Implications for primary migration, in Bjoroey, M., ed., Advances in organic geochemistry: New York, John Wiley and Sons, p. 129–135.

Uspenskii, V. A., 1962, The geochemistry of processes of primary oil migration: Geochemistry, no. 12, p. 1153–1178.

van Krevelen, D. W., 1950, Graphical-statistical method for the study of structure and reaction processes of coal: Fuel, v. 29, p. 269–284.

—— , 1961, Coal: Amsterdam, Elsevier, 514 p.

Waples, D. W., 1979, Simple method of source rock evaluation: American Association of Petroleum Geologists Bulletin, v. 63, p. 239–245.

—— , 1980, Time, temperature in petroleum formation; Application of Lopatin's method to petroleum exploration: American Association of Petroleum Geologists Bulletin, v. 64, p. 916–926.

—— , 1985, Geochemistry in petroleum exploration: Boston, Massachusetts, International Human Resources Development Corporation, 232 p.

Watts, H., 1963, The possible role of adsorbtion and diffusion in the accumulation of crude petroleum: Geochimica et Cosmochemica Acta, v. 27, p. 925–928.

Welte, D. H., Schaefer, R. G., Stoessinger, W., and Radke, M., 1984, Gas generation and migration in the deep basin of western Canada, in Masters, J. A., ed., Elmworth; Case study of a deep basin gas field: American Association of Petroleum Geologists Memoir 30, p. 35–47.

MANUSCRIPT ACCEPTED BY THE SOCIETY SEPTEMBER 20, 1989

Chapter 15

Pore system aspects of hydrocarbon trapping

William R. Almon
Anadarko Petroleum Corporation, Box 5050, Denver, Colorado 80217
John B. Thomas
Amoco Production Company, 1670 Broadway, Denver, Colorado 80202

The petroleum geologist attempts to understand the occurrence of petroleum accumulations and to find ways to predict the location of new accumulations. Aspects that need to be considered range in scale from the geometry and orientation of the reservoir (structural geology, sedimentology, stratigraphy) to the pore system (hydrology, geochemistry, and reservoir engineering). This chapter examines the role the pore system plays in hydrocarbon trapping. The selective transmission or retention of fluids depends upon the details of the interconnectedness of rock pore systems. The arrangement of pores makes the difference between a reservoir rock and a seal rock.

A hydrocarbon accumulation or trap may be visualized in terms of the Earth's potential energy field. A hydrocarbon trap is an underground region in which the potential energy of the hydrocarbon phase is at a minimum with respect to its surroundings. The boundaries on subsurface accumulations are usually (but not always) physical ones, with bounding media that impede the migration of hydrocarbons by virtue of their structure, geometry, or composition (Dahlberg, 1982). The typical accumulation consists of hydrocarbons trapped between water below and impermeable sedimentary rocks above. The overlying seal is usually a rock type with an extremely fine pore system relative to the surface tensions of the hydrocarbon and water. The capillary pressure, which strongly imbibes the wetting fluid into a fine pore system, produces a high potential energy gradient for hydrocarbons relative to coarser pore systems of the reservoir rock. Thus, the nature of the voids composing the pore system affects the distribution of hydrocarbon potential energy in a rock column.

Pore system structure directly affects four aspects of hydrocarbon trapping. (1) Migration pathways connecting source beds and reservoir rocks must have sufficient porosity or interconnectedness to transmit fluids, including hydrocarbons. (2) Reservoir rock must be present with a volume of porosity for significant hydrocarbon storage. The pores must be interconnected in order to transmit stored hydrocarbons to a wellbore. (3) The pore structure of a good seal rock must prevent or retard hydrocarbon migration on a geologically significant time scale. (4) Trap geometry requires an arrangement of seal and reservoir rocks that produces a closed potentiometric low relative to the hydrocarbon phase.

ROCK PROPERTIES

Pores develop as a series of voids within a solid framework of detrital grains, biogenic fragments, or crystals that have been cemented by authigenic minerals. It is the pore system that permits the storage and transmission of fluids, be they water or hydrocarbon. The size, shape, and pattern of this pore system is very difficult to specify because of its geometric variability and small size.

Porosity is a mass property that quantifies the relationship between the solid and void space of a material. Porosity does not indicate the degree to which the voids are either connected or isolated. Porosity is a scalar quantity expressed as a percent of rock volume.

At deposition, it is sorting, grain shape, and grain packing that most strongly influence porosity in clastic sediments (Clark, 1969). These controls carry over into lithified rocks because they strongly influence the course of diagenesis that may be the main control on porosity in rocks. Beard and Weyl (1973) studied the effects of grain size and sorting on the porosity of artificially mixed and packed sand samples and indicated that sorting is the key to porosity in sand packs. The same control can be observed in sandstone of the Permian Rotliegendes Formation in the Dutch sector of the North Sea. In these sands, porosity is virtually independent of grain size but is strongly related to sorting (Nagtegall, 1979).

At the time of deposition, coarser sands commonly have greater porosity than finer sands because of differences in sorting and grain shape. Fine grained sands usually include many platy and angular grains that produce lower porosity when packed together. On the other hand, coarser grained sands are usually better sorted as a result of the higher energy required to transport these grains; they are composed of more rounded grains and have higher initial porosity.

Once a rock is lithified as a result of burial, this relationship

Almon, W. R., and Thomas, J. B., 1991, Pore system aspects of hydrocarbon trapping, *in* Gluskoter, H. J., Rice, D. D., and Taylor, R. B., eds., Economic Geology, U.S.: Boulder, Colorado, Geological Society of America, The Geology of North America, v. P-2.

is frequently reversed by cementation and other processes. Fine grained sandstones in the subsurface frequently have higher porosity than the associated coarser grained sandstones. Von Englehardt (1960) attributes this to the greater number of grain contacts per unit volume in the fine sand, which causes greater resistance to compaction. However, Fuchtbauer (1967) attributes this relationship to grain shape. The greater percentage of rounded grains in coarse sands permits grains to slide past one another and to compact more than fine sands that have more angular grains and prohibit this slippage. An alternative explanation might lie in the role of permeability during diagenesis. Coarse grained sandstones have higher permeability than fine grained sandstones. Hence, they are exposed to more water flow, resulting in greater cement precipitation relative to finer sandstones.

Most research into the relationships between permeability and textural properties (packing, grain size, and shape) has been conducted on sandstones or sand packs. Similar trends can be expected for carbonate rocks, but pore geometry suggests that the relationship is not consistent with average particle size. The orientation and packing of sandstone framework grains appear to have only weak control on permeability in the plane of the bedding, although permeability is greatest in the direction of sand transport and is least perpendicular to bedding (Mast and Potter, 1963).

PORE SYSTEM CLASSIFICATION

The fundamental characteristics of pores in rocks—their shape, size, and interconnectedness—are commonly referred to as the pore geometry. Pittman (1979) provides a useful ternary diagram and nomenclature for textural pore geometries of sandstones. By extension, it may also be utilized in carbonate rocks with some modification. The three types of porosity on the Pittman diagram are (1) intergranular porosity that occurs between clastic grains or particles, (2) dissolution/moldic porosity developed through the complete removal of grains or as voids within grains, and (3) microporosity, which is any type of porosity having pore throat size of less than 0.5 microns. Intergranular porosity is equivalent to intercrystal porosity in coarsely crystalline rocks and probably represents the pore geometry having the best storage capacity and fluid deliverability.

Comparisons of particle size to pore and pore throat size are shown schematically in Figure 1. The diameters of the pores and pore throats decrease through the sand size ranges until they stabilize, in the silt and clay size ranges, with both pores and pore throats being the same size. In the silt and clay size ranges, the grains tend to be platy and 20 microns or less in diameter. The pores between such grains will be on the order of 0.5 to 5 microns in diameter, and the pore throats will be in the range of 0.1 to 1 micron in diameter. The trend to finer pore sizes is accompanied by an increase in pore system surface area, and pores change to a more tabular, rather than equidimensional configuration (Coalson and others, 1983).

Dissolution porosity includes vugs and moldic spaces developed by the dissolution of framework grains, such as biogenic

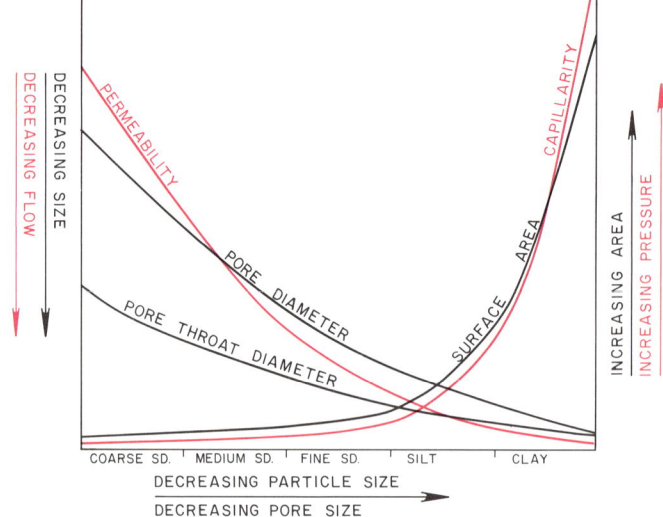

Figure 1. Pore size and pore throat diameter decrease with decreasing grain size. Pore system surface area changes only a small amount in the larger sand sizes but increases dramatically as grain size decreases in the silt and clay size range. As pore size decreases, permeability decreases slowly in the larger pore sizes and more rapidly in the progressively finer pore sizes. Capillarity increases slowly and then increases rapidly with a decrease in pore size for the smaller pores. (After Coalson and others, 1983).

fragments and unstable rock fragments, as well as fenestral pores that may be of primary origin. These pores have shapes and sizes that are not related to the size and shape of the framework constituents. Often these pore spaces are larger than the individual framework constituents. Rocks dominated by this pore type will have porosities and permeabilities that may be unrelated to rock fabric.

As in the case of intergranular pore space, the permeability will depend on the degree to which the dissolution pores are directly connected to one another. If solution pores are isolated (i.e., pinpoint vugs), fluid movement must occur through the rock matrix, and permeability is dependent on the nature of the matrix pore system. Connected dissolution pores are not strongly affected by the matrix pore system because fluid moves directly from one dissolution void to the next. If the matrix pore system has high porosity and permeability, total hydrocarbon storage is enhanced, but total permeability of the rock may not be significantly increased.

Microporosity is defined on the basis of capillary pressure test data; a pore throat of less than 0.5 microns in equivalent spherical radius is by definition the upper threshold of microporosity. Micropores tend to be tabular and are not much larger in diameter than the pore throat (Coalson and others, 1983). Microporosity is the dominant pore type in shales, chalks, sandstones that contain high percentages of detrital clay or authigenic clay, and finely crystalline carbonates. Micropore systems are generally impermeable to fluids that do not wet the pore system surface no matter how high their porosity. Seal rocks, therefore, have a higher component of microporosity.

Fracture porosity is generally not a textural porosity but a secondary feature that may enhance the other three porosity types. Fractures by themselves contribute little storage porosity to a reservoir rock, and unless coupled to a rock matrix with significant hydrocarbon storage capacity, they do not offer much reserve potential.

Porosity and permeability relationships

Porosity-permeability crossplots are very helpful for identification of pore system characteristics in sedimentary rocks (Coalson and others, 1983). Because permeability is controlled by pore throat size and pore system geometry, groupings of porosity and permeability data are often apparent on crossplots (Fig. 2). The general envelopes of porosity-permeability relate to fluid flow in a reservoir.

Capillary phenomena

Fluids in the subsurface usually flow through pore systems so small that movement and fluid distribution are determined by capillary size. This is especially true in systems that contain two immiscible fluids such as oil and water. Capillary effects in hydrocarbon accumulations control the original static distribution of fluids in an untapped reservoir and provide the mechanism whereby oil and gas move through reservoir pore spaces until they are barred from further movement. The static capillary pressure in a reservoir pore system is in part a function of the relative fluid saturations. The capillary pressure and the pore geometry combine to determine the distribution of fluids within the pore system.

Capillary phenomena arise when two immiscible fluids are in contact, because molecular attractions between similar molecules in each fluid are greater than the attraction between the different molecules of the two fluids. The boundary, or interface, between the two fluids has unique properties. Molecular attraction is greater on the side of the more dense fluid, causing the interface to be convex toward the more dense fluid. When one fluid is immersed in another fluid with which it is immiscible, it assumes a spherical shape, the shape that possesses the minimum surface area to volume ratio.

The force that acts on the interface of a gas and a liquid is called surface tension. This same force acting on the interface of two liquids is called interfacial tension. The result of this force is to produce a pressure difference across the contact surface. This pressure difference is capillary pressure. The pressure difference between the two immiscible fluids can be calculated as:

$$Pc = 2\gamma/r$$

Thus the capillary pressure is proportional to the interfacial

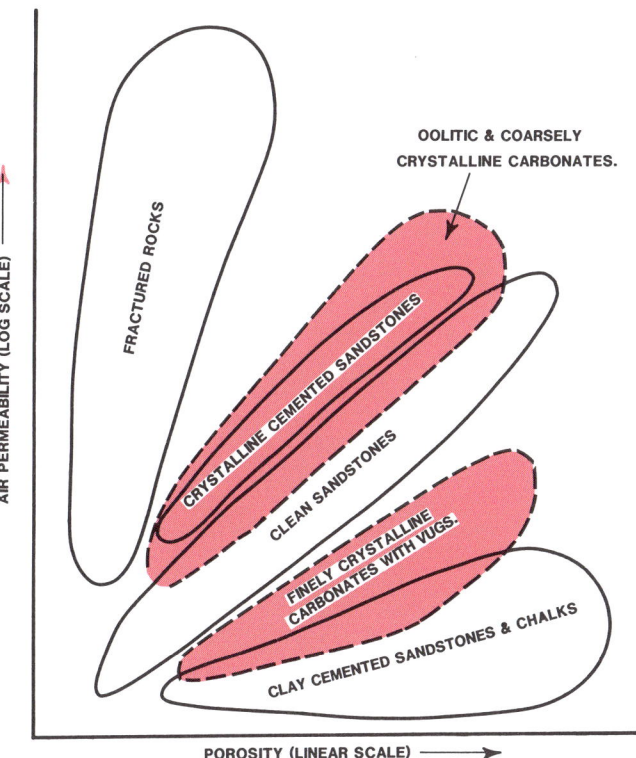

Figure 2. Crossplots of porosity versus permeability are useful for defining rock types with consistent pore system characteristics (Coalson and others, 1983).

tension between the two fluids (γ) and inversely proportional to the radius of curvature (r) of the drop (Berg, 1975).

In a pore system, a third material—the rock surface—is present. The mutual attraction between each of the liquids and the rock surface is called the force of wettability. This force is a surface effect and is one aspect of the energy of adhesion. For a fluid that is wetting, the adhesive attraction between the surface and the fluid is greater than the cohesive attraction of the fluid molecules for one another. This force causes the wetting fluid to spread over the solid surface in a thin layer to contact as much of the surface as possible. In the subsurface, water is usually the wetting fluid. The adhesive force, or the attraction of the wetting fluid to the solid in any hydrocarbon-water-rock system, is the result of the combined interfacial energies of the hydrocarbon-water, water-rock, and hydrocarbon-rock surface.

Wettability is expressed mathematically by the contact angle (θ) of the hydrocarbon-water interface and the rock surface, as measured through the aqueous phase. Contact angles (θ) of 0 to 90 degrees indicate a water-wet system. The angle θ approaches zero for most reservoir systems (Treiber and others, 1972).

If two sand grains (Fig. 3A) are in contact and the water saturation is higher than the oil saturation, the oil-water interface will occupy the position b. If the capillary pressure is increased,

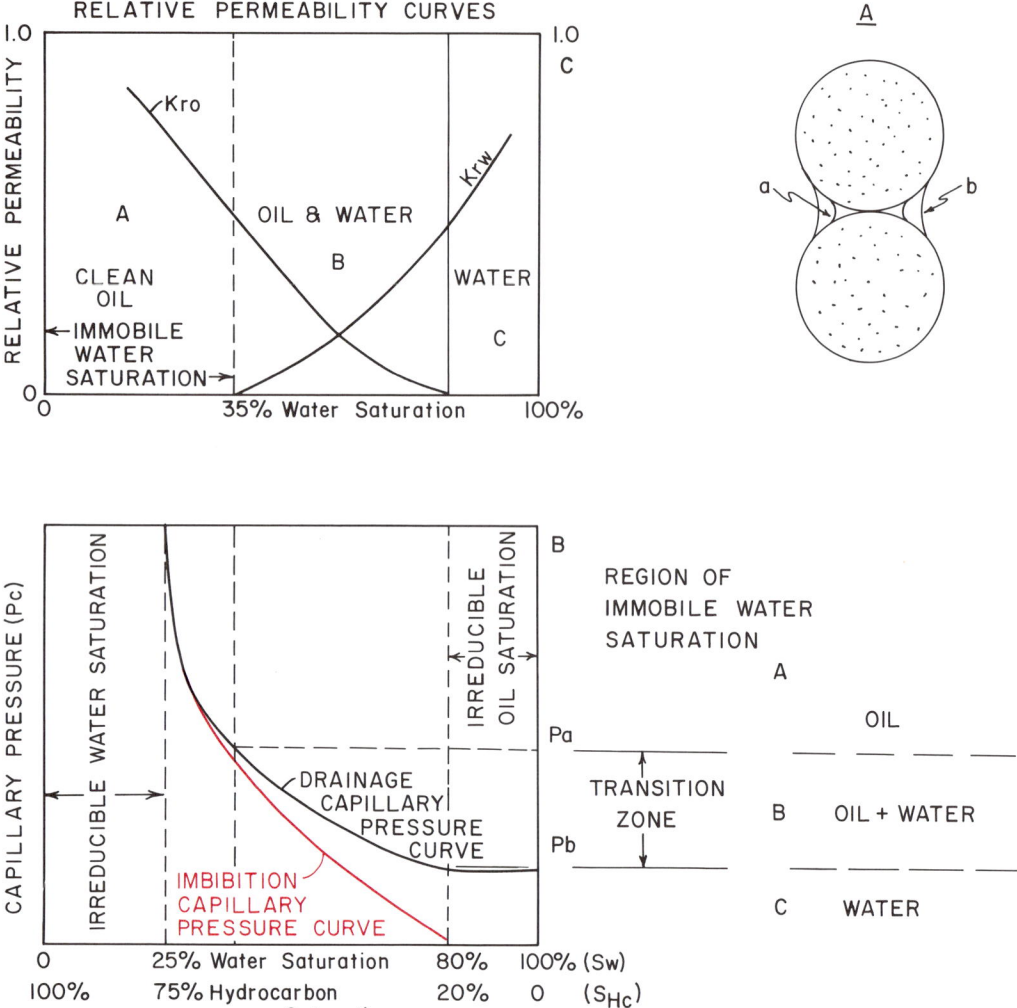

Figure 3. For a two phase, oil-water system, saturation, capillary pressure, and relative permeability are strongly related. In a pore (A), if the capillary pressure is low, the interfacial curvature will be low and the oil-water interface will occupy position *b*. If more oil is introduced, the oil-water interface will be forced to position *a*. The interfacial curvature will be high, as will be the capillary pressure (after Levorson, 1967). The related curves representing capillary pressure and saturation are shown in B. The corresponding pair of oil-water relative permeability curves are shown in C. These related properties determine the saturation distribution and productive properties of a reservoir rock (after Arps, 1964).

the oil saturation is increased; the water saturation will be correspondingly reduced, and the interface will be forced to occupy position a. The capillary pressure for such a system can be plotted as a fuction of the saturation as shown in Figure 3C. The capillary pressure required to increase oil saturation becomes progressively larger as the oil saturation increases.

Capillarity reflects the tenacity with which water is held in a pore. High capillarity indicates strongly held water in a pore, which requires greater energy to displace water with hydrocarbons. Therefore, high capillarity results in high immobile water saturation. Capillarity and immobile water saturation increase slowly with pore size until the smaller pore sizes are reached, and then they increase rapidly. Permeability decreases rapidly and more or less uniformly with decreasing pore size through the larger pore sizes. In the smaller pore sizes, the rate of permeability decrease slows (Coalson and others, 1983).

RELATIVE PERMEABILITY

Natural rock pore systems commonly contain more than one fluid. In such systems the permeability to any one fluid is less than would be measured if that fluid were the only fluid in the pore system. The effective permeability of a fluid is the ability of a pore system to conduct that fluid in the presence of other fluids. The ratio between the effective permeability for a given fluid at a partial saturation and the absolute permeability for the same fluid at 100 percent saturation is known as the relative permeability (Levorson, 1967).

The relative permeability of a pore system depends on the saturation, the nature of the fluids, and the geometry of the pore system. Together these factors determine the saturations of wetting and non-wetting phases in that particular rock system. The distribution and the dimensions of each phase at a particular saturation control the flow of that phase irrespective of any others that are present.

Relative permeabilities of two-phase systems (e.g., oil-water) are usually presented as curves of relative permeability for each phase as a function of pore volume saturation. Examination of such a curve for oil and water (Fig. 3C) indicates that there is no permeability to oil until some threshold value (here 10 percent) is exceeded. This is the residual oil saturation, which represents oil trapped as isolated droplets in pores that have been flushed with water. Residual oil saturation is a function of pore geometry. As the size difference between pores and their pore throats increases, the residual oil saturation increases (Wardlaw and Cassan, 1979).

Further examination of an oil-water relative permeability curve (Fig. 3C) indicates that a non-zero water saturation exists at which water ceases to be mobile. This is the immobile water saturation and represents the water that coats the surface of the water-wet pore system. At this saturation, the remainder of the pore system is saturated with oil, and only oil is mobile.

The oil-water relative permeability characteristics of reservoirs are strongly influenced by pore system geometry and surface area. Reservoirs with large, clean pore systems have low irreducible water saturations because pore walls have a relatively small surface area. Conversely, small pore systems with high surface area-to-volume ratios have large irreducible water saturations. The high immobile water saturation leaves little room for the flow of mobile fluids, and the end points of the relative permeability curves are depressed. It is easy to infer that rocks with high immobile water saturations serve as more effective seal rocks than those with low immobile water saturations.

A low initial Kro (Fig. 3C) suggests that very small pore throats control fluid flow in the sandstone. This control results from the fact that the wetting phase (water) takes up a large percentage of the space in small pores and leaves less space for oil to flow. In water-wet sandstones, final Krw values are believed (Morgan and Gordon, 1970) to be lower than initial Kro values (Fig. 4A) because the residual oil saturation, which occupies a portion of the larger pores, and the irreducible water saturation are both immobile.

Because pore system capillary phenomena and relative permeability are storngly related, curves plotting capillary pressure and saturation may be compared directly to relative permeability curves (Fig. 3B-C). The interaction between saturation, capillary

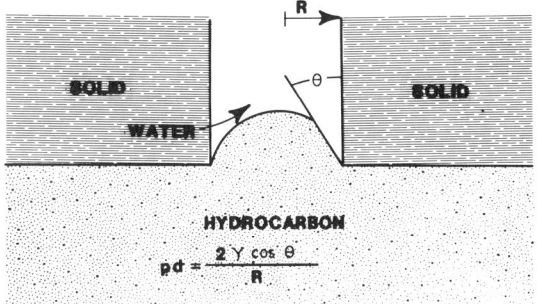

Figure 4. The hydrocarbon-water displacement pressure of a pore system is a function of hydrocarbon-water interfacial tension, wettability, and the radius of the pore throat (Purcell, 1949).

pressure, and relative permeability determines the producing characteristics of a water-wet, oil-water reservoir system by controlling the vertical distribution of interstitial water in that reservoir.

Arps (1964) demonstrated that there is not a single oil-water contact in a reservoir (Fig. 3B). Instead, there is a transition zone in which the water saturation varies gradually upward in the reservoir from 100 percent to the irreducible water saturation. This transition may be very sharp or extend over tens of meters. Such of transition zone may be divided into three parts. In the lowest part, the reservoir will produce only water because the capillary pressure causes the water saturation to exceed the saturation at which oil can flow. In the middle zone, the water saturation gradually decreases until the critical water saturation at which water ceases to be mobile is encountered. In this zone, both oil and water can be produced. The uppermost part of the transition zone produces only oil.

Analysis of oil-water transition zones indicates that the minimum trap closure required for water-free production is the vertical height above the free-water surface necessary to achieve a water saturation below the critical water saturation (Arps, 1964). This height is controlled by the capillary pressure characteristics of the reservoir pore system, being higher for pore systems containing small pore throats than for systems containing large pore throats. The capillary pressure required to reach the critical water saturation is roughly proportional to the square root of the ratio of porosity to permeability (Arps, 1964).

SEAL CHARACTERISTICS

Commercial hydrocarbon accumulations occur where barriers to further migration exist. Such barriers, or seals, may be described in terms of permeability, capillarity, or interconnectedness of pore systems.

Seals such as shales and siltstones retard hydrocarbons because of their wettability and entry pressure properties. The fluid that wets the surface of the mineral grains in the migration barrier can pass through the barrier at a rate determined by its permeability. Non-wetting fluids are trapped because they are unable to enter the fine pore system of the seal (Hill, 1956). Seal capacity is determined by the pore size distribution because for a migration barrier to exist, the entry pressure of the barrier must exceed the buoyant pressure of the oil column contained by the trap.

The minimum capillary pressure that will force the entry of a non-wetting fluid into a capillary opening saturated with a wetting fluid is known as the displacement or entry pressure. In the case of a water-wet rock, the displacement pressure is the resistance that must be overcome if a filament of non-wetting hydrocarbon is to move from one pore to the next through the intervening pore throat (Schowalter, 1979). The magnitude of this resistant force in any rock-water system is shown in Figure 4.

The displacement pressure of a pore system increases as the radius of the pore throats decreases and as the hydrocarbon-water interfacial tension increases. Displacement pressure decreases as the degree of hydrocarbon wetting increases. Techniques for estimating values of the hydrocarbon-water and gas-water interfacial tension are given by Schowalter (1979).

The displacement pressure can be measured in the laboratory by injecting a non-wetting fluid into a cleaned and dried sample plug of rock under increasing pressure. The pressure at which a connected filament of non-wetting fluid extends across the sample is measured. Schowalter (1979) determined that fluid breakthrough occurred at saturations ranging from 5 to 17 percent of the rock pore volume. The technique is generally analogous to the movement of hydrocarbons through a water-wet rock. Such movement is described as secondary migration.

Laboratory capillary pressure tests on rock samples can use almost any fluid for the wetting and non-wetting phases. Obviously, the best test for establishing the displacement pressure for a seal would use the hydrocarbon and water from the associated reservoir. Unfortunately these tests are difficult and time consuming. Almost all laboratory measurements of capillary pressure use air as the wetting fluid and mercury as the non-wetting fluid. The capillary pressure is measured as the non-wetting fluid saturation (mercury) is increased under pressure from a few pounds (psia) to a maximum of 2000 psia, resulting in a "drainage" (of wetting fluids) capillary pressure curve. In routine circumstances, the "imbibition" (withdrawal of mercury) curve is not measured.

If injection of the non-wetting phase is continued beyond the saturation at which a continuous filament of non-wetting fluid is established, the entire range of capillary properties can be measured and a pore size distribution can be determined for the pore system. A curve relating the non-wetting fluid saturation to the pore system capillary pressure is a "drainage" capillar pressure curve, usually referred to as a "capillary pressure curve." The capillary pressure of the system, as the non-wetting fluid is expelled and wetting fluid is imbibed, can be determined by reversing the previous process. The pressure on a pore system saturated with a non-wetting fluid is lowered in stages, and the decrease in non-wetting fluid saturation is monitored. This is called an "imbibition" capillary curve and is analogous to the production of hydrocarbons from a reservoir.

Air-mercury capillary pressure data do not in themselves represent reservoir seal behavior in rock systems. The quantitative application of mercury capillary pressure data to the analysis of reservoir seal capacity requires the conversion of the data to subsurface hydrocarbon-water capillary pressure values. The conversion is possible if the interfacial tension of the hydrocarbon-water system (γhw) and the contact angle of the hydrocarbon-water interface (θhw) are known at reservoir conditions. The mercury-air surface tension (γma) is 480 dynes/cm at laboratory conditions, and the contact angle between mercury and quartz is 40 degrees, making COS θma equal to 0.776. According to Purcell (1949), the capillary pressure for the hydrocarbon-water system at reservoir conditions, (Pc) hw, is:

$$(Pc)\ hw = \frac{\gamma hw \times COS\ \theta hw}{\gamma ma \times COS\ \theta ma} \times (Pc)\ ma$$

In the evaluation of this equation, the pore system is usually assumed to be completely water-wet, resulting in a contact angle of zero degrees for the oil-water interface with the pore surface. The equation simplifies, after the insertion of appropriate mercury-air values, to:

$$(Pc)\ hw = .02685 \times \gamma hw \times (Pc)\ ma$$

Once the mercury-air capillary pressure data have been converted to appropriate hydrocarbon-water data, the vertical height of the continuous pore to pore oil phase required for buoyant migration into the seal can be calculated by the equation:

$$H = \frac{PdB - PdR}{(\rho w - \rho h) \times .0433}$$

.0433 is a unit conversion factor. The length of the hydrocarbon column held by a seal can be substantially increased or decreased by hydrodynamic factors (Hubbert, 1953).

Calculations of seal capacity must be used with care. Downey (1984) points out the difficulty in extrapolating precise measurements made on a few core samples to the entire seal surface of a large hydrocarbon accumulation. The probability that the properties of a given sample are representative of a large area are vanishingly small. This is unfortunate because the weakest point in the seal determines the actual hydrocarbon column that can be retained. Average values or random values may have little relevance to those at the weakest point. One open fracture can undo the finest seal. The importance of seal integrity and the absence of open extension fractures are critical to hydrocarbon entrapment. In a modeled case, a single 0.1-millimeter-wide open fracture in a cap rock overlying a 150-meter oil column would leak at the rate of 150 million bbl/1,000 yr. (Downey, 1984).

The estimation of seal capacity can be useful in exploration if it is coupled with an assessment of the mechanical properties of the seal rock and its structural setting. If a rock is not likely to fail by brittle fracture in its structural setting, then seal capacity estimates may be made with more confidence. Such estimates of seal capacity are enhanced when hydrocarbons can be demonstrated to be present in the seal pore system (Schowalter, 1979).

TRAP CONSIDERATIONS

At the conceptual level, a hydrocarbon trap is a geometrical arrangement of reservoir rock juxtaposed with a seal rock in such a way as to produce containment. In this view, a trap is a passive container of hydrocarbons. Careful examination of the relationship of reservoirs, seals, and their role in the migration of hydrocarbons suggests that this concept is too passive.

Commonly a reservoir is a relatively coarse-textured rock interbedded between rocks of finer texture. Fluid migration within a reservoir bed is greater than in the surrounding beds and is generally parallel to its boundary with the finer textured beds. Expulsion from the source bed is thought to be more or less perpendicular to the bed boundaries (Hubbert, 1953).

The high transmissibility of reservoir beds, relative to surrounding finer textured rocks, causes these beds to be principal pathways for fluid movement in the subsurface. Reservoir beds serve to import hydrocarbons into regions of relative energy minima where they accumulate by buoyancy or capillary filtration while water escapes through the seal rock (Toth, 1980).

That the seals associated with hydrocarbon traps are permeable to water has long been recognized (Hubbert, 1953; Hill, 1956; Berg, 1975). Capillary action prevents the escape of hydrocarbons from the trap while inducing the escape of water through the fine pores of the seal. Oil is retained by capillary pressure differences between hydrocarbon and water in the small pores of the seal.

The passage of water, at significant rates, through a relatively thick shale seal has been demonstrated in the subsurface. In tests of the Gatesville Sandstone, Witherspoon and Neuman (1967) determined that a 5-meter-thick shale separating two sandstone beds had a permeability of 0.7×10^{-4} md. This permeability is not due to fracturing because it is less than the permeability (1.8×10^{-4} md) obtained from laboratory core analysis.

Thus a trap is not just a container but an apparatus for creating hydrocarbon accumulations. The thought of a trap as an active mecahnism for hydrocarbon accumulation fits in well with the concept of the "hydrocarbon machine" put forth by Meissner and others (1984). In this view, all the factors that affect the process of hydrocarbon generation, migration, and accumulation constitute a total system. The individual elements are interdependent and are in effect driven by the organic maturation process. Fluids are generated from organic-carbon rich rocks in the area of thermal maturity. Expulsion of the fluids is primarily vertical from the area of generation into the overlying or underlying carrier/reservoir beds. The high fluid pressures commonly associated with active hydrocarbon generation may initiate fracturing and allow the fluids to penetrate thick sequences of relatively impermeable rocks and reach a carrier/reservoir bed. Hydrocarbons then migrate laterally and updip to a site of entrapment. The hydrocarbons are retained by capillary action, and the associated waters are discharged upward (Meissner and others, 1984).

Diagenetic aspects of trapping

Diagenesis is an important factor in hydrocarbon trapping, as it alters the pore system geometry, creating reservoir quality rock by enlarging pore space through secondary porosity generation or creating seals by pore size reduction through cementation (Cant, 1986). The formation of a diagenetic trap requires that parts of a sedimentary deposit react differently from each other during burial. These differences can be caused by such factors as: (1) the rate of fluid through-flow, (2) slight initial differences in mineralogy, and (3) variations in early diagenesis.

Diagenetic reservoirs may be formed by the development of secondary porosity resulting from the enlargement of primary pores or from leaching of grains and cement (Schmidt and others, 1977). An excellent discussion by Surdam and others (1984) suggests that secondary porosity generation by alumnio-silicate or carbonate mineral dissolution is a natural consequence of the interaction of organic and inorganic reactions during progressive diagenesis.

The burial of sediment prisms eventually results in the maturation of organic matter. Organic maturation reactions result in the generation of significant amounts of organic acids and carbon dioxide (Surdam and others, 1984). These become components of migrating subsurface waters with or ahead of the migrating hydrocarbons and can attack the reservoir rock constituents to generate secondary porosity. The degree to which secondary porosity develops depends on a number of factors, including the amount and composition of the source rock organic material and the sedimentary geometry.

Diagenesis also may produce seal rock by progressive cementation of the initial pore space. Clay cementation can alter the capillary properties of sandstones and carbonates to such an extent that they serve as seals for hydrocarbon accumulations in less altered rocks. The Spirit River Formation of Alberta (Cant, 1983) is a good example. Sandstones in this formation have been cemented extensively by quartz and illitic clay. Laboratory measurements of Spirit River permeability range between 0.001 and 0.5 md, and overlap values measured for typical "seal quality" shales (10^{-3} to 10^{-7} md) by Freeze and Cherry (1979).

Trap classification

The history of attempts to classify hydrocarbon traps has

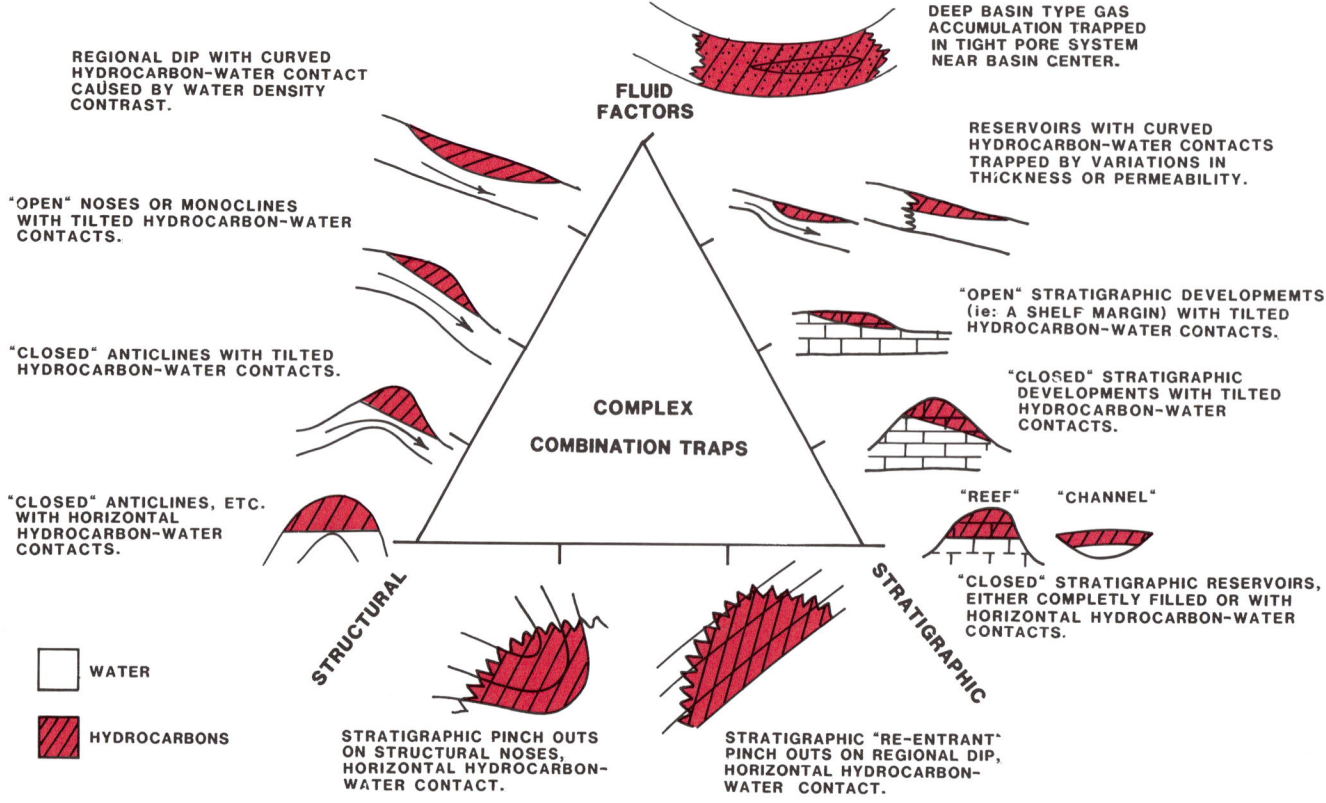

Figure 5. A schematic ternary classification of trap types illustrating the combination trap types produced by the interaction of different trapping controls (after Levorson, 1966).

been reviewed by Dott and Reynolds (1969). The earliest schemes were based on purely structural concepts. As knowledge of producing zones increased, the concept of the stratigraphic trap was included in classification schemes. More recent classification schemes have focused on combination structural-stratigraphic traps. The effect of fluids has yet to become a part of classification schemes, and the concept of structural trap and stratigraphic trap end members with a range of combination traps in between has become the operative classification scheme of most exploration.

Levorson (1966) states that we are "entering the era of the combination trap." He recognized that structure, stratigraphy, and fluid phenomena can act alone or in combination to produce a trap. He introduced a classification scheme in which these three components of a trap are end members on a ternary diagram (Fig. 5). The degree to which the various end members contribute to the formation of a particular trap can be reflected by plotting the trap as a point (or area) on the ternary diagram. All of the traps are plotted along the edges of the diagram, indicating that each is a combination of only two of the three trapping factors. The center of the ternary diagram is the realm of the subtle trap, influenced by all three factors. Though this classification scheme does not appear to have gained wide acceptance, it is very useful at a conceptual level when considering the mechanics of hydrocarbon accumulation.

If we consider the boundaries on hydrocarbon accumulations, traps can be classified into static or dynamic types. Static trap boundaries are provided by reservoir rock-seal geometry and contain static hydrocarbons at the end of a migration pathway. The traditional structural trap and most stratigraphic traps are static. Dynamic trap boundaries are provided by fluid pressure phenomena. They are formed along migration paths where rapid changes in potentiometric gradient occur or where pore system capillarity retards the migration of hydrocarbons. Such dynamic traps are transient. Examples of dynamic traps include deep basin gas accumulations and hydrodynamic traps. The hydrocarbons in dynamic traps have the potential for further migration.

If the concept of static versus dynamic traps is combined with Levorson's (1966) ternary classification, the stratigraphic and structural end members have physical boundaries and represent the end points of particular migration pathways. The third end point has boundaries that are not material but are created by

Pore system aspects of hydrocarbon trapping   249

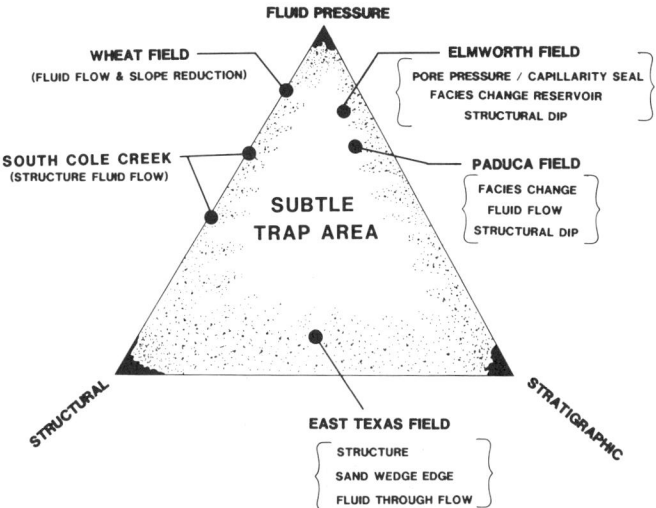

Figure 6. Ternary plot of example fields according to trapping mechanism (Levorson, 1966).

the interaction of pore geometry and pore fluid pressure. The impact of diagenesis is most strongly felt at this pole because of its strong influence on pore geometry (Fig. 5).

FIELD EXAMPLES OF TRAPPING MECHANISMS

The five field examples represent only a fraction of the possible array of hydrocarbon traps. If they are plotted on a ternary diagram (Fig. 6), some of the variability in trapping mechanisms can be seen. Each field requires that two or more trapping components come together to produce the trap. Three of these fields, Paduca, South Cole Creek, and Wheat, indicate the existence of fluid flow, which—combined with even minor changes in permeability or structural attitude—can provide a trap for hydrocarbons. In the Elmworth Field example, the fluid pressure aspect of trapping is seen in the capillarity with which water is held in the pore system. Water is so strongly held that gas permeability is sufficiently restricted in the tight sand on a geologically significant time scale. The East Texas Field is trapped by a combination of structure and stratigraphy, but the size of the accumulation may be related to the increased vertical permeability to water over the field.

Wheat Field

Wheat Field is located in the Delaware Basin of west Texas and is an oil accumulation in an unclosed structure (Fig. 7). According to Adams (1936), this accumulation occurs in a sandstone of the Delaware Mountain series. Regionally, the sand dips homoclinally to the east at an average rate of 20 meters/kilometer. Wheat Field is formed on a local structural terrace where dip

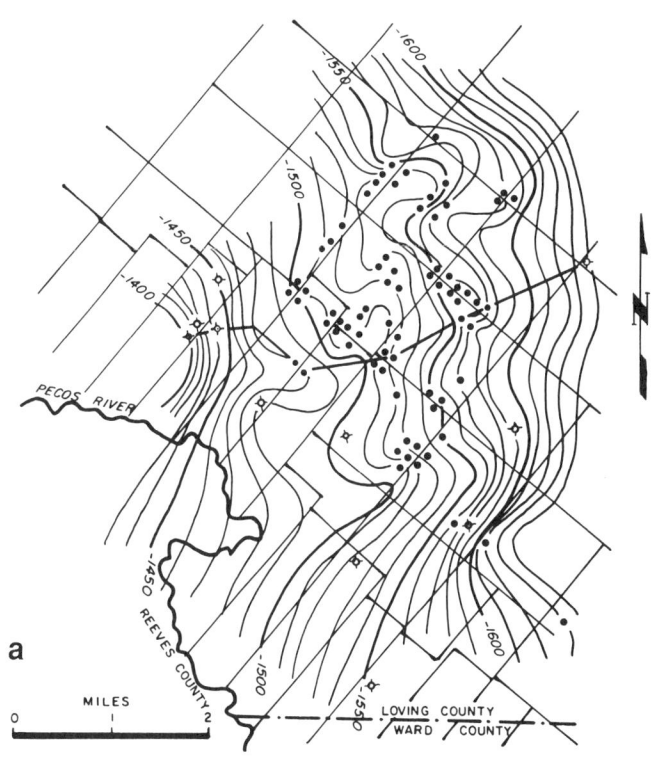

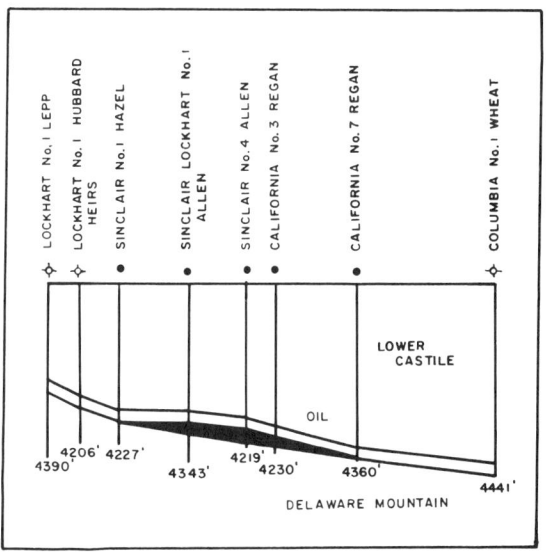

Figure 7(a). Structure map on the top of the Delaware Mountain Group at Wheat Field, Loving County, Texas (Contour interval—10 feet). (b) Diagrammatic cross section of Wheat Field showing the lack of a lithologic seal (Hubbert, 1953).

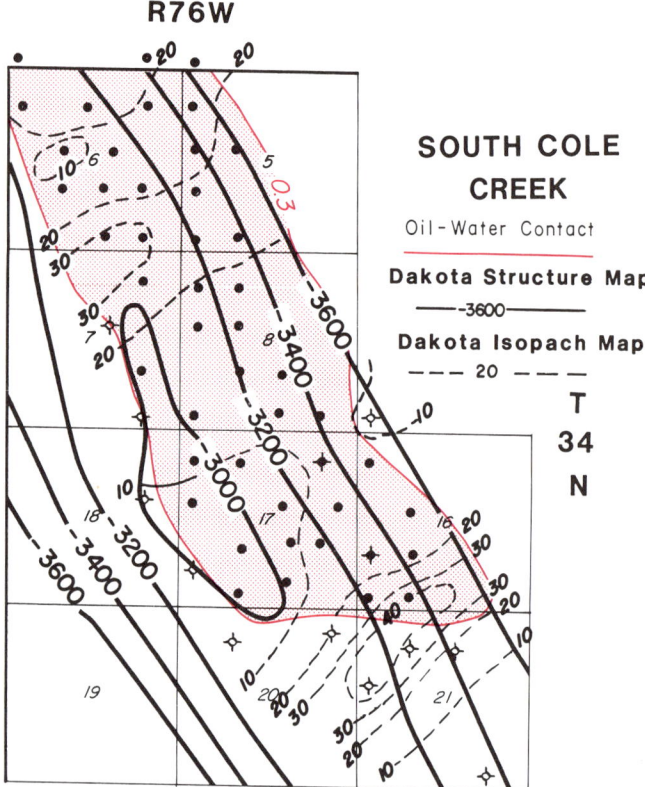

Figure 8. Dakota sandstone structure map (contour interval—20 feet) and isopach map (contour interval—10 feet) at South Cole Creek Field. A section is 1 mile wide. The location of the oil/water contacts shows the influence of a hydrodynamic flow (Moore, 1984).

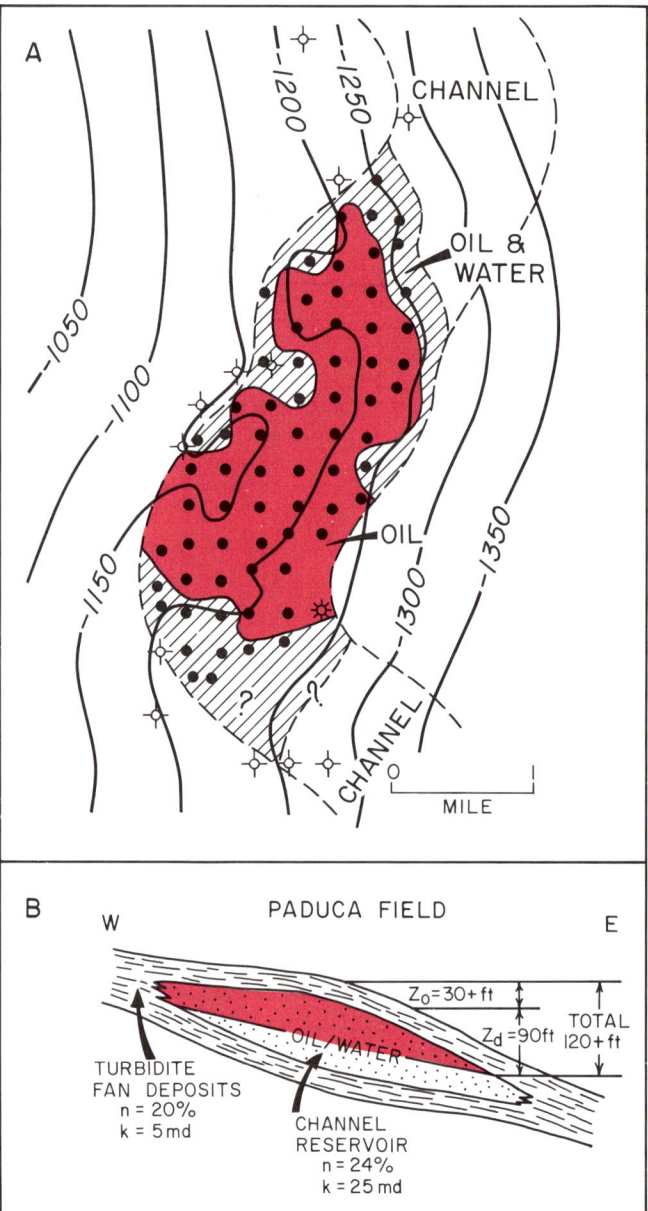

Figure 9(A). Structure on the Bell Canyon Sandstone, Paduca Field, New Mexico, showing the outline of turbidite channel reservoir facies. (B) Diagrammatic cross section of Paduca Field showing capillary and hydrostatic oil columns that total 120 feet or more (Berg, 1975).

decreases to 10 meters/kilometer. There is no evidence of faulting, and dry holes on the western side of the field are water productive and indicate good permeability. There is no evidence of the field being trapped against a facies change or permeability barrier. Hubbert (1953) concludes that Wheat Field is a hydrodynamic trap produced by water flowing eastward from the Delaware Mountains.

South Cole Creek Field

South Cole Creek Field is located in the southwestern Powder River Basin in Converse County, Wyoming (Moore, 1984). It is on a major south-plunging anticlinal trend that is generally regarded as the structural division between the Powder River Basin and the Casper Arch. At South Cole Creek, the Lower Cretaceous Dakota Sandstone has been folded into an anticline with approximately 180 meters of structural reversal (Fig. 8).

Thickness of the productive Lower Cretaceous Dakota Sandstone varies between 3 and 12 meters. The Dakota reservoir contains oil only on the east and north flanks of the structure. The south and west flanks are wet even though reservoir quality sand is present. Wells that tested small amounts of oil and abundant water from the Dakota are structurally higher than wells that produced oil and no water. These facts clearly show that structure and stratigraphy are not the controlling factors of the area extent of the oil accumulation at South Cole Creek.

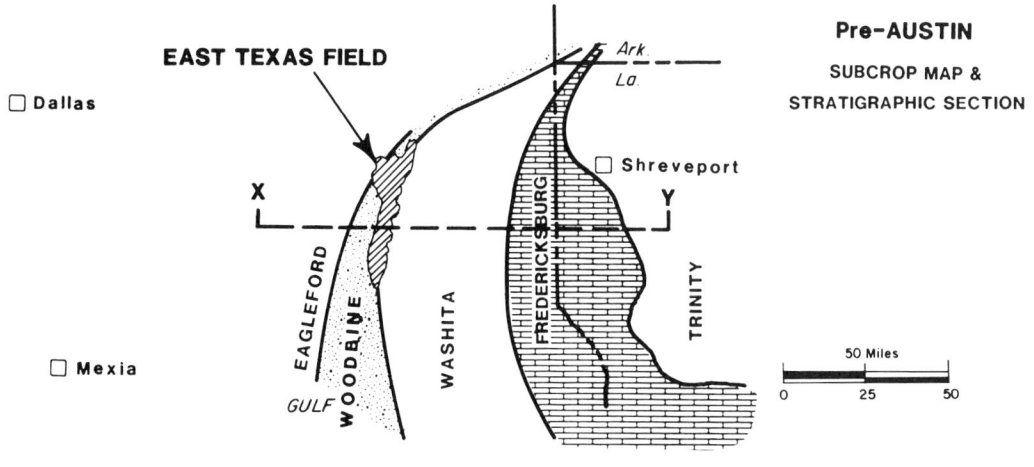

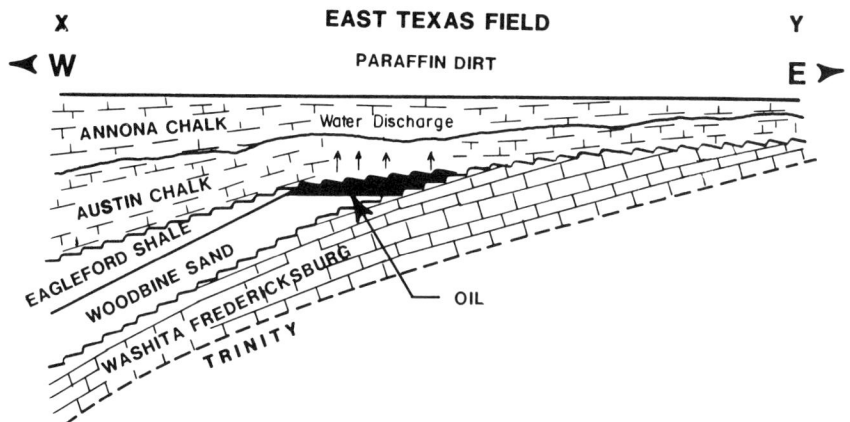

Figure 10. The geology of East Texas Field is quite simple. The trapping mechanism of this combination structural-stratigraphic trap is very subtle (after Levorson, 1967).

Paduca Field

An example of a hydrodynamic-stratigraphic trap is Paduca Field in the Delaware Basin of southeastern New Mexico. A structure map shows that the Guadalupian Bell Canyon Sandstone is dipping eastward at a rate of 20 m/km (Fig. 9) (Berg, 1975). The Bell Canyon reservoir facies is a massive, very fine grained turbidite channel sandstone with a maximum thickness of 10 meters. Reservoir porosity and permeability average 24 percent and 25 md, respectively. The seal facies is thin bedded siltstone turbidite fan deposits with 20 percent porosity and 5 md permeability.

If Paduca Field were a simple stratigraphic trap under hydrostatic conditions, the expected oil column would be about 11 m; however, oil production extends over a much greater height. The Bell Canyon Sandstone has a hydrodynamic gradient of about 5 m/km that is generally parallel to the structural dip. The hydrodynamic flow can be expected to enhance the seal capacity.

The hydrodynamic flow is sufficient to tilt the oil-water contact at a rate of about 14 m/km or a total of 35 m across the 2.5 km downdip length of Paduca Field. This tilt corresponds to the additional oil that can be trapped by hydrodynamic flow. The hydrodynamic component of the seal capacity is nearly three times the stratigraphic component.

East Texas Field

The East Texas Field is located in Gregg and Rusk counties, Texas, near the Louisiana state line (Fig. 10). This field has long been considered a classic example of a subtle combination structural-stratigraphic trap in which a stratigraphic element created a permeability limit within a reservoir rock, and this permeability limit was modified by a structural element that deformed the stratigraphic element (Levorson, 1966).

The Upper Cretaceous Woodbine aquifer of the East Texas Basin rises on the west flank of the Sabine Uplift. To the west of the field, the Woodbine Sand regionally is covered by the Eagleford Shale. At East Texas Field, the Woodbine Sand, however, is in direct contact with the overlying Austin Chalk. To the east of the field, the Austin Chalk lies directly on the pre-Woodbine Lower Cretaceous Washita Group. According to Levorson (1967), there is nothing in the surface or subsurface geology to suggest the huge accumulation of approximately 5,860 million barrels of recoverable oil. The angle of truncation on pre–Austin Chalk strata is very low, less than ¼ degree. This results in an erosional thinning on the order of 4 m/km. The structure on the top of the Woodbine Sand is very subtle (Fig. 11), with structural dip less than 5 m/km (Levorson, 1967).

Roberts (1980) believes that the key to the existence of East Texas Field lies in recognizing the shift in the overlying rock from the Eagleford Shale to the more permeable Austin Chalk. He states that the trapping of hydrocarbons is related to the increased vertical transmissibility over the trap. This seems reasonable if traps are viewed as dynamic mechanisms that focus and promote the flow of subsurface waters while retaining migrating hydrocarbons. Reports of paraffin dirt in the soils overlying the East Texas Field suggest the leakage of some hydrocarbons with the water.

Elmworth Gas Field

The largest gas reservoir in the Rocky Mountain region, Elmworth Gas Field, occurs in the most subtle trap that has been closely examined. At Elmworth and elsewhere in the Deep Basin of Alberta, the trap may be a combined stratigraphic diagenetic type. No discrete stratigraphic or lithologic unit is serving as a seal for the abnormally pressured gas accumulation in the Deep Basin of Alberta (Fig. 12).

In the area of the Elmworth Gas Field, the Wilrich Member of the Spirit River Formation comprises five transgressive and regressive cycles, with alternating marine and nonmarine deposits (Cant, 1983). Each cycle begins with laminated, fine grained shoreface and shallow-marine sandstones. These are succeeded by cross-bedded, pebbly conglomerates containing sandy matrix deposited in distributary channels that down-cut into the shoreface sandstones. The channel deposits are capped by beach deposits of well-sorted, matrix-free granule conglomerates. The sequence is locally completed by a thin coal.

The coarser sandstones and conglomerates of the beach zone are exceptionally good reservoirs (Fig. 12). Measured permeability may exceed one Darcy, and porosity values as high as 15 percent are common. The finer grained shoreface and shallow-marine sandstones exhibit permeabilities in the range of 0.001 to 0.5 md. These tight shoreface or shallow marine sands are the dominant facies and seals in the Spirit River Formation in the area between Elmworth Field and the water zone 20 kilometers to the east.

Porosity reduction in these tight sands is caused by grain crushing and cementation by illitic and kaolinitic clays and quartz carbonate. Secondary porosity, which enhances the remaining primary porosity, was formed as a result of chert and sedimentary rock fragment leaching (Cant, 1983). The seals in the Spirit River Formation are formed by the diagenetic redistribution of material leached from chert grains and sedimentary rock fragments. The permeability of these sealing sandstones approaches that of shales. Actual values may be lower because laboratory measurement of sandstones of this type is very sensitive to the drying technique

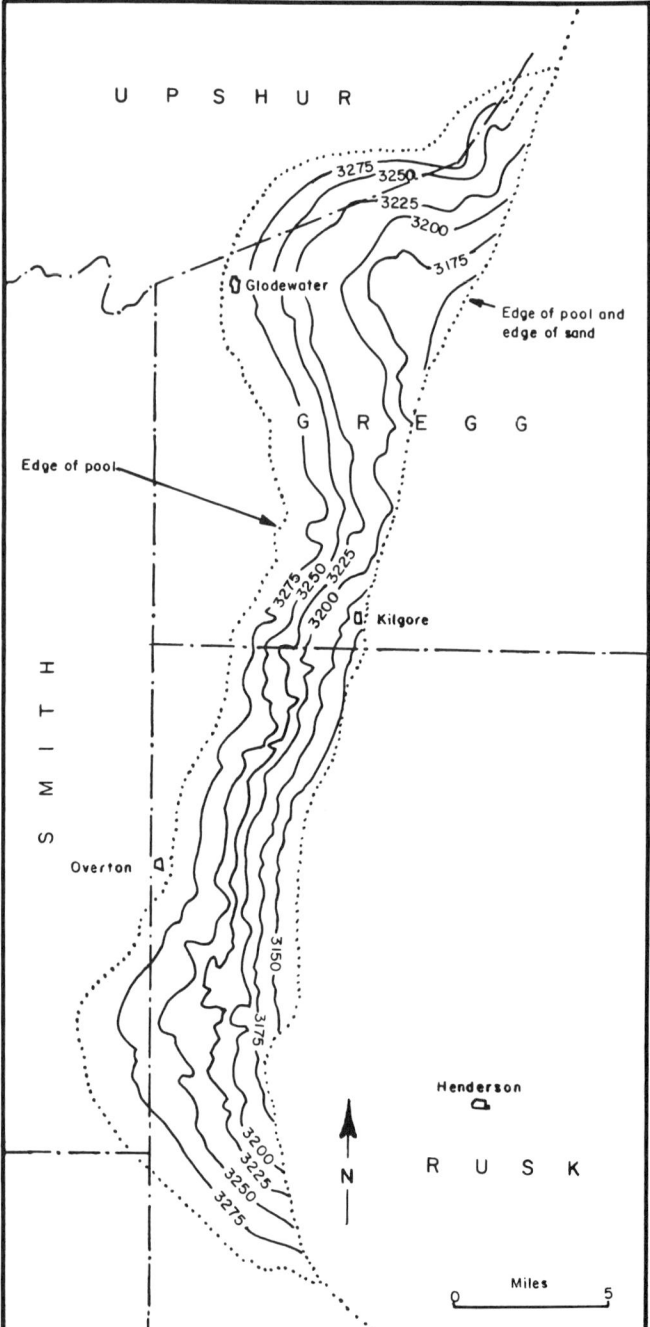

Figure 11. Structure on the top of the Woodbine (Upper Cretaceous) producing sand in the East Texas Field. Structural dip is generally less than 50 feet per mile (Levorson, 1967).

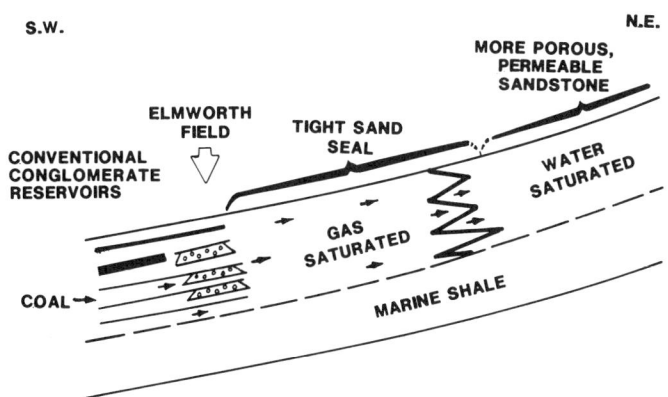

Figure 12. In the diagenetically mature Deep Basin of Alberta, the Spirit River Formation developed diagenetic seals during the geopressured, deep burial-organic maturation stage (Cant, 1983).

used to prepare the sample (Pallatt and others, 1984) and to the conditions of permeability determination (Wells, 1982).

The trap is a dynamic feature because the seals are allowing gas to leak out of the accumulation. The critical factors in this accumulation are the amount of gas and rate of gas generation and the rate at which gas can migrate through the tight sands. Currently or in the recent past, the gas generation rate was higher than the gas leakage rate, allowing the sands to develop a gas pressure greater than hydrostatic. Gas generation in the Spirit River Formation is presently ocurring in the interbedded coals (Tissot and Welte, 1978; Welte and others, 1982). Law and Dickinson (1985) point out that such overpressured gas accumulation types may eventually evolve into underpressured accumulations such as the San Juan Basin.

SUMMARY

These examples indicate the complexity of trapping mechanisms with the infinite number of variables affecting reservoir generation, filling, sealing, and location. The geologist must possess skills in a wide breadth of disciplines to understand such traps. Certainly knowledge of the role of pore systems on hydrocarbon entrapment is requisite. Ultimately, it is the nature of changes in the pore system in hydrocarbon source area, migration path, reservoir, and seal that controls the accumulation of hydrocarbons. Each piece of evidence from the disciplines of geology, fluid dynamics, organic and inorganic geochemistry, engineering, and physics must be blended to lead to the successful location and development of hydrocarbon traps.

REFERENCES

Adams, J. E., 1936, Oil pool of open reservoir type: American Association of Petroleum Geologists Bulletin, v. 20, p. 780–796.

Arps, J. J., 1964, Engineering concepts useful in oil finding: American Association of Petroleum Geologists Bulletin, v. 48, p. 157–165.

Beard, D. G., and Weyl, P. K., 1973, Influence of texture on porosity and permeability of unconsolidated sands: American Association of Petroleum Geologists Bulletin, v. 57, p. 349–369.

Berg, R. R., 1975, Capillary pressures in stratigraphic traps: American Association of Petroleum Geologists Bulletin, v. 59, p. 939–956.

Cant, D. J., 1983, Spirit River Formation; A stratigraphic diagenetic gas trap in the deep basin of Alberta: American Association of Petroleum Geologists Bulletin, v. 67, p. 577–587.

—— , 1986, Diagenetic traps in sandstones: American Association of Petroleum Geologists Bulletin, v. 70, p. 155–160.

Clark, N. J., 1969, Elements of petroleum reservoirs: Dallas, American Institute of Mining Metallurgical and Petroleum Engineering, 250 p.

Coalson, E. B., Hartmann, D. J., and Thomas, J. B., 1983, Applied petrophysics in exploration and exploitation: Denver, University of Colorado at Denver, 126 p. (private publication).

Dahlberg, E. C., 1982, Applied hydrodynamics in petroleum exploration: New York, Springer-Verlag, 161 p.

Dott, R. H., and Reynolds, M. J., 1969, Source book for petroleum geology: Tulsa, American Association of Petroleum Geologists, 469 p.

Downey, M. W., 1984, Evaluating seals for hydrocarbon accumulations: American Association of Petroleum Geologists Bulletin, v. 68, p. 1752–1763.

Freeze, R. A., and Cherry, J. A., 1979, Groundwater: Englewood Cliffs, New Jersey, Prentice Hall, 604 p.

Fuchtbauer, H., 1967, Influence of different types of diagenesis on sandstone porosity, in Proceedings of the 7th World Petroleum Congress, v. 2, p. 359–369.

Hill, G. A., 1956, Trap barriers; Sieves or seals? [abs.]: American Association of Petroleum Geologists, 41st Annual Meeting, Chicago, p. 35–36.

Hubbert, M. K., 1953, Entrapment of petroleum under hydrodynamic conditions: American Association of Petroleum Geologists Bulletin, v. 37, p. 1954–2026.

Law, B. E., and Dickinson, W. W., 1985, Conceptual model for origin of abnormally pressured gas accumulations in low permeability reservoirs: American Association of Petroleum Geologists Bulletin, v. 69, p. 1295–1304.

Levorson, A. I., 1966, The obscure and subtle trap: American Association of Petroleum Geologists Bulletin, v. 50, p. 2058–2067.

—— , 1967, Geology of petroleum (second edition): San Francisco, W. H. Freeman and Company, 724 p.

Mast, R. F., and Potter, P. E., 1963, Sedimentary structures, sand shape fabrics, and permeability, part II: Journal of Geology, v. 71, p. 548–565.

Meissner, F. F., Woodward, J., and Clayton, J. L., 1984, Stratigraphic relationships and distribution of source rocks in the greater Rocky Mountain Region, in Woodward, J., Meissner, F. F., and Clayton, J. L., eds., Hydrocarbon source rocks of the greater Rocky Mountain region: Denver, Rocky Mountain Association of Geologists, p. 1–34.

Moore, W. R., 1984, Hydrodynamic control on oil entrapment in channel sandstones of the Dakota sandstone, South Cole Creek Field, Converse County, Wyoming: The Mountain Geologist, v. 21, p. 105–113.

Morgan, J. T., and Gordon, D. T., 1970, Influence of pore geometry on oil-water relative permeability: Journal of Petroleum Technology, v. 22, p. 1199–1208.

Nagtegaal, P.J.C., 1979, Relationship of facies and reservoir quality in Rotliegendes Desert sandstones, southern North Sea region: Journal of Petroleum Geology, v. 2, p. 145–158.

Pallatt, N., Wilson, J., and McHardy, B., 1984, The relationship between permeability and the morphology of diagenetic illite in reservoir rocks: Journal of Petroleum Technology, v. 36, p. 2225–2227.

Pittman, E. D., 1979, Porosity, diagenesis, and productive capability of sandstone reservoir, *in* Scholle, P. A., and Schluger, P. R., eds., Aspects of diagenesis: Society of Economic Paleontologists and Mineralogists Special Publication 26, p. 159–173.

Purcell, W. R., 1949, Capillary pressures; Their measurement using mercury and the calculation of permeability therefrom: Petroleum Transactions, American Institute of Mining Engineers, v. 186, p. 39–48.

Roberts, W. H., 1980, Design and function of oil and gas traps, *in* Roberts, W. H., and Cordell, R. J., eds., Problems of petroleum migration: Tulsa, American Association of Petroleum Geologists, p. 217–240.

Schmidt, V., McDonald, D. A., and Platt, R. L., 1977, Pore geometry and reservoir aspects of secondary porosity in sandstones: Bulletin of Canadian Petroleum Geology, v. 25, p. 271–290.

Schowalter, T. T., 1979, Mechanics of secondary hydrocarbon migration and entrapment: American Association of Petroleum Geologists Bulletin, v. 63, p. 723–760.

Surdam, R. C., Boese, S. W., and Crossey, L. J., 1984, The chemistry of secondary porosity, *in* Surdam, R. C., and McDonald, M. A., eds., American Association of Petroleum Geologists Memoir 37: Tulsa, American Association of Petroleum Geologists, p. 127–149.

Tissot, B. P., and Welte, D. H., 1978, Petroleum formation and occurrence: New York, Springer-Verlag, 538 p.

Toth, J., 1980, Cross-formational gravity-flow of groundwater; A mechanism of the transport and accumulation of petroleum (the generalized hydraulic theory of petroleum migration), *in* Roberts, W. H., and Cardell, R. J., eds., Problems of petroleum migration: Tulsa, American Association of Petroleum Geologists, p. 121–167.

Treiber, L. E., Archer, D. L., and Owens, W. W., 1972, A laboratory evaluation of the wettability of fifty oil producing reservoirs: Society of Petroleum Engineers Journal, v. 12, p. 531–540.

Von Englehart, W., 1960, Der porenraum der sedimente: Berlin, Springer-Verlag, 207 p.

Wardlaw, N. C., and Cassan, J. P., 1979, Oil recovery efficiency and the rock-pore properties of some sandstone reservoirs: Bulletin of Canadian Petroleum Geology, v. 24, p. 117–138.

Wells, J. D., 1982, Tight gas sands-permeability, pore structure, and clay: Journal of Petroleum Technology, v. 34, p. 2708–2714.

Welte, D. H., Schaefer, R. G., Radke, M., and Weiss, H. M., 1982, Origin, migration, and entrapment of natural gas in Alberta Deep Basin [abs.]: American Association of Petroleum Geologists Bulletin, v. 66, p. 642.

Witherspoon, P. A., and Neuman, S. P., 1967, Evaluating a slightly permeable caprock in aquifer gas storage; 1. Caprock of infinite thickness: Journal of Petroleum Technology, v. 19, p. 959–965.

MANUSCRIPT ACCEPTED BY THE SOCIETY MAY 16, 1987

Chapter 16

Exploration techniques

Edward A. Beaumont
1444 S. Boulder, Tulsa, Oklahoma 74119
Norman H. Foster and Richard R. Vincelette
1625 Broadway, Suite 530, Denver, Colorado 80202
Graham R. Curtis
Gold Cup Exploration, Inc., 11880 Swadley Drive, Lakewood, Colorado 80215

INTRODUCTION

The ultimate goal of the explorationist is to find and develop commercial accumulations of hydrocarbons, and the focus of all efforts should always be directed toward this objective. North (1985, p. vii) states, "Geology is an integrative science and among its many constituents, petroleum geology is the most integrative of all." The successful petroleum geologist must be able to integrate and understand information from a wide variety of sources, including organic and inorganic geochemistry, geophysics, well logging, sedimentology, stratigraphy, paleontology, structural geology, and economics. As of yet, no direct method of precisely identifying the presence and location of a commercial oil and gas accumulation has been developed. Typically, most accumulations are hidden in the subsurface under thousands of meters of overlying sedimentary cover. Therefore, the explorationist must rely on a variety of indirect indicators to identify potential locations of hydrocarbon accumulations and if these indicators are judged to be strong enough, the drilling of an exploratory well will be required to prove or disprove the presence of the accumulation.

DEVELOPMENT OF EXPLORATION CONCEPTS

To find oil and gas we create concepts that lead to the discovery of oil and gas fields. An understanding of the techniques utilized in the creative thinking process are therefore very important. Foster (1989a, b) has summarized the stages involved in the creative thought process and how the process can be applied to petroleum exploration.

Well-planned oil and gas exploration projects involve the identification and study of three critical elements: exploration targets, exploration areas, and exploration techniques. Exploration targets are the type of oil or gas accumulation being sought, the exploration area is the geographic or geologic province selected in which to search for these targets, and the exploration techniques are those chosen to help locate the targets. Creative exploration involves the development of new ideas in any of these three critical elements. All three are closely interrelated, and the selection of one will have a strong influence on the selection of the others. For example, exploring for oil-bearing Tertiary reefs in the jungles of New Guinea will require much different techniques than will the search for methane-rich gas trapped in Paleozoic coal seams in the Black Warrior basin of Alabama.

Many areas that offer exploration opportunities have already been examined and in many cases condemned. Therefore, in order to find oil and gas in these areas, we need new insights, concepts, interpretations, techniques, data, applications, enthusiasm, and action.

Effective exploration projects progress through four basic stages of development: (1) identify objectives, (2) develop exploration concept, (3) test concept, (4) act on concept (Vincelette, 1989).

In stage one, the explorationist identifies exploration objectives. Should the targets be large fields? Is it best to explore in North America or somewhere else in the world? Do mature or frontier basins offer the best opportunities? Should targets be shallow, intermediate, or deep? Will the drilling program be limited to a few tests or will many wells be drilled? Insight is important in identifying objectives, and insight is generally a product of experience. Experienced explorationists intuitively recognize opportunities and therefore define objectives more effectively. In stage one the explorationist tries to assess the critical factors of the play by gathering data. Are source rocks present? Are they mature? Are reservoirs present? What is the nature of the traps?

In stage two the explorationist develops the exploration concept. At this stage he or she relies on the creative process of saturation, incubation, and illumination. Visualization skills, such as those described by Beaumont (1989), are critical. Data acquisition is a large part of stage two, and the acquired data are processed and examined in many ways. During stage two the explorationists continues to identify and focus on the critical elements. Models for a play are developed. Analog fields found locally or in other basins are important for developing play models. In stage two it is important to stay open-minded and not build conceptual blocks.

In stage three, the concepts developed in stage two must be rigorously tested. Critical data must be reexamined to verify play concepts. The explorationist must recognize negative factors and problems in the play and prospects because the "wilder" the play the more the uncertainty. While recognizing the problems, he or she must begin to analyze possible solutions. One effective method for overcoming problems is through the use of analogies. Risk evaluation is also part of stage three. Failure to test the concepts rigorously can result in unnecessary and premature failure.

Finally, in stage four the explorationist must find the courage to act on the play concepts. He or she makes recommendations for land purchases and drilling commitments. Persistence is critical because it is difficult to convince others they have overlooked the play, and in some cases a valid play concept may have to be presented to many different prospective buyers before someone agrees to test it with the drill. Many explorationists fail to carry out this stage. Stage four requires courage.

ANALOG STUDIES

A thorough understanding of the physical attributes of analog oil and gas accumulations, along with a knowledge of the typical habitat where these accumulations occur, enable the geologist to devise the most effective exploration techniques.

Detailed field studies typically will include an analysis of basin type in which the accumulation occurs (foredeep, regional arch), reservoir rock type (sandstone, carbonate, volcanic), reservoir environment of deposition (tidal flat, beach, reef), type of petroleum (gas, condensate, light or heavy oil), and the trap type (structural, stratigraphic, or combination) (Beaumont and Foster, 1990). These same studies will provide considerable detail on the location, exploration and production history, discovery method, local and regional structural setting, detailed stratigraphy from surface to basement, physical dimensions of the trap and reservoir, porosity and permeability, description of probable source rocks, and history of hydrocarbon migration into the trap.

The information provided by the field studies leads to the development of models or analogs from which the explorationist can devise exploration concepts and techniques to be applied in the search for similar accumulations elsewhere. In addition, known fields provide test sites where new exploration tools or techniques can be tested to determine their effectiveness.

SUBSURFACE TECHNIQUES

Most of the information utilized in subsurface studies is derived from boreholes. Even though many boreholes are "dry holes," they can provide valuable information that, in combination with other sources of data, can ultimately lead to the successful discovery of oil or gas. In addition, in frontier areas it is often necessary to drill a series of stratigraphic tests to gain information about the types and distribution of potential reservoirs, seals, and source rocks.

Borehole geophysical logs

Well logging with geophysical instruments has been one of the most widely used tools of the petroleum geologist since its invention in 1912. Geophysical logging instruments in general measure subsurface rock properties using electrical, nuclear, and sonic (or acoustic) methods (Asquith, 1982). New methods are continually in development.

Displays of the recorded information provide a detailed vertical profile of rocks penetrated by the borehole. Logs help to identify discrete lithologic units at their precise depth and position in the subsurface. Different formation and lithologic units typically have a unique log signature that can then be correlated with similar log characteristics from other wells in the general area. Using this information, the geologist can make a variety of maps interpreting the subsurface geology of the area under investigation, including maps showing the structural position and attitudes of formations of interest, as well as maps showing the thickness and distribution of potential reservoir rocks. In addition, these logs provide specific information on such important variables as porosity, permeability, hydrocarbon saturation, water salinity, lithology, formation temperature, and bedding attitudes.

Electric logs. The first geophysical well logging tool was the electric log. Today it is still the most widely used. When we run only one logging survey it is usually an electric-log survey. Generally, the electric log consists of three components: the spontaneous potential log, a shallow-reading resistivity log, and a deep-reading resistivity log (Fig. 1). Each sees a different aspect of the rocks penetrated by the borehole.

The spontaneous potential tool measures voltage variations in millivolts caused by differences in salinities between drilling-mud filtrate and formation-water resistivity within permeable beds.

Resistivity logs record differences in the resistive character of formations. The matrix and grains of rocks are nonconductive. The ability of a rock to conduct an electrical current is therefore a function of the amount of water in the pore space. The greater the porosity of a rock, the lower its resistivity. Hydrocarbons are nonconductive; because of this, as hydrocarbon content increases, resistivity of the rock containing the hydrocarbon increases. Resistivity logs can be used to estimate porosity, locate permeable zones, and determine hydrocarbon versus water-bearing zones.

Microresistivity devices, such as the dipmeter, measure the orientation of bedding planes and other detailed features in a borehole. Dipmeter logs can be used to interpret location and type of fault (Fig. 2), regional dip, unconformities, draping caused by differential compaction, and depositional environment.

Porosity logs. The sonic or acoustic log measures interval transit time of a compressional sound wave traveling through a formation in order to measure porosity. Interval transit time is a function of both lithology and porosity. Sonic porosity can be derived from a chart that determines matrix porosity.

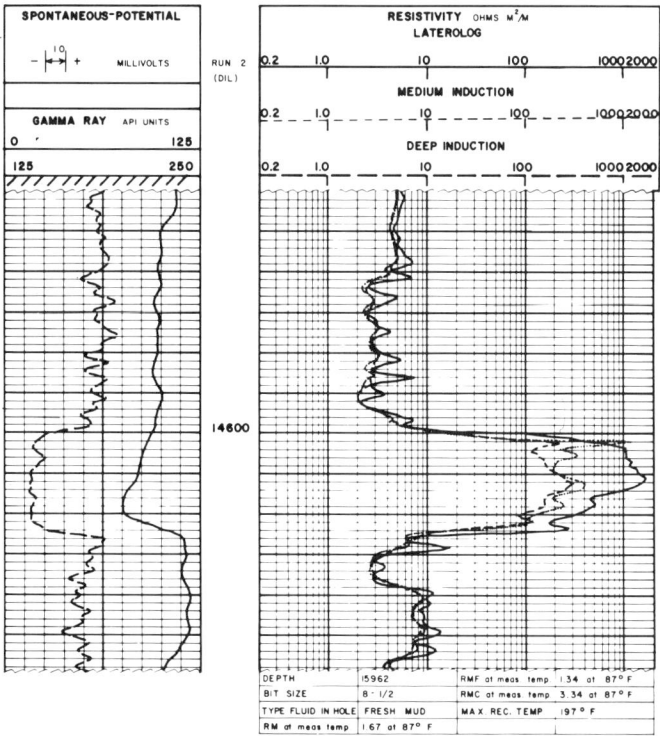

Figure 1. Resistivity log with spontaneous potential and gamma ray logs through a sandstone-shale interval. Sandstone interval is from 14,600 to 14,625 ft. Bold lines on vertical scale are in increments of 10 ft (from Asquith, 1982).

Using a gamma-ray source, the density tool measures the electron density of a formation. Gamma rays collide with electrons in the formation and scatter. A detector located a fixed distance from the source measures returning gamma rays, which are an indicator of formation density (Fig. 3). Matrix density, porosity, and density of the fluid in the pores determine the formation bulk density. The matrix density and type of fluid in the borehole must be known to calculate porosity. This kind of log is useful for evaluating shaly sand reservoirs and complex lithologies, determining hydrocarbon density, detecting gas-bearing zones, and identifying evaporite minerals (Asquith, 1982).

Neutron logs measure porosity because the tool responds to variations in the hydrogen-ion concentration of a formation. In formations with little clay matrix, the neutron log records the amount of liquid-filled porosity (water or oil). A chemical source creates neutrons that collide with nuclei in the formation. This process results in a net loss of energy in the neutrons. Maximum energy loss occurs when a neutron collides with a hydrogen atom, which is nearly the same mass as the neutron. The amount of energy lost by the neutrons emitted by the tool relates directly to the amount of fluid in the pore spaces; therefore, neutron-log measurements in rocks with water-filled pores can measure formation porosity (Fig. 3).

Sample examination

Sample examination includes looking at cores and cuttings obtained from boreholes. Sample description is the foundation of subsurface geology and is the petroleum geologist's closest link to the subsurface. Sample data give information that can be used to calibrate data from other sources to the rocks.

Examination of cuttings allows direct observation of rock fabric, diagenetic fabric, pore and pore-throat geometry, and shows of oil and gas. Porosity and permeability or rocks can be estimated. Cores also offer the opportunity to view primary sedimentary and tectonic structures and the sequence of lithologies. Sample data are the most reliable data that can be used to map facies in the subsurface. Microfossils obtained from cores and cutting can identify age and depositional environment. In addition, well cuttings and cores can be analyzed for hydrocarbon source-rock potential and maturity. Sample data are especially effective when they can be calibrated with other data.

Subsurface mapping

Subsurface maps are the manifestation of the petroleum geologist's visualization of subsurface geology. They are compiled for the purpose of locating traps that contain oil or gas. Just about every aspect of petroleum geology can be represented on a map; the only limit to mapping data is imagination. Most information comes from wells, but abundant information is also available from geophysical surveys, production history of fields, pressure and temperature surveys of wells, oil and gas shows, and downward projection of surface data.

Basic maps compiled in the process of studying an area for exploration depict either external geometry of rocks (structural attitude or thickness) or the internal character of the rocks in a particular interval (lithofacies or porosity).

Cross sections

Before constructing the first subsurface map, the geologist should make a grid of correlation cross sections to establish regional correlations. Good well-log quality and control are main considerations in locating cross sections. Correlation sections force the explorationist to "learn the habits of the rocks, their stratigraphic relations, the potential reservoir rocks, the position and nature of unconformities, the time of folding and faulting, the changes of folding with depth, and the facies changes in the section of rocks being mapped" (Levorsen, 1967, p. 592).

Next, detailed stratigraphic and structural cross sections should be made through selected fields to illustrate structural or stratigraphic features responsible for trapping. If possible, field cross sections should use wells that have core or sample information to aid in mapping facies or porosity. Test data should be posted on tested wells in the section.

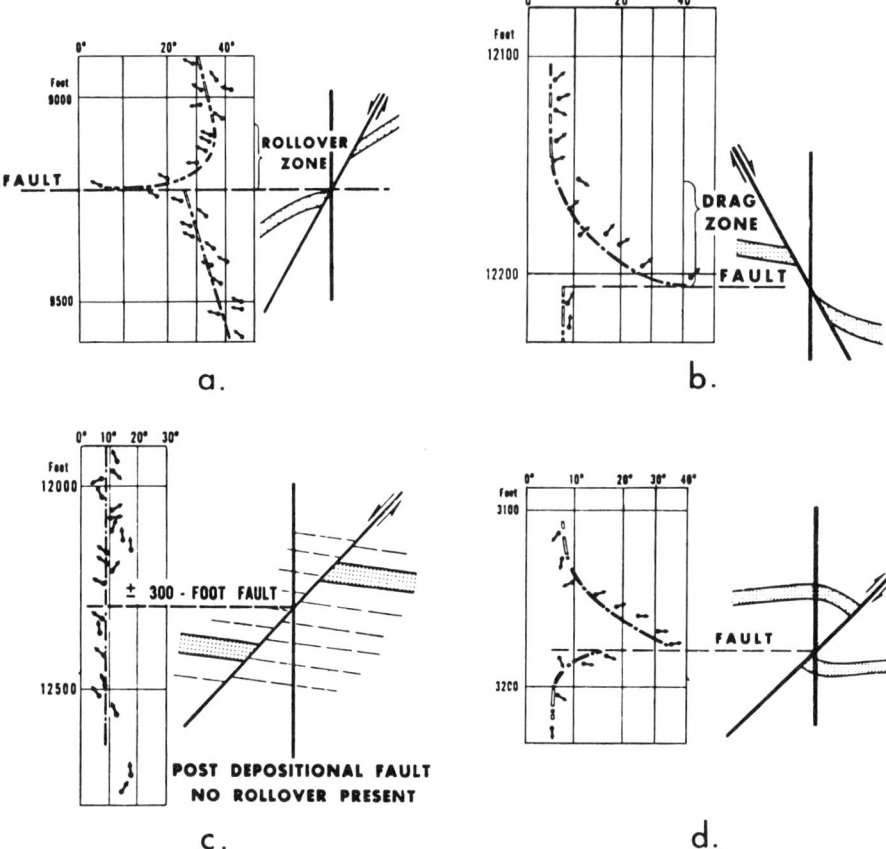

Figure 2. Dipmeter response patterns to different fault types (from Doveton, 1986).

Contour maps

The most frequently used exploration maps are subsurface structure and stratigraphic maps. Structure maps illustrate the present configuration of any correlatable surface, such as a formation boundary or unconformity. Stratigraphic or isopach maps illustrate the geometry of an interval between two correlatable horizons, such as the top and bottom surfaces of a reservoir or formation. Structure and isopach maps, together, are powerful prospecting tools; however, any data can be contoured. For example, maps can be made of subsurface temperatures, pressures, oil gravity, and production.

It takes considerable time to contour all available data sets. Therefore, it is necessary to determine which contoured data can enhance understanding of the critical elements of a play.

Before contouring, visualize the geology to be represented and mentally project it onto the data points of your map. If there is enough time, try contouring the data in several different ways with different interpretations in mind.

Pressure plots and maps

Fluid pressure plots and maps aid in locating discontinuities in reservoirs, determining lengths of trapped hydrocarbon columns, inferring migration pathways, and looking for tilted oil-water contacts.

The following procedure can be used to evaluate the subsurface fluid pressures. (1) Map pressure data from drill-stem tests run in water-saturated portions of reservoirs. If a fluid pressure change is found, mapping the top of the pressure variation may reveal a regional seal that could trap hydrocarbons. (2) Construct pressure-depth plots in order to evaluate the degree of continuity of "the number and nature of pressure systems indicated and the predicted stratigraphic positions of oil-water, gas-water, and oil-gas interfaces where appropriate." Use these plots to determine how the data might be mapped: "as discrete systems or single systems in which pressure differences reflect internal hydrodynamic gradients" (Dahlberg, 1982, p 39).

Show maps

Indications of the presence of hydrocarbons in wells or what are called "shows" can be important clues to discovering oil and gas. To evaluate a show: (1) determine if the show is from a continuous-phase hydrocarbon occurrence; (2) if it is from a continuous-phase hydrocarbon occurrence, determine if it is from a commercial accumulation; and (3) if it is a continuous-phase oil show from a suspected commercial accumulation, determine if the show is from the waste or transition zone.

Shows can be seen in drilling fluids, in core or cutting samples, and in formation or production tests, or they may be indicated by "anomalies" on geophysical logs. Hydrocarbons occur in the subsurface as (1) continuous-phase oil or gas, (2) isolated droplets of oil or gas, (3) molecular-scale dissolved hydrocarbons, or (4) minor shows associated with oil or gas source rocks (Schowalter and Hess, 1982). Of these, only shows from continuous-phase oil or gas indicate a possible commercial accumulation.

A continuous-phase hydrocarbon occurrence is a continuous connection of hydrocarbon through the pore network of a water-saturated porous rock. Continuous-phase shows can easily be distinguished from shows of molecular-scale dissolved hydrocarbons and shows associated with oil or gas source rocks. Shows from dissolved hydrocarbons are always gas, and the volume of gas produced from these shows is small and has a limited flow. Shows associated with source rocks come from kerogen within the source rock and not the reservoir. Kerogen dispersed within the reservoir is generally only found in small quantities. Shows from residual or isolated droplets of hydrocarbons are most difficult to distinguish from continuous-phase shows. This can be done by analysis of the percent saturation of hydrocarbons with respect to water, and by analysis of the type and amount of fluid recovered. To have a continuous hydrocarbon phase the saturation must be 10 percent or more. The minimum saturation for a residual show is about 5 to 10 percent (Schowalter and Hess, 1982). However, as a result of migration or drainage of hydrocarbons in the reservoir, residual hydrocarbon saturation can be as much as 35 percent for sandstones and 80 percent with an average of 55 percent for carbonates. In general, therefore, only hydrocarbon saturations of more than 35 percent for sandstones and 55 percent for carbonates are positive indications of continuous-phase hydrocarbons. The most reliable method of calculating in-situ saturations is by well-log measurements (Schowalter and Hess, 1982).

With regard to the type and amount of fluid recovered from a reservoir, continuous-phase shows have some relative permeability to oil, whereas residual shows have no relative permeability to oil. Thus, only continuous-phase shows will flow oil into the well bore during testing. Any significant recovery of free oil on drill-stem testing or production tests would positively identify a continuous-phase show from a residual phase (Schowalter and Hess, 1982).

Distinguishing continuous-phase gas shows is more difficult; however, gas flowing at sustained measurable rates is a positive indication of continuous-phase gas. Other occurrences of gas, such as gas dissolved in water, should not be able to produce at sustained rates.

Shows from continuous-phase hydrocarbon occurrences also may be used to estimate the oil or gas column needed to force oil or gas into the pores of a water-saturated reservoir. The hydrocarbon-water contact determines the downdip extent of the reservoir, which can be calculated if the capillary properties, hydrocarbon saturation, hydrocarbon-water interfacial tension, oil-water densities, and hydrodynamic conditions are known.

In many cases, the updip and downdip limits of oil reservoirs are sharp, but often they are transitional (Fig. 4). The updip zone is known as the waste zone, and the downdip zone is known as the transition zone. When testing a well, it is important to know which zone produced a show because both will yield oil and/or water on formation tests. The waste zone is distinguished from the transition zone by higher fluid pressure (which indicates a large oil column) and by higher productivity (Fig. 5).

SURFACE TECHNIQUES

Remote sensing

Sabins (1987, p. 1) gives a broad definition of remote sensing as "collecting and interpreting information about a target without being in contact with the object." Remote-sensing instruments record variations of intensity in electromagnetic energy caused by unique properties of the materials being imaged. Photogeologic/photogeomorphic approaches, which utilize an understanding of geomorphology to interpret geologic structure, are also a form of remote sensing.

The electromagnetic spectrum is the range of frequencies or wavelengths of energy that travels at the speed of light and propagates through a vacuum. The part of the spectrum of geologic interest extends from 0.4 micrometers to 50 cm and is divided into segments: ultraviolet, visible and near infrared, short wavelength infrared, midinfrared, and microwave. Different remote-sensing tools measure different parts of the electromagnetic spectrum. Some of these tools are described below.

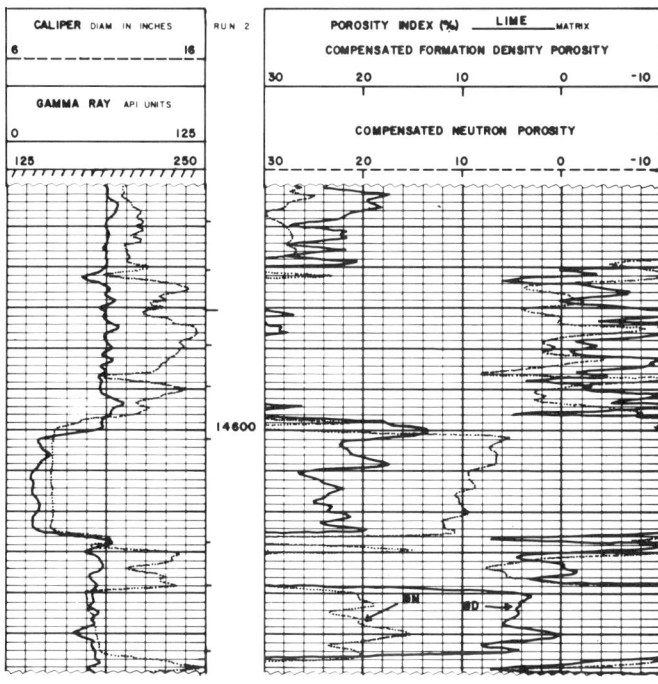

Figure 3. Gamma-ray and neutron log from same well and interval as log shown in Figure 1 (from Asquith, 1982).

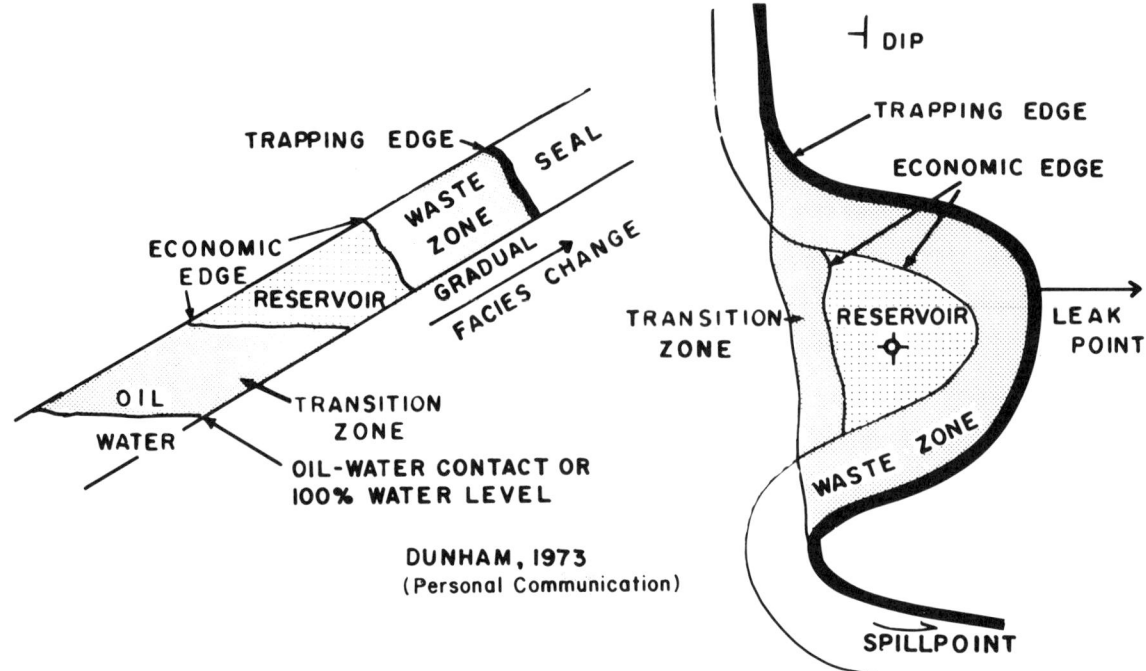

Figure 4. Transition and waste zones in a hypothetical stratigraphic trap (from Schowalter and Hess, 1982).

Aerial photography

In the 1930s, aerial photography developed into an exploration tool. At that time, only black-and-white photography was available; now normal color and infrared-color are commonly used. Aerial photography has the advantage of high spatial resolution, stereoscopic viewing, and vertical exaggeration. Color and infrared color contain much more information because the eye senses many more colors than shades of gray.

Satellite imagery

Landsat imagery became available in the early 1970s and is especially useful as a reconnaissance tool for revealing regional features such as basins, arches, and major fault systems. Landsat images have ground-resolution cells ranging from 79 by 79 m for Multispectral scanner (MSS) data to 30 by 30 m for Thematic mapper (TM) data and have the advantage of global and seasonal coverage. Computer processing enhances Landsat images to reveal anomalies by highlighting certain minerals, clay types, stressed vegetation, or thermal signatures.

Radar

Side-looking radar images have been acquired by aircraft since the 1950s. Radar provides its own radiation and can be operated at night if necessary or in any weather. The angle of illumination can be controlled to enhance subtle topographic features that are caused by subsurface structures.

Thermal infrared imagery

Thermal infrared (IR) imagery records the radiant temperature of minerals. IR became widely used in petroleum and mineral exploration when it was acquired by satellites in the 1970s. Different parts of the IR spectrum are effective for different purposes. Broadband IR is especially effective in terrain with low relief and thin soils because it can detect structure by mapping differences in thermal inertia or moisture content of minerals in soil or rocks. Multispectral thermal IR is useful for differentiating quartzose sandstones, carbonates, bedded gypsum/anhydrite, and shales. For petroleum exploration, multispectral thermal IR can detect microseepages associated with mineral alteration such as gypsum to calcite, the reduction and mobilization of iron oxide, and the recementation of clastic sedimentary rocks with calcite, pyrite, and montmorillinite (Goward and Taranik, 1986). Multiband works best in areas with less than an average 40 percent vegetative ground cover.

Use of remote sensing in exploration

Photogeology, photogeomorphology, and remote sensing can be one of the explorationist's most effective exploration methods when properly used. Many geologic features are expressed at the surface as distinctive geomorphic anomalies. For example, differential compaction of shales flanking a reef may cause beds overlying the reef to drape. The drape may extend to the surface, causing characteristic geomorphic patterns to be formed. The same type of geomorphic expression may develop

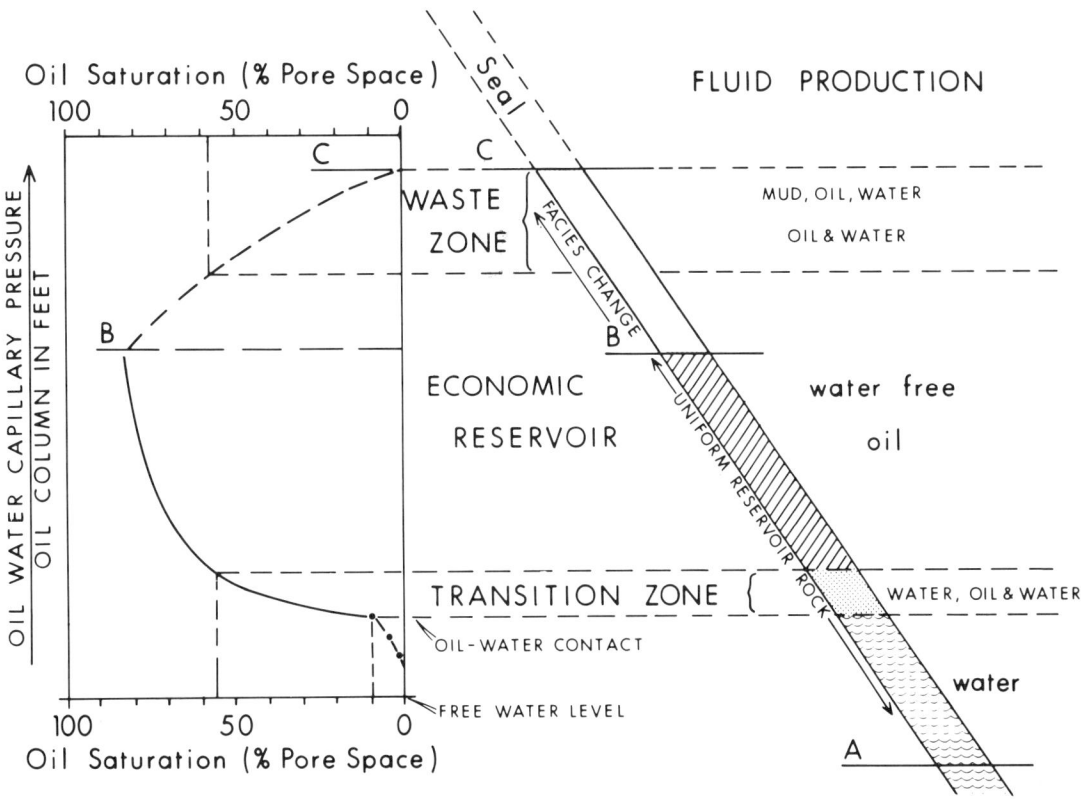

Figure 5. Relation of waste and transition zones in a complex stratigraphic trap to capillary pressure (from Schowalter and Hess, 1982).

over rising salt domes. Features might include an anomalously high topographic feature, with a rough or hummocky topography formed by erosion of fractures associated with the feature, drainage deflections, radial drainage, barbed tributaries, or other characteristic patterns.

Strike and dip can be mapped from standard aerial photographs, so that an accurate surface structural map can be made when adequate stereoscopic coverage is available. When using remotely obtained surface data to aid in exploration, thorough study of the surface expression of a known field that produces from a trap similar to the type of trap anticipated is recommended. The analog may be present anywhere on the Earth.

Surface mapping and field geology

The development and evolution of the anticlinal theory, in generl relation to hydrocarbon origin and migration from field studies by Sir William Logan (1844) and others, laid the foundations for the early beginnings of petroleum geology. The modern field geologist, with perhaps more knowledge and sophistication, is confronted with the same chores and approach to field geology as those early pioneers. The same principles of planning, going, observing, recording, thinking, extrapolating, and reporting are required.

GEOCHEMICAL TECHNIQUES

Source-rock analysis

Geochemistry lowers petroleum exploration risk in both frontier and mature basins. In frontier basins, geochemistry aids explorationists in the identification of source rocks and determination of the amount of hydrocarbon generation from the source rock in space and time. In mature basins, geochemistry is useful for determining areas with deeper or undeveloped remaining potential.

Assessment of source-rock quality, maturation of organic matter, and hydrocarbon migration (Cornford, 1986) requires determination of: (1) abundance of organic matter (sometimes measured by TOC or total organic carbon); (2) levels of maturation of organic matter; (3) types of organic matter (oil or gas prone, mixed, or inert); (4) areal extent, thickness, and lateral variation; and (5) relative amounts of interbedded sands and silts. Taken together, these analyses help explorationists answer the following questions (Tissot and Welte, 1984):

1. Which strata have potential as source rocks? What is their regional extent and what are their paleogeographic relations to time-equivalent nonsource rocks?

2. Is the basin oil prone, gas prone, or both?

3. When, at what depth, and where in the basin did source rocks generate petroleum?

4. How much petroleum was generated?

5. When and along which pathways did hydrocarbon migration occur?

6. What traps were present during hydrocarbon migration?

Ideally, by knowing the answers to these questions the explorationist can increase his or her chance of success by drilling prospects along pathways where the greatest volumes of petroleum migrated to preexisting traps. However, these questions are usually only answered after many wells have been drilled in a basin. Realistically, to use geochemistry effectively at an early stage of basin exploration, the geologist must fill in gaps in the data with his or her imagination and continually enter new data into the geochemical exploration model as they become available. Details about the origin and migration of oil and gas are presented by Meissner (this volume).

Surface geochemical prospecting

Surface geochemical prospecting in petroleum exploration is primarily based on finding and mapping hydrocarbons that leaked from oil and gas traps at depth. Early petroleum exploration began by drilling near visible seeps of oil and gas. Surface geochemical prospecting is really a modern extension of that well-worn technique. Just as additional geological data are required to understand the exploration significance of a surface seep, microseeps also require supplementary data before their exploration significance can be known.

Gas chromatography is the most widely used onshore and offshore direct detection technique. Onshore detection of these "invisible" seeps or microseeps is more problematic because of more complex migration mechanisms associated with near-surface water containing dissolved gas (Philp and Crisp, 1982). Offshore direct detection techniques use "sniffers" and have the advantage of minimal lateral movement of dissolved gas in the water column.

Sensitive gas chromatographs rapidly detect and measure gas concentrations in parts per billion, and gas chromatographs can separate and measure individual amounts of methane, ethane, propane, or heavier gases. Therefore, an explorationist can determine if the gas detected in surface anomalies is microbially produced gas or if it leaked to the surface from traps in the subsurface. In some cases it is possible to determine if gas in anomalies is related to gas found in nearby oil and gas reservoirs (McIver, 1985).

Other surface or near-surface geochemical prospecting techniques (McIver, 1985) involve the collection of soil samples and ascertaining secondary indicators such as: (1) radiogenic or inert gases; (2) radioactivity; (3) heavy hydrocarbons, including paraffin dirt; (4) methane-utilizing bacteria; (5) spectral fluoresence or luminescense of soil extracts; (6) trace metals in soils and plants; (7) inorganic salts or anions; and (8) carbonates, sometimes analyzed for ^{13}C isotope content.

GEOPHYSICAL EXPLORATION TECHNIQUES

Geophysical methods are fundamental tools of petroleum exploration. They can be used to interpret the geometric relations between basement and sedimentary rocks, structural configurations of rocks, the stratigraphy of a basin, facies distribution, the porosity and permeability of potential reservoirs, and in some cases, the identification of hydrocarbon-bearing zones in the subsurface.

The seismic method is especially well suited for unraveling the details of subsurface structure and stratigraphy. Other techniques such as gravity, magnetics, and magnetotellurics are normally utilized to gain information on regional subsurface conditions and in reconnaissance exploration surveys, but also provide valuable information in those areas where good seismic data are not obtainable. The most commonly used methods are reviewed below.

The seismic method

The most powerful and widely used geophysical exploration technique today is reflection seismic. Layered sedimentary rocks have a relatively narrow range of densities and only weak magnetic signatures. Therefore, gravity and magnetics are of little use when trying to unravel stratigraphic or structural details of sedimentary rocks. Reflection seismic mapping can reveal stratigraphic relations, structural geometries, facies, and occasionally porosity distributions.

The ultimate use of reflection seismic methods is in the identification of drilling targets that have a high probability of containing hydrocarbon accumulations. Oil and gas accumulations typically occupy a very small part of a geographic area under investigation. Therefore, the seismic data acquired must be carefully interpreted to identify those "anomalies" in the regional structural or stratigraphic patterns that could be the locis of hydrocarbon accumulations. Originally utilized to map structural closures such as anticlines or fault traps in the subsurface, advances in seismic acquisition and processing methods have enabled this technique to have increasing success and application in the identification of stratigraphic traps (such as carbonate reefs and channel sandstones).

In addition, gas can have a profound effect on the character or amplitude of the seismic signal. This effect has resulted in the use of seismic data to identify and map the distribution of gas-bearing sandstones ("bright spots") in the subsurface as well as identifying gas-water contacts in carbonate reefs.

Seismic theory

According to Anstey (1982, p. 2), the seismic method, "reduced to its essentials it is simply this: we make a bang, and we listen for echoes." The seismic method involves generating sound waves that spread spherically downward into the earth and echo off rock surfaces that they encounter.

Exploration techniques 263

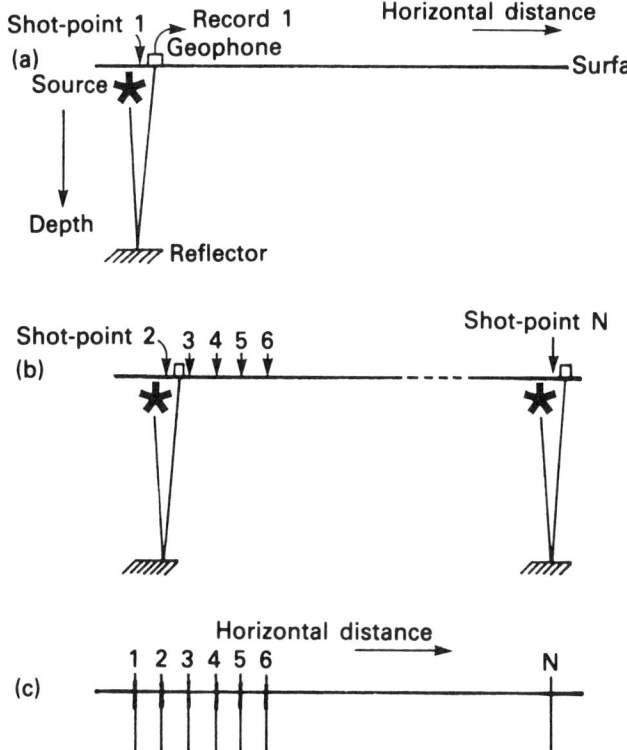

Figure 6. The basic concept of measuring reflecton times along a surveyed line to construct a seismic section (from Anstey, 1982).

Receivers and recording equipment at the surface record travel time to and from reflecting horizons inthe subsurface. The process of building a seismic section essentially involves a repetition of generating sound at different surveyed locations along lines and recording reflection times at each location (Fig. 6). Variations in travel time to a particular reflector along a seismic line define structure on that reflector. Variations in reflector amplitude, continuity, phase, and relations to other reflectors gives information about subsurface stratigraphy.

Structural application

A seismic survey usually consists of a grid of lines. An interpreter takes this grid of lines, picks continuous clean reflections, and correlates or "ties" stratigraphically similar reflectors from different lines at line intersections in the grid. Next, the interpreter chooses a reflector from a horizon of interest on which to create a structure contour map. At control points, usually shot points at a given interval, plots are made of values of either time, or depths converted from time, derived from the chosen reflector and values are contoured. All of this is simple, but as Anstey (1982) points out, there are at least three major pitfalls to be aware of when interpreting seismic data for structure.

1. In areas with flat-lying beds, ray paths for seismic energy are essentially straight up and down; points of reflectance are directly below the energy source and receiver. This situation results in an unambiguous portrayal of subsurface structure. However, dipping beds introduce ambiguity because the point of reflectance is not directly under the receivers but at some distance horizontally. A mechanical process called migration remedies this situation by moving reflections back to their proper location on the seismic section. Basically, migration involves drawing circles from shot points to corresponding reflectors. The circles converge at the actual location of the reflector (Fig. 7).

2. Seismic energy passes through the earth as spheres, not as circles (i.e., three dimensions versus two dimensions). Three-dimensional features in areas of complex geology cause spurious reflections to be included in the section. To use spheres to define the areas of possible reflection, three-dimensional migration is performed and reflectors are moved to their proper locations on the lines.

3. Velocity variations in the subsurface caused by the heterogeneous nature of rocks mean that a single value for velocity to measure depths is done when echo-ranging in air or water cannot be used. Instead, velocity values of different stratigraphic units at different depths must be known. Sonic logs or common depth

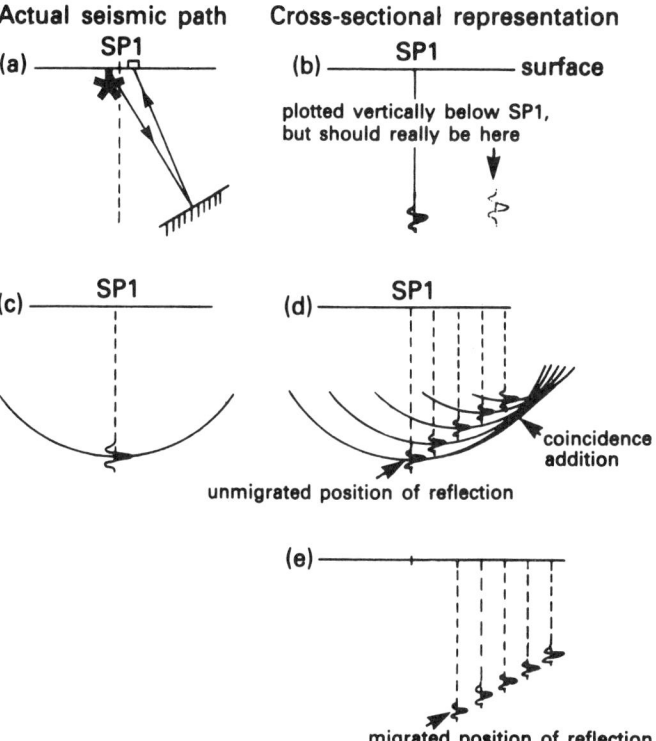

Figure 7. The process of migration (c to e) used to make corrections in seismic sectons for reflections from a dipping interface (ab) (from Anstey, 1982).

point gathers (CDP gathers) give information for determining subsurface velocity. With sonic logs, velocity values are read directly from the log. In CDP gathers, shot points and recorders are spaced symmetrically about a reflection point or common-depth point. In order to calculate velocity, the reflection times from a source-receiver pair near the common depth point and a source-receiver pair distant from the common depth point are necessary (Fig. 8). The extra distance that a seismic pulse travels between the close-in source receiver pair and the distant source receive pair, divided by the extra time to travel the extra distance gives the velocity.

Variations in velocity can lead to misinterpretation of seismic sections. "Pull-ups" result from higher velocities in a localized thickening of a rock type characterized by high velocity such as salt. This situation results in a pull-up of reflectors from strata beneath the thick high-velocity zone. A classic example of velocity pull-up is from salt domes where there appears to be an anticline under the dome (Fig. 9).

There are many other pitfalls to interpreting structure in seismic sections, including geometrical and processing problems. Knowing the geology and the pitfalls may help to correct the data and avoid a misinterpretation or to see something significant that might otherwise have been missed.

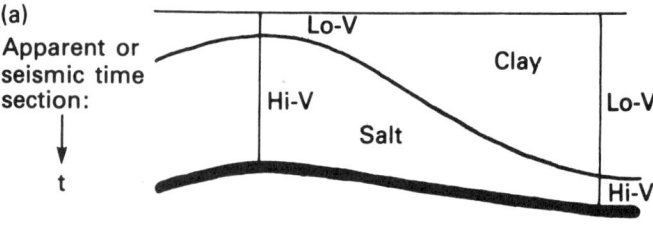

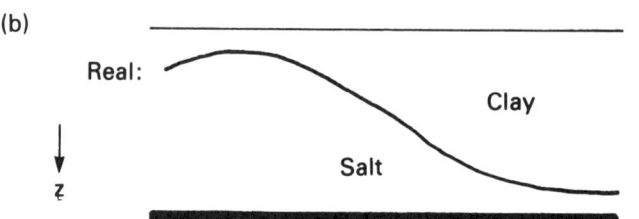

Figure 9. The classic velocity hoax: spurious structure introduced under a salt pillow (from Anstey, 1982).

Stratigraphic application

Explorationists first used the seismic method to locate subsurface structure. Lines were generally short, and it was difficult to interpret anything but structure. When explorationists began using the seismic reflection method offshore, lines became much longer. Vail and others (1977) observed that regional stratigraphy could easily be deciphered on these longer lines.

Stratigraphy can be interpreted from seismic in two ways: (1) by variations in reflector character, and (2) by the relations of reflectors to one another. Establishment of depositional sequences is the first step in a seismic stratigraphic analysis. A depositional sequence "is a stratigraphic unit composed of a relatively conformable succession of genetically related strata and bounded at its base by unconformities or correlative conformities" (Mitchum and others, 1977b, p. 53; see Fig. 10). Depositional sequences are defined through careful analysis of seismic sections and establishment of their equivalent seismic sequences. Correlation of seismic sequences throughout a basin builds the stratigraphic framework in the basin from which relative sea-level changes, tectonic history, and sedimentary history are interpreted.

Boundaries of depositional sequences and their equivalent seismic sequences are determined by discordant relations of strata to unconformities or, in the seismic stratigraphic sense, by finding discordant reflectors in a seismic section (Fig. 11). Once sequences are delineated, seismic facies can be identified within the sequence in order to decipher its depositional history. "Seismic facies units are mappable, three-dimensional seismic units composed of groups of reflections whose parameters differ from those of adjacent facies units" (Mitchum and others, 1977a, p. 117). Facies units can be interpreted in terms of environmental setting, depositional processes, and lithology after the internal reflection

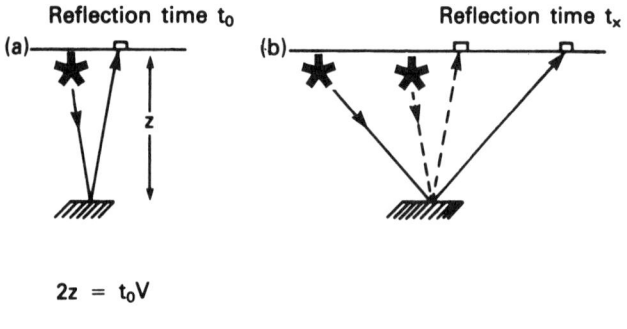

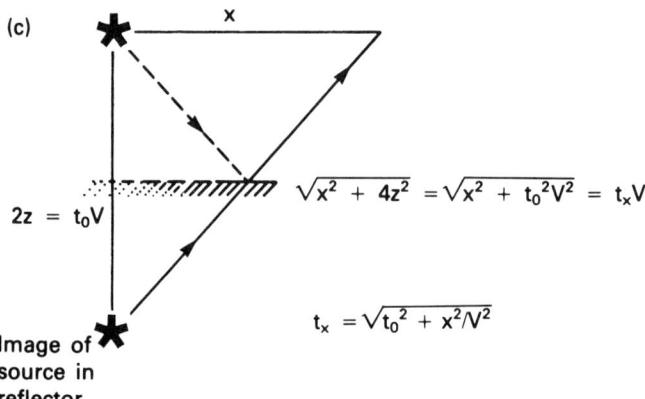

Figure 8. The concept of velocity analysis: the velocity is the extra distance divided by the extra time (from Anstey, 1982, p. 13).

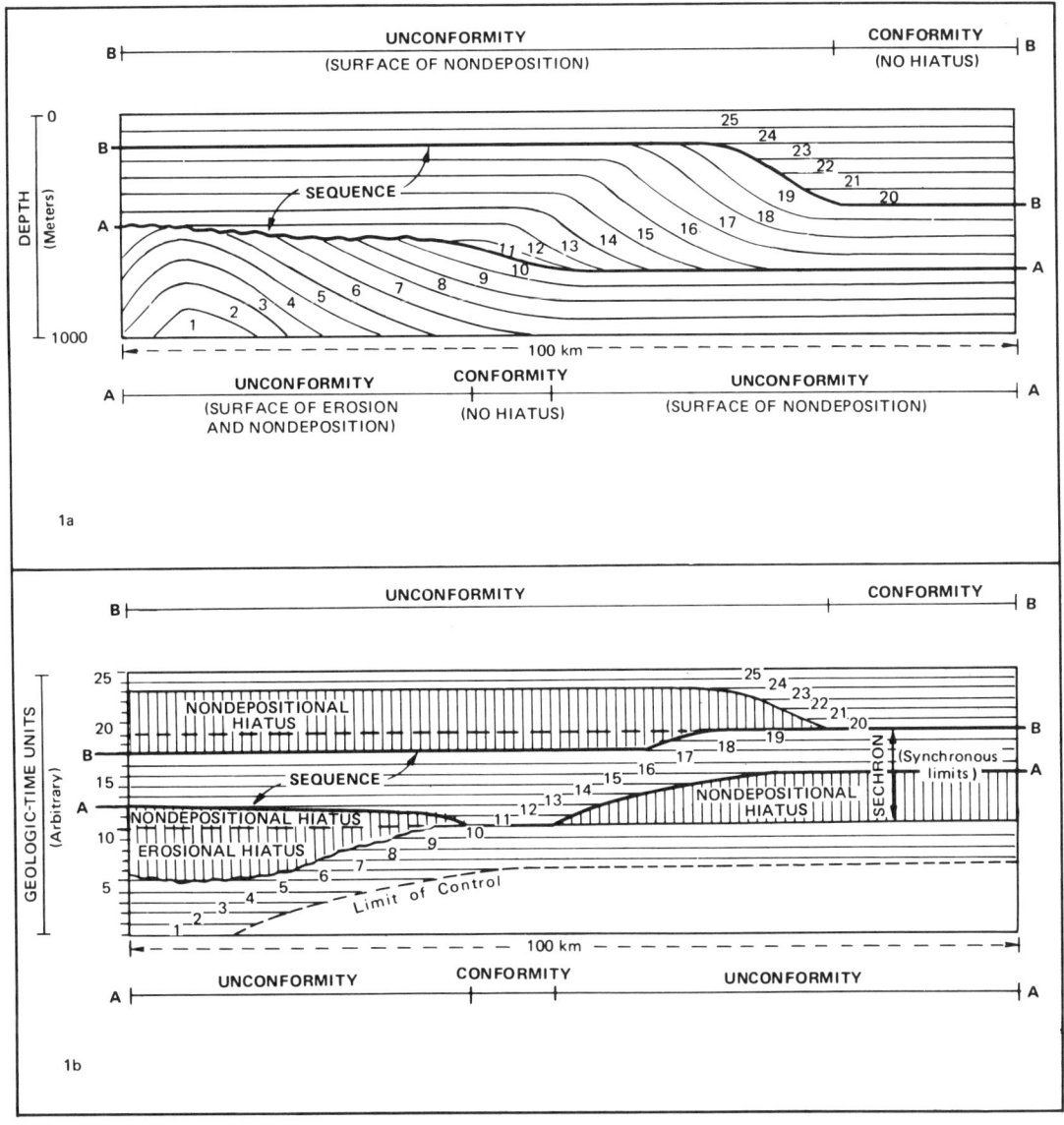

Figure 10. The basic concepts of depositional sequence represented by a stratigraphic section (A) and a chronostratigraphic section (B) (from Mitchum and others, 1977b).

parameters (Fig. 12), external geometry, and three-dimensional associations are delineated.

Resolution of reflection seismic data limits the amount of extractable stratigraphic detail available from a seismic section both horizontally and vertically. "Resolution is the ability to tell that more than one feature is contributing to a given effect" (Sheriff, 1985, p. 1). Velocity and frequency of the seismic energy determine the vertical resolution. The length of the wavelet recorded on the section is a product of velocity and frequency and a combination of many reflecting horizons (Fig. 13). The lower the velocity and the higher the frequency, the better the resolution. As seismic energy passes deeper into the earth, velocity increases and frequency lowers; therefore, resolution decreases. Widess (1973) observed that we can resolve down to a limit of one-fourth the length of a wavelet (Fig. 13). Using an example from Sheriff (1985), for shallow reflections of 1,600 m in an unconsolidated sand-shale section, with a velocity of 2,000 m and a frequency of 60 hertz, the one-fourth wavelength thickness is 8 m. When we consider a deeper reflection with a higher velocity of approximately 5,000 m/second and a lower frequency of approximately 15 hertz, the one-fourth wavelength resolution is 80 m.

The width of the first Fresnel zone determines the horizontal resolution (Fig. 14). Seismic energy passes downward into the subsurface as a sphere. When a spherical wave is incident on a plane reflector, instead of a reflection point there is a reflection area that is a system of circles and rings. The innermost circle is the first Fresnel zone. Outside the first Fresnel zone are a series of annular rings and circles of other Fresnel zones. Contributions to the total reflection from zones outside the first Fresnel zone alternatively cancel, resulting in no net effect (Sheriff, 1985). The dimension of the reflecting area is dependent on frequency (wavelength), velocity, and depth. It may be thousands of feet across for low-frequency reflections deep in a high-velocity section. Because higher frequencies attenuate with depth, the deeper the reflector the wider the first Fresnel zone. Horizontal gaps between reflectors are harder and harder to detect with depth.

Seismic modeling

Seismic modeling is an excellent interpretive tool for both exploration and development of reservoirs. This is essentially a mathematical simulation of a seismic response to a subsurface geologic model, and is especially useful for: (1) testing the mappability of a geological concept (for example, what does the expected trap look like in a reflection seismic section [such as, structural geometry, sand body shape]?); (2) analyzing the effect of expected geologic variability (porosity, thickness, faulting, folding etc.) on the seismic response; and (3) evaluating changes or anomalies in wavelet amplitude. These changes can be compared to variations in lithology, thickness, and fluid content (Meckel and Nath, 1977).

Vertical seismic profiling

A major advance in the seismic method has been vertical seismic profiling (VSP) or borehole seismic. Basically, VSP involves lowering a geophone (receiver) down a borehole to record the total upgoing and downgoing seismic wavefields propagating through the stratigraphic section penetrated by the borehole. Advantages of VSP include: (1) better vertical and lateral velocity control; (2) greater understanding of the origin of reflections (i.e., are reflections time-stratigraphic horizons or lithostratigraphic horizons, or are wavelet character anomalies due to local cementation, porosity, or the presence of hydrocarbons?); (3) looking ahead of the bit; and (4) detection of complex structure where reflections from complexly folded or faulted beds are not detectable at the surface (Anstey, 1982).

Three-dimensional seismic

A dense grid of seismic sections is a requisite for three-dimensional seismic sections. If the lines are spaced closely enough, a volume of seismic data points results, and instead of processing the individual lines, the data can be processed as a whole. True three-dimensional migration, not possible with two-dimensional seismic data, is then possible (Anstey, 1982). With three-dimensional seismic data, migration processing truly moves spurious reflections from complex geologic features to their proper positions in the seismic section regardless of the orientation of the seismic line.

Advantages of three-dimensional seismic are: (1) a seismic section can be produced in any direction regardless of the orienta-

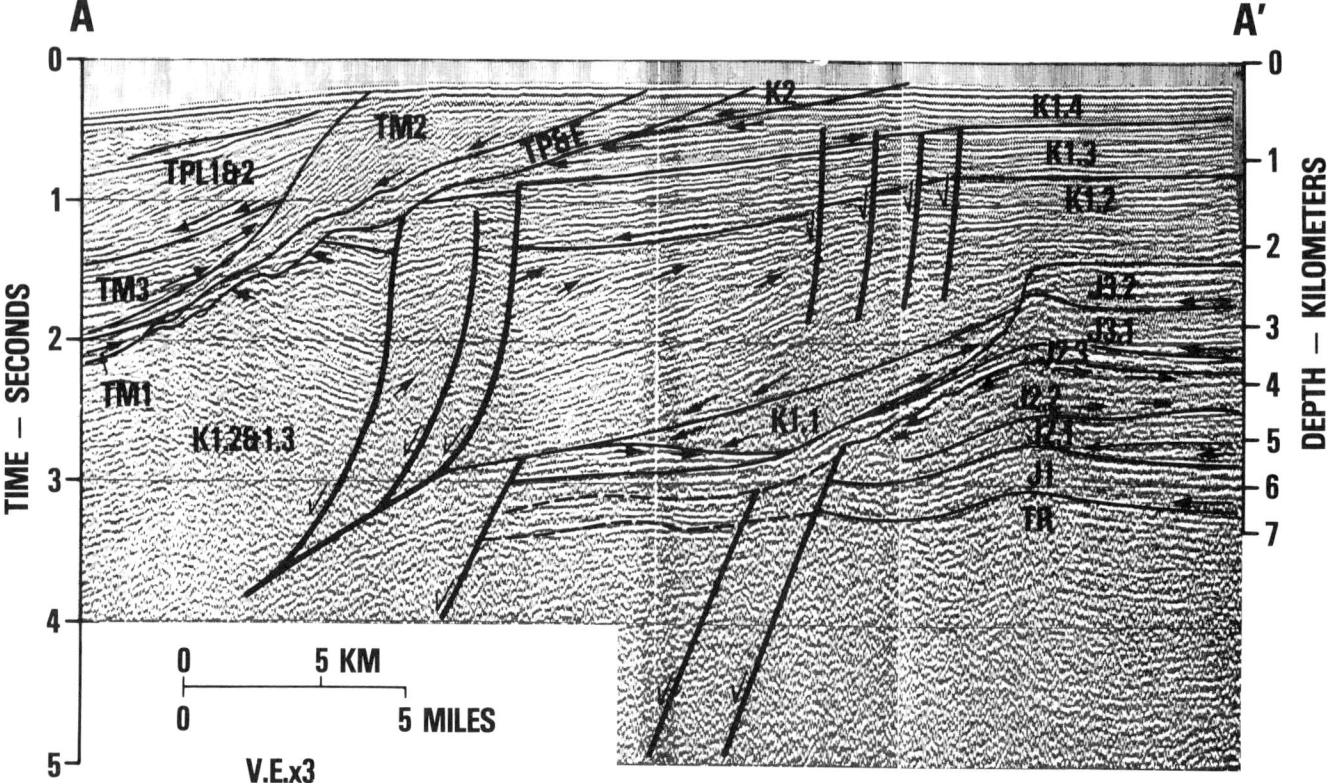

Figure 11. Depositional sequences represented by refections in a seismic line from offshore northwest Africa (from Mitchum and others, 1977b).

tion of the grid; (2) a seismic section can be generated in the zig-zag pattern that a well-log cross section would have; (3) a contour map of a surface can be produced from the data; and (4) a time slice can be made through any part of the data volume (Fig. 15; Anstey, 1982). With three-dimensional seismic interpretation, it becomes possible to look through a trap by generating a series of either horizontal or vertical seismic slices.

Gravity

Gravity anomalies in a basin are a function of horizontal variations in density of subsurface rocks. "The magnitude and form of the gravity effect depend on the details of the densities involved, their magnitudes, vertical relief, depth and horizontal extent" (Nettleton, 1971, p. 47). In petroleum exploration, gravity aids interpretation of the configuration and location of basement blocks and salt domes or, on a smaller scale, location or buried channels and reefs. The scale of an anomaly found with gravity analysis depends on removal of regional gravity effects.

Before the data gathered from a gravity survey can be interpreted, several corrections are made: (1) a latitude correction is made due to the increase of gravity from the poles to the equator; (2) an elevation or "free air" correction is made to account for the variations in gravity caused by elevation; (3) a Bouguer or terrain correction is made to allow for local topographic gravity effects; and (4) residual maps are then generally derived from the Bouguer map by removing the effect of the regional gravity. The resulting gravity anomalies help to identify potential subsurface drilling targets.

Magnetics

The magnetic method is similar to the gravity method in that both are applications of variations in potential fields. The mathematics of the magnetic field, however, are much more complex due to latitudinal and time variations in the Earth's magnetic field. Interpretations, however, are simpler because, in most cases, only basement rocks generate magnetic anomalies, not the entire lithologic section.

Magnetic disturbances caused by rocks are localized effects superimposed on the normal magnetic field of the Earth. The distribution of magnetite in rocks almost exclusively causes all variations observed in magnetic surveys. Magnetite is not the only magnetic mineral, but it is so much more magnetic and common than other minerals that it is the dominant cause of magnetic anomalies (Nettleton, 1962). The amount of magnetite contained in basement rocks is generally so much greater than the amount of magnetite contained in sedimentary rocks that only basement rocks are considered responsible for variations in the magnetic field. Because of this condition, depth to basement, thickness of the sedimentary section, and structural features of the basement within basins are considered.

Magnetic anomalies show different signatures depending on their latitude. Figure 16 shows examples of the total intensity

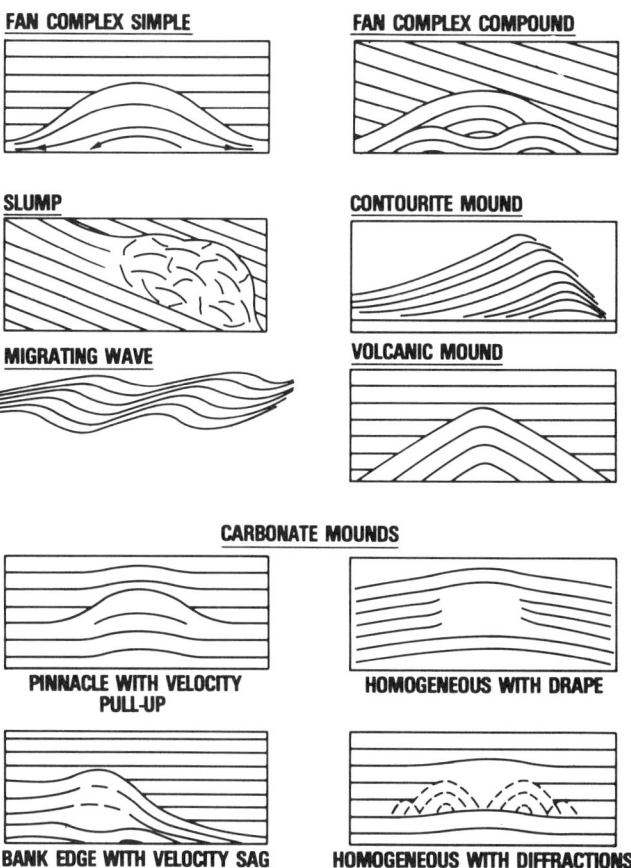

Figure 12. Examples of reflection patterns define seismic facies within a depositional sequence (from Mitchum and others, 1977).

anomaly for the same disturbing body at different latitudes. Near the north magnetic pole the magnetizing field is vertical, and the anomaly is a strong positive peak with weak negative zones on each side. Moving south where the magnetic inclination is 45°, the magnetic field shows a strong negative anomaly on the north side and a shift of the maximum to the south At 22.5°, the positive and negative anomalies are nearly equal. The negative anomaly is strongest at the equator, where it is flanked by two weak positive anomalies. In the Southern Hemisphere the anomaly patterns are opposite of those described above.

Figures 17 and 18 compare magnetic and gravity contour maps and measurement profiles for the same disturbing body.

Magnetotellurics

Fluctuations in the Earth's magnetic field induce electrical currents in rocks called telluric currents, which flow in vast sheets within the outermost layers of the Earth's crust. In a gross sense, telluric currents may be used to measure resistivity. The magnetotelluric method determines resistivity distribution in the subsurface by measuring potential differences at the Earth's surface caused by telluric currents.

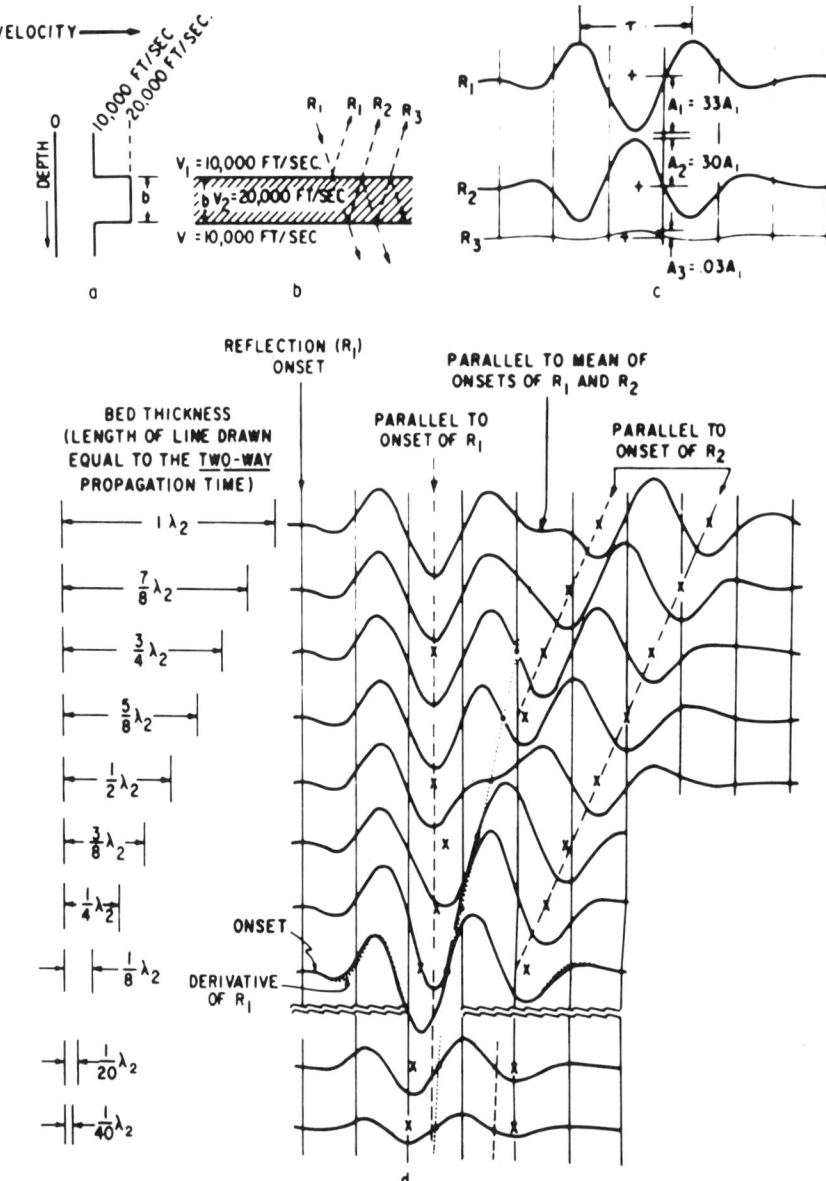

Figure 13. Symmetrical wavelet response to varying bed thickness (from Widess, 1973).

Time variations of the Earth's magnetic field are treated as the magnetic component of a plane electromagnetic wave. A simple relation exists between the amplitude of the magnetic field changs, the voltage gradients induced in the Earth, and the Earth's resistivity. Because the penetration depth of an electromagnetic wave is a function of frequency and the resistivity of the conductor, resistivity can be computed as a function of depth within the Earth using amplitude variations in the magnetic and electric fields recorded over a wide range of frequencies (Keller and Frischknecht, 1966).

The magnetotelluric (MT) method is used to interpret large-scale structural features of sedimentary basins by measuring resistivities to great depths within the Earth. It is also used to map broad features of porosity distribution in sedimentary basins and to evaluate areas with thick surface deposits of sand and gravel, or especially to map structure beneath areas with thick volcanic rock sequences where seismic reflection fails.

COMPUTER APPLICATIONS

The petroleum geologist should view the computer not only as a labor-saving device but also as a tool to define problems and formulate and test approaches to problems (Harbaugh and Merriam, 1968). Computers provide experimental approaches to both geological and economic exploration problems. Computers can reduce complex interrelations within large data bases to simpler interrelations using mathematical analysis. The effect of

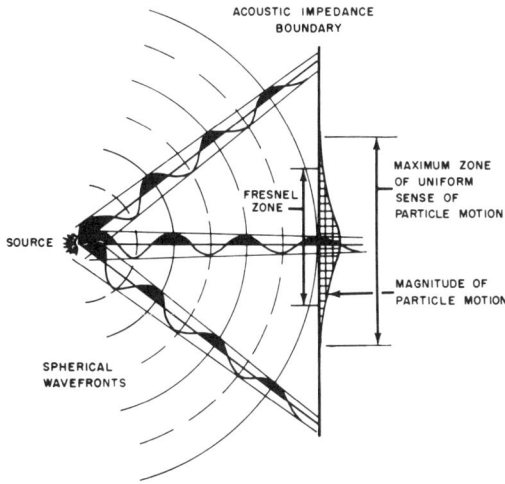

Figure 14. Concept or spherical wave motion of seismic energy generating Fresnel zone after encountering a flat reflector (from Neidell and Poggiagliolmi, 1977).

time, so important to geologic processes, can be simulated in mathematical models.

Applications of the computer can be broken down into at least three categories: (1) manipulation of large data bases, (2) modeling or simulating geology, and (3) artificial intelligence.

Data bases

A petroleum geologist can spend large amounts of time acquiring, storing, retrieving, processing, and displaying data. To be efficient the petroleum geologist must have data organized in files, catalogues, or libraries. Computers are extremely efficient information processors.

Using multivariate statistical techniques, data bases can be analyzed to ascertain whether proposed exploration models are supported by the data or not. Data bases can be used to build models that use the uncertain predrilling geological input about source rocks, migration, trap size reservoir, and seal for estimation of a predictive distribution of the volume of oil, gas, and gas liquids in an exploration prospect outlined by geophysical methods (Sluijk and Nederlof, 1984).

Modeling and simulation

Computers make possible modeling and simulation of geology in two, three, and four dimensions. Petroleum geologists can compile and manipulate extremely large data bases into experimental models to more fully comprehend both geology and responses of exploration tools to geology. A few examples follow.

1. Through application of massive data bases, processes that develop throughut the evolution of a basin can be simulated. As Welte and Yukler (1984, p. 27) explain, "Input data consist of heat flux, initial physical and thermal properties of sediments,

Figure 15. Horizontal slice from a three-dimensional seismic survey showing a meandering stream channel (from Anstey, 1982).

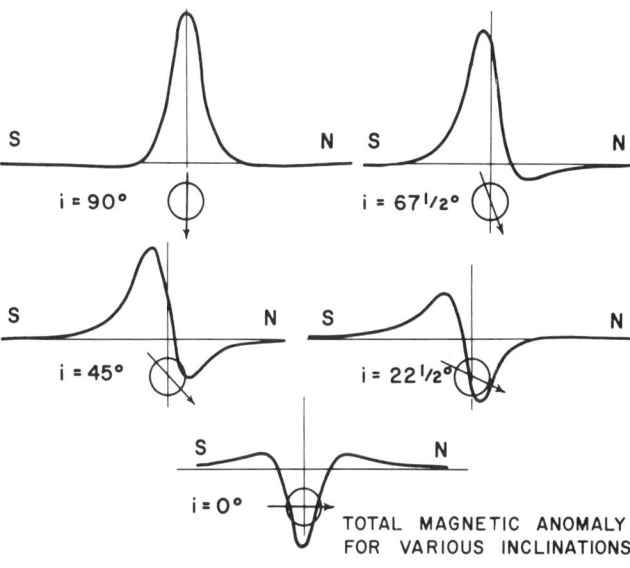

Figure 16. Variation in form of anomaly in total magnetic intensity with change in magnetic latitude (from Nettleton, 1962).

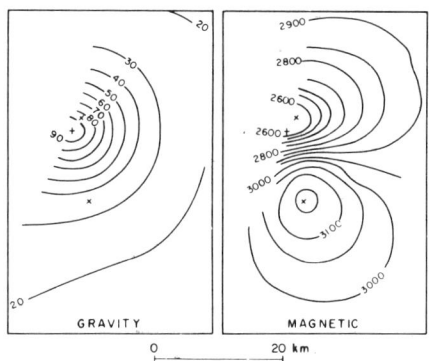

Figure 17. Comparison of gravity and magnetic contours for the same disturbing body (from Nettleton, 1962).

paleobathymetric estimates, sedimentation rate, and amount and type of organic matter. The model computes pressure, temperature, physical and thermal properties of sediments, maturity of organic matter, and the hydrocarbon potential of any source rock as a function of space and time. Thus the complex dynamic processes of petroleum formation and occurrence in a given sedimentary basin can be quantified." The model described above by Welte and Yukler is primarily a migration model and provides information on petroleum and water movement. Migration models can indicate which areas and strata have potential petroleum accumulation or which areas might have abnormal pressures. Migration models can also be used to estimate ultimate reserves of a sedimentary basin (Tissot and Welte, 1984).

2. Quantitative simulation models of plate tectonics are used to predict paleogeography and paleoclimates of basins through time, which in turn can be used to predict distribution and lithology of source beds (Barron, 1985) and reservoir units (Dickinson and Seely, 1979). Plate tectonic models are even more effective for predicting source and reservoir characteristics when used with knowledge of global sea-level changes.

3. The configuration of seismic reflectors in a seismic section can often be interpreted in many ways. Computer-generated seismic models aid delineation of stratigraphy and interpretation of structure in seismic sections. Based on geological models, they predict ray paths for downgoing seismic energy and thereby predict the appearance of resulting reflectors on the seismic display. With seismic models the limits of resolution of the seismic wavelets with respect to a concept of the geology or the target trap can be defined. Thus, computer modeling can narrow the possibilities for geological interpretation of seismic sections.

Artificial intelligence

Petroleum geologists need to know something about organic and inorganic geochemistry, geophysics, well logging, sedimentology, stratigraphy, paleontology, tectonics, structure, economics, and on and on. There is no way to become expert in all but a few fields. Instead of spending hours and hours searching for a possible solution to a problem in an unfamiliar field, expert systems (a subdiscipline of artificial intelligence) instantly put a massive amount of knowledge at the fingertips of the user. Thus, instead of calling in an authority, the problem can be approached by using an especially designed expert system. Expert systems also improve objectivity in making decisions by consistency in evaluation of problems and, in addition, operators of expert systems learn by doing; therefore, expert systems can be efficient teachers.

ECONOMIC ANALYSIS OF A PROSPECT

This simple formula describes the economic analysis of a prospect:

$$\text{Profit} = \text{Income} - (\text{Investment} + \text{Operating Cost} + \text{Taxes}).$$

Depending on whether the prospect is in a frontier basin or a mature basin, the economic analysis ranges from an educated guess to a well-documented calculation. In mature exploration areas (Wright, 1985), most of the variables in the analysis are known to within a predictable range, and the following formula applies:

$$\text{Income} = (\text{Estimated Recoverable Petroleum Reserve} \times \text{Petroleum Price}) - \text{Royalty}.$$

Subsurface and geophysical studies define the size of a target trap. Target size is defined as productive area multiplied by pay thickness. Nearby production from the target zone is used to estimate the recovery factor. These data combine to give estimated recoverable oil and/or gas in the formula:

$$\text{Estimated Recoverable Petroleum} = \text{Area} \times \text{Thickness} \times \text{Recovery Factor}.$$

Royalty is the share of income paid to the landowner—or, in some cases, prospect promoters—and is established by the terms of a lease. The landowner royalty amount generally ranges between 12.5 percent and 16 percent of the gross production

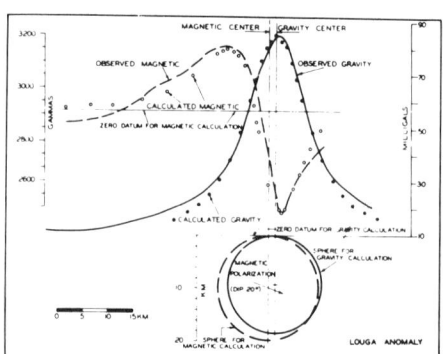

Figure 18. Comparison of gravity and magnetic profiles for the same disturbing body (from Nettleton, 1962).

income. In some lease contracts the prospect promoters will keep an overriding royalty that ranges from ap art of a percent to 10 percent or more. Royalties are open to negotiation.

Investment is the total capital expenditure for items such as costs for obtaining mineral leases; geological and geophysical expenses; drilling exploration and development wells; development facilities such as offshore platforms and treatment plants; and pipelines and transportation facilities.

Operating costs are the amount of money required to keep a field in operation on a daily basis. This varies with the type of petroleum, type of lift, and well-bore maintenance problems.

In a frontier prospect the volume of recoverable oil and gas is subjective and highly dependent on experience. Seismic mapping and other subsurface data are critical for estimation of trap size. Taxes, investment, and operating costs are dependent on the geography and government of the country where the frontier prospect is located.

In economic prospect analysis, investment timing and the time value of money play a crucial role in determining the economic success of a prospect. The amount of money required prior to drilling a prospect and the time delay between beginning to invest in a prospect and receiving income from a successful prospect can make a great difference in the economics. The discounted value of the money (in percent), variable petroleum prices, and risk must be factored in.

Risk is part of any business venture. Even putting money in a bank involves some risk. Risk is especially high in petroleum exploration. For example, based on data from 1974 to 1984, only one in 89 wildcat wells will find a field with 1 to 10 million barrels of oil, and only one in 5,962 wells will find a field with 50 or more million barrels of oil (Wright, 1985).

Estimation of geological risk is subjective. To determine a risk factor, which is used to discount the income estimated using the formula given at the beginning of this section, each critical geologic factor is considered and given a number in percent of its chance of being correctly predicted. For example, the chance of a seal being present might be 90 percent, the chance that a reservoir will be present might be 50 percent, the chance that a source is present might be 80 percent, and the chance that a trap is present might be 50 percent. These numbers are multiplied together to give the risk factor; in this example, the risk factor would indicate that there is an 18 percent chance that the geological predictions are correct. However, if any more factors are added, the risk factor could drop significantly. If there is a 75 percent chance that a trap formed in time to catch migrating hydrocarbons, the risk factor would change to 13.5 percent, assuming that the previous estimates for each geologic variable were correct. Risk analysis is an art.

Drilling many wells is the best way to offset the effects of risk. If only one well is drilled in a basin where the success rate is 10 percent, there is 10 percent chance that the well will be successful. However, if 10 wells are drilled in the same basin there is a 60 percent chance that at least one will be successful. If 20 wells are drilled there is a 90 percent chance one will be successful. The bottom line is that the value of the petroleum discovered by the successful wells has to offset the expenses of drilling all the exploration wells (including dry holes) plus the costs of developing and producing the petroleum.

CONCLUSION

A tremendous variety of sophisticated tools and techniques are currently available for use in the search for oil and gas accumulation. However, continued research and development is required to create new tools and methods of exploration. In addition, as our understanding of the nature of oil and gas accumulations increases, as well as our ability to identify the critical elements that control these accumulations, so will our success in finding additional fields improve. Finally, it is our firm belief that the ultimate realization of the full potential of the remaining undiscovered energy resources of the world will require the creative input of many individuals and companies, all generating different ideas, developing new concepts and techniques, and applying them to different objectives. When and if the proper environment is created or exists for this type of activity to flourish, then and only then will the true potential of the creative explorationist be realized.

REFERENCES CITED

Anstey, N. A., 1982, Simple seismics for the petroleum geologist, the reservoir engineer, the well-log analyst, the processing technician, and the man in the field: Boston, Massachusetts, International Human Resources Development Corp., 169 p.

Asquith, G. B., 1982, Basic well log analysis for geologists: American Association of Petroleum Geologists Methods in Exploration Series, 216 p.

Barron, E. J., 1985, Numerical climate modelling, a frontier in petroleum source rock prediction; Results based on Cretaceous simulations: American Association of Petroleum Geologists Bulletin, v. 69, no. 3, p. 448–459.

Beaumont, E. A., 1989, Creativity in petroleum exploration: American Association of Petroleum Geologists Continuing Education Course Notes, 186 p.

Beaumont, E. A., and Foster, N. H., eds., 1990, Atlas of oil and gas fields: American Association of Petroleum Geologists, Treatise of Petroleum Geology.

Cornford, C., 1986, Source rocks and hydrocarbons of the North Sea, in Glennie, K. W., ed., Introduction to the petroleum geology of the North Sea: Blackwell Scientific Publications, p. 197–236.

Dahlberg, E. C., 1982, Applied hydrodynamics in petroleum exploration: New York, Springer-Verlag, 161 p.

Dickinson, W. R., and Seely, D. R., 1979, Structure and stratigraphy of forearc regions: American Association of Petroleum Geologists Bulletin, v. 63, p. 2–31.

Doveton, J. H., 1986, Log analysis of subsurface geology: New York, John Wiley and Sons, 273 p.

Foster, N. H., 1989a, Creativity vital in exploration: American Association of Petroleum Geologists Explorer, April 15, Annual Meeting Special Edition, p. 1 and 11.

—— , 1989b, Creativity vital for successful exploration: American Association of

Petroleum Geologists Explorer, May, p. 1 and 8.

Goward, S. N., and Taranik, J. V., 1986, Commercial applications and scientific research requirements for thermal-infrared observations of terrestrial surfaces; A report of the Joint EOSAT/NASA Thermal Infrared Working Group: Washington, D.C., National Aeronautics and Space Administration Earth Science and Applications Division, 145 p.

Harbaugh, J. W., and Merriam, D. F., 1968, Computer applications in stratigraphic analysis: New York, John Wiley and Sons, 282 p.

Keller, G. V., and Frischknecht, F. C., 1966, Electrical methods in geophysical prospecting: New York, Pergamon Press Series of Monographs in Electromagnetic Waves, v. 10, 517 p.

Levorsen, A. I., 1967, Geology of petroleum: San Francisco, California, W. H. Freeman, 724 p.

Logan, W. E., 1844, Canada Geological Survey report of progress: Canada Geological Survey, 41 p.

McIver, R. D., 1985, Near-surface hydrocarbon surveys in oil and gas exploration: Oil and Gas Journal, v. 82, no. 39, p. 115–117.

Meckel, L. D., and Nath, A. K., 1977, Geologic considerations for stratigraphic modeling and interpretation, *in* Payton, C. E., ed., Seismic stratigraphy; Applications to hydrocarbon exploration: American Association of Petroleum Geologists Memoir 26, p. 417–438.

Mitchum, R. M., Jr., Vail, P. R., and Sangree, J. B., 1977a, Chronostratigraphic significance of seismic reflections, *in* Payton, C. E., ed., Seismic stratigraphy; Applications to hydrocarbon exploration: American Association of Petroleum Geologists Memoir 26, p. 117–134.

Mitchum, R. M., Jr., Vail, P. R., and Thompson, S., III, 1977b, The depositional sequence as a basic unit for stratigraphic analysis, *in* Payton, C. E., ed., Seismic stratigraphy; Applications to hydrocarbon exploration: American Association of Petroleum Geologists Memoir 26, p. 53–62.

Neidell, N. S., and Poggiagliolmi, E., 1977, Stratigraphic modeling and interpretation; Geophysical principles and techniques, *in* Payton, C. E., ed., Seismic stratigraphy; Applications to hydrocarbon exploration: American Association of Petroleum Geologists Memoir 26, p. 389–416.

Nettleton, L. L., 1962, Gravity and magnetics for geologists and seismologists: American Association of Petroleum Geologists Bulletin, v. 46, p. 1815–1835.

——— , 1971, Elementary gravity and magnetics for geologists and seismologists: Society of Exploration Geophysicists Monograph 1, 121 p.

North, F. K., 1985, Petroleum geology: London, Allen and Unwin, 607 p.

Philp, R. P., and Crisp, P. T., 1982, Surface geochemical methods used for oil and gas prospecting; A review: Journal of Geochemical Exploration, v. 17, p. 1–34.

Sabins, F. R., Jr., 1987, Remote sensing principles and interpretation, 2nd ed.: San Francisco, California, W. R. Freeman and Company, 449 p.

Schowalter, T. T., and Hess, P. D., 1982, Interpretation of subsurface hydrocarbon shows: American Association of Petroleum Geologists Bulletin, v. 66, p. 1302–1327.

Sheriff, R. E., 1985, Aspects of seismic resolution, *in* Berg, C. R., and Woolverton, D. G., eds., Seismic Stratigraphy II; An integrated approach to hydrocarbon exploration: American Association of Petroleum Geologists Memoir 39, p. 1–12.

Sluijk, D., and Nederlof, M. H., 1984, Worldwide geological experience as a systematic basis for prospect appraisal, *in* Demaison, G., ed., Petroleum geochemistry and basin evolution: American Association of Petroleum Geologists Memoir 35, p. 15–26.

Tissot, B. P., and Welte, D. H., 1984, Petroleum formation and occurrence, 2nd ed.: New York, Springer-Verlag, 699 p.

Vail, P. R., and others, 1977, Seismic stratigraphy and global changes of sea level, *in* Payton, C. E., ed., Seismic stratigraphy; Applications to hydrocarbon exploration: American Association of Petroleum Geologists Memoir 26, p. 49–212.

Vincelette, R. R., 1989, Practical application of creative thinking to hydrocarbon exploration, *in* Creativity in petroleum exploration: American Association of Petroleum Geologists Short Course Notes, p. 1–5.

Welte, D. H., and Yukler, M. A., 1984, Petroleum origin and accumulation in basin evolution; A quantitative model, *in* Demaison, G., ed., Petroleum geochemistry and basin evolution: American Association of Petroleum Geologists Memoir 35, p. 27–40.

Widess, M. B., 1973, How thin is a thin bed?: Geophysics, v. 38, p. 1176–1180.

Wright, T., 1985, Petroleum geology and oil exploration; Notes for California Offshore Petroleum Conference Short Course, Long Beach, California, March 26, 1985: Fullerton, California State University, Statewide Energy Consortium of California, 27 p.

MANUSCRIPT ACCEPTED BY THE SOCIETY OCTOBER 8, 1990

Chapter 17

Petroleum geology of the Appalachian basin

Wallace de Witt, Jr.
U.S. Geological Survey, 955 National Center, Reston, Virginia 22092
Robert C. Milici
Virginia Division of Mineral Resources, P.O. Box 3667, Charlottesville, Virginia 22903

INTRODUCTION

The Appalachian basin is a broad, elongate synclinorium that extends from the southern shore of Lake Ontario in New York generally southwestward about 1,500 km to the Gulf Coastal Plain in Alabama. It has a surface area of more than 500,000 km^2 and is filled by about 1,800,000 km^3 of Paleozoic rocks ranging in age from Early Cambrian to Early Permian. The western edge of the basin is at the crest of the Cincinnati arch (Fig. 1A). Paleozoic strata dip gently eastward from the arch, first beneath the Appalachian Plateaus and then into and beneath the complexly folded and faulted Valley and Ridge. Except where they were thrust westward along the flanks of the Blue Ridge during the Alleghanian Orogeny, easternmost Paleozoic strata of the basin lie concealed beneath Blue Ridge and Piedmont thrust sheets.

The Appalachian basin contains two conspicuous, smaller basins: the Dunkard in southwestern Pennsylvania and contiguous West Virginia and Ohio, and the Black Warrior in Alabama and Mississippi (Plate 6H). The Dunkard basin lies above the Rome trough, suggesting that it may be in part inherited from that early Paleozoic feature. The Dunkard basin is a shallow synclinal trough elongated parallel to the regional strike of the Appalachians. In contrast, the Black Warrior basin is a homocline that dips generally southwestwardly from the Nashville dome (Thomas, 1988) at a high angle to the strike of the Valley and Ridge Paleozoic rocks.

In general, the majority of the siliciclastic sediments of the Appalachian basin were derived from sources to the north and east of the basin, in particular from those tectonic highlands raised by the Taconic, Acadian, and Alleghanian Orogenies. As a result, shales, sandstones, and red beds dominate the Paleozoic sequences of eastern Pennsylvania, whereas marine carbonate strata dominate the Plateau regions of Tennessee, Georgia, and northern Alabama.

Parts of this chapter are condensed from de Witt and Milici (1989) and Milici and de Witt (1988). The reader is referred to those papers for a more detailed discussion of stratigraphy and a more detailed documentation of the geological literature.

STRUCTURE

Except for the Rome trough and the Dunkard basin, the Appalachian basin is a regionally extensive homocline. Autochthonous crystalline basement dips eastward from about 610 m subsea in western Ohio to as much as 15,240 m in eastern Pennsylvania. The eastern margin of the homocline was turned up along some of its length, however, when marginal tectonic lands were uplifted by the Taconic Orogeny at the end of the Ordovician.

The eastern segment of the Appalachian basin is partly buried beneath crystalline thrust sheets of the Blue Ridge and Piedmont. The Valley and Ridge to the west of the Blue Ridge is a belt of greatly deformed Paleozoic strata characterized by elongate folds and thrust sheets in the central Appalachians and by great imbricate thrust sheets in the southern Appalachians. The fold-and-thrust belt lies above a basal décollement in Cambrian siliciclastic strata. In general, thrusts rooted in this basal décollement extend diagonally upward along moderately dipping tectonic ramps, in places flattening in younger subhorizontal strata, so that multiple levels of décollement extend westward beneath the Appalachian Plateau. Superficial folds and faults have developed in strata shortened above these décollements. Their sizes and general characteristics are functions of the amount of tectonic shortening and the stratigraphic positions of the décollements.

The Cincinnati arch, at the western margin of the Appalachian basin, is warped by two major domal structures, the Nashville dome in Tennessee and the Jessamine dome in Kentucky, separated by the Cumberland saddle. The Nashville is the older of the two, and apparently was intermittently active since the Middle Ordovician. The Jessamine, however, shows no evidence of tectonic activity until after the Silurian. Both of the domes are today breached by erosion, exposing rocks as old as Middle Ordovician.

STRATIGRAPHY

Paleozoic strata of the Appalachian basin may be divided into broad depositional phases that are generally related to the

---

de Witt, W., Jr., and Milici, R. C., 1991, Petroleum geology of the Appalachian basin, *in* Gluskoter, H. J., Rice, D. D., and Taylor, R. B., eds., Economic Geology, U.S.: Boulder, Colorado, Geological Society of America, The Geology of North America, v. P-2.

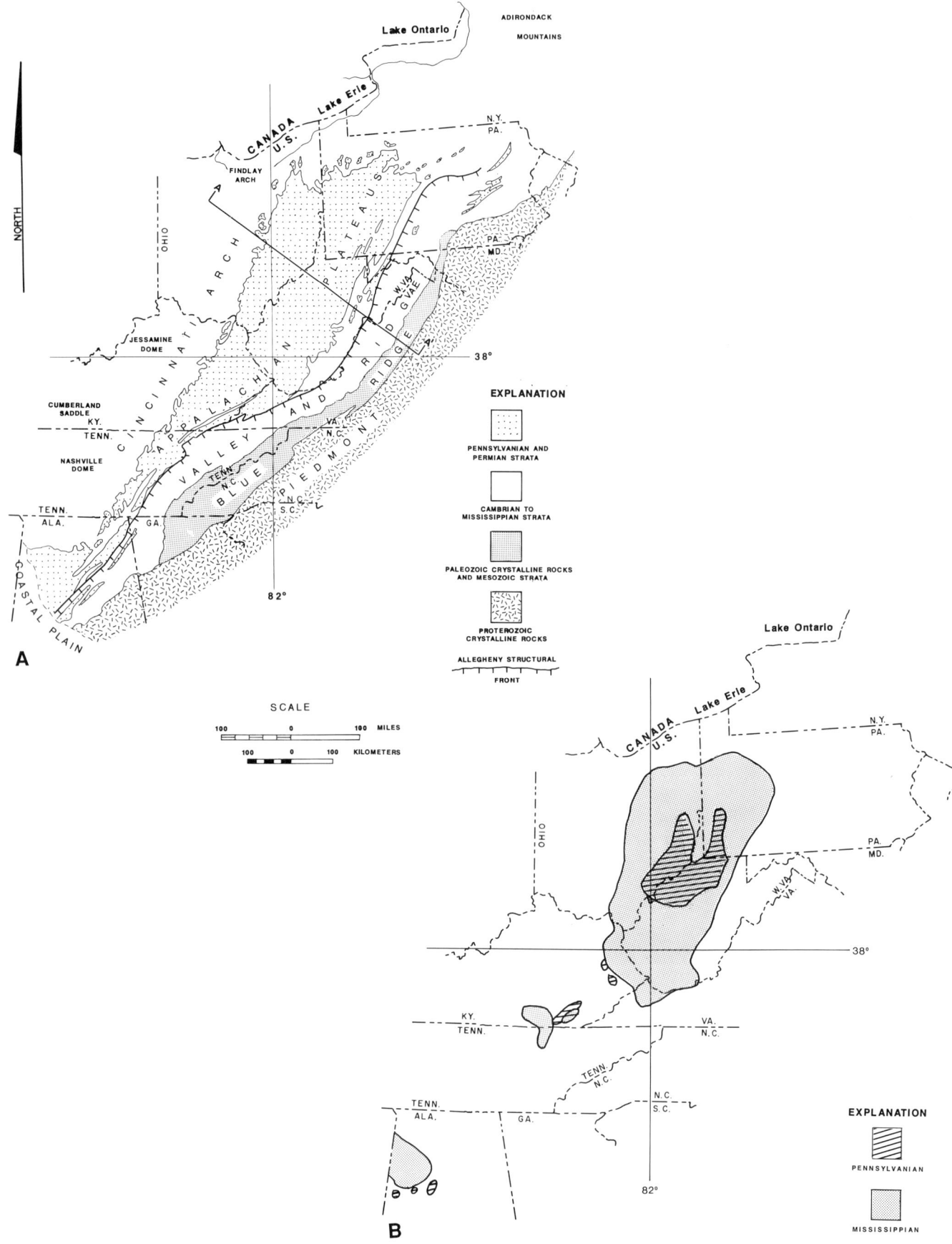

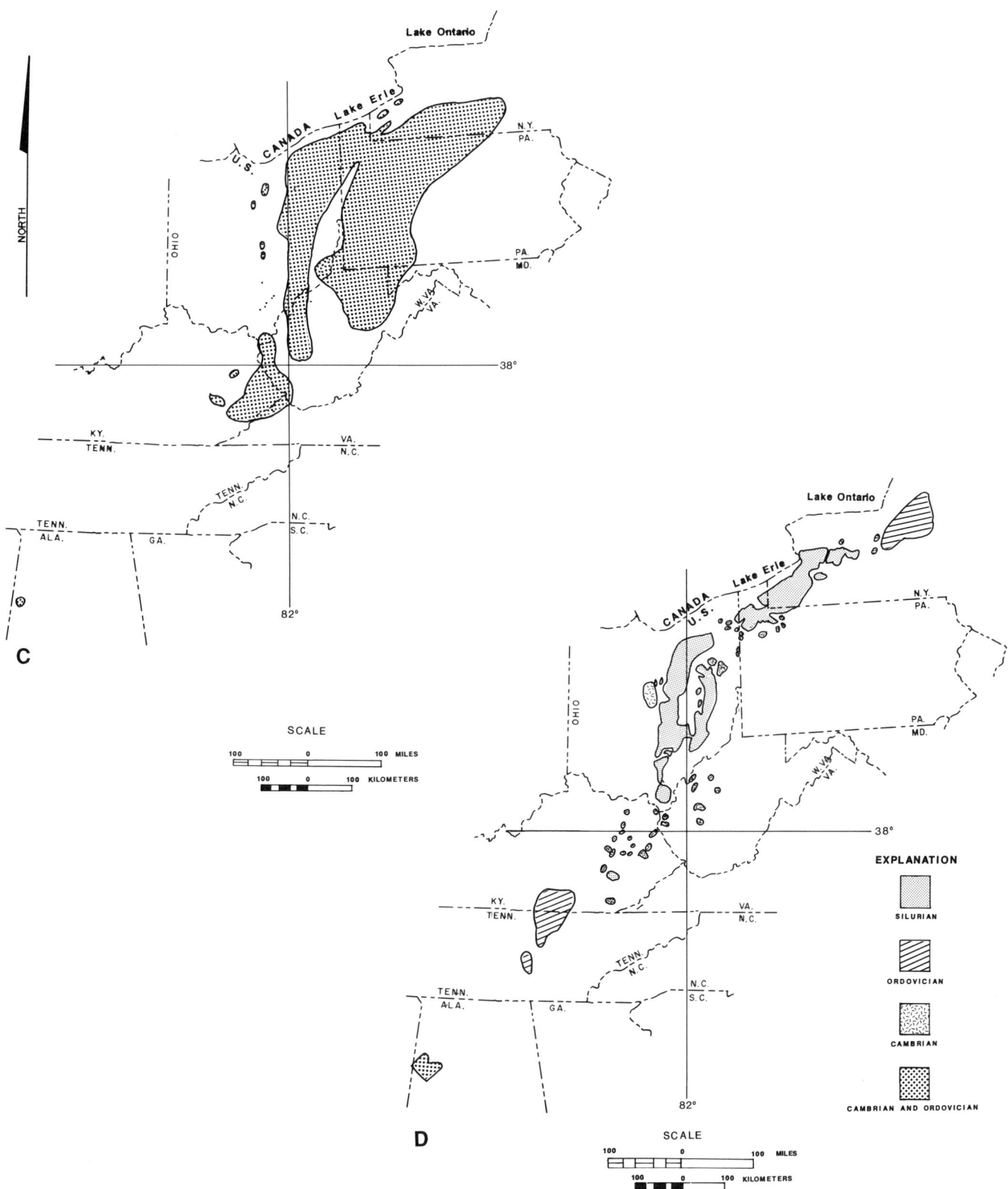

Figure 1. Geologic maps of the Appalachian basin, showing principal hydrocarbon-producing areas from Cambrian to Pennsylvanian strata. A. generalized geologic map; B. areas producing from Mississippian and Pennsylvanian strata; C. areas producing from Devonian strata; D. areas producing from Cambrian, Ordovician, and Silurian strata.

tectonic evolution of the Paleozoic continental margin. A generalized stratigraphic column for the central part of the basin is shown in Figure 2. The earliest phase began during late Precambrian time, when rifting of Grenville-age crust along the eastern edge of the continent resulted in the formation of a marginal ocean, Iapetus (Harland and Gayer, 1972). Thick sequences of siliciclastic strata, some containing volcanic rocks, were deposited both on the edge of the continent and in epicontinental grabens nearby. As the margin subsided, and transgressing seas inundated the Paleozoic craton, thicker, older sequences of Precambrian to Cambrian siliciclastics on the continental margin were replaced westward by thinner, younger Upper Cambrian epicontinental sequences. In general, the siliciclastic rocks are less mature at the base, where they overlie granitic terrane, and are more mature at the top, where they were reworked by transgressive seas. Unlike siliciclastic deposits related to the Taconic, Acadian, and Alleghanian Orogenies, these earliest siliciclastic rocks were derived from the cratonic interior under divergent, passive-margin tectonic conditions.

Basal siliciclastic sequences are overlain by a thick carbonate shelf deposit of Cambrian and Early Ordovician age. The carbonate strata extend basin wide, from the Canadian Shield and Adirondack crystalline basement on the north, southwestward into the subsurface of central Alabama and Mississippi. Indeed, they transcend the crest of the Cincinnati arch, and extend westward beneath the central plains. The carbonate beds are commonly dolomite or dolomitic limestone, reflecting either their deposition under restricted marine conditions or alteration by relatively fresh ground waters. The extent and thickness of these shelf deposits suggest both a quiet tectonic environment and a general lack of sources for siliciclastic sediment.

The eastern passive margin became active, or convergent, at the beginning of Middle Ordovician time. At first, thick sequences of Taconic flysch, black muds, and turbidites filled foreland basins along the continental margin. When these basins were filled, siliciclastic sediments overran the adjacent carbonate shelf, mixing with and diluting carbonate accumulations in a variety of peritidal environments. By the end of the Late Ordovician, a great blanket of siliciclastic strata, the Queenston delta, occupied much of the northern part of the Appalachian basin. Taconic flysch was succeeded by Silurian molasse, as the relatively coarse-grained siliciclastic rocks of the Taconic highlands were eroded, and the derived sediment spread westward to accumulate in a variety of fluvial, delta plain, beach, barrier, and marine environments.

Progressive lowering of the source areas and a diminishing supply of siliciclastic detritus led to the formation of restricted interior seas and the reestablishment of extensive carbonate bank deposits. Evaporitic conditions were ultimately established over much of the central Appalachian basin and in the adjacent Michigan basin near the close of the Silurian, resulting in the widespread accumulation of anhydrite and salt. These are major constituents of the interval beneath the Dunkard basin, whereas dolomite and anhydrite fill the Late Silurian interval to the east and west of that basin. Carbonate deposits, together with some relatively mature marine quartzose sandstones and bedded chert, persisted into the Early Devonian over much of the central and part of the southern Appalachian basin. These strata in turn were succeeded by the black shales, turbidites, and red beds of the great Catskill delta, during the beginning of the Acadian Orogeny.

In general, the Catskill delta is a thick basin-filling sequence, which consists of several tongues of black shales at the base that in turn grade upward and laterally into distal and proximal turbidites and marginal marine and deltaic deposits, and ultimately into continental red beds. Upper parts of the sequence, the Pocono and Price deltas, are of Mississippian age and consist of alluvial and deltaic facies that in places contain mineable coal beds. These siliciclastic sediments spread south and west from tectonic highlands in the northern Appalachians, which were created by collision between the ancient land masses of Armorica, Laurentia, and Baltica (Perroud and others, 1984). They are overlain by Mississippian carbonate sequences that are commonly fossiliferous, cherty, and oolitic, indicative of shallow-marine and littoral depositional environments. In contrast with the Devonian and Mississippian deltaic deposits, carbonate rocks are thick in the south, and thin into the northern part of the Appalachian basin.

Carbonate deposition was supplanted by red, green, and gray shale, mudstone, and sandstone in Late Mississippian time, the precursors of the epicontinental wedge of coal-bearing strata of Mississippian, Pennsylvanian, and Permian age that resulted from Alleghanian continental collision. Carboniferous siliciclastic sediment was derived in part from cratonic sources to the north of the Appalachian basin, but chiefly from tectonic sources along the eastern and southern margin of the continent.

The terminal Alleghanian Orogeny deformed the accumulated mass of Appalachian strata into the great folds and thrusts that characterize the mountain chain. Post-Paleozoic erosion has breached the Appalachian Mountain system, exposing its crystalline core and hinterland. Sediments derived by erosion of the Appalachians during the Mesozoic and Cenozoic were swept eastward and southward to produce the great sheets of siliciclastic molasse that blanket the present continental margin. Even today, turbidites are accumulating along the nearby Atlantic continental margin. Should the gross tectonic cycles initiated in the Paleozoic continue, the North American continent will subside again beneath ocean waters, and carbonate sediment now accumulating along its southern margin will be spread blanketlike into the interior.

THERMAL HISTORY

The thermal history of the rocks of the Appalachian basin can be grossly inferred from conodont color alterations and from carbon ratios of the contained coal beds. Isograds of conodont color alteration indices (CAI values 1 to 5) show broad regional trends in thermal maturation that are roughly parallel to the northeastern trend of the Appalachians (Fig. 3). In detail, these

Petroleum geology, Appalachian basin   277

Era	Period	Stratigraphic unit	Reservoir Rock	Lithology	Type of Trap	Source bed
PALEOZOIC	Permian	Dunkard Group				
	Pennsylvanian	Monongahala Group				coal, black shale
		Conemaugh Group	Cow Run Sand / Dunkard Sand	Cl / Cl	S / S	coal, black shale
		Allegheny Group	Burning Springs SS. / Gas sands	Cl / Cl	C / S	coal, black shale
		Pottsville Group	Salt sands	Cl	S	coal
	Mississippian	Mauch Chunk / Pennington Formations	Ravencliff sand / Maxon sand	Cl / Cl	S / S	
		Greenbrier Limestone	Blue Monday Sand / Big lime	Cl / Ca	S / C	
		Burgoon Mbr. of Pocono Fm.	Big Injun Sand / Squaw Sand / Weir Sand	Cl / Cl / Cl	C / C / S	
		Rockwell / Spechty Kopf Formations	Berea sand / Murrysville sand	Cl / Cl	S,C / C	Sunbury Shale
	Devonian	Catskill Formation	Venango sands / Bradford sands	Cl / Cl	S,C / S,C	Upper Devonian Black Shales
		Brallier Fm.	Elk sands	Cl	S	
		Harrell Fm.				Burket Member of Harrell Fm.
		Hamilton Group				Marcellus Shale
		Onondage Group	Huntersville Chert / "Corniferous" lime	Cl / Ca	St / S	
		Oriskany Ss.	Oriskany sand	Cl	S,C	
		Helderberg Group				
	Silurian	Bass Islands Dolomite				
		Salina Group				
		Lockport Dolomite	"Corniferous" lime	Ca	S	
		Clinton Group	Keefer or Big Six sand	Cl	S	Rose Hill Formation
		Tuscarora Formation	Tuscarora/Clinch/ Clinton/Medina sand	Cl	S,C / St	
	Ordovician	Juniata/Bald Eagle Fms.	Bald Eagle sand	Cl	S	
		Reedsville Shale				
		Utica / Antes Shale				Utica/Antes Shale
		Trenton Ls.	Trenton Ls.	Ca	S, St	Trenton Ls.
		Black River Ls.				
		Chazy Group				
		Beekmantown Group	Upper Knox / Rose Run sand	Ca / Cl	S / C	
	Cambrian	Gatesburg Formation	Trempeleau Dolo. / Gatesburg sand	Ca / Cl	S / C	
		Warrior Formation				
		Elbrook Dolomite				
		Rome Fm.	Rome Fm.	Cl	S	Rome Fm. (?)
		Shady Dolo.				
		Chilhowee Group				

EXPLANATION

LITHOLOGY
Cl : Clastic quartzose rock
Ca : Carbonate rock

TYPE OF TRAP
S : Stratigraphic trap
St : Structural trap
C : Combination of S and St, including fracture-porosity traps

Figure 2. Generalized stratigraphic sequence of the central Appalachian basin, showing position of some major oil and gas source beds and reservoir rocks.

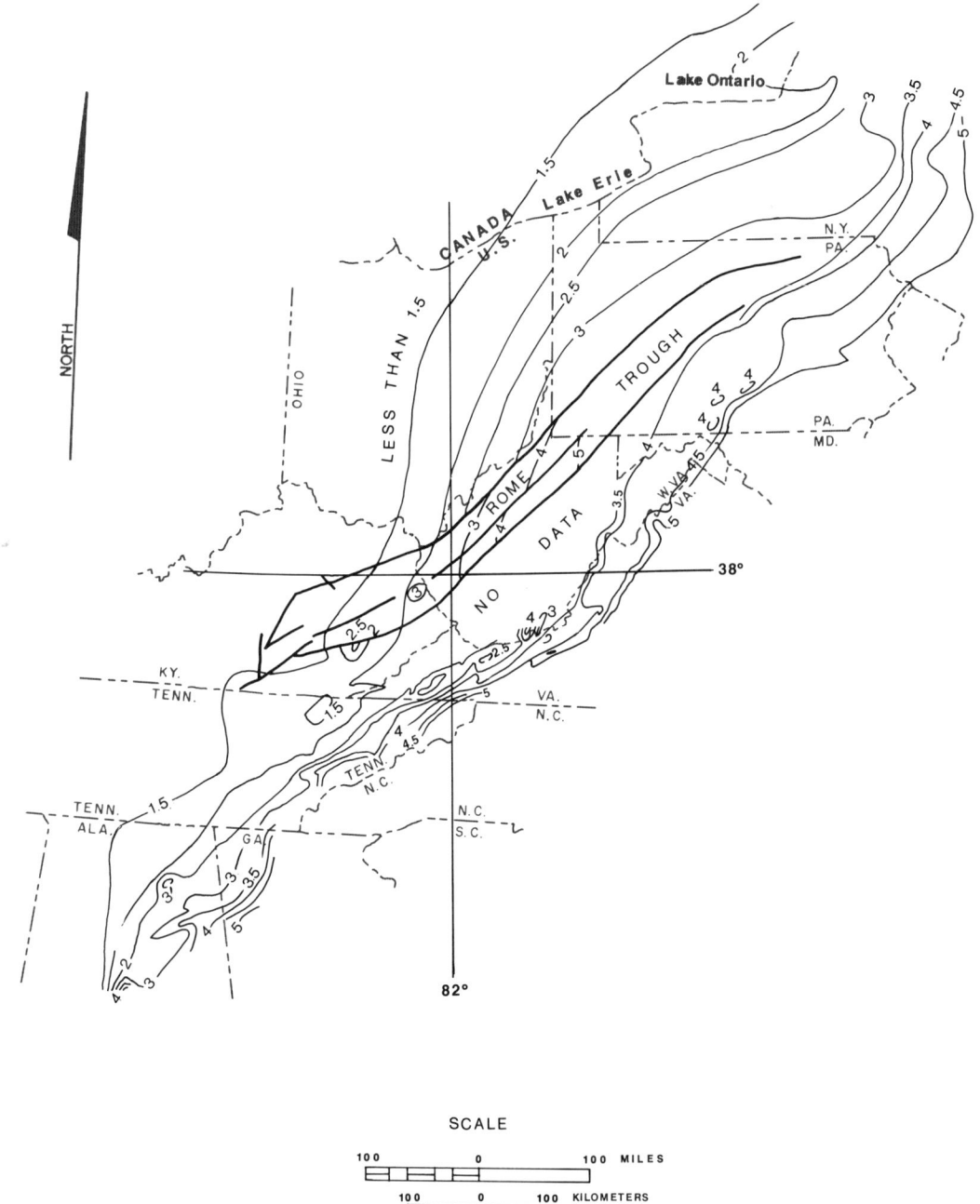

Figure 3. Conodont color-alteration index (CAI) isograds in Ordovician carbonate rocks (modified from Harris and others, 1978, sheet 1).

isograds trend in a more northerly direction than the Appalachian structural grain (Harris and others, 1978), more closely paralleling regional isopach trends. This divergence between isograds and structural grain, particularly in the southern Appalachians, and parallelism with isopachs indicates that the isograd pattern was largely developed by sediment loading before the Alleghanian Orogeny. Alleghanian thrusting telescoped the eastern part of the continent, covering a broad section of Paleozoic strata beneath crystalline rocks and juxtaposing rocks of high CAI index and rocks of lower CAI index across thrust faults. In general, rock temperature, which is a function of depth of burial and the existing geothermal gradient, increases eastward from low values (CAI 1, temperature less than 50° C) along the Cincinnati arch to relatively high values (CAI 5, temperatures 300° to 400°C) along the western toe of the Blue Ridge from Pennsylvania to Alabama.

The eastern limit of oil production appears to coincide with the CAI 2 isograd, which equates approximately with 65 percent fixed carbon and a depth of burial of 2,440 m (Harris and others, 1978, sheet 1, Table 1). Thus, in the absence of other information about geothermal gradients, strata buried less than 2,440 m may be considered mature with regard to oil generation. The lower limit of oil generation is at a lower temperature, however, than the 50° C required to produce a CAI of one. Temperatures in this range may be attained by depths of burial of several hundred meters.

Depth-to-basement maps of the Appalachian basin (Colton, 1970, Fig. 3; Harris, 1975, sheet 2) suggest that Cambrian and older strata lie at depths of 3,000 m or more in the eastern half of the basin from southern New York to Alabama. Apparently these older rocks have passed through the "oil window" in this area. Similarly, Ordovician shale source rocks are buried 2,700 m or more in a large area that extends from southern New York and eastern Ohio southward into Pennsylvania, West Virginia, Maryland, Virginia, and eastern Kentucky (Harris and others, 1978, sheet 3). These Middle to Upper Ordovician shales appear to have been the source of much oil and gas in Ordovician and possibly Silurian reservoir rocks in the Appalachian basin (Fig. 1D). Locally, in northern New York west of the Adirondacks, these mature shales produce only dry gas. A smaller area of thick Middle Ordovician to Pennsylvanian strata lies at the southern end of the Appalachians in the Valley and Ridge of Alabama and Black Warrior basin of Alabama and Mississippi.

In contrast, the several sequences of brown and black Devonian shales rich in organic matter, which underlie most of the western half of the western part of the basin, are presently in the oil-generating zone with their CAIs below 2. These rocks are the main source beds for most of the gas and oil in the Devonian to Pennsylvanian reservoir rocks in the basin (Figs. 1B, 1C, and 2) and are, by far, the most important source beds in the Appalachians. The CAI isograd maps of Devonian through Permian rocks (Harris and others, 1978, sheet 2) suggest that the Devonian rocks of eastern Pennsylvania, eastern West Virginia, Maryland, and most of Virginia have also passed the upper limit for oil generation. A small area in the fold-and-thrust belt of southwestern Virginia and adjacent eastern Tennessee is potentially a favorable region for oil generation in both Ordovician and Devonian shales.

The combined thickness of Carboniferous strata exceeds 2,700 m only in parts of the anthracite district of eastern Pennsylvania, southwestern Virginia, and southeastern Alabama. This indicates that almost all of the Carboniferous strata in the Plateaus segment of the Appalachian basin have not exceeded the uppermost limit for oil generation and are suitably mature. It is likely, however, that their contained kerogen is largely plant-derived and is more suitable for gas than for oil.

HISTORY AND EXPLORATION

Early developments: 1821 to 1900

The first productive wells deliberately drilled for gas or oil in the United States were drilled in the Appalachian basin during the nineteenth century. The oil drillers used percussion tools and techniques invented and perfected by Appalachian salt-well drillers during the early decades of that century. While drilling for brine, drillers often encountered both gas and oil in considerable amounts (Hildreth, 1833, p. 54). The hydrocarbons, however, were of little use at that time. Furthermore, they were considered a nuisance because they interfered with recovery of the brine.

In 1821, William Hart (Orton, 1899, p. 496) drilled the country's first gas well at Fredonia, Chautauqua County, New York, near a gas seep in Canadaway Creek. He demonstrated the utility of natural gas for heating, cooking, and illuminating to the villagers and to a number of visitors, including General Lafayette. Although Hart's shallow low-pressure gas wells from fractured Devonian black shales initiated the natural-gas industry, the large integrated gas company, organized to serve many communities, was many years in the future.

Although crude oil had been obtained in small quantities from oil seeps, springs, alluvial gravels, and salt wells, Edwin Drake's well, drilled to a depth of 21 m in the summer of 1859 along Oil Creek near Titusville, Pennsylvania, is credited as the first producing well deliberately drilled for oil in the United States. In drilling this well, Drake launched the modern petroleum business, which lubricated and illuminated America's industrial revolution.

Most early oil wells were generally drilled along stream valleys close to surface indications of gas or oil. Drilling in valleys was popular because the Drake well had been drilled close to an oil seep in Oil Creek. Because many of the early "gushers," 500 to 2,500 barrels per day, were drilled between Titusville and Oil City in the 1860s following the trends of creeks, "Creekology" (Galey, 1985, p. 423) and off-setting large producing wells were popular methods used by the "practical" oil men (Owen, 1975) to select well sites and to discover new pools. Another technique widely employed was to select well sites along lines projected between gushers or along lines extended from pools or fields. Drilling "on trend" became a much-favored technique for locating test wells in the early mythology of the Appalachian basin's oil fields, and was used later in many other basins in the country and worldwide. Later, with the benefit of much surface and subsurface data, petroleum geologists established the scientific rationale for exploration along specific trends in response to certain geologic criteria. In the 1860s and 1870s, however, scant surface and subsurface data were available, and because professional geologists little understood the conditions for generation, migration, and accumulation of petroleum, they were generally unable to assist the infant petroleum industry. During this time, a strong distinction was made between professional geologist and "practical" oil man in the Pennsylvania oil fields (Owen, 1975, p. 96). Generally, money went with the oil man, and considerable solid geological advice was ignored.

In 1861, T. Sterry Hunt of the Canadian Geological Survey (Owen, 1975, p. 61) proposed the concept of the anticlinal accumulation of oil and gas by gravity separation. Hunt's theory, which in later years would guide geologists and wildcatters to vast amounts of oil and gas, had little impact in the western part of the Appalachian basin, where most of the accumulations of oil in the Pennsylvania and Ohio fields were in stratigraphic traps, and local geologic structures are commonly poorly defined at the surface.

In the same year, E. B. Andrews related the presence of oil in the Burning Springs anticline of Virginia, later West Virginia, to zones of vertical fractures along the flanks of the fold. He noted that the more productive oil wells were located in the zones of most intense fracturing. He called these open fractures crevices, a term that was popular at the time. Because Billy Smith, Drake's driller, had reported encountering a crevice at the oil-producing level in the Drake well, crevices were much sought after by the early oil men. Unfortunately, Andrews' emphasis on the importance of fractures in the accumulation of oil at Burning Springs masked the importance of the anticlinal structure as the primary feature for the accumulation of oil (DeGolyer, 1961, p. 19).

In lucid reports of his careful and detailed study of the subsurface of parts of western Pennsylvania for the Pennsylvania Geological Survey in the 1870s, John F. Carll set forth many of the basic principles of petroleum geology. He recognized stratigraphic traps, and discussed the deposition and correlation of reservoir sandstones, and the depositional conditions conducive to stratigraphic accumulations of gas and oil. Carll named, defined, and explained the potential benefits of water flooding (Carll, 1880, p. 256–263) and speculated on its use in the enhanced recovery of oil. He clearly foresaw the advantages of unitization of pools for the maximum recovery of oil, and urged the saving of drill cuttings, well logs, and location plats. Most unfortunately for subsurface geologists in the Appalachian basin, Carll's astute advice was ignored in the hectic drilling booms of the late 1800s. Forty or fifty years later, many of Carll's ideas were resurrected and applied to enhance recovery of oil in the basin and elsewhere.

Drilling in the 1860s and 1870s was commonly shallow, generally less than 450 m, in the sequence of Ordovician to Pennsylvanian rocks along the west flank of the Appalachian basin. As drilling tools and rigs became larger and heavier, wells were drilled to test deeper reservoir rocks, and locations were sited farther to the east and south in the basin. Repeatedly during the last half of the century, wells deepened beneath old pools found more oil in older strata, particularly in the Pennsylvanian and Mississippian rocks of Ohio and West Virginia and in the Mississippian and Devonian rocks of Pennsylvania and West Virginia.

Beginning in the 1860s, many wells were stimulated by the ignition of a considerable volume of explosive, commonly nitroglycerin, in the oil-productive zone (Carll, 1880). The stimulation technique, commonly known as "torpedoing" or "shooting," generally increased the well's yield by fracturing the reservoir rock and increasing its permeability at the well bore. Nitroglycerin is, however, a very hazardous substance to manufacture, store, transport, and use in liquid form. It's touchy stuff! Consequently, numerous accidents, both destructive to property and fatal to well "shooters," occurred in the Appalachian oil and gas fields.

In the 1880s, I. C. White of West Virginia (White, 1885, p. 521–522) popularized and expanded Hunt's anticlinal accumulation theory by demonstrating that many large gas wells were located at or near the crests of anticlines in western Pennsylvania and contiguous West Virginia. Fortuitously for White, his reports coincided with the first great industrial demand for natural gas in western Pennsylvania, and White's presentation gained wide popular acceptance. The professional geologist began to make his presence known as detailed field mapping became essential to locating anticlines in the surface rocks along the western flank of the basin.

Much drilling on trend and on structure during the last decades of the nineteenth century outlined most of the major oil fields in Devonian strata from western New York to central West Virginia, including the Bradford field—the largest oil field in the Appalachian basin. Oil and gas were found at relatively shallow depths in stratigraphic traps in the basal Silurian sandstones, the Clinton in Ohio and the Medina in New York. In northwestern Ohio, along the Findlay segment (Plate 6H) of the Cincinnati arch, large amounts of oil and gas were found in extensively dolomitized zones in the upper part of the Trenton Limestone of Ordovician age. The wells, with oil or gas above salt water, were shallow, generally less than 460 m deep. Porosity and permeability were high in the dolomitized zones, and wells were prolific. Some wells produced as much as 10,000 to 40,000 barrels per day for short periods of time after being stimulated with explosives (Bownocker, 1903, p. 62, 75). Owen (1975, p. 129) noted that Orton, in his 1888 report on the Trenton oil fields, was the first geologist to use the term "trap" for the structural confinement of petroleum.

During the last two decades of the nineteenth century, geologists mapped many structures, particularly in the coal-bearing parts of the Appalachian basin, for both the federal and state geological surveys. Application of the anticlinal theory to their structure contour maps led to the discovery of many oil and gas pools in the basin and greatly enhanced the status of the professional geologist in the petroleum industry.

1900 to World War I

The early years of the twentieth century saw continued improvements in surface and subsurface mapping in the basin. Geologists, however, were commonly frustrated by the lack of subsurface data in old oil and gas fields, many of which had been abandoned for 20 or 30 years. Careful and detailed studies by geologists from the several state surveys and the U.S. Geological Survey led to the general acceptance of many of the geological principles developed and set forth by Carll, White, and Orton (Owen, 1975).

World War I to World War II

In response to the increased demand for both oil and gas after the first World War, drillers moved eastward deeper into the basin to extend known petroliferous zones and to evaluate the more deeply buried older rocks. Drilling was by cable tool, which was slow and required expert drillers to drill and complete the deeper wells. Small geological departments were established by the larger companies in the area, and company geologists were engaged in collecting data and choosing well sites. Drillers saved complete strings of well cuttings, prepared more complete well logs, and cooperated with company and government geologists in accumulating subsurface information. State surveys and other agencies began to require the filing of well locations and logs, particularly in the coal-bearing areas of Ohio and West Virginia. Regulations were promulgated for the plugging of abandoned wells, and efforts were instituted to protect the potable water supply from subsurface contamination.

By the late 1920s many of the larger fields had been discovered and delineated in the western part of the basin in the basal Silurian sandstones of Ohio, in the Devonian and Silurian carbonate rocks of eastern Kentucky, in the Upper Devonian sandstones of western Pennsylvania and contiguous New York or West Virginia, and most of the shallower sandstones of Pennsylvania and Mississippian age. In the 1920s, gas was found in fracture reservoirs in the dark gray to black Devonian shales rich in organic matter in parts of eastern Kentucky. Because these fractures could not be located from surface data, and because 40 percent of the wells had no indication of gas when drilled (Hunter, 1964, p. 25) and required extensive stimulation with several tons of explosive before production was obtained, development of Devonian shale gas in the Big Sandy field of eastern Kentucky and adjacent West Virginia took several decades. Wells producing from the Devonian shales, however, are long lived and yield gas relatively high in natural gasoline and associated hydrocarbons.

Enhanced recovery by water flooding, originally suggested by Carll in 1880, became legal in New York in 1919 and a year later in Pennsylvania. It was first applied on a large scale to the Upper Devonian sandstones of the great Bradford oil field, with considerable success. The need for much detailed data on porosity, permeability, and fluid saturation of the reservoir rocks was met with the development of a cable tool core barrel, which permitted recovery of short cores from the reservoir sandstones for the determination of many physical properties needed for successful water flooding. The Bradford sandstones responded well to water flooding, and the success at Bradford led to water floods in many other basins in the world (Owen, 1975, p. 625).

In the 1930s, the Lower Devonian Oriskany Sandstone became the prime target for exploration in the Appalachian basin. In contrast to the low permeability of the gas-productive Upper Devonian sandstones, the Oriskany is highly permeable and very porous near the western edge of the sheet of sandstone, where gas and some oil occur in stratigraphic traps of considerable extent. The largest of these traps, the Elk-Poca field near Charleston, West Virginia, has produced about 1 trillion cubic feet of gas to date (Cardwell and Avary, 1982, p. A-14), although at present much of the field has been converted to the storage of gas from other parts of the country.

Farther east and deeper under the central part of the basin, Oriskany gas has accumulated in structural traps, augmented by fracture porosity, along many of the major anticlines of the Alleghany Plateau (Plate 6H). Oriskany gas has also been found locally in structural traps in the more complexly faulted and folded rocks of the Valley and Ridge to the east of the Plateau. In Pennsylvania, the anticlines of the Alleghany Plateau are generally well defined, whereas in New York their surface expression is subdued, and detailed surface mapping is generally required to define the structure adequately for the selection of drill sites.

Drilling for anticlinal traps in the Oriskany began in south-central New York, where reservoirs at depths of 600 to 900 m yielded as much as 30 million cubic feet of gas per day per well. Drilling spread southwestward to the more deeply buried Oriskany in Pennsylvania and West Virginia. Drilling by cable tools with high-pressure gas reservoirs at depths of 1500 to 2400 m was slow and hazardous. The high yield of the Oriskany, however, produced a considerable play for the deep Oriskany gas.

In order to resolve the geometry of fault traps in the Oriskany along anticlinal trends, the reflection seismograph was introduced in the mid-1930s (Randall, 1938, p. 443–444). The first interpretations of Oriskany structures were not particularly accurate because of the difficult terrain, cumbersome equipment, inadequate data base, and the considerable length of time required to process the data by hand. In the following decades, geologists gained extensive knowledge of the characteristics of the stratigraphic sequence from samples and cuttings, thus permitting identification and correlation of conspicuous reflecting strata, and considerably enhancing the geophysicists' ability to delineate Oriskany structures. Equipment for collecting and processing the subsurface data was greatly improved, and reflection seismic profiles have become an important tool for working in the folded and faulted rocks of the basin.

The Oriskany play of the 1930s and 1940s produced a number of relatively deep wells, from which cuttings and a few short cores were obtained. Subsurface geologists, particularly C. R. Fettke in Pennsylvania and J.H.C. Martens in West Virginia, published detailed lithologic descriptions and identified stratigraphic units widely within the basin. Company geologists aided in establishing a general framework for the basin's Devonian subsurface stratigraphy by correlating drillers' logs, locally augmented by lithostratigraphic sample data.

A scattering of deep stratigraphic tests had been drilled to the Cambrian or Ordovician rocks in the western part of the basin. With the exception of Ohio's Trenton oil and some small oil pools in the Ordovician rocks of the Cumberland saddle of south-central Kentucky and contiguous Tennessee, however, most of these deep wells were not productive. The great depth to the older rocks in the central part of the basin and the considerable difficulty of drilling through the folded and faulted rocks in the Valley and Ridge restricted exploration to the shallower parts of the basin and depths generally less than 1,500 m.

During the 1930s, stimulation of oil and gas from carbonate reservoir rocks by acidization, the introduction of as much as several thousand gallons of acid into oil- or gas-productive zones (Kingston, 1936, p. 9; Leonardon, 1961, p. 598), became widespread in the Appalachian area. Acid treatment enlarges joints and fractures in carbonate rocks and dissolves carbonate cement in siliciclastic reservoir strata. By increasing permeability of reservoir rocks adjacent to the well bore, acidization augmented production in a number of relatively low permeability, tight, rocks.

Natural gas was first discovered in the Valley and Ridge in 1931, in the Early Grove gas field in the overthrust region of southwestern Virginia (Averitt, 1941). A decade later, oil was discovered in the Rose Hill district of Virginia, the first significant discovery east of the Appalachian Plateaus (Miller and Fuller, 1954). Oil in southwestern Virginia, chiefly from fractured Trenton limestones, has persisted as the only commercial production in the Appalachian Valley and Ridge to date.

World War II to the present

During the late 1940s and in the following decades, exploratory drilling continued to advance eastward and deeper into the basin. Mississippian carbonate and siliciclastic reservoir rocks were more closely defined and extensively drilled in parts of Ohio, Kentucky, Tennessee, West Virginia, and Virginia. Much drilling of the Upper Devonian and Oriskany sandstones was done in Pennsylvania and West Virginia, mainly seeking gas. Exploration in the Clinton sandstones spread eastward in Ohio to depths greater than 1,500 m, at which depths stimulation with explosives was no longer effective in increasing well productivity. Deep drilling by cable tool reached a maximum depth of 3,397 m in the New York State Natural Gas Corporation's #1 E. C. Kesselring, in Van Etten Township, Chemung County, New York; this well was abandoned June 3, 1953, in the Upper Cambrian Theresa Dolomite (Fettke, 1961), short of its intended Cambrian Potsdam Sandstone target.

Revitalization of the petroleum industry in the Appalachian basin took place in the mid-1950s with the introduction and widespread use of new techniques in drilling, logging, and stimulating wells. Rotary drilling methods were introduced to speed the completion of deep wells and to explore strata below the depths obtainable by cable tools. With rotary drilling, wells could be drilled and completed in a fraction of the time required by cable tool methods. Suites of electric logs yielding data on reservoir characteristics could be obtained from either a mud- or an air-filled well bore. The analysis of geophysical wire-line well logs enabled the subsurface geologist to select specific intervals in the well to test and evaluate for gas or oil.

Well stimulation by hydraulic fracturing (Leonardon, 1961, p. 600), rather than by the use of explosives, proved particularly effective in the low-permeability sandstones of the Upper Devonian sequence; in the basal Silurian sandstones of Ohio, Pennsylvania, and New York; and in the nonconventional source bed and reservoir rock of the black Devonian gas shales. Hydraulic fracturing induced production in areas where stimulation by explosives was ineffectual. Thus, large areas in the basin were effectively rejuvenated for exploration. At first oil or water, "river frac," was used as the fracturing fluid. Water, however, caused clays to swell in the reservoir rock by base exchange, and reduced permeability rather than enhancing fluid flow. Water also blocked pore throats and prevented gas from flowing to the well bore through shales or tight silty rocks. Consequently, low-density foam, gas, or liquified gas was substituted for water in the fracturing process, with encouraging results. Foam and gas fracs have been successful in many places in the basin where water fracs were only moderately effective. Stimulation by hydraulic fracturing opened large areas to exploration in low-permeability, tightly cemented sandstones and siltstones, thereby enlarging the areas from which gas and oil could be produced from the Devonian shale sequence in the western and central parts of the basin. In contrast to the past practice of shooting wells with nitroglycerin, at present most well stimulations are by some type of hydraulic fracturing, a process considerably less hazardous than well shooting.

Production was extended into the more southerly parts of the Appalachians, where carbonate reservoirs, chiefly of Mississippian age, supplant siliciclastic reservoirs to the north. The intergranular porosity of fossiliferous and fragmental limestones of the Fort Payne Chert yielded oil prolifically in some parts of the Tennessee Plateau. Interoolitic porosity in other Mississippian carbonate rocks has yielded gas in parts of the southern Appalachian Plateaus.

Common Depth Point (CDP) reflection seismic profiling has recently been used relatively extensively to define structures in the complexly folded and faulted parts of the Appalachian basin, that part in which thin-skinned décollement overthrusting is the dominant style of deformation. Deeply buried structures, particularly those in a series of stacked thrust sheets in which structures in the lower sheets are unrelated to the surface structures, can only be detected and defined by seismic profiling. Only recently, within the past decade, have geologists been able to resolve the geometry of subsurface structures in the overthrust belt satisfactorily to locate zones of fracture porosity in brittle-bed reservoir strata, using a meld of surface geology, CDP seismic profiles, and scattered sets of drill cuttings. As yet, little use has been made of three-dimensional seismic profiling in the Appalachian basin because of the unfavorable terrain, absence of closely spaced roads in target areas, and the high cost per unit of area surveyed.

In contrast to the early days of independents or small groups drilling on small lease blocks, at present, major petroleum companies in the basin employ well-staffed geological departments, with a cadre of geologists, geophysicists, and engineers. Working together, they seek out and evaluate prospects and select drill sites by using a combination of geological, geochemical, and geophysical surface and subsurface data in combination with an extensive data base of subsurface information, well history and production data, and pertinent geological literature. These data are commonly manipulated by computers using appropriate software programs. Small companies and independent operators drill

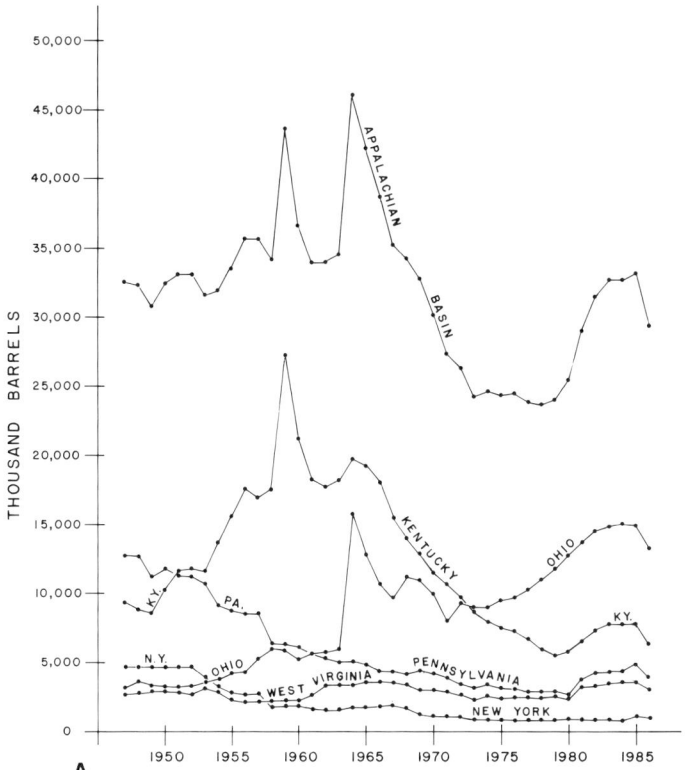

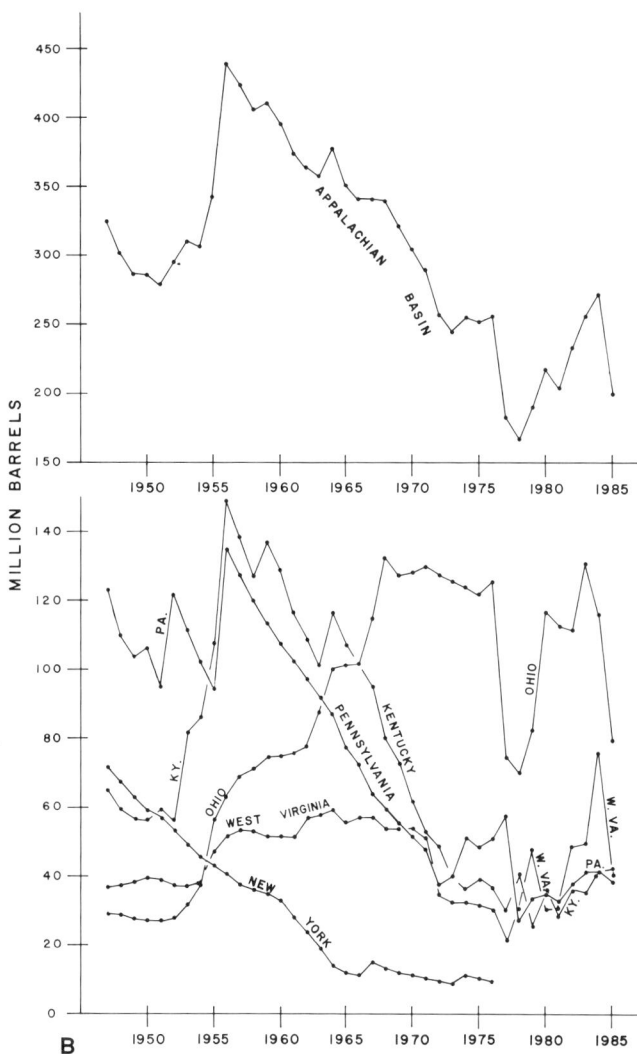

Figure 4. Appalachian basin crude oil production and reserves (includes western Kentucky and western Ohio). Data from American Petroleum Institute. A. Appalachian basin crude oil production, 1947 to 1986; B. Proved crude oil reserves of the Appalachian basin, 1947 to 1985.

many wells in the basin, using data generally available from the federal or state geological surveys, academia, and a scattering of service companies. Although much of the deeper drilling and drilling in rapidly developing areas is by rotary methods, about one quarter of the wells in the basin are as yet drilled with cable tools, particularly in the older areas and shallow fields where time is not essential and cost per foot is an important factor.

PRODUCTION AND POTENTIAL

In the discussion of oil and gas production in the Appalachian basin, the following volumetric terminology is used: MMBO = million barrels of oil ($N \times 10^6$), BBO = billion barrels of oil ($N \times 10^9$), BCF = billion cubic feet of gas ($N \times 10^9$), and TCF = trillion cubic feet of gas ($N \times 10^{12}$).

The Appalachian region (including adjacent parts of Ohio, Kentucky, and Tennessee) has produced more than 3 BBO and more than 35 TCF of gas, exclusive of waste, since 1860 (Miller, 1975; American Petroleum Institute, 1987). In general, oil production in the area declined from about 46 MMBO in 1964 to about 30 MMBO in 1985. Of the major producing Appalachian states, all except Ohio have declined significantly during the past several decades (Fig. 4a). Similarly, proven crude oil reserves of the area have declined generally, from about 440 MMBO in 1956 to about 200 MMBO in 1985 (Fig. 4b).

In contrast, annual production of natural gas in the Appalachian area has generally increased since 1947, from an average of about 425 BCF to almost 600 BCF in 1985 (Fig. 5a). Of the major producing states in the Appalachians, West Virginia and Kentucky have shown significant declines during the past few decades, whereas production has increased substantially in Pennsylvania and Ohio (Fig. 5a). Proved reserves of natural gas increased steadily from the late 1940s to the early 1970s (Fig. 5b). Subsequently, apparently resulting from unanticipated problems induced in the market by OPEC policies, natural gas reserves has fluctuated greatly but with an overall net increase (Fig. 5b).

A recent resource estimate of the undiscovered recoverable conventional oil and gas resources of the United States by the U.S. Geological Survey (Dolton and others, 1981) suggests that the Appalachians may contain between 100 MMBO and 1.5 BBO barrels of oil and between 6.4 TCF and 45.8 TCF of gas yet

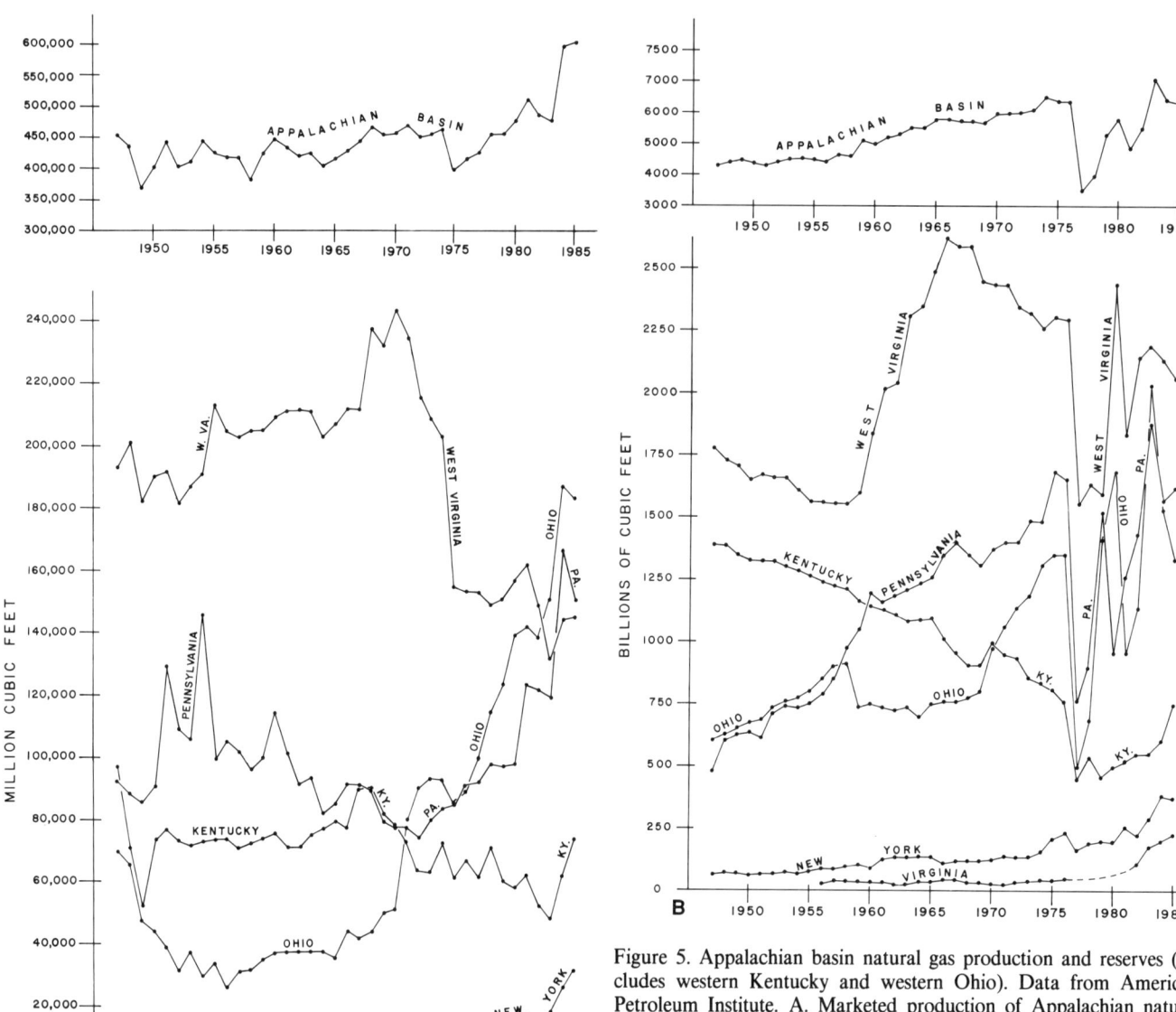

Figure 5. Appalachian basin natural gas production and reserves (includes western Kentucky and western Ohio). Data from American Petroleum Institute. A. Marketed production of Appalachian natural gas. 1947 to 1985; B. Proved natural gas resources of the Appalachian basin, 1947 to 1985.

to be recovered. If the high estimate for oil resources is correct, the Appalachians will ultimately have yielded about 4.3 BBO. Half of this amount was produced by the early 1940s, some 80 years after initial Appalachian production. Current production trends suggest that almost all of the remainder will be produced by 2020, with a sharp, irreversible downturn in production beginning at that time, if not sooner. New techniques in enhanced oil recovery, however, may sustain oil production considerably beyond 2020, because many fields were abandoned with a large percentage of the original in-place oil unrecovered.

Natural gas resources of the Appalachians are in a considerably less mature state of development than oil. Historically, production from the basin has been consistently maintained at slightly less than 5 TCF per decade. If the ultimate amount of recoverable natural gas is about 80 TCF, then half of the resource (40 TCF) will be produced a few years before the end of the twentieth century. Current trends indicate that this resource will be available during much, if not all, of the following century.

An enigma to the gas industry is the Devonian black shale, which contains an estimated 225 TCF (National Petroleum Council, 1980, low-estimate) to 1,131 TCF (Charpentier and others, 1982, high estimate) of in-place gas largely adsorbed on the organic matter in the shale. However, to date, only about 3.0 TCF of gas have been recovered from the gas-shale sequence, mainly in areas of considerable natural fracturing. A large potential resource of gas awaits new methods of well stimulation to free the adsorbed gas in the Devonian gas-shale sequence. Recovery of a significant percentage of the adsorbed gas would extend the gas supply in the Appalachian area well into the twenty-second century.

FUTURE EXPLORATION

The exploration trends of the past—drilling to deeper petroliferous zones and moving eastward into the less explored and structurally more complex segment of the basin—will continue to be followed in the future. Explorationists, using a full suite of geologic, geochemical, and geophysical data, will search for deeply buried, subtle structural traps in the overthrust belt or for stratigraphic traps defined and delineated from detailed basin analysis, or a combination of both sorts of traps in areas having the potential of producing hydrocarbons. A careful synthesis of source-bed potential, thermal maturation, and reservoir characteristics is essential if hydrocarbons are to be found in commercial quantities in the fold-and-thrust belt of the eastern Appalachians. Enhanced recovery of gas from nonconventional fractured shale reservoirs and from coal beds, as well as oil from carbonate and siliciclastic reservoir rocks will continue to be important factors in the recovery of additional oil and gas in the Appalachian basin.

Siliciclastic reservoir rocks in the Upper Ordovician to Mississippian sequence will continue to be drilling targets, particularly the widespread basal Silurian sandstones in the northern half of the basin. Stratigraphic traps and fracture porosity reservoirs in the carbonate rocks from the Mississippian to Cambrian sequence in the southern half of the basin—for example, in the vicinity of the Cumberland saddle—will yield both oil and gas. In general, the pools will be small to medium in size. Cambrian and Ordovician carbonate rocks have been little explored, except along the western and northwestern edge of the basin. Considerable thicknesses of these rocks remain to be tested, especially in the overthrust region of the Appalachians. Subtle structural and stratigraphic traps in the Ordovician and Cambrian rocks in the deeper parts of the basin will be drilling targets in the future. Oil and gas may be found in these traps in the western part of the basin, whereas dry gas will be found in the eastern part of the basin. The Rome trough (Plate 6H) has been little drilled, except in parts of Kentucky, and only small amounts of hydrocarbons have been recovered from 1,220 m to 2,440 m of sub-Knox rocks in this extensive complex graben.

The Appalachian basin presents a full spectrum, from the very maturely developed Late Devonian oil fields of western Pennsylvania, which were sites of the country's first oil boom and which have been redrilled several times in the past 125 years, to the relatively undrilled and untested structurally complex frontier area of the Valley and Ridge in the eastern part of the basin. Although the giant fields, Bradford and Elk-Poca, have been long discovered and exploited, many small- to medium-sized pools and fields remain to be found. More than 750,000 wells have been drilled in the 414,400 km^2 of the Appalachian Plateau, mainly in the shallower parts of the basin. In contrast, less than 500 wells have been drilled in the 129,800 km^2 of the Valley and Ridge, and most of the wells have been concentrated in local areas, such as in southwestern Virginia. Drilling will undoubtedly continue in the Appalachian basin for many years, particularly to recover oil that was trapped in siliciclastic reservoir rocks during the early development of the area when "blowing off the gas to get the oil," then an accepted production practice, effectively reduced reservoir pressures, thereby decreasing the movement of oil from the reservoir to the well bore.

REFERENCES CITED

American Petroleum Institute, 1987, Basic petroleum data book: American Petroleum Institute, v. 7, n. 3.

Averitt, P., 1941, The Early Grove gas field, Scott and Washington counties, Virginia: Virginia Geological Survey Bulletin 56, 50 p.

Bownocker, J. A., 1903, The occurrence and exploitation of petroleum and natural gas in Ohio: Ohio Geological Survey, 4th series, Bulletin no. 1, p. 62–75.

Cardwell, D. H., and Avary, K. L., 1982, Oil and gas fields of West Virginia: West Virginia Geological and Economic Survey Mineral Resources Series no. MRS-7B, p. A–14, B–33.

Carll, J. F., 1880, Geology of the oil regions of Warren, Venango, Clarion and Butler Counties [PA]: Pennsylvania Geological Survey, 2nd series, v. 3, p. 256–263, 325–329.

Charpentier, R. R., de Witt, W., Jr., Claypool, G. E., Harris, L. D., Mast, R. F., Megeath, J. D., Roen, J. B., and Schmoker, J. W., 1982, Estimates of unconventional natural-gas resources of the Devonian shale of the Appalachian basin: U.S. Geological Survey Open-File Report 82-474, 43 p.

Colton, G. W., 1970, The Appalachian basin—its depositional sequences and their geologic relationships, in Fischer, G. W., Pettijohn, F. J., Reed, J. C., Jr., and Weaver, K. N., eds., Studies of Appalachian Geology, Central and Southern: New York, Interscience Publishers, p. 5–47.

DeGolyer, E. L., 1961, Concepts on occurrence of oil and gas, in History of Petroleum Engineering: American Petroleum Institute Division of Production, p. 19.

de Witt, W., Jr., and Milici, R. C., 1989, Energy resources of the Appalachian orogen, in Hatcher, R. D., Thomas, W. A., and Viele, G. W., eds., The Appalachian-Ouachita Orogen in the United States; U.S.: Boulder, Colorado, Geological Society of America, The Geology of North America, v. F-2, p. 495–510.

Dolton, G. L., and 12 others, 1981, Estimates of the undiscovered recoverable conventional resources of oil and gas in the United States: U.S. Geological Survey Circular 860, 87 p.

Fettke, C. R., 1961, Well-sample descriptions in northwestern Pennsylvania and adjacent states: Pennsylvania Geological Survey, 4th series, Bulletin M-40, p. 567–606.

Galey, J. T., 1985, The anticlinal theory of oil and gas accumulation—Its role in the inception of the natural gas and modern oil industries in North America, in Drake, E. T., and Jordan, W. A., eds., Geologists and ideas—A history of North American Geology: Boulder, Colorado, Geological Society of America, Centennial Special Volume 1, p. 423–442.

Harland, W. B., and Gayer, R. A., 1972, The Arctic Caledonides and earlier oceans: Geology Magazine, v. 109, p. 289–314.

Harris, A. G., Harris, L. D., and Epstein, J. B., 1978, Oil and gas data from Paleozoic rocks in the Appalachian basin; Maps for assessing hydrocarbon potential and thermal maturity (Conodont Color Alteration Isograds and overburden isopachs): U.S. Geological Survey Miscellaneous Investigations Series Map I–917-E, scale 1:2,500,000, 4 sheets.

Harris, L. D., 1975, Oil and gas data from the Lower Ordovician and Cambrian rocks of the Appalachian basin: U.S. Geological Survey Miscellaneous Investigations Series Map I–917-D, scale 1:2,500,000, 3 sheets.

Hildreth, S. P., 1833, Observations on the saliferous rock formation, in the Valley of the Ohio: American Journal of Science, 1st series, v. 24, no. 1, p. 54–55.

Hunter, C. D., 1964, Gas development, production, and estimated ultimate recovery of Devonian shale in eastern Kentucky: Kentucky Geological Survey Special Publication 8, series 10, p. 21–29.

Kingston, B. M., 1936, Acidizing Handbook: Houston, Texas, Gulf Publishing Co., p. 9.

Leonardon, E. G., 1961, Logging, sampling, and testing, in History of Petroleum Engineering: American Petroleum Institute Division of Production, p. 598–600.

Milici, R. C., and de Witt, W., Jr., 1988, The Appalachian basin, in Sloss, L. L., ed., Sedimentary Cover—North American Craton; U.S.: Boulder, Colorado, Geological Society of America, The Geology of North America, v. D-2, p. 427–470.

Miller, B. M., 1975, A summary of oil and gas production and reserve histories of the Appalachian basin, 1859–1972: U.S. Geological Survey Bulletin 1409, 36 p.

Miller, R. L., and Fuller, J. O., 1954, Geology and oil resources of the Rose Hill district–the fenster area of the Cumberland overthrust block, Lee County, Virginia: Virginia Geological Survey Bulletin 71, 383 p.

National Petroleum Council, 1980, Unconventional gas sources—Devonian shale: U.S. Department of Energy, National Petroleum Council, v. 3, p. 21.

Orton, E., 1899, Petroleum and natural gas in New York: New York State Museum Bulletin, v. 6, no. 30, p. 496–497.

Owen, E. W., 1975, Trek of the oil finders—A history of exploration for petroleum: American Association of Petroleum Geologists Memoir 6, 1647 p.

Perroud, H., Van der Voo, R., and Bonhommet, N., 1984, Paleozoic evolution of the American plate on the basis of paleomagnetic data: Geology, v. 12, p. 579–582.

Randall, L. E., 1938, Use of geophysical methods in Oriskany prospecting [abs.]: American Petroleum Institute, Drilling and Production Practice, p. 443–444.

Thomas, W. A., 1988, Black Warrior basin, in Sloss, L. L., ed., Sedimentary Cover—North American Craton; U.S.: Boulder, Colorado, Geological Society of America, The Geology of North America, v. D-2, p. 471–492.

White, I. C., 1885, Geology of natural gas: Science, v. 5, no. 125, p. 521–522.

Manuscript Accepted by the Society October 27, 1988

ACKNOWLEDGMENTS

The manuscript was reviewed by Donald C. Le Van, Virginia Division of Mineral Resources, Charlottesville, Virginia.

Chapter 18

The Michigan basin

C. R. Reszka, Jr.
Geological Survey Division, Michigan Department of Natural Resources, Box 30028, Lansing, Michigan 48909

INTRODUCTION

The Michigan basin is a large, relatively deep structure of Paleozoic age centered in the lower peninsula of the state of Michigan. The province includes Michigan's entire lower peninsula, parts of northern Indiana, most of eastern Wisconsin, the eastern part of Michigan's upper peninsula, western portions of Ontario, and northwestern Ohio. It is defined geologically by a series of subtle structural highs: the post-Silurian Findlay arch forms the southeastern boundary, the Early Ordovician Kankakee arch forms the southwestern boundary, and the Wisconsin Highland and the Wisconsin arch form the western boundary. To the north and northeast is the Canadian Shield, and to the east is the Algonquin arch, a major Precambrian feature in Ontario (Fig. 1). For a more complete discussion of the physiography and age of the Michigan basin, see Fisher and others (1988).

In the past, the greatest amount of drilling in the Michigan Basin took place in the sectors in Indiana, Ohio, and Ontario. However, with modern production from the Ohio and Indiana sectors currently in decline, most drilling activity is confined to Michigan's lower peninsula. For this reason most of the statistical and geological data in this report, unless otherwise stated, will concern the lower peninsula.

GENERAL CHARACTERISTICS

The Michigan basin is a relatively deep, slightly elliptical, intracratonic basin occupying about 316,000 km^2 (Plate 7B). It contains a sedimentary column approximately 6,100 m thick with a volume of 558,000 km^3. These figures exclude Pleistocene glacial deposits, which almost entirely cover the lower peninsula of Michigan to a thickness of 425 m.

With the exception of the Permian, all systems of the Paleozoic Era are represented in the basin, with a remnant of Jurassic System deposits limited to the Central Basin area (Figs. 1 and 2). The pre-Pennsylvanian deposits are predominantly marine carbonates, deposits of the Pennsylvanian System are predominantly terrestrial sandstones with some marine carbonates, and sediments of the Jurassic System are entirely terrestrial sandstones.

The Michigan basin is classified as an interior sag basin (Kingston and others, 1983). An interior sag basin is characterized by being entirely intracratonic, several hundred kilometers wide, subcircular in outline, and containing several hundred thousand cubic kilometers of sediment. Formation of this type of basin is attributed to a simple downwarping of the crust. There may be small-scale, localized deformation, but no major structural contortions that deform large portions of the sediment within the basin.

There is no clear solution to the mechanism that generates interior sag basins. Mechanisms that have been proposed include sediment loading, rifting, stretching and loading, and thermal contraction following emplacement of denser material below the crust (Sleep and others, 1980). The major difficulty with all these theories is that they hypothesize a single event. The Michigan basin has undergone many sequences of subsidence and rebound during its formation, and none of these theories adequately account for multiple events. At present the thermal contraction hypothesis is the most widely accepted. However, there is no evidence of the proper thermal activity, at the appropriate times, which could initiate the uplift-subsidence sequence in the Michigan basin (Sloss, 1982).

Interior sag basins, regardless of origin, are important sources of petroleum reserves. The shallow marine and nonmarine environments that covered these basins facilitated source rock development by the formation of blanket reservoirs and evaporite seals. In addition, hydrocarbon plays are associated with features such as basement structures and stratigraphic traps, which are usually found within interior sag basins.

Most continents contain identifiable interior sag basins. However, those of approximately the same age but smaller in scale than the Michigan basin are generally too shallow and have too little burial depth for hydrocarbon maturation. Those that are morphologically similar, but younger, may have sufficient sedimentary deposit depth but insufficient burial time for hydrocarbon generation.

Although the shape and the major characteristics of the Michigan basin province have been known for 130 years, the exact age of the structure is still unknown. Newcombe (1933) suggested a mid-Proterozoic (Keweenawan) age. More recent

Reszka, C. R., Jr., 1991, The Michigan basin, *in* Gluskoter, H. J., Rice, D. D., and Taylor, R. B., eds., Economic Geology, U.S.: Boulder, Colorado, Geological Society of America, The Geology of North America, v. P-2.

authors suggest that a recognizable basin structure existed by Late Cambrian time (Catacosinos, 1973; Bricker and others, 1983; Fisher and others, 1988). Most authors, however, agree that the basin existed in its present form by at least Middle Ordovician (for example, Haxby and others, 1976; Sloss, 1982).

Nearly all continents have interior sag basins, although they may not be the same age or have the same history. In this volume the Michigan basin serves as a representative example for interior sag basins. Kingston, Dishroon, and Williams, in their "Global Basin Classification System" (1983), describe in great detail the various criteria used to designate basin types in their system. St. John and others (1984) list and locate, on a large-scale map, all the "known sedimentary accumulations greater than 1,000 meters thick." The author used this list and personal communication with D. R. Kingston and M. W. Leighton in an attempt to catalogue other North American interior sag basins like Michigan's.

The Williston basin is most similar to the Michigan basin;

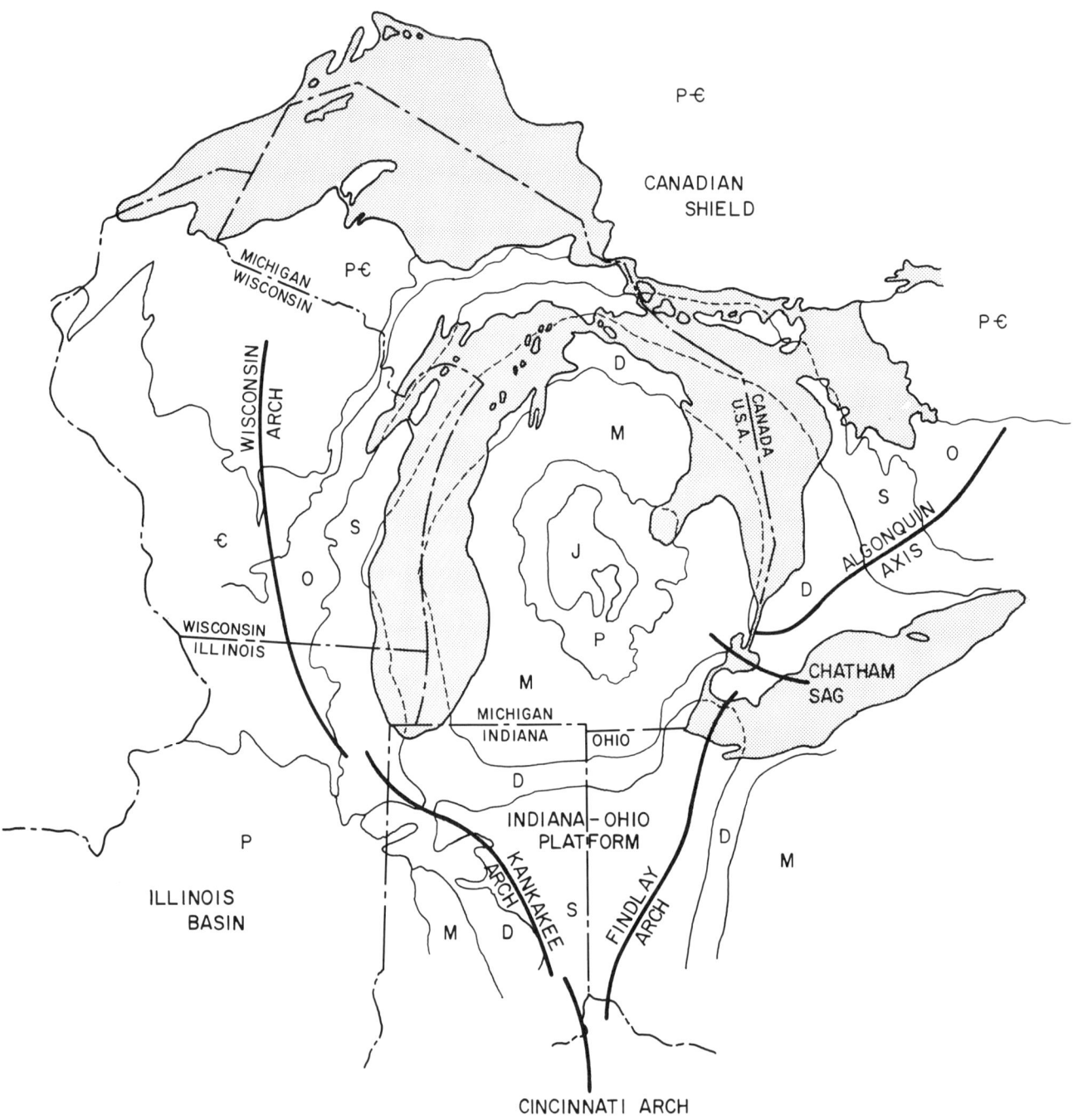

Figure 1. The Michigan basin and surrounding structure (adapted from Stonehouse, 1969, and Ells, 1969).

the Illinois basin, the Hudson Bay basin, and the Salina–Forest City basin are also interior sag basins (see Fig. 3). The Illinois and Williston basins have production from giant fields (500 million barrels [mbbls] or more of recoverable oil or at least 3 trillion cubic feet [tcf] of gas in a gas field). The Michigan and Salina–Forest City basins produce from fields classified as subgiant, and the Hudson Bay basin is nonproductive (St. John and others, 1984).

HISTORY

Early French settlers first noted gas seeps in southeastern Michigan, including some near the present location of Detroit. In neighboring Canada, seeps near Petrolia, in southwest Ontario, were used for "illuminating and lubricating purposes" (Cohee and Landes, 1955). Later, in 1648, Jesuits reported seeps from the Trenton Limestone on Manitoulin Island in the northeastern part of the Michigan basin. As early as 1836, near Findlay in northwest Ohio, oil was actively sought to heat homes. In 1858 a well was dug in the area of the Petrolia seeps, and oil was discovered in the glacial deposits. This was the first commercial exploration for hydrocarbons in the world and preceded Drake's oil discovery in Titusville, Pennsylvania, by one year. It was only after Drake's discovery that interest in exploring for oil spread to the rest of the Michigan basin, but commercial drilling did not begin in the U.S. sections of the Michigan basin until 1884. A well sunk at this time to the Middle Ordovician Trenton Group in Findlay began the extensive Lima-Indiana field that eventually covered much of the Indiana-Ohio Platform (see Fig. 6). By 1896, between 70,000 and 100,000 wells were drilled in the Ohio portion alone. Unfortunately, this was also the year in which exploration and production peaked in this field (Gray, 1983). In 1886 the first commercial well was drilled in the state of Michigan. This well, drilled in Saint Clair County near the Ontario-Michigan border, began the Port Huron field, the first significant field in Michigan. Production was from the Dundee Limestone at a depth of about 180 m.

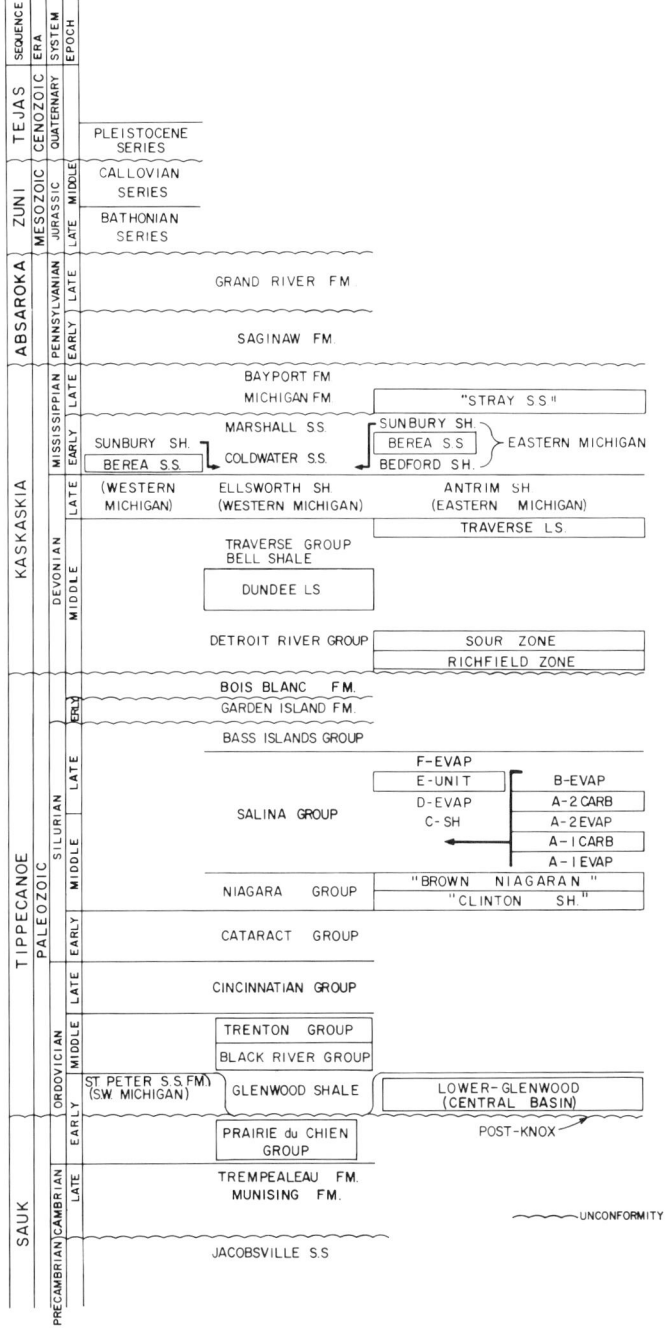

Figure 2. Stratigraphic succession in Michigan (adapted from Michigan Department of Natural Resources, 1964). The boxed units indicate major producing horizons.

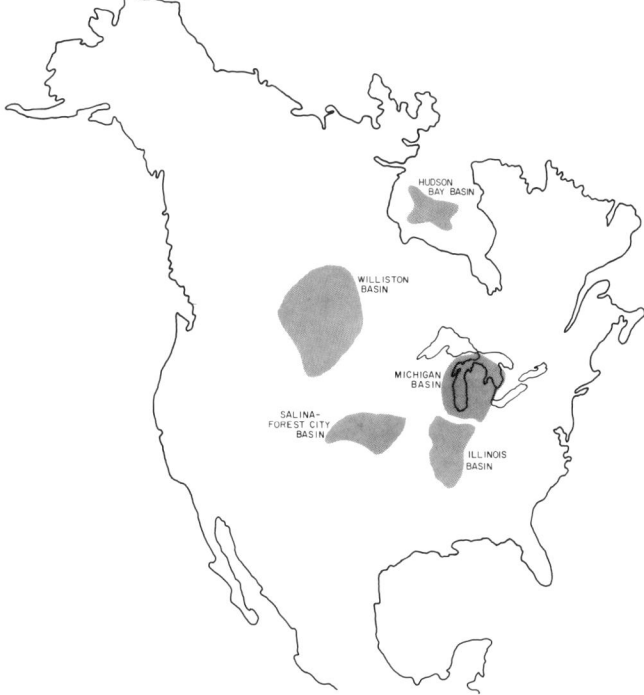

Figure 3. Interior Sag basins of North America (after Gardner and Bray, 1984).

Many oil and gas shows were reported in the southern and eastern parts of the Michigan basin, but it was not until the discovery in 1928 of the Muskegon field (in western Michigan) and the Mount Pleasant field (in the central basin area) that the real possibilities of the basin were demonstrated. Until this time it was thought that the hydrocarbons migrated to the basin edges, but the Mount Pleasant field was drilled on an anticlinal structure near the basin center, indicating that oil could be found anywhere within the basin. Parts of the Port Huron, Muskegon, and Mount Pleasant fields are still producing and undergoing continued development.

The greatest production, in the earliest years, was from the Stray Sandstone in Michigan and the Dundee Limestone and Trenton Group in the rest of the basin. In 1920, carbonate sediments of the Middle Ordovician Trenton and Black River Groups in Michigan near the Ohio-Michigan border were found to be oil productive, and in 1927 the Middle Silurian Niagara Group of southeastern Michigan was found to be productive. The discovery in 1957 of the large Albion-Scipio field of south-central Michigan made the Trenton and Black River Groups the production leaders for several years. Then in 1969 the discovery of a reef of Middle Silurian age in northeastern Michigan—coupled with an increased understanding of, and dependence on, seismic exploration technology—resulted in the discovery of several hundred "Niagaran" reservoirs. Silurain strata now dominate the oil and gas production in the basin.

GEOLOGICAL STRUCTURE

No large-scale tectonic activity has greatly disturbed the beds of the Michigan basin, but erosion, small-scale folding, faulting, and solution activity have influenced and modified the orderly arrangement of the layers. The folds are primarily anticlinal and predominantly occupy the central basin, but several large anticlinal structures are found in Michigan in the southeastern and eastern parts of the basin. A few large faults occur in these areas, but they are predominantly basement features, and even the largest extends only to the Mississippian System (Fisher and others, 1988). Deep seismic reflection studies suggest that the basement is highly faulted and may be a complex series of horsts and grabens. Sediment drape over these features may explain some of the anticlinal structures seen in the southeastern and eastern parts of the basin. It is not known if the folds in the center of the basin have the same origin, or if some later tectonic event, unrelated to basement faulting, was the cause. Prouty (1983) believes many of the structures in the basin are due to the action of extensive wrench and shear faulting. On the other hand, Fisher and Barratt (1985) attribute the folds in the central basin to reactivation and "essentially vertical movements of basement fault blocks."

GEOPHYSICAL FEATURES

The largest geophysical feature in the Michigan basin is the Mid-Michigan Gravity High. This is a large, sinuous, well-defined positive gravity anomaly that crosses the lower peninsula from northwest to southeast. Drill-hole data, outcrop studies, and gravity, magnetic, and seismic investigations show a connection with the Lake Superior Basin (Oray and others, 1973), which has been shown to be a part of the Mid-Continent Geophysical Anomaly (Craddock, 1972). Therefore, some authors (Craddock, 1972; Chase and Gilmer, 1973; Hinze and others, 1975) consider the structure in Michigan to be an extension of the Mid-Continent Geophysical Anomaly. The Mid-Michigan Gravity High was thought to be a Late Precambrian rift zone partially filled with Keweenawan Age clastic and volcanic rock (Hinze and others, 1975). A COCORP (Consortium for Continental Reflection Profiling) deep seismic line across the Mid-Michigan Gravity High supports the rift hypothesis (Brown and others, 1982).

The rift is not considered to have caused the basin subsidence. It does not conform to the surrounding basin geometry, and its age, although not known with certainty, suggests it was in place more than 100 m.y. before the earliest basin structure (Sloss, 1982).

THERMAL HISTORY

The thermal history of the Michigan basin is still a controversial subject. Organic geochemical studies have recently shown that organic-rich sediments throughout the basin have moderate to high maturity levels. These range from 2.7 to 3.9 on the Thermal Alteration Index scale with corresponding vitrinite reflectance values of 0.7 percent Ro to 3.0 percent Ro (Moyer, 1982; Cercone, 1984; Gardner and Bray, 1984). In Michigan the Thermal Alteration Index (TAI) and vitrinite reflectance (Ro) values show that the "oil window," a subsurface region where the proper temperature and pressure exist to foster the generation of liquid hydrocarbons, has TAI values between 2.65 (0.65% Ro) and 3.67 (2.0% Ro) (Plate 7B). Below 2.65, any potential source rocks will not have begun to generate oil, and above 3.67 any oil will be converted to gas. The upper physical limit of the modern "oil window" in the Michigan Basin is about 500 m deep, and the lower limit is about 2,500 m deep (Cercone, 1984).

To determine source beds for the producing horizons in the basin, Moyer (1982) tested samples from the Cambrian to the Pennsylvanian for total organic content and thermal maturity. Table 1 lists the Thermal Alteration Index values obtained from these samples. The values are indicators of the thermal maturity of the samples and show that source rocks in the central or northern parts of the basin, from the Ordovician System to the Early Pennsylvanian System, are mature enough to produce hydrocarbons. In the southern part of the basin, strata of the Mississippian System have low maturity levels and may not have been able to generate oil or gas, but units tested in the underlying Ordovician and Silurian Systems exhibit high maturity levels.

When these data are used in an attempt to determine the thermal history of the basin, a major problem becomes evident. If the modern average geothermal gradient of 25°C/km is applied, the observed maturity levels of the formations tested are too high

TABLE 1. OBSERVED ORGANIC MATURITIES IN
THE MICHIGAN BASIN*

	Observed Maturity	
Formation	TAI	(Ro)
Central Michigan Basin		
Saginaw	2.70	(0.70)
Michigan	2.75	(0.77)
Antrim	3.20	(1.30)
Bell	3.30	(1.36)
Salina C	3.50	(1.50)
Utica	3.90	(3.00)
Northern Michigan Basin		
Salina G	3.10	(1.22)
Salina C	3.20	(1.30)
Niagaran	3.40	(1.42)
Utica	3.50	(1.50)
Southern Michigan Basin		
Antrim	2.20	(0.40)
Salina C	3.60	(1.75)
Utica	3.70	(2.00)

*Modified from Cercone (1984).

for their depth of burial. In addition, whenever either burial depth or geothermal gradient is held constant to the modern value, the other factor must become unrealistically high to account for the observed maturities (Cercone, 1984). It is probable that both burial depth and geothermal gradient values for the past differed from the modern values.

Cercone (1984) used a variation of the Lopatin Method (Waples, 1980) to develop a thermal history that produced estimates very close to the observed maturity values and supported the multiple subsidence theory for the Michigan basin. Data included known periods of erosion in the basin, sporopollen maturity, the presence of about 1,700 m of Carboniferous System sediments in the adjacent Illinois and Appalachian basins, extrapolation of the surface of the Upper Mississippian System marine sediments from the basin center to the northern lower peninsula, and determination of the organic maturity of Pennsylvanian System coal. From this study, Cercone estimated that an additional 1,000 m of overburden was deposited at one time and then subsequently removed from the basin.

Using this additional burial depth and the observed organic maturity, Cercone calculated the thermal maturity and developed burial history curves for the northern, central, and southern basin (Fig. 4). Cercone also tested a range of geothermal gradients but found that no single gradient was consistent with the observed data and that the values 35°C/km and 45°C/km best suited the existing conditions. These values, however, each predict only half the sample, overestimating or underestimating the remainder.

In Figure 5 the dashed lines labeled 45°C/km and 35°C/km are the thermal maturity indicators that represent the top of the "oil window" for each of the proposed geothermal gradients. Throughout the history of the basin, any strata below these lines were capable of generating hydrocarbons. Using these values, Cercone determined that the "oil window" during the Paleozoic ranged from 1,900 to 2,300 m and stated that post-Pennsylvanian erosion and continued thermal maturation raised the "oil window" to its present-day level. The fact that the data are best explained by either of two geothermal gradient values, or the range between those values, suggests that no single thermal event was responsible for hydrocarbon generation and that there was a thermal history of at least two, or perhaps more, reheating events.

MAJOR PRODUCING SYSTEMS

There are twenty-five hydrocarbon producing zones in the Michigan basin with nearly every system of the Paleozoic Era having one or more zones. This chapter will concern itself with the fifteen most productive zones. The other ten have only been productive historically or are of very limited extent, as collectively they produce only 0.1 percent of the total oil and gas in the basin.

Some fields have been extensively studied, but information such as porosity and permeability values, temperature and pressure measurements, trapping mechanisms, and migration paths exists for a few isolated fields only. The variability of the available data makes it difficult to arrive at basinwide values for each

Figure 4. Regions of the Michigan basin with burial history curves.

producing horizon. The values used in describing the following producing systems are generalized and may not describe individual fields with complete accuracy.

It is important to understand that the lithological units that act as source rocks for the producing horizons have not been unequivocally defined. Total organic carbon and thermal maturity studies (Moyer, 1982) showed that some part of all the formations tested in the basin can generate hydrocarbons. These same data also show that some units may be considered source rocks in one part of the basin but not in others. Illich and Grizzle (1985) reached similar conclusions, although they approached the source rock problem through geochemical analyses of selected Michigan oils. Vogler and others (1981) and Nunn and others (1984) suggest that oil and gas for the producing formations were generated in the central basin from the rocks of the deeper Ordovician System. Then a series of postulated faults, acting as conduits to both shallower and more outlying reservoirs, allowed the hydrocarbon to migrate up column and updip. This hypothesis seems unlikely because the hydrocarbons would have had to traverse several hundred meters of evaporites and at least three unconformities to reach the shallower producing units in the central basin. To determine the source rocks for the producing horizons throughout the basin is a complex task that is far from complete; any references to source rocks must therefore be considered somewhat tentative.

Unless otherwise noted, the cumulative oil and gas statistics used in this chapter are for 1984 (Michigan Geological Survey, unpublished data) and are the most current available.

CAMBRO-ORDOVICIAN

Cambrian Age production currently is not very prominent. The only Cambrian production is from Ontario on or near the Algonquin Axis at the extreme eastern edge of the Michigan Basin. Production is from stratigraphic traps in dolomite and porosity pinchouts or fault blocks in a sandy dolomite. Cumulative production since 1925 is 2.7 mbbls (million barrels) of oil and 954 mmcf (million cubic feet) (Ontario Geological Survey, 1980).

The Ordovician System has production from the Early Ordovician Prairie du Chien Group and the Middle Ordovician Black River and Trenton Groups.

There has always been some controversy concerning the nomenclature of the Early Ordovician and deeper formations in the Michigan basin, primarily because of the lack of hard data. The recent discoveries and accelerated drilling program have not only added to the body of knowledge but have also fueled the controversy, with the Prairie du Chien Group as the focus of the dispute. Every stratigraphic unit in the group is controversial somewhere in the basin.

A large-scale unconformity, referred to as the Post-Knox unconformity (Syrjamaki, 1977; Bricker and others, 1983; Shaver, 1985), caps the Prairie du Chien Group and represents a hiatus as long as 35 m.y. (Shaver, 1985). Above this unconformity, different beds are encountered in different parts of the basin. In the southwest part of the lower peninsula the St. Peter Sandstone is immediately above the Prairie du Chien Group but is not present in the rest of the basin. In the center of the basin a unit consisting of sand, shales, and carbonates, stratigraphically equivalent to the St. Peter Sandstone, caps the Prairie du Chien Group. In the remainder of the basin the Glenwood Shale, the basal member of the Black River Group, immediately overlies the Prairie du Chien Group.

In its most complete form, the Prairie du Chien Group consists of two sandy dolomite units bracketing a sandstone unit. At the basin margins, only the lowest dolomite interval remains, but toward the central basin the character of the group changes. Below the top of the Glenwood Shale is the sandstone, shale, and carbonate unit variously called the Zone of Unconformity (Bricker and others, 1983) or Lower Glenwood (Fisher and Barratt, 1985; W. Harrison, personal communication, 1986). Below this unit the Prairie du Chien Group generally consists of an upper massive sand interval, up to 305 m thick, and a lower shaley dolomitic siltstone or a shaley, sandy dolomite, the thickness of which depends on the placing of the underlying Trempealeau Formation. The lithology change between the Trempealeau Formation and the overlying Prairie du Chien Group strata is not

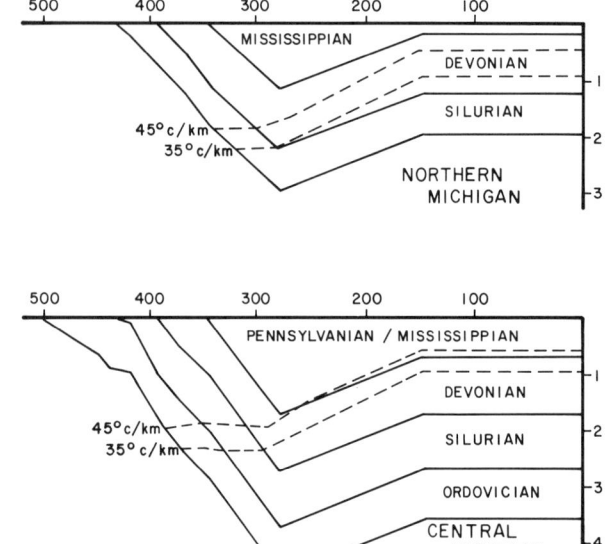

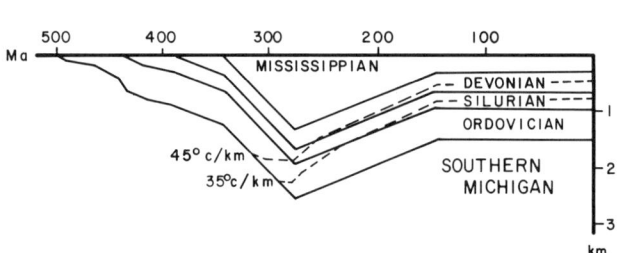

Figure 5. Burial history diagrams (after Cercone, 1984). Strata below the dashed lines are mature enough to generate hydrocarbons.

always clear. A new age determination of two wells in the central basin area (Fisher and Barratt, 1985) and an associated conodont study may initiate a renaming and recorrelation of many of the units below the Lower Glenwood.

Currently the Prairie du Chien Group, at 3,200 m, is the deepest producing unit in the Michigan basin. Historically it has produced oil only from the lowest dolomite unit at depths averaging 1,150 m in the southern part of the lower peninsula of Michigan. Deeper wells of about 2,850 m in the northern and central basin produce only gas and condensate from four zones within the group. The first zone is a thin sandstone stringer that is actually above the Prairie du Chien Group in the Lower Glenwood, but it is included in the Prairie du Chien Group until the nomenclatural debate is clarified. The second zone is located approximately 30 m below the top of the massive sand interval. The third producing zone is about two-thirds down the same unit but below a section of sandstone tightly cemented with carbonate or slightly shaley. The fourth zone in the Prairie du Chien Group is a relatively clean sand stringer about 30 m below the top of the siltstone/sandy dolomite described as underlying the massive sand unit. The Prairie du Chien Group in Michigan has produced 39,000 bbls of oil since 1960 and about 5 bcf (billion cubic feet)

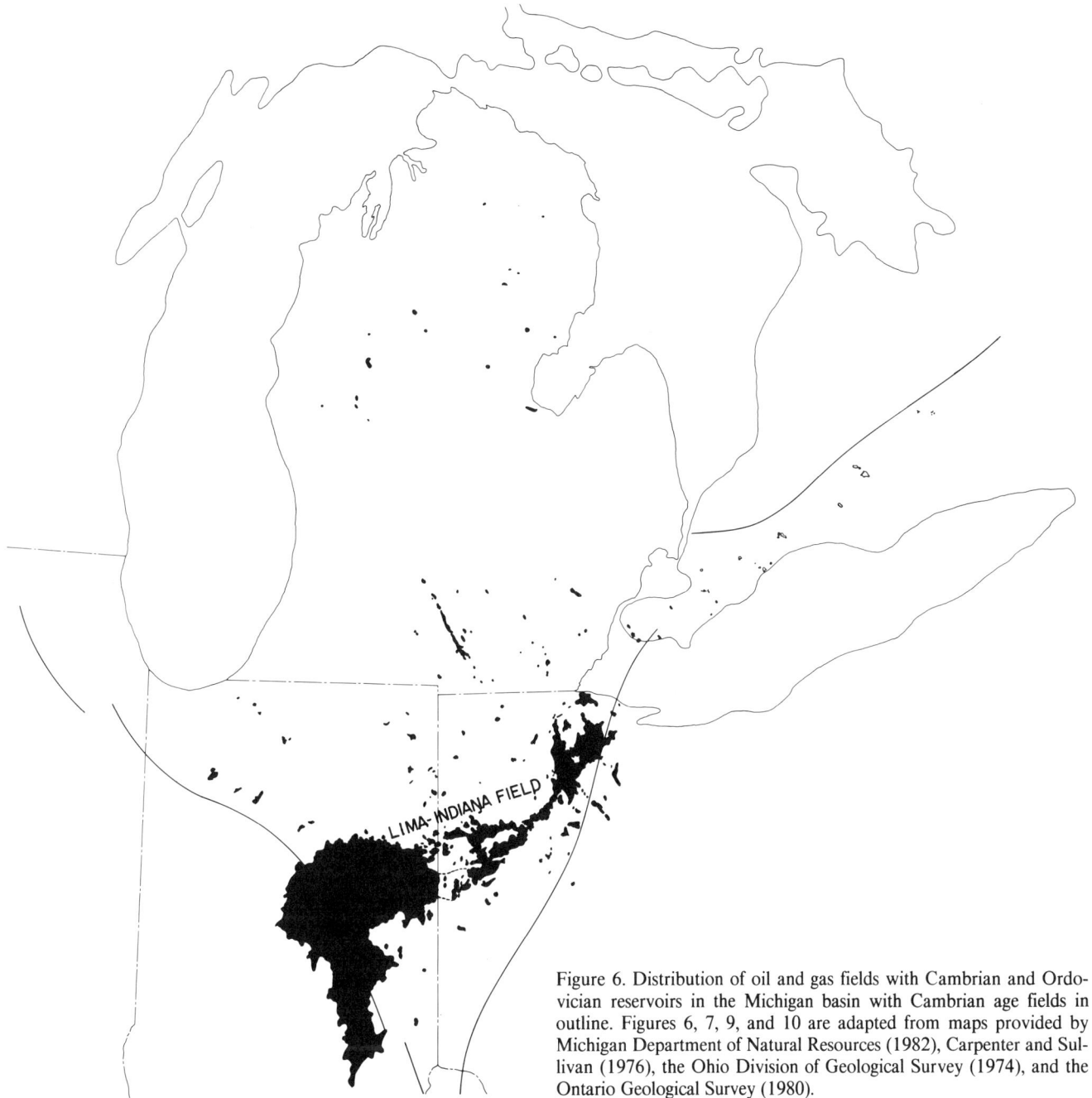

Figure 6. Distribution of oil and gas fields with Cambrian and Ordovician reservoirs in the Michigan basin with Cambrian age fields in outline. Figures 6, 7, 9, and 10 are adapted from maps provided by Michigan Department of Natural Resources (1982), Carpenter and Sullivan (1976), the Ohio Division of Geological Survey (1974), and the Ontario Geological Survey (1980).

of gas. The producing wells have been found by drilling slight structural highs or as deeper tests of shallower producing formations.

Trap mechanisms associated with the Prairie du Chien Group might have more to do with porosity and permeability variations than existing structures. Porosity is intergranular and varies from 0 to 20 percent, and permeabilities range from 1 to 300 md (millidarcies). Porosity and permeability values are greater in the western part of the Michigan basin but are reduced, probably because of poor sorting and compaction, in the deeper central parts of the basin. Although structures are targeted, their influence on porosity appears negligible. The reasons for the variable are not completely understood, so the presence of effective porosity remains somewhat unpredictable. However, Fowler and Barratt (1982) suggest that clay mineral inclusions between quartz grains inhibit cementation and that consequently the best porosity should be where the clay fraction is greater, as it is on the western margin of the basin.

The Trenton and Black River Groups are light brown to grey fossiliferous limestones with fracture related and dolomitized porosity. The Albion-Scipio field, which is the largest active field in the Michigan basin, produces from a postulated fault system more than 56 km long that extends vertically through these groups. Dolomite has replaced most of the limestone within the fractures in this field, with white dolomite crystals lining some vugs and fractures. Unaltered Trenton and Black River Group limestones, "tight" dolomite, and the Utica Shale act as lateral and vertical seals. Core measurements in the Albion-Scipio field show both intercrystalline porosity and vugular porosity. Intercrystalline porosity can range from 0.3 to 27.6 percent, and permeability from 0.1 to 8,000 md (Buehner and Davis, 1968). Some of the vugular porosity is cavernous, as was demonstrated when a drill bit dropped 19 m during drilling. The Trenton and Black River Groups in Michigan have produced more than 127 mbbls of oil and more than 235 bcf of gas since their initial discovery in 1920.

The carbonate bank that forms the huge Lima-Indiana field (Fig. 6) can reach 91 m in thickness (Gray, 1983). The upper 15 m has been redolomitized and forms the reservoir rocks of the field. Porosity is vugular and ranges from 0 to 15 percent, and the permeability ranges from 0.10 to 10 md (Gray, 1983). The hydrocarbons are trapped stratigraphically by porosity changes between limestone and dolomite. They are also trapped structurally by the Findlay arch and, in the northwest-southeast trends, by dolomization of faults such as the Bowling Green fault. Although in decline, there is still production from the Trenton fields of Ohio and Indiana. The Ohio data were unavailable, but cumulative Trenton production from the Indiana portion of the Michigan basin was 106 mbbls of oil through 1984 (Carpenter and Keller, 1984).

Trenton production in the Canadian section of the basin is predominantly from dolomitized fault zones. Cumulative oil production from "prior to 1900" (Ontario Geological Survey, 1980) to 1980 is 925,000 bbls of oil with gas production at 13 bcf.

Recent discovery and development of the Stoney Point field, east of and much like the Albion-Scipio field, has sparked new interest in the Trenton and Black River Groups as a producing horizon. Active exploration is primarily limited to the southern and southeastern part of the basin. However, given the fractured basement and the continued deep exploration, it is not unreasonable to assume that more producing targets in the Trenton and Black River Groups can be found in other parts of the basin.

H. A. Illich (personal communication, 1985) believes the thick Utica Shale, which caps the Trenton, is the source for all the hydrocarbons in the Ordovician System. Moyer (1982) does not believe this likely, as his studies show that the unit has low total organic carbon content. However, he feels there may have been some hydrocarbon generated from the Utica Shale in the southeastern or southwestern areas of the basin. Within the Black River Group is a member informally termed the "Black River Shale"; this may be a source rock, but no studies were completed on the unit.

SILURIAN

The Silurian System is the most prolific producer in the Michigan basin. Within the basin, the system produces from five different horizons; the Clinton Shale and Brown Niagaran units of the Niagara Group, the A1 and A2 Carbonates, and the E Unit of the Salina Group (Fig. 2). The most prolific producing horizon is the upper member of the Middle Silurian Niagara Group, the Brown Niagaran.

The first producing Silurian Age well in the Michigan basin was drilled in 1889 in Essex County, Ontario. The producing horizon was a dolomitized bioherm within the A1 Carbonate. Discoveries from Silurian reservoirs continued in Ontario, and in 1927 production spread to southern Michigan. In 1952, oil was found in reef structures in the northern part of the Lower Peninsula, although significant hydrocarbon production from Silurian Age reservoirs did not take place until the late 1950s in the southern basin and until 1970 in the northern basin. The hydrocarbons are trapped in numerous reef structures that occupy a band encircling the central basin and that may extend under Lakes Huron and Michigan (Fig. 7). Tall pinnacle reefs occupy the Northern Trend; the southern area structures are shorter and broader. More than 780 productive reefs have been found in Michigan, and approximately 100 in southwest Ontario (Petroleum Frontiers, 1984). They range in size from 50 to 200 acres and have individual reserves of about 0.5 mbbls of oil and 4.5 bcf gas (Michigan Geological Survey, unpublished data).

Porosity and permeability vary widely between reefs, as they depend on the presence of salt plugging and the amount of dolomitization of the reef. Salt plugging is more likely to be basinward in the Northern Trend, but its occurrence is still poorly understood. The dolomitized reefs are more porous, and although dolomitization generally increases toward the shelf bank, porosity development remains unpredictable.

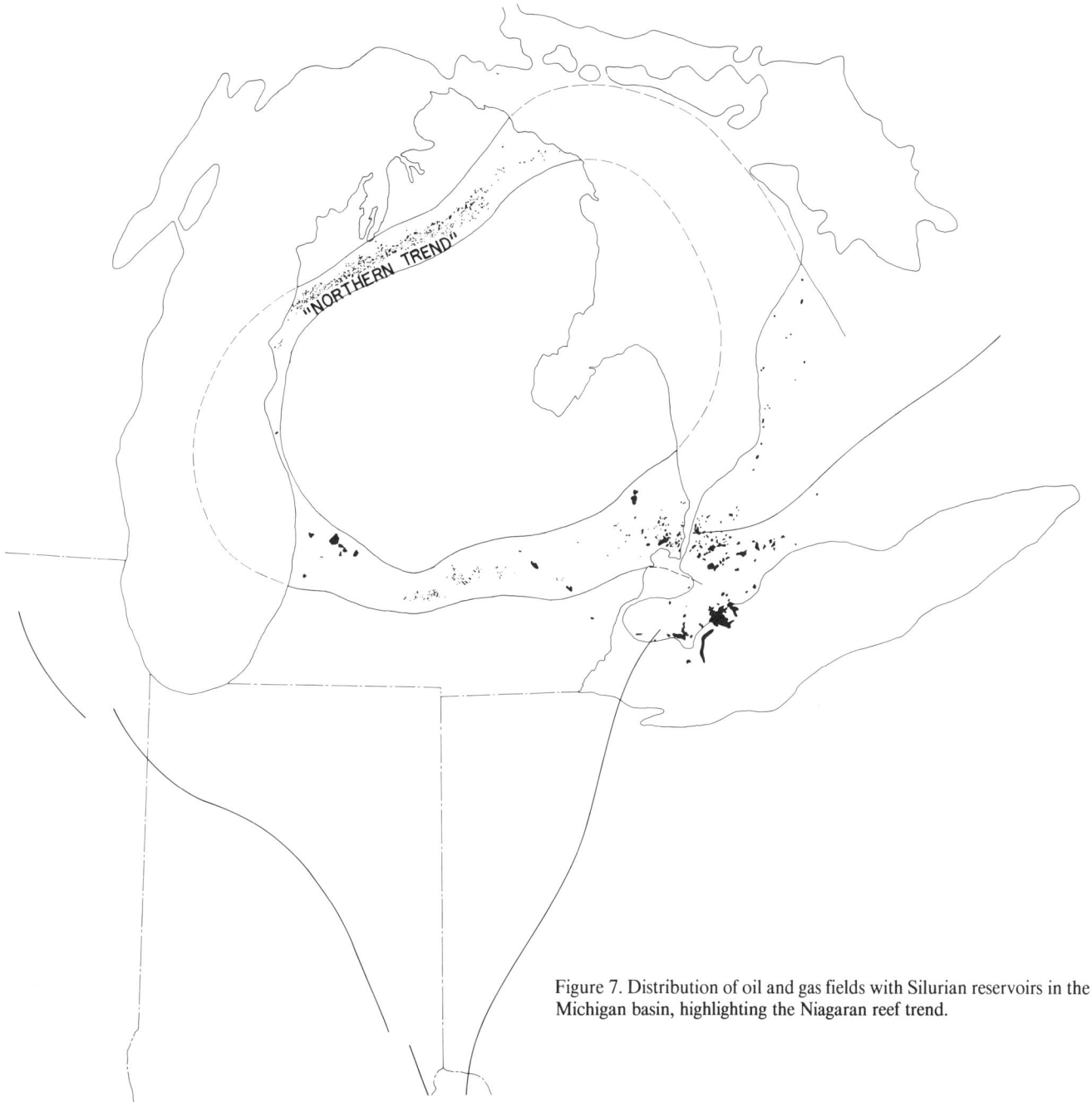

Figure 7. Distribution of oil and gas fields with Silurian reservoirs in the Michigan basin, highlighting the Niagaran reef trend.

Most authors (e.g., Illich and Grizzle, 1985; Vogler and others, 1981; Pruitt, 1983) agree that Silurian oils appear distinct from all other oils in the basin. This argues for a hydrocarbon source that is different from the others in the basin. Gardner and Bray (1984) believe that the source for the oil in the Brown Niagaran is most likely a series of carbonaceous muds in the A1 Carbonate overlying the Niagara Group. Regionally there is a layer of salt and anhydrite, the A1 Evaporite, between the A1 Carbonate and the Niagara Group. In the area of a Niagaran reef the evaporites of the A1 surround the reef and spread up its sides but do not cover it (Fig. 8). Therefore, the A1 Carbonate can be in direct contact with the Brown Niagaran of the reef itself.

Gardner and Bray (1984) also believe that the algal mats of the Brown Niagaran itself may act as a secondary source for the hydrocarbons in the Silurian System. Michigan's cumulative production from all sources in the Silurian System, but predominantly from the Niagara Group, is 290 mbbls oil and 1.7 tcf gas. Canadian production from the Michigan Basin, through 1982, totaled 10 mbbls of oil and 540 bcf of gas.

The possibility that any reefs discovered in the fairway beneath Lakes Huron and Michigan may contain oil presents the real possibility of oil spills in the Great Lakes. Because of this danger it is presently illegal to drill in the waters of Lake Michigan and those waters that are under U.S. jurisdiction in Lake

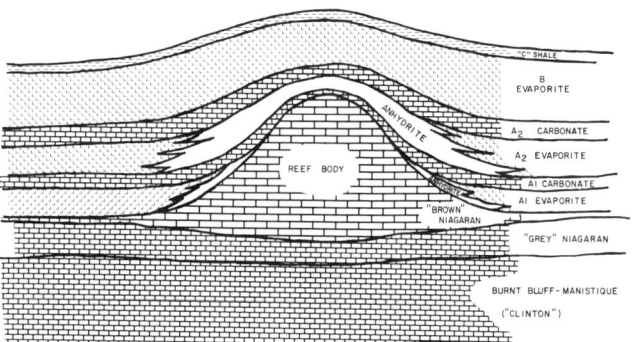

Figure 8. Generalized cross section of a Silurian pinnacle reef in the "Northern Trend."

Huron. Canada has drilled for Silurian Age hydrocarbons in the reef trend beneath Lake Erie since 1936, but only gas reservoirs may be produced.

In Michigan, "Clinton" is an informal term for the basal member of the Niagara Group that is underlain by shale members of the Cataract Group (Fig. 2). In southern Michigan the Clinton is a green, silty shale. Basinward, and north to its outcrop in the upper peninsula, the unit undergoes a facies change to a thick carbonate series. In the carbonate facies there are two distinct, mappable, lithologic units: the upper Manistique Group and the lower Burnt Bluff Group. The change is gradational, with the carbonates becoming more argillaceous and clastic to the south. As this unit shows promise as a producing horizon, studies are now being undertaken to determine the boundary between the shale and carbonate facies and the area where the Manistique and Burnt Bluff Groups first become distinct.

Hydrocarbon production from the Clinton is limited to the central and north-central parts of the basin where the carbonates are fairly porous; in the southern basin, where the same unit is a shale, it is nonproductive. Accumulation and entrapment are due to porosity pinchout along regional dip, structural highs, and in at least one case what is believed to be a reef.

Although it is not expected to approach the reef trend in magnitude, the Clinton has considerable potential. Exploration outside the central basin, wherever the unit is predominantly a carbonate, could prove productive. The same can also be said for the other producing horizons in the Silurian System. The A1 Carbonate has hydrocarbon-bearing zones in many areas of the Michigan Basin, but E Unit and A2 Carbonate production are usually targets only in the western and southwestern basin areas.

DEVONIAN

The Devonian System has several producing zones from Middle Devonian horizons. These include the Detroit River Group (Lucas Formation of Canada), the Dundee Limestone (Muscatatuck Group of Indiana), and the Traverse Group (Fig. 2). In Michigan, the Detroit River Group is a thick sequence of dolomite, limestone, and evaporites that produces sour crude from one zone, called the Sour Zone, and sweet crude from another zone, the Richfield Zone.

The Richfield Zone is composed of alternating layers of dolomite and anhydrite with a unit thickness of about 60 m. It underlies the massive anhydrite bed at the base of the Sour Zone and overlies the Amherstburg Formation, a black coralline limestone. Porosities of the Richfield average 16 percent, and permeabilities range from 4 md to 60 md (Upp, Jr., 1968; Sutton, 1968; Fugate, 1968).

The Sour Zone is 30 m to 45 m thick, and the pay horizons are thin dolomite stringers intercalated with limestone and anhydrite. The entire unit is usually overlain by the lowest massive salt in the Detroit River Group and underlain by a massive anhydrite bed. The Sour Zone is highly porous, averaging about 18 percent, but low in permeability, about 15 md (Upp, Jr., 1968; Sutton, 1968; Fugate, 1968). In both these units accumulation is generally attributed to anticlinal trends.

The Detroit River Group is found throughout most of the basin, but production has only been from the central basin area and through the Chatham sag in the southeast area of the basin (Figs. 1, 9). Production from the Detroit River Group has amounted to 86 mbbls of oil and 89 bcf of gas in Michigan since 1938 and 9 mbbls of oil from Ontario since 1898.

The Dundee Limestone directly overlies the Detroit River Group and is predominantly buff colored and fine to coarsely crystalline. In some areas, thick beds of dolomite and anhydrite are found in the Dundee Limestone, with the dolomite fraction increasing to the west and southwest where the unit can be entirely dolomitic. The Dundee Limestone is found over much of the basin, but is is most productive in the central basin where the formation is a combination of limestone, dolomite, and anhydrite beds. Hydrocarbons accumulate on the flanks and crests of a series of northwest trending anticlinal structures. In the Ontario part of the basin, primary Devonian Age production is from the Dundee from dolomitized zones in domal structures (Ontario Geological Survey, 1980).

The thickness of the Dundee Limestone ranges from 9 to 175 m, with the possibility of the productive zones being located in any level of the unit. In Michigan the average porosity is approximately 11 percent, and the average permeability is 18 millidarcies (Lundy, 1968; Upp, 1968). Lenticular zones of well-developed secondary porosity are the best productive horizons, but porosity can also be intergranular, vugular, fracture related, and primary (Champion, 1968).

Historically the Dundee Limestone was the largest and most productive play in the Michigan Basin. Since 1884, 349 mbbls of oil and 42 bcf of gas have been produced in Michigan, and 30 mbbls of oil in Ontario. The Dundee Limestone is still a major producer but is now exceeded by production from the Middle Silurian Brown Niagaran.

The Traverse Group is a thick sequence of alternating dark grey shales and brown limestones, with the inclusion of dolomite and anhydrite beds in the western part of the basin. The pay is

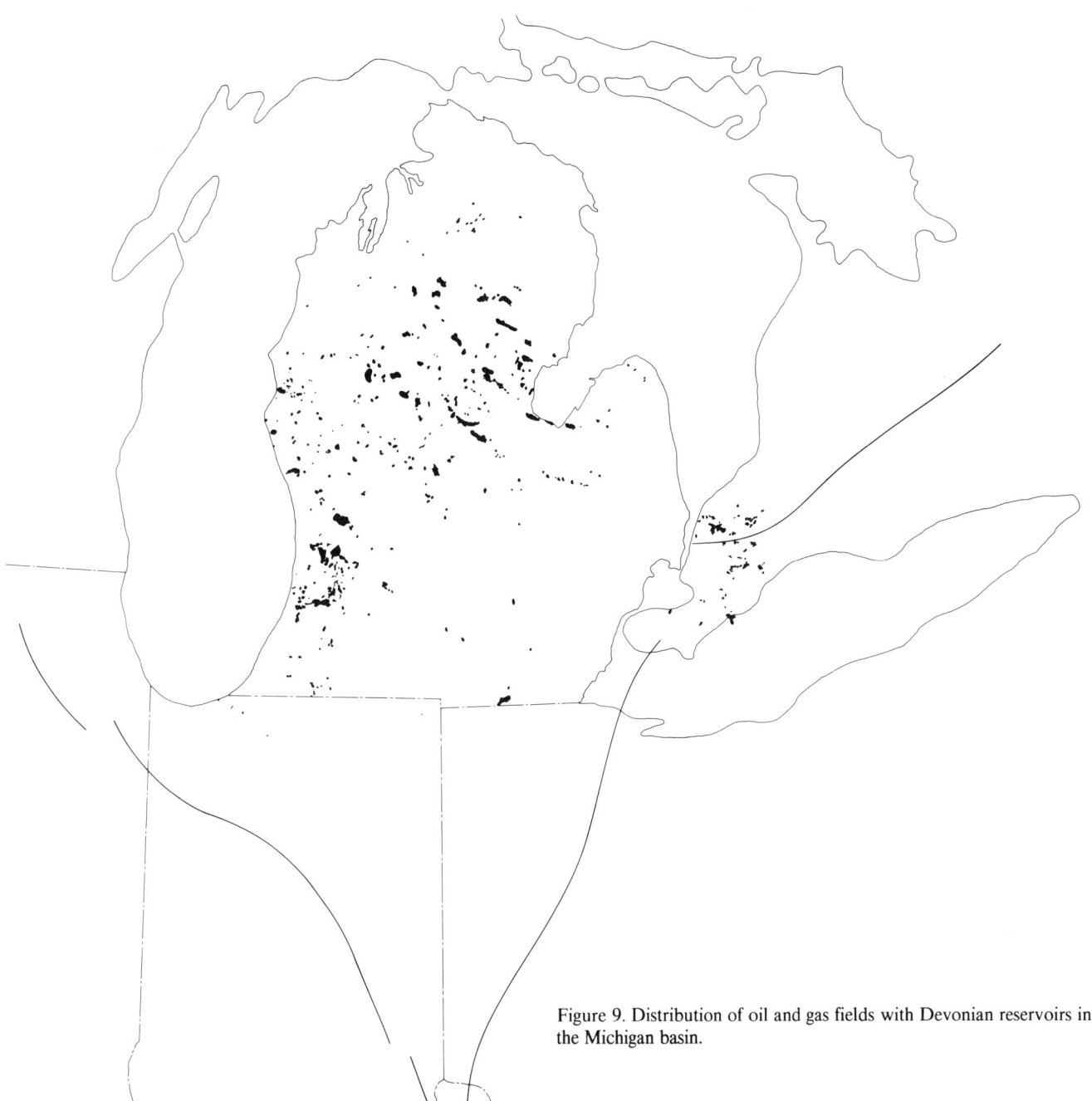

Figure 9. Distribution of oil and gas fields with Devonian reservoirs in the Michigan basin.

from several thin zones, each averaging somewhat over 1 m in thickness, collectively called the Traverse Limestone. They form the upper part of the Traverse Group. Most production is from fields in the west and southwest part of Michigan (Fig. 9), where the lithology is more than 80 percent carbonate (Petroleum Frontiers, 1984). Reefs are also observed and may act as reservoirs, although they are not as numerous as the Niagaran Reefs. In the anticlinal structures of eastern and central Michigan, porosity pinchouts and structural traps are common.

Another form of trap is formed by salt collapse. In the southwestern part of Michigan is a structurally positive feature called the Allegan platform. This structure may have been a shelf margin that created a depositional edge for Salina Group salts. Local leaching of these salts created solution caves, which cased the collapse of overlying formations. Masses adjacent to the collapsed areas remain elevated and formed salt-cored hills or domes (Ells, 1967).

Porosity in the Traverse Limestone is both intercrystalline and vugular and measures 9 to 10 percent, with a permeability maximum of about 100 md (Petroleum Frontiers, 1984). As with the other Devonian producing horizons, the Traverse Group can be found in most places in the lower peninsula of Michigan, but production is from the western, southwestern, and eastern areas of Michigan. The first producing well drilled in the Traverse Group was in 1927; since that time the unit has produced 109 mbbls of oil and 11 bcf of gas.

The oils of the Dundee Limestone appear to have as their source the rocks of the Ordovician System (Illich and Grizzle, 1985; Nunn and others, 1984; Pruitt, 1983; Vogler and others, 1981). However, the source for the oil in the Traverse and Detroit River Groups is disputed. Illich and Grizzle (1985) believe that the Devonian oils, other than those produced from the Dundee Limestone, have their source in the Devonian System itself.

Pruitt (1983) partially agrees but argues for a mix of Silurian and Ordovician oils to produce the oil in the Detroit River Group; Vogler and others (1981) believe this same mix produces the Traverse Group oil. They further suggest an entirely separate, unspecified source for the Detroit River oil. Nunn and others (1984) conclude that all the Devonian oil is generated from Ordovician System source rocks.

MISSISSIPPIAN

The Mississippian System has two major producing units, the Berea Sandstone and the "Stray" Sandstone of the Michigan

Figure 10. Distribution of oil and gas fields with Mississippian reservoirs in the Michigan basin.

Formation. In central and eastern Michigan, the Berea Sandstone is overlain by the Sunbury Shale. The Berea Sandstone is a light grey, fine grained, dolomitic, silty, shaley, pyritic sandstone where it can be distinguished from the underlying Bedford Shale. In the west and northwest the Berea becomes increasingly dolomitic, and although it occupies the same stratigraphic position as the eastern Berea, it is intercalated with the upper portion of the Ellsworth Shale.

The Berea Sandstone covers much of Michigan, but the hydrocarbon-bearing zones are limited to central and western areas of the state (Fig. 10). The reservoirs are primarily related to structural closures in the central basin and to porosity and permeability traps in the western basin. The Berea Sandstone has produced 3.5 mbbls of oil and 13 bcf of gas since 1925.

The source for the Berea oil has not been defined with any greater clarity than the sources for other producing horizons. Illich and Grizzle (1985) maintain the source is Devonian or younger.

The "Stray" Sandstone is an informal term for the basal member of the Michigan Formation. The youngest and shallowest major producing formation in the basin, it is primarily a white, fine-grained, well sorted sandstone underlain by the lithologically similar Marshall Sandstone. In the past the Stray Sandstone was an important producer, with 93,000 bbls of oil and 214 bcf of gas produced since 1928. Producing fields are located on structural highs and are limited to the central basin. Gypsum and anhydrite beds of the overlying Michigan Formation act as cap rocks, and its shales act as the source rocks for the hydrocarbon in the Stray (Moyer, 1982). The Mississippian System is no longer a major producer when compared with deeper formations. However, recent discovery of the Williams field in the central basin shows that exploration is continuing for shallow producing formations and that new fields can be found and developed in these shallower horizons.

FUTURE POTENTIAL

The Michigan basin has undergone exploration and development of its petroleum resources for more than 100 years, and for this reason the basin is generally classified as mature. However, this classification suggests a feeling of completed development, a designation that should not be applied to this basin. In a purely statistical sense the development shows no sign of declining. Drilling in Michigan has steadily increased from a low of 379 permits in 1969 to a level at or above the thousand permit mark in recent years. Interest in the shallower horizons fluctuates with the discovery of new and deeper fields, such as the Middle Silurian Niagaran reefs, but the development of existing fields continues.

Interest is increasing as the deeper horizons within the basin are explored. Devonian wildcats are still being drilled, and some of the large fields are being further developed. Silurian targets have generated the greatest amount of drilling since 1970. Most Silurian exploration is for Niagaran reefs, but the new Clinton target has evoked much interest. The other producing horizons in the Silurian System continue to be developed in the regions where they have proven successful in the past. Exploration for these targets in other areas, not previously successful, and the testing of untried horizons, as with the Clinton, indicate a continued interest in the Silurian System.

The deeper Ordovician targets represent what might be the greatest promise for new frontiers. Exploration of these horizons is increasing because of the relative shallowness of the gas when compared with many other "deep gas" targets outside the basin. The discoveries of the Stoney Point field, much like the giant Albion-Scipio in age, structure, and lithology, and the proliferating Prairie du Chien discoveries indicate a bright future for deep basin plays. Although the Michigan basin may be considered mature, it still exhibits continued development and rejuvenation.

REFERENCES

Bricker, D. M., Milstein, R. L., and Reszka, C. R., 1983, Selected studies of Cambro-Ordovician sediments within the Michigan Basin: Michigan Geological Survey, R.I. 26, p. 9–27.

Brown, L., Jensen, L., Oliver, J., Kaufman, S., and Steiner, D., 1982, Rift structure beneath the Michigan Basin from COCORP profiling: Geology, v. 10, p. 645–689.

Buehner, J. H., and Davis, S. H., Jr., 1968, Albion-Pulaski-Scipio trend field, in Proceedings, Michigan Basin Geological Society Oil and Gas Symposium: Michigan Basin Geological Society, p. 37–49.

Carpenter, G. L., and Keller, S. J., 1984, Oil development and production in Indiana during 1984: Indiana Department of Natural Resources, Geological Survey Mineral Economics Series 31, p. 17–39.

Carpenter, G. L., and Sullivan, D. M., 1976, Map showing oil, gas, and gas storage fields in Indiana: Indiana Geological Survey, scale 1:500,000.

Catacosinos, P. A., 1973, Cambrian lithostratigraphy of the Michigan Basin: American Association of Petroleum Geologists Bulletin, v. 57, p. 2404–2418.

Cercone, K. R., 1984, Thermal history of Michigan Basin: American Association of Petroleum Geologists Bulletin, v. 68, no. 2, p. 130–136.

Champion, B. L., 1968, Oil and gas producing zones in Michigan, in Proceedings, Michigan Basin Geological Society Oil and Gas Symposium: Michigan Basin Geological Society, p. 17–36.

Chase, C. G., and Gilmer, T. H., 1973, Precambrian plate tectonics, The Midcontinent gravity high: Earth and Planetary Science Letters, v. 21, p. 70–78.

Cohee, G. V., and Landes, K. K., 1955, Oil in the Michigan Basin, in Weeks, L. G., ed., Habitat of oil, American Association of Petroleum Geologists, p. 473–493.

Craddock, C., 1972, Late Precambrian regional geological setting, in Geology of Minnesota; A centennial volume: Minnesota Geological Survey, p. 281–291.

Ells, G. D., 1967, Michigan's Silurian oil and gas pools: Michigan Department of Natural Resources, Geological Survey Division, R.I. 2, p. 24–29.

——, 1969, Architecture of the Michigan Basin, in Studies of the Precambrian of the Michigan Basin, Michigan Basin Geological Society Annual Field Excursion: Michigan Basin Geological Society, p. 64.

Fisher, J. H., and Barratt, M. W., 1985, Exploration in Ordovician of central Michigan Basin: American Association of Petroleum Geologists Bulletin, v. 69, no. 12, p. 2065–2076.

Fisher, J. H., Barratt, M. W., Droste, J. B., and Shaver, R. H., 1988, Michigan Basin, in Sloss, L. S., Vail, P. R., and Mankin, C. J., eds., Sedimentary cover of the craton; U.S.: Boulder, Colorado, Geological Society of America, Geology of North America, v. D-2, p. 361–382.

Fugate, R. I., 1968, East Norwich Field, in Proceedings, Michigan Basin Geological Society Oil and Gas Symposium: Michigan Basin Geological Society, p. 69–87.

Gardner, W. C., and Bray, E. E., 1984, Oil and source rocks of Niagaran reefs (Silurian) in the Michigan Basin, in Petroleum geochemistry and source rock potential of carbonate rocks: American Association of Petroleum Geologists Studies in Geology 18, p. 33–44.

Gray, J. D., 1983, Oil and gas trends in Ohio: Ohio Division of Geological Survey, p. 2–6.

Haxby, W. F., Turcotte, D. L., and Bird, J. M., 1976, Thermal and mechanical evolution of the Michigan Basin: Tectonophysics, v. 36, p. 57–75.

Hinze, W. J., Kellogg, R. L., and O'Hara, N. W., 1975, Geophysical studies of basement geology of southern peninsula of Michigan: American Association of Petroleum Geologists Bulletin, v. 59, no. 9, p. 1562–1584.

Illich, H. A., and Grizzle, P. L., 1985, Discussion of Thermal subsidence and generation of hydrocarbons in Michigan Basin: American Association of Petroleum Geologists Bulletin, v. 69, no. 9, p. 1401–1404.

Kingston, D. R., Dishroon, C. P., and Williams, P. A., 1983, Global basin classification system: American Association of Petroleum Geologists Bulletin, v. 67, p. 2177–2178.

Lundy, C. L., 1968, Deep River field, in Proceedings, Michigan Basin Geological Society Oil and Gas Symposium: Michigan Basin Geological Society, p. 61–69.

Michigan Department of Natural Resources, 1964, Stratigraphic succession in Michigan: Geological Survey Division, Chart 1.

——, 1982, Michigan oil and gas fields, 1982: Geological Survey Division map, scale 1:380,000.

Moyer, R. B., 1982, Thermal maturity and organic content of selected Paleozoic formations; Michigan Basin [M.S. thesis]: Lansing, Michigan State University, 61 p.

Newcombe, R. B., 1933, Oil and gas fields of Michigan [Ph.D. thesis]: Ann Arbor, University of Michigan, 103 p.

Nunn, J. A., Sleep, N. H., and Moore, W. E., 1984, Thermal subsidence and generation of hydrocarbons in Michigan Basin: American Association of Petroleum Geologists Bulletin, v. 68, no. 3, p. 296–315.

Ohio Division of Geological Survey, 1974, Oil and gas fields of Ohio: map, scale 1:500,000.

Ontario Geological Survey, 1980, Oil and gas pools and pipelines of southwestern Ontario: Petroleum Resources Map P-2499, scale 1:250,000.

Oray, E., Hinze, W. J., and O'Hara, N. W., 1973, Gravity and magnetic evidence for the eastern termination of the Lake Superior syncline: Geological Society of America Bulletin, v. 84, p. 2763–2780.

Petroleum Frontiers, 1984, Michigan Basin; Expanding the deep frontier: v. 1, no. 3, p. 35–36.

Prouty, C. E., 1983, The tectonic development of the Michigan Basin intrastructures, in Tectonics, structure, and karst in northern lower Michigan, Michigan Basin Geological Society 1983 Field Conference: Michigan Basin Geological Society, p. 36–81.

Pruitt, J. D., 1983, Comment on Comparison of Michigan Basin crude oils: Geochimica et Cosmochimica Acta, v. 47, p. 1157–1159.

St. John, W., Bally, A. W., and Klemme, H. D., 1984, Sedimentary provinces of the world; Hydrocarbon productive and nonproductive: American Association of Petroleum Geologists, 1 map, scale 1:31,368,000 at equator, 35 p.

Shaver, R. H., coordinator, 1985, Midwestern basin and arches region: Correlation of Stratigraphic Units of North America project (COSUNA), American Association of Petroleum Geologists, Correlation chart MBA.

Sleep, N. H., Nunn, J. A., and Chou, L., 1980, Platform basins: Annual Review of Earth and Planetary Sciences, v. 8, p. 17–34.

Sloss, L. L., 1982, The Michigan Basin: University of Missouri–Rolla Journal, no. 3, p. 25–29.

Stonehouse, H. B., compiler, 1969, Studies of the Precambrian of the Michigan Basin: Michigan Geological Society Annual Field Excursion, inside front cover map.

Sutton, D. G., 1968, Headquarters field, in Proceedings, Michigan Basin Geological Society Oil and Gas Symposium: Michigan Basin Geological Society, p. 99–115.

Syrjamaki, R. M., 1977, The Prairie du Chien Group of the Michigan Basin [M.S. thesis]: Lansing, Michigan State University, 140 p.

Upp, J. E., Jr., 1968, Reed City field, in Proceedings, Michigan Basin Geological Society Oil and Gas Symposium: Michigan Basin Geological Society, p. 149–161.

Vogler, E. A., Meyers, P. A., and Moore, W. A., 1981, Comparison of Michigan Basin crude oils: Geochimica et Cosmochimica Acta, v. 45, p. 2287–2293.

Waples, D. W., 1980, Time and temperature in petroleum formation; Application of Lopatin's method to petroleum exploration: American Association of Petroleum Geologists Bulletin, v. 64, no. 6, p. 916–926.

Manuscript Accepted by the Society May 16, 1987

NOTE ADDED IN PROOF

When this manuscript was accepted, the best statistical information available was for 1984. Since then, two producing formations have increased dramatically in importance: production rates of the Middle Silurian Niagaran reefs have been almost completely replaced by those of the Upper Devonian Antrim Formation and the Lower Ordovician Prairie du Chien Group.

Wells drilled to the Antrim Formation presently account for more than 70 percent of the total drilling in Michigan. The Antrim occurs at shallow depths of approximately 457 m and consists of a dark gray to black, hard, thin-bedded, carbonaceous shale interbedded with some gray shales. It is found throughout the Michigan basin, but drilling activity is currently confined to Otsego and Antrim Counties, Michigan, in the northern part of the basin. The Antrim wells produce gas and have a success rate that is currently about 100 percent.

Most wells are perforated near the base of the Antrim Formation where it is most carbonaceous. The interval must be fractured to initiate production. Average production for wells in the current play is 60,000 ft^3/day and 80 barrels of water per day, and they will produce at nearly that rate for many years.

The Prairie du Chien Group has also become increasingly important in the Michigan basin. Since the central basin deep play began in 1981, 62 new discoveries have been made. Most were classified as deeper pool tests, having been drilled through shallower productive zones. In the Michigan basin, 1,764 wells penetrate the Prairie du Chien Group, including 543 in the deep play of the central basin. According to 1989 statistics, the exploratory success rate was 56 percent for the Prairie du Chien Group play, whereas development-well drilling had a 61 percent success rate. There is minor oil production in the eastern part of the basin, but it is predominantly a gas play; approximately 28 percent of the deep play wells are producers, and more than 90 percent of them are gas wells.

Chapter 19

The northern Gulf of Mexico basin

D. M. Curtis*
800 Anderson, Belaire, Texas 77401

INTRODUCTION

Location and geologic setting

The United States Gulf Coast hydrocarbon province includes the states that border the Gulf of Mexico, extending inland to the Cretaceous outcrops that overlie the Paleozoic fold belts (Fig. 1). The northern Gulf basin (NGB) province is the southern passive margin of North America, characterized since Triassic time by extensional and gravitational tectonics. It is underlain by Paleozoic continental crust approximately to the present continental slope (Buffler and Sawyer, 1985).

An early rifting stage took place in Late Triassic to mid-Jurassic time (Salvador, 1987) along the trend of the Ouachita fold belt, which became the landward margin of the subsiding basin. A broad continental platform developed during mid-Jurassic to Late Cretaceous subsidence. Increased subsidence and progradation followed, from Late Cretaceous to Holocene time. The hydrocarbon-bearing sequences of the province range in age from Late Jurassic to Quaternary.

Rocks of the province include Mesozoic and Cenozoic terrigenous clastics, carbonates, evaporites, and minor igneous intrusives and extrusives. Carbonates predominate in the Mesozoic. Platform limestones and shelf-edge carbonate buildups are prominent in Jurassic and Lower Cretaceous sequences. A composite Early Cretaceous reef trend defines the Mesozoic shelf margin. Upper Cretaceous sequences are a mixture of carbonate and terrigenous clastic facies. Cenozoic sequences are almost exclusively terrigenous clastics except in Florida where platform limestones characterize the entire section. Jurassic salt and Cenozoic low-density shale are responsible for the gravity tectonics that affect the entire province (Salvador, 1988).

The NGB strongly resembles many other basins developed on Mesozoic-Cenozoic passive margins, including the Niger Delta (Weber, 1986) and the Beaufort-Mackenzie Delta provinces (Curtis, 1986), and the Mahakem Delta (Durand and others, 1986). The similarities (except for the salt) include geologic history, regional and local gravity tectonics, abnormal formation pressures, age, and lithologies, as well as hydrocarbon reservoirs and traps.

*Deceased.

Exploration history and techniques

Although there had been shallow production from Cretaceous rocks in east Texas as early as 1895, the Gulf Coast petroleum story really began with the discovery in 1901 of oil in salt-dome caprock at Spindletop, in southeast Texas near the Texas-Louisiana border (Owen, 1975). The dome has slight but perceptible surface expression. This feature, along with gas seeps, led to the discovery and stimulated an intensive search for topographic expressions of domal structures. Within a year, Texas' daily production had risen from 2,300 barrels to more than 50,000 barrels.

Exploration techniques for the next two decades included surface and subsurface geologic mapping and "trendology." Gravity exploration (torsion balance and other gravimetric instruments) was first successful in locating a productive salt dome in 1924. By 1926, large parts of southeast Texas and southwest Louisiana had been explored by refraction seismograph. The first salt-dome discovery attributable to reflection seismology was recorded in 1929. Surface, near-surface (core hole), and subsurface mapping were responsible for many discoveries during the 1920s and 1930s. The discovery of the East Texas field (a giant stratigraphic trap, the largest field in the conterminous United States) in 1929 resulted from surface and near-surface mapping (and intuition?). Many fields discovered during that era are attributed to random drilling. By the late 1930s, emphasis had shifted to the use of the reflection seismograph and subsurface mapping, aided by biostratigraphy and wire-line logging. Since the 1950s, exploration offshore and for deep targets onshore has depended on reflection seismology, in which many advances in technology and processing have increased exploration success, and on subsurface mapping aided by biostratigraphy, which is invaluable for chronostratigraphic correlation and environmental interpretation and in the interpretation of seismic stratigraphic sequences.

STRUCTURE

The NGB is a gently dipping homocline within which are several subregional positive and negative elements that have

Curtis, D. M., 1991, The northern Gulf of Mexico basin, *in* Gluskoter, H. J., Rice, D. D., and Taylor, R. B., eds., Economic Geology, U.S.: Boulder, Colorado, Geological Society of America, The Geology of North America, v. P-2.

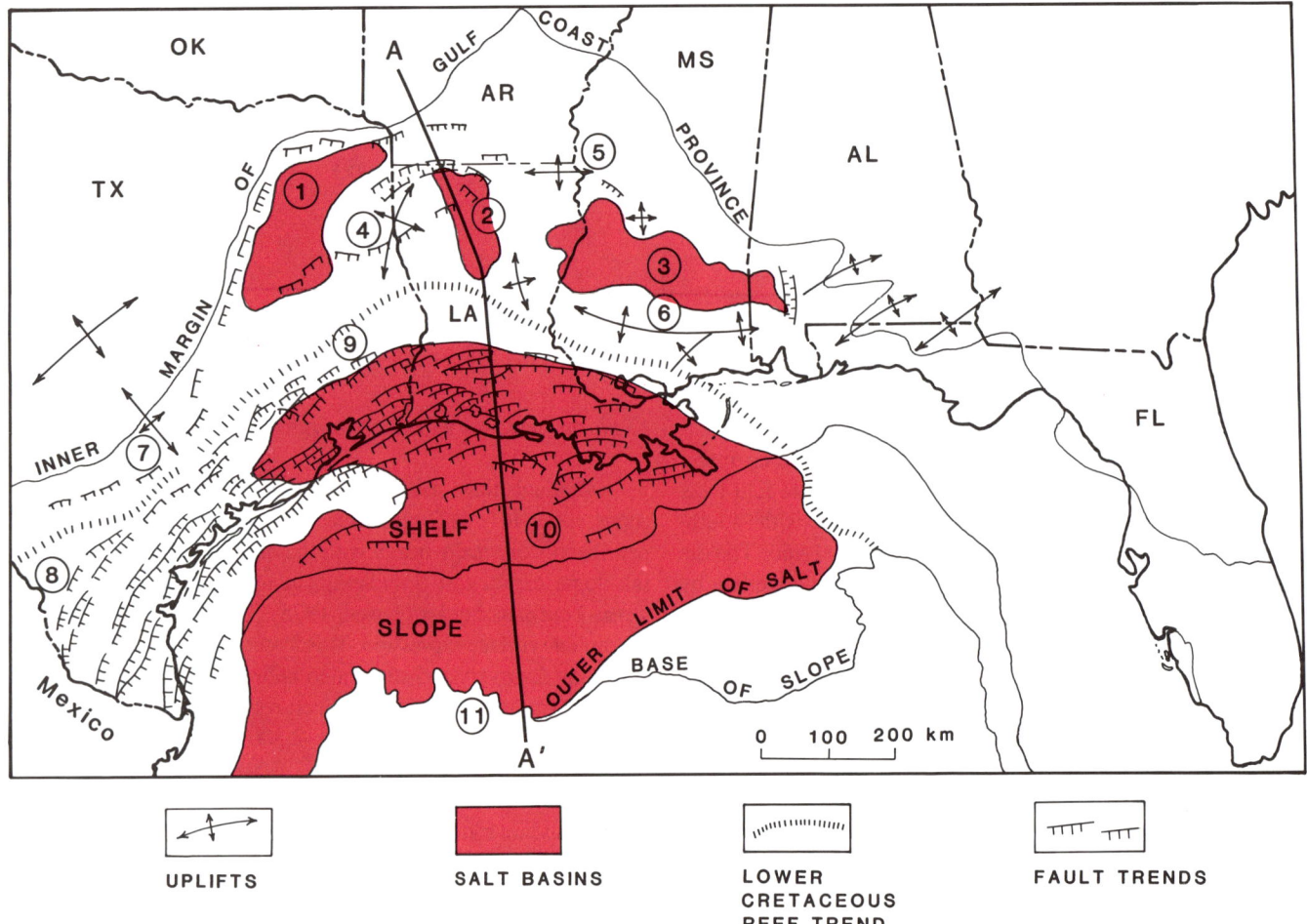

Figure 1. Index map of northern Gulf basin province. A–A′, line of regional cross section (Plate 7D); 1, East Texas salt basin; 2, North Louisiana salt basin; 3, Mississippi salt basin; 4, Sabine uplift; 5, Monroe uplift; 6, Wiggins anticline; 7, San Marcos arch; 8, Rio Grande embayment; 9, Houston embayment; 10, Gulf Coast salt basin (including continental slope salt basin); 11, Sigsbee Deep.

maintained their character through time (Murray, 1961; Martin, 1978) (Plate 7D). Along the northern edge of the province, graben structures represent the early rifting stage of the developing subsiding passive margin. Along a zone that is more or less coincident with the graben trend, a complex arcuate system of slightly younger normal faults bounds the northern rim of the continuously subsiding basin and determines the structural and depositional strike from south Texas along the Ouachita fold belt trend, across east Texas and north Louisiana, and southeastward across Mississippi and Alabama into the Gulf. Many of the faults in the basin-rim trends have grown intermittently and are synsedimentary, at least in part (Bishop, 1973).

Between the updip boundary faults and the Early Cretaceous shelf margin are major positive structures inherited from the Paleozoic: the San Marcos arch, Sabine uplift, Monroe uplift, and Wiggins arch. Between the uplifts are the "interior salt basins": the East Texas, North Louisiana, and Mississippi salt basins. They contain Mesozoic sediments and structures and are characterized by isolated salt diapirs and deep-seated salt anticlines (Seni and Jackson, 1983). Low-relief anticlines and normal faults are associated with the salt tectonics or with the updip boundary-fault systems. A few buried features that trend southwest are inherited from Appalachian trends in the eastern Gulf area.

Basinward from the Early Cretaceous shelf margin trend, extensional and gravity tectonics predominate in the Cenozoic sequences that have prograded some 400 km (250 mi) beyond the Cretaceous shelf in the rapidly subsiding basin. Areas of greater subsidence and sediment accumulation are the Rio Grande embayment, the Houston embayment in southeast Texas, and the Mississippi embayment. Contemporaneous faulting (syndepositional or growth faulting) and diapiric salt or shale structures characterize the region (Jackson and Galloway, 1984). As on similar passive margins, a low-density, overpressured, undercompacted marine shale facies that underlies the reservoir facies provides the mechanism for extensional and gravity tectonics in the progradational basin fill. This mobile substrate, and the older underlying salt layer, are intrinsic to the tectonics of the basin and provide a mechanism for the development of regional contemporaneous listric faults and diapiric structures (Bruce, 1973, 1984).

Growth faults (Curtis, 1970; Galloway, 1986; Shelton, 1984) originate at the shelf edge of each major depositional sequence, where vertical density imbalance results from sedimentary loading of sandstones that prograde over low-density shales, or where high pore pressures provide a mechanism for basinward gravitational sliding with a strong horizontal stress component. A fault is active during successive depositional sequences, so that downthrown intervals are thicker than upthrown. Displacement and growth ratio (downthrown: upthrown) are greatest in the oldest stratigraphic units and decrease upward. Some of the faults have surface expression, but most are buried. Counter-regional dip and "rollover" anticlines that result from "reverse drag" into the fault are common in downthrown blocks.

The growth faults occur in arcuate trends that are oriented roughly parallel with the Lower Cretaceous shelf margin and are progressively younger basinward. They are related in time and space to successive Cenozoic depocenters. These faults define the structural and depositional strike in the Cenozoic. Most of them are down-to-the-basin listric faults, with which may be associated up-to-the-basin (antithetic or counter-regional) normal faults. The listric faults flatten with depth and form a décollement surface in the overpressured shale or on the salt.

Salt underlies a large part of the province. The Gulf Coast salt basin occupies the area gulfward from the Early Cretaceous shelf margin under south Louisiana and southeast Texas, and under the continental shelf extending approximately to the present shelf edge. This area is characterized by diapiric salt structures and deep-seated, salt-related domal and anticlinal uplifts (Halbouty, 1979; Martin, 1978), as well as by growth-fault trends related to the gravity and extensional tectonics associated with the sedimentary progradation into the basin. The Texas-Louisiana slope salt basin, which extends from the present shelf edge to the Sigsbee Deep in the central Gulf, is characterized by broad, semi-continuous diapiric salt walls and ridges and associated shale masses (Humphris, 1979). Growth fault trends on the slope are more localized and more closely related to salt trends than on the shelf. South of the limits of the Gulf Coast salt basin, across the San Marcos arch, and in the Rio Grande embayment, growth faults and rollover anticlines, along with shale domes and anticlines, are the characteristic structures. A small salt basin is present in the Rio Grande embayment. In the eastern Gulf area, the DeSoto salt basin underlies the continental slope.

STRATIGRAPHY

Mesozoic and Cenozoic sediments rest unconformably on Precambrian and Paleozoic rocks (Vernon, 1971). Much of the stratigraphy is better developed in the subsurface than in outcrop. In general the strata are progressively younger gulfward. Overpressuring in deep sections is characteristic.

The *Mesozoic* rocks include both carbonates and terrigenous clastics (Fig. 2). The eastward increase in clastics, especially in the Lower Cretaceous strata, reflects an Appalachian provenance.

Eagle Mills (Upper Triassic–Lower Jurassic) red beds, 0 to 2 km (0 to ~7,000 ft) thick, which are intruded by basalt, fill rift basins in the Paleozoic "basement." Early Middle Jurassic time is represented by a post-rift unconformity, so that truncated Eagle Mills beds are overlain by upper Middle Jurassic to Upper Cretaceous rocks. Thick salt deposits (*Louann*), 0 to probably more than 3 km (10,000 ft) thick, and associated anhydrite (*Werner*) deposited during the late Middle Jurassic are overlain by Upper Jurassic *Norphlet* and *Smackover* Formations (Newkirk, 1971).

The *Norphlet* and *Smackover* Formations (Oxfordian) are extensively distributed in the subsurface in the northern part of the province (Moore, 1984; Presley, 1984). Together they reach a thickness of more than 600 m (>2,000 ft). The Norphlet is a basal clastic unit, predominantly nonmarine arkosic conglomerates and sandstones. The Smackover consists of dark calcareous sandstones and dense argillaceous limestones, gradationally overlain by oolitic and pelletal grainstones and packstones with sandstone interbeds.

Overlying the Smackover are sediments of Kimmeridgian age that are known as *Buckner* (anhydrite and red beds), *Haynesville* (limestone and sandstone), and *Bossier* (shale). Together they reach more than 750 m (>2,500 ft) in thickness).

Beds of Tithonian age (lower part of the *Cotton Valley* Group, *Schuler* Formation, and part of the *Bossier* Shale) form a thick progradational wedge of nonmarine and deltaic to marine terrigenous clastics that are widely distributed in the subsurface. They reach a thickness of more than 1 km (>4,000 ft) in north Louisiana.

The upper part of the *Cotton Valley* Group, earliest Cretaceous in age, includes nonmarine, deltaic, and near-shore sandstones and conglomerates, siltstones, and shales, that grade basinward and laterally to shelfal and basinal carbonates and shales. The strata thicken toward the shelf margin where their thickness reaches ~300 to ~600 m (~1,000 to ~2,000 ft).

A regional unconformity separates Cotton Valley strata from the overlying *Hosston* Formatin. *Hosston-Sligo* beds (Hauterivian–Barremian–early Aptian) up to 1 km (~3,500 ft) thick include nonmarine, deltaic, and near-shore sandstones and conglomerates with interbedded red shales in the Hosston, that grade upward and basinward into basinal shales and carbonates of the Sligo. At their updip edge, Hosston red beds overlap truncated Paleozoic rocks. Downdip, in north Louisiana, rudist bank facies form a prominent Hosston-Sligo outer shelf buildup that reaches more than 1 km (>3,500 ft) in thickness. The "Sligo reef trend" is the oldest of the Mesozoic shelf-margin carbonate buildups that are known in composite as the "Lower Cretaceous reef trend" (Bebout, 1974; McFarlan, 1977).

Rocks of Aptian to early Cenomanian age (Bebout and Loucks, 1977; McFarlan, 1988; Robertson Research, 1983) are known as *Trinity* (Aptian-Albian), *Fredericksburg* (late Albian), and *Washita* (early Cenomanian) groups. These strata crop out in

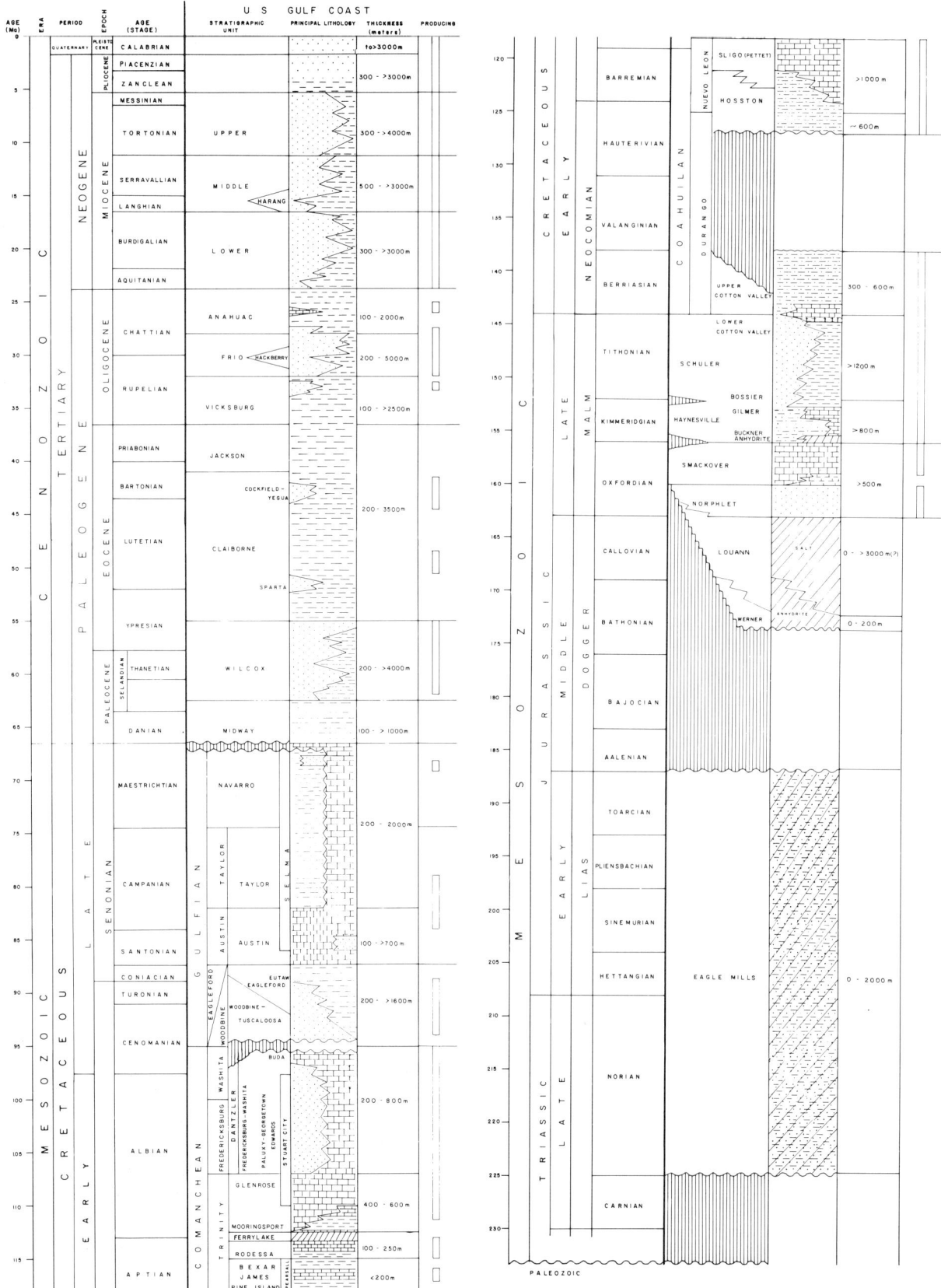

Figure 2. Columnar section. Terminology reflects Gulf Coast usage and may not conform with standard guidelines.

the northern Gulf area and are well known from the subsurface throughout the region.

A lower *Trinity* shale-limestone unit (*Pearsall*) is less than 200 m (<700 ft) thick. Sandstones increase eastward; shales increase gulfward. A clastic-carbonate-anhydrite sequence approximately 100 to 250 m (~400 to ~800 ft) thick includes *Rodessa* and *Ferry Lake* Formations (middle Trinity). Shallow marine carbonates predominate in south Texas. Clastics increase eastward and updip throughout the trend. From east Texas to Florida, the Ferry Lake, thin-bedded anhydrite and shale, forms a useful electric log correlation datum. A carbonate bank extends across the entire Gulf margin, and the shelf edge is rimmed with rudist limestone banks, part of the composite "Lower Cretaceous reef trend." In south Florida, Sunniland limestone is part of a carbonate-anhydrite sequence of equivalent age (Rainwater, 1971). Upper Trinity *Mooringsport–Glen Rose* strata as thick as 600 m (~2,000 ft), like the older sequences, include an updip terrigenous clastic facies that grades gulfward and westward to finer grained clastics and shelf carbonates. A shelf-edge rudist bank, sometimes referred to as the "Glen Rose reef trend," represents the start of the younger reef trend known in south Texas as the "Stuart City reef trend," which continued to develop through early Cenomanian time.

The *Paluxy* Formation (*Fredericksburg*) is a thick progradational deltaic wedge. Toward the shelf margin in south Texas the terrigenous clastics grade into shelf carbonates and discontinuous rudist banks. The terms *Edwards* reef, *Edwards* limestone, and *Edwards* Group are variously applied to this group of strata. The "Edwards reef" is partly synonymous with the late Albian part of the Stuart City.

Washita Group rocks in Texas (*Georgetown–Del Rio–Buda*) include carbonates and shales. The shelf-margin carbonates are part of the Stuart City reef trend. Terrigenous clastics increase eastward to become an undifferentiated deltaic sequence (Washita-Fredericksburg) known as the Dantzler Formation. Combined Washita-Fredericksburg thickness reaches about 1 km (>3,000 ft).

Upper Cretaceous (middle and upper Cenomanian to Maastrichtian) deposits (Holcomb, 1971; Sohl, 1988; Stehli and others, 1972) onlap a truncated mid-Cenomanian surface and unconformably overlap older Mesozoic beds and the boundary faults, to lie on Paleozoics. The development of this unconformity marks the change in depositional style from the broad stable-shelf deposition that characterized the Jurassic and Early Cretaceous, to the increasingly progradational filling of the more rapidly subsiding basin during the Tertiary. The Upper Cretaceous stratigraphy represents a transition between the two styles. The widespread Late Cretaceous transgression connected the Gulf of Mexico with the western interior seaway.

Woodbine-Eagleford-Tuscaloosa sandstones and shales and their equivalents are known from the San Marcos arch eastward to Florida. The sandstones onlap the mid-Cenomanian unconformity and grade laterally into dark marine shales, so that the shales lie unconformably on increasingly older beds. In south-central Louisiana, lower Tuscaloosa deltaic and turbidite sands have prograded across the Early Cretaceous shelf margin and across "Tertiary style" growth faults to reach a thickness of more than 3 km (>10,000 ft;) (Smith, 1985).

Austin strata are predominantly carbonates (limestone and chalk). In the eastern part of the province, terrigenous clastics grade downdip to marls and chalks. Intrusive and extrusive volcanics of the same age are common from central Texas to the Rio Grande embayment.

Beds of the *Taylor* Group are marly in the eastern Gulf area and become more calcareous westward. Fringing reef deposits are associated with volcanic cones in south Texas, the Jackson dome in Mississippi, and the Monroe uplift in Louisiana.

The *Navarro* Group consists of calcareous clays and marls, glauconitic and calcareous sands, and chalks. In south Texas it also includes sandstones and shales with coal and lignite.

After the Late Cretaceous inundation, extensive *Cenozoic* progradation of terrigenous clastics (Fig. 2) took place in sequences of offlapping deposits that extended beyond the Lower Cretaceous shelf edge to the present continental slope. Progradation and filling of the subsiding basin proceeded in a series of depocenters (Murray, 1961) that were progressively younger from southwest to northeast and basinward (Galloway, 1988; Robertson Research, 1980; Winker, 1984). Each depocenter contains repeated sequences of shifting and migrating delta complexes with marine shales at the base, overlain by prodelta shales and sands, deltaic sands and shales, and finally fluvial and alluvial sands and clays. Each regressive sequence is overlain by relatively thin transgressive marine shale. The lateral gradation is the same from basin to land. The sequence is repeated hundreds of times in each depocenter. Outside of the depocenters, contemporaneous strand-plain and shelf sandstones, silts, and shales are replaced basinward by marine shales. The resulting deposits are arranged in belts that are progressively younger basinward, along a depositional strike that is roughly governed by the Early Cretaceous shelf margin. Cenozoic sequences reach a combined thickness of 10 to 15 km (~35,000 to ~50,000 ft) under the Louisiana continental shelf.

Each stratigraphic unit increases in thickness across a series of growth faults that are progressively younger basinward, related to the age of the depocenter with which they are associated. The subsiding basin setting is one in which synsedimentary tectonics prevail, and sedimentation, stratigraphy, and tectonics are intimately interrelated with hydrocarbon migration and accumulation (Curtis, 1986; Fails, 1985; Humphris, 1985).

The deltaic deposits in the depocenters contain dip-oriented sand bodies. The strand-plain and shelf deposits contain strike-oriented sand bodies. Submarine-channel sands fill channels carved in a mid-Oligocene erosion surface in an area known as the "Hackberry embayment" that straddles the Texas-Louisiana border. Several other occurrences of sedimentary sequences that were deposited during low stands of sea level are known (Curtis and Echols, 1985).

Major sediment influx and rapid progradation began in the

southwest with sediments derived from the Laramide orogenic belt and took place in several episodes. Each episode is represented by a thick progradational wedge of terrigenous clastics: (a) Paleocene and early Eocene (*Wilcox*), principally in Texas and southwestern Louisiana (Bebout and others, 1979; Fisher and McGowan, 1967; Lofton and Adams, 1971); (b) mid- and late Oligocene (*Frio*), principally in Texas and southwest Louisiana (Galloway and others, 1982); (c) *early Miocene* in southwest Louisiana and southeast Texas (Tipsword and others, 1971); (d) *middle* and *late Miocene* in southeast Louisiana (Curtis, 1970; Shinn, 1971); and *Plio-Pleistocene* in offshore Louisiana (Lehner, 1969; Powell and Woodbury, 1971; Woodbury and others, 1973). The principal source of sediments had shifted from the west to the midcontinent and the Mississippi embayment by Miocene time. Intervals representing lower rates of sediment influx are in the early Paleocene (*Midway*), middle and late Eocene (*Claiborne* and *Jackson*) (Lofton and Adams, 1971), early Oligocene (*Vicksburg*) (Stanley, 1970), and late Oligocene (*Anahuac*) (Tipsword and others, 1971). The strata in these units are more shaley, with localized depocenters of aggradational deltaic deposits, and widespread strand-plain and shelf deposits.

The progradational facies patterns, in which sand content and sand-bed thickness increase upward, result in a diachronous magnafacies pattern for the entire Cenozoic; a massive sand unit is at the top and is underlain by mixed sand and shale, with thick marine shale at the bottom. The marine shales are low-density, overpressured, undercompacted sediments that form a mobile substrate (Curtis, 1970).

Overpressures (geopressures) are abnormally high interstitial fluid pressures, in which hydrostatic pressure gradients are exceeded, and trapped pore fluids support not only the overlying water column but also part of the geostatic load. Geopressure results from many mechanisms, several of which may interact (Barker, 1972; Bruce, 1984; Dickinson, 1953; Fertl, 1976; Gretener, 1981; Hedberg, 1974; Price, 1976). The mechanisms are different in different basins (e.g., Spencer, 1987). The mechanisms in the Gulf basin probably include undercompaction, hydrocarbon generation, aquathermal pressuring, and clay mineral diagenesis. Depths to the top of the geopressured section vary with the stratigraphy; they tend to be greater in the deltaic depocenters where sedimentation rates are high. Depth and age of the top of the geopressured section tend to change across growth faults.

THERMAL HISTORY

Subsidence in the Gulf basin has been continuous since Late Jurassic time except for a 15- to 20-m.y. interval in Early Cretaceous time and a brief (1 to 2 m.y.) interruption in mid-Cenomanian time. Thus, present-day geothermal gradients may be assumed to approximate geothermal gradients through time in this basin. Gradients in this province range from less than 2.3°C/100 m (1.2°F/100 ft) to more than 3.7°C/100 m (2°F/100 ft). In general, areas of rapid Neogene deposition (south Louisiana and offshore) have the lowest gradients. Thus, thermal maturity has been reached at shallower depths in the older sediments with the higher gradients, assuming it is valid to project present gradients to the past.

Lopatin plots, constructed for two wells with different geothermal gradients, are shown in Figures 3A and B. Figure 3A, which represents thermal history for a well in the middle to late Miocene depocenter with a geothermal gradient of 2.3°C/100 m (top of lower Miocene estimated at approximately 9,100 m, or ~30,000 ft), shows that the base of the lower Miocene section was buried to approximately 5,100 m (~17,000 ft) at 130°C (268°F) when thermal maturity (based on Time-Temperature Index of Waples, 1980) began at approximately 9 Ma. The top of the lower Miocene section reached depths needed for thermal maturation at about 5,800 m (~19,000 ft) at 140°C (284°F) at approximately 5 Ma and remained almost entirely in the oil window until the present burial depth of approximately 7,900 m (~26,000 ft). The section below the base of the Miocene is at depths suitable for condensate generation. Middle Miocene sediments remain thermally immature to present burial depths of approximately 5,100 m (~17,000 ft). Fields in the area are producing oil and gas from middle and upper Miocene reservoirs. No source-rock quality sediments have been identified.

Figure 3B represents the thermal history of a well in south-central Texas with a geothermal gradient of 3.7°C/100 m that penetrated most of the Lower Cretaceous section to a depth of 6,363 m (~21,000 ft). The base of the Lower Cretaceous (base of Hosston) entered the zone of thermal maturity at approximately 2,900 m (~9,500 ft) of burial at approximately 95 Ma (120°C; 248°F) and remained in the oil-generating zone to ~3,600 m (~12,000 ft) of burial at 65 Ma (150°C; 310°F). During continued burial to ~4,800 m (15,900 ft) (190°C; 372°F) at 48 Ma, the Lower Cretaceous sediments continued to generate oil and gas. At present depth the base of the Lower Cretaceous is just below the gas-generating zone. The top of the Lower Cretaceous is still in the oil-generating zone. Upper Cretaceous potential source rocks have not reached thermal maturity. Fields in this trend are producing from Lower Cretaceous, Upper Cretaceous, and Paleocene–lower Eocene reservoirs.

Nunn and Sassen (1986) have constructed burial history curves for Gulf of Mexico slope and Mississippi salt basin wells. The two curves reflect a difference in rates and times of sediment accumulation. More than half the sediment thickness on the continental slope is less than 30 m.y. old, whereas in the salt basin rapid sediment accumulation ceased almost 100 m.y. ago. Using a thermal model developed by Nunn (1984), Nunn and Sassen (1986) predicted temperature histories for selected time horizons in the two areas that showed the most rapid change in thermal regime since Late Cretaceous time on the Gulf slope, but in Jurassic and Cretaceous time in the salt basin. Their thermal history plots indicate that Cretaceous and early Tertiary sediments at depths between 6,000 and 9,000 m (20,000 and 30,000 ft) are likely sources for hydrocarbon accumulations in late Cenozoic reservoirs on the slope, and that Jurassic (Smackover) sediments at depths between 3,000 and 6,000 m (10,000 and 20,000 ft) are probable source rocks in the salt basin.

Using present-day gradients, depths to thermal maturation have been predicted for the onshore portion of Cretaceous to Pliocene deposits. Depths to vitrinite reflectance 0.6 R_0 and 1.2 R_0 from analyses by Dow (1980) and calculated depths based on geothermal gradients and time (Robertson Research, 1980) indicate the possibility of source rocks that range in age from Cretaceous to Miocene.

CHARACTERISTIC PRODUCING SYSTEMS

As of 1978 (Nehring, 1981; Murray and others, 1985), an estimated total of more than 125 billion barrels of oil and liquid equivalents (LE; converted at 6,000 cfg = 1 bbl) had been discovered in the province, of which 94 billion barrels LE had been produced. Seventy-five percent of the reserves are in terrigenous clastic Cenozoic reservoirs; the rest are in Mesozoic reservoirs, one-fourth of which are carbonates.

Jurassic

Producing trends and depths. Jurassic rocks are productive in an arcuate trend along the landward margin of the province extending from the San Marcos arch to offshore Alabama and the Florida "panhandle" (Fig. 4A). Producing depths range from less than 1,000 m to greater than 7,500 m (~3,000 to ~25,000 ft). Principal production is from Smackover, Norphlet, and Cotton Valley reservoirs (Pearcy and Ray, 1986; Ventress and others, 1984); both oil and gas are produced. A deep gas trend in southern Mississippi and the Alabama continental shelf produces from Norphlet reservoirs at depths to 7,500 m (Mancini and Mink, 1985).

Source rocks. Thermally mature organic-rich lower Smackover dense limestones and algal mudstones are recognized as the source rocks for hydrocarbons in Jurassic reservoirs from east Texas to Florida (Sassen and Moore, 1987). In east Texas, some Haynesville slope limestones are also thermally mature.

Migration. Regional diagenetic studies (Moore, 1984) indicate relatively early migration (Late Jurassic) in east Texas and later migration (Early Cretaceous) in Arkansas, north Louisiana, and Mississippi-Alabama. Thermal history profiles and modeling for the Mississippi salt basin (Nunn, 1984) also indicate that Smackover source rocks there began to generate oil in Early Cretaceous time after about 1,800 m (~6,000 ft) of burial. Apparently oil began to migrate into Jurassic traps as early as 15 m.y. to as late as 40 or 50 m.y. after the source rocks were buried. The timing of migration relative to trap formation is critical,

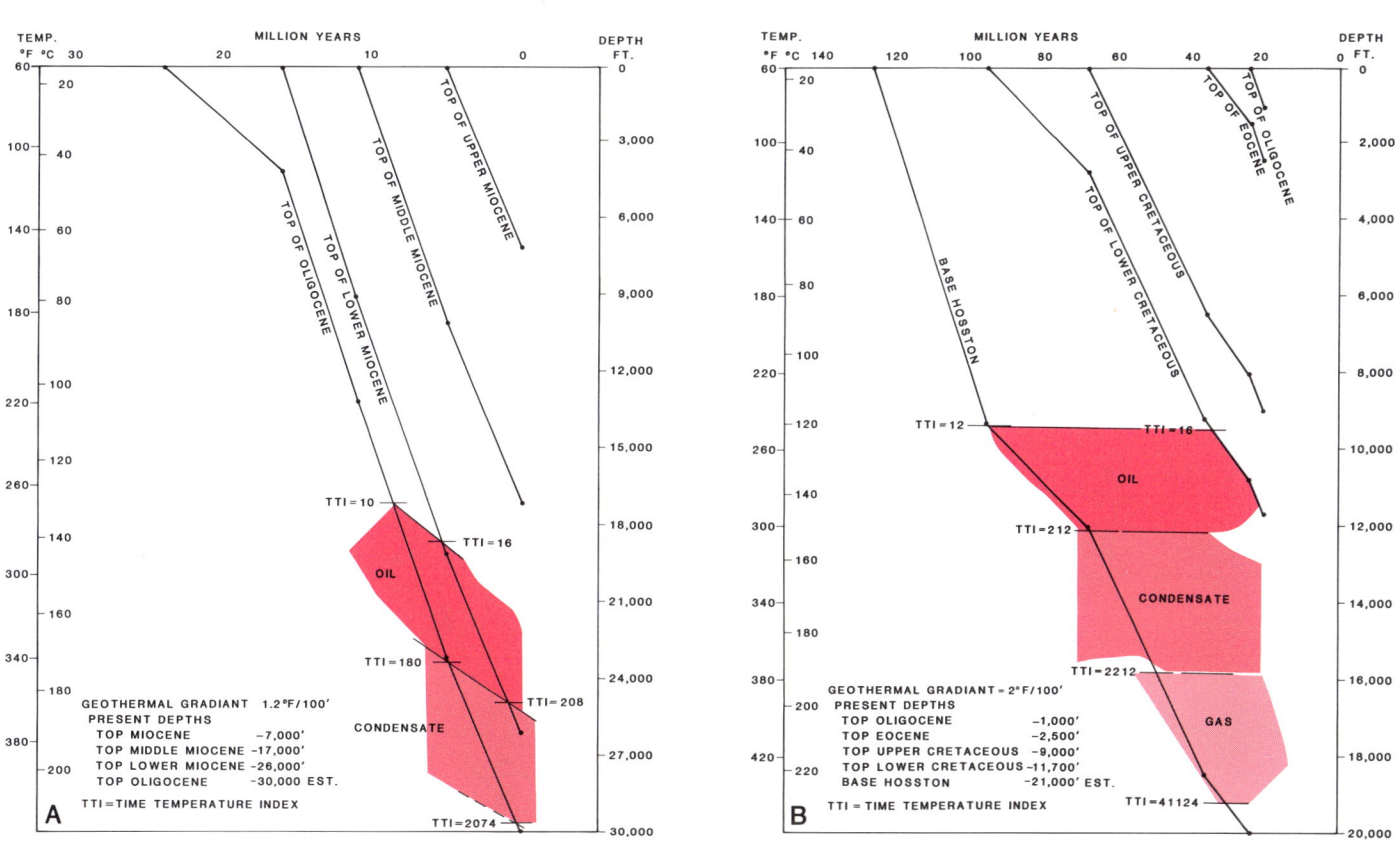

Figure 3. Thermal histories. (A) Thermal history for a deep Miocene well, south Louisiana; geothermal gradient 2.3°C/100 m (1.2°F/100 ft). (B) Thermal history for a deep Lower Cretaceous well, central Texas; geothermal gradient 3.7°C/100 m (2°F/100 ft).

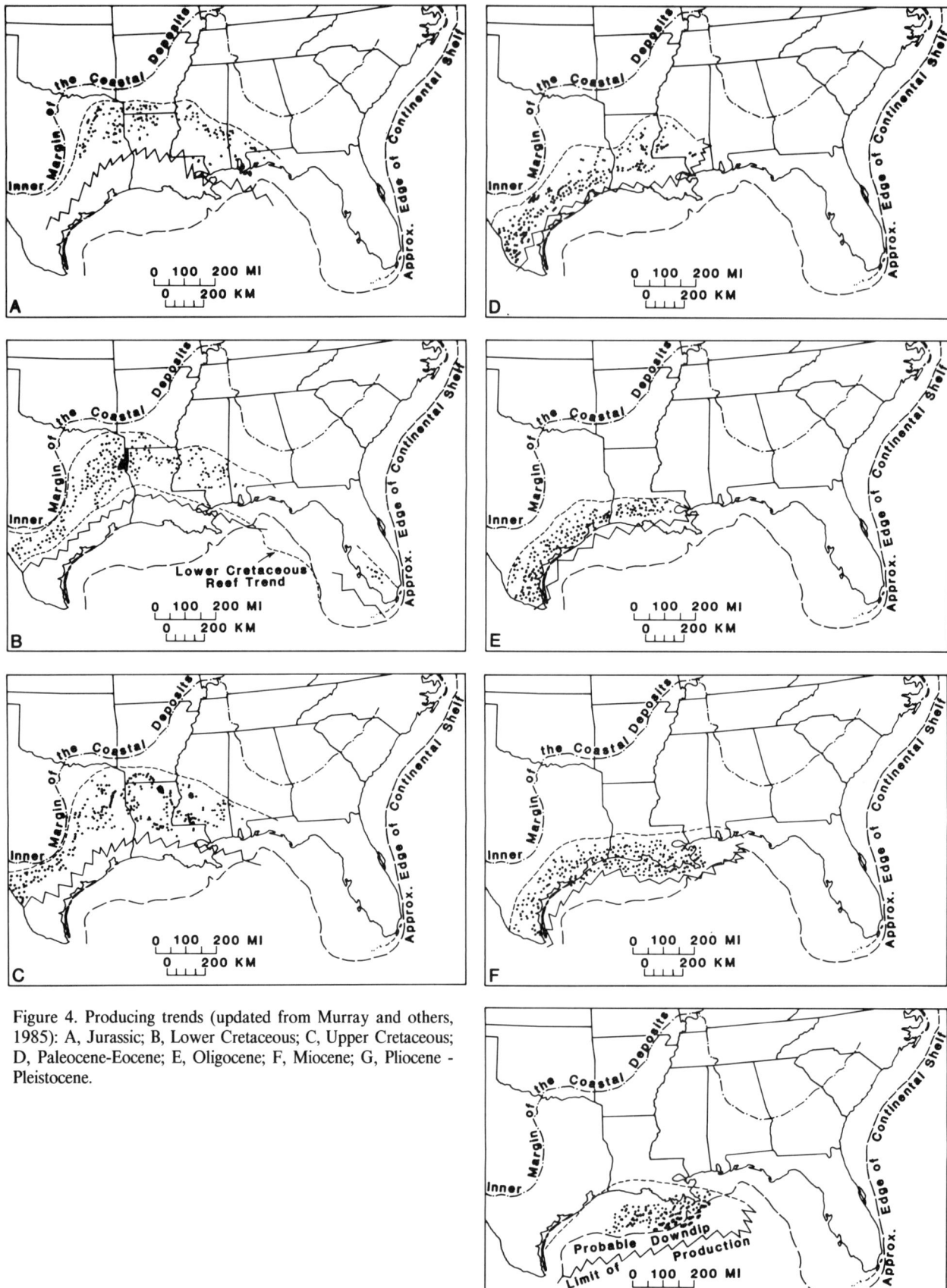

Figure 4. Producing trends (updated from Murray and others, 1985): A, Jurassic; B, Lower Cretaceous; C, Upper Cretaceous; D, Paleocene-Eocene; E, Oligocene; F, Miocene; G, Pliocene - Pleistocene.

especially in east Texas where migration appears to have begun earlier than in the rest of the trend, and where some traps probably formed after migration.

Short-distance vertical or lateral migration from source to adjacent reservoir facies is probable throughout much of the Smackover and Norphlet producing trends in the salt basins. Vertical migration paths from Smackover source rocks into Cotton Valley reservoirs can be permeable facies or faults, including Late Jurassic growth faults (Bishop, 1973). Pressure gradients from geopressured Smackover sections are probably also involved in vertical migration. Because most of the potential Smackover source rocks thin and pinch out updip, lateral migration from the basin margin across the shelf is also important.

Reservoirs. Smackover, Norphlet, and Cotton Valley reservoirs account for most of the Jurassic production. Principal Smackover reservoirs are high-energy shelf limestones and sandstones (Newkirk, 1971; Moore, 1984). Oolite bars are the most common reservoir facies. Reservoir characteristics vary widely, depending on their diagenetic history, with porosity ranges from 5 to 30 percent, permeability from 0.5 to 1,400 md, and "pay" thickness as great as 60 m (200 ft) (Moore, 1984; Benson and Mancini, 1984). Porosity is both primary intergranular and diagenetic; porosity trends are mappable and regionally predictable, based on diagenetic facies. In the deep subsurface, secondary porosity may be associated with hydrocarbon maturation (Moore and Druckman, 1981). Both oil and gas are produced from these reservoirs. The gas from deep geopressured Smackover reservoirs contains high percentages of H_2S and CO_2.

Norphlet subarkosic sandstone reservoirs are productive principally in southeastern Mississippi and offshore Alabama. Porosities range from 8 to 25 percent; permeabilities generally range from 10 to 197 md, but some producing reservoirs have permeabilities as low as 0.5 md. Thickness of reservoir facies is at least 30 m (Hartman, 1968). Production is mostly gas.

Cotton Valley reservoirs (Collins, 1980) are generally low-permeability sandstones that produce, or have potential for production, from east Texas to central Mississippi. Porosity ranges from 6 to 18 percent; permeability ranges from 10 to 750 md, and "pay" thickness may be as much as 60 m. Most production is gas.

Traps and seals. The bulk of Jurassic production comes from low-relief salt anticlines that formed in response to loading of the Louann salt by the overlying Jurassic sediments. These anticlines are elongate, with long axes parallel with regional strike. They average 6 to 10 km in length but can be as long as 30 km. Closure at the top of the Smackover is normally 15 to 120 m (50 to 400 ft).

Salt anticlines with high relief are less common. They may have several kilometers of closure affecting strata as young as Cretaceous. They are in the deeper parts of the salt basins where they formed in response to higher rates of sedimentary loading. In many cases the Smackover is missing from the crests and is productive on the flanks. Such structures are responsible for a significant portion of Cotton Valley production.

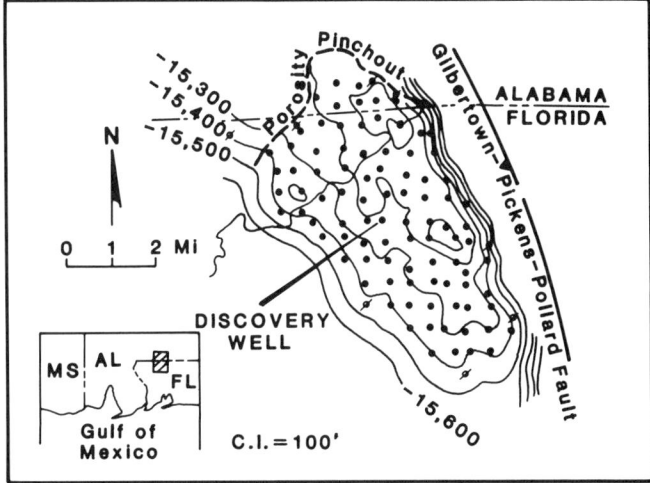

Figure 5. Jay field, northwest Florida-Alabama, producing from Jurassic Smackover limestone porosity pinchout on deep-seated salt anticline (after Ottman and others, 1976).

Traps also include (1) updip fault traps and complex graben faults along the updip margin of the province, (2) basement highs generally associated with the peripheral graben faulting, (3) residual anticlines related to salt withdrawal, and (4) stratigraphic traps (permeability pinchouts, truncation traps). True stratigraphic traps are rare. Permeability variations in combination with subtle structural anomalies are responsible for significant accumulations.

Cumulative production and reserves. The total volume of hydrocarbons discovered in the Jurassic as of 1978 (Nehring, 1981) was estimated at more than 5.1 billion barrels LE, about half of which is gas. To this should be added post-1978 high-volume Norphlet gas discoveries in southwest Alabama and offshore, where reserves have been estimated at at least 12 trillion cfg (cubic feet of gas). About 2.5 billion barrels LE had been produced by 1978, leaving reserves of more than 2.6 billion barrels LE yet to be produced. Although more than 200 fields produce from Jurassic reservoirs, about 80 percent of the production comes from approximately 70 fields. Seven major fields, with recoverable volumes of more than 100 million barrels LE each, are responsible for more than half the total.

Typical fields. Jay field (Fig. 5) illustrates a typical trapping style for Jurassic production. Additional fields that exemplify characteristics of the Jurassic trend are listed in Table 1. Their locations are shown in Figure 6.

Future potential for exploration. Potential is considerable in coastal Alabama and on the eastern continental shelf (Mancini and Mink, 1985) in all reservoir facies, especially the deep Norphlet. Most of the new production will be gas. Greatest potential for Smackover is in the south Texas basin-margin trend, south of the San Marcos arch, where suitable reservoir facies and source rocks appear to be present. Shelf-margin limestones are also prospective on the flanks of the east Texas basin, around the

TABLE 1. SELECTED MAJOR FIELDS REPRESENTING TYPICAL RESERVOIR AND TRAP TYPES

No. and Field Name	Area	Age	Producing Formation	Principal Reservoir	Principal Trap	Reference
1. Brantley-Jackson	NE Tex	Jur.	Smackover	Oolitic shelf limestone	Upthrown fault block in updip fault line	Newkirk, 1971
2. Bryans Mill	NE Tex	Jur.	Smackover	Shelf limestone	Diagenetic; faulted nose on basement high	Ventress and others, 1984
3. Dorcheat-Macedonia	Ark	Jur.	Smackover and Cotton Valley	Oolitic shelf limestone Sandstone	Anticline with salt swell	Newkirk, 1971
4. Jay	NW Fla/Ala	Jur.	Smackover	Shelf limestone	Diagenetic on deep-seated salt anticline	Ottman and others, 1976
5. Thomasville	Miss	Jur.	Smackover	Geopressured turbidite sandstone	Deep-seated anticline on salt swell	Olsen, 1982; Parker, 1974
6. Walker Creek	Ark	Jur.	Smackover	Calcarenite shelf limestone	Diagenetic	Chimene, 1976; Becher and Moore, 1976
7. Black Lake	N La	L Cret.	Sligo-Pettet	Reef limestone	Permeability closure on nose	Bailey, 1978; Spooner, 1968
8. Citronelle	Ala	L Cret.	Rodessa	Channel sandstone; fluvial	Dome on residual salt anticline	Eaves, 1976
9. Fairway	NE Tex	L Cret.	James	Reef limestone	Fault-bounded reef porosity on uplift	Galloway and others, 1983; Terriere, 1976
10. Luling	S Tex	L Cret.	Edwards Austin	Fractured limestone Fractured chalk	Sealing fault on nose	Galloway and others, 1983; Rainwater, 1971
11. Rodessa-Caddo Pine Island	NE Tex	L Cret.	Glen Rose	Oolitic and bioclastic limestone	Anticline w/ updip pinchout and sealing fault	Galloway and others, 1983
12. Sligo	N La N La	L Cret.	James Sligo-Hosston Rodessa-Paluxy Cotton Valley	Oolitic limestone Fossiliferous calcarenite Shoreface sandstone	Faulted anticline	Skillern, 1968
13. Delhi	N La	Jura. U Cret. L Cret.	Tuscaloosa Paluxy	Fluviodeltaic sandstone	Truncation on flank of Monroe Uplift	Powell, 1972
14. East Texas	NE Tex	U Cret.	Woodbine	Fluviodeltaic sandstone	Onlap and truncation on Sabine Uplift	Galloway and others, 1983; Murray, 1961; Halbouty and Halbouty, 1982
15. Gwinville	Miss	U Cret.	Tuscaloosa Rodessa	Fluviodeltaic sandstones Sandstones	Deep-seated faulted salt dome	Beebe, 1968
16. Hawkins	NE Tex	U Cret.	Woodbine	Deltaic and strand-plain sandstones	Faulted anticline on salt pillow	Galloway and others, 1983; Murray, 1961
17. Heidelberg	Miss	U Cret.	Austin-Eagleford	Sandstones	Deep-seated faulted dome on salt uplift	Murray, 1961
18. Kurten	NE Tex	U Cret.	Woodbine	Shelf sandstone	Reservoir pinchout	Galloway and others, 1983; Berg and Leethem, 1985
19. Mexia-Powell	NE Tex	U Cret.	Woodbine	Fluvial sandstone	Nose in updip fault trend	Galloway and others, 1983
20. Tinsley	Miss	U Cret.	Tuscaloosa	Shoreface and barrier sandstones	Deep-seated anticline w/ sand pinchout	Shelton, 1976
21. Van	NE Tex	U Cret.	Woodbine	Deltaic and strand-plain sandstones	Faulted anticline w/ stratigraphic	Galloway and others, 1983; Murray, 1961
22. Government Wells	SW Tex	Eocene	Jackson-Yegua	Strand-plain sandstones	Updip porosity pinchout	Galloway and others, 1983
23. NE Thompsonville	SW Tex	Eocene	Wilcox	Deltaic sandstones	Upthrown fault block	Lofton and Adams, 1971
24. Rosita	SW Tex	Eocene	Wilcox	Deltaic sandstones	Rollover anticline	Edwards, 1981
25. Seeligson-Borregas	SW Tex	Oligo.	Frio	Deltaic sandstones	Rollover anticline	Galloway and others, 1983
25a. McAllen Ranch	SW Tex	Oligo.	Vicksburg	Turbidite channel sandstone	Rollover anticline	Berg and others, 1979
26. Stratton-Agua Dulce	SW Tex	Oligo.	Frio	Deltaic sandstones	Rollover anticline	Galloway and others, 1983
27. Tom O'Connor-Greta	SW Tex	Oligo.	Frio + Anahuac	Strand-plain sandstones	Rollover anticline	Galloway and others, 1983
		Mio.	Lower	Fluvial sandstone		Mills, 1970; Murray, 1961
28. West Ranch	SW Tex	Oligo.	Frio	Strand-plain sandstone	Rollover anticline	Galloway and others, 1983
29. Conroe	SE Tex	Eocene	Yegua Wilcox	Deltaic sandstone	Deep-seated faulted dome	Galloway and others, 1983
30. Katy	SE Tex	Eocene	Yegua	Deltaic sandstone(?)	Deep-seated residual anticline	Galloway and others, 1983
31. Old Ocean			Frio	Strand-plain sandstones	Rollover anticline	Galloway and others, 1983; Halbouty, 1968

TABLE 1. SELECTED MAJOR FIELDS REPRESENTING TYPICAL RESERVOIR AND TRAP TYPES (continued)

No. and Field Name	Area	Age	Producing Formation	Principal Reservoir	Principal Trap	Reference
32. Pierce Junction	SE Tex	Mio.	Lower	Sandstones	Piercement salt dome	Galloway and others, 1983; Murray, 1961
		Oligo.	Frio, Vicksburg Anahuac	Sandstones Limestone		
33. Segno	SE Tex	Eocene	Jackson-Yegua	Sandstone	Rollover anticline	Galloway and others, 1983
34. Sheridan	SE Tex	Eocene	Wilcox	Deltaic sandstones	Fault-bounded anticline	Lofton, 1968
35. Sour Lake	SE Tex	Eocene	Wilcox	Deltaic sandstones	Piercement salt dome	Galloway and others, 1983
	SE Tex	Mio.	Lower	Deltaic sandstones		Galloway and others, 1983
	SE Tex	Oligo.	Frio Anahuac	Deltaic sandstones Sandstones and limestone		
35a. Spindletop	SE Tex	Eocene	Jackson-Yegua	Sandstones	Piercement salt dome	Galloway and others, 1983
	SE Tex	Mio.	Lower	Sandstones		
36. Webster	SE Tex	Oligo.	Frio	Sandstones	Deep-seated salt dome	Galloway and others, 1983
37. Bay Marchand-Caillou Island-Timbalier Bay	S La	Plio. Mio.	Upper	Deltaic sandstones Deltaic sandstones	Salt domes on salt ridge	Frey and Grimes, 1970
38. Black Bay complex	S La	Mio.	Upper & Middle	Deltaic sandstones	Rollover anticline	NOGS, 1965, 1983b
39. Deep Lake	S La	Mio.	Middle & Lower	Strand-plain sandstones	Rollover anticline	LGS, 1964
40. Delta Farms	S La	Mio.	Upper & Middle	Deltaic sandstones	Faulted low-relief dome	NOGS, 1983b
41. Fordoche	S La	Oligo.	Frio	Strand-plain sandstones	Faulted rollover anticline	LGS, 1970
		Eocene	Jackson-Yegua	Strand-plain sandstones		
42. Grand Bay	S La	Mio.	Upper & Middle	Deltaic sandstones	faulted anticline	NOGS, 1983b
43. Iowa	S La	Oligo.	Frio + Anahuac	Strand-plain sandstones	Piercement salt dome	LGS, 1964
44. Jennings	S La	Mio.	Lower	Fluvial sandstones	Piercement salt dome	NOGS, 1962
		Oligo.	Frio + Anahuac	Strand-plain sandstones		
45. Lake Pelto-Bay Ste. Elaine	S La	Pleisto. Plio.		Deltaic sandstones Deltaic sandstones	Piercement salt dome	NOGS, 1965; 1983a
46. Paradis	S La	Mio.	Upper	Deltaic sandstones	Deep-seated salt dome	Murray, 1961
47. Quarantine Bay	S La	Mio.	Upper & Middle	Deltaic sandstones	Low-relief faulted anticline	NOGS, 1983b
48. Venice	S La	Mio.	Upper & Middle	Deltaic sandstones	Piercement salt dome	NOGS, 1962, 1967
49. Vinton	S La	Mio.	Upper	Strand-plain sandstones	Piercement salt dome	NOGS, 1960, 1967
		Oligo.	Lower Frio + Anahuac	Strand-plain sandstones		
50. Weeks Island	S La	Mio.	Lower	Strand-plain and deltaic sandstones	Piercement salt dome	NOGS, 1962, 1983a
51. West Bay	S La	Pleisto. Plio.		Deltaic sandstones Deltaic sandstones	Deep-seated dome	NOGS, 1967; 1983b
52. Eugene Is Blk 126	La OCS	Mio. Plio.	Upper	Deltaic sandstones Deltaic sandstones	Piercement salt dome	NOGS, 1974; 1983a
53. Eugene Is Blk 330	La OCS	Pleisto. Plio.	Upper	Strand-plain sandstones(?) Deltaic sandstones	Faulted rollover anticline	Holland and others, 1980
54. Grand Isle Blk 16	La OCS	Mio.	Upper	Deltaic sandstones	Piercement salt dome	NOGS, 1974; Steiner, 1976
55. Grand Isle Blk 47	La OCS	Plio.	Upper	Deltaic sandstones	Faulted anticline	NOGS, 1974
56. Main Pass Blk 35	S La	Mio.	Upper & Middle	Deltaic sandstones	Deep-seated faulted dome	NOGS, 1974; Hartman, 1976
57. So Pass Blk 24	S La	Plio.	Upper	Deltaic sandstones	Faulted interdomal area	Hartman, 1976; Murray, 1961
58. So Pass Blk 27	S La	Mio. Plio.	Upper	Deltaic sandstones	Piercement salt dome	Smith, 1961; NOGS, 1983a
59. Vermilion Blk 14	La OCS	Mio.	Upper & Middle	Strand-plain sandstones(?)	Deep-seated faulted anticline	NOGS, 1974
60. Vermilion Blk 39	La OCS	Mio.	Upper & Middle	Strand-plain sandstones(?)	Deep-seated faulted dome	LGS, 1967

OCS = Outer Continental Shelf   NOGS = New Orleans Geological Society   LGS = Lafayette Geological Society

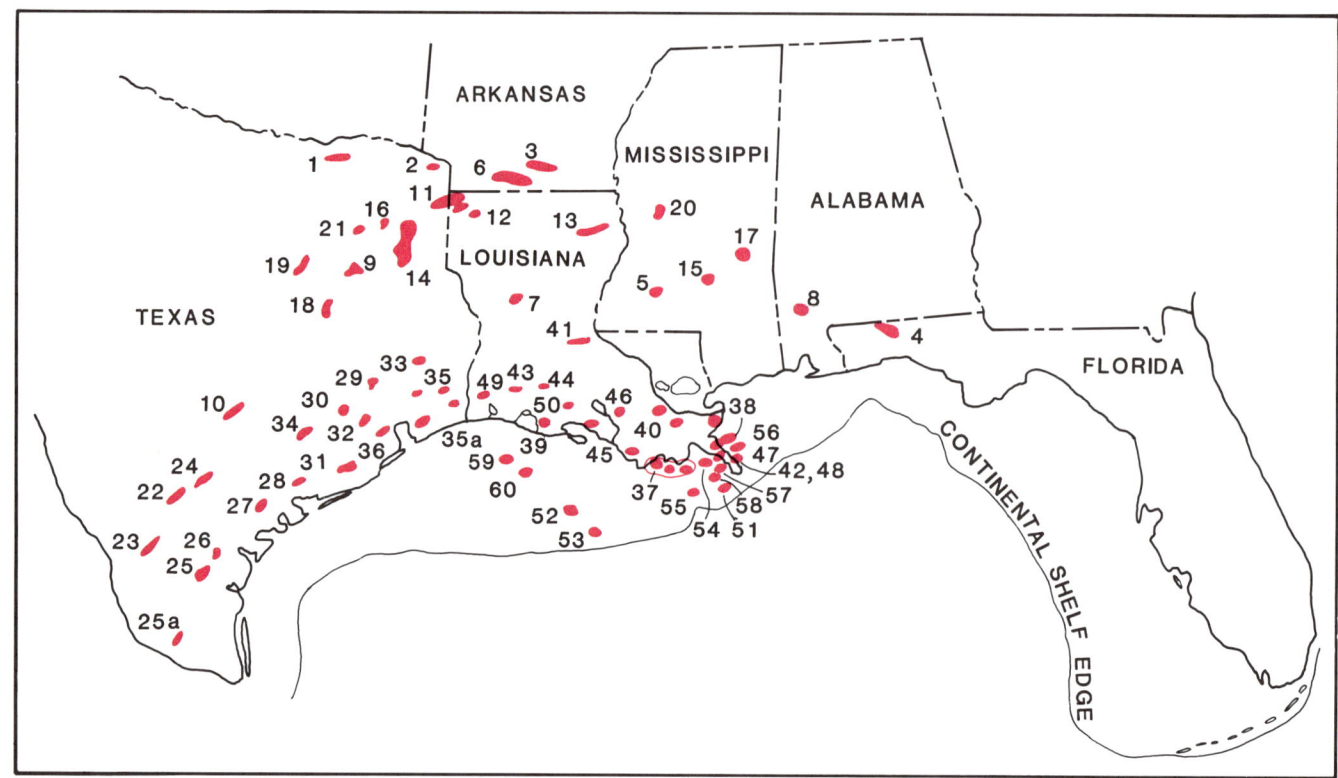

Figure 6. Reference map for fields listed in Table 1, which contains selected major fields representing typical reservoirs and traps. (Fields not numbered are not listed but are significant as part of producing trend.)

Sabine uplift, and in the Smackover reef facies in Louisiana and Mississippi (Baria and others, 1982), where additional major gas reserves in permeability traps and fractured reservoirs may be found. In the deeply buried carbonate basins, secondary porosity that may be associated with hydrocarbon maturation offers exploration potential. Deep gas will be found in low-permeability downdip Cotton Valley sandstones.

Cretaceous

Producing trends and depths. Principal Cretaceous production is in an arcuate trend that extends from south Texas to Florida, parallel with the updip boundary of the province and basinward to or beyond the composite Early Cretaceous shelf edge (Figs. 4B, C). Distribution of Cretaceous production coincides in part with that of the Jurassic. Producing depths range from less than 1,000 m to greater than 6,000 m (~3,000 to >20,000 ft). Each stratigraphic unit is productive in some area in the province. The most prolific sectors for Cretaceous production are east Texas and north Louisiana–Arkansas (the Ark-La-Tex area). East Texas contributes over 60 percent of the oil and 20 percent of the gas in the entire trend. North Louisiana–Arkansas contributes about 20 percent of the oil and more than 50 percent of the gas in the trend (Holcomb, 1971).

Hosston-Sligo rocks produce both oil and gas principally in the Ark-La-Tex area, with minor production in Mississippi. Trinity oil and gas production extends through the Ark-La-Tex area into Mississippi-Alabama-Florida. Fredericksburg-Washita production is found in a discontinuous trend throughout the province from south Texas to Alabama; both oil and gas are produced. Woodbine-Eagleford-Tuscaloosa production is concentrated principally in east Texas, Louisiana, and Mississippi. Younger Cretaceous fields produce oil and gas in south Texas, east Texas, north Louisiana, and Mississippi.

Source rocks. Upper Cretaceous Eagleford black shales are the major source for accumulations in the Upper Cretaceous and the upper part of the Lower Cretaceous section from south Texas to Mississippi (Surles, 1985). Lower Cretaceous source rock facies have also been identified in the Pearsall shales in south Texas, east Texas, and Florida. Geochemical and thermal maturation studies have shown that much of the oil in Lower Cretaceous reservoirs probably came from Jurassic (Smackover) source rocks.

Migration. Isotopic analyses of oil in Cretaceous reservoirs and oil-rock comparisons indicate the presence of two "families" of oils (Dow, 1983): one "family" from Cretaceous and the other from Jurassic source rocks. There are also mixed oils from both types of source rocks throughout the province. Indigenous oils (derived from Cretaceous source rocks) tend to be associated with simple unfaulted anticlinal and stratigraphic traps. Oils derived from Jurassic source rocks tend to be associated with faulted traps, which indicates vertical migration. This interpretation is consistent with the thermal history for the Jurassic previously described, in which expulsion from Smackover source rocks began early in Cretaceous time, after about 1,500 to 1,800 m (~5,000 to 6,000 ft) of burial. Where Eagleford source rocks are too shallow for thermal maturity in updip areas, lateral migration into updip Cretaceous reservoirs is indicated.

Reservoirs. In each Cretaceous stratigraphic unit, limestone and sandstone reservoir facies are related to depositional environment and diagenetic history (Rainwater, 1971; Robertson Research, 1983).

Lower Cretaceous carbonate reservoirs are found in the Sligo, Pearsall (James and Rodessa), and Stuart City (Glen Rose and Edwards). Upper Cretaceous carbonate reservoirs are present in Austin and Selma chalks. Several depositional/diagenetic facies are recognized: (a) shelf limestones, including high-energy deposits such as oolite bars, skeletal grainstones, and calcarenites in the Sligo and Rodessa of east Texas, with primary interparticle and secondary porosity that ranges from 0 to 30 percent; (b) shallow-marine and tidal-flat limestones from low-energy environments in the Edwards in south and central Texas, with secondary porosity; (c) shelf-margin facies, including reefs and banks, tidal bars and channels in the Stuart City–Edwards reef trend, and in the Sligo and Glen Rose (Adams and Watkins, 1985; Cook, 1979), with both primary and diagenetic porosity that is highly variable and ranges as high as 35 percent in the high-energy facies of the reef trend and as high as 20 percent in the rudist grainstone banks; (d) patch reefs and carbonate mounds in numerous fields, including Black Lake (Sligo), Fairway (James), and fields in the Glen Rose and Stuart City shelf trends, with variable primary and secondary porosity; (e) chalks and micritic limestones in the Austin and Selma chalks and Edwards limestone, with enhanced fracture porosity; (f) calcarenite reservoirs that onlap igneous plugs in south and central Texas.

Terrigenous sandstones also are important reservoirs in the Cretaceous. Porosity varies with facies, depth, and diagenetic effects. Several depositional systems contain reservoir facies: (a) fluviodeltaic systems with well-developed dip-oriented reservoir facies in Hosston, Sligo, Rodessa (e.g., Citronelle Field), Paluxy-Dantler, and Woodbine-Tuscaloosa; (b) strand-plain systems in the Woodbine-Tuscaloosa in east Texas (e.g., East Texas field; Halbouty and Halbouty, 1982); (c) marine-shelf sandstone reservoirs, including bars, shoreface deposits, and shallow-marine sheet sands in the Hosston in east Texas, Paluxy in the Ark-La-Tex area, Woodbine in southeast Texas, and in Upper Cretaceous sandstones in Mississippi and south Texas; (d) turbidite reservoirs in the deep lower Tuscaloosa gas trend in central Louisiana downdip from the Lower Cretaceous shelf edge.

Traps and seals. Structural, stratigraphic, and combination traps control accumulations in Cretaceous reservoirs. The most common trapping mechanisms in this trend are fault traps, deep domal or anticlinal traps, stratigraphic traps on subtle structural anomalies, and salt domes. Rollover anticlines are less common except in the deep lower Tuscaloosa gas trend.

Domes and anticlines that are related to deep-seated salt structures form traps in the interior salt basins, where they trap hydrocarbons in carbonate and sandstone reservoirs in most of the Cretaceous section. Fault traps are mostly in updip peripheral fault zones, generally upthrown on an antithetic fault. Accumulations of this type are common in both carbonate and sandstone reservoirs. Many updip Upper Cretaceous fields are in such traps.

Stratigraphic traps may result from porosity variations, reservoir pinchouts, truncation, or onlap on topographic highs. Many examples of such trapping mechanisms are well known. The famous East Texas field is a stratigraphic trap that formed by onlap of Woodbine sands on a truncated Lower Cretaceous surface on the west flank of the Sabine uplift, followed by truncation and deposition of impermeable Austin chalk. Black Lake field in north Louisiana is formed by a porous reef facies that extends landward from the shelf margin and is surrounded by tight carbonates. Linear calcarenite or sandstone bars and fractured chalk also form stratigraphic traps.

Several deep-seated faulted domes in the interior salt basins have significant accumulations in reservoirs that wedge out or are truncated against their flanks, as in the Van and Hawkins fields in east Texas and Heidelberg field in Mississippi. Rollover anticlines associated with growth faults are the trapping mechanism in the deep Tuscaloosa turbidite reservoirs in central Louisiana, where the section has expanded across growth faults along the Lower Cretaceous shelf edge.

Shales, tight limestones, chalks, and anhydrite are seals.

Cumulative production and reserves. Almost 72 billion barrels LE have been discovered in Cretaceous reservoirs. Cumulative production from Lower Cretaceous rocks to 1978 (Nehring, 1981; Murray and others, 1985) was approximately 5 billion barrels LE, with some 7 billion barrels yet to be produced of the estimated 12.1 billion barrels LE total reserves. Upper Cretaceous cumulative production to 1978 was approximately 11 billion barrels, of which more than 5 billion had come from the East Texas field. Total discovered reserves in Upper Cretaceous fields were estimated at 14.5 billion barrels, leaving some 3.5 billion barrels LE yet to be produced. Although more than 500 fields have significant production from the Cretaceous, only about 20 percent of them have produced more than 90 percent of the hydrocarbons. Seventy-five percent of the hydrocarbons that have been produced are from 26 giant fields (>100 million barrels each).

Typical fields. East Texas field (Fig. 7) illustrates one Cretaceous trapping type. Additional typical fields are listed in Table 1 and located in Figure 6.

Future potential. Cretaceous producing trends are maturely explored except on the Florida shelf (Faulkner and Applegate, 1986). Future potential lies in several directions. In the Hosston-Sligo, potential for discoveries exists in deep-basin gas sands below ~4,500 m (~15,000 ft), and in shelf-margin buildups and forereef debris piles in south Texas, and north Louisiana. There is also potential in porosity traps on the shelf in east Texas and north Louisiana. Potential discoveries are all gas.

Gas discoveries may also be made in Trinity reservoirs.

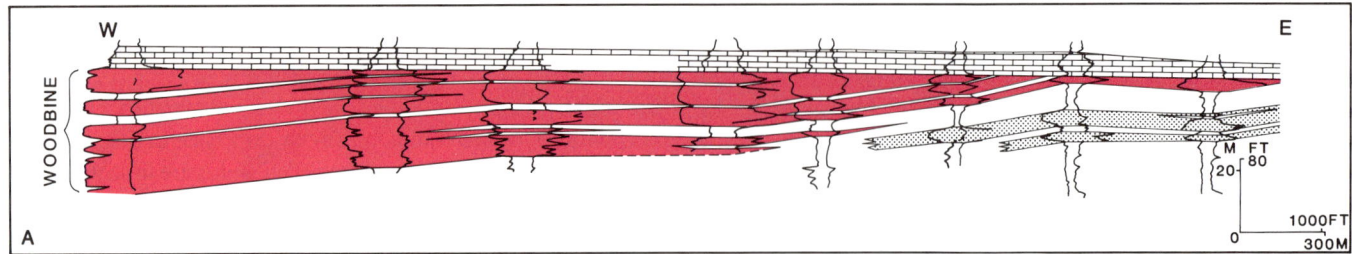

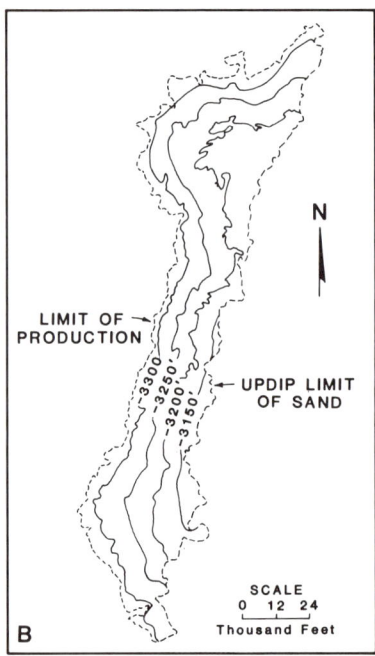

Figure 7. East Texas field, northeast Texas, producing from Upper Cretaceous Woodbine reservoir in truncation trap (after Galloway and others, 1983). (A) Cross section showing onlap of Woodbine sandstone, truncation, and Austin chalk seal. (B) Structure map of top of Woodbine.

Facies and diagenetic models suggest potential for the entire Lower Cretaceous shelf margin. New porosity trends may also be found on the shelf in limestone banks, fractured reservoirs, and linear bars. There is also potential for deep gas discoveries in forereef facies below 6,000 m (>20,000 ft). Potential for discoveries in sandstone reservoirs in Rodessa or Mooringsport delta-fringe and bar deposits in Mississippi and Alabama also exists.

Fredericksburg-Washita potential for gas discoveries is in the Stuart City reef trend, in linear banks and porosity trends on the shelf, in backreef deposits, and in forereef facies below 6,000 m (>20,000 ft). There is also further potential in bank-margin and linear-bank deposits in the Edwards.

Woodbine-Tuscaloosa sandstones in delta-fringe and barrier island trends have potential for gas production in east Texas and Mississippi-Alabama.

Austin chalks and limestones have some potential for gas production in fracture porosity trends in south Texas, east Texas, and north Louisiana south of the Sabine uplift. Navarro-Taylor production may be discovered in basinal sandstones downdip from the Stuart City reef trend in south Texas and in fractured limestones in east Texas and north Louisiana.

Tertiary and Quaternary

Producing trends and depths. Hydrocarbons are produced from Tertiary and Quaternary strata in trends that extend from the Rio Grande to east of the Mississippi Delta, offshore on the Texas-Louisiana continental shelf, and on the Louisiana continental slope (Figs. 4D to G) at depths that range from very close to the surface to more than 7,800 m (>26,000 ft). Principal producing systems are associated with major progradational wedges of terrigenous clastic sediments. Producing trends are progressively younger from south Texas eastward and gulfward. Major Paleogene trends are in Texas; major Neogene trends are in Louisiana and on the continental shelf and slope. A producing trend of each age is closely identified with each Cenozoic depocenter, and each contains deltaic and strand-plain facies tracts. Each producing system is a result of an intimate interdependence of sedimentation and tectonics in time and space that has produced a finely tuned relationship of geologic history with migration and accumulation of hydrocarbons (Bouma and others, 1978; Curtis, 1986).

Each trend extends from updip shallow reservoirs at 600 to 1,000 m depth (~2,000 to 3,000 ft) to downdip production at 4,800 to 8,000 m (~16,000 to >26,000 ft). Production is limited downdip and along strike by the extent of the reservoir sandstones. Depths of producing trends tend to be limited by porosity and extreme overpressuring. Most of the producing trends produce oil, or oil and gas, from the normally pressured section and gas from the geopressured section. Depths to the top of the geopressured section range from about 1,800 m to greater than 5,000 m (~6,000 to >18,000 ft). The geopressured section is deepest in south Louisiana in the middle and late Miocene depocenters in trends similar to geothermal gradient trends.

Source rocks. Many Gulf Coast geologists have assumed that the marine and prodelta shales that are interbedded with or adjacent to the reservoir sands must be the source rocks, but source rocks for Cenozoic accumulations have not been identified. With few exceptions (Laplante, 1974), source rocks are believed to be slope shales of Tertiary depositional systems (Dow, 1978) and Cretaceous Eagleford shales in some areas. No organic-rich, thermally mature shales have yet been identified in Cenozoic strata. Throughout the Cenozoic section, reservoirs and their associated shales are in thermally immature and organically poor intervals.

Based on geothermal gradient, and on age and depth, it can be shown that the distribution of possible thermally mature source rocks parallels the distribution of producing trends but is offset from the age of the producing trend to an older sequence (Dow, 1980, 1984). For example, source rocks for the Frio in Texas (Galloway and others, 1982) are probably in the underlying lower Oligocene and upper Eocene slope shales. Because of their great thickness and volume, the thermally mature slope shales that underlie the productive sequences in each producing trend can be source rocks, even with extremely low organic carbon content (Jones, 1981, 1986). Jones estimated that there are about 3,000 m (~10,000 ft) of low total organic carbon (TOC) shales underlying reservoir sequences. Dow (1984) believes the oil source beds may be localized in Tertiary marine slope shales in anoxic intraslope basins behind salt ridges, especially during times of rising sea level.

Using biomarkers, Walters and Cassa (1985) demonstrated that the Tertiary beds show increasing maturity from east to west on the Louisiana continental shelf. They concluded that early to middle Miocene source rocks generated oil on the eastern shelf, where production is from upper Miocene and Pliocene reservoirs, and that upper Miocene to Pliocene source rocks probably generated oil on the outer continental shelf farther west, where production is from Pleistocene reservoirs.

For accumulations in Plio-Pleistocene reservoirs on the continental slope, source rocks may be the thick sequence of Lower Cretaceous to upper Miocene sediments beneath the allochthonous salt tongue that extends to the Sigsbee Deep (Dinkelman, 1986; Dinkelman and Curry, 1987). Calculated thermal maturity models based on deep seismic stratigraphy indicate that Cretaceous and probably early Tertiary sediments on the slope are presently within the thermal maturation range for generation and migration of hydrocarbons into Plio-Pleistocene reservoirs. Younger sediments are immature. The potential Cretaceous and early Tertiary source rocks matured during the past 30 m.y. during rapid late Miocene to Pleistocene loading.

Recent studies of oils from the continental slope (Nunn and Sassen, 1986) indicate some oils have biomarkers that are consistent with carbonate source rocks. Although Mesozoic source rocks that underlie the shelf are overmature, Mesozoic sediments that underlie the slope have a burial history more favorable for relatively recent hydrocarbon generation and migration.

Migration. The dubious assumptions regarding the source-rock potential of the interbedded thermally immature shales derive from assumptions about early migration. Many years ago, Gulf Coast geologists demonstrated correctly that hydrocarbons migrated into traps that were formed shortly after burial of the reservoirs. From this they concluded that early migration from adjacent source rocks must have taken place. Geochemical constraints, however, require that source rocks in this setting must be buried many thousands of meters before they are thermally mature. This implies late expulsion and migration from the source

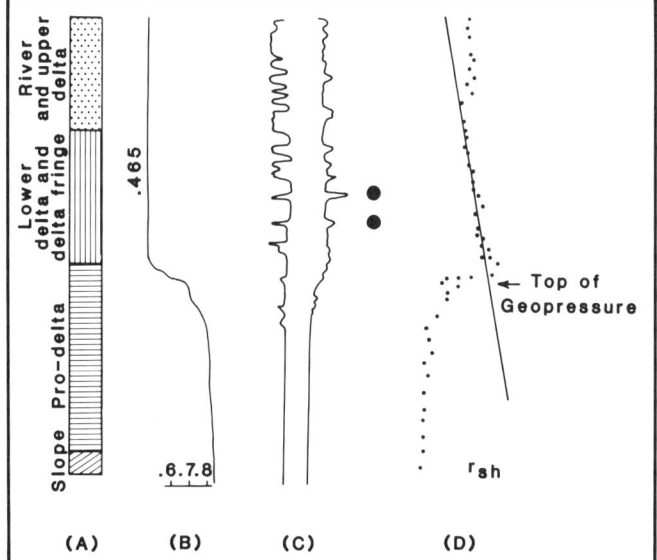

Figure 8. Relationship of (A) depositional environments in a deltaic sequence with (B) pressure gradients and top of geopressured section with gradients greater than 0.465 psi/ft; (C) producing interval; and (D) shale resistivity, as commonly observed in Tertiary trends (Curtis, 1986).

rock. The two concepts are compatible, however, if migration into the trap takes place long after the source rock is deposited ("late migration"), but soon after trap formation in a younger sequence ("early migration").

Given the relationship between possible source rocks and producing reservoirs, it is possible to postulate a migration concept. The Cenozoic source rocks are in the geopressured shale, through which migration must take place in order for hydrocarbons to reach the reservoir section. Pressure gradients may cause hydraulic fracturing and may facilitate primary migration (Jones, 1981; Durand and others, 1986). Generation and expulsion of hydrocarbons from Tertiary slope shales occurs too late for primary migration to be involved with compaction-water movement, but it may take place by processes such as diffusion through shale pore systems (Hinch, 1980) or microfractures in the geopressured shales (Hedberg, 1980; Magara, 1981). The source rocks are below the flat listric faults, but once the hydrocarbons have been expelled from the source rocks and have migrated through the geopressured shales, they find their way into the fault systems along which migration into the younger reservoir systems takes place. There is an intricate relation of pressures and fluid movement along faults; the faults serve as both migration paths and seals, depending on pressure gradients and permeability barriers (Allan, 1986; Downey, 1984; Smith, 1980; Weber, 1986).

Pleistocene reservoirs, in which temperature and burial time are both low, contain a considerable amount of biogenic methane as well as thermogenic gas that has migrated vertically from much more deeply buried source rocks (Rice, 1980; Rice and Claypool, 1981).

Calculated hydrocarbon ages of south Louisiana oils are

older than the ages of the reservoirs from which they were produced (Young and others, 1977). This indicates considerable vertical migration. Vertical migration of as much as 3,000 m (~11,000 ft) is suggested by the average difference in ages of the Neogene oils and reservoirs (Dow, 1984).

Reservoirs. Reservoirs are almost exclusively sandstones. The sand bodies in fluviodeltaic depositional systems have dip-oriented distribution patterns. Interdeltaic strand-plain and marine-shelf sands are strike-oriented. Turbidites and slump deposits are reservoirs in several settings. Submarine channel and fan deposits in low-stand wedges are significant. The ages of major producing reservoirs correspond closely with the distribution of depocenters previously discussed.

The major oil production from Cenozoic reservoirs onshore and on the continental shelf comes from lower delta-plain and delta-fringe sands and from linear bars, where the percentage of sand in the section is less than 30 percent. Many of the producing reservoirs are on the downthrown sides of major growth faults in the section just above the geopressured section. Gas is produced throughout the section, but geopressured reservoirs produce mostly gas. Reservoirs of each age produce both oil and gas, but the predominance of oil or gas depends on dip position, structural complexity, type of source rock, pressure regime, and depositional environment. In general, more gas is produced from Texas reservoirs and more oil is produced from Louisiana reservoirs.

Data from both Cretaceous and Tertiary sections show a

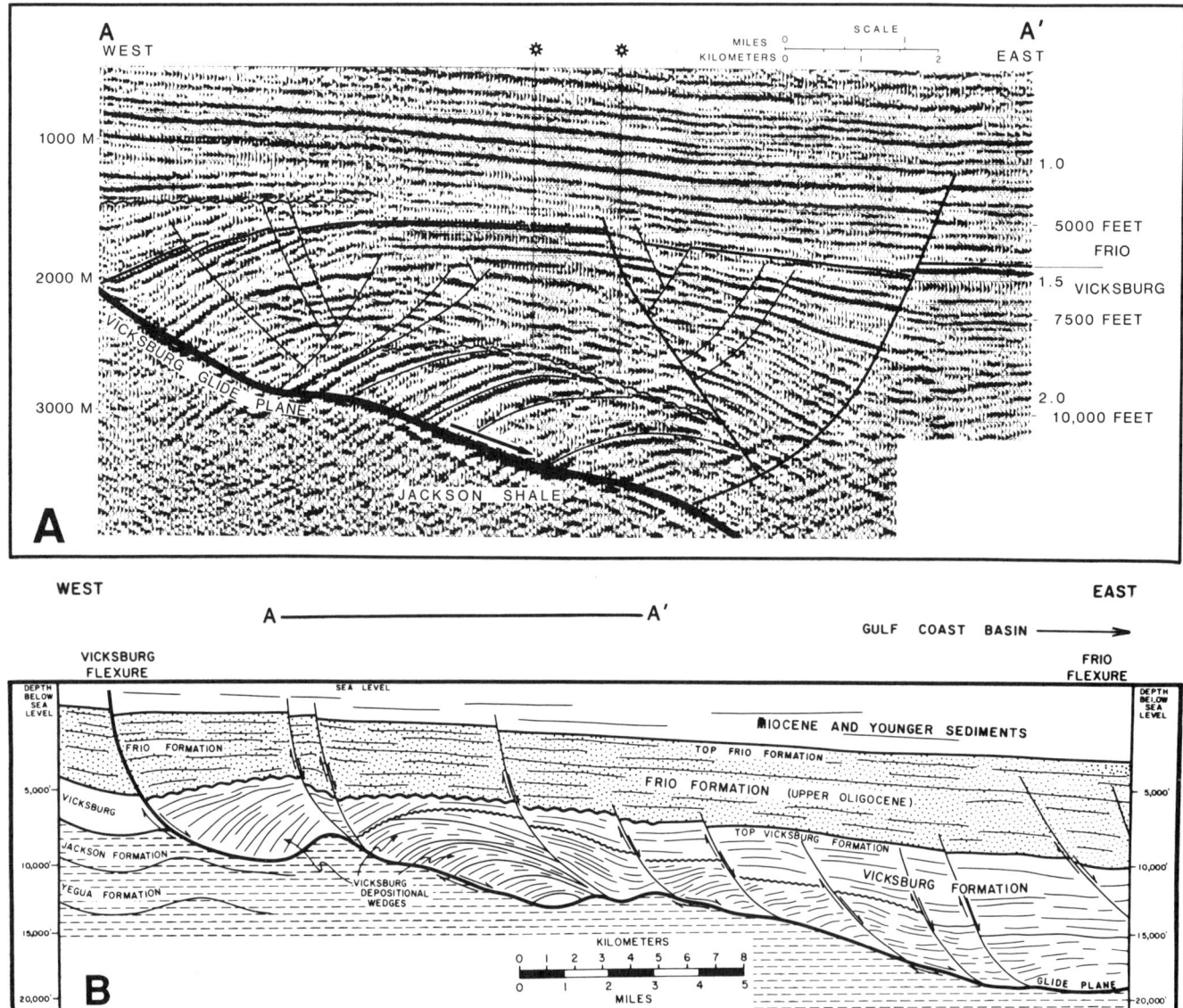

Figure 9. (A). Regional seismic line A–A' showing growth faults, rollover structures, antithetic faults, and glide plane (décollement surface) in south Texas Oligocene trend (after Bally, 1983, Fig. 2.3.1–26). (B). Regional structural and stratigraphic cross section based on seismic and subsurface interpretations, south Texas Oligocene trend. A–A' is the portion of this cross section represented in seismic line A–A', Figure 9A (after Bally, 1983, Fig. 2.3.1–23).

linear decrease in porosity and recovery per unit volume with depth for normally pressured sections (Atwater and others, 1986; Maxwell, 1964). Higher porosities are associated with excess pore pressures and lower geothermal gradients, and porosity may be either enhanced or decreased by diagenesis. The empirical relationship of geopressures, porosity, shale resistivity, and hydrocarbon accumulation in the reservoir section above the geopressure shale is shown in Figure 8. In general, reservoirs with formation pressures over 0.85 psi/ft do not contain commercial volumes of hydrocarbons, but a few producing zones with pressures as high as 0.975 psi/ft have been reported (Timko and Fertl, 1971).

In each producing trend, fields are associated with numerous shifting and migrating depositional systems. Characteristically, therefore, multiple reservoirs are stacked on a productive structure. Producing trends also overlap, so that multiple objectives of several ages are generally available. The major part of each trend is underlain by the more distal part of an older trend.

Traps and seals. Most traps for hydrocarbons in Cenozoic reservoirs are related to salt or shale structures or to rollover anticlines associated with growth faults. Salt-related structural traps include piercement and nonpiercement domes, salt ridges, and deep-seated domal uplifts, all with complex faulting. Structures that are not salt related are diapiric or residual shale domes and anticlines.

Salt domes have a complex growth history (Seni and Jackson, 1983). Most of the dome growth is contemporaneous with deposition, but is intermittent or continuous, depending on rates of deposition of sediments overlying the salt. This type of growth results in multiple reservoirs that thin toward the dome or are truncated against the flank, and that dip away from the dome. Hydrocarbons are trapped over the crests and on the flanks of salt domes in many blocks of the complex fault pattern associated with the domal growth. Many domes also have an early growth stage followed by truncation of uptilted beds; if the truncation surface is overlain by shale, truncation traps are formed.

In offshore portions of the salt basin the relative proportion of individual salt domes decreases, and huge elongate salt ridges and massifs become common under the continental slope. Traps are formed in associated fault blocks and in pinchouts and truncations on the flanks of salt ridges. Shale diapirism is common. As much as 6,000 m (~20,000 ft) of sediments fill the inter-ridge basins.

In areas outside the salt basins, particularly in the Texas Gulf Coast, the dominant trap types are rollover anticlines (Fig. 9a). Deep-seated shale diapirs and domal uplifts (Fig. 9b) at the toes of flat listric faults are also common traps. Growth-fault–related structures include rollover anticlines in downthrown blocks, deep low-relief domal uplifts related to gravity sliding, and uplifts at the toe-ends of listric faults.

Extensive areas of counter-regional dip, developed in connection with older parts of each growth fault system, characterize the area. In areas of counter-regional dip, stratigraphic traps may be present as gulfward pinchouts of distal deltaic sands. Upthrown fault traps also occur in this setting on antithetic faults.

TABLE 2. CENOZOIC RESERVES, BY AGE

Age of Reservoir	Total Discovered	Cumulative Production	Reserves Remaining
	(in billion barrels LE; 6,000 cfg = 1 bbl)		
Plio-Pleistocene	8.9	3.0	5.9
Miocene	47.5	33.2	14.3
Oligocene	28.8	11.7	17.1
Paleocene-Eocene	8.9	3.7	5.2

Local sand pinchouts on deep-seated uplifts form combination traps.

Differential pressures become a trapping mechanism where fault traps or stratigraphic traps are reinforced by pressure differentials that cause trapping in the lower pressure segment. The downdip pressure drop on the flanks of diapiric structures provides the necessary hydrostatic reinforcement to trap long hydrocarbon columns by influencing water levels (Myers, 1968). Pressure gradients are strongly displaced in adjacent fault blocks. Trapping is improved in downthrown blocks if reservoirs are juxtaposed against overpressured, low-permeability sediments across the fault. Similarly, the geopressured shale "sheath" that surrounds many piercement salt stocks may enhance the trapping capabilities of reservoirs that pinch out against the dome (Downey, 1984). Fault planes may act as seals or migration paths, depending on pressure differentials across them (Allan, 1986; Downey, 1984; Smith, 1980).

Cumulative production and reserves. Total discovered reserves in Cenozoic fields onshore and offshore as of 1978 approximate 94 billion barrels LE. About 51.6 billion barrels had been produced as of 1978, with 42.4 billion barrels reserves remaining. Conservative cumulative production and reserve estimates, broken down by trends (modified from Murray and others, 1985), are shown in Table 2 for onshore and the continental shelf.

Pleistocene reserves have been considerably augmented by recent "flexure trend" (shelf edge and upper slope) and continental slope discoveries.

Forrest (1986) estimated that an additional 600 million barrels had already been discovered by 1986 in trends in water depths of 300 to 3,000 m (1,000 to 10,000 ft). Miocene reserves in offshore Texas have been augmented by discoveries in downdip rollover structures on flattened listric faults like the Corsair trend (Vogler and Robison, 1987).

Cenozoic reserves are credited to thousands of small and large fields, but fewer than 1,000 fields are responsible for about 98 percent of the production. About 25 percent of these, with volumes greater than 100 million barrels each, are responsible for more than half the reserves. Paleocene-Eocene reserves are in about 150 fields, 18 of which have produced a total of more than 2 billion barrels. Oligocene production is largely from fewer than 200 fields, of which 72 have produced about 8 billion barrels. Most of the Miocene production comes from fewer than 500 fields. Two-thirds of the Miocene production is from about 20 percent of the fields. Plio-Pleistocene production is mostly from 128 fields, of which 23 have produced more than 75 percent of the hydrocarbons. Offshore fields must have larger volumes of commercially recoverable reserves because of economic constraints. Cumulative Cenozoic production from major fields by region is shown in Table 3.

Typical fields. Figures 10 through 12 illustrate typical trap types of Cenozoic fields. Additional fields are listed in Table 1 and located in Figure 6.

Future potential. Although Cenozoic trends have been maturely explored onshore and on the continental shelf, hydrocarbons remain to be found in deeper objectives and offshore, especially on the continental slope. Downdip from each producing trend are the distal ends of each trend that have not been intensively explored because they underlie younger, less deeply buried trends. The deeper, older trends may be extensions of distributary-channel sands, delta-fringe sands, or turbidite and fan deposits. The Pleistocene Mississippi fan has extensive, thick, reservoir-quality sands. Much of the potential in the downdip trends is in deep geopressured gas sands that were deposited as turbidites. Submarine-channel and fan deposits remain to be found and explored in low-stand wedges in settings where lowered sea level is suspected or known.

The "flexure trend" of Pleistocene exploration on the present continental slope has great potential as technology for deep-water drilling continues to develop. Forrest (1986) estimated undiscovered resources of 3 to 5 billion barrels of oil and 100 trillion cfg in upper Cenozoic reservoirs in waters 300 to 3,000 m deep.

Dolan (1986) predicted discoveries of considerable reserves in turbidites and in older Cenozoic and Mesozoic reservoirs under the lower continental slope and continental rise. He suggested that the U.S. Geological Survey estimates of 2.6 billion barrels of oil and 26.5 trillion cfg for the deep-water Gulf (to 3,000 m) may be low. The "Maritime Boundary Area" (from the 3,000-m water

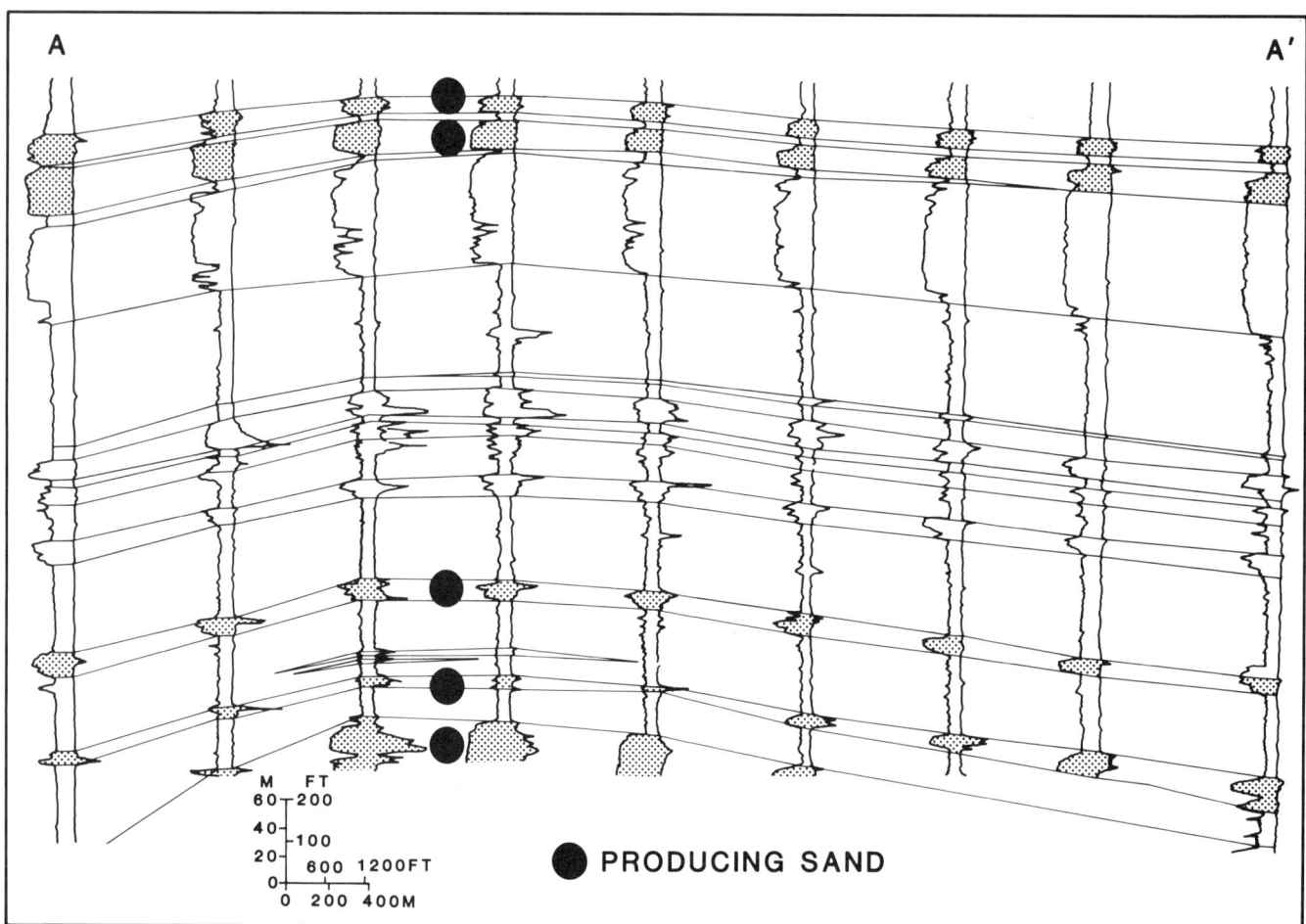

Figure 10. Tom O'Connor field, southwest Texas, producing from Oligocene Frio sandstone reservoirs on rollover anticline (after Galloway and others, 1983): cross section showing stacked barrier sands and producing intervals.

**TABLE 3. CUMULATIVE PRODUCTION FROM MAJOR FIELDS
(>100 MILLION BARRELS)
BY REGION**

Region	No. of Fields	Cumulative Production (billion bbl LE)	Trap Types
Louisiana onshore and offshore	43	10.14	60% salt domes + deep-seated domes; rollovers
Texas upper Gulf Coast	22	7.58	75% salt domes + deep-seated uplift; 25% rollover anticlines
Texas lower Gulf Coast	19	5.13	Rollover anticlines and combination

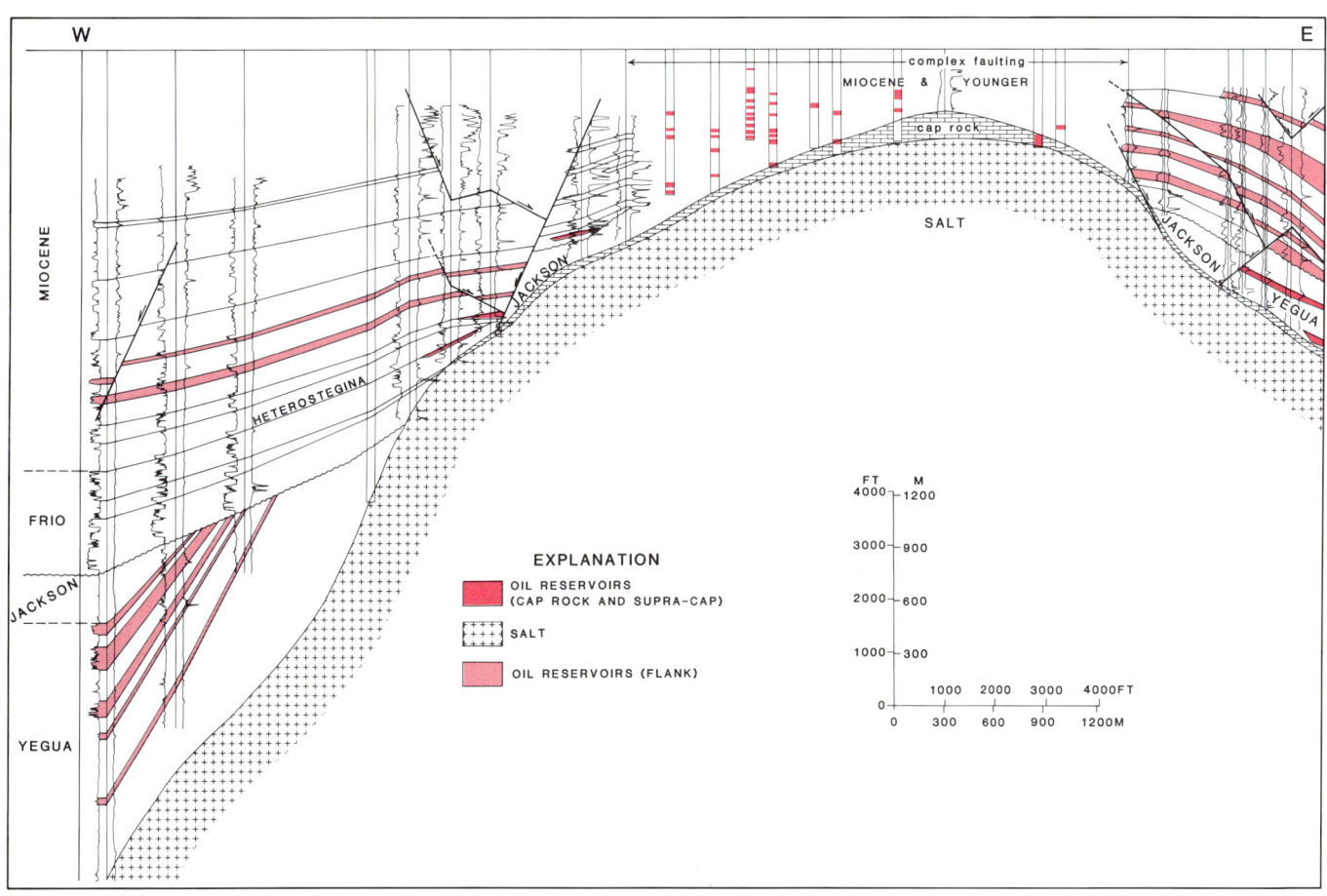

Figure 11. Sour Lake field, southeast Texas, producing from salt dome (after Galloway and others, 1983): cross section showing cap rock production; production from Miocene sands on dome crest, and from Miocene and Oligocene sands on the flanks; production from Eocene Jackson and Yegua sands in deep truncation traps. The piercement dome has slight surface expression.

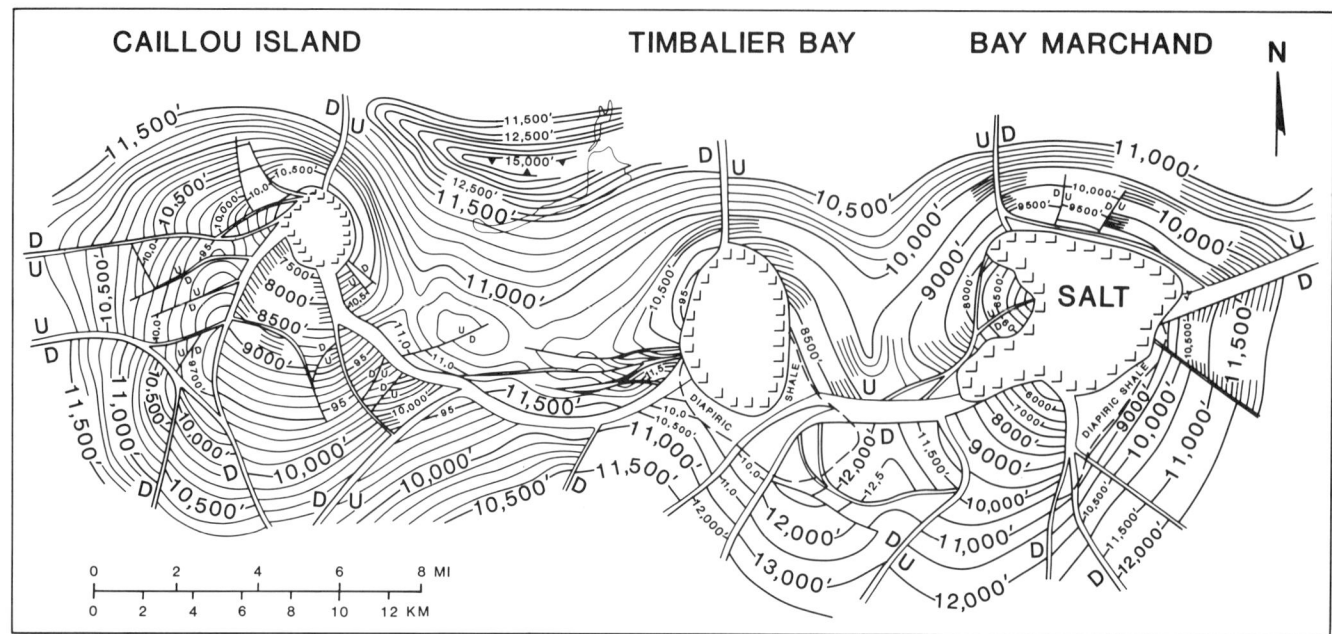

Figure 12. Bay Marchand–Timbalier Bay–Caillou Island fields, south Louisiana, producing from Miocene-Pliocene sands on salt domes on salt ridge (after Frey and Grimes, 1970): structure map of an upper Miocene sand. Contours are depths in feet below sea level.

depth to the Sigsbee Deep) may have large volumes of undiscovered resources (Foote and others, 1983). The U.S. Geological Survey has estimated that between 2.24 and 21.99 billion barrels of oil (median 9.11 billion barrels) and between 5.48 and 44.4 trillion cfg (median 18.77 trillion cfg) may be discovered in the deep Gulf of Mexico.

REFERENCES

Adams, G. S., and Wakins, G. B., 1985, Subsurface Glen Rose reef trend of east Texas, Louisiana, and Mississippi: American Association of Petroleum Geologists Bulletin, v. 69, p. 233.

Allan, U. S., 1986, Model for hydrocarbon migration and entrapment: American Association of Petroleum Geologists Bulletin, v. 70, p. 559.

Atwater, G. I., Miller, E. E., and Wiggins, G. B., 1986, Effect of decreased porosity with depth on oil and gas reserves in south Louisiana sandstone reservoirs: American Association of Petroleum Geologists Bulletin, v. 70, p. 561.

Bailey, J., 1978, Black Lake field, Natchitoches Parish, Louisiana; A review: Gulf Coast Association of Geological Societies Transactions, v. 28, p. 11–24.

Bally, A. W., ed., 1983, Seismic expression of structural styles; Vol. 2, Tectonics of extensional provinces: American Association of Petroleum Geologists Studies in Geology no. 15, v. 2, section 2.3, figs. 2.3.1-1 to 2.3.1-49.

Baria, L. R., Stoudt, D. L., Harris, P. M., and Crevello, P. D., 1982, Upper Jurassic reefs of Smackover Formation, United States Gulf Coast: American Association of Petroleum Geologists Bulletin, v. 66, p. 1449–1482.

Barker, C., 1972, Aquathermal pressuring, role of temperatures in development of abnormal-pressure zones: American Association of Petroleum Geologists Buletin, v. 56, p. 2068–2071.

Bebout, D. G., 1974, Lower Cretaceous Stuart City shelf margin of south Texas; Its depositional and diagenetic environments and their relation to porosity: Gulf Coast Association of Geological Societies Transactions, v. 24, p. 138–159.

Bebout, D. G., and Loucks, R. G., eds., 1977, Cretaceous carbonates of Texas and Mexico: University of Texas Bureau of Economic Geology Report of Investigations 89, 332 p.

Bebout, D. G., Weise, B. R., Gregory, A. R., and Edwards, M. B., 1979, Wilcox reservoirs in the deep subsurface along the Texas Gulf Coast; Their potential for production of geopressured geothermal energy: University of Texas Bureau of Economic Geology Report of Investigations 117, 125 p.

Becher, J. W., and Moore, C. H., 1976, The Walker Creek field, a Smackover diagenetic trap: Gulf Coast Association of Geological Societies Transactions, v. 26, p. 34–56.

Beebe, B. W., 1968, Natural gas in post-Paleozoic rocks of Mississippi, in Beebe, C. W., and Curtis, B. F., eds., Natural gases in rocks of Mesozoic age: American Association of Petroleum Geologists Memoir 9, v. 1, pt. 2, p. 1176–1226.

Benson, D. J., and Mancini, E. A., 1984, Porosity development and reservoir characteristics of the Smackover Formation in Alabama, in Ventress, W.P.S., Bebout, D. G., Perkins, B. S. and Moore, C. H., eds., The Jurassic of the Gulf rim: Society of Economic Paleontologists and Mineralogists, 3rd Annual Research Conference Proceedings, p. 1–18.

Berg, R. R., and Leethem, J. T., 1985, Origin of the Woodbine-Eagleford reservoir facies, Kurten Field, Brazos County, Texas: Gulf Coast Association of Geological Societies Transactions, v. 35, p. 11–18.

Berg, R. R., Marshall, W. D., and Shoemaker, P. W., 1979, Structural and depositional history, McAllen Ranch Field, Hidalgo County, Texas: Gulf Coast Association of Geological Societies Transactions, v. 29, p. 24–28.

Bishop, W. F., 1973, Late Jurassic contemporaneous faults in north Louisiana and south Arkansas: American Association of Petroleum Geologists Bulletin, v. 57, p. 858–877.

Bouma, A. H., Moore, G. T., and Coleman, J. M., eds., 1978, Framework, facies, and oil-trapping characteristics of the upper continental margin (Gulf of Mexico): American Association of Petroleum Geologists Studies in Geology 7, 326 p.

Bruce, C. H., 1973, Pressured shale and related sediment deformation, mechanism for development of regional contemporaneous faults: American Association of Petroleum Geologists Bulletin, v. 57, p. 878–886.

——, 1984, Smectite dehydration, its relation to structural development and hydrocarbon accumulation in northern Gulf of Mexico basin: American Association of Petroleum Geologists Bulletin, v. 68, p. 673–683.

Buffler, R. T., and Sawyer, D. S., 1985, Distribution of crust and early history, Gulf of Mexico basin: Gulf Coast Association of Geological Societies Transactions, v. 35, p. 333–344.

Chimene, C. A., 1976, Upper Smackover reservoirs, Walker Creek Field area, Lafayette and Columbia Counties, Arkansas, in Braunstein, J., ed., North American oil and gas fields: American Association of Petroleum Geologists Memoir 24, p. 177–204.

Collins, S. E., 1980, Jurassic Cotton Valley and Smackover reservoir trends, east Texas, north Louisiana, and south Arkansas: American Association of Petroleum Geologists Bulletin, v. 64, p. 1004–1013.

Cook, T. D., 1979, Exploration history of south Texas Lower Cretaceous carbonate platform: American Association of Petroleum Geologists Bulletin, v. 63, p. 32–49.

Curtis, D. M., 1970, Miocene deltaic sedimentation, Louisiana Gulf Coast, in Morgan, J. P., ed., Deltaic sedimentation modern and ancient: Society of Economic Paleontologists and Mineralogists Special Publication 15, p. 293–308.

——, 1986, Comparative Tertiary petroleum geology of the Gulf Coast, Niger delta, and Beaufort–Mackenzie delta areas: Geological Journal, v. 21, p. 225–255.

Curtis, D. M., and Echols, D. J., 1985, Habitat of oil and gas in the middle Frio (Oliogcene) Hackberry, in Perkins, B. F., and Martin, G. B., eds., Habitat of oil and gas in the Gulf Coast: Society of Economic Paleontologists and Mineralogists 4th Annual Research Conference Proceedings, p. 263–274.

Dickinson, G., 1953, Geological aspects of abnormal reservoir pressures in the Gulf Coast region of Louisiana: American Association of Petroleum Geologists Bulletin, v. 37, p. 410–432.

Dinkelman, M. G., 1986, Stratigraphic framework of the Louisiana continental margin—clues to hydrocarbon sourcing: Gulf Coast Association of Geological Societies Transactions, v. 36, p. 71.

Dinkelman, M. G., and Curry, D. J., 1987, Significance of anoxic slope basins to occurrence of hydrocarbons along the flexure trend, Gulf of Mexico; A reappraisal: American Association of Petroleum Geologists Bulletin, v. 71, p. 548.

Dolan, P., 1986, Deep water (200–1,800 meters) hydrocarbon potential of the U.S. Gulf of Mexico, in Halbouty, M. T., ed., Future petroleum provinces of the world: American Association of Petroleum Geologists Memoir 40, p. 243–268.

Dow, W. G., 1978, Petroleum source beds in continental slopes and rises: American Association of Petroleum Geologists Bulletin, v. 62, p. 1584–1606.

——, 1980, Applications of basic principles of geochemistry to the Gulf Coast Tertiary, in Framework for oil and gas occurrence in the Gulf Coast Tertiary: Houston, Texas, Robertson Research (U.S.) Incorporated, in association with Curtis, D. M. and Echols, D. J., v. 1, Report, Appendix D, p. D-1 to D-19.

——, 1983, Geochemistry, in Gulf Coast Cretaceous; A regional geological synthesis: Houston, Texas, Robertson Research (U.S.) Incorporated, p. 6-1 to 6-95; Appendix p. C-1 to C-90.

——, 1984, Oil source beds and oil prospect definition in upper Tertiary of the Gulf Coast: Gulf Coast Association of Geological Societies Transactions, v. 34, p. 329–340.

Downey, M. W., 1984, Evaluating seals for hydrocarbon accumulations: American Association of Petroleum Geologists Bulletin, v. 68, p. 1754–1763.

Durand, B., Bessereau, G., Ungerer, P. H., and Dudin, J. L., 1986, Mechanisms of hydrocarbon migration in Mahakem delta, Kalimantan, Indonesia: American Association of Petroleum Geologists Bulletin, v. 70, p. 584.

Eaves, E., 1976, Citronelle oil field, Mobile County, Alabama, in Braunstein, J., ed., North American oil and gas fields: American Association of Petroleum Geologists Memoir 24, p. 259–275.

Edwards, M. B., 1981, Upper Wilcox Rosita delta system of south Texas; Growth-faulted shelf-edge deltas: American Association of Petroleum Geologists Bulletin, v. 65, p. 54–73.

Fails, T. G., 1985, Intimate relationships of growth faulting and diapirism in south Louisiana: American Association of Petroleum Geologists Bulletin, v. 69, p. 254.

Faulkner, B. M., and Applegate, A. V., 1986, Hydrocarbon potential of offshore south Florida basin: American Association of Petroleum Geologists Bulletin, v. 70, p. 588.

Fertl, W. H., 1976, Abnormal formation pressures: American Elsevier, New York, p. 1–48.

Fisher, W. L., and McGowan, J. H., 1967, Depositional systems in the Wilcox Group of Texas and their relationship to occurrence of oil and gas: Gulf Coast Association of Geological Societies Transactions, v. 17, p. 105–125.

Foote, R. Q., Martin, R. G., and Powers, R. B., 1983, Oil and gas potential of the maritime boundary region in the central Gulf of Mexico: American Association of Petroleum Geologists Bulletin, v. 67, p. 1047–1065.

Forrest, M. C., 1986, Deep water Gulf of Mexico exploration, geology, hydrocarbons, and economics: Gulf Coast Association of Geological Societies Transactions, v. 36, p. xiv–xlvii.

Frey, M. G., and Grimes, W. I., 1970, Bay Marchand–Timbalier Bay–Caillou Island salt complex, in Halbouty, M. T., ed., Geology of giant petroleum fields: American Association of Petroleum Geologists Memoir 14, p. 277–291.

Galloway, W. E., 1986, Faults and fault-related structures of prograding terrigenous clastic continental margins: Gulf Coast Association of Geological Societies Transactions, v. 36, p. 121–128.

——, 1988, The Cenozoic, in Salvador, A., ed., The Gulf of Mexico Basin: Boulder, Colorado, Geological Society of America, The Geology of North America, v. J (in press).

Galloway, W. E., Hobday, D. K., and Magara, K., 1982, Frio Formation of Texas Gulf Coastal Plain; Depositional systems, structural framework, and hydrocarbon distribution: University of Texas Bureau of Economic Geology Report of Investigations 122, 78 p.

Galloway, W. E., Ewing, T. E., Garrett, C. M., Jr., Tyler, N., and Bebout, D. G., 1983, Atlas of major Texas oil reservoirs: University of Texas Bureau of Economic Geology, 139 p.

Gretener, P. E., 1981, Pore pressures; Fundamentals, general ramifications, and implications for structural geology: American Association of Petroleum Geologists Continuing Education Course Note Series 4, 81 p.

Halbouty, M. T., 1968, Old Ocean field, Brazoria and Matagorda Counties, Texas, in Beebe, C. W., and Curtis, B. F., eds., Natural gases in rocks of Cenozoic age: American Association of Petroleum Geologists Memoir 9, v. 1, pt. 1, p. 295–305.

——, 1979, Salt domes—Gulf region United States and Mexico (2nd edition): Houston, Gulf Publishing Company, 561 p.

Halbouty, M. T., and Halbouty, J. J., 1982, Relationship between East Texas Field region and Sabine Uplift in Texas: American Association of Petroleum Geologists Bulletin, v. 66, p. 1042–1054.

Hartman, J. A., 1968, The Norphlet sandstone, Pelahatchie Field, Rankin County, Mississippi: Gulf Coast Association of Geologica Societies Transactions, v. 18, p. 2–11.

——, 1976, "New" oil from old fields: Gulf Coast Association of Geological Societies Transactions, v. 26, p. 82–91.

Hedberg, H. D., 1974, Relation of methane generation to undercompacted shales, shale diapirs, and mud volcanoes: American Association of Petroleum Geologists Bulletin, v. 58, p. 661–673.

——, 1980, Methane generation and petroleum migration: American Association of Petroleum Geologists Studies in Geology 10, p. 179–206.

Hinch, H. H., 1980, The nature of shales and the dynamics of hydrocarbon expulsion in the Gulf Coast Tertiary section: American Association of Petroleum Geologists Studies in Geology 10, p. 1–18.

Holcomb, C. W., 1971, Hydrocarbon potential of Gulf (Upper Cretaceous) section of western Gulf Basin, in Cram, I. H., ed., Future petroleum provinces of the United States; Their geology and potential: American Association of Petroleum Geologists Memoir 15, p. 887–900.

Holland, D. S., Nunan, W. E., Lammlein, D. R., and Woodhams, R. L., 1980,

Eugene Island Block 330 field, offshore Louisiana, *in* Halbouty, M. T., ed., Giant oil and gas fields of the decade 1968–1978: American Association of Petroleum Geologists Memoir 30, p. 253–280.

Humphris, C. C., Jr., 1979, Salt movement on continental slope, northern Gulf of Mexico: American Association of Petroleum Geologists Bulletin, v. 63, p. 782–798.

——— , 1985, Relationship of structural development and Cenozoic sedimentation, northern Gulf of Mexico: American Association of Petroleum Geologists Bulletin, v. 69, p. 268.

Jackson, M.P.A., and Galloway, W. E., 1984, Structural and depositional styles of Gulf Coast Tertiary continental margins; Application to hydrocarbon exploration: American Association of Petroleum Geologists Continuing Education Course Note Series 25, 226 p.

Jones, R. W., 1981, Some mass balance and geological constraints on migration mechanisms: American Association of Petroleum Geologists Bulletin, v. 65, p. 103–122.

——— , 1986, Origin, migration, and accumulation of hydrocarbons in Gulf Coast Cenozoic: American Association of Petroleum Geologists Bulletin, v. 70, p. 605.

Lafayette Geological Society, 1964; 1967; 1970, Typical oil and gas fields of southwestern Louisiana: Lafayette Geological Society, v. 1 unpaginated; supplement unpaginated; v. 2 unpaginated.

Laplante, R. E., 1974, Hydrocarbon generation in Gulf Coast Tertiary sediments: American Association of Petroleum Geologists Bulletin, v. 58, p. 1281–1289.

Lehner, P., 1969, Salt tectonics and Pleistocene stratigraphy on the continental slope of northern Gulf of Mexico: American Association of Petroleum Geologists Bulletin, v. 53, p. 2431–2497.

Lofton, C. L., 1968, Sheridan field, Colorado County, Texas, *in* Beebe, C. W., and Curtis, B. F., eds., Natural gases in rocks of Cenozoic age: American Association of Petroleum Geologists Memoir 9, v. 1, pt. 1, p. 306–329.

Lofton, C. L., and Adams, W. M., 1971, Possible future petroleum provinces of Eocene and Paleocene, western Gulf basin, *in* Cram, I. H., ed., Future petroleum provinces of the United States; Their geology and potential: American Association of Petroleum Geologists Memoir 15, p. 855–886.

Magara, K., 1981, Mechanics of natural fracturing in sedimentary basins: American Association of Petroleum Geologists Bulletin, v. 65, p. 123–132.

Mancini, E. A., and Mink, R. M., 1985, Petroleum production and hydrocarbon potential of Alabama's coastal plain and territorial waters, *in* Perkins, B. F., and Martin, G. B., eds., Habitat of oil and gas in the Gulf Coast: Society of Economic Paleontologists and Mineralogists 4th Annual Research Conference Proceedings, p. 25–42.

Martin, R. G., 1978, Northern and eastern Gulf of Mexico continental margin: stratigraphic and structural framework, *in* Bouma, A. H., Moore, G. T., and Coleman, J. M., eds., Framework, facies, and oil-trapping characteristics of the upper continental margin: American Association of Petroleum Geologists Studies in Geology 7, p. 21–42.

Maxwell, J. C., 1964, Influence of depth, temperature, and geologic age on porosity of quartzose sandstone: American Association of Petroleum Geologists Bulletin, v. 48, p. 697–709.

McFarlan, E., Jr., 1977, Lower Cretaceous sedimentary facies and sea level changes, U.S. Gulf Coast, *in* Bebout, D. G., and Loucks, R. G., eds., Cretaceous carbonates of Texas and Mexico: University of Texas Bureau of Economic Geology Report of Investigations 89, p. 5–11.

——— , 1988, The Lower Cretaceous, *in* Salvador, A., ed., The Gulf of Mexico basin: Boulder, Colorado, Geological Society of America, The Geology of North America, v. J (in press).

Mills, H. G., 1970, Geology of Tom O'Connor field, Refugio County, Texas, *in* Halbouty, M. T., ed., Geology of giant petroleum fields: American Association of Petroleum Geologists Memoir 14, p. 292–300.

Moore, C. H., 1984, The upper Smackover of the Gulf rim; Depositional systems, diagenesis, porosity evolution, and hydrocarbon production, *in* Ventress, W.P.S., Bebout, D. G., Perkins, B. S., and Moore, C. H., eds., The Jurassic of the Gulf rim: Society of Economic Paleontologists and Mineralogists 3rd Annual Research Conference Proceedings, p. 283–307.

Moore, C. H., and Druckman, Y., 1981, Burial diagensis and porosity evolution, upper Smackover, Arkansas and Louisiana: American Association of Petroleum Geologists Bulletin, v. 65, p. 597–628.

Murray, G. E., 1961, Geology of the Atlantic and Gulf Coastal province of North America: New York, Harper and Brothers, 692 p.

Murray, G. E., Rahman, A., and Yarborough, H., 1985, Introduction to the habitat of petroleum, northern Gulf (of Mexico) coastal province, *in* Perkins, B. F., and Martin, G. B., eds., Habitat of oil and gas in the Gulf Coast: Society of Economic Paleontologists and Mineralogists 4th Annual Research Conference Proceedings, p. 1–24.

Myers, J. D., 1968, Differential pressures, a trapping mechanism in Gulf Coast oil and gas fields: Gulf Coast Association of Petroleum Geologists Transactions, v. 28, p. 56–59.

Nehring, R., 1981, The discovery of significant oil and gas fields in the United States (prepared for U.S. Geological Survey and U.S. Department of Energy) R 2654 = USGS/DOE: Santa Monica, Rand Corporation, 236 p., Appendices, 477 p.

Newkirk, T. F., 1971, Possible future petroleum potential of Jurassic, western Gulf basin, *in* Cram, I. H., ed., Future petroleum provinces of the United States; Their geology and potential: American Association of Petroloem Geologists Memoir 15, p. 927–953.

New Orleans Geological Society, 1960; 1962; 1983a, Salt domes of southeast Louisiana: New Orleans Geological Society, v. 1, 145 p.; v. 2, 107 p.; v. 3, 131 p.

New Orleans Geological Society, 1965; 1967; 1983b, Oil and gas fields of southeast Louisiana: New Orleans Geological Society, v. 1, 195 p.; v. 2, 189 p.; v. 3, 225 p.

New Orleans Geological Society, 1974, Offshore Louisiana oil and gas fields: New Orleans Geological Society, 123 p.

Nunn, J. A., 1984, Subsidence and temperature histories for Jurassic sediments in the northern Gulf Coast, *in* Ventress, W.P.S., Bebout, D. G., Perkins, B. S., and Moore, C. H., eds., The Jurassic of the Gulf rim: Society of Economic Paleontologists and Mineralogists 3rd Annual Research Conference Proceedings, p. 309–322.

Nunn, J. A. and Sassen, R., 1986, The framework of hydrocarbon generation and migration, Gulf of Mexico continental slope: Gulf Coast Association of Geological Societies Transactions, v. 36, p. 257–262.

Olsen, R. S., 1982, Depositional environment of Jurassic Smackover sandstone, Thomasville Field, Rankin County, Mississippi: Gulf Coast Association of Geological Societies Transactions, v. 32, p. 59–66.

Ottman, R. D., Keyes, P. L., and Ziegler, M. A., 1976, Jay field, Florida; A Jurassic stratigraphic trap, *in* Braunstein, J., ed., North American oil and gas fields: American Association of Petroleum Geologists Memoir 24, p. 276–286.

Owen, E. W., 1975, Trek of the oil finders; A history of exploration for petroleum: American Association of Petroleum Geologists Memoir 6, 1,674 p.

Parker, C. A., 1974, Geopressure and secondary porosity in the deep Jurassic of Mississippi: Gulf Coast Associatio of Geological Societies Transactions, v. 24, p. 69–80.

Pearcy, J. R., and Ray, P. K., 1986, Production trends of the Gulf of Mexico; Exploration and development: Gulf Coast Association of Geological Societies Transactions, v. 36, p. 263–273.

Powell, J. B., 1972, Exploration history of Delhi Field, northeastern Louisiana, *in* King, R. E., ed., Stratigraphic oil and gas fields; Classification, exploration methods, and case histories: American Association of Petroleum Geologists Memoir 16, p. 548–558.

Powell, L. C., and Woodbury, H. O., 1971, Possible future petroleum potential of Pleistocene, western Gulf basin, *in* Cram, I. H., ed., Future petroleum provinces of the United States; Their geology and potential: American Association of Petroleum Geologists Memoir 15, p. 813–823.

Presley, M. W., ed., 1984, The Jurassic of east Texas: East Texas Geological Society, 304 p.

Price, L. C., 1976, Aqueous solubility of petroleum as applied to its origin and migration: American Association of Petroleum Geologists Bulletin, v. 60, p. 213–244.

Rainwater, E. H., 1971, Possible future potential of Lower Cretaceous, western Gulf basin, in Cram, I. H., ed., Future petroleum provinces of the United States; Their geology and potential: American Association of Petroleum Geologists Memoir 15, p. 901–926.

Rice, D. D., 1980, Chemical and isotope evidence of the origins of natural gases in offshore Gulf of Mexico: Gulf Coast Association of Geological Societies Transactions, v. 30, p. 203–213.

Rice, D. D., and Claypool, G. E., 1981, Generation, migration, and resource potential of biogenic gas: American Association of Petroleum Geologists Bulletin, v. 65, p. 5–25.

Robertson Research, 1980, Framework for oil and gas occurrence in the Gulf Coast Tertiary: Houston, Texas, Robertson Research (U.S.) Incorporated, in association with Curtis, D. M., and Echols, D. J., Report 241 p., Appendices p. A-1 to D-19, Maps, Cross sections, Atlas.

——, 1983, Gulf Coast Cretaceous; A regional geological synthesis: Houston, Texas, Robertson Research (U.S.), Incorporated, Report P. 1-1 to 9-76, Appendices p. A-1 to D-58, Maps, cross sections, Atlas.

Salvador, Amos, 1987, Late Triassic–Jurassic paleogeography and origin of the Gulf of Mexico basin: American Association of Petroleum Geologists Bulletin, v. 71, p. 419–451.

——, ed., 1988, The Gulf of Mexico basin: Boulder, Colorado, Geological Society of America, The Geology of North America, v. J (in press).

Sassen, R., and Moore, C. H., 1987, Source rock study of Smackover Formation from east Texas to Florida: American Association of Petroleum Geologists Bulletin, v. 71, p. 610.

Seni, S. J., and Jackson, M.P.A., 1983, Evolution of salt structures, east Texas diapir province: American Association of Petroleum Geologists Bulletin, v. 67, p. 1219–1274.

Shelton, J. W., 1984, Listric normal faults; An illustrated summary: American Association of Petroleum Geologists Bulletin, v. 68, p. 801–815.

Shelton, M. F., Jr., 1976, Tinsley oil field, Yazoo County, Mississippi, in Braunstein, J., ed., North American oil and gas fields: American Association of Petroleum Geologists Memoir 24, p. 239–258.

Shinn, A. D., 1971, Possible future petroleum potential of upper Miocene and Pliocene, western Gulf basin, in Cram, I. H., ed., Future petroleum provinces of the United States; Their geology and potential: American Association of Petroleum Geologists Memoir 15, p. 824–835.

Skillern, I. E., 1968, Sligo Field, Bossier Parish, Louisiana, in Beebe, C. W., and Curtis, B. F., eds., Natural gases in rocks of Mesozoic age: American Association of Petroleum Geologists Memoir 9, v. 1, pt. 2, p. 1146–1152.

Smith, D. A., 1961, Geology of South Pass Block 27 Field, offshore Plaquemines Parish, Louisiana: American Association of Petroleum Geologists Bulletin, v. 45, p. 51–71.

——, 1980, Sealing and nonsealing faults in Louisiana Gulf Coast basin: American Association of Petroleum Geologists Bulletin, v. 64, p. 145–172.

Smith, G. W., 1985, Geology of the deep Tuscaloosa (Upper Cretaceous) gas trend in Louisiana, in Perkins, B. F., and Martin, G. B., eds., Habitat of oil and gas in the Gulf Coast: Society of Economic Paleontologists and Mineralogists 4th Annual Research Conference Proceedings, p. 153–190.

Sohl, N. F., 1988, The Upper Cretaceous, in Salvador, A., ed., The Gulf of Mexico basin: Geological Society of America, Geology of North America, v. J (in press).

Spencer, C. W., 1987, Hydrocarbon generation as a mechanism for overpressuring in the Rocky Mountain region: American Association of Petroleum Geologists Bulletin, v. 71, p. 368–388.

Spooner, H. V., 1968, Black Lake field, Natchitoches Parish, Louisiana, in Beebe, C. W., and Curtis, B. F., eds., Natural gases in rocks of Mesozoic age: American Association of Petroleum Geologists Memoir 9, v. 1, pt. 2, p. 1152–1156.

Stanley, T. B., 1970, Vicksburg fault zone, Texas, in Halbouty, M. T., ed., Geology of giant petroleum fields: American Association of Petroleum Geologists Memoir 14, p. 301–308.

Stehli, F. G., Creath, W. B., Upshaw, C. E., and Forgotson, J. M., Jr., 1972, Depositional history of Gulfian Cretaceous of East Texas Embayment: American Association of Petroleum Geologists Bulletin, v. 56, p. 38–67.

Steiner, R. J., 1976, Grand Isle Block 16 Field, offshore Louisiana, in Braunstein, J., ed., North American oil and gas fields: American Association of Petroleum Geologists Memoir 24, p. 229–238.

Surles, M. A., 1985, Petroleum and source rock potential of the Eagleford Group: American Association of Petroleum Geologists Bulletin, v. 69, p. 309.

Terriere, R. T., 1976, Geology of Fairway field, east Texas, in Braunstein, J., ed., North American oil and gas fields: American Association of Petroleum Geologists Memoir 24, p. 157–176.

Timko, D. J., and Fertl, W. H., 1971, Relationship between hydrocarbon accumulation and geopressures and its economic significance: Journal of Petroleum Technology, p. 923–933.

Tipsword, H. L., Fowler, W. A., and Sorrell, B. J., 1971, Possible future petroleum potential of lower Miocene–Oligocene, western Gulf basin, in Cram, I. H., ed., Future petroleum provinces of the United States; Their geology and potential: American Association of Petroleum Geologists Memoir 15, p. 836–854.

Ventress, W.P.S., Bebout, D. G., Perkins, B. S., and Moore, C. H., eds., 1984, The Jurassic of the Gulf rim: Society of Economic Paleontologists and Mineralogists, Gulf Coast Section Foundation, 3rd Annual Research Conference Proceedings, 408 p.

Vernon, R. C., 1971, Possible future petroleum potential of pre-Jurassic, western Gulf basin, in Cram, I. H., ed., Future petroleum provinces of the United States; Their geology and potential: American Association of Petroleum Geologists Memoir 15, p. 954–979.

Vogler, H. A., and Robison, B. A., 1987, Exploration for deep geopressured gas; Corsair trend, offshore Texas: American Association of Petroleum Geologists Bulletin, v. 71, p. 777–787.

Walters, C. C., and Cassa, M. R., 1985, Regional organic geochemistry of offshore Louisiana: Gulf Coast Association of Geological Societies Transactions, v. 35, p. 277–286.

Waples, D. W., 1980, Time and temperature in petroleum formation; Application of Lopatin's method to petroleum exploration: American Association of Petroleum Geologists Bulletin, v. 64, p. 916–926.

Weber, K. J., 1986, Hydrocarbon distribution patterns in Nigerian growth-fault structures controlled by structural style and stratigraphy: American Association of Petroleum Geologists Bulletin, v. 70, p. 661–662.

Winker, C. D., 1984, Clastic shelf margins of the post-Comanchean Gulf of Mexico; Implications for deep-water sedimentation, in Characteristics of Gulf basin deep-water sediments and their exploration potential: Society of Economic Paleontologists and Mineralogists, Gulf Coast Section Foundation, 5th Annual Research Conference Program and Abstracts, p. 109–120.

Woodbury, H. O., Murray, I. B., Jr., Pickford, P. J., and Akers, W. H., 1973, Pliocene and Pl
of hydrocarbons in oils; Physical chemistry applied to petroleum geochemistry I: American Association of Petroleum Geologists Bulletin, v. 61, p. 573–600.

MANUSCRIPT ACCEPTED BY THE SOCIETY JULY 11, 1988

NOTE ADDED IN PROOF

Since the preparation of the manuscript for this chapter, abundant high-resolution deep seismic reflection profiles have become available. These data show clearly that many concepts regarding the genesis, structure, distribution, and nature of the salt features in and around the Gulf of Mexico must be modified. Interpretation of these seismic reflection profiles has revealed the presence of allochthonous salt bodies (detached from the Jurassic regional salt layer) and has made it possible to distinguish between these and the more deeply rooted autochthonous salt structures (attached to the mother salt). In addition to the previously recognized autochthonous diapiric structures such as pillows, stocks, and walls, the newer profiles show the presence of allochthonous salt sheets and sills, and salt domes generated from these remobilized salt features (Brooks, 1989; Lopez, 1989; Worrall and Snelson, 1989; Wu and others, 1990).

The salt structures illustrated in the Gulf Coast cross section (Plate 7D) are all shown as autochthonous. A revised version of this cross section should show the presence of widespread allochthonous features as well. The presence of unexplored subsalt drilling objectives (Pratsch, 1989) is an obvious implication of these facts for petroleum exploration.

REFERENCES

Brooks, R. O., 1989, Horizontal component of Gulf of Mexico salt tectonics, in Gulf of Mexico salt tectonics, associated processes and exploration potential: Gulf Coast Section SEPM Foundation, Tenth Annual Research Conference Program and Extended Abstracts, p. 22–24.

Lopez, J. A., 1989, Distribution of structural styles in the northern Gulf of Mexico and Gulf Coast, in Gulf of Mexico salt tectonics, associated processes and exploration potential: Gulf Coast Section SEPM Foundation, Tenth Annual Research Conference Program and Extended Abstracts, p. 101–107.

Pratsch, J. C., 1989, Salt in oil and gas exploration offshore Gulf Coast region, U.S.A., in Gulf of Mexico salt tectonics, associated processes and exploration potential: Gulf Coast Section SEPM Foundation, Tenth Annual Research Conference Program and Extended Abstracts: p. 111–114.

Worrall, D. M., and Snelson, S., 1989, Evolution of the northern Gulf of Mexico with emphasis on Cenozoic growth faulting and the role of salt, in Bally, A. W., and Palmer, A. R., eds., Geology of North America; An Overview: Geological Society of America, The Geology of North America, vol. A, p. 97–138.

Wu, S., Bally, A. W., and Cremez, C., 1990, Allochthonous salt, structure and stratigraphy of the northeastern Gulf of Mexico, Part II, Structure: Marine and Petroleum Geology, v. 7, no. 4, p. 334–370.

Chapter 20

Anadarko basin

Herbert G. Davis
Independent and Consulting Geologist, P.O. Box 3728, Edmond, Oklahoma 73083
Robert A. Northcutt
Independent and Consulting Geologist, 11032 Quail Creek Road, Suite 201, Oklahoma City, Oklahoma 73120

INTRODUCTION

The Anadarko basin is located in northwest Oklahoma, the Oklahoma and northern Texas Panhandles, southwest Kansas, and southeast Colorado, an area of about 155,400 km^2. The basin is tectonically bounded on the south by the Amarillo-Wichita uplift, on the west by the Sierra Grande uplift, on the northwest by the Las Animas arch, on the north by the Cambridge arch–Central Kansas uplift, on the northeast by the Nemaha Range (Eardley, 1962), on the east by the central Oklahoma fault zone (Amsden, 1975), and on the southwest by a major WNW-trending fault zone that connects the northwest end of the Arbuckle Mountains with the east end of the Wichita Mountains. Figure 1 illustrates several subsidiary elements of the greater Anadarko basin: the Hugoton Embayment, the Pratt anticline, the Sedgwick basin, the Keyes dome, and the Dalhart basin, all of which are important to understanding the economic geology of the basin. This is a major oil and gas province because of its large size, thick sedimentary section, and the fact that significant accumulations of hydrocarbons have been found in each of the Paleozoic systems.

The name "Anadarko Basin" was first proposed by Gould (1924). He described it as ". . . a large synclinal basin, . . ." He stated in his summary that "A structural trough, known as the Anadarko basin, is recognized as extending northwest from a point near the west end of the Arbuckle Mountains for a distance of more than 150 miles." He further speculated that ". . . this structural trough continues across the state line into the panhandle of Texas."

GENERAL GEOLOGIC CHARACTERISTICS

The Anadarko basin is a large, asymmetrical, structural basin. The south side is defined by a mobile belt consisting of a buried mountain range and a geosynclinal trough. There is a broad shelf area and a stable foreland to the north (Plate 6E).

Huffman's (1959) study of the Midcontinent region demonstrated that autogeosynclinal (Kay, 1951) conditions prevailed in western Oklahoma and north Texas in early and middle Paleozoic times. Carbonate rocks interspersed with lesser quantities of shale and sandstones were the principle deposits and include the equivalents of the well-known Knox and Ellenburger suite and the unique Chattanooga Shale. The adjacent shelf areas remained stable until very Late Mississippian time. Major orogenic movements in Morrowan and Atokan time produced most of the well-known tectonic features of the central Midcontinent, for example, the Amarillo-Wichita-Criner uplift, the Anadarko basin, the Nemaha range, the Las Animas arch, and the Sierra Grande uplift.

Ham and others (1964) identified the basement rocks beneath the southern part of this intracratonic geosyncline as mostly igneous flows and intrusives of Cambrian age. The overlying Paleozoic sediments in southern Oklahoma are divided into four major chronostratigraphic units: (1) Late Cambrian to Early Devonian marine sediments, mostly carbonates; (2) Late Devonian and Mississippian dark shales; (3) Pennsylvanian shales, sandstones, thin marine limestones, and local conglomerates; and (4) Permian red shales, siltstones and sandstones, and halite-gypsum evaporites. "The fullest representation is in the Anadarko basin, or western deep segment of the geosyncline where Late Cambrian through Permian sediments are 38,800 feet thick and rest upon Middle Cambrian volcanic flows and tuffs, probably at least 7,000 feet thick" (Ham, 1969).

The Anadarko basin can best be described as a continental multicycle basin associated with the craton margin. The Alberta, Appalachian, Arkoma, Black Warrior, Crazy Mountains, Eagle, Fort Worth, and Peel basins were also classified as having similar architecture (St. John and others, 1984).

Tectonic events in the Pennsylvanian Period, utilizing preexisting zones of weakness in many places, produced the great structural and stratigraphic diversity that characterize this basin. "Distribution and character of conglomerates in the Criner segment of the uplift indicate that movement may have started as early as Springer time" (Tomlinson and McBee, 1959), and if so, the central block was a source area. "The oldest conglomerates near the Criner uplift are in strata of late Springer age, and

Davis, H. G., and Northcutt, R. A., 1991, Anadarko Basin, *in* Gluskoter, H. J., Rice, D. D., and Taylor, R. B., eds., Economic Geology, U.S.: Boulder, Colorado, Geological Society of America, The Geology of North America, v. P-2.

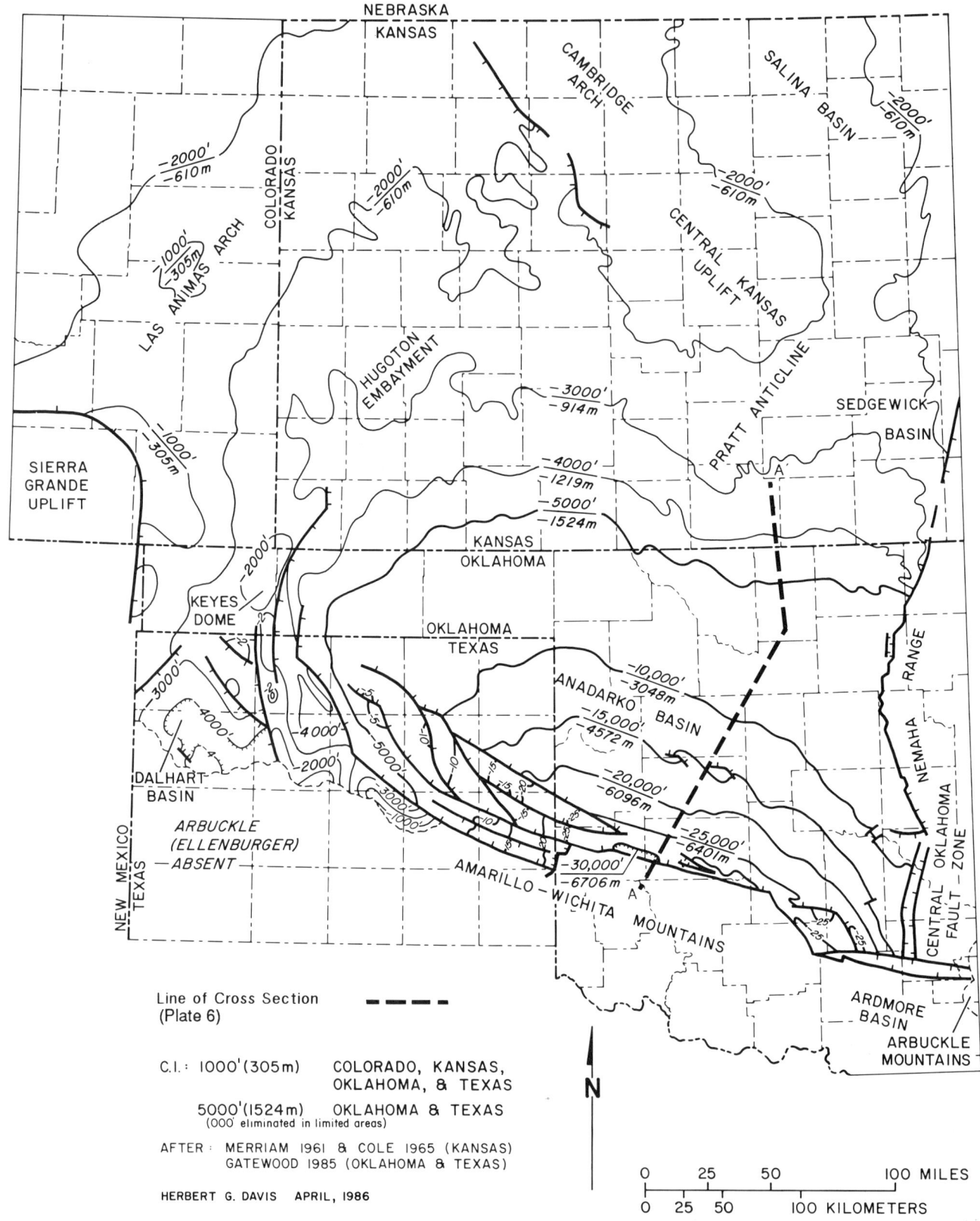

Figure 1. Structure contour map of the greater Anadarko basin, on top of the Arbuckle Group. Modified from Merriam (1963) and Cole (1976) in Kansas; modified from Gatewood (1985) in Oklahoma and Texas, also modified from personal maps. A-A' is line of cross section shown on Plate 6E.

overlying conglomerates of Morrow age contain cobbles and pebbles of Mississippian and Devonian rocks" (Frezon and Dixon, 1975).

Recent proprietary subsurface work on the Springer sandstones in the southeastern Anadarko basin (Cyril basin) indicates that no detritus from the Wichita or Criner uplifts is found in the massive Springer sandstones encountered in the deep tests drilled during the late 1970s or early 1980s (S. Takken, personal communication, 1986).

The area along the Amarillo-Wichita uplift to the west in the Texas Panhandle and the western portion of the deep basin in Oklahoma has detrital material in the upper Morrow, which indicates that the western part of the uplift was active and eroded before the eastern segment was uplifted during upper Morrowan time.

The Wichita orogeny (Tomlinson and McBee, 1959) resulted in structural uplift along the Amarillo-Wichita axis during Atokan time. The present configuration of the Anadarko basin was determined by faulting and folding during the Atokan. Major tectonic activity took place during Desmoinesian time: the Ouachita uplift east of the basin was formed, the Amarillo-Wichita mountains continued to rise, and the Sierra Grande uplift was rising to the west. During Missourian time, the Amarillo-Wichita uplift continued to be a positive area and was a source for a great volume of sediment. Tectonic features established in the basin were less active in Virgilian time than before.

EXPLORATION HISTORY AND TECHNIQUES

Exploration for petroleum began in the northeastern part of Indian Territory before Oklahoma became a state, focusing for the most part on oil seeps and hunches. By 1915, Carter County in southern Oklahoma was a major oil-producing area. The anticlinal theory of oil accumulation had been accepted, and geologists were mapping with plane table and alidade searching for "domes." A conspicuous fold in Permian rocks had been mapped in 1916 near the town of Cement in the southeastern part of the basin, and by the end of 1917, drilling had confirmed it as the first commercial oil field in the Anadarko basin.

The Panhandle gas field was discovered in 1918 in the Texas Panhandle. Ultimately this discovery led to the development of the largest known gas reserve in North America, a stratigraphic trap covering approximately 22,000 km^2 over three states. It is known today as the Panhandle-Hugoton gas field.

The introduction of seismic refraction surveys in 1922, together with new rotary drilling equipment, revolutionized oil exploration. Emphasis was still on the quest for structural traps. The wire-line logging tool introduced in 1929 made regional correlations of the Midcontinent stratigraphic section possible. The refinement of seismic capabilities, plus drilling and electric log technologies, continued during the 1930s and 1940s. Hydraulic fracturing was introduced in the 1950s and stimulated exploration by improving the economic performance of low-quality reservoirs.

Most of the large shallow structural traps had been found by 1950, and in the search for smaller ones, stratigraphic traps in the basin were being discovered. The Mocane-Laverne, Camrick, Putman, Postle, and Watonga fields are examples of stratigraphic traps that were developed in the 1950s and 1960s (Davis and Northcutt, 1989). Natural gas became a valuable commodity, and in the successful search for it, as well as for oil, the Anadarko basin came of age as a major hydrocarbon-bearing province.

Price controls on oil and gas depressed exploration from 1954 through 1973. The Arab oil embargo of 1973 resulted in dramatic increases in the prices for oil and gas at the wellhead by 1982. This economic incentive, combined with modern geological interpretations and exploration methods, and new drilling and producing techniques, created the greatest drilling boom in Anadarko basin history. The price structure for oil and gas collapsed in 1982 and then fell even further in 1986. Combined with changes in federal tax law in 1986, these factors had the effect of nullifying the profitability of oil and gas investments. As a result, the period from 1986 to 1990 has seen very little exploratory activity in the Anadarko basin.

If and when the economic incentives to find new oil and gas reserves are revived, future development of the Anadarko basin will require geologists to use sophisticated seismic techniques with emphasis on stratigraphic definition and detailed studies of rock properties.

STRUCTURE

The Anadarko basin is separated from the Ardmore basin by complicated faulting at its southeast end, and is composed of three distinct yet interconnected parts, as described below (Fig. 1; Plate 6E).

1. The complex frontal fault system extending from the buried Amarillo Mountains in the Texas Panhandle southeastward along the north side of the Wichita Mountains and thence to the Meers-Marlow-Doyle-Eola fault system that splays into the Mill Creek, Reagan, and Washita Valley system of faults in the Arbuckle Mountains (Harlton, 1963, 1972; Gatewood, 1985).

2. An area of relatively steep-dipping rocks between the frontal fault system and the hinge line of the basin. The hinge line appears to coincide with the regional –15,000-ft (4,572 m) Arbuckle structure contour (Fig. 1), the –13,000-ft (3,962 m) Hunton structure contour (Fig. 2), and the 2,000 ft (610 m) isopach line of the Morrow-Springer (Davis, 1974; Davis and Nondorf, 1974).

3. The northern shelf area extends northeast to the Nemaha range in Oklahoma and Kansas and north and northwest into Kansas, southeast Colorado, and the Texas Panhandle. This shelf area contains the major oil and gas producing stratigraphic traps in the basin.

STRATIGRAPHY

The Anadarko basin contains one of the most complete Paleozoic sections in the continental United States (Fig. 3). Logs from more than 70,000 wells provide the geologic data with which to study the stratigraphic architecture in three dimensions. One well, the Lone Star Production Company #1 Bertha Rogers, in Washita County, Oklahoma, is the deepest hole in North America and penetrated the West Spring Creek Formation of the Arbuckle Group at 9,520 m. If drilled to the basement, it would probably have reached the Cambrian Carlton rhyolite at approximately 11,765 m (Gatewood, 1985; Rowland, 1974b).

Granite and rhyolites of Early Cambrian age (525 to 535 Ma) that crop out in the Wichita Mountains are believed to floor the Anadarko basin as far north as the hinge line. These Early Cambrian igneous rocks are estimated to be 2,100 m thick in the deep basin (Ham and others, 1964; Ham and Wilson, 1967). North of the hinge, the central Oklahoma granite (1.36 Ga) is present beneath the Phanerozoic sediments (Denison and others, 1984).

The Phanerozoic section begins with the Late Cambrian Timbered Hills Group and the overlying Arbuckle Group; both are primarily carbonate rocks. The Timbered Hills and Arbuckle Groups together are the equivalent of the Sauk sequence of Sloss (1963), and are estimated to be 2,400 m thick in the deep Anadarko basin.

The Middle Ordovician Simpson Group ranges in thickness from 670 m in the deep basin to zero at its truncation in southwest Kansas. It consists of more or less equal amounts of clean quartz, and sheet-like sandstone units interbedded with limestone, shale, and siltstone (Huffman, 1959; Staler, 1965).

The Upper Ordovician Viola Group ranges in thickness from 460 m in the deep basin to zero in northwest Kansas. It consists mostly of finely crystalline and granular limestone and dolomite (Huffman, 1959; Chenoweth, 1966). The Sylvan Shale overlies the Viola Group and ranges in thickness from zero to 120 m. It is a brown, fissile dolomitic shale in the lower part, and is light green to gray-green and fossiliferous in the upper part. At its northern edge, near the Oklahoma-Kansas border, the stratigraphic equivalent is referred to as the Maquoketa Shale (Huffman, 1959; Ireland, 1966; Amsden, 1975).

The Silurian-Devonian Hunton Group is present in the deep basin and is truncated by the regional pre-Mississippian unconformity at its north and northwest limit on the shelf. It ranges in thickness from zero to 430 m. The lower Hunton, of Silurian age, is composed of cherty and fossiliferous limestone and dolomite. Where present, the upper Hunton of Devonian age, is fine to coarsely crystalline limestone with some local dolomitization and is locally cherty and fossiliferous (Rowland, 1974a; Amsden, 1975; S. E. Howery, personal communication, 1986).

The Late Devonian–Early Mississippian Woodford Shale is present throughout the Anadarko basin, but is absent on the Las

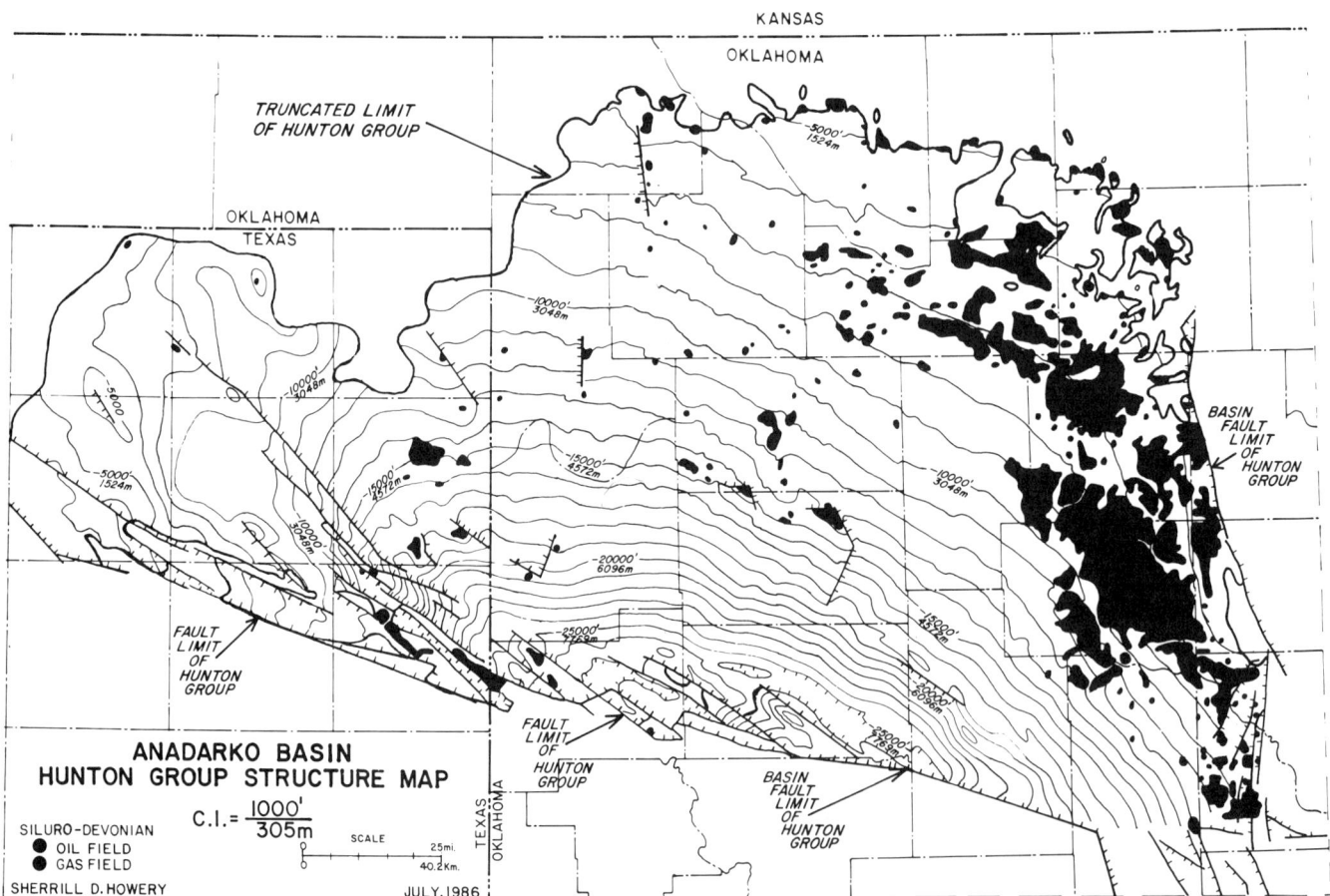

Figure 2. Structure contour map on the top of the Hunton Group, Anadarko basin.

ERA	SYSTEM	SERIES	Ma	GROUP	SUBSURFACE: FORMATION / MEMBER / LOCAL NAMES
CENOZOIC	QUATERNARY	PLEISTOCENE	2		Alluvium Colluvium Terrace Deposits Playa Deposits Pearlette Ash Windblown Deposits Odee
	TERTIARY	PLIOCENE	4.6 24		Black Mesa Basalt Ogallala
MESOZOIC	CRETACEOUS	GULFIAN	75	Montana	Pierre
				Colorado	Niobrara Carlile Greenhorn
		COMANCHEAN	100	Dakota	Graneros Purgatoire Kiowa Cheyenne
	JURASSIC		145		Morrison Ralston Creek
	TRIASSIC		210	Dockum	
PALEOZOIC	PERMIAN	OCHOAN	250	Quartermaster	Elk City Cloud Chief
		GUADALUPIAN	255	Whitehorse	
				El Reno	Blaine Glorieta
		LEONARDIAN	270	Enid - Clearfork - Sumner	Stone Corral Cimarron Hennessey Red Cave Tubb Ninnescah Wellington Panhandle
		WOLFCAMPIAN	275	Chase	Brown White Moore Herrington Krider Winfield Towanda Florence Wreford Ft. Riley
				Council Grove	Council Grove
				Admire	
	CARBONIFEROUS: PENNSYLVANIAN	VIRGILIAN	290	Wabaunsee	Hoover "Greenwood - Lansing"
				Shawnee	Heebner Topeka
				Douglas	Toronto Lovell Upper Wade
		MISSOURIAN		Pedee	Iatan Brown Lime
				Lansing	Tonkawa Avant Lower Wade
				Kansas City	Cottage Grove Marchand
				Pleasanton	Hogshooter Checkerboard Cleveland
		DES MOINESIAN	310	Marmaton	Big Lime Oswego
				Cherokee	Prue Verdigris Pink Deese Skinner Red Fork Inola Hart Osborne
		ATOKAN		Atoka	Atoka Lime "13 Finger Limestone" Novi Upper Dornick Hills
		MORROWAN		Morrow	UPPER Kearney Purdy Lower Dornick Hills Hollis Bradstreet Puryear MIDDLE Squaw Belly LOWER Keyes Primrose
		SPRINGERAN		Springer	Upper - Cunningham Lower - Spiers Middle - Britt Boatwright
	CARBONIFEROUS: MISSISSIPPIAN	CHESTERIAN	330	Chester	Kimball Okeene Parvin Manning Goddard Caney
		MERAMECIAN	340	Meramec	Ste. Genevieve St. Louis Salem Spergen Warsaw
		OSAGEAN	355	Osage	"Chat" "Mississippi Lime" Cowley "Mississippi Solid"
		KINDERHOOKIAN			Kinderhook
	DEVONIAN		365		Woodford Chattanooga Misener
			400	Hunton	Frisco Bois d'Arc Haragan
		CAYUGAN	405		Henryhouse
	SILURIAN	NIAGARAN			Chimneyhill
	ORDOVICIAN	CINCINNATIAN	425		Sylvan Maquoketa
				Viola	Fernvale Trenton
		CHAMPLAINIAN	455	Simpson	Bromide Tulip Creek McLish Oil Creek Joins
		CANADIAN	485	Arbuckle (Ellenberger)	West Spring Creek Kindblade Cool Creek McKenzie Hill Butterly
	CAMBRIAN	CROIXIAN	500		Signal Mountain Royer Ft. Sill
				Timbered Hills	Honey Creek Reagan
			525 ±	Carlton Rhyolite	Extrusive Igneous Rocks
PROTEROZOIC	PRECAMBRIAN		1100- 1400	Central Oklahoma Granite Group	Metasediments Granite

Vertical label alongside Group column (Wolfcampian to Morrowan): GRANITE · CARBONATE · WASHES - CHERT

References: Hills and Kottlowski
American Association of Petroleum Geologists
Correlation Chart Series 1983
COSUNA Chart Region - SSMC
CSD #360 and #450 G. Moore

Fay, R. O.
Oklahoma Geological Survey
1986 Personal Communication

Herbert G. Davis & Robert A. Northcutt, September 1986

Figure 3. Stratigraphic column for the Anadarko basin.

Animas arch in southeastern Colorado and west-central Kansas. It ranges in thickness from zero to 215 m and is the stratigraphic equivalent of the Chattanooga Shale and the Arkansas Novaculite. It is dark gray to black and is believed to be one of the primary sources of hydrocarbons in the basin (Huffman, 1959; Amsden, 1975; Cardott and Lambert, 1985; Comer and Hinch, 1987; Burruss and Hatch, 1989).

Rocks of Mississippian age are present throughout the basin and attain a thickness of over 1,500 m in the deep basin. They are predominantly shelf carbonates and interbedded shales; siliciclastic rocks account for a comparatively minor part.

The limestones, shales, and sandstones of Kinderhookian age are the thinnest and most areally restricted of the Mississippian rocks; they range from zero to 75 m and are widespread in the Hugoton Embayment and the Texas Panhandle (Frezon and Jordan, 1979). The possibility that they are present in the deep basin in Oklahoma is noted by Rowland (1974a), wherein he reported 60 m in the Lone Star Producing Company #1 Baden well in Beckham County, Oklahoma, an ultra-deep basin test drilled in 1972.

Rocks of Osagean age attain a thickness of more than 610 m in the deep basin and are composed of light-colored limestones and dolomites that are abundantly cherty. The rocks thin northward to zero on the flank of the Central Kansas uplift. A dark-colored shale facies known as the Cowley is encountered along the Oklahoma-Kansas line south of the Pratt anticline and in the shelf area of the Texas Panhandle. The Cowley facies consists of dark gray, argillaceous dolomites, and shales (Clair, 1948; Cunningham, 1969; Frezon and Jordan, 1979; Rascoe and Adler, 1983).

Rocks of Meramecian age are 395 m thick in the deep basin and thin northward to zero in northwestern Kansas (Cunningham, 1969; Frezon and Jordan, 1979). They consist of limestones and dolomites throughout the shelf portion of the basin. In the northern two-thirds of the basin, the rocks are known simply as the "Mississippi Lime" or the "Meramec Lime." In the deeper southeastern part of the basin a calcareous sandstone-to-shale facies overlies the carbonate units and is known as the Caney Shale. It includes rocks of both late Meramecian and early Chesterian age.

The Chesterian section ranges in thickness from zero to over 730 m in the deep basin and records a constriction of the Mississippian sea (Frezon and Jordan, 1979). It includes more shale than is found in the underlying Meramecian series and is limited in occurrence to the deep basin and portions of the shelf in northwest Oklahoma, the Texas Panhandle, and extreme southwest Kansas. The lower section of the Chester, in the northwestern shelf area of the basin, contains red, maroon, and green shales and some sandstones. Rocks of Chesterian age on the northeastern shelf of the basin, also called the Manning, consist of shales and limestones that are thin, light colored, chalky, fossiliferous, and oolitic. Southward, toward the deep basin, the limestones become darker colored, and shale increases to more than 50 percent of the section (Clair, 1948; Cunningham, 1969). In the southeastern part of the basin, the Caney is predominantly a medium-gray shale. It is overlain by the dark gray Goddard Shale, which was originally considered to be the lowermost part of the Springer Formation.

The Springer Formation is latest Mississippian and earliest Pennsylvanian in age. It is areally restricted to the deep and lower shelf portions of the Anadarko basin. The question of its age has been addressed by numerous authors, including Bennison (1956), Tomlinson and McBee (1959), Frezon and Dixon (1975), Frezon and Jordan (1979), and Sutherland (1982). The Springer is of special interest because its shales are believed to be one of the major gas-prone source rocks in the basin, its sandstones are some of the most important reservoirs, and it is commonly involved in structural dislocations because of the lubricity and thickness of its shales.

In the southeastern part of the Anadarko basin, the Springer consists of as much as 1,400 m of dark gray shale containing lenticular sandstones, which are usually concentrated in the upper third of the formation (Frezon and Dixon, 1975).

The top and bottom of the Springer, particularly in subsurface studies, are difficult to pick because the formation represents continuous deposition between the Caney Shale and the lower Morrowan section in the deep basin. Beyond the areas of continuous deposition, the top of the Springer is an unconformity that truncates the formation to a zero edge and bevels Early Pennsylvanian structures. The Springer sandstones are precursors of the great clastic wedges that manifest the structural growth of the Anadarko basin in Pennsylvanian time.

Because of the indeterminant nature of the Mississippian-Pennsylvanian boundary in the deep basin, the combined section is frequently referred to simply as the "Morrow-Springer" section, the top of which is the base of the Thirteen Finger Limestone (Atokan) and the bottom is usually picked at the top of the limestones of Chesterian age, or the Caney shale (Davis, 1971, 1974).

Rocks of Morrowan age are present throughout the deep Anadarko basin where they are predominantly dark gray shale with thick sandstones toward the base. Their northern limit occurs on the shelf area of northwest Oklahoma and the Hugoton Embayment. The series reaches a maximum thickness of 2,100 m in the deep basin (Frezon and Dixon, 1975).

Immediately north of the Amarillo-Wichita Mountains, the upper part of the section of Morrowan age consists primarily of chert conglomerate resulting from erosion of Lower Mississippian sediments exposed as the mountains were unroofed (Shelby, 1980). These conglomerate beds are rapidly replaced by gray marine shales into and across the axis of the basin. In the western part of the basin the upper Morrow rocks, consisting of quartz sandstones and gray shales derived from the Sioux uplift to the north, were deposited on a large deltaic plain (Swanson, 1979).

Rocks of Atokan age are more than 1,500 m thick (Frezon and Dixon, 1975) in the deep basin north of the Amarillo-Wichita Mountains, where they are dominantly carbonate, chert, and granite washes that evince the continued erosion of those mountains. These heterogeneous rocks wedge out toward the north, across the basin axis, where they interfinger with normal marine sediments of shale and limestone (Edwards, 1959; Lyday, 1985). The Atokan section thins northward, both by onlap of its base and truncation of its top, and goes to a zero edge along the northeastern shelf of the basin.

The Cherokee and Marmation Groups are the earliest Pennsylvanian units to extend over the entire basin. They reach maximum thickness of about 1,300 m in the Texas Panhandle (Frezon and Dixon, 1975) and in the extreme southeastern portion of the Anadarko basin, in the area referred to by Harlton (1972) as the Cyril basin.

The series consists of alternating thin limestones, lesser amounts of shale, and sandstones, which are locally important in the southeastern end of the basin and along the west flank of the Central Kansas uplift and into the Hugoton Embayment. Toward the basin the carbonates thicken at the expense of the shales, and in the deep basin the carbonates, in turn, give way to clastic sediments that were derived from the Ouachita fold belt in southeastern Oklahoma (Rascoe and Adler, 1983).

The Missourian Series is made up of the Pleasanton, Kansas City, and Lansing Groups in the subsurface of the Anadarko basin. It is more than 1,000 m thick in the deep basin area (Frezon and Dixon, 1975; Rowland, 1974a). Granite-wash and carbonate-wash clastics were eroded from the Amarillo-Wichita uplift and deposited on the flanks of that feature. Cyclic deposits of carbonates and shales were widely distributed over the shelf north of the deep Anadarko basin where the Missourian Series is represented by eight limestone-shale cyclothems of the Lansing–Kansas City Group (Rascoe and Adler, 1983).

The Virgilian Series comprises the Douglas, Shawnee, and Wabaunsee Groups and reaches a thickness of about 1,000 m in the eastern part of the Anadarko basin (Frezon and Dixon, 1975). The Douglas Group consists mostly of sandstones. The overlying Shawnee Group includes the Heebner shale, one of the more prominent marker beds used in mapping throughout the basin. The Heebner is approximately 1 m thick at the northern Kansas border and over most of the shelf area. It thickens southward from the basinal hinge line into the deeper basin where it is about 120 m thick. "Its great regional extent suggests that the Heebner Shale marked a major rise in sea level in the Mid-Continent and the time of greatest marine transgression during the Pennsylvanian. During this great rise in sea level the clastic sediments eroded from the Ouachita fold belt could no longer be delivered into the Anadarko basin by fluvial systems as before. As a result, a constructional shelf developed along the eastern margin of the basin" (Rascoe and Adler, 1983, p. 996). The uppermost Pennsylvanian Wabunsee Group is represented by alternating limestones and shales on the shelf changing to shale into the basin. Detritus from the rising Amarillo-Wichita Mountains predominates in rocks on the south flank of the basin.

Rocks of the Permian System attain a thickness of about 2,000 m in the Anadarko basin (McKee and others, 1967). The late Wolfcampian Chase Group changes from shelf carbonates in northern Oklahoma to semicontinental clastics updip in Texas and Kansas.

The Leonardian Series contains the first evidence of Permian aridity. It is composed of cyclic red beds, evaporites, and a few regionally extensive dolomite beds. The Guadalupian Series consists of evaporites, red shales, and dune sandstones.

Rocks of Mesozoic and Cenozoic ages are preserved in the extreme western part of the Anadarko basin. The Triassic section is 60 to 75 m thick; the Jurassic section is 75 to 150 m thick; and the Cretaceous section thins eastward from southeast Colorado across the Kansas part of the basin, ranging from 450 m in southeast Colorado to zero in south-central Kansas (Branson and Johnson, 1972; Merriam, 1963). The Pliocene Ogallala Formation is present in a broad north-south band across the Texas and Oklahoma panhandles and ranges in thickness from 15 to 150 m (Branson and Johnson, 1972). The Raton and Clayton lavas flowed on the Ogallala bajada in extreme northeastern New Mexico (Strower, 1972), beginning in latest Pliocene time and continuing intermittently nearly to the present.

THERMAL HISTORY

The thermal history of the Anadarko basin has been the subject of several theoretical studies. The source of heat and the timing and explanations for a gravity anomaly are some of the variables that are as yet poorly understood. Garner and Trucotte (1984) prepared a two-phase model in which the lithosphere beneath the deep basin was thinned and consequently heated during the interval of time between the onset of Springer deposition and the end of Desmoinesian deposition. Subsequently the rocks in the basin cooled slowly, became more rigid, and have supported themselves ever since. Building on that hypothesis, Schmoker (1986) stated, "Hydrocarbons have been generated in the Anadarko basin for >300 m.y., in an unusually long and continuous history that has contributed to the high oil and gas productivity of this Paleozoic province."

The degree of thermal maturation of the Upper Devonian–Lower Mississippian Woodford Shale in Oklahoma has been determined by vitrinite reflectance studies (Cardott and Lambert, 1985; Fig. 4). It is marginally mature with respect to oil generation (vitrinite reflectance, R_o values < 0.6 percent) along the shallow northern and eastern shelf areas and in the uplifted blocks of the frontal Wichita fault zone, and goes to post-mature with respect to oil generation (R_o values > 1.3 percent) in the deep Anadarko basin. The isoreflectance map (Fig. 4) is contoured to illustrate the change in vitrinite reflectance values from the shallow to the deep Anadarko basin. The contoured values are also illustrated on Plate 6E.

Plots of sediment thickness versus time (Dickinson and Yarborough, 1977; Donovan and others, 1983) indicate that there were two distinct episodes of subsidence in the basin. The rate of sediment accumulation was moderate during the Cambrian and Ordovician, slow from the Silurian to the Early Mississippian,

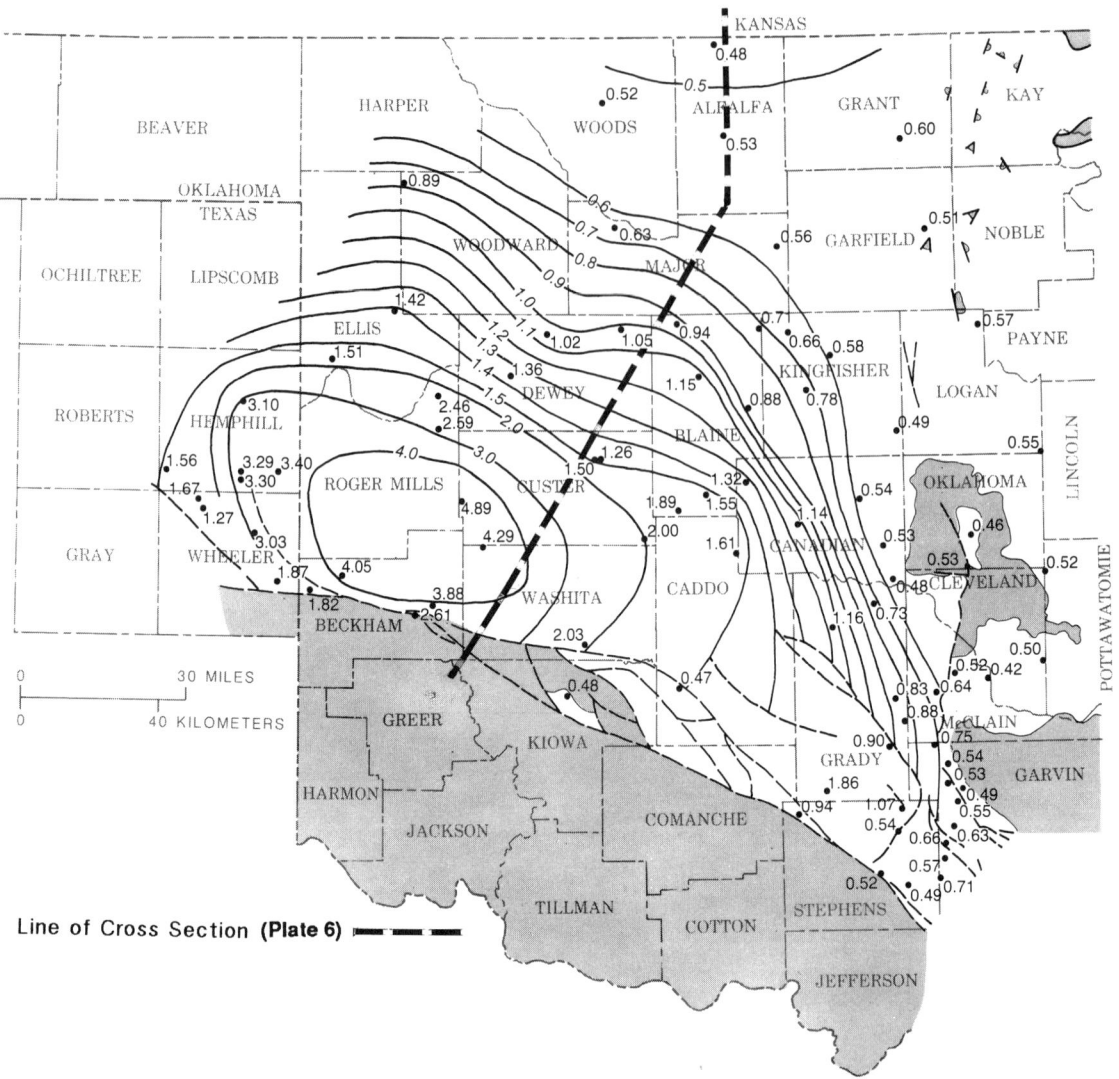

Figure 4. Isoreflectance map of the Woodford Shale in the Anadarko basin. Geology from Amsden (1975); faults indicated by fine dashed lines; shading represents areas where Woodford Shale is absent. Isoreflectance contour in percent. From Cardott (1989). A-A' is line of cross section shown on Plate 6E.

and very rapid during the Late Mississippian and Pennsylvanian. Feinstein (1981) showed that maximum thermal maturity of pre-Woodford strata in the Anadarko basin occurred during late Pennsylvanian and Early Permian times, corresponding to the time of greatest sediment accumulation. A localized high-temperature anomaly was postulated to explain the high vitrinite reflectance profile for that time. Present temperatures in the rock column are estimated to be cooler than those reached when the basin was forming in Pennsylvanian and Permian times (Cardott and Lambert, 1985).

CHARACTERISTIC PRODUCING SYSTEMS

Hydrocarbons are or have been produced from rocks of every Paleozoic system present in the Anadarko basin. As a generality, pre-Pennsylvanian rocks produce both oil and gas from structurally controlled traps, some of which formed prior to Woodford Shale deposition and were subsequently rejuvenated during the Pennsylvanian orogenies. The pre-Pennsylvanian reservoir rocks are predominantly carbonates, with the exception of the locally prodigious Bromide sandstones of late Ordovocian age. Reservoirs in stratigraphic traps are common in the Pennsylvanian sandstone-shale sequences. Pennsylvanian sandstones are also productive on many rejuvenated pre-Pennsylvanian structures. Permian carbonates produce petroleum in the giant Panhandle-Hugoton field, which is a stratigraphic trap. Some Permian sandstones produce on older, rejuvenated structures.

Hydrocarbons have become trapped in reservoir rocks wherever the source and reservoir sections are juxtaposed by faults, joints, or unconformities. Hydrocarbons have migrated from sources in the deep Anadarko basin since Late Mississippian time (Schmoker, 1986) and have charged reservoirs as far as 500 to 600 km away (Rich, 1931; Walters, 1958; Momper, 1975, 1978; Price, 1980).

Oils produced in the Anadarko basin can be characterized as having been generated in one of four primary sedimentary sections (Burruss and Hatch, 1989). The Upper Devonian and Lower Mississippian Woodford Shale is the source of the bulk of Anadarko basin oils. Pennsylvanian shales and some Ordovician carbonates and shales, are the other primary oil-generating units, but their combined output pales by comparison with that of the Woodford.

The tremendous volumes of gas that have been produced throughout the basin from Pennsylvanian rocks were most likely generated from the Upper Mississippian and Pennsylvanian shales. The great number of purely stratigraphic traps in Pennsylvanian sandstones indicates generation and entrapment much closer to the source.

Cumulative production of oil and gas for each of the producing systems cannot be accurately determined from the manner of record keeping in Oklahoma. Therefore, only total basin production can be obtained. As shown by Figure 5 the Anadarko basin, through 1985, has produced 82.4 trillion ft^3 of gas and 5.37 billion barrels of oil (Davis and Northcutt, 1989). This production has been obtained from hundreds of oil and gas fields located throughout the basin (Fig. 6).

Cambro-Ordovician producing system

Hydrocarbons are found in carbonates of the Arbuckle Group where diagenetic enhancement of porosity and intensive fracturing have created locally developed reservoirs. The most significant traps are structurally controlled and involve deep-seated faulting, early uplift, and possible early migration. Updip lateral seals are commonly provided by faulting of the reservoirs against an impervious rock. Top seals consist of shales of Woodford or of Pennsylvanian age.

Production from traps of these types is best exemplified by the Oklahoma City field on the east side of the basin where more than 815 million barrels of oil and 435 billion ft^3 of gas have been recovered since its discovery in 1928, of which 18 million barrels are from the Arbuckle carbonates (Gatewood, 1970). The West Mayfield and Mills Ranch fields located across the Oklahoma-Texas border immediately north of the Wichita Mountains, are structurally complex and thrust faulted, and produce from several different Paleozoic reservoirs, including those in the Arbuckle Group. Wells completed in the Arbuckle have produced more than 82 billion ft^3 of gas since 1975. The Chevron-Freeport 1 Ruth Ledbetter in Wheeler County, Texas, is the deepest producing well in the world; at 7,955 m from the Arbuckle Group (Jemison, 1979).

Significant volumes of oil and gas are produced from reservoirs in the Simpson Group and the Viola Formation, which are of Middle and Late Ordovician age. Massive Simpson sandstones are excellent reservoirs in the northeastern and eastern parts of the Anadarko basin where Gatewood (1970) attributed more than

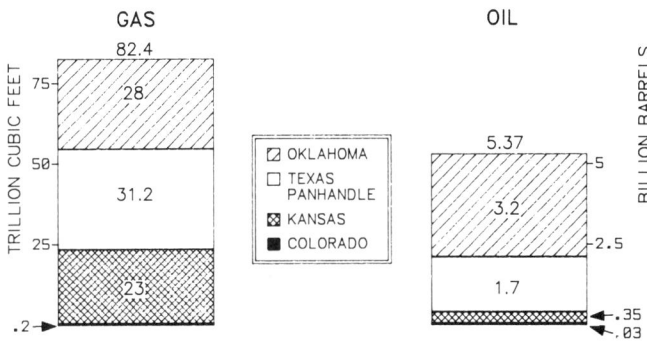

Figure 5. Oil and gas production of the greater Anadarko basin, 1917 to 1985.

90 percent of the reserves in the giant Oklahoma City field to them. Structural traps of Simpson sandstones in the Golden Trend and in several other fields have produced large volumes of oil per acre foot, but these are commonly of small areal extent; future Simpson discoveries are likely to be similar.

Carbonates of the Viola Group produce from both structural and stratigraphic traps, but the gross volume of such production is not large. The principal source of Viola oil seems to have been shales of the Woodford, Mississippian, and Pennsylvanian. Reservoirs are largely determined by locally developed porosity and permeability in conjunction with structural traps generally associated with early faulting. Viola reservoirs are usually sealed by Woodford shales. Wells completed in the Viola commonly come on streams with high initial production rates that soon decline to low sustained rates. The future economic viability of wells completed in Viola reservoirs will be more a question of oil prices than of geology.

Silurian and Devonian producing system

The principal reservoirs of Silurian and Devonian age in the Anadarko basin occur in the Hunton Group. Most of the Hunton oil was probably generated from the Wodford Shale, but Pennsylvanian shales may also have contributed to the total. These latter shales began expelling hydrocarbons into the migration pathways from the depocenter of the Anadarko basin in Middle Pennsylvanian time after the Hunton reservoirs had been enhanced by fracturing and diagenetically altered to dolomites.

Hydrocarbons accumulated in the Hunton reservoirs in both structural and stratigraphic traps. The West Mayfield–Mills Ranch field area is typical of complicated structural traps in the southern part of the deep basin. There, fracturing during Pennsylvanian structural movements greatly enhanced the Hunton reservoirs. Other productive structural traps on horst blocks include Mathers Ranch, Washita Creek, and Buffalo Wallow fields in the Texas Panhandle. Structural traps are also found at the shelf edge in western Oklahoma at North Custer City and Aledo fields

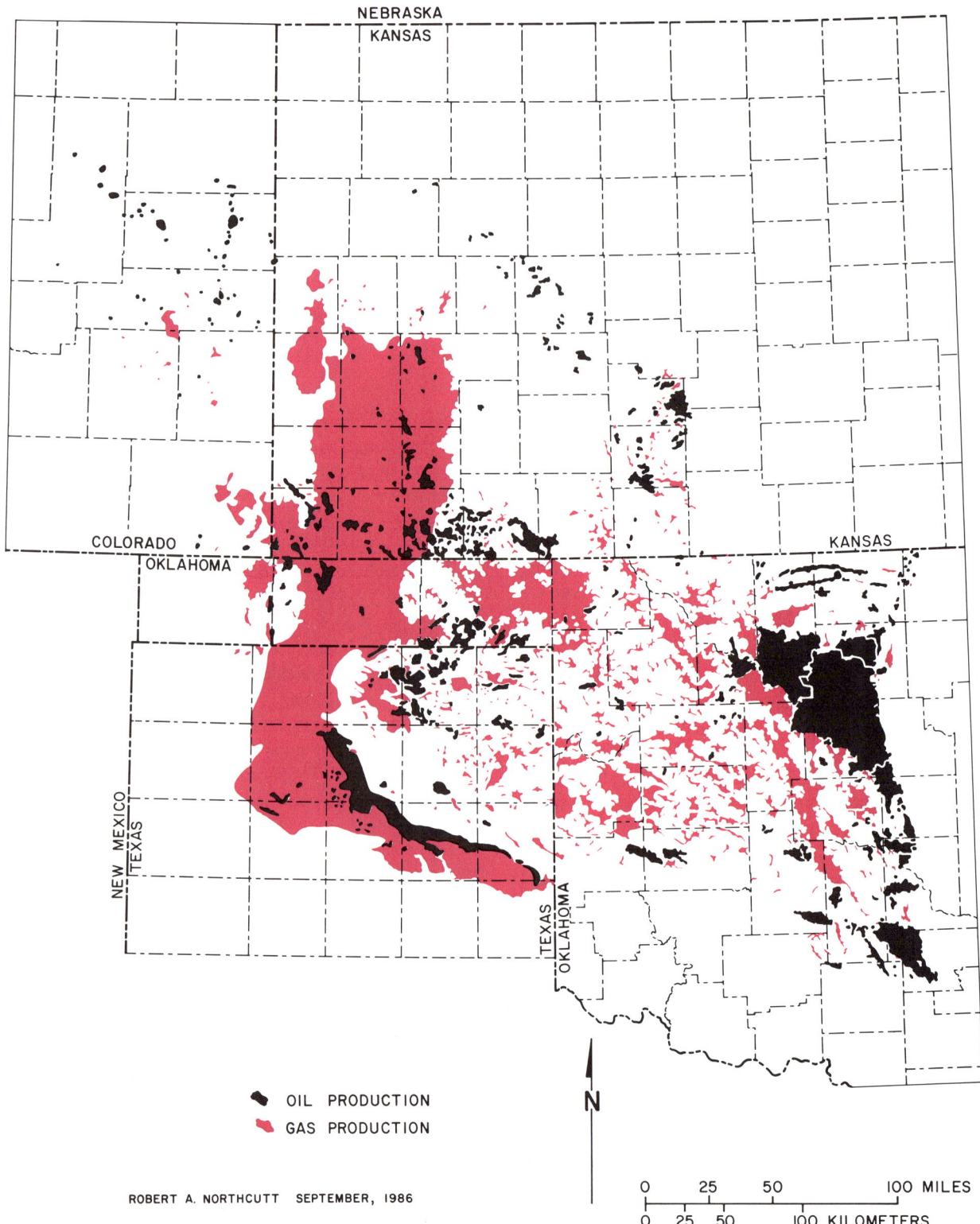

Figure 6. Oil and gas field map of the greater Anadarko basin. Many small fields are not shown, because of scale limitations. Field outlines are modified from maps as follows: Oklahoma, OGS Map GM-28 (Burchfield, 1985); Kansas KGS Map M-17 (Paul and others, 1982); Colorado, CGS Map Series 22 (Scanlon, 1983); Texas Panhandle (Petroleum Information Corporation, 1981); and from personal maps.

where up-to-the-basin faulting accounts for the lateral seal and the Woodford Shale seals the top.

Stratigraphic traps consisting of Hunton reservoirs overlain by shales at the post-Hunton unconformity have produced large volumes of petroleum along the eastern margin of the basin at the West Edmond and West Moore fields and in the Golden Trend in the southeastern part of the basin. Several small accumulations of Hunton oil have been found at the truncated updip limit of the lower Hunton on the shelf in western Oklahoma where diagenetic factors control the accumulations.

Cumulative oil production from the Hunton at West Edmond field exceeds 160 million barrels, and Hunton rocks are continuing to yield hydrocarbons from large, newly developed areas in the giant Golden Trend (Davis and Northcutt, 1989).

There is still potential for future discoveries in the Hunton. Most of them will be limited to subtle stratigraphic traps that are diagenetically controlled, possibly in conjunction with faulted structures.

The Misener sandstone is a locally prolific petroleum reservoir that was deposited in isolated pods and stringers on the eroded Hunton and older surface before the Upper Devonian Woodford Shale covered everything. Misener accumulations, usually found by accident, are often prodigious but of small areal extent in the northern shelf of the Anadarko basin.

Mississippian producing system

Source rocks for oil and gas trapped in Mississippian reservoirs are the Woodford Shale, Mississippian argillaceous carbonates, and shales of Late Mississippian and Early Pennsylvanian ages. Most of these rocks in the deep Anadarko basin began generating hydrocarbons in the Late Mississippian and continued to do so throughout Pennsylvanian time. Migration of hydrocarbons from the depocenter of the Anadarko basin led to early accumulations of oil and gas in Mississippian rocks.

Accumulations in carbonate reservoirs of Osagean and Meramecian ages are generally controlled by fractures and account for large areas of stratigraphic traps such as those found in the Sooner Trend along the northeastern margin of the Anadarko basin. This large stratigraphic trap is sealed by the overlying Mississippian shales and, where they are truncated to the northeast at the post-Mississippian unconformity, by Middle Pennsylvanian shales. Cumulative production from the giant Sooner Trend exceeds 300 million barrels of oil and 1 trillion ft^3 of gas (Davis and Northcutt, 1989). Average well reserves from these fractured carbonate reservoirs are quite low and in most cases do not provide favorable economics.

Meramec rocks in southwest Kansas produce at numerous small structural fields where porosity has been diagenetically enhanced by dolomitization. Updip pinchout of these porosity zones at the post-Mississippian unconformity on small structures at the Mississippian level are trapped by overlying Pennsylvanian shales. Larger structural features in southwest Kansas, as at the Pleasant Prairie field which has produced 20 million barrels of oil from porosity zones in the Meramec rocks, are worthwhile economic targets.

Rocks of Chesterian age are present throughout all but the outer limits of the Anadarko basin where they have been removed by erosion. Reservoirs in these rocks have a strong diagenetic imprint where solution and dolomitization have created favorable porosity and permeability in the carbonate units. Fracturing further enhances these reservoirs, which are stratigraphically trapped, with the Chesterian shales providing lateral and vertical seals and, where overlying Chester shales have been eroded, the Pennsylvanian shales providing top seals.

The large Ringwood field, discovered in 1945 in the northeastern part of the basin in Major County, Oklahoma, has produced nearly 90 million barrels of oil and 900 billion ft^3 of gas primarily from Chesterian carbonate rocks (Davis and Northcutt, 1989).

Potential for discovery and development of oil and gas reserves from Mississippian carbonate rocks in the Anadarko basin will depend on detailed stratigraphic studies to identify and isolate favorable porosity trends. At this mature stage of exploration, significant amounts of new production do not appear possible, but small areas of production may still be found.

The giant Watonga-Chickasha Trend and the Golden Trend oil field produce from the Springer sandstones, and several Springer gas fields that were found in the deep basin during the 1970s and 1980s will prove to be major accumulations (Davis, 1974; Frezon and Dixon, 1975; Rascoe and Adler, 1983; Walker, 1985).

Pennsylvanian producing system

Rocks of Pennsylvanian age are present throughout the Anadarko basin and account for the second-largest quantity of oil and gas produced there. Only the Permian section in the giant Panhandle-Hugoton gas field has produced more. Limestones and sandstones in every Pennsylvanian series produce on the shelf and in the deep basin from both structural and stratigraphic traps. The basal sandstones of Morrowan age in the Anadarko basin became well known as stratigraphic gas traps beginning in 1960. These lenticular sandstones, which were deposited in fluvial, fluvial-deltaic, deltaic, and pro-deltaic systems, produce oil and gas from southeastern Colorado across the entire basin and into the Ardmore basin (Davis, 1971); the lower sandstones are roughly equivalent to the sandstones referred to previously in the upper Springer section.

The Atokan section was found to contain several significant gas fields during the late 1970s and early 1980s (e.g., Berlin field; Lyday, 1985). For the most part, these fields are stratigraphic traps in arkosic graywackes.

Reservoirs in the Cherokee Group produce oil and gas from the Red Fork sandstones in Oklahoma and Cherokee sandstones in the Texas Panhandle and southwest Kansas; some Cherokee granite-wash fields are located immediately north of the Amarillo-Wichita Mountains in Oklahoma and Texas. Significant oil and gas production has been found in the Red Fork sandstone in the southeastern part of the basin, and several fields there will ultimately reach major size (e.g., East Clinton field; Pulling, 1985).

Reservoirs in the overlying Marmaton Group produce hydrocarbons from stratigraphically trapped limestones that are termed Marmaton in Kansas and Big Lime or Oswego in northwest Oklahoma and the Texas Panhandle. Putnam field, in northwest Oklahoma, is one of the most significant Oswego limestone fields in the Anadarko basin and is located at the approximate hinge line. It has produced more than 1,000 billion ft^3 of gas and 54 million barrels of oil (Brown, 1963; Northcutt, 1985).

Significant stratigraphic traps occur in sandstones of Missourian age where the fluvial-deltaic facies is trapped on the north and west by marine facies. The most prominent stratigraphic traps occur in the Tonkawa sandstone of northwest Oklahoma and the Oklahoma and Texas Panhandles. The reserves of gas in the Tonkawa of the Mocane-Laverne field are estimated to be as much as 800 billion ft^3 (Pate, 1968; Rascoe and Adler, 1983). Reservoirs of Missourian age in large structural traps occur in the deep basin. They include granite wash, which is the primary reservoir in the Elk City field (Edwards, 1959), and sandstones, which account for approximately 50 percent of the production at the Cement field (Hermann, 1961; Rascoe and Adler, 1983).

The Hoover sandstone in the Mocane-Laverne field (Pate, 1968) and the "Greenwood-Lansing" limestone in the Greenwood field (Wingerten, 1968) of southwest Kansas and southeast Colorado include major stratigraphic accumulations of gas in reservoirs involving the Wabaunsee Group.

Permian producing system

Wolfcampian rocks contain the largest accumulation of natural gas in the United States in the Panhandle-Hugoton field of Kansas, Oklahoma, and Texas. Figure 7 compares production from this giant gas field with the total Anadarko basin. The shelf carbonates of the Chase Group grade updip into semicontinental clastics to the west and north, creating a stratigraphic trap (Pippin, 1970; Rascoe and Adler, 1983). In addition, there is a complex hydrodynamic trap in the Kansas part of the field (Mason, 1968).

The Panhandle oil field is on the northeast flank and down dip of the Panhandle gas field in Texas and produces oil from dolomites of Wolfcampian and granite washes of Leonardian age. The oil occurs beneath the Panhandle gas and is trapped against the Amarillo uplift (Pippin, 1970).

The Red Cave sandstone of Leonardian age produces gas in the Interstate field located on the north flank of the Keyes dome in Morton County, Kansas (Rascoe and Adler, 1983).

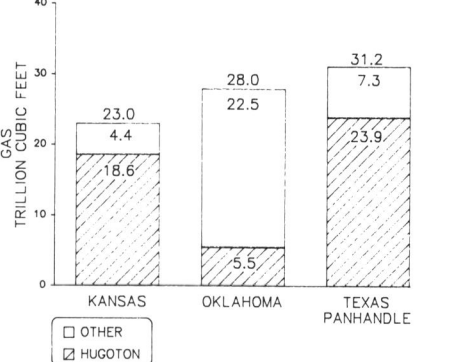
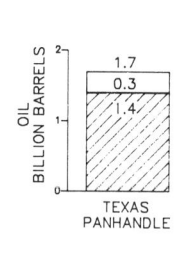

Figure 7. Comparison of oil and gas production from the greater Anadarko basin with that from the Panhandle-Hugoton field, 1971 to 1985.

POTENTIAL OF THE ANADARKO BASIN

As early as 1967, the oil reserves of the region were predicted to decline, and significant gas reserves were predicted to be added (Adler and others, 1971). It was stated that "significant stratigraphic accumulations of oil and gas will continue to be discovered in the Mid-Continent even though much of this region is now in a late mature stage of exploration and development" (Rascoe and Adler, 1983).

The deep drilling campaign conducted during the late 1970s and early 1980s determined that the entire basin affords structural and stratigraphic oil and gas possibilities to depths below 7,500 m. Gas will be the primary target, with extensions and re-drills on increased density patterns for the lenticular Pennsylvanian sandstones and Mississippian carbonates still being the primary targets.

The shelf area of the basin is in its late stage of exploration; however, the deep basin and mountain-front fault systems still afford opportunities for significant stratigraphic and structural traps. The thick sequence of Pennsylvanian siliciclastic sediments, and the Mississippian, Devonian, and Cambro-Ordovician carbonates, afford multiple stratigraphic plays as our knowledge of the stratigraphy unfolds by seismic studies and the drill bit.

Reinterpretation of the existing subsurface data and seismic coverage will undoubtedly allow exploration in the deep basin for years to come. Improvements in the sciences of geology and geophysics, and in petroleum engineering, coupled with the advancing use of computers and data processors, and both high-technology and horizontal drilling methods should lead to more efficient drilling, higher success, and better recoveries of reserves from reservoirs in the future.

REFERENCES CITED

Adler, F. J., and others, 1971, Future petroleum provinces of the Mid-Continent, region 7, *in* Cram, I. H., ed., Future petroleum provinces of the United States; Their geology and potential: American Association of Petroleum Geologists Memoir 14, v. 2, p. 985–1120.

Amsden, T. W., 1975, Hunton Group (Late Ordovician, Silurian, and Early Devonian) in the Anadarko basin of Oklahoma: Oklahoma Geological Survey Bulletin 121, 214 p.

Bennison, A. P., 1956, Springer and related rocks of Oklahoma: Tulsa Geological Society Digest, v. 24, p. 111–115.

Branson, C. C., and Johnson, K. S., compilers, 1972, Generalized geologic map of Oklahoma: Oklahoma Geological Survey Educational Publication 1 (updated 1979), 4 p., scale of 1:2,000,000.

Brown, D. P., 1963, Putnam field, *in* Oil and gas fields of Oklahoma: Oklahoma City Geological Society Reference Report, v. 1, p. 94A.

Burchfield, M. R., 1985, Oil and gas fields map of Oklahoma: Oklahoma Geological Survey Map GM-28, scale 1:500,000.

Burruss, R. C., and Hatch, J. R., 1989, Geochemistry of oils and hydrocarbon source rocks, greater Anadarko basin; Evidence for multiple sources of oils and long-distance oil migration: Oklahoma Geological Survey Circular 90, p. 53–64.

Cardott, B. J., 1989, Thermal maturation in the Woodford Shale in the Anadarko basin, *in* Johnson, K. S., ed., Anadarko Basin Symposium, 1988: Oklahoma Geological Survey Circular 90, p. 32–46.

Cardott, B. J., and Lambert, M. W., 1985, Thermal maturation by vitrinite reflectance of Woodford Shale, Anadarko basin, Oklahoma: American Association of Petroleum Geologists Bulletin, v. 69, p. 1982–1998.

Chenoweth, P. A., 1966, Viola oil and gas fields of the Mid-Continent: Tulsa Geological Society Digest, v. 34, p. 110–118.

Clair, J. R., 1948, Preliminary notes on lithologic criteria for identification and subdivision of the Mississippian rocks in western Kansas: Kansas Geological Society, 14 p.

Cole, V. B., 1976, Configuration of the top of the Precambrian rocks in Kansas: Kansas Geological Survey Map M-7, scale 1:500,000.

Comer, J. B., and Hinch, H. H., 1987, Recognizing and quantifying expulsion of oil from the Woodford Formation and age-equivalent rocks in Oklahoma and Arkansas: American Association of Petroleum Geologists Bulletin, v. 71, p. 844–858.

Cunningham, B. J., 1969, Identification and division of the pre-Pennsylvanian sediments, western Anadarko basin, *in* Pre-Pennsylvanian geology of western Anadarko basin: Panhandle Geological Society, p. 7–15.

Davis, H. G., 1971, The Morrow-Springer trend, Anadarko basin, target for 70s: Oklahoma City Geological Society Shale Shaker, v. 22, no. 3, p. 64–73.

——— , 1974, High pressure Morrow-Springer gas trend, Blaine and Canadian Counties, Oklahoma: Oklahoma City Geological Society Shale Shaker, v. 24, no. 6, p. 104–118.

Davis, H. G., and Nondorf, J. L., 1974, Morrow-Springer pressures of the Anadarko Basin: Society of Petroleum Engineers of American Institute of Mining, Metallurgical, and Petroleum Engineers, Inc., Deep Drilling and Production Symposium, Amarillo, Texas, September 8–10, 1974 paper 5174, p. 59–64.

Davis, H. G., and Northcutt, R. A., 1989, The greater Anadarko Basin: An overview of petroleum exploration and development: Oklahoma Geological Survey Circular 90, p. 13–24.

Denison, R. E., Lidiak, E. G., Bickford, M. E., and Kisvarsanyi, E. B., 1984, Geology and geochronology of Precambrian rocks in the central interior region of the United States: U.S. Geological Survey Professional Paper 1241-C, 20 p.

Dickinson, W. R., and Yarborough, H., 1977, Plate tectonics and hydrocarbon accumulation: American Association of Petroleum Geologists Continuing Education Course Note Series 1, 108 p.

Donovan, R. N., and 8 others, 1983, Subsidence rates in Oklahoma during the Paleozoic: Oklahoma City Geological Society Shale Shaker, v. 33, no. 8, p. 86–88.

Eardley, A. J., 1962, Structural growth of North America: New York, Harper and Row, 743 p.

Edwards, A. R., 1959, Facies changes in Pennsylvanian rocks along north flank of Wichita Mountains, *in* Petroleum geology of southern Oklahoma: American Association of Petroleum Geologists, v. 2, p. 142–155.

Feinstein, S., 1981, Subsidence and thermal history of southern Oklahoma aulacogen; Implications for petroleum exploration: American Association of Petroleum Geologists Bulletin, v. 65, p. 2521–2533.

Frezon, S. E., and Dixon, G. H., 1975, Texas Panhandle and Oklahoma; Paleotectonic investigations of the Pennsylvanian system in the United States; Part 1, Introduction and regional analyses of the Pennsylvanian system: U.S. Geological Survey Professional Paper 853-J, p. 177–195.

Frezon, S. E., and Jordan, L., 1979, Oklahoma, *in* Craig, L. C., and others, eds., Paleotectonic investigations of the Mississippian system in the United States; Part 1, Introduction and regional analysis of the Mississippian system: U.S. Geological Survey Professional Paper 1010-I, p. 147–159.

Garner, D. L., and Turcotte, D. L., 1984, The thermal and mechanical evolution of the Anadarko basin: Tectonophysics, v. 107, p. 1–24.

Gatewood, L. E., 1970, Oklahoma City field; Anatomy of a giant, *in* Halbouty, M. T., ed., Geology of giant petroleum fields: American Association of Petroleum Geologists Memoir 14, p. 223–254.

——— , 1984, Arbuckle structure map: Oklahoma City, L. E. Gatewood, scale 1:384,000.

Gould, C. N., 1924, A new classification of the Permian Redbeds of southwestern Oklahoma: American Association of Petroleum Geologists Bulletin, v. 8, p. 322–341.

Ham, W. E., 1969, Regional geology of the Arbuckle Mountains, Oklahoma: Oklahoma Geological Survey Guide Book 17, 52 p.

Ham, W. E., and Wilson, J. L., 1967, Paleozoic epeirogeny and orogeny in the central United States: American Journal of Science, v. 265, p. 332–407.

Ham, W. E., Denison, R. E., and Merritt, C. A., 1964, Basement rocks and structural evolution of southern Oklahoma: Oklahoma Geological Survey Bulletin 95, 302 p.

Harlton, B. H., 1963, Frontal Wichita fault system of southwestern Oklahoma: American Association of Petroleum Geologists Bulletin, v. 47, p. 1552–1580.

——— , 1972, Faulted fold belts of southern Anadarko Basin adjacent to frontal Wichitas: American Association of Petroleum Geologists Bulletin, v. 56, p. 1544–1551.

Herrmann, L. A., 1961, Structural geology of Cement-Chickasha area, Caddo and Grady Counties, Oklahoma: American Association of Petroleum Geologists Bulletin, v. 45, p. 1971–1993.

Huffman, G. G., 1959, Pre-Desmoinesian isopachous and paleogeologic studies in central Mid-Continent region: American Association of Petroleum Geologists Bulletin, v. 43, p. 2541–2574.

Ireland, H. A., 1966, Resume and setting of middle and upper Ordovician stratigraphy, Mid-Continent and adjacent regions: Tulsa Geological Society Digest, v. 34, p. 26–40.

Jemison, R. M., Jr., 1979, Geology and development of Mills Ranch complex; World's deepest field: American Association of Petroleum Geologists Bulletin, v. 63, p. 804–808.

Kay, M., 1951, North American geosyncline: Geological Society of America Memoir 48, 143 p.

Lyday, J. R., 1985, Atokan (Pennsylvanian) Berlin field; Genesis of recycled detrital dolomite reservoir, deep Anadarko basin, Oklahoma: American Association of Petroleum Geologists Bulletin, v. 69, p. 1931–1949.

McKee, D. A., and others, 1967, Paleotectonic maps of the Permian system: U.S. Geological Survey Miscellaneous Geological Investigations Map I-450, 164 p.

Mason, J. W., 1968, Hugoton Panhandle field, Kansas, Oklahoma, and Texas, *in* Natural gases of North America; Part 3, Natural gases in rocks of Paleozoic age: American Association of Petroleum Geologists Memoir 9, v. 2, p. 1539–1547.

Merriam, D. F., 1963, The geologic history of Kansas: Kansas Geological Survey Bulletin 162, 137 p.

Momper, J. A., 1975, Time and temperature relations reflecting the origin, expulsion, and preservation of oil and gas, *in* Proceedings of the 9th World Petroleum Congress: London, Applied Science Publishers, Geology, v. 2, p. 199–205.

——, 1978, Oil migration limitations suggested by geological and geochemical considerations, *in* Physical and chemical constraints on petroleum migration: American Association of Petroleum Geologists Course Note Series 8, p. B-1–B-60.

Northcutt, R. A., 1985, Oil and gas development in Oklahoma, 1891–1984; Oklahoma City Geological Society Shale Shaker, v. 35, no. 6, p. 123–132.

Pate, J. D., 1968, Laverne gas area, Beaver and Harper Counties, Oklahoma, *in* Natural gases of North America; Part 3, Natural gases in rocks of Paleozoic age: American Association of Petroleum Geologists Memoir 9, v. 2, p. 1509–1524.

Paul, S. E., Chang, L. H., and Burt, S., 1982, Oil and gas fields in Kansas: Kansas Geological Survey Map M-17, scale 1:500,000.

Petroleum Information Corporation, 1981, Midcontinent oil and gas production map: Denver, Colorado, Petroleum Information Corporation, scale 1:1,000,000.

Pippin, L., 1970, Panhandle-Hugoton field, Texas, Oklahoma, Kansas: The first fifty years, *in* Halbouty, M. T., ed., Geology of giant petroleum fields: American Association of Petroleum Geologists Memoir 14, p. 204–222.

Price, L., 1980, Shelf and shallow basin oil as related to hot-deep origin of petroleum: Journal of Petroleum Geology, v. 3, no. 1, p. 91–116.

Pulling, D. M., 1985, Clinton gas field; A significant stratigraphic discovery [abs.]: American Association of Petroleum Geologists Bulletin, v. 69, p. 1319.

Rasco, B., Jr., and Adler, F. J., 1983, Permo-Carboniferous hydrocarbon accumulations, Mid-Continent, U.S.A.: American Association of Petroleum Geologists Bulletin, v. 67, p. 979–1001.

Rich, J. L., 1931, Function of carrier beds in long distance migration of oil: American Association of Petroleum Geologists Bulletin, v. 15, p. 911–924.

Rowland, T. L., 1974a, The historic 1 Baden Unit and a brief look at exploration in the Anadarko basin: Oklahoma Geology Notes, v. 34, p. 3–9.

——, 1974b, Lone Star 1 Rogers Unit captures world depth record: Oklahoma Geology Notes, v. 34, p. 185–189.

Scanlon, A. H., 1983, Oil and gas fields map of Colorado: Colorado Geological Survey Map Series, No. 22, scale 1:500,000.

Schmoker, J. W., 1986, Oil generation in the Anadarko Basin, Oklahoma and Texas; Modeling using Lopatin's method: Oklahoma Geological Survey Special Publication 86-3, p. 2.

Shelby, J. M., 1980, Geologic and economic significance of the upper Morrow chert conglomerate reservoir of the Anadarko Basin: Journal of Petroleum Technology, v. 32, no. 3, p. 489–495.

Sloss, L. L., 1963, Sequences in the cratonic interior of North America: Geological Society of America Bulletin, v. 74, p. 93–113.

St. John, B., Bally, A. W., and Klemme, H. D., 1984, Sedimentary provinces of the World; Hydrocarbon productive and nonproductive: American Association of Petroleum Geologists Map Series, 36 p., 1 sheet.

Staler, A. T., 1965, Stratigraphy of the Simpson group in Oklahoma: Tulsa Geological Society Digest, v. 33, p. 162–211.

Strower, J. C., 1972, Age and nature of volcanic activity on the southern High Plains, New Mexico and Colorado: Geological Society of America Bulletin, v. 83, p. 2443–2448.

Sutherland, P. K., ed., 1982, Lower and Middle Pennsylvanian stratigraphy in south-central Oklahoma: Oklahoma Geological Survey Guidebook 20, 44 p.

Swanson, D. C., 1979, Deltaic deposits in the Pennsylvanian upper Morrow Formation of the Anadarko basin, *in* Hyne, N. J., ed., Pennsylvanian sandstones in the Mid-Continent: Tulsa Geological Society Special Publication 1, p. 115–168.

Tomlinson, C. W., and McBee, W., 1959, Pennsylvanian sediments and orogenies of Ardmore district, Oklahoma, *in* Petroleum geology of southern Oklahoma: American Association of Petroleum Geologists, v. 2, p. 3–52.

Walker, J. R., 1985, Development and economic significance of Springer-Britt sandstone, Eakly field, Caddo, Custer, and Washita Counties, Oklahoma [abs.]: American Association of Petroleum Geologists Bulletin, v. 69, p. 1320.

Walters, R. F., 1958, Differential entrapment of oil and gas in Arbuckle dolomite of central Kansas: American Association of Petroleum Geologists Bulletin, v. 42, p. 2133–2173.

Wingerten, H. R., 1968, Greenwood gas field, Kansas, Colorado, and Oklahoma, *in* Natural gases of North America; Part 3, Natural gases in rocks of Paleozoic age: American Association of Petroleum Geologists Memoir 9, v. 2, p. 1557–1566.

Manuscript Accepted by the Society February 12, 1991

ACKNOWLEDGMENTS

The authors thank the following individuals for significant help with review, materials and data for illustrations, and general support during the preparation of this manuscript: R. H. Espach, Jr., L. E. Gatewood, S. D. Howery, B. J. Cardott, C. J. Mankin, and M. Summers.

Chapter 21

The Permian basin

Bernold M. Hanson
P.O. Drawer 1269, Midland, Texas 79702
Brian K. Powers
4100 Dyer Circle, Midland, Texas 79707
Chester M. Garrett, Jr.
Bureau of Economic Geology, University of Texas, Box X, University Station, Austin, Texas 78713
Donald E. McGookey
310 West Illinois, Suite 314, Midland, Texas 79701
Ed H. McGlasson
Mobil Producing Texas-New Mexico, Box 633, Midland, Texas 79702
Ralph L. Horak
802 Twilight Circle, Richardson, Texas 75080
S. J. Mazzullo
S. J. Mazzullo and Associates, 401 East Illinois, Suite 301, Midland, Texas 79701
Alastair M. Reid
1000 West Storey, Midland, Texas 79701
Gerald G. Calhoun
2606 Terrace, Midland, Texas 79705
John Clendening and Brenda Claxton
Amoco Production Company, P.O. Box 3092, Houston, Texas 77253

INTRODUCTION

The Permian basin of southeast New Mexico and western Texas occupies 297,850 km^2 in the southwestern part of the United States craton. It comprises the area south of the Matador uplift, west of the Concho Arch, north of the Marathon-Ouachita thrust belt, and east of the Diablo Platform (Fig. 1). Major structural elements within the greater Permian basin include the Midland basin on the east, the Delaware basin on the west, and between these basins, an uplift called the Central Basin Platform. The structural bowl of the region is the Delaware basin. Maximum sediment thickness is on the order of 8,200 m (Plate 7C).

STRUCTURE AND SEDIMENTATION

The tectonic history of the Permian basin was comparatively stable for long stretches of time. Rocks in the Permian basin range in age from Precambrian through recent (Table 1). The greatest thickness of the sedimentary section was deposited in the Paleozoic Era. During the early and middle Paleozoic, west Texas was the site of the Tobosa basin (Galley, 1958), which was a shallow marine basin occupying approximately the same area as the later Permian basin. Sediments of the Tobosa basin were deposited during the early phase of the Paleozoic Era. A period of tectonic activity culminating in the pulsations of the Marathon-Ouachita orogeny occurred during late Middle Pennsylvanian, and the final pulsation occurred during the Early Permian. The present structural expression was developed in Pennsylvanian and Early Permian time (Fig. 2). Vertical or near-vertical faults characterize the structural style, and Henderson and others (1983) attribute vertical or near-vertical faults to strike-slip or wrench tectonics. Although Laramide structures are found in west Texas along the Rio Grande River, evidence of tectonic movements during the Laramide in the Permian basin is scanty or nonexistent. At the western extremity of the Permian basin, basin-and-range type structures occur.

The rocks throughout the Permian basin are equally divided between clastics and carbonates. Early Paleozoic time was characterized by carbonate development on a stable cratonic margin. In Pennsylvanian and Permian time, uplifts were capped by carbonate reefs and banks, and basins were filled with clastics and evaporites.

Hanson, B. M., Powers, B. K., Garrett, C. M., Jr., McGookey, D. E., McGlasson, E. H., Horak, R. L., Mazzullo, S. J., Reid, A. M., Calhoun, G. G., Clendening, J., and Claxton, B., 1991, The Permian basin, *in* Gluskoter, H. J., Rice, D. D., and Taylor, R. B., eds., Economic Geology, U.S.: Boulder, Colorado, Geological Society of America, The Geology of North America, v. P-2.

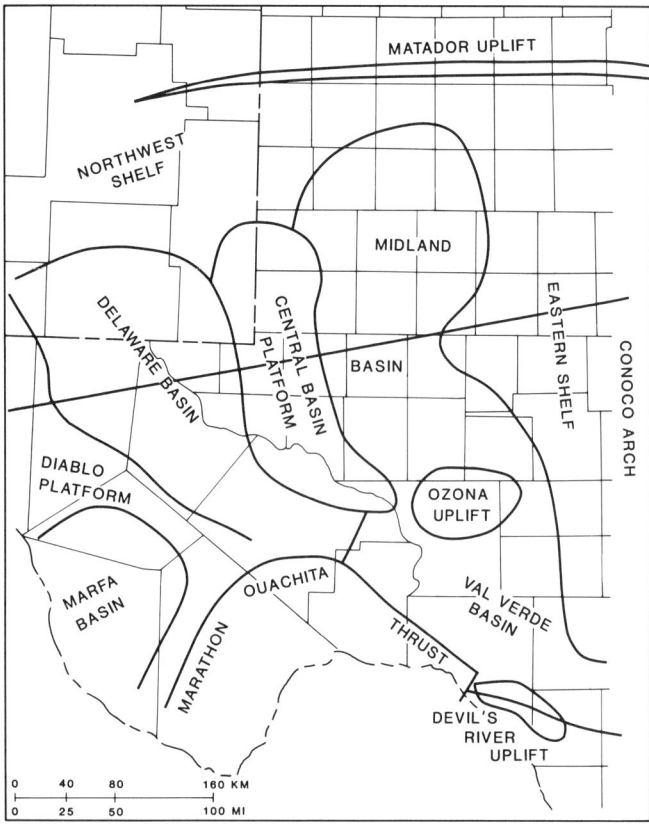

Figure 1. Outline of major structural elements of the greater Permian basin with partial line of east-west cross section as shown on Plate 7C.

The discussion in this chapter is confined to the sedimentary producing section (Table 2). The thermal history of these rocks is illustrated in a scale found on Plate 7C.

ORDOVICIAN, SILURIAN, DEVONIAN, AND MISSISSIPPIAN PRODUCING SYSTEMS

Introduction

Early and middle Paleozoic sediments were deposited in the Tobosa basin. This dish-shaped cratonic basin formed in western Texas and southeastern New Mexico during the Ordovician Period and persisted until the Early Pennsylvanian. It was bounded on the northeast, north, and west by low-relief land areas, and it was probably connected toward the southeast with deeper waters (Fig. 2). Although sea level oscillated considerably during the middle Paleozoic, there were no orogenic or taphrogenic tectonics in the area of the Tobosa basin.

Ordovician reservoirs comprise the Lower Ordovician Ellenburger Group, the Middle Ordovician Simpson Group, and the Upper Ordovician Montoya/Sylvan Formation. These rocks are the oldest important reservoirs in the Permian basin. The Ellenburger accounts for more than half of the pre-Pennsylvanian structurally controlled oil and gas production and reserves in the Permian basin and 7.5 percent of the basin's cumulative production.

The middle Paleozoic reservoirs are the Fusselman Formation, the Upper Silurian carbonate facies, and the Lower Devonian Thirtyone Formation (Hills and Hoenig, 1979). The Late Devonian to Early Mississippian Woodford Shale is considered the source of most of the hydrocarbons in Silurian and Devonian reservoirs.

Reservoirs

The Ellenburger carbonates are present in the entire Permian basin (and extend well beyond its limits), except locally where they were removed during the late Paleozoic. The maximum thickness of the group is between 490 and 520 m in the Val Verde basin and southern Delaware basin (Fig. 3). The section is diachronous, so exposures at the type locality in the Ellenburger Hills of central Texas are older than exposures of the approximately equivalent El Paso Group in the Franklin Mountains at El Paso. The top of the Ellenburger is regarded to be an erosional unconformity everywhere.

Lithologically the Ellenburger varies from dominantly dark, finely crystalline dolomite in the subsurface to interbedded limestone and dolomite at outcrops in central Texas (Barnes and Cloud, 1946; Barnes and others, 1959). The Ellenburger equivalent is principally limestone in outcrops in the region of the Marathon-Ouachita thrust belt (Marathon Limestone), the Sierra Diablo, the Hueco Mountains, and the Franklin Mountains (El Paso Group).

The Ellenburger was deposited in a variety of platform and strandline environments, which probably include, in part, restricted (evaporitic) shelves. In the lower Ellenburger, shelf facies include laminated mudstone, burrowed mudstone, interclast packstone, algal boundstone, and oolite grainstone. Loucks and Anderson (1980) inferred that the lower Ellenburger was deposited on an evaporitic tidal flat. Retained blocky lath-like textures suggest that the original evaporites were removed by regional dissolution. A wider diversity of middle and upper Ellenburger lithologies suggests deposition in subtidal and intertidal/supratidal channel belts (Loucks and Anderson, 1980).

Probably the most conspicuous attribute of the Ellenburger is that it is fractured and brecciated regionally. Brecciation is thought to have been related to removal of the original evaporites by regional dissolution, a relationship that has been recognized in many other carbonate basins. J. H. Halsey (personal communication, 1986) suggests that the *storage* capacity of the Ellenburger as a hydrocarbon reservoir is related to the localized breccia systems. The *drainage* capacity of the reservoir is related to the systematic fractures and joints, which are regional in scope. Fracture occurrence may be coincident with major episodes of orogenic stress and activity during the late Paleozoic and Laramide time.

TABLE 1. STRATIGRAPHIC NOMENCLATURE OF THE PERMIAN BASIN
AND IDENTIFICATION OF RESERVOIR ROCKS, SOURCE ROCKS, AND SEALS

SYSTEM	SERIES	NORTHWEST SHELF	CENTRAL BASIN PLATFORM	MIDLAND BASIN & EASTERN SHELF	DELAWARE BASIN	VAL VERDE BASIN			
PERMIAN	OCHOA	DEWEY LAKE RUSTLER SALADO	DEWEY LAKE RUSTLER SALADO	DEWEY LAKE RUSTLER SALADO	DEWEY LAKE RUSTLER SALADO CASTILE	RUSTLER SALADO	●		□
	GUADALUPE	TANSILL YATES SEVEN RIVERS QUEEN GRAYBURG SAN ANDRES ─── GLORIETA ───	TANSILL YATES SEVEN RIVERS QUEEN GRAYBURG SAN ANDRES ─── GLORIETA ───	TANSILL YATES SEVEN RIVERS QUEEN GRAYBURG SAN ANDRES ─── SAN ANGELO ───	DELAWARE MT GROUP BELL CANYON CHERRY CANYON BRUSHY CANYON	TANSILL YATES SEVEN RIVERS QUEEN GRAYBURG SAN ANDRES	●	△	□
	LEONARD	ABO YESO CLEAR FORK WICHITA	CLEAR FORK WICHITA	LEONARD SPRABERRY, DEAN	BONE SPRING	LEONARD	●	△	□
	WOLFCAMP	WOLFCAMP	WOLFCAMP	WOLFCAMP	WOLFCAMP	WOLFCAMP	●	△	□
PENNSYLVANIAN	VIRGIL	CISCO	CISCO	(ABSENT OR THIN) / CISCO	(ABSENT OR THIN)	(ABSENT OR THIN)	●	△	□
	MISSOURI	CANYON	CANYON	CANYON			●		□
	DES MOINES	STRAWN	STRAWN	STRAWN	STRAWN	STRAWN	●	△	□
	ATOKA	ATOKA ─ BEND ─	ATOKA ─ BEND ─	ATOKA ─ BEND ─	ATOKA ─ BEND ─	ATOKA ─ BEND ─	●	△	□
	MORROW	MORROW	(ABSENT)	(ABSENT)	MORROW	(ABSENT)	●	△	□
MISSISSIPPIAN		CHESTER MEREMEC OSAGE KINDERHOOK	CHESTER MEREMEC-OSAGE •BARNETT•	CHESTER MEREMEC-OSAGE •BARNETT•	CHESTER MEREMEC-OSAGE •BARNETT•	MEREMEC-OSAGE •BARNETT•	●	△	
			KINDERHOOK	KINDERHOOK	KINDERHOOK		●	△	□
DEVONIAN		─── WOODFORD ─── DEVONIAN	─── WOODFORD ─── DEVONIAN	─── WOODFORD ─── DEVONIAN	─── WOODFORD ─── DEVONIAN	─── WOODFORD ─── DEVONIAN	●		
SILURIAN		SILURIAN (UNDIFFERENTIATED)	SILURIAN SH FUSSELMAN	SILURIAN SH FUSSELMAN	MID. SILURIAN FUSSELMAN	MID. SILURIAN FUSSELMAN	●	△	□
ORDOVICIAN	UPPER	MONTOYA	MONTOYA	SYLVAN MONTOYA	SYLVAN MONTOYA	SYLVAN MONTOYA	●		
	MIDDLE	SIMPSON	SIMPSON	SIMPSON	SIMPSON	SIMPSON	●	△	□
	LOWER	ELLENBURGER	ELLENBURGER	ELLENBURGER	ELLENBURGER	ELLENBURGER	●		
CAMBRIAN	UPPER	CAMBRIAN	CAMBRIAN	CAMBRIAN	CAMBRIAN	CAMBRIAN			
PRECAMBRIAN									

● -RESERVOIR ROCK   △ -SOURCE ROCK   □ -SEAL

Throughout the Permian basin, the Simpson Group unconformably overlies the Ellenburger, except where it has been removed by later tectonism or was never deposited. Thickness of the Simpson ranges up to 700 m within the basin (Fig. 4), and the lithology is predominantly green to gray shales interbedded with limestone and occasional sandstone. The Simpson can be readily divided into formations that are persistent and correlative over great distances; these divisions are shown on Table 3.

The three basal sands are the principal reservoirs for oil and gas and are the best developed on the southern Central Basin Platform. These sands of the Delaware basin, which is also the western extent of Simpson deposition, exhibit good reservoir development and the same zonation characteristics as the Central Basin Platform. The sands thin west and south in the interior Delaware basin. The section as a whole thickens and changes into an allochthonous "Marathon-type" lithology.

Lateral thinning of each of the Simpson formations away from the area of maximum sediment accumulation in the basin (as shown in the stratigraphic cross section, Fig. 4) indicates continued subsidence of the basin depocenter and perhaps gentle uplift of the basin margins throughout Middle Ordovician time. Therefore, the youngest units, the Bromide and Tulip Creek Formations, are absent from the basin flanks, but the older units are present. The close of the Middle Ordovician is marked by a regional unconformity.

The Montoya Formation unconformably overlies the Simpson and transgresses it on the west, north, and northeast. Ranging in thickness up to almost 180 m within the basin, the Montoya contains Simpson-derived quartz grains throughout its basal member. The rest of the Montoya consists of limestone, dolomite, and chert in the eastern part of the basin. The lithology grades north and west into predominantly dolomite with chert west of the Central Basin Platform.

The Montoya is not a prime hydrocarbon objective in the Permian basin, but it is productive in a number of fields, particularly where it is vertically continuous with the Silurian and Devonian as one reservoir. Generally, the formation has very little porosity, but fracture and vuggy porosity are locally developed on the largest structures.

Perhaps the most important factor that limits Montoya production is that the section is not overlain by a good shale seal or source rock. Therefore, its productive possibilities depend primarily upon its being vertically continuous with overlying Silurian and/or Devonian reservoirs. This reasoning holds true for the Midland basin and the Central Basin Platform; there, the Montoya produces only where the overlying Silurian Fusselman also produces. However, Montoya updip truncation traps occur on the platform.

The oldest Silurian formation in the subsurface of western Texas and southeastern New Mexico is the Fusselman Forma-

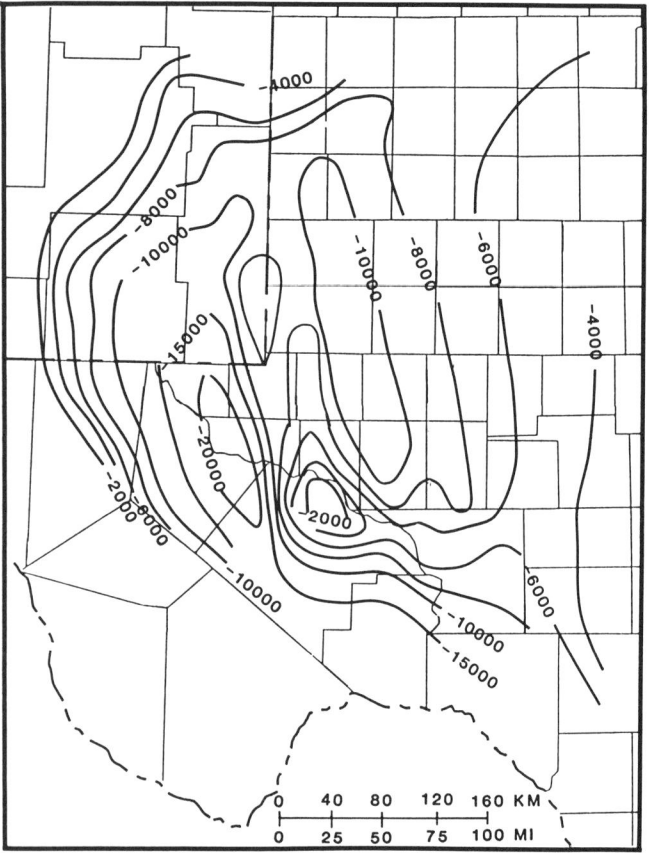

Figure 2. Structure map on top of the Ellenburger Group in the Permian basin. Contour intervals are 2,000 ft and 5,000 ft (after Galley, 1958).

TABLE 2. CUMULATIVE HYDROCARBON PRODUCTION BY AGE, PERMIAN BASIN*

	Oil (Oil + condensate in billions of barrels)	Gas (Dry + casinghead in trillions of ft^3)
Permian	15.90	26.00
Pennsylvanian	4.00	9.00
Mississippian	0.03	0.08
Silurian and Devonian†	1.82	3.57
Ordovician	1.65	9.00

*Data compiled from International Oil and Gas Development (1985) and USGS Circular 828 (1980) and should be considered estimates.

†Silurian (includes Fusselman) and Devonian are used synonymously in reserves and reports in the chapter, and are therefore combined for purposes of accuracy.

tion, a shallow-water carbonate unit resting on the Late Ordovician Montoya Formation and on the equivalent Sylvan Shale toward the east. The Fusselman consists of two major lithofacies, and both contain productive reservoirs. The most basinal facies is a limestone from less than 15 m to about 60 m thick. Toward the north and west the Fusselman thickens to more than 250 m and changes to a coarsely crystalline dolomite facies. The limestone facies of the Fusselman Formation produces primarily from reservoirs developed in biogenic banks.

The limestone facies of the Fusselman is overlain conformably by the Wristen Formation of Late Silurian age (Hills and Hoenig, 1979). It consists of calcareous mudstone and green, dark brown, and gray shale. This rock unit ranges from less than 30 cm to more than 90 m thick and is present across the southern Central Basin Platform and southern Midland basin. Toward the north and west the Upper Silurian thickens sharply to over 460 m and changes to a shallow-water carbonate facies that includes limestone and dolomite with coral and stromatoporoid reefs (McGlasson, 1967; Fig. 5).

The Thirtyone Formation (Hills and Hoenig, 1979), of lower Devonian age, conformably overlies the Wristen Formation. It consists of chert in the west grading eastward to chert and limestone, then to mostly limestone in the eastern side of the basin. The Thirtyone Formation ranges from 0 to more than 300 m, and it is present in approximately the same area as the Wristen Formation (Figs. 5 and 6). The Thirtyone Formation produces from three principal types of reservoirs. In the west, the unit consists entirely of chert, which is generally tectonically brecciated on structures. These chert breccias are the Devonian reservoirs in the Delaware basin. Toward the east, in the central and southern part of the Central Basin Platform, the Thirtyone Formation consists of chert and limestone intervals, and most of the production is from an unusual reservoir type called "tripolite." Tripolite nearly always has calcite and dolomite mixed with a siliceous framework, but the rock may vary from siliceous dolomite or limestone to slightly calcareous, dolomitic chert. Tripolite is usually characterized by very high porosity, low permeability, and very small pores. The tripolite reservoirs are found at various intervals beneath the top of the Thirtyone Formation. Farther east, in the southern Midland basin, the Devonian is mostly limestone, and many of the reservoirs are in biogenic bank facies in fossiliferous calcarenites.

The Fusselman, Wristen, and Thirtyone Formations were deposited in a deepening basin (McGlasson, 1967). They are beveled by a major Middle Devonian unconformity. The Late Devonian and Early Mississippian Woodford Shale is a transgressive marine unit that overlaps all of the Silurian and Devonian rocks (Fig. 6). The transgressive sequence continued through the Mississippian.

The Woodford is overlain by Middle Mississippian limestones and dark shales of the Upper Mississippian. The Upper Mississippian (Chester) Chappel Formation is an oil reservoir in

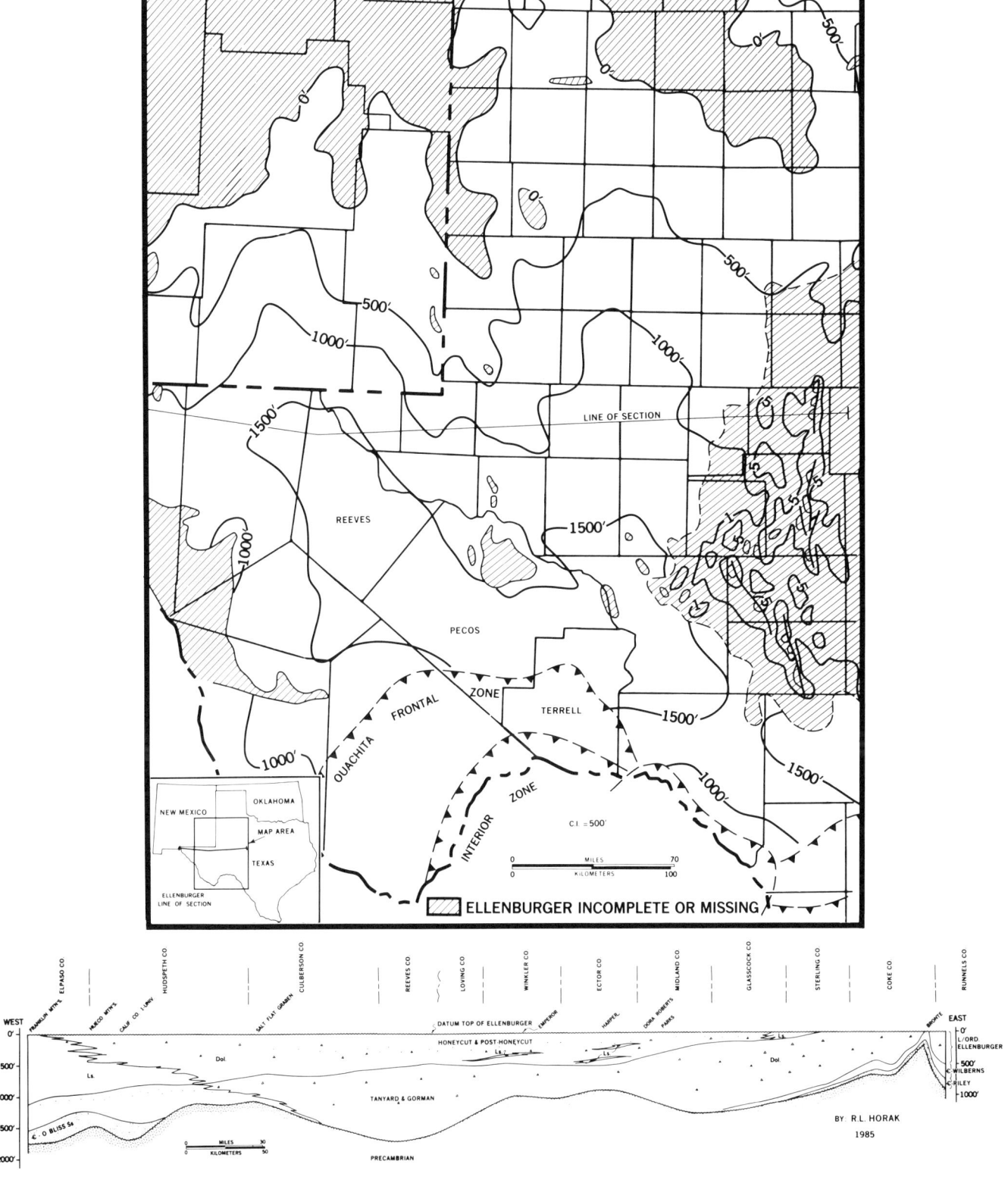

Figure 3. Isopach map of the Lower Ordovician Ellenburger Group and east-west stratigraphic section. Datum is top of Lower Ellenburger.

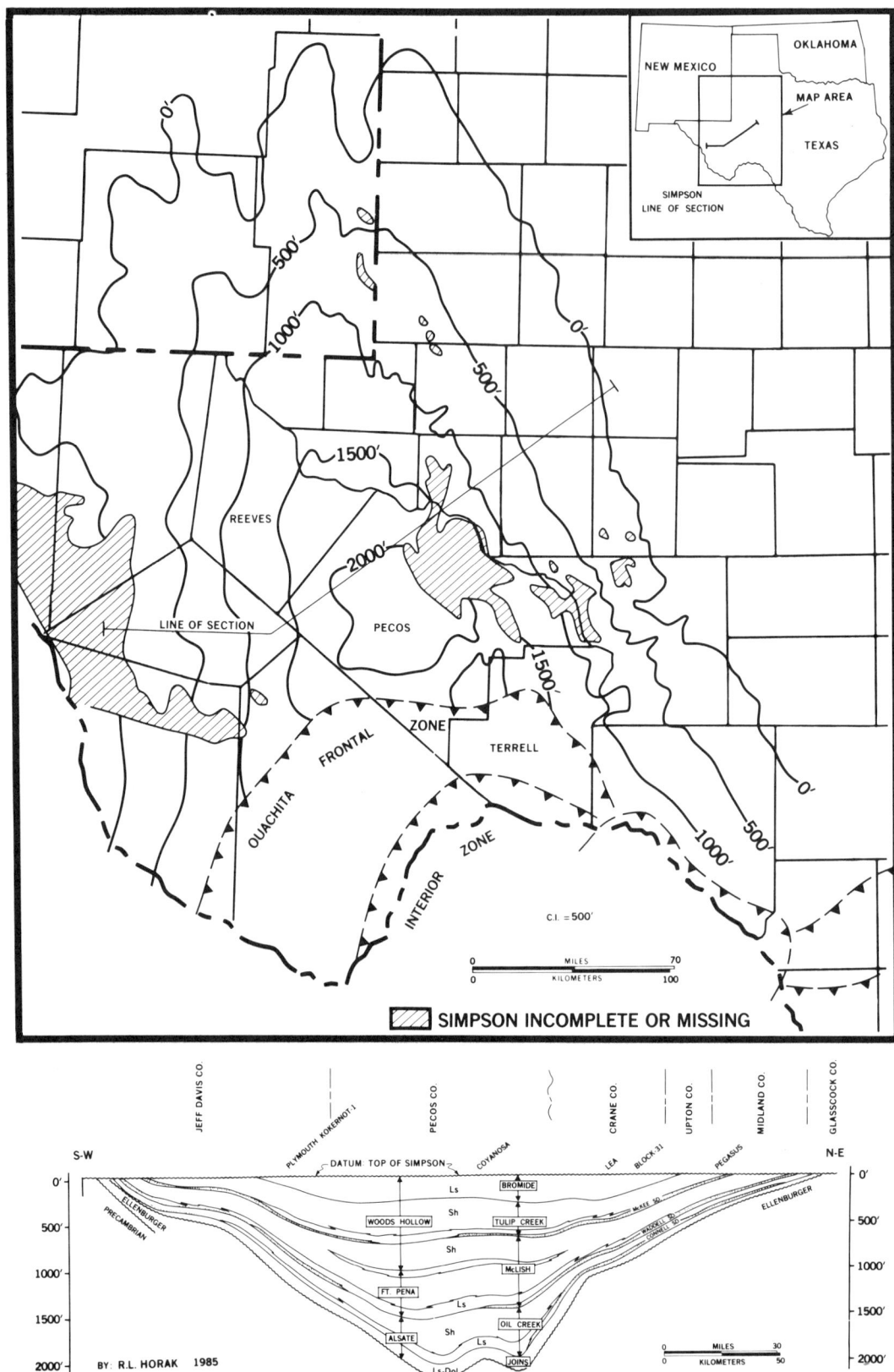

Figure 4. Isopach map of the Middle Ordovician Simpson Group and east-west stratigraphic section. Datum is top of Simpson.

TABLE 3. COMPARATIVE STRATIGRAPHIC NOMENCLATURE FOR DIVISIONS OF THE SIMPSON GROUP FROM AUTOCHTHONOUS AND ALLOCHTHONOUS AREAS, PERMIAN BASIN REGION

Permian Basin Terminology (Autochthonous)	Marathon Terminology (Allochthonous)
Montoya Formation	
——————————— Unconformity ———————————	
Bromide Limestone	
	Woods Hollow Shale and Limestone
Tulip Creek Shale	
Basal McKee Sand	
McLish Shale and Limestone	
Basal Waddell Sand	Ft. Pena Limestone
Oil Creek Shale	Alsate Shale
Basal Connell Sand	
Joins Limestone or Dolomite	
——————————— Unconformity ———————————	
Ellenburger Group	Marathon Limestone

the northeastern part of the Permian basin, and the Middle Mississippian (Meramec-Osage) Limestone is productive in various places in the basin.

Source

The Middle Ordovician Simpson shales are regarded as the primary source rock for Ellenburger oil and gas. Eastward, beyond the limits of the Simpson, Late Devonian (Woodford) Mississippian, and Upper Pennsylvanian shales are thought to be source rocks for Ellenburger oil on the Eastern Shelf where these units directly overlie the Ellenburger. In this region, Ellenburger, Mississippian, and Pennsylvanian oils are generally indistinguishable from each other (Kvenvolden and Squires, 1967). In the greater part of the Permian basin, however, where the Ellenburger underlies the Simpson, Ellenburger oils have uniquely low ^{13}C values (Kvenvolden and Squires, 1967).

Simpson shales are also considered to be the source rocks for Simpson oil and gas. Because the three principal producing sands (McKee, Waddell, and Connell) act as separate reservoirs, it is thought that hydrocarbons produced from the Simpson originated within the Simpson Group. Some Montoya reservoirs are continuous across the Ordovician-Silurian boundary and extend upward into the Silurian Fusselman. This condition suggests a possible Silurian shale source for these Montoya reservoirs.

The Woodford Shale is generally considered to be the source of virtually all of the hydrocarbons trapped in Devonian reservoirs and much of those in Silurian reservoirs except for some high structures on the Central Basin Platform where the Devonian, Silurian, and older formations have been truncated by Pennsylvanian and Early Permian erosion and buried by Permian or Pennsylvanian sediments. Although little work has been published chemically relating the oils of the Permian basin area to the source rocks, the oils themselves are chemically identifiable, and oils found beneath Woodford cover are chemically distinct from oils in the same age reservoirs beneath Permian or Pennsylvanian cover (where the Woodford has been removed by erosion; Williams and Coester, 1968).

Traps and Seals

Almost all the traps found in Ordovician rocks to date are structurally controlled, and these traps are associated with foreland deformation temporally linked to the Ouachita Orogenic Belt. Some of the low-relief Eastern Shelf Ellenburger fields may be erosional highs and would thus be primarily stratigraphic in origin.

Large, basement-controlled asymmetric fault blocks and anticlines with up to 900 m of closure are characteristic of the Delaware and Val Verde basins and the Central Basin Platform. In the Midland basin, the structures are similar but relief is much less. Structural relief decreases eastward so Eastern Shelf structures have only 15 to 30 m of relief at the Ellenburger level and contain only modest volumes of oil.

On the Central Basin Platform, along a 130-km north-south trend, many Ellenburger and Simpson fields produce from modified structural traps. In these settings the faulted asymmetric anticlines have been severely truncated and are overlain by Permian rocks. The updip edge of the reservoir terminates against the unconformity, thus forming a wedge-out type of trap.

Seals for Ellenburger fields over most of the area are the overlying ductile Simpson shales, which are draped over the reservoir and deformed plastically. Where the producing interval is well beneath the top of the Ellenburger, tight upper Ellenburger dolomite seems to provide the seal. Major bounding faults may also provide impermeable barriers laterally. In updip wedge-out traps, impermeable dense carbonates overlying the unconformity provide the seal. On the Eastern Shelf, the Ellenburger is overlain by either the Upper Ordovician Sylvan Shale, the Devonian Woodford Shale, Mississippian shale, or Pennsylvanian shale, depending upon the location. All these shales serve as seals. Simpson fields are sealed by the Tulip Creek, McLish, or Oil Creek shales, depending on which of the three principal sands (McKee, Waddell, Connell, respectively) are productive. Montoya fields are generally considered to be sealed by tight carbonates in the upper part of the section. Where the reservoir is continuous upward into the Silurian and Devonian sections, the Silurian shale or Woodford Shale provides the seal.

Traps in the Fusselman Formation are of three basic types: (1) structural closures, (2) truncation by Permian or Pennsylvanian erosion on high uplifts on the Central Basin Platform, and (3)

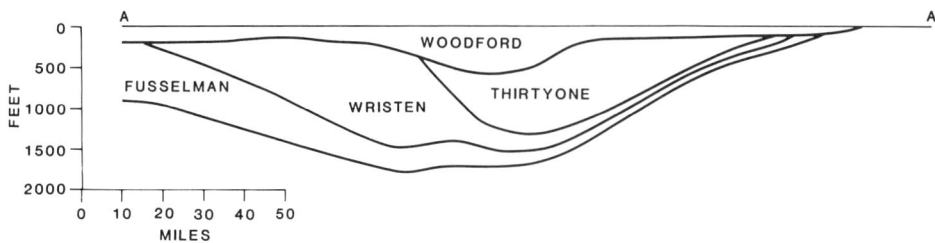

Figure 5. Diagrammatic east-west stratigraphic section of Silurian through Late Devonian. Datum is top of the Woodford Shale. See Figure 6 for location of line of section.

truncation by the Woodford Shale in the eastern eroded margin of the formation. The shaly facies of the overlying Wristen Formation is thought to be the source of much of the hydrocarbons in the limestone facies of the Fusselman, and it is the seal for many of the traps. The Woodford is the most likely source of hydrocarbons trapped immediately beneath it, and it is the seal where the Fusselman is directly overlain by Woodford. The extensive dolomite facies of the Fusselman appears to be very porous and permeable everywhere it is found, but it contains hydrocarbons only near the limestone facies, where it is overlain by a sealing facies of the Upper Silurian and at its eroded margin, where it is overlain and sealed by the Woodford Shale. Over most of its extent, it is overlain by a nonsealing, permeable carbonate facies of the Upper Silurian.

The shaly facies of the Upper Silurian has no reservoirs, but furnishes the seal and probable source rock for Fusselman reservoirs beneath it. The carbonate facies, however, does contain prolific reservoirs, primarily in dolomite. Hydrocarbon production is almost always from near the top of the unit where it is sealed by the Woodford Shale, which is also the likely source of the hydrocarbons. Upper Silurian traps are mostly structural, but some are in bioherrmal reefs that have considerable depositional and/or erosional relief.

Traps of the Thirtyone Formation in the Delaware basin are structural, mainly in horsts and faulted anticlines, and they are sealed by the Woodford Shale. Seals of reservoirs in the southern and central parts of the Central Basin Platform appear to be tight limestone intervals within the formation, as well as the Woodford Shale where the reservoir is developed near the top of the formation. Permian or Pennsylvanian sediments seal some Devonian tripolite reservoirs where the Woodford has been eroded. Traps in the southern Midland basin are both structural and in the beveled margins of beds where they are truncated by the erosion surface beneath the Woodford Shale.

Migration

Little is known about migration paths in the Permian basin. Systematic fractures, caused by late Paleozoic and Laramide orogenic stress, may provide the major pathways from mature depocenters to the available traps.

It is unlikely that any significant quantity of hydrocarbons was generated prior to Permian time, but very likely that middle Paleozoic source beds did begin to generate oils during the Permian Period. The carbonaceous material probably began to reach thermal maturity in the deeper part of the Val Verde and Delaware basins by the end of Leonardian time and over most of the Permian basin area by the end of Guadalupian time.

Hydrocarbon maturation history provides some insight into the *timing* of migration. Calculations to determine the onset of hydrocarbon generation in the Delaware basin, using a modified Lopatin (1971) method, suggest that the onset of oil and gas generation from Simpson rocks occurred in the Permian Leonardian. Appreciable gas generation occurred in the Triassic and has continued to the present.

There is no record of multiple oil (or gas)/water contacts within the Ellenburger. This observation, and the fact that hydrocarbons usually occur immediately below the unconformity at the top of the Ellenburger or in the lowest porous and permeable zone beneath the Simpson, suggest that Ellenburger hydrocarbons migrated stratigraphically downward from the overlying shale rather than being indigenous to the Ellenburger.

Since the structural traps of the middle Paleozoic reservoirs were already formed by the time of the late Paleozoic orogeny, the oil and gas generated in the Middle and Late Permian found a home waiting for them. As in the Ellenburger, downward migration to middle Paleozoic reservoirs may be inferred where a reservoir rock is covered by a Woodford, Permian, or Pennsylvanian source.

PENNSYLVANIAN PRODUCING SYSTEM

Introduction

As the Permian basin evolved from the precursor Tobosa basin in Middle Mississippian to Early Permian time, there was more or less continuous subsidence and deepening of the Midland and Delaware basins and intermittent uplift of the surrounding platform and shelves (Galley, 1958; Hills, 1970, 1984). South of the Permian basin proper, a foredeep trough received flysch sediments derived from the ancestral Ouachita-Marathon area dur-

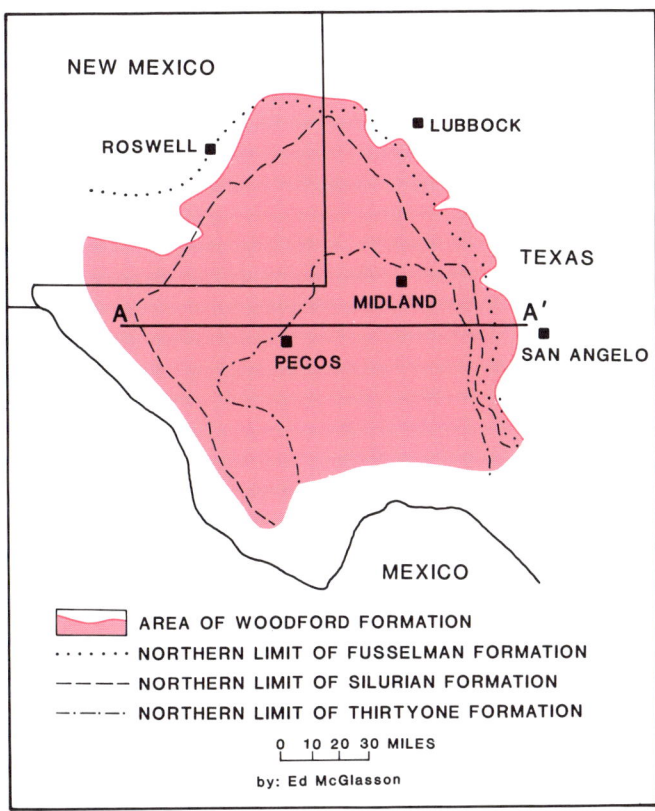

Figure 6. Paleoerosional limits of the Silurian and Devonian rock units showing the offlapping relationship of the pre-Woodford units.

ing most of the Pennsylvanian, until it was buried beneath thrust sheets during the Late Pennsylvanian (Virgilian) and Early Permian (Wolfcampian).

Reservoirs

Changing structural conditions in the Pennsylvanian produced a complex variety of reservoir facies.

Representative Morrowan sediments are recognized only on the Northwest Shelf of the Delaware basin in southeastern New Mexico and adjacent Texas (Fig. 7). The Morrow Formation is a thick section of relatively thin individual units of repetitive terrigenous to nearshore siliciclastics and subordinate carbonates that were deposited on a gently southward-sloping shelf. The fine-to-coarse sandstones and shales were derived from the erosion of the Pedernal massif of central New Mexico, the Matador uplift, and the Central Basin Platform. The principal hydrocarbon reservoirs include fluvio-deltaic, tidal-channel, nearshore-bar, and strandline sandstones. The associated carbonates are principally oolite grainstones and micritic limestones deposited in areas away from active clastic input. The shelf facies pass farther south into basinal shales of the Delaware basin.

An active phase of subsidence caused the bipartite division of the Permian basin into the component Delaware and Midland basins in Atokan time. Facies are varied and complex; cyclic marine and terrestrial deposits are recognized throughout the entire region. From the Atokan through the Pennsylvanian, the Delaware basin persisted as a deeper marine embayment than the Midland basin.

On the Northwest Shelf, the Atokan section consists of many time-transgressive carbonate shelves that prograded southward into the Delaware basin (James, 1985; Fig. 8). Hydrocarbon reservoirs in this area are most notably developed in porous, phylloid algal/*Donezella* reefs and associated oolites and biograinstones (Mazzullo, 1981).

On the Central Basin Platform, Ozona uplift, and portions of the northern Midland basin, the Atokan is represented by a relatively thin, cyclic sequence of shallow-marine carbonates (phylloid algal/*Donezella* reefs and high-energy oolite and skeletal sand shoals) and nearshore-marine to terrestrial siliciclastics and coals. These facies pass gradually via ramps (i.e., no pronounced shelf margins as in the Delaware basin) into dark-colored shales and argillaceous, micritic limestones in deeper portions of the Midland basin. In the northern part of the Eastern Shelf, a variable thickness of Bend Conglomerate is present. Bend carbonate-clast conglomerates and minor siliciclastics were deposited in terrestrial to nearshore-marine environments.

The Desmoinesian or "Strawn" section in the Permian basin is separated into two informal stratigraphic units based on fusulinid zonations and lithologic correlations. The early Desmoinesian was a time of widespread shallow-marine and terrestrial sedimentation over much of the Midland basin and Northwest Shelf of the Delaware basin (Fig. 9). The lower Strawn section over most of the Central Basin uplift, Ozona arch, and North Basin platform includes cyclic fluvial and nearshore-marine siliciclastics, coals, and subordinate marine limestones. Thick alluvial fan chert conglomerates adjoin major block-faulted terrains along parts of the southern Central Basin Platform. As a result of periodic tectonism, the more elevated portions of these positive features remained subaerially exposed well into the Permian System. The marginal facies grade via gentle ramps into a relatively thin sequence of shallow-basinal shales and argillaceous lime micrites in the deeper, offshore-marine environments of the Midland basin. In places along the Eastern Shelf and Northwest Shelf, thick phylloid algal-biograinstone banks defined distally steepened ramps into the Midland Basin.

Upper Desmoinesian sedimentation was profoundly influenced by numerous eustatic sea-level fluctuations and only minor tectonism. This period heralded the beginning of rapid and prolonged subsidence of the Midland and Delaware basins. Accordingly, many of the former nonmarine realms on the platforms and arches were inundated, and both basins steadily became deep-water, starved environments that persisted throughout the remainder of the Pennsylvanian (Adams and others, 1951). Shelf facies peripheral to the basins include cyclic, back-reef micritic limestones and accessory shales, and shelf-marginal phylloid algal

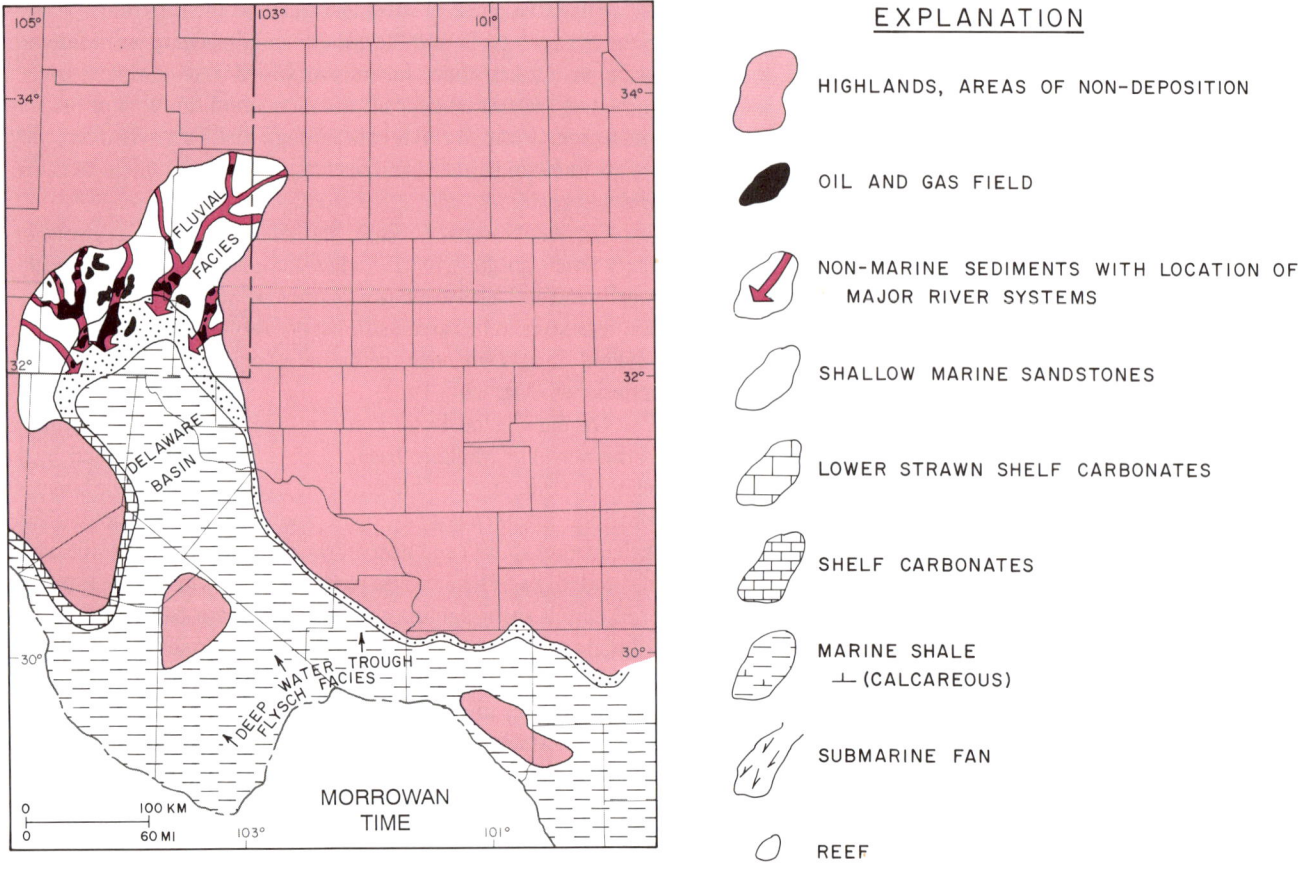

Figure 7. Morrowan paleogeographic map showing pretectonic Pennsylvanian facies in subsiding Delaware basin. Map is in part after James (1985).

reefs and biograinstones. Along major tracts of the Eastern Shelf, north flanks of the Midland basin, and Northwest Shelf, the concurrent deposition of shelf-margin buildups and rapid basin subsidence resulted in the establishment of steep shelf-to-basin profiles. Generally, loci of thick limestone deposition in the upper Strawn coincide with those of the subjacent late Cherokee part of the lower Strawn.

A 290-km north-south trend of thick, shelf-marginal limestone reefs, consisting primarily of biograinstones and phylloid algal facies, occurs in the upper Strawn on the Eastern Shelf. Many of these buildups served as the foundations for continued reef development in the Upper Pennsylvanian. Large oil fields along this trend typically are characterized by stacked reservoirs (Galloway and others, 1983).

In contrast to the dominant pattern of carbonate deposition in the Permian Basin during the late Desmoinesian, some areas in the northeastern part of the Midland basin were the sites of deltaic sedimentation. Large limestone banks grew between and west of these deltaic lobes (Cleaves and Erxleben, 1982). Locally, the foreset beds of these deltas appear to have had more than 120 m of depositional relief into the Midland basin. Shallow-water carbonates that onlap disconformities developed at the base of these slopes suggest that there may have been fluctuations of sea level of that magnitude.

The most prominent feature in the northern Midland basin is the Garza Platform (Fig. 9), which served as the foundation for the Horseshoe Atoll of late Desmoinesian to early Wolfcampian age. Considered as an entity, the Horseshoe Atoll includes the largest Pennsylvanian hydrocarbon accumulation in the United States, and the second largest trap in the conterminuous United States. The largest field is the 1.3 billion-barrel Kelly-Snyder Field. Total production from the atoll is over 2.1 billion barrels of oil. In the thickest area along the southwest part, total relief over the surrounding basin exceeds 770 m.

In late Desmoinesian (upper Strawn) time, all but the northern portions of the Garza platform were the site of deposition of marginal grainstone shoals, islands, and leeward algal wackestone facies. These thick upper Strawn sections are the sites where the Canyon and younger buildups of the Horseshoe Atoll were localized. The bulk of the Atoll is Missourian to Virgilian in age. On the east side, lower and middle Cisco (Virgilian) limestone is present on pinnacles. In the southern part, carbonate deposition

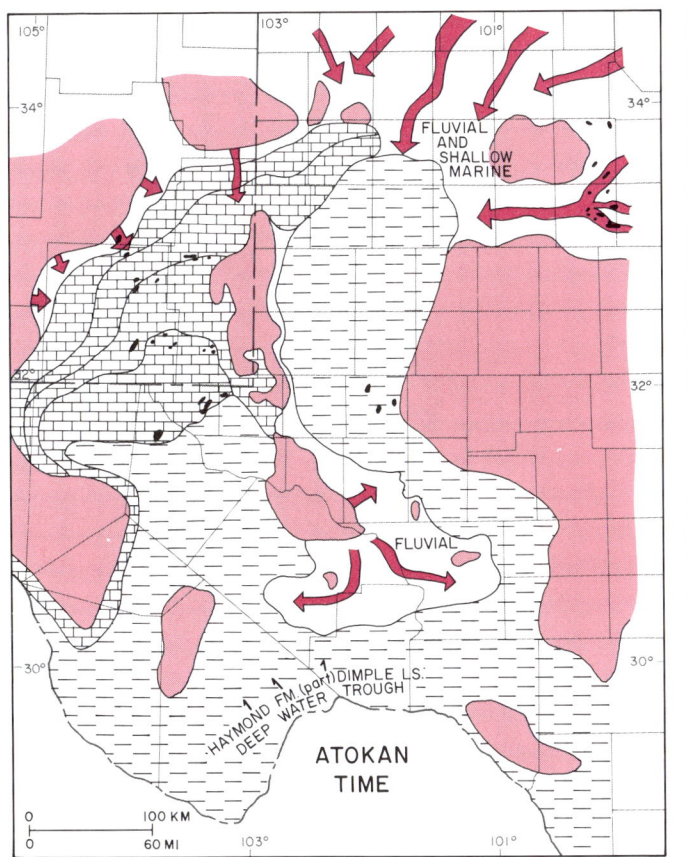

Figure 8. Atokan paleogeographic map showing opening of Midland and Delaware basins. Map is in part after Mazzullo (1981) and James (1985). See Figure 7 for explanation of symbols.

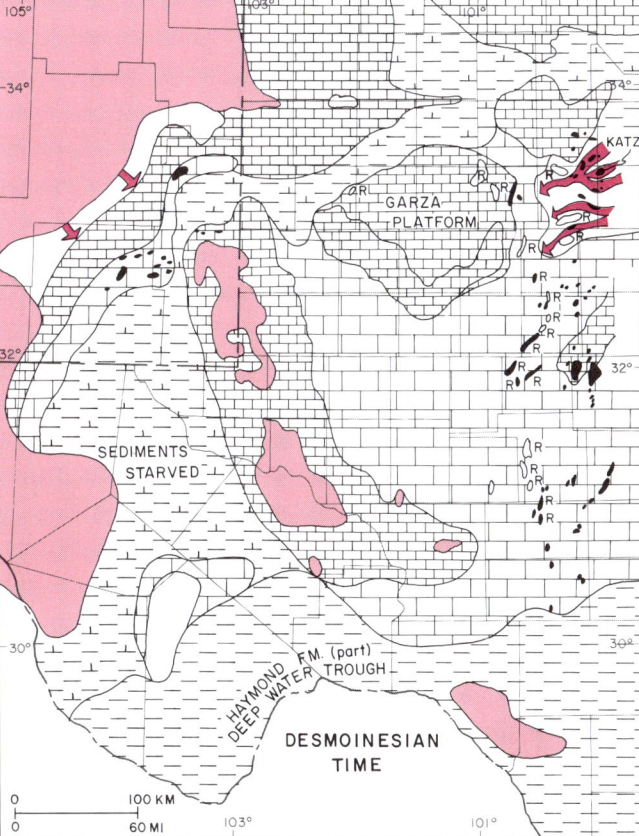

Figure 9. Composite of early and late Desmoinesian (upper and lower Strawn) paleogeographic map showing basinwide establishment of platform and shelf margins. See Figure 7 for explanation of symbols.

continued into late Virgilian. In the southwestern part, it continued into the early Wolfcampian (Vest, 1970). By late Virgilian the eastern part of the Atoll was being buried by slope muds and sands from prograding shelf sediments to the east. Due to westward tilting as the Midland basin continued to subside throughout the Permian, much of the oil migrated into traps along the eastern edge of the Atoll. The reefs of the Horseshoe Atoll are primarily high-energy bioclastic grainstones and oolite shoals deposited along the leading edge of the platform. Solution during these erosional intervals is responsible for much of the porosity as well as erosional topography.

Because of an overall continuous deepening of the basins and the resulting high stands of sea level, the Missourian (Canyon) sections are almost entirely shelf limestones (Fig. 10). Along the eastern side of the Midland basin, there was continued reef, bank, and shelf-edge accumulation over many upper Strawn carbonate buildups, resulting in the development of thick, stacked limestone deposits. The central parts of the Midland and Delaware basins received only thin sections of basinal shales that have been referred to as "sediment starved" areas (Adams and others, 1951).

The sediments deposited during Virgilian time are collectively referred to as the Cisco section. Over most of the basin the same types of deposition as that of the upper Strawn and Canyon sections continued through the Cisco section (Fig. 11). However, along the eastern side of the Midland basin, there was a drastic change in depositional style. The basins to the east were filled by this time, and shelf edges prograded into the eastern part of the Midland basin. The cyclical deposits of the upper part of the Cisco section involved sequences that had progradation of deltas to the shelf edge during low sea-level stands, usually followed by limestone deposition during the high sea-level stands, usually followed by a hiatus, then a repeat of the cycle. Beyond the shelf edges there were thick wedges of slope muds, and sand was deposited in the basin. The bottom topography was irregular in many areas because of the presence of Strawn, Canyon, and early Cisco carbonate buildups. The slope sediments wrapped around these features. Where deep-water sand deposition occurred, usu-

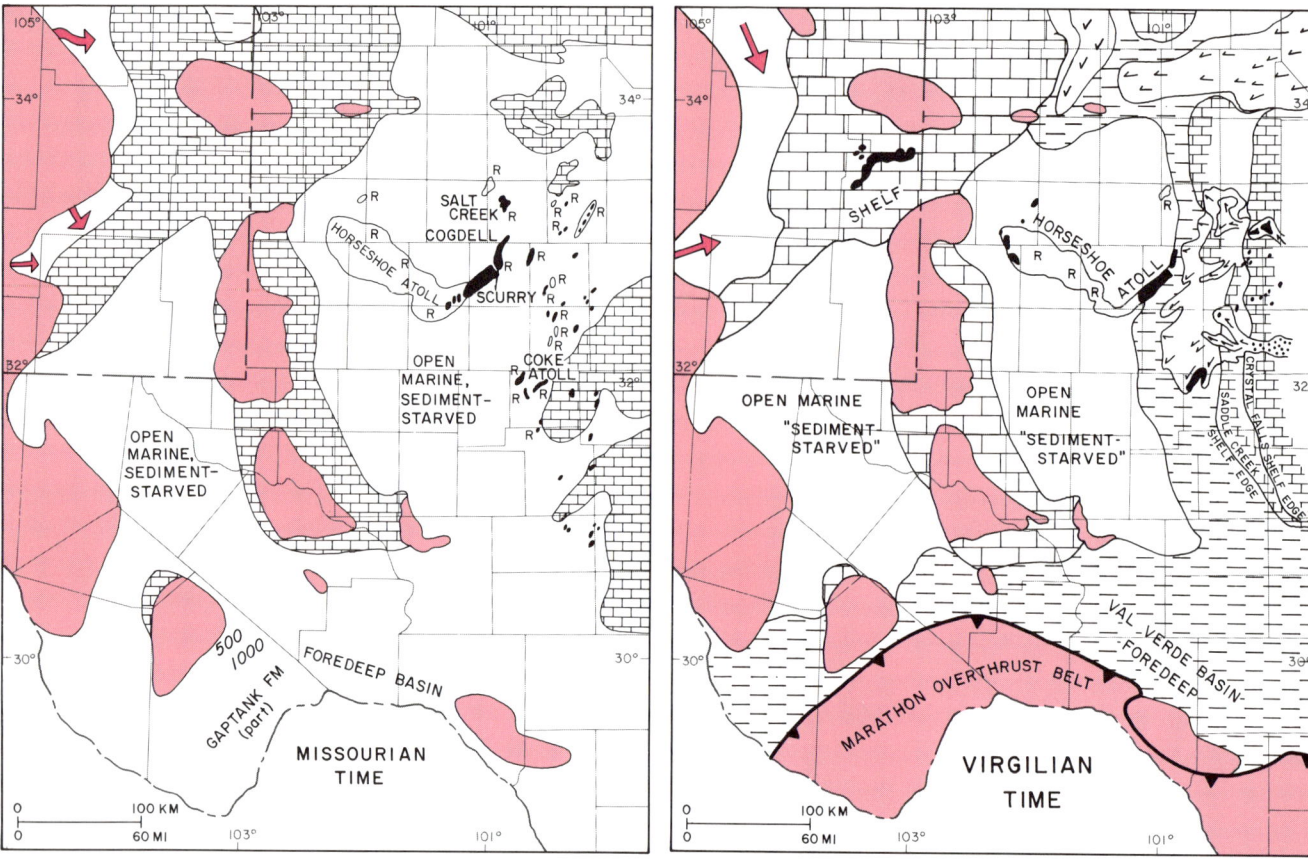

Figure 10. Missourian (Canyon) paleogeographic map displays continuous deepening of basins. See Figure 7 for explanation of symbols.

Figure 11. Virgilian (Cisco) paleogeographic map showing platform and basin relationships that persisted through Permian time. Filling of "sediment-starved" basins began in earliest Permian and closed in Permian Ochoan time. See Figure 7 for explanation of symbols.

ally in large submarine fans, the sandstones have significant oil and gas reserves. Because the sandstones are lateral to the Canyon-age part of the reefs, they were mistakenly identified as Canyon sediments by early operators. This is a common problem with the nomenclature throughout the Midland and Val Verde basins. In both areas, almost all of the so-called "Canyon" sandstones are either of Virgilian or Wolfcampian age.

Another prolific type of Cisco reservoir is the widespread algal limestones on the Northwest Shelf of New Mexico. These extensive limestones were deposited at a high stand of sea level that flooded a broad, shallow shelf. This environment was ideal for the growth of green algae because the water was clear, shallow, and warm.

On the Central Basin Platform, Missourian-Virgilian facies are mainly platform-margin carbonates (phylloid algal reefs and biograinstones). Inner-platform facies are mainly red beds and, locally, coals.

The Marathon-Ouachita mountain belt developed along the south side of the Permian basin during Virgilian and into middle Wolfcampian time. Other than shedding muds and sands into the adjacent basins, the emplacement of these overthrust sheets had little effect on the Permian basin. Both the submarine fan deposits north of and under the sheets and the fractured chert reservoirs in the thrust sheets are exploration targets, but to date only minor oil reserves have been found.

Source migration and traps

The primary source beds for Pennsylvanian oil in the Permian Basin are the Morrow and Atoka basinal shales, thin Strawn shales, and the shales of the Cisco and Wolfcamp. In the Delaware basin the older source beds have been buried deep enough that significant gas has been generated. In the Midland basin, only small amounts of gas have been generated; the primary hydrocarbon is oil. In both areas, the migration has been from the basins to surrounding uplifts and shelves. The variety of traps in limestone and sandstone reservoirs is diverse. The vast majority of Pennsylvanian production has been from stratigraphic and combination structural-stratigraphic traps, including the prolific reefs. The seals for these traps are, in all cases, lateral and overlying shales.

PERMIAN PRODUCING SYSTEM

Introduction

Permian reservoirs have dominated the production of oil and gas from the Permian basin. More than two-thirds of the total oil production from all reservoirs in the basin and one-third of the total gas production from all reservoirs are from Permian reservoirs (Table 1; USGS, 1980).

Oil and gas production from Permian reservoirs was discussed by Galley (1958), and the framework established by him has been used for this chapter, except that the San Andres and Grayburg reservoirs have been considered together rather than separating them into lower and upper Guadalupian. Galloway and others (1983) grouped Permian reservoirs into plays on the basis of common depositional and/or diagenetic facies, and that grouping has been followed for this report.

Permian reservoirs discussed herein include those in the Wolfcampian (Wolfcamp), Leonardian (Glorieta, Clearfork, Tubb, Abo, Wichita, Wichita-Albany, Spraberry/Dean, and Bone Spring), and lower and lower-middle Guadalupian (Queen, Seven Rivers, Yates, Tansill, and Formations of the Delaware Mountain Group).

The general facies relationships of the Permian units are displayed in Plate 7C. In Permian time the Pennsylvanian tectonics and Pre-Penn structure were masked by platform-to-basin deposition. Upper Permian structure (Fig. 12) displays subtle indication of Pre-Permian section.

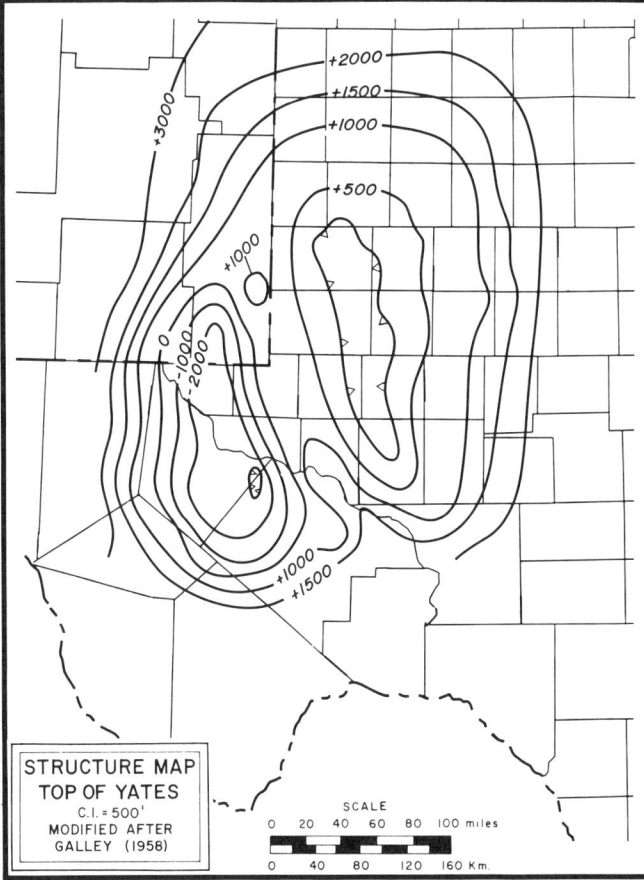

Figure 12. Structure map on top of Permian Yates Formation showing structural expression near close of the Permian. Map is modified from Galley (1958). Contour intervals are 500 ft and 1,000 ft.

Reservoirs

Carbonate shelves and margins developed around the Midland and Delaware basins through Wolfcampian time. The largest production has been from reef-associated sediments. The western portion of the Horseshoe Atoll reef continued growth into the lower Wolfcamp. Phylloid-algal bafflestones are important hydrocarbon reservoirs on the Northwest Shelf in New Mexico (Frenzel and others, 1988) and along the Eastern Shelf in Texas (M. L. VanderLoop, personal communication, 1984). More recently, A. M. Reid (personal communication, 1985) has suggested that Wolfcamp reservoirs of the Eastern Shelf may have originated as debris flows and were deposited basinward of reefal development. The northeastern portion of the Central Basin Platform is another area where depositional topography (organic buildup) has enhanced structural closure.

The nearly four billion barrels of oil produced from Leonardian rocks came from a large number of reservoirs that range from basinal sand deposits (Spraberry and Dean) to reefal limestones (Wichita-Albany, "Abo," Fig. 13). Oil has been produced from the Delaware basin Bone Spring by concentrating on debris-flow deposits near the northeast margin of the basin. The Central Basin Platform was the site for the production of shallow-water platform carbonates (Clear Fork) and associated eolian clastics (Tubb and Glorieta). On the northern flanks of the Midland basin and the Northwest Shelf, shallow-marine as well as supratidal and intertidal facies of the shelf carbonates of the Clear Fork, intermixed with eolian sand deposits of the Tubb and Glorieta, provide reservoirs for oil and gas accumulations. In New Mexico, other names have been applied for these platform and shelf deposits (i.e., Yeso, Blinebry, Drinkard; Table 1).

Sandstones of the Lower Permian Abo (Leonardian/Wolfcampian) were deposited in meandering-stream channels in this fluvial-deltaic environment on the Northwest Shelf. Dolomitized organic buildups of the Wichita-Albany (Abo) are important hydrocarbon reservoirs along the northern margin of the Delaware basin and northwest margin of the Midland basin (Fig. 13; Cys and Gibson, 1988).

On the Eastern Shelf, Clearfork and Glorieta reservoirs account for an important portion of the total Leonardian produc-

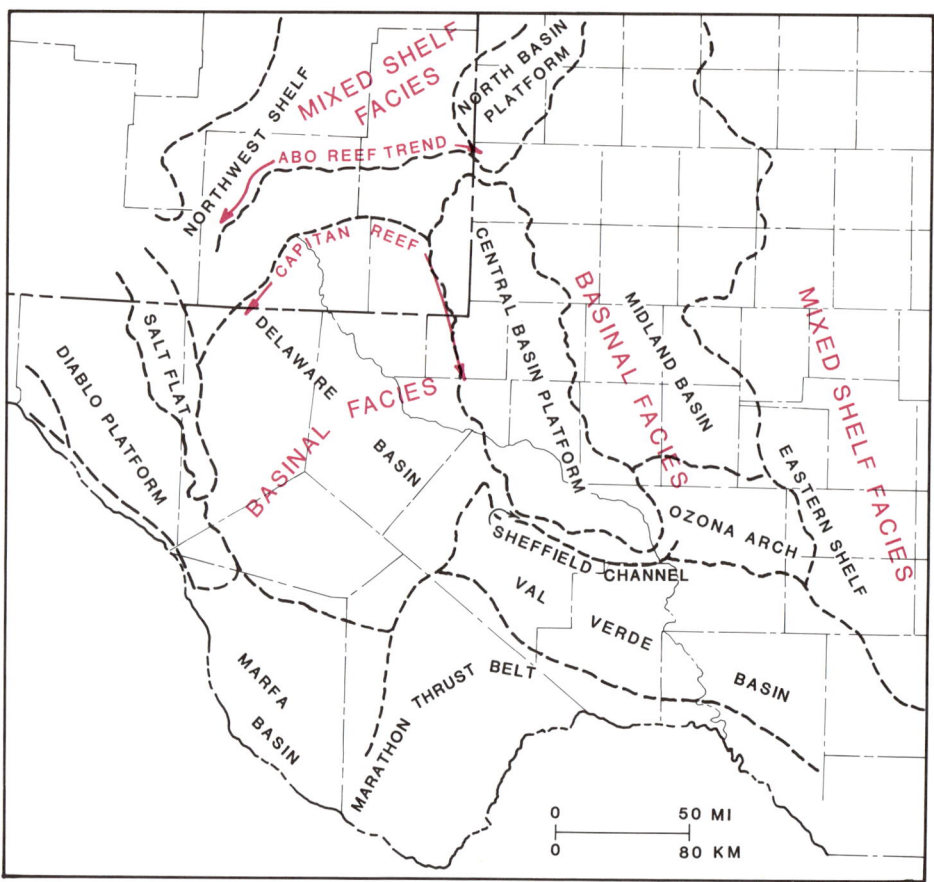

Figure 13. Map of generalized facies of the Permian showing location of principal margin reef units. Map is modified from Vertrees and others (1959).

tion. Supratidal dolomites and wind-blown sand deposits of the tidal flat were intermixed with subtidal dolomites as the shelf migrated westward through time (Van Siclen, 1958).

Distribution of the San Andres and Grayburg reservoirs illustrates the structurally dominant position of the Central Basin Platform. Conditions favored the development of extensive shallowing-upward cycles of shelf and platform carbonates in the lower Guadalupian San Andres grading upward to the more restricted shallow-water lagoonal and supratidal sabkha deposits in the middle Guadalupian Grayburg. Reservoirs occur in rocks deposited in all these environments; however, shallow-water and lagoonal deposits are favored. Anhydrite, common throughout the Permian deposits, is particularly abundant in the San Andres and Grayburg. Much of the original porosity has been filled; therefore, the better reservoirs are in the muddier sections that were not invaded by the calcium-sulphate–laden brines.

Reservoirs adjacent to the northern Midland basin and Northwest Shelf are mostly in porous zones in the lower San Andres. Intercrystalline dolomite porosity is best developed in mudstone or wackestone facies in extensive restricted-shelf deposits that formed basinward of the sabkhas in which interbedded dolomite, anhydrite, salt, and siltstone were being deposited. These transgressions in the lower San Andres are marked by widespread subtidal, porous dolomites interbedded with intertidal to supratidal nonporous dolomite and anhydrite. Progradation was complete over most of the platform by middle San Andres time (Ramondetta, 1982b).

The Yates field at the extreme southern end of the Central Basin Platform is unique. Widespread emergence allowed meteoric waters to create vugs, cavities, and caverns in the skeletal grainstone deposits. The shallow depth to the pay horizon, 300 to 450 m, and the large capacity of some of the wells, combine to make this a truly remarkable oil accumulation. Remaining reserves of a billion barrels confirm the importance of the field.

San Andres and Grayburg reservoirs on the Ozona uplift are on broad structures whose productive limits are controlled by depositional and diagenetic facies. Reservoirs are primarily in restricted-platform dolowackestones, commonly in cyclical sequences.

On the Eastern Shelf, dolowackestones and dolomicrites of

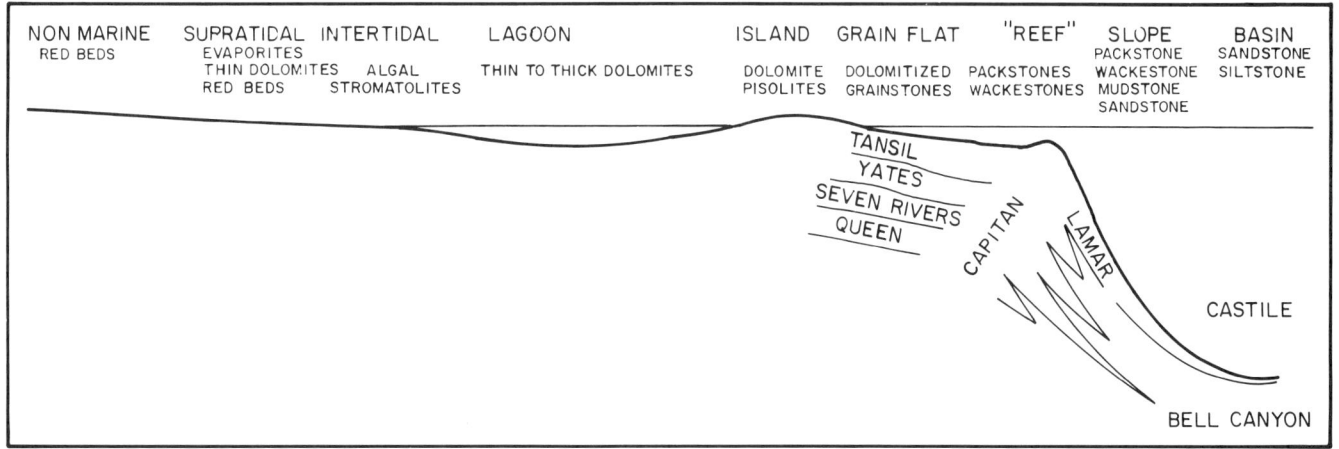

Figure 14. Diagrammatic profile of the late Guadalupian (Capitan Reef). "Restricted-shelf facies" with occasional clastic deposits characterize the back-reef environment. Saline density currents transported significant quantities of clastics into the basins. Diagram is after McGlasson (1982).

the San Andres are reservoirs, whereas dolograinstones and dolopackestones have become tightly cemented. This reversal from original depositional porosity is the result of early deposition of anhydrite in the available pore space in the grainier rocks, while the low-permeability mud-supported rocks resisted the precipitation of anhydrite (Chuber and Pusey, 1969).

Reservoirs of the upper Guadalupian Queen, Seven-Rivers, Yates, and Tansill form a nearly continuous productive trend along the entire western margin of the Central Basin Platform, whereas the Capitan Reef itself is most often water filled. Sandstones, derived from a northwest source, accumulated on the shelfward side of the Capitan Reef that grew to encircle the Delaware basin during the period (Fig. 13, Fig. 14). Upper Guadalupian strandplain sandstone deposits, particularly in the Queen, are also productive in the northern part of the Midland basin, and along the Northwest Shelf of the Delaware basin.

Reservoirs in the upper part of the Delaware Mountain Group (Bell Canyon sandstones) are well-sorted, very fine-grained sandstones interbedded with siltstones and organic-rich shales. Saline-density currents transported significant quantities of these clastics from broad tidal flats across the reef and down into the depths of the basin. These sediments were deposited in broad, internally braided, anastomosing channels along the lower slope and floor of the deep Delaware basin (Williamson, 1977).

The Ochoan Series contains predominantly evaporites that were laid down in great thickness in the Delaware basin and were spread more thinly over the balance of the shrinking Permian basin. In the eastern part of the area, the proportion of carbonate rocks is small. Except for the south edge of the basin, there is no evidence of large reefs or carbonate shales, which were characteristic in all the earlier Permian environments. Noncarbonate deposits contain the highest percentage of clastic strata in the eastern part of the basin, but these volumes are indeed small because the total interval is relatively thin. The clastics increase eastward at the expense of the evaporites. The Ochoan evaporites contain a large rock-salt section, except in the southwest portion of the basin where they are predominantly anhydrite.

Small amounts of oil have been found in the Ochoan strata in the Delaware basin, and one oil field has been found on the south flank of the Central Basin Platform. The ultimate recovery of the small pools has been less than half a million barrels of oil. The crude oil is sour, like that in the back-reef reservoirs of the Guadalupe series, and the gravities range from 18° to 28° API. Hydrocarbons probably migrated into the Ochoan reservoirs from the underlying strata because the Ochoan environment was too saline for organisms to exist.

The Ochoan environments are similar to the back-reef environments of Early Permian epochs, and this evaporitic environment had encroached upon the entire Permian basin. The sea during Late Permian time was rapidly disappearing, and no doubt a broad regional uplift occurred, which closed the Ochoan epoch and simultaneously the Paleozoic Era.

Source

Wolfcampian basinal facies, especially in the northern Midland basin, have been shown to be good source beds for hydrocarbons (Ramondetta, 1982a). Total organic carbon (TOC) averaged 2.8 percent (max. 4.4) in a 300-m section in Lynn County, Texas (Dutton, 1980). Pennsylvanian shales should also be included for consideration as source beds for the Permian reservoirs, as they have attained or passed through the hydrocarbon generation window in many parts of the basin. Leonardian basinal facies in the Midland and Delaware basins are considered to be adequate source beds. In the Midland basin, for example, many thin but widespread black shales of the Spraberry commonly contain 1 to 3 percent TOC, most of which is oil prone, algal, and amorphous (Ramondetta, 1982a).

Basinal facies of the San Andres also contain organic-rich deposits that have potential as source beds; however, the lesser depth of burial and, consequently, lower temperatures tend to minimize the contribution from these beds. In the Delaware basin, the Bone Spring is seen as having significance as a hydrocarbon-generating source rock as well. These deposits comprise a large percentage of basinal carbonates as well as the siliciclastic shales and silts. Younger (shallower) deposits in the middle and upper Guadalupian and Ochoan have a considerably lower potential as a source for Permian oils.

Traps and seals

Depositional topography associated with reefal buildups along the Horseshoe Atoll characterizes Wolfcampian accumulations in Texas. The reservoir seals for these hydrocarbon traps are the later Wolfcampian shales that finally killed the reef. The various phylloid-algae–dominated mounds that developed on the carbonate shelves surrounding the Delaware and Midland basins became traps due to depositional and diagenetic facies changes that provided updip stratigraphic closure. Wolfcampian reservoirs on the Central Basin Platform exhibit principally structural control with an assist from depositional topography. Reservoir seals are provided by anhydrite and/or impermeable carbonates.

Significant volumes of hydrocarbon reserves were trapped where low-porosity sandstones of the Spraberry and Dean thin updip and pinch out, and the submarine-fan deposits onlap the toes of the surrounding shelves in the Midland basin. Seals for these accumulations are the associated tight siltstones and organic-rich shales.

Natural gas is produced from naturally fractured red-bed sandstones in the Lower Permian Abo of New Mexico. The trap here is stratigraphic, depending on the presence of mudstone seals and natural fracturing of the meandering-stream, channel-sand deposits of the Lower Permian. Hydrocarbon traps for Clear Fork, Glorieta, and Tubb reservoirs on the Central Basin Platform are principally structural; however, stratigraphy is important in determining porosity development, which controls productive limits on the structures. Seals for these accumulations are, most commonly, impermeable carbonates, though evaporites also play a part. However, depositional topography is considered the principal trapping factor for the Wichita-Albany reservoirs along the Northwest Shelf margin, although drape over the Wichita-Albany reef trend provides part of the structural closure for Clear Fork reservoirs along the northern Midland basin and Northwest Shelf. In addition, the facies change resulting from the introduction of clastics and evaporites (supratidal deposits) in an updip (shelfward) direction provides a stratigraphic-trapping element and seal for accumulations. The combination of structural and stratigraphic elements accounts for the traps along the Eastern Shelf as well. Facies changes in the prograding shelf deposits provide a marked permeability barrier to the east.

Hydrocarbon traps for San Andres and Grayburg reservoirs involve, in most cases, both structural and stratigraphic elements; however, the principal controlling factor is stratigraphically determined porosity-permeability development. As a general rule, structures that serve as traps are large in areal extent and have relatively low relief, with poorly defined oil-water contacts. Seals for the accumulations consist of impermeable anhydrite beds, tight, fine-grained silts and/or well-cemented dolomites.

Hydrocarbon traps in the San Andres along the buried Wichita-Albany (probable Abo sediments) shelf-margin reef trend of the northern Midland basin and Northwest Shelf result from depositional drape over the reef trend, which also has localized fractures. Reservoir traps north and northwest of this trend are greatly assisted by the depositional and diagenetic facies changes that form updip seals for the accumulations. Traps along the eastern and central parts of the Central Basin Platform are associated with large anticlinal closures, often partially productive due to stratigraphic control of porosity and permeability. Poorly defined oil-water contacts for many accumulations are provided by tightly cemented, anhydritic dolomites or siltstones. Traps on the Ozona arch are often associated with low-relief domal structures with productive limits controlled by facies. Anhydrites play an important role in sealing accumulations in this area as well. Eastern Shelf traps are related to the prograding shelf margin and not primarily to the underlying structure. Facies changes related to the progradation of the shelf provide a marked permeability barrier to the east, updip from reservoir development in the carbonates.

Hydrocarbon traps in these upper Guadalupian reservoirs depend upon the updip pinchout of permeable facies shelfward from the Capitan Reef trend. The seal for the accumulations is provided by the associated evaporite deposits.

Hydrocarbons in the basinal sand units, the Delaware Mountain Group, occur in stratigraphic traps. The channel deposits that occur in broadly lenticular belts pinch out laterally and updip. Although the lack of a definite impermeable seal in the updip direction has suggested to some (Williamson, 1977) that hydrodynamics may play a role in the trap formation, depositional control seems primary.

Migration

The presence of fractures (Finley and Gustavson, 1981) and the probability that the fractures existed at the time of migration provide a means for the movement of fluids. Again, the generation and migration of hydrocarbons in the deeper parts of the Permian basin may have occurred prior to and during Permian time (Hills, 1984). In the shallower Midland basin, generation and expulsion of hydrocarbons from the Wolfcampian and Leonardian source beds may not have occurred until later in the Permian or even into the Triassic or later.

CONCLUSIONS

The general framework of the Permian basin was delineated by shallow drilling early in the exploration history of the region. As deeper drilling became prevalent, the general concept of a Permian-age basin remained valid. The presence of an early Paleozoic section and the complex structure beneath Permian cover were defined. Structure, stratigraphy, and diagenesis of the basin combine to determine the nature in which reservoir rocks form hydrocarbon traps. Future hydrocarbon potential (Table 4) of the Permian basin exists throughout the geologic section. As in other mature hydrocarbon provinces, detailed stratigraphic studies are necessary to understand the areal extent of these reservoirs and to extract the unfound reserves.

TABLE 4. FUTURE HYDROCARBON POTENTIAL BY AGE, PERMIAN BASIN

	Oil (Oil + condensate in billions of barrels)	Gas (Dry + casinghead in trillions of ft^3)
Permian	4.00	10.00
Pennsylvanian	2.00	3.00
Mississippian	0.04	0.41
Silurian and Devonian	0.42	10.93
Ordovician	0.20 - 0.30	5.0 - 10

REFERENCES

Adams, J. E., Frenzel, H. N., Rhodes, M. L., and Johnson, D. P., 1951, Starved Pennsylvanian Midland basin: American Association of Petroleum Geologists Bulletin, v. 35, p. 2600–2607.

Barnes, V. E., and Cloud, P. E., 1946, The Ellenberger Group of central Texas: University of Texas Publication 4612, p. 30–75.

Barnes, V. E., and 6 others, 1959, Stratigraphy of the pre-Simpson Paleozoic subsurface rocks of Texas and southeast New Mexico: University of Texas Publication 5924, v. 1, p. 30–75.

Cleaves, A. W., and Erxleben, A. W., 1982, Upper Strawn and Canyon (Pennsylvanian) depositional systems, surface and subsurface, north-central Texas, *in* Cromwell, D. W., ed., Middle and Upper Pennsylvanian system of north-central and west Texas: Permian Basin Section, Society of Economic Paleontologists and Mineralogists 1982 Symposium and Field Conference Guidebook, p. 49–86.

Chuber, S., and Pusey, W. C., 1969, Cyclic San Andres facies and their relationship to diagenesis, porosity, and permeability in the Reeves Field, Yoakum County, Texas, *in* Elam, J. G., and Chuber, S., eds., Cyclic sedimentation in the Permian Basin: West Texas Geological Society, p. 135–150.

Dutton, S. P., 1980, Petroleum source rock potential and thermal maturity, Palo Duro Basin, Texas: Texas, Bureau of Economic Geology Geological Circular 80-10, 48 p.

Finley, W. L., and Gustavson, T. C., 1981, Lineament and analysis based on Landsat imagery, Texas Panhandle: University of Texas at Austin, Bureau of Economic Geology Geological Circular 81-5, 37 p.

Frenzel, H. N., and 13 others, 1988, The Permian Basin region, *in* Sloss, L. L., ed., Sedimentary Cover—North American Craton, U.S.: Boulder, Colorado, Geological Society of America, The Geology of North America, v. D-2, p. 261–306.

Galley, J. E., 1958, Oil and geology in the Permian Basin of Texas and New Mexico, *in* Weeks, L. G., ed., Habitat of oil; A symposium: American Association of Petroleum Geologists, p. 395–446.

Galloway, W. E., Ewing, T. E., Garrett, C. M., Tyler, N., and Bebout, D. G., 1983, Atlas of major Texas reservoirs: Texas Bureau of Economic Geology, p. 86–93, 139.

Henderson, G. J., Lake, E. A., and Douglas, G., 1983, Langley Deep Field discovery and interpretation: Oil and Gas Journal, Oct. 31, p. 151–160.

Hills, J. M., 1970, Late Paleozoic structural directions in southern Permian Basin, west Texas and Southeastern New Mexico: American Association of Petroleum Geologists Bulletin, v. 54, p. 1809–1827.

—— , 1984, Sedimentation, tectonism, and hydrocarbon generation in Delaware Basin, west Texas and southeastern New Mexico: American Association of Petroleum Geologists Bulletin, v. 68, p. 250–267.

Hills, J. A., and Hoenig, M. A., 1979, Proposed type sections for Upper Silurian and Lower Devonian subsurface units in Permian Basin, west Texas: American Association of Petroleum Geologists Bulletin, v. 63, p. 1510–1521.

International Oil and Gas Development, 1985, Texas West RRC Districts 7C, 8, 8A, *in* Lockstedt, B., and Lockstedt, D., eds., Yearbook 1982: International Oil Scouts Association, v. 5-52, p. 681–734.

James, A. D., 1985, Producing characteristics and depositional environments of Lower Pennsylvanian reservoirs, Parkway–Empire South area, Eddy County, New Mexico: American Association of Petroleum Geologists Bulletin, v. 69, p. 1043–1063.

Kvenvolden, K. A., and Squires, R. M., 1967, Carbon isotopic composition of crude oils from Ellenburger Group (Lower Ordovician), Permian Basin, west Texas and eastern New Mexico: American Association of Petroleum Geologists Bulletin, v. 51, p. 1293–1303.

Lopatin, N. V., 1971, Temperature and geologic time as factors in coalification: Akademiya Nauk SSSR Izvestiya, Seriya Geograficheskaya, no. 3, p. 95–106.

Loucks, R. G., and Anderson, J. H., 1980, Depositional facies and porosity development in Lower Ordovician Ellenburger Dolomite, Puckett field, Pecos County, Texas, *in* Halley, R. B., and Loucks, R. G., eds., Carbonate reservoir rocks: Society of Economic Paleontologists and Mineralogists Notes for Core Workshop no. 1, p. 1–31.

Mazzullo, S. J., 1981, Facies and burial diagenesis of a carbonate reservoir; Chapman Deep (Atoka) Field, Delaware Basin, Texas: American Association of Petroleum Geologists Bulletin, v. 65, p. 850–865.

McGlasson, E. H., 1967, The Siluro-Devonian of west Texas and southeastern New Mexico, *in* Silurian-Devonian rocks of Oklahoma and environs; A symposium: Tulsa Geological Society Digest, v. 35, p. 148–164.

—— , 1982, Introduction to the Delaware basin, *in* Delaware Basin Guidebook: West Texas Geological Society, p. 111–112.

Ramondetta, P. J., 1982a, Genesis and emplacement of oil in the San Andres Formation, northern shelf of the Midland Basin, Texas: University of Texas at Austin, Bureau of Economic Geology Report of Investigations 116, 39 p.

—— , 1982b, Facies and stratigraphy of the San Andres Formation, northern and northwestern shelves of the Midland Basin, Texas and New Mexico: University of Texas at Austin, Bureau of Economic Geology Report of Investigations 128, 56 p.

U.S. Geological Survey, 1980, Future supply of oil and gas from the Permian Basin of west Texas and southeast New Mexico: U.S. Geological Survey Circular 828, 57 p.

Van Siclen, D. C., 1958, Depositional topography; Examples and theory: American Association of Petroleum Geologists Bulletin, v. 42, p. 1897–1913.

Vertrees, C., Atchison, C. H., and Evans, G. L., 1959, Paleozoic geology of the Delaware basin, *in* Geology of the Val Verde Basin and Field Trip Guidebook: West Texas Geological Society, p. 64–73.

Vest, E. L., Jr., 1970, Oil fields of Pennsylvanian-Permian Horseshoe Atoll, west Texas, *in* Halbouty, M. T., ed., Geology of giant petroleum fields: American Association of Petroleum Geologists Memoir 14, p. 185–203.

Williams, J. L., and Coester, B. B., 1968, Relationships of oil composition and stratigraphy in multipay fields, *in* Holmquest, H. J., ed., The compositional and stratigraphic relationships of Permian Basin oil, Texas and New Mexico: West Texas Geological Society Oil Study Committee, p. 136–174.

Williamson, C. R., 1977, Deep-sea channels of the Bell Canyon Formation (Guadalupian), Delaware basin, Texas and New Mexico, *in* Upper Guadalupian facies, Permian Reef Complex, Guadalupe Mountains, New Mexico and west Texas: Society of Economic Paleontologists and Mineralogists Permian Basin Section Publication 77-16, v. 1, p. 409–431.

Manuscript Accepted by the Society December 4, 1987

ACKNOWLEDGMENTS

As chairman for this chapter, I subdivided the Permian Basin into the geologic systems. Ralph Horak, then with Mobil Oil Corporation, prepared the Ordovician Producing System; E. H. McGlasson, Mobil Producing Texas and New Mexico, prepared the Silurian, Devonian, and Mississippian Producing Systems; Donald P. McGookey, S. J. Mazzullo, and Alastair M. Reid, all independents, wrote on the Pennsylvanian Producing System; Chester M. Garrett, Jr., Bureau of Economic Geology at the University of Texas at Austin, wrote on the Permian Producing System; John Clendening and Brenda Claxton, both with Amoco in Houston, prepared the vitrinite reflectance data; Gerald G. Calhoun, then with Pennzoil Company, prepared the cross section; and Brian K. Powers, then with Harper Oil Company, and I prepared the remainder of the paper.

Bernold M. Hanson

Printed in U.S.A.

Chapter 22

Oil and gas resources of the San Juan basin, New Mexico and Colorado

James E. Fassett
U.S. Geological Survey, MS 939, Box 25046, Denver Federal Center, Denver, Colorado 80225

INTRODUCTION

The San Juan basin of northwestern New Mexico and southwestern Colorado contains the second-largest gas field in the conterminous United States, second in total estimated gas reserves only to the Hugoton field of Texas, Oklahoma, and Nebraska. The basin is in the Four Corners area, near the common corners of New Mexico, Arizona, Utah, and Colorado. The major tectonic element of the basin is the monocline bounding the central basin on the east, north, and west sides (Fig. 1). The central basin has no southern structural boundary; its southern limit on Figure 1 is drawn roughly along the outcrop of the Pictured Cliffs Sandstone. Outside the monocline, rocks generally dip less steeply toward the middle of the basin. In this chapter, the San Juan basin comprises three elements: the central basin, Chaco slope, and Four Corners platform. All of the oil and gas fields discussed herein are within the San Juan basin, so defined, and the production statistics for the San Juan basin are for fields within this area.

The structural axis, or deepest part of the central basin (Fig. 2), is arcuate and generally trends northwest in the northern part of the basin. Except along the monoclinal rim of the central basin, dips are gentle and range from less than 1° to commonly less than 2°. These structural relations are also shown on the north-trending geologic cross section on Plate 6D. Precambrian rocks crop out north of the San Juan basin on the San Juan uplift, to the east on the Nacimiento uplift, to the south on the Zuni uplift, and to the southwest on the Defiance uplift (Fig. 1).

Oil and gas production in the San Juan basin through 1987 has been from 313 fields or reservoirs in New Mexico and Colorado. Most production has come from Upper Cretaceous rocks. Hydrocarbons in these fields are contained in stratigraphic, structural, and stratigraphic-structural traps. Papers on each of the oil and gas fields in the basin known through 1978 are in Fassett (1978a, 1983a). Those volumes also contain papers on the geologic history and stratigraphy of the producing rocks, and the oil and gas production history of the basin. The first comprehensive study of the oil and gas resources of the San Juan basin was by Peterson and others (1965).

STRATIGRAPHY AND GEOLOGIC HISTORY

General discussion

The San Juan basin contains sedimentary rocks of Cambrian, Devonian, Mississippian, Pennsylvanian, Permian, Triassic, Jurassic, Cretaceous, Tertiary, and Quaternary age (Fig. 3; Plate 6D). The maximum known thickness of sedimentary rocks in the basin was penetrated in the central basin by the El Paso Natural Gas Company San Juan 29-5 Unit 50 well (Fig. 1). The well was drilled to a total depth of 4,396 m and penetrated Precambrian rocks at 4,276 m. The thickness of Paleozoic rocks in this well is 1,340 m, Mesozoic rocks are 2,130 m thick, and Cenozoic rocks are 790 m thick. Upper Cretaceous rocks are 1,650 m thick in this well.

Knowledge of the stratigraphy of San Juan basin rocks is excellent for most of the Cretaceous, fair for the Mesozoic, and poor for the Paleozoic part of the section. This wide range of knowledge results from three factors:

1. Only 10 drill holes have penetrated the entire sequence of sedimentary rocks within the central basin area; the rest of the 19,000 wells in the basin have stopped either in the Upper Cretaceous or have barely penetrated the top of the Jurassic.

2. Outcrops of Paleozoic rocks are limited to small areas on the perimeter of the basin. Middle Mesozoic rocks are fairly well exposed and more limited on the east and north sides.

3. All of the sedimentary rocks exhibit facies changes across the basin, and some of the pre-Cretaceous rocks of the same age have been assigned different stratigraphic names in various parts of the basin.

Tertiary rocks (except for the Ojo Alamo Sandstone), although well exposed and penetrated by a large number of wells, are still poorly understood basinwide. The primary reason for this lack of knowledge is that these rocks do not contain significant economic mineral deposits; thus, there has been little economic incentive to study them in detail. In addition, the rocks have not been consistently differentiated in the subsurface of the central basin area because their formation boundaries are difficult to locate on geophysical logs.

Fassett, J. E., 1991, Oil and gas resources of the San Juan basin, New Mexico and Colorado, *in* Gluskoter, H. J., Rice, D. D., and Taylor, R. B., eds., Economic Geology, U.S.: Boulder, Colorado, Geological Society of America, The Geology of North America, v. P-2.

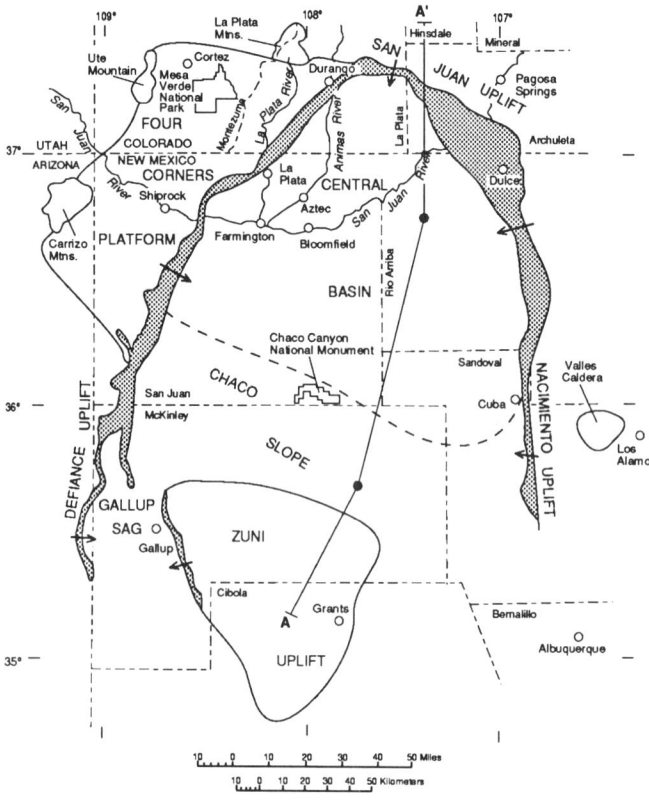

Figure 1. Index map showing the geographic and structural setting of the San Juan basin. Structural elements were generlaized from a structure map of the basin area by Thaden and Zech (1984). Areas of steeper dip (monoclines) are patterned; arrows indicate the direction of dip. The dashed line separating the central basin from the Chaco slope is drawn approximately along the outcrop of the Pictured Cliffs Sandstone. A-A' is the line of the cross section for the San Juan basin shown on Plate 6D; the drill hole shown on the Chaco slope is the Great Western Hospah 1 well (sec.1,T.17N.,R.9W.), the drill hole in the central basin is the El Paso Natural Gas Co. San Juan 29-5 Unit 50 well (sec.7, T.29N.,R.5W.).

Paleozoic rocks (Fig. 3; Plate 6D) in the San Juan basin consist of Cambrian quartzite; Devonian limestone, dolomite, black shale, and glauconitic sandstone; Mississippian rocks, principally limestone; Pennsylvanian limestone, black shale, and sandstone; and Permian rocks, mostly continental sandstone. These rocks are discussed by Armstrong and Mamet (1977), Baars and Stevenson (1977), Jentgen (1977), Stevenson and Baars (1977), Fassett (1983b), Stevenson (1983a, 1983b), and Huffman (1989).

Triassic rocks in the basin consist of continental sandstone, siltstone, and mudstone facies (O'Sullivan, 1977). Jurassic rocks consist of a variety of continental mudstone, siltstone, and sandstone beds, and marine limestone and anhydrite deposits. These rocks were discussed by Condon and Peterson (1986), Condon and Huffman (1988), and in other papers in Turner-Peterson and others (1986).

Cretaceous rocks. The Lower Cretaceous Burro Canyon Formation overlies Jurassic rocks (Morrison Formation) in the northern San Juan basin. Although the age of the lower part of the formation has not been determined, Craig (1981, p. 200) wrote that "it seems possible that [the Burro Canyon] represent[s] much of Early Cretaceous time." Craig further stated that deposition of the Burro Canyon appeared "to represent a continuation of Morrison deposition" although Burro Canyon deposition "may reflect a distinct period of uplift in the source areas and may have been accompanied by slight increase in gradients across the depositional plain." The Burro Canyon is conglomeratic sandstone, especially at its base, and was probably deposited by north-flowing braided streams (Ridgley, 1977; Craig, 1981). Harr (1988, Fig. 6) showed the expression of the Burro Canyon on a geophysical log in the northern (Colorado) part of the basin and stated that the unit ranges from 0 to more than 30 m thick. The Burro Canyon thins southward across the basin and is not present in the southern part.

The Upper Cretaceous rocks of the San Juan basin range from 0 to 1,800 m thick and consist of a series of alternating continental and marine rocks (Fig. 3). These rocks represent a remarkably complete record of a series of transgressions and regressions of the western shoreline of the Western Interior seaway into and out of the San Juan basin area. Indeed, it was these rocks that provided the first model for transgressive and regressive shoreface deposits described in the now-classic paper by Sears and others (1941). That model, simply stated, supposed a continuously subsiding trough receiving clastic sediment at a varying rate from a southwestern source area. A high rate of sediment influx resulted in an outbuilding of the shoreline to the northeast (shoreline regression), and a low rate of sediment influx resulted in landward advance of the shoreline to the southwest (shoreline transgression). This engine operated for about 25 m.y. with streams bringing sediment into the basin area from the southwest and the northwest-trending shoreline rhythmically shifting back and forth across the basin area in response to the varying rate of sediment supply.

Many studies of the stratigraphy and depositional history of Upper Cretaceous rocks in the San Juan basin have been published; therefore, a detailed exposition on those subjects is not included in this chapter. Some of the more relevant publications include: Fassett and Hinds (1971), Fassett (1976, 1977, 1978b, 1983b), Peterson and Kirk (1977), Molenaar (1977, 1983, 1988), and Huffman (1989). Discussions of the geometry and lithology of Cretaceous and older, oil- and gas-bearing rocks is included in the "Oil and gas" section of this chapter.

Tertiary rocks and structural evolution of the San Juan basin. The following discussion is almost entirely from Fassett (1985). In latest Cretaceous time, and probably continuing into earliest Paleocene time, the basin area was uplifted and tilted, resulting in widespread erosion and removal of as much as 650 m of Upper Cretaceous rock in the east-central part. Following this erosional episode, the basin area again began to subside and collect sediment. This time, however, the sediment source was from the north, a radical change from the southwest source during nearly all of Late Cretaceous time. The first unit deposited in Paleocene time was the Ojo Alamo Sandstone. The Ojo Alamo is a complex unit consisting of stacked multiple layers of sandstone and conglomeratic sandstone and mudstone. The sandstone beds are poorly sorted, medium- to very coarse-grained quartzose sandstone. On the west side of the basin, conglomeratic sandstone is abundant; conglomerates are rare to nonexistent on the eastern side. The Ojo Alamo represents a braided fluvial sandstone complex; the sandstone and conglomerate beds were deposited in river channels, and the mudstone interbeds represent overbank deposits. Streams in Ojo Alamo time flowed southeast (Powell, 1973) or south (Sikkink, 1986) in response to the first major pulse of Tertiary Laramide uplift in the San Juan and La Plata Mountain areas in southern Colorado.

The Ojo Alamo ranges from 0 to 150 m thick and covers all except the northern part of the central basin where it was eroded as the result of an early pulse of uplift there. The Ojo Alamo is early Paleocene in age and may be time-transgressive, becoming slightly younger eastward across the basin. The structural San Juan basin had still not begun to form in Ojo Alamo time.

Conformably overlying the Ojo Alamo Sandstone in the southern part of the basin and unconformably overlying the Kirtland Shale in the northern part are the upper part of the Animas and Nacimiento Formations of late Paleocene age. The upper part of the Animas is present only in the northern part of the basin and grades southward into the laterally equivalent Nacimiento. The upper Animas is a coarse-grained to conglomeratic siliceous sandstone containing abundant volcaniclastic rock fragments, especially in the north. The boundary between the upper Animas and Nacimiento Formations is drawn at the southern limit of conglomerates and macroscopic volcaniclastic rock fragments in the Animas. The Nacimiento Formation consists of interbedded mudstone, claystone, and sandstone beds, with sandstone geneally becoming less abundant southward.

These rock units were deposited in middle to late Paleocene time as a result of widespread volcanic eruptions in the San Juan volcanic center, northeast of the San Juan basin. These eruptions produced a rapid influx of volcaniclastic material into the northern part of the basin, forming what is now the upper part of the Animas Formation. Southward, the Nacimiento Formation was probably deposited by the same south-flowing streams that deposited the upper part of the Animas to the north. Deposition of the 825-m-thick Animas Formation in the northern San Juan basin area probably partly reflected subsidence on the south flank of the San Juan Mountains eruptive center and the beginning of the formation of the San Juan basin.

The basin continued to subside into early Eocene time, allowing deposition of the mostly fluvial San Jose Formation across the basin area. The San Jose is more than 600 m thick in the northern part of the basin and consists of interbedded sandstone, siltstone, mudstone, and claystone beds. The San Jose was also deposited by mostly south-flowing streams, but some sediment influx may have come from the rising Nacimiento uplift to

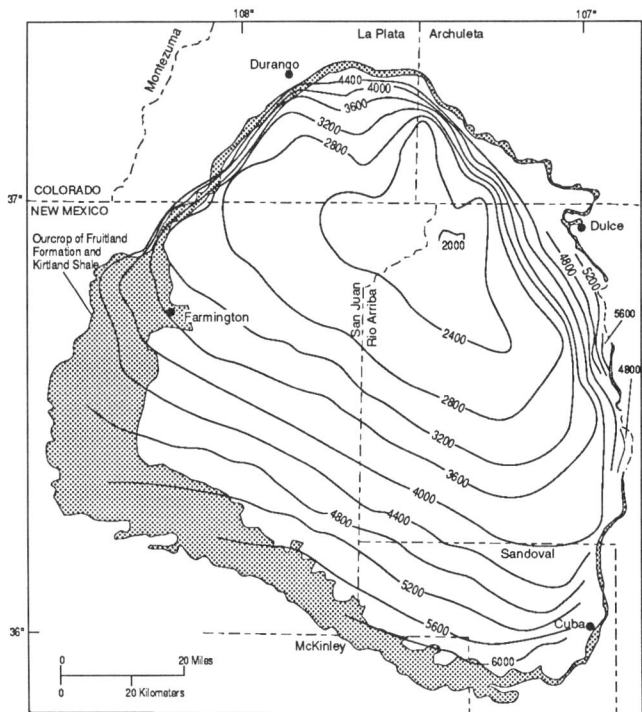

Figure 2. Structure contour map of the central San Juan basin. Contours (in feet; 3.28 ft equal 1 m) are on the Huerfanito Bentonite Bed (Fig. 3) of the Lewis Shale, contour interval is 400 ft, datum is mean sea level. (Modified from Fassett and Hinds, 1971.)

the east. The youngest San Jose in the basin is late early to early middle Eocene in age (Smith and others, 1985).

The basin continued subsiding, probably into middle Eocene time, as evidenced by tilted San Jose rocks around the north and east sides of the basin. On the west side of the basin the Chuska Sandstone is preserved in an area straddling the New Mexico–Arizona border. The Chuska outcrop is southwest of the Four Corners platform; its southeasternmost part barely extends into the western part of the Chaco slope (Fig. 1). The Chuska is essentially flat-lying on an erosion surface that truncates rocks that dip eastward into the San Juan basin. It reaches a thickness of more than 600 m and consists of a lower fluvial unit of conglomerate, sandstone, and mudstone and an upper massive unit of eolian sandstone containing layered volcanic ash beds. Thus, it seems cler that the San Juan basin had ceased to subside, an episode of uplift and erosion followed, and finally the Chuska was deposited. The age of the Chuska is as yet unknown; the only constraint is that it is no younger than 25 Ma (late Oligocene), the age of the igneous rocks that intrude it (Fassett, 1985). Smith and others (1985, Fig. 2) show the Chuska as late Eocene to Oligocene in age. It seems likely that rocks laterally equivalent to the Chuska once existed across the present basin area. If that is true, then based on present relief in the basin, an additional 760 m of rock must have been present in the central basin area (above the

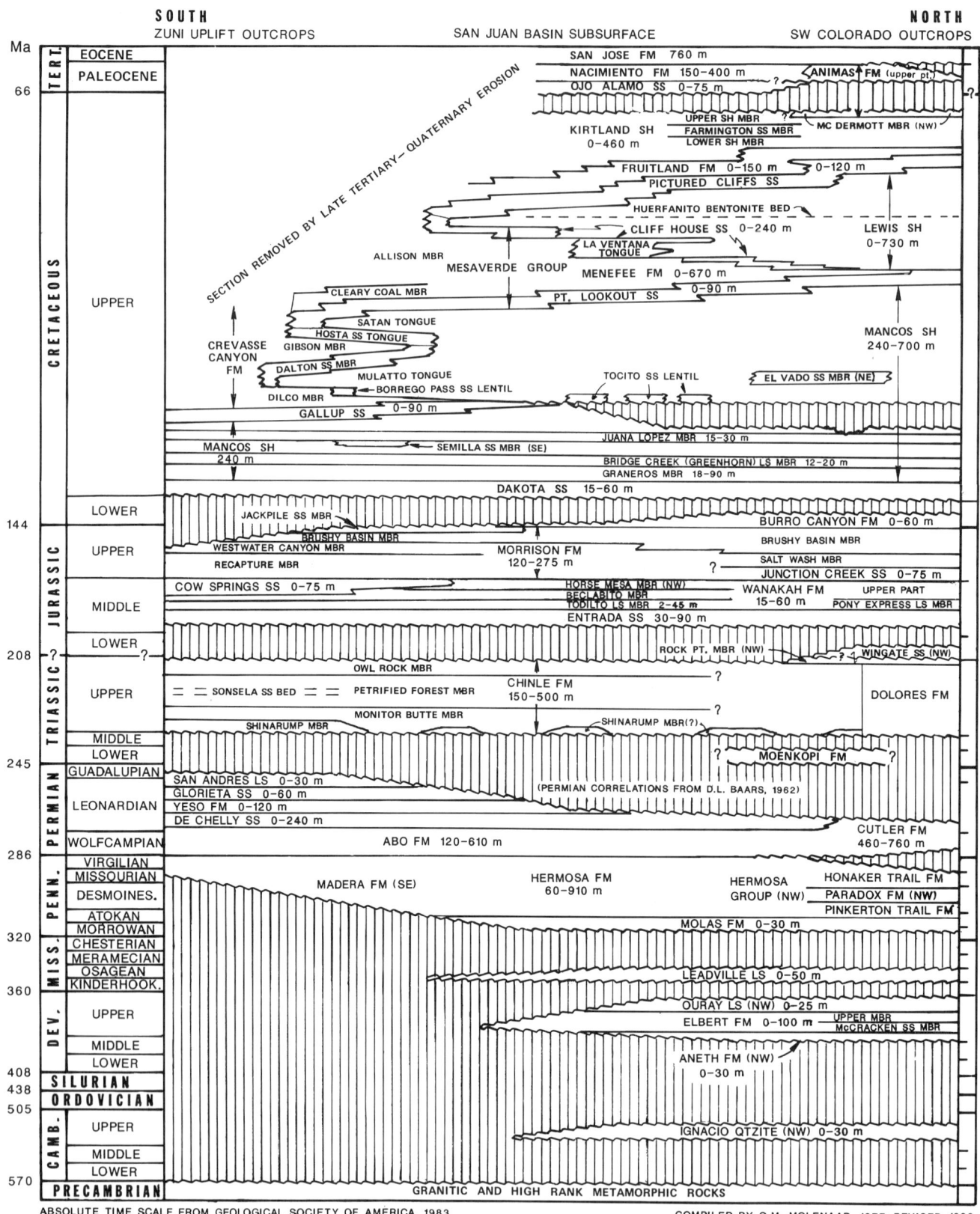

Figure 3. Correlation chart of sedimentary rocks in the San Juan basin (modified from Molenaar, 1989).

present top of the San Jose Formation) in Oligocene time (Fassett, 1985).

After uplift of the Colorado Plateau and an extended period of erosion, a relatively brief aggradation episode occurred during Pleistocene time, resulting in deposition of a thin layer (a few tens of meters thick) of alluvial material over most of the present basin area. Following that episode, erosion resumed in the basin and continues today. Present erosion is uncovering the pre-Pleistocene erosion surface as the Pleistocene alluvium is stripped away.

OIL AND GAS

Stratigraphic nomenclature

Any discussion of oil and gas production in the San Juan basin must be prefaced with a few words about the nomenclature of the producing rocks. The nomenclature is confusing because rather arbitrary designations and definitions were given to many of the producing rock intervals in the basin by the New Mexico Energy and Minerals Division before most of the detailed stratigraphic work had been completed. These definitions do not always coincide with formal stratigraphic nomenclature conventions as prescribed by the North American Commission on Stratigraphic Nomenclature. A detailed discussion of this subject is beyond the scope of this chapter; however, it is comprehensively discussed in Fassett (1978b, 1983b). Here the rock units are discussed in terms of their formally accepted rock-stratigraphic names, unless otherwise noted.

History

The following history of oil and gas development in the basin is abstracted from Dugan (1977), Matheny and Ulrich (1983), Matheny and Talley (1983), and Dugan and Williams (1988). The first well-documented oil discovery in the San Juan Basin, and in New Mexico, was made in 1911 at the Seven Lakes field in the south-central part of the Chaco slope in a well drilled for water. The well produced about 12 barrels of oil (BO) per day from the Menefee Formation at a depth of about 100 m. The first discovery of gas in the basin was made in October 1921, in a well about 1.6 km south of Aztec, New Mexico. The well was completed in the Farmington Sandstone Member of the Kirtland Shale (Fig. 3) at a depth of about 300 m, and a pipeline was built to Aztec where the first natural gas in New Mexico was marketed. (It is ironic that oil and gas production from the rock units that produced these first discoveries today represents but a tiny fraction of a percent of the total oil and gas production from the San Juan basin.)

In September 1922, oil was discovered in the Dakota Sandstone (Hogback field) at a depth of 245 m on the Four Corners platform, about 32 km west of Farmington. This discovery triggered a flurry of exploration that continued throughout the 1920s. During this period, many of the surface structures on the Four Corners platform were drilled and found productive from Paleozoic and Cretaceous rocks. During this same era, gas was discovered in the central basin in the Pictured Cliffs Sandstone and the Mesaverde Group. In 1930, the first pipelines out of the basin were built to Albuquerque and Sante Fe, providing a greatly expanded market for San Juan basin gas.

Over the next 20 years, modest development of oil and gas continued in the basin. The major discoveries during this period were two large Pennsylvanian gas fields on anticlinal structures on the Four Corners platform: the Barker dome and Ute dome fields (Fig. 4). These discoveries, plus the continuing gas discoveries in the Upper Cretaceous shoreface sandstone beds in the central basin, spurred interest in expanding the market for San Juan basin gas even wider; in 1951, a 24-in pipeline to California was completed. With an outlet for large volumes of gas now available, the great drilling boom of the 1950s began, resulting in delineation of the three major gas-producing sandstone units in the central basin: the Dakota Sandstone, Mesaverde Group, and the Pictured Cliffs Sandstone. During this decade, the two largest oil fields to produce from the Tocito Sandstone Lentil of the Mancos Shale were discovered: the Bisti and Horseshoe fields.

Since the 1960s, drilling in the basin has fluctuated, controlled by local events, such as changes in spacing of orders, and global events, such as changes in prices for oil and gas, resulting from both political and supply-and-demand influences. No new

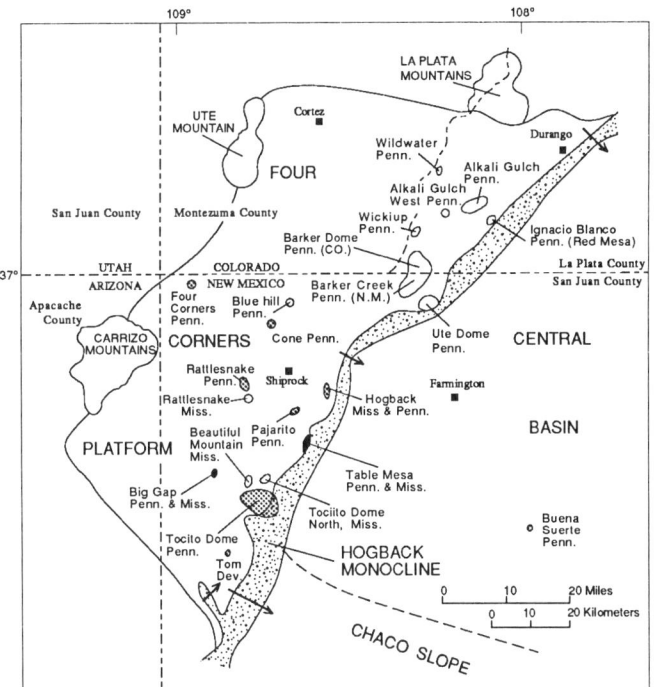

Figure 4. Map showing Paleozoic oil and gas fields in the San Juan basin. New Mexico and Colorado recognize 28 Paleozoic fields. Only 22 named fields are shown because some fields contain more than one named reservoir or pool within a named field. Oil fields are shown crosshatched, gas fields are white, and oil and gas fields are black. Producing systems are abbreviated: Dev. = Devonian, Miss. = Mississippian, and Penn. = Pennsylvanian.

large discoveries were made during this time until quite recently, when the large coal-bed methane resources in the Fruitland Formation were recognized. The rate drilling for Fruitland gas has been accelerating rapidly for the past few years. Because of the relatively low demand for natural gas at this time, it is unlikely that there will be another boom in the San Juan basin in the foreseeable future, but rather, a long, slow, steady development of the basin's tremendous gas resources is anticipated.

Production

Oil and gas have been produced from 313 fields in the San Juan basin. These fields produce (or have produced) from Devonian, Mississippian, Pennsylvanian, Jurassic, Cretaceous, and Tertiary rocks. Production statistics for these fields through 1987 were compiled from annual reports of the New Mexico Oil and Gas Engineering Committee (1988) and the Colorado Oil and Gas Conservation Commission (1988) (Table 1). New Mexico lists named fields and undesignated fields; field totals and production totals include the New Mexico undesignated fields. All fields that have ever been known to produce are included. New Mexico uses the term "pools" for producing reservoirs; Colorado uses the term "reservoirs." In this chapter, I refer to producing reservoirs as "fields."

Cretaceous rocks produced 93 percent of the oil and 98 percent of the gas (Table 1), and 86 percent of the fields in the San Juan basin are in these rocks. Colorado produced 984,767,761 thousand cubic feet (MCF) of gas and 1,786,667 barrels of oil (BO) from 53 fields, whereas New Mexico produced 15,174,405,028 MCF of gas and 286,646,079 BO from 260 fields. The number of producing wells in the basin was 19,174; 1,056 in Colorado and 18,118 in New Mexico. New Mexico has produced 93 percent of the gas and 99 percent of the oil from the San Juan basin.

Oil and gas production in New Mexico has come from 108 gas fields and 152 oil fields; in Colorado, from 38 gas fields and 15 oil fields. Oil-production totals for the basin (Table 1) include both oil from oil wells and condensate and oil from gas wells. Oil production from oil fields totals 220,869,334 BO; condensate and minor oil production from gas wells (mostly from New Mexico) is 67,563,412 barrels. It should be noted that the New Mexico Oil and Gas Engineering Committee (1988) listed all liquid-hydrocarbon production from gas wells as "oil." Most of these liquids were condensate; however, a small amount was oil. Gas produced from gas fields totals 15,570,634,458 MCF, and gas from oil fields totals 588,538,331 MCF.

Natural-gas production in the San Juan basin comes mostly from three geologic units: the Dakota Sandstone, the Mesaverde Group, and the Pictured Cliffs Sandstone. Most of the gas produced from the Mesaverde Group has come from the Point Lookout Sandstone (Fig. 3). The Dakota Sandstone has produced 4.4 trillion cubic feet (TCF) of gas from 15 gas fields, the Mesaverde Group 7.4 TCF of gas from 13 gas fields, and the Pictured Cliffs Sandstone 3.0 TCF of gas from 26 gas fields. Total

TABLE 1. CUMULATIVE OIL AND GAS PRODUCTION FOR THE SAN JUAN BASIN BY GEOLOGIC SYSTEM*

System	Oil† (barrels)	Gas§ (thousand cubic feet)	No. of Fields
Tertiary	659	431,290	4
Cretaceous	268,954,102	15,793,218,942	270
Jurassic	5,130,454	37	11
Pennsylvanian	14,239,956	359,039,228	22
Mississippian	106,123	6,475,109	5
Devonian	1,605	8,183	1
Total	288,432,899	16,159,172,789	313
Non-Cretaceous	19,478,797	365,953,847	43
Percent Non-Cretaceous	7	2	14

*Through December 31, 1987.
†Includes condensate and oil produced from gas wells.
§Includes casinghead gas produced from oil wells.

gas production from gas fields in these three rock units is 14.8 TCF, about 14 TCF from New Mexico and 0.8 TCF from Colorado. The three largest gas fields in New Mexico, the Basin Dakota, Blanco Mesaverde, and the Blanco Pictured Cliffs South, have produced nearly 12 TCF of gas. These rock units have also produced relatively small amounts of casinghead gas from oil fields.

Most condensate production is from the Dakota and the Point Lookout Sandstones. Indeed, nearly 66 million barrels of the more than 68 million barrels of condensate produced from the entire San Juan basin comes from two New Mexico fields, the Basin Dakota and the Blanco Mesaverde.

The largest percentage of the oil produced in the basin comes from the Tocito Sandstone Lentil and the El Vado Sandstone Member of the Mancos Shale. Together, these rock units have produced more than 150 million barrels of oil (MBO) of the 219 MBO recovered from oil fields in the basin. This production is from 71 fields. The two largest fields, the Bisti and the Horseshoe Gallup, have produced 74.7 MBO. The only statistically significant hydrocarbon production from non-Cretaceous rocks in the San Juan basin is the 14-MBO production from the Pennsylvanian system; nearly 12.8 MBO is from the Tocito dome field.

Oil and gas field characteristics

Paleozoic fields. Oil and gas have been produced from Paleozoic rocks in 28 fields in the San Juan basin; 12 are oil fields and 16 are gas fields (Fig. 4). With one exception, all of these fields are on the Four Corners platform. One field produced from the Devonian, 5 from the Mississippian, and 22 from the Pennsylvanian; 2 of the Pennsylvanian fields produced minor amounts of oil or gas from the Mississippian. The Devonian field produced oil from a sandstone reservoir on an anticlinal structure; the Mississippian fields produced nonflammable, high-nitrogen, helium-bearing gas and a small amount of oil from porous limestone and dolomite beds on anticlines; and the Pennsylvanian rocks produced from carbonate beds containing zones of porosity developed within algal mounds (sometimes referred to as "carbonate buildups") mostly on structural features. A few Pennsylvanian fields produce from stratigraphic traps. Devonian and Mississippian production is confined to the southwestern part of the Paleozoic productive trend on the Four Corners platform.

Pennsylvanian rocks produce only gas in the northeastern part of the Four Corners platform, and oil and gas in the southwestern part (Fig. 4). In the northeastern part of the platform, some zones in the Pennsylvanian produce sulfur-rich gas; H_2S from those zones ranges from 1 to more than 12 percent. The Alkali Gulch field in Colorado produces about 25 tons of sulfur per day from Pennsylvanian sour gas; the sulfur is sold locally. Some of these Pennsylvanian wells also produce gas containing as much as 21 percent CO_2.

Most of the Paleozoic fields on the Four Corners platform are aligned in a northeast-trending band that parallels the Hogback monocline (Fig. 4). The structural features on the platform were probably created during the time the structural basin was forming, from middle to late Paleocene time and continuing into early Oligocene time (Fassett, 1985). The lone Paleozoic field (Buena Suerte) within the central basin produced oil from Pennsylvanian rocks at a depth of 3,350 m. This one-well field is about 65 km east of the Four Corners platform.

Paleozoic fields on the Four Corners platform generally become deeper from southwest to northeast; depths to producing zones range from 1,220 m to 3,050 m. All of the gas-producing fields in the northeastern part of the platform produce from depths in excess of 2,320 m; the oil fields to the southwest produce from depths less than 2,320 m.

Jurassic fields. Eight named and three undesignated oil fields produce from the Entrada Sandstone of Jurassic age. All of these fields are in the southeastern part of the basin in a 70-km-long, northwest-trending belt (Fig. 5). Production is from the eolian sandstone facies (upper part) of the Entrada. All of the Entrada fields produce from stratigraphic traps, with closure reported to be from relict subaerial dune topography (Vincelette and Chittum, 1981); they suggested that the oil is trapped in the dune crests. The seal for the traps is the overlying Todilto Limestone Member of the Wanakah Formation. The thickness of the Todilto has an inverse relation to the Entrada paleo-topography; it thins over the Entrada highs and thickens over the Entrada lows. The Todilto, which is a fetid, organic-rich limestone and anhydrite deposit, is also thought to be the source rock of the Entrada oil. One of the fields (Media) is on a surface structural nose.

All of the Entrada fields are semicircular with a slightly longer north-to-northeast dimension. Vincelette and Chittum (1981, Fig. 16, p. 2558) presented a map of Entrada seismic anomalies in the southeastern San Juan Basin and stated that it showed a strong northeast trend to the "Entrada sand thicks or topographic highs."

All of the fields have similar reservoir characteristics, with an average porosity of about 23 percent and permeability of about 300 milidarcies. The oil has a 29 to 36° API gravity and high pour point (50° to 90°F) that seasonally necessitates the use of insulated production equipment The amount of water produced with the oil has increased to 50 percent in a few months in all fields and to 95 percent or more after a year of production. Disposal of this water has been a problem. Three Entrada fields—Eagle Mesa, Media, and Papers Wash—have produced three-fourths of the total Entrada production in the basin, from 1 to 1.5 MBO each. The number of productive wells in each field ranges from 1 to 6.

A few wells produce natural gas from the Brushy Basin Member of the Morrison Formation in the Ignacio Blanco field in the northern part of the central basin in Colorado and in the Red Mesa field on the Four Corners platform in Colorado. According to Harr (1988), this production comes from fluvial channel sandstone beds. All of these wells are reported as Dakota-Morrison producers by the Colorado Oil and Gas Commission; thus, Morrison gas production cannot be quantified. Dakota-Morrison production totals 630 million cubic feet (MMCF) of gas from 16 wells; Morrison production is thus relatively small at this time.

A small amount of oil has been produced from the Morrison Formation in the Red Mesa field from two Dakota-Morrison wells and one Morrison well. Total production is 18,000 BO.

Cretaceous fields. Upper Cretaceous sandstone beds and silty to sandy mudstone beds are the primary producers of oil and gas in the San Juan basin. Coal beds have produced a relatively small amount of gas, but are becoming increasingly large gas producers. Upper Cretaceous production has come from 270 fields: 127 gas fields and 143 oil fields. Most of the natural gas has been produced from the Dakota, Point Lookout, and Pictured Cliffs Sandstones. More than 80 percent of this gas comes from three New Mexico fields: Basin Dakota, Blanco Mesaverde, and Blanco Pictured Cliffs South. Most of the Cretaceous oil comes from the shelf-sandstone lenses of the Tocito Sandstone Lentil of the Mancos Shale (designated "Gallup" producing interval by New Mexico and Colorado) and from fractured shale reservoirs in the El Vado Sandstone Member of the Mancos. (New Mexico has classified some El Vado fields as "Mancos" fields and some as "Gallup" fields; Colorado calls them "Gallup" fields). Most of this production has come from the central basin area; a few, relatively small fields produce Cretaceous oil from structural traps outside the central basin on the Chaco slope and on the Four Corners platform (Figs. 1 and 5).

Oil. Most of the Upper Cretaceous oil (not including condensate) produced in the San Juan basin has come from reservoirs within the lower part of the Upper Cretaceous section. Most oil production from this part of the section is from the Dakota Sandstone (39 fields), Gallup Sandstone (6 fields), Tocito Sandstone Lentil (30 fields), and the El Vado Sandstone Member of

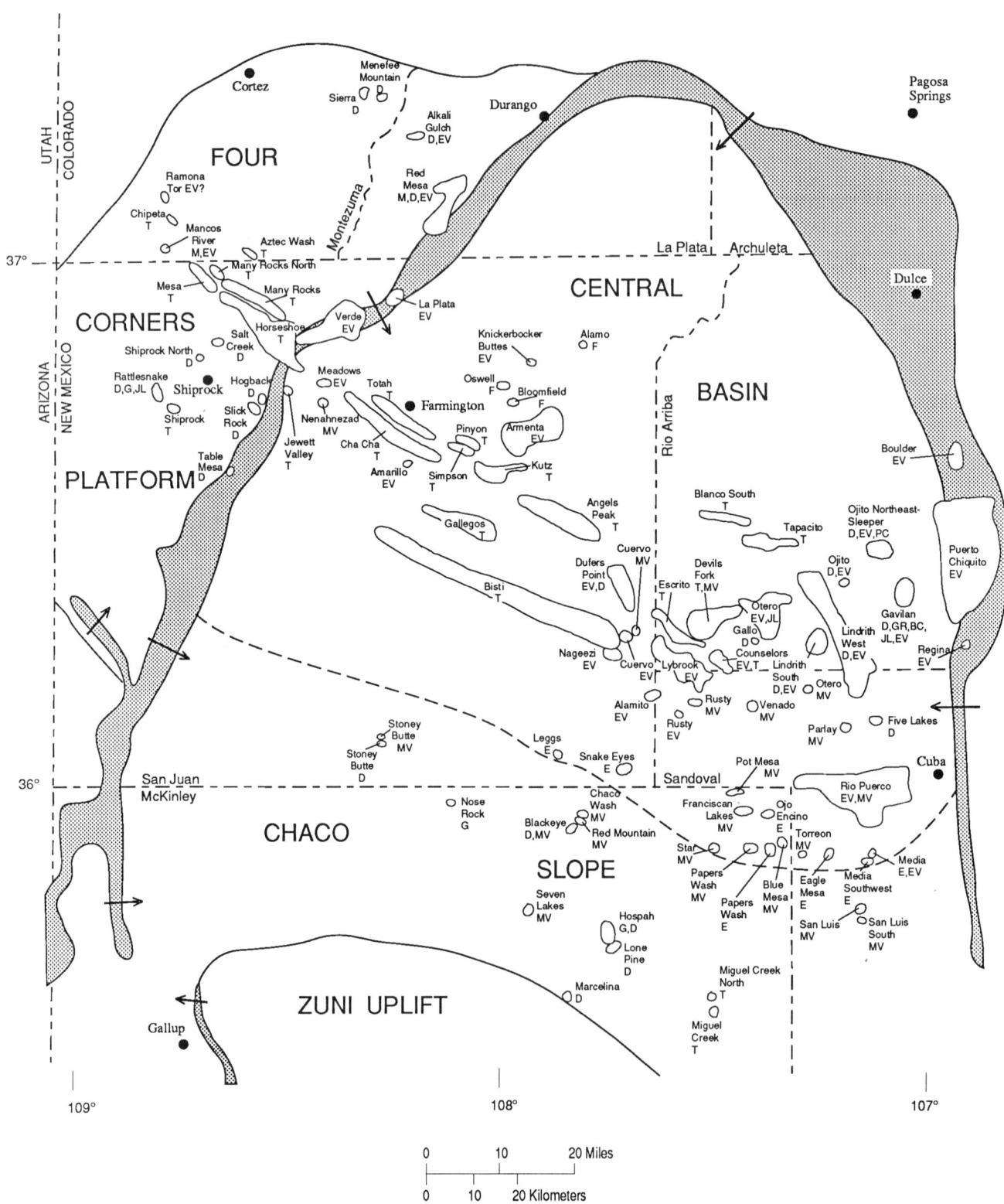

the Mancos Shale (39 fields). About 10 percent of the oil has come from Dakota fields, 10 percent from Gallup fields, and 80 percent from Tocito and El Vado fields. The source rock for this oil was most certainly the very black, organic-rich, lower Mancos marine shales. For a discussion of correlation of San Juan basin oils to source rock, see Ross (1980).

The Dakota Sandstone produces oil from 39 fields, including structural and stratigraphic traps on the south flank of the basin, and structural traps on the Four Corners platform (Fig. 5). The Hogback, Rattlesnake, and Table Mesa fields are the largest Dakota fields on the Four Corners platform, having produced 5.6, 4.8, and 1.4 MBO, respectively. In the central basin, the only large Dakota producer is the Lindrith West field. This field has produced over 13 MBO. Oil from this field represents commingled production from the El Vado Sandstone Member of the Mancos Shale (designated "Gallup" by the state of New Mexico) and the Dakota Sandstone; thus, Dakota production from this field cannot be precisely quantified. The Chacon field, which was merged with the Lindrith West field in 1984, had produced 3.2 MBO from the Dakota through 1983. At that time the Lindrith West field had produced 3.7 MBO from the Dakota and the El Vado. On the Chaco slope, Dakota oil has been produced from the Lone Pine field (2.6 MBO) and the Hospah field (nearly 0.2 MBO). The largest Dakota oil field in the Colorado part of the basin is the Red Mesa field; it is on the Four Corners platform and has produced a little more than 0.8 MBO.

Five Gallup Sandstone fields produce oil on the Chaco slope (Fig. 5): 3 on the Hospah structure (faulted anticline) have produced 19.5 MBO; 1 (Marcelina) on a faulted anticlinal nose about 16 km southwest of Hospah has produced 51,000 BO; and 1 (Nose Rock) on an updip-pointing, paleo-stream-channel meander on a structural nose (Bruce Black, personal communication, 1989) has produced 19,000 BO. All of these fields apparently produce from the Torrivio Sandstone Member of the Gallup Sandstone. The Torrivio is a fine- to coarse-grained fluvial sandstone facies of the mostly marine to marginal marine Gallup sandstone and is the upper part of the Gallup. One Gallup field produces from the Rattlesnake anticline (Matheny, 1983) on the Four Corners platform. The Gallup Sandstone in the Chaco slope oil fields has porosities of 25 percent and greater, and permeabilities in the hundreds of millidarcies (Luce, 1978a, b; Bircher, 1978; Edmister, 1983).

Tocito oil production comes from 30 fields in a northwest-trending, slightly arcuate band across the southern part of the central basin and onto the Four Corners platform (Fig. 5). The Tocito has produced more than 109 MBO. Tocito fields are stratigraphic traps consisting of shelf-sandstone lenses enclosed in marine shale. Most of these fields are within the central basin. Three New Mexico fields, the Many Rocks, Many Rocks North, and Mesa fields, are on the Four Corners platform. The Horseshoe field is mostly on the Four Corners platform, but its southeast end is draped over the northwestern monoclinal rim of the basin (Fig. 5). Trapping of the oil in these fields is both structural and stratigraphic (Matheny, 1978a). In Colorado, the Aztec Wash, Ramona, and Chipeta fields produce Tocito oil from sandstone lenses on the Four Corners platform.

Tocito sandstone beds are generally the best Cretaceous reservoir rocks in the central basin; they range from fine to coarse grained with porosities averaging around 15 percent and permeabilities ranging from 25 to well over 100 millidarcies. Many fields produce from two and sometimes three stacked sandstone beds, sometimes interconnected and sometimes not. A few Tocito fields contain fractured, low-permeability sandstone beds in which the fractures have created the reservoir. Generally, the Tocito reservoirs in the central basin are oil saturated and contain negligible water. The two largest Tocito fields are the Horseshoe, which has produced 38.3 MBO, and the Bisti, which has produced 36.4 MBO.

The El Vado Sandstone Member of the Mancos Shale has produced more than 41 MBO from 39 fields (Fig. 5). Most of this production is from "fractured shale" reservoirs. Gorham and others (1977) postulated that fractured shale reservoirs formed in the San Juan Basin in the carbonate-cemented, brittle, El Vado sandstone and siltstone beds where they were tectonically stressed, both vertically and horizontally, especially in places where the monoclinal rim of the basin makes sharp bends. The El Vado, where it is not fractured, generally has insufficient inherent porosity or permeability to make a reservoir. The El Vado is present throughout much of the central basin and is as much as 180 m thick. The reference log for the El Vado in the subsurface of the San Juan basin is shown in Fassett and Jentgen (1978, p. 238).

Fractured El Vado reservoirs on the east edge of the basin are the Puerto Chiquito and Boulder fields; these fields have produced nearly 16 MBO (14 MBO and 1.8 MBO, respectively) and are classified by New Mexico as producing from the Mancos Shale.

Fractured El Vado reservoirs on the northwestern edge of the central basin in New Mexico are the Verde, La Plata, and Meadows fields. These fields have produced nearly 9 MBO (7.9 MBO, 0.6 MBO, and 0.2 MBO, respectively). New Mexico classifies these fields as "Gallup" producers. In Colorado, the Red

Figure 5. Map showing Jurassic and Cretaceous oil fields in the San Juan basin. Undesignated New Mexico fields are not shown. Some fields produce from more than one formation and thus contain more than one designated pool or reservoir. Field names in New Mexico and Colorado always include the name of the producing interval, e.g., "Horseshoe Gallup" field. On this map, the names of the producing intervals are not included in the field names due to space limitations. Producing formations for each field are indicated by the following abbreviations: E, Entrada Sandstone; D, Dakota Sandstone; M, Mancos Shale; GR, Graneros Shale; BC, Bridge Creek Limestone Member of the Mancos Shale (formerly Greenhorn Limestone Member); JL, Juana Lopez Member of the Mancos Shale; G, Gallup Sandstone; T, Tocito Sandstone Lentil of the Mancos Shale; EV, El Vado Sandstone Member of the Mancos Shale; MV, Mesaverde Group; PC, Pictured Cliffs Sandstone; F, Farmington Sandstone Member of the Kirtland Shale.

Mesa field has produced more than 0.5 MBO from a fractured El Vado reservoir located mostly on the Four Corners platform. The Red Mesa field is located on an anticlinal flexure (Lauth, 1983); this structure may account for the fracturing of the El Vado. Two other small fields have produced minor amounts of oil from the El Vado on the Four Corners platform in Colorado. Colorado classifies this production as coming from the "Gallup."

Most of the rest of the El Vado fields are clustered in the southeastern part of the central basin (Fig. 5). Many of these fields are associated with structural noses, which probably accounts for fracturing of the El Vado. A few fields, however, seem to have no apparent local structural control and appear to be fractured along narrow, northwest-trending lineaments.

Twenty-four small fields have produced nearly 1.5 MBO from the Mesaverde Group. Except for the Nenahnezad field in the west-central part of the central basin, all of the Mesaverde fields are in the southeastern part of the central basin and on the Chaco slope (Fig. 5). Most of these fields are stratigraphic traps consisting of channel sandstone beds enclosed by impermeable finer-grained overbank facies in the lowermost part of the Menefee Formation. Many of these fields were discovered by mapping surface structures. A few of the fields produce from upper-shoreface sandstone beds in the upper part of the Point Lookout Sandstone. Most of the fields are quite small; the largest, Franciscan Lake, has produced 0.5 MBO. The source of this oil is probably the organic-rich carbonaceous rocks of the lowermost Menefee Formation.

One Pictured Cliffs field and one undesignated Pictured Cliffs well have produced a little more than 100,000 BO. Four fields have produced slightly more than 100,000 BO from the Farmington Sandstone Member of the Kirtland Shale (Fig. 5). According to Matheny (1978b), the Pictured Cliffs oil is probably condensate that collected in the low part of a stratigraphic trap. The Farmington oil is high-gravity (56 to 59° API) and, according to D. D. Rice (personal communication, 1989), is probably condensate that has migrated into the Farmington from older rocks. The Farmington oil is trapped in channel-sandstone stratigraphic traps.

Gas. Natural gas in the three major gas-producing sandstone units, the Dakota Sandstone, Mesaverde Group, and Pictured Cliffs Sandstone, is concentrated largely in broad, overlapping, northwest-trending stratigraphic traps in the south-central part of the basin (Fig. 6). Each of these gas-bearing sandstone units is present throughout the basin, and each crops out around its periphery. The Pictured Cliffs Sandstone is missing in two narrow areas on the east side of the basin. A relatively small amount of gas is present in these three units in smaller stratigraphic and structural traps in the northern and southeastern parts of the central basin. With no apparent trap, why hasn't all of the gas migrated updip through the sandstone beds to the outcrop and escaped from this 45-m.y.-old structural basin?

Berry (1959) was the first worker to offer a solution to this conundrum; he suggested that the gas in these sandstone units had somehow been prevented from escaping by hydrodynamic forces. The argument was that these gas-bearing rocks were underpressured, principally as the result of uplift of the Colorado Plateau in Miocene time and the rapid erosion of a few thousand meters of overlying rock. The resultant pressure release on the gas-bearing sandstone beds caused dilation of the pore space in these rocks and a decrease in temperature. With this pressure drop, meteoric water moved centripetally down the pressure gradient through the three gas-bearing sandstone units, sweeping the gas toward the deepest part of the basin. Berry's (1959) hydrodynamic-trap model was embraced by most subsequent workers. The hydrostatic-gas-trap hypothesis was recently evaluated by Cumella (1981, p. 168–175). Cumella discounted the hypothesis to some extent but did not completely dismiss it; further, he suggested that secondary kaolinite cement in the southwestern part of the basin, along with stratigraphic trapping of gas in the stair-steps of the Point Lookout and Pictured Cliffs Sandstones (Fig. 3), are also factors contributing to the trapping of the gas in these rocks.

A comprehensive review of all available publications on the stratigraphy of the three sandstone units, personal observations of these rock units on the outcrop and throughout the subsurface on geophysical logs, and my own detailed studies of the Pictured Cliffs Sandstone throughout the basin lead me to the conclusion that most of the gas trapped in the Dakota, Point Lookout, and Pictured Cliffs is stratigraphically trapped and that the hydrodynamic gas-trap hypothesis, however elegant in a theoretical sense, is a chimera. On the basis of all available information, it seems apparent that the gas is trapped in the three sandstone units by permeability barriers. Reservoir characteristics of these formations are clearly structurally enhanced by fracturing, but little has been published on the subsurface fracture pattern of the basin.

Even though these three gas-producing sandstone bodies were considered in the past to be sheet-like sandstone layers with interconnected permeability basinwide, subsequent studies have shown that each of them consists of a complex of individual sandstone beds separated by impervious mudstone layers. Berry's (1959) concept of a hydrodynamic trap for these sandstone units was proposed before the stratigraphy of these rock units had been worked out in detail. More recent work on these rocks, discussed below, does not support the concept of interconnected permeability of these rock units throughout the basin.

The Dakota Sandstone consists of a complex of fluvial and offshore marine sandstone bodies (Deischl, 1973; Hoppe, 1978). In the New Mexico part of the basin, most of the Dakota gas is trapped within the upper, marine facies; the underlying fluvial facies is generally water saturated (Hoppe, 1978). Most Dakota gas, therefore, appears to be trapped in several offshore marine sandstone lenses surrounded by impervious marine shale. The Dakota is a tight (low-permeability) sandstone having an average porosity of 5 to 15 percent and permeabilities ranging from 0.1 to 0.25 millidarcy. "Fracturing, either natural or induced, is required to obtain commercial flow rates from the low-permeability [Dakota] reservoirs" (Rice, 1983, p. 1202). The major areas of gas-productive Dakota Sandstone are shown on Figure 6.

The Point Lookout and Pictured Cliffs Sandstones are sim-

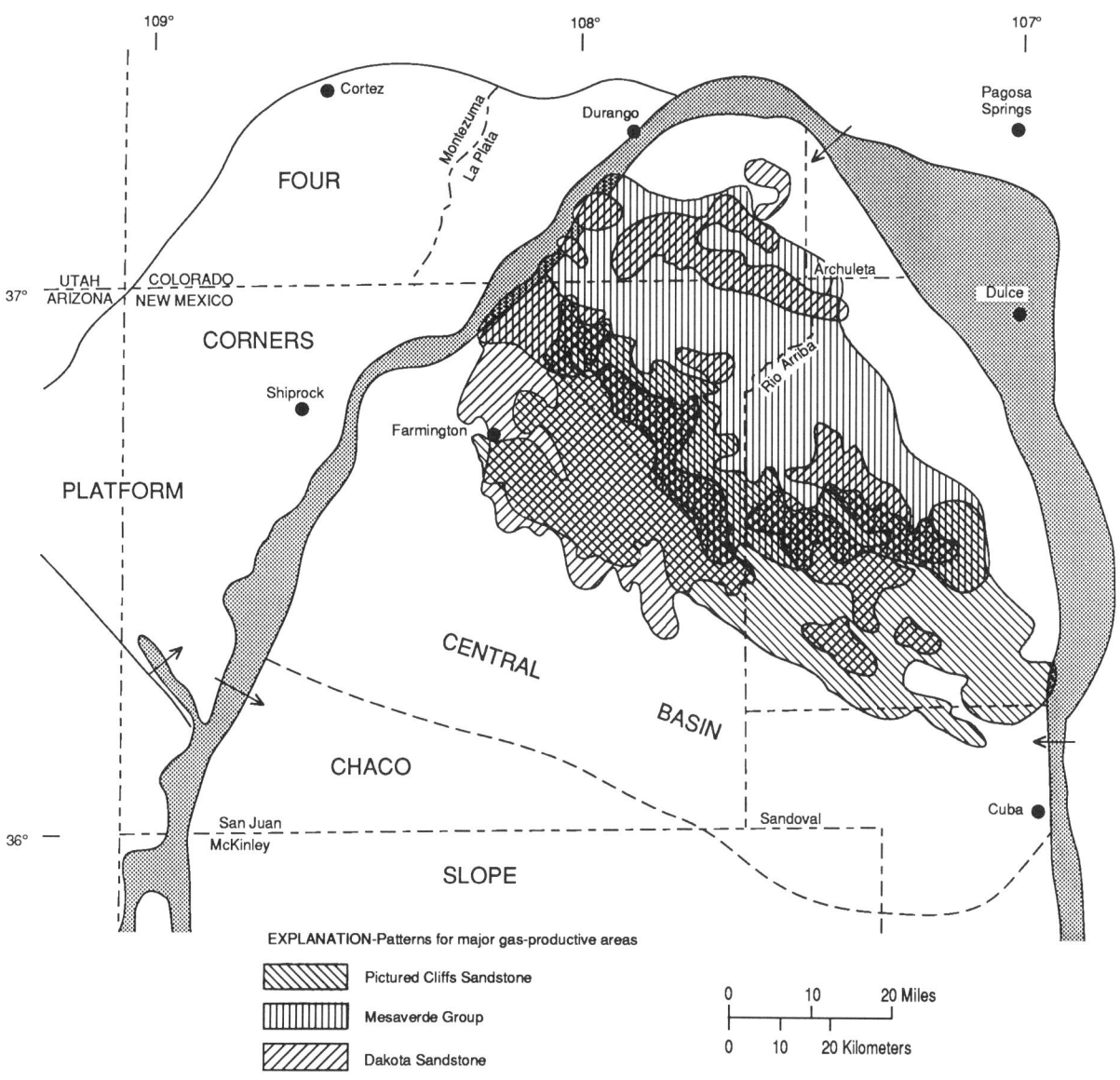

Figure 6. Map showing major gas-producing areas for the Upper Cretaceous Dakota Sandstone, Mesaverde Group, and Pictured Cliffs Sandstone in the San Juan basin.

ilar in their environments of deposition and geometry (Fassett, 1977, 1986), and thus are here discussed together. Both of these units represent regressive-marine shoreface deposits laid down as the shoreline shifted across the basin from southwest to northeast. When viewed in northeast-oriented cross sections, these formations are seen to consist of a series of stairsteps that rise stratigraphically northeastward across the basin (Fig. 3). Detailed surface observations of these units at right angles to their paleo-shoreline orientations show that they consist of a series of overlapping sandstone beds, commonly separated by impervious mudstone layers. Along the shoreline trend, shoreface sandstone beds are punctuated by distributary channel deposits and in places contain complexes of storm-caused cut-and-fill features frequently containing mudstone drapes. Only relatively recently have detailed studies of these kinds of features been discussed in the literature (Flores and Erpenbeck, 1978; Wright, 1986).

Thus, rather than being homogeneous sheet sandstone beds with interconnected permeability throughout the basin, the Dakota, Point Lookout, and Pictured Cliffs are complexes of individual sandstone bodies separated by impervious mudstone or claystone barriers. It is these permeability barriers that have trapped the natural gas in these sandstone beds. Where there is interconnected permeability in these rock units to the outcrop, the gas has, indeed, long ago escaped, thus explaining the gas-barren aureole around the shallower part of the central basin (Fig. 6). Water has subsequently moved downdip through these sandstones as far as their interconnected permeability existed, leaving them water-saturated around much of their periphery.

The Point Lookout and Pictured Cliffs Sandstones both contain water in the southwestern part of the basin and are more tightly cemented in the northeastern part of the basin. Both formations are tight, with average porosities of 10 to 15 percent and permeabilities of from 0 to 5.5 millidarcies (Pritchard, 1978; C. F. Brown, 1978). Production of gas from these two formations is enhanced or may be controlled, to a large extent, by natural fractures. If this is true, the fractures are apparently confined to the more brittle sandstones and did not form or have not remained open in the finer-grained mudstones and siltstones, thus maintaining the integrity of the impermeable seals trapping the gas. Cumella (1981, Fig. 28) presented a map showing the distribution of the initial potential for natural gas for wells completed in the Pictured Cliffs Sandstone, and suggested that the resulting pattern reflected shoreline trends for the Pictured Cliffs. An alternative interpretation of that map is that the pattern reflects not shoreline trends, but the orientation of natural fractures in the Pictured Cliffs.

In the northern (Colorado) part of the San Juan basin, gas has been trapped in the Dakota and Point Lookout Sandstones on a broad, northwest-trending structural feature called the Ignacio Blanco anticline (Harr, 1988). This structure is probably more complex than a simple structural anticlinal trap. Because of the inherently tight nature of the Dakota and Point Lookout, natural fractures in these rock units, created parallel to the axis of this anticline as it was formed in early Tertiary time, have probably greatly enhanced or, indeed, may have even created these reservoirs. The Pictured Cliffs Sandstone has produced minor amounts of gas from stratigraphic traps along a northwest trend in the Colorado part of the basin. (For a comprehensive discussion of the gas production in the northern San Juan Basin, see Harr, 1988.)

Other, smaller Cretaceous fields produce gas from the El Vado Sandstone Member and Tocito Sandstone Lentil of the Mancos Shale (22 fields) and an interval in the Cliff House Sandstone (designated "Chacra" by New Mexico) (9 fields). El Vado production comes from areas where the commonly tight, fine-grained sandstone and siltstone beds of the El Vado are fractured or contain slightly more porous or permeable sandstone or siltstone. "Chacra" production comes from offshore marine sandstone beds of the La Ventana Tongue of the Cliff House (Fassett, 1978b, Fig. 4; 1983b).

Fruitland Formation coal beds and channel-sandstone deposits have produced a modest amont of gas from 28 fields. Most of this gas has been produced from coal beds. However, due to intermingled production in many of these fields, the amount of coal-bed methane produced cannot be determined exactly. Fassett (1989a) estimated that the coal-bed gas produced from these 28 fields totaled 85.1 billion cubic feet (BCF) through 1987. Most of the Fruitland coal-bed gas produced to date has come from the Colorado part of the San Juan basin, from within the Southern Ute Indian Reservation. The largest Fruitland coal-bed methane fields are the Ignacio Blanco Fruitland–Pictured Cliffs (estimated coal-bed methane production, 40 BCF), the Cedar Hill Basal Coal (9.5 BCF of gas), and the Basin Fruitland (6.6 BCF of gas) (Fassett, 1989a). Fruitland coal-bed methane is a resource of enormous potential and is discussed in more detail below.

The Farmington Sandstone Member of the Kirtland Shale has produced gas from four very small fields from channel sandstone deposits. All of these fields depleted rapidly, indicating limited reservoir size. The origin of this gas is unknown.

Tertiary rocks have produced gas from the Paleocene Nacimiento Formation in four very small fields, three in the north-central part of the basin and one in the east-central part. Gas in each field was trapped stratigraphically in channel-sandstone deposits of limited lateral extent. Even though the potential for Tertiary rocks appears to be minimal, these fields are of interest because the gas in them appears to have been generated from older, more mature, organic-rich continental rocks, and the gas has migrated upward (D. D. Rice, written communication, 1989). A single well has produced gas from the Tertiary Ojo Alamo Sandstone in the southeastern part of the basin on the Schmitz anticline (Needham, 1978). This gas was probably generated in the marine Lewis Shale (only 60 m below), charging the Ojo Alamo channel-sandstone bed.

THERMAL MATURITY

The thermal maturity of the sedimentary rocks in the San Juan basin is discussed in several recent papers (Rice, 1983; Meissner, 1984; Bond, 1984; Rice and others, 1988; Clarkson and Reiter, 1988). These studies used essentially the same data base, which consists of two elements: the distribution of the fixed carbon and volatile content of Fruitland Formation coal beds as reported in Fassett and Hinds (1971), and the pattern of vitrinite reflectance values reported in Rice (1983). These data sets are in good agreement. Bond (1984) attempted to portray the thermal history of the basin on the basis of a reconstruction of the burial history of the geologic section penetrated in a well in the north-central part of the basin.

The general consensus of these writers is that thermal maturity increases toward the present structural axis of the basin, but that the area of maximum maturity may be slightly offset north of the basin axis. This offset has been attributed to a heat source to the north (the San Juan volcanic center), but this concept is controversial because of questions regarding the thermal conductivity of rock (Clarkson and Reiter, 1988). A simpler solution may be that the greatest depth of burial during the time between the Oligocene and the present was slightly north of the present axis of the bsain.

On the southern limb of the San Juan basin, where most of the oil fields are located, the greatest depths of oil production from the following rock units are (number in parentheses is elevation relative to mean sea level): Entrada Sandstone, 1,800 m (280 m); Dakota Sandstone, 2,290 m (−40 m); Tocito Sandstone Lentil 2,070 m (−90 m), and Mesaverde Group, 1,400 m (670 m). On the east rim of the basin, the deepest wells producing from fractured El Vado Sandstone Member wells are 2,040 m (400 m). On the Four Corners platform, the deepest oil production from Pennsylvanian rocks is from around 2,130 m (–520 m); at depths greater than 2,500 m (–580 m); only gas is produced from Pennsylvanian rocks. An empirical conclusion based on this data is that oil is present in rocks in the south-central part of the basin down to at least 90 m below sea level; on the Four Corners platform, the maximum depth for Pennsylvanian oil production is between 520 and 580 m below sea level.

The difference of 460 m for maximum oil depth (–90 versus –550) between the central basin area and the Four Corners platform is probably the result of different burial histories for these two areas. Rice (1983) showed the downdip limit of oil for the Pennsylvanian on the Four Corners platform and for the Dakota Sandstone and Mancos Shale (including the Tocito and El Vado) in the central basin area. He also stated that the shallowest oil in the southern San Juan basin is in rocks having vitrinite reflectance values less than 0.6 percent, and suggested that this oil migrated upward from where it was generated in more mature rocks in a deeper part of the basin.

An anomalous well has produced a little more than 5,000 BO from Pennsylvanian rocks at a depth of nearly 3,360 m in the south-central part of the basin. This depth equates to 1,400 m below sea level. This well, which constitutes the Buena Suerta oil field, is now abandoned. The vital statistics of the well were presented by H. H. Brown (1978). No oil should be present at this depth according to most theories of oil generation and thermal maturity. There is presently no explanation for the presence of oil at this depth in this well. The downdip limit of wet gases (containing condensate) is discussed and depicted in detail in the Rice (1983) report.

POTENTIAL TARGETS

Conventional targets

Despite the more than 19,000 wells that have produced hydrocarbons in the San Juan basin, the basin is still far from being completely explored. Most of the drilling has been concentrated in the gas-producing fairways (Fig. 6) that trend northwest across the central part of the basin. Outside of those fairways, large areas of low drilling density still exist. Furthermore, only a few basement tests exist in the central basin area; most wells stop in Cretaceous rocks and, in a few places, in the Jurassic Entrada Sandstone.

The most promising plays in the future wil be the virtually untested Paleozoic carbonate rocks in the central basin at depths where probably only gas can be expected. The tests drilled to date have not found reservoir rocks, and future plays will be based on an attempt to find algal mound porosity or possibly tectonic-dolomite porosity over deep structures. Seismic-reflection analysis will be the exploration tool. There is still some potential for undiscovered Paleozoic oil or gas on the Four Corners platform, but all of the known surface structures there have been drilled and stratigraphic traps will be difficult to locate.

Smaller plays will probably involve the search for additional oil-bearing Tocito Sandstone Lentil reservoirs along the main productive trend. In addition, small Dakota oil fields may be found associated with subtle surface structures in the southern part of the basin. Very small Menefee Formation channel-sandstone oil accumulations are almost certainly present along the southwestern flank of the basin, but these accumulations will be difficult to find because of their small size and absence of surface expression. Fractured reservoirs in the El Vado Sandstone Member have potential for oil near the monoclinal rim of the basin wherever structural conditions may have stessed the El Vado. El Vado oil and gas could also be present in the central basin area away from the basin's monoclinal rim where northwest-trending linear fracture zones may be present.

Harr (1988) discussed recent discoveries of limited gas resources in the Jurassic Morrison Formation and the Lower Cretaceous Burro Canyon Formation in the northern San Juan Basin in Colorado. He suggested that these rock units have further potential in that part of the basin.

Coal-bed methane

The real sleeping giant with enormous potential in the San Juan basin is coal-bed methane in Fruitland Formation coal beds. Fruitland coals have been known since before the turn of the century, and beginning in the 1930s and continuing through the 1950s, surface exposures of these coal beds were mapped. A detailed subsurface study of these coals by Fassett and Hinds (1971) estimated Fruitland coal resources of 200 billion tons and presented a model for deposition of the coal.

When the discovery was made in the mid-1970s that Fruitland coals contain adsorbed (or absorbed) natural gas that could be produced by pumping the water out of the coal (thereby lowering the pressure on the coal and allowing the desorbed gas to flow into the bore hole), interest in this newly recognized resource began to grow. Some recently discovered Fruitland coal-gas fields in the north-central part of the basin in New Mexico are overpressured and flow gas at rates greater than 10 MMCF per day. In the late 1980s, additional studies of Fruitland coals beds (Kelso and others, 1987; Kelso and Wicks, 1988) confirmed that the total Fruitland Formation coal resource in the San Juan basin exceeded 200 billion tons and estimated that these coals contained nearly 50 TCF of gas. If these estimates are correct and half of this gas is recoverable, previous estimates of the basin's ultimate production of natural gas (which did not include coal-bed methane) would more than double.

The Fruitland gas play is only just beginning, and the geologic factors controlling coal-bed methane production are still not completely understood. On the basis of present experience, however, Fruitland coal beds apparently produce more gas in places where the coal is more highly fractured. The coal seems to contain some natural fractures or cleats everywhere, but in places where tectonic features exist, such as faults, folds, or lineaments, the coal-fracture density is increased, and these areas seem to be the prime targets for Fruitland coal-bed methane exploration. The development and production of Fruitland coal gas is presently concentrated in the northern part of the basin where total coal thicknesses are the greatest, but exploration is beginning to be extended southward.

One of the primary attractions of Fruitland coal gas is its shallow depth; the Fruitland at its deepest is only slightly over 1,200 m and averages around 600 to 900 m. In addition, because the thickness, distribution, and physical and chemical characteristics of Fruitland coals are so well known, little risk is involved in locating prime target areas. Questions still remain unanswered, however, such as: Can coal beds outside areas of enhanced fracture permeability produce commercial coal-bed methane? What is the minimum depth at which Fruitland coal beds contain commercial gas? And, why are some Fruitland coal reservoirs overpressured?

The major problems in the production of coal-bed methane are controlling the influx of coal fines into the wellbore and disposing of the water produced with the gas. The most promising method for water disposal appears to be underground injection; various well-completion methods are still being tried to minimize the production of coal fines. For a comprehensive collection of papers on all aspects of coal-bed methane production and coal geology in the San Juan Basin, see Fassett (1988); for a review of coal-bed methane production in the San Juan Basin through 1987, see Fassett (1989a).

The Menefee and Crevasse Canyon Formations also contain significant coal deposits in the San Juan basin, but their resources have not yet been assessed in detail throughout the basin area. Fassett (1989b) estimated the non-Fruitland coal resource in the basin to be about 15 billion tons. Non-Fruitland coal beds are relatively thin and discontinuous (Fassett, 1986) and, thus, will present a much more challenging exploration target, if they indeed are found to contain commercial quantities of coal-bed methane.

REFERENCES CITED

Armstrong, A. K., and Mamet, B. L., 1977, Biostratigraphy and paleogeography of the Mississippian system in northern New Mexico and adjacent San Juan Mountains of southwestern Colorado, *in* San Juan Basin III: New Mexico Geological Society 28th Field Conference Guidebook, p. 111–127.

Baars, D. L., and Stevenson, G M., 1977, Permian rocks of the San Juan Basin, *in* San Juan Basin III: New Mexico Geological Society 28th Field Conference Guidebook, p. 133–138.

Berry, F.A.F., 1959, Hydrodynamics and geochemistry of the Jurassic and Cretaceous systems in the San Juan basin, northwestern New Mexico and southwestern Colorado [Ph.D. thesis]: Stanford, California, Stanford University, 192 p.

Bircher, J. E., 1978, Hospah Upper Sand oil field, *in* Fassett, J E., ed., Oil and gas fields of the Four Corners area, v. I: Four Corners Geological Society, p. 344–346.

Bond, W. A., 1984, Application of Lopatin's method to determine burial history, evolution of the geothermal gradient, and timing of hydrocarbon generation in Cretaceous source rocks in the San Juan Basin, northwestern New Mexico and southwestern Colorado, *in* Woodward, J., Meissner, F. J., and Clayton, J. L., eds., Hydrocarbon source rocks of the greater Rocky Mountain region: Rocky Mountain Association of Geologists Guidebook, p. 433–447.

Brown, C. F., 1978, Blanco Pictured Cliffs gas field, *in* Fassett, J. E., ed., Oil and gas fields of the Four Corners area, v. I: Four Corners Geological Society, p. 225–227.

Brown, H. H., 1978, Buena Suerte Pennsylvanian oil field, *in* Fassett, J. E. ed., Oil and gas fields of the Four Corners area, v. I: Four Corners Geological Society, p. 254–255.

Clarkson, G., and Reiter, M., 1988, An overview of geothermal studies in the San Juan Basin, New Mexico and Colorado, *in* Fassett, J. E., ed., Geology and coal-bed methane resources of the northern San Juan Basin, Colorado and New Mexico: Rocky Mountain Association of Geologists Guidebook, p. 285–291.

Colorado Oil and Gas Conservation Commission, 1988, 1987 oil and gas statistics for the State of Colorado: Denver, Colorado, Oil and Gas Conservation Commission, 538 p.

Condon, S. M., and Huffman, A. C., Jr., 1988, Revisions in nomenclature of the Middle Jurassic Wanakah Formation, northwestern New Mexico and northeastern Arizona: U.S. Geological Survey Bulletin 1633-A, p. 1–12.

Condon, S. M., and Peterson, F., 1986, Stratigraphy of Middle and Upper Jurassic rocks of the San Juan Basin; Historical perspective, current ideas, and remaining problems, *in* Turner-Peterson, C. E., Santos, E. S., and Fishman, N. S., eds., A basin analysis case study; The Morrison Formation, Grants uranium region, New Mexico: American Association of Petroleum Geologists Studies in Geology 22, p. 7–26.

Craig, L. C., 1981, Lower Cretaceous rocks, southwestern Colorado and southeastern Utah, *in* Wiegand, D. L., ed., Geology of the Paradox Basin: Rocky Mountain Association of Geologists Guidebook, p. 195–200.

Cumella, S. P., 1981, Sedimentary history and diagenesis of the Pictured Cliffs Sandstone, San Juan Basin, New Mexico and Colorado: Texas Petroleum Research Committee Report UT 81–1, 219 p.

Deischl, D. G., 1973, The characteristics, history, and development of the Basin Dakota gas field, *in* Fassett, J. E., ed., Cretaceous and Tertiary rocks of the southern Colorado Plateau: Four Corners Geological Society Memoir, p. 168–173.

Dugan, T. A., 1977, The San Juan Basin; Episodes and aspirations, *in* San Juan Basin III: New Mexico Geological Society 28th Field Conference Guidebook, p. 83–89.

Dugan, T. A., and Williams, B. L., 1988, History of gas produced from coal seams in the San Juan Basin, *in* Fassett, J. E., ed., Geology and coal-bed methane resources of the northern San Juan Basin, Colorado and New Mexico: Rocky Mountain Association of Geologists Guidebook, p. 1–9.

Edmister, J. A., 1983, Marcelina Gallup oil field, *in* Fassett, J. E., ed., Oil and gas fields of the Four Corners area, v. III: Four Corners Geological Society, p. 987–988.

Fassett, J. E., 1976, What happened during Late Cretaceous time in the Raton and San Juan basins–With some thoughts about the area in between, *in* Vermejo Park: New Mexico Geological Society 27th Field Conference Guidebook, p. 185–190.

——, 1977, Geology of the Point Lookout, Cliff House, and Pictured Cliffs Sandstones of the San Juan Basin, New Mexico and Colorado, *in* San Juan Basin III: New Mexico Geological Society 28th Field Conference Guidebook, p. 193–197.

——, ed., 1978a, Oil and gas fields of the Four Corners area (v. I and II): Four Corners Geological Society, 727 p.

——, 1978b, Stratigraphy and oil and gas production of northwest New Mexico, *in* Fassett, J. E., ed., Oil and gas fields of the Four Corners area: Four Corners Geological Society, p. 46–61.

——, ed., 1983a, Oil and gas fields of the Four Corners area (vol. III): Four Corners Geological Society, p. 729–1143.

——, 1983b, Stratigraphy and oil and gas production of northwest New Mexico updated through 1983, in Fassett, J. E., ed., Oil and gas fields of the Four Corners area, vol. III: Four Corners Geological Society, p. 849–863.

——, 1985, Early Tertiary paleogeography and paleotectonics of the San Juan Basin area, New Mexico and Colorado, in Flores, R. M., and Kaplan, S. S., eds., Cenozoic paleogeography of the west-central United States: Denver, Colorado, Rocky Mountain Section of the Society of Exploration Paleontologists and Mineralogists, p. 317–334.

——, 1986, The non-transferability of a Cretaceous coal model in the San Juan Basin of New Mexico and Colorado, in Lyons, P. C. and Rice, C. I., eds., Paleoenvironmental and tectonic controls in coal-forming basins in the United States: Geological Society of America Special Paper 210, p. 155–171.

——, ed., 1988, Geology and coal-bed methane resources of the northern San Juan Basin, Colorado and New Mexico: Rocky Mountain Association of Geologists Guidebook, 351 p.

——, 1989a, Coal-bed methane; A contumacious, free-spirited bride, the geologic handmaiden of coal beds, in Lorenz, J. C., and Lucas, S. G., eds., Energy frontiers in the Rockies: Albuquerque Geological Society, Albuquerque, New Mexico (Transactions/Summary volume, American Association of Petroleum Geologists Rocky Mountain Section Meeting, Albuquerque, New Mexico), p. 131–146.

——, 1989b, Coal resources of the San Juan Basin, in Finch, W. I., Huffman, A. C., Jr., and Fassett, J. E., eds., Coal, uranium, and oil and gas in Mesozoic rocks of the San Juan Basin; Anatomy of a giant energy-rich basin: 28th International Geological Congress Field Trip Guidebook, Field Trip T120, p. 19–26.

Fassett, J. E., and Hinds, J. S., 1971, Geology and fuel resources of the Fruitland Formation and Kirtland Shale of the San Juan Basin, New Mexico and Colorado: U.S. Geological Survey Professional Paper 676, 76 p.

Fassett, J. E., and Jentgen, R. W., 1978, Blanco Tocito South oil field, in Fassett, J. E., ed., Oil and gas fields of the Four Corners area: Four Corners Geological Society, p. 233–240.

Flores, R. M., and Erpenbeck, M. F., 1978, Differentiation of delta-front and barrier lithofacies of the Upper Cretaceous Pictured Cliffs Sandstone, southwestern San Juan Basin, New Mexico: Mountain Geologist, v. 18, p. 23–34.

Gorham, F. D., Jr., Woodward, L. A., Callender, J. F., and Greer, A. R., 1977, Fracture permeability in Cretaceous rocks of the San Juan Basin, in San Juan Basin III: New Mexico Geological Society 28th Field Conference Guidebook, p 235–241.

Harr, C. L., 1988, The Ignacio Blanco gas field, northern San Juan Basin, Colorado, in Fassett, J. E., ed., Geology and coal-bed methane resources of the northern San Juan Basin, Colorado and New Mexico: Rocky Mountain Association of Geologists Guidebook, p. 205–219.

Hoppe, W. F., 1978, Basin Dakota gas field, in Fassett, J. E., ed., Oil and gas fields of the Four Corners area, v. I: Four Corners Geological Society, p. 204–206.

Huffman, A. C., Jr., 1989, Petroleum geology of the San Juan Basin, in Finch, W. I., Huffman, A. C., Jr., and Fassett, J. E., eds., Coal, uranium, and oil and gas in Mesozoic rocks of the San Juan Basin; Anatomy of a giant energy-rich basin: 28th International Geological Congress Field Trip Guidebook, Trip T120, p. 33–38.

Jentgen, R. W., 1977, Pennsylvanian rocks in the San Juan Basin, New Mexico and Colorado, in San Juan Basin III: New Mexico Geological Society 28th Field Conference Guidebook, p. 129–132.

Kelso, B. S., and Wicks, D. E., 1988, A geologic analysis of the Fruitland Formation coal and coal-bed methane resources of the San Juan Basin, southwestern Colorado and northwestern New Mexico, in Fassett, J. E., ed., Geology and coal-bed methane resources of the northern San Juan Basin, Colorado and New Mexico: Rocky Mountain Association of Geologists Guidebook, p. 69–79.

Kelso, B. S., Decker, A. D., Wicks, D. E., and Horner, D. M., 1987, GRI geologic and economic appraisal of coal-bed methane in the San Juan Basin, in Proceedings, The 1987 Coal-bed Methane Symposium, November 16–19, 1987, The University of Alabama, Tuscaloosa, Alabama: Gas Research Institute, The University of Alabama, and U.S. Department of Labor, sponsors, p. 119–130.

Lauth, R. E., 1983, Red Mesa oil and gas field, in Fassett, J. E., ed., Oil and gas fields of the Four Corners area, v. III: Four Corners Geological Society, p. 899–900.

Luce, R. M., 1978a, Hospah Lower Sand, South oil field, in Fassett, J. E., ed., Oil and gas fields of the Four Corners area, v. I: Four Corners Geological Society, p. 341–343.

——, 1978b, Hospah Upper Sand, South oil field, in Fassett, J. E., ed., Oil and gas fields of the Four Corners area, v. I: Four Corners Geological Society, p. 347–349.

Matheny, J. P., 1983, Rattlesnake Gallup oil field, in Fassett, J. E., ed., Oil and gas fields of the Four Corners area, v. III: Four Corners Geological Society, p. 1002–1005.

Matheny, M. L., 1978a, Many Rocks Gallup oil field, in Fassett, J. E., ed., Oil and gas fields of the Four Corners area, v. II: Four Corners Geological Society, p. 398–402.

——, 1978b, Sleeper Pictured Cliffs gas field, in Fassett, J. E., ed., Oil and gas fields of the Four Corners area, v. II: Four Corners Geological Society, p. 495–496.

Matheny, M. L., and Talley, D. A., 1983, Oil and gas field discovery lists and index maps, in Fassett, J. E., ed., Oil and gas fields of the Four Corners area, v. III: Four Corners Geological Society, p. 811–829.

Matheny, M. L., and Ulrich, R. A., 1983, A history of the petroleum industry in the Four Corners area, in Fassett, J. E., ed., Oil and gas fields of the Four Corners area, v. III: Four Corners Geological Society, p. 804–810.

Meissner, F. F., 1984, Cretaceous and lower Tertiary coals as sources for gas accumulations in the Rocky Mountain area, in Woodward, J., Meissner, F. F., and Clayton, J. L., eds., Hydrocarbon source rocks of the greater Rocky Mountain region: Rocky Mountain Association of Geologists Guidebook, p. 401–431.

Molenaar, C. M., 1977, Stratigraphy and depositional history of Upper Cretaceous rocks of the San Juan Basin area, New Mexico and Colorado, with a note on economic resources, in San Juan Basin III: New Mexico Geological Society 28th Field Conference Guidebook, p. 159–166.

——, 1983, Major depositional cycles and regional correlations of Upper Cretaceous rocks, southern Colorado Plateau and adjacent areas, in Reynolds, M. W., and Dolly, E. D., eds., Mesozoic paleogeography of west-central United States: Denver, Colorado, Rocky Mountain Section, Society of Economic Paleontologists and Mineralogists, p. 201–224.

——, 1988, Cretaceous and Tertiary rocks of the San Juan Basin, in Baars, D. L., ed., Basins of the Rocky Mountain region, in Sloss, L. L., ed., Sedimentary cover–North American craton; U.S.: Geological Society of America, The Geology of North America, v. D-2, p. 129–134.

——, 1989, San Juan Basin stratigraphic correlation chart, in Finch, W. I., Huffman, A. C., Jr., and Fassett, J. E., eds., Coal, uranium, and oil and gas in Mesozoic rocks of the San Juan Basin; Anatomy of a giant energy-rich basin, 28th International Geological Conference Field Trip Guidebook T 120: Washington, D.C., American Geophysical Union, p. xi (Plate 2).

Needham, C. N., 1978, Schmitz Torreon–Puerco gas field, in Fassett, J. E., ed., Oil and gas fields of the Four Corners area, vol. II: Four Corners Geological Society, p. 482–485.

New Mexico Oil and Gas Engineering Committee, 1988, Annual Report, 1987, Volume II, Northwest New Mexico: Hobbs, New Mexico, New Mexico Oil and Gas Engineering Committee, 11 p.

O'Sullivan, R. B., 1977, Triassic rocks in the San Juan Basin of New Mexico and adjacent areas, in San Juan Basin III: New Mexico Geological Society 28th Field Conference Guidebook, p. 139–146.

Peterson, F., and Kirk, A. R., 1977, Correlation of the Cretaceous rocks in the San Juan, Black Mesa, Kaiparowits, and Henry basins, southern Colorado

Plateau, *in* San Juan Basin III: New Mexico Geological Society 28th Field Conference Guidebook, p. 167–178.

Peterson, J. A., Loleit, A. J., Spencer, C. W., and Ullrich, R. A., 1965, Sedimentary history and economic geology of San Juan Basin: American Association of Petroleum Geologists Bulletin, v. 49, p. 2076–2119.

Powell, J. S., 1973, Paleontology and sedimentation of the Kimbeto Member of the Ojo Alamo Sandstone, *in* Fassett, J. E., ed., Cretaceous and Tertiary of the southern Colorado plateau: Four Corners Geological Society Memoir, p. 11–122.

Pritchard, R. L., 1978, Blanco Mesaverde gas field, *in* Fassett, J. E., ed., Oil and gas fields of the Four Corners area, vol. I: Four Corners Geological Society, p. 222–224.

Rice, D. D., 1983, Relation of natural gas composition to thermal maturity and source rock type in San Juan Basin, northwestern New Mexico and southwestern Colorado: American Association of Petroleum Geologists Bulletin, v. 67, p. 1199–1218.

Rice, D. D., Threlkeld, C. N., Vuletich, A. K., and Pawlewicz, M. J., 1988, Identification and significance of coal-bed gas, San Juan Basin, northwestern New Mexico and southwestern Colorado, *in* Fassett, J. E., ed., Geology and coal-bed methane resources of the northern San Juan Basin, Colorado and New Mexico: Rocky Mountain Association of Geologists Guidebook, p. 55–59.

Ridgley, J. L., 1977, Stratigraphy and depositional environments of Jurassic–Cretaceous rocks in the southwest part of the Chama basin, New Mexico, *in* San Juan Basin III: New Mexico Geological Society 28th Field Conference Guidebook, p. 153–158.

Ross, L. M., 1980, Geochemical correlation of San Juan Basin oils; A study: Oil and Gas Journal, v. 78, p. 102–110.

Sears, J. D., Hunt, C. B., and Hendricks, T. A., 1941, Transgressive and regressive Cretaceous deposits in southern San Juan Basin, New Mexico: U.S. Geological Survey Professional Paper 134, 70 p.

Sikkink, P.G.L., 1986, Lithofacies relationships and depositional environment of the Tertiary Ojo Alamo Sandstone and related strata, San Juan Basin, New Mexico and Colorado, *in* Fassett, J. E., and Rigby, J. K., Jr., eds., The Cretaceous–Tertiary boundary in the San Juan and Raton Basins, New Mexico and Colorado: Geological Society of America Special Paper 209, p. 81–104.

Smith, L. N., Lucas, S. G., and Elston, W. E., 1985, Paleogene stratigraphy, sedimentation, and volcanism of New Mexico, *in* Flores, R. M., and Kaplan, S. S., eds., Cenozoic paleogeography of the west-central United States: Rocky Mountain Section of the Society of Exploration Paleontologists and Mineralogists, p. 293–315.

Stevenson, G. M., 1983a, Paleozoic rocks of the San Juan Basin; An exploration frontier, *in* Fassett, J. E., ed., Oil and gas fields of the Four Corners area, v. III: Four Corners Geological Society, p. 780–788.

——— , 1983b, Stratigraphy and gas production of southwest Colorado, *in* Fassett, J. E., ed., Oil and gas fields of the Four Corners area, v. III: Four Corners Geological Society, p. 844–848.

Stevenson, G. M., and Baars, D. L., 1977, Pre-Carboniferous paleotectonics of the San Juan Basin, *in* San Juan Basin III: New Mexico Geological Society 28th Field Conference Guidebook, p. 99–110.

Thaden, R. E., and Zech, R. S., 1984, Preliminary structure contour map of the San Juan Bsain: U.S. Geological Survey Miscellaneous Field Studies Map MF–1673, scale 1:500,000.

Turner-Peterson, C. E., Santos, E. S., and Fishman, N. S., eds., 1986, A basin analysis case study; The Morrison Formation, Grants uranium region, New Mexico: American Association of Petroleum Geologists Studies in Geology 22, 391 p.

Vincellete, R. R., and Chittum, W. E., 1981, Exploration for oil accumulations in Entrada Sandstone, San Juan Basin, New Mexico: American Association of Petroleum Geologists Bulletin, v. 65, p. 2546–2570.

Wright, R., 1986, Cycle stratigraphy as a paleogeographic tool; Point Lookout Sandstone, southeastern San Juan Basin, New Mexico: Geological Society of America Bulletin, v. 97, p. 661–673.

Manuscript Accepted by the Society September 20, 1989

ACKNOWLEDGMENTS

The quality and accuracy of this chapter was greatly improved as the result of colleague reviews by A. C. Huffman, Jr., C. M., Molenaar, and D. D. Rice, all with the U.S. Geological Survey, Denver, Colorado. I also gratefully acknowledge the insight, knowledge, and wisdom shared by many oil-patch geologists in the Farmington, New Mexico–Durango, Colorado area; without their contributions to my continuing education about the oil business over the past 28 years, this chapter could not have been written.

Chapter 23

Powder River basin

James E. Fox
Department of Geology and Geological Engineering, South Dakota School of Mines and Technology, Rapid City, South Dakota 57701
Gordon L. Dolton and Jerry L. Clayton
U.S. Geological Survey, Box 25046, Denver Federal Center, Denver, Colorado 80225

INTRODUCTION

Location and general geologic characteristics

The Powder River basin is the largest intermontane basin of Laramide origin in the northern Rocky Mountains of the United States. The basin occupies a large part of northeastern Wyoming and a small part of southeastern Montana (Fig. 1).

This deep, northerly trending, asymmetric, mildly deformed basin is approximately 370 km long and 160 km wide. The basin axis is close to the western margin, which is defined by hogbacks with steeply dipping and overturned strata along the flanks of the Bighorn Mountains, the Casper arch, and the northern end of the Laramie Mountains. It is bordered on the south by the Hartville uplift and on the east by the Black Hills uplift. Strata along the eastern margin are folded and faulted along monoclines. The northern margin is topographically less well defined but is along the subtle northwest-trending Miles City arch.

The present form of the Powder River basin is primarily the result of Laramide deformation. Major thrusting along the west and south sides of the basin began during the late Paleocene followed by renewed uplift, downwarping, and regional tilting as late as the Miocene.

Strata in the basin are dominantly clastic in composition, reflecting, in part, the large quantity of Upper Cretaceous and Tertiary detritus deposited in association with the Laramide orogeny. Carbonate rocks are abundant in Cambrian, Ordovician, Devonian, Mississippian, Pennsylvanian, and Permian strata. Minor amounts of carbonate rocks are present in Triassic, Jurassic, and Cretaceous strata.

Petroleum exploration history

The Powder River basin is one of the major petroleum producing areas in the United States and one of the largest in the Rocky Mountain region. Although this basin contains large quantities of oil, it has relatively minor amounts of gas. Exploration

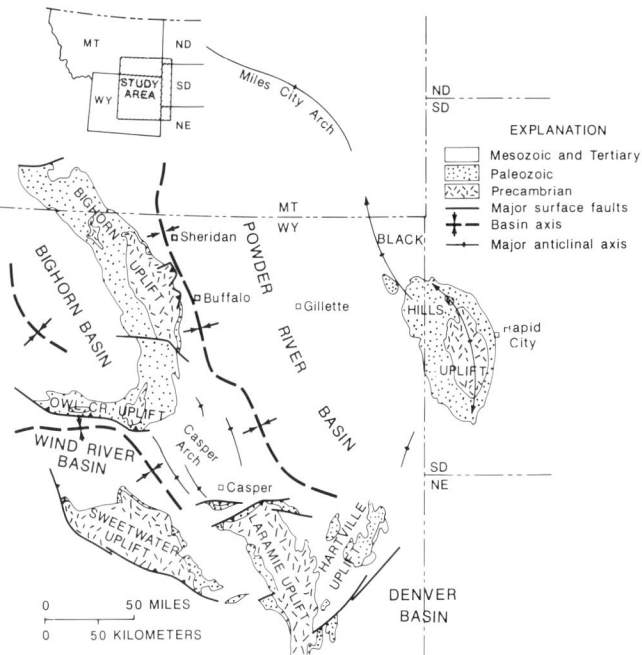

Figure 1. Index map showing location of Powder River basin and surrounding uplifts.

targets are relatively shallow, ranging from a few hundred meters on the east side to more than 4,575 m in the deep axial portion of the basin. The distribution of oil fields throughout the basin is shown on Figure 2, and stratigraphy is shown on Figure 3.

The history of petroleum exploration in the Powder River basin dates back to the late 1800s, as discussed by Strickland (1958). Numerous oil seeps were known to exist along the margins of the basin in territorial days, and by 1887 at least 60 wells had been drilled on the eastern side of the basin near Moorcroft, Wyoming. Several of these wells, ranging in depth from about 122 to 500 m, were successful. The first oil field in the Powder

Fox, J. E., Dolton, G. L., and Clayton, J. L., 1991, Powder River basin, *in* Gluskoter, H. J., Rice, D. D., and Taylor, R. B., eds., Economic Geology, U.S.: Boulder, Colorado, Geological Society of America, The Geology of North America, v. P-2.

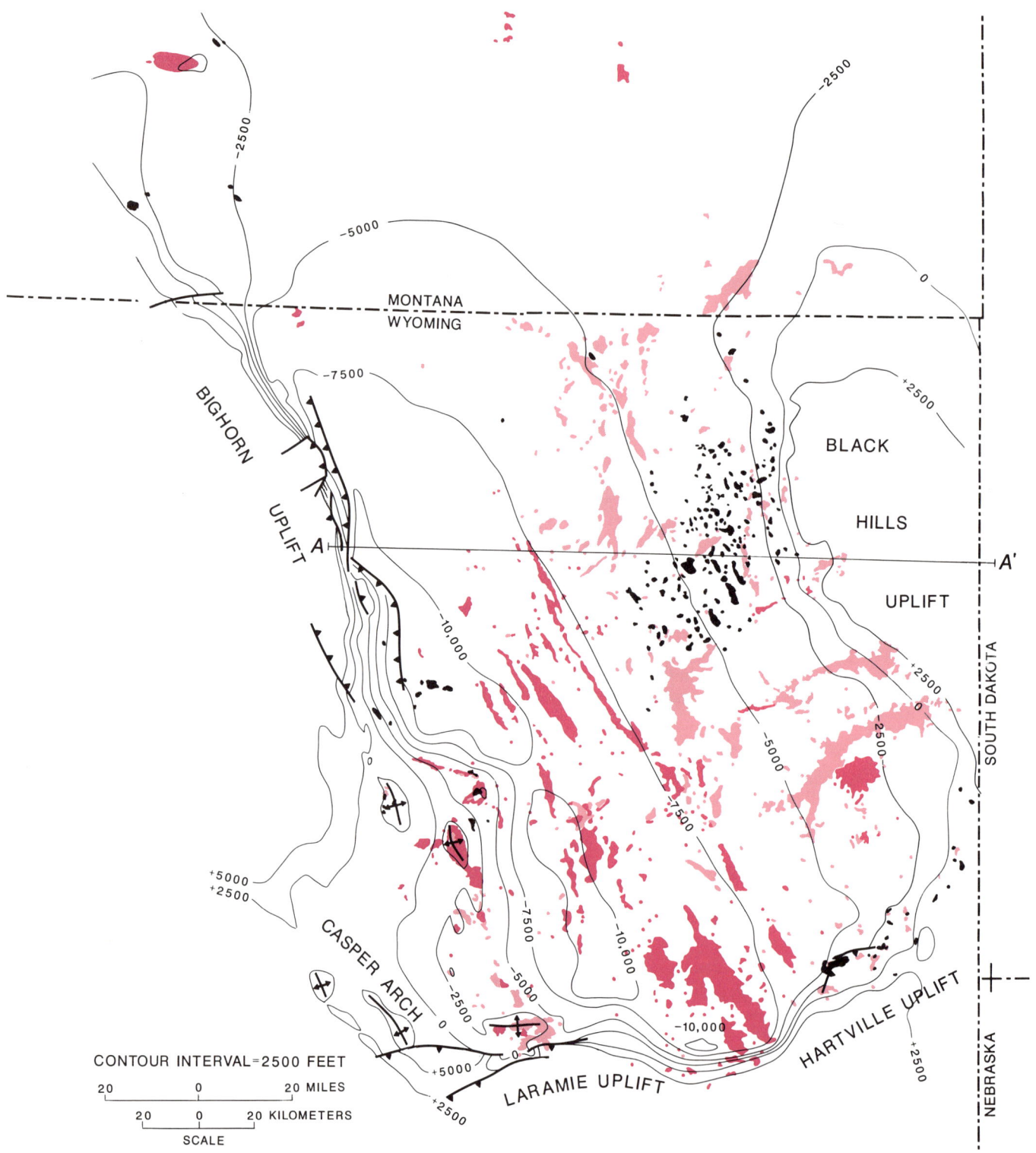

Figure 2. Oil fields and structure contours on top of the Minnelusa Formation (Pennsylvanian and Permian), Powder River basin. Cross-section A-A' shown on Plate 6C. Red, black, and pink colors show ages of petroleum reservoirs; red = Upper Cretaceous, pink = Lower Cretaceous and Jurassic, and black = Pennsylvanian and Permian.

River basin was established in this area and produced 22-degree API gravity green oil from a stratigraphic trap in the Lower Cretaceous Newcastle Sandstone. In 1889 a well was drilled into the Upper Cretaceous Shannon Sandstone Member of the Steele Shale on a fault closure off the north end of the huge Salt Creek anticline. This well produced a small amount of green 24-degree API gravity oil at a depth of 212 m. The real "boom," however, did not start until 1908 when a major oil well, flowing up to 600 barrels of oil daily, was drilled on the crest of this anticline. Production was from sandstone in the Upper Cretaceous Frontier Formation, locally identified as the "first Wall Creek sand," and later (1917) from a lower sandstone in the Frontier Formation that came to be known as the "second Wall Creek sand." Other oil-bearing sandstone formations at Salt Creek include the "third Wall Creek sand" and the Lakota Formation, sandstone in the Jurassic Sundance Formation, and Permian sandstone in the upper part of the Minnelusa Formation (Tensleep Sandstone). Minor production was also discovered in fractured shales of the Cretaceous Carlile and Lakota Formations and in sandstone of the Jurassic Morrison Formation.

Subsequent drilling of surface structures along the flanks of the basin resulted in the discovery of additional anticlinal oil fields over the next 25 years, including Big Muddy, Lance Creek, and Mule Creek.

During the late 1940s, exploration was renewed, augmented with geophysical techniques, and from 1948 to 1952, ten large oil fields were found at depths of about 2,700 m or less. These include South Cole Creek, Fiddler Creek, Sussex, Sage Spring Creek, South Glenrock, Meadow Creek, North Meadow Creek, Clareton, East Salt Creek, and West Sussex. They produced oil from a wide range of reservoirs in both structural and stratigraphic traps.

In 1953 a well produced about 1,500 barrels of oil per day from the Lower Cretaceous Newcastle Sandstone (Muddy Sandstone equivalent) in the Clareton trend on the basin's east flank. Major activity centered around this area and extended the area of production. A variety of smaller and more subtle stratigraphic traps as well as the large Bell Creek field were discovered in the Muddy Sandstone in the late 1950s and into the 1960s. These include fields ranging from the large Recluse and Rozet fields to smaller ones such as West Kitty. Discoveries continue to be made in the Muddy and Newcastle as exploration continues.

Exploration began in the 1950s for stratigraphic traps in the Fall River Sandstone, a Lower Cretaceous formation deeper than the Muddy or Newcastle. In 1957 the discovery of oil in the Upper Cretaceous Parkman Sandstone Member of the Mesaverde Formation at Dead Horse Creek field began expansion of the "fairway" in the central part of the basin. By 1975, exploration in the central part of the basin was extended to stratigraphic traps in Upper Cretaceous sandstones of the Teapot, Teckla, Sussex, and Shannon Sandstone Members; the Frontier Formation; and finally to the Lower Cretaceous sandstones of the Fall River and Muddy Sandstones.

Exploration for stratigraphic traps on the broad east flank of

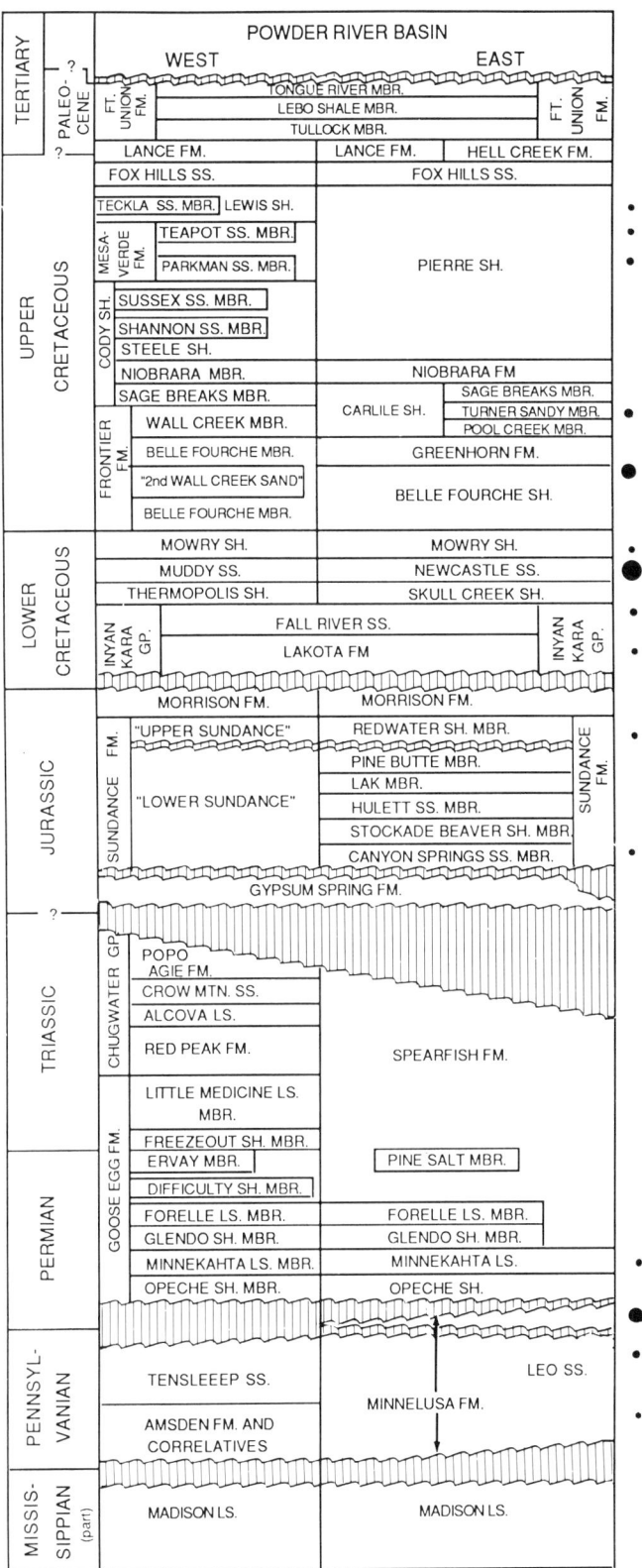

Figure 3. Stratigraphic columns of the western and eastern flanks of the Powder River basin, showing principal petroleum producing intervals (black dots). Size of dots denotes relative volume of oil.

the basin has continued to the present, with a general expansion into the deeper parts of the basin and emphasis on the Pennsylvanian and Permian Minnelusa Formation as well as on Cretaceous sandstones.

STRUCTURAL GEOLOGY

Introduction

The Powder River basin is asymmetric, with a thickness of at least 5,500 m of sedimentary rocks in the deep axial portion, which lies very close to its western margin (Fig. 2 and Plate 6C). At least 2,440 m of these strata are nonmarine Upper Cretaceous and lower Tertiary clastics associated with Laramide orogenesis.

The tectonic style of the Laramide uplifts surrounding the basin is variable. Deformation was most intense along the western and southern margins of the basin where tectonic activity formed anticlinal structures, some of which are highly productive hydrocarbon traps. Recent exploratory drilling, geophysical surveys, and field mapping have revealed the presence of thrust faults (Gries, 1983; Blackstone, 1981; Jenkins, 1986). Although the lateral extent of these faults is not completely known at present, their existence is documented.

Structural history

Curry (1971) addressed the topic of Laramide structural history of the Powder River basin. He states that structural movements are known to post-date deposition of the Lewis Shale (post-early Maestrichtian) because these marine sediments thicken to the south in the basin with no indication of Laramide tectonic activity. It is also known that strong deformation ended before deposition of the Oligocene White River Formation. Thrusting along the basin margin faults probably began during the late Paleocene, with renewed uplift, downwarping, and regional tilting as late as Miocene time.

In his summary, Curry (1971, p. 60) presents the following sequence of structural events during the Laramide orogeny in the Powder River basin area: (1) upper Maestrichtian Lance Formation, southward basin subsidence but no local basin subsidence; (2) lower Paleocene Tullock Member, slight subsidence in southern and southwestern parts of the basin but no prominent subsidence along the basin axis; (3) middle Paleocene Lebo Shale Member, first strong subsidence along the axis of the Powder River Basin but no evidence of influx of coarse clastics from adjacent uplifts. The first strong evidence of the Laramide orogeny is indicated; (4) upper Paleocene Tongue River Member, strong subsidence continued along the axis of the basin. The influx of sandstone is the first indication of uplift and erosion of adjacent mountain uplifts; (5) upper Paleocene and Eocene Wasatch Formation, mountains continued to be uplifted and eroded to their Precambrian core as evidenced in conglomerate fans that were deposited and subsequently deformed along the flank of the Bighorn Mountains; (6) Oligocene White River Formation, orogenic movements stopped before the Oligocene, and White River volcaniclastic rocks filled the basin and buried the Precambrian cores of the mountains; (7) post-Oligocene, moderate subsidence of the uplifts along faults caused local tilting of White River strata toward the sinking mountain blocks.

Structural style

The style of Laramide structural deformation has been variously interpreted. Perhaps one of the most seriously debated questions has been whether the mountain-front fault boundaries are high-angle reverse or thrust faults. Berg (1962) proposed that the uplifts were thrust and/or fold-thrusted as a result of crustal compression. Stearns (1978) interpreted them to be forced folds resulting from dominantly vertical uplift. In the past few years, drilling through the edges of six of these Laramide thrusts has revealed that the angle of the thrust plane ranges from 20 to 45 degrees (Gries, 1983). Recent work by Jenkins (1986) along the northeastern Bighorn Mountain front has substantiated the low-angle thrust planes. His investigation has also revealed that the structural style along the east side of the Bighorn Mountains is highly complex and that modifications of both Berg's and Stearns' models are necessary to interpret the structural geology of this area.

Jenkins (1986), Blackstone (1981), and other geologists have used angular discordances in Tertiary rocks near Buffalo and Story, Wyoming, to estimate timing and intensity of the folding and faulting. In these areas, displacement along thrust faults ranges from 3 to 10 km. South of this area, a system of faults occurs along a broad structural terrace. Here, almost half of all the oil that the basin has ever produced has occurred in folds developed along this structural terrace forming a major belt from Sussex oil field southward to Big Muddy oil field (Fig. 2).

The steep south flank of the basin adjacent to the Laramie and Hartville uplifts may be structurally similar to the west flank, discussed above. Faults may also have influenced the emplacement of oil in structural traps in this area.

The Stearns model may apply to the less-deformed eastern side of the basin where Phanerozoic strata are drape folded over basement blocks that were uplifted during the Laramide orogeny. It is not known, however, whether the faults in this area are high-angle normal or reverse. A large part of this eastern basin flank is characterized by monoclines that formed adjacent to the Black Hills uplift as discussed by Lisenbee (1979; 1985).

Structural deformation in the interior of the Powder River basin may have influenced patterns of sedimentation and may have had a minor control on oil entrapment. Chamberlin (1945) was one of the first geologists to postulate that the major folds in the northern Great Plains were intimately related to Precambrian basement fracturing of wide geographical extent. Slack (1981) suggested that virtually all of the stratigraphic production of petroleum in the Powder River basin can be directly related to subtle repetitive movements along linear structural trends representing faults extending upward from zones of basement weakness. According to Slack, the Belle Fourche arch extends southwest diagonally across the basin and is the major uplifted block in the basin. He states that movements along basement faults have been taking place perhaps since the Precambrian and have affected the occurrence of reservoir-quality sandstones in strata as old as the Pennsylvanian and Permian Minnelusa Formation and perhaps even older. Weimer and others (1982) also present evidence in support of recurrent movement of basement faults and their influence on development of reservoir-quality sandstone bodies in Lower Cretaceous strata of the east flank of the Powder River basin. Rasmussen and Bean (1984) present evidence suggesting a process of dissolution of Permian salt, which may have influenced the growth and development of some of these Mesozoic syndepositional sandstone bodies.

STRATIGRAPHY

Introduction

The Powder River basin is filled with a sequence of Phanerozoic strata that is predominantly clastic in composition and is thickest in the area of the basin axis (in excess of 5,500 m) (Fig. 2 and Plate 6C). The regional post-Mississippian stratigraphy of the Powder River basin is illustrated on a series of 22 geophysical well-log cross sections by Fox (1986; 1987) and is summarized in Figure 3. Rocks younger than Cretaceous or older than Pennsylvanian are not discussed in this chapter because they have not yielded significant amounts of petroleum in the Powder River basin nor have they contributed petroleum to producing units.

Pennsylvanian and Permian

Pennsylvanian time was characterized by periodic uplift of the Ancestral Rocky Mountains and associated subsidence of surrounding areas (Agatston, 1954; Mallory, 1967). Also, worldwide eustatic sea-level fluctuations in response to glaciation have been reported during this time. As a result, highly variable cyclothemic strata are typical of the Pennsylvanian of the region. The Pennsylvanian-age Amsden Formation and Tensleep Sandstone, present along the western side of the basin, typify these facies. Here the Amsden Formation, with about 60 m of red mudstone, sandstone, and cherty limestone, is overlain by the Tensleep Sandstone, which is approximately 30 to 150 m thick and is composed of white-to-tan cross-bedded sandstone and sandy dolomite (Mallory, 1975).

At Casper Mountain there is no Amsden equivalent, and the Pennsylvanian is represented by the Casper Formation, which rests unconformably on the Mississippian Madison Limestone. Morrowan, Atokan, and most of the Desmoinesian strata are locally absent; most of the Casper Formation in this area is of Wolfcampian age. It is about 150 to 180 m of interbedded sandstone and limestone, with cross-bedded sandstone composing the upper 60 m (E. K. Maughan, personal communication, 1986).

A marked thickness and facies change takes place between Casper Mountain and the Hartville uplift region where about 365 m of carbonate and evaporite rocks are equivalent to the coarser clastics of Casper Mountain. These strata are part of the Pennsylvanian Hartville Formation and are very similar to the Minnelusa Formation. The Minnelusa Formation comprises equivalent Pennsylvanian and Permian strata in the Black Hills, and the name is widely used for the subsurface part of the Powder River basin and in some adjacent areas. In the Hartville area, Condra and Reed (1950) divided the Hartville Formation to include, from bottom to top, the Fairbank Formation and the Reclamation, Roundtop, Hayden, Meek, Wendover, Broom Creek, and Cassa Groups. Except for the Fairbank, these names have not been widely adopted. Rather, two regional unconformities within the Minnelusa and Hartville provide boundaries for lower, middle, and upper members of these formations (Lageson and others, 1979); the middle member is informally called "Leo."

The Minnelusa Formation was deposited during the Pennsylvanian and part of the Permian Periods and is up to at least 365 m in thickness. The lower member of the Minnelusa is Morrowan and Atokan in age, the middle member is Desmoinesian, Missourian, and Virgilian, and the upper member of the Minnelusa is Early Permian Wolfcampian (Maughan, 1975). On the western side of the basin, the Amsden is Morrowan and Atokan, and the relatively thin Tensleep section is of Desmoinesian age but may include Missourian and Virgilian strata locally. Upper Pennsylvanian rocks in the northern Big Horn Mountains and adjacent areas in the Powder River basin are, however, generally absent due to truncation.

The Permian-Triassic boundary is paraconformable within the Goose Egg Formation. Most of the Guadalupian and the Ochoan are missing (Schock and others, 1981). The Permian and Lower Triassic Goose Egg Formation contains several interbedded tongues of shale, siltstone, and limestone. These include, from bottom to top, the Opeche Shale, Minnekahta Limestone, Glendo Shale, Forelle Limestone, Difficulty Shale, and Ervey Members of Permian age, and the Freezeout Shale and Little Medicine Limestone Members of Early Triassic age. The carbonate beds are tongues of the Permian Park City Formation and the Triassic Dinwoody Formation of western Wyoming. In the Powder River basin these carbonate rock members thin to a northerly to northwesterly direction (Maughan, 1967).

Triassic

The name Chugwater Group is applied to Triassic strata in the western and southern portion of the basin, and equivalent strata of the eastern portion are included in the upper part of the Spearfish Formation. Dominant lithotypes are red mudstones, siltstones, and silty sandstones and thin beds of gypsum and limestone.

Strata of the Chugwater Group are subdivided into several formations (Pipiringos, 1968). From bottom to top they are the Red Peak Formation. Alcova Limestone, Jelm Formation (or its equivalent, the Crow Mountain Sandstone), and the Popo Agie Formation. The younger units are recognized only in the extreme southwestern part of the basin, having been removed by pre-Jurassic erosion in other areas. At some localities the Alcova is overlain by the Jelm Formation or its equivalent, the Crow Mountain Sandstone. Originally the Jelm, Crow Mountain, and Popo Agie probably had a more widespread distribution. The part of the Spearfish Formation that is equivalent to the Red Peak Formation is also beveled at the top in the northern part of the basin. Triassic strata in the basin are about 245 m thick. The thin, upper two members of the Goose Egg Formation discussed above (Freezeout Shale and Little Medicine Limestone Members), are also Triassic.

Jurassic

Jurassic strata rarely exceed about 215 m in thickness in the Powder River basin; Lower Jurassic rocks are absent. Middle Jurassic rocks include the Gypsum Spring Formation, which occurs only in the northern one-third of the basin, and the overlying lower part of the Sundance Formation of Callovian age (Lower Sundance Formation of subsurface usage). An unconformity at the top of this sequence is overlain by Oxfordian Upper Jurassic rocks of the upper part of the Sundance Formation (Upper Sundance Formation of subsurface usage). This is overlain by the Morrison Formation. Rocks of the Sundance Formation are present throughout the basin and are almost entirely clastic, consisting of gray and greenish gray shale, glauconitic quartzose sandstone, dark gray and greenish gray shale, and minor amounts of thin-bedded oolitic limestone.

The Sundance has been subdivided by Imlay (1947; 1980) into members, based on strata cropping out along the eastern side of the Powder River basin. They include, in ascending order, Canyon Springs Sandstone, Stockade Beaver Shale, Hulett Sandstone, Lak, Pine Butte, and Redwater Shale Members. The lower five members are equivalent to the Callovian Lower Sundance, and the upper member is equivalent to the Oxfordian Upper Sundance of subsurface usage in the Powder River basin. Strata of the overlying Morrison Formation are present throughout the basin and are composed of highly lenticular varigated shales and sandstones. The contact between the Jurassic Morrison Formation and Lower Cretaceous Inyan Kara Group is unconformable.

Lower Cretaceous

Thickness of Lower Cretaceous strata is generally less than about 300 m, which is much less than that of the Upper Cretaceous. Continental and marine Lower Cretaceous strata were deposited across the area of the present Powder River basin during the initial transgression of the craton by a boreal sea that occupied the Western Interior region throughout most of the Cretaceous. Wedges of continental and marginal marine strata (Lakota, Fall River, and Muddy Sandstones) intertongue with marine shales (Skull Creek, Thermopolis, and Mowry Shales) to the north and west. The initially deposited Lakota of this sequence is fluvial in origin, with discontinuous sandstone bodies within varicolored shales. The Fall River is variously interpreted as fluvial channel, deltaic, and shoreline marine sandstone and may comprise elements of all three of these depositional facies. Deposition of the Fall River generally marks the beginning of a prolonged period of marine deposition, which prevailed with few minor interruptions until the seas finally withdrew near the end of the Cretaceous Period. The name "Dakota silt" is used in the subsurface for a sequence of silty shale above the Fall River Sandstone. This sequence is a part of the Thermopolis and Skull Creek Shales.

The contact between Lower and Upper Cretaceous strata in the Powder River Basin is between the Lower Cretaceous Mowry Shale and the Upper Cretaceous Frontier Formation or Belle Fourche Shale. At this stratigraphic position is a prominent and widespread stratigraphic marker called the Clay Spur Bentonite Bed.

Upper Cretaceous

Upper Cretaceous rocks thicken from about 1,525 m at the northern margin of the basin to about 2,745 m at the southern margin. They are more coarsely clastic in the southwestern part of the basin near their sediment source, reflecting their general distribution as a clastic wedge that prograded eastward into the Western Interior seaway. Within this clastic wedge, major transgressive and regressive cycles are indicated that provide a basis for mapping (Weimer, 1961; Gill and Cobban, 1966, 1973; McGookey and others, 1972). Deltaic, strandline, and shelf sandstones intertongue with marine shales, reflecting deposition in a seaway with an oscillating shoreline. This oscillation was due, in part, to episodic local tectonic movements, shifting deltaic depocenters, and eustatic sea level changes.

Upper Cretaceous strata of the Powder River basin are primarily shale and sandstone. Minor volumes of coal, limestone, chalky marl, and bentonite are also present. Marine shales are predominant in the eastern part of the basin with only one significant sandstone unit, the Turner Sandy Member of the Carlile Shale. Small sandstone bodies of the Shannon Sandstone Member of the Pierre Shale extend quite far east, as reported by Rice and Shurr (1983). In contrast to the eastern part of the basin, the southwestern part of the basin contains numerous sandstone beds in the Frontier Formation ("1st," "2nd," and "3rd Wall Creek sands"), Steele Shale (Shannon and Sussex Sandstone Members), Mesaverde Formation (Parkman and Teapot Sandstone Members), and Lewis Shale (Teckla Sandstone Member).

The contact between the Upper Cretaceous and Tertiary is gradational in most areas and difficult to identify. This boundary may be at the contact between the Lance and Fort Union Formations or within the Lance Formation itself.

MAJOR PETROLEUM PLAYS

Introduction

Oil and gas production in the Powder River basin is limited to Pennsylvanian, Permian, and Cretaceous strata except for some minor occurrences in rocks as old as Mississippian and as young as Paleocene and Oligocene. Figure 3 shows the relative importance of principal petroleum-producing intervals.

Petroleum plays in this basin are both structural and stratigraphic. Structural plays are basin margin anticlines or basin margin subthrusts. Stratigraphic plays are segregated into three major petroleum source and producing systems which involve, respectively, Pennsylvanian and Permian, Lower Cretaceous, and Upper Cretaceous rocks. The following discussion is limited to the characteristics of the traps and reservoirs within the plays, while their source rocks are discussed separately later.

Pennsylvanian and Permian stratigraphic plays

One of the major oil-producing formations in the basin is the Pennsylvanian and Permian Minnelusa Formation. Production is from sandstones of two distinct stratigraphic intervals in different areas of the basin. The oldest petroleum-productive interval in the Minnelusa is termed the "Leo sandstone" in the Pennsylvanian middle part of the Minnelusa Formation. This play is limited to the southern part of the basin, although production from equivalent rocks extends beyond it.

The youngest interval is the upper part of the Minnelusa Formation above the "red marker zone." This zone of oxidized red claystone and siltstone formed from subaerial exposure of strata during the time marking the systemic boundary between the Pennsylvanian and Permian. Petroleum production from the upper part of the Minnelusa occurs across the central part of the basin on a trend southwest from the northern Black Hills uplift.

Middle Minnelusa ("Leo") and upper Minnelusa sandstone reservoirs. The Minnelusa consists of several lithologies including sandstones carbonates, and evaporites, deposited in a variety of environments associated with offshore-prograding eolian sand dunes (Fig. 4; Tromp and others, 1981; Fryberger and others, 1983; Fryberger, 1984; George, 1984; and Motes, 1984). Quartzose sandstones provide the principal reservoirs.

Middle Minnelusa ("Leo") and upper Minnelusa petroleum traps and seals. As discussed by Van West (1972), the largest portion of in-place Minnelusa oil has been trapped in paleotopographic highs or erosional remnants at the top of the Minnelusa, upon which the overlying Opeche Shale was deposited (Fig. 5). Other less-significant traps include updip permeability pinchouts and structural closure. The majority of oil found to date is in areas with maximum paleotopographic relief. Moore (1983) has substantiated this in the Minnelusa strata of the Halverson field on the east flank of the basin. Here the Opeche Shale is thick, filling a low area on the irregular unconformable west-sloping surface of the Minnelusa Formation. Oil is trapped along the updip side of the erosional remnant of the Minnelusa by the impermeable Opeche Shale.

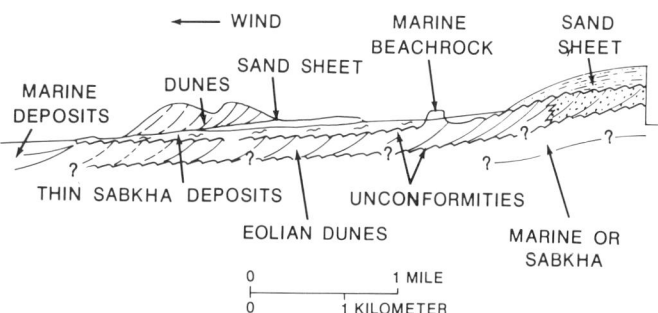

Figure 4. Diagrammatic depositional model of coastal dune and associated facies, used as a model to interpret similar facies of the Minnelusa Formation, Powder River basin (after Fryberger and others, 1983).

Traps and seals within the Leo sandstone in the southern part of the basin are mostly subtle stratigraphic traps, including sandstone pinchouts into nonreservoir impermeable facies. These impermeable facies may act as both traps and seals. The depth of the Leo-equivalent sandstones in this play extends to about 4,575 m in the deep western side of the basin.

Lower Cretaceous stratigraphic plays

These plays account for much of the petroleum production in the basin. Reservoir rocks are primarily sandstones of the Inyan Kara Group (Lakota Formation and Fall River Sandstone) and the Muddy and Newcastle Sandstones.

Inyan Kara Group sandstone reservoirs. Inyan Kara reservoir sandstones occur in two formations, the Lakota Formation and Fall River Sandstone. The lower alluvial part of the Inyan Kara sequence consists of the lithologically variable Lakota Formation. Reservoirs are fine to coarse sandstones which are locally pebbly or conglomeratic. Most contain chert and lithic fragments. These fluvial channel and associated sandstones are highly variable in scale and geometry but are usually small-scale, discontinuous, and lenticular. Reservoir-quality sandstones, although discontinuous, are widespread throughout the basin. There is only minor production from the Lakota, however.

The overlying Fall River Sandstone contains oil and gas in stratigraphic traps within a major regressive clastic wedge. Following the initial Cretaceous marine invasion of the region, this widespread thin clastic wedge prograded into the Cordilleran seaway from the south and east (MacKenzie and Poole, 1962; MacKenzie and Ryan, 1962; Dondanville, 1963). In the area of the Powder River basin it is composed of a widespread marine, deltaic, and alluvial complex that developed along the eastern margin of the seaway. These strata become progressively more marine to the west, consisting entirely of marine shales and siltstones on the west side of the basin.

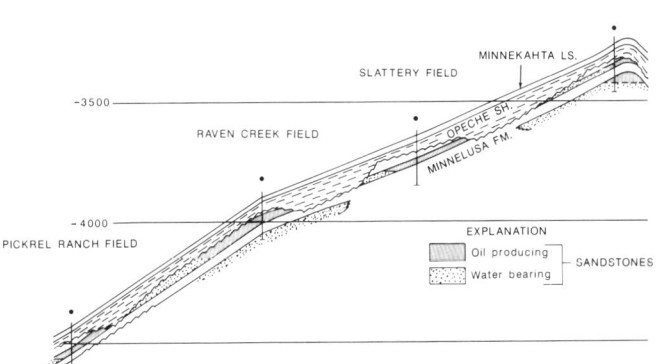

Figure 5. Diagrammatic representation of trap types in the upper part of the Minnelusa Formation, east flank of the Powder River basin. The unconformity is on top of the Minnelusa Formation (after Van West, 1972).

Most oil and gas occurs in the structurally uncomplicated east flank of the basin. Here, a number of fields are developed in discrete, essentially single-channel systems. To the west, near the edge of the clastic wedge, several fields have been discovered in combination traps on large structural closures or plunging anticlinal noses. These traps, although excluded from the play because of the dominance of structure, indicate the importance of stratigraphy for entrapment of hydrocarbons.

Inyan Kara Group petroleum traps and seals. Traps in the Lakota are invariably within discrete or composite alluvial channel sandstones and are sealed by fine-grained alluvial deposits. Although these traps in some cases occur in combination with structural traps, the essential trapping mechanism of the play is stratigraphic.

Seals for traps within the Lakota are fine-grained alluvial and deltaic rocks of the Inyan Kara Group and overlying Skull Creek Shale. The underlying Jurassic Morrison Formation may also provide a lateral seal where sandstones are entrenched into it.

Traps in the Fall River Sandstone play are often point bar deposits, formed in channels that have prograded through older marine or strandline sediments. Typically they are closed updip by fine-grained abandoned channel deposits within or at the edges of meander belts (Mettler, 1968; Berg, 1968; Harris, 1976). Marine bar sandstone traps resulting from pinchouts are also considered prospective, although they are not as well documented. In a few instances, structure plays a minor role in providing additional closure.

The Fall River wedge is sealed top and bottom, respectively, by enclosing shale of the Skull Creek and Fall River Formations. However, lateral and vertical seals for individual traps within the unit are provided by fine-grained rocks, such as those found in abandoned channel oxbow fills and lower-energy marine sequences.

Muddy and Newcastle Sandstone reservoirs. Lower Cretaceous Muddy and Newcastle Sandstones are composed of sediment transported into the Cretaceous seaway from the east, accompanying or following subareal erosion over much of the area (Stone, 1972; Mitchell, 1976). This was succeeded by a gradual transgressive phase interrupted by periodic regressive pulses in which considerable sand was supplied. The result is a compound wedge consisting of a variety of depositional environments, which produces stratigraphic traps. These include marine bar and strandline, channel, estuarine, alluvial, and lower delta plain sandstones (Haun and Barlow, 1962; Stone, 1972; Waring, 1976).

Muddy and Newcastle petroleum traps and seals. The Muddy and Newcastle sequence contains abundant shale as well as sandstone and rests upon an unconformity of regional extent cut into the Cretaceous Thermopolis and Skull Creek Shales. In many cases the thicker sandstones accumulated within more deeply dissected troughs or valleys. The approximately northeast-striking estuarine and alluvial Fiddler Creek and Clareton trends are examples. The dominantly marine bar facies traps, such as Hilight, Kitty, and Recluse, are generally nearly north trending, except for Bell Creek. They may represent still-stands during the transgressive phase of Muddy deposition, when more extensive high-energy marine bar or barrier sandstone deposits accumulated. Often a single field is a composite of overlapping separate traps and oil pools, as documented by Berg (1976). Seals for the various traps are provided by the enclosing Skull Creek and Mowry Shales.

Upper Cretaceous stratigraphic plays

Strata involved in these plays are part of a largely western-derived regressive clastic wedge. They include deltaic and marine shelf sandstones that grade into siltstone or shale, commonly resulting in stratigraphic petroleum entrapment. Marine shelf sandstones are the Frontier Formation and Shannon and Sussex Sandstone Members of the Steele Shale. Deltaic sandstones include the Teapot and Parkman Members of the Mesaverde Formation and the Teckla Sandstone Member of the Lewis Shale.

Frontier marine shelf sandstone reservoirs. Sandstones in the upper part of the Frontier, known variously as "First Wall Creek" or "First Frontier," are considered the principal objectives in this play, although genetically similar sandstones lower in the formation on the west side of the basin are also considered prospective. Petroleum occurs in stratigraphic traps of offshore marine shelf sandstones in a depositional setting similar to the Sussex and Shannon Sandstone Members (Fig. 6). These sandstone units in the Frontier are a part of a regressive sequence with clastics derived from the west (Haun, 1958; Barlow and Haun, 1970; Goodell, 1962). These shelf sandstones are located primarily in the deeper parts of the basin (Winn and others, 1983). Here the sandstones trend generally northwest but locally coalesce into less-regular trends and are productive in oil and gas fields such as the Powell and Spearhead Ranch fields.

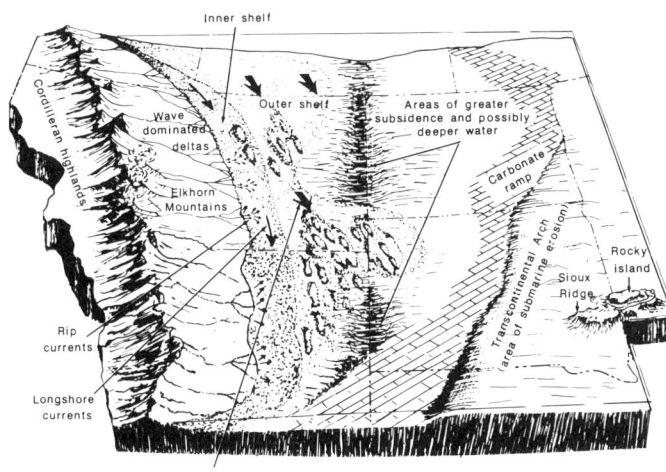

Figure 6. Diagrammatic depositional model of Upper Cretaceous shelf sandstones and delta complexes (modified from Rice and Shurr, 1983).

To the west of these shelf sandstones, deltaic facies of equivalent age have been identified. To the east, shelf sandstones become thinner, and some grade into widespread offshore sandstones of the Turner Sandy Member (Merewether and others, 1979).

Frontier petroleum traps and seals. Traps are stratigraphic, with accumulations resulting from trapping at the margins of individual bars or bar complexes, or from internal porosity loss within the sandstones. Seals are the enclosing shales and siltstones of the Frontier Formation and overlying shale of the Carlile and Niobrara Members.

Shannon and Sussex marine shelf sandstone reservoirs. Marine shelf sandstone plays include the Shannon and Sussex Sandstone Members of the Steele Shale. These sandstones are interpreted as having been deposited on a marine shelf seaward from the marine shoreline (Fig. 6).

Shannon Sandstone Member. Facies of the Shannon have characteristics of a wide-shelf model of deposition (Asquith, 1974). Asquith has reconstructed the shelf, slope, and basin topography using electric log correlations of bentonite marker beds in the Cody Shale and has shown the Shannon to be situated at the top of this progradational shelf sequence of predominantly offshore muds. Tillman and Martinsen (1984) and Spearing (1976) presented evidence from outcrops and subsurface strata to support the interpretation of the Shannon as an offshore shallow-marine sandstone transported by marine currents in a longshore direction. Detailed subsurface facies analysis by Crews and others (1976) has indicated that the shales underlying the Shannon have been scoured locally and beveled regionally, suggesting that the same marine currents that transported the Shannon sands may also have shaped the pre-Shannon surface.

Sussex Sandstone Member. Hobson and others (1982) interpreted the Sussex at House Creek field to have been deposited as part of an offshore bar complex many tens of kilometers from the shoreline in an outer shelf location. They invoke a mechanism whereby, as a rapid transgression took place, rapid shoreface retreat left residual sand sheets on the shelf. These were later transported and formed into broad elongate sand complexes or offshore bars by marine currents. The productive limits of the fields suggest a relatively narrow and sinuous sandstone distribution. However, the sand bodies are much broader with very low relief on the order of tens of meters over tens of kilometers. Brenner (1978) attributes the lobate lateral edges characteristic of the sand bodies to periodic breaching by storm-generated currents. Washover fans and/or tidal deltas formed at the mouths of channels cut through the sand bodies. Detailed mapping by Crews and others (1976) revealed at least 12 well-sorted sandstone bodies, imbricated, with the youngest to the east. They trend generally N30° to 40°W and are separated by areas of siltstone and mudstone. These sandstone bodies are about 1 to 3 km wide, 8 to 48 km long, and up to 10 m thick.

Shannon and Sussex petroleum traps and seals. These plays are limited in distribution to the deeper parts of the basin. According to Brenner (1978), petroleum in the Sussex Sandstone Member accumulated in classic updip pinchout traps where porous and permeable shelf sandstone bars pinch out into shale. House Creek field in the south-central part of the basin produces from the Sussex. It trends N40° to 45°W, which is nearly parallel to the shelf trend and is an updip stratigraphic trap surrounded by Steele Shale. Reservoir-quality rocks do not generally exceed about 10 m in thickness and may extend 30 or more km along strike and up to 5 km down dip (Sabel, 1985). Similar traps have been found in the Shannon Sandstone Member.

Mesaverde and Lewis deltaic reservoirs. Mesaverde and Lewis deltaic reservoir sandstones were deposited in a wave-dominated high-destructive shoreline and were locally modified into offshore bars (Fig. 6). These bars include porous sands that pinch out seaward and have formed traps for hydrocarbons.

Curry (1976) highlights stratigraphic studies that have led to current sedimentological interpretations of Upper Cretaceous strata in the Powder River basin. Gill and Cobban (1966) studied the relationships of marine and nonmarine sandstones within the Teapot and Parkman Members. Weimer (1961) interpreted interbedded marine and nonmarine Mesaverde strata to be deltaic rather than shifting marine shoreline sands. Asquith (1970) also interpreted deltaic deposition for these Upper Cretaceous strata and illustrated inclined surfaces of deposition.

Parkman Sandstone Member. The oldest of the Mesaverde and Lewis cycles of deltaic deposition in the Powder River basin is the coarsening-upward progradational Parkman Sandstone Member. Sandstone was derived from the Cretaceous cordillera in eastern Idaho and southwestern Montana (Hubert and others, 1972). Deposition took place on a southeasterly prograding, wave-dominated, high-destructive delta. The Parkman delta prograded over upper prodelta slope deposits of the Cody Shale. As indicated by the orientation of pillow structures at the base of delta-front sandstone beds, subaqueous parts of the delta surface sloped to the southeast. Upper delta plain facies that comprise the upper part of the Parkman include carbonaceous mudstone, lenses of channel sandstone, and a few thin lignite beds deposited in floodplains, meandering streams, lakes, swamps, coastal marshes, and tidal creeks. During the subsequent destructional phase, caused by compaction of thick prodelta mud over which the deltas prograded, marine shoreline sandstone was deposited, followed by marine shale deposition. This marked the termination of delta growth.

Curry (1973) has correlated outcrops with nearby subsurface well logs and has interpreted facies changes into the subsurface. Barrier island sandstone is thought to be present seaward of the delta plain sandstones. Still farther to the east, these barrier island sandstones grade into finer-grained marine sandstones, siltstones, and shale.

Teapot Sandstone Member. During a second progradational event, the Teapot Sandstone Member was deposited in a fashion similar to that of the Parkman. Curry (1976), using electric logs, mapped six stages of imbricate prodelta slope deposition. As these prodelta facies were being deposited, sandstones of the Teapot were being deposited in a variety of environments. These included subaerial delta plains, marine bars seaward from the delta plain, as well as deeper marine slopes. A destructive deltaic phase followed, with deposition of the marine Lewis Shale.

Teckla Sandstone Member. During deposition of the Lewis Shale, a third and more local deltaic progradation resulted in deposition of the Teckla Sandstone Member. The inferred locations of the Teckla delta systems from subsurface mapping coincide approximately with similar systems in the Teapot Sandstone Member. Deposition of deltaic strata was followed by transgression of the Lewis Sea once again and renewed deposition of finer-grained marine clastics.

Because the Teckla does not crop out, Runge and others (1973) defined a subsurface type section with two, delta-front, shoreline sandstone beds separated by a thin shale. These facies grade westward into delta plain facies that include thin swamp and marsh coals and lignites, and distributary channel sandstones with east-west porosity trends.

Mesaverde and Lewis petroleum traps and seals. Traps are produced in these plays by updip pinchout of shallow-marine sandstones into finer-grained prodelta facies, which also act as seals. Using the Teckla as an example, Runge and others (1973) described the occurrence of oil at Poison Draw, the largest oil field with Teckla production. Here, the reservoirs are a complex of strandline sandstones in which oil is trapped by updip loss of porosity due to increasing siltstone and shale content. The complexity of the sandstone bodies is attested to by the presence of multiple oil-water contacts within the reservoir.

Isabell and others (1976) reported on the petroleum geology of the Well Draw field. In this field, production is from marine sandstones in the Teapot, representing a large northwest-trending stratigraphic trap formed by an updip facies change from porous shallow-water marine sandstone into tight offshore siltstone and shale. They also note that the productive Teapot trend continues northward to include Mikes Draw and Don Draw fields. The depth of productive sandstones in these plays ranges to about 3,050 m in the western part of the basin.

Structural plays

Basin margin anticlinal play. The Powder River basin has a long history of oil and gas production from structural traps, mostly large anticlinal features around the margins of the basin, formed during the Laramide orogeny. The majority of these traps are relatively simple anticlinal closures that are often reverse-faulted at depth and contain extensional faults on their crests. However, also included are plunging anticlinal noses that have been faulted to form traps. Often subsidiary and sometimes complex anticlinal features and combination traps are associated with the large structures. Salt Creek field is an example of a large faulted anticlinal structure (Fig. 7B).

Strata with reservoirs in these structures range in age from Pennsylvanian to Upper Cretaceous. Most important reservoirs are sandstone in the Frontier, Tensleep, and Minnelusa Formations, Inyan Kara Group and Muddy Sandstone, and various Jurassic sandstones. Many fields contain multiple producing horizons. Petroleum is sealed in these reservoirs by numerous shales interbedded with the sandstones. Major seals occur at the base of the Permian and Triassic red beds as well as within Lower and Upper Cretaceous shale sequences.

Basin margin subthrust play. As discussed by Gries (1983), nearly all of the Laramide basins of the Rocky Mountain foreland have thrust faults along at least part of their margins. In this play, petroleum is trapped in deformed Phanerozoic strata below the thrust (Fig. 7C). The overthrust wedge of impermeable Precambrian rocks may have acted as a trap and seal of fluids in the underlying Phanerozoic sedimentary rocks. Traps may also have been formed in folds beneath the thrusts.

The first sub-Precambrian production in the Rocky Mountains was from a development well drilled by True Oil in 1971 in South Glenrock field at the southern end of the Powder River Basin (Fig. 7A). This well recovered 70 barrels of oil per day from the Muddy Sandstone at a depth of 2,340 m. This oil occurs below the Precambrian, which was thrust northward along the north side of the Laramie Range (Gries, 1983).

THERMAL HISTORY AND PETROLEUM SOURCE ROCKS

The thermal history of the Powder River basin is directly related to the burial history of the basin. Owing to burial beneath a thick Upper Cretaceous and Tertiary sedimentary sequence, petroleum source rocks of Early and Late Cretaceous age began expelling oil in early Eocene time (Momper and Williams, 1979). Thermal maturity measurements of Momper and Williams (1979, 1984) show that the pattern of thermally mature Cretaceous source rocks parallels the structural contours of the basin

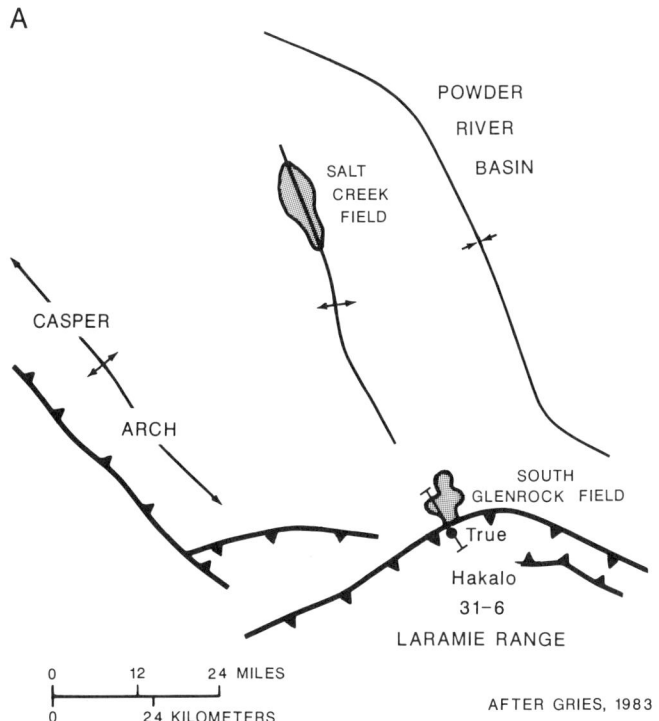

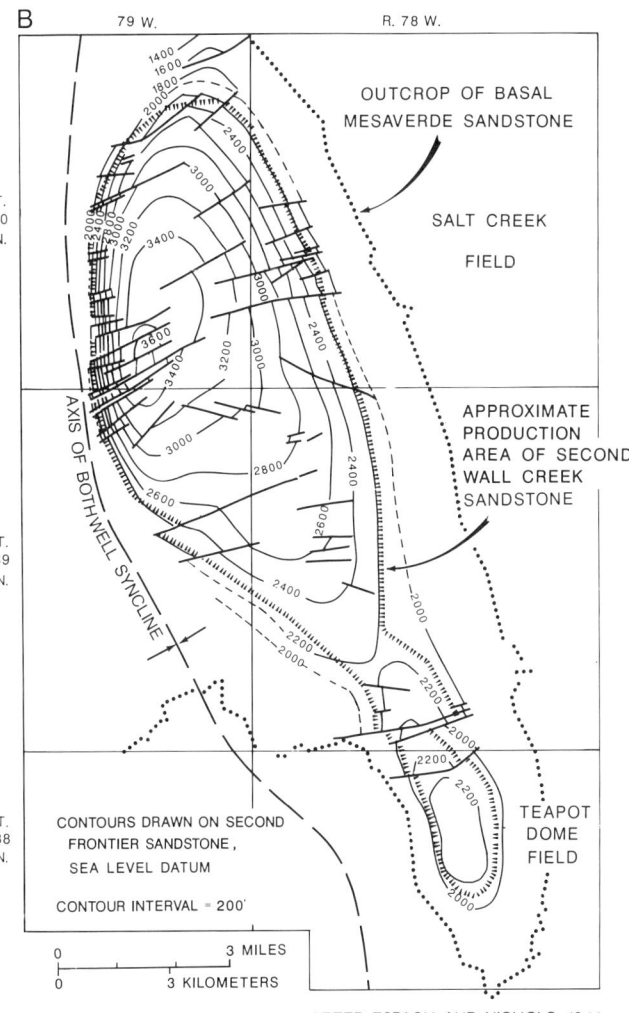

(Fig. 8). Oil expulsion usually occurred above about 190°F (88°C), corresponding to present burial depths between about 2,438 m and 3,658 m. Cooling probably occurred in some parts of the basin in the late Cenozoic because of uplift and erosion, invasion of relatively cool meteoric water, and possibly climatic cooling (Momper and Williams, 1984). Cooling by invasion of meteoric water is a possible explanation for the anomalously low present-day geothermal gradients observed in the deep part of the basin near the Big Horn Mountains (Fig. 9). Movement of water into the west side of the basin along fault-plane conduits on the east flank of the Big Horn uplift could have occurred more or less continuously since the Laramide orogeny. This would account for moderate thermal maturities of Cretaceous source rocks in the area, despite relatively great burial depths (generally greater than 3,660 m). The existence of relatively low geothermal gradients in this area throughout much of the basin's history is also indicated by the occurrence of oil (instead of gas) at approximately 4,570 m in Paleozoic (Minnelusa Formation) reservoirs at Reno field and several other fields (Fig. 9). A higher geothermal gradient would have resulted in thermal cracking of oil to form mostly gas and condensate at these depths, especially in Paleozoic-age source rock/reservoir systems. A similar structural and hydrodynamic setting associated with the Black Hills uplift may be the cause of the low geothermal gradients (1.4°F/100 ft; 2.4°C/100 m) along the west side of the Black Hills.

Another noteworthy feature of the present geothermal gradient distribution in the basin is the occurrence of relatively high

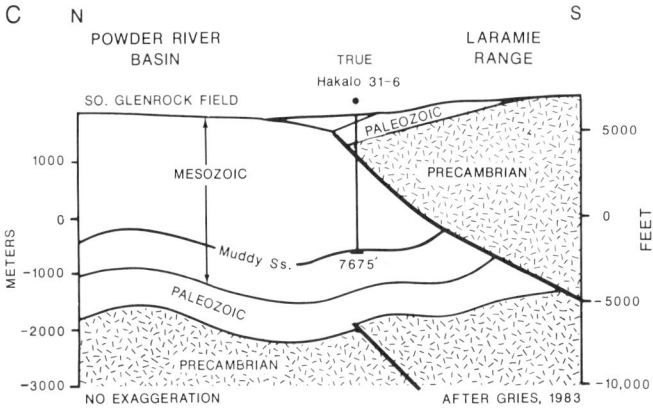

Figure 7. Salt Creek and South Glenrock fields, Powder River basin. (A) Map showing location of fields and generalized structures (after Gries, 1983). (B) Structure map of Salt Creek field, a basin margin anticlinal structure on the east flank of the Casper arch (after Espach and Nichols, 1941). (C) Cross section showing relation of South Glenrock field to a basin margin subthrust structure along the north flank of the Laramie Range. The cross-section traverse is shown on Figure 7A (after Gries, 1983).

gradients (approximately 2.0°F/100 ft; 3.5°C/100 m) in the southeastern and northeastern areas. In the southeastern part of the basin, high heat flux coincides with the Hartville uplift and is probably related to loss of sedimentary cover through erosion during and after the uplift. Because of the small scale used in Figure 9, the detailed pattern of geothermal gradients related to the Hartville uplift is not shown. However, an abrupt increase in geothermal gradients from less than 1.8°F/100 ft (3.1°C/100 m) to greater than 2.0°F/100 ft (3.5°C/100 m) occurs across the fault system bounding the north side of the Hartville uplift, with the higher gradients occurring on the uplift side of the faulted area. No explanation has been generally agreed upon for the area of high geothermal gradients in the northeastern part of the basin (greater than 2.0°F/100 ft; 3.5°C/100 m).

Figure 10 shows reconstruction of the burial history and calculation of time-temperature index (TTI) values according to the methods of Lopatin (1971) and Waples (1980) for the Government-Tracy #1 well (F. F. Meissner, personal communication, 1986). Located near the basin axis in the area where thermally mature Cretaceous source rocks occur, this well illustrates the relationship between thermal history in the basin and generation of oil. The TTI calculation of Figure 10 is based on the present geothermal gradient and does not consider possible compactional effects or temporal changes in the geothermal gradient. Some minor erosion (approximately 305 m) is estimated to have occurred. As the burial history diagram and TTI calculations show, the episode of Late Cretaceous–early Tertiary sedimentation and attendant burial heating was critical for maturation of source rocks in the basin. The calculated TTI and measured vitrinite reflectance data show that the Minnelusa Formation is in the wet gas zone of thermal maturity at this location. This observation supports the hypothesis discussed previously that a lower geothermal gradient must have existed throughout much, if not all, of the post-Laramide period in the area east of the Big Horn uplift in order for oil to have been preserved in Minnelusa Formation reservoirs at burial depths in excess of 4,500 m.

Burtner and Warner (1984) studied the potential for hydrocarbon generation from the Lower Cretaceous Mowry and Skull Creek Shales in this region. The Mowry and Skull Creek Shales were found to contain a mixture of types II and III organic matter. Because it has a generally higher proportion of type II material, the Mowry is a better petroleum source than the Skull Creek Shale. Momper and Williams (1984) estimate that the Mowry Shale may have expelled about 11.9 billion barrels of oil.

According to Momper and Williams (1984), the Niobrara Formation and Carlile Shale, collectively, are also a major source of oil found in Upper Cretaceous reservoirs. However, the areal extent of effective source rocks in these formations is less extensive than the deeper Mowry Shale. Also, data of Momper and Williams (1984) indicate that the Lower Cretaceous Fuson Shale and Upper Cretaceous shale in the Frontier and Steele Formations have expelled oil in amounts secondary to the major Cretaceous source rocks. Based on limited data, they estimate the Frontier may have expelled 0.8 billion barrels of oil.

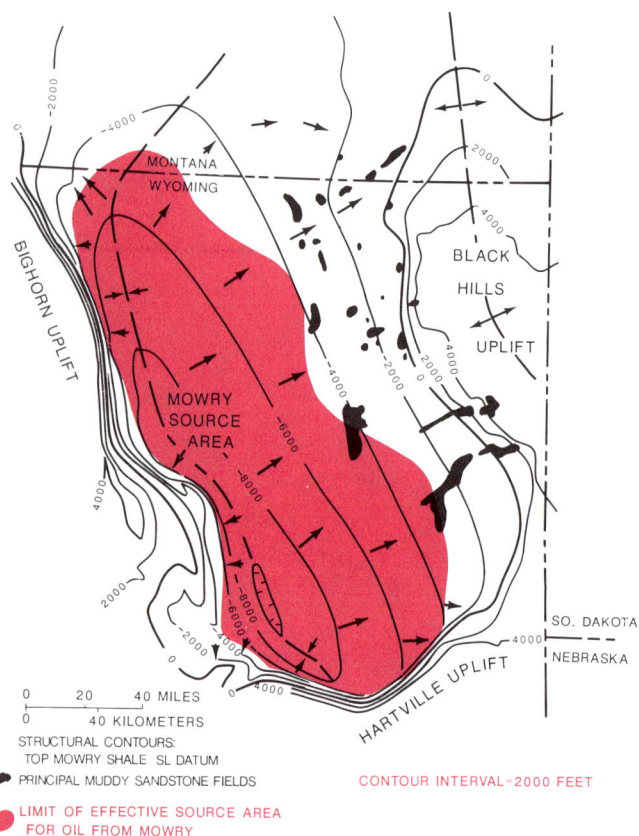

Figure 8. Basin structure map showing areas of oil generation and expulsion from the Mowry Shale (in pink). Arrows show migration paths to Muddy oil fields (in black). After Momper and Williams (1984).

Merewether and Claypool (1980) analyzed lowermost Upper Cretaceous potential hydrocarbon source rocks from well bores in three areas. The Belle Fourche Shale, Greenhorn Formation, and Carlile Shale from the shallow eastern side of the basin contain calcareous shale of offshore and open-marine origin with abundant, hydrogen-rich, organic material, derived largely from marine organisms. These rocks are in immature stages of thermal alteration with respect to oil generation. The Frontier Formation and most of the Sage Breaks Member of the Cody Shale along the western flank of the basin at a depth of about 318 m contain noncalcareous shale of nearshore-marine origin with abundant hydrogen-deficient organic matter derived mainly from land plants. These rocks are potential sources of gas at optimum levels of thermal maturity, but are currently thermally immature with respect to hydrocarbon generation. The Frontier Formation and Cody Shale near the axis of the basin are noncalcareous and of nearshore-marine origin. They contain low amounts of hydrogen-deficient organic matter. The organic matter in these rocks has undergone sufficient thermal alteration to have generated appre-

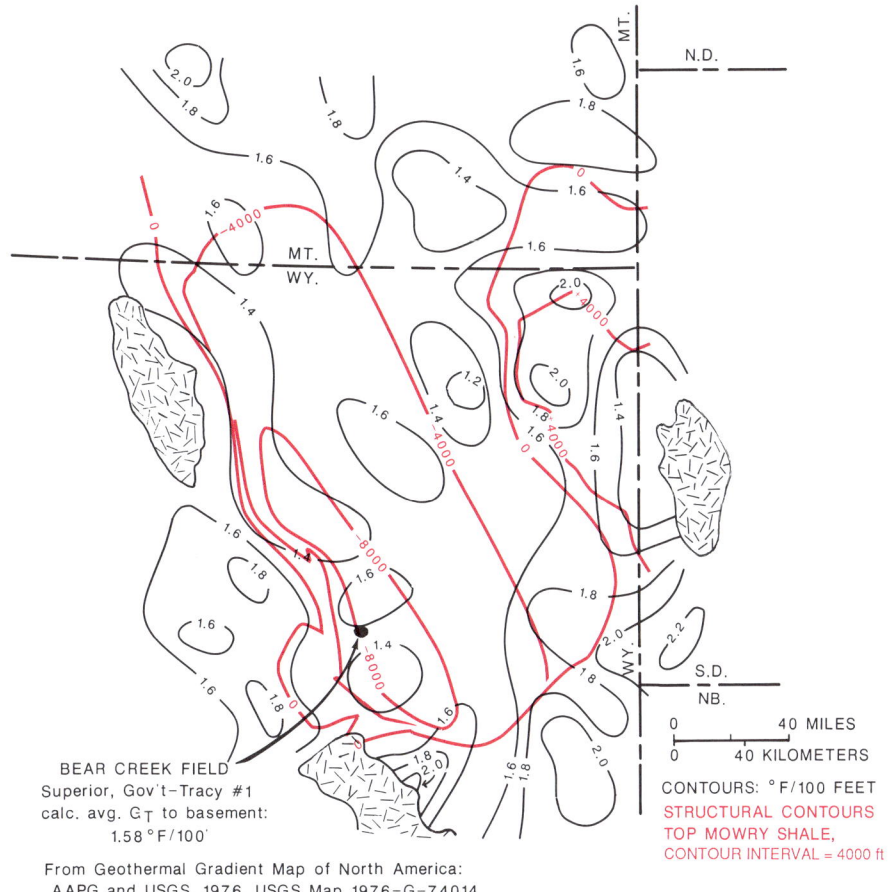

Figure 9. Isograms showing geothermal gradients in the Powder River basin. Modified from the Geothermal Gradient Map of North America (1976).

ciable quantities of hydrocarbons, especially if the original (before thermal alteration) organic matter was predominantly hydrogen rich.

The source rocks for petroleum in the middle ("Leo") and upper members or parts of the Minnelusa Formation may have been Permian shales in the present areas of western Wyoming and eastern Idaho, or the source rocks may be indigenous to the Minnelusa.

Permian black shales in the area of western Wyoming and eastern Idaho would probably have been buried deeply enough to generate hydrocarbons by Jurassic time. Organic carbon-rich facies of the Minnelusa Formation on the west side of the basin had sufficiently high thermal histories to produce hydrocarbons contemporaneously with or subsequent to the Laramide orogeny. If petroleum was supplied from both distant and local sources, some of it could have moved into the area of the present basin during the Jurassic. It would have been trapped until the Laramide orogeny, when surrounding mountain blocks were uplifted. At this time, the oil may have moved again and been redistributed into new structures forming around the flanks of the basin or into newly formed updip pinchout stratigraphic traps.

Clayton and Ryder (1984) found that black shales in the middle member ("Leo") of the Minnelusa Formation and equivalent rocks of Desmoinesian age in Nebraska are excellent source rocks. Organic carbon content ranges from less than 1 to 26 weight percent and averages 5.4 percent. Further, the black shales have reached sufficient thermal maturity to generate substantial quantities of liquid hydrocarbons. Momper and Williams (1984) also consider black shales in the middle member to be a source of oil in reservoirs of the middle member in the southeastern part of the basin. They believe, however, that long-distance migration from Permian Phosphoria Formation source rocks accounts for oil in the upper member.

PRODUCTION HISTORY AND FUTURE POTENTIAL OF PROVINCE

To date, approximately 2.5 billion barrels of oil have been discovered in the basin. About 75 percent of the oil in the basin is

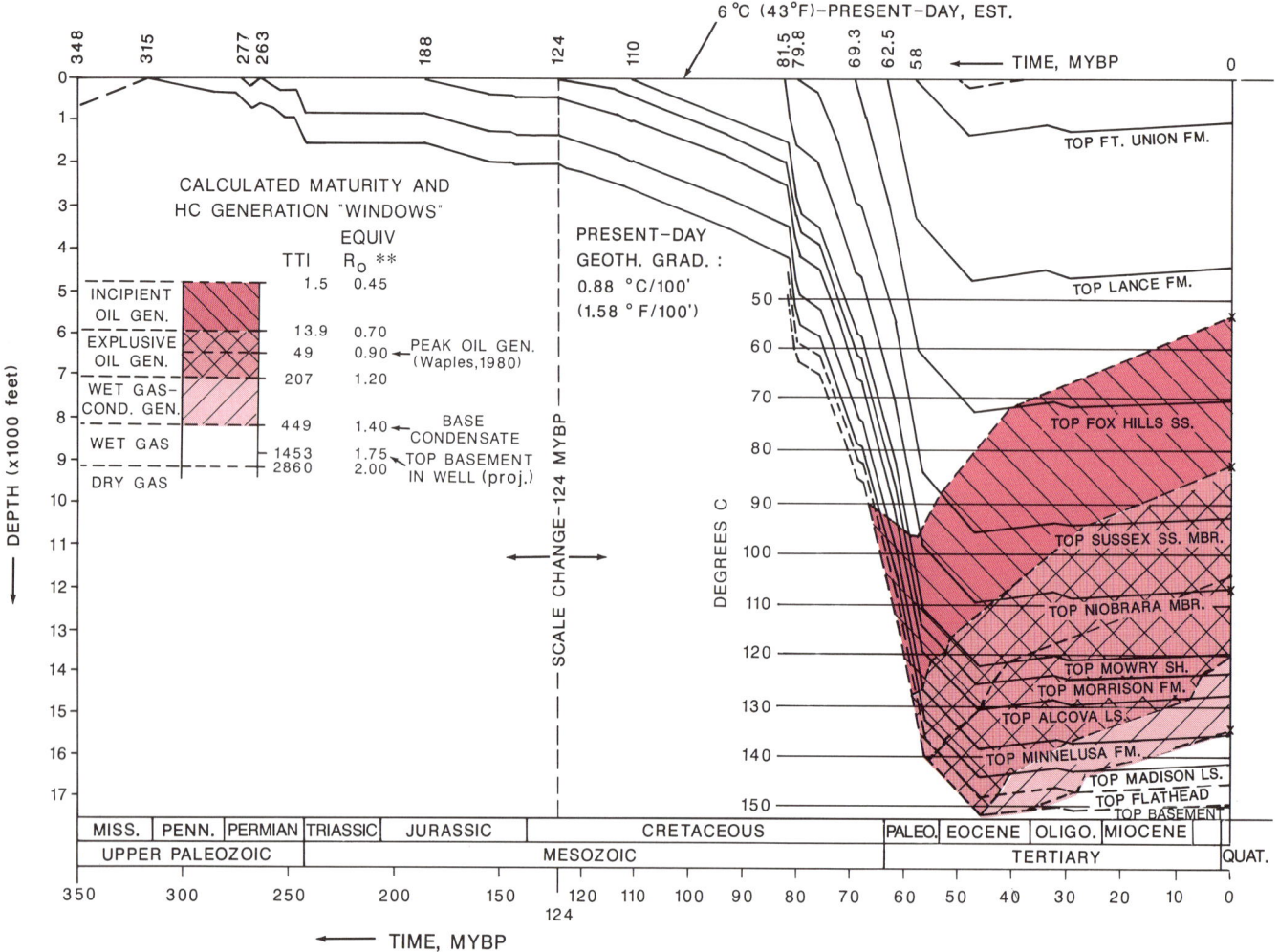

Figure 10 (this and facing page). Lopatin diagrams illustrating the thermal history of the Powder River basin based on data from Superior Government Tracy #1 well (SE¼,NW¼,Sec.26,T.38N.,R.75W.) in Bear Creek field. See Figure 9 for location. Burial profiles are based on present-day thickness and interpreted erosion. No correction was made for compaction. The temperature profile is based on present-day temperatures measured during log runs, with cooling corrections. Lopatin diagrams after data from F. F. Meissner, written communication, 1986.

from Cretaceous reservoirs; most of the remainder is from the Pennsylvanian and Permian. Almost half the oil has been found in large anticlinal structures around the basin margin (more than 700 million barrels in Salt Creek field alone), while stratigraphic traps in sandstone beds of Pennsylvanian, Permian, and Early Cretaceous age account for most of the remainder. Gas in the basin is found principally as associated-dissolved gas, with minor amounts occurring in nonassociated gas fields.

Pennsylvanian and Permian stratigraphic plays

The play in the upper part of the Minnelusa is well established, with an exploration history of about 30 years. Exploration ranges from well explored over a significant part of the eastern flank of the Powder River basin to very lightly explored in the deeper parts. Over 160 fields have been discovered through 1983; they are estimated to contain more than 350 million barrels of recoverable oil. The largest field discovered, Raven Creek, has slightly more than 50 million barrels, while mean size is approximately 2.5 million barrels. Sizes of undiscovered accumulations are expected to be similar.

The "Leo" play is in the category of lightly explored, with only a few fields in the category of greater than 1 million barrels of oil equivalent and only a few others in the category of less than

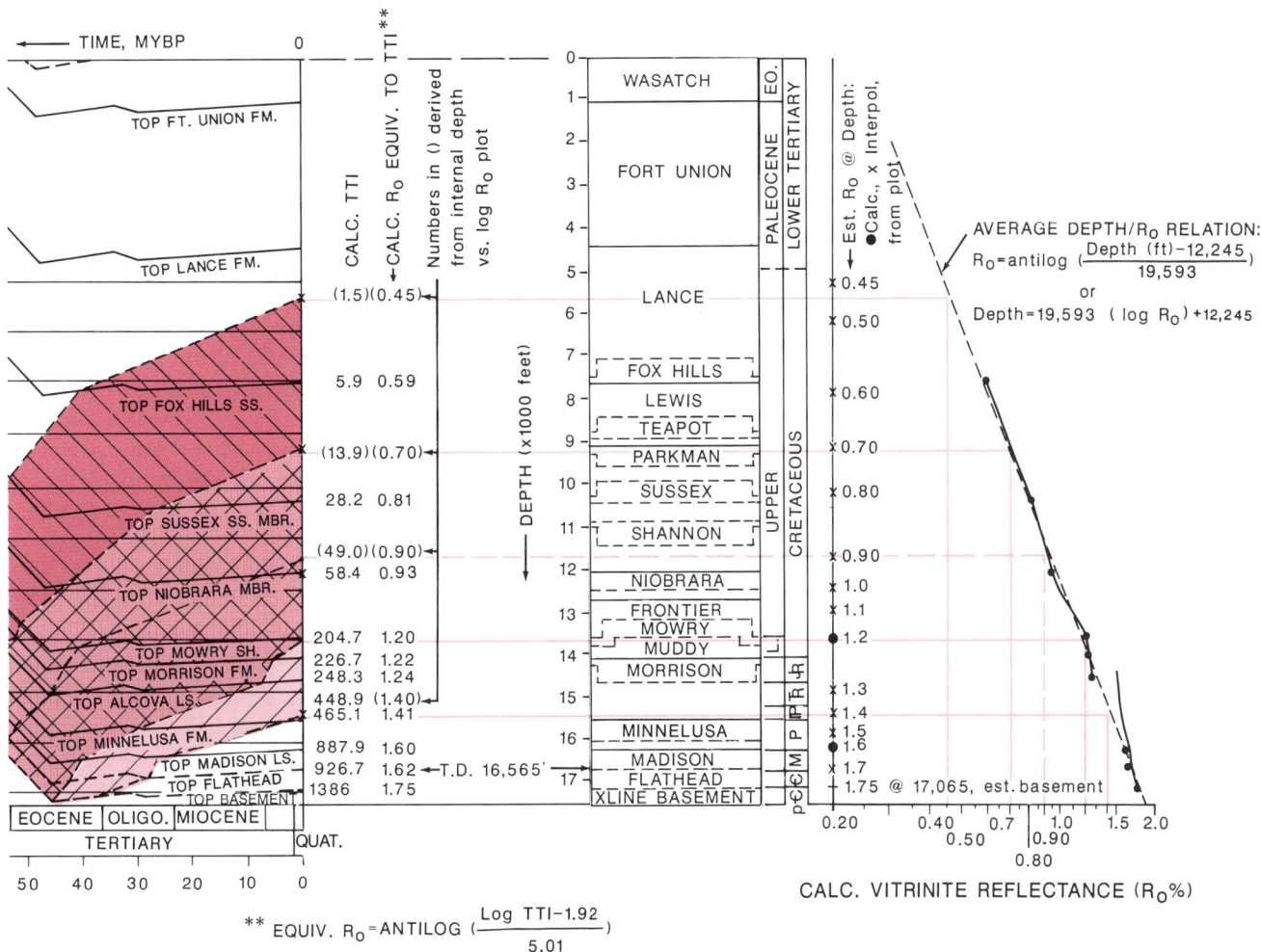

1 million barrels of oil equivalent. It is anticipated that future discoveries will probably total less than 10 million barrels of oil or 10 billion cubic feet of gas.

Lower Cretaceous stratigraphic plays

The Lakota Sandstone play is generally lightly explored due to the small size, unpredictability, and difficulty of detecting accumulations in channel sandstones. These stratigraphic traps also occur in combination with structural noses or anticlinal closures. Through 1982, about a dozen fields have been discovered in the unstructured east flank of the basin, accounting for about 10 million barrels of oil. Sizes of undiscovered accumulations will tend to be small (averaging less than 1 million barrels), although they may occur in large number. A few larger traps (5 to 10 million barrels range) may also be found. The sizes of Lakota pools in large Laramide structures peripheral to or within the play area suggest the potential of these reservoirs, even though they are not representative of sizes of stratigraphic traps in the play.

Exploration in the Fall River Sandstone play has continued in the shallower east flank of the basin for approximately 30 years and has resulted in discovery of more than 30 individual pools or fields, aggregating between 60 and 100 million barrels of oil. The largest accumulation, Coyote Creek, is approximately 20 million barrels. Pool sizes to be discovered will probably be similar to those discovered in the past, mostly small accumulations of less than 10 million barrels.

The Muddy and Newcastle Sandstone play is well established, with a history of more than 60 years. Exploration ranges from intensive over the shallower parts of the east flank of the basin, to lightly explored in the deeper parts. More than 180 recognized pools and fields have been discovered through 1982, and discovered quantities exceed 550 million barrels of recoverable oil. The largest field, Bell Creek in Montana, is estimated to contain over 139 million barrels, and average pool size is approximately 3 million barrels. Sizes of undiscovered accumulations are expected to be somewhat smaller than those already discovered, with an abundance of small accumulations plus a few of substantial size.

Upper Cretaceous shelf sandstone stratigraphic plays

The Frontier Formation is the major producing formation of the Powder River basin. However, excluding production from several major anticlinal traps around the margin of the basin, stratigraphic traps in sandstone in the Frontier Formation and Turner Sandy Member of the Carlile Shale have accounted for little more than 30 million barrels of oil and a small amount of gas. Nevertheless, Barlow and Haun (1970) have attributed the giant oil pool in the "Second Frontier sandstone" at Salt Creek field (more than 330 million barrels) to remigration of oil out of pre-existent stratigraphic traps.

Exploration in the Frontier stratigraphic play is generally restricted to the deep western part of the basin where exploration began in the early 1970s, stimulated by discovery of Spearhead Ranch field. Field sizes are not well documented, but the largest field discovered to date is probably Powell-Ross with more than 10 million barrels. There is evidence that several of the approximately 20 discovered pools in this play will eventually coalesce. Fields remaining to be discovered in the play will very likely be small, with the largest on the order of about 10 to 20 million barrels. They will probably average between 1 and 4 million barrels.

About 75 million barrels of oil and 60 billion ft^3 of gas have been produced from the Shannon and Sussex Sandstone Members. About 10 fields are in the category of greater than 1 million barrels of oil equivalent, and another 10 fields are in the category of less than 1 million barrels of oil equivalent. House Creek field has produced about 19 million barrels of oil from an estimated 160 million barrels of oil originally in place in the Sussex Member. In the analogous Shannon Member play, Hartzog Draw field is the largest, having produced about 33 million barrels of oil and about 14 billion ft^3 of gas.

The southern part of the basin is well explored, but the northern part is lightly explored. The size of fields remaining to be discovered is thought to be in the same range as those discovered to date, 10 million barrels or less of oil and 10 billion ft^3 or less of gas.

Upper Cretaceous deltaic sandstone stratigraphic plays

Approximately 50 million barrels of oil and 90 billion ft^3 of gas have been produced from these traps. About 12 fields in the category of greater than 1 million barrels of oil equivalent and 45 fields with less than 1 million barrels of oil equivalent have been discovered. The largest field is Well Draw field with ultimate recoverable oil of about 35 million barrels and gas of about 100 billion ft^3.

A very large area in the northern part of the basin remains to be evaluated. Fields to be discovered in the future will probably be of the size range of those discovered to date, with ultimate recoverable reserves of less than 10 million barrels of oil and less than 5 billion ft^3 of gas.

Structural plays

Exploration in basin margin anticlines has proceeded for approximately 100 years and has resulted in discovery of most of the major oil fields in the basin. These discoveries, which include Salt Creek, Teapot Dome, Lance Creek, and many others, were made early in its exploratory history. The largest field is Salt Creek, with almost three-quarter billion barrels of oil. Total discovered resources in structural traps exceed one billion barrels and include substantial dissolved-associated gas.

Exploration in the play is nearing its conclusion; little potential remains. Most future discoveries will probably be made in small and more subtle traps associated with some of the larger structures or trends.

Basin margin subthrust traps, either documented or hypothetical, may be explorable along flanks of the basin. Geologic data are very limited and do not allow accurate prediction of future reserves or field sizes.

REFERENCES CITED

Agatston, R. S., 1954, Pennsylvanian and lower Permian of northern and eastern Wyoming: American Association of Petroleum Geologists Bulletin, v. 38, no. 4, p. 504–583.

Asquith, D. O., 1970, Depositional topography and major marine environments, Late Cretaceous, Wyoming: American Association of Petroleum Geologists Bulletin, v. 54, no. 7, p. 1184–1224.

—— , 1974, Sedimentary models, cycles, and deltas, Upper Cretaceous, Wyoming: American Association of Petroleum Geologists Bulletin, v. 58, no. 11, p. 2274–2283.

Barlow, J. H., and Haun, J. D., 1970, Regional stratigraphy of the Frontier Formation and its relation to Salt Creek Field, in Halbouty, M. T., ed., Geology of Giant Petroleum Fields: American Association of Petroleum Geologists Memoir 14, p. 147–157.

Berg, R. R., 1962, Mountain flank thrusting in Rocky Mountain foreland, Wyoming and Colorado: American Association of Petroleum Geologists Bulletin, v. 46, no. 11, p. 2019–2032.

—— , 1968, Point-bar origin of Fall River sandstone reservoirs, northeastern Wyoming: American Association of Petroleum Geologists Bulletin, v. 52, no. 11, p. 2116–2122.

—— , 1976, Hilight Muddy Field–Lower Cretaceous transgressive deposits in the Powder River basin, Wyoming: The Mountain Geologist, v. 13, no. 2, p. 33–46.

Blackstone, D. L., Jr., 1981, Compression as an agent in deformation of the east-central flank of the Bighorn Mountains, Sheridan and Johnson Counties, Wyoming: Laramie, University of Wyoming Contributions to Geology, v. 19, no. 2, p. 105–122.

Brenner, R. L., 1978, Sussex Sandstone of Wyoming; Example of Cretaceous offshore sedimentation: American Association of Petroleum Geologists Bulletin, v. 62, no. 2, p. 181–200.

Burtner, R. L., and Warner, M. A., 1984, Hydrocarbon generation in Lower Cretaceous Mowry and Skull Creek shales of the northern Rocky Mountain area: Rocky Mountain Association of Geologists 1984 Symposium, p. 449–467.

Chamberlin, R. T., 1945, Basement control on Rocky Mountain deformation: American Journal of Science, v. 243-A, p. 98–116.

Clayton, J. L., and Ryder, R. T., 1984, Organic geochemistry of black shales and oils in the Minnelusa Formation (Permian and Pennsylvanian), Powder River Basin, Wyoming: Rocky Mountain Association of Geologists 1984 Symposium, p. 231–253.

Condra, G. E., and Reed, E. C., 1950, Correlation of the formations of the Laramie Range, Hartville Uplift, Black Hills, and Western Nebraska: Nebraska Geological Survey Bulletin, no. 13-A, 52 p.

Crews, G. C., Barlow, J. A., Jr., and Haun, J. D., 1976, Upper Cretaceous Gammon, Shannon, and Sussex sandstones, central Powder River basin, Wyoming: Wyoming Geological Association 28th Annual Field Conference Guidebook, p. 9–20.

Curry, W. H., III, 1971, Laramide structural history of the Powder River Basin, Wyoming: Wyoming Geological Association 23rd Annual Field Conference Guidebook, p. 49–60.

—— , 1973, Parkman delta in central Wyoming: Earth Science Bulletin, v. 6, no. 4, p. 5–18.

—— , 1976, Late Cretaceous Teapot Delta of southern Powder River Basin, Wyoming: Wyoming Geological Association 28th Annual Field Conference Guidebook, p. 21–28.

Dondanville, R. F., 1963, The Fall River Formation, northwestern Black Hills; Lithology and geologic history: Wyoming Geological Association and Billings Geological Society 1st Joint Field Conference Guidebook, p. 87–99.

Espach, R. H., and Nichols, H. D., 1941, Petroleum and natural-gas fields in Wyoming: U.S. Bureau of Mines Bulletin 418, 185 p.

Fox, J. E., 1986, Stratigraphic cross-sections A-A' through F-F' showing electric logs of Upper Cretaceous and older rocks, Powder River Basin, Wyoming and Montana: U.S. Geological Survey Open-File Report OF 86-465 A-F.

—— , 1987, Stratigraphic cross-sections G-G' through V-V' showing electric logs of Upper Cretaceous and older rocks, Powder River Basin, Wyoming: U.S. Geological Survey Open-File Report OF 86-465 G-V.

Fryberger, S. G., 1984, The Permian Upper Minnelusa Formation, Wyoming: Ancient example of an offshore-prograding eolian sand sea with geomorphic facies, and system-boundary traps for petroleum: Wyoming Geological Association 35th Annual Field Conference Guidebook, p. 241–271.

Fryberger, S. G., Al-Sari, A. M., and Clisham, T. J., 1983, Eolian dune, interdune, sand sheet, and siliciclastic sabkha sediments of an offshore prograding sand sea, Dhahran area, Saudi Arabia: American Association of Petroleum Geologists Bulletin, v. 67, no. 2, p. 280–312.

George, G. R., 1984, Cyclic sedimentation and depositional environments of the upper Minnelusa Formation, central Campbell County, Wyoming: Wyoming Geological Association 35th Annual Field Conference Guidebook, p. 75–95.

Geothermal Survey of North America Subcommittee of the American Association of Petroleum Geologists' Research Committee, 1976, Geothermal gradient map of North America: American Association of Petroleum Geologists, scale 1:5,000,000, two sheets.

Gill, J. R., and Cobban, W. A., 1966, The Red Bird section of the Upper Cretaceous Pierre Shale in Wyoming: U.S. Geological Survey Professional Paper 339-A, 73 p.

—— , 1973, Stratigraphy and geologic history of the Montana Group and equivalent rocks, Montana, Wyoming, and North and South Dakota: U.S. Geological Survey Professional Paper 776, 37 p.

Goodell, H. G., 1962, The stratigraphy and petrology of the Frontier Formation of Wyoming: Wyoming Geological Association 17th Annual Field Conference Guidebook, p. 173–210.

Gries, R., 1983, Oil and gas prospecting beneath Precambrian of foreland thrust plates in Rocky Mountains: American Association of Petroleum Geologists Bulletin, v. 67, no. 1, p. 1–28.

Harris, S. A., 1976, Fall River ("Dakota") oil entrapment, Powder River Basin: Wyoming Geological Association 28th Annual Field Conference Guidebook, p. 147–164.

Haun, J. D., 1958, Early Upper Cretaceous stratigraphy, Powder River Basin, Wyoming: Wyoming Geological Association 13th Annual Field Conference Guidebook, p. 84–89.

Haun, J. D., and Barlow, J. A., Jr., 1962, Lower Cretaceous stratigraphy of Wyoming: Wyoming Geological Association 17th Annual Field Conference Guidebook, p. 15–22.

Hobson, J. P., Jr., Fowler, M. L., Beaumont, E. A., 1982, Depositional and statistical exploration models, Upper Cretaceous offshore sandstone complex, Sussex Member, House Creek Field, Wyoming: American Association of Petroleum Geologists Bulletin, v. 66, no. 6, p. 689–707.

Hubert, J. F., Butera, J. G., and Rice, R. F., 1972, Sedimentology of Upper Cretaceous Cody-Parkman delta, southwestern Powder River Basin, Wyoming: Geological Society of America Bulletin, v. 83, no. 6, p. 1649–1670.

Imlay, R. W., 1947, Marine Jurassic of Black Hills area, South Dakota and Wyoming: American Association of Petroleum Geologists Bulletin, v. 31, p. 227–273.

—— , 1980, Jurassic paleobiogeography of the conterminous United States in its continental setting: U.S. Geological Survey Professional Paper 1062, 134 p.

Isbell, E. B., Spencer, C. W., and Seitz, T., 1976, Petroleum geology of the Well Draw Field, Converse County, Wyoming: Wyoming Geological Association 28th Annual Field Conference Guidebook, p. 165–174.

Jenkins, C. D., 1986, Tectonic analysis of the northeastern flank of the Bighorn Mountains; Buffalo to Dayton, Wyoming [M.S. thesis]: Rapid City, South Dakota School of Mines and Technology, 121 p.

Lageson, D. R., Maughan, E. K., and Sando, W. J., 1979, The Mississippian and Pennsylvanian (Carboniferous) Systems in the United States; Wyoming: U.S. Geological Survey Professional Paper 1110-U, p. U1–U38.

Lisenbee, A. L., 1979, Laramide structure of the Black Hills uplift, South Dakota-Wyoming-Montana: Geological Society of America Memoir 151, p. 165–196.

—— , 1985, Tectonic map of the Black Hills uplift, Montana, Wyoming, and South Dakota: Geological Survey of Wyoming, Map Series 13, scale 1:250,000.

Lopatin, N. V., 1971, Temperature and geologic time as factors in coalification: Akademiia Nauk SSSR. Izvestia Seriia Geologicheskaia, no. 3, p. 95–106 (in Russian); English translation by N. H. Bostick, Illinois State Geological Survey, 1972.

MacKenzie, D. B., and Poole, D. M., 1962, Provenance of Dakota Group sandstones of the Western Interior: Wyoming Geological Association 17th Annual Field Conference Guidebook, p. 62–71.

MacKenzie, F. T., and Ryan, J. D., 1962, Cloverly-Lakota and Fall River paleocurrents in the Wyoming Rockies: Wyoming Geological Association 17th Annual Field Conference Guidebook, p. 44–61.

Mallory, W. W., 1967, Pennsylvanian and associated rocks in Wyoming: U.S. Geological Survey Professional Paper 554-G, 31 p.

—— , 1975, Middle and southern Rocky Mountains, northern Colorado Plateau, and eastern Great Basin region, in McKee, E. D., and Crosby, F. J., co-ordinators, Introduction and regional analyses of the Pennsylvanian system, Pt. 1; Paleotectonic investigations of the Pennsylvanian system in the United States: U.S. Geological Survey Professional Paper 853-N, p. 265–278.

Maughan, E. K., 1967, Eastern Wyoming, eastern Montana, and the Dakotas, in McKee, E. D., Oriel, S. S., and others, coordinators, Paleotectonic investigations of the Permian system in the United States: U.S. Geological Survey Professional Paper 515-G, p. 125–152.

—— , 1975, Montana, North Dakota, northeastern Wyoming, and northern South Dakota, in McKee, E. D., and Crosby, E. J., coordinators, Introduction and regional analyses of the Pennsylvanian system, Pt. 1; Paleotectonic investigations of the Pennsylvanian System in the United States: U.S. Geological Survey Professional Paper 853-O, p. 279-293.

McGookey, D. P., Haun, J. D., Hale, L. A., Goodell, H. G., McCubbin, D. G., Weimer, R. J., and Wulf, G. R., 1972, Cretaceous System, *in* Mallory, W. W., and others, eds., Geologic Atlas of the Rocky Mountain Region: Denver, Colorado, Rocky Mountain Association of Geologists, p. 190-228.

Merewether, E. A., and Claypool, G. E., 1980, Organic composition of some Upper Cretaceous shale, Powder River basin, Wyoming: American Association of Petroleum Geologists Bulletin, v. 64, no. 4, p. 488-500.

Merewether, E. A., Cobban, W. A., and Cavanaugh, E. T., 1979, Frontier Formation and equivalent rocks in eastern Wyoming: The Mountain Geologist, v. 6, no. 3, p. 67-102.

Mettler, D. E., 1968, West Moorcroft and Wood Dakota Fields, Crook County, Wyoming: Wyoming Geological Association 20th Annual Field Conference Guidebook, p. 89-94.

Mitchell, G. C., 1976, Grieve Oil Field, Wyoming; A Lower Cretaceous estuarine deposit: The Mountain Geologist, v. 13, no. 3, p. 71-88.

Momper, J. A., and Williams, J. A., 1979, Geochemical exploration in the Powder River basin: Oil and Gas Journal, v. 77, p. 129-134.

—— , 1984, Geochemical exploration in the Powder River basin, *in* Demaison, G., and Morris, R. J., eds., Petroleum Geochemistry and Basin Evaluation: American Association of Petroleum Geologists Memoir 35, p. 181-191.

Moore, W. R., 1983, The nature of the Minnelusa-Opeche contact in the Halverson Field area, Powder River Basin, Wyoming: The Mountain Geologist, v. 20, no. 3, p. 113-120.

Motes, A. G., III, 1984, A sedimentologic study of the middle and upper Minnelusa Formation, Crook County, Wyoming [M.S. thesis]: Rapid City, South Dakota School of Mines and Technology, 129 p.

Pipiringos, G. N., 1968, Correlation and nomenclature of some Triassic and Jurassic rocks in south-central Wyoming: U.S. Geological Survey Professional Paper 594-D, 26 p.

Rasmussen, D. L., and Bean, D. W., 1984, Dissolution of Permian salt and Mesozoic syndepositional trends, central Powder River Basin, Wyoming: Wyoming Geological Association 35th Annual Field Conference Guidebook, p. 281-294.

Rice, D. D., and Shurr, G. W., 1983, Patterns of sedimentation and paleogeography across the Western Interior seaway during the time of deposition of Upper Cretaceous Eagle Sandstone and equivalent rocks, northern Great Plains, *in* Reynolds, M. W., and Dolly, E. D., eds., Mesozoic paleogeography of west-central United States: Rocky Mountain Section of the Society of Economic Paleontologists and Mineralogists, Rocky Mountain Paleogeography Symposium 2, p. 337-358.

Runge, J. S., Wicker, W. L., and Eckelberg, D. J., 1973, A subsurface type section of the Teckla Sand Member of the Lewis Shale Formation: Earth Science Bulletin, v. 6, no. 3, p. 3-18.

Sabel, J. M., 1985, House Creek field; Past and future: Wyoming Geological Association 36th Annual Field Conference Guidebook, p. 45-50.

Schock, W. W., Maughan, E. K., and Wardlaw, B. R., 1981, Permian-Triassic boundary in southwestern Montana and western Wyoming: Montana Geological Society Field Conference and Symposium Guidebook to southwest Montana, p. 59-69.

Slack, P. B., 1981, Paleotectonics and hydrocarbon accumulation, Powder River Basin, Wyoming: American Association of Petroleum Geologists Bulletin, v. 65, no. 4, p. 730-743.

Spearing, D. R., 1976, Upper Cretaceous Shannon Sandstone; An offshore, shallow-marine sand body: Wyoming Geological Association 28th Annual Field Conference Guidebook, p. 65-72.

Stearns, D. W., 1978, Faulting and forced folding in the Rocky Mountain foreland, *in* Matthews, V., III, ed., Laramide folding associated with basement block faulting in the western United States: Geological Society of America Memoir 151, p. 1-36.

Stone, W. D., 1972, Stratigraphy and exploration of the Lower Cretaceous Muddy Formation, northern Powder River Basin, Wyoming and Montana: The Mountain Geologist, v. 9, no. 4, p. 355-378.

Strickland, J. W., 1958, Habitat of oil in the Powder River Basin: Wyoming Geological Association 13th Annual Field Conference Guidebook, p. 132-147.

Tillman, R. W., and Martinsen, R. S., 1984, The Shannon shelf-ridge sandstone complex, Salt Creek Anticline area, Powder River Basin, Wyoming, *in* Tillman, R. W., and Siemers, C. T., eds., Siliciclastic shelf sediments: Society of Economic Paleontologists and Mineralogists Special Publication 34, p. 85-142.

Tromp, P. L., Cardinal, D. F., and Steidtmann, J. R., 1981, Stratigraphy and depositional environments of the "Leo sands" in the Minnelusa Formation, Wyoming and South Dakota: Wyoming Geological Association 32nd Annual Field Conference Guidebook, p. 11-22.

Van West, F. P., 1972, Trapping mechanisms of Minnelusa oil accumulations, northeastern Powder River Basin, Wyoming: The Mountain Geologist, v. 9, no. 1, p. 3-20.

Waples, D. W., 1980, Time, temperature in petroleum formation: American Association of Petroleum Geologists Bulletin, v. 64, p. 916-926.

Waring, J., 1976, Regional environments of the Muddy Sandstone, southeastern Montana: Wyoming Geological Association 28th Annual Field Conference Guidebook, p. 83-96.

Weimer, R. J., 1961, Uppermost Cretaceous rocks in central and southern Wyoming and northwest Colorado: Wyoming Geological Association 16th Annual Field Conference Guidebook, p. 17-33.

Weimer, R. J., Emme, J. J., Farmer, C. L., Anna, L. O., Davis, T. L., and Kidney, R. L., 1982, Tectonic influence on sedimentation, Early Cretaceous, east flank Powder River Basin, Wyoming and South Dakota: Colorado School of Mines Quarterly, v. 77, no. 4, 61 p.

Winn, R. D., Stonecipher, S. A., and Bishop, M. G., 1983, Depositional environments and diagenesis of offshore sand ridges, Frontier Formation, Spearhead Ranch Field, Wyoming: The Mountain Geologist, v. 20, no. 2, p. 41-58.

MANUSCRIPT ACCEPTED BY THE SOCIETY SEPTEMBER 9, 1987

ACKNOWLEDGMENTS

F. F. Meissner, Bird Oil Corporation, kindly provided diagrams, which are the basis for Figures 9 and 10. T. Kostick, U.S. Geological Survey, drafted the figures.

Chapter 24

Geologic controls on hydrocarbon occurrence, Fossil basin area, Cordilleran thrust belt

M. A. Warner
13799B East Marina Drive, Aurora, Colorado 80014

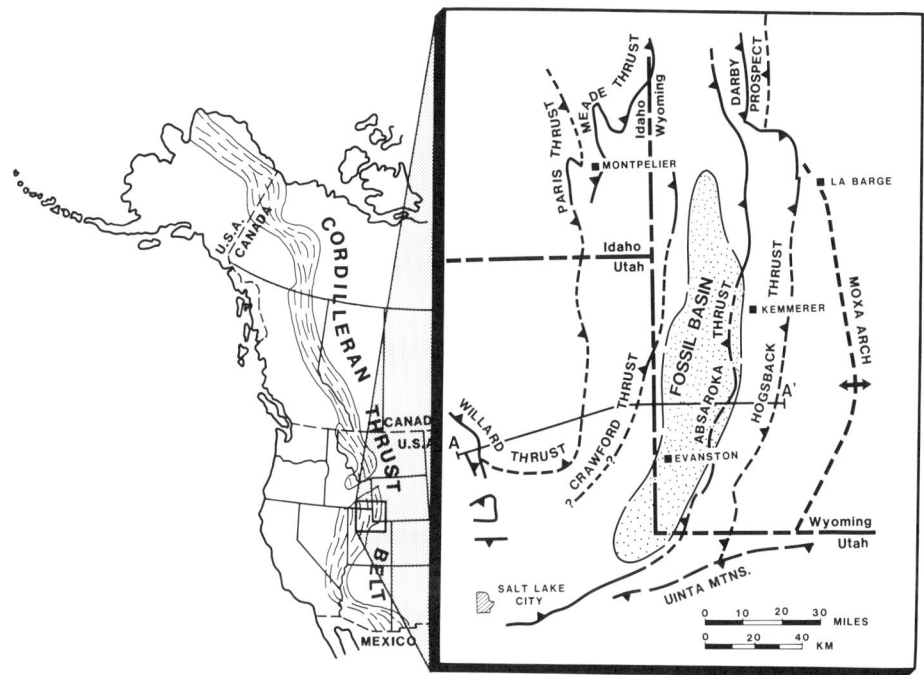

Figure 1. Index map of the Cordilleran thrust belt of western North America showing the location and major tectonic elements of the Fossil basin and nearby parts of the thrust belt. Structural cross section A-A′ is shown on Plate 7A.

INTRODUCTION

The Cordilleran thrust belt extends for 4,800 km or more along the foreland margin of the Cordilleran orogen and is one of the major structural features of western North America (Fig. 1). The belt is the structural expression of extensive horizontal shortening or telescoping of the supracrustal sedimentary wedge in response to generally easterly and northeasterly directed compressive forces. Throughout its length the thrust belt is characterized by low-angle thrust faults and elongate, thrust-associated concentric folds. The belt is segmented by a variety of transverse elements, of which some predate, and others postdate the main periods of thrusting. Although thrust faults are common to all parts, distinct differences in structural style are found in various segments of the belt. Such differences result primarily from changes in thickness and stratigraphic composition of the sedimentary wedge along the length of the deformed belt. Differences in structure combine with regional and subregional changes in source rock quality and maturation history to create a variety of environments for hydrocarbon generation and accumulation in the different segments of the Cordilleran thrust belt. An area that has been proven to have a particularly favorable environment for hydrocarbons is the part of the thrust belt that extends across the southwest corner of Wyoming into northern Utah. Commonly

Warner, M. A., 1991, Geologic controls on hydrocarbon occurrence, Fossil basin area, Cordilleran thrust belt, *in* Gluskoter, H. J., Rice, D. D., and Taylor, R. B., eds., Economic Geology, U.S.: Boulder, Colorado, Geological Society of America, The Geology of North America, v. P-2.

referred to as the Fossil basin, this area contains the largest and most significant oil and gas fields in the U.S. portion of the Cordilleran thrust belt and for that reason will be the main subject of the discussion that follows.

The Fossil basin (Fig. 1) is a rather loosely defined area that was a site of lacustrine and associated fluvial deposition in a depression formed on the eroded Absaroka thrust plate during early Tertiary time. The name is derived from the richly fossiliferous lacustrine beds of the Eocene Green River Formation. The sediments deposited in the Eocene lake have no genetic association with the hydrocarbons found in the area and serve only to add to the burial depth of the pre-Eocene strata that are the source, reservoir, and seal for the oil and gas accumulations.

GEOLOGIC HISTORY

The geologic history of the Wyoming-Idaho-Utah segment of the Cordilleran thrust belt involves three distinct phases: (1) a pre-orogenic depositional phase, (2) an orogenic phase of compressional deformation and synorogenic deposition, and (3) a post-orogenic phase of extensional faulting accompanied by fluvial and lacustrine deposition. During the pre-orogenic phase, from late Precambrian to late Jurassic, a westward-thickening wedge of mostly shallow water marine sediments was deposited in the Cordilleran miogeosyncline along the western border of the North American craton. To the west in northern Utah and southern Idaho, Proterozoic sediments are present in the miogeosynclinal wedge, which had a total thickness in excess of 18 km. The thickness decreased eastward to about 3 km near the east margin of the thrust belt in western Wyoming (Armstrong and Oriel, 1965; Royse and others, 1975). The stratigraphic column of the wedge is dominated by limestone, dolomite, and quartzose sandstone or quartzite, which occur in thick units separated by much thinner shaley units (Fig. 2). Although western source areas created by the Antler and Sonoma orogenies shed large volumes of debris eastward in Late Devonian–Early Mississippian and Permo-Triassic times, most of the clastic material deposited in eastern Idaho and western Wyoming was derived from the eastern craton (Churkin, 1962; Armstrong and Oriel, 1965). Deposition along the cratonic margin was interrupted several times during the pre-orogenic phase by regional crustal warping and eustatic sea level changes, with the result that the stratigraphic succession is punctuated by a number of regional and subregional unconformities.

Deposition of miogeosynclinal sediments along the cratonic margin was halted in late Jurassic time by the onset of the Sevier orogeny (Rubey, 1955; Armstrong, 1968). The sedimentary wedge was disrupted and subjected to large-scale eastward translation on a series of low-angle faults. Throughout most of the region, deformation is of the epidermal or thin-skinned type, in which the deformed sedimentary wedge is structurally detached from the relatively undisturbed crystalline basement. The Wasatch Mountain uplift in northern Utah shown on the western end

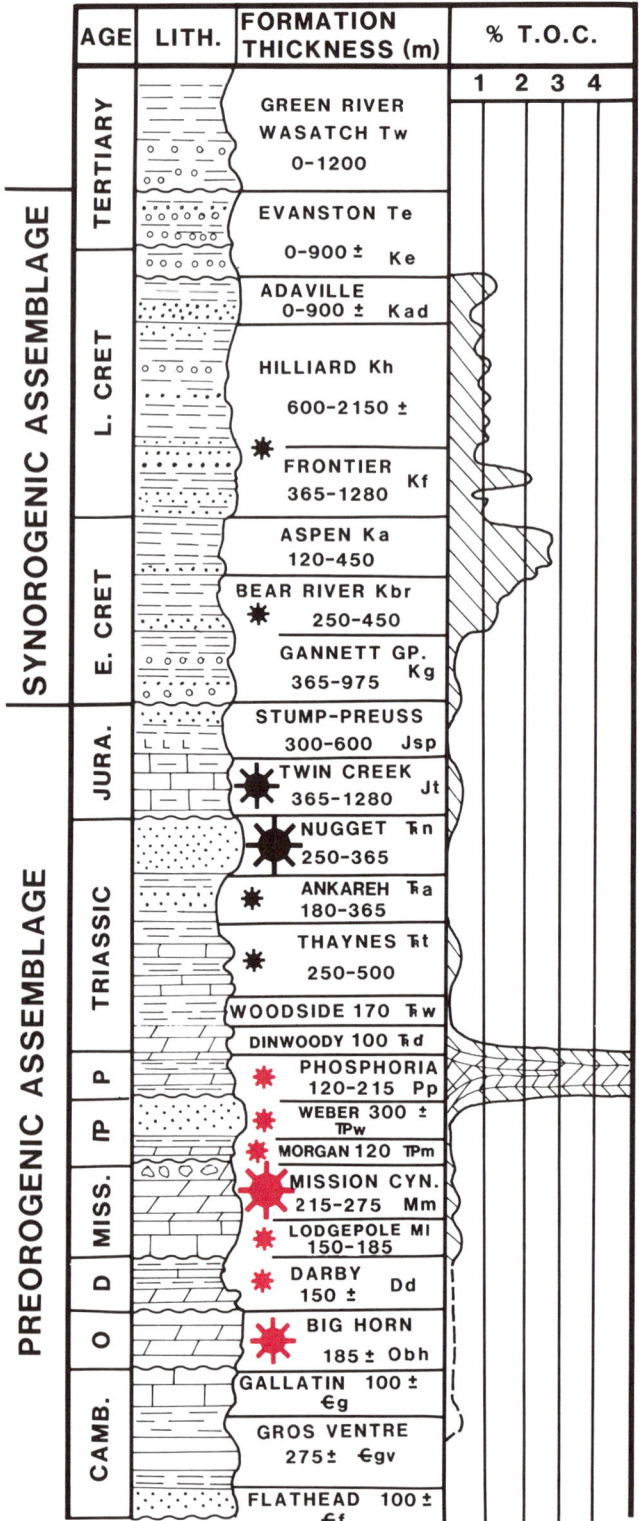

Figure 2. Generalized stratigraphic column of the Fossil basin area with gross lithologies and range of thickness of formations. Right side of figure shows generalized curve of total organic carbon (TOC) content as determined by analysis of representative samples of the formations. Productive intervals are denoted by well symbols that are red where production is primarily wet gas and condensate and black where production is primarily oil with associated gas.

of the cross section in Plate 7A is one of the few areas where crystalline basement is involved in thrusting.

Thrust faulting was episodic during a period of about 90 m.y. between late Jurassic and early Eocene times. Although a variety of names has been applied to individual fault traces, regional studies allow the thrust faults of the Wyoming-Idaho-Utah segment of the thrust belt to be grouped chronologically into four major fault systems: (1) the Paris-Willard system, (2) the Meade-Crawford system, (3) the Absaroka system, and (4) the Darby-Prospect-Hogsback system. The westernmost system (Paris-Willard) was formed first, and the major faults become progressively younger to the east.

The original width of the terrain involved in the thrust belt appears to undergo a shortening of approximately 50 percent as a result of thrusting. Shortening of this magnitude has been calculated by Bally and others (1966) and Price and Mountjoy (1970) for the Canadian thrust belt and by Rubey and Hubbert (1959) and Royse and others (1975) for parts of the Wyoming-Idaho-Utah thrust belt. Total shortening is about 97 km at the latitude of Whitney Canyon near the north-south midpoint of the Fossil basin. Shortening amounts to about 80 km near the southern end of the Fossil basin and increases to 105 km at the north end of the basin. Although the shortening results from movements that are primarily horizontal, a significant vertical component of motion has been involved. As the thrust plates moved eastward relative to the foreland, they rode upward over ramps in thrust planes to create uplifts that were subjected to immediate erosion and denudation.

Detritus derived from the rising thrust plates was deposited in a subsiding trough or basin along the foreland margin of the zone of thrusting. Sediments deposited in the marginal trough reached a maximum thickness of about 6.1 km (Armstrong and Oriel, 1965; Royse and others, 1975). The axis of the depositional trough migrated eastward during the long period of deformation in company with the eastward progression of faulting. Successively younger thrust faults cut and deformed foreland basin sediments, which were derived from uplifts created by preceding thrusts as the thrust-basin couplet migrated eastward during Cretaceous and early Tertiary time. Paleontological dating of preserved syntectonic deposits provides a sound basis for a refined dating of the several periods of thrusting. The Wyoming-Idaho-Utah segment of the thrust belt is somewhat unique in having numerous datable deposits that are derived from, are cut by, or overlap major thrusts. As a result, the history of thrusting in this area probably is the most precisely dated of any thrust system in the world (Armstrong and Oriel, 1965; Royse and others, 1975; Wiltschko and Dorr, 1983).

The stratigraphic assemblage of the synorogenic foreland basin is compositionally as well as genetically distinct from the assemblage of the pre-Jurassic miogeosynclinal wedge. The sedimentary section of the foreland basin is dominated by mudstone and shale with interbedded immature sands and synorogenic conglomerates (Fig. 2). Whereas the miogeosynclinal wedge is made up almost entirely of marine sediments, a significant part of the deposition in the foreland basin occurred under nonmarine conditions. Although much of the material deposited in the foreland basin came from reworking sediments of the older miogeoclinal wedge, an active magmatic arc that formed west of the foreland thrust belt (Burchfiel and Davis, 1975) was also a source of sediment during deposition of the Aspen, Frontier, and Hilliard formations.

A period of crustal extension started in late Eocene time following the main orogenic phase and has continued to the present. In the thrust belt, the extensional deformation occurs primarily in the form of west dipping listric normal faults that generally form half-grabens. The half-graben basins are filled with alluvial deposits derived from nearby source areas. Displacement on the larger of the normal faults exceeds 3 km, and in places such as the Wasatch Front near Salt Lake City, the displacement may be more than 6 km (Plate 7A). Normal faults commonly developed above ramps in previously formed thrust fault plates, and they sole in the plane of the older thrust fault. In many cases the older strata dip into the normal faults as do the alluvial sediments deposited in the half-grabens, so that rollover anticlinal structures were formed at the hinge along the west side of the basins.

EXPLORATION HISTORY

The presence of hydrocarbons in the form of seeps at various places along the Cordilleran thrust belt was known to explorers as early as the mid-1800s. Petroleum exploration activity during the late 1800s and early 1900s was largely confined to drilling shallow wells in the vicinity of the known seeps. The first oil field in Montana resulted from the drilling of several shallow wells along Swiftcurrent Creek in 1901 in an area that is now part of Glacier National Park (Darrow, 1955). Small amounts of oil were produced from folded and faulted Cretaceous strata below the surface trace of the Lewis thrust. At about the same time an oil boom of sorts developed in southwestern Wyoming. Several oil camps were built, and about 180 shallow wells were drilled near known seeps along the toe of the Absaroka thrust (Veatch, 1907). This activity resulted in the discovery of several small oil pools at depths of less than 450 m. Discovery of the first significant oil field in the Cordilleran fold and thrust belt occurred in 1918 when production was established in the LaBarge field in the central part of the Wyoming segment of the belt. Oil and gas were found at depths of 180 to 300 m in deformed Tertiary strata that lie structurally below and adjacent to the exposed Darby (Hogsback) thrust plate. Although early records are spotty, the LaBarge field is credited with a cumulative production of more than 25 million barrels (mbbl) of oil and 26 billion cubic feet (bcf) of gas through mid-1985 and was still producing approximately 1,000 bbl of oil per day at that time.

With the popularization of the anticlinal theory of hydrocarbon accumulation and the increased application of geology in the search for oil during the early decades of the twentieth century, emphasis shifted to testing the many anticlinal folds known from surface mapping of various parts of the thrust belt. The results of this activity were disappointing because the surface maps did not offer a valid picture of subsurface structure in this structural environment where the folds were highly asymmetrical and/or detached at relatively shallow depths. Some success was achieved in the LaBarge area, but the success ratio in drilling surface anticlines in the thrust belt was quite low compared with that achieved by testing anticlines in other areas.

Early efforts at seismic mapping of subsurface structure also yielded disappointing results in the U.S. part of the Cordilleran thrust belt, although considerable success was achieved in the Canadian thrust belt. Complex horizontal as well as vertical variations in velocity, steep structural dips, and rugged topography combined to make the acquisition of reliable seismic data a formidable task during the early stages of development of the seismic tool. A number of unsuccessful wells were drilled to test seismically mapped structural features that proved to be velocity anomalies rather than buried folds. By the 1970s, seismic technology had advanced to the point where reliable mapping could be done under the adverse conditions found in the Wyoming-Idaho-Utah segment of the thrust belt. Technological developments such as common depth point (CDP) digital recording and sophisticated computer processing of digitally recorded data provided the tools for making sufficiently accurate subsurface structural maps.

Concurrent with the advances in seismic technology, geological studies in the Canadian and Appalachian thrust belts provided information necessary for the formulation of a set of empirical rules of thrust belt structural behavior (Rich, 1934; Bally and others, 1966; Dahlstrom, 1970; Gretener, 1972). Structural models based on these rules provided a means of projecting surface data into an interpretation of subsurface structural form and supplied the foundation for making reliable interpretations of seismic data. As summarized by Royse and others (1975), the rules state that: (1) Folding is of the concentric type in which plastic flow and cleavage folding do not occur to any significant degree; (2) thrust faults cut up section in the direction of tectonic transport; (3) thrust faults are commonly parallel to bedding in incompetent strata and cut obliquely across bedding in competent strata; and (4) within a given belt of deformation, *major* thrusts are successively younger in the direction of tectonic transport.

By the mid-1970s the geological and geophysical tools necessary for a successful exploration program in the highly complex structural setting of the Fossil basin were at hand, and the needed economic stimuli were supplied by the anticipated shortages and rapidly escalating prices of oil and gas. The latest phase of exploratory activity was ushered in by the discovery of a major oil field at Pineview, Utah, by American Quasar Petroleum in early 1975. An intensive industry exploratory effort followed this discovery and led to the discovery of 23 additional fields by 1982. Of these, five are classed as giant fields (fields having recoverable reserves of at least 100 mbbl of oil or oil-equivalent gas). Hydrocarbons recoverable from the 24 fields by conventional means are estimated to total 650 to 700 mbbl of oil and 7.5 to 8.5 trillion cubic feet (tcf) of gas.

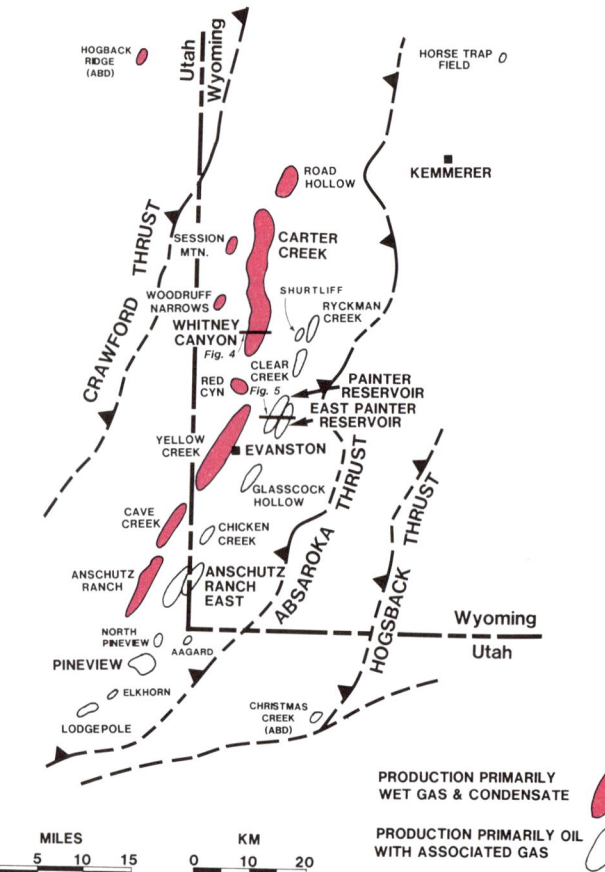

Figure 3. Location map of oil and gas fields in the Fossil basin area of southwestern Wyoming and northern Utah. The Nugget sandstone (Triassic) is the primary reservoir in the eastern line of fields. Bulk of gas reserves in western line of fields are in Paleozoic age reservoirs.

Activity in the Fossil basin as well as in other parts of the thrust belt declined markedly from 1982 to 1986, and there were no significant discoveries during that period. The decline resulted not only from the general overall reduction in industry activity but also from the fact that the prime prospects identified in the Fossil basin producing trends had been tested. Various types of structural leads in parts of the thrust belt outside the Fossil basin have been drilled in the post-1975 period with marginal results. Although all significant discoveries of the post-Pineview period of exploration are in the southern part of the Fossil basin, the presence of oil and gas shows in other areas indicates that opportunities for finding economically attractive reserves still exist in the Cordilleran thrust belt province.

HYDROCARBON ACCUMULATIONS IN THE FOSSIL BASIN

All of the significant oil and gas accumulations in the Fossil basin area shown in Figure 3 are trapped in anticlinal culminations along two subparallel lines of folding within the Absaroka thrust plate. The folds were formed at the time the thrust sheet was emplaced and in response to the same compressive forces responsible for the thrusting. The now abandoned Hogback Ridge gas field is the only field discovered to date (mid-1986) on the Crawford thrust plate, and only two minor fields (Horse Trap and Christmas Creek) have been found on the Hogsback plate.

The eastern line of folds extends 80 km in a general northeasterly direction from the Lodgepole field in northern Utah to the Ryckman Creek field in southwestern Wyoming (Fig. 3). Hydrocarbons are trapped in a number of anticlinal culminations aligned along the major fold trend. Most of the anticlines are asymmetrical or overturned to the east, and all are broken by numerous faults. Low sulfur, high gravity oil and gas are produced from a number of reservoir rocks of Mesozoic age, but the Nugget sandstone of Triassic age (Fig. 2) is by far the most important and contains more than 90 percent of the total reserves found along the eastern line of fields. Other productive zones are found in the Triassic Ankareh and Thaynes formations, which lie below the Nugget sandstone, in the Jurassic Twin Creek Formation, and in the Cretaceous Kelvin Formation. As shown in Plate 7A, the eastern line of folds lies east of the hanging wall cutoff of Paleozoic strata in the Absaroka thrust plate, so no Paleozoic rocks are present in these folds. Four of the fields along the eastern trend are considered to be giants. These are Pineview, Anschutz Ranch East, Painter Reservoir, and East Painter Reservoir (Lamerson, 1982).

Paleozoic rocks are involved in the folding and provide the main reservoirs for the hydrocarbon accumulations along the western anticlinal trend, which extends 61 km northeast from Anschutz Ranch field to Road Hollow field (Fig. 3). The hydrocarbons are trapped in a series of culminations along a large ramp anticline associated with the hanging wall cutoff of the Paleozoic section in the Absaroka thrust plate. Structure in Paleozoic rocks in the core of this large concentric fold is quite complex as a result of the imbricate faulting of less competent units as the fold tightens downward. As pointed out by Lamerson (1982), the structural relief of the northern part of the western anticlinal trend results not only from the initial folding over the fault ramp but also from the subsequent development of a "pillow" or duplex structure created by imbrication of the Cretaceous strata beneath the main Absaroka plate. Both features are illustrated in the cross section through the Whitney Canyon area in Figure 4 and in section A-A', Plate 7A. The Whitney Canyon/Carter Creek Field is the largest field along the western trend and has recoverable reserves well in excess of the 100 million barrels of oil or oil-equivalent gas required for classification as a giant field.

Productive capacity has been demonstrated in 10 different formations in the fields along the western line of folding. The reservoir rocks range from Jurassic to Ordovician in age and are found over a vertical interval of more than 1,800 m. The Paleozoic reservoirs, which contain the bulk of the hydrocarbons, produce sour gas and condensate; the relatively minor production from reservoirs in Triassic and Jurassic age strata is sweet gas and condensate.

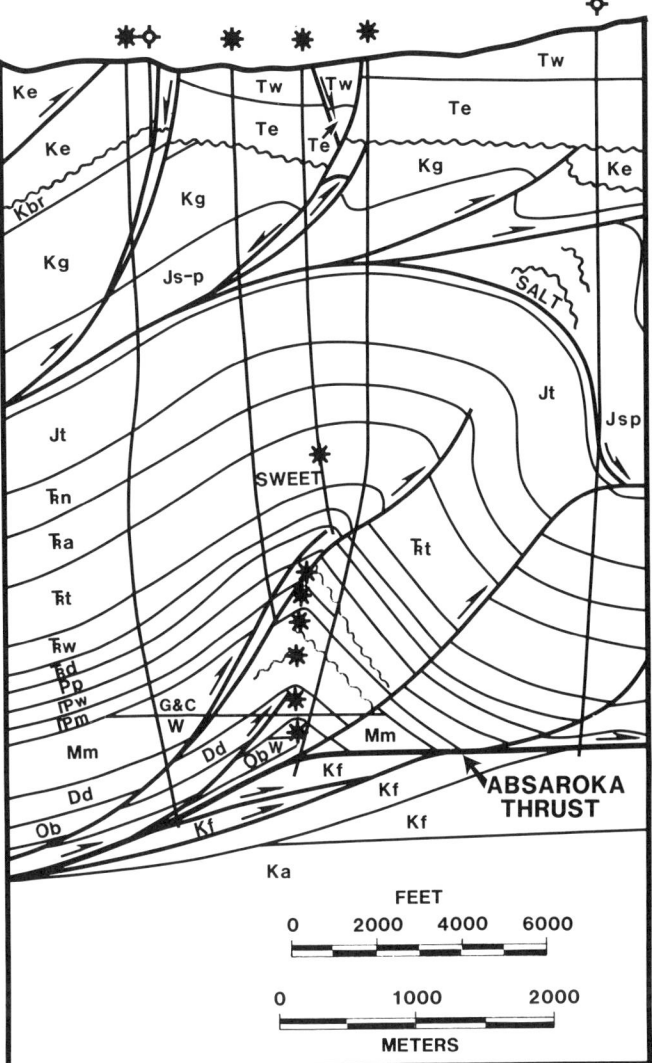

Figure 4. East-west structural cross section through Whitney Canyon field (modified from Lamerson, 1982). Formations involved are listed in Figure 2. Line of section is shown on Figure 3. The letters G, C, and W indicate gas, condensate, and water respectively.

GEOLOGICAL CONTROLS OF HYDROCARBON ACCUMULATION

The giant hydrocarbon accumulations of the Fossil basin result from the favorable interaction of a number of geological factors. Although the commonly considered elements of source, reservoir, and seal are as important in the Fossil basin as elsewhere, the time factor is unusually critical because of the complex tectonic and depositional history of the area. Creation of the anticlinal folds that serve to trap the hydrocarbons and the generation of these hydrocarbons are specific geologic events, which in this case are not genetically related. Hydrocarbons generated before thrusting were either lost for lack of suitable structural traps or were destroyed by later tectonic disruption of whatever traps might have existed. In some parts of the thrust belt, excellent structural traps are barren because the folds were formed after generation and expulsion of hydrocarbons from rocks that are in a position to serve as the source. The time of hydrocarbon generation in a specific source rock unit depends not only on the depth and rate of sedimentary or tectonic burial but also on the paleogeothermal gradient. Because the potential source rock units in the sedimentary column of the Fossil basin area are limited in number, the geologic history of each unit is important in evaluating the potential of the area.

SOURCE ROCKS

Analysis of a large number of samples from the various formations making up the lithologic column of the Fossil basin indicates that the shale members of the Permian Phosphoria Formation and the several thick shale units in the Cretaceous section are the only significant source rocks in this part of the thrust belt. Other stratigraphic units do not contain enough organic matter to be considered likely sources of hydrocarbons, as demonstrated by the representative values for total organic carbon (TOC) content plotted next to the lithologic column in Figure 2. Except for the shales of the Phosphoria Formation, which were deposited under reducing conditions (Maughan, 1975; Tisoncik, 1984), the pre-Cretaceous sediments of the miogeosynclinal wedge were deposited in a more or less open shelf environment where oxidizing conditions were not conducive to the preservation of organic matter. Although organic-rich sediments were deposited under more basinal conditions that existed farther west in Utah and Idaho during several other Paleozoic periods (Sandberg, 1975; Sandberg and Gutschick, 1984), equivalent strata deposited on the shelf area of western Wyoming contain little organic carbon.

The Meade Peak and Retort Phosphatic Shale members of the Phosphoria Formation are geographically widespread and exceptionally rich in organic carbon in the Wyoming-Idaho-Utah segment of the thrust belt. The richest beds contain as much as 30 wt. percent total organic carbon (TOC), and the average TOC of each member is as high as 10 wt. percent in the richest sections (Maughan, 1984). Claypool and others (1978) estimated that the shale members of the Phosphoria Formation in Wyoming, Idaho, and Utah contained 132×10^9 metric tons of organic carbon. This estimate was made without allowance for the duplication of section in the thrust belt where the thickest and organically richest shales are found, so the true value for the amount of organic carbon present in the Phosphoria Formation before thrusting may have been twice as large.

A tremendous volume of dark shale containing sufficient organic carbon to be considered an adequate source of hydrocarbons was deposited in the thrust belt during Cretaceous time. Where the full section is preserved, the thickness of the dark shales exceeds 1.8 km. Organic shales are present throughout the Cretaceous section, with the richest being found in the Bear River and Aspen formations of Early Cretaceous age (Fig. 2). Marine shales in the Late Cretaceous Frontier and Hilliard formations also contain enough organic carbon to be considered potential source rocks. The Phosphoria shales are considerably richer in organic carbon than the shales of the Cretaceous section, where individual samples rarely contain more than 4 wt. percent TOC, and shale units have an average TOC content between 1 and 2.5 wt. percent. Humic material, which is generally not found in the Phosphoria, is common in the organic matter of the Cretaceous shales, so that the Cretaceous rocks are somewhat more gas prone than the Phosphoria shales.

Although both the Permian (Phosphoria) and Cretaceous shales contain sufficient organic matter to have supplied the hydrocarbons trapped in the Fossil basin area, two different types of evidence indicate that the oil and presumably the associated gas were derived from Cretaceous source rocks. First and most important, studies by Rosenfeld and others (1980), Seifert and Moldowan (1981), and Warner (1982) yield conclusive geochemical evidence that all liquid hydrocarbons studied in the Fossil basin fields were generated in and migrated from the Cretaceous source rocks—not from the Phosphoria. Second, as will be discussed in more detail in a following section, the period of peak oil generation in the Phosphoria took place before the Absaroka thrust and associated fold traps were formed, so the Phosphoria Formation is not a likely contributor to the accumulations found in these traps.

Rosenfeld and others (1980) used mass spectral, gas chromatographic, and carbon isotope analyses to make oil-to-oil and oil-to-source comparisons at Pineview field. They concluded that oil from the Twin Creek and Nugget reservoirs at Pineview was from the same source and that the source was in the Cretaceous section rather than the Permian Phosphoria. Seifert and Moldowan (1981) established biomarker fingerprints for a number of oils and shale extracts by separating individual biomarker sterane compounds using computer-assisted gas chromatograph–mass spectrometry techniques. They concluded that (1) the Cretaceous and Phosphoria shales had distinctly different biomarker fingerprints, as did the oils derived from each, and (2) oil from the Nugget reservoir at Pineview field had the biomarker fingerprint of the Cretaceous source rocks. Warner (1982) identified characteristic differences in peak height ratios of gasoline range hydrocarbons in oils linked to either Cretaceous or Permian (Phosphoria) source rocks by the biomarker fingerprinting techniques of Seifert and Moldowan (1981) or by geological constraints. The chromatographic signagures of the known oils were compared with the signatures of oils and condensates from nine of the fields in the Fossil basin. Chromatograms of all of the 16 Fossil basin samples taken from 10 different formations ranging from late Jurassic to Cambrian in age yielded signatures typical of the Cretaceous sourced oils. None of the oils or condensates from the Fossil basin resembled the Phosphoria sourced oils used in the study. Furthermore, no other geochemical evidence has been presented to indicate that any of the liquid hydrocarbons in the Fossil basin came from a source in the Phosphoria Formation.

The geometric relationships between the reservoir rocks and the possible source rocks would make a strong argument for considering the footwall Cretaceous shales to be the source of the reservoired hydrocarbons even if no geochemical data were available. For an east-west distance of about 24 km beneath the Fossil basin, the Absaroka thrust rides on shales of the Bear River, Aspen, and Frontier formations, which are cut at a very low angle (Plate 7A). Reservoir rocks in the thrust plate are in direct contact with the organic-rich subthrust Cretaceous shales across much of this distance. Hydrocarbons generated in the Cretaceous shales moved along permeable strata of the Absaroka thrust hanging wall, as well as along numerous faults associated with the Absaroka thrust that provide migration pathways to the anticlinal traps.

The hydrocarbon accumulations along the western line of folds have a much higher gas content than those along the eastern line of folds. This difference in composition of the hydrocarbons is accounted for by differences in organic facies and differences in burial depth of the Cretaceous source rocks beneath the two lines of folding. The Bear River, Aspen, and Frontier formations were deposited under marginal marine conditions and become more nonmarine in a westward direction. The preserved organic material in these units is only moderately oil prone at best and becomes more gas prone to the west as the proportion of humic material increases. The observed change in organic facies from east to west may be the primary reason for the preponderance of gas along the western trend, although the maturation level of the source rocks probably has some influence on the composition of the trapped hydrocarbons. The Cretaceous units beneath the Absaroka thrust were buried somewhat deeper before, during, and after thrusting along the western line of folds than were equivalent strata along the eastern trend. As a result of the somewhat higher temperatures, the maturation level of the organic matter has advanced to the wet gas–condensate stage at most places along the western trend, but only to the oil stage along the eastern trend.

The fact that the wet gas and condensate found in the Paleozoic reservoirs along the western line of fields in the Fossil basin is sour, whereas that in the eastern trend is not, was taken as evidence of a Phosphoria source for the western fields by some workers because Phosphoria sourced oils are typically sour throughout Wyoming. However, chromatographic evidence indicates that the *liquid* hydrocarbons in the various Paleozoic reservoirs are derived from Cretaceous source rocks. Because the source of the *gaseous* hydrocarbons cannot be identified, it has been argued that the hydrogen sulfide (H_2S) and other sulfur compounds were introduced into the Paleozoic reservoirs along with the gas that came from a downdip source in the Phosphoria. Although such an argument cannot be disproven by geochemical data, the idea of a dual source for each of the many Paleozoic reservoirs seems unnecessary. Sulfur-bearing minerals such as anhydrite, which are abundant in some of the Paleozoic units, are a more probable source of the sulfur associated with the hydrocarbons. The large amounts of hydrogen sulfide and sulfur-bearing organic compounds found here are formed by thermocatalytic reactions between mineral phases of the reservoir rocks and trapped hydrocarbons at the elevated temperatures reached during thermal maturation of the oils (Orr, 1974; Ho and others, 1974). Reduction of the mineral sulfate is part of a complex series of reactions that involve dehydrogenation of hydrocarbons and formation of hydrogen sulfide. A relationship between the hydrogen sulfide and the sulfur-bearing mineral phase is demonstrated by the high hydrogen sulfide content of the gas from those reservoir rocks containing the greatest amount of anhydrite. For example, anhydrite is abundant in parts of the Mississippian and Pennsylvanian sections in the Whitney Canyon/Carter Creek area, and gas from reservoirs in these units generally contains 15 to 21 mol percent hydrogen sulfide. Only traces of anhydrite have been found in the Triassic and Ordovician strata. No hydrogen sulfide is present in gas produced from the Triassic at Whitney Canyon, and gas from the Ordovician at Carter Creek contains less than 1 mol percent hydrogen sulfide.

RESERVOIR ROCKS

The bulk of the hydrocarbon reserves in the Fossil basin area are trapped in reservoir rocks in the Nugget Sandstone (Triassic) and the Mission Canyon Formation (Mississippian). Although production has either been established or is indicated in 12 other formations ranging in age from late Cretaceous to Ordovician, the combined ultimate production from these units in the fields discovered to date will probably amount to less than 15 percent of the total production. Both the Nugget and Mission Canyon formations have thick zones of significant porosity and permeability. Rates of production may be enhanced by fracturing, especially in the Mission Canyon, but fracturing is not essential for production. In the other productive units, porosity and permeability are generally poor or confined to thin zones. Production from these units is frequently enhanced by extensive natural fracturing.

Nugget

The Nugget Sandstone is a very fine- to medium-grained subarkose or quartz arenite of eolian origin. The unit is made up of a multitude of cross-bedded dune sets interspersed with laminated, horizontally bedded interdune deposits. Thickness of the individual dune sets averages about 9 m and ranges from less than 0.3 m to more than 18 m. Bimodal grain size distribution is common in the dune deposits, and well-rounded, frosted quartz grains are abundant. In the Fossil basin area the Nugget ranges from about 245 m to 365 m in thickness. The mineralogy of the Nugget sand is relatively simple; monocrystalline quartz is the most abundant detrital constituent. Feldspar, mostly orthoclase, makes up 5 to 10 percent of the average sample, and heavy minerals and rock fragments are present in trace amounts. Authigenic minerals are quartz, as druzy grain coatings and quartz overgrowths; carbonate in the form of poikilotopic calcite and rhombic dolomite; illite and mixed-layer illite-smectite clay; chlorite; feldspar; and iron oxide (hematite?).

Porosity measured on cores of the Nugget sand ranges from 2 to 23 percent. The average porosity is 12. 5 percent in the Painter Reservoir/Clear Creek area (Frank and others, 1982) and is 10 percent or less in the Anschutz Ranch East field (Bergosh and others, 1982). Throughout the producing area the measured porosity is largely preserved, primary intergranular porosity. Both fracture porosity and porosity resulting from dissolution of detrital grains or cements have been reported (Bergosh and others, 1982), but neither makes a significant contribution to the overall porosity. Reduction of porosity from the original values at the time of deposition to the present lower values results primarily from the reorganization of quartz through solution and autocementation. Authigenic cements other than quartz are not present in sufficient volume to have caused the large loss of porosity.

Permeability measurements in the Nugget sandstone show a great range of values from less than 0.1 to more than 2,000 md (Frank and others, 1982). The very large values are rare and may reflect the presence of unobserved fractures in the whole core used for analysis. Permeability is strongly influenced by the grain size, sorting, and composition of detrital grains in the individual beds making up cross-bed sets. The original permeability was relatively lower in the finer-grained and more poorly sorted units because of the small size of pore throats and greater tortuosity of the pore system. Feldspars, which are presumed to be the source for cations included in the authigenic clays and carbonates, are concentrated in the finer grained layers. As a result, the concentration of authigenic clay and carbonate is higher in the finer layers, and the effects on permeability are relatively great because of the small primary pore throats.

The makeup and geometry of the cross-bed sets cause the fluid transmissibility of the Nugget reservoirs to have a pronounced directional character that may not be evident from published permeability data. Permeability measured in a direction parallel with the plane of bedding and reported as horizontal permeability may be more than 100 times greater than the permeability measured in a direction transverse to bedding in the thinly bedded or laminated parts of the dune sets and reported as vertical permeability. Most recently published values are for horizontal permeability measurements, which represent the maximum permeability for that sample. Where horizontal permeability is measured in inclined, thinly bedded strata, the values obtained apply only in the direction of strike of the cross beds. Permeability in any other horizontal direction is lower. The better reservoir rocks are found in the thicker, more uniform beds where horizontal and vertical permeability are approximately equal and average 10 to 30 md.

Fluid transmissibility is also influenced by fracturing of the Nugget reservoir rock. Open fractures that enhance permeability are present in the Nugget but are not abundant. Much more common are fractures characterized by zones of pulverized quartz or gouge. The gouge zones range from less than 1 mm to 1 m or more in thickness and are most commonly a few millimeters thick. The gouge material has very low porosity and permeability, so that gouge zones act as permeability barriers; this characteristic may serve to compartmentalize the Nugget reservoir in some cases.

Mission Canyon

The Mission Canyon Formation is a dominantly shallow-water carbonate and evaporite unit deposited along the shelf margin of the Cordilleran miogeosyncline during a Mississippian (Osasgian-Meramecian) regression (Rose, 1976). In the Whitney Canyon/Carter Creek area the Mission Canyon is about 255 m thick. Hoffman and Balcells-Baldwin (1982) subdivided the formation into three lithofacies: (1) an upper unit 106± m thick composed of dolomite, limey dolomite, and nodular anhydrite deposited under evaporitic conditions in a restricted shelf or sabkha environment; (2) a middle unit approximately 112 m thick consisting of dolomitized grainstone deposited in a higher energy environment along the shelf margin; and (3) a lower unit 37± thick of shaley, micritic limestone and dolomitic limestone deposited in a deeper water environment along the lower shelf. The upper surface of the Mission Canyon Formation is a widespread unconformity along which extensive solution breccias have developed.

The principal reservoirs of the Mission Canyon Formation are in the dolomitized grainstone unit where several discrete zones of intercrystalline and vuggy porosity are developed (Hoffman and Balcells-Baldwin, 1982). Matrix porosity in these zones ranges from 2 to 20 percent as measured by porosity logs. The Mission Canyon Formation has not been extensively cored, so direct permeability measurements are not available for most of the productive area, but production tests show that matrix permeability is sufficient for commercial rates of flow of gas and condensate. Thickness of effective pay (defined as rock with greater than 2 percent porosity) in the Mission Canyon has a wide range in the Fossil basin area. In the giant Whitney Canyon/Carter Creek producing complex, the Mission Canyon pay section has an average thickness of 102 m and an average porosity of 6.6 percent (Hoffman and Balcells-Baldwin, 1982).

TRAPS

The anticlinal structures of the Fossil basin are complexly folded and faulted. The cross sections through Whitney Canyon and Painter Reservoir fields shown in Figures 4 and 5 are more or less typical and give an idea of the complexities involved. Faulting plays an important role in trap definition. In the more competent or brittle strata, faults may act as conduits, and the hydrocarbon column may be restricted by the position of a fault intersection with the reservoir unit below fold closure. Fracture systems formed in conjunction with folding and faulting not only act as pathways for oil migration from source rocks into hanging wall reservoirs but also serve to form "leaky" traps at places such as Pineview field where the seal is disrupted by faulting.

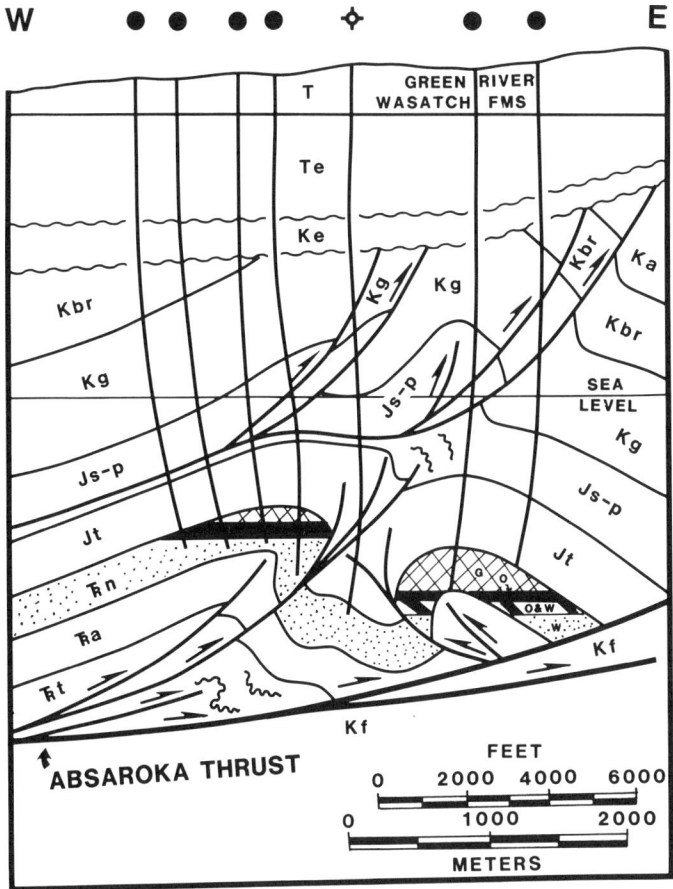

Figure 5. East-west structural cross section through Painter Reservoir and East Painter Reservoir fields (modified from Lamerson, 1982). Formations involved are listed on Figure 2. Line of section is shown on Figure 3.

Anhydrite deposited in the Jurassic Twin Creek Formation provides an effective barrier to fluid migration and acts as the cap rock for several of the large accumulations in the Nugget Sandstone, while thick shales in the Triassic section act as a seal for Paleozoic reservoirs. Salt in the Jurassic Preuss Formation serves as a regional glide plane and a zone of structural detachment. Structure above the detachment surface is complicated by salt flowage and extensive imbrication of overlying strata, as shown in Figures 4 and 5. In many places, deeper structure is effectively masked by the structural discontinuity at the level of the Preuss salts.

TIME OF HYDROCARBON GENERATION

The several periods of thrust faulting that took place between latest Jurassic and early Tertiary time had a profound influence on the process of catagenesis of organic matter in source rocks of the thrust belt. Cooling of rocks brought to or near the surface by uplift and erosion of a thrust plate brought the process of catagenesis of organic matter in these rocks to a halt. Continued heating of equivalent rocks that remained buried beneath the thrust plate or beneath clastic deposits derived from the thrust plate allowed catagenesis to continue. Measurements of the level of thermal maturation of the source rocks involved in the several closely dated periods of thrusting can be used in conjunction with stratigraphic data to construct a reasonably accurate history of hydrocarbon generation in this area.

Geochemical data gathered from analysis of a large number of outcrop and shallow subsurface samples of Phosphoria shales from the Fossil basin and adjacent parts of the thrust belt show an advanced stage of thermal maturity for the Phosphoria throughout this area. Samples from the Crawford-Meade, Absaroka, and Hogsback thrust plates analyzed in laboratories of Chevron U.S.A., Inc. show a level of maturation well into the dry gas stage. Comparable maturation levels were found in Phosphoria samples from thrust belt localities north and west of the Fossil basin reported by Claypool and others (1978); kerogen color, atomic hydrogen-carbon (H/C) ratio of the kerogen, and ratios of hydrocarbon compounds to total organic carbon were used to determine the maturation level. In the samples analyzed by Chevron, the H/C ratios were used in conjunction with a kerogen color scale calibrated to H/C ratios and Rock Eval pyrolysis data.

A reconstruction of the burial (thermal) history shows that the high level of thermal maturity observed in these Phosphoria samples was reached during deep burial before the rocks became involved in thrust faulting. The process of catagenesis was arrested by removal of overburden during thrust uplift and erosion so that the level of maturation is about the same now as it was at the time the thrust plates were formed. The geochemical data together with information on the burial history demonstrate that the hydrocarbon generative capacity of the Phosphoria shales was virtually depleted before the Absaroka thrust was formed. The difference in timing between hydrocarbon generation and trap formation during thrusting makes the Phosphoria Formation an unlikely contributor of significant amounts of hydrocarbons, especially oil, to the Fossil basin accumulations.

The rather peculiar nature of the organic matter in the Phosphoria shales causes problems in measuring the true degree of catagenesis. Vitrinite and spores or pollen are the types of particulate matter most frequently used in measurements of the level of the thermal maturation of kerogen. Neither of these materials is common in the rather structureless sapropelic material of the Phosphoria, so other analytical methods must be employed. Solidified bitumen (pyrobitumen?) is found in many samples of the Phosphoria and has been used by some investigators in measuring reflectance values (Edman and Surdam, 1984). Problems arise in interpreting reflectance values of bitumen because bitumen, unlike kerogen, has the capacity to migrate. Once sediment is consolidated, particulate matter such as vitrinite is fixed in place, and subsequent measurements of the level of thermal maturation will reflect the total thermal stress applied to the rock. Bitumen, however, may migrate at any stage of the burial or tectonic history of the rock, so reflectance values of bitumen need not be a true measure of total thermal stress on the rock, even though bitumen responds to thermal stress in the same manner as vitrinite, especially in the more advanced stages of catagenesis. The degree of thermal matural of kerogen may be higher than indicated by reflectance measurements on bitumen if the bitumen has migrated into an environment of high premigration thermal stress. As a consequence, any reflectance measurement indicating a relatively low level of thermal maturation of the Phosphoria in the Wyoming-Idaho-Utah segment of the thrust belt is open to question and should be confirmed by other types of analyses.

Data gathered through analysis of a large number of samples of Cretaceous shales from both hanging wall and footwall positions provide convincing evidence that the peak period of hydrocarbon generation in the organic-rich Cretaceous shales preserved beneath the Absaroka thrust occurred after emplacement of the thrust (Warner, 1982; Warner and Royse, 1987). The abundance of vitrinite, spores, and pollen in most Cretaceous shale samples makes determination of the level of thermal maturity a rather simple and straightforward process. Kerogen in samples of the Bear River Formation from outcrops and shallow core holes on the Absaroka and Hogsback plates in the vicinity of the Fossil basin has a maturation level below that commonly considered necessary for peak hydrocarbon generation. Measurements of vitirinite reflectance and thermal alteration index (TAI) yield values that fall in the immature to incipient oil generation phases of catagenesis. Because these values represent the highest level of maturation reached in the Cretaceous section during prefault burial, no significant volume of hydrocarbons could have been generated in these or stratigraphically higher rocks before uplift on the thrust faults. At the time faulting started, the level of thermal maturity in the Bear River Formation was the same on the two sides of the incipient fault. In other words, both hanging wall and footwall sections of the Bear River Formation were in the immature stage at that time. However, analysis of samples of the Cretaceous footwall section taken from a number of deep wells drilled through the Absaroka plate in the Fossil basin show that at the present time the process of catagenesis has advanced at least to the stage of peak oil generation at all localities sampled, and has reached the dry gas stage in many places. Clearly the Cretaceous age source beds beneath the Absaroka thrust have passed into or through the oil generative "window" after being overridden by the fault.

Application of the Lopatin method (Waples, 1980) of calculating the time-temperature index of maturity (TTI) of Cretaceous footwall strata at various places in the Fossil basin strengthens the conclusion that the period of peak hydrocarbon generation in these rocks occurred after the Absaroka thrust was emplaced. An example is shown in Figure 6, a reconstruction of the time-depth history of footwall Cretaceous source rocks at Ryckman Creek field. The regional setting and generalized structure at Ryckman Creek are shown in Plate 7A. The present geothermal gradient of 22°C/km was used to calculate the position

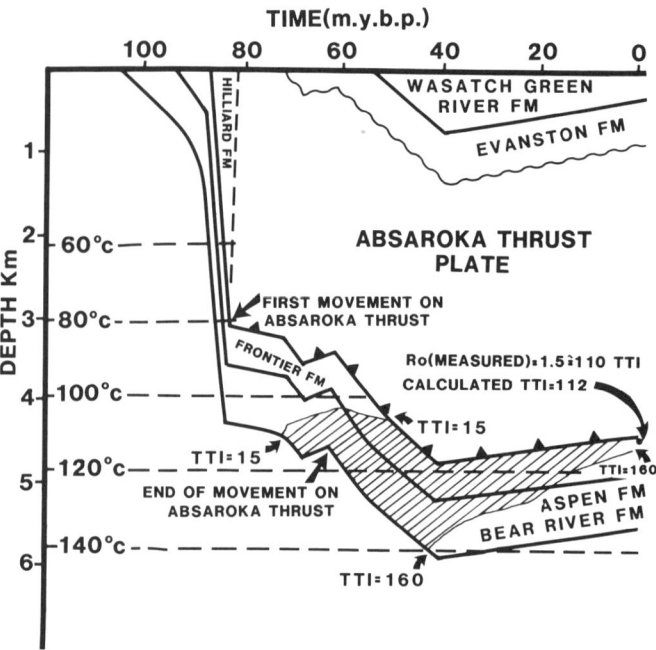

Figure 6. Lopatin diagram for Cretaceous source rock sequence preserved beneath the Absaroka thrust at Ryckman Creek field. A geothermal gradient of 22°C/km and a surface temperature of 15°C were used to calculate the position of isotherms. Hachured area defines the oil generative window between TTI values of 15 and 160 as defined by Waples (1980).

of isotherms in the diagram, because studies of maturation and coalification of organic material at different stratigraphic levels in the Cretaceous sequences indicate that the gradient has not changed significantly in the last 100 m.y. The Bear River, Aspen, and Frontier formations, which contain the best source shales of the Cretaceous section (Fig. 3), are preserved beneath the thrust. At the time of first motion on the Absaroka thrust the top of this preserved footwall section had been buried to a depth of about 3,050 km, and the base was more than 4,115 m deep. Despite this deep burial, no part of the source sequence had advanced to the stage of oil generation, which occurs in the range of TTI values from 15 to 160, according to Waples (1980). Figure 6 shows that the base of the preserved source rock section reached the threshold of oil generation (TTI = 15) about 9 m.y. after initiation of motion on the Absaroka thrust system, and part of the section has remained in the oil window until the present.

The remarkably good agreement between the level of maturity calculated from the Lopatin diagram for the Ryckman Creek locality and the level measured in footwall samples from a wellbore demonstrates that the burial history and temperature gradient used in the diagram are not grossly in error. The calculated TTI value for footwall strata a short distance below the fault at a depth of 4,359 m is 112. Vitrinite in a core sample at 4,375 m has a measured reflectance of 1.15 percent, which equates to a TTI of 110, according to Waples' (1980) calibration.

ROLE OF THRUSTING IN HYDROCARBON GENERATION

In the Fossil basin area, thrust faulting did not play an active role in the generation of the hydrocarbons trapped in thrust-related structures. No evidence has been presented to support ideas that thrust faulting creates large-scale thermal events that are in turn responsible for rapid generation of hydrocarbons in the affected source beds. To the contrary, the evidence available from the Fossil basin and other parts of the Wyoming-Idaho-Utah thrust belt indicates that the thermal effects of thrust faulting were negligible (Warner and Royse, 1987). Analysis of organic matter in samples taken from close proximity to major thrust faults shows that little if any frictional heat was developed through movement on the faults. Likewise the analytical data show that there was no sudden large increase in temperature of footwall rocks from deeper burial by the thrust plate or from the thermal effects of an overriding mass of hotter thrust plate material as visualized in some thrust models.

Stratigraphic information gathered from intensive geologic studies of the Fossil basin reveals only a minor increase in burial depth of the footwall section as a result of emplacement of the Absaroka thrust. Before the fault was formed, the space now occupied by the thrust plate in the Fossil basin was occupied by a thick section of Upper Cretaceous sediments that was transported eastward and replaced by Paleozoic and older Mesozoic rocks during thrusting. Rapid erosion of uplifted hanging wall sediments reduced the thickness of the thrust plate to its present value during or very shortly after thrusting. The burial reconstruction in Figure 6 illustrates both situations. At the time of initiation of motion on the thrust, the Frontier Formation beneath the thrust is shown to have been buried under 3,050 m of younger sediment, the upper Frontier and Hilliard formations, which was replaced by hanging wall material. The thrust uplift was reduced to base level and overlapped by the Evanston Formation soon after thrusting began. Subsequent deeper burial of the footwall section results from the deposition of younger sediments in the latest Cretaceous and Tertiary. Generation of hydrocarbons would have occurred in the Lower Cretaceous source rocks through normal burial even if no thrusting had taken place.

The primary role of thrust faulting in the Fossil basin oil and gas scene was in the creation of traps, not in the generation of hydrocarbons. Faulting placed the thick reservoirs of Paleozoic and early Mesozoic age above the source rocks in the proper position to receive the hydrocarbons subsequently generated. Folding associated with thrust motion created the anticlinal structures at the critical time before expulsion of hydrocarbons from the source rocks. Oil and gas would have been generated in the deeply buried Cretaceous source rocks with or without emplacement of the Absaroka thrust. However, without folding and translation of reservoir rocks by fault motion, the large structures and thick reservoirs would not have been available to trap the hydrocarbons.

REFERENCES CITED

Armstrong, F. C., and Oriel, S. S., 1965, Tectonic development of Idaho-Wyoming thrust belt: American Association of Petroleum Geologists Bulletin, v. 49, p. 1847–1866.

Armstrong, R. L., 1968, Sevier orogenic belt in Nevada and Utah: Geological Society of America Bulletin, v. 79, p. 429–458.

Bally, A. W., Gordy, P. L., and Stewart, G. A., 1966, Structure, seismic data, and orogenic evolution of southern Canadian Rocky Mountains: Bulletin of Canadian Petroleum Geology, v. 14, p. 337–381.

Bergosh, J. L., Good, J. R., Hillman, J. T., and Kolodzie, S., 1982, Geological characterization of the Nugget sandstone, Anschutz Ranch East, in Symposium and Field Conference Overthrust Belt of Utah: Utah Geological Association Special Publication no. 10, p. 253–265.

Burchfiel, B. C., and Davis, G. A., 1975, Nature and controls of Cordilleran orogenesis, western United States; Exensions of an earlier hypothesis: American Journal of Science, v. 275-A, p. 363–396.

Churkin, M., Jr., 1962, Facies across Paleozoic miogeosynclinal margin of central Idaho: American Association Petroleum Geologists Bulletin, v. 46, p. 569–591.

Claypool, G. E., Love, A. H., and Maughan, E. K., 1978, Organic geochemistry, incipient metamorphism, and oil generation in black shale members of Phosphoria Formation, western United States: American Association of Petroleum Geologists Bulletin, v. 62, p. 98–120.

Dahlstrom, C.D.A., 1970, Structural geology in the eastern margin of the Canadian Rocky Mountains: Bulletin of Canadian Petroleum Geology, v. 18, p. 332–406.

Darrow, G., 1955, The history of oil exploration in northwestern Montana, 1892–1950, in Billings Geological Society Guidebook Sixth Annual Field Conference, Sweetgrass Arch–Disturbed Belt, Montana: Billings Geological Society, p. 225–232.

Edman, J. D., and Surdam, R. C., 1984, Influence of overthrusting on maturation of hydrocarbons in Phosphoria Formation, Wyoming-Idaho-Utah overthrust belt: American Association of Petroluem Geologists Bulletin, v. 68, p. 1803–1817.

Frank, J. R., Cluff, S., and Bauman, J. M., 1982, Painter Reservoir, East Painter Reservoir, and Clear Creek Fields, Uinta County, Wyoming, in Powers, R. B., ed., Geological studies of the Cordilleran thrust belt: Denver, Rocky Mountain Association of Geologists, p. 601–611.

Gretener, P. E., 1972, Thoughts on overthrust faulting in a layered sequence: Bulletin of Canadian Petroleum Geology, v. 20, p. 583–607.

Ho, T. Y., Rogers, M. A., Drushel, H. V., and Koons, C. B., 1974, Evolution of sulfur compounds in oil: American Association of Petroleum Geologists Bulletin, v. 58, p. 2338–2348.

Hoffman, M. E., and Balcells-Baldwin, R. N., 1982, Gas giant of the Wyoming thrust belt; Whitney Canyon–Carter Creek field, in Powers, R. B., ed., Geological studies of the Cordilleran thrust belt: Denver, Rocky Mountain Association of Geologists, p. 613–618.

Lamerson, P. R., 1982, The Fossil basin and its relationship to the Absaroka thrust system, Wyoming and Utah, in Powers, R. B., ed., Geologic studies of the Cordilleran thrust belt: Denver, Rocky Mountain Association of Geologists, p. 279–340.

Maughan, E. K., 1975, Organic carbon in shale beds of the Permian Phosphoria Formation of eastern Idaho and adjacent states; A summary report, in Wyoming Geological Association Guidebook 27th Annual Field Conference, Big Horn Basin: Wyoming Geological Association, p. 107–115.

——, 1984, Geological setting of petroleum source rocks in Permian Phosphoria Formation [abs]: American Association of Petroleum Geologists Bulletin, v. 68, p. 942.

Orr, W. L., 1974, Changes in sulfur content and isotopic ratios of sulfur during petroleum maturation; Study of Big Horn Basin Paleozoic oils: American Association of Petroleum Geologists Bulletin, v. 58, p. 2295–2318.

Price, R. A., and Mountjoy, E. W., 1970, Geologic structure of the Canadian Rocky Mountains between Bow and Athabasca Rivers; A progress report: Geological Association of Canada Special Paper 6, p. 7–25.

Rich, J. L., 1934, Mechanics of low-angle overthrust faulting illustrated by Cumberland thrust block, Virginia, Kentucky, and Tennessee: American Association of Petroleum Geologists Bulletin, v. 18, p. 1584–1596.

Rose, P. R., 1976, Mississippian carbonate shelf margins, western United States, in Hill, J. G., ed., Symposium on geology of the Cordilleran hingeline: Denver, Rocky Mountain Association of Geologists, p. 135–152.

Rosenfeld, J. K., Ho, T.T.Y., and Dembicki, H., Jr., 1980, Oil-to-source rock correlation; Pineview field, overthrust belt, Utah [abs]: American Association of Petroleum Geologists Bulletin, v. 64, p. 776.

Royse, F., Jr., Warner, M. A., and Reese, D. L., 1975, Thrust belt structural geometry and related stratigraphic problems, Wyoming, Idaho, and northern Utah, in Bolyard, D. W., ed., Symposium on deep drilling frontiers of the Central Rocky Mountains: Denver, Rocky Mountain Association of Geologists, p. 41–54.

Rubey, W. W., 1955, Early structural history of the overthrust belt of western Wyoming, in Wyoming Geological Association Guidebook 10th Annual Field Conference: Wyoming Geological Association, p. 125–126.

Rubey, W. W., and Hubbert, M. K., 1959, Role of fluid pressure in mechanics of overthrust faulting, part II: Geological Society of America Bulletin, v. 70, p. 167–206.

Sandberg, C. A., 1975, Petroleum geology of Paleozoic rocks of Cordilleran miogeosyncline: U.S. Geological Survey Open-File Report 75-96, 7 p.

Sandberg, C. A., and Gutschick, R. C., 1984, Distribution, microfauna, and source-rock potential of Mississippian Delle Phosphatic Member of Woodman Formation and equivalents, Utah and adjacent states, in Woodard, J., and others, eds., Symposium on hydrocarbon source rocks of the Greater Rocky Mountain Region: Denver, Rocky Mountain Association of Geologists, p. 135–178.

Seifert, W. K., and Moldowan, J. M., 1981, Paleoreconstruction by biomarkers: Geochimica et Cosmochimica Acta, v. 45, p. 783–794.

Tisoncik, D., 1984, Regional lithostratigraphy of Permian Phosphoria Formation, western overthrust belt [abs.]: American Association of Petroleum Geologists Bulletin, v. 68, p. 952.

Veatch, A. C., 1907, Geography and geology of a portion of southwestern Wyoming: U.S. Geological Survey Professional Paper 56, 178 p.

Waples, D. W., 1980, Time and temperature in petroleum formation; Application of Lopatin's method to petroleum exploration: American Association of Petroleum Geologists Bulletin, v. 64, p. 916–926.

Warner, M. A., 1982, Source and time of generation of hydrocarbons in the Fossil basin, western Wyoming thrust belt, in Powers, R. B., ed., Geological studies of the Cordilleran thrust belt: Denver, Rocky Mountain Association of Geologists, p. 805–815.

Warner, M. A., and Royce, F., 1987, Thrust faulting and hydrocarbon generation; A discussion: American Association of Petroleum Geologists Bulletin, v. 70, p. 890–896.

Wiltschko, D. V., and Dorr, J. A., Jr., 1983, Timing of deformation in the overthrust belt and foreland of Idaho, Wyoming, and Utah: American Association of Petroleum Geologists Bulletin, v. 67, p. 1304–1322.

MANUSCRIPT ACCEPTED BY THE SOCIETY MAY 16, 1987

ACKNOWLEDGMENTS

Most of the information on which this paper is based was gathered while the author was an employee of Chevron U.S.A. Inc. The cooperation of Chevron management in granting permission to publish proprietary information and in supplying facilities for drafting and manuscript preparation is gratefully acknowledged. Special thanks are due Frank Royse, who provided many helpful ideas and critically reviewed the manuscript.

Chapter 25

Petroleum potential of the Great Basin

Norman H. Foster and Richard R. Vincelette
1625 Broadway, Suite 530, Denver, Colorado 80202

INTRODUCTION

Great Basin is a term used to describe a portion of the Basin and Range Physiographic Province that has internal surface drainage (Fig. 1). The entire basin is characterized by horst and graben blocks, generally oriented in a north-south direction, formed by Cenozoic extensional faulting. The major grabens form present-day valleys, and the large horst blocks form the intervening mountain ranges.

This chapter describes the oil accumulations that have been discovered in the area to data, and discusses the potential for finding additional hydrocarbon accumulations.

Only truncation fault-block traps located within grabens are productive at this time. Potential exists, however, for commercial accumulations in folded structures, some of which are thrusted, in both horst and graben blocks. In addition, regional subcrop traps also have high potential in the area, as do local stratigraphic traps, especially in the Diamond Peak and Chainman sandstones of Mississippian age.

The truncation fault-block traps discovered to date range in size up to 15 to 20 million barrels of oil reserves. Because excellent mature source rocks and reservoirs, along with good potential traps, are present in some parts of the Great Basin, the possibility exists for undiscovered resources.

HISTORY OF EXPLORATION

A number of oil and natural gas seeps and other surface shows have been documented in the Great Basin by Osmond and Elias (1971), Foster and others (1979), and Bortz (1983). These are indicated along with other surface and subsurface shows of hydrocarbons and oil fields, on Figure 2. Oil fields in the eastern Great Basin, include the Rozel Point field (discovered near the Rozel Point oil seeps, circa 1904), and West Rozel field (discovered in 1978). Both are located offshore in the Great Salt Lake, Utah (Bortz, 1983; Bortz and others, 1985).

In Nevada's Railroad Valley, oil fields that are currently producing (Fig. 2) include the Eagle Springs field, the first commercial discovery in Nevada (1954), the Trap Spring field (1976), the Bacon Flat field (1981), the Grant Canyon field (1983), and the Kate Spring field (1985 to 1986). The Bacon Flat and Grant Canyon fields were identified as a geomorphic anomaly in the original exploration project (Foster, 1979; Dolly, 1979; Foster and others, 1985), and were confirmed by low-quality seismic data obtained over the anomaly. Grant Canyon is a two-well field, but one of the wells has flowed more than 4,000 barrels of oil per day since it was completed and has produced a cumulative volume of oil in excess of 8,000,000 barrels. The other well has flowed more than 2,000 barrels of oil per day. Both wells produce from the Devonian Guilmette Formation, which has matrix porosity but is also highly fractured. The first well has one of the highest rates of daily production for any well in the onshore United States, which has created a strong level of interest in the area for additional potential.

In Pine Valley, Nevada, several fields have been brought into production, most notably the Blackburn field, discovered in 1982. The Tomera Ranch and North Willow Creek fields were found in 1987 and 1988, respectively. So far they are both relatively small accumulations.

A wildcat well, located in Pine Valley several kilometers northwest of Blackburn, has reportedly recovered free oil from Paleozoic, Mesozoic, and Tertiary rocks and may turn out to be a discovery.

All accumulations found to date have been from the same type of trap, a truncation fault block located within a graben. Reservoirs range in age from Devonian to Oligocene; most are highly fractured, and some have matrix porosity. Source rocks range in age from Ordovocian to Eocene, and are both marine and freshwater deposits.

GEOLOGIC HISTORY

A brief summary of the sedimentation and tectonics of the Great Basin is given here. More complete details are found in Hintze (1973), Hunt (1979), Eaton (1979), Stewart and others (1977), Stewart (1980), Cook (1988), and Peterson (1988).

During late Precambrian time, some 9,100 m of sediment accumulated in a north-south band along the western edge of the North American craton. A thick wedge of Cambrian through Devonian sediments, ranging from 9,100 m in Central Nevada to only about 900 m in central Utah, was deposited in shallow water

---

Foster, N. H., and Vincelette, R. R., 1991, Petroleum potential of the Great Basin, *in* Gluskoter, H. J., Rice, D. D., and Taylor, R. B., eds., Economic Geology, U.S.: Boulder, Colorado, Geological Society of America, The Geology of North America, v. P-2.

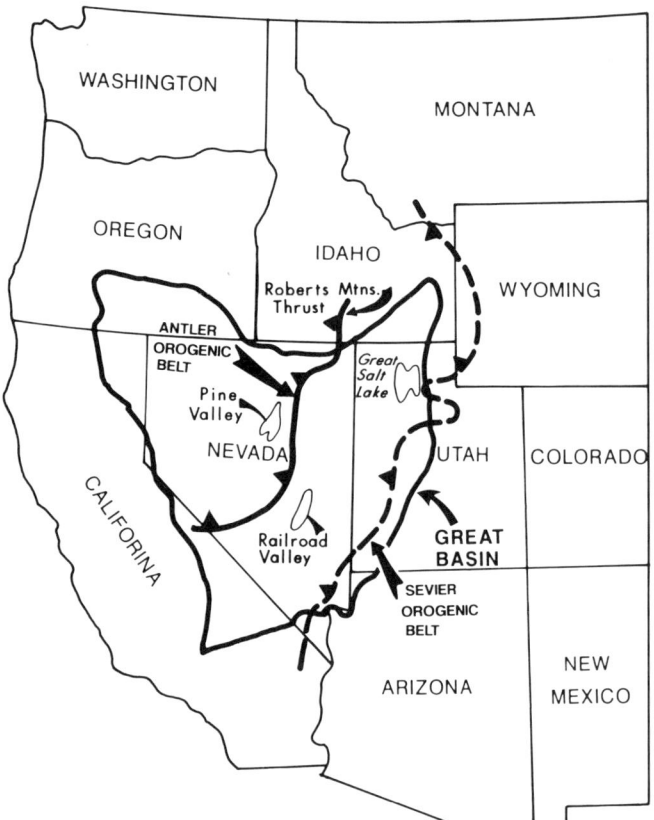

Figure 1. Index map showing Great Basin of the Basin and Range Province. The locations of the Antler and Sevier orogenic belts are shown along with the positions of the Pine, Railroad, and Great Salt Lake grabens.

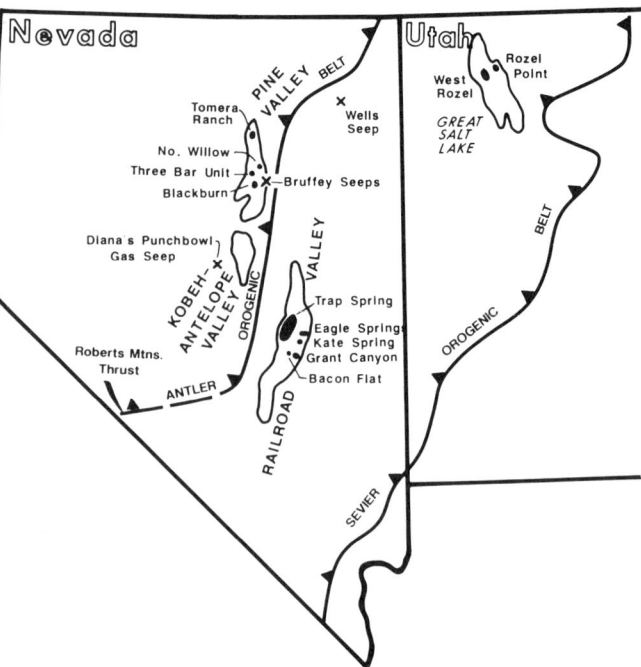

Figure 2. Map of western Utah and Nevada showing the Pine, Railroad, Kobeh-Antelope, and Great Salt Lake grabens. The locations of oil fields are also indicated. Several oil and gas seeps are shown along with the Antler and Sevier orogenic belts.

on the broad shelf bordering the craton. The eastern-facies rocks from central Nevada to central Utah are terrigenous clastics and carbonates, whereas the rocks of the western facies, deposited in western Nevada, are deeper-water shales, radiolarian cherts, and pillow lavas (Stewart, 1980). Potential reservoirs for hydrocarbons accumulated during this time, including some sandstones within the Cambrian, the Eureka Quartzite (which in some places has excellent matrix porosity), and the Silurian and Devonian carbonates (which are reefal in some areas). Most of the units, including quartzites, are potential fractured reservoirs. For example, high-gravity crude oil has been recovered on a drill stem test and production tests of Cambrian fractured quartzites from a well in Railroad Valley, Nevada.

The Antler orogeny (Roberts, 1968) occurred in Late Devonian and Early Mississippian time (Fig. 1). Deeper-water sediments in the west were thrust as much as 145 km eastward (Stewart, 1980) over miogeosynclinal eastern-facies rocks of equivalent age. The orogeny produced the Antler highland in central Nevada, from which terrigenous clastics were shed eastward into a foreland basin, forming the sandstones of the Mississippian Diamond Peak Formation and within the Chainman Shale. These sandstones thin toward the east and finger into shale.

This facies change of sandstone into oil source rocks forms one of the most prospective potential stratigraphic trap plays in the Great Basin. The chances for major stratigraphic accumulations of hydrocarbons within the Diamond Peak–Chainman sequence are probably very good. The truncation of these sandstones beneath valley fill in a favorable structural setting is also prospective. In eastern Nevada and western Utah, the Chainman Shale is believed to be one of the most important oil source rocks in the Great Basin (Poole and Claypool, 1984).

Stewart (1980, p. 5) describes the sedimentary and tectonic provinces of Nevada, during Late Paleozoic time, from east to west, as: "1. a shallow-water carbonate shelf; 2. a foreland basin containing coarse detrital material derived from the west as well as more widespread shallow-water carbonates; 3. the Antler highland, overlapped in Pennsylvanian and Permian time by thin coarse detrital marine sediments; 4. a western deep-water basin containing fine to coarse detrital rocks, radiolarian chert, silty limestone and mafic lava; and 5. a Permian magmatic arc terrane, largely of mafic lava." The provinces are basically the same for the entire Great Basin, with the carbonate shelf extending across western Utah.

The Sonoma orogeny described by Silberling and Roberts (1962) occurred in Late Permian and Early Triassic time (Silberling, 1973). Deep-water facies were overthrust was much as 97 km eastward on the Golconda thrust, onto thin eastern shelf facies rocks of equivalent age (Stewart, 1980).

Triassic and Lower Jurassic rocks occur in Utah and eastern Nevada, are missing in central and east-central Nevada, and are present again in western Nevada. The western Nevada assemblage is shallow-water marine carbonate and deeper-water marine mudstone (Stewart, 1980). Volcanic sediments, lavas, and fine-grained clastics occur farther to the west. Of interest to the petroleum geologist is the Triassic Favret Formation, which is a possible hydrocarbon source rock where it crops out east of Dixie Valley in west-central Nevada (Bortz, 1983). Cretaceous rocks are fairly sparse and consist mainly of the early portions of freshwater lake deposits known as the King Lear Formation and the Newark Canyon Formation and possibly the Sheep Pass Formation (Fouch and others, 1979). Deposition in these freshwater lakes continued into the Paleocene and even Eocene when similar deposits accumulated in other lakes such as the Elko and Kinsey Canyon. These deposits, believed to be the source of some of the petroleum discovered in Nevada, are similar in age and lithology to the oil-rich Green River Shales of Utah and Wyoming.

Folding and east-verging thrusting continued in the Mesozoic, culminating in the Sevier orogeny of Late Jurassic and Cretaceous time (Fig. 1). Bartley and others (1985), Bartley and Martin (1986), Martin and Bartley (1986), and Lund (1986) have documented folds and thrusts of this age in central Nevada. Some low-angle detachment faults, which Coney (1980) has associated with metamorphic core complexes, are mapped in various areas of east-central Nevada, including the Ruby Range of Elko County and the Snake Range in White Pine County; they also occur in the Grant Range east of Railroad Valley.

McKee and others (1970) and McKee (1971) have related the change from acidic to basic volcanic rocks near the end of Oligocene time (about 17 Ma) to the onset of Basin and Range extensional faulting. Ekren and others (1968) have shown, based on fault evidence in the field, that in southern Nevada the Basin and Range started to develop between 17 Ma and 14 Ma. The change from compressional to tensional forces in the Great Basin may be related to a change from subduction of the Pacific Plate under the North American Plate, to right-lateral movement between the plates, along the San Andreas fault, Walker Lane fault zone, and other faults just to the west of the Basin and Range area (Atwater, 1970).

Back-arc spreading has also been used to explain Basin and Range structure (Thompson and Burke, 1974). Other theories include that of Menard (1960) that the East Pacific Rise has subducted western North America and is spreading beneath the Basin and Range area. Also, Matthews and Anderson (1973) have suggested that mantle plumes are responsible for the spreading.

Substantial volcanic activity, mainly during Oligocene time (ignimbrite deposits) just prior to the beginning of the development of the Basin and Range, resulted in the formation of some excellent reservoirs for hydrocarbons. The ignimbrite sequence contains welded tuffs that have excellent columnar jointing and are highly fractured. Welded tuffs form the main reservoir at Eagle Springs and Trap Spring oil fields in Railroad Valley.

Valley-fill sequences, ranging in age from Miocene to Recent, fill the valley grabens and form the top seals for the oil fields found to date in Nevada. Fanglomerates along the basin margins grade basinward into fine-grained sandstones, siltstones, and shales. Basaltic volcanic extrusions have occurred during Miocene to Pliocene time. Basalt flows often occur within the valley-fill sequence. These basalts have proven to be reservoirs in the valley fill of the Great Salt Lake graben, where they have produced oil at the Rozel Point and West Rozel fields. The oil is locally sourced in the Pliocene and is immature.

THE INITIAL EXPLORATION MODEL

One of the critical elements in any successful exploration program is an understanding of the types of potential traps that may be present in an area. Many exploration programs fail because the exploration target being searched for has little likelihood of occurring in the area under investigation. Therefore, an accurate determination of the critical elements that control existing hydrocarbon accumulations in an area is an important first step in any exploration program.

Eagle Springs oil field

Dolly (1979) and Foster (1979) undertook a study of Eagle Springs oil field to understand the trap and find geologic clues that could lead to the discovery of similar traps in both Railroad Valley and White River Valley to the east. This field currently produces about 130 barrels of oil per day and had a cumulative production of 3,960,269 barrels of oil to the end of April 1990. Murray and Bortz (1967) demonstrated that Eagle Springs is a truncation fault-block trap (Fig. 3). The valley fill provides the top seal, and low-permeability Paleozoic rocks or faults provide the side seals for a truncated wedge of reservoir rocks consisting of Oligocene Garret Ranch volcanics and Eocene Sheep Pass lake deposits. The fault block is in a down-faulted position in the graben—on the downthrown side of the major bounding fault of the graben to the east, and on the down-faulted side of a high graben block located to the south. The oil is trapped in the truncated and faulted wedge of Tertiary on the high structural part of the down-faulted block. Foster (1979) and Dolly (1979) theorized that other similar fault blocks could occur within this and other basins where mature source rocks and reservoir rocks were present (Fig. 4). The truncated wedge of Tertiary rocks formed because of erosion of the Tertiary beds at the monocline that developed where graben boundary faulting would eventually occur (Fig. 5).

One of the keys to exploration was the recognition that the basin-margin faults could be mapped on aerial photographs and that high blocks within the graben could also be delineated on aerial photos (Foster, 1979). This led to the photogeomorphic interpretation of Railroad Valley (Fig. 6A). Seismic lines were shot across Eagle Springs oil field, the basin, and the large geomorphic anomaly on the west side of Railroad Valley. These lines

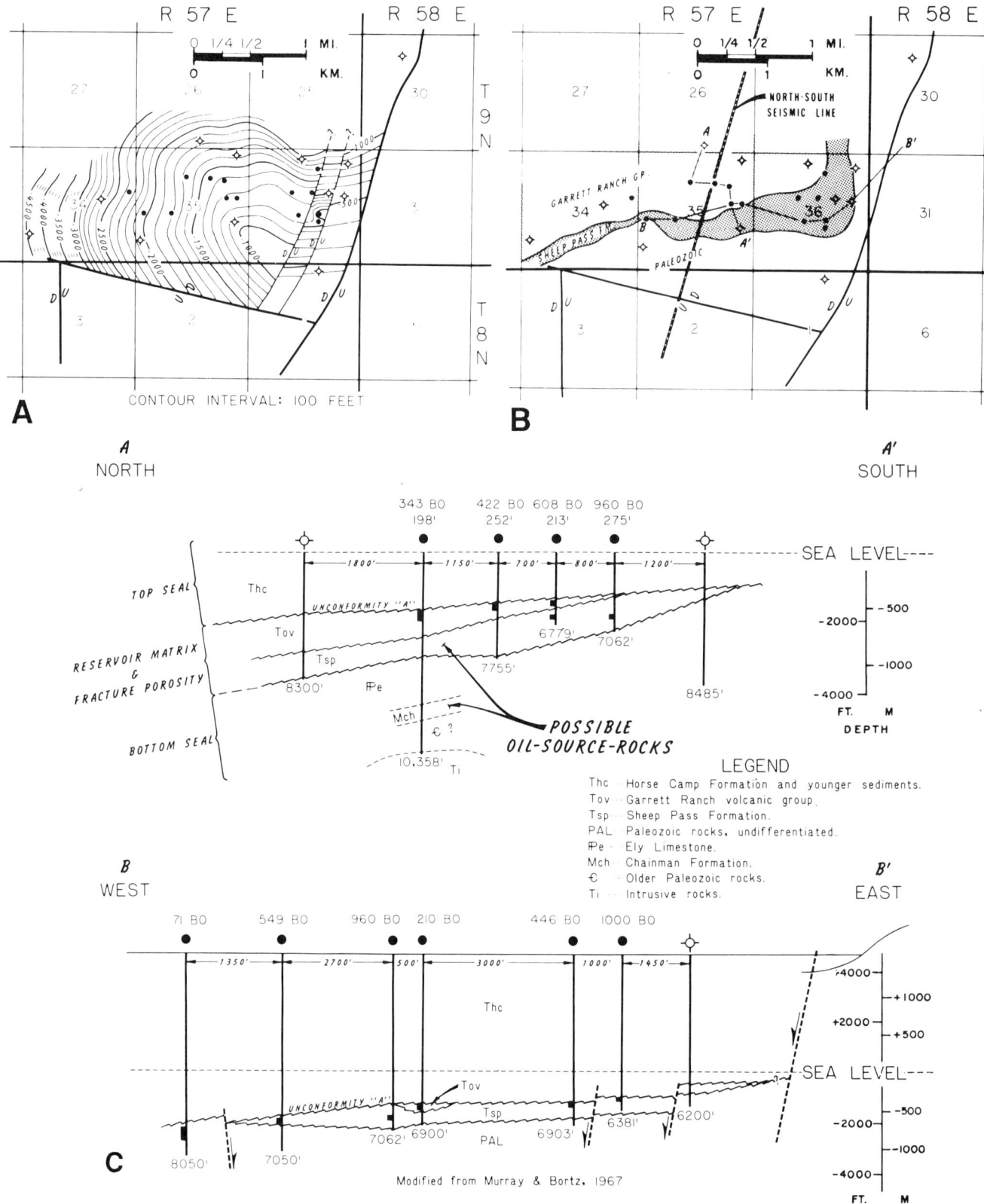

Figure 3. A. Structure map constructed on the base of the valley fill at Eagle Springs oil field (from Dolly, 1979). B. Paleogeographic map of rocks subcropping beneath the valley fill at Eagle Springs oil field (from Dolly, 1979). C. North-south and east-west cross sections at Eagle Springs oil field (modified from Murray and Bortz, 1967).

confirmed that a truncation fault-block trap was present on the west side of the graben (Fig. 7). Subsequently, the discovery well at Trap Springs oil field was drilled in 1976 (Duey, 1979).

Trap Spring oil field

Trap Spring oil field is located on the downthrown side of the main boundary fault on the west side of Railroad Valley. The top seal is the valley fill, and side seals are faults. The reservoir is in fractured volcanics of the Garret Ranch Group. Since its discovery in 1976, this field, which is still under active development, has produced more than 8,600,000 barrels of oil, and in December of 1989 was producing at a rate of 2,300 barrels of oil per day. Individual wells have produced at rates as high as 3,000 barrels of oil per day. This field is the largest in areal extent of the fields found to date in the Great Basin, and currently extends 8 km along the west margin of Railroad Valley.

EVOLUTION OF THE TRUNCATION FAULT-BLOCK EXPLORATION MODEL

After a period of time, it became clear that within the fault-block configurations in the valley grabens, not only the low fault blocks like Eagle Springs and Trap Spring held accumulations, but also the high fault blocks could produce (Fig. 4).

Grant Canyon and Bacon Flat oil fields

The Grant Canyon and Bacon Flat fields are located on a high fault block, from which the Tertiary sediments have been stripped by erosion prior to deposition of the valley fill sequence (Plate 6G). At Bacon Flat and Grant Canyon, the Devonian Guilmette Formation is encountered directly beneath the valley-fill sequence. The Guilmette is an excellent reservoir with matrix porosity and fracturing. The top seal is again the valley fill with faults as side seals (Veal and others, 1988a).

This field, which is currently producing at a little over 6,000 barrels of oil per day, had a cumulative production of 11,838,587 barrels of oil through December 1989.

Kate Spring oil field

Development drilling is continuing at the Kate Spring field, located less than 2 km south of the Eagle Springs field in Railroad Valley (Plate 6G). Production of low-gravity (10.5° API) crude oil at rates in excess of 1,500 barrels of oil per day has been obtained from Devonian Guilmette and Pennsylvanian Ely Formation carbonates beneath the valley-fill unconformity at an average depth of 1,300 to 1,350 m (Flanagan, 1988). Kate Spring field

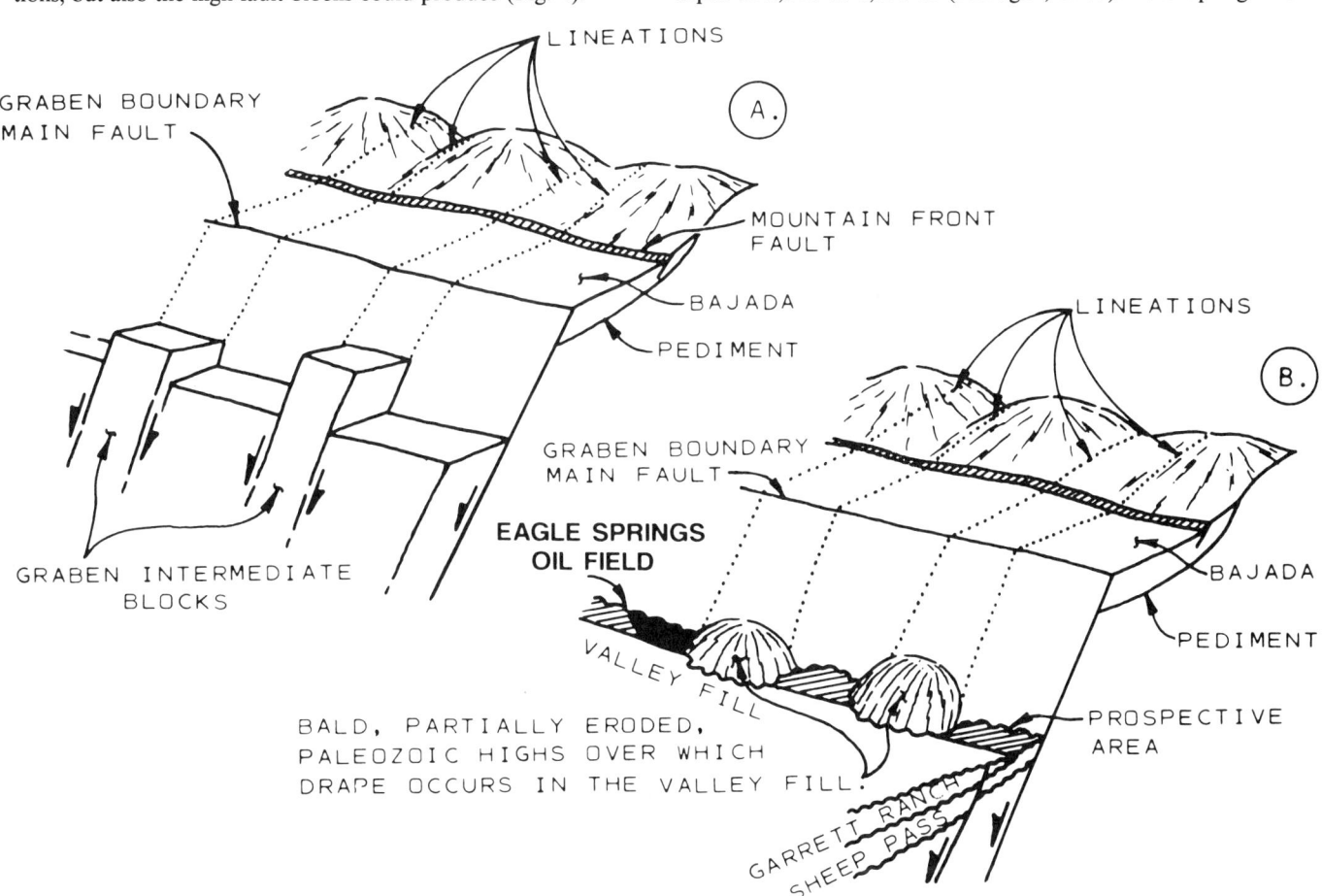

Figure 4. Initial exploration models that focused on downfaulted blocks similar to the Eagle Springs setting.

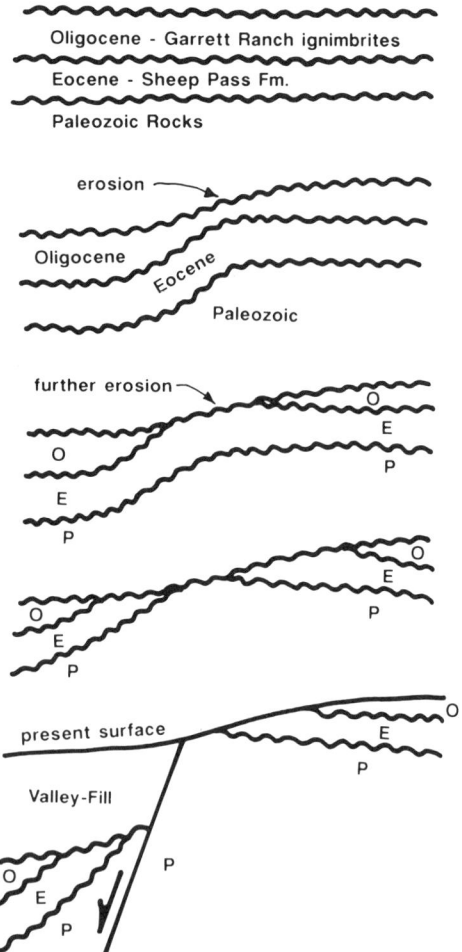

Figure 5. Schematic cross sections illustrating the development of the lower Tertiary truncated wedge sediments.

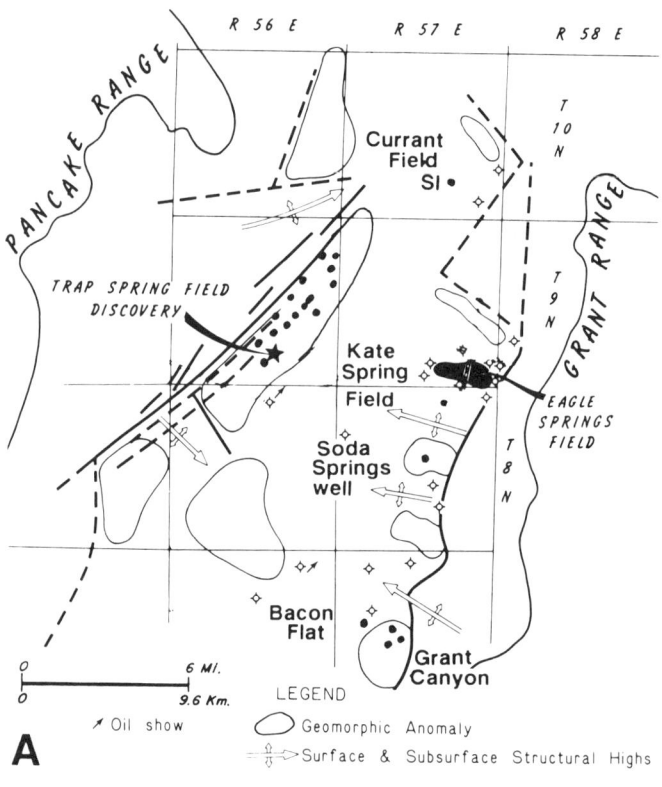

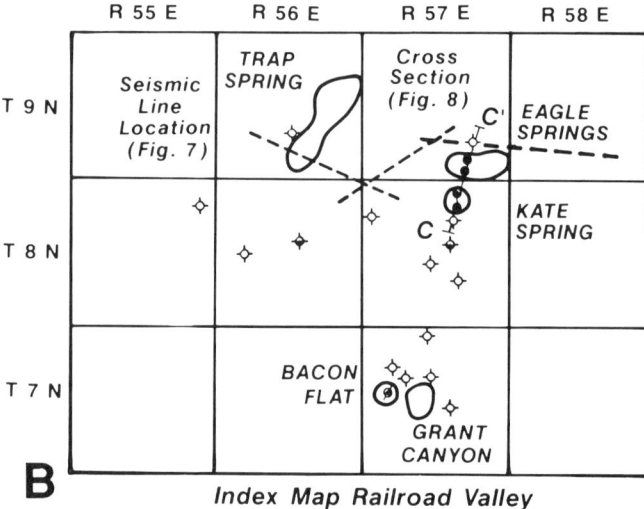

Figure 6. A. Geomorphic anomaly map showing locations of oil discoveries made through 1987 (from Veal and others, 1988a). B. Index map showing location of cross sections in Railroad Valley, Nevada.

in December 1989 averaged 1,300 barrels of oil per day, and had a cumulative production of 271,532 barrels of oil. The field also produces gas (45,891 thousand cubic feet from November 1988 through December 1989).

This field is located on a high fault block separated from the Eagle Springs field by a normal fault with at least 610 m of vertical throw (Fig. 8). High gravity oil (35° API) was tested at a depth of 2,350 m from a Cambrian quartzite in a well located in a separate lower fault block approximately 3.2 km south of the Kate Spring accumulation, attesting to the wide variety of oil types and reservoir changes that occur in short distances in this tectonic province.

Blackburn oil field

Blackburn field is located in Pine Valley some 160 km northwest of Railroad Valley. Oil is trapped on a high faulted block within the valley graben (Fig. 9). The top seal again is valley fill, but there are also internal seals provided by the Tertiary volcanics and Mississippian shales, and faults provide the side seals. Oil occurs in low-permeability volcanic rocks, but little production has come from the volcanic reservoirs. Oil has also filled Paleozoic-age reservoirs in the same oil column. The sandstones within the Chainman Shale are productive, as well as the Devonian Nevada Dolomite. In December 1989, this field was producing at a rate of 750 barrels of oil per day, and had cumulative production of 1,905,504 barrels of oil.

Gravity data show that Blackburn is located on a cross-basin high in Pine Valley. A new discovery is apparently also located on this cross-basin feature. The accumulation at Blackburn demonstrates that oil fills any porous and permeable reservoir that is structurally well-located on a fault block, and is sealed by valley fill and faulting.

EXPLORATION TECHNIQUES

The complex structural geology of the Great Basin has created a challenging environment in which to explore for oil and gas. In the present-day valleys where most of the exploration effort has been focused, the potential oil accumulations are hidden and masked by hundreds to thousands of meters of Tertiary valley-fill sediments shed from the surrounding mountains. These sediments, which unconformably overlie the underlying Paleozoic through early Tertiary reservoir beds, also provide the ultimate top seal for most of the accumulations found to date. Unfortunately, in most of the active exploration areas it has proven difficult to predict either the structural attitude or the age and type of potential reservoir rock beneath the valley-fill unconformity prior to drilling a well. A successful exploration effort requires an integration of all available exploration techniques to minimize risk. Perhaps the most fundamental, but often overlooked, exploration technique involves detailed surface geologic mapping and geomorphic analysis, which has proven very effective in this province. Mapping of basin boundary faults and projection of structural highs and reentrants along the basin margins based on geomorphic expression has been very effective in outlining potential traps in the basins (Foster, 1979).

Reflection seismic shooting has been the tool of choice in

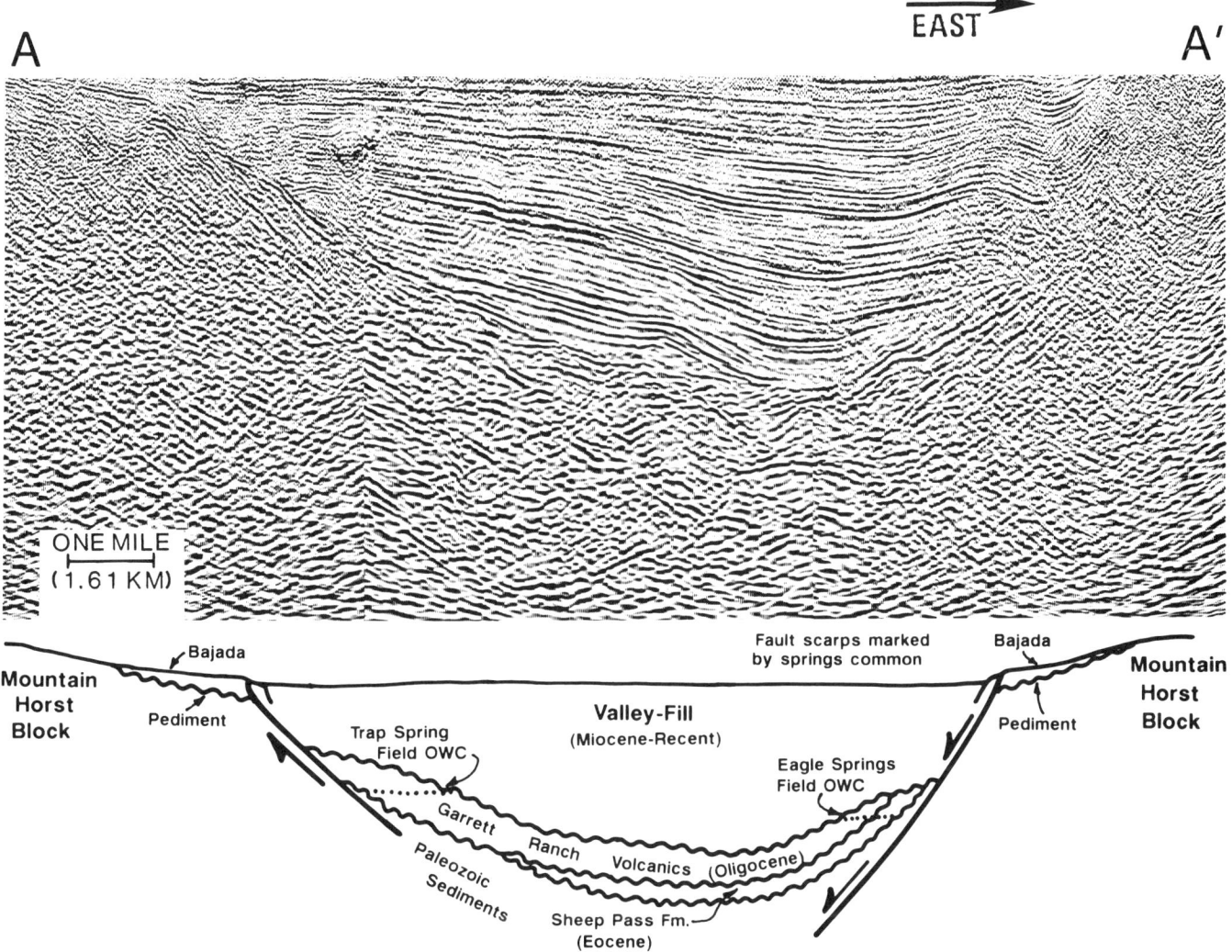

Figure 7. Diagrammatic east-west cross section of Railroad Valley, and an interpreted seismic section through the Eagle Springs oil field on the east and Trap Spring oil field on the west.

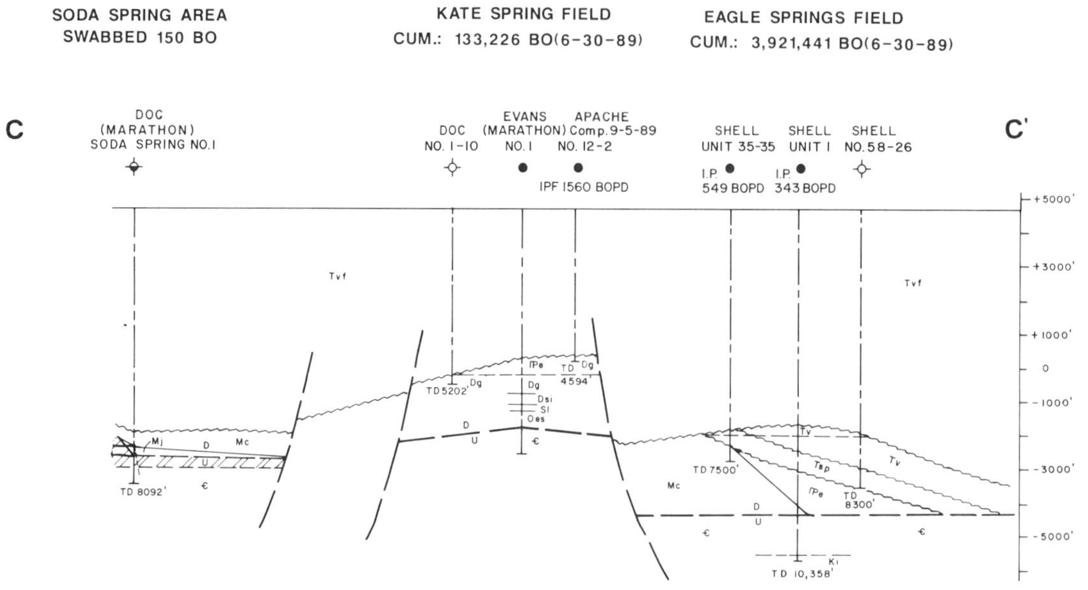

Figure 8. North-south cross section through the Kate Spring and Eagle Springs fields, Railroad Valley, Nevada.

mapping structural closure and fault-block boundaries at the base of the Tertiary unconformity. Limitations on this tool are caused by the presence of locally thick fanglomerates near the basin margins, which cause a deterioration in the seismic signal in areas where many of the most prospective fault block traps are thought to occur. In addition, the structural attitude and the nature of the rock beneath the unconformity (i.e., reservoir or nonreservoir) commonly cannot be determined from seismic data.

Gravity is a useful adjunct to seismic and surface geology in the area, and often can provide leads in areas of poor seismic definition. The high structural blocks and intervening lows can usually be mapped on the gravity, as can the major boundary faults. In addition, depth estimates to basement in unexplored areas can also be made with the gravity. Gravity data also provide a better interpretation of where the basement or Paleozoic reservoirs are closest to the surface beneath the unconformity, because the Tertiary volcanics and lake sediments beneath the unconformity may have the same density as the overlying valley fill, whereas the Paleozoic carbonates and underlying basement rocks have higher denisities. Thus, areas of bald-headed highs can be located more precisely with gravity than with seismic data.

Soil-gas geochemistry may also provide valuable information to help determine whether or not a prospective trap defined by other techniques may be hydrocarbon-bearing, because surface gas and oil seeps are present along surface faults in some areas (Foster and others, 1979), and leakage of oil up into the Tertiary valley fill has been observed over several of the known oil fields.

Another technique that appears to have potential in the area is the use of controlled source audio-frequency magnetotellurics (CSAMT). Several surveys over known accumulations appear to be able to define general positions of boundary faults where seismic data are poor.

Successful exploration in the Basin and Range province will require integration of a wide variety of techniques. Improvements in the data quality and interpretations made from these techniques will undoubtedly play a major role in the discovery of additional accumulations in this area.

From an exploration standpoint, perhaps the most frustrating element in this area is the knowledge that a potential exists for the discovery of additional hydrocarbon reserves, because all of the elements of source rocks, reservoirs, and traps are widely distributed. The complexity of the geology and masking of the reservoirs and traps by the overlying valley fill requires the use of sophisticated exploration techniques; however, many of these, in their present state of development, cannot fully resolve the problem of identifying precisely the most likely position in the subsurface where potential traps and reservoirs should occur. Development of accurate theoretical models is extremely critical in determining where these traps and reservoirs occur, and a good deal of interpretation is required in analyzing available data. Therefore, the more accurate the model, the more likely will be the successful discovery of a hydrocarbon-bearing trap.

REMAINING POTENTIAL

Since the discovery of the Trap Spring oil field in 1976, a modern cycle of exploration activity has taken place in the Basin and Range Province. As a result of additional drilling, geophysical and geological work in the Great Basin has continued to add to our knowledge of the critical factors that control oil and gas accumulations. Detailed studies of source rocks, migration histories, thermal gradients, reservoir rock distribution and geologic

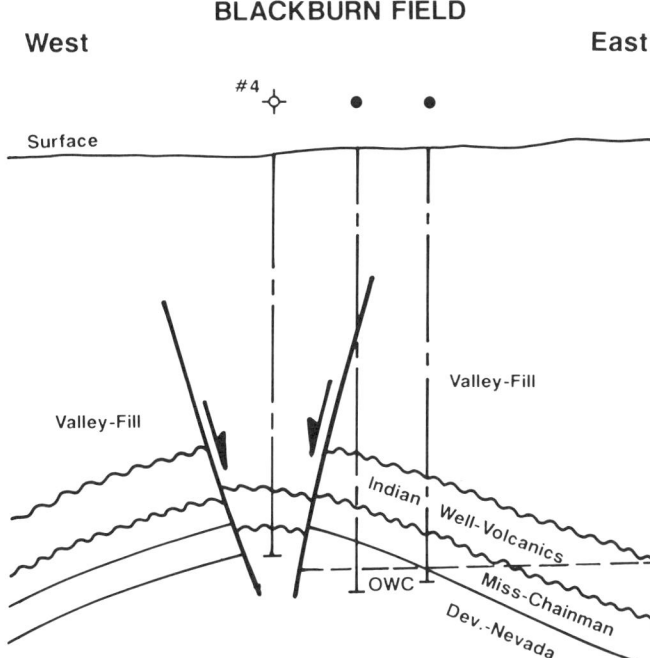

Figure 9. Cross section through the Blackburn oil field in Pine Valley, Nevada.

history, tectonic style, and a better understanding of the trap types make possible evaluation and location of those areas that have remaining potential for significant oil and gas accumulations.

Source rocks, maturation, and migration

Published analyses of source-rock potential and paleotemperature (Harris and others, 1980; Poole and others, 1983; Poole and Claypool, 1984) enable a general evaluation of areas in the Great Basin that will most likely contain mature source rocks in a position to charge potential reservoirs.

The oldest confirmed source rocks consist of siliceous shales of the Ordovician Vinini Formation and the Devonian Woodruff Formation deposited west of the Roberts Mountains thrust (Figs. 10 and 11). Locally these deposits are organic rich, with as much as 25 percent total organic carbon (TOC), and are capable of yielding as much as 30 gallons per ton of oil from outcrop samples (Poole and Claypool, 1984). Excellent oil shows have been reported from these shales in wells drilled in the Kobeh–Antelope Valley area, and ultimately, significant production could be obtained from reservoirs charged by these shales as well as potential oil production from fractures in the shales themselves. Based on conodont alteration indices, vitrinite refectance studies, and pyrolysis data, these source-rock units probably become overmature at distances of from 16 to 40 km west of the leading edge of the Roberts Mountains thrust. Whether these overmature areas are due to abnormally high geothermal gradients caused by high heat flows during Tertiary igneous activity or are due to greater depths of burial prior to uplift and overthrusting during development of the Roberts Mountains thrust is not apparent at the present time. Nonetheless, the paleotemperature history appears to provide a zone where good source-rock potential is present in these western eugeosynclinal assemblage rocks. To the west of these areas, source-rock potential may be provided by Triassic or younger sediments present in local depositional basins described by Bortz (1983) in the Dixie Valley area.

Perhaps the most important and widespread source rock in the Basin and Range Province is the Chainman Shale of Mississippian age. Distribution maps showing relative organic richness and maturity values outline a source-rock fairway in east-central Nevada extending from a few kilometers to the west of the Roberts Mountains thrust to approximately the Nevada-Utah border (Fig. 11). Significant variations in both organic richness and level of maturity occur within this broad belt, related to both depositional environment and thermal history. Perhaps the most significant element indicated by published data is that local hot spots exist, within which the Chainman is overmature on outcrop, but nearby areas are still immature and capable of generating hydrocarbons. These local hot spots may possibly be controlled by local igneous activity. One of the unique aspects of oil generation and migration in the Great Basin area is the close association between Tertiary volcanic activity, its associated elevated heat flow, and the maturation of potential source rocks. A good case can be made that if it were not for the high heat flows created by the Tertiary igneous activity, many of the potential source rocks would never have reached sufficient thermal maturity to have generated hydrocarbons. In addition, one of the major reservoir rocks in the area is provided by highly jointed and fractured igneous volcanic flow units (ignimbrites). Also, evidence has been presented that local igneous intrusions at both the Grant Canyon and Blackburn fields may have caused thermal hydrofracturing of the overlying sediments, thus enhancing reservoir permeability as well as localized source rock maturity (Hulen and others, 1990).

Because of the complex fault-block distribution in the Basin and Range Province, the migration paths from source rock to reservoir were probably limited to fairly short distances; that is, long-range migration would be extremely difficult for oil and gas presently trapped in individual truncation fault-block traps and is at least limited to the typical 10- to 20-km widths of the valleys or Neogene grabens separated by the intervening mountain ranges. In many cases, available migration paths to potential traps involve even smaller distances. As a consequence, some of the dry holes drilled on apparent favorable structural blocks may have more to do with the absence of a nearby mature source rock than with lack of reservoir, trap, or seal.

Source rock–crude oil correlations indicate that the Chainman Shale is probably the major source for the oils found in the Trap Spring, Bacon Flat, and Grant Canyon fields in Railroad Valley, and also for oil in the Blackburn field in Pine Valley (Poole and Claypool, 1984). The Vinini Shale may also ultimately prove to be a significant source in Pine Valley and other areas to the west of the Roberts Mountains thrust.

Superimposed on the regional Chainman and Vinini-

Woodward Paleozoic source rocks, much younger source rocks of more localized areal extent have been identified in the Mesozoic and Cenozoic deposits in the Great Basin. In western Nevada, oil shows have been reported in fossil ammonites in Triassic sediments (Bortz, 1983). In east-central Nevada, Late Cretaceous to early Tertiary lake beds that developed locally in the Great Basin prior to subsequent breakup of the area into isolated horsts and grabens are local source rocks (Poole and Claypool, 1984). These units, which make up the Newark Canyon Formation, the King Lear Formation, and the Elko Shale, though generally submature on the outcrops, are capable of generating significant quantities of hydrocarbons where buried to a sufficient depth or subjected to a high enough heat flow to become mature. Lacustrine shales in the Sheep Pass Formation are thought to have provided the main source for oil in the Eagle Springs field as well as having generated other noncommercial accumulations (Poole and Claypool, 1984). These lacustrine source rocks typically produce a high pour-point paraffinic crude oil that has less desirable production characteristics than the marine-sourced oils of the Chainman Shale.

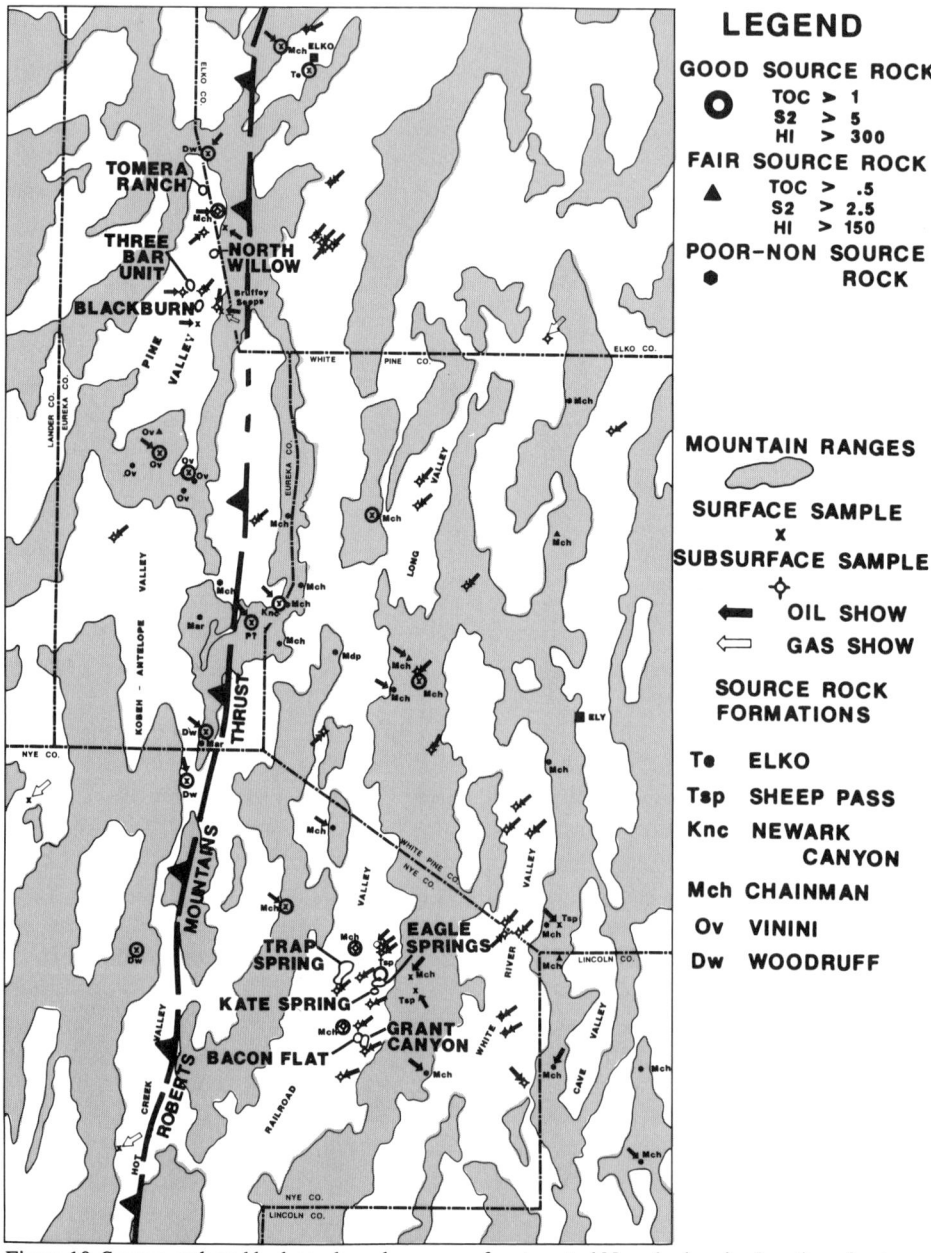

Figure 10. Source-rock and hydrocarbon-show map of east-central Nevada showing location of outcrop and subsurface samples identified as having source rock potential, surface and subsurface hydrocarbon shows, producing oil fields, and major tectonic elements. Major sources of published information include Harris and others (1980), Bortz (1983), Poole and others (1983), and Poole and Claypool (1984).

Reservoirs

Oil has been produced from rocks ranging in age from Cambrian to Tertiary. The lithologies vary widely. Oil apparently accumulates in lithologies of any age that are porous and/or permeable and are favorably located structurally within a trap.

Oil is produced from ignimbrite flows or ash-flow tuffs of the Oligocene Garrett Ranch Group at Eagle Springs and Trap Spring fields in Railroad Valley, Nevada, and at Blackburn field in Pine Valley, Nevada. Good descriptions of ignimbrites can be obtained by referring to French and Freeman (1979) and Dolly (1979). Ignimbrites are zones from bottom to top as follows: a thin basal vitric-clastic zone, a thin glass zone, a thick crystal-vitric zone (welded tuff), and a thin top zone of pumice. The welded tuff zones, which have excellent columnar joints and fracture easily due to folding and faulting, form the main producing intervals. The zones may be 75 to 100 m thick. The entire volcanic sequence of ignimbrites may be more than 1,000 m thick. At Trap Spring field there is a 670-m oil column—oil is produced from a stacked sequence of welded tuffs interbedded with the thinner zones de-

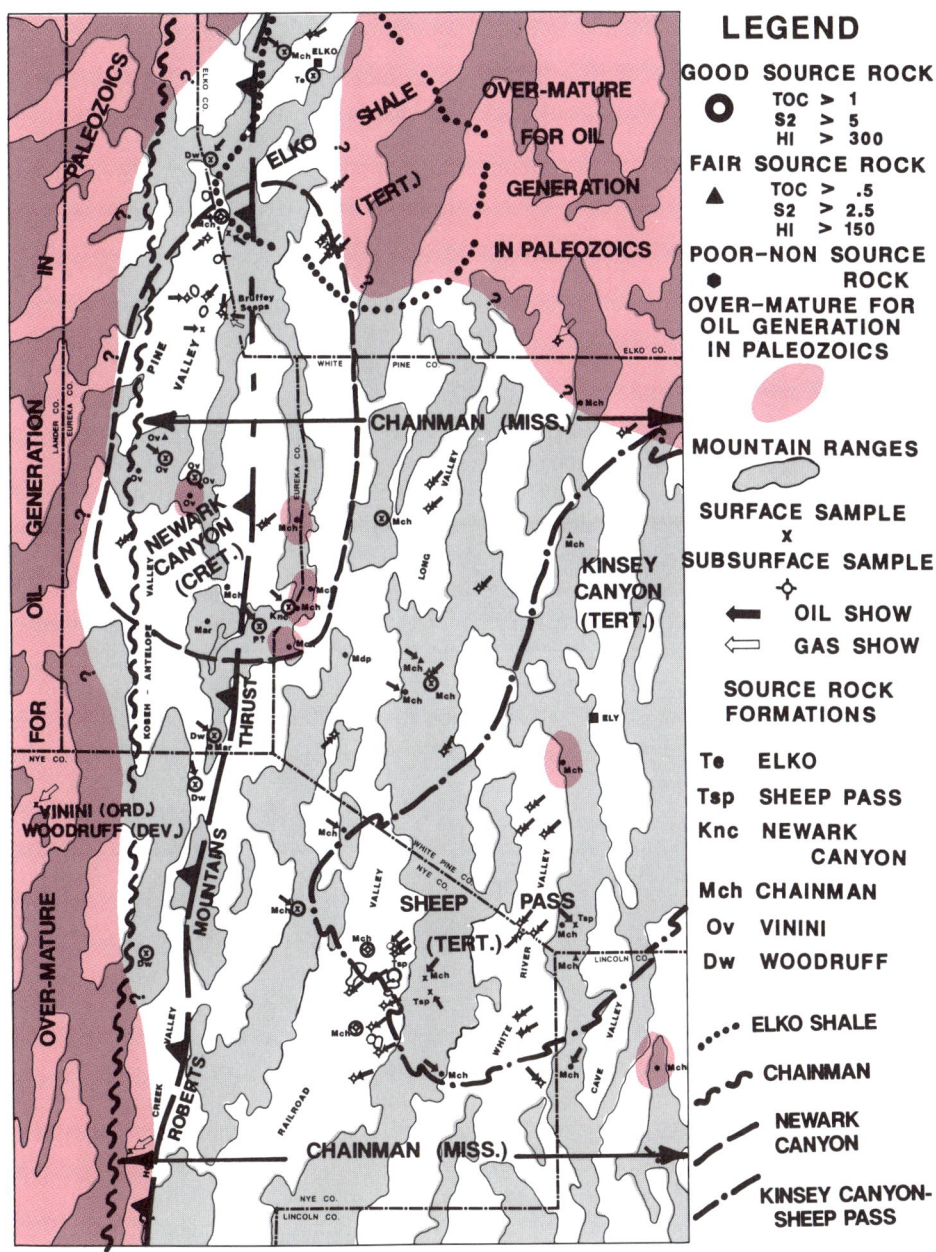

Figure 11. Source-rock distribution and maturity map of east-central Nevada showing distribution and thermal maturation of major source rock units. Generalized stratigraphic column of east-central Nevada shows major reservoirs and source-rock units. Major sources of published information include Harris and others (1980), Poole and others (1983), Bortz (1983), Poole and Claypool (1984).

scribed above. Welded tuffs have been truncated by erosion and subcrop beneath the valley-fill sequence. Effective porosity and permeability is provided by fracturing or columnar jointing. Wells completed in volcanic reservoirs generally produce oil at high rates due to the fracturing. For example, a well at Trap Springs flowed at a sustained rate of 3,000 barrels of oil per day for a period of time.

The Eocene Sheep Pass Formation has produced some oil at the Eagle Springs field from sandstones and limestones within the lacustrine sequence. The production is a million barrels of oil or less. Good porosity and permeability exist within the sequence so that these types of Tertiary beds could provide good reservoirs if found in a favorable structural position.

The Permian Ely Formation has produced less than 10,000 barrels of oil at two wells in Eagle Springs field from low-permeability, slightly fractured limestone (Dolly, 1979). In addition, production has recently been obtained from the Ely Formation in the Kate Spring field (Flanagan, 1988).

Oil is produced from interbedded fractured sandstones and siltstones in the Chainman Shale of Mississippian age at the Blackburn field. These sands have locally good porosity and permeability, and have had good oil shows in several other areas, both in the valley grabens and mountain horst blocks.

The Devonian Guilmette and Simonson Formations have provided spectacular production in the Grant Canyon field. Two wells have been flowing steadily at rates of about 2,000 and 4,000 barrels of oil per day since about 1983. The reservoir at Grant Canyon is composed of highly fractured, vuggy dolomite. Read and Zogg (1988, p. 25) have described three types of porosity from core studies: "(1) moldic porosity which resulted from early dissolution of stromotoporoid colonies and other fossil material, (2) quartz-lined fracture porosity, and (3) open-fracture porosity. Intercrystalline porosity in the dolomites is rare." Fracturing is caused by Basin and Range extensional faulting, and Read and Zogg (1988) have cited possible early and middle Tertiary solution collapse during erosion and karst development. The oil column at Grant Canyon is about 168 m.

At Blackburn field in Pine Valley, which produces oil from the Nevada dolomite, Hulen and others (1990) have shown that hydrothermal fracturing of the dolomites is important in producing fractures. They believe nearby intrusives provided the heat source. The reservoirs at Grant Canyon and Bacon Flat fields (which also produces from fractured Guilmette) have probably experienced hydrothermal fracturing. The Troy Canyon intrusive of Cretaceous age has been penetrated in nearby wells, and the reservoir temperature is abnormally high at Grant Canyon and Bacon Flat (Veal and others, 1988b). However, Zogg (written communication, 1990) states that, "There is no evidence of any connection between these two. The intrusive is Cretaceous and was emplaced at approximately 12.5 km depth (Fryxell, 1988). The reservoir and trap can be no older than mid-Miocene, by which time the Troy Granite had to be nearly exposed near the surface. By the time it had been uplifted and had 12.5 km of overlying rock stripped off, it was certainly cool and not providing "abnormal" temperatures to the reservoir. Now, the volcanic activity in the Pancake Range is another story. . . ."

Many other porous and/or permeable units are potential oil reservoirs in the Great Basin. For example, the Soda Springs well (Fig. 6B) recovered 35° API gravity oil from highly fractured Cambrian quartzite overlain by chloritic schists at a depth of about 2,380 m on both drillstem and production tests. About 150 barrels of oil were produced along with water from this zone before the well was abandoned.

Traps

The productive fields found to date in the Great Basin are truncation fault-block traps. Similar fields of this type will probably continue to be the major trap type found in this province. However, the potential for a wide variety of other trap types exists in the area. Of significance may be unconformity subcrop traps in which one or more porous reservoir units, separated by tight nonreservoir beds that can provide effective bottom seals, are truncated updip along the regional valley-fill unconformity. Obviously, this unconformity surface is of great significance, because most of the oil accumulations found to date rely on this surface to provide the ultimate top seal for the accumulations. Stratigraphic units that may lend themselves to this type of subcrop wedge trap include fractured ignimbrite reservoirs such as those producing at Eagle Springs and at Trap Spring. These are typically separated from overlying and underlying brittle units by softer tuffaceous units and volcanic clastic sediments, which are not fractured. In addition, individual units of lacustrine sandstones and oolitic carbonate shoals, developed in the Sheep Pass and in similar-age lacustrine sediments along the margins of the lakes in which they were deposited, are typically separated from one another by sealing units of shale. Sandstones developed locally in the Chainman Shale section are also obvious targets for subcrop traps beneath the Tertiary unconformity. Similar packages of intercalated porous and low-permeability reservoir units can be expected in other stratigraphic intervals as well. At both Eagle Springs and Trap Spring fields, different oil-water contacts have been observed in different parts of the fields in different stratigraphic units beneath the unconformity. Some of these variations are undoubtedly caused by fault separation within the overall field, and perhaps by hydrodynamic tilting of the oil-water contacts, but subcrops of individual units beneath the unconformity, each with their own individual oil-water contact, could also play an important role in these accumulations.

Drilling either updip from a tight well beneath the unconformity or downdip from a mapped oil-water contact may be a high risk, but could result in the discovery of hydrocarbon reserves on the flanks of subsurface high blocks where progressively younger reservoir units might be expected to subcrop beneath the unconformity, and in structurally low reentrants where the subcropping units could be trapped in a favorable updip position. In addition, regional subcrops of reservoir units beneath the unconformity on the gentle flanks of these asymmetric basins, which typically are in the form of tilted half-grabens, could result in areally significant accumulations.

Another potential trap type associated with the fault-block traps may be structural closure within deeper units where interbedded shales or other low-permeability lithologies provide separate top seals at a considerable distance below the unconformity. For example, in a fault-block trap there is no reason why the Chainman Shales shouldn't provide a top seal for interbedded sandstones of the underlying Joanna limestone, or the underlying Devonian carbonates such as the Guilmette Formation. Boundary faults could provide the requisite side seals.

Numerous folded and thrusted structures are present in outcrop within the horst blocks. The geologic history of the Great Basin had several episodes of folding and thrusting during late Paleozoic time and throughout the Mesozoic. The Antler, Golconda, and Sevier orogenic belts represent culminations of this compressional tectonic activity. Folded and thrusted structures that are masked by valley-fill sediments are also present in the graben blocks as well. Much exploration from the earliest days to the present time has focused on drilling folded structures in the horst blocks, so far without success. Some wells have also been drilled for folded structures in the valley grabens, also without success. On horst blocks, surface geological mapping of structures combined with seismic data has been the principal exploration approach; in valley grabens, seismic data combined with gravity data and subsurface well control is the usual approach. Exploration for both kinds of structures has been hampered by the relatively poor quality of seismic data obtained on horst blocks and below the valley fill in the grabens. The overprint of Basin and Range normal faulting has also complicated interpretation of these structures, and has probably caused breaching and destruction of some of the older closures. Even so, the potential for future petroleum accumulations is present in folded and thrusted structures in the Basin and Range.

CONCLUSION

The Great Basin is one of the few remaining frontier exploration provinces onshore in the continental United States. Large areas with hydrocarbon potential remain largely undrilled. Even though complex geology, remoteness from markets, lack of reliable data, and inability of existing exploration techniques to precisely identify exploration targets combine to create somewhat formidable exploration and economic hurdles, continued research and exploration activity should ultimately result in the further development of hydrocarbon resources in this area.

REFERENCES CITED

Bartley, J. M., and Martin, M. W., 1986, Structural geology of the Freiberg thrust, Worthington Mountains and Quinn Canyon Range, Nevada: Geological Society of America Abstracts with Programs, v. 18, p. 83.

Bartley, J. M., Murray, M. E., and Wright, S. D., 1985, Mesozoic thrusts and folds in the northern Quinn Canyon Range, Nye County, Nevada: Geological Society of America Abstracts with Programs, v. 17, p. 340.

Bortz, L. C., 1983, Hydrocarbons in the northern Basin and Range, Nevada and Utah: Geothermal Resource Council Special Report 13, p. 179–197.

Bortz, L. C., Cooke, S. A., and Morrison, O. J., 1985, Great Salt Lake area, Utah, in Gries, R. R., and Dyer, R. C., eds., Seismic exploration of the Rocky Mountain region: Rocky Mountain Association of Geologists and Denver Geophysical Society, p. 275–282.

Coney, P. J., 1980, Cordilleran metamorphic core complexes; An overview, in Crittenden, M. D., Jr., Coney, P. J., and Davis, G. K., eds., Cordilleran metamorphic core complexes: Geological Society of America Memoir 153, p. 7–34.

Cook, H. E., 1988, Overview; Geologic history and carbonate petroleum reservoirs of the Basin and Range Province, western United States in Carbonate Symposium: Rocky Mountain Association of Geologists, p. 213–228.

Dolly, E. D., 1979, Geological techniques utilized in Trap Spring field discovery, Railroad Valley, Nye County, Nevada, in Newman, G. W., and Goode, H. D., eds., Basin and Range Symposium: Rocky Mountain Association of Geologists and Utah Geological Association, p. 455–467.

Duey, H. D., 1979, Trap Spring oil field, Nye County, Nevada, in Newman, G. W., and Goode, H. D., eds., Basin and Range Symposium: Rocky Mountain Association of Geologists and Utah Geological Association, p. 469–476.

Eaton, G. P., 1979, Regional geophysics, Cenozoic tectonics, and geologic resources of the Basin and Range Province and adjoining regions, in Newman, G. W., and Goode, H. D., eds., Basin and Range Symposium: Rocky Mountain Association of Geologists and Utah Geological Association, p. 11–39.

Ekren, E. B., Rogers, C. L., Anderson, R. E., and Orkild, P. P., 1968, Age of Basin and Range normal faults in Nevada Test Site and Nellis Air Force Range, Nevada, in Eckel, E. B., eds., Nevada Test Site: Geological Society of America Memoir 110, p. 247–250.

Flanagan, D. M., 1988, Kate Spring Field discovery, Nevada Basin and Range: The Mountain Geologist, v. 25, p. 159–169.

Foster, N. H., 1979, Geomorphic exploration used in the discovery of Trap Spring oil field, Nye County, Nevada, in Newman, G. W., and Goode, H. D., eds., Basin and Range Symposium: Rocky Mountain Association of Geologists and Utah Geological Association, p. 478–486.

Foster, N. H., Howard, E. L., Meissner, F. F., and Veal, H. K., 1979, The Bruffey oil and gas seeps, Pine Valley, Nevada, in Newman, G. W., and Goode, H. D., eds., Basin and Range Symposium: Rocky Mountain Association of Geologists and Utah Geological Association, p. 531–540.

Foster, N. H., Vreeland, J. H., and Dolly, E. D., 1985, Seismic profiles in Railroad Valley, Nye County, Nevada, in Gries, R. R., and Dyer, R. C., eds., Seismic exploration of the Rocky Mountains: Rocky Mountain Association of Geologists, p. 283–288.

Fouch, T. D., Hanley, J. H., and Forester, R. M., 1979, Preliminary correlation of Cretaceous and Paleogene lacustrine and related nonmarine sedimentary and volcanic rocks in parts of the eastern Great Basin of Nevada and Utah, in Newman, G. W., and Goode, H. D., eds., Basin and Range Symposium: Rocky Mountain Association of Geologists and Utah Geological Association, p. 305–312.

French, D. E., and Freeman, K. J., 1979, Tertiary volcanic stratigraphy of Trap Spring field, Nye County, Nevada, in Newman, G. W., and Goode, H. D., eds., Basin and Range Symposium: Rocky Mountain Association of Geologists and Utah Geological Association, p. 487–502.

Fryxell, J. E., 1988, Geologic map and descriptions of stratigraphy and structure of the west-central Grant Range, Nye County, Nevada: Geological Society of America Map Chart Series MCH-064, 16 p.

Harris, A. G., Wardlaw, B. R., Rust, C. C., and Merrill, G. K., 1980, Maps for assessing thermal maturity (conodont color alteration index maps) in Ordovician through Triassic rocks in Nevada and Utah and adjacent parts of Idaho and California: U.S. Geological Survey Miscellaneous Investigations Series Map I-1249, scale 1:2,500,000.

Hintze, L. F., 1973, Geologic history of Utah: Provo, Utah, Brigham Young University Geological Studies, v. 20, pt. 3, Studies for Students 8, 181 p.

Hulen, J. B., Bereskin, S. R., and Bortz, L. C., 1990, High-temperature hydrothermal origin for fractured carbonate reservoirs in the Blackburn oil field, Nevada: American Association of Petroleum Geologists Bulletin, v. 74, p. 1262-1272.

Hunt, C. B., 1979, The Great Basin; An overview and hypothesis of its history, *in* Newman, G. W., and Goode, H. D., eds., Basin and Range Symposium: Rocky Mountain Association of Geologists and Utah Geological Association, p. 1-9.

Lund, K., 1986, Structural history of the northern Grant Range, east-central Nevada; Overprinting of structural styles: Geological Society of America Abstracts with Programs, v. 18, p. 392.

Martin, M. W., and Bartley, J. M., 1986, Brittle-ductile transition associated with syntectonic intrusion, Worthington Mountains, Nevada: Geological Society of America Abstracts with Programs, v. 18, p. 154.

Matthews, V., III, and Anderson, C. E., 1973, Yellowstone convection plume and break-up of the western United States: Nature, v. 243, p. 158-159.

McKee, E. H., 1971, Tertiary igneous chronology of the Great Basin of western United States; Implications for tectonic models: Geological Society of America Bulletin, v. 82, p. 3497-3502.

McKee, E. H., Noble, D. C., and Silberling, M. L., 1970, Middle Miocene hiatus in volcanic activity in the Great Basin area of the western United States: Earth and Planetary Science Letters, v. 8, p. 93-96.

Menard, H. W., Jr., 1960, The East Pacific Rise: Science, v. 132, no. 3441, p. 1737-1746.

Murray, D. K., and Bortz, L. C., 1967, Eagle Springs oil field, Railroad Valley, Nye County, Nevada: American Association of Petroleum Geologists Bulletin, v. 51, p. 2133-2145.

Osmond, J. C., and Elias, D. W., 1971, Possible future petroleum resources of Great Basin; Nevada and western Utah, *in* Cram, I. H., ed., Future petroleum provinces of the United States; Their geology and potential: American Association of Petroleum Geologists, v. 1, p. 413-430.

Peterson, J. A., 1988, Eastern Great Basin and Snake River downwarp; Geology and petroleum resources: U.S. Geological Survey Open-File Report 88-50, 51 p.

Poole, F. G., and Claypool, G. E., 1984, Petroleum source-rock potential and crude-oil correlation in the Great Basin, *in* Woodward, J., Meissner, F. F., and Clayton, J. L., eds., Hydrocarbon source rocks of the greater Rocky Mountain region: Rocky Mountain Association of Geologists, p. 179-229.

Poole, F. G., Claypool, G. E., and Fouch, T. D., 1983, Major episodes of petroleum generation in part of the northern Great Basin: Geothermal Resources Council Special Report 13, p. 207-213.

Read, D. L., and Zogg, W. D., 1988, Description and origin of the Devonian dolomite oil reservoir, Grant Canyon field, Nye County, Nevada, *in* Goulsby, S. M., and Longman, M. W., eds., The occurrence and petrophysical properties of carbonate reservoirs in the Rocky Mountain region: Rocky Mountain Association of Geologists, p. 22-32.

Roberts, R. J., 1968, Tectonic framework of the Great Basin, *in* V. H. McNutt Geology Department Colloquim Series 1, A coast to coast tectonic study of the United States: Rolla, University of Missouri Journal 1, p. 101-119.

Silberling, N. J., 1973, Geologic events during Permian-Triassic time along the Pacific margin of the United States, *in* Logan, A., and Hills, L. V., eds., The Permian and Triassic Systems and their mutual boundary: Alberta Society of Petroleum Geologists, p. 345-362.

Silberling, N. J., and Roberts, R. J., 1962, Pre-Tertiary stratigraphy and structure of northwestern Nevada: Geological Society of America Special Paper 72, 53 p.

Stewart, J. H., 1980, Geology of Nevada; A discussion to accompany the geologic map of Nevada: Nevada Bureau of Mines and Geology Special Publication 4, 136 p.

Stewart, J. H., Stevens, C. H., and Fritsche, A. E., eds., 1977, Paleozoic paleogeography of the western United States: Pacific Section, Society of Economic Paleontologists and Mineralogists Pacific Coast Paleogeography Symposium 1, 502 p.

Thompson, G. A., and Burke, D. B., 1974, Regional geophysics of the Basin and Range Province: Earth and Planetary Science Annual Review, v. 2, p. 213-238.

Veal, H. K., Duey, H. D., Bortz, L. C., and Foster, N. H., 1988a, Grant Canyon and Bacon Flat oil fields, Nye County, Nevada: Oil and Gas Journal, Part 1, March 28, p. 67-70, Part 2, April 4, p. 56-60.

—— , 1988b, Grant Canyon and Bacon Flat oil fields, Nye County, Nevada: The Mountain Geologist, v. 25, p. 193-209.

MANUSCRIPT ACCEPTED BY THE SOCIETY NOVEMBER 29, 1990

Chapter 26

San Joaquin basin, California

David C. Callaway and Ernest W. Rennie, Jr.*
ARCO Oil and Gas Company, Box 1346, Houston, Texas 77251

INTRODUCTION

Location

The San Joaquin basin is located in the south half of the topographically expressed Great Valley of California and is separated from the Sacramento Basin, which occupies the north half of the Great Valley, by the Stockton Arch fault (Fig. 1). The San Joaquin basin is a northwest trending basin 380 km long and 65–80 km wide. It is bounded on the north by the Stockton Arch fault, on the east by the Sierra Nevada Range, on the south by the Tehachapi Mountains and San Emigdio Mountains, and on the west by the Temblor Range, San Andreas Fault Zone, and Diablo Range.

General Geologic Characteristics

The Sacramento and San Joaquin basins, developed during late Mesozoic and early Cenozoic time as marine forearc basins (Ingersoll, 1979). During late Cenozoic time, subduction along the continental margin was replaced by right lateral transform shear along the San Andreas Fault Zone, which isolated the basin from the ocean to the west. Consequently, the forearc region evolved from a terraced forearc, with a deep forearc trough, to a shelved forearc (Dickinson and Seely, 1979). According to Klemme's (1984) classification, the San Joaquin is a continental rifted basin with a transform rifted margin (III Bb), developed under regional stress conditions of extension plus wrench compression.

As defined by today's basement configuration, the basin is sharply asymmetrical, with the major synclinal axis parallel and close to the western basin margin. The east side dips gently southwestward and is characterized by high-angle extension faults. The west side is highly folded and faulted, responding to compressional forces set up by shear along the San Andreas Fault Zone (Harding, 1976). The north end of the basin is terminated by the Stockton Arch fault (Fig. 1), a high angle reverse fault with 500 m of down to the north separation and 3,200 m of lateral displacement (Teitsworth, 1964). The fault is dated as post Eocene, but growth south of the fault, beginning in Paleocene time, is postulated (W. F. Edmondson, personal communication, 1986). Cherven (1983) suggests that an opposite sense of displacement (down to the south) was present during Cretaceous deposition.

Early workers interpreted the Sacramento and San Joaquin Basins as being separate and distinct. However, as more subsurface control was accumulated and an improved understanding of early and imprecise paleotime and paleoenvironmental interpretations was gained, it became obvious that the Stockton Arch fault was not a barrier to Cretaceous deposition (Edmondson, 1967). Consequently, this paper will deal with both the Sacramento and San Joaquin basins when discussing the Cretaceous.

An additional feature in the center of the basin on the western periphery is called the Vallecitos Trough. It is thought to be a seaway connecting the basin with the open ocean from Late Cretaceous to at least early Miocene time.

"Basement" is used here as an exploration term signifying the base of the hydrocarbon prospective sediments. Thus, "basement" is lithologically heterogeneous. Underlying the eastern part of the basin, it is pre–Upper Cretaceous Sierran granite with some schistose rocks, both meta-sedimentary and meta-igneous, sometimes interpreted as inliers or roof pendants in the magmatic material. Beneath the west side of the basin, the basement is Franciscan subduction (ophiolite) complex and a melange of Jurassic and Cretaceous sandstone, shale, and chert.

Basin Development. From the Cenomanian through the Campanian, Mesozoic basinal sediments prograded eastward onto the subsiding west facing edge of the Sierra (Ingersoll, 1979), filling the Sacramento basin and the northern half of the San Joaquin basin. The southern half remained emergent. Toward the end of the Campanian, sediment supply overcame basin subsidence, allowing prograding shelf clastic sediments to fill the basin from the north and the east while coeval basinal deposition continued in the west. More than 7,500 m of Cretaceous sedi-

---

*Present address: 674 Country Square Drive, Suite 103, Ventura, California 93003.

Callaway, D. C., and Rennie, E. W., Jr., 1991, San Joaquin basin, California, *in* Gluskoter, H. J., Rice, D. D., and Taylor, R. B., eds., Economic Geology, U.S.: Boulder, Colorado, Geological Society of America, The Geology of North America, v. P-2.

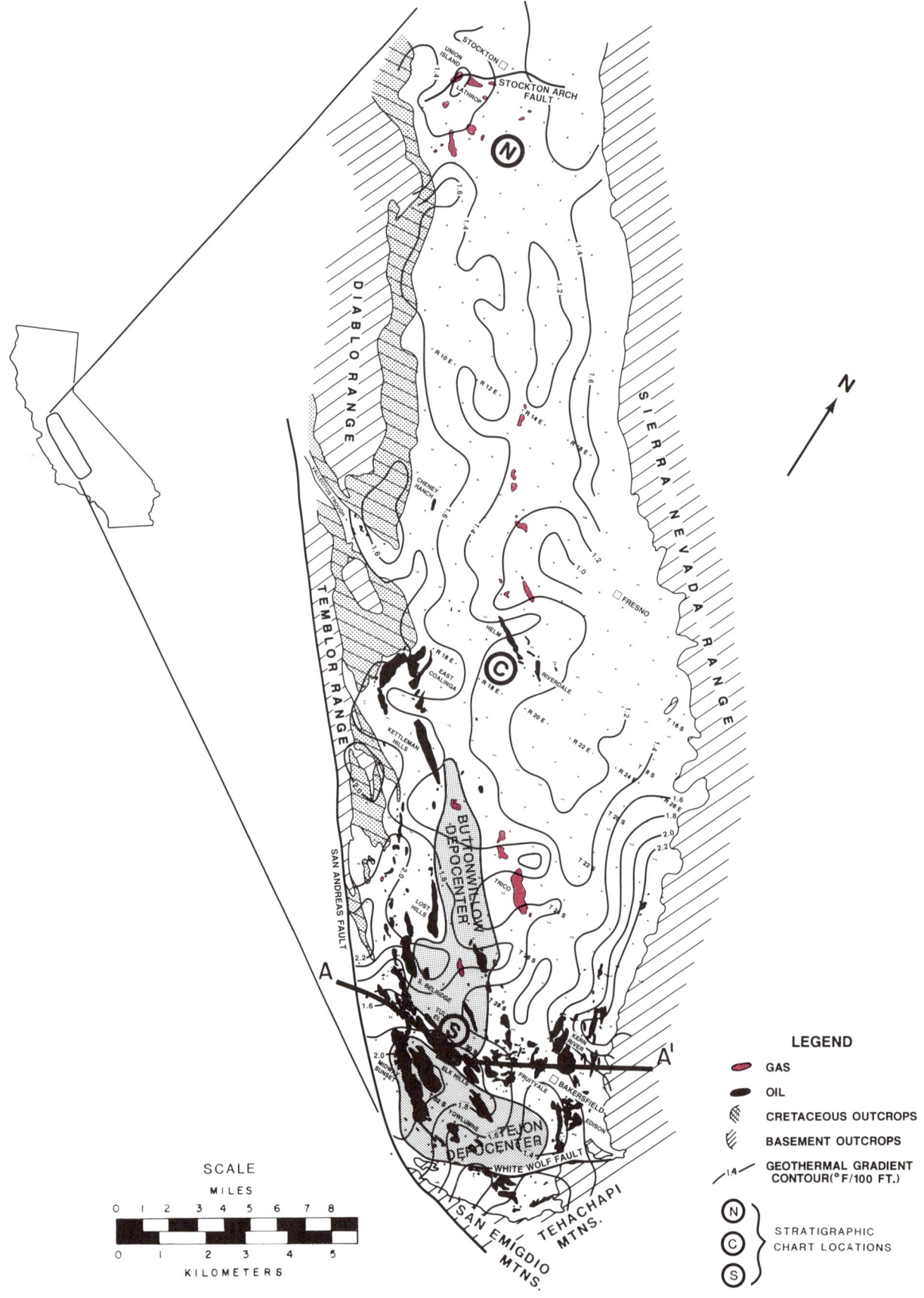

ments were deposited in the Sacramento basin and the northern part of the San Joaquin basin (Hackel, 1966).

Uplift south of the Stockton Arch Fault, extending as far south as Fresno, separated the two loci for Tertiary deposition: the Sacramento Basin and the southern San Joaquin basin. In the Sacramento Basin, post-Cretaceous sediments reached a maximum thickness of 3,500 m. In the southern San Joaquin basin, two depocenters developed—the Buttonwillow and Tejon (Fig. 1)—separated by the cross-basin Bakersfield Arch. The latter is a structurally high, cross-basin feature located in the approximate position of the line of section A-A' (Fig. 1; Plate 6F). Maximum post-Cretaceous deposition for the Buttonwillow and Tejon depocenters was 5,000 m and 9,000 m, respectively (Zieglar and Spotts, 1978). By combining more than 6,000 m of Cretaceous sediments (Hackel, 1966) with the post-Cretaceous in the Buttonwillow depocenter, the San Joaquin Basin can be described as filled entirely by clastic sediments to a thickness exceeding 11,000 m.

Production History. The Sacramento basin is almost exclusively a gas province; the San Joaquin basin, on the other hand, is largely an oil province, with the oil production being localized predominantly in the south half of the basin. The north half of the basin has minor dry gas production, most of which is adjacent to the Stockton Arch fault.

As of January 1, 1986, 117 oil fields had produced a cumulative total of 9.2 billion barrels (bbbl) of oil and 9.7 trillion cubic feet (tcf) of associated gas. Proven reserves are estimated to be 3.1 bbbl of oil and 1.4 tcf of gas. Nineteen dry gas fields have produced 0.9 tcf of gas with an estimated proven reserve of 0.134 tcf (California Division of Oil and Gas, 1986). Nineteen of the 117 oil fields are classified as giants (more than 100 mbbl), with two having produced more than 1 billion barrels. The deepest oil production at present is from 4,430 m at Rio Viejo, located 11 km east of Youlumne. The deepest oil production (now abandoned) was from 5,500 m at Semitropic, located 19 km east of Lost Hills. Part of one field (McKittrick) has been in pilot study for production by open pit mining (California Division of Oil and Gas, 1985).

Although the San Joaquin basin is an important oil province, it is gaining even greater recognition for its *heavy* oil production. Oil gravities (in degrees API) range from a low of 8° to a high of 63°, with the average being 27°. A breakdown of 432 individual reservoirs shows oil gravities grouped as follows: 138 reservoirs with gravities between 10° and 20°, 100 with gravities between 20° and 30°, 146 with gravities between 30° and 40°, 39 with gravities between 40° and 50°, and 9 with gravities over 50°

Figure 1. Index map of the San Joaquin basin showing geothermal gradient contours (modified from U.S. Geological Survey, 1975), oil and gas fields, mountain ranges, major faults, basement and Cretaceous outcrops, the location of the Buttonwillow and Tejon depocenters (modified from Zieglar and Spotts, 1978), location of cross section A-A' (Plate 6F), and location of stratigraphic sections N, C, and S shown in Figure 2.

(California Oil and Gas Fields, 1982). Higher gravity oil is usually from older reservoirs—i.e., Miocene, Oligocene, and Eocene. The low gravity oil, the result of bacterial degradation, usually occurs in beds shallower than 600 m of Late Miocene, Pliocene, and Pleistocene age.

Much of the oil produced and reserves added since the 1960s have been the result of improved production technology. Before the 1960s, many "heavy oil" wells produced less than 10 barrels per day. These fields could expect to ultimately recover 10 to 15 percent or less of their oil in place. Early stimulation methods utilized bottom hole heaters and fire flooding to reduce oil viscosity and improve oil recoveries. These were partially effective, but the discovery by Shell Oil of the "huff and puff," or the steam soak method of reservoir stimulation, was a vast improvement that revived many of the old, low gravity oil fields.

Enhanced oil recovery (EOR) projects include: (1) *Combustion* in the Midway Sunset, South Belridge, and Lost Hills fields; (2) CO_2 *miscible* at North Coles Levee field; (3) *Micellar-polymer* at East Coalinga field; (4) *Hot water* in the Kern River, Midway Sunset, and McKittrick fields; and (5) *Steam* in the North Belridge, South Belridge, Blackwells Corner, Coalinga, Cymric, Edison, Fruitvale, Kern Bluff, Kern Front, Kern River, Lost Hills, Midway Sunset, Poso Creek, and Tejon-Grapevine fields (135 individual steam projects) (Leonard, 1986). To emphasize the impact of steam EOR on the basin's production, oil fields of the San Joaquin Basin produced 256.5 mbbl of oil (8 percent of the nation's total in 1984). Steam injection was responsible for 59 percent of that production.

Exploration History. Oil was first discovered in the Great Valley at the Coalinga oil field in 1890. Early exploration centered near oil seeps and then moved on to surface expressed anticlines. Utilizing subsurface well control and surface geologic data, the search continued for the hidden anticlinal and fault trap fields, an approach still used effectively today. With the advent of reflection seismology in the mid 1930s, a number of large anticlines were defined, features that had no surface expression. Although major oil companies relied heavily on seismic data from the inception of the technique, severe surface weathering problems, rapid subsurface stratigraphic changes with commensurate strong velocity gradients, and heavy multiple masking of weak real events reduced the effectiveness of the seismic method. Persistent pursuit of prospects by subsurface correlation following up seismic "misses" found most fields before common depth point (CDP) seismic methods came into their own in the mid-1960s.

STRUCTURE

The San Joaquin basin has two basic structural styles. The east side of the basin is characterized by extension high angle normal faults. Conversely, the west side of the basin, deformed by compression and shear, is characterized by large, usually asymmetric anticlines and fault types that range from thrusts near the basin edge to high angle reverse to high angle normal faults. The difference in structural style may be a result in part of a change in basement type, from ductile Franciscan rocks on the west to rigid, stable granite on the east (Kuespert, 1983).

Fold axes on the west side trend northwest. Those folds nearest the basin edge are en echelon, are more westerly oriented, and impinge onto the San Andreas Fault Zone. Basinward anticlines are oriented parallel to the longitudinal axis of the basin. This structural style is considered by Harding (1976) to be caused by the right lateral strike slip movement of the San Andreas Fault.

Generally, the timing of structural deformation on the west side of the basin, as described by Harding (1976), is older near the basin edge and younger toward the basin center. Earliest folding (post–late Eocene to pre-Oligocene) affected the Belgian Anticline Field area closest to the San Andreas Fault. Early Miocene growth affected some nearby, slightly basinward fields and was repeated at Belgian Anticline. The same pattern of structural growth on basinward fields and repeated growth on previously affected, more westward fields continued in late Miocene and Pliocene time and into Pleistocene time.

From a different point of view, recent work utilizing the progression, internal stratigraphy, and lithologic characteristics of tectonostratigraphic sequences, combined with the kinematic analysis of regional cross sections, favors a fold and thrust belt model for the development of west side structure and casts doubt on the wrench fault tectonic model (Davis and Lagoe, 1984; Namson and Davis, 1984).

Namson and Davis (1984) describe the structures of the southwest border of the basin as two-phased and compressional: (1) Mid-Tertiary (30–17 Ma) development of a west-verging fold-thrust belt, resulting in the transition of the basin from west-facing continental slope to an asymmetric, semi-closed basin, and (2) Plio-Pleistocene to Recent (less than 3 Ma) development of an east-verging fold-thrust belt coincident with a Pacific/North American plate motion change.

The south end of the basin is a combination of both west- and east-side structural styles. Here six deformational pulses have been proposed by Davis and Lagoe (1984): (1) apparent south-verging Oligocene (30 Ma) thrusting and folding, (2) late Oligocene to early Miocene (18–28 Ma) volcanism and normal faulting, (3, 4, and 5) three Miocene and Pliocene (16–18 Ma, 6–11 Ma, and 0.6–5 Ma) pulses of down-to-the-north normal faulting, and finally (6) localized north-vergent thrusting and folding that began in late Miocene (5–8 Ma) and continued until late Pliocene to Recent (less than 3 Ma).

STRATIGRAPHY

Figure 2 is a stratigraphic chart depicting west- and east-side stratigraphy for three representative areas in the San Joaquin basin. The general locations for these representative areas are indicated on Figure 1.

Nomenclature

The stratigraphy and the stratigraphic nomenclature in the San Joaquin basin are extraordinarily complex. Early defined formal formation names were originally established on the basis of molluscan stages. Though occasionally still used, they have been modified, subdivided, and sometimes replaced by informal names accepted by common usage in the petroleum industry. For example, Cretaceous outcrops on the west side of the basin were mapped early by the U.S. Geological Survey (USGS) (e.g., Anderson and Pack, 1915). Later, oil industry geologists redescribed and informally named the Upper Cretaceous units (F. M. Anderson, 1941; Bennison, 1940; Payne, 1941; J. Q. Anderson, 1941). These informal names have become commonly accepted by the entire oil industry. Still later, work by the USGS created additional formal and different names, most of which have never been fully accepted by oil industry workers. Consequently, publications by the industry (and this chapter) continue to use the informal names, while USGS publications use the formal nomenclature.

In California, formation and member names are often applied by the petroleum industry to time-rock units and have strong paleoenvironmental implications. For example, Santa Margarita implies Upper Miocene, shallow marine sandstone, whereas Stevens implies Upper Miocene, deep marine (turbidite) sandstone. There are exceptions to this custom. The late Oligocene to middle Miocene Temblor Formation is a sand-shale sequence usually considered to have a transitional marine to shallow shelf paleoenvironment (Addicott, 1973; Cooley, 1982), but the Temblor also consists of deep basin turbidites (Kuespert, 1983).

The distribution of named time-rock units also is not well defined in California. For example, the "Phacoides" (formally, the Wygal Sandstone Member of the Temblor Formation) is an upper Oligocene shallow-water marine sandstone (Addicott, 1973) areally restricted to the west side of the San Joaquin basin. Another shallow water marine sandstone, the Santa Margarita Sandstone, was originally applied to rocks in the Salinas Valley west of the San Andreas Fault but has been commonly applied by the petroleum industry to many separate upper Miocene shallow water sandstones in parts of the San Joaquin, Cuyama, Santa Maria, and Ventura basins.

The extremely complex stratigraphy of the San Joaquin basin is caused by three main factors: (1) basin configuration, i.e., a tectonically active west flank and a shallow dipping east flank; (2) four distinct, somewhat separated source areas that simultaneously contributed shelf and basinal sediments; and (3) sporadic uplift and erosion as well as downwarp and subsidence in various places at various times. This complexity is exacerbated as the dating of rock units employs mainly benthic foraminiferal stages, some of which are time transgressive. Additionally, in certain areas the rocks are sparsely fossiliferous while in others the sediments have undergone diagenetic alteration that has destroyed most or all of the fossils.

Mesozoic Sediments

From pre-Coniacian through Campanian time, the north half of the basin was filled by Sierran (east) sourced deep marine

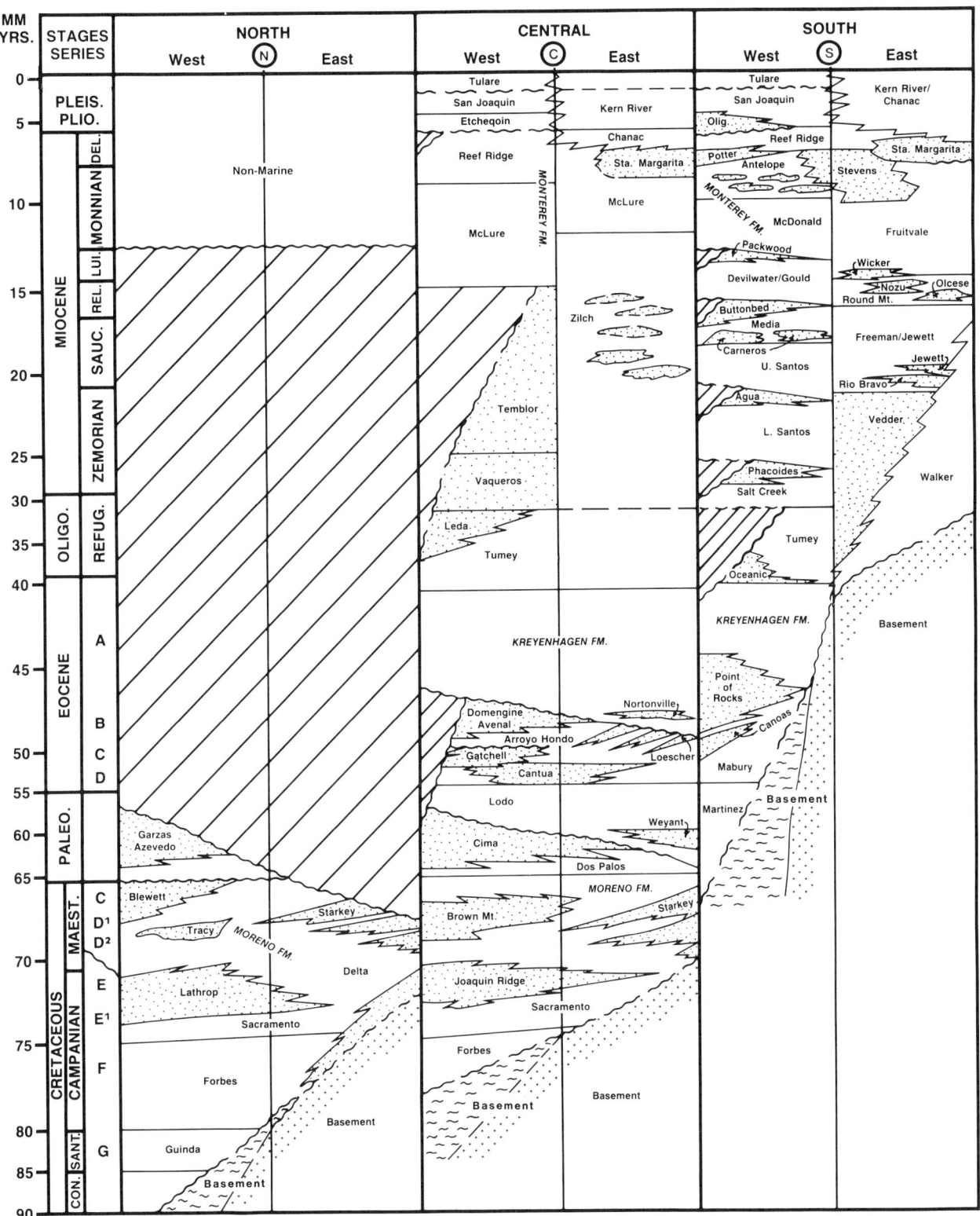

Figure 2. Stratigraphic chart for the north (N), central (C), and south (S) areas of the San Joaquin basin, each area relating the rock nomenclature of the west side to the east side of the basin. Capitalized shale formations are considered source shales. All sands except the Cima and Cantua are hydrocarbon reservoirs.

sediments deposited in a transgressive manner, onlapping Franciscan basement on the west and granitic basement on the east, while the south half was emergent. In the Maastrictian, sediment supply overwhelmed basin subsidence, resulting in the development of an easterly fringe of prograding shelf sands as well as the continued deposition of basinal sediments with numerous, complex submarine fans (Cherven, 1983). The shelf, as determined by the presence of shelf sands, close to and parallel to the east side of the basin, can be traced southward from the Sacramento Basin into the San Joaquin basin as far as the Helm and Riverdale fields, south of Fresno. As the southern half of the basin was emergent during the Cretaceous, it follows that the shelf should swing westerly somewhere between Coalinga and Bakersfield, but its exact location is still not known.

Cenozoic Sediments

Basin filling continued to the close of the Paleocene, with marine sediments restricted to an area near the Panoche Hills, located approximately 10 km southwest of the Cheney Ranch field on the west side of the basin.

At the beginning of the Eocene, the southern San Joaquin basin experienced rapid subsidence relative to the northern half of the basin. During the Eocene, subsidence progressed southward, leaving the major Buttonwillow depocenter—initially a salt filled playa (see Plate 6F) beneath the Elk Hills oil field—to be inundated with coarse clastics. These clastics were derived from the northeast, southeast, and a new western source, west of Devil's Den/Antelope Hills, located in the general area of small oil fields 20 km northwest and 5 km west of the Lost Hills field (Kappeler and others, 1984). The westerly sourced sands were shallow marine. The northerly sourced sands were primarily deep marine and possibly coeval with an areally small group of prograding shelf sands deposited on the north edge of the subsiding basin, south of Helm. The southerly and easterly sourced sands were mainly nonmarine. A rapid rise of sea level toward the close of the Eocene filled the basin with a thick blanket of organic rich, biogenic, siliceous shale: the Kreyenhagen (Issacson and Blueford, 1984).

At the beginning of the Oligocene, either falling sea level or renewed local uplift on the basin margins (more likely the latter, competing with a slowly rising sea level) restarted the cycle of prograding coarse clastic deposition. In the southern San Joaquin Basin, western, eastern, and southern sources contributed shallow marine sands and deep basin turbidites to the basin. These processes continued, with occasional minor interruptions, to the close of the Miocene.

Shales deposited in the late Miocene (coincident with a high stand of sea level) had an important role for hydrocarbon potential, as they served as both rich source and favorable reservoir rocks. Central basin shales of this type are Reef Ridge and McLure; the southern basin shales, in addition to the Reef Ridge, include the Antelope, McDonald, and Devilwater/Gould (Fig. 2). These local stratigraphic unit names are preferred by many geologists rather than the more familiar and inclusive formation name, the Monterey. These shales exceed a thickness of 1,800 m. They are biogenic siliceous rocks deposited in low energy, nutrient-rich waters. The Monterey formation is more calcareous at the base, grading upward to become more siliceous. Late Miocene cooling produced increased upwelling and abundant diatom productivity. Some areas have almost pure diatomite still preserved, whereas in other areas, diagenesis has altered the diatomite to procelanites and cherts (Graham and Williams, 1985).

The Pliocene and Pleistocene sediments represent the gradual, final fill of the basin. They are mainly nonmarine sediments with the area of marine deposition confined to the Buttonwillow and Tejon depocenters.

THERMAL HISTORY

The thermal gradient of the San Joaquin basin rocks ranges from a low of 2.2°C per 100 m to a high of 3.6°C per 100 m (Zieglar and Spotts, 1978). One well in the Trico field, reported to have a thermal gradient of 4.2°C per km based on vitrinite reflectance (Clark and Clark, 1982), is developed from data probably not accurate enough to distinguish between 3.6°C per 100 m and 4.2°C per 100 m. Figure 3 shows Lopatin diagrams (burial history–hydrocarbon generation diagrams) for three areas of the Great Valley of California: the northerly Delta depocenter in the Sacramento basin and the Buttonwillow and the Tejon depocenters in the southern San Joaquin basin (Zieglar and Spotts, 1978). Thermal gradient contours for the entire San Joaquin basin (USGS, 1975) are shown in Figure 1, which also locates the two major depocenters within the basin. The third, the Delta depocenter, is to the north, outside of the area shown in Figure 1.

A generalized west-east cross section in the south end of the San Joaquin basin (in the approximate location of the Bakersfield Arch) (Plate 6F) shows the thermal maturity of the sediments by employing two R_O value contours (heavily dashed). The contours were derived by combining data from the Zieglar and Spotts (1978) Lopatin diagrams for the Buttonwillow and Tejon depocenters with the thermal gradient contours from Figure 1. Though not defined, the writers assume the top of the oil generative window is at an R_O value of 0.6 (at approximately 120°C), and the top of the gas generative window is at an R_O value of 1.4 (at approximately 150°C). The depths to the window temperatures were then computed from the geothermal gradient profile along the section.

PRODUCING SYSTEMS

Source Rocks

Northern San Joaquin Basin. Compared to the southern San Joaquin basin, which is dominated by numerous oil fields, the northern San Joaquin basin has few fields, and they produce

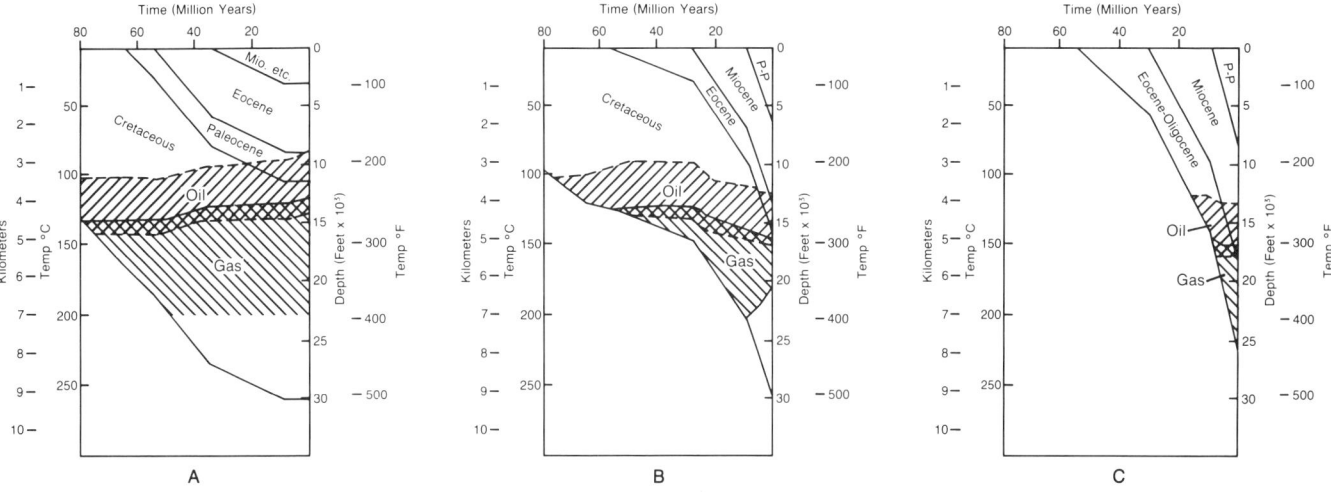

Figure 3. Lopatin diagrams for the thickest part of the following depocenters: A, Delta depocenter, Sacramento basin; B, Buttonwillow depocenter, central San Joaquin basin; and C, Tejon depocenter, southern San Joaquin basin (from Zieglar and Spotts, 1978).

mainly dry gas. Eight, at the north edge of the basin, account for 58 percent of the San Joaquin basin's total dry gas production. The remaining four, centrally located, account for an additional 12 percent of the basin's gas. A hypothesis to account for the lack of hydrocarbons in the northern San Joaquin basin can be made.

Zieglar and Spotts (1978) suggest that the Delta depocenter, 65 km north of the Stockton Arch Fault, is the only place where an adequate quantity of potential source beds is buried to depths greater than 3,000 m, where adequate thermal maturity for generation is possible. The source beds are most likely Cretaceous and possibly lower Eocene shales. Cretaceous beds generally contain less than 1 percent by weight of organic carbon, most of which is humic or nonsapropelic (Type III kerogen), and are considered gas generative. This is supported by the fact that almost all the fields in the Sacramento and northern San Joaquin basins are mainly dry gas fields. From the Delta depocenter, long distance migration, possibly along fault planes, would explain the placement of the gas fields on the northern edge of the San Joaquin basin, near the Stockton Arch Fault.

The lack of hydrocarbons in the remainder of the northern San Joaquin basin, between Stockton and Fresno, could be explained by the combination of the basin's high sand/shale ratio (67/33 on the east side of the basin and 45/55 on the west) and the low thermal gradient (Fig. 1), which results in an unfortunate coincidence of poor source rock quality, quantity, and maturity level.

Central and Southern San Joaquin Basin. Although easterly sourced sediments contain more land-derived kerogen than the westerly sourced rocks, the central and southern parts of the San Joaquin basin are largely dominated by marine amorphous, sapropelic organic material (Type I and Type II kerogen) and are therefore highly oil prone. Cenozoic shale beds usually have more than 1 percent by weight organic carbon, with some richer zones containing more than 5 percent (Zieglar and Spotts, 1978).

Monterey Shale. This thick sequence of deep basinal biogenic, siliceous shale was deposited in anoxic conditions. These fine-grained sediments were partly diluted in some places by relatively coarse (sand size) clastic input from turbiditic flows.

The Monterey Shale generally has more than 2 percent by weight organic carbon, with some zones containing more than 7 percent where penetrated by four wells on the periphery of the Buttonwillow depocenter. The maturation level in three wells shallower than 2,800 m is immature. The deepest well, at 3,470 m, found the maturation level at and slightly into the oil generation window (Graham and Williams, 1985). The most kerogen-rich lithotype in the Monterey contains laminations of quasi-stromatolitic, organic mats that have as much as 10 percent by weight organic carbon (Kruge and Williams, 1982).

The very rich, oil prone source potential of the Monterey is indicated from Rock Eval pyrolysis data (Kruge, 1983). Zones enriched in free hydrocarbons (migrated) have S_1 yields of 10 to 30 mg hydrocarbons per gram of rock. Pyrolysis yields from the conversion of kerogen to hydrocarbons (S_2, generation potential) range from 10 to 50 mg hydrocarbons per gram of rock on a bitumen-free basis. The S_2 yields are very high and indicate excellent oil source potential.

Kuespert's (1983) study of the Temblor Formation, which underlies the Monterey at North Kettleman Hills, found thermal alteration indices (TAI) that range from 0.5 to 1.5, indicating thermal immaturity, and an average Tmax from Rock-Eval pyrolysis of 428°C, which is near the very top of the oil generation window. He suggests either the Monterey or Kreyenhagen as potential source rocks for the Temblor oil.

Kreyenhagen Shale. The Eocene Kreyenhagen Shale is a biocalcareous and biosiliceous shale with thick phosphatic laminated intervals suggestive of an anoxic environment. With total organic carbon (TOC) commonly in the 2 to 5 percent range and an abundance of Type I and II kerogen, this 300- to 400-m-thick shale is considered the most likely oil source for Eocene sand reservoirs adjacent to the Buttonwillow depocenter.

Moreno Shale. The upper Cretaceous–lower Tertiary Marca Shale and Dos Palos Shale members of the Moreno Formation are laminated diatomaceous, phosphatic, and radiolarian-rich sediments deposited under anoxic conditions (McGuire and Ingle, 1984). The Marca has TOC values as high as 7.25 wt percent. Pyrolysis and vitrinite reflectance data indicate marginal maturity at relatively shallow depths for Marca samples from a well near Coalinga. McGuire (1986) suggests that where the Marca is more deeply buried to the northeast, it would be at a level of generative maturity and be particularly favorable as a potential petroleum source rock.

Reservoirs: Sandstone

The San Joaquin basin contains many sandstone units, most of which produce hydrocarbons. Zieglar and Spotts (1978) studied 165 productive sandstone reservoirs in the San Joaquin basin, ranging in age from late Cretaceous to Pleistocene and determined that initial sandstone porosity of 35 to 40 percent decreased an average of 0.56 percent per 100 m of burial. Data suggest porosity is independent of age. A minimum rate of decrease (best possible case) was 0.46 percent per 100 m, which is quite severe when compared to Gulf Coast sandstone reservoirs. The main cause of this rapid loss of porosity with depth in the San Joaquin basin is suggested to be the fact that San Joaquin rocks are more poorly sorted and more mineralogically immature than Gulf Coast rocks and are therefore more susceptible to burial diagenesis at any temperature.

Data from 54 reservoirs in 19 fields on the west side of the San Joaquin basin demonstrate an above average porosity loss rate of 0.73 percent per 100 m. A rate this high throws doubt on the possibility of having more than 15 percent porosity (a number generally considered to be the minimum producible porosity when lower viscosity oil is considered) at depths greater than 3,600 m.

Data from reservoirs in the Sacramento basin and the east side of the San Joaquin basin show a less severe rate of porosity loss with depth, about 0.42 percent per 100 m, allowing the exception of having 15 percent porosity as deep as 6,000 m.

The higher rate of porosity loss found on the west side of the San Joaquin basin was primarily attributed to postburial uplift that affects the western oil fields to a great degree. Less emphasis was put on the effects of a higher geothermal gradient and deformation (Zieglar and Spotts, 1978). Because west-side reservoirs include more metamorphic and volcanic debris than eastern reservoirs, which have a Sierran granitic provenance, we would give the role of provenance as well as the diagenetic effects of the higher geothermal gradient more importance to account for the higher rate of porosity loss in west-side reservoirs.

Clay is an important component of most sands in the basin, so much so that Schlumberger developed special logging tools for California specifically designed to handle the "dirty sand" problem.

From a study of the Yowlumne and Rio Viejo oil fields on the south end of the basin, Brasher and others (1982) found that the upper Miocene Stevens channel turbidite sands are arkoses with a quartz/feldspar ratio of 2:1. Differences in amounts of K-feldspar and plagioclase are considered important for the preservation of good porosity. K-feldspar alteration yielded mainly kaolinite clay, whereas plagioclase alteration yielded expandable clays that have a negative effect on porosity. K-feldspar altered zones at 3,400 to 4,500 m have porosities that range from 14 to 20 percent with permeabilities up to 200 md. Intervals with high plagioclase alteration have decreased reservoir quality with respect to both porosity and permeability.

Porosity vs. Permeability. In a sixteen-field study by Dixon and Kirkland (1985) which included some fields in the San Joaquin basin, data indicated a direct relationship between porosity gradient (rate of porosity loss) and the present thermal gradient. As the geothermal gradient increases, so does the porosity gradient. A similar relationship was found between permeability loss with depth and the geothermal gradient.

As seen on the porosity versus permeability plot from the Elk Hills oil field on the west side of the basin (Zieglar and Spotts, 1978), the "best reservoir limit" line indicates that as porosity decreases by 7 percent, permeability decreases by a factor of ten. The plot itself shows considerable scatter, suggesting that one would be hard put to predict permeability from porosity. The Elk Hills example shows that for a porosity of 35 percent, permeabilities range from greater than 5,000 millidarcies to less than 20 millidarcies.

The extreme variability of porosity and permeability within specific units was illustrated by a core analysis study of samples of the Miocene Temblor Formation taken from a well at the north end of the Buttonwillow depocenter (Estill, 1980). On inspection of the electric-log, a geologist probably would have assumed that the section consisted of alternating sands and shales. However, continuous cores from 3,870 to 4,050 m indicated that the zone was 100 percent sand. Porosities ranged from 0 to 22 percent, and permeabilities from 0 to 1,700 millidarcies. The porosity-permeability cross plot was definitely bimodal, with good porosity corresponding to favorable and unfavorable permeability in different parts of the unit. Good permeability in this case was tied directly to grain size and provenance.

Reservoirs: Shale and Diatomite

Mention has been made of the diatomaceous nature of the Monterey Formation. Almost pure diatomite (greater than 80 wt.

percent) has up to 62 percent porosity, but only 1 md. of permeability (Schwartz and others, 1981). It exists as biogenous opal-A (diatom frustules) (Williams, 1982). With increased depth of burial and compaction, the opal-A transforms to opal-CT accompanied by a commensurate loss of porosity to about 35 percent. A further porosity loss to about 10 percent occurs when opal-CT is transformed to quartz (Isaacs, 1981).

Highly porous, unaltered diatomite is a reservoir rock for oil in the South Belridge and McKittrick fields. In the latter case, the 832 mbbl of probable in-place oil are recoverable only by open pit mining (Mulhern and others, 1983).

The Monterey, in its brittle, quartz form, is productive as a fractured reservoir in a number of oil fields—e.g., Elk Hills, the Antelope Shale pool of the Midway Sunset field (there pure diatomite is also the seal), and Lost Hills. Although the Monterey is considered as both a source rock and a reservoir rock, it is not assumed to be both in the same place. Where it is productive, either as a diatomite or in its brittle, quartz state, it is probably too shallow to be in the oil generative window and for the oil to have been locally formed. Both Lost Hills and South Belridge fields are adjacent to the large Buttonwillow depocenter where deep Monterey rocks would be oil generative (Graham and others, 1982). Oil may have migrated to these fields through faults and fractures from depths where thermal maturation was possible (Kruge, 1983). In spite of this, the phosphatic aspect of the Monterey at Lost Hills intrigued Kruge (1983), who speculated that biogenic or early diagenetic bitumen may be the source of a portion of the oil in Lost Hills reservoirs.

The late Oligocene–early Miocene Santos Shale, a highly fractured, silty, siliceous shale with oil saturations that range from 20 to 50 percent, is a fractured shale reservoir in the Belgian Anticline field (Wilson, 1979).

Trap Types

In the San Joaquin basin, stratigraphy plays the major trapping role, with a strong structural component. To demonstrate this, we have analyzed every field and pool in the basin, zone by zone. Although there were some difficult and perhaps arbitrary decisions made, it was determined that the percentage of cumulative production found in stratigraphic, anticlinal, and fault traps was 58, 28, and 14 percent, respectively.

Stratigraphic Traps. On the west side of the basin, unconformities have created numerous truncation and buttress traps, usually in association with major marine transgressions after short periods of sand progradation. These features are found, for example, at the base of Oligocene, base of lower Miocene, base of middle Miocene, base of upper Miocene (Reef Ridge), and base of Pliocene (Etchegoin). Facies trapping was especially prevalent in the Late Miocene, occurring where sand-filled channels have crossed structural noses and anticlines—e.g., the 555 sand at Buena Vista Hills, the Asphalto field, and the 26-R and 24-Z sands at Elk Hills (Webb, 1981).

On the east side of the basin, the nonmarine Kern River, Chanac, and Zilch formations, as well as the transitional marine Etchegoin Formation, have stratigraphic traps of lenticular pods and lenses of sand that either shale out completely or become so tight as to create permeability traps.

On the Bakersfield Arch, in the southern San Joaquin basin, the upper Miocene Stevens turbidite sands were deposited on a homocline. Differential compaction of shales created highs over early channel-filling sand lobes, leaving the intervening low areas to be filled with sand later, an example of compensational deposition (MacPherson, 1978). Stratigraphic trapping relative to this style of deposition is represented by a number of smaller oil accumulations—e.g., the English Colony and Rosedale fields.

In the northern San Joaquin basin, Cretaceous turbidites provide stratigraphic trapping. Submarine fan reservoirs, found usually in midfan position, are exemplified by the Lathrop field and the adjacent Union Island field in the Sacramento Basin (Cherven, 1983). Farther south, the Cheney Ranch field is an example of a distal turbidite margin trap.

Submarine Channel Traps. In the Sacramento basin, a number of shale-filled submarine channels ("gorges") are important hydrocarbon-trapping features. Stratigraphic trapping of truncated anticlines and fault blocks, as well as erosional and channel fill topography, is developed by "gorge" shale seals. To date, there has been only one similar feature identified in the northern San Joaquin basin: the Chowchilla gorge (Callaway, 1964). In this case, however, no hydrocarbons have been trapped. MacPherson (1978) describes the Fruitvale oilfield as a submarine-canyon infill composed of nonmarine Chanac Formation sands and shales, completely sealed by overlying shales.

Anticlinal traps. Since such heavy emphasis has been put on the dominant role of stratigraphy, even on anticlinal folds, it should be stated that anticlines in the San Joaquin basin trap hydrocarbons and that they are important contributors to the basin's total hydrocarbon production. Many of these tectonic features are in the major field category: e.g., Kettleman Hills North Dome, Elk Hills, Buena Vista Hills, Ten Section, and Greeley. On the Bakersfield Arch and at the northern end of the basin, some anticlinal traps include smaller structures developed by differential compaction over sand lobes and channels: e.g., Strand and Canfield Ranch (Webb, 1981) and McMullin Ranch (Cherven, 1983).

Fault Traps. On the west side of the basin, where fault style changes from low angle overthrust close to the basin edge, to high angle reverse, and then to high angle normal nearer the basin center, traps have been found where reservoir sands are faulted against sealing shales.

Faulting on the Bakersfield Arch is related to the deposition of the upper Miocene Stevens Sand. Nearly vertical, syndepositional down-to-the-basin faults, similar to the growth faults of the Gulf Coast, are described by MacPherson (1978), who relates them to the midfan areas of the turbidite complexes. Like those in the Gulf Coast, these faults migrate basinward in response to the addition of prograding younger sediments and have accompanying downthrown (down-to-the-coast) antithetic faulting. An upper Miocene Stevens growth fault on the west side of the basin is described by Webb (1981). It is similar to the Bakersfield Arch faults, except it is down-to-the-east in response to westerly sourced sedimentation. At the confluence of two separate sand channels, the downthrown side of the fault has a threefold increase in sand thickness relative to the upthrown side. It is the site of the Tule Elk field, probably a growth faulted anticline.

The east side of the basin is gently west dipping and characterized by high angle extension faults. Hydrocarbon trapping usually occurs when down-to-the-east faults and transecting downthrown faults put westerly dipping sands in contact with downthrown shales.

Fractures. Fracture trapping has been previously commented on with regard to the Monterey and Santos shales. It should be noted that other shales, though not so important as the Monterey, are fractured reservoirs. They include the Tumey (late Eocene) and the Kreyenhagen (middle Eocene).

A variation of fault and fracture trapping is found in the Edison oil field, which has a nonmarine, nonsedimentary reservoir. Schistose basement rocks that had been faulted and fractured were expressed as a topographic hill, exposed to weathering and erosion, and later surrounded by and finally covered with marine sediments. Being structurally high with buttressing peripheral sand beds and capping marine shales, the feature became a locus for oil migration, with the fractures and faults becoming the reservoir (Bruer, 1965). This "basement type" reservoir, though unique in the San Joaquin basin, is present in other California basins.

Diagenetic Traps. Some oil fields, usually explained or described as being trapped by updip permeability barriers, have been reanalyzed and re-evaluated as diagenetic traps. The Pleasant Valley oil field (Schneeflock, 1978) is described as an accumulation in the Eocene Gatchell Sand that was positioned on an existing, gentle anticlinal high. The formation waters beneath the oil pool reacted with feldspars and micas to precipitate kaolinite, which effectively plugged the formation below the oil. Later, structural deformation eliminated the early fold, leaving the diagenetic trap frozen in position off-structure. The oil reserves in the Gatchell sand in the nearby East Coalinga, Guijarral Hills, and Kettleman North Dome oil fields are similarly categorized.

McCulloh and Stewart (1980) noted tilted oil reservoirs above diagenetic laumontite plugged formations and implied that the tilting is caused not by structural readjustment but by a steeply sloping diagenetic front that crosses gently plunging folds. The interface between the zeolitized strata and the overlying less altered beds ranges in depth from 2 to 6 km. Laumontite crystallizes from interstitial water of immature sandstone when (1) geothermal gradients equal or exceed the range set by 59°C at 1,100 m to 180°C at 4,150 m, (2) fluid pressure gradients are near 113 bar/km, and (3) the solutions have exceptionally low salinities and are depleted in dissolved carbonate species. Laumontite has been identified in 25 wells in the south end of the San Joaquin Valley, and active laumontite development in Eocene sands has been observed in selected wells on the west side of the basin at Cymric, North Belridge, and Northeast McKittrick (T. H. McCulloh, personal communication, 1986).

Hydrodynamic Traps. Normally, hydrodynamic traps are thought of as being displaced from anticlinal crests and having tilted oil/water interfaces. Three such examples are the Willmax pool, the Republic anticline pool of the Midway Sunset field, and the East Coalinga field. Another type of hydrodynamic trap, the fluid level trap, is described by Foss (1972) as an oil accumulation trapped by the plane of zero hydrostatic pressure or the potentiometric surface of the water table. This shallow phenomenon is a trapping device attributed to the Kern River field (approximately 1 billion barrels of cumulative oil production), the Tulare pool of the South Belridge field, and the Coalinga field. Similar, but a variant, is the Tulare of the Cymric-McKittrick area, where oil originally filled a stratigraphic-structural trap below the paleo regional groundwater table (RGT). The reservoir was subsequently uplifted above the RGT, resulting in gravity drainage into structural lows, the present-day reservoir. The highs are no longer reservoirs, as they are undersaturated (Chamberlain and Madrid, 1986).

FUTURE OPPORTUNITIES

With the combination of a high density of exploratory wells on the shallower flanks of the basin and an appreciable number of deep tests on the productive anticlines, it would appear that the better potential plays in the future will be mainly stratigraphic and will be found in the deeper parts of the basin.

One of the major producing horizons in the San Joaquin basin, and a candidate for continued and future exploration, is the late Miocene Stevens Formation in the south half of the southern San Joaquin. The Stevens' debut as a producing horizon was in 1936 with the discovery by Shell Oil Company of the Ten Section field, the first oil to be found in the center of the basin. As indicated in Figure 2, the Stevens is a Monterey-age sand. In the late 1950s, the Stevens began to be recognized as a turbidite deposit, but it was not until the late 1970s and early 1980s that real understanding began to develop (MacPherson, 1978; Webb, 1981). Three different source/input areas on the east, one on the south, and a possible five on the west contributed massive amounts of sand to the basin at various times throughout the late Miocene. Some structures were already in place, so the sand detoured around the folds. Some structures grew contemporaneously with sand deposition, and the successive sand flows offlapped the structures. In other instances, the sand geometry (lobe, channel, or fan) was buried, and the sequence was repeated elsewhere. This potential horizon has the fortunate circumstance of being perhaps within, or at least adjacent to, the Buttonwillow and the Tejon depocenters where the organic-rich Monterey Formation is in the oil generating window.

The middle Miocene Temblor Formation, in the center of the north half of the southern San Joaquin basin, is another target for future exploration. Internal unconformity traps in its shallow marine facies, porosity and permeability changes in its turbidite facies, and shale-outs where turbidite sands are juxtaposed or terminated against slope shale facies all present potential trapping situations. These will present a challenge for both detailed rock/log and stratigraphic-seismic analysis. The Temblor sands are overlain by Middle and Upper Miocene shales (McLure and Monterey) and in turn overlie the Kreyenhagen Shale, the two prime source rocks of the basin. The relatively recent discovery of the Tulare Lake field (1983), located approximately 19 km east of Kettleman Hills, has renewed exploration activity for this objective in the central part of the basin.

The Eocene Gatchell Sand, underlying the organic-rich Kreyenhagen Shale within and adjacent to the Buttonwillow depocenter, is another highly potential objective for future exploration for many of the same reasons cited previously for the Temblor. It has been only lightly explored and in the basin center would be found at considerably greater depths than the Temblor. Diagenesis probably would have more adverse impact on this formation and would be a critical factor in reservoir preservation.

The Monterey and Kreyenhagen Shales are two formations we consider high on the list for future exploration. These brittle, siliceous, and organic-rich source rocks, when buried deeply enough to be in the oil generating window, could develop into a combination fractured source and reservoir.

Although there has been minor production of light-gravity oil and dry gas from the Cretaceous reservoirs in the area, those rocks on the west side of the southern San Joaquin basin have been largely ignored and sparingly explored. Several large, Cretaceous exposed anticlines have been tested no deeper than 1,100 m in spite of the fact that there are 4,900 to 6,100 m of outcropping Cretaceous beds on the extreme western edge of the basin, described by Marsh (1960) as containing both impermeable and permeable marine sands.

Recent surface geochemical prospecting by Jones and Drozd (1983), in the Lost Hills area, demonstrated that the chemical compositions of near-surface hydrocarbon soil gases, measured by flame ionization chromatography, are largely controlled by the hydrocarbons found in nearby, underlying reservoirs. Surface geochemical data, in conjunction with geological and geophysical work, could possibly be used as a predictive tool to reduce some element of risk in future hydrocarbon prospecting.

OPERATIONAL FACTORS

The southern San Joaquin Basin can be divided into three distinct, east to west, linear (basin parallel) regions based on drilling problems related to mineralogy and pressure regimes. The most easterly, from the Sierra Nevada Range to a line just west of the Fruitvale field on the Bakersfield Arch, is dominated by loosely consolidated sands and silts in normal to subnormal pressured intervals and has no major drilling problems other than keeping the hole to gauge.

The central region, encompassing the middle of the valley, is dominated by mud-making claystones and shales with overpressuring in many intervals. The pressure development, however, is not necessarily consistent and definitely not wholly related to depth. For example, in the Tulare Lake area, 16 to 24 km east of Kettleman Hills, minor overpressuring in the Pliocene San Joaquin and Etchegoin formations decreases to normal pressure in the upper part of the Miocene Reef Ridge Shale. In the lower part of the Reef Ridge the pressure builds until it peaks at as much as 6.7 kg. per liter mud weight equivalent in the Miocene McLure Shale. The pressure declines to normal in the lower Temblor Sand. In the Eocene Kreyenhagen Shale the pressure abruptly and significantly increases (up to 7.25^+ kg per liter mud weight equivalent). Below the Kreyenhagen, pressure again returns to normal. Drilling problems are associated with controlling pressure and clay swelling due to high bentonite content in the sediments.

The western region is on the extreme west side of the basin where formations are more folded, faulted, and steeply dipping and are generally older and more compacted (and less water sensitive). Pressure here has been mostly relieved, but there are areas that have anomalously high pressures. Because this region generally has higher temperature gradients than the other regions, the drilling concerns here are for long-term hole integrity with good temperature stability (Clifford, personal communication, 1986).

Overbalanced mud weight, mud fluids incompatible with formation clays, thin beds, freshwater sands, and low resistivity mineralogy are additional factors throughout the San Joaquin Basin that continue to thwart the best efforts of the log analysts and production engineers.

Two aspects of land management in the state of California tend to plague oil operations. First, an inordinate amount of mineral fee is held by major oil companies. The bulk of this acreage is on the west and south sides of the basin and across the Bakersfield Arch, the prime producing areas in the basin (Callaway, 1971). A farmout large enough to support the drilling of a relatively deep well is not easy to obtain. Occasionally a large dollar/work commitment will tie up enough acreage for a sustained exploration program to function effectively, but with deep wells and a costly commitment, the smaller operators are usually excluded from this type of operation. Second, county and federal permitting, although not too large a concern in this basin, is becoming more and more time consuming.

On the plus side, the California Department of Conservation has a Division of Oil and Gas, supported by a special tax on the oil industry, that publishes an annual "Summary of Oil and Gas Operations" that provides production, reserve, and exploration statistics. The division also publishes separate oil and/or gas field studies including maps, cross sections, and reservoir data. This work is excellent and is very helpful to the industry. The state will, on request, hold all information from exploratory wells confidential for two years. Confidentiality can then be extended, upon approval, for six month periods, but extension is limited to four years.

There is an extremely functional well sample repository at California State College at Bakersfield which has the support of the entire oil industry. Contributions of dollars, cores, and service by many members of the petroleum community, both individual and organizational, have built this into a highly respected facility. Its usage goes far beyond California boundaries and the repository will be a basic aid to explorationists for years to come.

REFERENCES

Addicott, W. O., 1973, Oligocene molluscan biostratigraphy and paleontology of the lower part of the type Temblor formation, California: U.S. Geological Survey Professional Paper 791, 51 p.

Anderson, F. M., 1941, Synopsis of the Later Mesozoic in California: California Division of Mines Bulletin 118, pt. 2, p. 183–186.

Anderson, J. Q., 1941, Generalized columnar section of the Cretaceous rocks of the Alcalde Hills: Talk before the Pacific Section, Society of Economic Paleontologists and Mineralogists at Bakersfield, June 6, 1941.

Anderson, R., and Pack, R. W., 1915, Geology and oil resources of the west border of the San Joaquin Valley, north of Coalinga, California: U.S. Geological Survey Bulletin 603, 220 p.

Bennison, A., 1940, Late Cretaceous of the Diablo Range: Read at the meeting of the Le Conte Club, Stanford University, March 2, 1940.

Brasher, J. E., Pike, J. D., Tieh, T. T., and Berg, R. R., 1982, Effects of arkosic sandstone diagenesis on reservoir rock properties [abs.]: American Association of Petroleum Geologists Bulletin, v. 66, p. 552–553.

Bruer, W. G., 1965, Edison oil field abstract, in Pierce, R. L., ed., Geology of southern San Joaquin Valley, California, Kern River to Grapevine Canyon: Pacific Section American Association of Petroleum Geologists Guidebook, p. 17–19.

California Division of Oil and Gas, 1985, Seventieth annual report of the State Oil and Gas Supervisor, 1984: California Division of Oil and Gas Publication PR 06, 151 p.

——, 1986, California oil and gas statistics and new well operations for 1985: Publication PR 03, 18 p.

California Oil and Gas Fields, 1982, Central California map and data sheets: California Division of Oil and Gas, TR11-R, scale 1:500,000.

Callaway, D. C., 1964, Distribution of uppermost Cretaceous sands in the Sacramento–northern San Joaquin basins of California: San Joaquin Geological Society Selected Papers, v. 2, p. 5–18.

——, 1971, Petroleum potential of San Joaquin Basin, California, in Cram, I. H., ed., Future petroleum provinces of the United States: American Association of Petroleum Geologists Memoir 15, p. 239–253.

Chamberlain, E. R., and Madrid, V. M., 1986, Influence of uplift on oil migration; Tulare heavy oil accumulations, west side San Joaquin Valley, California [abs.]: American Association of Petroleum Geologists Bulletin, v. 70, p. 940.

Cherven, V. B., 1983, A delta-slope-submarine model for Maestrichtian part of Great Valley sequence, Sacramento and San Joaquin basins, California: American Association of Petroleum Geologists Bulletin, v. 67, p. 772–816.

Clark, J. L., and Clark, M. N., 1982, Vitrinite reflectance and kerogen analysis of Miocene Monterey formation rocks of the San Joaquin Valley, in Williams, L. A., and Graham, S. A., eds., Monterey formation and associated coarse clastic rocks: Pacific Section Society of Economic Paleontologists and Mineralogists Guidebook, no. 25, p. 43–54.

Cooley, S. A., 1982, Depositional environments of the lower and middle Miocene Temblor formation of Reef Ridge, Fresno, and Kings counties, California, in Williams, L. A., and Graham, S. A., eds., Monterey formation and associated coarse clastic rocks: Pacific Section Society of Economic Paleontologists and Mineralogists Guidebook, no. 25, p. 55–72.

Davis, T. L., and Lagoe, M., 1984, Cenozoic structural development of the north-central Transverse Ranges of southern margin of the San Joaquin Valley: Geological Society of America Abstracts with Programs, v. 16, no. 6, p. 484.

Dickinson, W. R., and Seely, D. R., 1979, Structure and stratigraphy of forearc regions: American Association of Petroleum Geologists Bulletin, v. 63, p. 2–31.

Dixon, S. A., and Kirkland, D. W., 1985, Relationships of temperature to reservoir quality for feldspathic sandstones of southern California [abs.]: American Association of Petroleum Geologists Bulletin, v. 69, p. 250.

Edmondson, W. F., 1967, Sacramento Valley, Winters to Modesto: Pacific Section American Association of Petroleum Geologists, Correlation Section 16.

Estill, W. D., 1980, Kettleman City "When you core enough you get the very best": San Joaquin Geological Society Selected Papers, v. 5, p. 58–66.

Foss, C. D., 1972, A note on the fluid level traps in the San Joaquin Valley, in Rennie, E. W., Jr., ed., Geology and oil fields, west side central San Joaquin Valley: Pacific Section American Association of Petroleum Geologists Guidebook, p. 15.

Graham, S. A., Williams, L. A., Bate, M., and Weber, L. S., 1982, Stratigraphic and depositional framework of the Monterey formation and associated course clastics of the central San Joaquin Basin, in Williams, L. A., and Graham, S. A., eds., Monterey formation and associated coarse clastic rocks: Pacific Section Society of Economic Paleontologists and Mineralogists Guidebook, no. 25, p. 3–16.

Graham, S. A., and Williams, L. A., 1985, Tectonic, depositional, and diagenetic history of Monterey formation (Miocene), central San Joaquin Basin, California: American Association of Petroleum Geologists Bulletin, v. 69, p. 385–411.

Hackel, O., 1966, Summary of the geology of the Great Valley, in Bailey, E. H., ed., Geology of northern California: California Division of Mines and Geology Bulletin 190, p. 217–238.

Harding, T. P., 1976, Tectonic significance and hydrocarbon trapping consequences of sequential folding synchronous with San Andreas faulting, San Joaquin Valley, California: American Association of Petroleum Geologists Bulletin, v. 60, p. 356–378.

Ingersoll, R. V., 1979, Evolution of the late Cretaceous forearc basin, northern and central California: Geological Society of America Bulletin, pt. 1, v. 90, p. 813–826.

Isaacs, C. M., 1981, Porosity reduction during diagenesis of the Monterey formation, Santa Barbara coastal area, California, in Garrison, R. E., and others, eds., The Monterey formation and related siliceous rocks of California: Pacific Section Society of Economic Paleontologists and Mineralogists Special Publication no. 15, p. 257–283.

Issacson, K. A., and Blueford, J. R., 1984, Kreyenhagen formation and related rocks; A history, in Blueford, J. R., ed., Kreyenhagen formation and related rocks: Pacific Section Society of Economic Paleontologists and Mineralogists Guidebook, no. 37, p. 1–7.

Jones, V. T., and Drozd, R. J., 1983, Predictions of oil and gas potential by near-surface geochemistry: American Association of Petroleum Geologists Bulletin, v. 67, p. 932–952.

Kappeler, K. A., Squires, R. L., and Fritsche, A. E., 1984, Transgressive marginal-marine deposits of the Avenal sandstone, Reef Ridge, central California, in Blueford, J. R., ed., Kreyenhagen formation and related rocks: Pacific Section Society of Economic Paleontologists and Mineralogists Guidebook, no. 37, p. 9–27.

Klemme, H. D., 1984, Basin classification chart, in St. John, B., Bally, A. W., and Klemme, H. D., eds., Sedimentary provinces of the world; Hydrocarbon productive and nonproductive: American Association of Petroleum Geologists, p. 5.

Kruge, M. A., 1983, Diagenesis of Miocene biogenic sediments in Lost Hills oil field, San Joaquin Basin, California, in Issacs, C. M., and Garrison, R. E., eds., Petroleum generation and occurrence in the Miocene Monterey formation, California: Pacific Section Society of Economic Paleontologists and Mineralogists, p. 39–51.

Kruge, M. A., and Williams, L. A., 1982, Silica diagenesis in the Monterey formation (Miocene) in the Lost Hills oil field, San Joaquin Valley, California, in Williams, L. A., and Graham, S. A., eds., Monterey formation and associated coarse clastic rocks: Pacific Section Society of Economic Paleontologists and Mineralogists Guidebook, no. 25, p. 37–42.

Kuespert, J. G., 1983, The depositional environment and provenance of the Miocene Temblor formation and associated Oligo-Miocene units in the vicinity of Kettleman north dome, San Joaquin Valley, California [M.S. thesis]: Stanford University, 105 p.

Leonard, J. E., 1986, Increased rate of EOR brightens outlook: Oil and Gas Journal, v. 84, no. 15, p. 71–101.

McCulloh, T. H., and Stewart, R. J., 1980, Subsurface interface between zeolitized and overlying less-altered rocks, southern San Joaquin Valley, California; Configuration and implications for petroleum entrapment [abs.]: American Association of Petroleum Geologists Bulletin, v. 64, no. 3, p. 446.

McGuire, D. J., 1986, Source rock potential of upper Cretaceous–lower Tertiary (Maestrichtian-Danian) Moreno Formation, west-central San Joaquin Valley, California [abs.]: American Association of Petroleum Geologists Bulletin, v. 7, p. 472.

McGuire, J. D., and Ingle, J. C., 1984, Late Cretaceous–early Tertiary anoxic facies of the central California margin; The Marca–Dos Palos shale sequence: Geological Society of America Abstracts with Programs, v. 16, no. 6, p. 588.

MacPherson, B. A., 1978, Sedimentation and trapping mechanism in upper Miocene Stevens and older turbidite fans of southeastern San Joaquin Valley, California: American Association of Petroleum Geologists Bulletin, v. 62, p. 2243–2274.

March, O. T., 1960, Geology of the Orchard Peak area, California: California Division of Mines Special Report 62.

Mulhern, M. E., Eachman, J. C., Jr., and Lester, G. K., 1983, Geology and oil occurrence of displaced diatomite member, Monterey formation–McKittrick oil field, *in* Issacs, C. M., and Garrison, R. E., eds., Petroleum generation and occurrence in the Miocene Monterey formation, California: Pacific Section Society of Economic Paleontologists and Mineralogists, p. 17–37.

Namson, J. S., and Davis, T. L., 1984, Deformational history, thrust-belt structural styles, and plate tectonic origins of Coast Range structures along the San Joaquin Valley, California: Geological Society of America Abstracts with Programs, v. 16, no. 6, p. 607.

Payne, M. B., 1941, Moreno Shale, Panoche Hills, Fresno County, California [abs.]: Geological Society of America, Cordilleran Section Program, p. 19.

Schneeflock, T. R., 1978, Permeability traps in Gatchell (Eocene) sand of California: American Association of Petroleum Geologists Bulletin, v. 62, p. 848–853.

Schwartz, D. E., Hottman, W. E., and Sears, S. O., 1981, Geology and diagenesis of Belridge diatomite and brown shale, San Joaquin Valley, California [abs.]: American Association of Petroleum Geologists Bulletin, v. 65, p. 988–989.

Teitsworth, R. A., 1964, Geology and development of the Lathrop gas field, San Joaquin County, California: San Joaquin Geological Society Selected Papers, v. 2, p. 19–29.

U.S. Geological Survey, 1975, Geothermal gradient maps: American Association of Petroleum Geologists and United States Geological Survey, nos. 22, 23, and 24.

Webb, G. W., 1981, Stevens and earlier Miocene turbidite sandstones, southern San Joaquin Valley, California: American Association of Petroleum Geologists Bulletin, v. 65, no. 3, p. 438–465.

Williams, L. A., 1982, Lithology of the Monterey formation (Miocene) in the San Joaquin Valley of California, *in* Williams, L. A., and Graham, S. A., eds., Monterey formation and associated coarse clastic rocks: Pacific Section Society of Economic Paleontologists and Mineralogists Guidebook, no. 25, p. 17–35.

Wilson, M. J., 1979, The Santos; A case history of fractured shale development: Proceedings of the California regional meeting of the Society of Petroleum Engineers, Contribution SPE 7978, 8 p.

Zieglar, D. L., and Spotts, J. H., 1978, Reservoir and source-bed history of Great Valley, California: American Association of Petroleum Geologists Bulletin, v. 62, p. 813–826.

MANUSCRIPT ACCEPTED BY THE SOCIETY MAY 18, 1987

NOTE ADDED IN PROOF

Because this paper was submitted for publication in 1986, an update on production is included, as is a discussion of the impact of a most important new concept in exploration—sequence stratigraphy—and discussion of the potential for the newly developed "horizontal drilling" technology.

Production History

As of January 1, 1988, 123 oil fields have produced a cumulative total of 9.8 billion barrels of oil and 10.3 trillion cubic feet of associated gas. Proven reserves are estimated to be 3.1 billion barrels and 1.4 trillion cubic feet, respectively. Twenty dry gas fields have produced 0.9 trillion cubic feet of gas with an estimated proven reserve of 0.134 trillion cubic feet (California Division of Oil and Gas, 1989).

Stratigraphy

Nomenclature. Previously, we discussed the complexity of the lithic units in the San Joaquin basin and the confusion related to the rather undisciplined nomenclatural usage that prevails. Charles Foss (Foss and Blaisdell, 1968; Foss, 1972a; and Foss, 1972b), came closest to understanding what is now termed "sequence stratigraphy." Sequence stratigraphy, a depositional concept that involves its own nomenclature, terminology, and philosophy, was initiated by Frazier (1974) and refined by Vail and others (1977). The concept relates deposition to eustatic changes in sea level and commensurate changes in accommodation space within a basin. It allows similarly named sands to be uniquely identified by depositional process and environment and therefore to be considered for individual reservoir potential.

Callaway (1990), utilizing these concepts with some minor variations, compiled a comprehensive reference list for all the lithostratigraphic units (names most commonly used in the published literature) found in the San Joaquin basin. Identified were 13 low-stand sequence boundaries and internal transgressive surfaces with age control adapted to the DNAG time scale from the chronology of Haq and others (1987). Also included were five semiregional sections, one hypothetical cross section, and a chart that identifies and chronologically positions unique, depositionally different lithostratigraphic units (numbered) as: transgressive sandstones (46), high-stand deltaic sandstones (34), low-stand deltaic sandstones (23), incised "channel fill" sandstones (6), "slope bypass" fan sandstones (29), low-stand fan sandstones (23), and slump sandstones (1), as well as shales (96) and non-marine units (48).

The concept of sequence stratigraphy is based primarily on the identification of unconformities (sequence boundaries) in outcrop, correlated to sequence boundaries and system tracts interpreted from seismic reflector terminations. As yet, no work of this nature has been published for the San Joaquin basin. Three studies combining well logs and outcrop work are adaptable to sequence stratigraphic analysis: Bloch and others (1990) describe sequence stratigraphy for a selected interval from late Zemorrian to Luisian on the "East Side" of the basin, tying outcrop studies to well logs; Kuespert (1985) describes the Coalinga/Kettleman Hills area; and DeCelles (1988) describes the Tejon area at the south end of the basin.

Future opportunities

Horizontal drilling is a newly developing technology considered primarily for fractured reservoirs. Horizontal wells can be drilled to intersect a maximum number of vertically oriented fractures, compared to conventional wells that rarely intersect vertical fractures parallel or sub-parallel to the well bore. This technology was immediately successful in the fractured Austin Chalk of the Texas Gulf Coast and the Bakken Formation in the Williston basin where initial production rates increased on the order of several magnitudes over vertically drilled wells. Many other formations are being considered throughout the world for this technology.

In the San Joaquin basin, the most obvious candidate for horizontal drilling technology is the Monterey Formation, a shale often found to be fractured. As yet, no horizontal wells have been drilled in the Monterey.

In 1988, one horizontal well was drilled into the 26R Sand in Elk Hills field. This zone has been producing for many years; the gas/oil interface is close to the oil/water interface, and gravity drainage is the primary recovery mechanism. Vertical wells are subject to water coning. The new well was drilled horizontally for 243 m, approximately 20 m above the oil/water interface. The results were extremely encouraging because of the high production rate with less gas and water coning. The advantages of horizontal drilling in this reservoir are that fewer wells are required, higher production rates are obtained, less draw-down occurs, and in steeply dipping beds, more individual reservoir beds are encountered (Hart, 1990).

REFERENCES

Bloch, R. B., Olson, H. C., 1990, Stratigraphy and structural history of the lower and middle Miocene section, east side San Joaquin Valley, *in* Kuespert, J. G., and Reid, S. A., eds., Structure, stratigraphy and hydrocarbon occurrences of the San Joaquin basin, California: Pacific Sections Society of Economic Paleontologists and Mineralogists and American Association of Petroleum Geologists Guidebook, no. 64 and 65, p. 287–291.

California Division of Oil and Gas, 1989, Seventy-Fourth annual report of the State Oil and Gas Supervisor, 1988: California Division of Oil and Gas Publication PR 06, 155 p.

Callaway, D. C., 1990, Organization of stratigraphic nomenclature for the San Joaquin Basin, California, *in* Kuespert, J. G., and Reid, S. A., eds., Structure, Stratigraphy and Hydrocarbon Occurrences of the San Joaquin Basin, California: Pacific Sections Society of Economic Paleontologists and Mineralogists and American Association of Petroleum Geologists Guidebook, no. 64 and 65, p. 5–21.

DeCelles, P. G., 1988, Middle Cenozoic depositional, tectonic and sea level history of southern San Joaquin Basin, California: American Association of Petroleum Geologists Bulletin, v. 72, p. 1297–1322.

Foss, C. D., and Blaisdell, R., 1968, Stratigraphy of the west side of the San Joaquin Valley, *in* Karp, S. E., ed., Guidebook to geology and oil fields of westside southern San Joaquin Valley: American Association of Petroleum Geologists, Society of Engineering Geologists, Society of Economic Paleontologists and Mineralogists Pacific Sections Annual Field Trip Guidebook, p. 33–43.

Foss, C. D., 1972a, Lower Eocene to upper Cretaceous stratigraphy, central San Joaquin Valley, *in* Rennie, E. W., Jr., ed., West side central San Joaquin Valley, geology and oil fields: Pacific Sections American Association of Petroleum Geologists, Society of Engineering Geologists, Society of Economic Paleontologists and Mineralogists Field Trip Guidebook, p. 59–63.

Foss, C. D., 1972b, A preliminary sketch of the San Joaquin Valley stratigraphic framework, *in* Rennie, E. W., Jr., ed., West side central San Joaquin Valley, geology and oil fields: Pacific Sections American Association of Petroleum Geologists, Society of Engineering Geologists, Society of Economic Paleontologists and Mineralogists Field Trip Guidebook, p. 40–50.

Frazier, D. E., 1974, Depositional episodes; Their relationship to the Quaternary stratigraphic framework in the northwestern portion of the Gulf Basin: Texas Bureau of Economic Geology Circular 74-1, 28 p.

Haq, B. U., Hardenbol, J., and Vail, P. R., 1987, Chronology of fluctuating sea levels since the Jurassic: Science, v. 235, p. 1156–1166.

Hart, O., 1990, Elk Hills medium radius horizontal well, *in* Kuespert, J. G., and Reid, S. A., eds., Structure, stratigraphy and hydrocarbon occurrences of the San Joaquin basin, California: Pacific Sections Society of Economic Paleontologists and Mineralogists and American Association of Petroleum Geologists Guidebook, no. 64 and 65, p. 169–172.

Kuespert, J. G., 1985, Depositional environments and sedimentary history of the Miocene Temblor Formation and associated Oligo-Miocene units in the vicinity of Kettleman North Dome, San Joaquin Valley, California, *in* Graham, S. A., ed., Geology of the Temblor Formation, western San Joaquin basin, California: Pacific Sections Society of Economic Paleontologists and Mineralogists Guidebook, v. 44, p. 53–67.

Vail, P. R., Mitchum, R. M., Jr., and Thompson, S., III, 1977, Seismic stratigraphy and global changes of sea level, part 3; Relative changes of sea level from coastal onlap, and part 4; Global cycles of relative changes of sea level, *in* Payton, C. E., ed., Seismic stratigraphy—Applications to hydrocarbon exploration: American Association of Petroleum Geologists Memoir 26, p. 63–97.

Chapter 27

Geologic controls on hydrocarbon occurrence within the Santa Maria basin of western California

J. B. Dunham, B. W. Bromely, and V. J. Rosato
Unocal Center, Room 312, Unocal Corporation, P.O. Box 7600, Los Angeles, California 90051

INTRODUCTION

Giant oil fields have been discovered within several Tertiary basins of western California. Subsidence, accumulation of source and reservoir lithofacies, and subsequent growth of structural traps within these basins were controlled by plate-tectonic interactions that deformed the western edge of the North American plate throughout late Tertiary (Neogene) time. Oil present in the Santa Maria basin (Figs. 1 and 2) reflects an accordance of tectonics and sedimentation in that rapid tectonic subsidence during middle Miocene time produced a deep marine basin that was isolated from sources of terrigenous detritus and thus starved of coarse-clastic sediment. However, oceanographic conditions in surface waters above the basin were conducive to proliferation of marine-plant plankton such that a thick section of biogenic organic-rich sediment (diatom ooze, coccolith ooze) was deposited within the basin. Continued subsidence led to burial diagenesis that transformed soft, low-density ooze into brittle, high-density rock. An abrupt change in tectonic style occurring in the early Pliocene brought about a sudden compression of the sediment-filled basin that resulted in the growth of a series of broad folds as well as tectonic fracturing of brittle rocks. Tens of billions of barrels of oil are presently trapped in fractured reservoirs within these structures; the ultimate recovery from the Santa Maria basin may exceed several billion barrels.

Location and regional setting

The Santa Maria basin lies within the southern Coast Ranges Province of California (Fig. 1). Although dissected by major faults, the Neogene rocks in the onshore basin are continuous offshore to the Santa Lucia Bank to the west (Fig. 2), where the Neogene sequence onlaps and laps out against an uplift of Mesozoic basement named the Santa Lucia High (Payne and others, 1979). The southern margin of the offshore basin (Fig. 2) is delineated by a northeast-trending basement uplift termed the Amberjack High (Crain and others, 1985), where Neogene sediments abruptly thin and lap out against basement. The northern and eastern boundaries of the basin are defined by Mesozoic basement rocks exposed onshore in the San Rafael Mountains; the basement rocks are composed largely of trench-related Franciscan Complex lithologies of Jurassic-Cretaceous age, and similar rocks (metasediments, altered basic igneous rocks) have been dredged from acoustic basement on the Santa Lucia High (McCullough and others, 1977). Subsurface data demonstrate that the Franciscan Complex constitutes the basement of the basin both onshore and offshore. Coastal outcrops at the physiographic boundary between the onshore and offshore basins consist primarily of the late Tertiary Monterey and Sisquoc Formations; Monterey outcrops are similar to oil-bearing strata present in the offshore basin (Dunham and Blake, 1987). The trend of structures in the Santa Maria region gradually changes from east-west near the Santa Barbara Channel to northwest-southeast at the northern edge of the basin (Fig. 2).

GEOLOGIC HISTORY

Compilation of the history

The present understanding of the basin and its sediments has developed through a compilation of detailed descriptive field studies (Arnold and Anderson, 1907; Bramlette, 1946; Woodring and Bramlette, 1950; Dibblee, 1950; Hall and Corbato, 1967; Pisciotto, 1978; Isaacs, 1980; Grivetti, 1982), combined with conceptual advances in appreciation of regional and global tectonics (Atwater, 1970; Crowell, 1974; Blake and others, 1978; Hall, 1978; Crouch and others, 1984; Hornafius, 1985). The stratigraphy, sedimentology, and structure of the basin are so dominated by its tectonic setting that advances in understanding of its geology have coincided with improved recognition of global-tectonic influences on the western North American margin. Specifically, interpretation of western California geology in a plate-tectonic light (Atwater, 1970) led to development of the general concept of the pull-apart basin (Crowell, 1974) and to the eventual proposal of a wrench-tectonic origin for the Tertiary basins of coastal California (Blake and others, 1978). Hall (1978) recognized and described the field relations indicating that the

---

Dunham, J. B., Bromley, B. W., and Rosato, V. J., 1991, Geologic controls on hydrocarbon occurrence within the Santa Maria basin of western California, *in* Gluskoter, H. J., Rice, D. D., and Taylor, R. B., eds., Economic Geology, U.S.: Boulder, Colorado, Geological Society of America, The Geology of North America, v. P-2.

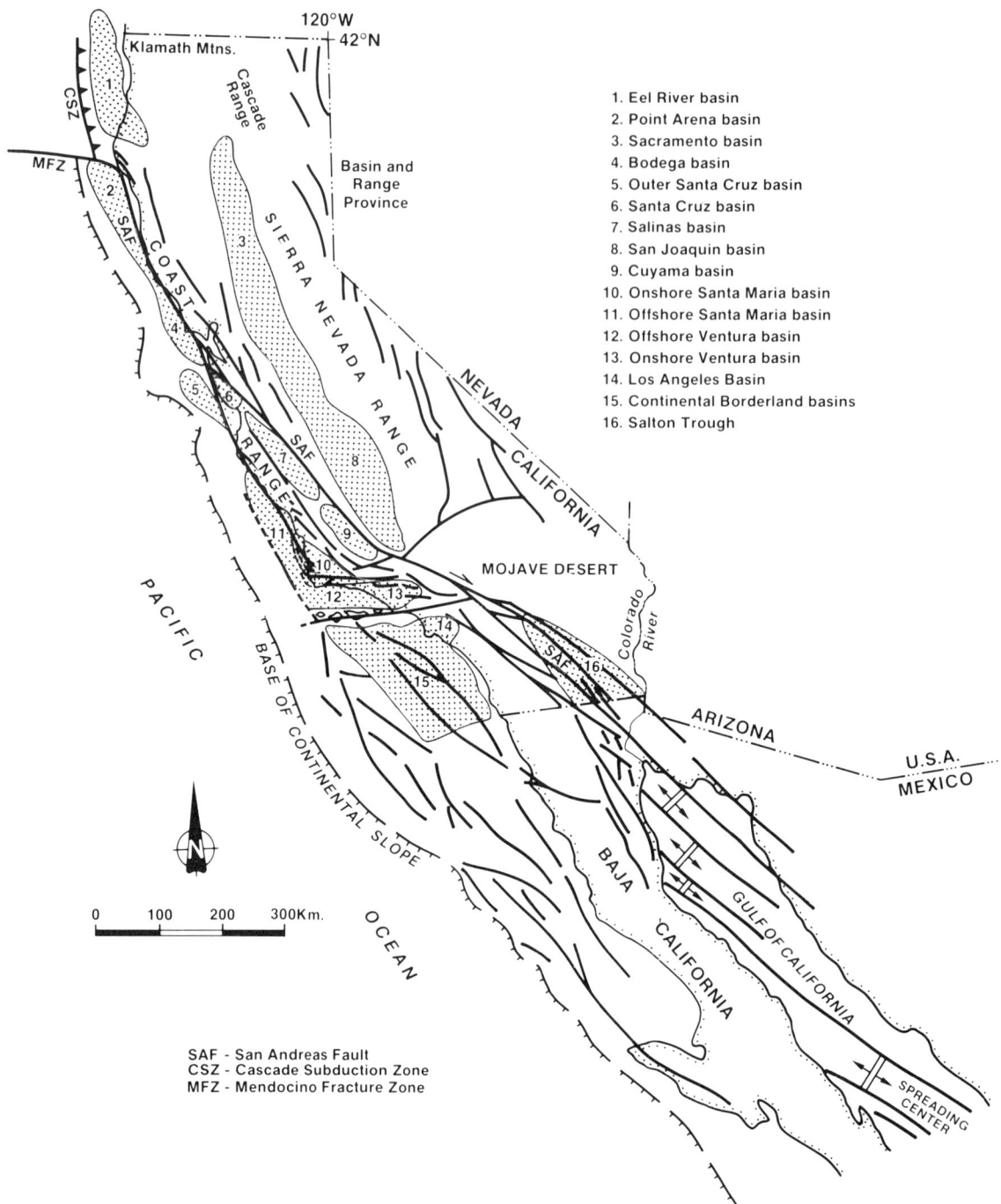

Figure 1. Generalized geologic map of California showing the locations of major Tertiary basins (adapted from Crowell, 1979; Miall, 1984).

Santa Maria basin did in fact originate as a result of right-slip tectonics; while Crouch and others (1984) called attention to the effects of compressional tectonics on the post-Miocene history of the basin. The synthesis of Hornafius (1985), based on paleomagnetic data, integrates the wrench-tectonic and compressional-tectonic models into a plate-tectonic–driven model that explains basin subsidence and later structural growth in the context of tectonic rotation of crustal blocks along the western California margin. Appreciation of the geologic history of the Santa Maria basin has followed an evolutionary path that has been influenced by advances in comprehension of global tectonics. Since it is unlikely that this evolution has attained its highest grade, continued study of California geology will crystallize further insights into the natural history of the Santa Maria basin.

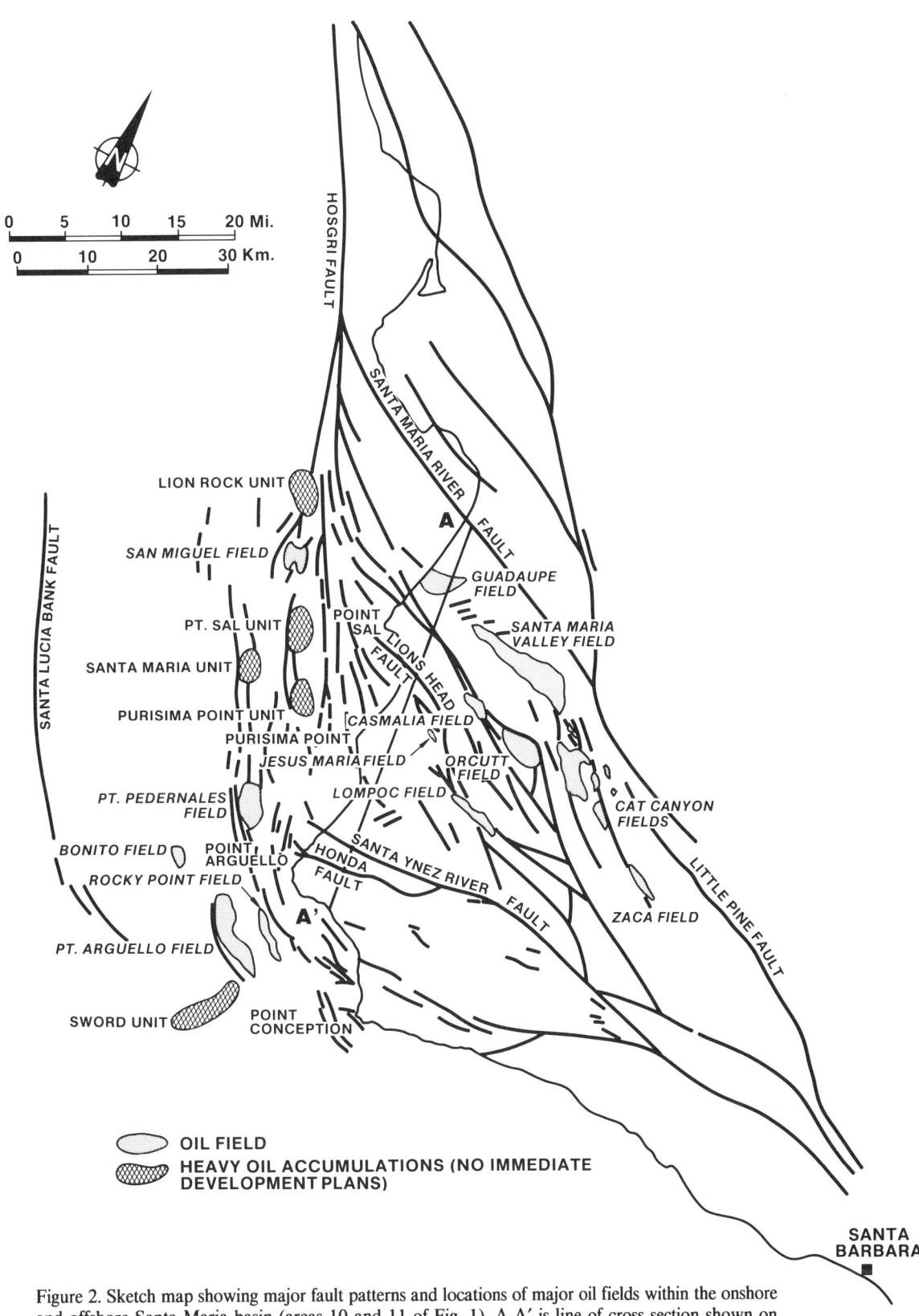

Figure 2. Sketch map showing major fault patterns and locations of major oil fields within the onshore and offshore Santa Maria basin (areas 10 and 11 of Fig. 1). A-A' is line of cross section shown on Plate 6B.

General geologic history of the Santa Maria basin

From the late Mesozoic into middle Tertiary time, the continental margin of western North America was subjected to Andean-type subduction and resultant development of the Cordilleran arc-trench system, reflected by trench-related Franciscan Complex metasediments and arc-related plutons of the Sierra Nevada batholith (Dickinson, 1981). At about 30 Ma, subduction of ocean crust brought the East Pacific Rise directly in contact with the edge of western North America, with the result that relative motion along the coast changed from subduction to right-slip displacement (Atwater, 1970; Crowell, 1979). The Neogene basins of western California and offshore Baja California developed in response to right-lateral shearing of the continental margin (Blake and others, 1978).

Initial subsidence in the Santa Maria basin was triggered by inception of right-slip motion along the Santa Maria River–Little Pine fault system, and the associated Santa Ynez River fault to the south (Fig. 2), which resulted in pull-apart, extension, and foundering of Franciscan basement rocks (Hall, 1977, 1978). Coarse alluvial conglomerates of the late Oligocene–early Miocene Lospe Formation record the initial rifting (Figs. 3 and 4); however, subsequent subsidence took place very rapidly, such that a deep marine basin developed within the region. Coincidentally, climatic and oceanographic events combined to produce conditions favorable for high plankton productivity in surface waters above this deep basin (Pisciotto and Garrison, 1981), with the result that the basin rapidly filled with a mix of organic-rich pelagic and hemipelagic sediments. However, the history of the basin does not end with Miocene subsidence. Despite the importance of wrench tectonics to basin subsidence, the present geometry of folds and faults within the Santa Maria basin chiefly reflects post-Miocene northeast-southwest–directed compression (Crouch and others, 1984). Based on interpretation of seismic reflection profiles, Crouch and others (1984) concluded that the Hosgri, and other major northwest-trending faults within the basin, appear to be predominantly thrusts rather than strike-slip faults. In fact, compressional structures are ubiquitous, and form some of the major oil-producing anticlines of the region (Fig. 5). Nearly every field in the Santa Maria basin is bounded or cut by reverse faults. Thus, there is evidence for a change in tectonic style, from right-slip rifting to horizontal compression, within the Santa Maria basin.

The transition from wrench to compressional tectonics was brought about by changes in fault geometries associated with clockwise rotation of fault-bounded blocks at the southern and eastern boundaries of the Santa Maria basin (Hornafius, 1985; Hornafius and others, 1986; Luyendyk and Hornafius, 1987). Right-slip displacement on northwest-trending faults in the Coast Ranges and Peninsular ranges was accommodated by clockwise rotation of the Transverse Ranges during Neogene time (Luyendyk and others, 1980); the changes in tectonic style affecting the Santa Maria region reflect various stages of this rotation. Specifically, the change from divergent rifting to tectonic compression

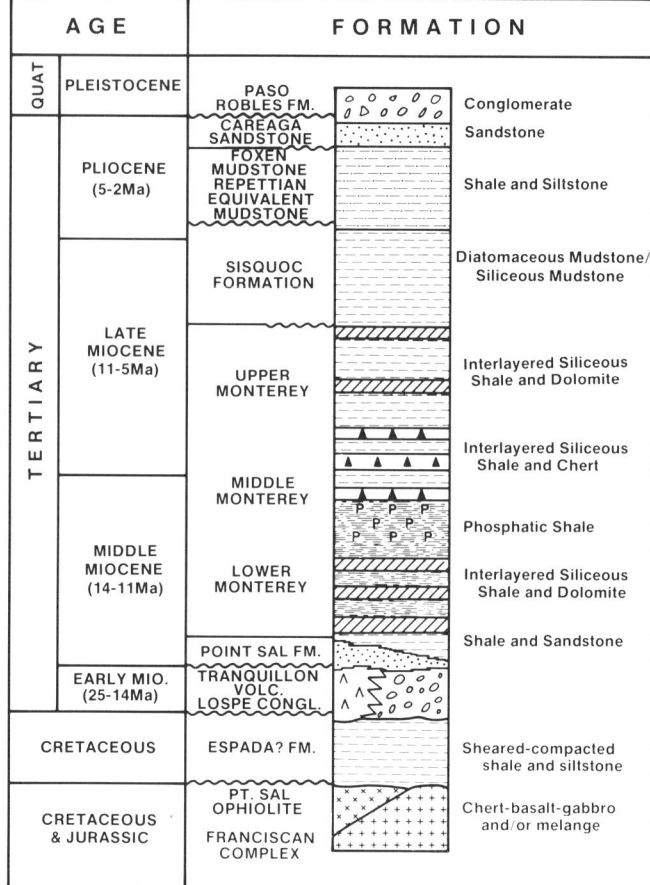

Figure 3. Stratigraphy of the onshore Santa Maria basin. Facies change to deeper-water clastics takes place in the Plio-Pleistocene interval of the offshore basin.

(Hornafius, 1985) appears to correlate with the eastward jump in the position of the Pacific–North American transform boundary from the southern California borderland to the Gulf of California at the end of the Miocene (Atwater and Molnar, 1973).

Note that by this model, faults that originated through right-slip rifting would have experienced normal as well as right-slip displacement during the middle Miocene, prior to becoming reactivated as reverse faults in the Plio-Pleistocene. The first phase of fault translation took place coincident with subsidence and sedimentation of the Monterey and pre-Monterey Formations. As sedimentation continued coincident with strike-slip movement, original depositional facies patterns would have been disrupted by continuous lateral displacement. Consequently, early Monterey sedimentation patterns necessarily became offset from one another by later Monterey-age strike-slip fault movements. The maximum offset of sedimentation patterns has affected pre-Monterey and early Monterey strata, while later Monterey and post-Monterey sedimentation patterns have undergone proportionally less offset. Major thickness and facies changes occur over short distances within the Lospe, Point Sal, and lower Monterey Formations, while lateral correlations within the uppermost

Monterey, Sisquoc, and Foxen Formations are much easier to establish. These observations reflect a pattern of syn-sedimentary strike-slip tectonics.

As for the relative importance of wrench tectonics versus compressional tectonics to the geologic history of the Santa Maria basin, it is evident that both tectonic styles have had a significant influence on basin evolution. Initial basin subsidence was related to strike-slip faulting and wrench tectonics, while later stages of fault movement were produced by compression normal to the present San Andreas system; the compression resulted in reactivation of former right-slip faults rather than in development of a second set of faults oriented with regard to the new stress field. Older fault planes are commonly reactivated within changing stress fields (Aki and Richards, 1980).

It has been suggested that the high-angle reverse faults cutting the Santa Maria basin, which dip steeply near the surface (Fig. 2), may flatten or sole out into a décollement at depth (Crouch and others, 1984). However, gravity data seem to indicate that the faults remain steeply dipping down to at least 2.4 km (approximately 7,900 ft) within the onshore Santa Maria basin (Kieniewicz and Luyendyk, 1986), although the decrease in dip and the detachment surface would likely lie at greater depth. If the faults flatten, they do so beneath the depth of resolution of the gravity data. On the whole, the gravity data seem consistent with a two-layer model of the basin; that is, a high-density basement layer overlain by a lower-density layer of Neogene rocks.

The Santa Maria basin clearly has been a site of evolving structural character throughout its history. Consequently, when considering structural type, it must be noted that the basin is a hybrid of distinctly different tectonic styles. The Santa Maria basin evolved from an initial pull-apart structural aspect, which characterized it during its period of early Neogene basin subsidence, into an area that underwent strong horizontal compression, which resulted in reactivation of faults in a new stress regime and consequent growth of large anticlines that presently localize major oil fields. Tectonism resulting from NE-SW–directed compression continues to affect the Santa Maria basin to the present day (Hamilton and Hall, 1987); anticlines are still growing, synclines are still subsiding, and oil-water contacts continue to shift within the great closed structures of the basin.

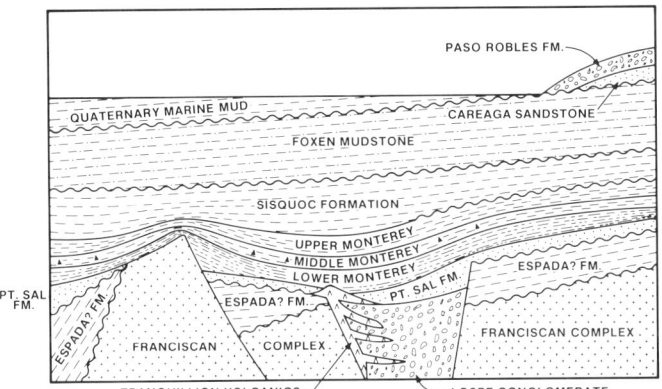

Figure 4. Sketch showing the stratigraphy of the Santa Maria basin.

EXPLORATION HISTORY

The Santa Maria basin is one of the oldest oil-producing regions of California (Jenkins, 1943). Prospecting was underway by the late 1890s, with interest focused on the well-known oil seeps of the region; the first commercial well was brought in by the Western Union Oil Company in 1901 (Arnold and Anderson, 1907, p. 9). The discovery sparked a boom, and by December, 1903, 21 wells were producing oil from the Santa Maria basin, with oil gravities ranging from 27 to 16 degrees API, and at a well-head price of as much as 80 cents per barrel (Aubury, 1904). Arnold and Anderson (1907) noted the high productivity of wells from the Santa Maria district, ranging from 300 to 3,000 barrels per day, and suggested that the region had the promise of becoming one of the most productive oil districts in the western United States.

The first gusher in California was located within the Santa Maria district, in what would later be designated the Orcutt field (Fig. 5). The Union Oil Company Hartnell No. 1 well, completed in 1904, flowed at a rate of 10,000 barrels per day for several months (Woodring and Bramlette, 1950). By 1908, drilling had defined the locations of major oil fields that are productive even to the present day; including Orcutt field, Lompoc field, and Cat Canyon field. By 1913, most of the oil was recognized as being localized in anticlines, produced from the Monterey Formation, and as for the nature of oil storage within the Monterey, ". . . undoubtedly in fracture zones" (McLaughlin and Waring, 1914, p. 403).

The early success of oil companies working in the Santa Maria district stemmed from their application of a number of fundamental concepts, including prospecting near known seeps, the use of cross sections to illustrate subsurface structure and stratigraphy (McLaughlin and Waring, 1914, p. 403), and an early appreciation of the anticlinal theory of oil accumulation, coupled with the fortunate circumstance that the structures were so young that they had well-developed surface expression. Exploration strategy within the onshore Santa Maria basin has changed little since these early times; onshore discoveries continue to be made by correlating known producing horizons into structural traps, although the methods of analysis, including refined stratigraphic terminology and improved cross sections, have evolved since the early days.

Exploration in the offshore basin

Exploration in the offshore Santa Maria basin has been underway since the 1950s (McCulloch and others, 1977, p. 13), although only one well had been drilled in the offshore Santa Maria basin prior to 1981 (Humble P-060-1 Oceano well), and no significant shows of hydrocarbons were identified in that well (McCulloch and others, 1977, p. 36). In the late 1970s and early 1980s, preparation for Federal Offshore Sale 53 provided an opportunity for reevaluation of the offshore basin. Ignoring the

ORCUTT OIL FIELD

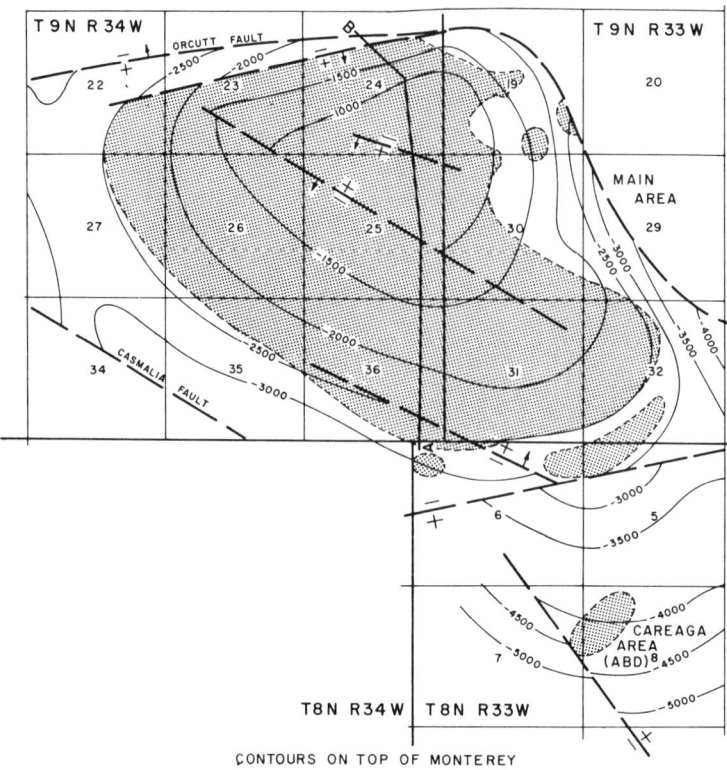
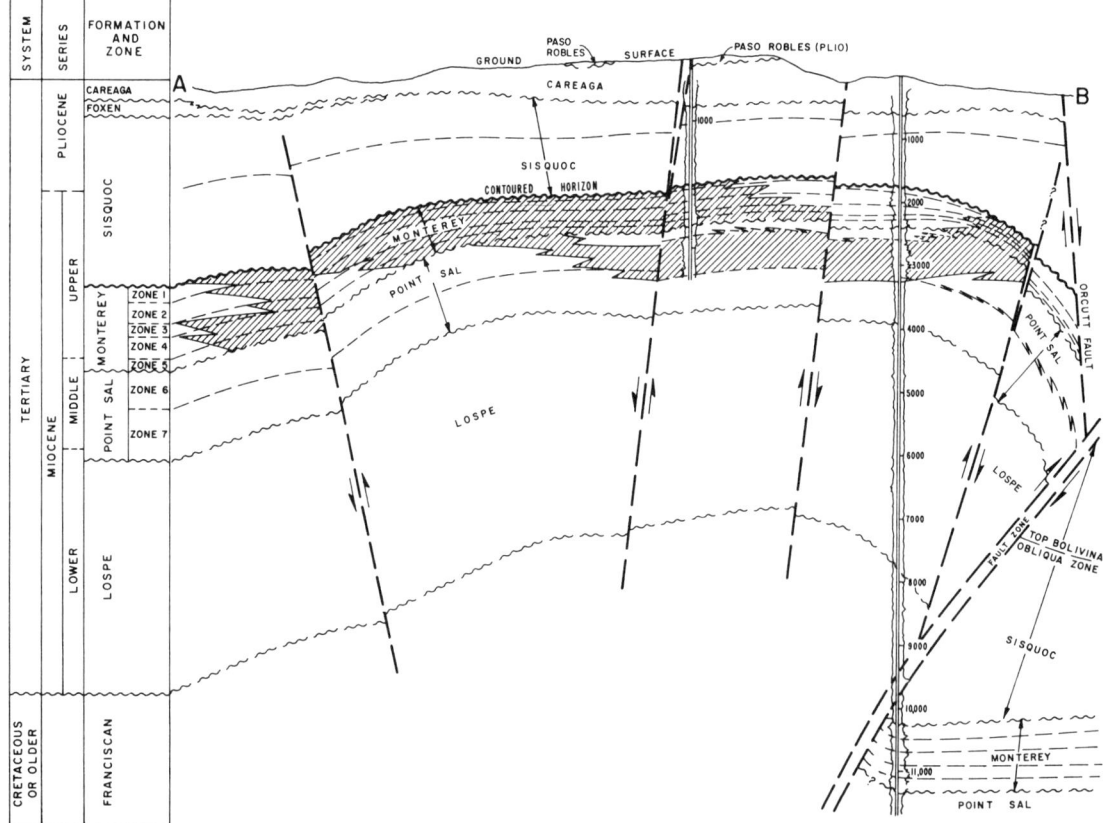

Figure 5. Map of, and cross section through, Orcutt Field (from California Division of Oil and Gas, 1974). See Fig. 2 for location of field.

results of the single dry hole, more than 20 companies, individually or in partnership, posted bids in Sale 53 in June, 1981; subsequent drilling resulted in the discovery of notable oil fields.

Point Arguello field (Fig. 2) is the largest U.S. find since Prudhoe Bay (Rintoul, 1985). Chevron USA Inc. made the initial discovery in 1981 (Crain and others, 1987); recoverable reserves are estimated to be on the order of 300 million barrels, representing approximately 17 percent of actual oil in place (Rintoul, 1986). Evaluation of exploration wells on the border of the Sale 53 area led Chevron and partner Phillips Petroleum to post a record bid of $333.6 million for lease OCS-P 0450. Production rates for delineation wells drilled in Point Arguello field tested at rates as great as 4,000 barrels per day (Rintoul, 1987). Projections of peak production at Point Arguello field lie in the range of several tens of thousands of barrels per day.

Point Pedernales field (Fig. 2), discovered in January 1983, by Union Oil Company (now Unocal) and partners Gulf (now Chevron) and Superior (now Mobil), was the first oil field to go on production from acreage sold at OCS Sale 53. Platform Irene, operated by Unocal Corporation in 74 m of water on tract OCS-P 0441, was the first platform installed in the offshore Santa Maria basin; platform Irene presently produces approximately 20,000 barrels per day of 16 degrees API gravity oil. Plans are under consideration for location of a second platform within the Point Pedernales unit.

San Miguel field, which lies 48 km north of Point Arguello field (Fig. 2), was discovered by Occidental Petroleum in 1983. Delineation wells within the field have recorded rates of as much as 3,780 barrels per day of 12 degree API gravity oil. Shell has purchased the controlling interest in the field and is now the operator. Shell is preparing for the installation of platform Julius in 146 m of water on tract 409, and projects production at a peak of 40,000 barrels per day of 12 degrees API gravity oil (Rintoul, 1987).

Development plans have yet to be revealed for several other announced discoveries, including Bonito, a new field discovery by Chevron of 22 degrees API gravity oil, offsetting the Point Arguello field to the northwest; as well as Rocky Point, another Chevron discovery of 25 to 35 degrees API gravity oil located 3 km northeast of Point Arguella field. Chevron is considering the installation of platform Hacienda for development of Rocky Point field (Rintoul, 1985).

Plans for Conoco's Sword discovery (Fig. 2) await research and development of economic methods for recovery of the heavy oil (Rintoul, 1987). Similar low-gravity oil (less than 12 degrees API) has been discovered in several other areas of the offshore Santa Maria basin, but the commerciality of the finds remains to be determined; discoveries have been announced from the Point Sal, Lion Rock, and Purisima Point units (Fig. 2), although the current price for heavy oil does not permit development of these resources (Rintoul, 1986). Low-gravity oil underlies major portions of the offshore basin; should technology and oil price reach a favorable concurrence, then billions of barrels of heavy oil may yet be recovered from the offshore Santa Maria basin.

HYDROCARBON ACCUMULATIONS IN THE SANTA MARIA BASIN

Most of the oil and gas accumulations within the Santa Maria basin (Fig. 2) are trapped in anticlines that are bounded by major reverse faults (Figs. 2 and 5). The only significant nonstructural trap discovered to date within the basin lies within the western Santa Maria Valley field, where a trap is formed by the lateral pinchout of the Monterey Formation against basement, with low-permeability, Sisquoc Formation, diatomaceous shales forming the top seal. Undoubtedly other stratigraphic traps remain to be discovered within the basin.

Oil has been produced from several formations within the basin, as shown on Figure 3, but naturally fractured rocks in Monterey Formation have by far accounted for the greatest production. Other productive zones are associated with fractures present in Mesozoic basement rocks, but these fracture developments are minor relative to those of the Monterey. Oil is also trapped within conventional sandstone interparticle porosity present in the pre-Monterey Tertiary clastic sequence of the Lospe and Point Sal Formations. In the section above the Monterey, the Tinaquaic Sandstone Member of the basal Sisquoc Formation is a reservoir zone, while locally, very low-gravity oil present within high-porosity diatomaceous rocks of the Sisquoc Formation forms tar-diatomite deposits.

Regan and Hughes (1949) noted that most reservoirs within the Santa Maria district were limited to zones containing rocks that were susceptible to fracturing. "These rocks include in order of importance: (1) chert, (2) calcareous shales, (3) platy siliceous shale, and (4) sandstones" (Regan and Hughes, 1949, p. 33). Chert-bearing zones were recognized as being confined almost exclusively to intervals of Mohnian (late Miocene) age within the Monterey Formation, while the underlying Luisian and late Relizian Monterey strata consist largely of carbonates and limey shales. Based on variations in distribution of chert within the basin, Regan and Hughes (1949) concluded that chert appears to be a depositional facies; that is, some factor inherited from the depositional environment controlled the present distribution of chert. Fractures were recognized as being tectonic in origin and extending beyond the crests of structures; that is, "rocks sufficiently brittle to be susceptible to fracturing are regionally fractured and broken and are potential reservoirs" (Regan and Hughes, 1949, p. 47). Regan and Hughes (1949) noted that the favorable coincidence of brittle rocks combined with a good structural trap capped by ductile sealing shale facies, provided the key to successful hydrocarbon exploration in the Santa Maria district.

Since the time of Regan and Hughes, it has been recognized that many of the carbonate layers within the Monterey Formation of the Santa Maria basin actually consist of dolomite (Garrison and others, 1984). Dolomite formed as a result of diagenetic replacement of original calcium carbonate sediment, which consisted largely of nannofossil-foraminiferal ooze. Highly variable carbon isotopes within these dolomites reflect the influence of bacterial metabolism on dolomite formation (Pisciotto, 1981a). Dolomitization of calcium carbonate is inhibited by high concentration of dissolved sulfate in sea water (Baker and Kastner, 1981). However, due to the fact that anoxic conditions persisted on the sea floor during Monterey deposition, sulfate-reducing bacteria were prolific within the near-surface sediment, and these bacteria consumed all available sulfate from sediment-pore waters. The sulfate-free, bacterially modified sea water brought about the rapid early-diagenetic alteration of fine-grained calcium carbonate sediment into lithified dolomite rock. Because dolomite is a brittle rock, it too constitutes an important fractured-reservoir lithology within the Santa Maria basin (Roehl and Weinbrandt, 1985). Although the dolomite is diagenetic in origin, its distribution within the basin was ultimately controlled by the original distribution of nannofossil ooze within the depositional basin.

Depositional-facies controls on hydrocarbon accumulation

Two separate facies controls affect the distribution of recoverable oil within the basin. The first control involves the distribution of source rocks, as organic-rich shales within limited intervals of the Monterey Formation appear to constitute the only volumetrically significant source rocks of the basin. The second control involves factors affecting the distribution of brittle rocks within the basin, as fractured rocks make up the principal reservoir zones of the basin. Regan and Hughes (1949) noted considerable variation in the actual number of fractured-reservoir beds across the basin, due to differences in percentage of chert and carbonate beds within the Monterey Formation across the basin. Regan and Hughes perceived that these reservoir-quality variations were facies controlled in the sense that in some areas, "... the elements necessary for chert were deposited, while they were not deposited elsewhere in the basin" (Regan and Hughes, 1949, p. 46).

An explanation for variations in chert content within the basin had been offered by Bramlette (1946), who noted that chert within the Santa Maria basin formed through alteration of original diatomaceous sediment. Mining operations had shown that unusually pure diatomite beds were confined to distinct beds, seldom more than a few meters thick, which were interbedded with thicker zones of less pure diatomaceous mudstone that consisted of a mix of diatoms and clay (Bramlette, 1946, p. 14). Bramlette (1946, p. 50) recognized that chert formed as a result of alteration of originally pure diatomaceous sediment; the interbedded cherts and shales present within the Monterey were the products of diagenetic alteration of an originally interlayered sequence of pure diatomite and less pure diatomaceous mudstone. The brittle cherts and ductile shales present within reservoir intervals followed the original depositional distributions of pure-diatom sediment and less pure diatomaceous mudstone within the basin. Bramlette suggested that upwelling of nutrient-rich waters might be a factor favoring prolific growth of diatoms, and factors such as variation in upwelling coupled with variations in terrigenous influx might be responsible for interlayering of pure diatomites with impure diatomaceous mudstones. Although the process of alteration was not well understood, it was apparent that silica diagenesis proceeded through dissolution of diatoms followed by immediate reprecipitation of silica during the course of burial diagenesis (Bramlette, 1946, p. 54).

Recent research has refined the understanding of silica diagenesis within the Monterey Formation. Starting with the work of Murata and Larson (1975), the evolution of biogenic silica from opal-A (original amorphous-silica shells such as diatoms or radiolaria), through the intermediate stage of opal-CT (opal-CT is a mixture of cryptocrystalline cristobalite and tridymite) and eventually into quartz, became recognized as the method by which Monterey diatomites and diatomaceous mudstones were converted into cherts and siliceous shales. The process involves dissolution of biogenic opal (opal-A), as envisioned by Bramlette (1946), and is related to burial; but the relation to burial is indirect. In fact, the principal control on alteration of opal-A into opal-CT and quartz is temperature (Murata and Larson, 1975); overburden pressure in itself is not a factor. The relation to burial depth involves geothermal heating with burial, such that the burial depth necessary to convert opal-A to opal-CT may vary from hundreds to thousands of meters from place to place, depending on regional variations in geothermal gradient. Over the years, estimates of temperature of silica phase change have been refined, such that the relatively broad, early estimates have been narrowed to the following ranges: opal-A to opal-CT transition at about 45° to 50°C, and opal-CT to quartz transition at about 75° to 85°C (Isaacs, 1984). Factors such as detrital content of the sediment (Keller and Isaacs, 1985) may affect the precise temperature at which the phase changes will occur, thus resulting in interlayering of opal-A and opal-CT phase rock (or opal-CT and quartz-phase rock) over a broad interval of many hundreds of meters.

Pisciotto (1978, 1981b) studied in detail the evolution of biogenic silica within the Santa Maria basin, and showed that silica-phase-change boundaries were independent of formation contacts; that is, the phase changes cut across stratigraphy. The phase-change boundaries were related to differences in burial history from place to place within the basin; they have nothing to do with defining formation contacts. Specifically, the contact between the Monterey Formation and the overlying Sisquoc Formation does not correspond with the location of a silica-phase-change boundary. Both the Monterey and Sisquoc Formations are observed to lie within the opal-A, opal-CT or quartz phase at different locations within the Santa Maria district.

The contact between the Monterey and Sisquoc Formations is a lithofacies change that is unrelated to silica diagenesis. This contact often defines the vertical transformation from reservoir to cap rock within the Santa Maria basin. The contact is marked by sediment-composition differences that are independent of silica diagenetic grade. The Monterey Formation is characterized by a rhythmic interlayering of more pure and less pure siliceous rocks, while the Sisquoc Formation consists of massive, less pure siliceous rocks. The contact between the Monterey and the Sisquoc Formations was produced by an increase in influx of terrigenous clastics into the basin, and the consequent disappearance of pure-diatom facies. The Monterey-Sisquoc contact is apparent by an upsection change from thin-bedded, rhythmically interlayered, siliceous rocks to massive siliceous mudstones. During Monterey time, rhythmic pulses of high and low terrigenous influx affected the basin; during Sisquoc time, terrigenous influx remained relatively high throughout deposition. This lithofacies change (formation contact) is visible both in outcrops and on well logs. Barron (1986) has correlated the Monterey-Sisquoc contact with a major global sea-level drop at about 6 Ma, suggesting that increased terrigenous influx into the Santa Maria basin was related to increased efficiency of clastics transport across exposed shelf areas, hence directly into the basin, during this low stand. The change from interlayered brittle and ductile rock in the Monterey to entirely ductile rock in the Sisquoc is the factor that forms the seal atop the prolific Monterey reservoirs of the Santa Maria basin; this vertical facies change marks a critical control on hydrocarbon accumulation.

Diagenetic controls on hydrocarbon accumulation

Certainly, silica diagenesis has played a role in the evolution of soft, highly porous, diatomaceous sediment into dense, brittle, highly fractured chert. Snyder (1987) discusses the manner by which silica diagenesis, responsible for transformation of soft sediment to brittle rock, coupled with compressional tectonics such that fracture formation is the natural result of folding and faulting of brittle rock. Siliceous rocks must have attained at least opal-CT-grade diagenesis before they are brittle enough to fracture. Consequently, for maximum production of reservoir-quality fractures, tectonic folding should follow, and not precede, silica diagenesis.

However, it must be emphasized that diagenesis in itself was not the only factor responsible for formation of brittle siliceous rock within the Santa Maria basin. Diagenesis has formed very brittle, glassy, opal-CT chert as a result of alteration of original pure-silica/pure-diatom sediment; in contrast, less pure diatomaceous mudstone consisting of a mix of diatoms and terrigenous clay has been transformed into ductile, non-glassy opal-CT siliceous shale. A similar evolution marks the alteration of opal-CT rocks into quartz-phase siliceous rocks. Glassy quartz chert formed only from diagenetic replacement of original, pure-silica, opal-CT rock. In contrast, noncherty, opal-CT-phase siliceous shale has evolved into noncherty, quartz-phase siliceous shale. It is critical to understand that silica diagenesis has not in itself resulted in the formation of glassy chert. Rather, it was a coincidence of proper original-sediment composition and proper burial heating that brought about the diagenetic generation of brittle, glassy-chert reservoir rocks at certain localities within the basin.

This transition process has been observed in outcrops (Dunham and Blake, 1987) as well as in the subsurface. Thus, although chert is of diagenetic origin, the distribution of chert within the Santa Maria basin was controlled ultimately by original sedimentation patterns of diatoms and clay across the basin. Consequently, as a result of depositional-facies changes, brittle reservoir rocks are not uniformly distributed within the Monterey Formation across the Santa Maria basin. There are some areas where the Monterey Formation contains thick intervals of fractured-reservoir rock, whereas in other areas the Monterey consists largely of nonfractured siliceous shale.

Origin of fractures within the Monterey Formation

The origin of fractures within the Monterey Formation is related to the effects of sediment composition and diagenesis, but the relation is indirect: diagenesis must be followed by still another event. Natural fractures within the Monterey Formation have formed as a result of tectonic folding of brittle rock (Grivetti, 1982; Belfield and others, 1983; Snyder, 1987). Whenever a rock like chert is deformed into a fold, the rock will become fractured, simply because chert is too brittle to undergo plastic deformation. Natural fracturing of the Monterey Formation is a result of horizontal compression and tectonic folding of rocks that have been made brittle by burial diagenesis of pure biogenic sediment.

Reservoir rocks outside the Monterey Formation

Although the Monterey Formation accounts by far for most of the oil trapped within the Santa Maria basin, there are pre- and post-Monterey sandstones and conglomerates that do produce significant quantities of oil from primary sandstone porosity. Coarse conglomerates of the Lospe Formation are productive in the Casmalia field, whereas the Point Sal Formation accounts for major oil production from Orcutt field, where it is known as the "Third Zone Sand." Both the Lospe and the Point Sal Formations represent basal clastic facies that were shed from Mesozoic and early Tertiary basement highs; these conglomerates and sands were then deposited in rapidly subsiding troughs during the first stages of Santa Maria basin subsidence. Lospe alluvial conglomerates record the earliest phase of basin subsidence, whereas Point Sal turbidite sands record progression to deep-water marine sedimentation (J. M. Casey, written communication, 1982). These clastic formations are much more limited in areal extent than the Monterey Formation, and they undergo rapid lateral facies changes that strongly affect sandstone reservoir quality over short distances. The most important control on reservoir quality within the Lospe and Point Sal Formations is depositional facies, with the best reservoir quality occurring in the better sorted, more porous, clastic facies.

Sandstones do not make up important reservoir intervals within the Monterey Formation of the Santa Maria basin. In the Santa Maria basin, the Monterey generally contains only fine-grained terrigenous clastics in the form of silt and clay; thus, the effective porosity within the Monterey is restricted to fracture porosity developed in nonclastic biogenic rocks. In contrast, sandstone locally represents an important reservoir facies in the post-Monterey section. The basal Sisquoc Formation contains sandstones that are productive in a number of fields within the Santa Maria basin. These sands were derived from erosion of fault blocks of Tertiary and Mesozoic-basement rocks that were uplifted as a result of the onset of compressional tectonics in the region during the late Miocene to early Pliocene. These sands generally overlie angular unconformities at the base of the Sisquoc, and rather than representing laterally continuous sheets, they seem to represent local accumulations developed adjacent to individual uplifted blocks.

The basal Sisquoc, referred to as the Tinaquaic Sandstone Member is productive in the Santa Maria Valley field, where it is capped by impermeable diatomaceous mudstones of the upper Sisquoc. The Tinaquaic consists of eroded Mesozoic basement, as well as eroded, opal-CT phase Monterey chert clasts; thus, it represents a late Miocene–early Pliocene pulse of tectonism that resulted in uplift and erosion of Monterey and pre-Monterey rocks. Where the Tinaquaic is present, it overlies the Monterey along an angular unconformity. Based on the presence of opal-CT phase clasts in the Tinaquaic, it is estimated that the eroded Monterey must have been buried to the depth of the opal-A to opal-CT transition prior to uplift, erosion, and deposition of the Tinaquaic in the western Santa Maria Valley field. The estimated present depth of the transition in this area ranges from 3 to 5 km, thus suggesting considerable uplift and erosion of Monterey rocks at the end of the Miocene. Basal Sisquoc sandstones are also productive in Cat Canyon field and Guadalupe field.

The basal Foxen Formation produces from sandstones within the Santa Maria Valley field, but the lack of sealing facies at other localities has precluded production. Also, although sandstones are present in the Pliocene Careaga Formation and the Plio-Pleistocene Paso Robles Formation, there are no sealing facies in the section (Fig.3); thus, economic hydrocarbon accumulations within the Santa Maria basin are largely limited to the rocks older than early Pliocene, due to absence of cap rocks younger than this age. The distribution of cap rocks represents a most fundamental facies control on hydrocarbon distribution within the Santa Maria basin. Although massive tar accumulations are present in both the Sisquoc and Foxen Formations, these have not proven to be economic.

Facies controls on source rocks within the Santa Maria basin

Up to this point, this discussion of geologic controls on hydrocarbon distribution within the basin has focused on reservoir rocks and cap rocks. Naturally, another control on hydrocarbon distribution within the basin involves the geology of source rocks. Within the Santa Maria basin, the principal source rocks are made up of the organic-rich shale facies of the Monterey Formation.

Analysis of a large number of samples, coupled with examination of well logs for qualitative indications of high organic content (low density, high uranium radioactivity), indicate that organic-rich rocks are limited to 0.5 to 2-m-thick shale intervals that are interlayered with thin dolomite beds within the lower to middle members of the Monterey Formation (Fig. 3). Kerogen contents commonly exceed 5 percent and locally exceed 18 percent within certain organic-rich Monterey shale beds. Other stratigraphic units, with the exception of thin shale layers within the upper Point Sal Formation, lack sufficient organic matter to constitute source rocks for hydrocarbons.

Organic-rich shales reflect a combination of effects related to the depositional environment, including prolific biogenic productivity that supplied the original biomass, a relatively slow sediment-accumulation rate that permitted concentration of organic matter relative to mineral matter in the sediments, and anoxic conditions at the sediment-water interface that allowed preservation of the organic matter. Organic-rich shales of the Monterey Formation typically contain nodules and thin laminae of calcium phosphate (carbonate fluorapatite), and are commonly referred to as phosphatic shales or organic-phosphatic shales. The presence of phosphate allows the inference of a number of unusual conditions within the depositional environment of the sediment. Garrison and others (1987) concluded that three main factors were responsible for controlling the distribution of the phosphatic facies within the Monterey Formation; these include: (1) deposition of calcareous sediments dominated by coccoliths during times when oceanographic conditions did not favor high diatom productivity; (2) slow sedimentation of dominantly pelagic components with relatively little terrigenous sediment; and (3) low-oxygen water masses, which led to widespread accumulation of laminated, organic-rich sediments. Each of these factors also has affected the accumulation and preservation of organic matter within these same Monterey rocks. As a contrast to the Santa Maria basin, Garrison and others (1987) attributed the absence of phosphatic-marlstone facies within the Monterey Formation of the Cuyama basin, as described by Lagoe (1987), to the fact that it was situated proximal to terrestrial sources of detritus; consequently, the high detrital input into the proximal basin inhibited phosphate formation.

The very organic-rich facies of the Monterey is largely restricted to shales of late Relizian, Lusian, and earliest Mohnian age, which are widely distributed within the basin, occurring stratigraphically beneath the principal chert-bearing interval of the formation throughout the basin. In contrast, cherts within the Monterey are largely Mohnian in age. Originating as the replacement products of pure diatom sediments, these cherts are related to climatic and paleoceanographic conditions that favored high diatom productivity and high rates of sediment accumulation; conditions that likely were brought on by major polar cooling and associated increased upwelling at the beginning of Mohnian time (Barron, 1986).

In contrast to the chert facies of the Monterey, the organic-rich shale facies was deposited prior to the global cooling that led to high diatom productivity; in fact, the rapid extraction of carbon from the global ocean-atmosphere system actually may have triggered the global cooling that led to subsequent diatom productivity. Extraction of carbon dioxide (CO_2) from the atmosphere by marine plankton, and the subsequent failure to return this CO_2, as a result of preservation of organic carbon within Monterey sediments, may have precipitated a reverse greenhouse effect that affected the entire planet (Vincent and Berger, 1985). The fact that pure-chert sediments succeed organic-rich sediments within the Monterey may be no coincidence at all; this stratigraphic succession may reflect a cause-and-effect relation that operated on a global scale.

Above the phosphatic-shale sequence (Fig. 3), the siliceous shales that are interlaminated with chert and dolomite beds of the Monterey of Mohnian age, do not have a high organic content. In fact, Isaacs (1983) noted an inverse correlation between abundance of biogenic silica and total organic-carbon content. This relation likely reflects the sedimentation-rate control on organic content as discussed above. Rocks with high silica content are interpreted as representing high diatom productivity (Isaacs, 1983); consequently, these rocks represent rapid accumulation of diatomaceous sediment, with a resultant high ratio of inorganic (silica shells) to organic matter within the sediment. Thus, Monterey siliceous shales are observed to have a low kerogen content due to dilution of organic matter by biogenic silica. Note that thin laminae of organic-rich phosphatic marl occasionally do occur interlaminated with chert and siliceous shale; these organic shales are interpreted as representing short-term, climatically related decreases in diatom productivity in surface waters (Dunham and Blake, 1987, p. 24).

The distribution of source rocks, as well as reservoir rocks, within the Santa Maria basin is controlled primarily by depositional facies. The base of the Monterey Formation was marked by a decrease in influx of terrigenous clastics into the rapidly subsiding Santa Maria basin that likely resulted from a latest early Miocene, global sea-level rise, which trapped clastics on shallow shelves and thus isolated the basin from clastic input (Isaacs, 1983). Barron (1986) identifies this basal Monterey event as correlative with the onset of rising sea level at about 17.5 Ma. Organic-rich sediments were deposited only during this time of reduced sedimentation rate, during the late Relezian, Lusian, and brief intervals of Mohnian time, as a result of the coincidence of very slow sediment accumulation within an anoxic water mass. An increase in sediment accumulation rate as a result of enhanced diatom productivity in Mohnian time ended the prolonged period of organic-shale deposition within the Santa Maria basin, and led to deposition of pure-diatom sediments that would eventually evolve into fractured cherts. Near the end of the Miocene, and onward throughout the remaining history of the Santa Maria basin, high rates of terrigenous-sediment influx effectively diluted the concentration of organic matter within the shales, and at the same time prevented the deposition of pure-biogenic sediment.

HYDROCARBON GENERATION WITHIN THE SANTA MARIA BASIN

Organic maturity and thermal history of the Santa Maria basin

The kerogen present within Monterey Formation organic-rich shales is Type II amorphous kerogen (Tissot and others, 1987) derived from bacterial alteration of marine planktonic algae (phytoplankton). When total organic carbon (TOC) values of the Monterey shales are reported, they may record both indigeneous organic matter (kerogen) and migrated organic matter (oil) that commonly fills the high matrix porosity (usually 20 to 30 percent). As a result, a shale reported as having a 10 percent TOC, may in fact consist of 3 percent inextractable organic matter in the form of kerogen, and 7 percent extractable organic matter in the form of oil. The presence of both in situ and migrated organic matter has led to confusion regarding thermal history, source-rock maturity, and timing of hydrocarbon generation within Monterey shales. For example, both extractable and inextractable organic matter are present in the Monterey Formation in the Point Conception Coastal Offshore Stratigraphic Test (COST) well, located in the southwestern corner of the offshore Santa Maria basin (King and Claypool, 1983). The extractable component consists of thermally mature hydrocarbons, while the inextractable component of the shales is immature kerogen. The mature hydrocarbons within reservoir beds of the well could not have originated from the immature source rocks that are present in the well; rather, the oil was generated from more deeply buried source rocks of higher maturity that were located in generative depressions down the flanks of the oil-bearing structure. Following generation, the oil migrated updip and was trapped at a much shallower depth than where it was generated, thus becoming associated with organic-rich but immature shales with which it had no genetic affinity (King and Claypool, 1983).

Many workers cite the Monterey as being its own source rock and reservoir; a true statement that, nevertheless, has been misinterpreted with regard to short-distance (primary) or long-distance (secondary) migration. The oil within a Monterey reservoir may not have been generated from the very same shales that are interbedded with the fractured-reservoir rocks. Rather, the oil may have migrated a considerable distance updip within fractures before becoming structurally trapped. Ogle and others (1987) have clearly distinguished between the source-reservoir concept and the migration concept; these authors envision significant updip migration of oil within the Monterey from deeply buried source rocks in the troughs of synclines, on into fractured reservoir rocks at the crests of adjacent anticlines.

Further complicating the evaluation of thermal history is evidence that classic indicators of thermal maturity may not behave in a normal manner within the unusual lithologies of the Monterey Formation. Walker and others (1983) have noted that organic matter dominated by amorphous algal material, as in the Modelo and Monterey Formations, may mature at a significantly faster rate than structured organic debris such as vitrinite. Additionally, vitrinite dispersed in predominately amorphous organic matter may fail to develop the reflectance reached under equal conditions in vitrinite-rich kerogen; therefore, vitrinite reflectance "... is an unreliable indicator of maturity in the most oil-prone source rocks" (Walker and others, 1983, p. 185).

Peterson and Hickey (1984) also report vitrinite-reflectance values of 0.3 percent from areas of the Santa Maria basin where oil generation has obviously taken place; they suggest that indirect estimates of maturity, such as vitrinite reflectance, should be evaluated in conjunction with more direct chemical methods for estimating maturity of Monterey kerogen. However, Peterson and Hickey (1984) introduce a topic that will be expanded on below; they suggest the possibility that Monterey oil may have been generated at unusually low temperature as a consequence of the unusual chemistry of Monterey kerogen.

Not only do thermal indicators appear to be anomalous within the Santa Maria basin, but the basin itself may have had an anomalous thermal history. Isaacs (1984) has suggested that unconventional patterns of compaction and thermal conductivity may have affected the thermal history of sediments within the Santa Maria basin. Isaacs noted that diatomaceous sediments retain very high porosities even after considerable burial; porosities as much as 70 percent are observed in diatomites buried to more than 450 m. Such extraordinary porosity values, relative to typical clastic sediment, result in the thermal conductivity of diatomite being extremely low relative to virtually all other sediment types (Isaacs, 1984). Furthermore, the porosity reduction associated with silica-phase changes results in major compaction of original bed thicknesses, such that an original 10-cm-thick layer of surface sediment may be reduced to less than 3 cm of quartz-phase siliceous rock. Since the thermal conductivity of opal-A sediments is low, the geothermal gradient over much of the Monterey's burial history may have been as much as twice as high as present. And, since the original sedimentary thickness of the Monterey has likely been reduced by a factor of three, temperature differences across the thickness of the Monterey may have been up to six times greater during early burial than presently observed based on present strata thickness and Recent geothermal gradient (Isaacs, 1984). A more realistic way to model geothermal heating is not simply to multiply Holocene geothermal gradient by Holocene stratal thickness, but rather to model the varying thermal conductivities and thicknesses of strata during burial and silica-phase change. For example, Isaccs noted that, based on recent thickness and thermal-gradient data from the vicinity of the Point Conception COST well, the base of the Monterey is estimated to have reached a temperature of only about 55°C by the end of deposition of the Sisquoc Formation in early mid-Pliocene time (ca. 3.5 Ma). On the other hand, if a more realistic model is used, the base of the Monterey may actually have reached a temperature of 100°C by the end of Sisquoc deposition at 3.5 Ma (Isaacs, 1984, p. 80). Clearly, this model has significant implications for modeling timing of hydrocarbon generation.

Another factor with implications regarding hydrocarbon generation involves the unusual tectonic history of the basin as described above, including migration of triple junctions and volcanism associated with wrench tectonics. Heasler and Surdam (1985) speculate that crustal extension associated with pull-apart basin subsidence resulted in a higher than normal thermal regime for the region associated in part with middle Miocene emplacement and extrusion of high-temperature volcanic material of the Tranquillon and Obispo Formations of the Santa Maria and Pismo basins. High heat flow, accompanied by rapid basin subsidence, would accelerate the entry of Monterey organic-rich shales into the zone of oil generation, thus affecting the time portion of time-temperature index function. For example, Tissot and others (1987) point out that applying the Lopatin-Waples method of calculating a time-temperature index (TTI) (see Waples, 1980) to basins with a rapid burial rate tends to result in an underestimate of maturation. Tissot and others noted that when temperatures are low, the influence of time on TTI is important, but when temperatures are high, the effect of time is slight, resulting in a TTI calculation that may not accurately reflect true maturity. As noted above, the special tectonic history of the Monterey has resulted in a situation where temperature, and not time, has had the most significant effect on oil generation; consequently, TTI calculations that suggest immaturity must be considered in error. That is, which line of evidence is most compelling regarding Monterey oil generation: the fact that TTI is low, or the fact that billions of barrels of mature hydrocarbon are present in the basin?

Timing of hydrocarbon generation

The question is reduced to when and where did the Monterey Formation subside to a depth sufficient to carry it to a temperature level adequate for oil generation, given the local geothermal gradient? In fact, there are many areas of the basin where the Monterey presently is situated at depths where temperatures exceed 120°C; at these locations the Monterey is situated within the classic oil window. However, it has been suggested that significant oil generation could take place at relatively low temperature. Orr (1986) has described Monterey kerogen as being different from typical Type II kerogen; he has proposed that Monterey kerogen be described as a separate category, designated Type II-S, which comprises amorphous algal kerogen with an unusually high sulfur content. Orr (1986) suggests that Type II-S may in fact generate oil at significantly lower temperatures than typical Type II kerogens, due to breakage of weak sulfur bonds within complex organic molecules of high-sulfur kerogen. The suggestion has yet to be documented; it remains to be proven that the sulfur present within high-sulfur kerogen is actually an integral part of the structure of oil-generating molecules (S. R. Larter, personal communication, 1987).

It is not necessary to appeal to unusually low-temperature generation to account for the oil of the Santa Maria basin. There are many areas of the basin where organic-rich shales have been carried to depths sufficient to heat them to at least 120°C. In contrast, it is apparent that the Monterey has never been deeply buried within other areas of the basin. For example, where biogenic silica still exists as opal-A, it is clear that the Monterey has never been heated to more than 55°C; it is unlikely that it could have generated oil from any type of kerogen at such low temperature. In fact, there is considerable regional variation in burial depth of the Monterey within the basin; there are areas where the Monterey has resided within the zone of oil generation (hotter than 120°C) since the late Miocene, while there are other areas where the Monterey has never been buried deeply enough to generate oil. Thus, there is no single Monterey oil-generation event in the Santa Maria basin; oil generation began in the late Miocene and likely continues to the present in tectonically subsiding regions of the basin where immature Monterey shales are only now being carried into the oil window.

Oil gravity variation within the Santa Maria basin

One of the most significant factors affecting hydrocarbon production from the Santa Maria basin is oil gravity. Historically, the Santa Maria basin has been considered a heavy-oil province. Because classic indicators of thermal maturity appeared to demonstrate that Monterey kerogen was immature, it seemed to follow that thermal immaturity was responsible for the occurrence of heavy oil within the basin. Oil generation at low temperature and shallow depths of burial was considered the most likely explanation for the preponderance of 6 to 12 degree API gravity oil, all too commonly found within the basin. Nevertheless, it is an observation that oil gravity within the Santa Maria basin is not exclusively heavy; in fact, it varies considerably, ranging from less than 5 to as light as 40 degrees API. To regard the basin as entirely a heavy-oil province is a misconception.

Intrabasinal variation in the nature of the kerogen within source rocks could be a factor affecting oil-gravity diversity, although Bromley and Senftle (1987) report little variation in the nature of the Type II kerogen present within the Monterey, based on petrographic and chemical analysis of kerogens from a range of Monterey samples from within and outside the Santa Maria basin. They concluded that variation in kerogen alone cannot account for the observed range of oil qualities present within the Monterey Formation.

Another model suggests that identical kerogens, associated with rocks of different mineralogy, may have produced different oils. Tannenbaum and others (1986) report that pyrolysis experiments performed on Monterey kerogen with and without associated calcite, illite, and montmorillonite, resulted in generation of different oils. Both illite and montmorillonite were found to selectively adsorb a considerable portion of high-molecular-weight compounds, while calcite exhibited no such adsorptive effect. Therefore, it was suggested that organic diagenesis of source rocks associated with carbonate minerals would result in generation mainly of heavy oils, whereas light oils would be the main products from source rocks that contain high montmorillonite contents. Since the detrital content of the Monterey Formation does vary within the Santa Maria basin, then it is possible that variations in clay content could have had an effect on oil gravity. It must be noted, however, that these experiments were not performed on actual Monterey rocks; rather, the samples consisted of chemical reagents that were artificially prepared in a laboratory. It has not been shown that the above effect actually occurs within water-saturated sediment.

Additionally, it has been suggested that the high-sulfur nature of Monterey kerogen results in the generation of heavy oil due to the presence of weak carbon-sulfur bonds, which cause the kerogen to fragment into large, heavy molecules such as asphaltene and sulfur-rich aromatics (Orr, 1986, p. 515). Although intuitively appealing, this suggestion requires further research and documentation. The occurrence of oil in the 20 to 30^+ degree gravity range in both the onshore and offshore Santa Maria basin indicates that Monterey kerogen can generate more than just heavy oil.

A factor that very likely has had a major influence on oil-gravity variation within the basin involves bacterial alteration of migrated oil within Monterey reservoirs. Magoon and Isaacs (1983) have documented the presence of biodegraded oils within the Santa Maria basin that are heavier than normal oils. Magoon and Isaacs (1983) note that three conditions are necessary for petroleum biodegradation, including: (1) a subsurface reservoir temperature low enough to allow petroleum-consuming microbes to exist, (2) availability of nutrients at the oil-water interface to feed the microbes, and (3) a means of introducing microbes into the oil pool. Both microbe infection and nutrient supply are accommodated by downdip migration of surface-derived water into the reservoir, while the reservoir temperature is a function of geothermal gradient and depth. Magoon and Isaacs (citing Philippi, 1977) suggest an average cutoff of 73°C for microbial biodegradation of oil. Thus, when considering thermal history of the Monterey, it is just as important to consider the low-temperature range of the reservoir history as it is the oil-generation temperature. For, if oil has migrated into a trap that has a low reservoir temperature, then biodegradation may have converted the original oil into noncommercial tar.

In summary, billions of barrels of oil are present in the Santa Maria basin despite the conventional indicators of maturity, which suggest that source rocks have not reached peak oil generation. Furthermore, oil lighter than 30 degrees API gravity is present within the basin; thus, the Monterey kerogen is capable of generating light oil.

SIMILAR BASINS

Tertiary basins in California similar to Santa Maria include the Los Angeles, Ventura–Santa Barbara, San Joaquin, Pismo, Huasna, Cuyama, Salinis, Santa Cruz, Bodega, Point Arena, and Eel River basins (Fig. 1; also see McCulloch and others, 1977; and Blake and others, 1978, for descriptions). Each of these basins has had a similar history of Neogene tectonic subsidence, deposition of biogenic-siliceous sediments, and late Neogene tectonic compression. While the Los Angeles, Ventura–Santa Barbara, and San Joaquin basins are known to contain billions of barrels of oil sourced by the Monterey, the remaining basins have not been thoroughly explored. Although reconaissance drilling has been carried out in some portion of each of these Santa Maria–like basins, these few wells can not have resulted in a thorough evaluation.

Neogene diatomaceous sediments are not limited to the California coast. Rather, they are present around virtually the entire rim of the North Pacific Ocean, including the Gulf of California, Bering Sea, and Sea of Japan; ". . . the ubiquitous occurrence of these siliceous deposits is indeed not coincidential but in fact represents the combined product of major Neogene paleoceanographic and tectonic events affecting the entire Pacific margin" (Ingle, 1981, p. 159). Modern analogs may be found in the California continental borderland (Blake, 1981), the Gulf of California (Donegan and others, 1981), the Panama Basin (Van Andel, 1973), offshore Peru (Soutar and others, 1981), offshore West Africa (Thomson and others, 1984), and in the Antarctic (Brewster, 1980).

SUMMARY

Tectonics and sedimentation have combined to localize major hydrocarbon accumulations within large anticlines of the Santa Maria basin. Structure is the key to anticlinal trap development in giant onshore and offshore fields; however, depositional facies provides the ultimate control on source-rock, reservoir-rock, and cap-rock quality within these structures. Economic oil accumulations are present only at localities where the proper depositional facies, at the appropriate diagenetic grade, have been deformed into structures that are capped by seals that have not been broken by major faults. This coincidence of sedimentology, diagenesis, and structural growth, combined in the proper order, constitutes the set of necessary conditions sufficient for development of economic hydrocarbon accumulations within the Santa Maria basin of western California.

REFERENCES CITED

Aki, K., and Richards, P. G., 1980, Quantitative seismology; Theory and methods: San Francisco, California, W. H. Freeman and Company, 557 p.

Arnold, R., and Anderson, R., 1907, Preliminary report on the Santa Maria oil district: U.S. Geological Survey Bulletin 317, 66 p.

Atwater, T., 1970, Implications of plate tectonics for the Cenozoic tectonic evolution of western North America: Geological Society of America Bulletin, v. 81, p. 3513–3536.

Atwater, T., and Molnar, P., 1973, Relative motion of the Pacific and North American plates deduced from sea-floor spreading in the Atlantic, Indian, and South Pacific Oceans: Stanford, California, Stanford University Publications in the Geological Sciences, v. 13, p. 136–148.

Aubury, L. E., 1904, Production and use of petroleum in California: California State Mining Bureau Bulletin 32, 230 p.

Baker, P. A., and Kastner, M., 1981, Constraints on the formation of sedimentary dolomite: Science, v. 213, p. 214–216.

Barron, J. A., 1986, Paleoceanographic and tectonic controls on deposition of the Monterey Formation and related siliceous rocks in California: Palaeogeography, Palaeoclimatology, Palaeoecology, v. 53, p. 27–45.

Belfield, W. C., Helwig, J., La Pointe, P., and Dahleen, W. K., 1983, South Ellwood oil field, Santa Barbara Channel, California, a Monterey Formation fractured reservoir, in Isaacs, C. M., and Garrison, R. E., eds., Petroleum generation and occurrence in the Miocene Monterey Formation, California: Pacific Section Society of Economic Paleontologists and Mineralogists Special Publication, p. 213–221.

Blake, G. H., 1981, Biostratigraphic relationship of Neogene benthic foraminifera from the southern California outer continental borderland to the Monterey Formation, in Garrison, R. E., and others, eds., The Monterey Formation and related siliceous rocks of California: Pacific Section Society of Economic Paleontologists and Mineralogists Special Publication, p. 1–14.

Blake, M. C., Jr., and 7 others, 1978, Neogene basin formation in relation to plate-tectonic evolution of San Andreas fault system, California: American Association of Petroleum Geologists Bulletin, v. 62, p. 344–372.

Bramlette, M. N., 1946, The Monterey Formation of California and the origin of its siliceous rocks: U.S. Geological Survey Professional Paper 212, 57 p.

Brewster, N. A., 1980, Cenozoic biogenic silica sedimentation in the Antarctic Ocean: Geological Society of America Bulletin, part I, v. 91, p. 337–347.

Bromley, B. W., and Senftle, J. T., 1987, Quality variations in middle Miocene oils of southern California; Source organic matter controls: Abstracts of Papers; 193rd American Chemical Society national meeting, Anaheim, California, paper GEOC-97.

California Division of Oil and Gas, 1974, California oil and gas fields, south-central Coastal and offshore California, Volume 2; Sacramento, California Division of Oil and Gas, unpaginated.

Crain, W. E., Mero, W. E., and Patterson, D., 1985, Geology of the Point Arguello discovery: American Association of Petroleum Geologists Bulletin, v. 69, p. 537–545.

—— , 1987, Geology of the Point Arguello field, in Ingersoll, R. V., and Ernst, W. G., eds., Cenozoic basin development of coastal California, Rubey Volume 6: Englewood Cliffs, New Jersey, Prentice-Hall, Inc., p. 407–426.

Crouch, J. D., Bachman, S. B., and Shay, J. T., 1984, Post-Miocene compressional tectonics along the central California margin, in Crouch, J. K., and Bachman, S. B., eds., Tectonics and sedimentation along the California margin: Pacific Section Society of Economic Paleontologists and Mineralogists Volume 38, p. 37–54.

Crowell, J. C., 1974, Origin of Late Cenozoic basins in southern California, in Dickinson, W. R., ed., Tectonics and sedimentation: Society of Economic Paleontologists and Mineralogists Special Publication 22, p. 190–204.

—— , 1979, The San Andreas fault system through time: Journal of the Geological Society of London, v. 136, p. 293–302.

Dibblee, T. W., Jr., 1950, Geology of southwestern Santa Barbara County, California: California Division of Mines Bulletin 150, 95 p.

Dickinson, W. R., 1981, Plate tectonics and the continental margin of California, in Ernst, W. G., ed., The geotectonic development of California, Rubey Volume 1: Englewood Cliffs, New Jersey, Prentice-Hall, Inc., p. 1–28.

Donegan, D., and Schrader, H., 1981, Modern analogues of the Miocene diatomaceous Monterey Shale of California; Evidence from sedimentologic and micropaleontologic study, in Garrison, R. E., and others, eds., The Monterey Formation and related siliceous rocks of California: Pacific Section Society of Economic Paleontologists and Mineralogists Special Publication, p. 149–158.

Dunham, J. B., and Blake, G. H., 1987, Guide to coastal outcrops of the Monterey Formation of western Santa Barbara County, California: Pacific Section Society of Economic Paleontologists and Mineralogists Book 53, 36 p.

Garrison, R. E., Kastner, M., and Zenger, D. H., eds., 1984, Dolomites of the Monterey Formation and other organic-rich units: Pacific Section Society of Economic Paleontologists and Mineralogists, v. 41, 215 p.

Garrison, R. E., Kastner, M., and Kolodny, Y., 1987, Phosphorites and phosphatic rocks in the Monterey Formation and related Miocene units, coastal California, *in* Ingersoll, R. V., and Ernst, W. G., eds., Cenozoic basin development of coastal California, Rubey Volume 6: Englewood Cliffs, New Jersey, Prentice-Hall, Inc., p. 348–381.

Grivetti, M. C., 1982, Aspects of stratigraphy, diagenesis, and deformation in the Monterey Formation near Santa Maria–Lompoc, California [M.A. thesis]: Santa Barbara, University of California, 154 p.

Hall, C. A., 1977, Origin and development of the Lompoc–Santa Maria pull-apart basin and its relation to the San Simeon–Hosgri fault, California: Geological Society of America Abstracts with Programs, v. 9, p. 428.

——, 1978, Origin and development of the Lompoc–Santa Maria pull-apart basin and its relation to the San Simeon–Hosgri strike-slip fault, western California: California Division of Mines and Geology Special Report 137, p. 25–31.

Hall, C. A., and Corbato, C. E., 1967, Stratigraphy and structure of Mesozoic and Cenozoic rocks, Nipomo Quadrangle, southern Coast Ranges, California: Geological Society of America Bulletin, v. 78, p. 559–582.

Hamilton, D. H., and Hall, N. T., 1987, Structure and tectonics of the San Luis–Pismo–Santa Maria region, coastal central California: Geological Society of America Abstracts with Programs, v. 19, p. 386.

Heasler, H. P., and Surdam, R. C., 1985, Thermal evolution of coastal California with application to hydrocarbon maturation: American Association of Petroleum Geologists Bulletin, v. 69, p. 1386–1400.

Hornafius, J. S., 1985, Neogene tectonic rotation of the Santa Ynez Range, western Transverse Ranges, California, suggested by paleomagnetic investigation of the Monterey Formation: Journal of Geophysical Research, v. 90, no. B14, p. 12503–12522.

Hornafius, J. S., Luyendyk, B. P., Terres, R. R., and Kamerling, M. J., 1986, Timing and extent of Neogene tectonic rotation in the western Transverse Ranges, California: Geological Society of America Bulletin, v. 97, p. 1476–1487.

Ingle, J. C., 1981, Origin of Neogene diatomites around the North Pacific rim, *in* Garrison, R. E., and others, eds., The Monterey Formation and related siliceous rocks of California: Pacific Section Society of Economic Paleontologists and Mineralogists Special Publication, p. 159–179.

Isaacs, C. M., 1980, Diagenesis in the Monterey Formation examined laterally along the coast near Santa Barbara, California [Ph.D. thesis]: Stanford, California, Stanford University, 329 p.

——, 1983, Compositional variation and sequence in the Miocene Monterey Formation, Santa Barbara coastal area, California, *in* Larue, D. K., and Steel, R. J., eds., Cenozoic marine sedimentation, Pacific Margin, U.S.A.: Pacific Section Society of Economic Paleontologists and Mineralogists Special Publication, p. 117–132.

——, 1984, The Monterey; Key to offshore California boom: Oil and Gas Journal, v. 82, no. 2, p. 75–81.

Jenkins, O. P., 1943, Geologic formations and economic development of the oil and gas fields of California: California Division of Mines Bulletin 118, 773 p.

Keller, M. A., and Isaacs, C. M., 1985, An evaluation of temperature scales for silica diagenesis in diatomaceous sequences including a new approach based on the Miocene Monterey Formation, California: Geo-Marine Letters, v. 5, p. 31–35.

Kieniewicz, P. M., and Luyendyk, B. P., 1986, A gravity model of the basement structure in the Santa Maria basin, California: Geophysics, v. 51, p. 1127–1140.

King, J. D., and Claypool, G. E., 1983, Biological marker compounds and implications for generation and migration of petroleum in rocks of the Point Conception deep-stratigraphic test well, OCS–CAL 78–164 No. 1, offshore California, *in* Isaacs, C. M., and Garrison, R. E., eds., Petroleum generation and occurrence in the Miocene Monterey Formation, California: Pacific Section Society of Economic Paleontologists and Mineralogists Special Publication, p. 191–200.

Lagoe, M. B., 1987, Middle Cenozoic basin development, Cuyama Basin, California, *in* Ingersoll, R. V., and Ernst, W. G., eds., Cenozoic basin development of coastal California, Rubey Volume 6: Englewood Cliffs, New Jersey, Prentice-Hall, Inc., p. 172–206.

Luyendyk, B. P., and Hornafius, J. S., 1987, Neogene crustal rotations, fault slip, and basin development in southern California, *in* Ingersoll, R. V., and Ernst, W. G., eds., Cenozoic basin development of coastal California, Rubey Volume 6: Englewood Cliffs, New Jersey, Prentice-Hall, Inc., p. 259–283.

Luyendyk, B. P., Kamerling, M. J., and Terres, R., 1980, Geometric model for Neogene crustal rotations in southern California: Geological Society of America Bulletin, v. 91, part 1, p. 211–217.

Magoon, L. B., and Isaacs, C. M., 1983, Chemical characteristics of some crude oils from the Santa Maria basin, California, *in* Isaacs, C. M., and Garrison, R. E., eds., Petroleum generation and occurrence in the Miocene Monterey Formation, California: Pacific Section Society of Economic Paleontologists and Mineralogists Special Publication, p. 201–211.

McCullough, D. S., Clarke, S. H., Field, M. E., Scott, E. W., and Utter, P. M., 1977, A summary report on the regional geology, petroleum potential, and environmental geology of the southern proposed lease sale 53, central and northern California outer continental shelf: U.S. Geological Survey Open-File Report 77–593, 57 p.

McLaughlin, R. P., and Waring, C. A., 1914, Petroleum industry of California: California State Mining Bureau Bulletin 69, 519 p.

Miall, A. D., 1984, Principals of sedimentary basin analysis: New York, Springer-Verlag, 490 p.

Murata, K. J., and Larson, R. R., 1975, Diagenesis of Miocene siliceous shales, Temblor Range, California: U.S. Geological Survey Journal of Research, v. 3, p. 553–566.

Ogle, B. A., Wallis, W. S., Heck, R. G., and Edwards, E. B., 1987, Petroleum geology of the Monterey Formation in the offshore Santa Maria/Santa Barbara areas, *in* Ingersoll, R. V., and Ernst, W. G., eds., Cenozoic basin development of coastal California, Rubey Volume 6: Englewood Cliffs, New Jersey, Prentice-Hall, Inc., p. 348–381.

Orr, W. L., 1986, Kerogen/asphaltene/sulfur relationship in sulfur-rich Monterey oils, *in* Leythaeuser, D., and Rullkotter, J., eds., Advances in organic geochemistry 1985: Oxford, Pergamon, p. 499–516.

Payne, C. M., Swanson, O. E., and Schell, B. A., 1979, Investigation of the Hosgri fault offshore southern California, Point Sal to Point Conception: U.S. Geological Survey Open-File Report 79–1199, 17 p.

Peterson, N. F., and Hickey, P. J., 1984, California Plio-Miocene oils; Evidence of early generation, *in* Meyer, R. F., and others, eds., Exploration for heavy crude oil and bitumen: American Association of Petroleum Geologists Conference Proceedings, October 28–November 2, 1984, Santa Maria, California, v. 2, unpaginated.

Philippi, G. T., 1977, On the depth, time, and mechanism of origin of the heavy to medium-gravity naphthenic crude oils: Geochimica et Cosmochimica Acta, v. 41, p. 33–52.

Pisciotto, K. A., 1978, Basinal sedimentary facies and diagenetic aspects of the Monterey Shale, California [Ph.D. thesis]: Santa Cruz, University of California, 450 p.

——, 1981a, Review of secondary carbonates in the Monterey Formation, California, *in* Garrison, R. E., and others, eds., The Monterey Formation and related siliceous rocks of California: Pacific Section Society of Economic Paleontologists and Mineralogists Special Publication, p. 273–284.

——, 1981b, Diagenetic trends in the siliceous facies of the Monterey Shale in the Santa Maria region, California: Sedimentology, v. 28, p. 547–571.

Pisciotto, K. A., and Garrison, R. E., 1981, Lithofacies and depositional environments of the Monterey Formation, California, *in* Garrison, R. E., and others, eds., The Monterey Formation and related siliceous rocks of California: Pacific Section Society of Economic Paleontologists and Mineralogists Special Publication, p. 97–122.

Regan, L. J., and Hughes, A. W., 1949, Fractured reservoirs of Santa Maria district, California: American Association of Petroleum Geologists Bulletin, v. 33, p. 32–51.

Rintoul, B., 1985, California offshore: Pacific Oil World, v. 77, no. 11, p. 7–12.

——, 1986, California offshore update: Pacific Oil World, v. 78, no. 5, p. 4–8.

——, 1987, California offshore: Pacific Oil World, v. 79, no. 1, p. 50–61.

Roehl, P. O., and Weinbrandt, R. M., 1985, Geology and production characteristics of fractured reservoirs in the Miocene Monterey Formation, West Cat Canyon oilfield, Santa Maria Valley, California, in Rohel, P. O., and Choquette, P. W., eds., Carbonate petroleum reservoirs: New York, Springer-Verlag, p. 525–545.

Snyder, W. S., 1987, Structure of the Monterey Formation; Stratigraphic, diagenetic, and tectonic influences on style and timing, in Ingersoll, R. V., and Ernst, W. G., eds., Cenozoic basin development of coastal California, Rubey Volume 6: Englewood Cliffs, New Jersey, Prentice-Hall, Inc., p. 321–347.

Soutar, A., Johnson, S. R., and Baumgartner, T. R., 1981, In search of modern depositional analogs to the Monterey Formation, in Garrison, R. E., and others, eds., The Monterey Formation and related siliceous rocks of California: Pacific Section Society of Economic Paleontologists and Mineralogists Special Publication, p. 123–148.

Tannenbaum, E., Huizinga, B. J., and Kaplan, I. R., 1986, Role of minerals in thermal alteration of organic matter; 2, A material balance: American Association of Petroleum Geologists Bulletin, v. 70, p. 1156–1165.

Thomson, J., Calvert, S. E., Mukherjee, S., Burnett, W. C., and Bremner, J. M., 1984, Further studies of the nature, composition, and ages of contemporary phosphorite from the Namibian Shelf: Earth and Planetary Science Letters, v. 69, p. 341–353.

Tissot, B. P., Pelet, R., and Ungerer, P., 1987, Thermal history of sedimentary basins, maturation indices, and kinetics of oil and gas generation: American Association of Petroleum Geologists Bulletin, v. 71, p. 1445–1466.

Van Andel, T. H., 1973, Texture and dispersal of sediments in the Panama Basin: Journal of Geology, v. 81, p. 434–457.

Vincent, E., and Berger, W. H., 1985, Carbon dioxide and polar cooling in the Miocene; The Monterey hypothesis, in Sundquist, E. T., and Broecker, W. S., eds., Natural variations in carbon dioxide and the carbon cycle: American Geophysical Union, Monograph Series, v. 32, p. 455–467.

Walker, A. L., McCulloh, T. H., Petersen, N. F., and Stewart, R. J., 1983, Anomalously low reflectance of vitrinite, in comparison with other petroleum source-rock maturation indices, from the Miocene Modelo Formation in the Los Angeles basin, California, in Isaacs, C. M., and Garrison, R. E., eds., Petroleum generation and occurrence in the Miocene Monterey Formation, California: Pacific Section Society of Economic Paleontologists and Mineralogists Special Publication, p. 185–190.

Waples, D. W., 1980, Time and temperature in petroleum formation; Application of Lopatin's method to petroleum exploration: American Association of Petroleum Geologists Bulletin, v. 64, p. 916–926.

Woodring, W. P., and Bramlette, N. M., 1950, Geology and paleontology of the Santa Maria district, California: U.S. Geological Survey Professional Paper 222, 142 p.

Manuscript Accepted by the Society September 20, 1989

ACKNOWLEDGMENTS

We thank those who read our early drafts, including G. A. Crawford, J. R. Odermatt, J. F. Kowalski, and D. H. Zenger, and we thank the management of Unocal Corporation for permission to publish this paper. Our manuscript has benefited from the scientific criticisms and technical editing of GSA reviewers R. E. Garrison, C. A. Hall Jr., and A. R. Palmer, although we alone are responsible for the above interpretations. We note that over the years, our understanding of the Santa Maria basin has developed through interaction with many colleagues, including G. H. Blake, N. A. Brewster, R. T. Budden, S. M. Carey, J. M. Casey, J. P. Chauvel, M. L. Cotton-Thornton, R. G. Hickman, C. M. Isaacs, K. W. Northrup, T. A. Redin, G. H. Smith, R. B. Tallyn, R. E. Williams, and many others.

Chapter 28

North Slope of Alaska

Kenneth J. Bird
U.S. Geological Survey, 345 Middlefield Road, Menlo Park, California 94025

INTRODUCTION

The North Slope petroleum province occurs at the northern extension of the Rocky Mountains and Great Plains provinces. It is one of the six most prolific petroleum producing areas of North America (Bird, 1989) and currently produces about 25 percent of U.S. oil production. The province spans the entire width of northern Alaska (Fig. 1). Total area is about 400,000 km^2, nearly one-half of which lies offshore. The onshore area includes the northern part of the Brooks Range, the Northern Foothills, and the Coastal Plain physiographic provinces; offshore the area includes the continental shelves of the Chukchi Sea on the west—one of the broadest shelves in the world—and the Beaufort Sea on the north. The name North Slope is interchangeable with Arctic Slope.

The North Slope petroleum province is primarily a composite basin of Paleozoic to Cenozoic age. The northern edge of the composite basin includes Cretaceous and Tertiary passive margin deposits, which form the southern flank of the modern Canada basin (Fig. 2). The composite basin includes late Paleozoic and Mesozoic south-facing continental margin deposits overlain by late Mesozoic and Cenozoic north-facing foreland basin deposits. Beginning in the Late Jurassic, north-verging compressional tectonism resulted in the uplift of the ancestral Brooks Range as a fold and thrust belt with a foredeep basin (Colville basin) to the north. Since the Early Cretaceous opening of the Canada basin, crustal shortening in the south has been occurring simultaneously with crustal extension in the north. The Chukchi and Arctic platforms, relatively stable areas characterized by nearly flat-lying strata, lie north of the fold and thrust belt. The Barrow arch, a broad, subsurface basement ridge, marks the boundary between the downwarped and downfaulted northern rifted margin and the gently south-dipping continental margin deposits.

The petroleum-prospective rocks of the North Slope province are predominantly clastic deposits. The Colville basin and laterally equivalent passive margin deposits are composed of clastic sedimentary rocks derived from a southern source, the Brooks Range orogen. These rocks, known as the Brookian sequence, overlie Mississippian to Early Cretaceous continental margin deposits known as the Ellesmerian sequence. The Ellesmerian is

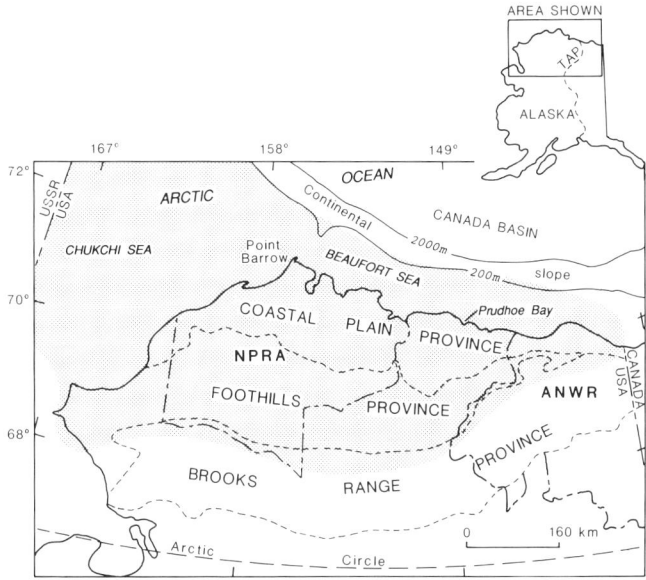

Figure 1. Index map showing the location of the North Slope petroleum province (shaded area). NPRA, National Petroleum Reserve in Alaska; ANWR, Arctic National Wildlife Refuge; TAP, trans-Alaska pipeline.

composed of both clastic (~75 percent) and carbonate (~25 percent) rocks that were deposited on the south-facing margin (in present-day coordinates) of a stable continental land mass. Together, the Brookian and Ellesmerian sequences, names coined by Lerand (1973), make up the composite basin deposits; they unconformably overlie metamorphosed and deformed sedimentary and igneous rocks of Proterozoic to Devonian age that form the basement for petroleum exploration in this region.

With one critical difference, the North Slope basin is similar to the numerous intracontinental composite basins located along the east side of the American Cordillera (Klemme, 1981). Whereas these basins display an asymmetric basin profile, with the longer, tapering limb toward the continental interior (the foreland), the Paleozoic through Mesozoic section of the North Slope is unique in that the foreland side of the basin is truncated by rifting, and the expected continental interior is replaced by the Arctic Ocean basin. Pre-rift rocks of the North Slope, in fact,

Bird, K. J., 1991, North Slope of Alaska, *in* Gluskoter, H. J., Rice, D. D., and Taylor, R. B., eds., Economic Geology, U.S.: Boulder, Colorado, Geological Society of America, The Geology of North America, v. P-2.

show that a continent once lay to the north (in present-day coordinates). The continental connection was severed in Early Cretaceous time, and the whereabouts of that connection is a point of ongoing debate and one of the main reasons that northern Alaska is considered a suspect terrane (Coney and Jones, 1985).

This report summarizes the petroleum geology of the North Slope province, including its exploration history and future petroleum potential. The treatment of the structure, stratigraphy, and tectonic development of this region is generalized from the more detailed presentations by Moore and others (1990) and Grantz and others (1990).

EXPLORATION HISTORY AND TECHNIQUES

Intermittent petroleum exploration of the North Slope can be documented for 50 of the last 70 years, and the role of government in these efforts makes this region unique. Details of North Slope petroleum exploration history provided by Reed (1958), Gyrc (1970, 1988), Jamison and others (1980), Bird (1981), and Tailleur and Weimer (1987) show that government exploration was conducted intermittently beginning in 1923 and that industry exploration has been continuous since 1958.

Standard petroleum exploration methods are employed on the North Slope. However, its Arctic environment and frontier setting dictate modification of some techniques. For example, the near-roadless nature of the region often necessitates air transport of equipment and supplies for exploratory drilling at remote sites. Furthermore, the marshy terrain of the coastal plain makes summertime surface activities difficult and potentially disruptive of the tundra; therefore, most onshore seismic surveys and exploratory drilling are conducted during the winter months when the ground surface is frozen and snow covered, thus minimizing the environmental impact of these activities. Temporary (single season) airstrips and roads are usually constructed of ice, and offshore, at shallow-water locations, islands have been constructed of either ice or gravel to serve as drilling platforms for one-season exploratory wells (Oil and Gas Journal, 1989). Schindler (1988) describes many of the logistical and engineering details related to Arctic exploration.

Government exploration

Government exploration of the North Slope was conducted in three phases. The initial phase, following World War I, was

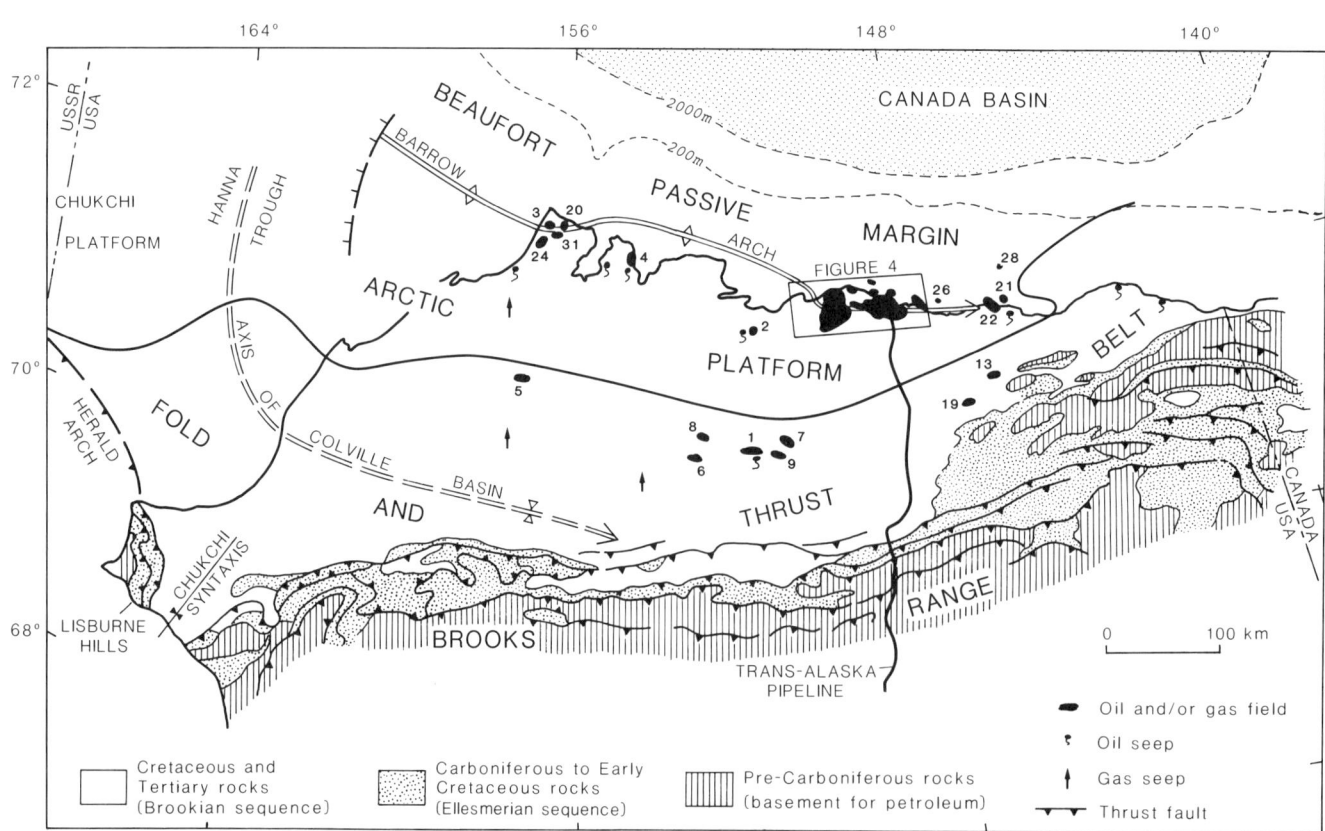

Figure 2. Generalized geologic map of North Slope petroleum province showing locations of oil and gas fields, hydrocarbon seepages, and major tectonic elements. The northern edge of the Brooks Range generally coincides with the limit of pre-Cretaceous rocks, while the foothills–coastal plain boundary coincides with the northern edge of the fold and thrust belt. Numbers coincide with oil and gas fields or accumulations listed in Table 1.

prompted by a perceived shortage of oil for the Navy. About one-half of the prospective area of the North Slope was set aside by the U.S. government in 1923, under the jurisdiction of the Navy. This area, initially known as Naval Petroleum Reserve No. 4 (NPR-4), is now known as the National Petroleum Reserve in Alaska (NPRA; Fig. 1). Exploration was conducted here during four consecutive seasons (1923 to 1926) by geologic-topographic parties from the U.S. Geological Survey (Smith and Mertie, 1930). Between 1926 and 1943, no further exploration of the North Slope was undertaken either by the government or by private industry.

The second phase of exploration began during World War II (1944), when the U.S. government mounted a full-scale petroleum exploration program known as Pet-4. It was focused on NPR-4 but was not limited to it. This program, described in detail by Reed (1958), lasted ten years; it utilized all of the petroleum exploration techniques then available and introduced new ones. Extensive and far-ranging geologic studies were conducted by the U.S. Geological Survey (USGS). A full complement of geophysical surveys was conducted, including seismic, gravity, and aeromagnetic. A total of 36 test wells and 45 shallow core tests were drilled, resulting in the discovery of three oil fields and five gas fields (Table 1, Fig. 2). All petroleum discoveries were subeconomic, although accumulation sizes are only approximately known. The South Barrow gas field was developed for local usage. Most discoveries were made on anticlinal structures in Cretaceous sandstone reservoirs, although two of the three oil accumulations (Fish Creek and Simpson, nos. 2 and 4 on Fig. 2) were discovered by drilling near surface oil seepages in the absence of anticlinal structures.

The third and final phase of government exploration was stimulated by the discovery of the Prudhoe Bay oil field (1967), and the Arab oil embargo (1973). Initiated in 1974 and concluded in 1982, this program was limited to the NPRA. It was a comprehensive exploration effort, including geologic, geochemical, and geophysical surveys as well as exploratory drilling. Nearly 24,000 line-kilometers of reflection seismic data was collected and interpreted. A total of 28 test wells were drilled, primarily for pre-Cretaceous objectives similar to those developed at Prudhoe Bay. Two subcommercial gas accumulations were discovered (East Barrow and Walakpa, numbers 20 and 24 on Fig. 2), and indications of oil and/or gas were found in nearly all the test wells. Additional exploratory and development wells were drilled in the Barrow area to supplement the local gas supply. A summary of this program and results of studies may be found in Gryc (1988) and in various contractor reports that are available through the National Geophysical Data Center, NOAA, Boulder, Colorado 80303.

Industry exploration

Industry exploration of the North Slope dates from 1958, when the government lifted a land freeze and offered acreage for

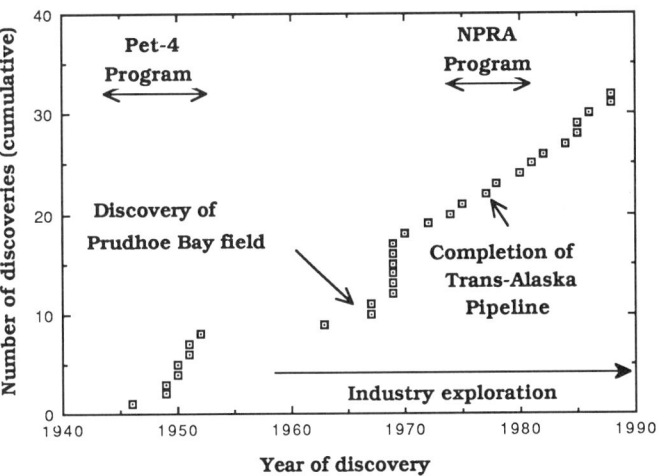

Figure 3. Summary of North Slope petroleum exploration history in terms of oil and gas discoveries tabulated since the beginning of exploratory drilling in 1944. Most of the discoveries are not economic or are not producing. Data from Table 1. Pet-4, Petroleum Exploration Program; NPRA, National Petroleum Reserve in Alaska.

lease. Until 1979, when the first offshore lease sale was held, industry exploration was restricted by land availability to the area between the NPRA and the Arctic National Wildlife Refuge (ANWR, Fig. 1). Since then, additional sales have been held, both offshore and onshore, including four sales in the NPRA and, most recently, the first sale in the Chukchi Sea. Limited exploration, including seismic surveys and surface geologic studies, was allowed on a part of the ANWR coastal plain in 1984 and 1985. Results of these studies, described in Bird and Magoon (1987), indicate that this is the most promising onshore area for petroleum exploration remaining on the North Slope. A decision about whether or not to allow leasing and drilling in the coastal plain of ANWR is now pending in Congress.

Industry drilling began in the foothills, testing Cretaceous anticlinal objectives, similar to those tested by the Navy in NPR-4; eight wells were drilled from 1964 to 1967. Although only one subcommercial gas field (East Umiat) was discovered, indications of oil and gas were encountered in every well. Exploration activity then shifted northward to the coastal plain, where pre-Cretaceous objectives could be tested at shallower depths than in the foothills, and where there was better opportunity to acquire solid lease-blocks on state-owned acreage. The third coastal plain well was drilled in 1967, when exploration activity on the North Slope was winding down toward a virtual standstill. This well, the eleventh industry attempt on the North Slope, resulted in the discovery of the Prudhoe Bay field, the largest commercial oil accumulation in North America. In the year or so following the Prudhoe Bay discovery, North Slope exploration activity blossomed, and the rate of oil and gas discovery shot upward. Since then, the rate has tapered off to one discovery every two years (Fig. 3).

TABLE 1. SUMMARY OF SELECTED DETAILS FOR NORTH SLOPE OIL AND GAS ACCUMULATIONS*

Map No	Field or Accumulation Name	Discovery Date	Reservoir	In Place Oil BBO	In Place Gas TCF	Cumulative Production Oil MBO	Cumulative Production Gas BCF	Reserves Oil MBO	Reserves Gas BCF	Data Source
1	Umiat	1946	Nanushuk	<1	<<1	....	....	70	?	1
2	Fish Creek	1949	Nanushuk	<<1	....	....	....	?	?	2
3	South Barrow	1949	Barrow	....	<<1	....	18	....	7	3
4	Simpson	1950	Nanushuk	<<1	<<1	....	....	12	?	2
5	Meade	1950	Nanushuk	....	<<1	....	....	....	20	2
6	Wolf Creek	1951	Nanushuk	....	<<1	....	....	....	?	2
7	Gubik	1951	Colville	....	<1	....	....	....	295	2
8	Square Lake	1952	Colville	....	<<1	....	....	....	58	2
9	East Umiat	1963	Nanushuk	....	<1	....	....	....	?	4
10	Prudhoe Bay	1967	Sadlerochit	23	27	5,507	7,652	4,219	23,441	5
11	Lisburne	1967	Lisburne	3	3	22	76	143	406	6
12	Kuparuk River	1969	Kuparuk	~4	~2	398	470	1,105	634	7
13	Kavik	1969	Sadlerochit	....	<1	....	....	....	?	8
14	West Sak	1969	Sagavanirktok	20†	<<1	....	....	....	....	9
15	Ugnu	1969	Sagavanirktok	15†	....	....	....	....	....	9
16	Milne Point	1969	Kuparuk	<1	<<1	5	2	95	?	8
17	Gwydyr Bay	1969	Sadlerochit	<1	<<1	....	....	60	?	10
18	North Prudhoe	1970	Sadlerochit	<1	<<1	....	....	75	?	10
19	Kemik	1972	Shublik	....	<1	....	....	....	?	8
20	East Barrow	1974	Barrow	....	<<1	....	4	....	8	3
21	Flaxman Island	1975	Canning	?	?	....	....	?	?	4
22	Point Thomson	1977	Thomson	<1	6	....	....	350	5,000	11
23	Endicott	1978	Kekiktuk	1	<2	9	8	366	907	12
24	Walakpa	1980	Walakpa	....	<<1	....	....	....	?	13
25	Niakuk	1981	Kuparuk	<1	<<1	....	....	58	30	14
26	Tern Island	1982	Kekiktuk	?	?	....	....	?	?	15
27	Seal Island	1984	Sadlerochit	<1	<1	....	....	150	?	16
28	Hammerhead	1985	Sagavanirktok	?	?	....	....	?	?	15
29	Colville Delta	1985	Kuparuk?	?	?	....	....	?	?	16
30	Sandpiper	1986	Sadlerochit	?	?	....	....	?	?	15
31	Sikulik	1988	Barrow	....	<<1	....	....	....	?	4
32	Point McIntyre	1988	Kuparuk	1	?	....	....	~300	?	17
	Totals			~69	~40	5,941	8.230	7,003	30,791	

*For additional details see table in Magoon, 1990.
Map numbers correspond to locations shown in Figures 2 and 4; Reservoir, main hydrocarbon-bearing rock unit; see Figure 6 for age and lithology; In Place, reported values or rounded estimates based on assumed recovery factors of about 30 percent (oil) and 80 percent (gas); Oil, reported in BBO (10^9 barrels, 159×10^8 m^3) or MBO (10^6 barrels, 159×10^3 m^3)(barrel = 42 U.S. gallons = 0.159 m^3); Gas, reported in TCF (10^{12} ft^3, 283×10^8 m^3) or BCF (10^9 ft^3, 283×10^5 m^3); Cumulative Production through 12/31/87; Reserves, remaining reserves as of 12/31/87; ? = amount of resource unknown; .... = resource not present; † = midpoint of range.
Data Sources: 1. Molenaar, 1982; 2. Collins and Robinson, 1967; Reed, 1958; 3. Lantz, 1981; 4. Well file; 5. State of Alaska, 1977; 6. State of Alaska, 1984; 7. Carman and Hardwick, 1983; van Poollen and Associates and State of Alaska, 1978; 8. State of Alaska, 1985; 9. Werner, 1987; 10. Van Dyke, 1980; 11. Oil and Gas Journal, 1984; Bird and Magoon, 1987; Craig and others, 1985; 12. Woidneck and others, 1987; Harris, 1987; 13. Gryc, 1988; 14. Harris, 1988; 15. Minerals Management Service, 1988; 16. Alaska Report, 1989; 17. Wall Street Journal, 1989.

Exploratory drilling

Exploratory drilling by government and industry has been conducted for 45 years on the North Slope, resulting in the discovery of 32 oil and gas accumulations, which encompass both onshore and offshore areas. The rate of discovery is plotted in Figure 3, based on the tabulation in Table 1. The oil and gas accumulations are located in Figures 2 and 4. Although most of the oil and gas accumulations are noncommercial by current North Slope standards, five are now producing. Their total ultimate recovery (reserves plus produced) is estimated to be nearly 12 billion barrels (1.9×10^9 m^3) of oil. Even more indicative of the richness of this province are the total in-place resources shown in Table 1: nearly 70 billion barrels (11×10^9 m^3) of oil and 40 trillion cubic ft (1.13×10^{12} m^3) of gas, more than 95 percent of which are concentrated within a 65-km radius of Prudhoe Bay (Fig. 4).

Hydrocarbon production

Hydrocarbon production on the North Slope began in 1949, when the government produced natural gas for local use at Barrow. Commercial oil production began in 1977 from the Prudhoe

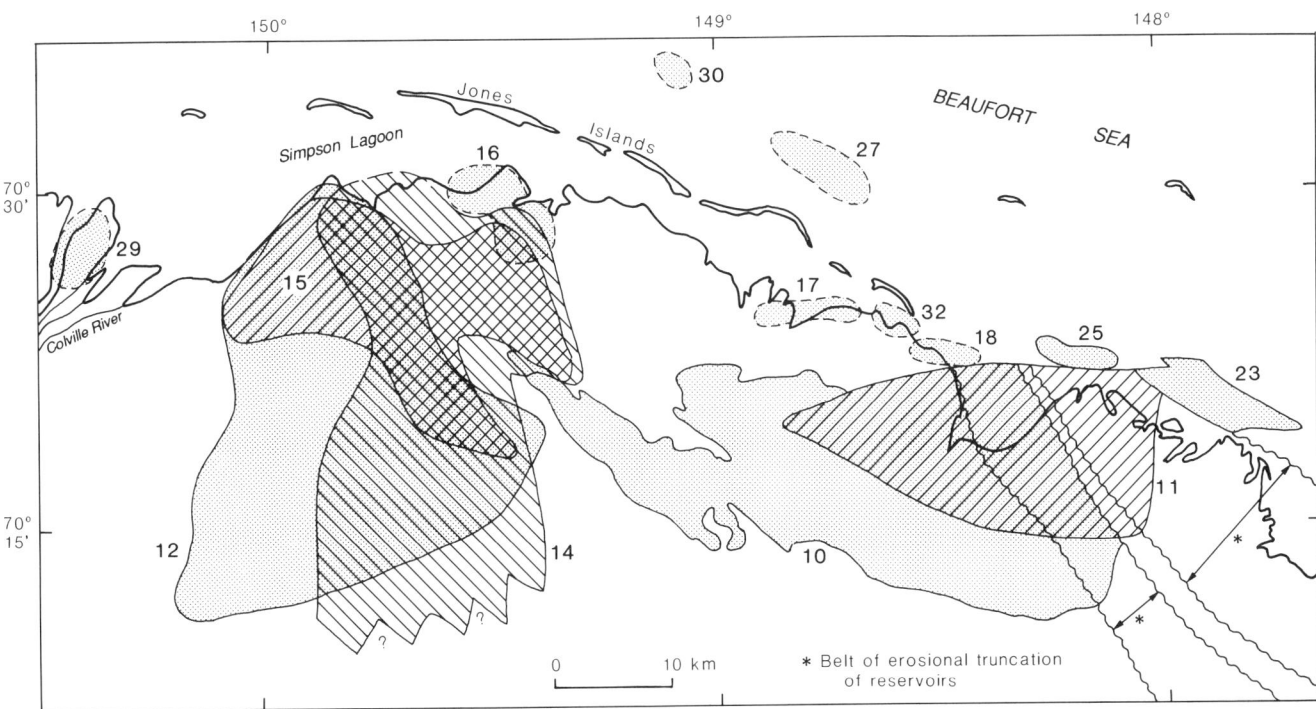

Figure 4. Map of the greater Prudhoe Bay area, where most North Slope oil and gas resources occur, showing the location and areal extent of fields and the truncation trends of producing reservoir units beneath the Lower Cretaceous unconformity (LCU). Numbers coincide with fields listed in Table 1.

Bay field when the trans-Alaska pipeline was completed—almost ten years after the field was discovered. Production from the Prudhoe Bay and Kuparuk River fields, currently the two leading U.S. oil producers, combined with that from three other North Slope fields (Endicott, Lisburne, and Milne Point), supplied about 2 million barrels (3.2×10^5 m^3) of oil per day in 1988, or about 25 percent of total U.S. production. North Slope production rates on a per-well basis are typical of frontier regions (Table 2). This tabulation of the United States' most important oil-producing states shows Alaskan production rates average nearly 2,000 barrels per day per well, about two orders of magnitude greater than any other state. Alaska's rates reflect the remarkable characteristics of the Sadlerochit reservoir at the Prudhoe Bay field—some wells produce nearly 20,000 barrels per day (Jamison and others, 1980)—and the economics of Arctic petroleum development. At least in part for lack of a transportation system, natural gas is not presently economic, even though large volumes of gas have been discovered. Some gas, however, is produced for use in the production facilities at Prudhoe Bay and nearby oil fields.

STRUCTURAL SETTING

The complex history of basin evolution, foreland to rift-type basins, results in a complex structure contour pattern on the basement surface in the North Slope province (Fig. 5). Burial depth to pre-Carboniferous basement exceeds 10 km beneath the axis of the Colville basin and along the passive margin; minimum burial depths, some as shallow as 1 km, occur along the Barrow arch and on the Chukchi platform. Brooks Range thrusting and uplift interrupt these trends and bring basement rocks to the surface. As summarized below and in the north-south cross section (Plate 6A), most preserved North Slope structures formed in Cretaceous and Tertiary time in response to compressional forces in the south and extensional forces in the north.

The Brooks Range and the deformed rocks of the Foothills province compose a fold and thrust belt (Fig. 2). The Brooks Range is a continental (A-type) subduction orogenic belt more than 1,000 km long and as much as 300 km wide (Bally and Snelson, 1980). In the western part of the North Slope, this orogenic belt makes a sharp bend (the Chukchi syntaxis), virtually doubling back on itself, and continues northward through the Lisburne Hills and northwestward offshore along the Herald arch (Fig. 2). Except for the area of the syntaxis, where structural trends oppose each other, the orogen is characterized by east-trending, north-vergent structures. Important differences in structural style and time of deformation occur along the northern margin of the Brooks Range; faults are more steeply dipping and imbrications less numerous in the northeastern part of the range compared to the western part of the range (Moore and others, 1990). West of 148°W, the northern margin of the range is made up of a stack of far-traveled thrust packages (allochthons) that are defined by distinctive sequences of stratified Late Devonian to Early Cretaceous rocks and characterized by complex folds and imbricate thrust faults. East of 148°W, where the front of the range extends far to the north and displays northeast-trending

TABLE 2. RANKING OF THE TEN TOP OIL-PRODUCING STATES AS OF 1984 SHOWING ALASKA'S EXTREMELY HIGH PER-WELL PRDUCTION RATE AND ITS RELATIVELY SMALL NUMBER OF PRODUCING WELLS, REFLECTIONS PRIMARILY OF THE ECONOMICS OF ARCTIC OIL PRODUCTION*

Rank	State	Daily Rate of Production		Number of Producing Wells
		Total (10^3 bbl)	Per Well (barrels)	
1	Texas	2,435	12	209,040
2	Alaska	1,825	1,868	977
3	Louisiana	1,392	54	25,823
4	California	1,161	23	49,847
5	Oklahoma	446	4	103,000
6	Wyoming	352	29	12,038
7	New Mexico	215	12	18,697
8	Kansas	207	4	51,888
9	North Dakota	139	38	3,697
10	Utah	112	58	1,944

*Adapted from Gerhard and others, 1988.

structures, the range is dominated by anticlinoria cored by pre-Carboniferous rocks, reflecting regional north-vergent duplex structures. Basal thrusts lie in pre-Carboniferous basement rocks, and roof thrusts cut various stratigraphic levels in younger stratigraphic units. Two phases of thrusting are recognized in the Brooks Range: (1) an early phase of displacement during Late Jurassic to Neocomian time, and (2) a late phase of displacements from mid-Cretaceous to Tertiary time. The latter deformational episode resulted in uplift of the range and development of many of its conspicuous structures, as well as deformation of foredeep deposits in the Foothills province. Patterns of development and filling of the foredeep basin suggest that the latter deformational event was diachronous, being older in the west and becoming progressively younger in the east. Earthquake epicenters and deformed terrace deposits suggest that deformation continues in the northern ANWR and offshore in the eastern Beaufort Sea (Grantz and others, 1988).

The foothills structural province consists of deformed foredeep deposits and, along its southern margin, thrust blocks of older Brooks Range rocks (Plate 6A). This province is characterized by detachment folds. Thrust faults occur at many levels and generally cut stratigraphically upsection to the north. Fault displacements and fold amplitudes diminish northward; their northern extent generally coincides with the northern limit of the Foothills province. Structures characteristic of this province extend offshore to the west beneath the southern part of the Chukchi Sea (north of the Herald arch) and to the northeast beneath the Beaufort shelf east 146°W, offshore from the ANWR (Craig and others, 1985; Thurston and Theiss, 1987; Grantz and May, 1988; Grantz and others, 1988; Fig. 2).

North of the region affected by the Brooks Range orogeny, the North Slope petroleum province consists mostly of nearly flat-lying Mississippian and younger strata and east-trending extensional faults. Prominent structural features of this area are the (1) Arctic platform, (2) Barrow arch, and (3) Beaufort passive margin. An exception to the generalization above is the region of extensive north-trending wrench faults and associated structures in the western part of the United States Chukchi shelf, the Chukchi platform and Hanna trough (Fig. 2).

The Arctic platform, that region beveled by pre-Carboniferous erosion and upon which the Ellesmerian and Brookian sequences were deposited, has remained fairly stable from Mississippian to Early Cretaceous time. Its northern margin was disrupted by rifting in Early Cretaceous time and its western edge is marked by the Hanna trough. Its gentle southward tilt during Ellesmerian deposition was accentuated by Brooks Range thrust loading and deposition of Brookian foredeep deposits beginning in Late Jurassic time. On the Arctic platform, several local basins developed in Mississippian time as sags or partially fault-bounded basins (half-grabens) that are inferred to be filled largely with nonmarine, coal-bearing clastic deposits. The combined Ikpikpuk and Umiat basins, with an areal extent of about 14,000 km^2 and more than 3 km of fill, are the largest of these basins (Fig. 5; Plate 6A). High-angle normal and reverse faults (with possible strike-slip motion) that cut Pennsylvanian to Jurassic rocks extend eastward across the central part of the NPRA, and probably beyond. The Inigok-1 well, located in the eastern NPRA (Fig. 5; Plate 6A), tested a prominent anticlinal fold related to this faulting.

In the Chukchi Sea, the north-trending Hanna trough (Central Chukchi basin of Thurston and Theiss, 1987) is a down-bowed and downfaulted feature with a long and complex history of subsidence that separates the Arctic platform from the Chukchi platform (Fig. 2). Ellesmerian and Brookian deposits in the Hanna trough are more than 12 km thick; these deposits accumulated in three areally coincident structural basins that formed at different times in highly contrasting settings, according to Thurston and Theiss (1987). These authors report numerous structural features commonly associated with wrench tectonics in a north-trending zone more than 160 km wide (the Hanna wrench-fault zone) that crosses the Hanna trough and Chukchi platform. Southward, the Hanna trough merges with the axis of the Colville basin.

The Chukchi platform is a long-standing positive feature beneath the western Chukchi shelf characterized by westward-shallowing basement, onlapping Ellesmerian rocks, and thin Brookian rocks. Lengthy, north-trending faults segment the Chukchi platform. From seismic stratigraphic evidence, these faults are Ellesmerian in age, and in the northern part of the platform, they were reactivated in Late Cretaceous(?) to early Tertiary time as wrench faults. Diapiric structures are also identified in the northern part of the platform (Thurston and Theiss, 1987; Grantz and May, 1988). The southern part of the Chukchi platform is bounded by the Herald arch, the offshore extension of the Brooks Range thrust belt. Northward across the Chukchi platform, Brookian deposits thicken toward the tectonic hinge that marks the southern boundary of the North Chukchi basin, a part of the passive margin sequence (Fig. 5).

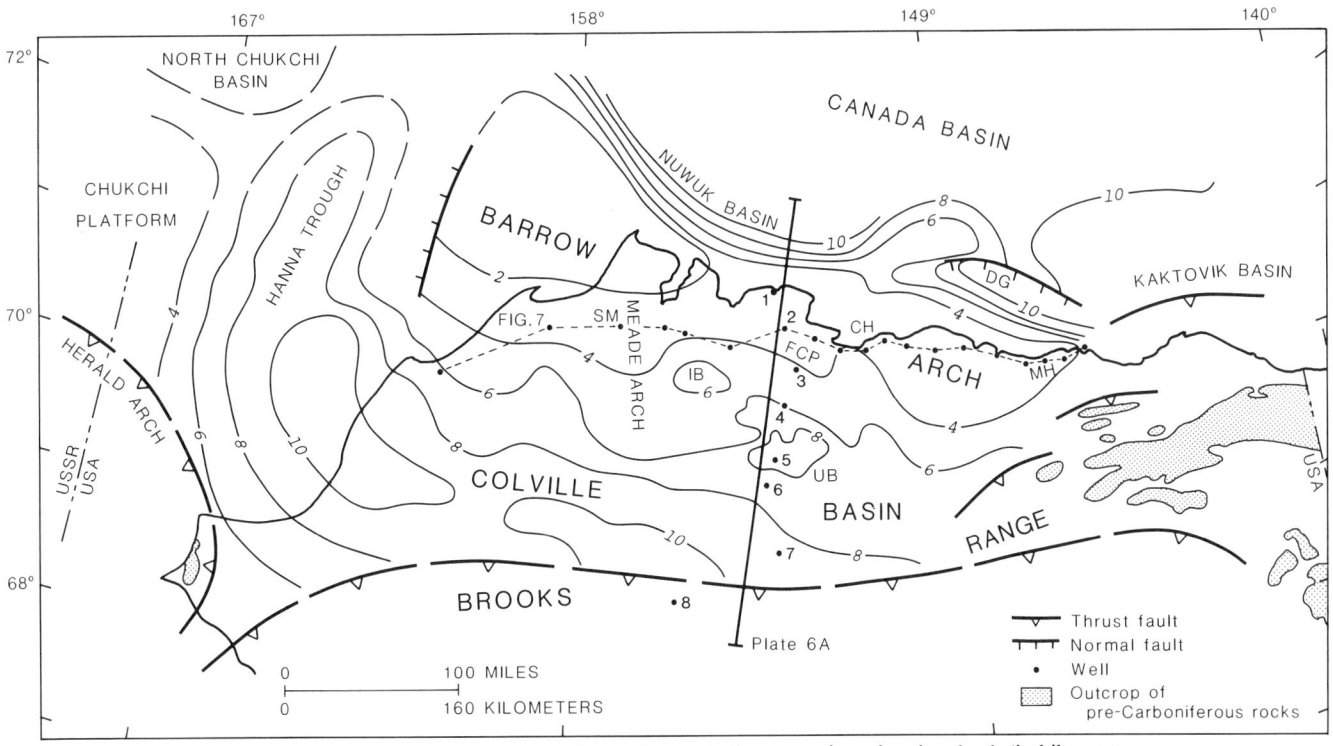

Figure 5. Structure contour map of the North Slope petroleum province showing depth (in kilometers subsea) to pre-Carboniferous basement. Offshore contours generalized from Grantz and May (1988) and Grantz and others (1988). Geologic features: CH, Colville high; DG, Dinkum graben; FCP, Fish Creek platform; IB, Ikpikpuk basin; MH, Mikkelsen high; SM, South Meade; UB, Umiat basin. Wells: 1. J. W. Dalton-1; 2. East Teshekpuk-1; 3. North Inigok-1; 4. Inigok-1; 5. Square Lake-1. 6. Wolf Creek-3; 7. East Kurupa-1; 8. Lisburne-1. Line of cross section (Fig. 7) is shown.

The Barrow arch, a broad, east-trending, ridge-like feature, underlies the coast of northern Alaska and the northeastern Chukchi Sea and separates the Colville basin from the Beaufort passive margin and the Canada basin (Fig. 2). Most North Slope oil and gas accumulations occur along or near this arch. Flank dips on the Barrow arch are generally less than 2°, and its axis plunges eastward at about 0.5°. Defined as the line along which previously south-dipping beds (the Ellesmerian sequence) begin to dip northward as a result of sagging of the new (rifted) continental margin (Rickwood, 1970), the Barrow arch is not a simple, symmetrical anticlinal feature. Its eastward plunge results from uplift in the Point Barrow area, and subsidence to the east, probably related to tectonic and sedimentary loading by the northeast Brooks Range and adjacent foredeep deposits (Fig. 5). The concentration of oil and gas accumulations on the Barrow arch in the general area of Prudhoe Bay is explained by Hubbard and others (1987) as the result of four-way closure that developed in a favorable up-dip position that trapped migrating hydrocarbons generated in nearby Brookian depocenters. The four-way closure resulted from the intersection of the high-standing rift margin trend and lithospheric flexural bulges that developed in front of the Brooks Range thrust belt along the northern edge of the foreland basin.

The continental margin of northern Alaska is an Atlantic-type passive margin, called the Beaufort passive margin by Hubbard and others (1987). In this margin, Grantz and May (1988) and Grantz and others (1988) recognize three regions of contrasting structure and stratigraphy. Typically, the margin consists of 8- to 10-km-thick prism of Cretaceous and Tertiary sediments (North Chukchi, Nuwok, and Kaktovik basins) overlying horst and graben structures developed during a 60-m.y. period (Middle Jurassic to Early Cretaceous) of rift-related faulting. Structures in the passive margin deposits trend subparallel to the coastline and consist of growth faults, related rollover anticlines, and diapiric ridges. The earlier episode of extensional tectonism produced sediment-filled grabens, a regional unconformity (the LCU), the Barrow arch, and the oceanic Canada basin.

STRATIGRAPHY

The stratigraphic record of the North Slope province probably spans a billion years of geologic time, but rocks with petroleum potential are mostly younger than Devonian (Figs. 6, 7). Traditionally, the petroleum-prospective rocks of the North Slope have been grouped into two sequences (Ellesmerian and Brookian) to emphasize source areas and genetic relations. Both sequences contain important petroleum source and reservoir rocks, although present oil extraction is entirely from Ellesmerian reservoirs. Pre-Carboniferous (basement) rocks, especially carbonate rocks, that have been buried and metamorphosed beyond the

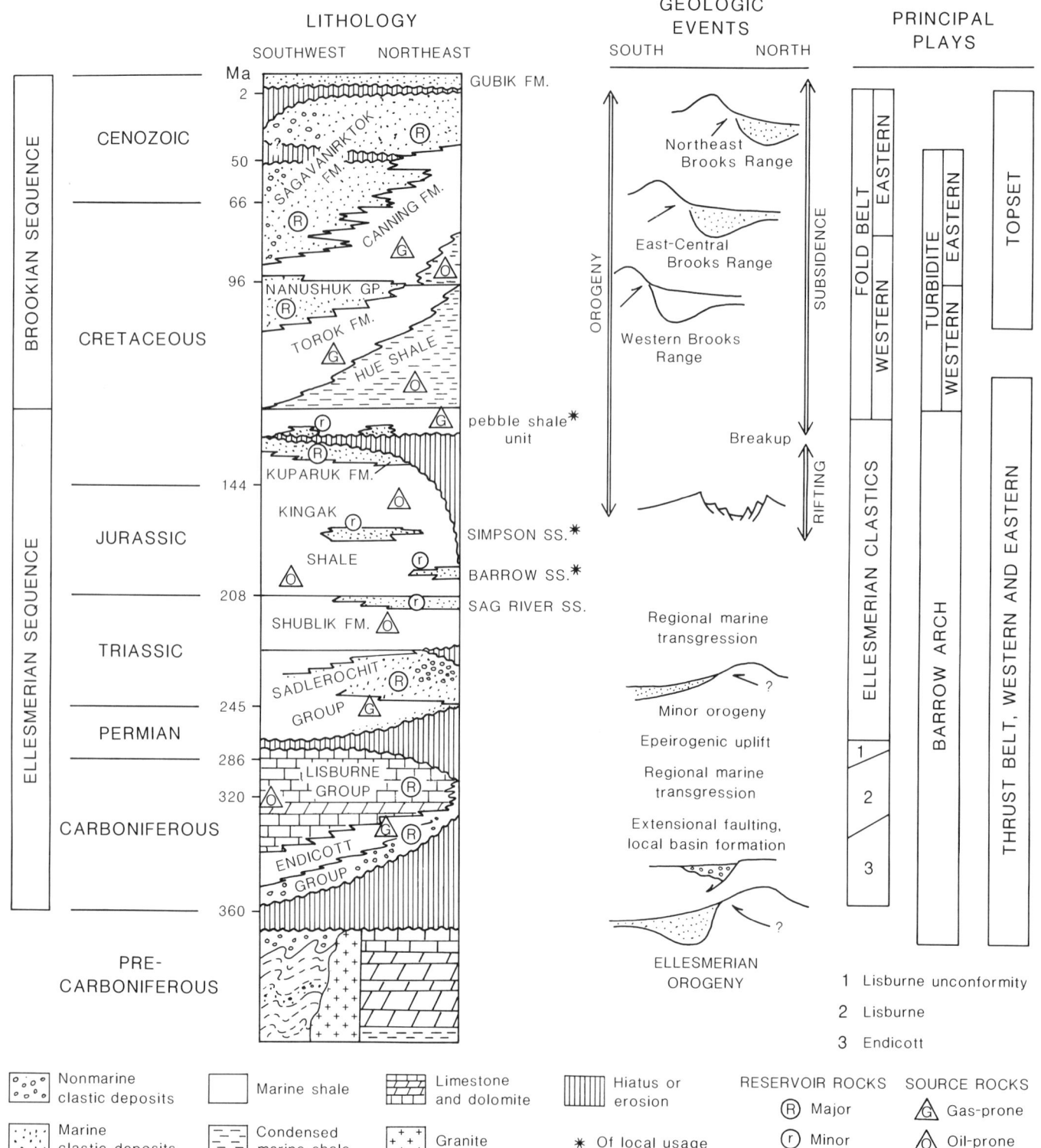

Figure 6. Generalized stratigraphic column for the North Slope petroleum province showing petroleum source and reservoir rocks, major geologic events, and the stratigraphic intervals encompassed by principal petroleum plays. Petroleum plays are described in Bird (1990). Absolute time scale from Palmer (1983).

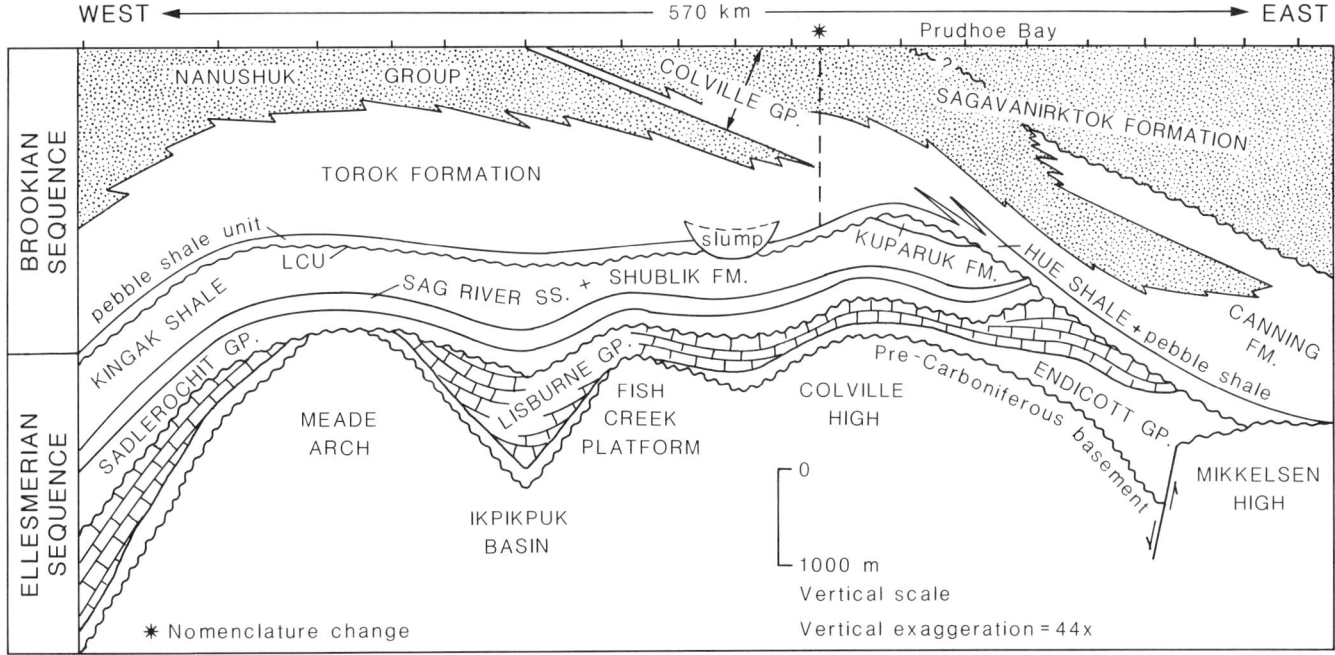

Figure 7. West-to-east cross section of the North Slope Coastal Plain province illustrating major structural and stratigraphic features of the Ellesmerian and Brookian sequences. Tick marks along top edge of figure are locations of wells providing subsurface control. Figure is generalized from Molenaar and others (1986). See Figure 5 for location. LCU, Lower Cretaceous unconformity.

thermal stage for oil, may, in favorable circumstances, provide reservoirs for oil or gas.

Pre-Carboniferous

Pre-Carboniferous rocks of the North Slope province consist of a complex assemblage of slightly to moderately metamorphosed Proterozoic to Devonian sedimentary and igneous rocks; Precambrian crystalline rocks typical of the Canadian Shield are not known to occur in this area. In the subsurface, pre-Carboniferous rocks (known mostly along the Barrow arch) consist of steeply dipping and slightly metamorphosed argillite of Ordovician and Silurian age (Carter and Laufeld, 1975). Their fine-grained character and apparent great thickness suggest a continental slope-depositional environment. Near Point Barrow, these rocks have an organic carbon content of as much as 12 percent and may have generated oil or gas prior to Mississippian time (Magoon and Bird, 1988). Most primary porosity in pre-Carboniferous rocks was destroyed during their long history of deep burial, heating, and deformation. Secondary porosity, however, may occur as a result of tectonic fracturing or selective leaching of grains or cements. Oil and condensate have been recovered from fractured(?) pre-Carboniferous sandstone and carbonate rocks beneath the Early Cretaceous Thomson sand of local usage in the Point Thomson field (no. 22 on Fig. 2; Bird and others, 1987).

Ellesmerian sequence

The Ellesmerian sequence records a major northward advance of the sea following the Ellesmerian orogeny. The sequence consists of marine carbonate rocks, marine and nonmarine clastic rocks, and scarce igneous rocks representing about 220 m.y. (Mississippian to Early Cretaceous) of continental margin sedimentation (Fig. 6). Northward-directed stratigraphic features such as onlap, convergence, truncation, increasing grain size, and marine to nonmarine facies changes indicate that the ancient shoreline lay near the present coast and that the open ocean was to the south. Total Ellesmerian thickness is generally less than 2 km, but in local areas, such as the Umiat basin, may exceed 6 km. Ellesmerian sandstones, typically composed of quartz and chert grains (van de Kamp, 1988), generally make better reservoirs than the relatively immature Brookian sandstones that contain more feldspar and rock fragments (Bartsch-Winkler and Huffman, 1988). Because the Ellesmerian sequence is generally less than 2 km thick, rocks rich in organic material deposited during this time were not capable of generating oil until buried by the much thicker Brookian deposits.

The Mississippian to Early Permian part of the Ellesmerian constitutes a transgressive megacycle, as much as 4 km thick in some areas. It consists of nonmarine coal-bearing sandstone, shale, and conglomerate (Kekiktuk Conglomerate) that is succeeded by shallow marine black shale (Kayak Shale) or, along the northern basin margin, by red and green shale (Itkilyariak Formation). These shale units grade upward and laterally into an areally extensive carbonate platform sequence composed of limestone and dolomite (Lisburne Group). Epirogenic movements in Late Pennsylvanian(?) and Early Permian time caused a withdrawal of the sea and development of a regional unconformity at the top of the Lisburne. Significant oil accumulations occur in the

Kekiktuk Conglomerate (Endicott field) and in limestone and dolomite of the Lisburne Group (Lisburne field).

Advance of the sea over the eroded Lisburne platform resulted in the deposition of the next megacycle, the clastic deposits (~300 m thick) of the Permian and Triassic Sadlerochit Group. The lower part of the Sadlerochit is a northward-thinning, transgressive marine sandstone and siltstone unit (Echooka Formation). The Echooka is abruptly overlain by prodelta shale and siltstone (Kavik Member of the Ivishak Formation) that grades upward into a southward prograding clastic wedge of marine and nonmarine sandstone and conglomerate (Ledge Sandstone Member of the Ivishak Formation). The Ledge is the main reservoir of the Prudhoe Bay oil field; there it consists of alluvial fan and deltaic facies (Melvin and Knight, 1984). The uppermost part of the Sadlerochit Group consists of a transgressive upward-fining and northward-thinning marine siltstone and argillaceous sandstone (Fire Creek Siltstone Member of the Ivishak Formation).

The transgression that began with deposition of the Fire Creek Siltstone continued in Middle and Late Triassic time with deposition of as much as 100 m of richly fossiliferous shale, siltstone, mudstone, and limestone (Shublik Formation). This unit, believed to represent deposition under oceanic, upwelling conditions (Parrish, 1987), is an important petroleum source rock. A thin (<30 m), regressive marine sandstone and siltstone (Sag River Sandstone, Karen Creek Sandstone) gradationally overlies the Shublik Formation along the northern basin margin and marks the end of the Shublik transgressive cycle.

Jurassic to Early Cretaceous (Neocomian) strata consist of as much as 1.5 km of marine shale and siltstone (Kingak Shale and pebble shale unit) with locally developed, predominantly marine sandstones (e.g., Kuparuk Formation). The Kingak Shale, an important petroleum source rock composed of a complex of southward prograding, offlapping and downlapping wedges of shale and siltstone, was deposited during an episode of crustal extension and represents the last complete megacycle of the Ellesmerian sequence. Normal faulting and development of sediment-filled grabens and half-grabens, some with as much as 3 km of fill (e.g., Dinkum graben, Fig. 5; Grantz and others, 1988), occurred mainly north of the present-day coastline; rock types of the graben-fill are unknown because of a lack of well penetrations.

Uplift along the rift margin in Early Cretaceous (Valanginian) time resulted in the formation of a northwesterly elongate land mass that, at its maximum, occupied the area beneath most of the present-day coastal plain and the inner continental shelf. Erosion of this landmass produced the regional Lower Cretaceous unconformity (LCU). Subsidence of the rift margin resulted in a marine transgression and deposition of local sandstones, such as the Kemik Sandstone. A blanket-like marine shale (pebble shale unit) that forms the final deposit of the Ellesmerian sequence constitutes an important petroleum source rock, and provides the seal for many of the petroleum accumulations in the Prudhoe Bay area (Fig. 4). The pebble shale unit is conformably overlain by distal, condensed marine shale deposits (Hue Shale) representing the basal unit of the Brookian sequence.

Brookian sequence

The Brookian sequence includes all of the sediments that were shed across northern Alaska from the Brooks Range orogenic belt. This sequence spans 180 to 150 m.y. (Middle or Late Jurassic to the present) and mostly represents foreland basin deposits developed ahead of northward advancing thrust sheets. The oldest Brookian strata (Jurassic and earliest Cretaceous), probably deposited several hundred kilometers to the south of the present Brooks Range, were tectonically transported in Early Cretaceous time with the Brooks Range allochthons, and are now discontinuously preserved in the Brooks Range and adjacent foothills. The majority of Colville basin fill (>8 km thick) is Aptian(?) and younger in age. Filling of the basin occurred from southwest to northeast in a series of migrating depocenters. Because Brookian deposits provided the overburden necessary for petroleum source rocks to be heated to reach maturity, Brookian depocenters provided changing sites of maturity and changing directions of petroleum migration. Shifting depocenters may also reflect geographically and temporally distinct orogenic pulses. Brookian deposits thin northward over the Barrow arch and grade into the passive margin sequence, where they again thicken to as much as 8 km (Plate 6A).

The earliest preserved Brookian depocenter developed during Aptian(?) to Cenomanian time in the western and central part of the North Slope (beneath the Chukchi shelf and NPRA). During Cenomanian to Eocene time, the depocenter shifted to the central part of the North Slope (approximately the area between the NPRA and the ANWR). From Eocene time to the present, the depocenter has been located in the northern ANWR and adjacent offshore (Moore and others, 1990).

Seismic reflection profiles show that the Brookian sequence is characterized by well-developed sigmoidal reflectors. These reflectors, interpreted as time lines, record successive basin profiles (Molenaar, 1988). Well and outcrop studies show that (1) topset reflections represent deltaic coal-bearing sandstone, conglomerate, and shelf shale deposits; (2) foreset reflections depict slope shale and turbidite sandstone deposits; and (3) bottomset reflections include basin-plain shale and turbidite sandstone deposits and distal, condensed shale with interbedded volcanic ash (Molenaar and others, 1987; Molenaar, 1988). Brookian deltaic deposits include the Nanushuk Group and Sagavanirktok Formation; marine shelf, slope, and basin-plain shale and interbedded sandstone include the Torok and Canning Formations; and distal, condensed basin-plain shale and bentonites are represented by the Hue Shale (Figs. 6, 7).

THERMAL HISTORY

Compared to most basins, the broad outlines of the thermal history are reasonably well known for the onshore North Slope; for most parts offshore, there are no direct observations. Undoubtedly this region has been affected by regional thermal events related to rifting and thrusting, as well as localized events related to magma intrusion, contrasts in radioactive heat production, fluid flow, and thermal conductivity variations. However, the contribution and relative importance of these factors at any particular locality are poorly known. Our knowledge of North Slope thermal history is based on measurements of organic metamorphism for surface and well samples (mainly vitrinite reflectance [R_0], thermal alteration index, conodont alteration index), present-day thermal profiles, and most recently, apatite fission-track analysis. Trends of organic metamorphic thresholds for the onset of oil (R_0 = 0.6 percent), condensate (R_0 = 1.3 percent), and dry gas (R_0 = 2.0 percent) generation are plotted on the north-south structure section (Plate 6A). These trends and other features are discussed below.

Organic metamorphic grade of rocks at the surface and at similar subsurface depths generally decreases northward in the North Slope province; this is a reflection of uplift and erosion in the orogenic belt and subsidence and sedimentation in the adjacent foredeep basin and passive margin. Data from Brosge and others (1981), Harris and others (1987), and Magoon and Bird (1987, 1988) show that the 2-percent R_0 value is generally found at the surface within the northern margin of the Brooks Range and at depths of 4 km or more beneath the Colville basin. The 0.6-percent R_0 value intersects the surface in the foothills and plunges northward and eastward into the subsurface; along the coastline it occurs at a depth of about 600 m near Point Barrow and about 4,000 m at the western edge of the ANWR. The distance (vertical separation) between R_0 values of 0.6 and 2.0 percent tends to increase from about 1.5 km in the north to about 4.5 km in the south, a reflection of higher thermal gradients to the north and lower gradients to the south.

The Lisburne-1 well, located near the northern edge of the Brooks Range (Fig. 5), shows a progressive vitrinite reflectance increase with depth through five thrust repetitions (Magoon and others, 1988, pl. 19.38). This relation appears to indicate that thermal maturity was achieved after thrusting and before uplift. At this site, more than 3 km of uplift and erosion are indicated.

For the offshore Beaufort passive margin, Grantz and others (1988) utilized heat flow measurements from the Canada basin, combined with a model of passive-margin thermal history, to estimate that the top of the oil-generating zone occurs about 3 km below the sea floor, and the top of the thermal gas generation zone occurs at about 6 km.

Thermal history relative to hydrocarbon maturation at the 6,127-m-deep Inigok well in the eastern part of the NPRA is illustrated in Figure 8. Burial history curves for this well show continuous subsidence and sedimentation from Mississippian to middle Tertiary time, followed by an estimated 300 m of uplift and erosion. Lopatin's method of integrating time and temperature (Waples, 1980) shows that thermal conditions favorable for oil generation (TTI value of 10) in the Endicott Group and lower part of the Lisburne Group occurred here as early as Triassic time. As subsidence and burial continued, the zone of oil genera-

tion (TTI 10 to 1,000) effectively moved upward through the stratigraphic section to its present position, encompassing rocks of Jurassic to mid-Cretaceous age between the depths of 1,980 to 3,690 m. The zone of oil generation is expected to have migrated northeastward across the North Slope in concert with the filling of the Colville basin.

The present-day North Slope thermal regime is characterized by thick permafrost and variable geothermal gradients (American Association of Petroleum Geologists and U.S. Geological Survey, 1976; Blanchard and Tailleur, 1982; Lachenbruch and others, 1988). The long-term mean surface temperature systematically varies from –12°C along the northern coast to –4°C inland near the Brooks Range. Depth to base of permafrost, the 0°C isotherm, generally ranges from 200 to 400 m, but in the area of the Prudhoe Bay oil field it reaches a depth of 630 m. Thermal gradients within the permafrost range from 15 to 50 °C/km, while gradients below the permafrost range from 24 to 47 °C/km. Although a simple pattern of gradient variation has yet to emerge from these data, higher gradients are generally found in coastal plain wells (area of Barrow arch and northern flank of Colville basin) and lower gradients in foothill wells (area of Colville basin and the fold and thrust belt). Anomalous thermal characteristics—high geothermal gradients, thin permafrost, and a shallow occurrence of the oil window—have been identified in the area of the South Meade well and in wells on the Fish Creek platform (Fig. 5). The origin of these anomalies is unknown.

The present-day North Slope thermal regime provides conditions suitable for the occurrence of natural gas hydrates (solids composed of light gases caged in an ice crystal lattice). Gas hydrates may occur offshore where water depths exceed about 400 m or onshore where permafrost thickness exceeds 240 m. Evidence for the presence of gas hydrate was found on about 75 percent of the seismic lines collected in areas along the continental slope north of Alaska where water depths exceed 400 m (Grantz and others, 1988). A comprehensive study of onshore North Slope hydrates (Collet and others, 1988) shows that multiple hydrate-bearing reservoirs underlie the western part of the Prudhoe Bay oil field and the eastern part of the Kuparuk River oil field (Fig. 4). The amount of gas contained in these hydrates is estimated at 10 to 12×10^{12} ft^3 (28 to 34×10^{10} m^3), or about half the volume of gas in the Prudhoe Bay oil field.

OILS AND SOURCE ROCKS

Chemical analyses of North Slope oils and source rocks indicate that there are multiple oil types that have been generated by multiple source rocks. Prior to the discovery of the Prudhoe Bay field, investigations of North Slope oils from wells and seeps (McKinney and others, 1959) focused on the quality of products that could be derived from these oils rather than their relations to source rocks or to other oils. This emphasis changed, however, with the discovery of multiple oil-bearing reservoirs in the Prudhoe area (Fig. 4). Analysis indicated that oils from widely separated Prudhoe-area reservoirs—Sadlerochit, Kuparuk, and Sagavanirktok—were similar and thus, commonly sourced (Jones and Speers, 1976). The most comprehensive study aimed at discovering related types of North Slope oils is that of Magoon and Claypool (1981). From their analyses of 40 samples collected from seeps and wells all across the North Slope, two groups of oil were identified: (1) the Barrow-Prudhoe (from the Barrow and Prudhoe Bay fields), and (2) the Simpson-Umiat (from the Simpson seeps and Umiat field). The Barrow-Prudhoe group, volumetrically the predominant North Slope oil, occurs in reservoirs of Mississippian to Tertiary age. It is characterized by high sulfur content, medium API gravity, light isotopic composition, and a pristane-phytane ratio less than 1.5. The Simpson-Umiat group occurs in Cretaceous and Tertiary reservoirs and, surprisingly, is the only oil found in seeps. This group is characterized by low sulfur content, high API gravity, heavy isotopic composition, and a pristane-phytane ratio greater than 1.5. Other studies of the same and newly discovered oils (e.g., Seifert and others, 1980; Magoon and Claypool, 1985, 1988; Curiale, 1987; Sedivy and others, 1987; Anders and others, 1987) reveal various oil types within the two groups. Many investigators now regard the Simpson and Umiat oils as being derived from different terrigenous source-rock facies. Additional oil types identified within the Simpson-Umiat group include the Manning (from the Manning Point seep), the Jago (from oil-stained rocks along the Jago River), and the Kavik (from oil-stained rocks near the Kavik field) in the ANWR area (Anders and others, 1987) and the pebble shale in the Barrow area (Magoon and Claypool, 1988). The Kingak oil in the Prudhoe area, a type within the Barrow-Prudhoe oil group, is locally derived from the marine Kingak Shale (Seifert and others, 1980).

The organic-carbon content of most North Slope rock-units exceeds the threshold value of 0.5 weight-percent of potential petroleum source rocks (Fig. 9); although their hydrogen content, an indicator of propensity to generate oil or gas, varies considera-

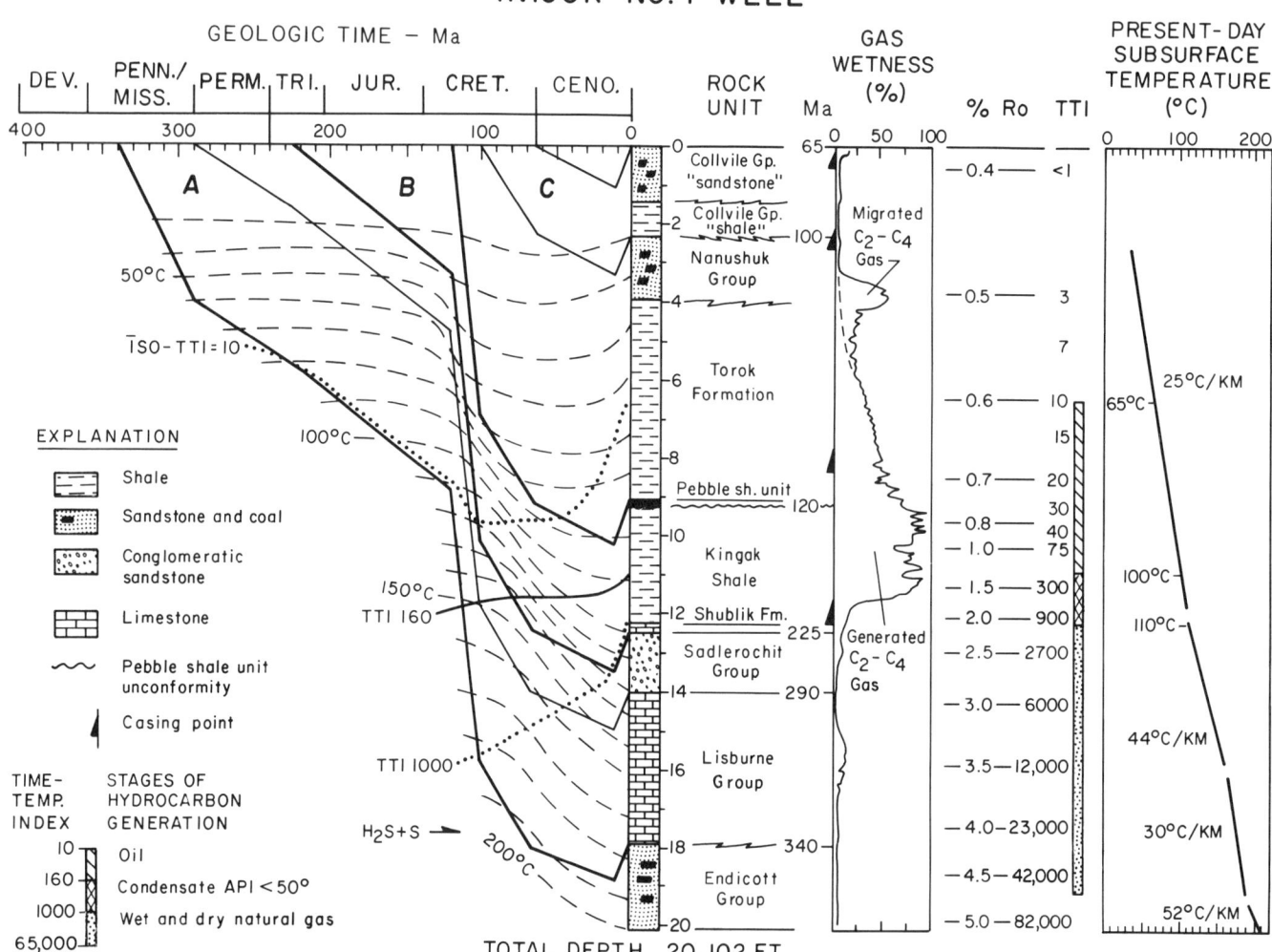

Figure 8. Burial and thermal history of the second deepest well in Alaska; Inigok-1, central North Slope (from Magoon and Claypool, 1983). Zone of oil (and condensate) generation, which corresponds to interval between time-temperature index (TTI) values of 10 and 1000, is confirmed by the gas wetness curve, based on analysis of drill cuttings. See Figure 5 for well location.

bly. Generally, the deltaic and prodeltaic units (Endicott Group, Sadlerochit Group, Nanushuk Group, Torok Formation, Colville Group, Sagavanirktok Formation, and Canning Formation) that have relatively high organic-carbon content but low hydrogen content are considered gas-prone source rocks. Marine units, such as the Shublik Formation, Kingak Shale, Hue Shale, and parts of the pebble shale unit, which have both high organic-carbon and hydrogen content, are considered oil-prone source rocks (Fig. 6). The inferred North Slope thermal history indicates that all but the youngest of these rock units are mature to overmature somewhere on the North Slope (Magoon and Bird, 1988).

Considerable effort has been devoted to matching North Slope oils with specific source rocks. The earliest efforts (Morgridge and Smith, 1972) identified Cretaceous shales above the LCU and the Kingak Shale as the most probable source rocks for the Prudhoe Bay oils, based on geologic relations and bulk geochemical characteristics of the proposed source rocks. Later, comparison of biomarker compounds from rocks and oils (Seifert and others, 1980) suggested that these oils were sourced from an assemblage of rocks including the Shublik Formation, Kingak Shale, and deeply buried HRZ shale (Hue Shale). Isotopic correlations of Sedivy and others (1987, Fig. 10) complement the biomarker results by showing an excellent correlation between source rocks (Shublik Formation and Kingak Shale) and the Prudhoe oils. The landmark USGS-sponsored oil-rock correlation study (Magoon and Claypool, 1985) was a multilaboratory effort that focused on a common set of oil and candidate source-rock samples, mostly from outside of the Prudhoe Bay area. The majority (17 of 30 laboratories) agreed that the Shublik Formation and, to a lesser extent (eight laboratories), the Kingak Shale are source rocks for the Prudhoe oils. There was also general agreement (14 laboratories) that the pebble shale unit and, to a lesser extent (seven laboratories), the Torok Formation are source rocks for the Umiat oil type (Claypool and Magoon, 1985).

Geologic relations, source-rock geochemistry, and oil analyses indicate that the most important petroleum source rocks for the North Slope are marine shales of Mesozoic age. These rocks began to generate oil as early as mid-Cretaceous time, and some may continue to generate oil today. Magoon (1988) recognizes three separate North Slope petroleum systems—a petroleum system is that set of geologic elements (such as source rocks and reservoir rocks) and processes (such as petroleum generation, migration, and entrapment) necessary to form oil and gas deposits. The Ellesmerian petroleum system is responsible for most North Slope oil (Barrow-Prudhoe group). Most of this oil originated from Triassic and Jurassic source rocks and accumulated in early-formed traps on the Barrow arch in the Prudhoe Bay area during Late Cretaceous and early Tertiary time. Eastward regional tilting in middle to late Tertiary time redistributed this oil, changed the size and shape of accumulations, and even created new accumulations (Jones and Speers, 1976; Carman and Hardwick, 1983; Bird, 1985; Werner, 1987). The Simpson-Umiat oil group originated from two petroleum systems. Those oils of this group from west of the Prudhoe Bay area are of the Torok-

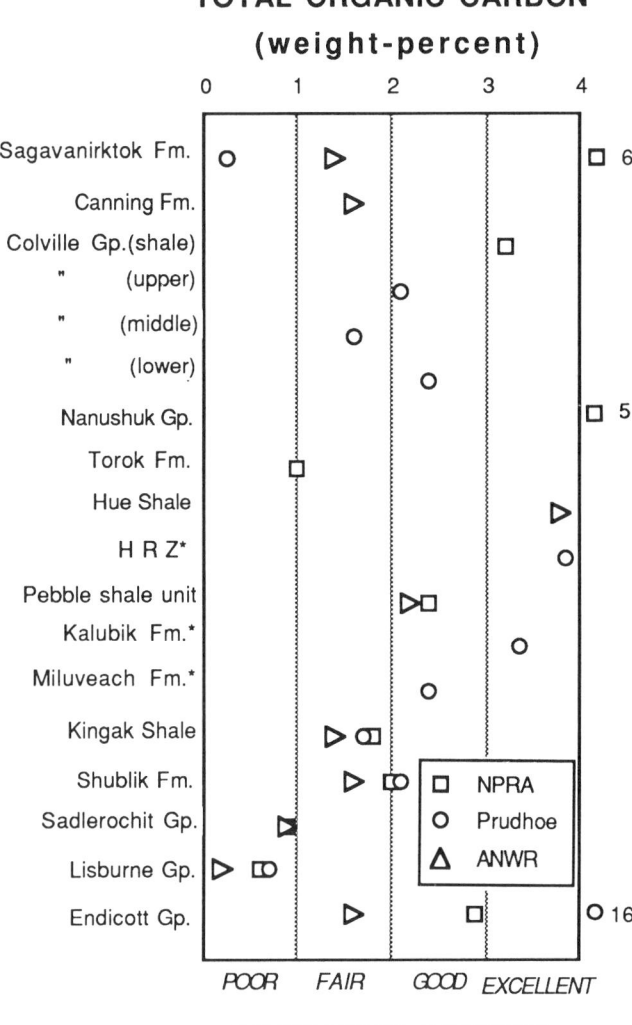

Figure 9. Average organic-carbon content of North Slope rock units. NPRA area values modified from Magoon and Bird (1988); Prudhoe Bay area values from Sedivy and others (1987); ANWR area values from Magoon and others (1987). Rock units with values greater than 4 percent are coal bearing.

Nanushu, petroleum system, while those oils east of Prudhoe Bay are of the Hue-Sagavanirktok petroleum system (Magoon, 1989, 1990).

FUTURE POTENTIAL OF THE PROVINCE

The future petroleum potential of the North Slope province is considerable, as indicated by the results of the latest assessment of undiscovered oil and gas resources conducted by Mast and others (1989). These results, summarized in Table 3, show that the range of uncertainty (F_{05} and F_{95} values) is broad, and the mean values indicate that about the same amounts of oil and gas

TABLE 3. ESTIMATED UNDISCOVERED RECOVERABLE OIL AND GAS RESOURCES OF THE NORTH SLOPE PETROLEUM PROVINCE*

Area	Oil† (in billion barrels)			Gas (in trillion ft^3)		
	F_{95}	F_5	Mean	F_{95}	F_5	Mean
Onshore and State offshore	3.69	32.56	13.64	14.07	131.21	54.09
Beaufort Shelf	0.49	3.74	1.27	2.14	12.81	8.26
Chukchi Sea	0.00	7.19	2.22	0.00	16.87	6.33
Totals§	5.78	36.94	17.13	23.16	148.15	68.68

*Adapted from Mast and others, 1989, Tables A1 and A2. These estimates include resources in undiscovered accumulations analogous to those in existing fields that are producible with current recovery technology and efficiency, but without reference to economic viability. Estimates of economically recoverable oil and gas resources for this area are considerably less than those listed; they can be found in the source document mentioned above.
†Oil values include natural gas liquids.
§Fractile values (F_{95}, F_5) are not additive.

remain to be found onshore as have already been found. Offshore potential, both state and federal, is also estimated to be significant.

The recently completed assessment of U.S. undiscovered oil and gas resources utilized a play-analysis method (Mast and others, 1989). In petroleum exploration and assessment, a play is an area characterized by geologically similar petroleum prospects. Twelve plays were identified and assessed for the North Slope. The stratigraphic interval encompassed by each play is shown in Figure 6. Maps, descriptions, and estimated oil and gas resources for each play are provided elsewhere (Bird, 1990).

Onshore, the area that makes up the foothills and coastal plain of the ANWR, which is presently closed to exploration, is considered the most promising region for future petroleum discoveries (Bird and Magoon, 1987). Greatest potential in this area is believed to occur in the thrust belt play. West of the ANWR, the coastal plain is considered highly prospective for oil and gas in combination structural-stratigraphic traps similar to known accumulations in the Prudhoe Bay field and nearby areas. Although nearly 40 separate structures have been tested in the Foothills province, resulting in the discovery of one oil and six gas accumulations, several times as many structures remain to be tested.

Offshore, the extension of the Barrow arch is considered highly prospective, as are large areas of the Chukchi and Beaufort shelves that have a variety of structural and stratigraphic traps. However, those areas are just now in an early stage of exploration (Craig and others, 1985; Grantz and May, 1988; Grantz and others, 1988; Thurston and Theiss, 1987).

SUMMARY

The North Slope petroleum province, one of North America's most prolific petroleum producing areas, currently provides 25 percent of total U.S. oil production; it is also the site of large but presently uneconomic gas reserves. This is a large, geologically complex province composed of a composite basin of late Paleozoic to Cenozoic age, flanked on its north by a Cretaceous to Tertiary passive margin. The composite basin consists of a relatively thin sequence of northern-sourced continental margin deposits of late Paleozoic to Early Cretaceous age overlain by very thick, southern-sourced Cretaceous and Tertiary foredeep deposits. The majority of oil and gas accumulations, and all commercial oil deposits, lie along a major subsurface positive trend, the Barrow arch, that separates the foredeep basin deposits from the passive margin deposits. Geochemical signatures identify Triassic and Jurassic marine shales as the primary source rocks for North Slope oils, although other potential source rocks are also present in the region. Major oil accumulations occur in nonmarine to shallow-marine sandstone reservoirs of Mississippian, Triassic, Cretaceous, and early Tertiary ages. Oil generation, migration, and entrapment date from mid-Cretaceous and Tertiary time. Recent estimates show that considerable oil and gas remain to be discovered in both the onshore and offshore parts of this province.

REFERENCES CITED

Alaska Report, 1989, Consortium seeks extension of Beaufort Sea leases: Alaska Report, v. 35, no. 42, p. 1.

American Association of Petroleum Geologists and U.S. Geological Survey, 1976, Geothermal gradient map of North America: Tulsa, Oklahoma, American Association of Petroleum Geologists Geothermal Survey of North America, scale 1:5,000,000, 2 sheets.

Anders, D. E., Magoon, L. B., and Lubeck, S. C., 1987, Geochemistry of surface oil shows and potential source rocks, in Bird, K. J., and Magoon, L. B., eds., Petroleum geology of the northern part of the Arctic National Wildlife Refuge, northeastern Alaska: U.S. Geological Survey Bulletin 1778, p. 181–198.

Bally, A. W., and Snelson, S., 1980, Realms of subsidence, in Miall, A. D., ed., Facts and principles of world petroleum occurrence: Canadian Society of Petroleum Geologists Memoir 6, p. 9–94.

Bartsch-Winkler, S., and Huffman, A. C., Jr., 1988, Sandstone petrography of the Nanushuk Group and Torok Formation, in Gryc, G., ed., Geology and exploration of the National Petroleum Reserve in Alaska, 1974 to 1982: U.S. Geological Survey Professional Paper 1399, p. 801–831.

Bird, K. J., 1981, Petroleum exploration of the North Slope, Alaska, in Mason, J. F., ed., Petroleum geology in China: Tulsa, Oklahoma, Pennwell Publishing Company, p. 233–248.

—— , 1985, The framework geology of the North Slope of Alaska as related to oil-source rock correlations, in Magoon, L. B., and Claypool, G. E., eds., Alaska North Slope oil/rock correlation study: American Association of Petroleum Geologists Studies in Geology 20, p. 3–29.

—— , 1989, North American fossil fuels, in Bally, A. W., and Palmer, A. R., eds., The geology of North America; An overview: Boulder, Colorado, Geological Society of America, The Geology of North America, v. A, p. 555–573.

——, 1990, Petroleum geology, play descriptions, and oil and gas resources of the Alaskan North Slope: U.S. Geological Survey Open-File Report (in press).

Bird, K. J., and Magoon, L. B., eds., 1987, Petroleum geology of the northern part of the Arctic National Wildlife Refuge, northeastern Alaska: U.S. Geological Survey Bulletin 1778, 329 p.

Bird, K. J., Griscom, S. B., Bartsch-Winkler, S., and Giovannetti, D. M., 1987, Petroleum reservoir rocks, in Bird, K. J., and Magoon, L. B., eds., Petroleum geology of the northern part of the Arctic National Wildlife Refuge, northeastern Alaska: U.S. Geological Survey Bulletin 1778, p. 79–99.

Blanchard, D. C., and Tailleur, I. L., 1982, Preliminary geothermal isograd map, NPRA, in Coonrad, W. L., ed., The United States Geological Survey in Alaska; Accomplishments during 1980: U.S. Geological Survey Circular 844, p. 47–48.

Brosge, W. P., Reiser, H. N., Dutro, J. T., Jr., and Detterman, R. L., 1981, Organic geochemical data for Mesozoic and Paleozoic sheets, central and eastern Brooks Range, Alaska: U.S. Geological Survey Open-File Report 81-551, 17 p.

Carman, G. J., and Hardwick, P., 1983, Geology and regional setting of the Kuparuk oil field, Alaska: American Association of Petroleum Geologists Bulletin, v. 67, no. 6, p. 1014–1031.

Carter, C., and Laufeld, S., 1975, Ordovician and Silurian fossils in well cores from North Slope of Alaska: American Association of Petroleum Geologists Bulletin, v. 59, no. 3, p. 457–464.

Claypool, G. E., and Magoon, L. B., 1985, Comparison of oil-source rock correlation data for Alaskan North Slope; Techniques, results, and conclusions, in Magoon, L. B., and Claypool, G. E., eds., Alaska North Slope oil/rock correlation study: American Association of Petroleum Geologists Studies in Geology 20, p. 49–81.

Collett, T. S., Bird, K. J., Kvenvolden, K. A., and Magoon, L. B., 1988, Geologic interrelations relative to gas hydrates within the North Slope of Alaska: U.S. Geological Survey Open-File Report 88-389, 150 p.

Collins, F. R., and Robinson, F. M., 1967, Subsurface stratigraphic, structural, and economic geology, northern Alaska: U.S. Geological Survey Open-File Report 287, 252 p.

Coney, P. J., and Jones, D. L., 1985, Accretion tectonics and crustal structure in Alaska: Tectonophysics, v. 119, p. 265–283.

Craig, J. D., Sherwood, K. W., and Johnson, P. P., 1985, Geologic report for the Beaufort Sea planning area, Alaska; Regional geology, petroleum geology, environmental geology: Minerals Management Service OCS Report MMS 85-0111, 192 p.

Curiale, J. A., 1987, Crude oil chemistry and classification, Alaska North Slope, in Tailleur, I., and Weimer, P., eds., Alaskan North Slope geology: Pacific Section Society of Economic Paleontologists and Mineralogists, v. 50, p. 161–167.

Gerhard, L. C., Graber, L. A., and Brostuen, E. A., 1988, A look at the status of U.S. petroleum: Oil and Gas Journal, June 20, 1988, p. 73–78.

Grantz, A., and May, S. D., 1988, Regional geology and petroleum potential of the United States Chukchi shelf north of Point Hope, in Gryc, G., ed., Geology and exploration of the National Petroleum Reserve in Alaska, 1974 to 1982: U.S. Geological Survey Professional Paper 1399, p. 209–229.

Grantz, A., May, S. D., and Dinter, D. A., 1988, Geologic framework, petroleum potential, and environmental geology of the United States Beaufort and northeasternmost Chukchi Seas, in Gryc, G., ed., Geology and exploration of the National Petroleum Reserve in Alaska, 1974 to 1982: U.S. Geological Survey Professional Paper 1399, p. 231–255.

Grantz, A., May, S. D., and Hart, P. E., 1990, Geology of the Arctic continental margin of Alaska, in Grantz, A., Johnson, G. L., and Sweeney, J. F., eds., Geology of the Arctic region: Boulder, Colorado, Geological Society of America, The Geology of North America, v. L, p. 257–288.

Gryc, G., 1970, History of petroleum exploration in Alaska, in Adkison, W. L., and Brosge, M. M., eds., Proceedings of the geological seminar on the North Slope of Alaska: Pacific Section, American Association of Petroleum Geologists, p. C1–C8.

——, ed., 1988, Geology and exploration of the National Petroleum Reserve in Alaska, 1974 to 1982: U.S. Geological Survey Professional Paper 1399, 940 p.

Harris, A. G., Lane, H. R., Tailleur, I. L., and Ellersieck, I., 1987, Conodont thermal maturation patterns in Paleozoic and Triassic rocks, northern Alaska; Geologic and exploration implications, in Tailleur, I., and Weimer, P., eds., Alaskan North Slope geology: Pacific Section Society of Economic Paleontologists and Mineralogists, v. 50, p. 181–191.

Harris, M., 1987, Endicott benefits from lessons learned: Alaska Construction and Oil, October, 1987, p. 15–16.

——, 1988, Beaufort causeways: Alaska Construction, v. 29, no. 12, p. 12–16.

Hubbard, R. J., Edrich, S. P., and Rattey, R. P., 1987, Geologic evolution and hydrocarbon habitat of the "Arctic Alaska microplate", in Tailleur, I., and Weimer, P., eds., Alaskan North Slope geology: Pacific Section Society of Economic Paleontologists and Mineralogists, v. 50, p. 797–830.

Jamison, H. C., Brockett, L. D., and McIntosh, R. A., 1980, Prudhoe Bay; A 10-year perspective, in Halbouty, M. T., ed., Giant oil fields of the decade 1968–1978: American Association of Petroleum Geologists Memoir 30, p. 289–314.

Jones, H. P., and Speers, R. G., 1976, Permo–Triassic reservoirs of Prudhoe Bay field, North Slope, Alaska, in Braunstein, J., ed., North American oil and gas fields: American Association of Petroleum Geologists Memoir 24, p. 23–50.

Klemme, D. H., 1981, Types of petroliferous basins, in Mason, J. F., ed., Petroleum geology in China: Tulsa, Oklahoma, Pennwell Publishing Company, p. 101–115.

Lachenbruch, A. H., and 8 others, 1988, Temperature and depth of permafrost on the Arctic Slope of Alaska, in Gryc, G., ed., Geology and exploration of the National Petroleum Reserve in Alaska, 1974 to 1982: U.S. Geological Survey Professional Paper 1399, p. 645–656.

Lantz, R. J., 1981, Barrow gas fields, N. Slope, Alaska: Oil and Gas Journal, v. 79, no. 13, p. 197–200.

Lerand, M., 1973, Beaufort Sea, in McCrossan, R. G., ed., The future petroleum provinces of Canada; Their geology and potential: Canadian Society of Petroleum Geologists Memoir 1, p. 315–386.

Magoon, L. B., ed., 1989, The petroleum system; Status of research and methods, 1990: U.S. Geological Survey Bulletin 1912, 88 p.

——, 1990, The geology of known oil and gas resources by petroleum system; Onshore Alaska, in Plafker, G., and Berg, H. C., eds., The Cordilleran orogen; Alaska: Boulder, Colorado, Geological Society of America, The Geology of North America, v. G-1 (in press).

Magoon, L. B., and Bird, K. J., 1987, Alaskan North Slope petroleum geochemistry for the Shublik Formation, Kingak Shale, pebble shale unit, and Torok Formation, in Tailleur, I., and Weimer, P., eds., Alaskan North Slope geology: Pacific Section Society of Economic Paleontologists and Mineralogists, p. 145–160.

——, 1988, Evaluation of petroleum source rocks in the National Petroleum Reserve in Alaska, using organic-carbon content, hydrocarbon content, visual kerogen, and vitrinite reflectance, in Gryc, G., ed., Geology and exploration of the National Petroleum Reserve in Alaska, 1974 to 1982: U.S. Geological Survey Professional Paper 1399, p. 381–450.

Magoon, L. B., and Claypool, G. E., 1981, Two oil types on North Slope of Alaska; Implications for exploration: American Association of Petroleum Geologists Bulletin, v. 65, no. 4, p. 644–652.

——, 1983, Petroleum geochemistry of the North Slope of Alaska; Time and degree of thermal maturity, in Byoroy, M., and others, eds., Advances in organic geochemistry 1981: Chichester, United Kingdom, Wiley Heyden, p. 28–38.

——, eds., 1985, Alaska North Slope oil-rock correlation study: American Association of Petroleum Geologists Studies in Geology 20, 682 p.

——, 1988, Geochemistry of oil occurrences, National Petroleum Reserve in Alaska, in Gryc, G., ed., Geology and exploration of the National Petroleum Reserve in Alaska, 1974 to 1982: U.S. Geological Survey Professional Paper 1399, p. 519–549.

Magoon, L. B., Woodward, P. V., Banet, A. C., Griscom, S. B., and Daws, T.,

1987, Thermal maturity, richness, and type of organic matter of source rocks, *in* Bird, K. J., and Magoon, L. B., eds., Petroleum geology of the northern part of the Arctic National Wildlife Refuge, northeastern Alaska: U.S. Geological Survey Bulletin 1778, p. 127–179.

Magoon, L. B., Bird, K. J., Claypool, G. E., Weitzmann, D. E., and Thompson, R. H., 1988, Organic geochemistry, hydrocarbon occurrence, and stratigraphy in government-drilled wells, North Slope, Alaska, *in* Gryc, G., ed., Geology and exploration of the National Petroleum Reserve in Alaska, 1974 to 1982: U.S. Geological Survey Professional Paper 1399, p. 483–487.

Mast, R. F., and nine others, 1989, Estimates of undiscovered conventional oil and gas resources in the United States; A part of the nation's energy endowment: U.S. Department of the Interior, 44 p.

McKinney, C. M., Garton, E. L., and Schwartz, F. G., 1959, Analyses of some crude oils from Alaska: U.S. Bureau of Mines Report of Investigations 5447, 19 p.

Melvin, J., and Knight, A. S., 1984, Lithofacies, diagenesis and porosity of the Ivishak Formation, Prudhoe Bay area, Alaska, *in* McDonald, D. A., and Surdam, R. C., eds., Clastic diagenesis: American Association of Petroleum Geologists Memoir 37, p. 347–365.

Minerals Management Service, 1988, Alaska update; January 1987–August 1988: Minerals Management Service OCS Information Report MMS 88-0073, 44 p.

Molenaar, C. M., 1982, Umiat field, an oil accumulation in a thrust-faulted anticline, North Slope, Alaska, *in* Powers, R. B., ed., Geologic studies of the Cordilleran thrust belt: Denver, Colorado, Rocky Mountain Association of Geologists, p. 537–548.

—— , 1988, Depositional history and seismic stratigraphy of Lower Cretaceous rocks in the National Petroleum Reserve in Alaska and adjacent areas, *in* Gryc, G., ed., Geology and exploration of the National Petroleum Reserve in Alaska, 1974 to 1982: U.S. Geological Survey Professional Paper 1399, p. 593–621.

Molenaar, C. M., Bird, K. J., and Collett, T. S., 1986, Regional correlation sections across the North Slope of Alaska: U.S. Geological Survey Miscellaneous Field Studies Map MF-1907.

Molenaar, C. M., Bird, K. J., and Kirk, A. R., 1987, Cretaceous and Tertiary stratigraphy of northeastern Alaska, *in* Tailleur, I., and Weimer, P., eds., Alaskan North Slope geology: Pacific Section, Society of Economic Paleontologists and Mineralogists, v. 50, p. 513–528.

Moore, T. E., and 5 others, 1990, Geology of northern Alaska, *in* Plafker, G., and Berg, H. C., eds., The Cordilleran orogen; Alaska: Boulder, Colorado, Geological Society of America, The Geology of North America, v. G-1 (in press).

Morgridge, D. L., and Smith, W. B., 1972, Geology and discovery of Prudhoe Bay field, eastern Arctic Slope, *in* King, R. E., ed., Stratigraphic oil and gas fields: American Association of Petroleum Geologists Memoir 16, p. 489–501.

Oil and Gas Journal, 1984, Exxon; N. Slope gas/condensate field is a giant: Oil and Gas Journal, v. 82, no. 11, p. 30.

—— , 1989, Chevron Beaufort test to use spray ice island: Oil and Gas Journal, v. 87, no. 3, p. 23.

Palmer, A. R., 1983, The decade of North American geology 1983 geologic time scale: Geology, v. 11, p. 503–504.

Parrish, J. T., 1987, Lithology, geochemistry, and depositional environment of the Triassic Shublik Formation, northern Alaska, *in* Tailleur, I., and Weimer, P., eds., Alaskan North Slope geology: Pacific Section, Society of Economic Paleontologists and Mineralogists, v. 50, p. 391–396.

Reed, J. C., 1958, Exploration of Naval Petroleum Reserve No. 4 and adjacent areas, northern Alaska, 1944–53; Part 1, History of the exploration: U.S. Geological Survey Professional Paper 301, 192 p.

Rickwood, F. K., 1970, The Prudhoe Bay field, *in* Adkison, W. L., and Brosge, M. M., eds., Proceedings of the Geological Seminar on the North Slope of Alaska: Pacific Section, American Association of Petroleum Geologists, p. L1–L11.

Schindler, J. F., 1988, History of exploration in the National Petroleum Reserve in Alaska, with emphasis on the period from 1975 to 1982, *in* Gryc, G., ed., Geology and exploration of the National Petroleum Reserve in Alaska, 1974 to 1982: U.S. Geological Survey Professional Paper 1399, p. 645–656.

Sedivy, R. A., and 5 others, 1987, Investigation of source rock-crude oil relationships in the northern Alaska hydrocarbon habitat, *in* Tailleur, I., and Weimer, P., eds., Alaskan North Slope geology: Pacific Section, Society of Economic Paleontologists and Mineralogists, v. 50, p. 169–179.

Seifert, W. K., Moldowan, J. M., and Jones, R. W., 1980, Appliation of biological marker chemistry to petroleum exploration: Proceedings of the 10th World Petroleum Congress, Bucharest, p. 425–440.

Smith, P. S., and Mertie, J. B., Jr., 1930, Geology and mineral resources of northwestern Alaska: U.S. Geological Survey Bulletin 815, 351 p.

State of Alaska, 1977, Prudhoe Bay Unit operating plan: Anchorage, Oil and Gas Conservation Commission, May 5, 1977, Conservation Hearing No. 145, Exhibit No. 8.

—— , 1984, Lisburne field rules: Anchorage, Oil and Gas Conservation Commission, Proceedings of the November 29, 1984 public hearing, 86 p.

—— , 1985, 1984 statistical report: Anchorage, Oil and Gas Conservation Commission, 177 p.

Tailleur, I., and Weimer, P., eds., 1987, Alaskan North Slope geology: Pacific Section, Society of Economic Paleontologists and Mineralogists, v. 50, 874 p.

Thurston, D. K., and Theiss, L. A., 1987, Geologic report for the Chukchi Sea planning area, Alaska; Regional geology, petroleum geology, and environmental geology: Minerals Management Service OCS Report MMS 87-0046, 193 p.

van de Kamp, P. C., 1988, Stratigraphy and diagenetic alteration of Ellesmerian sequence siliciclastic rocks, North Slope, Alaska, *in* Gryc, G., ed., Geology and exploration of the National Petroleum Reserve in Alaska, 1974 to 1982: U.S. Geological Survey Professional Paper 1399, p. 833–854.

Van Dyke, W. D., 1980, Proven and probable oil and gas reserves, North Slope, Alaska: Anchorage, Alaska Department of Natural Resources, Division of Minerals and Energy Management, 11 p.

van Poollen, H. K., and Associates, Inc., and Alaska Division of Oil and Gas, 1978, In-place hydrocarbons determination Kuparuk River Formation Prudhoe Bay area, Alaska: Anchorage, State of Alaska Department of Natural Resources, 13 p.

Wall Street Journal, 1989, ARCO test of well in Alaska suggests big oil discovery: Wall Street Journal, August 7, 1989, p. A4.

Waples, D. W., 1980, Time and temperature in petroleum formation; Application of Lopatin's method to petroleum exploration: American Association of Petroleum Geologists Bulletin, v. 64, p. 916–926.

Werner, M. R., 1987, West Sak and Ugnu sands; Low-gravity oil zones of the Kuparuk River area, Alaskan North Slope, *in* Tailleur, I., and Weimer, P., eds., Alaskan North Slope geology: Pacific Section, Society of Economic Paleontologists and Mineralogists, p. 109–118.

Woidneck, K., Behrman, P., Soule, C., and Wu, J., 1987, Reservoir description of the Endicott field, North Slope, Alaska, *in* Tailleur, I., and Weimer, P., eds., Alaskan North Slope geology: Pacific Section, Society of Economic Paleontologists and Mineralogists, p. 43–59.

MANUSCRIPT ACCEPTED BY THE SOCIETY JULY 26, 1990

ACKNOWLEDGMENTS

This manuscript was reviewed by J. E. Eason, R. L. Foland, D. G. Howell, L. B. Magoon, C. M. Molenaar, and R. B. Powers; the estimates of undiscovered oil and gas resources were reaggregated by R. A. Crovelli. Many of the reviewers' suggestions were incorporated. The efforts of these individuals are gratefully acknowledged; however, the opinions and conclusions presented here remain the responsibility of the author.

Printed in U.S.A.

Chapter 29

Coal; A brief overview

Hal J. Gluskoter
U.S. Geological Survey, MS 956, National Center, Reston, Virginia 22092

INTRODUCTION

More than 40 years ago, in a paper entitled "Coal Geology: An Opportunity for Research and Study," the preeminent American coal geologist of this century, Gilbert H. Cady, described coal geology in the following cogent terms:

Coal geology may be conveniently partitioned into those fields of knowledge and inquiry that concern the geology of the coal beds and those that concern the geology of the coal material itself. The geology of the coal beds consists of the knowledge relating to the discovery and delineation of coal resources, the stratigraphic and structural conditions, and the geologic and economic conditions that affect the mining operation. The geology of the coal material, on the other hand, consists of knowledge relating to the nature and origin of its physical heterogeneity in composition and properties, of its variations in chemical composition and properties, and of those geological conditions whereby these several variations were produced (Cady, 1949, p. 2).

Implicit in equating coal geology with a body of knowledge is the concept that once the natural history of coal beds is understood, we will have the ability to more accurately predict the characteristics and the technologic behavior (coal quality) of coal.

Advances in geologic research have been extensive since Cady's discussion. The instrumentation available to the modern researcher is significantly advanced relative to that which existed a generation ago. Few geoscientists in the world, and almost none on this continent, were aware of the mobility of the plates on which they resided and the ramifications to coal geology and to all other fields of geological investigations that the mobility signified. Nevertheless, the basic framework of Cady's (1949) discussion of the nature of coal science remains valid today. The section of this volume concerned with coal is divided into the two aspects of coal geology as he discussed them, and the topics he listed as necessary to formulate a complete discussion of coal geology are addressed. Chapters 30 to 32 are discussions of the natural history of coal, and Chapters 33 to 37 are concerned with the nature of the major coal deposits of the United States. Although it is true that the two fields of knowledge that make up coal geology may be separated for purposes of discussion, all aspects of coal science must be integrated to understand the coal resources of a specific region. Therefore, the reader will find much geologic interpretation in the discussions of the major coal deposits in the United States that follow, and as much heterogeneity will be found in those chapters as is inherent in the organic sedimentary rock under consideration—coal.

One last reference to the 1949 paper by Cady gives testimony to the enduring nature of his observations—the running head, which is the text that appears at the top of alternate pages of Cady's article, is "Coal Geology, A Neglected Field of Study."

COAL MAP OF NORTH AMERICA

Plates 8 and 9, which accompany this volume, are maps of the coal deposits of the United States and nearly all of Canada. The two plates are slightly modified versions of the U.S. Geological Survey (USGS) Special Geologic Map, *Coal Map of North America* (Wood and Bour, 1988), which is reproduced here at its original scale (1:5,000,000). The reproduced versions of the coal map differ from the original in that they do not include much of Mexico or any of Central America on the southern sheet; the coals of Newfoundland and the North Atlantic islands are excluded from the northern sheet; and the coal deposits of Nova Scotia and New Brunswick are to be found on the southern sheet. These changes and omissions were the necessary result of reducing the total size of the map sheets to meet the requirements of this volume.

On the maps, the locations of coal deposits are color-coded based on geologic ages of the coal beds. Each coal field or occurrence is assigned a number referencing additional information provided in the 44-page pamphlet that accompanies the original version of the maps (Wood and Bour, 1988). In addition to the name and age of the coal deposit, the following information is included in the pamphlet: coal rank, maximum overburden, number of beds identified in the coal field, an indication of the coking characteristics of the coal, maximum sulfur and ash content, whether the region has been mined, and resource tonnages by several reliability categories. The pamphlet also includes a discussion of the definitions and the criteria used to construct the map, the summary statistics by geologic age and rank of coal, and a general bibliography.

The coal map is referred to by several of the authors of the

Gluskoter, H. J., 1991, Coal; A brief overview, *in* Gluskoter, H. J., Rice, D. D., and Taylor, R. B., eds., Economic Geology, U.S.: Boulder, Colorado, Geological Society of America, The Geology of North America, v. P-2.

chapters that follow. The map is most noteworthy for its geologic content and depiction of entire basins underlain by coal-bearing strata. It is one of the first maps of its type that shows the Gulf Coast as an extensive and continuous coal basin.

THE NATURAL HISTORY OF COAL

Although Chapters 30 to 32 are not site specific, nor do they describe the occurrence and distribution of coal in the United States, they do discuss the genesis of coal from several different, important approaches. Peter McCabe (Chapter 30), discusses the environments of deposition of coal precursors, including a variety of environments (mires) in which peats can accumulate, and the several clastic sedimentary environments that were hosts for the major coal deposits of the United States.

The most basic nature of coal (that is, the plants from which it is derived), is addressed by Tom L. Phillips and Aureal T. Cross in Chapter 31. The authors have distilled from their own extensive research and from the literature an impressive, concentrated body of information. Plate 4, which is a reconstruction of a Middle Pennsylvanian peat swamp, is an original contribution in which the flora and the ecological relations during that coal-forming time are accurately depicted in a detailed and aesthetically pleasing manner.

Coal itself is not deposited; rather, it evolves from organic material that accumulates in mires. Under favorable sedimentary conditions, the peat may be preserved. The coalification process, which proceeds over geologic time, results in coal formation. The complex physical and chemical alterations of these processes and the methods by which they are interpreted are summarized by Heinz H. Damberger in Chapter 32. Much of the discussion of coalification is concerned with thermal history; the lessons of coalification can provide the key to the effective interpretation of the thermal history of a sedimentary basin.

COAL RESOURCES

The authors of chapters 33 to 37 have placed coal in its appropriate geologic context, and in each chapter the reader will find discussions of tectonism, sedimentation, and climates. The five chapters are approximately in the order of age of the coal sequences described, beginning with the oldest. Alan C. Donaldson and Cortland Eble discuss the Pennsylvanian coal of the central and the eastern United States in Chapter 33; the Cretaceous and Tertiary coal geology of the Rocky Mountains and the Northern Great Plains is summarized by Romeo M. Flores and Timothy A. Cross in Chapter 34; Tertiary coal of the Gulf Coast is reviewed by John A. Breyer in Chapter 35; and the disparate coal deposits of the western United States are examined by Aureal T. Cross in Chapter 36.

The coal deposits of Alaska have a brief chapter of their own (Chapter 37). This seems appropriate in that there is certainly more coal in Alaska than in any other state in the United States, although most would be classified as hypothetical resources, and there is much less known about the distribution, the quantity, and the quality of coal in Alaska than anywhere else in the country. Gary D. Stricker summarizes the coal resources of Alaska at approximately the rate of 1 trillion tons of coal per printed page.

A caveat should be placed here. An effort such as this volume, which has required the coordination of more than 50 authors and several editors, may have a long gestastion period between conception and publication. In this case it has been much longer than that of the largest mammals and it approaches the life cycle of members of the cicada family. Several authors completed their manuscripts well in advance of others and deserve credit for doing so. They certainly should not receive criticism for omitting significant references to works published subsequent to the submission of their manuscripts.

COAL: PAST, PRESENT, AND FUTURE (?)

In the middle of the nineteenth century, railroads expanded across the United States, and the use of coal surpassed wood as a source of heat and energy. By the beginning of this century, more than 200 million tons of coal were consumed annually in the United States, and during the first half of the 1900s, coal was the dominant source of energy.

The total energy in the United States produced from coal has declined from this dominant position to the current level of 23 percent. However, the absolute amount of coal mined has increased at a rather steady rate since 1961 (Fig. 1). United States coal production in 1990 is expected to reach 1 billion tons, an amount larger than that produced in any prior year. Significantly, the price paid for a ton of coal in the United States has not followed the same trend as production during the last 40 years (Fig. 2). From 1945 to 1969 the price of coal remained nearly constant at $4.50 to $5.00 per ton. The values shown in Figure 2 are not adjusted for inflation but are in dollars paid in each year. With adjustment for inflation, the value of a ton of coal in 1969 had decreased to 60 percent of the 1949 value. Increases in mechanization and in the amount of surface mining (the absolute amount and the amount relative to underground mining) and a trend toward fewer, larger mines allowed the coal industry to keep prices low and to maintain consumer cost below that of petroleum or other competing fuels.

Crude oil prices at the wellhead are shown in Figure 3. The curve of the price history of a ton of coal is similar to that for the price of crude oil for the period of 1949 to 1975. At the time of the 1973 oil embargo, the price structure of fossil fuels changed significantly, keeping ahead of inflation for the first time in several decades. The coal and oil price curves (Figs. 2 and 3, respectively), which diverged in the 1980's, show coal retaining more of the price gain relative to petroleum, although both indicate significant decreases.

Currently, 57 percent of the electricity in this country is produced from coal. In 1989, of all coal consumed in the United States, 86 percent was used by electric utilities (Energy Informa-

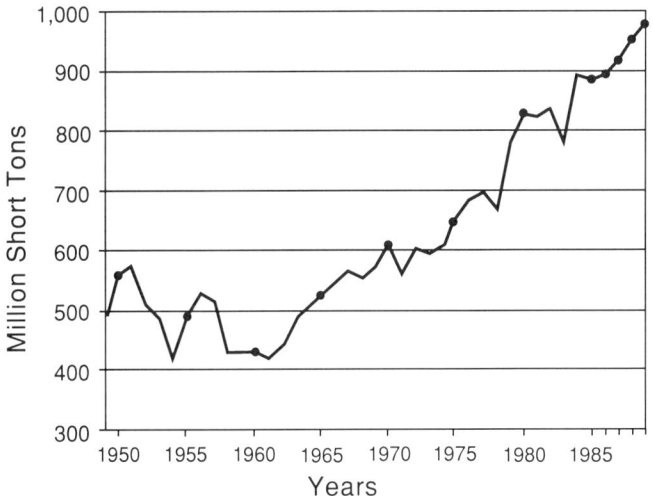

Figure 1. Production of coal in the United States, 1949 to 1989 (data from Energy Information Agency, 1990).

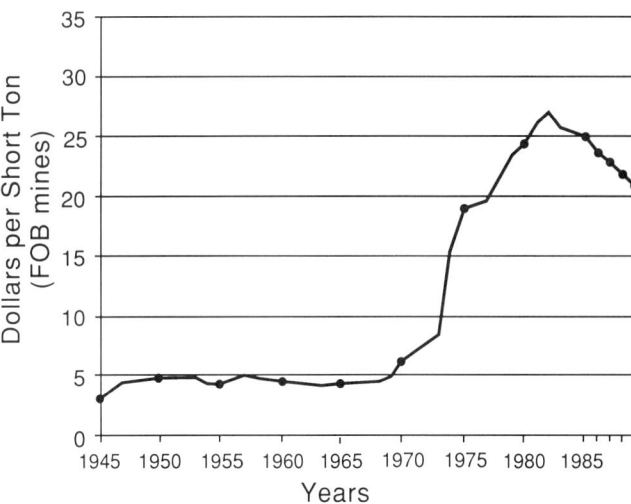

Figure 2. Price of coal in the United States, 1945 to 1989, not adjusted for inflation. Prices shown in current-year dollars (data from Energy Information Agency, 1990).

tion Administration, 1990); the remaining 14 percent was used in the manufacture of coke and in industrial and residential boilers. Approximately 10 percent of the U.S. coal production was exported.

In any field, risks are inherent in predicting the future. When making such predictions, one should be aware that many more incorrect answers will be found than correct ones. In a talk presented in 1980, Gluskoter (1982) based projections on energy usage to the end of the twentieth century, a period of only 20 years, on analyses published by a major petroleum company. We now know that he and the oil company were well off the mark, and current trends make it clear that it will be impossible to achieve more than half of the rate of increase in coal utilization that was predicted 10 years ago.

Recognizing the ease with which one can err in trying to predict the future use of coal and other energy sources, what can be forecast with some degree of confidence? If one looks ahead a relatively short time, a decade or so, then it is possible to determine the type of fuel that will be used to produce electricity. The life expectancy of a power plant is in the tens of years, and therefore, most of the electricity that will be produced in the year 2000 will be generated in plants currently in existence or under construction. Coal will continue to be a significant source, and likely the major source, of electric energy for some time to come. Looking far into the future and recognizing that fossil fuels are present on this planet in finite amounts and that they are not renewable on the human scale of existence, a shift to other sources of energy and to new methods of utilizing energy eventually will occur.

The uncertainty about the future use of coal is mostly concerned with an interim period that separates the current fossil fuel economy from the next generation of fuels; this period of time is likely to be from 30 to 300 years. The equation that must be solved in order to know the future of coal has too many variables (unknowns) at present. However, we can consider many factors, positive and negative. A large number negatively affect coal usage, and those, considered in the aggregate, suggest to some observers that the future of coal utilization is shorter rather than longer.

Coal is a dusty and bulky commodity that requires expensive handling. Removing it from the ground may cause environmental problems and hazards to those who mine it. It is an impure, heterogeneous material and contains significant amounts of inorganic matter that produces ash when the coal is burned. A reasonable estimate of "average" ash content is 10 percent. Therefore, approximately 100 million tons of waste products, which must be disposed of, will be produced by burning the 1 billion tons of coal mined in the United States in 1990.

Efforts to lessen environmental deterioration could have a great impact on the future use of coal. Emissions of oxides of sulfur and nitrogen and the correlation of those oxides with acid rain have resulted in a series of legislative actions. More recently, the problem of acid rain has been pushed from the environmental center stage by global warming and the greenhouse effect. The public perception is that there is a direct correlation between the combusion of fossil fuels and the warming of the Earth and an expected rise in sea level. Coal produces significantly more of the greenhouse gas, CO_2, per unit of heat than does oil or gas, and therefore is considered to be the least desirable of the fossil fuels by those who wish to limit the use of fossil fuels in order to reduce CO_2 emissions. The environmental concerns associated with coal combustion, acid emissions, sulfur and nitrogen oxides in the air we breathe, and carbon dioxide in the atmosphere are all factors that can adversely affect the future utilization of coal and offset the benefits inherent in using coal.

The positive (beneficial) factors are principally that coal is

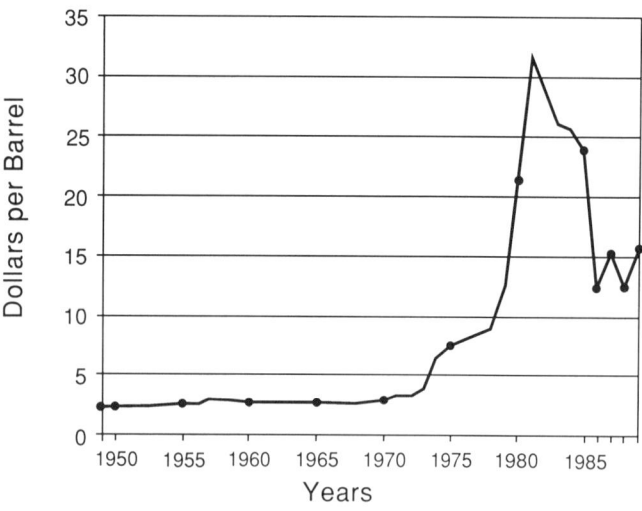

Figure 3. Crude oil prices in the United States, 1949 to 1989, not adjusted for inflation. Prices shown in current-year dollars (data from Energy Information Agency, 1990).

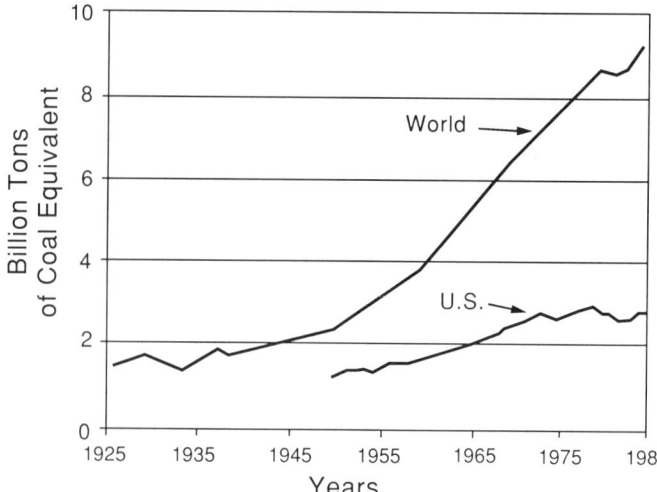

Figure 4. World and United States energy consumption (source: Speth, 1988).

available, it is relatively inexpensive, and the technologies for its use are well developed. The amount of the resource that could be produced in the United States is not a constraint on the potential, near-term utilization of coal. Although it is extremely difficult to arrive at precise values for coal resources and reserves and even more difficult to compare the coal resources of different countries, the United States has a large resource and is certainly one of the top three coal-rich countries of the world (Averitt, 1975; Energy Information Administration, 1990); the others are China and the USSR. There are regions within the United States where the coal has been extensively mined over a long period of time and where the bulk of the reserves may have been depleted, but the nation as a whole has vast resources.

Some alternatives to an increase in or even the continued use of coal for the generation of electricity in this country are currently available. The most obvious is nuclear; with that potential alternative in mind, a recent article was entitled "Will Nuclear Power Recover in a Greenhouse?" (Ahearne, 1989). The people of the United States may decide that, for our best national purposes, nuclear energy, with all its associated problems, is the lesser of several evils. To date, a trend in that direction has not made itself evident.

Another series of alternatives that would allow the continued use of coal involves technological advances resulting in the more efficient, cleaner burning of coal. This is not an answer to all the objections to coal burning, but it is environmentally more acceptable.

Predicting the future use of coal, especially worldwide use, is further complicated by the wide discrepancy in per capita energy use between developed and developing countries. Energy consumption in the United States and other highly industrialized and developed countries has leveled off, primarily due to the use of energy-efficient technologies (Fig. 4). However, the energy consumption of a developing country rises with the economic growth of that country, and while U.S. energy consumption has remained fairly level for the past 10 to 15 years, world energy consumption has risen dramatically. The global nature of the environmental problems is emphasized by Landsberg (1989, p. 10), ". . . the U.S. contribution to emissions is relatively small: its coal-burning power plants are responsible for 3 percent of worldwide emissions of all greenhouse gases at most." Similarly, the trend in sulfur dioxide emissions in the United States has turned downward during that same period of time, while the world sulfur dioxide emissions have increased. Müller (1984, p. 23) estimated that the annual sulfur dioxide emissions increased by slightly more than 50 percent from 1969 to 1985.

The factors that will affect the future of coal, and therefore the fate of coal geology and of the practitioners of the science, are in part economic, in part the perception of the concerned public as to its best environmental interests, and in part political. National security interests are best served by adequate, dependable supplies of critical minerals and energy resources. Recently, the more significant upward moves in the price structure of coal and oil (Figs. 2, 3) were the results of major worldwide political events and not technological factors or "normal" market reactions. Of all the disparate factors that influence the world's energy markets, the most difficult to predict are the national paroxysms of the major oil-producing regions. The resulting instability in the energy economy of the world has a greater impact on coal science and coal scientists than any other of the technological or environmental factors mentioned previously.

The more sophisticated the method whereby coal is to be utilized, the more in demand is information about coal quality and the availability of coal with certain desired characteristics. The coal geologist integrates knowledge of the genesis of peat, coalification, and the subsequent geologic history of coal into predictive geologic models. The ability to interpret these models and to make accurate predictions as to the occurrence, distribution, and quality of coals will be in demand to some degree for the near term, and if the synfuels-from-coal pendulum continues to oscillate as it has in the past, interest in synfuels will be heightened, and this predictive ability will be in great demand.

The chapters that follow indicate that the field of coal geology and the knowledge of the coal-bearing strata in the United States are sophisticated and extensive. Additionally, the chapters make us aware of the recent major advances that have taken place in this corner of the geosciences. It can be expected that by building upon an increasingly large and solid base, the next decade(s) will see an acceleration in the understanding of the coal-forming process and a concomitant increase in the ability to predict coal occurrence and coal quality.

REFERENCES CITED

Ahearne, J. F., 1989, Will nuclear power recover in a greenhouse?: Washington, D.C., Resources for the Future, Resources, Winter, 1989, no. 4, p. 14–17.

Averitt, P., 1975, Coal resources of the United States, January 1, 1974: U.S. Geological Survey Bulletin 1412, 131 p.

Cady, G. H., 1949, Coal geology; An opportunity for research and study: Economic Geology, v. 44, no. 1, p. 1–12.

Energy Information Administration, 1990, Annual energy review, 1989: Department of Energy, Energy Information Administration DOE/EIA-0384 (89), 313 p.

Gluskoter, H. J., 1982, Coal geology; Who needs it?, *in* Perspectives in geology: Illinois State Geological Survey Circular 525, p. 7–11.

Landsberg, H. H., 1989, Coal revisited: Washington, D.C., Resources for the Future, Resources, Winter 1989, no. 4, p. 8–11.

Müller, D., 1984, Estimation of the global man-made sulfur emission: Atmospheric Environment, v. 18, no. 1, p. 19–27.

Speth, J. G., 1988, Environmental pollution, *in* Earth 1988; Changing Geographic Perspectives, Proceedings of the Centennial Symposium: National Geographic Society, p. 262–282.

Wood, G. H., Jr., and Bour, W. V., III, 1988, Coal map of North America: U.S. Geological Survey Special Geologic Map, text 44 p., scale 1:5 million.

Manuscript Accepted by the Society November 29, 1990

Printed in U.S.A.

Chapter 30

Geology of coal; Environments of deposition

Peter J. McCabe*
Alberta Research Council, P.O. Box 8330, Postal Station F, Edmonton, Alberta T6H 5X2, Canada

INTRODUCTION

The environments of deposition of a coal and its surrounding sediments control the seam's geometry (thickness and lateral extent) and determine the composition of roof and floor rocks. Major aspects of the quality of the coal, including its maceral, ash, and sulfur content, are also determined in large part by the depositional setting. Predictive models of facies distribution can be useful for both coal exploration and mine development.

Geologists from the United States have played an important role in developing our present understanding of depositional environments of coal. Early work concentrated on the apparent cyclicity exhibited by many coal-bearing sequences. Udden (1912) recognized cycles in the Pennsylvanian of Illinois that he attributed to repeated transgression and regression. Theories on the origin of cyclothems were later explored by Weller (1930) and Wanless (Wanless and Weller, 1932). The work of Ferm and Horne (1979) and their students has been very important in developing an awareness that the understanding of modern clastic environments is critical to a sound interpretation of coal-bearing strata. U.S. geologists have also been at the forefront of studies of modern peat deposits as analogues to ancient coal-forming environments. Notable in this field have been workers at the Pennsylvania State University (Spackman and others, 1976).

This chapter reviews the ecosystems in which peat accumulates. It discusses the nature of coal facies and the relationships between coal and clastic sediments. Major environments of deposition are also reviewed. Examples are given throughout from coal-bearing strata in the United States. Canadian examples are used if they are particularly illustrative or if there are no good U.S. examples.

The term *mire* is used in this paper as a generic word to cover nonsaline wetlands in which peat accumulates (Gore, 1983). The term encompasses all ecosystems described as bog, fen, moor, muskeg, peatland, and swamp.

ENVIRONMENTS OF PEAT ACCUMULATION

Most peat forms in mires where the plant material accumulates close to its place of growth. The mires in which such autochthonous peats develop can be divided into three types: raised, low-lying, and floating (Fig. 1). Organic material may, however, be transported some distance before deposition and, if deposited with relatively little inorganic material, will form allochthonous peats. Such peats have been described from beaches and lake floors. This section describes the various environments of peat accumulation and discusses the factors controlling the type of peat in each environment.

It should be remembered that the environments are not mutually exclusive. Many mires show more than one type of morphology and contain streams and lakes. Figure 2, for example, shows the complex terrain of part of the Okefenokee Swamp. Mires may evolve through time (Romanov, 1968). Lakes may be infilled by sediment accumulating on the bottom of the lake or from the buildup of floating peat at the lake's surface. Low-lying mires (low moor) form in areas of poor drainage, including infilled lakes. If the climatic conditions are favorable, a raised mire (high moor) may develop from the low-lying mire (Fig. 1).

Peat composition is controlled by the type of organic matter that is added to the peat and the amount and type of degradation that occurs. The type of organic matter is determined by the variety of plants growing on the mire surface and/or the variety of plant parts added to the peat. Major factors controlling the type of organic matter in autochthonous peats are (1) the presence or absence of woody material and (2) the relative importance in contribution of roots and surface litter. The transportation mechanism is important in determining the content of allochthonous peats.

A peat profile can be divided into two parts: the catotelm and the acrotelm (Ingram and Bragg, 1984). The catotelm is the main body of peat. It is perennially waterlogged and permanently anaerobic except in the close vicinity of live roots. Permeability is low and relatively constant through the catotelm. The acrotelm is the zone of water table fluctuation, which extends up to the mire surface. Most living matter is confined to the acrotelm, which is

---

*Present address: U.S. Geological Survey, MS 972, Box 25046, Federal Center, Denver, Colorado 80225.

McCabe, P. J., 1991, Geology of coal; Environments of deposition, *in* Gluskoter, H. J., Rice, D. D., and Taylor, R. B., eds., Economic Geology, U.S.: Boulder, Colorado, Geological Society of America, The Geology of North America, v. P-2.

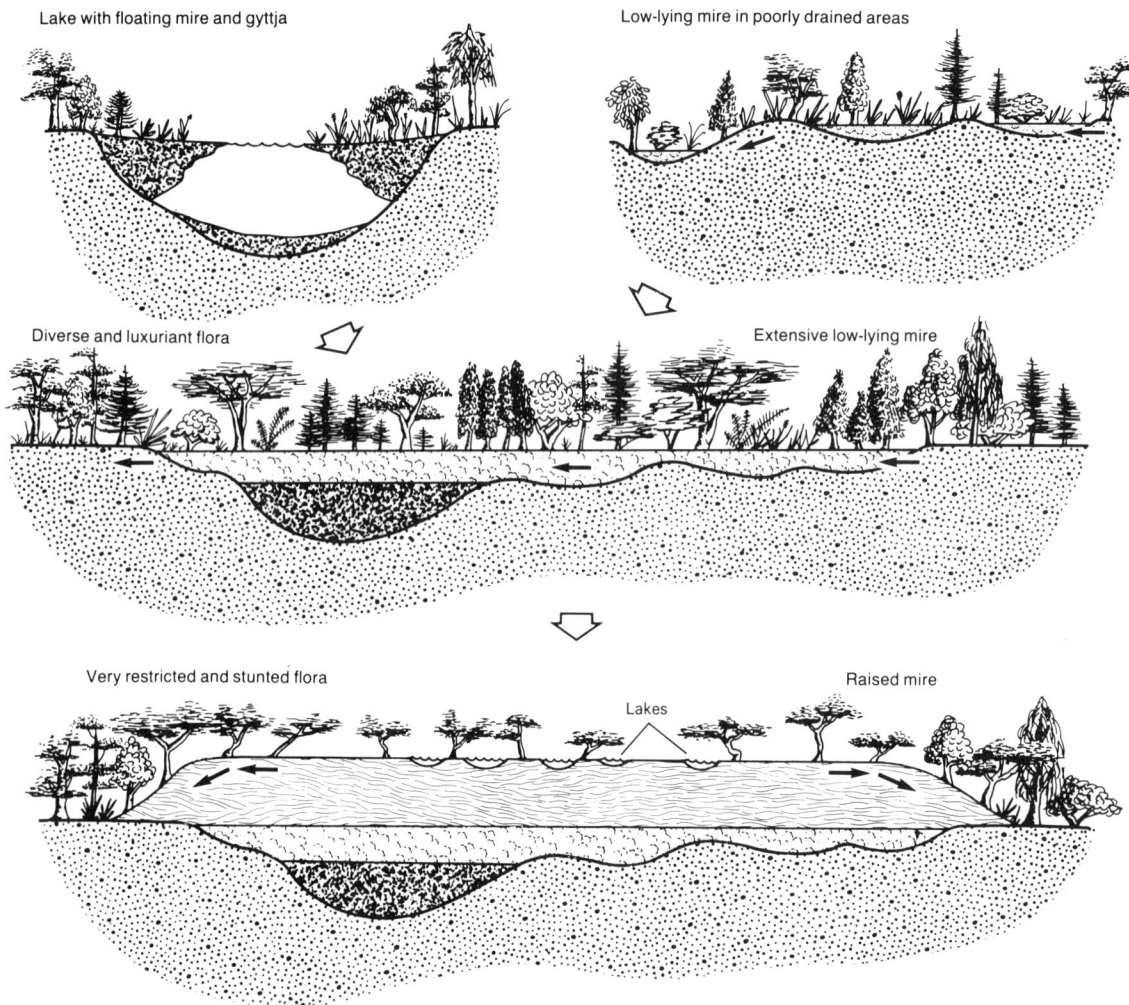

Figure 1. Types of mires showing evolutionary sequence of Romanov (1968).

periodically aerated. Permeability is variable and is highest near the surface. An active microflora of fungi and bacteria in the acrotelm degrades organic material, transforming plant material to peat. The degree of degradation is a function of the length of time the organic material lies within the acrotelm and the rate of microbial oxidation. This is dependent on the thickness of the acrotelm, the rate of peat accumulation, and the proportion of surface litter to roots in its composition. Higher temperatures facilitate more intense microfaunal activity.

Raised mires

Raised mires have convex upper surfaces that do not reflect a pre-existing topography (Fig. 1). They develop in areas where annual precipitation is greater than annual evaporation and where there are normally no long dry periods. They are most common in areas with a maritime climate in tropical and cool temperate regimes. Extensive raised mires are present in southeast Asia (Anderson, 1983) and northwest Europe (Moore, 1984). From a geological standpoint, the best-described raised mires in North America are those from the Fraser River delta in British Columbia (Styan and Bustin, 1983a, b). A cross section through one of them is shown in Figure 3.

Raised mires are able to build upward because they maintain their own water table. Mires in southeast Asia are raised up to 7 m above adjacent floodplains (Anderson, 1983), whereas those in temperate areas are less elevated. Margins of raised mires are typically steep and convex upward. Central parts are generally flat, and some show a concentric ridge and pool topography. Lakes may be present on the surface of mature raised swamps. Their formation may be related to fires burning through the acrotelm during unusually dry periods, creating a depression which then fills with water when normal precipitation resumes. The acrotelm is generally well developed in raised mires and is thickest in the central parts of the mire. There is little surface run-off due to the high permeability of the acrotelm, and raised mires have few streams. Groundwater in raised mires is generally oligotrophic but may be mesotrophic at margins. This lack of nutrients is due to the supply, which is limited to recycling from underlying peats and material blown into the mire or dropped by flying fauna. The groundwater is very acidic (usually pH < 4) because of the retention in the system of humic acid created during the breakdown of organic matter. The paucity of nutrients is reflected in the vegetation of raised mires, many of which show a strong concentric zonation of floral ecosystems. In the tropics, raised mires are densely forested, but the central areas have a low number of tree species and trees are stunted. Central areas of temperate mires are dominated by low, herbaceous flora with *Sphagnum* moss. A distinct peat succession, reflecting the concentric floral zonation, may develop as a mire becomes raised (Fig. 1).

Plant degradation in raised mires is facilitated by the relatively thick acrotelm, but this is counterbalanced by the acidic conditions which inhibit microfloral activity. Present-day peats in tropical raised mires are high in lignin and low in cellulose as a result of the woody vegetation and the high amount of degradation. In contrast, peats from temperate raised mires are low in lignin and high in cellulose as a consequence of the sparse woody flora and relatively low rates of degradation.

Low-lying mires

Low-lying mires are those that infill underlying topography and aggrade to a near-horizontal surface (Fig. 1). They may form in any waterlogged area and are not as dependent on climatic conditions as are raised mires. Worldwide, low-lying mires are much more common than raised mires. The well-described mires of the Okefenokee and Everglades in the southeastern U.S. (Spackman and others, 1976; Cohen, 1984; Cohen and others, 1984) are of this type. The surface relief of these mires is, in part, inherited from the underlying topography. Streams and lakes are common within the mires (see Fig. 2). Streams are usually sluggish, many are anastomosed, and in some cases, water flow through the mire is unchannelized. Lakes can form in a variety of ways. Some may be inherited from underlying relief; others may result from the blocking of drainage by the growth of vegetation or the burning of the acrotelm during droughts.

There is a through-flow of ground water in low-lying mires, with water draining down into the mire and water from the mire in many cases feeding a drainage system. In large mires the outflow may be considerably larger than the inflow because of the addition of precipitation over the mire area. The thickness of the acrotelm is dependent on variations in regional ground-water levels. The acrotelm in low-lying mires is probably thinnner than in raised mires. Substantial variations in water levels have, however, been described from low-lying mires (Spackman and others, 1976).

The through-flow of water to most parts of low-lying mires increases the nutrient supply and decreases the acidity. Ground water is generally eutrophic, with some areas being mesotrophic. Conditions are usually slightly acidic (pH between 4.8 and 6.5). These conditions allow a varied flora in these mires. Most are forested, except in areas where the water level is too high or the climate is too cold. Relative to raised mires, plant degradation may be enhanced in low-lying mires because of the less acidic conditions. The thinner acrotelm may, however, balance this effect.

Floating mires

In some places, mats of floating peat form the substrate for vegetation growth (Fig. 1). Peat mats may be free floating or, more commonly, may be attached to the shore of a lake. The most extensive floating peats reported in the United States are those in lakes of the Louisiana coastal plain (Kolb and Van Lopik, 1966; Russel, 1967). Most floating peats probably form by gradual build out from a shoreline. Spackman and others (1976), however, reported that water level fluctuations may create floating peat. In the Okefenokee Swamp, drowned peat mats may be buoyant enough to break away from underlying sediment. These peat mats may become reattached to the substrate by the downward extension of roots.

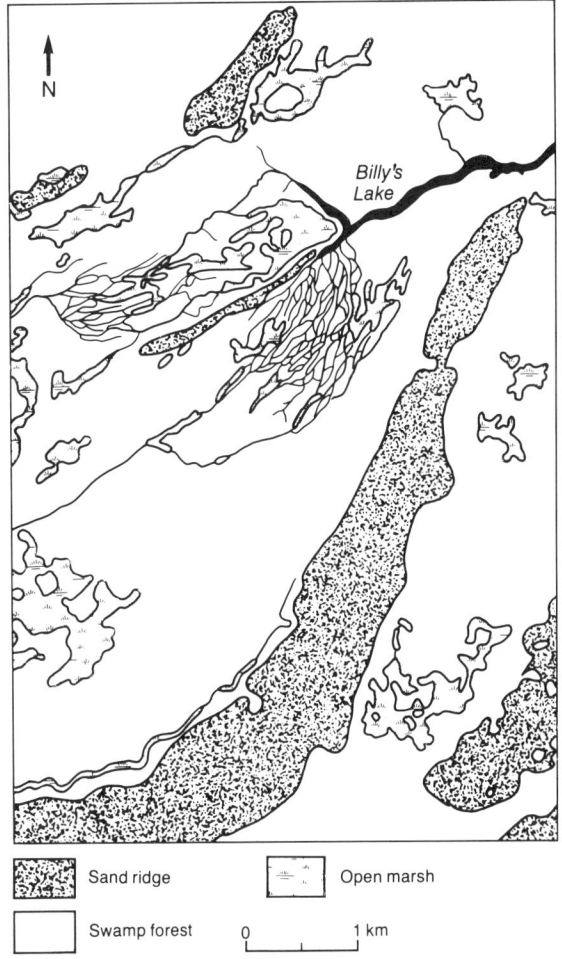

Figure 2. Map of portion of Okefenokee Swamp showing varied relief and ecosystems (based on Spackman and others, 1976). Sand ridges are remnants of beach-barriers that are topographically higher than the mire. Note the complex of anastomosed streams.

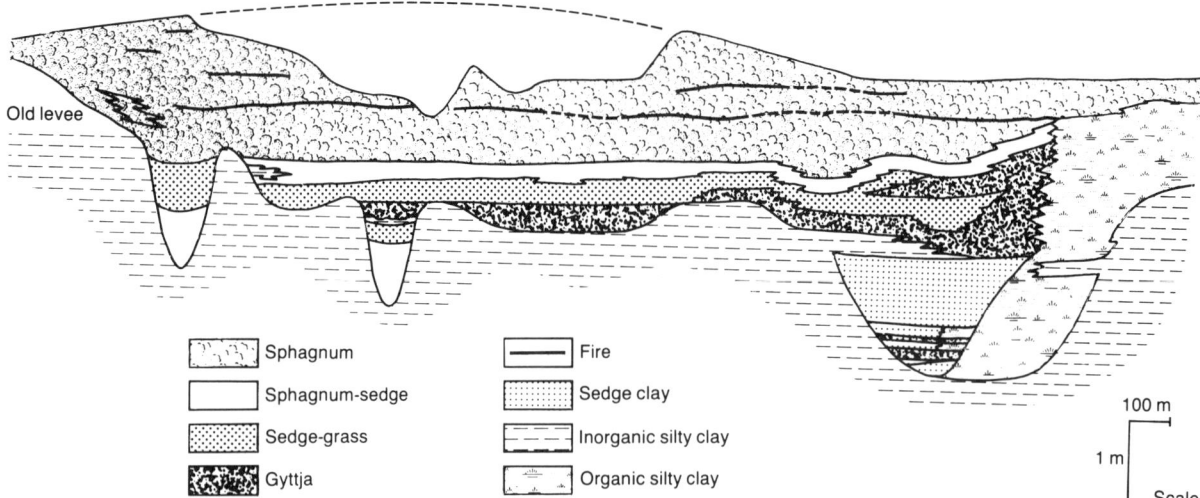

Figure 3. Cross section of raised mire on the Fraser River Delta, British Columbia (based on Styan and Bustin, 1983b). Note peat infill of abandoned channels and buildup of raised mire topography. Removal of peat in upper part of mire was by human beings, not by natural causes.

Little information is available on the nature of the porewater in floating mires. However, it is reasonable to assume that, because of the proximity of lake water, conditions are eutrophic with near neutral pH. The thickness of the peat mat is an important factor controlling the type of vegetation growth. Thick mats can be stable enough to support trees.

For peats of a floating mire to be preserved in the geologic record, the floating stage must be temporary. The peat mat eventually needs to join up with the lake bottom sediments. This may take place by increase in thickness of the peat mat or by sedimentation on the lake floor. Although there are few studies of floating peats, it is presumed that vegetation growth must be rapid for the peat mat to grow or even be maintained, because the mat is open to an oxidizing environment at both its upper and lower surface.

Lake bottom peats

Organic matter deposited at the bottom of a lake (Fig. 1) is usually fine grained. Such organic muds (gyttjae) have been well documented from many glacially formed lakes in the U.S. (see, for example, Treese and Wilkinson, 1982) but are also found in many lakes associated with other types of mires. Some organic muds are rich in particular types of material that are selectively transported to the lake (e.g., spores) or that grow within the lake waters (e.g., algae).

The rate of degradation of organic material on a lake floor may be quite high; if water movement is sluggish, however, conditions may become partly anaerobic, allowing sapropelitic muds to accumulate. The character of the organic material may be significantly altered with the selective preservation of those constituents most resistant to biological degradation.

Sapropelitic muds are the precursors of boghead and cannel coals. Boghead coals are rich in alginite. Cannel coals are composed of fine maceral particles, usually dominated by sporinite. Differential degradation is probably an important factor controlling the original maceral composition of most cannel coals.

Beach ridges of allochthonous peat

Because of its low density, detrital organic matter may accumulate as beach ridges at the edge of a lake or ocean. Spectacular examples of organic beach ridges are present on the Mahakam Delta in Indonesia (Allen and others, 1979). Organic beach ridges are present on the Mississippi Delta and along the coastline of the Everglades. Plant material of any size may be washed up onto a shoreline, but the beach ridges typically consist of small fragments that resemble coffee grounds. The amount of degradation in beach ridges is probably high, but the organic material has a chance of being preserved if buried relatively quickly.

CONTROLS ON ASH CONTENT

Origins of minerals in coal

Minerals in coal may be of organic, detrital, or authigenic origin. In this chapter only those minerals that are related to the environment of deposition are discussed.

Minerals are precipitated within plants as a result of the enrichment of dissolved minerals in the transpiration process. Silica, for example, is precipitated as opal, often as needles a few microns long. Many of the precipitated salts, however, are liable to be redissolved during the peatification process. It has been suggested that organically derived minerals may compose a considerable percentage of the ash content of coals (Renton and Cecil, 1979). Finkelman (1982) argued that this may be true for low ash coals, but in coals with over 5 weight percent ash, petrographic studies indicate that the majority of nonepigenetic minerals are detrital in origin. Ruppert and others (1985) recently used petrographic evidence to suggest that quartz in the Upper Freeport coal of Pennsylvania is authigenic in origin and was derived from phytoclasts.

Detrital minerals may be washed or blown into a mire. Because peat accumulates at relatively slow rates, even low rates of detrital sedimentation can be significant. For example, wind-transported mineral matter makes up between 3.5 and 15 dry weight percent of peat from Minnesota (Finney and Farnham, 1968). Many coals contain considerable amounts of volcanic ash even through deposited many hundreds of kilometers from active volcanoes. Finkelman and Stanton (1978) were even able to recognize an extraterrestrial component in the ash of some coals.

Syngenetic mineralization can be important in mires because of their unique chemistry compared to surrounding environments. Biochemical processes may be particularly important during the peatification stage. Mineralization may also take place shortly after a peat is buried or drowned. Early cementation within primary pore space is indicated by coal balls (Eggert and Phillips, 1982) and silicified peat (Ting, 1972a, b).

Sulfur

The origin of sulfur in coal has been the subject of a considerable amount of research. Sulfur is probably enriched and redistributed in most coals after burial. It is generally believed, however, that the sulfur content of a coal is a reflection of its depositional environment.

Casagrande and others (1977) reported that freshwater peats in the Okefenokee Swamp have between 0.02 and 0.21 percent sulfur on a dry weight basis, whereas sulfur in marine peats formed in mangrove swamps ranges from 1.29 to 10.11 percent. Bustin and others (1985) found that peats of the lower delta plain of the Fraser River have sulfur contents up to 6 percent, while those of the upper delta plain have less than 0.5 percent. Much of the sulfur is organic (either carbon-bonded or as an ester-sulfate), though other forms of sulfur found in coal are also present in the peat. While vascular plants serve as a sink for sulfur, isotope studies (Nissenbaum and Kaplan, 1972) indicate that microorganisms play an important role in concentrating sulfur. In deeper zones of the peat profile, the sulfur is converted to a pyritic form (Altschuler and others, 1983).

It appears that reduction of sulfate by microorganisms can also take place after drowning of a freshwater peat by marine water. Cohen and others (1983) noted that freshwater peats that have been buried by marine or brackish sediments are enriched in sulfur. This process may explain the high sulfur content of many Pennsylvanian coals. In the Herrin (no. 6) Coal of Illinois (Fig. 4), sulfur values are generally between 3 and 5 percent where the coal is overlain by marine sediments. In contrast, where there is at least 5.5 m of nonmarine shale over the coal, there is usually less than 2 percent sulfur (Gluskoter and Simon, 1968; Gluskoter and Hopkins, 1970). The higher values of total sulfur are related to increases in both organic and pyritic sulfur. This suggests that the peats were not derived from plants in contact with marine water, but rather, that sulfate reduction took place within the peat after a marine transgression. In the Pennsylvanian coals of the U.S. there is a good correlation between marine roof rocks and high sulfur contents (Williams and Keith, 1963; Davies and Raymond, 1983).

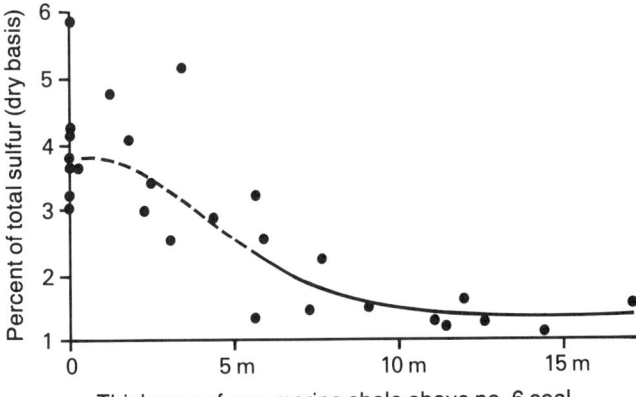

Figure 4. Relationship of total sulfur content of the Herrin (no. 6) Coal with thickness of nonmarine strata separating the coal from overlying marine strata (after Gluskoter and Hopkins, 1970).

Partings

Layers of mudstone, or more rarely sandstone, occur within most coal seams. Most can be explained by infrequent events, such as major floods, that introduced clastic sediments to the swamps. However, this may not be the only explanation. Degradation of peat could conceivably be sufficient to concentrate inorganic material into a discrete layer, creating a parting (Renton and Cecil, 1979). Such a mechanism may explain some thin partings that are enigmatic in that they are relatively constant in thickness. Sediment that is introduced into a vegetated area should be unevenly distributed; for example, there should be no deposition at the sites of plant stems and trunks, and there should be eddying around vegetation if there is any current flow. Furthermore, thin layers of sediment are likely to be thoroughly bioturbated into the peat by both plants and animals.

Some partings are tonsteins. These are kaolinitic mudrocks, most of which are interpreted as cinerites (i.e., altered volcanic tuffs). Tonsteins are present in most coal fields of the U.S. The transformation of tuff to cinerite is a slow process. Tertiary coals of Alaska have volcanic ash layers that show only partial alteration (Triplehorn, 1983). In older coals, the tonsteins' clay is usually monomineralic kaolinite. Many tonsteins are of wide lateral extent. The 70- to 100-mm-thick flint clay parting in the Hazard No. 4 coal of eastern Kentucky can be traced more than 2,000 km^2 (Bohor and Triplehorn, 1981). Tonsteins have proven valuable in correlating coal seams (Burger and Damberger, 1985; Ryer and others, 1980; Spears and Kanaris-Sotiriou, 1979). Evidence of the volcanic origin of tonsteins, in addition to their wide lateral extent, may include the presence of euhedral apatite, sanadine, zircon, β-form quartz or other crystals (Bohor and Triplehorn, 1981), and accretionary lapilli (Bohor and Triplehorn, 1984). Although most tonsteins are interpreted to be cinerites, it is possible that some tonsteins were formed by the degradation of peat with concentration of mineral matter (Moore, 1965).

COAL FACIES

Variable environments within a mire produce peats of differing characters. As the environment changes through time, the composition of successive peat layers may change. It is possible, therefore, to divide a peat bed into a variety of facies: bodies of peat with a set of specified characteristics. Perhaps the best demonstration of the application of Walther's (1894) Law of Facies to peat deposits is that described by Spackman and others (1976) from the Everglades (Fig. 5). A marine transgression in this area has produced a peat sequence showing an upward transition from fresh to marine facies.

The facies concept has so far been applied to peats and coals in limited ways. Most studies have been concerned with only one characteristic, such as biofacies, palynofacies, or petrofacies. There are, however, several studies that have taken a more integrated approach.

Microtome sections of peat have been used to define peat types within the Everglades (Cohen and Spackman, 1972) and Okefenokee (Cohen, 1974). The Everglades study indicates that there is a good relationship between pollen associations and peat types. Studies on coal seams have also shown a correlation between petrographic composition and palynoflora.

Smith (1962) examined the vertical variation in microlithotype and miospore content at 20- to 80-mm intervals through Pennsylvanian coal seams. He distinguished four miospore assemblages that were associated with coals of distinctive petrographic types. Smith observed coal sequences that he interpreted as upward transitions from wetter to drier conditions in the environment of deposition. He ascribed these changes either to climatic variation or to the development of raised mires.

During the last 20 years, several other workers have made detailed petrographic analyses of seam profiles, especially in Canada (Bustin and others, 1983). In a classic study of Pennsylvanian coal seams of Nova Scotia, Hacquebard and Donaldson (1969) defined facies on petrographic characteristics. They related these facies to the types of vegetation and relative water levels at the time of deposition. These facies also showed a strong relationship between petrographic composition and miospore distribution. The facies distributions were mapped from cores over a 50-km-long section.

A few studies of Recent mires have looked at several parameters in defining facies. For example, peats from the Fraser River delta of British Columbia were divided into a series of organic facies by Styan and Bustin (1983b) on the basis of type of plant material, degree of humification, ash content, and peat texture. Styan and Bustin showed the distribution of these facies in detailed cross sections through mires on the delta top (Fig. 3).

A similar approach to facies analysis of coals is that of Ting and Spackman (1975). They defined "lithotypes" on the basis of

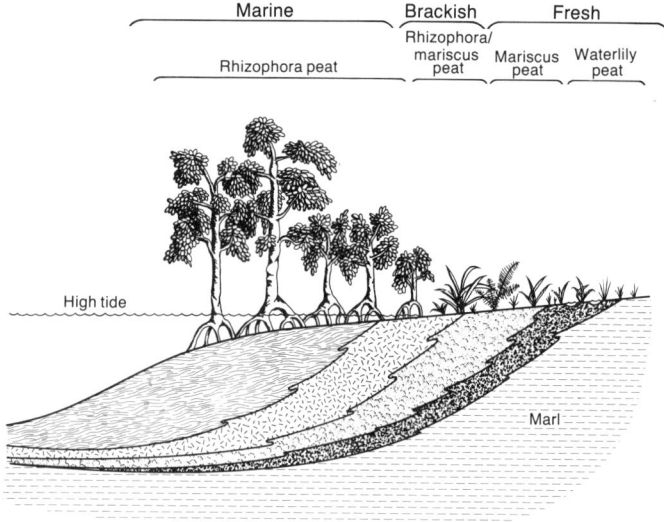

Figure 5. Generalized cross section of peat stratigraphy in the Everglades coastal area (after Spackman and others, 1976). Sequence is the result of a marine transgression in the area over the last 3,600 years.

maceral content, petrographic texture, spore content, and mineral content. Coal facies defined by several parameters are probably the most useful in interpreting the original environment of deposition. In other branches of sedimentary geology it has generally been found necessary to develop a separate classification scheme for the purpose of each study. The type of detailed facies classification proposed by Ting and Spackman could probably prove most useful if separate schemes are developed for different basins and coal zones. Such facies schemes may also be more readily accepted if facies are given less formidable names.

ASSOCIATION OF COALS WITH CLASTIC SEDIMENTS

Rates of peat accumulation are difficult to precisely quantify because so much compaction takes place within a peat in the mire environment. Recent peat accumulation rates are generally calculated by dividing the thickness of a peat profile by the age of basal deposits. Calculations for Recent peats vary from 2.3 mm/yr in the tropics to 0.1 mm/yr in Arctic regions (McCabe, 1984). Assuming a peat:coal compaction ratio of 10:1, 1 mm of coal represents peat accumulation over a period of between 4 and 100 years. Such slow rates of accumulation suggest that low-ash coals must have formed in areas that were well removed from active clastic deposition. Mudrock partings in coals probably represent events with a periodicity of thousands or even tens of thousands of years. Sediment from more frequent events (e.g., 50-year floods) should be an integral part of a coal's ash content.

Most facies models of coal-bearing environments show peat accumulating in close proximity to active clastic depositional environments. This appears to be an unlikely scenario if the mires are of the low-lying variety, as the sediment introduced by floods, storm surges, or exceptionally high tides would give the peat a very high ash content.

Many workers have suggested the Mississippi River delta as a modern analogue for coal-bearing strata. Recent work on peats of the abandoned lobes of the delta (Kosters, 1983; Kosters and Bailey, 1983) shows that the majority of organic sediments have a very high ash content. Only 9 of 215 samples from the Barataria and Sale-Cypremort areas, west of the modern delta, had more than 80 percent organic matter on a moisture free basis (Kosters, 1983). No peats were recorded with less than 17 percent ash. True peats, with less than 25 percent ash, exceed 1 m in thickness in only a few places. Kosters and Bailey (1986) showed that leaching of true peat can decrease the ash content by approximately 33 percent. They suggested (Kosters and Bailey, 1983) that leaching during early diagenesis could upgrade many peats to be potential precursors of commercial coal. There is, however, little evidence that such leaching processes occur. In contrast, there is good evidence for epigenetic mineralization in most coals. A careful study of Kosters' (1983) cross sections through Mississippi peats suggests that, after compaction, they would form only carbonaceous shales with thin (up to 0.15 m thick) coaly stringers, even if a significant amount of leaching occurred.

Most other deltas are also not sites of significant true peat accumulation (McCabe, 1984). Coastal and floodplain mires have been frequently suggested as "coal-forming environments," but a review of the literature on modern deposits in these areas indicates that peat precursors of coal are generally not present in such environments either (McCabe, 1984; Breyer and McCabe, 1986). The only exceptions are in areas where raised mires have developed.

In southeast Asia, raised mires are present close to some shorelines and major river systems and on some deltas (Anderson, 1964, 1983; Coleman and others, 1970). Margins of the raised mires are narrow and steep. The surfaces of the central parts of the mires are generally flat and are between 3 and 7.5 m above flood or high-tide levels. Ash content of these peats is usually less than 5 percent and over large areas of mires is less than 1 or 2 percent (Anderson, 1983). Raised mires have also been described in delta-top and shoreline environments from temperate climates. Peats from raised mires of the Fraser River delta in British Columbia have ash contents as low as 0.5 to 1.5 percent (Styan and Bustin, 1983a, b).

Although some low ash coals may have originated as peats in raised mires, it is unlikely that this is true for all of them. The paleoclimate under which many coals are thought to have accumulated would probably be unsuitable for the extensive development of raised mires.

The base of many coal seams may represent a considerable hiatus in deposition. Underlying sediments were probably deposited long before the mire was established, at which time the area could have been well removed from active clastic depositional environments. A considerable period of nondeposition preceded the initiation of peat accumulation in some of the mires of the southeastern United States. The peats of the Okefenokee (Cohen, 1984) and Snuggedy swamps (Staub and Cohen, 1979) are developed over Pleistocene beach ridges. These mires now lie 75 and 20 km, respectively, from the present Atlantic shoreline.

Care should be taken, therefore, before interpreting the depositional environment of a coal from an examination of enclosing clastic sediments. Detailed studies are required to determine the nature of the association.

COALS ASSOCIATED WITH FLUVIAL ENVIRONMENTS

Many coals are interbedded with sedimentary rocks of alluvial plain origin. Major examples from the U.S. include the Paleocene and Eocene strata of the Powder River Basin (Ethridge and others, 1981; Flores, 1981; Flores and Hanley, 1984), the Paleocene of northwest Colorado (Beaumont, 1979), the Cretaceous of the Raton and San Juan basins of Colorado and New Mexico (Flores, 1984), the Dunkard and Monongahela groups of the Appalachian region (Donaldson, 1974), and parts of the Wilcox Group of Texas (Fisher and McGowen, 1969; Kaiser and others, 1980). In Canada, coals are associated with very thick alluvial sequences in both the Pennsylvanian of Maritime Canada (Gersib and McCabe, 1981; Rust and others, 1984) and the Paleocene of Alberta (Gibson, 1977).

Coal seams associated with fluvial sediments have been reported to be laterally continuous over wide areas. Splits in seams are common. Coal seams may be thick; in the Powder River Basin, seams are up to 30 m thick (Ayers and Kaiser, 1984). However, not all coal-bearing alluvial sequences have thick seams; Pennsylvanian coals in New Brunswick have a maximum thickness of only 0.76 m (Ball and Gemmell, 1985). In addition to the depositional setting, factors such as subsidence rates and climatic regime control peat thickness.

The ash content of coals in alluvial strata can be relatively low. In the Powder River Basin, coals have between 4.3 and 11.3 percent ash (Flores, 1981), and in the Alberta Paleocene, ash contents range from 5.2 to 24.9 percent (Nurkowski, 1985). Raised mires have been suggested by Flores (1981) and Ethridge and others (1981) to explain the low ash coals of the Powder River Basin. Coals in alluvial sequences characteristically have low sulfur values, presumably because of the lack of any marine influence. In the Paleocene of the Powder River Basin, sulfur ranges between 0.3 and 1.6 percent (Flores, 1981); in Alberta, it is between 0.15 and 1.79 percent (Nurkowsi, 1985). A notable exception are the coals of Maritime Canada, which typically have up to 10 percent sulfur.

Sediments usually found in alluvial sequences include fining-upward channel deposits, overbank muds, crevasse splay sands, and thin lacustrine deposits (Gersib and McCabe, 1981). The latter are deposited in shallow floodplain lakes and in abandoned channels. Most mires are located in areas of low gradient, and consequently most fluvial sediments associated with coals appear to have been deposited in meandering systems. They can, however, be associated with alluvial fan and braided river deposits (e.g., in late Cretaceous and Tertiary intermontane basins of the Canadian Cordillera [Long, 1981]) and with anastomosed river deposits (e.g., in the Paleocene of the Powder River Basin [Flores, 1984] and the Pennsylvanian of Nova Scotia [Rust and others, 1984]).

The geometry of alluvial strata below and above a coal may control a seam's thickness. Graphic examples of coal thickening over underlying abandoned channels and thinning due to scouring from overlying channels have been described by Padgett and Ehrlich (1978) from the mid-Carboniferous of West Virginia. Mine maps of individual seams that show dendritic patterns of mine out areas or unmined areas indicate the importance of fluvial geometry in controlling the location of coals of economic thickness. Donaldson and others (1985) demonstrated that the geometry of the Waynesburg Coal of West Virginia is controlled by both underlying and overlying alluvial strata. They suggested that during the life of a mire there is decreasing importance of initial relief and increasing importance of differential compaction of underlying sediment in controlling thickness of accumulated peat. They also showed the variable nature of roof rock conditions due to the complex geometry of overlying fluvial channel sandstones.

Ribbonlike bodies of clastic sediment split or immediately overlie some coals. Spectacular examples occur in some Pennsylvanian coals of Illinois and Indiana. The Walshville channel cuts the Herrin Coal and underlying and overlying strata (Krausse and others, 1979; Treworgy and Jacobson, 1985). The channel is over 370 km long (Fig. 6), up to 3 km wide, and up to 30 m

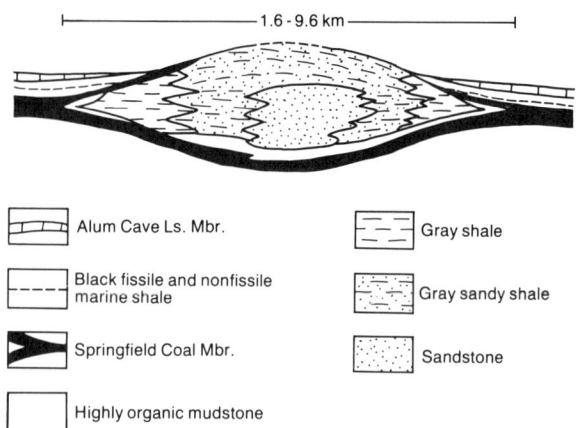

Figure 7. Diagrammatic cross section through ribbonlike body overlying the Springfield Coal of Indiana (after Eggert, 1984).

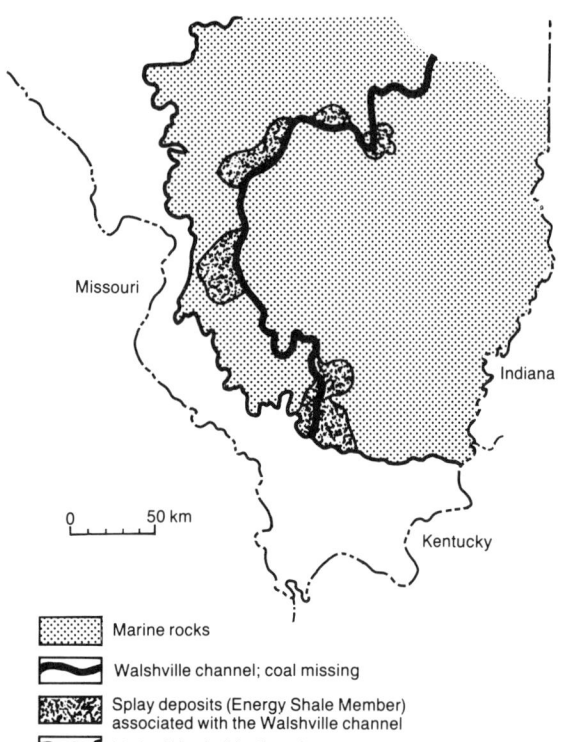

Figure 6. Distribution of rock types overlying the Herrin (no. 6) Coal in Illinois (after Treworgy and Jacobson, 1985).

thick. Coal adjacent to the channel is thin or contains splits, which suggests that the channel was in part contemporaneous with peat development. The Walshville channel affected both the maceral content (Harvey and Dillon, 1985) and the paleoflora (Phillips and others, 1985) of the surrounding coals. Overlying the coal and adjacent to the channel are lobate wedges of shales, siltstones, and sandstones, termed the Energy Shale, which have been interpreted as crevasse-splay and lacustrine sediments (Palmer and others, 1985). These clastic wedges range in area from 13 to 520 km^2 and are up to 30 m thick. After cessation of peat growth, the river appears to have been less confined, and large crevasse splay complexes were deposited over the dying mire. This preceded a marine transgression that deposited limestones and shales over the entire area.

A ribbon body, the Galatia channel, also overlies the Springfield and Harrisburg Coal of Illinois and Indiana (Eggert, 1984; Eggert and Adams, 1985; Treworgy and Jacobson, 1985). This channel can be traced for over 160 km and, in its upstream reaches, shows an anastamosed tributary pattern. Fluvial activity appears to have been in part contemporaneous with the upper part of the coal seam, as the seam thickens and splits toward the channel (Fig. 7). As with the Walshville channel, the coarse sediments of the central channel complex are flanked by finer sediments, the Dykersburg Shale, in a belt up to 24 km wide. It would appear that the rivers that deposited these ribbon bodies were restricted to relatively narrow paths for long periods of time. Whether this was because of the nature of the mire, compaction of underlying sediments, or another control is not known.

COALS ASSOCIATED WITH FLUVIAL-DOMINATED SHORELINES

Where the input of fluvial sediment exceeds the capacity of tides and waves to transport the sediment alongshore or offshore, a high-constructive delta develops (Fisher and others, 1969). Such deltas are either lobate or elongate. The past and present lobes of the Mississippi Delta exemplify this type of delta. Many coal-bearing sequences have been interpreted as having been deposited by high-constructive deltas. This is partly for historical reasons, as deltas were the first modern environment to be thoroughly studied by sedimentologists. The coarsening-upward sequence of the classic Pennsylvanian cyclothem appear to be easily explained by prograding deltas. Coals were interpreted as deposits of delta-top mires. As discussed earlier, however, high-constructive deltas do not appear to be the favorable sites for peat development as has long been assumed. Many sequences once interpreted as deltaic have been reinterpreted in light of recent advances in facies studies. A critical reappraisal of some other deposits may be warranted.

Much of the Pennsylvanian of the Appalachian region has been interpreted as delta deposits and has been compared to the modern Mississippi delta complex (Ferm, 1970, 1974, 1975; Donaldson, 1974; Baganz and others, 1975; Horne and others, 1978). Fluvial-dominated shoreline sequences have also been interpreted from coal-bearing strata in the Upper Cretaceous of Utah (Cotter, 1975; Ryer, 1981), the Paleocene of North Dakota (Belt and others, 1984), and the Eocene of Texas (Fisher and McGowen, 1969; Kaiser and others, 1980). All aspects of deltaic sediments—including prodelta, delta front, mouth bar, interdistributary bay, distributary channels, and fluvial channels—have been recognized in these studies.

Ferm and Williams (1965) suggested that the Pennsylvanian strata were controlled by the interaction of subsidence and variation in sediment supply. They suggested that coal is a product of transgression and that there may have been lateral migration of mires during transgressive episodes. Coal was recognized, therefore, as an abandonment facies of deltas. Ferm (1970) suggested that a major cause of variation in sediment supply was delta switching. He envisaged coal originating in mires developed on abandoned delta lobes.

Ferm (1974) suggested that the geometry of coal seams is determined by their position within the deltaic sequence. He suggested that lower delta plain coals were relatively thin but widespread, whereas upper delta plain coals were relatively thick but lenticular. Horne and others (1978) suggested that the best developed coals originated as peats in mires developed on infilled interdistributary bays transitional between the upper and lower delta plain. Here coals are thick and widespread. Coals interpreted as lower delta plain in origin are laterally continuous in the direction of depositional dip but discontinuous parallel with the depositional strike. It is suggested that the peats developed in mires on the narrow, poorly developed levees of distributary channels. Coals of the upper delta plain also tend to parallel depositional dip but are more pod-shaped, exhibiting abrupt variations in thickness over short distances. Numerous splits are present near strata interpreted as levees of meandering channels. If the Appalachian coals were deposited in as close proximity to active depositional systems as Horne and others (1978) suggest, the mires were presumably of the raised type to exclude the input of clastic sediments.

Ferm and Staub (1984) have recently indicated, however, that the geometry of coal seams in the Appalachians is not as strongly related to the deltaic setting as had previously been thought. They consider the major control on peat geometry to have been the topography on which the mire developed. They suggest that the topography was mainly controlled by subsidence rate—in large part, a function of differential compaction of underlying sediments.

In a detailed study of the coals associated with the Ferron Sandstone, Ryer (1981) showed that coal seams overlie regressive sandstone sequences interpreted as the deposits of lobate deltas. Coal seams continue landward of the sandstone sequences and reach their maximum thickness, up to 10 m, some 10 km landward of the landward pinchout of each regressive sequence (Fig. 8). Seams are traceable much farther parallel to the depositional strike than parallel to the depositional dip. Seams are, however, pod-like because the area of mire development was cut by streams flowing toward the coast (Fig. 8).

Sulfur contents of fluvially influenced shorelines are variable. As a general rule, coals interpreted as upper delta plain in origin have lower percentages of framboidal pyrite and pyritic sulfur than lower delta plain coals (Carruccio and Geidel, 1979), suggesting less marine influence. There are, however, some notable exceptions.

COALS ASSOCIATED WITH WAVE-DOMINATED SHORELINES

Beaches and barrier systems are developed where wave action is important in reworking shoreline sediment. Progradation of such shorelines results in beach ridges, which may be separated by lagoonal sediments. If the shoreline is fed by a river with a large supply of sediment, the coastline may build out as a wave-dominated delta.

Wave-dominated shoreline sediments have been interpreted from some coal-bearing sequences in the Appalachian Carboniferous (e.g., Hobday, 1974; Hobday and Horne, 1977), in the Cretaceous of the western United States (Cavaroc and Flores, 1984; Flores and others, 1984; Johnson, 1978; Ryer, 1977; Balsley, 1980), and in the Eocene of south Texas (Snedden and Kersey, 1981). Coals may overlie back barrier sediments and associated channel and overbank deposits or, in many cases, may sit directly on the beach-barrier sandstones. Modern analogues of such relationships are provided by the Okefenokee (Cohen, 1984) and Snuggedy swamps (Staub and Cohen, 1979).

Coals associated with wave-dominated shoreline sequences tend to be elongate parallel to the paleoshoreline. Isopach maps of the Hiawatha coal of Utah (Flores and others, 1984), the San Miguel lignite of Texas (Snedden and Kersey, 1981), and the coals of the Menefee Formation of New Mexico (Tabet and others, 1985) show coals subparallel to associated beach-barrier deposits. Both show local dip-trending segments interpreted as effects of channels cut through the ancient mires. In a detailed study of coals of the Cretaceous Rock Springs Formation in Wyoming, Levey (1985) showed that seams extend up to 24 km along depositional dip and up to 58 km along strike. Seams are thick in pods, which are 2 to 3 km wide where there are few splits or channel washouts. Levey also showed that the coals overlie regressive sandstone sequences, but unlike the fluvial-dominated shoreline deposits of the Ferron (Ryer, 1981), the thickest parts of the seam overlie the sandstones. Because of their association with marine strata, coals associated with wave-dominated shorelines could be expected to have relatively high sulfur contents. The San Miguel lignite, for example, has considerably more sulfur (1.47 percent) than other Texas coals (Snedden and Kersey, 1981). The Hiawatha coal of Utah has between 0.7 and 1.6 percent sulfur, in contrast to 0.4 to 0.45 percent in the closely associated Muddy No. 1 coal, which is interpreted as having formed in a more landward setting. In contrast, the coals of the Menefee Formation are relatively low in sulfur (Tabet and others, 1985).

COALS ASSOCIATED WITH TIDAL SEDIMENTS

Evidence of tidal currents may be found in sediments of open shelves, shorefaces, estuaries, and tidal channels. In areas of raised mire development, such as southeast Asia and northwest Europe, low-ash peats may be developed even where tidal ranges are high (e.g., Coleman and others, 1970; Wilks, 1979). Although some evidence of tidal currents is often found in barrier and deltaic sediments (e.g., Hobday and Horne, 1977), there are only a few coals that lie within sediments interpreted as having been deposited dominantly by tidal currents.

Breyer and McCabe (1986) have interpreted clastic sediments in the lower part of the coal-bearing Wilcox Group in south Texas as being predominantly tidal. A 50-m-thick sequence in the Chacon Creek area consists of six coal seams that are interbedded and cut by clastic sediments with streaky, lenticular, wavy, and flaser bedding. The sediments form fining- and coarsening-upward sequences up to 14 m thick. These are interpreted as tidal channel deposits because they commonly overlie a thin sandstone with mudclasts, interpreted as a lag, and because they apparently cut out seams. Breyer and McCabe suggested that the Chacon Creek sediments formed in a setting somewhat similar to the modern Snuggedy Swamp of South Carolina, which is

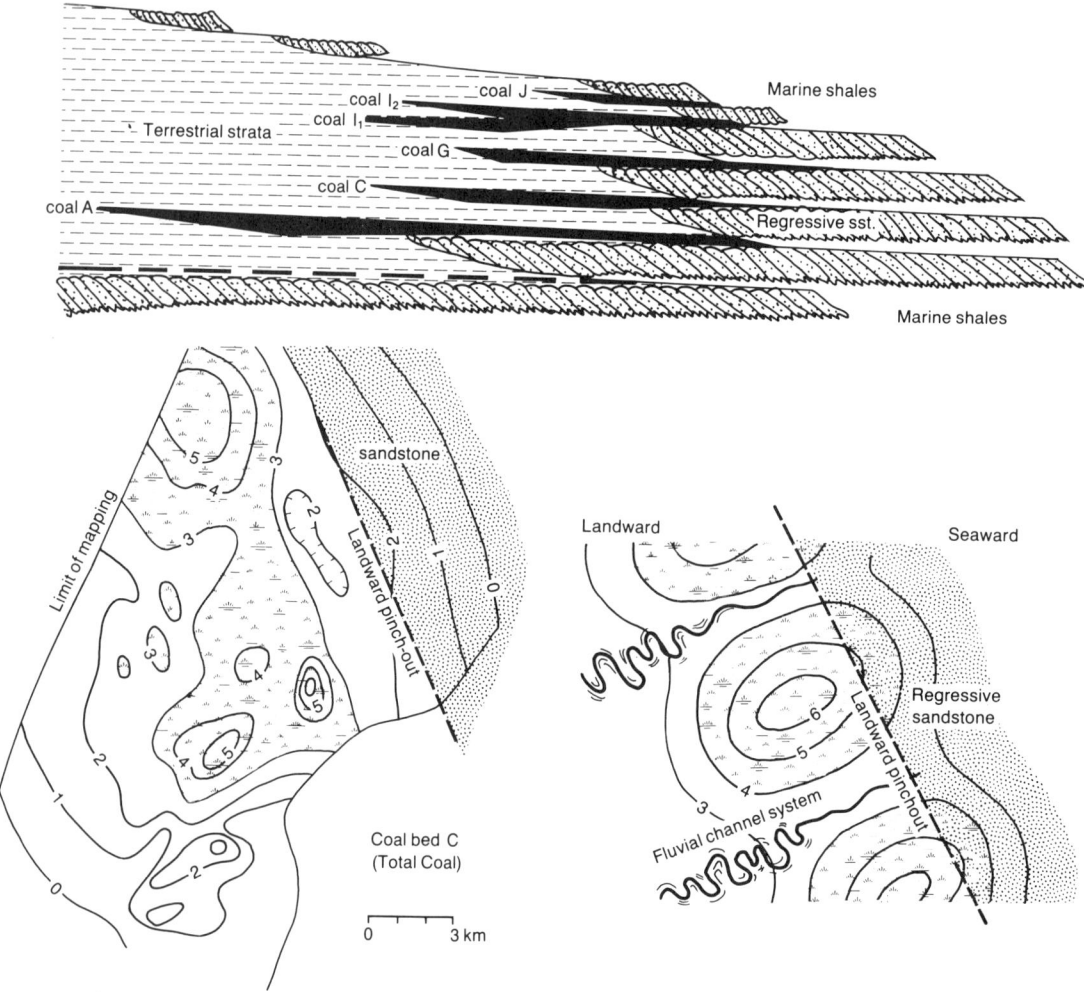

Figure 8. Distribution of coals associated with Ferron Sandstone (after Ryer, 1981). Top: diagrammatic cross section showing coal overlying and thickening landward of regressive sandstone units. Lower left: isopach of C coal bed showing maximum thickness landward of landward pinch-out of regressive sandstone unit. Lower right: model to explain podlike form of coal seams.

bounded by two small tributaries of a large estuarine complex. In such an environment, further regression would result in cleaner and more extensive peat development while a relative rise in sea level would cause encroachment on the mire by estuarine sediments.

Channel sandstones of the Upper Cretaceous Horseshoe Canyon Formation of Alberta show good evidence of deposition under tidal conditions (Rahmani, 1983). The thickest coals of this formation occur close to the contact zone of interfingering marine and terrestrial sediments, suggesting that the coals formed on a coastal plain incised in part by estuaries or tidal channels.

COALS ASSOCIATED WITH LACUSTRINE SEDIMENTS

Most lakes are intimately associated with at least one of the environments already discussed. They will therefore be dealt with only briefly here. Lakes may be developed in most environments as a result of local subsidence, channel abandonment, or the development of relief (e.g., levees) that impedes free drainage. A lake may be infilled by the progradation of a wave-influenced or fluvial-influenced shoreline or by crevasse splays from a nearby river.

Lacustrine sediments are common in sequences dominated by alluvial sediments (e.g., Belt and others, 1984; Flores and Hanley, 1984; Gersib and McCabe, 1981). These sediments are usually thin, rarely exceeding a few meters. In some coal basins, however, lacustrine sediments make up a large percentage of the strata (Falini, 1965; Gibling and others, 1985). Hacquebard and Donaldson (1969) have suggested that the Carboniferous Pictou coalfield of Nova Scotia was deposited in a narrow intermontane basin. Their lithofacies map shows lacustrine sediments in the central part of the basin with coal seams up to 13.4 m thick. Fluvial deposits are more common at the margin of the basin. It has recently been suggested by Ayers and Kaiser (1984) that there is a substantial lacustrine component to the Paleocene strata of the Powder River Basin, an interpretation disputed by Flores and Hanley (1984).

CONCLUSIONS

During the last 20 years, a much better understanding of the depositional environments of coal and coal-bearing strata has developed. In some cases, application of facies models to detailed studies of mine sites, such as those described by Horne and others (1978), has helped optimize production and lower costs. Models have also been developed to show the areas of better coal development in certain regional depositional settings (e.g., Ryer, 1981).

What directions should future sedimentological studies take? The following are suggested:

1. More detailed facies studies. Coal-bearing strata have generally not been subjected to the same rigorous analyses as oil- and gas-bearing strata. Detailed facies descriptions, including measured sections, need to be published to facilitate comparison between different formations.

2. Multidisciplinary studies. Cooperation betweeen clastic sedimentologists, coal petrographers, palynologists, paleobotanists, and geochemists is needed for a comprehensive understanding of the relationships between depositional environments and coal properties.

3. Economic applicability. Models need to be predictive, not just explanatory. It is also important that models predict something that needs to be predicted. The general location of most of the coals likely to be exploited in North America through the twenty-first century is already known. Prediction of the variations in quality within these coals could prove very useful.

REFERENCES

Allen, G. P., Laurier, D., and Thouvenin, J., 1979, Étude sédimentologique du delta de la Mahakam: Total, Compagnie Française des Pétroles, Paris, Notes et Mémoires no. 15, 156 p.

Altschuler, Z. S., Schnepfe, M. M., Silber, C. C., and Simon, F. O., 1983, Sulfur diagenesis in Everglades peat and origin of pyrite in coal: Science, v. 221, p. 221-227.

Anderson, J.A.R., 1964, The structure and development of the peat swamps of Sarawak and Brunei: Journal of Tropical Geography, v. 18, p. 7-16.

—— , 1983, The tropical peat swamps of western Malesia, in Gore, A.J.P., ed., Ecosystems of the World, v. 4B, Mires: Swamp, bog, fen, and moor, regional studies: Amsterdam, The Netherlands, Elsevier Scientific Publishing Company, p. 181-199.

Ayers, W. B., and Kaiser, W. R., 1984, Lacustrine-interdeltaic coal in the Fort Union Formation (Palaeocene), Powder River Basin, Wyoming and Montana, U.S.A., in Rahmani, R. A., and Flores, R. M., eds., Sedimentology of coal and coal-bearing sequences: International Association of Sedimentologists Special Publication no. 7, p. 61-84.

Baganz, B. P., Horne, J. C., and Ferm, J. C., 1975, Carboniferous and recent Mississippi lower delta plains; A comparison: Gulf Coast Association of Geological Societies Transactions, v. 25, p. 183-191.

Ball, F. D., and Gemmell, D. E., 1985, The New Brunswick coal resource, in Patching, T. H., ed., Coal in Canada: Canadian Institute of Mining and Metallurgy Special Volume 31, p. 70-77.

Balsley, J. K., 1980, Cretaceous wave-dominated delta systems; Book Cliffs, east-central Utah: American Association of Petroleum Geologists Continuing Education Course Field Guide, 163 p.

Beaumont, E. A., 1979, Depositional environments of Fort Union sediments (Tertiary, northwest Colorado) and their relation to coal: American Association of Petroleum Geologists Bulletin, v. 63, p. 194-217.

Belt, E. S., Flores, R. M., Warwick, P. D., Conway, K. M., Johnson, K. R., and Waskowitz, R. S., 1984, Relationship of fluviodeltaic facies to coal deposition in the lower Fort Union Formation (Palaeocene), south-western North Dakota, in Rahmani, R. A., and Flores, R. M., eds., Sedimentology of coal and coal-bearing sequences: International Association of Sedimentologists Special Publication no. 7, p. 177-195.

Bohor, B. F., and Triplehorn, D. M., 1981, Volcanic origin of the flint clay parting in the Hazard No. 4 (Fire Clay) coal bed of the Breathitt Formation in eastern Kentucky, in Cobb, J. C., Chestnut, D. R., and Hester, N. C., eds., Coal and coal-bearing rocks of eastern Kentucky: Geological Society of America Coal Division Fieldtrip, November 5-8, 1981, Guidebook, p. 49-54.

—— , 1984, Accretionary lapilli in altered tuffs associated with coal beds: Journal of Sedimentary Petrology, v. 54, p. 317-325.

Breyer, J. A., and McCabe, P. J., 1986, Coals associated with tidal sediments in the Wilcox Group (Paleogene), south Texas: Journal of Sedimentary Petrol-

ogy, v. 56, p. 510–519.
Burger, K., and Damberger, H. H., 1985, Tonsteins in the coalfields of western Europe and North America: Compte Rendu, Neuvième Congrès International de Stratigraphie et de Géologie du Carbonifère, Washington and Champaign-Urbana, 1979, v. 4, p. 433–448.
Bustin, R. M., Cameron, A. R., Grieve, D. A., and Kalkreuth, W. D., 1983, Coal petrology, its principles, methods, and applications: Geological Association of Canada, Short Course Notes 3, 273 p.
Bustin, R. M., Styan, W. S., and Lowe, L. E., 1985, Variability of sulphur and ash in humid-temperate peats of the Fraser River delta, British Columbia: Compte Rendu, Dixième Congrès International de Stratigraphie et de Géologie du Carbonifère, Madrid, 1983, v. 2, p. 79–94.
Carruccio, F. T., and Geidel, G., 1979, Using the paleoenvironment of strata to characterize mine drainage quality, in Ferm, J. C., and Horne, J. C., eds., Carboniferous depositional environments in the Appalachian region: Columbia, University of South Carolina, Department of Geology, Carolina Coal Group, p. 587–596.
Casagrande, D. J., Siefert, K., Berschinski, C., and Sutton, N., 1977, Sulphur in peat-forming systems of the Okefenokee Swamp and Florida Everglades; Origins of sulphur in coal: Geochimica et Cosmochimica Acta, v. 41, p. 161–167.
Cavaroc, V. V., and Flores, R. M., 1984, Lithologic relationships of the Upper Cretaceous Gibson-Cleary stratigraphic interval; Gallup coal field, New Mexico, U.S.A., in Rahmani, R. A., and Flores, R. M., eds., Sedimentology of coal and coal-bearing sequences: International Association of Sedimentologists Special Publication no. 7, p. 197–215.
Cohen, A. D., 1974, Petrography and palaeoecology of Holocene peats from the Okefenokee swamp-marsh complex of Georgia: Journal of Sedimentary Petrology, v. 44, p. 716–726.
——— , 1984, The Okefenokee Swamp; A low sulphur end-member of a shoreline-related depositional model for coastal plain coals, in Rahmani, R. A., and Flores, R. M., eds., Sedimentology of coal and coal-bearing sequences: International Association of Sedimentologists Special Publication no. 7, p. 231–240.
Cohen, A. D., and Spackman, W., 1972, Methods in peat petrology and their application to reconstruction of paleoenvironments: Geological Society of America Bulletin, v. 83, p. 129–142.
Cohen, A. D., Spackman, W., and Dolsen, P., 1983, Occurrence and distribution of sulfur in peat-forming environments of southern Florida, in Raymond, R., and Andrejko, M. J., eds., Mineral matter in peat; Its occurrence, form, and distribution: Los Alamos National Laboratory, Proceedings of workshop, 1983, p. 87–112.
Cohen, A. D., Casagrande, D. J., Andrejko, M. H., and Best, G. R., 1984, The Okefenokee Swamp; Its natural history, geology, and geochemistry: Los Alamos, Wetland Surveys, 709 p.
Coleman, J. M., Gagliano, S. M., and Smith, W. G., 1970, Sedimentation in a Malaysian high tide tropical delta, in Morgan, J. P., ed., Deltaic sedimentation, modern and ancient: Society of Economic Paleontologists and Mineralogists Special Publication, v. 15, p. 185–197.
Cotter, E., 1975, Deltaic deposits in the Upper Cretaceous Ferron Sandstone, Utah, in Broussard, M. L., ed., Deltas, models for exploration: Houston Geological Society, p. 471–484.
Davies, T. D., and Raymond, R., 1983, Sulfur as a reflection of depositional environments in peats and coals, in Raymond, R., and Andrejko, M. J., eds., Mineral matter in peat; Its occurrence, form, and distribution: Los Alamos National Laboratory, Proceedings of workshop, 1983, p. 123–139.
Donaldson, A. C., 1974, Pennsylvanian sedimentation of central Appalachians, in Briggs, G., ed., Carboniferous of the southeastern United States: Geological Society of America Special Paper 148, p. 47–78.
Donaldson, A. C., Moyer, C. B., and Renton, J. J., 1985, Factors affecting thickness and quality of Waynesburg coal, West Virginia: Compte Rendu, Neuvième Congrès International de Stratigraphie et de Géologie du Carbonifère, Washington and Champaign-Urbana, 1979, v. 4, p. 308–320.
Eggert, D. L., 1984, The Leslie Cemetery and Francisco distributary fluvial channels in the Petersburg Formation (Pennsylvanian) of Gibson County, Indiana, U.S.A., in Rahmani, R. A., and Flores, R. M., eds., Sedimentology of coal and coal-bearing sequences: International Association of Sedimentologists Special Publication, v. 7, p. 309–315.
Eggert, D. L., and Adams, S. C., 1985, Distribution of fluvial channel systems contemporaneous with the Springfield Coal Member (middle Pennsylvanian) in southwestern Indiana: Compte Rendu, Neuvième Congrès International de Stratigraphie et de Géologie du Carbonifère, Washington and Champaign-Urbana, 1979, v. 4, p. 342–348.
Eggert, D. L., and Phillips, T. L., 1982, Environments of deposition; Coal balls, cuticular shale, and gray-shale floras in Fountain and Parke counties, Indiana: Indiana Geological Survey Special Report 30, 43 p.
Ethridge, F. G., Jackson, T. J., and Youngberg, A. D., 1981, Floodbasin sequence of a fine-grained meander belt subsystem; The coal-bearing Lower Wasatch and Upper Fort Union Formations, southern Powder River Basin, Wyoming, in Ethridge, F. G., and Flores, R. M., eds., Recent and ancient nonmarine depositional environments; Models for exploration: Society of Economic Paleontologists and Mineralogists Special Publication 31, p. 191–209.
Falini, F., 1965, On the formation of coal deposits of lacustrine origin: Geological Society of America Bulletin, v. 76, p. 1317–1346.
Ferm, J. C., 1970, Allegheny deltaic deposits, in Morgan, J. P., ed., Deltaic sedimentation, modern and ancient: Society of Economic Paleontologists and Mineralogists Special Publication no. 15, p. 246–255.
——— , 1974, Carboniferous environmental models in eastern United States and their significance, in Briggs, G., ed., Carboniferous of the southeastern United States: Geological Society of America Special Paper 148, p. 79–95.
——— , 1975, Pennsylvanian cyclothems of the Appalachian plateau; A retrospective view: U.S. Geological Survey Professional Paper 853, part II, p. 57–64.
Ferm, J. C., and Horne, J. C., 1979, Carboniferous depositional environments in the Appalachian region: Columbia, University of South Carolina Department of Geology, Carolina Coal Group, 760 p.
Ferm, J. C., and Staub, J. R., 1984, Depositional controls of mineable coal bodies, in Rahmani, R. A., and Flores, R. M., eds., Sedimentology of coal and coal-bearing sequences: International Association of Sedimentologists Special Publication no. 7, p. 275–289.
Ferm, J. C., and Williams, E. G., 1965, Characteristics of a Carboniferous marine invasion in western Pennsylvania: Journal of Sedimentary Petrology, v. 35, p. 319–330.
Finkelman, R. B., 1982, Modes of occurrence of trace elements and minerals in coal, in Filby, R. H., Carpenter, B. S., and Ragaini, R. C., eds., Atomic and nuclear methods in fossil energy research: New York, Plenum, p. 141–149.
Finkelman, R. B., and Stanton, R. W., 1978, Identification and significance of accessory minerals from a bituminous coal: Fuel, v. 57, p. 763–768.
Finney, H. R., and Farnham, R. S., 1968, Mineralogy of the inorganic fraction of peat from two raised bogs in northern Minnesota: Proceedings of Third International Peat Congress, Quebec, 1968, p. 102–108.
Fisher, W. L., and McGowen, J. H., 1969, Depositional systems in Wilcox Group (Eocene) of Texas and their relation to occurrence of oil and gas: American Association of Petroleum Geologists Bulletin, v. 53, p. 30–54.
Fisher, W. L., Brown, L. F., Jr., Scott, A. J., and McGowen, J. H., 1969, Delta systems in the exploration for oil and gas: Texas Bureau of Economic Geology, Special Publication, 212 p.
Flores, R. M., 1981, Coal deposition in fluvial paleoenvironments of the Paleocene Tongue River Member of the Fort Union Formation, Powder River area, Powder River Basin, Wyoming and Montana, in Ethridge, F. M., and Flores, R. M., eds., Recent and ancient nonmarine depositional environments; Models for exploration: Society of Economic Paleontologists and Mineralogists Special Publication 31, p. 169–190.
——— , 1984, Comparative analysis of coal accumulation in Cretaceous alluvial deposits, southern United States Rocky Mountain basins, in Stott, D. F., and Glass, D. J., eds., The Mesozoic of middle North America: Canadian Society of Petroleum Geologists Memoir 9, p. 373–385.
Flores, R. M., and Hanley, J. H., 1984, Anastomosed and associated coal-bearing

fluvial deposits; Upper Tongue River Member, Palaeocene Fort Union Formation, northern Powder River Basin, Wyoming, U.S.A., *in* Rahmani, R. A., and Flores, R. M., eds., Sedimentology of coal and coal-bearing sequences: International Association of Sedimentologists Special Publication no. 7, p. 85–103.

Flores, R. M., Blanchard, L. F., Sanchez, J. D., Marley, W. E., and Muldoon, W. J., 1984, Paleogeographic controls of coal accumulation, Cretaceous Blackhawk Formation and Star Point Sandstone, Wasatch Plateau, Utah: Geological Society of America Bulletin, v. 95, p. 540–550.

Gersib, G. A., and McCabe, P. J., 1981, Continental coal-bearing sediments of the Port Hood Formation (Carboniferous), Cape Linzee, Nova Scotia, Canada, *in* Ethridge, F. G., and Flores, R. M., eds., Recent and ancient nonmarine depositional environments; Models for exploration: Society of Economic Paleontologists and Mineralogists Special Publication 31, p. 95–108.

Gibling, M. R., Ukakimaphan, Y., and Srisuk, S., 1985, Oil shale and coal in intermontane basins of Thailand: American Association of Petroleum Geologists Bulletin, v. 69, p. 760–766.

Gibson, D. W., 1977, Upper Cretaceous and Tertiary coal-bearing strata in the Drumheller-Ardley region, Red Deer River valley, Alberta: Geological Survey of Canada Paper 76–35, 41 p.

Gluskoter, H. J., and Hopkins, M. E., 1970, Distribution of sulfur in Illinois coals, *in* Smith, W. H., Nance, R. B., Hopkins, M. E., Johnson, R. G., and Shabica, C. W., eds., Depositional environments in parts of the Carbondale Formation, western and northern Illinois: Illinois State Geological Survey Guidebook Series 8, p. 89–95.

Gluskoter, H. J., and Simon, J. A., 1968, Sulfur in Illinois coals: Illinois State Geological Survey Circular 432, 28 p.

Gore, A.J.P., 1983, Introduction, *in* Gore, A.J.P., eds., Ecosystems of the world; 4A, Mires: Swamp, bog, fen and moor; General studies: Amsterdam, Elsevier Scientific Publishing Company, p. 1–34.

Hacquebard, P. A., and Donaldson, J. R., 1969, Carboniferous coal deposition associated with flood-plain and limnic environments in Nova Scotia, *in* Dapples, E. C., and Hopkins, M. E., eds., Environments of coal deposition: Geological Society of America Special Paper 114, p. 143–191.

Harvey, R. D., and Dillon, J. W., 1985, Maceral distributions in Illinois coals and their paleoenvironmental implications: International Journal of Coal Geology, v. 5, p. 141–165.

Hobday, D. K., 1974, Beach- and barrier-island facies in the Upper Carboniferous of northern Alabama, *in* Briggs, G., ed., Carboniferous of the southeastern United States: Geological Society of America Special Paper 148, p. 209–223.

Hobday, D. K., and Horne, J. C., 1977, Tidally influenced barrier island and estuarine sedimentation in the Upper Carboniferous of southern West Virginia: Sedimentary Geology, v. 18, p. 97–122.

Horne, J. C., Ferm, J. C., Caruccio, F. T., and Baganz, B. P., 1978, Depositional models in coal exploration and mine planning in Appalachian region: American Association of Petroleum Geologists Bulletin, v. 62, p. 2379–2411.

Ingram, H.A.P., and Bragg, O. M., 1984, The diplotelmic mire; Some hydrologic consequences reviewed: 7th International Peat Congress, Dublin, 1984, Proceedings, v. I, p. 220–235.

Johnson, J. L., 1978, Stratigraphy of the coal-bearing Blackhawk Formation on North Horn Mountain, Wasatch Plateau, Utah: Utah Geology, v. 5, p. 57–77.

Kaiser, W. R., Ayers, W. B., and La Brie, L. W., 1980, Lignite resources in Texas: Austin, Texas, Bureau of Economic Geology Report of Investigations no. 104, 52 p.

Kolb, C. R., and Van Lopik, J. R., 1966, Depositional environments of the Mississippi River deltaic plain; Southeastern Louisiana, *in* Shirley, M. L., ed., Deltas in their geologic framework: Houston Geological Society, p. 17–61.

Kosters, E. C., 1983, Louisiana peat resources: Louisiana Geological Survey Technical Report DOE/FE/05113, 63 p.

Kosters, E. C., and Bailey, A., 1983, Characteristics of peat deposits in the Mississippi River deltaic plain: Gulf Coast Association of Geological Societies Transactions, v. 33, p. 311–325.

——, 1986, A reassessment of Louisiana peat resources based on leaching experiments: Louisiana Geological Survey Coastal Geology Technical Report No. 2, 108 p.

Krausse, H.-F., and 6 others, 1979, Engineering study of structural geologic features of the Herrin (No. 6) Coal and associated rock in Illinois: Vol. 2; Illinois State Geological Survey, Final Report to U.S. Bureau of Mines, Contract No. H0242017, 205 p.

Levey, R. A., 1985, Depositional model for understanding geometry of Cretaceous coals; Major coal seams, Rock Springs Formation, Green River Basin, Wyoming: American Association of Petroleum Geologists Bulletin, v. 69, p. 1359–1380.

Long, D.G.F., 1981, Dextral strike slip faults in the Canadian Cordillera and depositional environments of related fresh-water intermontane coal basins, *in* Miall, A. D., ed., Sedimentation and tectonics in alluvial basins: Geological Association of Canada Special Paper 23, p. 153–186.

McCabe, P. J., 1984, Depositional environments of coal and coal-bearing strata, *in* Rahmani, R. A., and Flores, R. M., eds., Sedimentology of coal and coal-bearing sequences: International Association of Sedimentologists Special Publication, v. 7, p. 13–42.

Moore, L. R., 1964, Microbiology, mineralogy, and genesis of a tonstein: Proceedings of the Yorkshire Geological Society, v. 34, p. 235–292.

Moore, P. D., 1984, European mires: London, Academic Press, 367 p.

Nissenbaum, A., and Kaplan, I. R., 1972, Chemical and isotopic evidence for the in situ origin of marine humic substances: Limnology and Oceanography, v. 17, p. 570–582.

Nurkowski, J. R., 1985, Coal quality and rank variation within Upper Cretaceous and Tertiary sediments, Alberta plains region: Edmonton, Alberta Research Council Earth Science Report 85-1, 39 p.

Padgett, G., and Ehrlich, R., 1978, An analysis of two tectonically controlled integrated drainage nets of mid-Carboniferous age in southern West Virginia, *in* Miall, A. D., ed., Fluvial sedimentology: Canadian Society of Petroleum Geologists Memoir 5, p. 789–799.

Palmer, J. E., Trask, C. B., and Jacobson, R. J., 1985, Depositional environments of strata of late Desmoinesian age overlying the Herrin (No. 6) Coal in southwestern Illinois and the occurrence of low-sulfur coal: Compte Rendu, Neuvième Congrès International de Stratigraphie et de Géologie du Carbonifère, Washington and Champaign-Urbana, 1979, v. 4, p. 329–341.

Phillips, T. L., Peppers, R. A., and Dimichele, W. A., 1985, Stratigraphic and interregional changes in Pennsylvanian coal-swamp vegetation; Environmental inferences: International Journal of Coal Geology, v. 5, p. 43–109.

Rahmani, R. A., 1983, Facies relationships and paleoenvironments of a late Cretaceous tide-dominated delta, Drumheller, Alberta; The Mesozoic of middle North America, fieldtrip guidebook no. 2: Canadian Society of Petroleum Geologists Conference, May 8–11, 1983, Calgary, 66 p.

Renton, J. J., and Cecil, C. B., 1979, The origin of mineral matter in coal, *in* Donaldson, A. C., Presley, M. W., and Renton, J. J., eds., Carboniferous coal guidebook, vol. 1: West Virginia Geological and Economic Survey Bulletin B-37-1, p. 206–223.

Romanov, V. V., 1968, Hydrophysics of bogs: Jerusalem, Israel Program for Scientific Translations, 299 p.

Ruppert, L. F., Cecil, C. B., Stanton, R. W., and Christian, R. P., 1985, Authigenic quartz in the Upper Freeport coal bed, west-central Pennsylvania: Journal of Sedimentary Petrology, v. 55, p. 334–339.

Russel, R. J., 1967, River and delta morphology: Louisiana State University Press, Coastal Studies Series no. 20, 55 p.

Rust, B. R., Gibling, M. R., and Legun, A. S., 1984, Coal deposition in an anastomosing-fluvial system; The Pennsylvanian Cumberland Group south of Joggins, Nova Scotia, Canada, *in* Rahmani, R. A., and Flores, R. M., eds., Sedimentology of coal and coal-bearing sequences: International Association of Sedimentologists Special Publication no. 7, p. 105–120.

Ryer, T. A., 1977, Patterns of Cretaceous shallow-marine sedimentation, Coalville and Rockport areas, Utah: Geological Society of America Bulletin, v. 88, p. 177–188.

——, 1981, Deltaic coals of Ferron Sandstone Member of Mancos Shale; Predictive model for Cretaceous coal-bearing strata of Western Interior: American Association of Petroleum Geologists Bulletin, v. 65, p. 2323–2340.

Ryer, T. A., Phillips, R. E., Bohor, B. F., and Pollastro, R. M., 1980, Use of altered volcanic ash falls in stratigraphic studies of coal-bearing sequences; An example from the Upper Cretaceous Ferron Sandstone Member of the Mancos Shale in central Utah: Geological Society of America Bulletin, v. 91, p. 579–586.

Smith, A.H.V., 1962, The palaeoecology of Carboniferous peats based on the miospores and petrography of bituminous coals: Proceedings of Yorkshire Geological Society, v. 33, p. 423–474.

Snedden, J. W., and Kersey, D. G., 1981, Origin of San Miguel lignite deposit and associated lithofacies, Jackson Group, south Texas: American Association of Petroleum Geologists Bulletin, v. 65, p. 1099–1109.

Spackman, W., Cohen, A. D., Given, P. H., and Casagrande, D. J., 1976, The comparative study of the Okefenokee Swamp and the Everglades-mangrove swamp-marsh complex of southern Florida: A short course presentation of the Pennsylvania State University, 403 p.

Spears, D. A., and Kanaris-Sotiriou, R., 1979, A geochemical and mineralogical investigation of some British and other European tonsteins: Sedimentology, v. 26, p. 407–425.

Staub, J. R., and Cohen, A. D., 1979, The Snuggedy Swamp of South Carolina; A back barrier estuarine coal forming environment: Journal of Sedimentary Petrology, v. 48, p. 203–210.

Styan, W. B., and Bustin, R. M., 1983a, Petrography of some Fraser River Delta peat deposits; Coal maceral and microlithotype precursors in temperate-climate peats: International Journal of Coal Geology, v. 2, p. 321–370.

——, 1983b, Sedimentology of Fraser River Delta peat deposits; A modern analogue for some deltaic coals: International Journal of Coal Geology, v. 3, p. 101–143.

Tabet, D. E., Frost, S. J., and Kottlowski, F. E., 1985, Depositional environments for Menefee Formation low-sulfur coals in southeast San Juan Basin of New Mexico: Compte Rendu, Neuvième Congrès International de Stratigraphie et de Géologie du Carbonifère, Washington and Champaign-Urbana, 1979, v. 4, p. 321–328.

Ting, F.T.C., 1972a, Petrified peat from a Paleocene lignite in North Dakota: Science, v. 177, p. 165–166.

——, 1972b, Depositional environments of the lignite-bearing strata in western North Dakota: North Dakota Geological Survey Miscellaneous Series no. 50, 134 p.

NOTE ADDED IN PROOF

The last four years have seen an explosion in the publication of papers on the environments of deposition of coal. In particular, the reader is referred to the volumes resulting from the international coal geology meetings in London, England, in 1986 (published in Scott [1987] and The Journal of the Geological Society, Vol. 144, Part 3, 1987; see also Warwick and Stanton [1988]) and Orleans, France, in 1989 (published as no. 2 of the Bulletin de la Société Géologique de France, 8e Série, tome 162, 1991) and at the International Geological Congress in Washington (published as Volume 12 of the International Journal of Coal Geology). An excellent review of coal-forming environments has also been made by Haszeldine (1989).

Perhaps the most significant change in view over this period has been the increased realization that major properties of coals (bed thickness, maceral content, ash content, sulfur content) may be more dependent on allogenic controls (tectonism and climate) than on autogenic controls (environments of deposition). Recent papers espousing this viewpoint from different perspectives include Fielding (1987), Ferm and Weisenfluh (1989), Cecil (1990), and McCabe (1991).

Ting, F.T.C., and Spackman, W., 1975, The coal lithotype concept and seam profile: Compte Rendu, Septième Congres International de Stratigraphie et de Geologie du Carbonifere, Krefeld, 1971, v. 4, p. 307–311.

Treese, K. L., and Wilkinson, B. H., 1982, Peat-marl deposition in a Holocene paludal-lacustrine basin, Sucker Lake, Michigan: Sedimentology, v. 29, p. 375–390.

Treworgy, C. G., and Jacobson, R. J., 1985, Paleoenvironments and distribution of low-sulfur coal in Illinois: Compte Rendu, Neuvième Congrès International de Stratigraphie et de Géologie du Carbonifère, Washington and Champaign-Urbana, 1979, v. 4, p. 349–359.

Triplehorn, D., 1983, Cinerites and tonsteins from the Kenai Peninsula, Alaska: X Congreso Internacional de Estratigrafia y Geologia del Carbonifero, Madrid, 12–17 September 1983, Abstracts, p. 334.

Udden, J. A., 1912, Geology and mineral resources of the Peoria quadrangle, Illinois: U.S. Geological Survey Bulletin 506, 103 p.

Walther, J., 1894, Einleitung in die geologie als historische wissenschaft: Jena, Germany, Fischer Verlag, 3 volumes, 1055 p.

Wanless, H. R., and Weller, J. M., 1932, Correlation and extent of Pennsylvanian cyclothems: Geological Society of America Bulletin, v. 43, p. 1003–1016.

Weller, J. M., 1930, Cyclic sedimentation in the Pennsylvanian and its significance: Journal of Geology, v. 38, p. 97–135.

Wilks, P. J., 1979, Mid-Holocene sea-level and sedimentation interactions in the Dovey Estuary area, Wales: Palaeogeography, Palaeoclimatology, Palaeoecology, v. 26, p. 17–26.

Williams, E. G., and Keith, M. L., 1963, Relationship between sulfur in coals and the occurrence of marine roof beds: Economic Geology, v. 58, p. 720–729.

Manuscript Accepted by the Society May 18, 1987

ACKNOWLEDGMENTS

I wish to thank Drs. J. A. Boon, J. C. Cobb, T. A. Cross, J. C. Horne, and A. R. Palmer for constructive comments on a first draft of the manuscript. I also thank the Alberta Research Council's Graphics Department for drafting the figures and Maureen FitzGerald for typing the manuscript.

Cecil, C. B., 1990, Paleoclimate controls on stratigraphic repetition of chemical and siliciclastic rocks: Geology, v. 18, p. 533–536.

Ferm, J. C., and Weisenfluh, G. A., 1989, Evolution of some depositional models in Late Carboniferous rocks of the Appalachian coal fields: International Journal of Coal Geology, v. 12, p. 259–292.

Fielding, C. R., 1987, Coal depositional models for deltaic and alluvial plain sequences: Geology, v. 15, p. 661–664.

Haszeldine, R. S., 1989, Coal reviewed: depositional controls, modern analogues and ancient climates, in Whateley, M.K.G., and Pickering, K. T., eds., Deltas: Sites and traps for fossil fuels: Geological Society of London Special Publication No. 41, p. 289–308.

McCabe, P. J., 1991, Tectonic controls on coal accumulation: Bulletin de la Société Géologique de France, 8e Série, tome 162, no. 2, p. 133–138.

Scott, A. C., ed., 1987, Coal and coal-bearing strata: Recent advances: Geological Society of London Special Publication No. 32, 332 p.

Warwick, P. D., and Stanton, R. W., 1988, Depositional models for two Tertiary coal-bearing sequences in the Powder River Basin, Wyoming, USA: Journal of the Geological Society of London, v. 145, p. 613–620.

Chapter 31

Paleobotany and paleoecology of coal

Tom L. Phillips
Department of Plant Biology, University of Illinois, Urbana, Illinois 61801
Aureal T. Cross
Department of Geological Sciences, Michigan State University, East Lansing, Michigan 48823

INTRODUCTION

Coal is an environmentally controlled product of both biologic and geologic origins. Paleoecological reconstructions aid in establishing plant sources and constituents of coal and in interpreting environmental history. The two principal sets of environmental conditions are: interactions of once-living plants and their habitats (ecology) and those attendant to death, burial, and fossilization (taphonomy).

Most economically important coals were derived from vascular plant peat, accumulated in situ (autochthonous). Thus, coal is usually a transformed peat deposit—essentially a buried organic soil with litter stratified in succession and penetrated by roots. This simplified taphonomy aids significantly in paleoecological reconstructions; most terrestrial fossil-plant assemblages are more or less transported.

Some principal limitations in the taphonomy of peat deposits are conditions of preservation; plant debris may be too severely degraded or the coal rank too high for adequate fossil data. The links of fossil assemblages of coal to ecology are dependent on knowledge of the plant sources, and, in turn, their biology. Opportunities and constraints in paleoecological studies differ markedly in the North American coals of the Pennsylvanian, Cretaceous, and Tertiary, as did the plants, paleoclimates, and paleogeography.

FOSSIL-PLANT SOURCES

Botanic origins of coal are inferred from a combination of preservational states (Figs. 1 to 7): layered compressions within the coal (White and Thiessen, 1913); anatomy of tissues within the coal macerals, especially vitrinite (Winston, 1986); anatomy and morphology of relatively uncompacted plant assemblages in coal-ball concretions (DiMichele and others, 1986); and microfossil macerates yielding cuticles, resinlike bodies, amber, algae, fungi, pollen, and spores (Winslow, 1959).

Coal balls

The key means of estimating botanic constituents and reconstructing plants, where available, are anatomically preserved "peatlike stages" in coal-ball concretions. Anatomical preservation permits identification of the source plants because of early permineralization and minimal compaction. The so-called "peat stage" is coalified to about the same rank as the surrounding coal (Lyons and others, 1985). Coal balls or other permineralized plant material occur within lignite, bituminous, and anthracite coals; anatomical detail may be retained even to anthracitic rank. In North America, coal balls are most abundant in the Pennsylvanian (Fig. 8) and have been reported from more than 35 coals and over 130 localities. Permineralized plants are known in coals of the Upper Devonian of West Virginia (Scheckler, 1986), in the Upper Cretaceous of Alberta (Peter McCabe, personal communication, 1985), in the Paleocene of North Dakota (Ting, 1972), and in the upper Miocene to Pliocene of Alaska (Knoll, 1985).

Coal-ball concretions are generally composed primarily of calcium carbonate, with varying amounts of pyrite and dolomite, or occasionally of silica. The entombing carbonates generally have a δC^{13} signature, indicating origin of CO_2 partly or largely from degradation of the peat. Coal-ball deposits may occupy entire seam thicknesses, known up to 4 m, and massively occupy areas of football-field size. Deposits tend to be semilinear, usually as discontinuous pods. Coal-ball deposits most commonly occur below black marine roof-shale strata, but they also occur below marine limestone, gray shale, sandstone, and complex transitional sequences. For calcareous coal-ball origins, the crucial questions relate to calcium or magnesium supply sources and to circumstances of pH change for carbonate precipitation (Mamay and Yochelson, 1962; DeMaris and others, 1983). The larger question as to their abundant occurrence only in upper Paleozoic coals remains unanswered.

Tissues in coal

The use of thin-sectioning or etching techniques to identify tissue components in coal was established early in American coal studies (Thiessen, 1920; Thiessen and Sprunk, 1937). Only recently have they been reapplied with a reference to plant anat-

Phillips, T. L., and Cross, A. T., 1991, Paleobotany and paleoecology of coal, *in* Gluskoter, H. J., Rice, D. D., and Taylor, R. B., eds., Economic Geology, U.S.: Boulder, Colorado, Geological Society of America, The Geology of North America, v. P-2.

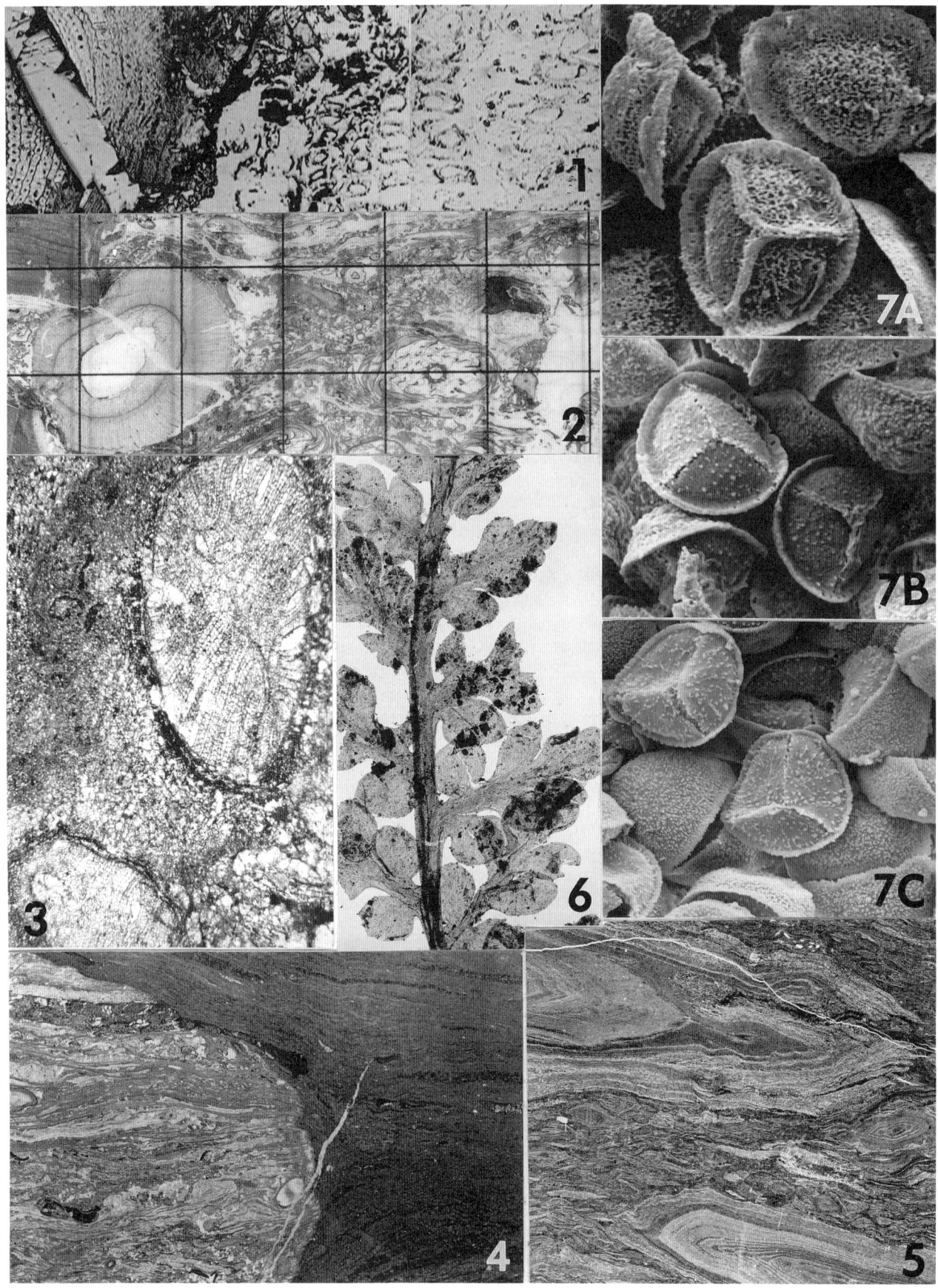

omy in adjacent coal balls (Winston, 1986). This may be the most direct source for determining the botanic composition of many coals, and the method is applicable up through low-volatile bituminous rank.

Palynology of coal

Coal pollen and spores and other macerated plant detritus provide useful means of sampling the floras. Much of the palynological data were generated originally for biostratigraphic purposes (Peppers, 1985). For paleoecological purposes, the utility and resolution of coal palynology depend on determining the source plants of spores, pollen, and cuticles, and on relating spore-pollen abundances to more direct quantitative estimates of the vegetation, such as biomass, ground cover, or crown area. Quite significant components of many macerates are the cuticles from leaves and stems. These permit rather specific identification of plant sources (DiMichele and others, 1984). In the transition from high- to low-volatile bituminous coal, spores and cuticles are markedly altered and usually are less accessible for paleobotanical purposes.

Compressions and other fossils in clastic deposits

It was generally assumed that roof-shale floras were indicative of coal-seam assemblages. On this premise, most of the classic reconstructions of upper Carboniferous swamp forests were

---

Figure 1. Transverse anatomy of lycopod bark in coal ball (right) and adjacent derived coal indicating compaction of 8:1. Etched, polished surface, Herrin Coal, ×150. Courtesy R. B. Winston.

Figure 2. *Paralycopodites* (lycopod) aerial assemblage of wood, bark, and cones, Herrin coal ball, with overlay of cm^2 grid system used to quantify botanical constituents. Etched surface, ×2.

Figure 3. Transverse anatomy of *Rhacophyton* wood in pyritic coal ball, polished surface. Upper Devonian, West Virginia. Courtesy, S. E. Scheckler.

Figure 4. Structural "peat" (left) and subsequent "lignitic" stage in Iowa coal ball traversed by a fusain band. Etched surface, ×5.

Figure 5. Structural "peat" of conifer wood in matrix of stems and leaves. Silicified coal ball, Upper Cretaceous, Alberta. Polished thin section, ×5. Courtesy P. J. McCabe.

Figure 6. *Karinopteris* (seed fern) frond cuticle with intact pinnules. Indiana "paper coal". Macerate, ×5. Courtesy W. A. DiMichele.

Figure 7. *Lycospora* species of dominant lycopod trees in Middle Pennsylvanian. A. *L. pellucida* from *Lepidophloios harcourtii*; B. *L. pusilla* putatively from *Lepidodendron hickii*; and C. *L. granulata* from *Lepidophloios hallii*. Spores extracted from cones in coal balls. SEM, ×1,000. Courtesy D. A. Willard.

made. In general, plant assemblages (compressions, spores) from clastic sediments proximal to coal are representative of other habitats. Use of such fossils as coal-flora indicators needs to be corroborated directly from the coal. Some forest assemblages at the top of coal beds, anchored in the peat and exhibiting large trunk bases, are different from the typical coal-swamp vegetation (DiMichele and Demaris, 1987).

PEAT-SWAMP FORESTS

The low diversity, freshwater forests of peat swamps are the principal contributors to repetitive, massive peat accumulation. Such forests exhibited a combination of productivity, decay-resistant litter, and buried root growth that enhanced the potential of preservation of the peat accumulation. The kinds of dominant plants differed in successive coal ages, and numerous changes occurred in extent of diversity, principal tissue types (biomass), sizes and longevity of trees or shrubs, reproductive biology, types of litter accumulation, and preservational qualities.

Evolution, migration, and extinction have resulted in a progression of dominant tree types in peat swamps, from lower vascular plants (spore-bearing) in the Pennsylvanian, to gymnosperms (conifers) in the Cretaceous, to angiosperms (flowering plants) along with conifers in the Tertiary. This pattern strongly reflects the evolution of reproductive systems as well as vegetative characteristics. The tree habit has evolved independently in numerous plant groups and first appears in Middle to Late Devonian time. Trees of most lower vascular plants were paratropical in origin and confined to that paleoclimatic zone. Trees of seed plants inhabited a broad range of climatic zones and subsequently dominated many dryland and wetland habitats.

Main evolutionary trends in vascular plants include morphologic and physiologic developments that allow conquest and survival in dryland habitats—escape from ephemeral and standing-water environments. Trees represent the zenith of structural complexity and longevity of growth and productivity. In view of the terrestrial adaptations of trees, it is not surprising that a rather limited number of species are tolerant of or adapted to peat-forming environments.

Similar patterns of stasis and change

Despite differences in the biologies of dominant trees in peat swamps of different coal ages, many shared similar kinds of distributional patterns. Genera and, to a lesser extent, species of the major tree types usually exhibit very long stratigraphic ranges and wide paleogeographic distribution. This stasis contrasts sharply to the rate and nature of evolutionary changes in tree forms outside of swamps. With passing geologic time, peat-swamp floras and especially the tree types, become relict. Peat-swamp floras seem inherently less responsive to the oscillating influences of climate than those of "typical" terrestrial habitats where the short-term effect of moisture and greater thermal variability exert stronger selective pressures. Consequently, peat-swamp floras are rather good average indicators of paleoclimate on a geologic time scale.

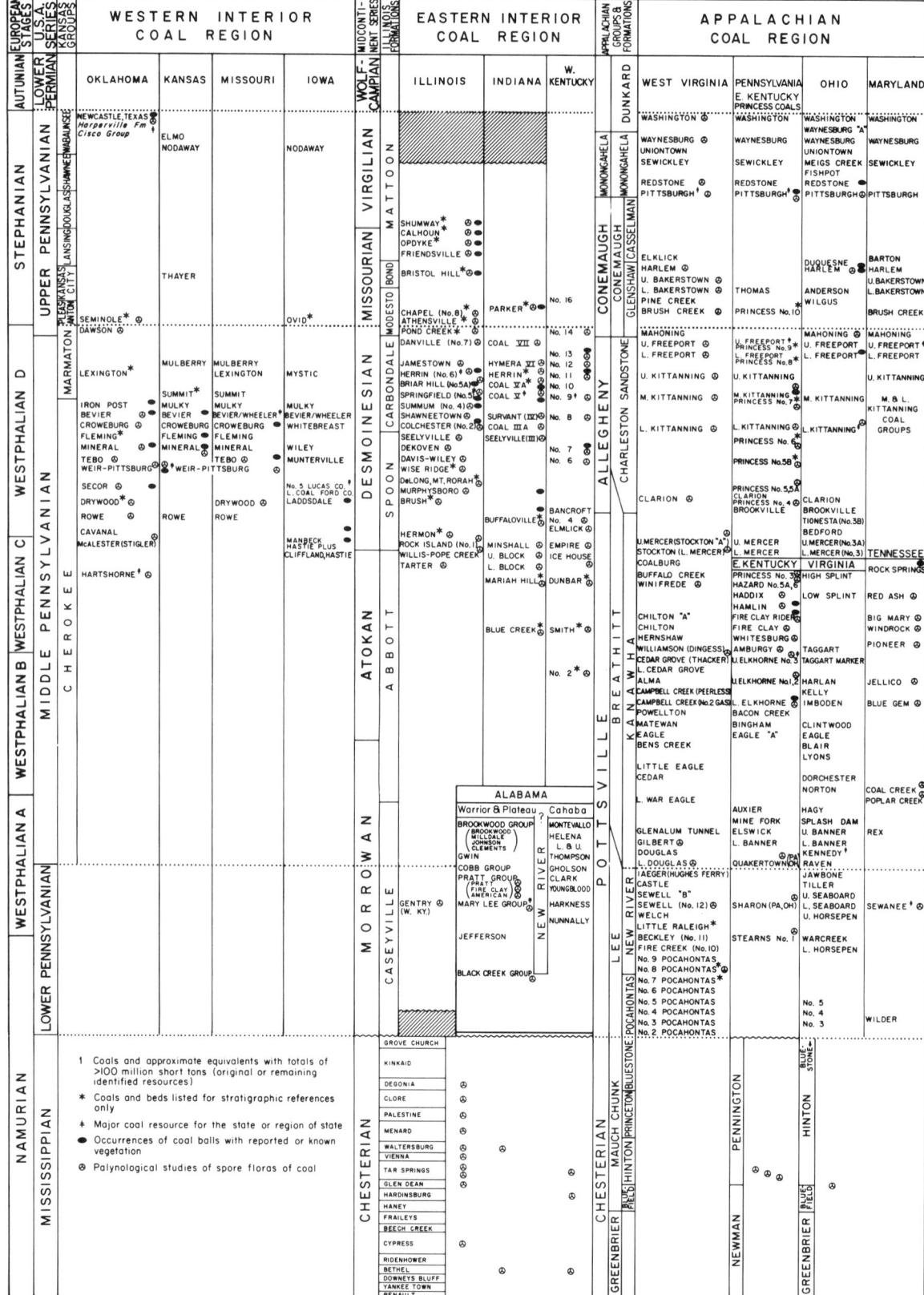

Figure 8. Stratigraphic distribution of major[1] bituminous coal deposits in the Pennsylvanian System of the United States with symbols indicating coal-ball occurrences and reported spore floras, including those of Chesterian strata. Revised from Phillips and others (1985).

Despite the many changes in population sizes of tree species during the contraction and expansion of peat swamps, such floras retained the characteristic array of tree types over and over again. Each termination of a peat swamp, even if time transgressive, was a case of mass extermination in large areas. However, there were usually adequate refugia for maintenance of plant populations between swamp-forming episodes as well as for access to colonize new swamp areas as they appeared.

The essentially closed nature of such an ecosystem suggests that floristic change took place in limited refugial habitats where severe perturbation resulted in "opening" them to various magnitudes of alteration: speciation, introduction of trees from nearby habitats, and/or extinctions. Extinction of swamp-centered genera was more likely to take place in long time spans between coal ages as habitats dwindled or became inaccessible for dispersal and occupancy. However, this can be relatively abrupt, geologically, after severe changes within a coal age. Following major changes in peat-swamp composition and structure (also in the early developmental stages of such communities), the floras are expected to more closely resemble surrounding vegetation from which components were derived.

Most changes in peat-swamp vegetation from one coal level to another in a stratigraphic interval are likely to be quantitative fluctuations in response to local and regional environmental parameters—not the evolutionary or permanent shifts in the floras. Major long-term changes in composition and structure are the exception, not the rule. They appear to have been geologically abrupt and divide stratigraphic intervals (or regions) into subpatterns characterized by particular dominant trees.

Environmental constraints and paleoecology

Distribution of terrestrial vegetation of the geologic past was controlled by evolution, paleoclimate, paleogeography, edaphic factors, and biotic interactions. Particularly in a coal age, paleoclimate is the most important. In peat-forming ecosystems leading to coal formation, the controls are largely telescoped. They are mainly mediated by resulting edaphic conditions that constrain diversity, evolutionary change, and productivity, and combine the chief variables: the water table (net rainfall, runoff, evapotranspiration), peat accumulation (net productivity versus degradation), and burial (submergence, compaction, subsidence). Peat will accumulate in any environment where plants can grow and reproduce, and organic matter accumulates and can escape total decay. Peat-accumulating environments are usually characterized by low nutrients and low oxygen availability, low pH, and low temperatures. The environmental variables affecting peat accumulation are emphasized in chapters by Ziegler and Cross and by McCabe (this volume). The emphasis here is on peat-swamp tree types and paleoecological implications.

Hydrologic limitations. Coal-forming ecosystems begin and end in water, and indigenous peat accumulation is mainly in a freshwater milieu. The succession of plant assemblages within a peat swamp is largely abiotically controlled by level of the water table, water quality, and frequency of disruption, such as fire or flooding by fresh or brackish water. Consequently, hydrologic succession patterns vary significantly within and between contemporaneous swamps. In order to interpret patterns of plant distribution, it is necessary to relate the plants to known indicators of the physical environment. For example, in the Taxodiaceae (bald cypress family), it seems likely that the moist to wettest series was *Metasequoia, Sequoia, Glyptostrobus,* and *Taxodium* (Fig. 12); in the lycopods it was probably *Sigillaria, Paralycopodites, Diaphorodendron,* and *Lepidodendron-Lepidophloios* (Plate 4). If these prove to be fairly reliable, there is potential to link plant dominance patterns with environmental controls on some peat-forming characteristics.

Typical water regimes of peat swamps range between intermittently exposed peat to long-term flooding of peat throughout most years or growing seasons. The flooding regime has considerable successional consequences, influencing potential for invasion by other tree types and for development of marshes. The approximate depth separating palustrine vegetation from deep inland water (aquatic) habitats is 2 m. This is typically beyond limits for trees, but it coincides with limits of emergent, herbaceous plants. Inundation depths and durations are usually quite restrictive for trees; bald cypress is a notably tolerant exception.

Mineral matter and nutrient impact. Transitions from peat swamps to clastic swamps are indicated by elevated mineral-matter content, often increased diversity, and usually different floras, depending on kind and distribution of clastic sediment. In general, the distinctions between peat swamps and clastic swamps are not well defined when impurity limits of coal are less than 50 percent mineral matter by weight or when soil designations as clastic histosols are more than 55 percent (Schopf, 1956; USDA, 1981). Most economic coals would be expected to have less than 20 percent mineral matter; it is mostly within that lower range that relationships between nutrient levels and vegetation are sought. In the North Carolina wetlands, the 75 percent organic/25 percent mineral content is considered a natural sedimentological break between true peat and peaty mineral sediment (Otte and others, 1987). Diagnostic coal floras may help delineate gradients within topogenous (planar) swamps as well as their distinctions from ombrogenous (domed) ones. High-ash coals enriched in mineral matter probably formed from floras quite distinct from those of economic coals of similar age. They probably represent ecological windows on other kinds of vegetation. The Oligocene Brandon Lignite of Vermont is an example (Barghoorn and Spackman, 1950; Traverse, 1955a).

The constraints of low pH (less than 5.5), water tables, and nutrients are particularly evident in peat lands, as in north temperate North America today (Jeglum, 1971). Low pH substrates occur in many wetlands such as spruce, pine, and fir forests of the northern Great Lakes region, as well as in tropical rainforests. It is the low pH of peat-water, mostly generated by degradative processes and limited oxygen availability, that impedes the decay process and affects nutrient supply and uptake.

Structure and botanic constituents of tree types

The botanic constituents of peat differ significantly according to the kinds and abundances of tree types. There were three main kinds represented among the lower vascular plants (lycopods, ferns, and calamites) and three among the seed plants (conifers, cycadophytes, and flowering dicots).

There are considerable differences among North American coal-forming ecosystems of the Pennsylvanian, Cretaceous, and Tertiary in the biology and taphonomy of dominant plants. Some tree types have environmental implications for growth conditions that closely relate to those of peat accumulation. The architecture of some trees aids in interpreting litter represented by plant constituents but not necessarily environments of peat accumulation.

Lower vascular plants. Lycopod trees originated in the Devonian and diversified in the Mississippian. The stigmarian lycopods apparently originated in swamps and were confined to them. They diminished during a mass extinction in North America at the end of the Middle Pennsylvanian when the broad coastal plain/delta-swamp environments were greatly reduced. The main scale trees, *Lepidophloios, Lepidodendron, Diaphorodendron, Paralycopodites,* and *Sigillaria,* were anchored by peculiar root systems called *Stigmaria* (Plate 4). They exhibited a determinate growth habit (fixed growth period) in both root system and aerial trunk-crown. The polelike architecture developed either lateral rows of deciduous branches with cones or, near maturity, dichotomously branched crowns with cones. Narrow, linear leaves were produced on scalelike, elliptical cushions, and both were progressively smaller toward the tree top. The trunk and larger branches were supported mostly by decay-resistant, barklike tissues around very small wood cylinders.

Tree ferns first appeared in abundance in Middle Pennsylvanian time. *Psaronius* tree ferns were unbranched, up to 7.5 m tall, and supported largely by a thick root mantle extending to near the crown of large fronds. The mantle of air-chambered, adventitious roots formed a basal buttress, with trunk diameters attaining as much as a meter. Cheap construction of root anatomy (sometimes the stem and leaf bases, too) and lack of secondary growth suggest that *Psaronius* had low nutrient requirements for relatively rapid development. These attributes preadapted *Psaronius* for swamp environments with the broadest ecologic range of any genus; tree ferns were not restricted to such habitats.

Calamites were more important in clastic environments of accumulation than those of peat swamps. Trees were composed mostly of wood and exhibited determinate growth and vegetative propagation. The arrangement of branches, leaves, and cone-units was whorled with a distinctly jointed anatomy (Plate 4). Erect trunks developed from buried prostrate stems; adventitious roots grew from underground portions.

Seed plants. Two kinds of wood occur in gymnosperms. Dense (pycnoxylic) and massive wood is indicative of coniferlike trees and those gymnosperms that grew in temperate regions. Bark is thin compared to wood and is sloughed off regularly as diameter increases. Weaker, more spongy wood construction, known as manoxylic, is present in cycad-like plants in the tropics. Such trees gain additional support from thickened tissues around the wood. They had unbranched trunks of limited diameter with relatively large fronds. The medullosan seed ferns of the Pennsylvanian tropics typify this construction (Plate 4).

Coniferlike cordaite trees were represented in tropical Pennsylvanian peat swamps and dominated those of the north temperate Permian (Angaran Realm). True conifers appeared in the paleotropics in the Middle Pennsylvanian and were not major peat-swamp components until much later. Both cordaites and conifers apparently had origins in dry, moisture-stressed regions and became more widespread in eastern North America as the humid Pennsylvanian coal age diminished and Permian semiaridity ensued.

The xeric origins of conifers, including survival in the dry Permian-Triassic transition, was probably important in preadapting some for the environments they dominate today in north temperate regions. During the Mesozoic, modern conifer families evolved, and among them, the Taxodiaceae (bald cypress) was particularly important in the temperate Cretaceous and Tertiary peat swamps. Surviving genera, such as *Taxodium* (bald cypress), *Metasequoia* (dawn redwood), and *Glyptostrobus* (East Asian water pine), have long stratigraphic ranges and formerly very wide paleogeographic distributions. Their present occurrences are refugial, with *Metasequoia* and *Glyptostrobus* near extinction in southeast China (prior to man's intervention), and *Taxodium* along the Mississippi Embayment, and Atlantic and Gulf coastal plains. Unlike most living conifers, which are typically evergreen, these genera usually have total seasonal fall of leaves or twigs (deciduous) in temperate zones.

Angiosperm peat-swamp trees of the latest Cretaceous and Tertiary are far more diverse than earlier types of plants and too numerous to refer to generically in detail here. Flowering plants have developed a much greater variety of architectures and ecological amplitudes than earlier plant groups, resulting in an array of woody habits, including trees, shrubs, and vines as well as herbs. Angiosperm dicot trees and shrubs exhibit a woody construction similar to conifers but with more varied wood characteristics and foliage types. Some peat-forming swamp genera, such as *Nyssa* (black and tupelo gums), are known to extend back as far as the Paleocene.

CURRENT STUDIES

Much of the quantitative paleobotanical information on coal origins and especially peat-swamp paleoecology is derived from research of the last decade or so.

Pennsylvanian

Stratigraphic changes. Vegetational patterns of peat swamps are stratigraphically divisible into five intervals coinciding largely with recognized series or groups. The most obvious time of change is at the Middle–Upper Pennsylvanian boundary, separating general lycopod dominance from that of tree ferns thereafter (Fig. 9). Peat swamps change from as much as 95 percent lycopod biomass in the Early Pennsylvanian to about 70 percent toward the end of the Middle Pennsylvanian, with shifting importance of genera and species; all of the abundant lycopod tree genera except *Sigillaria* became extinct in North America near the boundary. The main earlier changes began at the onset of Middle Pennsylvanian time with fluctuations in the dominant lycopod tree types. Palynological studies indicate sporadic densospore marshlands, increased abundances of cordaites and calamites, and a rise in number and diversity of tree-fern spores; the latter is inconsistent so far with available peat data. Vegetationally, the early Middle Pennsylvanian was, in part, a lycopod-cordaite interval, a time of many marshlands, a sequence of increasing species diversity, and a net diminution of lycopod dominance.

The mid–Middle Pennsylvanian peat swamps differed vegetationally across east-west gradients, reflecting apparently earlier accrued shifts as well as significant environmental change of the time. This interval coincides broadly with boundary transitions of the Atokan-Desmoinesian and Pottsville-Allegheny and with quite significant patterns of changing forests and coal characteristics. In the Appalachians there was an abundance of lycopod-cordaite-calamite vegetation, coinciding with the "splint coal interval" (Sprunk and others, 1940) and fluctuating increases in ash content of coal (Cecil and others, 1985). In the Western Interior Coal Region (Arkoma, Cherokee, Forest City Basins) there were cordaitean-rich swamps, diminishing eastward into lycopod-cordaitean communities in the Illinois Basin Coal Field. *Psaronius* tree ferns and *Medullosa* seed ferns attained subdominant levels at this time. As the largest swamps of the Pennsylvanian developed in the Allegheny and Desmoinesian, lycopods dominated with *Psaronius* and *Medullosa* secondary; cordaites became minor components. The late Middle Pennsylvanian shows a further increase in tree ferns in both major coal regions. Following mass extinction of most lycopod trees, the Late Pennsylvanian peat swamps were *Psaronius* forests with patchy distribution of calamites, cordaites, *Sigillaria,* and especially *Medullosa,* which was usually a subdominant.

Composition of coal-ball peats. The main botanic components of coal balls are summarized from selected seam averages (Fig. 9) and indicate that about 95 percent of the identifiable biomass was formed by trees. The unidentifiable debris (taxa, organs, tissues) varies from 5 to 33 percent and is usually <20 percent. Shoot/root ratios change from ≤1 (Westphalian A) to >1 (Middle Pennsylvanian) to <1 (Upper Pennsylvanian). With a few exceptions, the range of ratios in the Desmoinesian from the Western Interior Coal Region is 0.96 to 1.9, and from the Illinois Basin, 0.89 to 2.97. Those of the Missourian-age coals are 0.42 to 0.6. Use of 4/1 ratios for lycopod-dominated peats for comparison of expected to observed shoot/root ratios (estimates of minimal litter loss) suggests degradation of >50 to 87.5 percent aerial biomass for ratios of 2.0 to 0.5. Principal biases in estimates are the following: severe degradation resulting in enrichment of aerial bark; greater compaction of root systems or, usually, lack of compaction of air-chambered root systems, which alter the expected aerial biomass.

Roots usually form 30 to 55 percent of relatively uncompacted peat biovolumes. Woody root peats occur in the mid–Middle Pennsylvanian. Cordaites and calamites in such deposits resulted in 35 to 40 percent wood, of which almost half was root derived. Coals with subdominant cordaites commonly have 10

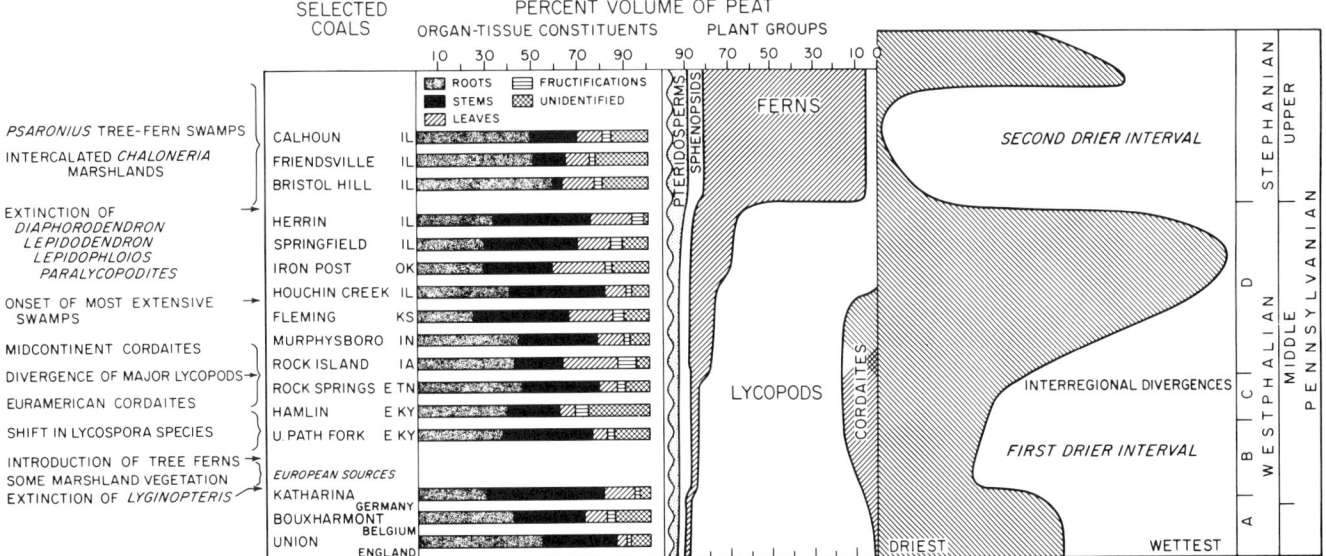

Figure 9. Stratigraphic patterns of Pennsylvanian peat swamps (from left) indicating major changes in plant genera, relative abundances of botanical constituents from selected coals and of the five major plant groups. At the right is a relative wetness curve based on a log plot of identified bituminous coal resources of the United States. Modified from Phillips and Peppers (1984) and Phillips and others (1985).

percent or more biomass of wood; those above or below the cordaitean intervals (Fig. 9) usually have about 5 percent wood composition.

Bark is a principal tissue in lycopod-dominated peat swamps, composing 20 to 45 percent of the biomass, with few exceptions. Foliage and cortical tissues of seed ferns and tree ferns vary considerably in the upper Middle and Upper Pennsylvanian and are indicators of good litter preservation.

Fusain content of coal-ball deposits is: 3 to 10 percent in the Westphalian A, up to 9 percent, mid–Middle Pennsylvanian; Desmoinesian, 2.6 to 5.8 percent in the Western Interior Coal Region and 2.4 to 6.9 percent in the Illinois Basin; and 1.4 to 4.1 percent, Missourian of the Illinois Basin. While there are some higher fusain contents in specific coals, no striking stratigraphic or interregional differences are observed. Notable increases in fusain level occur near paleochannels of the largeset coal seams: 8.6 percent near the Galatia paleochannel in the Springfield Coal and 8.1 to 10.4 percent near the Walshville in the Herrin Coal.

Profile and regional analyses of vegetation. Three kinds of succession patterns are encountered among multiple profiles in areas distant from paleochannels: very low-diversity vegetation varying little from one zone to another or laterally over a 100-m distance (small seams of irregular distribution or local parts of larger seams); high-diversity swamps showing distinctive vegetational changes within profiles and some similar patterns matched within a mine and in other mines many kilometers away (very large coal seams); and, in the Upper Pennsylvanian, small, floristically depauperate swamps dominated by *Psaronius* with marked variability in abundances of other trees, from meter to meter.

An ecological analysis of the Herrin Coal (Fig. 10) of Illinois based on coal-ball assemblages (Phillips and DiMichele, 1981), provides an example of vegetational succession with recurring *Lepidophloios* dominance among the other communities. This basic pattern is related to changes of the edaphic environment indicated in part by mineral-rich bands or disruption intervals.

The bottom zone is composed mostly of *Lepidophloios* stigmarian roots and some *Medullosa* litter; the combination of a stigmarian lycopod and *Medullosa* litter is a recurrent basal assemblage at other sites and in some other coals. From below to above each mineral-rich band there is a decrease in the abundance of *Lepidophloios,* an increase of other lycopod trees, and an increase or continued high abundance of *Medullosa* (except for the bottommost band). Maximum abundances of *Psaronius* tree ferns approximately alternate with those of *Lepidophloios*. The environmental interpretations from this and other profiles (Phillips and others, 1977; DiMichele and Phillips, 1988) suggest the following: *Lepidophloios* characteristically grew in standing water (Plate 4) with prolonged stability; abundant *Medullosa* assemblages were consistently associated with mineral-rich bands— higher nutrient availability and exposed peat from dry-downs (some following clastic influx from floods); *Lepidodendron* is probably indicative of standing water also, but with higher mineral matter content (clastic swamp); *Sigillaria* and *Paralycopodites* were abundant only in transitional fluctuating environments of low water table, hence associated with *Medullosa*; *Diaphorodendron* was ecologically intermediate between wet and drier extremes; and *Psaronius* overlapped significantly with most tree types.

In large peat swamps, as represented by Springfield and Herrin Coals, regional vegetational differences are evident. There is a greater abundance of lycopods (principally *Lepidophloios*) near the major paleochannels and in the thickest coal deposits (Phillips and Peppers, 1984; Phillips and others, 1985). Maximum levels of *Lycospora* (mainly *Lepidophloios*) differ: 35 to 60 percent in the Springfield and 65 to 85 percent in the Herrin Coal. Progressively away from such paleochannels, tree-fern spores tend to increase but not in simple patterns.

Regional patterns of the Desmoinesian-Allegheny indicate that cordaites were initially more abundant in the Western Interior Coal Region than in the Illinois Basin Coal Field and in both synchronously dropped to <1 percent of the biomass near the time of the Iron Post and Springfield peat swamps. The main lycopod trees were *Diaphorodendron* and then *Lepidophloios.* In the Appalachians, *Psaronius* was more abundant than in the Midcontinent, and *Lepidodendron* was probably as or more abundant than *Lepidophloios.* Despite the diverging dominance patterns, abrupt diminutions of lycopods in North America reduced the main pattern in Late Pennsylvanian peat swamps to tree ferns in all regions. It is noted that the Mahoning Coal of the basal Conemaugh was lycopod dominated in some environments, and the extinction break takes place above it (Phillips and Peppers, 1984).

Environmental implications. Patterns of Pennsylvanian peat-swamp vegetation in eastern North America suggest a changing relative wetness of climate and important times of disturbances. The plot of identified coal resources (Fig. 9) is generally indicative of wetness conveyed by interpretations of the dominance patterns (Phillips and others, 1985). The Middle Pennsylvanian was the wettest, with a gradual rise in the early part and a drastic decline near the Middle–Upper boundary. Early Pennsylvanian time was moderately wet, the Conemaugh was the driest interval, and the Monongahela was very wet seasonally in the Dunkard Basin. The basic pattern is viewed as three progressively more severe pulses of drier climate in the earliest Middle, and early Late Pennsylvanian and onset of the Permian. Studies by Cecil and others (1985) suggest a somewhat different pattern for the Appalachians based on mineral-matter content of coal and their geochemical model: wettest in the Early and early Middle Pennsylvanian, declining some in the late Middle and then also dropping to the driest in the Conemaugh. The abrupt decline of relative wetness near the Middle–Upper Pennsylvanian boundary is a key point of most analyses. The relative climatic changes are important in testing models that may predict peat-swamp types (topogenous versus ombrogenous), mineral-matter sources, and relevant petrologic compositions.

The vegetational changes noted near the onset of the early Middle Pennsylvanian, with some extinctions, and subsequent increase in diversity of peat swamps have been attributed to a somewhat drier paleoclimate (Phillips and Peppers, 1984). Those of the mid–Middle Pennsylvanian are attributed to brackish influences, especially in the Western Interior Coal Region, diminishing eastward across the Illinois Basin; in the Appalachians, Cecil and others (1985) have suggested that the mineral-matter content of coal increases stratigraphically in a fluctuating pattern because of a seasonally drier climate. Questions exist as to mineral sources in the "splint coal interval" (Sprunk and others, 1940), which could in part be volcanic ash (Burger and Damberger, 1985; Burger, 1985), as well as the result of changes in sedimentary regimes, such as influx of clastics (Donaldson and others, 1985).

Late Jurassic–Early Cretaceous

Coals of this time are prominent in Montana (Great Falls area) and Alberta, in the Morrison and Kootenay Formations, respectively, of the Northern and Canadian Rocky Mountains and Great Plains. These coals and associated clastics are characterized by conifers, cycadophytes (pineapple-like gymnosperms), ginkgophytes (ancestral maidenhair trees), and ferns. A basic dif-

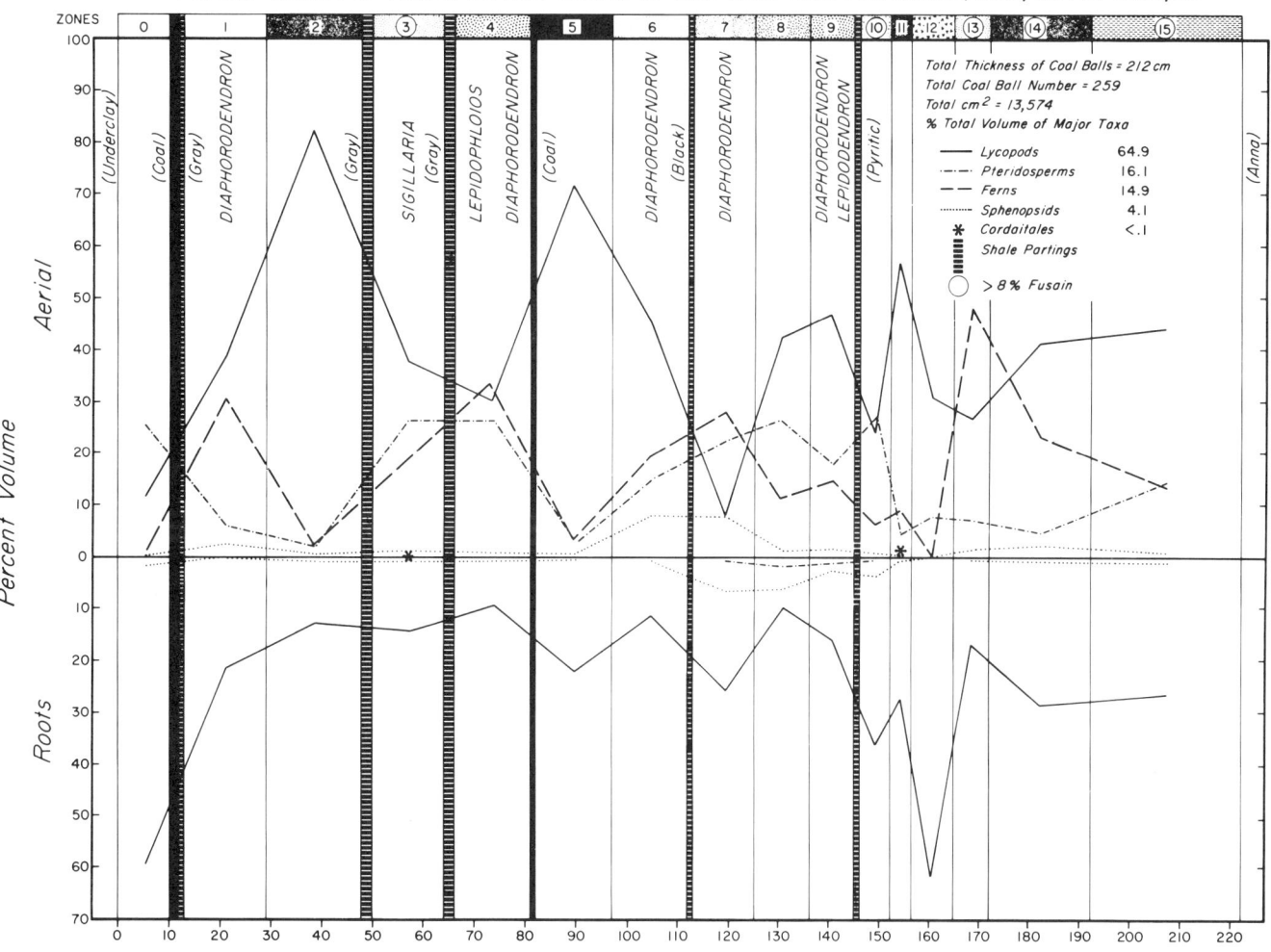

Figure 10. Percent biomass plots of major plant groups in a coal-ball profile of the Herrin Coal in southern Illinois. Aerial plant litter (stems, leaves, cones, and seeds) and all fern assemblages are plotted above the midline and roots (except ferns) below. Principal patterns of successional change are mainly related to the five mineral-rich or shale bands. *Lepidophloios* dominates the profile (Zones 1, 2, 5, 8, 11, 14, 15) with a profile total of 49 percent of the peat and declines in abundance from below to above each shale band. *Psaronius* tree ferns alternate in abundance with the *Lepidophloios* peaks and *Medullosa* seed ferns are most abundant in proximity to shale bands. See text for further explanation. Modified from Phillips and DiMichele (1981).

ference in the parent floras of the two areas is indicated by the plants in associated clastics of the Morrison coals, characterized by conifers, and of the Kootenay coals, in which cycadophytes are dominant. Whether this was due to climate or a difference in sedimentary environment has not been demonstrated. Conifer wood is predominant in the coals, but little coal petrographic work has been done except in the Alberta deposit. Most information on the plants constituting the coals is from palynological analyses, which have been extensive (Pocock, 1962; Fensome, 1983). Coniferous bisaccate pollen, *Classopollis* (Cheirolepidiaceae), and araucarian pollen indicate that these coals accumulated in coastal areas of a warm-arid paleolatitudinal belt.

The best known Early Cretaceous megaflora in North America occurs with coals of the Kootenai Formation of Montana (late Neocomian–Aptian). Microfossil floras from adjacent western Canada demonstrate great floral diversity (Singh, 1964). LaPasha and Miller (1984) provided an excellent evaluation of both megaflora and microfossils of associated rocks. They correlated this flora with the lower Blairmore of Alberta, Mannville of Alberta and Saskatchewan plains, Lakota of the Dakotas, Cloverly of Wyoming, and the lower portion of the Potomac of the Atlantic Coast. This flora includes two conifers, *Arthrotaxites* and *Elatides* (Taxodiaceae), which were not present in the underlying Morrison Formation. These and other conifers are dominant in the Kootenai Formation and lived in low-lying, coastal or delta-plain swamps and lakes of temperate climate. Only one cycadophyte and one ginkgophyte have been identified, but there are several ferns, pteridosperms, bryophytes, and sphenopsids in the flora. LaPasha and Miller (1985, p. 129) have postulated that, "the climate was seasonally wet with periodic flooding. Dry periods were not sufficient to cause subaerial exposure of the swamp sediments. The presence of tree ferns (Dicksoniaceae) and other ferns with delicate fronds, suggest moderate temperatures with rare if any freezing temperature."

Significantly lacking is any angiosperm, which seems strange considering the correlation with the Blairmore of Canada (Aptian to Albian age), which has a few angiosperms and numerous cycadophytes. The first angiosperm pollen (*Clavatipollenites*) appears in North America in the upper Barremian (Doyle and Robbins, 1977). In western Canada it is found first in middle Albian strata along with reticulate tricolpate pollen. Tricolpate pollen first appeared worldwide in the Albian (Singh, 1971, p. 25). Tricolporate pollen first appeared in latest Albian in western North America, but not until early Cenomanian in the eastern United States (Tschudy and others, 1984, p. 10). The first consistent occurrence is near the Early–Late Cretaceous boundary, generally represented by a relative of the tupelo gum tree, *Nyssapollenites*. Three genera, based on permineralized angiosperm wood of Albian age (Cedar Mountain and Burro Canyon Formations, Utah and Colorado [Tidwell and others, 1976]) correlate with the earliest angiosperm pollen from western United States.

Middle Cretaceous

In the "Dakota," a prolific and diverse angiosperm leaf flora of Albian to Cenomanian age is distributed widely through the central High Plains and parts of the Rocky Mountains, both east and west of the great interior seaway. Extensive minable coals and megafloras are present in many places on western shores of that epeiric seaway, but coal is very limited on the eastern margin. Current paleocontinental reconstructions of latitudinal positions place this interior area in the warm-temperate to subtropical belt. On the west side of the seaway, few megafloras of Cenomanian age have been studied. Many of the angiosperms (based on leaf morphology) and several of the ferns indicate very warm, humid climates. The Dakota flora is considered equivalent to the Woodbine of Texas and Oklahoma, which has been extensively studied palynologically (Hedlund, 1966). Hedlund's palynologic assemblage of 71 species in 45 genera closely approximates the palynoflora of the Dakota Group of Minnesota. It differs from the Cenomanian of the eastern Gulf Coast in lacking tricolporate and triporate pollen, probably due to climatic differences. Remarkable diversity is indicated by some mid–Gulf Coast Albian-Cenomanian palynofloras of 96 species in 62 genera described by Phillips and Felix (1971a, 1971b).

Great, largely undeveloped coal fields of northwestern Alaska in the Colville, Chandler, and Kuk River valleys, and in the Cape Lisburne area, are characterized by rich, gymnospermic floras of coastal-margin origin. These range in age from middle Albian to Cenomanian. The coals are thicker to the west, ranging from an aggregate thickness of about 6 m, in the Umiat Delta region of the Colville River valley, to over 60 m with an excellent megaflora in the Corwin area, near Cape Lisburne. A recent estimate (Sable and others, 1986) indicates that coals deposited in the river-dominated Umiat and Corwin deltas of the Nanushuk Group (Lower and Upper Cretaceous) may contain as much as one-third of all coal resources in the United States (2 to 3 trillion tons). Scott and Smiley (1979) recognized six zones from early or middle Albian to Santonian-Campanian. Marginally interdigitated marine beds in several areas have made possible rather precise dating. May and Shane (1985) summarized evidence from palynomorphs and leaf floras in the Umiat Delta area along the Colville River, by concluding that a diverse warm-temperate to subtropical flora of ferns, lycopods, and conifers dominated the nearshore and delta wetlands that produced the peat swamps. The cycadophytes and some conifers appear to have occupied drier sites.

Current studies of the Nanushuk Group along the Colville River (Spicer and Parrish, 1986) provide much additional information on the large-leaved angiosperms and conifers. They note that the large-leaved conifer, *Podozamites,* and *"Sequoia"* were ubiquitous elements in the younger coals of the Niakogon tonuges. At least one Cenomanian coal was composed entirely of *Podozamites.* They suggested that the swamp floras were mainly coniferous. From evidence of deciduousness of the floras, angiosperm leaf characteristics, and fossil woods, Spicer and Parrish (1986) interpreted the mid-Cretaceous climate of northern Alaska to have been cool.

Late Cretaceous–Paleogene

The principal coal floras of the Late Cretaceous in the United States are associated with structural basins on the margins of the interior seaway that gradually retreated north to the Arctic and south to the Gulf, after reaching maximum distribution during the late Albian-Cenomanian interval (Colorado Group). The principal zones of fossil plants are associated with coal seams that were deposited in various environments associated with the lower alluvial valleys of streams headed in the episodically rising, mesocordilleran highlands to the west, and in the prograding delta systems and behind long-shore coastal barriers. Major transgressions of the sea were followed by long-term, oscillating regressions during which most of the coal-bearing strata and fossil plants accumulated (see Flores and Cross, this volume). Numerous early floristic studies have been supplemented and modified in recent years by more than 150 papers on palynology relating directly to the resolution of floras of coals and associated rocks of the Cenomanian-Maastrichtian sequences in the Rocky Mountains and High Plains (Jameossanaie, 1987).

Only a few relatively recent papers have reported on the megafloras. In Parker's (1976) analysis of the environments of distribution of the leaf floras associated with the commercial coals of the Black Hawk Formation (Campanian) of Utah, three paleoenvironmentally controlled assemblages were proposed. The peat-forming swamps were characterized by conifer trees, including *Sequoia* and angiospermous trees (a buckthorn, *Rhamnites,* codominant with *Sequoia,* grape, and sycamore). A shrubby understory was limited, but small palms and cycadophytes were abundant. The herbaceous understory was dominated by two kinds of ferns. Water lily and Chinese water chestnut dominated open-water sites in the freshwater swamps. The surrounding bottomland communities are similarly subdivided into stratified zonation by Parker. Plants of the point-bar communities were dominated by *Araucaria.* Climatic inferences made on the basis of both representative species and leaf physiognomy indicate relatively humid, warm-temperate to subtropical conditions for the coal swamps. Seasonal growth is indicated by growth rings in wood and by layers of leaf mats intermittently interbedded with overbank muds. The coals include vitrain bands—mostly coniferalean—and occasional permineralized wood. Parker and Tidwell (1987) have also demonstrated minor amounts of vitrinized angiosperm tissue in some coals. Balsley (1980) and Parker prepared an unusual map of the stumps of gymnosperms and palms, along with fallen logs, interspersed by tracks of at least 5 kinds of saurian reptiles, including immature individuals (Fig. 11). The tracks and tree bases at the top of the coal probably represent communities present, following inundation of the swamp by a layer (10 to 30 cm) of overbank mud. Roots of the trees penetrated the upper meter of peat, and reptile tracks are compacted down as much as 30 cm into the upper peat layers. The vegetation represents the final peat-swamp flora and may differ due to changed edaphic conditions.

The phytogeographic distribution of floras of the Albian to Maastrichtian reflects the importance of the epeiric sea dividing North America and the relationship of some of the western floras with those in east Asia and Siberia. Extensive coal deposits of the Late Cretaceous and Paleogene strata have been mined or explored in much of the area from Manitoba and Alberta south to New Mexico and Arizona. These activities resulted in discovery of a number of early coal floras and led to later works that considerably expanded the knowledge of both the floras and the areas studied. However, major new studies are rare. One example of a relatively complete synthesis, particularly of the stratigraphy, sedimentation, and paleobotany, is the study of the Golden Valley Formation (Hickey, 1977) of transitional Paleocene-Eocene age near the top of the stratigraphic sequences containing the commercially important and widespread Williston Basin coal fields. This synthesis demonstrated the necessity of using modern sedimentological techniques to identify the type of surface on which the plants grew. It also stressed some of the problems of using modern analogs for interpreting fossil suites, particularly those older than Neogene.

The understanding of the extensive latest Cretaceous and Early Tertiary vegetational arrays of the Gulf Coast, both associated with and peripheral to the Gulf Coast lignites (see Breyer, this volume), is aided very little by the earlier studies of the megafloras. During the past few decades, the development of a broad spectrum of palynological information (Elsik, 1978, 1986; Frederiksen, 1980) and detailed analysis of the leaves on the basis of architecture and cuticular anatomy (Dilcher and Dolph, 1970), and of flowers and fruits (Crepet and Dilcher, 1977), have demonstrated that most earlier systematic assignments of the fossil plants are erroneous. Therefore, earlier paleoecological interpretations are unreliable.

The nonmarine lignites of the Gulf Coast are associated with leaf-bearing clays and siltstones, as well as with occasional interdigitated marine zones on the seaward margin of the coastal plain. Elsik (1978) characterized the following stratigraphic sequence of vegetational change: Calvert Bluff lignites (Paleocene–early Eocene) were derived from birch, myrica, and *Nyssa*-type trees and the source plant of *Thompsonipollis*; Yegua and Manning strata (middle to late Eocene) had abundant trees of *Nyssa* and *Engelhardia* (walnut family), and in some places *Amanoa* pollen (Euphorb family). All lignites lack abundant *Taxodium* (bald cypress) but have good fern floras. The Calvert Bluff also contains a *Sphagnum* moss component. The early marshes (Eocene) lack grasses, but there is a predominance of mangrove-type palms and some lilies and sedges; later lignites may have any of the above-named plants as major components along with some tree pollen and ferns. On the basis of an intensive study of late Eocene sporomorphs (166 types) by Frederiksen (1980), a winter-dry tropical climate was present in the southeastern

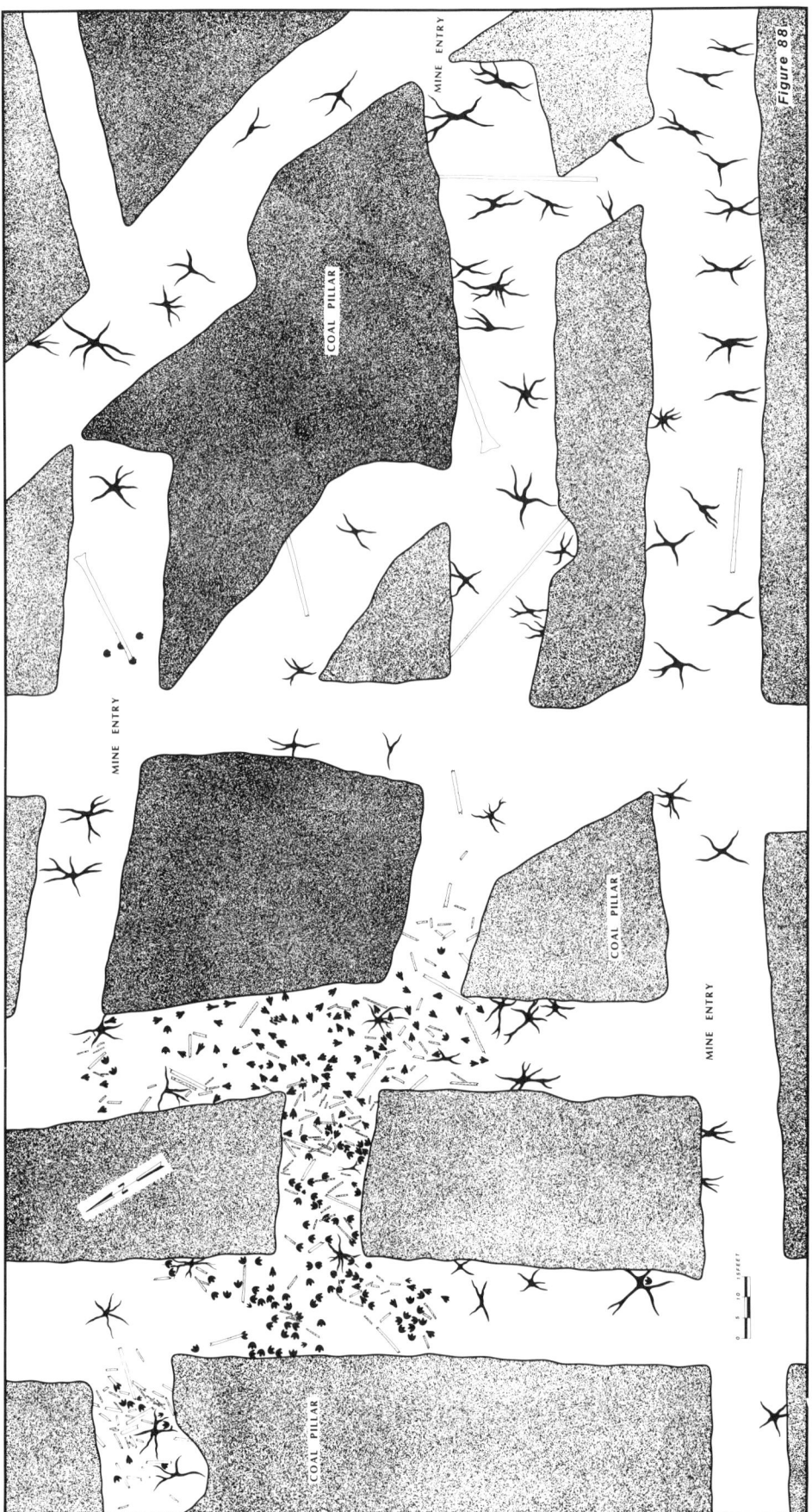

Figure 11. Map of coal roof in Kenilworth Mine, Utah (Black Hawk Formation) showing position, orientation, and size of tree bases and fallen logs interspersed with five types of dinosaur tracks. All trees and tracks are plotted at left; only principal trees shown at right. Parker (personal communication, 1987) calculated 400 trees per hectare with an unexplained northeast to southwest mean orientation of tree bases, mostly conifers. Many tracks face inward at tree bases, some in pairs, some on spreading roots, perhaps indicating feeding. Courtesy L. R. Parker; see Balsley (1980, p. 135).

United States, somewhat similar to the climate of the Florida Keys today. On the upper coastal plain, the climate was marginally humid and subtropical. Following the Eocene the climate rapidly became cooler.

The general nature of plant contributions to some thick Upper Cretaceous coals of the Powder River Basin has been interpreted by Satchell (1984; Fig. 12). She suggested that most of these swamps were dominated by *Glyptostrobus* rather than by *Taxodium.* Some peat accumulations were probably ombrogenous, which would account for great thicknesses of coal uninterrupted by clastic partings. Such raised swamps indicate high levels of rainfall year round rather than seasonally dry *Taxodium* swamps. Pocknall (1986) has suggested that some of the swamps were taxodiaceous, flood-basin swamps of an anastomosed fluvial system. The drier sites were dominated by *Nyssa* and hickory as in some of the Gulf Coast Paleogene swamps.

RESEARCH APPROACHES TO COAL AGES AND COMPARISONS OF PATTERNS

All tree genera that occurred in Pennsylvanian peat swamps are extinct. Ecological reconstructions are developed necessarily from an almost independent data base. Fortunately, probably more is known about the peat-swamp trees of the Pennsylvanian than any other group of extinct plants. Anatomically preserved plants in coal-balls and coal-spore floras are the bases for paleoecological analyses of the floras. Quantitative analysis from coalball deposits permits estimates of biomass constituents with resolution of preservational states. Succession is determined from coal-ball profiles that exhibit litter and rooting patterns. The combined use of palynology, coalified tissue (vitrinite and fusinite), and petrography ideally are more direct for most coal-seam studies, but there is a dearth of studies of coalified anatomy and plant sources of spores, pollen, and cuticles in every coal age.

Some genera that occurred within or near peat swamps in the Tertiary are living today in wetlands. Consequently, a biased research focus is firmly anchored to the living plants and their ecology. The Tertiary problem is the obverse of the Pennsylvanian; whereas much is known about living plants and their habitats, there is no vegetational reference base comparable to that of Pennsylvanian coal balls. Also, one of the most difficult tasks is selection of relevant modern peat-swamp environments for analyses and, in turn, model construction. While similarities of modern and fossil floras allow extrapolations of associational models for Neogene paleocommunities and some environments, these become increasingly tenuous when applied to the Early Tertiary and Late Cretaceous.

Dominance, diversity, and botanic constituents

General patterns of dominance and diversity are clearer in Pennsylvanian peat swamps, despite gaps, than in those of the Cretaceous and Tertiary. The Pennsylvanian data base is drawn widely from the eastern North American coal basins from one tropical paleoclimatic belt. Cretaceous and Tertiary peat swamps of western North America and the Gulf Coastal Region span greater paleoclimatic and paleogeographic regions. This complicates recognition of north-south patterns concurrent with angiosperm radiation. The poor resolution results in an outline that is apt to be misleading by its simplicity.

Pennsylvanian peat swamps were the most unusual of any coal age because the dominant trees were mostly lower vascular plants, generally lycopods through two-thirds of the period and then tree ferns after the Middle–Upper boundary extinctions. The unique aspect of diversity is the presence of all five major kinds of trees in the swamps. Apart from the generic diversity of lycopod trees, other groups were mainly represented by one genus. Probably ten tree genera account for 90 percent of the economic coal of the Pennsylvanian. Their patterns of tissue production and root-system architecture are divisible into four general types: lycopods, tree ferns, seed ferns, and the cordaites-calamites.

In general, the Cretaceous peat swamps were dominated by conifers, especially genera of the Taxodiaceae and some broadleafed trees of uncertain family affinities. In some uppermost Cretaceous coals of the Gulf Coastal Region and southern Western Interior Coal Region, angiosperms were at least locally important. The apparent time trend through the early Tertiary was a northward shift of conifer peat swamps and a southern development of angiosperm or mixed peat swamps. The angiosperm families of importance are only partially known and not well documented in some cases; compression records of the families only indicate association with depositional environments as opposed to their being indigenous to the peat swamps. In most cases the literature does not distinguish between clastic and peat-swamp habitats.

In general, the two most distinctive patterns of peat accumulation are represented by the lycopod-fern forests of the Pennsylvanian and those of woody seed plants of the Cretaceous and Tertiary. The patterns include differences in root systems, litter, and community turnover.

Root systems. Basic components of in situ peat accumulations are root systems, forming an intertwining matrix by shallow self burial. Fibrous and woody root systems can support large trees and compose one-third to most of the peat accumulated in some habitats (e.g., mangroves, Cohen and Spackman, 1977, 1980). The development of adventitious roots, prop roots, pneumatophores, knees, and varied spreading buttresses may create litter entrapments, provide additional habitats, impede water and comminuted-debris flow, or even constrain rapid burial of logs, depending on the water table.

Practically all swamp trees exhibit some kind of aerating or air-chambering system (aerenchyma, lacunae) in the roots. Some extend above the water and some within the peat matrix, allowing CO_2 and O_2 diffusion. Aerating structures of lower vascular plants usually had little wood, consisted mostly of air chambers, and were rather fragile. These features indicate short-term, herbaceouslike growth with low biomass allocation. In Pennsylvanian peats, lycopod stigmarian root systems and those of tree ferns (part was aerial support mantle) generated most of the root vol-

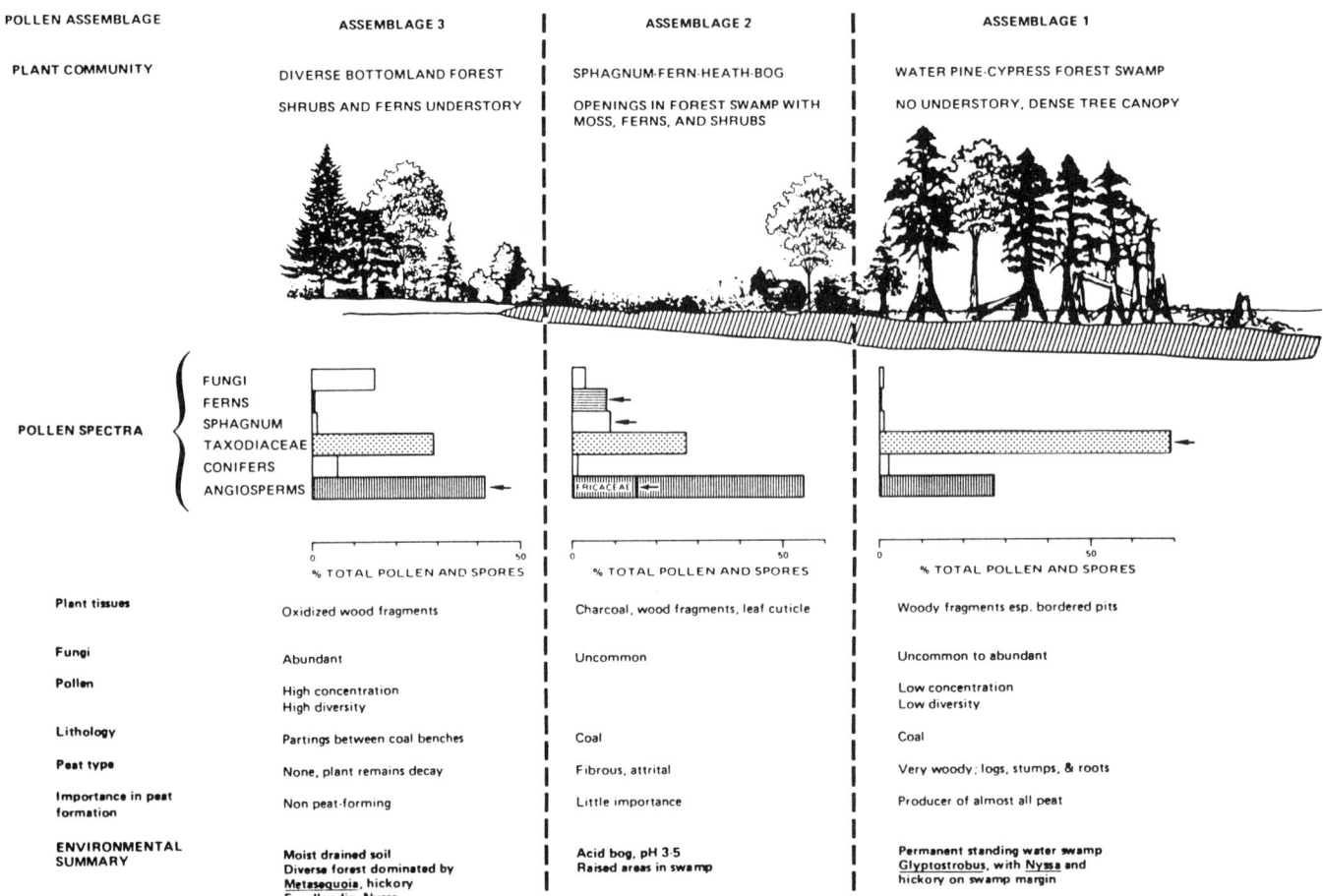

Figure 12. Reconstruction of vegetation, community structure and environments of some Paleocene coals of the Powder River Basin based on palynology and paleobotany. Courtesy L. S. Satchell and Exxon Production Research (see Satchell, 1984).

ume and often are the least compacted component in coal balls due to later growth entry into accumulated litter. Estimated volume of root systems from these trees forms a smaller proportion of the actual peat-to-coal than is apparent from coal-ball measurements. Such root peat was important in biomass accumulation in the Pennsylvanian but probably less so than that of woody trees with central root systems in the Cretaceous and Tertiary.

The aerating systems of roots of woody seed plants were usually secondary in development and continued to expand with prolonged growth. The biomass provided by woody root systems was a significant component of Cretaceous and Tertiary peats. Based on living conifers and woody angiosperm trees, whole-plant shoot/root ratios are generally about 4/1, indicative of root systems forming about 20 percent of mature tree biomass (Andersson, 1970; Grier and Logan, 1977; Santantonio and others, 1977).

Litter patterns. The woody habit of relatively large, long-lived conifers and angiosperm swamp trees of the Cretaceous-Tertiary, versus the smaller, short-lived trees of the Pennsylvanian, results in general differences in litter patterns. The botanic constituents produced by the dominant trees differed, as did the sizes of the typical litter debris. Furthermore, the rates of community turnover were faster in the Pennsylvanian where dominant trees were less woody and more cheaply constructed.

The largest lycopod trees in coal-ball deposits were probably 15 to 25 m tall and 30 to 40 cm in breast-high-diameter (poles). However, *Lepidodendron* trunks in top surfaces of coal are commonly 45 to 60 cm in diameter, with some >1 m in diameter. Some *Sigillaria* trunks, more than 1 m in diameter, also occur in coal mine roofs. In general, those in clastic swamps and at the top of the peat were larger than those typically in the peat. Holocene tropical angiosperms of Sarawak domed-peat swamps range up to 8 m in diameter with heights 45 to 60 m in the outer rings, to "stunted" (4 to 20 m tall and less than 90 cm in diameter) in the inner low-nutrient center of the catena. *Taxodium* in the Okefenokee and Dismal Swamps occasionally attains an age of more than 500 years and breast-high-diameter of a meter or more (Duever and Riopelle, 1984).

Large woody trees exhibit long-term "rain" of leaves (seasonally massive for deciduous habits), woody branches, bark, seeds, and fruits, terminally punctuated by tree fall. The bulk of biomass generated in such forests is wood, and trees can remain standing for hundreds of years, depending on events. Such woody tree litter presents a broad range of potential for peat preservation; tree falls could be largely buried or left exposed to almost total decay. The massiveness of such wood accumulations enhances potential peat thicknesses enormously, according to decay resistance of the wood and levels of tannins and lignins, which also inhibited degradation.

The main trees of the Pennsylvanian peat swamps were relatively short-lived (tens of years) and with a few exceptions were nonwoody (tree ferns) or contained little wood (lycopods). Different architectural forms of support, mostly soft or fibrous tissues in ferns and seed ferns, or "living bark" in the lycopods, suggest that these trees had cheaper biomass construction, as well as more rapid growth, than the truly woody cordaites of the same swamps. The combination of small tree size, determinate growth in some, weak tree structure, and short life spans is consistent with rapid community turnover compared to communities of woody trees. Litter from the five tree types was diverse: large foliage (ferns and seed ferns), bark sheets and logs (lycopods), deciduous branches (lycopods), root mantle (ferns), and wood (cordaites and calamites). According to dominance patterns, one or more of these components, along with buried roots, were characteristic of particular stratigraphic intervals.

The surviving litter in coal balls suggests dismemberment of plants into numerous chunky pieces, consistent with "break away" cheap construction; lycopod bark was the most abundant decay-resistant tissue. Those trees that had significant woody composition, allowing potentially long-term structural support, were either typically small and minor elements (calamites), stunted or often shrubby (cordaites), or unbranched, vine-like, and reinforced by cortical fibers (seed ferns). All three groups usually contributed disproportionately more fusain for their relative peat-biomass contribution to the coal.

Community turnover—Implications

There were several types of fairly distinctive Pennsylvanian peat-swamp communities. Disruptions in edaphic factors resulted in short-term fluctuations in characteristic forests and litter composition, and, in turn, a layering of the peat and banding in the coal seam. This suggests both rapid burial, considering the quality of preservation in coal balls, and close environmental links among disruptive events, litter patterns, and vegetational responses. Consequently, edaphically controlled environments of growth are probably much more closely keyed to depositional environments (and peat preservation) than are those occupied by long-lived trees.

Some trees living hundreds of years, and some swamps possibly existing an order of magnitude longer in the Cretaceous and Tertiary than swamps in the Pennsylvanian, suggest a much slower turnover and likely less evidence of vegetational response to environmental fluctuations below severe levels of damage or death and replacement. Environmental fluctuations would affect the preservation of upper peat layers and litter fall without significantly changing the vegetation. Rates of peat accumulation vary widely. Estimated rates for a 30-m Paleocene coal seam in the Powder River Basin in Montana and Wyoming were extrapolated as 88,663 years (Ayers and Kaiser, 1984). They noted this was a long time span for a fluvial system to maintain stability. Indeed, an old-growth bald cypress stand in the Okefenokee, an order of magnitude less in age than that estimated for the Powder River example, is presently giving way to a succession of *Gordonia* (loblolly bay) and secondarily to *Magnolia* and *Nyssa* (tupelo gum). This indicates, in particular, that *Taxodium* could not be perpetuated in such an environment without major favorable changes in the edaphic or climatic controls (Best, 1984) and that long-lived tree stands may survive considerable environmental fluctuations without reflecting them.

It has been suggested that aerobic exposure of Cretaceous and Tertiary peat may have been much more frequent, persistent, or to greater depths (greater fluctuation of water tables) than that in the Pennsylvanian (Schopf, 1952). Saprophytic fungi are mostly aerobic, and there is a remarkably greater abundance of hyphae, sclerotia, and spores in younger coals and a lack of bona fide sclerotia in older coals (Schopf, 1952). This may be due to evolutionary changes in fungi, chemistry of wood substrates, environments of deposition, and/or litter patterns.

The decay resistance of *Taxodium* wood is well-known commercially, yet even *Taxodium* wood may not survive cellular preservation in some swamp-marsh environments as well as nonwoody plants such as *Nymphaea* (water lilies) (Corvinus and Cohen, 1984). Holocene succession of *Nymphaea* and *Taxodium* is indicative of hydrologic extremes; the taxodiaceous swamps of the present may be more indicative of refugial habitats than good models for economic coals. In the Lower Sunnyside coal of the Upper Cretaceous of Utah, most of the conifer wood was highly degraded, and wood, bark, and leaves contributed about equal amounts to the coal mass (Thiessen and Sprunk, 1937). Although wood was the most important root and litter component in the Cretaceous and Tertiary, anatomical preservability of some angiosperm woods seems to have been much less than that of conifers (Thiessen and Sprunk, 1937, p. 11; Teichmüller and Teichmüller, 1982, p. 21). Litter patterns and decay of woody trees are more complex than generally assumed in evaluating the relationship between peat and coal (Conner and Day, 1976; Day, 1979).

The preservational fate of wood in limited cordaitean and calamitean-rich peats of the Pennsylvanian suggests that much of the trunk wood was degraded by fungi and detritivores. Fungi are also well represented in Pennsylvanian coal balls (Baxter, 1975), and fecal evidence of detritivores (Taylor and Scott, 1983) is widespread in swamps, being prominent in mid–Middle Pennsylvanian peats, which have the most wood. Much litter in Pennsylvanian peat swamps was preserved anatomically due to relatively rapid burial under low pH and anaerobic conditions (Schopf,

1952). This is consistent with structural evidence of tissue in coals (Winston, 1986) and in coal-ball peats (Phillips and others, 1985). It was earlier assumed that coal-ball peats exhibited better preservation only because they were permineralized at an "early stage". Most was entombed prior to much compaction, but the state of degradation is apparently the same as in the adjacent coal (Winston, 1986).

Peat marshes and aquatic environments

Peat-forming marshes characterized by emergent-herb vegetation, and aquatic habitats with submerged or floating plants, vary in their paleoecological implications and significance in different coal ages. Some lycopod marshes were probably indicative of ephemeral, shallow-water habitats with basally rooted plants, as represented by *Chaloneria,* in thin and bony Upper Pennsylvanian coals (DiMichele and others, 1979). Uncertainties exist about water tables and even salinity associated with the so-called *Densosporites* (lycopod) peat marshes (Smith, 1962), which occurred within peat-swamp sequences and dominated some coal-spore floras. Within temperate and tropical climates, marsh and aquatic vegetation usually indicate conditions unsuitable for peat-forming forests because of excess water depth or amplitudes of fluctuations. In the Pennsylvanian, no submergent or floating vascular plants are known for such habitats. The aquatic environments were the domains of algae until the advent of floating and submerged angiosperms as well as some aquatic ferns in the Cretaceous.

Open-marsh angiosperms such as the grasses, sedges, reeds, and others of the Tertiary opened new vegetational possibilities for peat formation in deeper water and marginal wetlands. Many grasses and sedges have low shoot/root ratios of biomass and contribute considerable root material to the peat. Although peat marshes are pervasive in so many wetlands, it is not clear that they had as much importance in the origins of economic coals as their present diversity and areal extent suggest in temperate regions. A key problem in interpreting the paleoecology and evaluating the importance of peat marshes in coal origins is the difficulty of accurately identifying their facies from fragmented, dispersed, and variably altered plant detritus *within* the coal beds.

Cannel and boghead coal-forming environments (Ashley, 1918; Thiessen, 1925; Parks and O'Donnell, 1956) have paleoecological significance, although they were widely scattered and account for only small amounts of coal resources. Their importance resides as indicators of a range of depositional environments distinct from the ecology of the plants that contributed to organic accumulations; they provide inferences of lacustrine or hydrologic regimes not conducive to in situ vascular plant-peat preservation. Such sapropelic coal deposits or facies become depositional components (open-water areas within domed peats) in models of humic coal formation (Esterle and Ferm, 1986), as well as habitats and/or death traps for vertebrates (Hook and Ferm, 1985).

Cannel coals are derived mostly from vascular plants that have been subjected to severe decay and usually to variable amounts of transport. Cuticles, spores, and pollen are particularly abundant, and algae are often minor components. Cannel coals grade into boghead coals and, usually being of high ash content, into carbonaceous shales. Some high-ash facies of humic coals contain concentrated cuticular layers that weather near the surface to form "paper coal", as in Indiana Lower and Upper Block coals of the Middle Pennsylvanian (Guennel and Neavel, 1959).

Boghead coals were derived mostly from true algae, usually fresh-water. From the Mississippian onward, *Botryococcus* is perhaps the most prevalent and interesting. It is a colonial green alga with cell sheaths rich in lipids and chitin-like polymers (Niklas, 1976), and exhibits a polymorphism of cell and colonial size and shape. Fossil colonies of *Botryococcus* are morphologically within the observed variation of living colonies, which vary according to light intensity, ammonium-ion concentration, and agitation (Niklas, 1974). Living *Botryococcus* inhabits both fresh-water and some brackish lakes; fossil specimens are found in a wide range of deposits (Traverse, 1955b).

RESEARCH PERSPECTIVES

It is important to assess what peat-swamp (and marsh) ecosystems have in common and what similarities and problems characterize different coal ages. Although such comparisons are seldom made, they do emphasize important concepts of biological and ecological origins of coal that are beginning to be recognized. The kinds of dominant trees in North American coal ages differ significantly, but there is a general paucity in diversity of trees adapted to or tolerant of peat-swamp habitats; they typically had broad geographic distribution with long stratigraphic ranges. Most economically important coal deposits were formed from autochthonous peats of fresh-water forest communities; some tree types, such as the swamp-centered lycopods, are probably good indicators of changing environments of peat accumulation, while some very long-lived tree types may provide less resolution. Peat marshes apparently differed in their relative importance in different coal ages and in their environmental implications. Distinctions need to be made in each coal age among the peat-forming communities associated with mineral-rich deposits and those from which economic coals are derived. In turn, mineral-rich bands and their origins in coal deposits have considerable ecological significance, representing disruptive conditions affecting the communities as well as tests of depositional models of peat accumulation.

Ecological differences among North American coal ages are interwoven historically with the biologies of the tree types as well as with depositional environments, paleogeography, and paleoclimates. Comparisons of the Pennsylvanian peat swamps with those of the Cretaceous and Tertiary yield only a first approximation of evident or likely differences. Our information and inferences about plant communities and environmental controls differ greatly both within and among coal ages. Of the many combined approaches to understanding coal origins and derived properties, peat-swamp paleoecology has lagged considerably behind. During the past decade, interest in plant paleoecology, particularly in ancient peat swamps, has increased. There is no dearth of problems or opportunities for research on coals of any age. If anything, it may be the abundance of research materials (e.g., tons of coal balls, vast thicknesses, and areal extent of coal seams) that presents a logistical challenge to researchers—a challenge to ask the right questions, choose the best available sites, and address the enormous gaps and misconceptions in our present views of peat-swamp paleoecology.

REFERENCES CITED

Andersson, F., 1970, Ecological studies in a Scanian woodland and meadow area, southern Sweden; II, Plant biomass, primary production, and turnover of organic matter: Botaniska Notiser, v. 123, p. 9–51.

Ashley, G. H., 1918, Cannel coal in the United States: U.S. Geological Survey Bulletin 659, 127 p.

Ayers, W. B., and Kaiser, W. R., 1984, Lacustrine-interdeltaic coal in the Fort Union Formation (Palaeocene), Powder River Basin, Wyoming and Montana, U.S.A., in Rahmani, R. A., and Flores, R. M., eds., Sedimentology of coal and coal-bearing sequences: International Association of Sedimentologists Special Publication 7, p. 61–84.

Balsley, J. K., 1980, Cretaceous wave-dominated delta systems; Book Cliffs, east-central Utah, a field guide: Denver, Amoco Production Company, 163 p.

Barghoorn, E. S., and Spackman, W., 1950, Geological and botanical study of the Brandon lignite and its significance in coal petrology: Economic Geology, v. 45, p. 344–357.

Baxter, R. W., 1975, Fossil fungi from American Pennsylvanian coal balls: Lawrence, University of Kansas Paleontological Contributions, v. 77, p. 1–6.

Best, G. R., 1984, An old growth cypress stand in Okefenokee Swamp, in Cohen, A. D., Casagrande, D. J., Andrejko, M. J., and Best, G. R., eds., The Okefenokee Swamp; Its natural history, geology, and geochemistry: Los Alamos, New Mexico, Wetland Surveys, p. 132–143.

Burger, K., 1985, Petrography and chemistry of tonsteins of the coal basins of Western Europe and North America, in Cross, A. T., ed., Economic geology; Coal, oil, and gas, Compte Rendu, Neuvieme Congres International de Stratigraphie et de Geologie du Carbonifere, Washington and Champaign-Urbana, May, 1979, Volume 4: Carbondale, Southern Illinois University Press, p. 449–466.

Burger, K., and Damberger, H. H., 1985, Tonsteins in the coalfields of Western Europe and North America, in Cross, A. T., ed., Economic geology; Coal, oil and gas, Compte Rendu, Neuvieme Congres International de Stratigraphie et de Geologie du Carbonifere, Washington and Champaign-Urbana, May, 1979, Volume 4: Carbondale, Southern Illinois University Press, p. 433–443.

Cecil, C. B., Stanton, R. W., Neuzil, S. G., Dulong, F. T., Ruppert, L. F., and Pierce, B. S., 1985, Paleoclimatic controls on late Paleozoic sedimentation and peat formation in the central Appalachian Basin (U.S.A.): International Journal of Coal Geology, v. 5, p. 195–203.

Cohen, A. D., and Spackman, W., 1977, Phytogenic organic sediments and sedimentary environments in the everglades-mangrove complex; Part II, The origin, description, and classification of the peats of southern Florida: Palaeontographica, Abt. B, v. 162, p. 71–114.

——— , 1980, Phytogenic organic sediments and sedimentary environments in the everglades-mangrove complex of Florida; Part III, The alteration of plant material in peats and the origin of coal macerals: Palaeontographica, Abt. B, v. 172, p. 125–149.

Conner, W. H., and Day, J. W., Jr., 1976, Productivity and composition of a baldcypress–water tupelo site and a bottomland hardwood site in a Louisiana swamp: American Journal of Botany, v. 63, p. 1354–1364.

Corvinus, D. A., and Cohen, A. D., 1984, Micropetrographic characteristics of peats from the Okefenokee Swamp; The origin of coal macerals, in Cohen, A. D., Casagrande, D. J., Andrejko, M. J., and Best, G. R., eds., The Okefenokee Swamp; Its natural history, geology, and geochemistry: Los Alamos, New Mexico, Wetland Surveys, p. 651–667.

Crepet, W. L., and Dilcher, D. L., 1977, Investigations of angiosperms from the Eocene of North America; A mimosoid inflorescence: American Journal of Botany, v. 64, p. 714–725.

Day, F. P., Jr., 1979, Litter accumulation in four plant communities in the Dismal Swamp, Virginia: American Midland Naturalist, v. 102, p. 281–289.

Demaris, P. J., Bauer, R. A., Cahill, R. A., and Damberger, H. H., 1983, Geologic investigation of roof and floor strata; Longwall demonstration, Old Ben Mine No. 24, Prediction of coal balls in the Herrin Coal: Illinois State Geological Survey Contract/Grant Report 1983-2, 69 p.

Dilcher, D. L., and Dolph, G. E., 1970, Fossil leaves of *Dendropanax* from Eocene sediments of southeastern North America: American Journal of Botany, v. 57, p. 153–160.

DiMichele, W. A., and DeMaris, P. J., 1987, Structure and dynamics of a Pennsylvanian-age *Lepidodendron* forest; Colonizers of a disturbed swamp habit in the Herrin (No. 6) Coal of Illinois: Palaios, v. 2, p. 146–157.

DiMichele, W. A., and Phillips, T. L., 1988, Paleoecology of the Middle Pennsylvanian Herrin coal swamp (Illinois) near a contemporaneous river system, the Walshville paleochannel: Review of Palaeobotany and Palynology, v. 52 (in press).

DiMichele, W. A., Mahaffy, J. F., and Phillips, T. L., 1979, Lycopods of Pennsylvanian age coals; *Polysporia*: Canadian Journal of Botany, v. 57, p. 1740–1753.

DiMichele, W. A., Rischbieter, M. O., Eggert, D. L., and Gastaldo, R. A., 1984, Stem and leaf cuticle of *Karinopteris;* Source of cuticles from the Indiana "paper" coal: American Journal of Botany, v. 71, p. 626–637.

DiMichele, W. A., Phillips, T. L., and Willard, D. A., 1986, Morphology and paleoecology of Pennsylvanian-age coal-swamp plants, in Broadhead, T. W., ed., Land plants; Notes for a short course: Knoxville, University of Tennessee Studies in Geology 15, p. 97–114.

Donaldson, A. C., Renton, J. J., and Presley, M. W., 1985, Pennsylvanian deposystems and paleoclimates of the Appalachians: International Journal of Coal Geology, v. 5, p. 167–193.

Doyle, J. A., and Robbins, E. I., 1977, Angiosperm pollen zonation of the continental Cretaceous of the Atlantic Coastal Plain and its application to deep wells in the Salisbury Embayment: Palynology, v. 1, p. 43–78.

Duever, M. J., and Riopelle, L. A., 1984, Successional patterns and rates on Okefenokee tree islands, in Cohen, A. D., Casagrande, D. J., Andrejko, M. J., and Best, G. R., eds., The Okefenokee Swamp; Its natural history, geology, and geochemistry: Los Alamos, New Mexico, Wetland Surveys, p. 112–131.

Elsik, W. C., 1978, Palynology of Gulf Coast lignites; The stratigraphic framework and depositional environment, in Kaiser, W. R., ed., Proceedings of the Gulf Coast Lignite Conference; Geology, Utilization, and Environmental Aspects: Texas University Bureau of Economic Geology Report of Investigations, v. 90, p. 21–32.

——— , 1986, Palynologic studies of Gulf Coast lignites, in Finkelman, R. B., ed., Geology of Gulf Coast lignites: Houston, Environment and Coal Association, p. 146–155.

Esterle, J., and Ferm, J. C., 1986, Relationship between petrographic and chemical characteristics and a geometry in the Hance Seam, Breathitt Formation, southeastern Kentucky: International Journal of Coal Geology, v. 6, p. 199–214.

Fensome, R. A., 1983, Miospores from the Jurassic–Cretaceous boundary beds, Aklavik Range, Northwest Territories, Canada [Ph.D. thesis]: Saskatoon, University of Saskatchewan, 761 p.

Frederiksen, N. O., 1980, Paleogene sporomorphs from South Carolina and quantitative correlations with the Gulf Coast: Palynology, v. 4, p. 125–179.

Grier, C. C., and Logan, R. S., 1977, Old-growth *Pseudotsuga menziesii* communities of a western Oregon watershed; Biomass distribution and production budgets: Ecological Monographs, v. 47, p. 373–400.

Guennel, G. K., and Neavel, R. C., 1959, Paper coal in Indiana: Science, v. 129, p. 1671–1672.

Hedlund, R. W., 1966, Palynology of the Red Branch Member of the Woodbine Formation (Cenomanian), Bryan County, Oklahoma: Oklahoma Geological Survey Bulletin 112, 69 p.

Hickey, L. J., 1977, Stratigraphy and paleobotany of the Golden Valley Formation (early Tertiary) of western North Dakota: Geological Society of America Memoir 150, 183 p.

Hook, R. W., and Ferm, J. C., 1985, Depositional model for the Linton tetrapod assemblage (Westphalian D, upper Carboniferous) and its palaeoenvironmental significance: Philosophical Transactions of the Royal Society of London, v. B311, p. 101–109.

Jameossanaie, A., 1987, Palynology and age of South Hospah coal-bearing deposits, McKinley County, New Mexico: New Mexico Bureau of Mines and Mineral Resources Bulletin 112, 65 p.

Jeglum, J. K., 1971, Plant indicators of pH and water level in peatlands at Candle Lake, Saskatchewan: Canadian Journal of Botany, v. 49, p. 1661–1676.

Knoll, A. H., 1985, Exceptional preservation of photosynthetic organisms in silicified carbonates and silicified peats: Philosophical Transactions of the Royal Society of London, v. B311, p. 111–122.

LaPasha, C. A., and Miller, C. N., Jr., 1984, Flora of the Early Cretaceous Kootenai Formation in Montana; Paleoecology: Palaeontographica, Abt. B, v. 194, p. 109–130.

—— , 1985, Flora of the Early Cretaceous Kootenai Formation in Montana; Bryophytes and Tracheophytes excluding conifers: Palaeontographica, Abt. B, v. 196, p. 111–145.

Lyons, P. C., Hatcher, P. G., Brown, F. W., Thompson, C. L., and Millay, M. A., 1985, Coalification of organic matter in a coal ball from the Calhoun coal bed (Upper Pennsyl;vanian), Illinois Basin, United States of America: Compte Rendu, Dixieme Congres International de Stratigraphie et de Geologie du Carbonifere, Madrid, September, 1983, v. 1, p. 155–159.

Mamay, S. H., and Yochelson, E. L., 1962, Occurrence and significance of marine animal remains in American coal balls: U.S. Geological Survey Professional Paper 354-I, p. 193–224.

May, F. E., and Shane, J. D., 1985, An analysis of the Umiat Delta using palynology and other data, North Slope, Alaska, *in* Huffman, A. C., Jr., ed., Geology of the Nanushuk Group and related rocks, North Slope, Alaska: U.S. Geological Survey Bulletin 1614, p. 97–120.

Niklas, K. J., 1974, Some problematic algae of the Paleozoic [Ph.D. thesis]: Urbana, University of Illinois, 511 p.

—— , 1976, Chemical examinations of some non-vascular Paleozoic plants: Brittonia, v. 28, p. 113–137.

Otte, L. J., Saunders, C. L., Mallison, D. J., and Purser, M. T., 1987, Controls over peat deposition on the North Carolina Coastal Plain, *in* Whittecar, G. R., ed., Geological excursions in Virginia and North Carolina: Norfolk, Virginia, Old Dominion University, p. 145–177.

Parker, L. R., 1976, The paleoecology of the fluvial coal-forming swamps and associated floodplain environments in the Blackhawk Formation (Upper Cretaceous) of central Utah, *in* Cross, A. T., and Maxfield, E. B., eds., Aspects of coal geology, northwest Colorado Plateau; Some geologic aspects of coal accumulation, alteration, and mining in western North America; A symposium, Coal Geology Division, Geological Society of America, Salt Lake City, 1975: Provo, Utah, Brigham Young University Geology Studies, v. 22, part 3, p. 99–116.

Parker, L. R., and Tidwell, W. D., 1987, Angiosperm and conifer xylem features preserved in vitrain: Botanical Society of America Abstracts, v. 74, p. 694.

Parks, B. C., and O'Donnell, H. J., 1956, Petrography of American coals: U.S. Bureau of Mines Bulletin 550, 193 p.

Peppers, R. A., 1985, Comparison of miospore assemblages in the Pennsylvanian System of the Illinois Basin with those in the upper Carboniferous of western Europe, *in* Sutherland, P. K., and Manger, W. L., eds., Biostratigraphy, Compte Rendu, Ninth International Congress of Carboniferous Stratigraphy and Geology, Washington and Champaign-Urbana, May, 1979, Volume 2: Carbondale, Southern Illinois University Press, p. 483–502.

Phillips, P. P., and Felix, C. J., 1971a, A study of Lower and Middle Cretaceous spores and pollen from the southeastern United States; I, Spores: Pollen et Spores, v. 13, p. 279–348.

—— , 1971b, A study of Lower and Middle Cretaceous spores and pollen from the southeastern United States; II, Pollen: Pollen et Spores, v. 13, p. 447–473.

Phillips, T. L., and DiMichele, W. A., 1981, Paleoecology of Middle Pennsylvanian age coal swamps in southern Illinois; Herrin Coal Member at Sahara Mine No. 6, *in* Niklas, K. J., ed., Paleobotany, paleoecology, and evolution: New York, Praeger Publishers, p. 231–284.

Phillips, T. L., and Peppers, R. A., 1984, Changing patterns of Pennsylvanian coal-swamp vegetation and implications of climatic control on coal occurrence: International Journal of Coal Geology, v. 3, p. 205–255.

Phillips, T. L., Kunz, A. B., and Mickish, D. J., 1977, Paleobotany of permineralized peat (coal balls) from the Herrin (No. 6) Coal Member of the Illinois Basin, *in* Given, P. N., and Cohen, A. D., eds., Interdisciplinary studies of peat and coal origins: Geological Society of America Microform Publication 7, p. 18–49.

Phillips, T. L., Peppers, R. A., and DiMichele, W. A., 1985, Stratigraphic and interregional changes in Pennsylvanian coal-swamp vegetation; Environmental inferences: International Journal of Coal Geology, v. 5, p. 43–109.

Pocknall, D. T., 1986, Composition of plant communities in and near late Paleocene coal swamps, Fort Union Formation, Powder River Basin, Wyoming–Montana [abs.]: Palynology, v. 10, p. 256–257.

Pocock, S.A.J., 1962, Microfloral analysis and age determination of strata at the Jurassic–Cretaceous boundary in the western Canada plains: Palaeontographica, Abt. B, v. 111, p. 1–95.

Sable, E. G., Stricker, G. D., and Affolter, R. H., 1986, Nanushuk Group coal investigations; North Slope of Alaska, *in* Carter, L.M.H., ed., USGS research on energy resources; 1986 program and abstracts: U.S. Geological Survey Circular 974, p. 59–60.

Santantonio, D., Hermann, R. K., and Overton, W. S., 1977, Root biomass studies in forest ecosystems: Pedobiologia, v. 17, p. 1–31.

Satchell, L. S., 1984, Reconstruction of vegetation and environments from palynology of Paleocene coals from Wyoming, U.S.A. [abs.]: Sixth International Palynological Conference, Volume of Abstracts, Calgary, p. 146.

Scheckler, S. E., 1986, Geology, floristics, and paleoecology of Late Devonian coal swamps from Appalachian Laurentia (U.S.A.): Annales de la Societe geologique de Belgique, v. 109, p. 209–222.

Schopf, J. M., 1952, Was decay important in origin of coal?: Journal of Sedimentary Petrology, v. 22, p. 61–69.

—— , 1956, A definition of coal: Economic Geology, v. 51, p. 521–527.

Scott, R. A., and Smiley, C. J., 1979, Some Cretaceous plant megafossils and microfossils from the Nanushuk Group, northern Alaska; A preliminary report, *in* Ahlbrandt, T. S., ed., Preliminary geologic, petrologic, and paleontologic results of the study of Nanushuk Group rocks, North Slope, Alaska: U.S. Geological Circular 794, p. 89–111.

Singh, C., 1964, Microflora of the Lower Cretaceous Mannville Group, east-central Alberta: Research Council of Alberta Bulletin 15, 238 p.

—— , 1971, Lower Cretaceous microfloras of the Peace River area, northwestern Alberta: Research Council of Alberta Bulletin 28, v. 1, 299 p.

Smith, A.H.V., 1962, The palaeoecology of Carboniferous peats based on the miospores and petrography of bituminous coals: Proceedings of the Yorkshire Geological Society, v. 33, p. 423–474.

Spicer, R. A., and Parrish, J. T., 1986, Paleobotanical evidence for cool north polar climates in middle Cretaceous (Albian–Cenomanian) time: Geology, v. 14, p. 703–706.

Sprunk, G. C., Ode, W. H., Selvig, W. A., and O'Donnell, H. J., 1940, Splint coals of the Appalachian region; Their occurrence, petrography, and comparison of chemical and physical properties with associated bright coals: U.S. Bureau of Mines Technical Paper 615, p. 1–59.

Taylor, T. N., and Scott, A. C., 1983, Interactions of plants and animals during the Carboniferous: Bioscience, v. 33, p. 488–493.

Teichmüller, M., and Teichmüller, R., 1982, The geological basis of coal formation, in Stach, E., Mackowsky, M.-Th., Teichmüller, M., Taylor, G. H., Chandra, D., and Teichmüller, R., eds., Stach's textbook of coal petrology: Berlin, Gebruder Borntraeger, p. 5–86.

Thiessen, R., 1920, Structure in Paleozoic bituminous coals: U.S. Bureau of Mines Bulletin 117, 296 p.

—— , 1925, Origin of the boghead coals: U.S. Geological Survey Professional Paper 132I, p. 121–138.

Thiessen, R., and Sprunk, G. C., 1937, Origin and petrographic composition of the lower Sunnyside coal of Utah: U.S. Bureau of Mines Technical Paper 573, p. 1–34.

Tidwell, W. D., Thayn, G. F., and Roth, J. L., 1976, Cretaceous and Early Tertiary floras of the Intermountain area; A summary, in Cross, A. T., and Maxfield, E. B., eds., Aspects of coal geology, northwest Colorado Plateau; Some geological aspects of coal accumulation, alteration, and mining in western North America; A symposium, Coal Geology Division, Geological Society of America, Salt Lake City, 1975: Provo, Utah, Brigham Young University Geology Studies, v. 22, part 3, p. 77–98.

Ting, F.T.C., 1972, Petrified peat from a Paleocene lignite in North Dakota: Science, v. 177, p. 165–166.

Traverse, A., 1955a, Pollen analysis of the Brandon lignite of Vermont: U.S. Bureau of Mines Report of Investigations 5151, 107 p.

—— , 1955b, Occurrence of the oil-forming alga *Botryococcus* in lignites and other Tertiary sediments: Micropaleontology, v. 1, p. 343–350.

Tschudy, R. H., Tschudy, B. D., and Craig, L. C., 1984, Palynological evaluation of Cedar Mountain and Burro Canyon formations, Colorado Plateau: U.S. Geological Survey Professional Paper 1281, 21 p.

U.S. Department of Agriculture, 1981, Soil series of the United States, Puerto Rico, and the Virgin Islands; Their taxonomic classification: U.S. Department of Agriculture, p. 1–13.

White, D., and Thiessen, R., 1913, The origin of coal: U.S. Bureau of Mines Bulletin 38, 390 p.

Winslow, M. R., 1959, Upper Mississippian and Pennsylvanian megaspores and other plant microfossils from Illinois: Illinois State Geological Survey Bulletin 86, 135 p.

Winston, R. B., 1986, Characteristic features and compaction of plant tissues traced from permineralized peat to coal in Pennsylvanian coals (Desmoinesian) from the Illinois Basin: International Journal of Coal Geology, v. 6, p. 21–41.

MANUSCRIPT ACCEPTED BY THE SOCIETY JUNE 1, 1988

ACKNOWLEDGMENTS

We thank the following for generous help in information retrieval, manuscript preparation, and reviews: William A. DiMichele, Joan Esterle, Debbie Gains, Judith Gennett, Sheila Hunt, Robert M. Kosanke, Carol Kubitz, Richard Leary, Peter J. McCabe, Russel A. Peppers, Alice Prickett, Debra A. Willard, Scott Wing, Gary Upchurch, and Richard B. Winston. We thank Lee R. Parker and Loretta Satchell for generously making available original figures.

NOTE ADDED IN PROOF

Since submitting the manuscript in January 1988, several papers have been published that are significant to the paleobotany and paleoecology of coals of the Carboniferous, Jurassic, Cretaceous, and Paleogene, mainly of North America. At the symposium, "Peat and Coal: Origin, Facies, and Depositional Models," at the 28th International Geological Congress, Washington, D.C., in 1989, more than 50 papers were presented (Lyons and Alpern, 1989a, b, 1990).

Particularly pertinent are reports on ancient and modern peat-forming systems and general overviews of the role of plants in peat accumulation, and their alteration and identification in coal. One comprehensive overview (Teichmuller, 1989a) summarized a broad spectrum of data on coal-forming plants, genesis of macerals, microlithotypes and lithotypes, and reconstructions of various types of peat swamps. In another paper (Cross and Phillips, 1990), we expanded on the discussions in the preceding text concerning the contribution of different plants to peat accumulation from Devonian to Tertiary, including some of the less important coal periods. We emphasized the importance of algae and herbaceous or shrubby plants to peat accumulation in pre-Carboniferous coals, and the evolutionary shifts to the main types of vascular plants that contributed to the peat of Carboniferous coals (lycopods, tree-ferns, and pteridosperms). Those Carboniferous plants were succeeded by conifers and other gymnosperms as the principal contributors to peat accumulations in the early and middle Mesozoic. These were succeeded by a combination of different conifers and angiosperms in the Late Cretaceous and Paleogene.

Winston (1988) has amplified earlier contributions on the identification of source plants from coalified tissues in coal seams from a Pennsylvanian coal. He was able to identify macerals derived from four major plant groups, several to the genus level. In some cases he differentiated the plants further into their origin as roots, stems, or other organs. He also was able to calculate the ratios of variable degrees of compaction for different plant organs, to determine relative contribution by volume, by comparing analyses of volume of permineralized tissues and organs in coal balls to laterally adjacent coalbed sections. He concluded that the identity and volume of the original plants could generally be determined from coal.

In a palynological study of an Illinois coal, Mahaffy (1988) demonstrated two distinctive patterns of plant distribution: (1) correlation of abundance of certain species to the areal extent of widespread coalbeds, i.e., those controlled by regional factors; and (2) the correlation of plant abundance to more restricted or local areas of the coal swamp, i.e., local depositional environments that are represented by certain bands in the coals. This diversity of microhabitats, as represented by diverse plants contributing to the Springfield coal, is in contrast to greater uniformity of habitats over wide areas in the Herrin coal above.

The Middle Pennsylvanian Hernshaw coal of West Virginia and correlative Fire Clay coal of Kentucky show vertical stratification into four floral groups, based on palynology (Eble and Grady, 1990). They suggest that these may correspond to floral groups found in modern domed peats. Arborescent lycopods at the base show progressive seral shift to ferns and herbaceous lycopod dominance in younger layers. Interruption of the peat-swamp floral succession by volcanic ash deposition resulted in an increase in cordaites and calamites in association with mineral soil.

In a study of the Smith and Anderson coalbeds in the Powder River basin, Wyoming (Moore and others, 1990), important plants in the associated floristic communities were not always principal contributors to peat accumulation. *Glyptostrobus* was dominant in the two coal-forming swamps in both peat accumulation and palynology. However, the important plants that initially characterized the peat flora in the Smith swamp were pine and spruce (conifers that formed woody peat), plus sycamore and elm. Later, a wider variety of angiosperms was more important in the flora. Peat of the Anderson swamp formed on a more stable platform (thick sands of a meander-belt complex). Though dominated by *Glyptostrobus,* the deposit was characterized by an important, broad-leaved angiosperm flora.

REFERENCES CITED

Cross, A. T., and Phillips, T. L., 1990, Coal-forming plants through time in North America, *in* Lyons, P. C., Callcott, T. G., and Alpern, B., eds., Peat and coal: Origin, facies and coalification: International Journal of Coal Geology, v. 16, p. 1–46.

Eble, C. F., and Grady, W. C., 1990, Paleoecological interpretation of a Middle Pennsylvanian coalbed in the central Appalachian basin, U.S.A.: International Journal of Coal Geology, v. 16, p. 255–286.

Lyons, P. C., and Alpern, B., eds., 1989a, Peat and coal: Origin, facies and depositional models: International Journal of Coal Geology, v. 12, 798 p.

Lyons, P. C., and Alpern, B., eds., 1989b, Coal: Classification, coalification, mineralogy, trace-element chemistry, and oil and gas potential: International Journal of Coal Geology, v. 13, 626 p.

Lyons, P. C., Callcott, T. G., and Alpern, B., eds., 1990, Peat and coal: Origin, facies, and coalification: International Journal of Coal Geology, v. 16, 237 p.

Mahaffy, J. F., 1988, Vegetational history of the Springfield coal (Middle Pennsylvanian of Illinois) and distribution patterns of a tree-fern miospore, *Thymospora pseudothiessenii,* based on miospore profiles: International Journal of Coal Geology, v. 10, p. 239–260.

Moore, T. A., Stanton, R. W., Pocknall, D. T., and Flores, R. M., 1990, Maceral and palynomorph facies from two Tertiary peat-forming environments in the Powder River Basin, U.S.A.: International Journal of Coal Geology, v. 15, p. 293–317.

Teichmüller, M., 1989, The genesis of coal from the viewpoint of coal petrology, *in* Lyons, P. C., and Alpern, B., eds., Peat and coal: Origin, facies, and depositional models: International Journal of Coal Geology, v. 12, p. 1–87.

Winston, R. B., 1988, Paleoecology of Middle Pennsylvanian-age peat-swamp plants in Herrin coal, Kentucky, U.S.A.: International Journal of Coal Geology, v. 10, p. 203–288.

Chapter 32

Coalification in North American coal fields

Heinz H. Damberger
Illinois State Geological Survey, Natural Resources Building, 615 East Peabody Drive, Champaign, Illinois 61820

INTRODUCTION

Coalification is the process that—over geologic time and as a result of rising pressure and temperature during increasingly deeper burial—matures peat into lignitic, subbituminous, bituminous, and anthracitic coal (*coalification series*). Both physical (pressure, heat) and chemical (bio-, thermo-) factors are influential in a complex way that is difficult to discriminate. The boundaries between the members of the coalification series are transitional, and any specific physical or chemical boundary values of a given classification scheme are arbitrary to a certain degree and vary between countries and researchers. The terminology of the American Society for Testing and Materials (ASTM) Standard Specification D388 "Classification of Coals by Rank" is used here; this standard is widely recognized around the world (Table 1).

A large number of physical and chemical parameters have been used to rank coals. No single parameter is useful over the full range of the coalification series. The best coalification parameters permit a high degree of certainty in discrimination between coals of only slightly different rank. Ideal places to observe systematic changes in coal rank are vertical drill holes that penetrate undisturbed, flat-lying strata. There, present absolute and former relative depths of burial can be measured accurately, and changes of physical and chemical properties of the organic matter can be linked to present and probable past increases of pressure, temperature, and duration of heating. The best parameters of rank change rapidly with depth, while exhibiting little scatter about the average trend line.

VERTICAL COALIFICATION PATTERNS AND THEIR CAUSES

Until fairly recently, systematic changes in coal rank with depth have received relatively little attention in North America, where mining has been limited to a single or a few coal beds separated only by a few tens of meters. In European coal fields with hundreds of meters of vertical separation between coal beds, numerous investigations of coal rank in relation to depth have been conducted during the past 100 years (Hilt, 1873).

Pressure and temperature induce changes in coal properties during burial, following mostly microbial degradation during the peat stage near the surface. Peat-forming environments of the North American Atlantic and Gulf Coastal Plains, particularly the Everglades and the Okefenokee Swamp, have been studied extensively during the past 25 years (Spackman and others, 1974; Altschuler and others, 1983; Cohen and others, 1984). Many factors have been identified that contribute to the specific nature of a peat deposit at any given site and to the vertical, local, and regional variations of these deposits. Due to the complexity of the processes involved and the heterogeneity of the resulting peats, no common starting point can be identified from which changes in material composition of "coal" can be measured. Rather, during coalification, the various components of peat follow their own paths of development. The integration of the chemical and physical changes of the individual components of coal results in the observed change of the "average" coal.

Generally, overburden pressure should primarily affect the physical properties of coal (e.g., porosity), whereas temperature and duration of heating should primarily affect chemical composition (e.g., C, H, and O contents) and molecular structure of coal. In reality, the effects of pressure and heating are difficult to separate.

Moisture content and heating value

The natural or inherent bed moisture content of coal on the ash-free or mineral-matter–free basis, which is a direct reflection of the porosity of the coal substance, has long been recognized as a useful indicator of coal rank (Schürmann, 1927; Patteisky and Teichmüller, 1960; Damberger and others, 1964). However, its decrease with depth has only occasionally been investigated in North American coal fields, either directly or indirectly, through the use of the heating value on a mineral-matter–free, moist basis (Damberger, 1971, 1974; Gilchrist, 1974; Hacquebard, 1977; Walsh and Phillips, 1982; Nurkowski, 1985).

Teichmüller (1968) sampled a 196-m-thick peat deposit in Greece and reported a 10.4 percent per 100 m decrease of the moisture content (weight percent), from 89 percent near the surface to 69 percent at 193 m. Edwards (1945) reported a 4.0

Damberger, H. H., 1991, Coalification in North American coal fields, *in* Gluskoter, H. J., Rice, D. D., and Taylor, R. B., eds., Economic Geology, U.S.: Boulder, Colorado, Geological Society of America, The Geology of North America, v. P-2.

TABLE 1. ASTM CLASSIFICATION OF COALS BY RANK (Standard D388, in box) AND CORRESPONDING OTHER COMMON RANK PARAMETERS*

ASTM Class	ASTM Group	ASTM Abbr.	Heating Value (Mineral-matter-free, moist)			Agglom-erating	% VM (mmf Dry)**	% Reflectance		% Moist. (mmf, Moist)	Ultimate Analysis (Min. matter free, dry)			Free Swelling Index
			1000 Btu/lb	MJ/kg	1000 kcal/kg			Oil, max. (Vitrinite)	Oil, random (Vitrinite)		% C	% O	% H	
Lignite	Peat		3.0-4.0†	7.0-9.3	1.67-2.22	No	72-62	0.2-0.4	0.2-0.4	95-50	50-65	42-30		
	Lignite B	ligB	Undefined -6.3†	Undefined -14.6	Undefined -3.50	No	65-40	0.2-0.4	0.2-0.4	70-45†	55-73	35-23	7-5	
	Lignite A	ligA	6.3-8.3†	14.6-19.3	3.50-4.61					50-31†				
Subbitu-minous Coal	Subbituminous C	subC	8.3-9.5†	19.3-22.1	4.61-5.28	No	55-35	0.3-0.7	0.3-0.7	38-25†	60-80§	28-15§		
	Subbituminous B	subB	9.5-10.5†	22.1-24.4	5.28-5.83					30-20†				
	Subbituminous A	subA	10.5-11.5†	24.4-26.7	5.83-6.39					25-18†				
Bitumi-nous Coal	High vol. C bit.	hvCb	10.5-13.0†	24.4-30.2	5.83-7.22	Yes	55-35	0.4-0.7	0.4-0.7	25-10†	76-83§	18-8§	6-4.5	1-5
	High vol. B bit.	hvBb	13.0-14.0†	30.2-32.6	7.22-7.78		50-35	0.5-0.8§	0.5-0.8§	12-5†	80-84§	12-7§		2-7
	High vol. A bit.	hvAb	>14.0	>32.6	>7.78		45-31	0.6-1.2§	0.6-1.1§	7-1†	78-88	10-6§		4-9
	Med. vol. bit.	mvb					31-22†	1.0-1.7†	0.9-1.6†	<1.5	84-91	9-4		9-7
	Low vol. bit.	lvb					22-14†	1.5-2.0†	1.4-1.9†	<1.5	87-92	5-3		9-1
Anthra-cite	Semianthracite	sa	>14.0	>32.6	>7.78	No	14-8†	1.8-2.8†	1.6-2.5†	<1.5	89-93§	5-3§	5-3†	
	Anthracite	an					8-2†	2.6-6.0†	2.4-5.1†	0.5-2	90-97§	4-2§	4-2†	
	Meta-anthracite	ma					<2	>5.5†	>5.0	1-3	>94§	2-1§	2-1†	

*Modified from Damberger and others, 1964.
†Well suited for rank discrimination.
§Moderately well suited for rank discrimination.
**VM = Volatile matter, on mineral-matter-free (mmf), dry basis; fixed (FC) content on that basis is supplement to 100%.

Note: Since coal rank is defined by heating value (mineral-matter-free, moist basis) and volatile matter content (mineral-matter-free, dry basis), respectively, the ranges of the equivalent values (outside box) overlap more or less depending on how wide the correlation bands between heating value and volatile matter contents and the equivalent values are. The values for the lower rank are consistently listed first.

percent per 100 m decrease of the moisture content of Australian brown coals (lignites) from about 70 percent to about 58 percent over a 300-m depth range, and Kutzner (in Teichmüller and Teichmüller, 1968a) showed a drop of 4.9 percent per 100 m in the average moisture content from 65 to 48 percent over a depth interval of 350 m for the brown coal deposits near Cologne, Germany.

By combining and extrapolating these observations (Fig. 1a), one may conclude that lignites that are at or near the surface of the Gulf Coast and the Northern Great Plains and have moisture contents ranging from about 50 to 30 percent (Energy Resources Company, 1980) may once have been buried beneath 600 to 1,100 m of overburden; and the subbituminous coals of the Rocky Mountain coal basins (e.g., Powder River basin), with moisture contents of 30 to 20 percent, may have been buried beneath about 1,100 to 1,600 m. Systematic studies of the change with depth in moisture content and heating value (moist basis) are still lacking for these coal basins. This is particularly unfortunate because the 45 to 20 percent moisture interval is not well represented in other major coal regions of the world. (Note "no information" gap in Fig. 1a).

In the Illinois Basin, with its high-volatile bituminous coals, the moisture content drops from somewhat more than 20 percent to approximately 1 percent over a depth interval of 1,830 m. The moisture-versus-depth gradient decreases from about 2.5 percent per 100 m in high-moisture coals to about 0.4 percent per 100 m in low-moisture coals (Damberger, 1971). Moisture-versus-depth gradients in other coal basins around the world are remarkably similar (Fig. 1a). The observed moisture reduction with increasing depth apparently stems primarily from decreasing porosity due to increasing compaction; the curve looks much like the compaction curves for mud (Weller, 1959; Altschaeffl and Harrison, 1959). The composite "world curve" for moisture versus depth in coal (Fig. 1a) thus is useful in estimating the order of magnitude of the former depth of burial of a given coal in the lignite to high-volatile A bituminous rank range. Based on this composite curve, coals now near the surface of the northwestern part of the Illinois Basin coal field with about 20 percent moisture may once (probably during late Paleozoic) have been buried under about 1.6 km of sediments, and the higher rank coals of southeastern Illinois, with moisture contents less than 5 percent, may have been buried under about 3 km of sediments, long since removed by erosion (Damberger, 1971). The moisture-versus-depth gradient has been shown to reflect the geothermal gradient, like other coalification gradients, but only in high-volatile bituminous-rank coals (Damberger, 1966, 1968; Teichmüller and Teichmüller, 1968a). The average curve of Figure 1a probably is representative for normal geothermal gradients only.

Hacquebard (1977) and Nurkowski (1985) used the gradients for moisture content and heating value (mineral-matter-free, moist basis) reported from several coal basins as the basis for estimating the amount of erosion and former depth of burial for the coals of the Alberta Plains region (Fig. 1a). Nurkowski constructed maps showing the probable thickness of eroded sediments and the depths of the subbituminous/bituminous boundary.

With increasing burial, a minimum of less than 1 percent moisture content is reached at about the boundary of the high-volatile/medium-volatile bituminous rank; the moisture content remains low through semianthracite. In anthracites a microporosity develops with increasing rank, and moisture contents as much as 5 percent have been observed (Fig. 1a). This porosity increase is accompanied by a significant increase in internal surface area mostly in micropores less than 10 Å in diameter (Thomas and Damberger, 1976).

Reflectance of vitrinite

Microscopic reflectance measurements on polished vitrinite under oil immersion (R_0) have become the most common method to determine the maturity of coaly material (McCartney, 1952; Kötter, 1960). A large body of literature exists that relates vitrinite reflectance data to the generation and occurrence of liquid and gaseous hydrocarbons, to present and past geothermics, and to other phenomena (see section on oil and gas in this volume). The obvious advantage of microscopic over routine chemical methods is that the analyzed sample spot is small, typically a few to a few tens of micrometers across, and the material measured is identifiable. Chemical analyses do not provide such material-specific information unless the components of coal can first be separated, a difficult task with coal because of the intimate mixture of its maceral constituents and their commonly small size. Reflectance measurements on polished coal samples generally are made on vitrinite (telocollinite type) in coal and in dispersed organic matter. Its uniform appearance, recognizable under the microscope, and the relatively rapid increase of R_0 during coalification make it a nearly ideal material for coalification studies.

Methods of measurement have been standardized at the national (ASTM, D2798) and international levels (ICCP, 1963; ISO, 1984). Repeatability (same laboratory) and reproducibility (different laboratories; Riepe and Wolff-Fischer, 1982) are generally satisfactory; however, some caution in the use of vitrinite reflectance data is in order. The observed scatter of reflectance data relates to differences between vitrinite submacerals (which are derived from different plants and plant tissues) and to variable bio- and geochemical conditions in the peat and later coalification stages. Goodarzi and others (1988) demonstrated the influence of facies (mineral and maceral association) on reflectance in the Hat Creek coal field of British Columbia. Vitrinites associated with large amounts of liptinites commonly have lower reflectivities than do vitrinites in common coals (Kalkreuth, 1982; Wolf and Wolff-Fischer, 1984) and some vitrinites in marine-influenced (high-sulfur) coal seams (Teichmüller, 1982; Littke, 1987). Most vitrinites in oil shales have lower reflectance compared to those in associated common coals or dispersed in other clastic sediments; thus the reflectance of vitrinites in black shales does not provide a directly comparable measure of

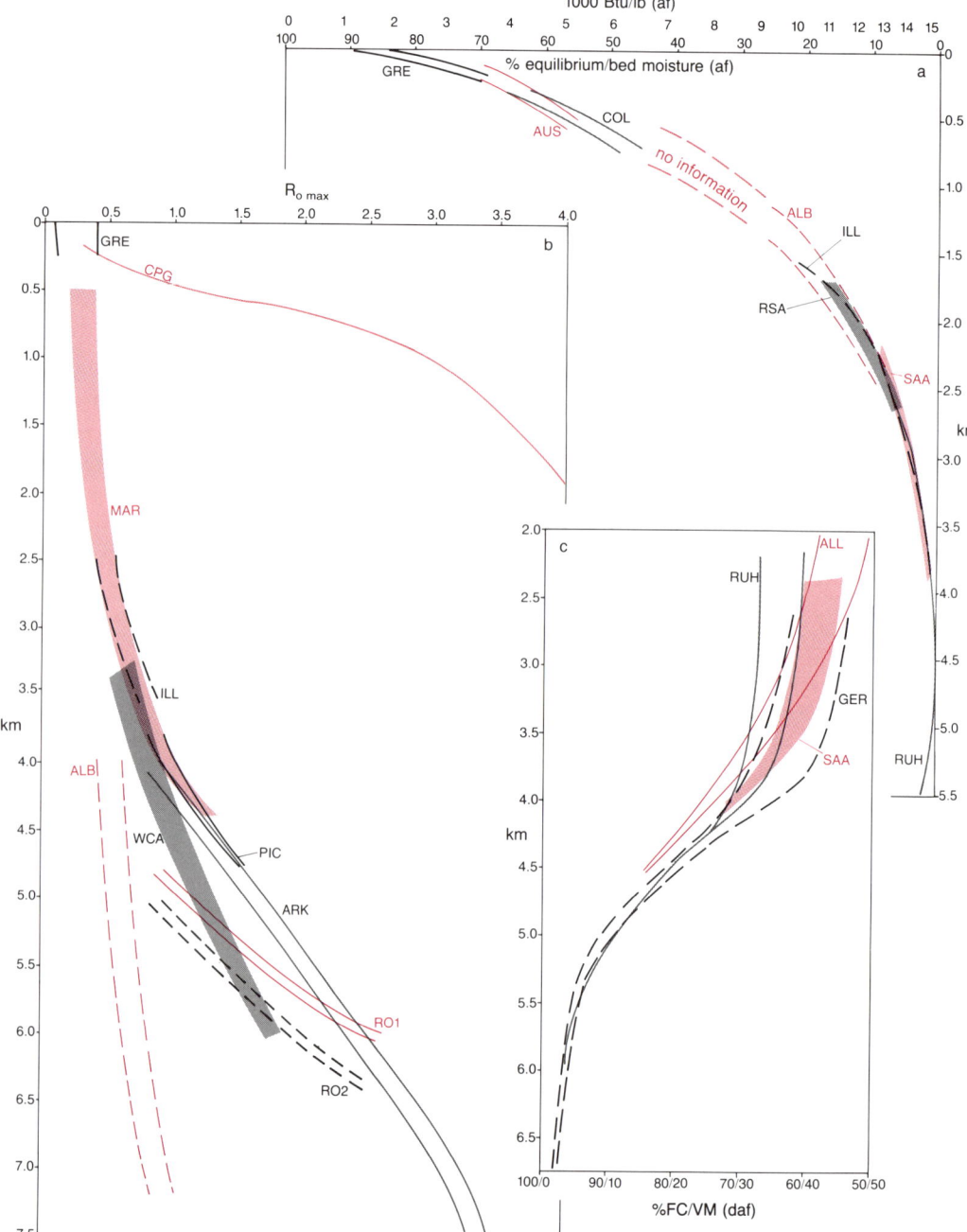

Figure 1. Coalification curves of inherent moisture (and heating value) (a), reflectance of vitrinite (b), and volatile matter (or fixed carbon) (c) versus relative depth. Individual curves for various coal basins are positioned at their inferred approximate former maximum depth of burial. Published data from North American coal fields is limited and was supplemented by data from overseas to show the general trends as well as possible ranges from region to region. ALB, Alberta Plains (England and Bustin, 1986b); ALL, Allegheny Basin (based on Reeves, 1928); ARK, Arkoma Basin (Houseknecht, 1987); AUS, Australia, lignite (Edwards, 1945); CPG, Cerro Prieto geothermal field (Barker and Elders, 1981); COL, Cologne lignite, West Germany (Kutzner in Teichmüller and Teichmüller, 1968a); GER, Germany (NW), Carboniferous (Teichmüller and Teichmüller, 1984); GRE, Greece, peat (Teichmüller, 1968); ILL, Illinois Basin (Damberger, 1971); MAR, Maritime Region, Canada (Hacquebard, 1974); PIC, Pictou coal field (Kalkreuth and Macauley, 1987); R01, Rocky Mountains, Inner Foothills (Hacquebard and Donaldson, 1974); R02, Rocky Mountains, Inner Foothills and Front Ranges (Kalkreuth and McMehan, 1984); RSA, Ruhr Basin, Upper Silesia, Aachen coal fields (Patteisky and Teichmüller, 1960); RUH, Ruhr coal field (Patteisky and Teichmüller, 1960); SAA, Saar coal field (Damberger, 1966); WCA, western Canada Basin (Radke and others, 1982a).

maturity and must be compared with caution to those in common humic coal (Kalkreuth and Macauley, 1987).

Figure 1b includes published North American R_0-versus-depth curves from coal-bearing strata that were plotted according to their estimated maximum depths of burial. Several curves or parts thereof are based on dispersed vitrinite particles in clastic sediments (but not oil shales or black shales) rather than vitrinites from coal beds. The R_0-versus-depth curves generally are extremely steep—in the 0.2 to 0.6 percent range. Reflectance values exhibit a wide scatter about the average trend; samples hundreds and even more than 1,000 m apart vertically do not necessarily yield significantly different R_0 values, even though they were exposed to different burial conditions. The relative depth uncertainty in this rank averages about 1,000 to 1,500 m, compared to about 250 m for the moisture (or heating value) curve in the same rank range. Beyond about 0.6 percent R_0, the values start to increase somewhat more rapidly with increasing relative depth until the gradient stabilizes beyond about 1.0 percent R_0. Concurrently, the scatter of individual values about the average curve diminishes considerably to within a narrow band, indicating that reflectance of vitrinite is an excellent rank parameter at 1.0 percent or more; the relative depth uncertainty is only about 50 to 100 m above about 1 percent R_0.

Vitrinite is bireflectant (Hower and Davis, 1981b), measurable in high-volatile bituminous coals, and increasingly more bireflectant with increasing rank. North American coal petrographers prefer to use maximum reflectance (R_0 max), which requires rotation of the microscope stage through 360° for each measurement (ASTM, D2798). Random reflectance (R_0 ran) is commonly used in western Europe and is almost universally used in oil and gas exploration because of the small size of grains to be measured and the inaccuracy due to optical alignment on such grains during rotation.

Reflectance versus depth gradients vary geographically, mirroring regional variations of paleogeothermal gradients. Hacquebard and Donaldson (1974), for instance, describe the variation of reflectance gradients in the Inner Foothills Belt of the Jurassic-Cretaceous coal-bearing sections of the Canadian Rocky Mountains of Alberta and British Columbia. Because of the variable gradients, they compare reflectance gradients from ten different stratigraphic sections to those of comparable rank in the Peel Region of the Netherlands. The observed gradients vary between 62 and 174 percent of the Peel "standard." Hacquebard and Donaldson attribute these variable deviations to variations in the paleogeothermal gradient. They conclude that the three areas with above-normal coalification gradients and high rank in the Inner Foothills Belt were subjected to a relatively large paleogeothermal gradient rather than deep burial.

The R_0-versus-depth curve for the Maritime Region of the Canadian Atlantic Provinces published by Hacquebard (1974, 1983) is a particularly useful curve because the sequence of Cretaceous and Upper Jurassic strata that overlie the Carboniferous represents a stratigraphic section with only minor discontinuities and little erosion of latest Tertiary sediments. Thus, the current depth of burial probably is very near the maximum depth of burial. Hacquebard found that the gradients in the R_0 range between 0.63 and 0.95 percent—in which the Mesozoic and Carboniferous overlap—were nearly identical. Consequently, he felt justified in joining the two curves into one composite coalification curve for the Maritime Region (Fig. 1b). The curve extends upward to the lignite stage (0.28 percent R_0) and thus permits estimation of the amount of strata that have been removed by erosion at any given bedrock surface location with known R_0. For instance, to reach about 1.0 percent R_0, burial to about 4 km is required in the Maritime Region. This is only about 15 percent more than what the composite moisture-versus-depth curve (Fig. 1a) predicts for the approximate equivalent point in the moisture curve, at about 1 to 2 percent moisture content.

Beyond about 1.5 percent R_0, published data from North American coal fields are still relatively scarce. Houseknecht (1987) published three depth-versus-R_0 profiles from the Atoka Formation of the Arkoma basin in Arkansas; they cover the range between 0.7 and 3.7 percent (Fig. 1b). A relatively small paleogeothermal gradient is suggested by these R_0-versus-depth gradients, which are significantly smaller than those reported from western Canada and northwestern Germany. One possible reason is that the Atoka Formation is rich in sandstones, the relatively good thermal conductivity of which causes reduced geothermal and consequently reduced coalification gradients (Damberger, 1966, 1968). Houseknecht hypothesizes, though, that the observed coalification patterns in the Arkoma basin are strongly influenced by the movement of hot fluids laterally outward from the basin.

How much coalification gradients are influenced by the geothermal regime is dramatically illustrated by R_0 data from boreholes in the Cerro Prieto geothermal field of Baja California, Mexico (Fig. 1b), where reflectance increases as steeply as temperature does with depth (about five times the normal geothermal gradient; Barker and Elders, 1981). But even within the same coal basin, significant variations of coalification gradients can be observed for a given rank (Kalkreuth and McMechan, 1988).

Volatile matter and fixed carbon

Volatile matter (VM) and fixed carbon (FC), both on a dry, ash-free or dry, mineral-matter-free basis have long been used to determine rank of coal (Hilt, 1873). "Hilt's rule" of rising rank in successive coal seams with increasing depth has been reconfirmed many times in coal basins around the world. Although White (1913, p. 129), in his classic book with Thiessen, clearly states that "coal, on account of its sensitive reaction to dynamic influences, offers a delicate medium for the observation of earth metamorphism" and its variation with depth, he offered no specific information on gradients. He considered both temperature and thrust pressure important factors in the devolatilization of coal. Reeves (1928) discussed Hilt's rule at length, emphasized its significance, and presented data sets with different coal ranks (between 17 and 48 percent VM) from 17 locations in the Appalachian coal field; the coal samples had been taken between 25 and 300 m apart vertically. This is still the best available published set of data on change of volatile matter against depth for any North American coal field, certainly for coal that is medium-volatile bituminous and higher in rank (<31 percent VM, >69 percent FC). Hacquebard and Donaldson (1970) analyzed coal samples from the Canadian Atlantic Provinces, but their VM values from drill holes were derived from reflectance measurements, using a R_0-VM cross plot.

The outstanding characteristic of VM coalification curves (Fig. 1c) is the rapid change in slope and in scatter of data points about the average curve at around 30 percent VM. From the surface down to that value, the curves are steep; individual values scatter within a wide band. Below about 30 percent VM the curves are less steep, and data points fall within a narrow band. This conspicuous turning point has been chosen in all coal classification schemes as an important boundary between major classes of coal rank. Stach (1953) referred to it as the "Inkohlungssprung," sometimes translated as "coalification jump," but more appropriately referred to as coalification break in reference to the break in the slope of the coalification curve. Because the "break" refers to a rapid but not abrupt change in the VM/depth slope, the boundary may still be chosen between the beginning of the gradient change at about 31 percent VM (used in ASTM Standard D388) and the completion of the rapid gradient change at about 28 percent VM (Germany).

The rapid devolatilization of coal below the break is particularly strong in the liptinite group of macerals. They quickly lose their distinctive dark appearance in polished sections, apparently due to the generation of methane (meaning high relative loss of H and corresponding increase in C contents). Liptinites approach the reflectance of vitrinite at about 1.5 to 1.6 percent R_0 of vitrinite or about 20 to 24 percent VM (Alpern, 1984).

VM contents continue to drop rapidly with increasing relative depth down to about 10 percent, below which the curves steepen until they are almost vertical below 4 to 5 percent VM. This less pronounced but still significant break in the coalification curve at about 10 percent VM marks the boundary between bituminous coals and anthracites in most coal classification systems. Most volatiles have been split off at this point, and the curve approaches the 0 percent value asymptotically. Again, different countries and researchers have chosen slightly different boundary values. The American standard sandwiches semianthracite with 14 to 8 percent VM between bituminous coal and anthracite (Table 1).

Above about 30 percent VM, only averages of a sufficiently large number of samples will show a regular but slow decrease in VM (or increase in FC) with depth. "Exceptions" to Hilt's rule are common between any pair of individual samples. Occasionally, such exceptions have been used to question the validity of Hilt's rule.

Another potential pitfall of VM values lies in the natural variability of the petrographic makeup of coal seams. Liptinite-rich coals, especially coals with over 30 percent VM, have above average VM contents while inertinite-rich coals have below average VM contents, due to the relatively high VM contents of liptinite macerals and the low VM contents of inertinites. Such differences in petrographic composition must be considered when comparing samples. Of course this is one of the biggest advantages of reflectance measurements that can be performed on nearly the same material (vitrinite), regardless of the petrographic composition of the coal. Patteisky and Teichmüller (1960), Patteisky and others (1962), and Damberger and others (1964) used hand-picked vitrain samples to limit petrographic variability and make chemically analyzed samples used in coalification studies more comparable.

Organic elements of coal (C, O, H)

Of the elements determined by ultimate analyses, C, O, and H contents (dry ash-free basis) exhibit useful changes with relative depth, each in a certain rank range. However, no published data on changes with depth for these elements are available from North American coal basins. Based on observations in coal-bearing strata from other continents, particularly from Europe (e.g., Patteisky and others, 1962), one can expect an almost linear increase in C content between 60 and 90 percent. In this range (high- and medium-volatile bituminous coals), the scatter band is fairly narrow and C is a reasonably good parameter for determining rank. The curve steepens around 88 to 90 percent, within the medium-volatile bituminous coal rank, and increases only slowly into the anthracites where a slight pickup in the rate of change is noticeable. In these coals of higher rank, the C content should therefore not be used to indicate maturity.

Generally, the O content drops in mirror fashion to the increase in C content. The H content exhibits very little change from lignites down into the medium-volatile bituminous coal range. Downward through the anthracites, the H content decreases gradually and regularly from more than 5 percent to less than 3 percent, within a fairly narrow coalification scatter band, indicating its usefulness as rank parameter in this range.

The atomic H/C ratio decreases through the entire coalification series from about 1.5 to less than 0.2, initially due to the increasing C content, then due to the decreasing H content in high-rank coals. Published H/C versus depth curves are rare and nonexistent for North America.

Other coal properties

Many other coal properties change systematically with maturation and can be used in coalification studies. Samples of coking coals are commonly analyzed by the Gieseler plastometer, dilatometer, and the free swelling test (FSI). Geologists and geochemists of the oil and gas industry have looked extensively at various organic geochemical parameters obtained using different organic solvents to extract coal (and sediments containing dispersed organic matter, kerogens), by means of controlled heating experiments (e.g., RockEval analysis), and by other techniques (Hatcher and others, 1982). The soluble and pyrolizable organic matter thus obtained is analyzed by a multitude of methods, especially gas chromatography and mass spectrometry (Tissot and Welte, 1984), and maturity parameters are derived from such analyses (see section on oil and gas in this volume). Fundamental properties of coal such as porosity, density, internal surface area, and aromaticity have also been investigated and correlated with coal maturation. Coal petrographers have demonstrated the usefulness of a number of optical parameters to indicate coal rank such as reflectance of macerals other than vitrinite, bireflectance, refractive index, and fluorescence-related parameters (especially for low-rank coals).

CORRELATIONS BETWEEN COALIFICATION PARAMETERS

Cross plots of maturation-related parameters provide another useful way to look at the changes that occur during coalification. Correlations between the most important coalification parameters are shown on Figures 2a to 2i. The fields of data for the various coal basins indicate the rank range and thereby compositional properties of the mineral-free coal represented in these basins. The width of the bands in the cross plots results from the normal scatter of both parameters (compared to only one in depth plots of Figure 1). The charts allow conversion between parameters that is often desirable in comparisons of rank. They also help in the selection of natural boundaries at places of rapid trend change in the coalification series.

Correlations for different coal fields (Fig. 2) generally exhibit the same basic trends, but band widths are by no means identical. These differences between basins probably reflect differences in environments of deposition (and thus differences in coal compositions), subsequent geologic evolution of the coals, and also the availability of samples and their distribution within the coal basin. The analyses were all converted to the mineral-matter-free (or ash-free), moist or dry bases, whichever is normally used in coalification studies.

The atomic ratio H/C versus O/C diagram (Van Krevelen, 1961) deserves special attention (Fig. 3). The curve for the main component of most coals (wood through vitrinite) takes a sharp turn at approximately 0.75 H/C and 0.05 O/C (equivalent to about 87 percent C); this is the coalification break between high- and medium-volatile bituminous coals. In considering the chemical basis for the changes shown in the Van Krevelen diagram, dehydration, decarboxylation and demethanation, as well as dehydrogenation, oxidation, and hydrogenation (removal of H_2O, CO_2, CH_4, and H, and addition of O and H, respectively) can be represented as straight lines (red in Fig. 3). Figure 3 can thus be easily interpreted in terms of these important chemical reactions during coalification.

DIFFERENT COURSES OF COALIFICATION OF ORGANIC CONSTITUENTS OF COAL

The bulk of the material of most coals is derived from wood and bark of stems, trunks, and roots (Phillips and Cross, this volume). This material is called huminite in lignites and vitrinite in bituminous and anthracite coals. Most North American coals are composed of about 70 to 90 percent vitrinite. For this reason, chemical parameters work well in ranking most coals; that is, the process of coalification reflected in chemical analyses of whole coal samples is mainly that of the vitrinite (huminite in low-rank coals) macerals. Coals rich in inertinite or liptinite macerals deviate significantly in chemical composition from common vitrinite-rich coals. Consequently, these coals cannot be classified by the standard schemes for ranking coal (e.g., ASTM, D388).

The Van Krevelen diagram illustrates the compositional changes that occur in the principal petrographically recognizable organic constituents of coal (macerals) during the coalification process (Fig. 3). The huminites/vitrinites follow a coalification path intermediate between those of exinites/liptinites and of the inertinites. Dehydration is an important process early in their coalification; then decarboxylation becomes more significant. Demethanation obviously dominates after the curve makes its sharp turn at the coalification break. The coalification path for liptinites and similar materials initially indicates a high dehydration and a significant demethanation behavior; the path then turns and almost parallels the demethanation lines. Alginites, resinites, and similar materials have the highest H/C ratio, and the coalification band has the highest demethanation component of all coal constituents. The coalification paths of vitrinites, exinites, and alginites become inseparable at an H/C atomic ratio of about 0.5 and an O/C atomic ratio of about 0.02 (within the low-volatile bituminous rank).

In reflected light, liptinites brighten quickly once they have reached medium-volatile bituminous rank; with increasing rank, they become increasingly difficult to separate from vitrinites. Inertinites have the lowest H/C ratios among macerals (Fig. 3). Fusinite undergoes little change through much of the coalification process and retains its distinctly high reflectance relative to other macerals. Only in anthracite does fusinite become difficult to distinguish from other macerals under the microscope (Lemos de Sousa, 1979).

BYPRODUCTS OF COALIFICATION

The principal byproducts of the coalification process of coals are H_2O, CO_2, CH_4, H_2S, N_2, and small quantities of both parafinic and aromatic hydrocarbons (Jüntgen and Klein, 1975). Figure 4 presents the coalification process during burial, in terms of changes of the components of coal as well as the processes and products involved.

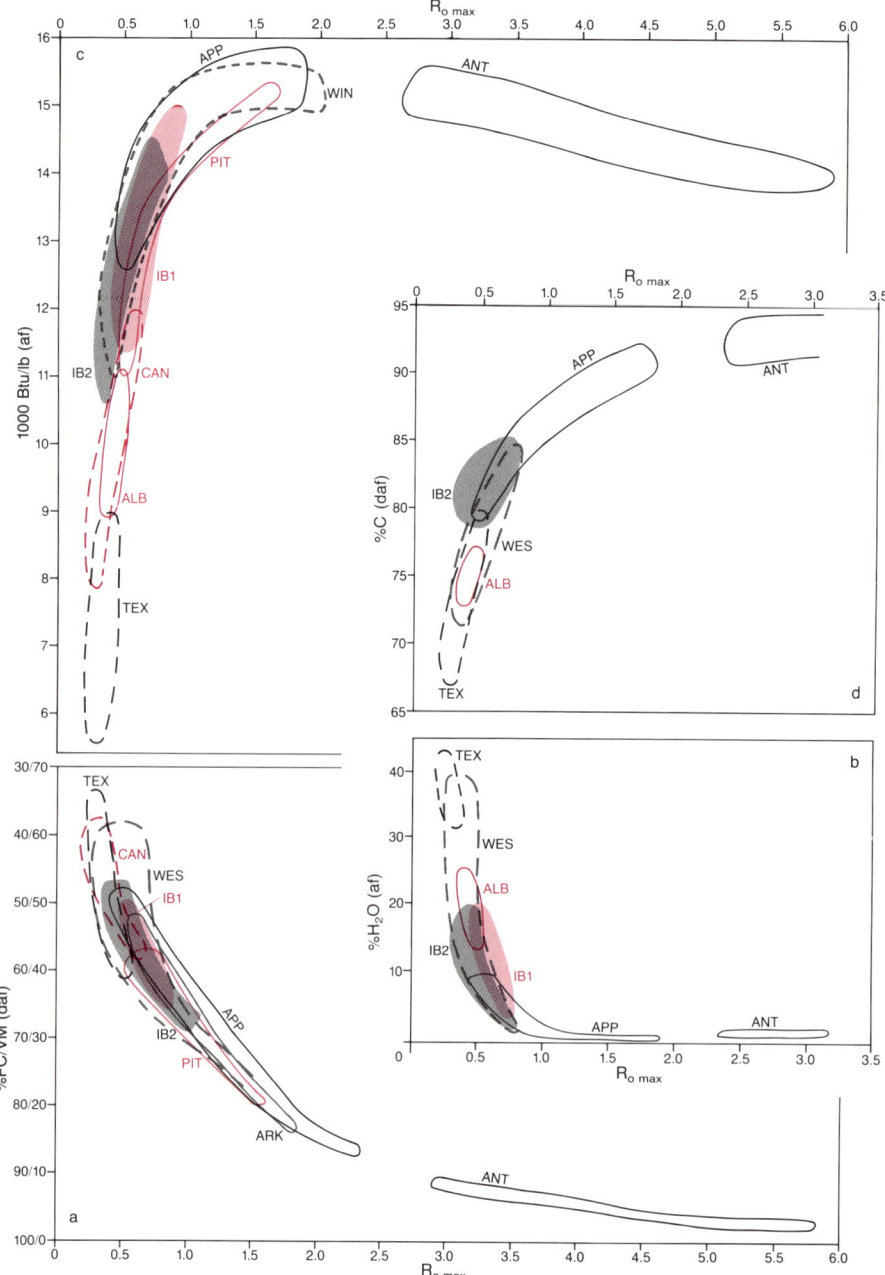

Figure 2. Correlation of reflectance versus fixed carbon (FC)/volatile matter (VM) (a), moisture (b), calorific value (c), and C (d); of moisture versus fixed carbon (FC)/volatile matter (VM) (e), calorific value (f), and C (g); and of C versus fixed carbon (FC)/volatile matter (VM) (h) and versus O (i). Sources: National Coal Resources Data System (NCRDS) of the USGS; Information System on Chemistry of Illinois Coal (ISCIC) of Illinois State Geological Survey; Bustin (1986); Chyi and others (1987); Damberger (1971); Energy Resources (1980); Hower and others (1985); International Committee for Coal Petrology (1963); Mukhopadhyay (1987); Nurkowski (1985); Parkash and Chakrabartty (1986); Tewalt and others (1986); Waddell and others (1978). ALA, Alabama lignites; ALB, Alberta Plains; ANT, Anthracite fields, Pennsylvania; APN, APP, APS, Appalachian Basin, north, entire, south; ARK, Arkoma Basin; BIG, Big Horn Basin; CAN, Canadian Arctic; DEN, Denver Region; FTU, Fort Union Basin; GRR, Green River Basin; IB1, Illinois Basin, Illinois Geological Survey; IB2, Illinois Basin (Penn State University); PIT, Pittsburgh Coal, Northern Appalachian Basin; PRB, Powder River Basin; TEX, Texas lignites; UIN, Uinta Basin; WES, Western United States coal fields combined (Penn State University); WIN, Western Interior coal basin.

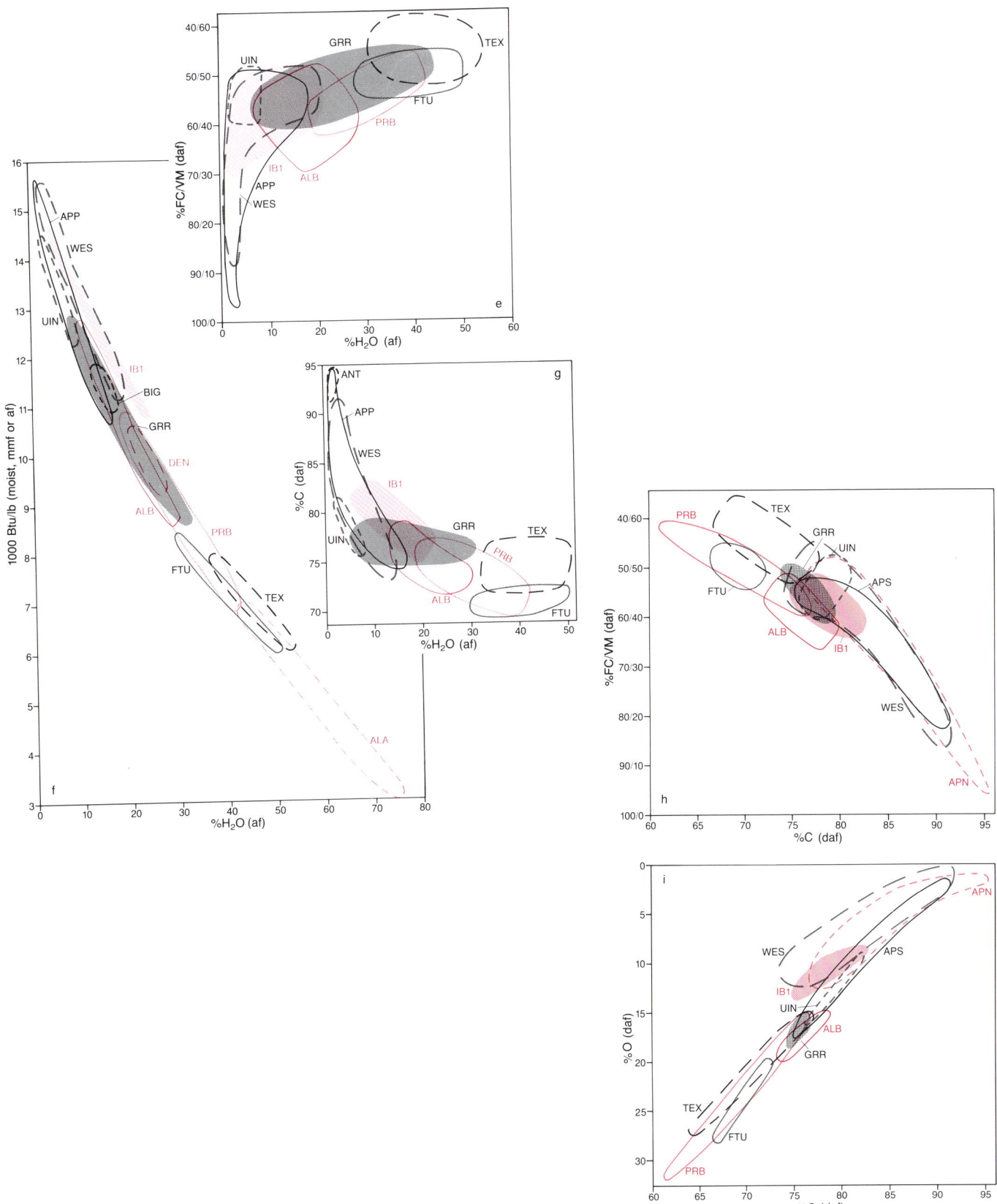

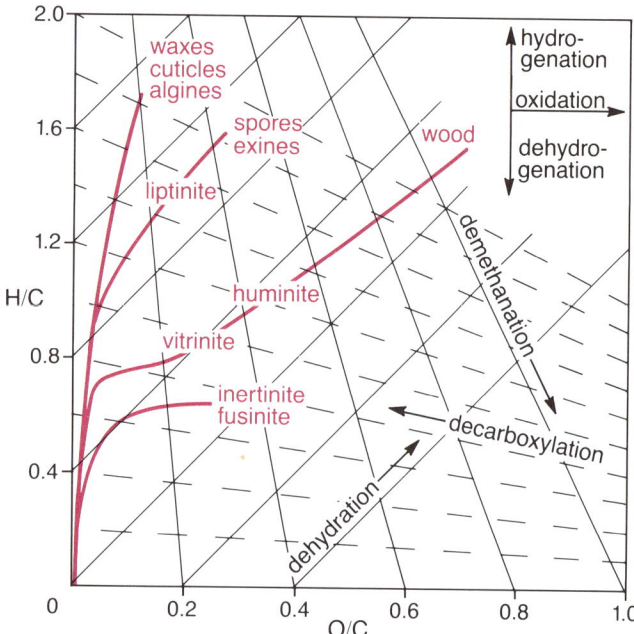

Figure 3. Van Krevelen diagram (H/C versus O/C atomic ratios) for the main components of coal and their predecessors with lines of dehydration, decarboxylation, demethanation, dehydrogenation, oxidation, and hydrogenation (modified after Van Krevelen, 1961; Tissot and Welte, 1984).

Methane is generated in quantity especially in medium-volatile bituminous to anthracitic ranges of rank. Even though the large internal surface area of coal (Thomas and Damberger, 1976) permits storage of substantial quantities of methane in the coal itself, the amount of methane produced per unit volume coal begins to exceed the coal's capacity to adsorb the gas on its internal surface area or to store it in pores of high- to medium-volatile bituminous coals (Fig. 5). Therefore, medium- and low-volatile bituminous coals generally are accompanied by methane, both within the coal and in associated strata (Rightmire and others, 1984). In recent years this coal gas has become the target of exploration and exploitation, in particular in the Raton Mesa, San Juan, and Black Warrior basins.

Besides methane, parafinic and aromatic hydrocarbons (HC) are also generated, particularly from the liptinite components of coal. Leythaeuser and Welte (1969) and Radke and others (1982b) demonstrated that such HC generation reaches a maximum in the range between about 0.8 and 1.05 percent R_0. Radke and others (1982b) observed a sudden increase in the proportion of fluorescing vitrinite at a little more than 0.8 percent R_0 and a sharp dropoff at more than about 1.4 percent R_0. They attributed the rise to generation of new bitumens and low-boiling HCs, and the drop to their destruction. They also observed a rather sudden occurrence of fluorescing fillings of pores (e.g., in fusinite) and fissures at about 0.9 percent R_0, and interpreted

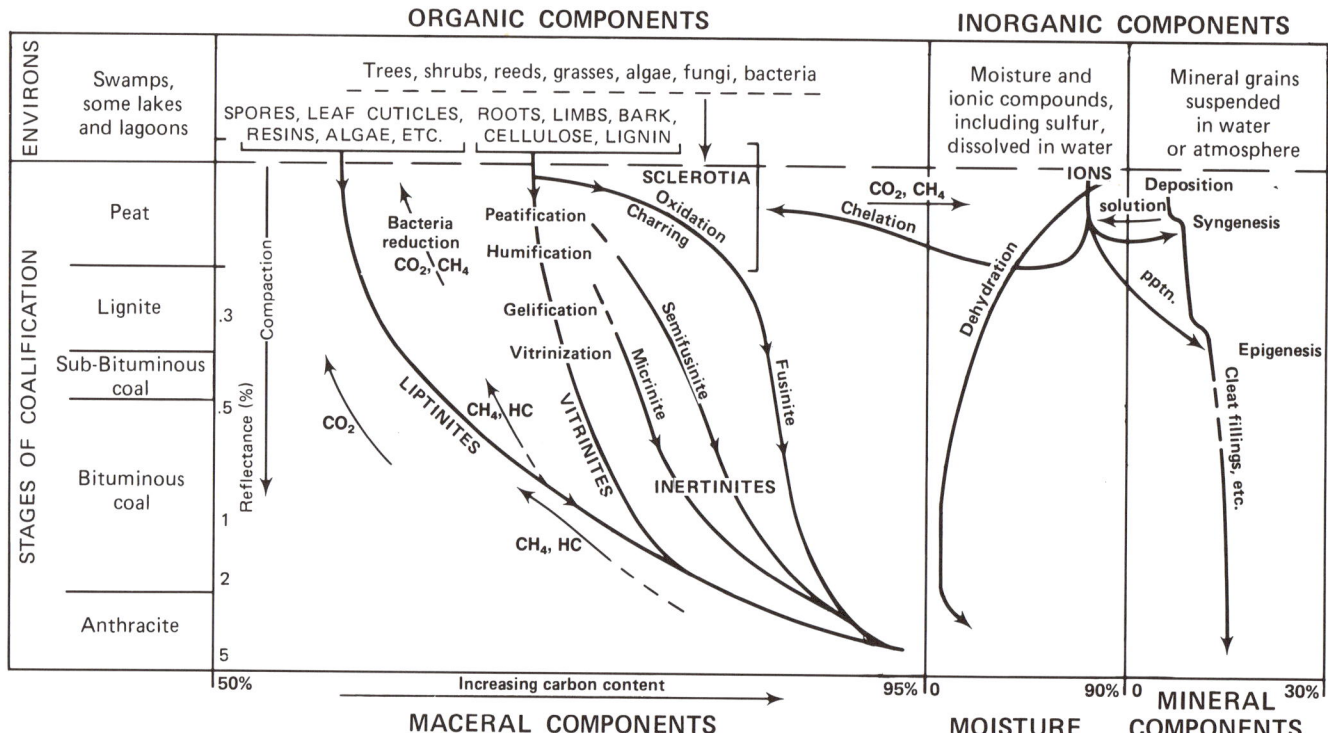

Figure 4. Courses of change of the principal constituents of coal during burial and coalification; depth scale is approximately linear (from Harvey and Ruch, 1986, slightly modified). V.M., volatile matter.

these as coalification products. Thomas and Damberger (1976) suggested that low-boiling parafins may contribute to the well-known reduction in porosity and internal surface area (Van Krevelen, 1961) from lignites through subbituminous and high-volatile bituminous coals by the deposition of parafins in pores and by clogging of pore throats. An increase in porosity and internal surface area in high-rank coals may subsequently occur due to volatilization of these parafins as the temperature rises during coalification in the low-volatile bituminous to anthracitic range.

Mukhopadhyay (1987) discusses the potential of Texas lignites as source rocks for both liquid HC and methane. The lignites are too immature near the surface, but gulfward—where they are deeply buried—they could well be a significant source of HCs.

The CO_2 generated during the early stages of coalification may end up in part in various carbonates precipitated from pore water as cement in sediments associated with coals, and in part as cleat (joint) filling in coals. Most of the CO_2 generated probably escapes as gas or becomes dissolved in pore water.

Micrinite is commonly considered a solid byproduct of the coalification of liptinitic macerals, including bituminite (Teichmüller, 1974) and exsudatinite (Shibaoka, 1978); however, some micrinite may originate in the peat stage through degradation of cell walls (Spackman and Barghoorn, 1966; Cohen, 1968; Cohen and Spackman, 1980). Exsudatinite itself is an early byproduct of the coalification process, primarily derived from liptinitic macerals in the subbituminous to high-volatile bituminous range at the onset of petroleum generation (Teichmüller, 1974, 1982). Exsudatinite is observed as a secondary product that fills cracks around liptinites and associated vitrinites.

REGIONAL COALIFICATION PATTERNS AND THEIR CAUSES

Pressure, temperature, and duration of heating must be considered as the main causes of coalification. Pressure is exerted by the overburden and by tectonic force that produces structural deformation (folding, faulting). Temperature is related to depth of burial and heat input, which can be strongly influenced regionally by deep-seated plutonic intrusions and movement of basin fluids or locally by vulcanism or igneous intrusion. The duration of exposure to specific pressures and temperatures can vary considerably due to regional and local variations in depth of burial, deformation, and geothermics.

Predeformational coalification patterns

White (1915) theorized that thrust pressure during the Appalachian orogenesis caused the devolatilization of coal because the coal rank of the affected strata increases systematically eastward into areas of more intense folding and faulting (Fig. 6).

Coal rank also increases toward the Inner Foothills of the Rocky Mountains of western Canada, where folding and thrust faulting are intense. Nevertheless, coal rank is independent of the degree of folding and thrusting, except for local effects apparently related to intense shearing along thrust faults (Kalkreuth and Langenberg, 1986). Surfaces of equal coal rank (isoranks) enclose a small angle with the coal beds, which is related to the regional increase in burial depth toward the Rocky Mountains (Kalkreuth and McMechan, 1988). The isorank surfaces are deformed with the coals (Fig. 7). Ting (1984) reports that the orientations of the optical axes of vitrinites remain the same relative to the bedding regardless of the dip of the coal beds in some western Canadian coal fields. In contrast to White's theory, which dominated thinking among coal and petroleum geologists for many years, these observations prove that the coalification process was completed before deformation took place.

Predeformational coalification patterns are preserved in a surprisingly large number of cases, even in Paleozoic basins. Damberger (1974) showed that original coalification patterns, probably established during maximum burial sometime in the late Paleozoic, are still preserved in cratonic basins and over arches of the Midcontinent. At any given stratigraphic horizon, coal rank decreases systematically northward toward the Canadian Shield. The pattern is modified by the basins and arches in such a way that along any east-west section, the rank of a chosen seam is higher in the basins than on the adjacent arches. Consequently, areas of higher coal rank extend farther north within basins than on adjacent arches, especially within a single coal bed (Fig. 6).

Variations in heat flow and thus geothermal gradients lead to different coal rank at comparable depths. If the average rank-

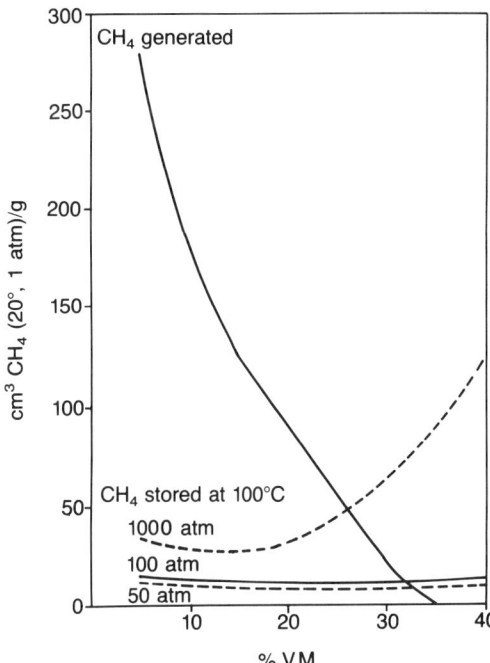

Figure 5. Cumulative amounts of methane (CH_4) generated during coalification, compared to storage capacity of coal (principally to internal surface area) at 100°C and 50, 100, and 1,000 atm gas pressure (from Jüntgen and Klein, 1975).

versus-depth curves of Figure 1 reflect the average effects of burial in a regime of normal geothermal gradients (a reasonable assumption), then the lines of equal rank in Figure 6 can be translated into a first approximation of maximum former depths of burial. For instance, the 15 percent moisture line (hvCb/hvBb boundary) would be equivalent to a former maximum burial depth of about 1.9 km. The equivalent maximum temperature would be about 80 to 90°C, assuming normal geothermal gradients. Maximum burial probably was attained during late Paleozoic (or Mesozoic?) time. These maximum depths are higher than those suggested by Chyi and others (1987) for coals of comparable rank in the Dunkard basin of the Appalachians, primarily because they assumed a 50 percent more than normal paleogeothermal gradient in the Late Permian and a gradual decrease of the gradient since then. The high geothermal gradient could be caused by elevated heat flow or a relatively low thermal conductivity of the now mostly eroded overburden, or by a combination of both.

Syndeformational coalification patterns

Syndeformational coalification takes place when folding or faulting and coalification occur simultaneously. Isoranks are folded and displaced the same way beds are, but less severely. The result is that a coal seam on major anticlines or on upthrown fault blocks is lower in rank than the same coal seam in adjacent synclines or downthrown fault blocks. Thus, in syndeformational coalification the total amount of deformation is only partially reflected in the deformation of isorank surfaces (less folded and/or displaced). England and Bustin (1986a, b) suspect that

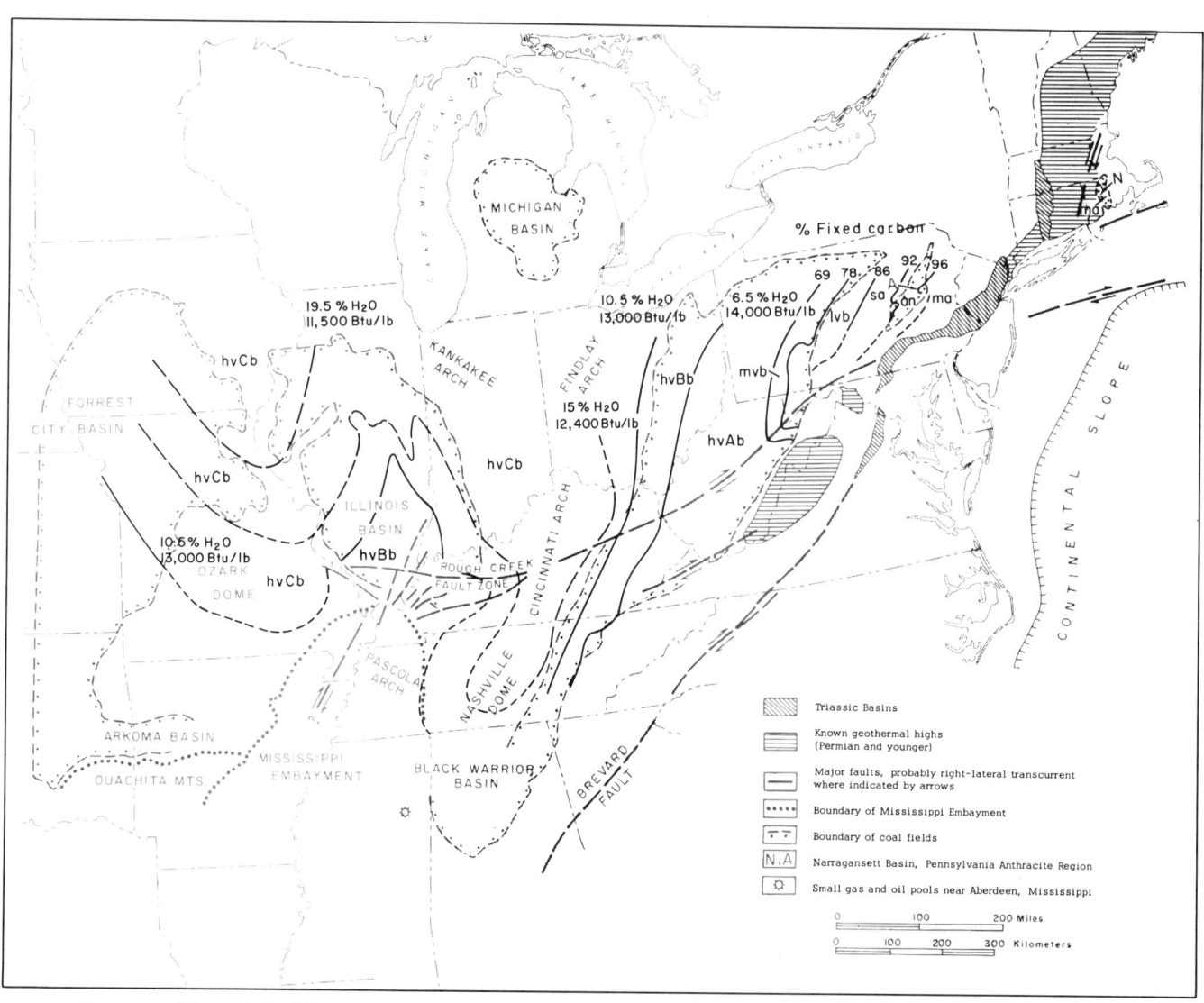

Figure 6. Coalification pattern of Pennsylvanian coal basins of the eastern United States pictured at the Illinois Herrin (No. 6) coal horizon and its equivalents, the Middle Kittanning coal bed of the Appalachian basin, and the Mystic of the Forest City basin (from Damberger, 1974).

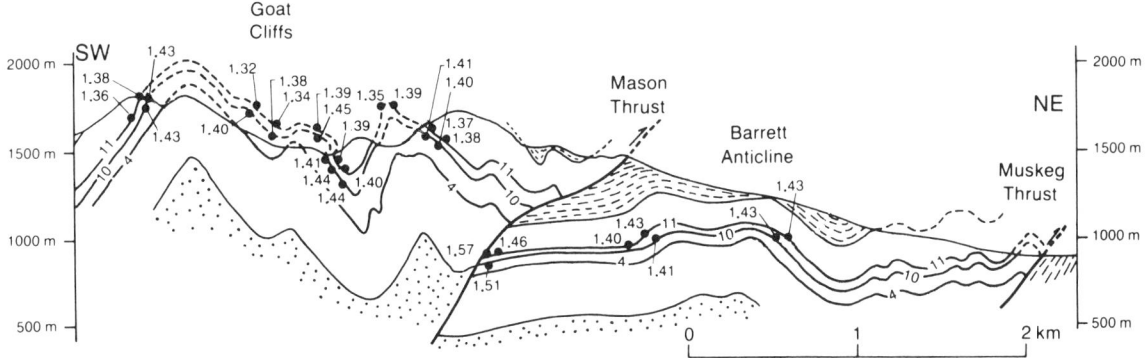

Figure 7. Predeformational coalification patterns in a cross section in the Canadian Rockies (Grande Cache area, Alberta): note that over a distance of about 4 km the reflectance of the no. 4 seam remains essentially unchanged up and down folds and across a major thrust (from Kalkreuth and Langenberg, 1986).

the origin of deformation in the Disturbed Belt of the Canadian Rocky Mountains is in part syndeformational, caused by large-scale overthrusting while coalification was ongoing.

Postdeformational coalification patterns

In the Sydney coal basin of Nova Scotia, the isoranks lie close to horizontal and are unrelated to the variable dip of the folded strata (Fig. 8; Hacquebard, 1975, 1983). Coalification must have taken place well after deformation, indicating that deepest burial occurred (or the paleogeothermal gradient increased significantly) subsequent to deformation. Hacquebard estimated that about 3.7 km of former overburden was eroded over the Sydney coals. Figure 1 would suggest removal of about 2.5 to 4.0 km by erosion (depth to 0.7 percent R_0). The approximate equivalent moisture content of 5 percent (Fig. 2a) leads to an estimate of about 3 km (Fig. 1a).

Along the southern closure of the Illinois basin, isorank surfaces dip more steeply than do coal seams (Fig. 9)—a condition contrary to what one expects to see along the margin of a cratonic basin and contrary to what one finds along the margins of other Midcontinent basins. Damberger (1971, 1974) suggests that this represents a postdeformational coalification pattern related to a late Paleozoic major geothermal event in the region.

Direct observations of the effects of heating on coal can be made along igneous dikes (Clegg, 1955; Dutcher and others, 1966; Bostick, 1974) or in laboratory experiments (Bostick, 1971). Apparently, many of the high-rank coals associated with the Rocky Mountains owe their high rank to heating during plutonism or vulcanism rather than to greater depth of burial (Kottlowski, 1986; Johnson, 1976; Johnson and others, 1963); but other, more detailed studies are needed for the Rocky Mountain coal fields.

The dominant influence of temperature on the coalification process was well demonstrated by the observation that the coalification gradient mirrors the geothermal gradient in several boreholes. At a given flow of heat, geothermal gradients are smaller in rocks with relatively high thermal conductivity (sandstone, limestone) than in rocks with a low thermal conductivity (shales, and especially coals). Due to the natural scatter in coalification data collected from drill holes, the direct correlation between coalification and geothermal gradients can only be detected in boreholes where thick rock sequences of highly contrasting heat conductivity occur (Damberger, 1966, 1968; Teichmüller and Teichmüller, 1966a, b). Hacquebard and Donaldson (1974) suggested that some of the variability in coalification gradients that they observed in the Rocky Mountain Inner Foothills of Alberta was caused by variable paleogeothermal gradients due to differences in the lithologic makeup of the rock sequences.

APPLICATIONS OF COALIFICATION STUDIES

The most widely used and economically important application of organic maturation studies is in exploration for oil and gas (see section on oil and gas, this volume), but there are many other applications.

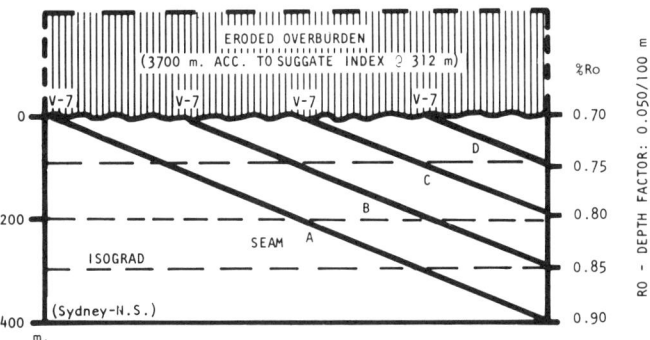

Figure 8. Typical postdeformational coalification pattern in a cross section from Nova Scotia (after Hacquebard, 1983; Hacquebard and Donaldson, 1970).

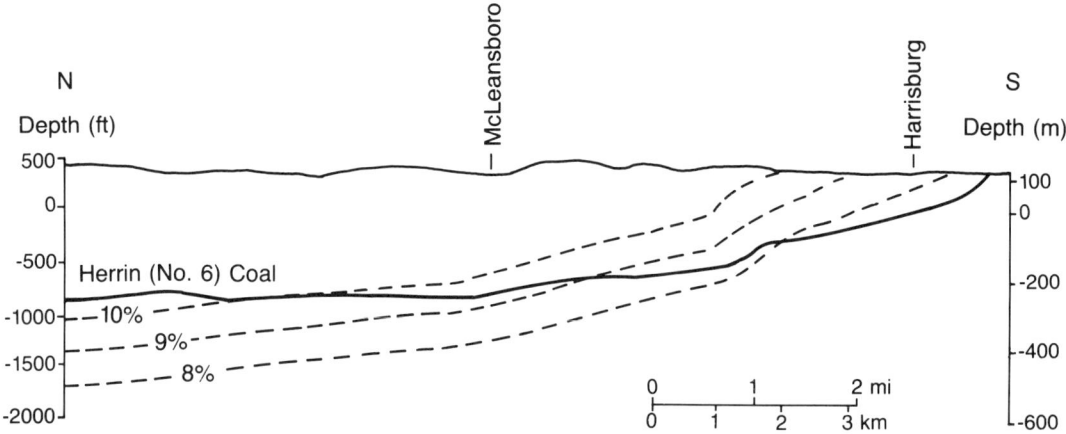

Figure 9. Cross section along the southern closure of the Illinois basin showing isorank surfaces dipping more steeply than strata as evidence for a postdeformational coalification pattern (from Damberger, 1971, 1974).

Structural analyses

Closely spaced sampling will define the position and deformation of isoranks in sections and maps, and determine the gradients. Oriented samples permit measurement of the orientation vectors of maximum and minimum reflectance values relative to structural features. Local and regional patterns of coalification that emerge from systematic sampling, analysis, and mapping can be interpreted in terms of relative ages of deformation and coal maturation processes. Knowledge of the regional pattern can be applied in several ways. For instance, if the pattern is predeformational, rank measurements can be used to correlate coal beds up and down folds or across faults. In the case of a fault penetrated in a drill hole or along a cross section, equivalent segments of the standard stratigraphic column and the composite coalification curve for the region will be missing or repeated at the intersection of the fault with the hole or section, thus permitting determination of displacement (Kalkreuth and Langenberg, 1986; Teichmüller and Teichmüller, 1958, 1966a, b; Damberger and others, 1964). Because of the normal scatter of rank data, closely spaced sampling is required, and displacements must exceed certain minimum values to be detectable (generally a few tens of meters for the best rank parameters).

In the case of postdeformational coalification, folding and/or faulting terminated well before the last phase of the coalification process was completed. A postdeformational pattern might result from renewed subsidence and burial under younger sediments long after folding and/or faulting and subsequent erosion took place. Such a case was described by Patijn (1964) for eastern Holland where Paleozoic coal-bearing strata are now buried much deeper than they were in late Paleozoic times when the main (initial) phase of coalification took place. The renewal of the coalification process obliterated the previous coalification pattern and resulted in the generation of large quantities of methane from maturing coal, which gave rise to one of the largest gas fields (Groningen) in the world.

Plutonism, which heats regions over long periods of time, is another possible cause for postdeformational coalification. It alters or completely obliterates any previous coalification pattern (Bartenstein and others, 1971; Stadler and Teichmüller, 1971), rendering coal rank useless for correlation across faults or folds. On the other hand, two separate deformational events—one predating, the other postdating coalification—could be easily distinguished.

The only well-documented example for postdeformational coalification in North America was described by Hacquebard (1975) referring to the Sydney coal field of Nova Scotia, Canada (Fig. 8), where isorank surfaces run parallel to a postdeformational erosional unconformity. Damberger (1971, 1974) pointed out that the coalification pattern along the southern closure of the Illinois basin may represent a postdeformational pattern related to deep-seated plutonism (Fig. 9). Based on broad coalification patterns rather than detailed studies, he also suggested that potential candidates for the detection of postdeformation coalification related to regional geothermal events include the Mississippi Embayment and the adjacent eastern part of the Arkoma basin; an area of low- and medium-volatile bituminous coals in southern West Virginia; a region of low- to medium-volatile coals in western Pennsylvania, Maryland, and northern West Virginia; the anthracite coal fields of eastern Pennsylvania; and the metaanthracite region of Rhode Island. Comprehensive studies of the interrelations between isoranks and structure need to be undertaken to fully evaluate the origin and significance of coalification patterns in these areas.

Alan Davis of Pennsylvania State University and his students applied the study of anisotropies of vitrinite reflectance to the analysis of coal metamorphism and the structural evolution of coal basins (Hower and Davis, 1981a, b; Levine, 1983, 1986; Levine and Davis, 1984; Paxton, 1983). The basic premise is that aromatic clusters of C-O-H functional groups developing during burial metamorphism in vitrinites and other macerals become arranged in response to the prevailing overburden pressure; and the degree of anisotropy is a measure of the depth of burial. Consequently, the maximum anisotropy is observed looking normal to the beding plane (Ting, 1984). Major differences in temperature at the same overburden pressure (and vice versa) should result in measurably different degrees of anisotropy. Furthermore, strain during folding could modify or completely obliterate anisotropy caused by depth of burial; or reflectance anisotropic axes could be folded passively with the bedding.

For instance, Levine (1983, 1986) reports that maximum reflectance axes of sporinite in strongly folded anthracite beds of the Western Middle Anthracite field of Pennsylvania are still aligned parallel to the bedding regardless of dip; therefore, they must have been folded passively with the bedding. Levine concludes that folding took place after the sporinites had passed through the stage of rapid devolatilization in the medium-volatile bituminous rank during basin subsidence. Deep burial to at least 5 km, and possibly as deep as 9 km at normal geothermal gradients (about 33° C/km) is considered the likely cause for attainment of the high rank. On the basis of paleomagnetic data, only 10 to 15 m.y. are available to accomplish such deep burial. To avoid having to invoke unusually high sedimentation (subsidence) rates, Levine (1986) suggests that part of the required deep burial of the anthracite-bearing strata was accomplished through emplacement of thick (several kilometers), large-scale overthrust sheets similar to those known from the southern Appalachians and other fold belts. Levine's coalification map shows a pattern that seems independent of the intense structural deformation of the region, except that the coals with the highest rank occur in the "Anthracite Depression," structurally a major synclinorium (Fig. 10). Such independence between structural and coalification patterns suggests postdeformational coalification. A much clearer understanding of the relative ages of deformation and coalification would emerge if a coalification map for a single stratigraphic horizon could be constructed.

Levine (1983, 1986) also reports that the minimum reflectance axes of vitrinites from the Broad Top coal field of south-central Pennsylvania in the Valley and Ridge Province generally plunge steeply toward the North American craton to the west, regardless of the dip of strata. Levine concludes that here the reflectance fabric was imposed postdeformationally, possibly by a combined burial and compressional stress field related to Alleghanian folding.

Clearly, the study of reflectance anisotropies provides potential additional insights into the relative timing of coalification and deformational events and an additional means to unravel the geologic history of complexly deformed terrains.

Paleogeothermal analyses

The dominant influence of temperature and duration of heating on the coalification process is no longer disputed as it was for many years following White's paper (1915) in which he related increased coalification to increased folding intensity (thrust pressure). Chemical reaction rates increase exponentially

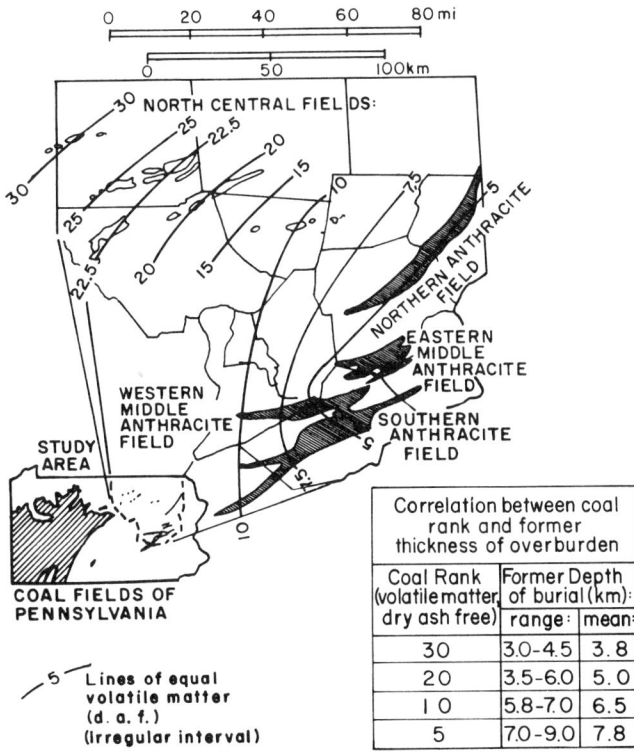

Figure 10. Coalification pattern of anthracite region in Pennsylvania depicted by lines of equal VM (daf) of coals at or near the surface regardless of their age (from Levine, 1986).

with increasing temperature (approximately doubling with every 10°C temperature increase). Therefore, the maximum temperature to which a coal was exposed during burial and the length of time it remained subjected to the maximum or nearly maximum temperature (within about 15°C) are considered the most important factors in determining coal maturity (Karweil, 1956; Teichmüller and Teichmüller, 1966b; Bostick, 1971, 1974; Lopatin, 1971; Tissot and Espitalie, 1975; Geologisches Landesamt Nordrhein-Westfalen, 1979; Buntebarth, 1982; Lerche and others, 1984; Waples, 1984). Karweil's (1956) often reproduced and modified nomograph correlates coal rank with the length of time coal was subjected to the maximum coalification temperature (Fig. 11a). The nomograph is based on reaction kinetics (Arrhenius equation) and was calibrated by geologic data from the Ruhr coal basin. It predicts, for instance, that for a coal to reach medium-volatile bituminous rank (30 percent VM) requires about 110 m.y. of exposure to 100°C, or 10 m.y. to 200°C; anthracite of 10 percent VM requires more than 300 m.y. of exposure to 100°C, but only about 30 m.y. exposure to 200°C. These selected time-temperature pairs illustrate how temperature and duration of heating can substitute for each other. A similar nomograph by Bostick and others (1979) (Fig. 11b) predicts slightly lower ranks for the same time-temperature pairs.

Tissot and Espitalie (1975) developed a comprehensive mathematical model that simulates the kinetics of kerogen and

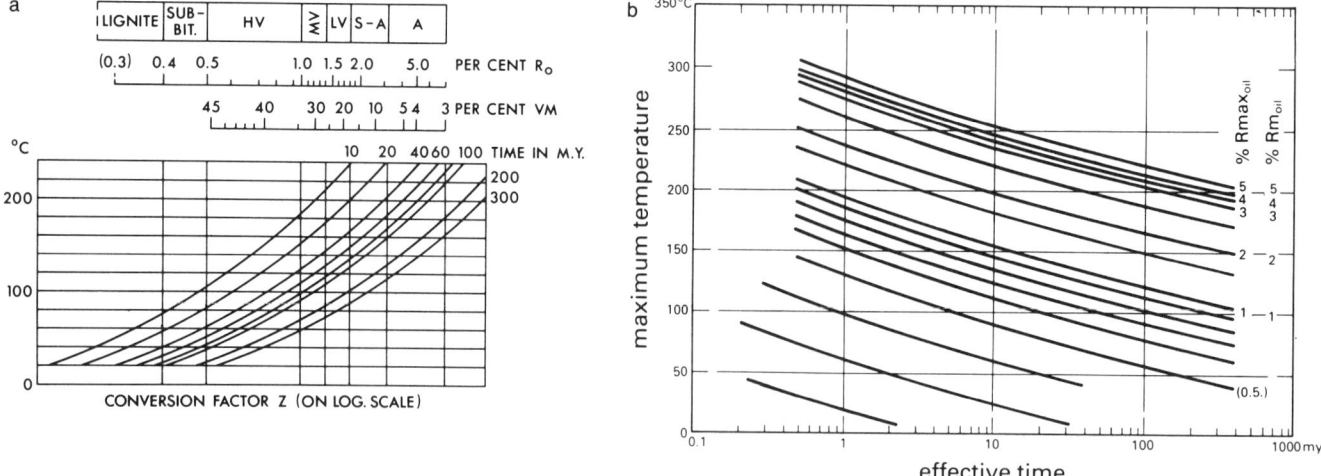

Figure 11. Karweil's nomograph (a), which relates coal rank to temperature at various exposure times (from Hacquebard, 1977, after Karweil, 1956), and a similar nomograph by Bostick (b), which relates vitrinite reflectance to maximum coalification temperature and effective coalification time, that is, the time exposed to within 15°C of maximum temperature (from Teichmüller, 1987; after Bostick and others, 1979).

coal alteration due to gradually rising temperatures during burial. Their formulas consider temperature, time, and character of organic material. Adjustments in exposure time and temperature are made until the model predicts the measured maturation values (geologic calibration). Lopatin's (1971) widely used and commonly computerized method uses stepwise integration of the effects of temperature and time, through use of a time-temperature index, TTI (Lopatin and Bostick, 1973).

Calibration is difficult in these mathematical models, particularly for maximum temperature in old sedimentary basins. If such basins have been subjected to higher temperatures in the past, the influence of duration of heating will be too great in the model. For example, Karweil (1956) assumed a paleogeothermal gradient of 40°C/km in constructing his nomograph; if the gradient had been much higher at some past time, his calibration would be in error. When the models are properly calibrated, they permit estimation of paleogeothermics (maximum temperature reached during burial, paleogeothermal gradient), given that burial history and rank gradients are known, or estimation of burial history, given that geothermal gradients are known or can be estimated with some degree of confidence.

Buntebarth's (1979, 1982, 1984) method estimates paleogeothermal gradients from vitrinite reflectance gradients. Paleogeothermal gradients thus estimated are often significantly different from today's geothermal gradients for the same region (Buntebarth and others, 1982). Maps and sections that show variations in coalification gradients of a region thus can now be interpreted as maps depicting paleogeothermal gradients. Such maps and sections have been published for several coal basins, including the western Canadian basin of the Alberta Plains (England and Bustin, 1986a, b, based in part on data published by Hacquebard and Donaldson, 1974). Kalkreuth and McMechan (1988) discuss the probable influence of burial depth and geothermal gradients on the regional coalification pattern between the Rocky Mountain Front Ranges and foreland of British Columbia and Alberta of western Canada.

Prediction of coal quality

Coal rank is also used to predict many important technological properties of coal, including behavior of coal in coking, combustion, and liquefaction and gasification processes. Facies is the other basic genetic factor to be considered (Elliott, 1986).

Local and regional trends of coal rank and relations to technological properties of coal, in addition to data on seam thickness, depth, and facies-related quality, must be known to properly assess coal reserves and resources. Knowledge of vertical coalification gradients is equally as important as knowledge of the lateral extent of the various members of the coalification series, either in a chosen coal bed or at selected elevation levels. For instance, rapid change of coal rank with depth (large gradient) reduces a region's resources of prime coking coals, while slow change would increase them. On the other hand, a large gradient may put coals of desirable characteristics within closer reach of the surface than a small coalification gradient.

Application in mineral exploration

Many important mineral deposits are related to plutonism, vulcanism, and basin evolution. Coalification studies can assist in delineating areas subjected to heating by deep-seated plutons that may have produced significant mineralization in their vicinity. Aureoles of high-rank coal around plutons help define areas of prospective mineralization: the Bramsche and Vlotho Massifs of northwestern Germany (Teichmüller and Teichmüller, 1968b); the Erkelenz anomaly of the western Rhine-Ruhr coal district

(Patteisky and others, 1962; Teichmüller and Teichmüller, 1968a); and the Alston Block of northern England (Trotter, 1954). The present erosional surface is often not a good reference horizon for delineating an anomaly, for instance, in the case of faulting that followed emplacement of the plutonic intrusion—a common occurrence.

OUTLOOK

Most coalification maps of North American coal fields depict coal rank at the present erosion surface, regardless of the age of the sampled coal seams. Such maps are difficult to interpret in terms of timing of the coalification process relative to deformational and geothermal events. New maps showing both structure and rank of a single coal seam, and cross sections illustrating coalification gradients and the relation between deformation of strata and isorank surfaces, should respond to questions raised in this review. Equilibrium moisture data are needed, especially for low-rank coals in western North America.

Coalification maps should be constructed to depict the elevation (structure, depth) of selected isorank surfaces, such as the top and base of prime coking coals (medium- and low-volatile bituminous ranks) and the top and bottom of the "oil window." Maps of coal rank at selected elevations (e.g., at sea level) would also have obvious applications.

The search for a reliable microscopic parameter of rank for coals with a vitrinite reflectance less than about 1.0 percent R_0 must continue. These coals are most heterogeneous, and microscopic methods are most likely to permit standardization for comparable macerals. Most chemical and physical parameters are normally determined on mixtures of the macerals of coal and thus tend to be more or less strongly influenced by the heterogeneous composition of coals. Porosity and inherent moisture content (or heating value, on mineral-matter-free, moist basis) exhibit a rapid change with depth for these low-rank coals and thus provide useful rank parameters. However, coals with the same moisture content may exhibit significantly different "chemical rank" (represented by C, O, VM contents, reflectance, and some technological and geochemical properties).

Information on coalification gradients is still limited for North American coal fields. It is needed for better coal resource assessments and to develop a better understanding of geothermal regimes through the geologic history of our coal basins, most of which also contain petroleum and mineral deposits whose areal distribution was strongly influenced by paleogeothermics.

REFERENCES CITED

Alpern, B., 1984, Petrographie des charbons et gazefication in situ: Bulletin Societe Geologique France, v. 26, p. 739-756.

Altschaeffl, A. G., and Harrison, W., 1959, Estimation of a minimum depth of burial for a Pennsylvanian underclay: Journal of Sedimentary Petrology, v. 29, p. 178-185.

Altschuler, Z. S., Schnepfe, M. M., Silber, C. C., and Simon, F. O., 1983, Sulfur diagenesis in Everglades peat and origin of pyrite in coal: Science, v. 221, p. 221-227.

American Society of Testing and Materials (ASTM), D388, Standard specification for classification of coals by rank; D2798, Standard method for microscopical determination of the reflectance of the organic components in a polished specimen of coal, in Annual Book of ASTM Standards: American Society of Testing and Materials, v. 29.

Barker, C. E., and Elders, W. A., 1981, Vitrinite reflectance geothermometry and apparent heating duration in the Cerro Prieto geothermal field: Geothermics, v. 10, p. 207-223.

Bartenstein, H., Teichmüller, M., and Teichmüller, R., 1971, Die Umwandlung der organischen Substanz im Dach des Bramscher Massivs: Fortschritte der Geologie von Rheinland und Westfalen, v. 18, p. 501-538.

Bostick, N. H., 1971, Thermal alteration of clastic organic particles as an indicator of contact and burial metamorphism in sedimentary rocks: Geoscience and Man, v. 3, p. 83-92.

——— , 1974, Phytoclasts as indicators of thermal metamorphism, Franciscan Assemblage and Great Valley Sequence (upper Mesozoic), California, in Dutcher, R. R., and others, eds., Carbonaceous materials as indicators of metamorphism: Geological Society of America Special Paper 153, p. 1-17.

Bostick, N. H., Cashman, S. M., McCulloch, T. H., and Waddel, C. T., 1979, Gradients of vitrinite reflectance and present temperature in Los Angeles and Ventura Basins, California, in Oltz, D. F., ed., Low temperature metamorphism of kerogen and clay minerals: Pacific Section, Society of Economic Paleontologists and Mineralogists, p. 65-96.

Buntebarth, G., 1979, Eine empirische Methode zur Berechnung von paläogeothermischen Gradienten aus dem Inkohlungsgrad organischer Einlagerungen in Sedimentgesteinen mit Anwendung auf den mittleren Oberrhein-Graben: Fortschritte der Geologie von Rheinland und Westfalen, v. 27, p. 97-108.

——— , 1982, Geothermal history estimated from the coalification of organic matter: Tectonophysics, v. 83, p. 101-108.

——— , 1984, Geothermics; An introduction: Berlin, Heidelberg, New York, Springer-Verlag, 144 p.

Buntebarth, G., Koppe, I., and Teichmüller, M., 1982, Paleogeothermics in the Ruhr Basin, in Cermak, V., and Hanel, R., eds., Geothermics and geothermal energy: Stuttgart, Schweizerbart, p. 45-55.

Bustin, R. M., 1986, Organic maturity of late Cretaceous and Tertiary coal measures, Canadian Arctic Archipelago: International Journal of Coal Geology, v. 6, p. 71-106.

Chyi, L. L., Barnett, R. G., Burfort, A. E., Quick, T. J., and Gray, R. J., 1987, Coalification patterns of the Pittsburgh coal; Their origin and bearing on hydrocarbon maturation: International Journal of Coal Geology, v. 7, p. 69-83.

Clegg, K., 1955, Metamorphism of coal by peridotite dikes in southern Illinois: Illinois Geological Survey Report of Investigation 178, 18 p.

Cohen, A. D., 1968, The petrology of some peats of southern Florida (with special reference to the origin of coal) [Ph.D. thesis]: University Park, Pennsylvania State University, 352 p.

Cohen, A. D., and Spackman, W., 1980, Phytogenetic organic sediments and sedimentary environments in the Everglades-Mangrove Complex of Florida; Part 3, The alteration of plant material in peats and the origin of coal macerals: Paleontographica Abteilung B, v. 172, p. 125-149.

Cohen, A. D., Casagrande, D. J., Andrejko, M. J., and Best, G. R., eds., 1984, The Okefenokee Swamp; Its natural history, geology, and geochemistry: Los Alamos, Wetland Surveys, 709 p.

Damberger, H., 1966, Die Abhängigkeit des Inkohlungsgradienten vom Gesteinsaufbau: Zeitschrift der deutschen geologischen Gesellschaft, Jahrgang 1965, v. 117, p. 8.

——— , 1968, Ein Nachweis der Abhängigkeit der Inkohlung von der Temperatur: Brennstoff-Chemie, v. 49, p. 73-77.

Damberger, H. H., 1971, Coalification pattern of the Illinois Basin: Economic Geology, v. 66, p. 488–494.

——, 1974, Coalification patterns of Pennsylvanian basins of the eastern United States, in Dutcher, R. R., and others, eds., Carbonaceous materials as indicators of metamorphism: Geological Society of America Special Paper 153, p. 53–74.

Damberger, H. H., Harvey, R. D., Ruch, R. R., and Thomas, J., Jr., 1984, Coal characterization: in Cooper, B. R., and Ellingson, W. A., eds., The science and technology of coal and coal utilization: New York, Plenum Press, p. 7–45.

Damberger, H., Kneuper, G., Teichmüller, M., and Teichmüller, R., 1964, Das Inkohlungsbild des Saarkarbons: Glückauf, v. 100, p. 209–217.

Dutcher, R. R., Campbell, D. L., and Thornton, C. P., 1966, Coal metamorphism and igneous intrusives in Colorado: American Chemical Society, Advancements in Chemistry Series, no. 55, p. 708–723.

Edwards, A. B., 1945, The composition of Victorian brown coals: Proceedings of the Australian Institute of Mineralogy and Metallurgy, no. 140, p. 252.

Elliott, M. A., ed., 1986, Chemistry of coal utilization, 2nd supplementary volume: New York, John Wiley and Sons, 2374 p.

Energy Resources Company, 1980, Low rank coal study; Vol. 2, Resource characterization: Prepared for U.S. Department of Energy, Contract DE-AC18-79FC10066, 262 p.

England, T.D.J., and Bustin, R. M., 1986a, Effect of thrust faulting on orogenic maturation in the southeastern Canadian Cordillera; Advances in organic geochemistry 1985: Organic Geochemistry, v. 10, p. 609–616.

——, 1986b, Thermal maturation of the western Canadian sedimentary basin south of the Red Deer River; 1, Alberta Plains: Bulletin of Canadian Petroleum Geology, v. 34, p. 71–90.

Geologisches Landesamt Nordrhein-Westfalen, 1979, Inkohlung und Geothermik: Fortschritte der Geologie von Rheinland und Westfalen, v. 27, 372 p.

Gilchrist, R., 1974, Principles of coal systematics, in Fryer, J. F., Campbell, J. D., and Speight, J. G., eds., Symposium on coal evaluation: Alberta Research Council Information Series 76, p. 31–39.

Goodarzi, F., Gentzis, T., Feinstein, S., and Snowdon, L., 1988, Effect of maceral subtypes and mineral matrix on measured reflectance of subbituminous coals and dispersed organic matter: International Journal of Coal Geology, v. 10, p. 383–398.

Hacquebard, P. A., 1974, A composite coalification curve of the Maritime Region and its value for petroleum exploration: Geological Survey of Canada Paper 74-1, Part B, p. 21–23.

——, 1975, Pre- and postdeformational coalification and its significance for oil and gas exploration, in Alpern, B., ed., Petrographie de la matiere organique des sediments, relations avec la paleotemperature et le potentiel petrolier: Editions Centre Nat. Rech. Sci. (CERCHAR), Colloquium in September 1973, in Paris, France, p. 225–241.

——, 1977, Rank of coal as an index of organic metamorphism for oil and gas in Alberta: Geological Survey of Canada Bulletin 262, p. 11–22.

——, 1983, Geological development and economic evaluation of the Sydney coal basin, Nova Scotia, in Current research, Part A: Geological Survey of Canada Paper 83-1A, p. 71–81.

Hacquebard, P. A., and Donaldson, I. R., 1970, Coal metamorphism and hydrocarbon potential in the upper Paleozoic of the Atlantic Provinces, Canada: Canadian Journal of Earth Sciences, v. 7, p. 1139–1163.

——, 1974, Rank studies of coals in the Rocky Mountains and Inner Foothills belt, Canada, in Dutcher, R. R., and others, eds., Carbonaceous materials as indicators of metamorphism: Geological Society of America Special Paper 153, p. 75–94.

Harvey, R. D., and Ruch, R. R., 1986, Mineral matter in Illinois and other U.S. coals, in Vorres, K. S., ed., Mineral matter and ash in coal: American Chemical Society Symposium Series 301, p. 10–40.

Hatcher, P. G., Breger, I. A., Szeverenyi, N., and Maciel, G. E., 1982, Nuclear magnetic resonance studies of ancient buried wood; 2, Observations on the origin of coal from lignite to bituminous coal: Organic Geochemistry, v. 4, p. 9–18.

Hilt, C., 1873, Die Beziehungen zwischen der Zusammensetzung und den technischen Eigenschaften der Steinkohle: Verein Deutscher Ingenieure, Zeitschrift des VDI, Sitzungsberichte des Aachener Bezirks Vereins, 2/12/73, v. 17, Heft 4, p. 193–202.

Houseknecht, D. W., 1987, The Atoka Formation of the Arkoma Basin; Tectonics, sedimentology, thermal maturity, sandstone petrology: Tulsa Geological Society Short Course, p. 19–24.

Hower, J. C., and Davis, A., 1981a, Application of vitrinite reflectance anisotropy in the evaluation of coal metamorphism: Geological Society of America Bulletin, v. 92, p. 350–366.

——, 1981b, Vitrinite reflectance anisotropy as a tectonic fabric element: Geology, v. 9, p. 165–168.

Hower, J. C., Trinkle, E. J., and Bland, A. E., 1985, Coal geoscience, petrology, geochemistry, and mineralogy of the Springfield (No. 9) and Herrin (No. 11) Coals in western Kentucky: Lexington, Kentucky Center for Energy Research Laboratory, KCERL/85-138, 277 p.

International Committee for Coal Petrology (ICCP), 1963, International handbook of coal petrography, 2nd ed. and supplement to 2nd ed. (1971): Centre National de Recherche Scientifique, Paris.

International Standards Organization (ISO), 1984, International Standard 7404/5, Methods for the petrographic analysis of bituminous coal and anthracite; Part 5, Method of determinating microscopically the reflectance of vitrinite: International Standards Organization 7404/5-1984 (E).

Johnson, V. H., 1976, Metamorphic patterns in western Cretaceous coals and their geoenvironmental implications: Provo, Utah, Brigham Young University Geology Studies, v. 22, pt. 3, p. 45–58.

Johnson, V. H., Gray, R. J., and Schapiro, N., 1963, Effect of igneous intrusives on the chemical, physical, and optical properties of Somerset coal: American Chemical Society, 145th National Meeting, Division of Fuel Chemistry, v. 7, no. 2, p. 110–124.

Jüntgen, H., and Klein, J., 1975, Entstehung von Erdgas aus kohligen Sedimenten: Erdöl und Kohle, v. 28, p. 65–73.

Kalkreuth, W., 1982, Rank and petrographic composition of selected Jurassic–Lower Cretaceous coals of British Columbia, Canada: Bulletin of Canadian Petroleum Geology, v. 30, p. 113–139.

Kalkreuth, W., and Langenberg, C. W., 1986, The timing in coalification in relation to structural events in the Grande Cache area, Alberta, Canada: Canadian Journal of Earth Sciences, v. 23, p. 1103–1116.

Kalkreuth, W., and Macauley, G., 1987, Organic petrology and geochemical (Rock-Eval) studies on oil shales and coals from the Pictou and Antigonish areas, Nova Scotia, Canada: Bulletin of Canadian Petroleum Geology, v. 35, p. 263–295.

Kalkreuth, W., and McMechan, M. E., 1984, Regional pattern of thermal maturation as determined from coal rank studies, Rocky Mountain Foothills and Front Ranges north of Grande Cache, Alberta; Implications for petroleum exploration: Bulletin of Canadian Petroleum Geology, v. 32, p. 249–271.

——, 1988, Burial history and thermal maturity, Rocky Mountain Front Ranges, Foothills and foreland, east-central British Columbia and adjacent Alberta, Canada: American Association of Petroleum Geologists Bulletin, v. 72, p. 1395–1410.

Karweil, J., 1956, Die Metamorphose der Kohlen vom Standpunkt der physikalischen Chemie: Zeitschrift der deutschen geologischen Gesellschaft, v. 107, Jahrgang 1955, p. 132–139.

Kötter, K., 1960, Die mikroskopische Reflexionsmessung mit dem Photomultiplier und ihre Anwendung auf die Kohlenuntersuchung: Brennstoff-Chemie, v. 41, p. 263–272.

Kottlowski, F., 1986, Coal seams and fields, New Mexico; 1987 Keystone Coal Industry Manual: New York, McGraw-Hill, p. 479–488.

Lemos de Sousa, M. J., 1979, Contribuicao do estudo das perantracites durienses para o conhecimento das curvas gerais de incarbonizacao dos carvoes norte-atlanticos: Publ. Mus. Labor. miner. geol. Fac. Ci. Porto, v. 91, 4. Ser., p. 253–265.

Lerche, I., Yarzab, R. E., and Kendall, C.G.St.C., 1984, Determination of paleoheat flux from vitrinite reflectance data: American Association of Petroleum Geologists Bulletin, v. 68, p. 1704–1717.

Levine, J. R., 1983, Tectonic history of coal-bearing sediments in eastern Pennsylvania using coal reflectance anisotropy [Ph.D. thesis]: University Park, Pennsylvania State University, 314 p.
——, 1986, Deep burial of coal-bearing strata, Anthracite region, Pennsylvania; Sedimentation or tectonics: Geology, v. 14, p. 577–580.
Levine, J. R., and Davis, A., 1984, Optical anisotropy of coals as an indicator of tectonic deformation, Broad Top coal field, Pennsylvania: Geological Society of America Bulletin, v. 95, p. 100–108.
Leythaeuser, D., and Welte, D. H., 1969, Relation between distribution of heavy n-parafins and coalification in Carboniferous coals from the Saar district, Germany, in Schenck, P. A., and Havenaar, I., eds., Advances in organic geochemistry 1968: Oxford, Pergamon Press, p. 429–442.
Littke, R., 1987, Petrology and genesis of upper Carboniferous seams from the Ruhr region, West Germany: International Journal of Coal Geology, v. 7, p. 147–184.
Lopatin, N. V., 1971, Temperature and geologic time as factors in coalification: Akad. Nauk. SSSR, Ser. Geol. Izvestiya, v. 3, p. 95–106 (translated into English by N. H. Bostick).
Lopatin, N. V., and Bostick, N. H., 1973, The geologic factors in coal catagenesis: Akad. Nauk. SSSR Otdeleniye Geologii, Geofiziki, Geochimi, Kom. Osad. Porodam, Moscow, Nauka Press, p. 79–90.
McCartney, J. T., 1952, A study of the Seyler theory of coal reflectance: Economic Geology, v. 47, p. 202–210.
Mukhopadhyay, P. K., 1987, Characterization of Tertiary coals from Texas on the basis of organic petrographic, chemical, and organic geochemical properties: Austin, University of Texas Bureau of Economic Geology contract report for the U.S. Geological Survey, 87 p.
Nurkowski, J. R., 1985, Coal quality and rank variation within Upper Cretaceous and Tertiary sediments, Alberta Plains region: Alberta Research Council Earth-Science Report 85-1, 39 p.
Parkash, S., and Chakrabartty, S. K., 1986, Microporosity in Alberta Plains coals: International Journal of Coal Geology, v. 6, p. 55–70.
Patijn, R.J.H., 1964, Die Entstehung von Erdgas infolge der Nachinkohlung im Nordosten der Niederlande: Erdöl und Kohle, v. 17, p. 2–9.
Patteisky, K., and Teichmüller, M., 1960, Inkohlungs-Verlauf, Inkohlungs-Maszstäbe und Klassifikation der Kohlen auf Grund von Vitrit-Analysen: Brennstoff-Chemie, v. 41, p. 79–84, 97–104, 133–137.
Patteisky, K., Teichmüller, M., and Teichmüller, R., 1962, Das Inkohlungsbild des Steinkohlengebirges an Rhein und Ruhr, dargestellt im Niveau von Flöz Sonnenschein: Fortschritte der Geologie von Rheinland und Westfalen, v. 3, p. 687–700.
Paxton, S. T., 1983, Relationships between Pennsylvanian-age lithic sandstone and mudrock diagenesis and coal rank in the central Appalachians [Ph.D. thesis]: University Press, Pennsylvania State University, 503 p.
Radke, M., Welte, D. H., and Willsch, H., 1982a, Geochemical study on a well in the western Canada Basin; Relation of the aromatic distribution pattern to maturity of organic matter: Geochimica et Cosmochimica Acta, v. 46, p. 1–10.
Radke, M., Willsch, H., Leythaeuser, D., and Teichmüller, M., 1982b, Aromatic components of coal; Relation of distribution pattern to rank: Geochimica et Cosmochimica Acta, v. 46, p. 1831–1848.
Reeves, F., 1928, The carbon-ratio theory in the light of Hilt's law: American Association of Petroleum Geologists Bulletin, v. 12, p. 795–823.
Riepe, W., and Wolff-Fischer, E., 1982, Die Zuverlässigkeit der Bestimmung des Reflexionsvermögens von Vitriniten: Glückauf-Forschungshefte, v. 43, p. 50–52.
Rightmire, C. T., Eddy, G. E., and Kirr, J. N., 1984, Coalbed methane resources of the United States: American Association of Petroleum Geologists Studies in Geology Series 17, 378 p.
Schürmann, H.M.C., 1927, Über jungtertiäre Braunkohlen in Ostborneo: Braunkohle, p. 609–641.
Shibaoka, M., 1978, Micrinite and exsudatinite in some Australian coals, and their relation to the generation of petroleum: Fuel, v. 57, p. 73–78.
Spackman, W., and Barghoorn, E. S., 1966, Coalification of woody tissue as deduced from a petrographic study of Brandon lignite: American Chemical Society Advances in Chemistry Series, v. 55, p. 695–707.
Spackman, W., Cohen, A. D., Given, P. H., and Casagrande, D. J., 1974, The comparative study of the Okefenokee Swamp and the Everglades-Mangrove Swamp-Marsh Complex of southern Florida, in A field guidebook for the Geological Society of America, field trip No. 6, Nov. 15–17, 1974: 265 p.
Stach, E., 1953, Der Inkohlungssprung im Ruhrkarbon: Brennstoff-Chemie, v. 34, p. 353–355.
Stach, E., and 5 others, 1982, Stach's textbook of coal petrology, 3rd edition: Stuttgart, Gebrüder Borntraeger, 535 p.
Stadler, G., and Teichmüller, R., 1971, Zusammenfassender Überblick über die Entwicklung des Bramscher Massivs und des Niedersächsischen Tektogens: Fortschritte der Geologie von Rheinland und Westfalen, v. 18, p. 547–564.
Teichmüller, M., 1968, Zur Petrographie und Diagenese eines fast 200 m mächtigen Torfprofils (mit Übergängen zur Weichbraunkohle?) im Quartär von Phillippi (Mazedonien): Geologische Mitteilungen, v. 8, p. 65–110.
——, 1974, Über neue Macerale der Liptinit-Gruppe und die Entstehung des Micrinits: Fortschritte der Geologie von Rheinland und Westfalen, v. 24, p. 37–64.
——, 1982, Origin of petrographic constituents of coal, in Stach, E., and 5 others, eds., Stach's textbook of coal petrology, 3rd ed.: Stuttgart, Gebrüder Borntraeger, p. 223–294.
——, 1987, Recent advances in coalification studies and their application to geology, in Scott, A. C., ed., Coal and coal-bearing strata; Recent advances; Geological Society of London Special Publication 32: London, Blackwell Scientific Publications, p. 127–169.
Teichmüller, M., and Teichmüller, R., 1958, Inkohlungsuntersuchungen und ihre Nutzanwendung: Geologie en Mijnbouw, n.s., v. 20, p. 41–66.
——, 1966a, Die Inkohlung im saarlothringischen Karbon, verglichen mit der im Ruhrkarbon: Zeitschrift der deutschen geologischen Gesellschaft, Jahrgang 1965, v. 117, p. 243–279.
——, 1966b, Geological causes of coalification: American Chemical Society Advances in Chemistry Series 55, p. 133–155.
——, 1968a, Geological aspects of coal metamorphism, in Murchison, D. G., and Westoll, T. S., eds., Coal and coal-bearing strata: Edinburgh, Oliver and Boyd, p. 233–267.
——, 1968b, Cainozoic and Mesozoic coal deposits of Germany, in Murchison, D. G., and Westoll, T. S., eds., Coal and coal-bearing strata: Edinburgh, Oliver and Boyd, p. 347–379.
——, 1984, Verbreitung und Eigenschaften tiefliegender Steinkohlen in der Bundesrepublik Deutschland: Glückauf-Forschungshefte, Jahrgang 45, Heft 3, p. 140–153.
Tewalt, S. J., 1986, Chemical characterization of Texas lignite: Austin, Texas, Bureau of Economic Geology Geological Circular 86-1, 54 p.
Thomas, J., Jr., and Damberger, H. H., 1976, Internal surface area, moisture content, and porosity of Illinois coals; Variations with rank: Illinois Geological Survey Circular 493, 38 p.
Ting, F.T.C., 1984, Paragenetic relationship of thermal maturation (coalification) and tectonic framework of some Canadian Rocky Mountain coals: Organic Geochemistry, v. 5, p. 279–281.
Tissot, B., and Espitalie, J., 1975, L'evolution thermique de la matiere organique des sediments; Applications d'une simulation mathematique: Rev. Inst. Franc. Petrol., v. 30, p. 743–777.
Tissot, B. P., and Welte, D. H., 1984, Petroleum formation and occurrence; A new approach to oil and gas exploration, 2nd ed.: Berlin, Heidelberg, New York, Springer-Verlag, 699 p.
Trotter, F. M., 1954, The genesis of the high rank coals: Proceedings of the Yorkshire Geological Society, v. 29, p. 267–303.
Van Krevelen, D. W., 1961, Coal; Typology-chemistry-physics-constitution: Amsterdam, Elsevier, 514 p.
Waddell, C., Davis, A., Spackman, W., and Griffiths, J. C., 1978, Study of the

interrelationships among chemical and petrographic variables of United States coals: University Park, Pennsylvanian State University Coal Research Section, PSV-TR 9, 240 p.

Walsh, T. J., and Phillips, W. M., 1982, Coal rank and thermal maturation in King County, Washington: Washington Geological Newsletter, v. 10, no. 1, p. 9–19.

Waples, D. W., 1984, Thermal models for oil generation, *in* Brooks, J., and Welte, D., eds., Advances in petroleum geochemistry: London, Academic Press, v. 1, p. 7–67.

Weller, J. M., 1959, Compaction of sediments: American Association of Petroleum Geologists Bulletin, v. 43, p. 273–310.

White, C. D., 1915, Some relations in the origin between coal and petroleum: Washington Academy of Sciences Journal, v. 5, p. 189–212.

White, D., 1913, Regional metamorphism of coal, *in* White, D., and Thiessen, R., eds., The origin of coal: U.S. Bureau of Mines Bulletin 38, p. 91–130.

Wolf, M., and Wolff-Fischer, E., 1984, Alginit in Humuskohlen karbonischen Alters und sein Einflusz auf die optischen Eigenschaften des begleitenden Vitrinits: Glückauf-Forshungshefte, v. 45, p. 243–246.

Manuscript Accepted by the Society September 30, 1989

ACKNOWLEDGMENTS

I greatly appreciate the constructive reviews of my original manuscript by Neely Bostick and Paul Lyons of the U.S. Geological Survey, Pete Palmer of the Geological Society of America, and Richard D. Harvey and Donald F. Oltz of the Illinois State Geological Survey. Linda Bragg of the U.S. Geological Survey, Branch of Coal Resources, prepared a floppy disk with basin-by-basin data sets on the chemistry of coals from the National Coal Resources Data System. I used these and other data to construct the correlation charts of Figure 2. Also I gratefully acknowledge the assistance of many others, who provided me with confidential or as yet unpublished data sets and copies of their publications.

Printed in U.S.A.

Chapter 33

Pennsylvanian coals of central and eastern United States

Alan C. Donaldson and Cortland Eble
Department of Geology and Geography, West Virginia University, Morgantown, West Virginia 26506

INTRODUCTION

Pennsylvanian coal measures

Geologic conditions during the Pennsylvanian created events necessary for coal-forming that are relatively rare in geologic history. This mix of conditions involved (1) tectonics accompanied by sedimentation; (2) depositional environments that favored poorly drained, clear-water, acidic swamps; (3) climates that encouraged peat formation and preservation; (4) plant evolution that produced sufficient biomass to keep pace with a changing base level; and (5) repeated cycles of marine and nonmarine sedimentation, alternating between detrital influx and "starvation" in response to eustatic, tectonic, and sediment-supply changes. Factors important in determining the quality and thickness of coal beds are the relative rates of subsidence, depositional environments where ancient swamps existed, climate, and the frequency and extent of shoreline shifts. Integration of the above conditions and factors creates a complex explanation for origin of the Pennsylvanian coal measures in the eastern and central United States.

Excellent regional stratigraphic synopses of the Pennsylvanian coal measures indicate that the Upper Carboniferous (Pennsylvanian) coal measures constitute a large part of the total United States resources (Branson, 1962; Arndt and others, 1968; U.S. Bureau of Mines, 1974; McKee and others, 1975; Craig and others, 1979; Rightmire and others, 1984; Lyons and Rice, 1986; Sloss, 1988). Pennsylvanian coal beds are mainly preserved in structural basins located in the central and eastern United States (Fig. 1). The quality and thickness of coal beds, greatest in the Appalachian basin (anthracite and bituminous extending from Pennsylvania to Alabama/Mississippi), decreases westward across the Eastern Interior basin (Illinois, Indiana, western Kentucky), the Western Interior basin (Arkansas, Oklahoma, Kansas, Missouri, Iowa, Nebraska), and the Southwestern Interior basin (Texas). Table 1 summarizes characteristics of these coal basins.

In the Appalachian basin, Carboniferous coal beds occur within the upper part of the Acadian clastic wedge (Lower Mississippian), the upper part of the Ouachita clastic wedge (Lower Pennsylvanian) in the Black Warrior basin of Alabama, and the Alleghany clastic wedge (Lower, Middle, and Upper Pennsylvanian). Coal measures in the foreland troughs are associated with molasse sedimentation derived from adjacent orogenic belts. The Eastern and Western Interior basins exhibit coal measures mainly of Middle and Late Pennsylvanian age, but these intracratonic basins are infilled by siliciclastics derived mostly from the Canadian Shield source area. Figure 1 shows the relative thicknesses of Pennsylvanian coal measures for the different basins.

Except for the brief overview below of Mississippian coal beds, this chapter emphasizes commercially important coal beds of the Pennsylvanian System in the lower part of the Absaroka Sequence (Sloss and others, 1949).

Mississippian coal measures

A major unconformity between Mississippian and Pennsylvanian rocks extends throughout the intracratonic basins, and from the craton across most of the foreland trough of the Appalachian basin, and for a lesser distance into the Ouachita foreland trough from Alabama to Texas.

Mississippian (Lower Carboniferous) coal beds are best developed in the Price Formation of Virginia and in the Valley Coal Fields of the central Appalachians (Englund and others, 1983; Simon and Englund, 1983), but these beds are not presently economic targets. Coaly beds also occur in the Black Warrior and Eastern Interior basins, and foreshadow the abundant coals of the Upper Carboniferous (Pennsylvanian System). In the Appalachians, these Lower Carboniferous coal beds are within the upper part of the Acadian clastic wedge. This wedge consists of the Catskill deltaic complex (late Devonian) in the lower part and the Price deltaic complex (early Mississippian) in the upper part. An extensive marine transgression separates the Catskill and Price progradational sequences. Price coal beds occur in the upper part of the progradational sequence, where nonmarine facies replace marine facies.

In the Black Warrior basin in Alabama and Mississippi, the lower Pottsville Formation contains noncommercial late Mississippian coal beds. The upper Pottsville (Lower Pennsylvanian) contains the productive coal measures that are best developed in the eastern part of the basin. Northeastward progradation of the upper Pottsville indicates a southwestern source area related to the Ouachita orogenic belt (Thomas, 1988). Horsey (1981) re-

Donaldson, A. C., and Eble, C., 1991, Pennsylvanian coals of central and eastern United States, *in* Gluskoter, H. J., Rice, D. D., and Taylor, R. B., eds., Economic Geology, U.S.: Boulder, Colorado, Geological Society of America, The Geology of North America, v. P-2.

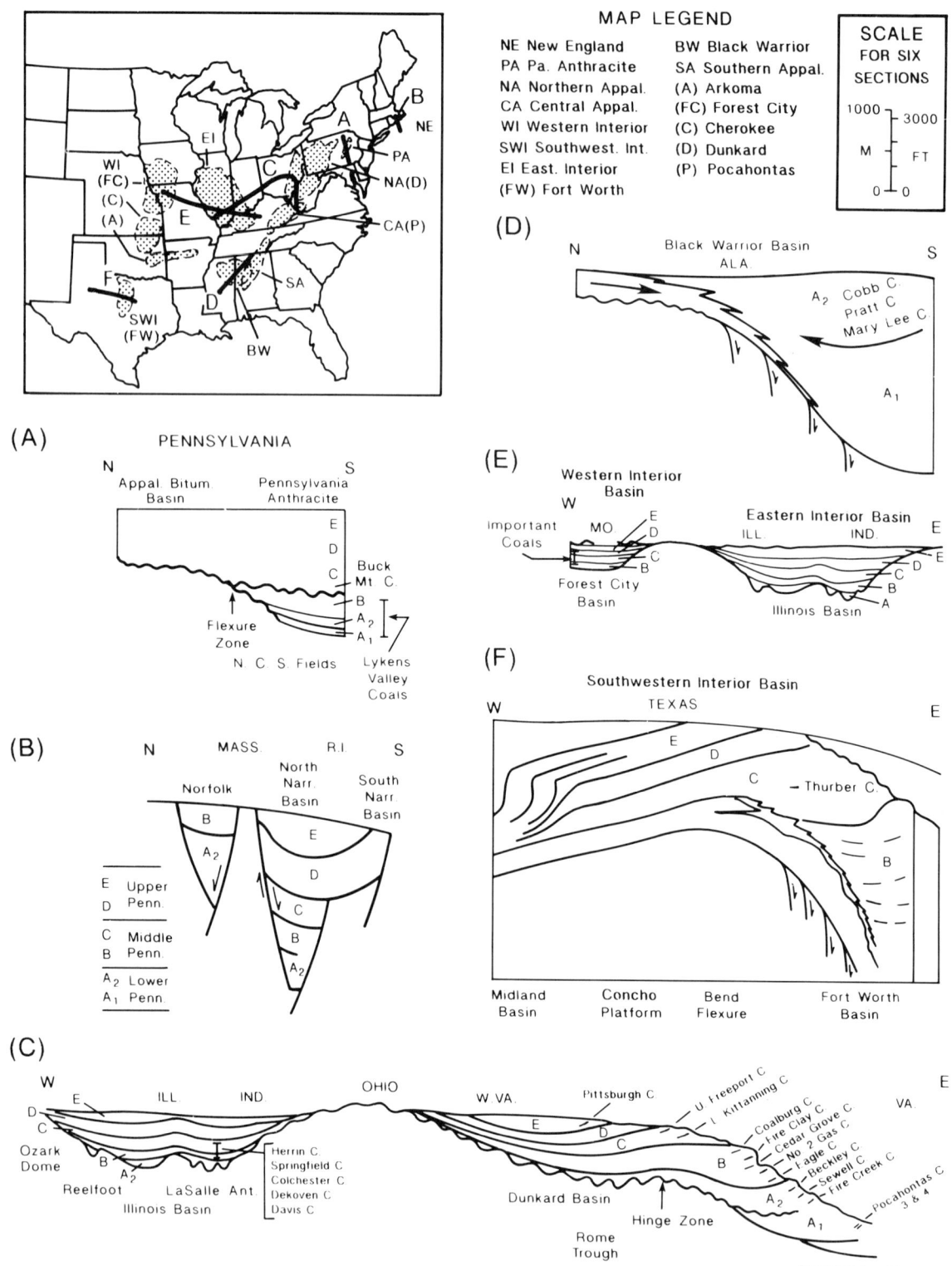

Figure 1. Generalized stratigraphic cross sections of Pennsylvanian coal basins in United States showing thicknesses and ages of coal measures and positions of most minable coal beds. USGS letter designations for time-stratigraphic units are from McKee and others (1975) and are also presented in Figure 2.

TABLE 1. COMPARISON OF SOME PROPERTIES OF COAL BEDS AND COAL MEASURES FROM PENNSYLVANIAN COAL BASINS OF THE UNITED STATES

Coal Basins, Regions	Location		Coal Measure Thickness		Number of Coal Beds	Accumulated Thickness (m)	Rank (BTU)	Coal Bed Properties			Methane Reserves (trillion ft³)	Coal Reserves (billion tons)
	States	Extent of Preserved Area (km²)	Maximum Thickness Preserved (m)	Estimated Thickness of Eroded Strata (m)				Coal Quality				
								Average Ash (%)	Average Sulfur (%)			
I Southern New England Meta-anthracite	MA, RI	2,460	3,000 to 3,700	?	28	14	13,000 to 14,000 meta-anthracite	>10 moderate	>1 low		?	?
II Pennsylvania Anthracite	PA	1,254	3,000	5,000–7,000	62	105	14,700 to 15,800 anthracite	13 moderate to high	<1		?	?
III Appalachian Bituminous		186,480										
A Northern (Dunkard)	PA, WV, OH	?	900	2,400	33	11	12,000 to 15,000 medium to low volatile	13 moderate to high	2.9–3.4		61	578
B. Central (Pocahontas)	WV, KY, VA	?	1,200	?	100	30		8 low	0.9		10–48 (Poc #3, 4)	
C Southern (Black Warrior, Coosa, Cahaba)	Al, GA, MISS, TN	?	2,800	?	152	38	13,000 mostly high volatile; some low volatile	9 low	<1–3 range		10	35
IV Eastern Interior	IN, IL, KY	137,270	900	1,000–3,000	75	20	8,600–11,000 high volatile A, B, C	8–15 mostly moderate to high	3–5%		25–33	365
V Western Interior (Forest City, Arkoma)	IA, KS, MO, OK, AR	174,825	450 (FC) 3,000 (A)	?	29	12	high volatile A, B, C [FC M-low volatile and Anthracite (A)]	>10 moderate to high	>2.5 high 1–5 range		? ?	7.8 (A) (FC)

ported progradation of a subordinate clastic wedge northwestward from the southern Appalachian orogenic source area.

Although the Mississippian is the greatest hydrocarbon-producing system in the Eastern Interior basin (Treworgy and Norby, 1989), no coal beds of economic importance are known. During late Mississippian, siliciclastics derived from the eastern Canadian Shield were deposited in the basin by the large ancestral Michigan River system, which prograded southwestward while shifting laterally to influence a fluvial-deltaic area of approximately 320 km along the coastal plain's northwest-trending shoreline (Swann, 1963).

TECTONISM AND SEDIMENTATION

Carboniferous coal beds of the Appalachian, Arkoma, and Fort Worth basins are associated with clastic wedges produced during the Alleghany and Ouachita orogenies when lithospheric plates collided along the southeastern margins of the North American continental crust. The orogenies created thrust sheets that loaded and downwarped the continental margins into foreland troughs (Quinlan and Beaumont, 1984). Sediment from erosion of these sheets first filled the foreland basins and then extended onto the adjacent craton.

Broad flexing of the basement probably occurred during the thrust-sheet loading and development of peripheral bulges (Quinlan and Beaumont, 1984; Hines and Thomas, 1987). The peripheral bulge is identified mainly with arches and domes bordering the foreland and interior basins (Cincinnati-Findley Arch, Nashville Dome, etc.), but satellite flexures on their flanks also are recognized. For example, a broad flexure called the *hingeline zone* subdivides the Appalachian foreland trough in West Virginia into two parts: the proximal (Pocahontas basin, also called the Central Appalachian basin), adjacent to the orogenic belt; and the distal (Dunkard basin), alongside the craton (Fig. 1, Section C).

Quinlan and Beaumont (1984) and Beaumont and others (1988) also relate events in the craton to isostatic adjustments reflecting downwarp at the continental margins by thrust-sheet loading on a viscoelastic crust during successive Paleozoic orogenies. The Paleozoic cratonic basins show a common history (Shumaker, 1986), starting with a rifting phase during the Precambrian and Cambrian, followed by crustal sagging of parts of the lengthy rift systems into single oval-shaped basins by mid-Paleozoic time, and subsequently modified in Pennsylvanian-Permian time by upthrusts caused by compressive stresses from the Ouachita and Alleghany orogenies. This created yoked basins in the Southwestern Interior basin and intrabasinal arches within the Western and Eastern Interior basins (Fig. 1).

Deformation was more intense in the southwestern craton, where yoked basins evolved and nearby source areas were created in Texas by Ouachita deformation associated with crustal collisions. The same compressive stresses caused upthrusts in the Western and Eastern Interior basins, but their relief was too low to form source areas during Pennsylvanian time. The Lasalle anticline formed in the Eastern Interior basin (Fig. 1C), and the Nemaha Ridge split the Western Interior basin into the Salina and Forest City basins.

Mississippian-Pennsylvanian unconformity

Withdrawal of epeiric seas from the North American craton and most of the Appalachian foreland trough during late Mississippian coincided with global lowering of sea level (Saunders and Ramsbottom, 1986), probably due to major glaciation and/or lithospheric plate conditions that increased ocean basin volume. In the Eastern Interior basin, Jennings and Fraunfelter (1986) interpret a regional unconformity throughout, with Lower Pennsylvanian deposits immediately overlying the unconformity in the southeast and progressing to Middle Pennsylvanian rocks northward. Rocks underlying the unconformity increase in age northward from late Mississippian to Ordovician, indicating regional uplift and erosion.

The entire basin was exposed to subaerial erosion, and the ancient Michigan River system etched the gently southwestward-sloping surface into dendritic-to-anastomosing paleovalleys (Siever, 1951; Potter and Siever, 1956; Shawe and Gildersleeve, 1969; Bristol and Howard, 1971; Davis and others, 1974; Howard, 1979; Damberger, 1989a). The largest paleovalleys are 32 km wide and 137 m deep, with steeply sloping walls of indurated Chesterian sediments that occur as large rotated slump blocks widely buried in valley fill (Bristol and Howard, 1974). Limestone beds capping terraces between the paleovalleys suggest a dry climate at the time of their existence (Shawe and Gildersleeve, 1969; Howard, 1979).

Garner (1974) interpreted from sedimentary and erosional patterns that, during Early Pennsylvanian time, the ancient Michigan River system flowed southwestward across the Michigan and Illinois basins into a gulf (Cane Hill Sea) in Arkansas as the paleoclimate changed from arid and semiarid to humid. Greb (1989) documented structural control of the paleovalleys in the southern part of the basin, particularly at the ancient aulacogen of the Mississippi embayment (Reelfoot rift and its northeastward Wabash Valley fault system extension, intersected by eastward-trending Rough Creek graben) and its adjacent satellite faults. In Missouri's Ozarks, cannel coal up to 30 m thick has been reported in karst sinkholes in Mississippian limestones (Branson, 1962).

The Sharon-Brownsville paleovalley system (Rice, 1984; 1985), a large tributary of the ancient Michigan River system in the Appalachian basin, occurs along a southwestward trend east of the Cincinnati arch in western Pennsylvania, Ohio, and the Cumberland saddle area of Kentucky. Paleovalleys associated with the Mississippian-Pennsylvanian unconformity also have been reported in northwestern West Virginia (Rice, 1984; Beuthin, 1989) and interpreted as tributaries of a drainage system flowing southeastward into a trunk stream draining the Pocahontas basin and trending southwestward toward the Ouachita foreland trough in Alabama.

The paleoslope of the sub-Pennsylvanian surface of the craton and Appalachian basin reflected a southern tilt, probably associated with crustal collisions along the southeastern and southern margins of the North American lithospheric plate during the late Mississippian (Pryor and Sable, 1974; Donaldson and Shumaker, 1981; Slingerland and Beaumont, 1989). The gentle up-arching of the continental interior apparently occurred during worldwide lowering of sea level and enhanced the subaerial exposure of the extensive surface represented by the Mississippian-Pennsylvanian unconformity.

Lower Pennsylvanian sedimentation, tectonism, and minable coal beds

In Figures 1 and 2, Lower Pennsylvanian strata are designated "A" (A_1 and A_2) in the coal basins of the eastern and central United States. Continuous sedimentation occurred in the foreland troughs adjacent to the orogenic belts, and Lower Pennsylvanian rocks are thickest in these areas.

Appalachian basin. In its component basins, preserved thicknesses of Lower Pennsylvanian strata are 600 to 825 m in the Black Warrior basin (Thomas, 1988), 150 to 600 m in the Pocahontas basin (Arkle, 1974), and 15 to 450 m in the Pennsylvania anthracite region (Wood and others, 1986). In these widely separated areas of the Appalachians, coal beds may have accumulated in coastal plain settings inland from (1) a wave-dominant shoreline (lower Pottsville of the Black Warrior basin; Thomas, 1988; Epsman and others, 1988); (2) a fluvial-tidal dominant shoreline (upper Pottsville of the Black Warrior basin; Thomas, 1988; Epsman and others, 1988; as well as Pocahontas Formation of the Pocahontas basin; Englund and others, 1986); (3) a tidal-dominant shoreline (New River Formation of Pocahontas basin; Cecil and Englund, 1989); and (4) an upland braidplain (Pennsylvania Anthracite region; Wood and others, 1986; Slingerland and Beaumont, 1989).

Black Warrior basin. Important minable coal beds occur in seven different coal zones in the upper Pottsville. The Mary Lee, Pratt, and Cobb coal groups contain the best-quality coals (<1 percent sulfur), with lateral variation controlled by growth structures, which have caused the greatest thickness and least ash content or coal splits on the upthrown blocks of normal faults (Weisenfluh and Ferm, 1984; Epsman and others, 1988). These three coal groups are equivalent to New River age (Westphalian A, A_2 in Fig. 1) and are younger than the Pocahontas Formation of the Pocahontas basin. Epsman and others (1988) believe the minable coals to have been planar-to-very-low domed peat swamps because of the splits and partings in the coal beds, which resulted from crevasse-splay influx of siliciclastics. These inland swamps in the Black Warrior basin apparently were slightly farther from the paleoequator than the Pocahontas basin, developing perhaps in a slightly drier climate. The interpreted elevations of the ancient peat swamps may be an indication of this, although the excessive tectonic activity occurring in the Black Warrior basin (resulting in a very thick sequence of Lower Pennsylvanian strata) must be considered.

Pocahontas basin. Pocahontas Formation coal beds are associated with lithic sandstones (Englund, 1974). The minable Pocahontas No. 3 and No. 4 beds display the thickest development overlying northwest-trending deltaic lobes. Younger New River Formation coal beds, commonly associated with quartz-rich sandstones (controversial in origin), include the Fire Creek, Beckley, and Sewell coals, which are also low in sulfur content (<1 percent) and ash yield (<10 percent).

Using coal-bed geometry and coal quality, Cecil and others (1985a, 1985b) interpreted the Pocahontas No. 2 and No. 3 coal beds to represent ancient analogs of raised peat swamps (domed) that developed along a tidal-influenced shoreline, similar to the modern swamps of Sumatra. Cecil and Englund (1989) cite the coal quality, compositional maturity of the sandstones, occasional trace fossils, and sedimentary structures as examples of a similar depositional setting, but interpret higher-energy tidal currents for the quartz-rich sandstones associated with younger New River coals. Others find abundant evidence with which to interpret fluvial deposition for the quartz-rich sandstones (Rice and others, 1979, 1984; Chesnut, 1989).

Pennsylvania anthracite region. Pottsville Formation anthracite beds are associated with conglomerates, lithic sandstones, quartz-rich sandstones, and shales. These may have accumulated in an alluvial plain (Edmunds and others, 1979; Wood and others, 1986; Edmunds, 1988). Although orogenic uplift traditionally has explained the dominant conglomerate content (50 to 60 percent) of the Pottsville Formation, noted by previous workers (Meckel, 1967; Edmunds, 1988), the relatively slow sediment accumulation and its different paleosols, compared to the underlying Mauch Chunk red beds, led Slingerland and Beaumont (1989) to suggest a climatic change from arid to humid. The eight important minable coal beds of the Lykens Valley coal zone are <1 to 3 m thick, with low sulfur, moderately high ash, and high Btu. Strata in the anthracite fields range from late Early Pennsylvanian (early New River; W. Gillespie, personal communications, 1990) to Late Pennsylvanian in age.

Southeastern New England. These coal-bearing basins may contain facies deposited in intermontane basins (Skehan and others, 1986), with graben-and-horst architecture. The basins were initiated at different times. Sedimentary rocks date the sequence of basin initiation as late Early Pennsylvanian time for the Norfolk basin, Middle Pennsylvanian for the northern Narragansett basin, and Late Pennsylvanian for the southern Narragansett basin. Alluvial fans dominated the basin fill (Skehan and others, 1986). Within this orogenic belt, mainly bed-load streams dispersed sediment, and peat accumulated in discontinuous swampy lowlands adjacent to these fluvial systems mostly during Middle to Late Pennsylvanian time.

The flora (Lyons, 1984; Skehan and others, 1986) and low sulfur content of the coal beds (Cecil and others, 1985b) suggest a tropical-to-subtropical climate. The original lenticular shapes of these discontinuous alluvial-fan coals (Skehan and others, 1986) were accentuated by structural changes associated with the Alleghany orogeny. The effect of increased temperatures produced

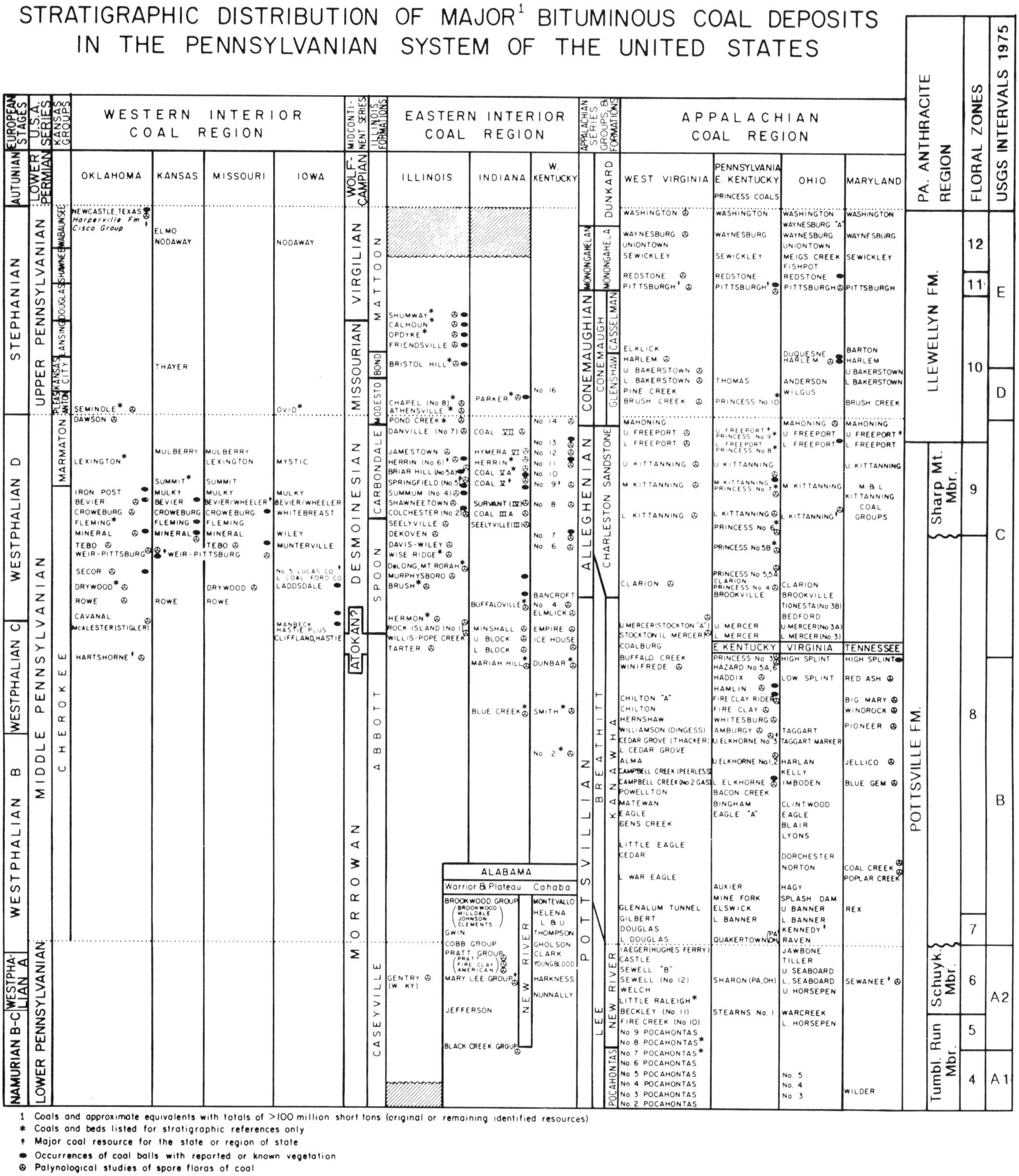

Figure 2. Stratigraphic distribution of major bituminous coal deposits in Pennsylvanian System of United States (modified after Phillips and Pepper, 1984).

coals of anthracite to meta-anthracite rank (Skehan and others, 1986).

Eastern Interior basin. Lower Pennsylvanian rocks (Caseyville Formation) are confined to the southeast part of the basin. They are thickest overlying the Rough Creek graben area, 60 to 90 m in interfluvial areas to 180 m in a paleochannel fill. Differential subsidence was greater prior to Lower Pennsylvanian sedimentation where this earlier rift zone had developed (Siever, 1951; Shawe and Gildersleeve, 1969; Bristol and Howard, 1971, 1974; Collinson and others, 1988; Damberger, 1989b; Greb, 1989; Chesnut and Cobb, 1989). Northeastward, these rocks are increasingly confined to the paleovalleys. Lower units of the Caseyville Formation pinch out and are overlapped by younger units in southwestern Illinois (Sonnefield, 1981). Progressive onlapping of Pennsylvanian rocks exists northward across the regional unconformity.

Another ancient graben, the Reelfoot, trends north-south and marks the northern limit of the Mississippi embayment, which attracted the southwestward-flowing Michigan River drainage system during shoreline retreat as well as the embayment front of a shallow sea transgressing northeastward onto the craton. Lower Pennsylvanian strata are characterized by quartzose pebbly sandstone (84 to 99 percent quartz), shale, and thin lenticular coal beds (Kosanke and others, 1960).

Origin of quartz-rich sandstones. Lower Pennsylvanian rocks are known for their coarse-grained siliciclastic and quartz-rich content. Conglomerates and sandstones derived from the northern Appalachian orogenic source area (Baltimore and New England) are less quartz-rich in the Pottsville Formation of the Pennsylvania anthracite region and in the New River Formation in the northeastern Pocahontas basin (Meckel, 1967; Houseknecht, 1980) than are correlatives from the Canadian Shield source area (Meckel, 1967; Potter and Siever, 1956; Rice, 1984, 1985; Rice and others, 1979, 1987; Krissek and others, 1986; Arkle and others, 1979; Englund, 1979). This is true for the Olean Conglomerate (New York), Sharon Conglomerate (Ohio), Lee Sandstone (eastern Kentucky), Nuttall Sandstone of the New River Formation (West Virginia), Parma Sandstone Member of the Saginaw Formation (Michigan), and Caseyville Formation (Illinois).

For depositional sites near the paleoequator (Fig. 3), the quartz-rich sandstones and gravels may have been derived from the Canadian Shield and northern Appalachian orogenic belt, which has an earlier origin than the southern Appalachian orogenic belt, during the unroofing of abundant sedimentary cover when the paleoclimate changed from arid to wet (Garner, 1974; Howard, 1979; Donaldson and Shumaker, 1981; Slingerland and Beaumont, 1989). They were then deposited in broad, extensive coastal plains by fluvial processes.

The drainage systems followed the paleoslope southwestward, eventually depositing sediments into the Ouachita foreland trough, according to evidence of (1) paleovalley trends in the Eastern Interior basin (Bristol and Howard, 1971, 1974; Rice, 1984), (2) decreasing grain size of the siliciclastics, and (3) paleoflow directions (Fuller, 1955; Potter and Siever, 1956; Edmunds and others, 1979; Rice and others, 1979). The extent of transportation of the quartz-rich sand and pebbles is indicated by the Jackford Sandstone in the northern margin, as well as within the Ouachita trough (Morris, 1974; Glick, 1975).

Although fluvial depositional environments are cited as dominant for sandstones and conglomerates of the Eastern Interior basin and the central and northern Appalachian basin, intercalated marine beds have been recognized, particularly in the southernmost areas, and tidal influence has been interpreted (Kvale and others, 1989; Cecil and Englund, 1989). Shallow seas presumably transgressed northward onto the craton and along the axis of the Appalachian foreland basin from the Ouachita foreland trough (Pryor and Sable, 1974; Donaldson and Shumaker, 1981; Rice and Schwietering, 1988).

Mountains of the southeastern and southern margins of the North American lithospheric plate, associated with the Alleghany orogeny crustal collisions, were responsible for mainly lithic sandstones and shales in the southern part of the Pocahontas and Black Warrior Basins. Where quartz-rich sandstones occur in these basins, marine currents are credited for their accumulation in coastal barrier islands and estuaries (Englund, 1974; Cecil and Englund, 1989; Hobday, 1974; Thomas, 1988).

Western Interior Basin. By Early Pennsylvanian time, the Nemaha anticline developed from southeastern Nebraska to Oklahoma City, dividing the Western Interior coal basin into the Salina basin on the west (lacking commercial coal beds), and the Forest City basin on the east (lacking definitely Lower Pennsylvanian rocks but containing minable coals in the Middle Pennsylvanian).

Middle Pennsylvanian sedimentation, tectonism, and minable coal beds

In Figures 1 and 2, Middle Pennsylvanian strata are designated "B" and "C" in the coal basins of the eastern and central United States.

Basal lithologies of the Middle Pennsylvanian are transitional from the Lower Pennsylvanian in coarseness and quartz content of the siliciclastics. In the Illinois basin, the Abbot Formation (Illinois), upper Mansfield Formation (Indiana), and Tradewater Formation (Kentucky) contain more shale and marine beds than do the underlying Caseyville Formation and its lateral equivalents. The Corbin Sandstone, an upper unit of the Lee Member in the Pocahontas basin (Rice, 1984), also indicates the transitional nature of basal Middle Pennsylvanian strata.

Appalachian basin. Although some marine or near-marine rocks occur throughout the Breathitt Formation of Kentucky (Chesnut, 1981), extensive mappable marine zones characterize it and the partially equivalent Kanawha Formation (West Virginia) of the Pocahontas basin. Extensive marine zones occur at the boundary between the Lower and Middle Pennsylvanian and at four stratigraphically higher intervals (Chesnut, 1989). In ascending order, these zones within the Breathitt are called the Betsie Shale, the Kendrick Shale, the Magoffin

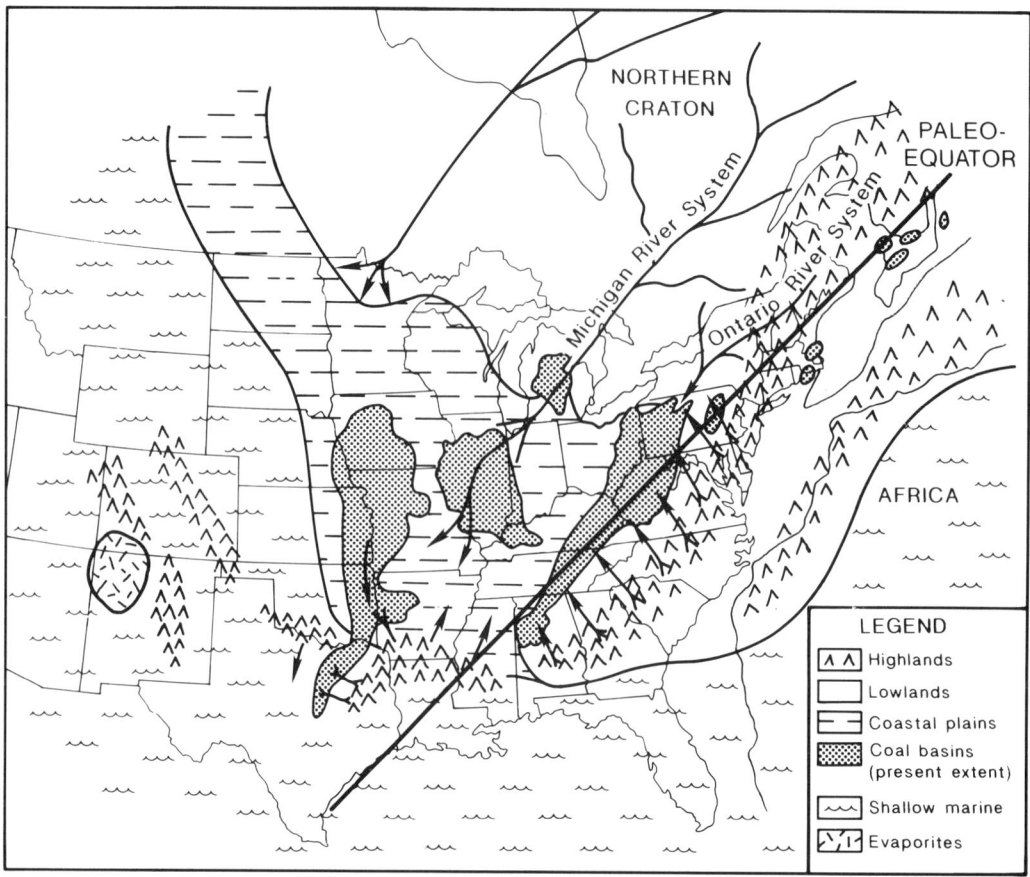

Figure 3. Relation of Pennsylvanian coal basins to suggested Desmoinesian-Virgilian paleoequator (from Heckel, 1977, 1980) and paleogeography, including interpreted ancient rivers. Coastal plains and shallow-marine environments alternated periodically over much of the midcontinent during major shifts of shoreline.

Member, and the Stoney Fork Member. The zones, which represent periodic transgressions of midcontinent seas, separate sandstones, shales, and coals that were deposited in fluvial-deltaic depositional environments of a coastal plain (Horne and others, 1978; Donaldson and Shumaker, 1981; Chesnut, 1989; Cobb and Chesnut, 1989). Slingerland and Beaumont (1989) interpret the increased marineness in the Middle Pennsylvanian strata as a response to increased thrust-sheet loading at the continental crust margin, which depressed the crust of the foreland basin.

Pedogenic flint clays are associated with coal beds in the upper part of the Middle Pennsylvanian of West Virginia and Pennsylvania (Donaldson and others, 1985; Cecil and others, 1985). The stratigraphic occurrence of the flint clays has been explained by Williams and others (1968; Bragonier, 1989) to have been the result of precipitation of colloidal clays from acidic swamps that were influenced by nearby alkaline environments (e.g., transgressive marine waters). Alternatively, Cecil (1990) interprets these flint clays to represent paleosol formation in a paleoclimate transitional between tropical everwet and savannah.

The Middle Pennsylvanian of the Appalachian, Eastern, and Western Interior basins is known for its cyclothems and minable coal beds (Fig. 2). The lack of preserved coal measures in the Black Warrior basin and the progressive shifting of marine sequences northward in the Appalachians into Late Pennsylvanian time shows the effect of southeastern uplift of the North American crust and the migration of the depressed crust into the foreland basin.

Beaumont and others (1988) interpret migration of the peripheral bulges outboard of the foreland trough as a result of relaxation of the viscoelastic crust with loading, as well as unloading due to erosion of the thrust-sheet pile. This condition is reflected by migration of sedimentation depocenters from the Pocahontas to Dunkard basins during Early to Late Pennsylvanian time (Slingerland and Beaumont, 1989). On a smaller scale, differential subsidence in the Appalachian foreland trough of Kentucky, West Virginia, and Pennsylvania produced growing structures that localized narrow seaways and fluvial drainage systems for the interval involving the Fire Clay coal bed of Kentucky (Haney and others, 1985) and for the Allegheny Group in western Pennsylvania (Williams and Bragonier, 1974).

Pennsylvania anthracite region. Upwarping occurred during Atokan time (Edmunds, 1988; Edwards and Eggleston,

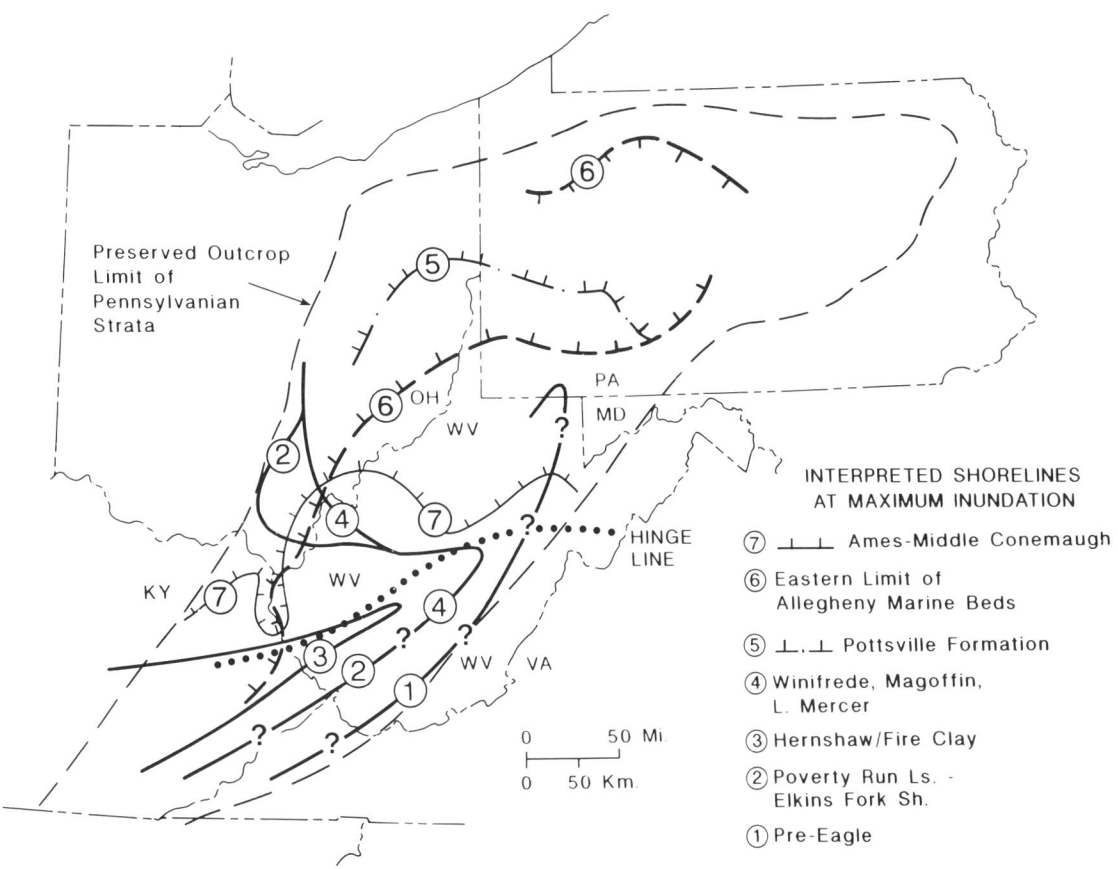

Figure 4. Interpreted shorelines for the central and northern Appalachians, showing a northward shift of marine deposits from the Pocahontas to Dunkard basin during Pennsylvanian time.

1989). Correlations (Donaldson and Eble, 1989) suggest that the eroded sediment from this eastern Pennsylvania area was deposited by bed-load fluvial systems westward in northwestern West Virginia and southwestern Pennsylvania as the pebbly, quartz-rich Connoquenessing sandstones. Figure 4 shows the northward shift in seaway positions in the central Appalachians during Early Pennsylvanian and early Middle to late Middle Pennsylvanian.

Eastern Interior basin. The southern portion of the Eastern Interior basin subsided the most; northward Middle Pennsylvanian strata overlap Lower Pennsylvanian strata and the sub-Pennsylvanian erosional surface. Along trends more parallel to the basin axis, differential subsidence produced intermittent warping (initially greatest in the DuQuoin monocline), separating the Western shelf from the Fairfield basin, but later including the LaSalle anticline belt and Salem and Louden anticlines (Treworgy and Bargh, 1982). Lithologies exhibit finer-grained clastics than Lower Pennsylvanian, and minable coal beds mainly are in the upper Middle Pennsylvanian (Desmoinesian Series, approximately equivalent to Westphalian D; Phillips and Peppers, 1984; Nelson and others, 1991). Damberger (1989b) refers to the Lower Pennsylvanian as the "lower sandstone-dominated sequence" and the Middle Pennsylvanian as the "middle cyclothem-dominated sequence."

The previously mentioned growth structures influenced the position of the large river systems of the coastal plain (e.g., the Walshville Channel). Fluvial-deltaic systems were large and widely spaced compared with the smaller multiple drainage systems originating in the orogenic belts alongside the Appalachian and Ouachita foreland troughs (Fig. 3). The Walshville fluvial channel also flowed through vast coastal swamps as thick peat deposits were being accumulated (based on crevasse-splay deposits of the paleochannel that split the commercial Herrin No. 6 coal bed; Palmer and Dutcher, 1979; Damberger, 1989b).

The repetitious, cyclic nature of diverse lithologies was first described by Udden (1912). Weller (1930) and Wanless and Weller (1932) later proposed the "ideal" Illinois cyclothems. Coal and marine beds of these cycles have served as key beds for correlation over extensive areas of the basin. Early workers recognized the cyclic sequences to be relatively uniform in thickness, very widespread, and representing alternating regressive-transgressive migrations of the shoreline hundreds of miles southwestward across the marine shelf and then northeastward across the coastal plain of the Eastern Interior basin. Although the shallow seas commonly transgressed and retreated across the entire length of the Eastern Interior basin while a cyclothem was being deposited, the progradational phase created a broad coastal plain where alluvial channels and flood plains predominated in the north while fluvial-delta and tidal-plain environments were deposited in the south. The most commercial coal beds in the Eastern Interior basin occur in these cyclothems (Table 2).

Western Interior Basin. Shallow seas transgressed from the Ouachita trough to the north. Repetitious cycles (maximum thickness 10 m) occur in the Forest City basin; consisting of thin marine beds of black shales, gray shales, and limestones, capped by thin to thick nonmarine siliciclastics and paleosols with thin underclays and coals (Wanless, 1975). During Middle Pennsylvanian time, this basin alternated between being a shelf and a coastal plain with extensive swamps developed due to fluctuations in sea level.

The minable coal beds are the Rowe, Bluejacket, Weir-Pittsburg, Tebo, Mineral, Fleming, Croweburg, Whitebreast, Wheeler, Bevier, Iron Post, Mulky, Lexington, Mystic, and Mulberry. Although the multiple thin coal beds have high sulfur contents and ash yields, the thin rock intervals between the lower coals have encouraged surface mining.

In the foreland trough of the Arkoma coal field (Branson, 1962), the minable coals are the Lower Hartshorne, Upper Hartshorne, McAlester, Stigler, Charleston, Paris, Secor, Morris, and Henryetta, which are lower in sulfur and ash content, and are also thicker than the Middle Pennsylvanian coal beds on the platform (Forest City basin). The Arkoma foreland trough (Johnson and others, 1988) infilled with sediment derived initially from the Michigan River system, dispersed from the Canadian Shield via the Mississippi embayment. Later, when the Ouachita Mountains developed and uplifted the eastern part, some sediment was probably transported through the Forest City basin. Increasingly with time, much sediment was derived from the emerging Ouachita Mountains so that vast fluvial-deltaic and tidal-deltaic plains provided a suitable setting for peat swamps.

Southwest Interior basin. The coal-bearing part of this basin, the Fort Worth basin of Texas, is a southward-trending part of the Ouachita foreland trough. Sediment infilling of this basin was similar to that of the Black Warrior basin at the other end of the trough, with similar depositional environments and sediment source mainly from the orogenic belt. However, its broad coastal plains with coal-forming swamps developed later (Middle rather than Early Pennsylvanian) with thinner, poor-quality coal beds. The Thurber coal bed is the only commercial coal, and it is marginal at that.

Upper Pennsylvanian sedimentation, tectonism, and minable coal beds

In Figures 1 and 2, Upper Pennsylvanian strata are designated "D" and "E" in the coal basins of the eastern and central United States.

Cyclothems also characterize the Upper Pennsylvanian rocks of United States coal basins. Generally, Upper Pennsylvanian coal measures have fewer minable coals and more limestone beds and calcareous sandstones than the Middle Pennsylvanian sequence. The coal beds typically have higher sulfur contents and

TABLE 2. COMMERCIAL COAL BEDS IN THE EASTERN INTERIOR BASIN*

Indiana	Kentucky	Illinois
Hymera (VI)		
	Herrin (No. 11)	Herrin (No. 6)
Springfield (V)	Springfield (No. 9)	Springfield (No. 5)
		Colchester (no. 2)
Seelyville (No. 3)	Dekoven (No. 7) Davis (No. 6)	DeKoven† Davis†

*In stratigraphic order.
†Southeast Illinois only.

ash yields, although specific amounts and lateral distribution are highly variable. In the Appalachians, red beds are common, with red claystones exhibiting pseudoanticlinal structures and carbonate nodules oriented orthogonal to bedding; this indicates ancient soil formation in a paleoclimate that alternated between heavy precipitation and drought (Donaldson and others, 1985). Phillips and Peppers (1984) concluded that the Late Pennsylvanian was more arid than Middle Pennsylvanian time, based on plant characteristics in the Illinois basin.

Upper Pennsylvanian deposits are preserved mainly in the Appalachian foreland trough in the western and northern part of the Appalachian basin—the Dunkard basin and Pennsylvania anthracite region.

Orogenic activity in the southeastern United States expanded the Alleghanian clastic wedge so that deposits from this source area, which first infilled the Appalachian basin, were dispersed into the Illinois basin (Nelson and others, 1990) for the first time during the late Missourian. This event coincided with the deposition of the Millersville/Livingston Limestone. To a lesser extent, the Ouachita orogenic belt supplied sediments to the southern Eastern Interior basin from the southwest (Horne, 1968; Giffin, 1978).

Four source areas—the Canadian Shield and to a lesser extent the Wisconsin dome, Ozark dome (covered by cherty Mississippian carbonates at this time), and Transcontinental arch—rimmed the northern Eastern Interior basin and fed southerly prograding deltas during the Early and Middle Pennsylvanian. For the first time in the Pennsylvanian, streams from emerging mountains to the southeast and south periodically extended across foreland trough areas and supplied deltas prograding northward and westward into the intracratonic basins.

From the Eastern Interior basin to the Western Interior basin, the abundance of extensive marine black phosphatic shales enclosed by marine limestones of the cyclothems suggests that the Illinois area was intermediate between dominant marine conditions in Kansas and the dominant coastal plain setting of the

Appalachians (Heckel, 1977, 1980; Klein and Willard, 1989). Heckel (1986) and Wanless and others (1970) concurred that the shoreline shifted hundreds of kilometers during a major cyclothem within (and between) the Appalachian and Interior basins during the Late Pennsylvanian. The interior sea mainly occupied the area of the midcontinent basins during the Late Pennsylvanian, but periodically flooded eastward into the northern Appalachian basin (Ohio, West Virginia, and Pennsylvania) during early Late Pennsylvanian time. Preceding these inundations, the shoreline shifted from the Ouachita foreland trough to the Western and Eastern Interior basins, providing environments for thin coals beneath the major marine cyclothems.

Heckel (1986) studied 55 cycles in the midcontinent area and estimated a 235,000- to 400,000-year duration for a major cyclothem of the Western Interior basin. From this short-term periodicity, he reasoned that sea-level fluctuations in the stable midcontinent cratonic area resulted from eustatic rather than tectonic causes. Boardman and Heckel (1989) used fossils to correlate 17 of these cycles to the shelf developed west of the foreland basin in north-central Texas. They concluded that the interpreted glacioeustatic explanation for the midcontinent sequence was not obscured by the tectonic overprint that is recognized in the foreland sequence in the adjacent Ouachita orogenic belt.

During Late Pennsylvanian time, the Appalachian basin continued to be uplifted from the southeast. This caused the northern Dunkard basin and Pennsylvania anthracite region to subside and be occupied first by periodic eastward extensions of the interior sea (Fig. 4), and later (when the sea retreated from the Appalachian basin) by a large lake within its extensive coastal plain (Fig. 5). This lake lay over the northern Rome trough, which evidently experienced reactivation, creating a small-scale starved-basin setting.

Previous chemical sediment buildups in this area had occurred during the Late Silurian (Salina Salt deposits) and Middle Devonian (Huntersville Chert deposits). During the Late Pennsylvanian, the Pittsburgh coal bed, one of the largest and most valuable mineral deposits in the world (McCulloch and others, 1975), accumulated in this same area.

The thickest preserved Monongahela coals (Pittsburgh, Redstone, Sewickley, and Waynesburg) are vertically stacked at the West Virginia–Pennsylvania border over the Rome trough and along the Dunkard basin axis. Their maximum development is between the main axes of northwest-directed fluvial sandstones (Fig. 5). Relatively high sulfur and ash in these Monongahela coals contrasts with low sulfur content and ash yield in Middle and especially Lower Pennsylvanian coal beds. Increased aridity during the Late Pennsylvanian produced planar rather than domed peat beds (Cecil and others, 1985b); the resulting influence of more-alkaline ground waters during peat formation and early burial is evidenced by higher sulfur and ash in these coal beds. Associated with these coal beds are abundant freshwater limestones and red-bed paleosols with carbonate nodules, adding credence to the interpretation of a climate with seasonal rainfall (Donaldson and others, 1985).

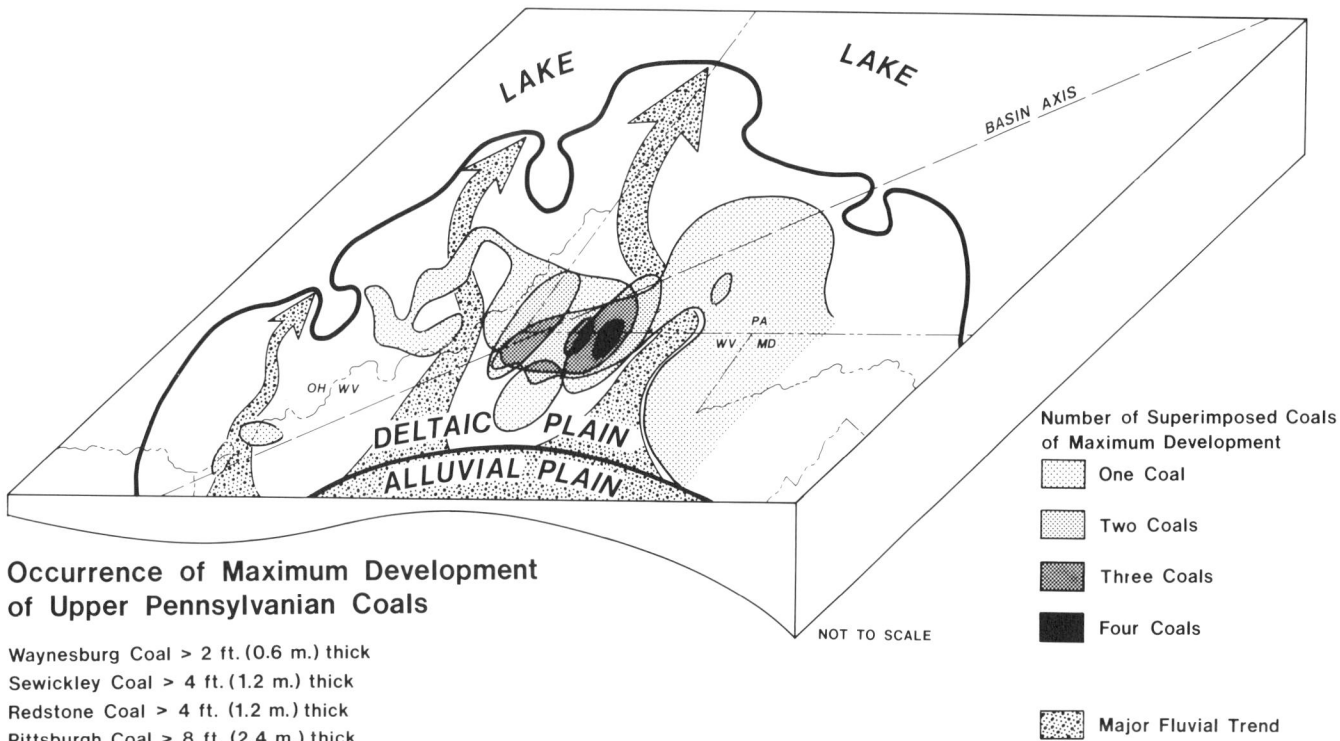

Figure 5. Occurrence of maximum development of Upper Pennsylvanian coals along Dunkard basin axis (also overlying Rome trough) located between adjacent major fluvial drainage systems.

DEPOSITIONAL ENVIRONMENTS AND CYCLES

The general depositional setting for the Pennsylvanian coal measures of the midcontinent and Appalachian bituminous coal basins is a wide coastal plain. Relatively abundant marine beds in the paralic sequences prompted early workers to label Western Interior basin cyclothems neritic, Eastern Interior basin cyclothems deltaic, and the Appalachian cycles piedmont (particularly the anthracite region; Wanless, 1947; Krumbein and Sloss, 1951).

Fluvial-deltaic depositional environments are common to all three basins. Differences among them reflect not only previously recognized degrees of marineness, but also the relative rate of subsidence (cratonic versus foreland-type deltas), water depth of prograding deltas (shallow versus deeper water), drainage basin size (single large river versus many moderate-size rivers), fluvial load (suspended versus bedload), and relative dominance of shoreline processes (marine tidal/waves versus fluvial). Minable coal beds commonly are correlated with regional transgressive events when the fluvial supply was greatly reduced on the coastal plain.

Rahmani and Flores (1984) described the "cyclothemic era" as beginning with Udden's description (1912) of cyclic sedimentation in Illinois coal measures. The concept became widely accepted when Weller (1930) and Wanless and Weller (1932) advocated this powerful method for regional correlation of Pennsylvanian rocks. Coal-measure cyclicity was attributed to regressive-transgressive couplets caused by tectonic or eustatic (particularly glacioeustatic) controls (Wanless and Weller, 1932; Wanless and Shepard, 1936; Weller and others, 1942). Many geologists have demonstrated the value of recognizing cycles in stratigraphic analysis (Krumbein and Sloss, 1951; Merriam, 1964; Duff and others, 1967).

Sedimentary cycles also develop in coastal areas where deltas shift laterally due to river avulsion upstream; thus transgression and regression of the shoreline can occur simultaneously within the same drainage basin. Such localized shoreline cycles prompted by sediment supply shifts are autogenic and called *autocycles* (Beerbower, 1964). Regionally, where an entire shoreline fronting several drainage basins exhibits transgression and the controls are probably tectonic and/or eustatic, this mechanism is *allogenic,* and the accumulated sediment of a regional transgression and regression is an *allocycle* (Beerbower, 1964).

According to Rahmani and Flores (1984), the "post-cyclothem innovations" began when cycles for the Pennsylvanian coal measures of the Appalachian and Illinois basins were described in terms of deltaic environments, initially by Williams and Ferm (1964) and later by Wanless and others (1970). Later, Ferm (1970) stressed the autocyclic control on sedimentation, whereas Wanless interpreted autocyclic changes within an overall allocyclic mechanism. Emphasis on depositional environments and their determining processes resulted in various deltaic models (Donaldson, 1969, 1974; Donaldson and others, 1970; Brown and others, 1973; Hobday, 1974; Horne and others, 1978) and a growing emphasis on autocyclic control. Renewed interpretation of allocycles for the Pennsylvanian coal measures in the eastern United States (Heckel, 1977, 1980, 1986; Busch and Rollins, 1984; Cecil and Englund, 1985) followed publications by Vail and others (1977) and Galloway (1981).

Galloway (1981) mapped Texas coastal plain Cenozoic units, as the scale of those river systems seems appropriate for Appalachian Basin stratigraphic analysis (Donaldson and others, 1979). Recognition that the Texas coastal plain also had been subjected to glacioeustatic sea-level changes during the Pleistocene-Holocene allowed for comparison with Carboniferous coal-measure rocks.

The Upper Pennsylvanian of the central Appalachians shows large fluvial-deltaic aprons (Galloway's 1981 terminology) of 160 to 320 km length and 100 to 140 km shoreline width (Fig. 6). Autocycles are recognized within an interpreted extrabasinal drainage system (fluvial-deltaic apron), and are smaller than the allocycles, which commonly are 15 to 30 m thick for the entire preserved Dunkard basin (Donaldson and others, 1985). This smallest-scale regional cycle is the minor allocycle of this text and is equivalent to the fifth-order allocycle of Busch and Rollins (1984) and to the major cycle of Heckel (1986).

Appalachian basin allocycles are compared in Figure 7 with the global sea-level curve of Veevers and Powell (1987) and their inferred ice volume for Gondwanaland. Allocycles correlated from three different foreland basin sites (Pennsylvania anthracite region, Pocahontas basin, Dunkard basin) show only the interpreted relative magnitude of shoreline shifts for illustrative purposes, because precise measurements are questionable. Although the allocycles can be grouped into three discrete classes (minor, intermediate, major), each class varies depending on whether it was deposited in the more rapidly subsiding Pocahontas basin or the more stable Dunkard basin. Pocahontas allocycles exhibit less shoreline shift and are thicker when compared to the Dunkard basin.

In the minor allocycle, laterally extensive chemical deposits (paleosol interval, coal beds, marine beds, limestones) are key beds. These represent times when rates of detrital sedimentation were low and are accompanied by transgressions, whereas the thicker clastic sequence developed during the dominantly regressive phase. Intermediate allocycles include several minor allocycles bounded by widespread marine zones such as the Betsie, Kendrick, Magoffin, and Kanawha Black Flint, all of which occurred during Middle Pennsylvanian time. Chesnut (1989), recognizing the importance of intermediate-scale allocycles in the Lower and lower Middle Pennsylvanian of the Pocahontas basin, estimated a 2.5-m.y. average duration for them. Major allocycles are bounded by maximum transgressive events within the basin, which for the Appalachian basin coincides with the separation of the Pennsylvanian System into the Lower, Middle, and Upper Series.

Regional differences in tectonic subsidence and/or orogenic uplift and sediment supply can cause skewed shorelines extending across several drainage systems and their delta complexes. The

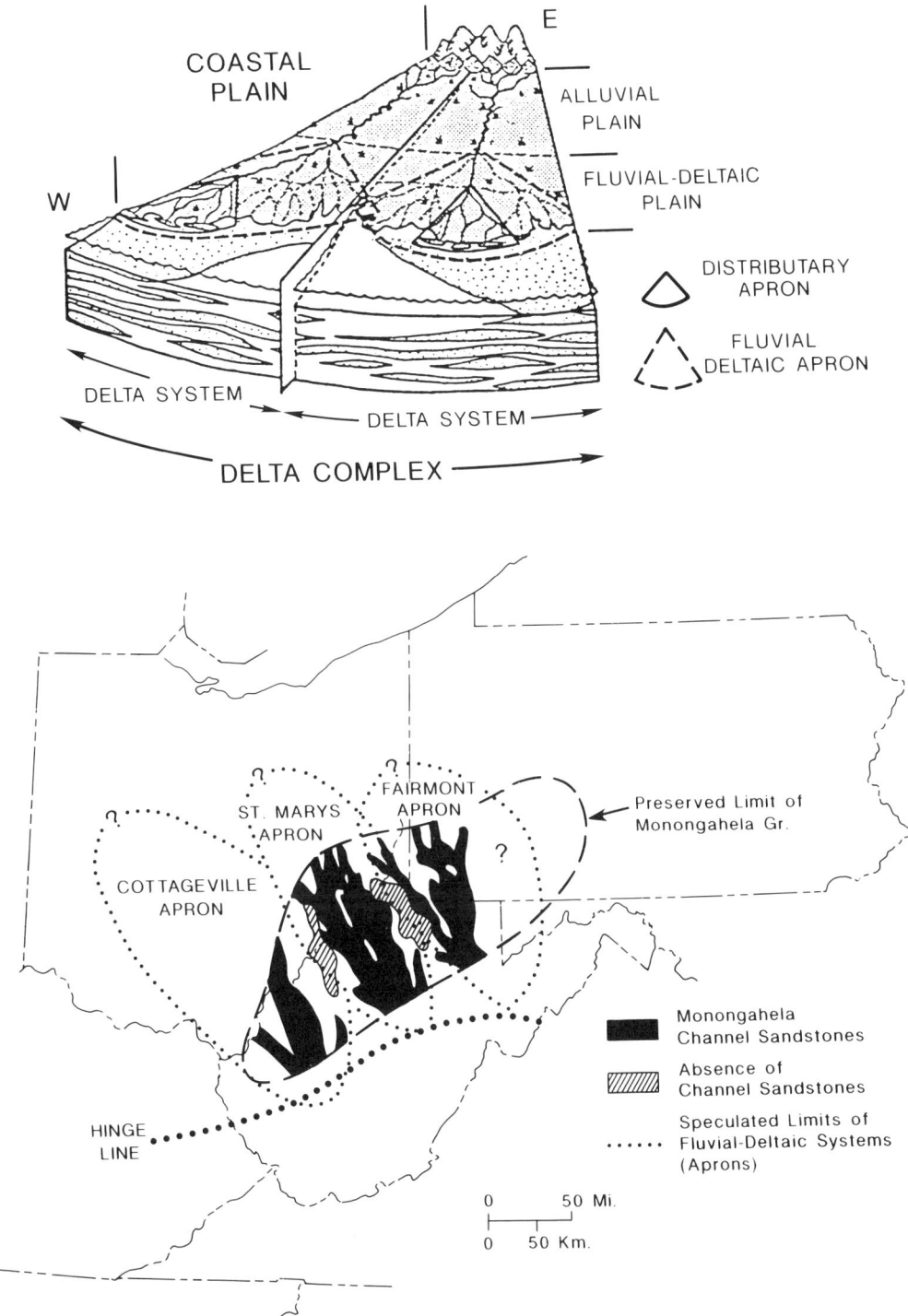

Figure 6. Interpreted fluvial-deltaic systems are shown for the Dunkard basin (northern Appalachians) during Late Pennsylvanian time. Nomenclature for the coastal plain depositional environments, as depicted in the block diagram, is from Boswell and Donaldson (1988).

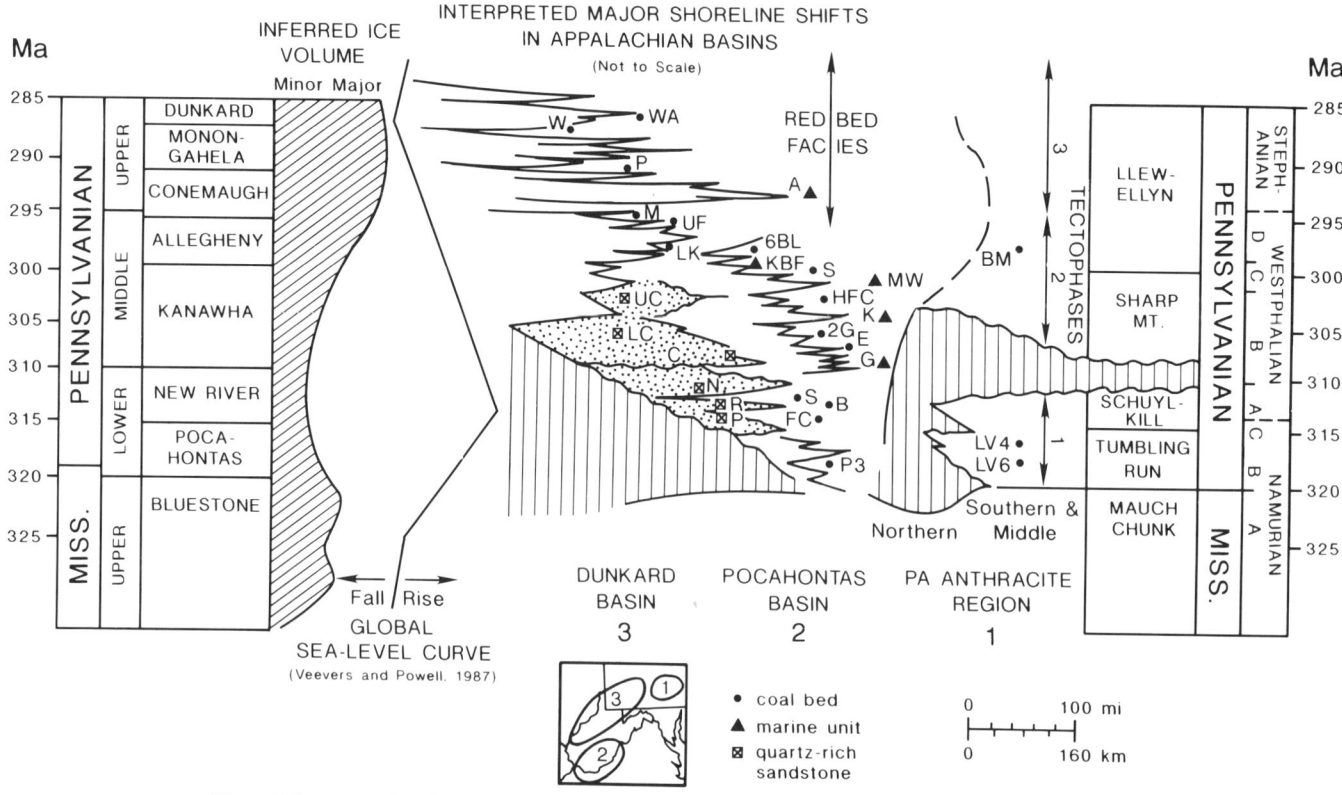

Figure 7. Interpreted major shoreline shifts and unconformities in the central and northern Appalachian basins compared with global sea-level curve and inferred ice volume in Gondwanaland for Late Mississippian and Pennsylvanian time. Source areas for quartz-rich sandstones are the Canadian Shield and orogenic mountains to northeast. Major cycles were controlled mainly by tectonism rather than eustacy and are called tectophases. Explanation for letters is as follows: LV, Lykens Valley; BM, Bucks Mountain; P3, Pocahontas #3; FC, Fire Creek; B, Beckley; S, Sewell; G, Gilbert; E, Eagle; 2G, #2 Gas; K, Kendrick; HFC, Hernshaw/ Fire Clay; MW, Magoffin/Winefrede; S, Stockton; KBF, Kanawha Black Flint; 6BL, #6 Block; P, Pineville; R (Raleigh); N, Nuttall; C, Corbin; LC, Lower Connequenessing; UC, Upper Connequenessing; LK, Lower Kittanning; UF, Upper Freeport; M, Mahoning; A, Ames; P, Pittsburgh; W, Waynesburg; WA, Washington.

major allocycles are skewed between the central (Pocahontas) and northern (Dunkard) Appalachian regions, suggesting that tectonic control during Pennsylvanian System sedimentation was not uniform throughout the foreland basin. The Tennessee salient (Thomas, 1977) shows progressive cratonward movement throughout Early and Middle Pennsylvanian time, whereas the Pennsylvania salient similarly, but to a lesser degree, loaded its crustal segment during Late Pennsylvanian time. This caused extensive marine inundation (lower Conemaugh) of the coastal plain in Pennsylvania and was followed by starved-basin infilling with thick chemical deposits in large lakes and swamps (Monongahela).

Because the major allocycles also do not correspond to the global sea-level curves of Veevers and Powell (1987), they are interpreted as tectonically controlled and called *tectophases* by Ettensohn (1985). The basal unit of each tectophase probably represents crustal loading by thrust sheets, downwarping of the depositional basin, and rapid transgression by epeiric seas.

Table 3 shows characteristics of the three scales of Appalachian basin allocycles. During the Alleghany orogeny and late Pennsylvanian time, a number of westward shoreline shifts of approximately 1,600 km occurred from the foreland basin to the continental interior (Schutter and Heckel, 1985). Minor allocycles probably reflect glacioeustatic controls (Wanless and Shepard, 1936; Busch and Rollins, 1984; Heckel, 1986; Veevers and Powell, 1987). The very exaggerated shoreline shifts probably happened during maximum glaciation on Gondwanaland (Veevers and Powell, 1987) and were associated with minimum crustal subsidence.

Boardman and Heckel (1989) correlated 17 cycles of the Upper Pennsylvanian in Texas with 17 cycles they previously had interpreted as glacioeustatic cycles of similar magnitude in the Western Interior basin of Oklahoma, Kansas, Missouri, Iowa, and Nebraska. They reasoned that glacial eustacy was the basic control for these cycles because tectonic masking was not evident. Correlation was based on biostratigraphic criteria of first, last, sole, or acme occurrence of ammonoid, conodont, and fusulinid taxa. These midcontinent allocycles consisted of maximum regressive deposits of paleosol mudstone and fluvial sandstones, and maximum transgressive deposits of widespread, ammonoid-bearing, conodont-rich, dark phosphatic shales. They also interpreted smaller cycles of persistent fossiliferous shales or limestones overlying terrestrial deposits also to have been dependent on glacioeustatic control rather than tectonic or delta switching.

Heckel (personal communication, 1987) claims that detailed biostratigraphic correlation among the Western Interior, Eastern Interior, and Appalachian basins is necessary before the importance of tectonic controls in the Appalachian basin can be evaluated accurately. However, the similarity in the minor allocycles (Table 3) of these areas in terms of facies, bounding lithologies, estimated duration, magnitude of shoreline shifts, and coincidence with known glacial activity supports a glacioeustatic explanation for allocycles of this scale.

A minor allocycle represents the basic model (Fig. 8), exhibiting facies formed regionally during shoreline regression followed by transgression. Transgressions are interpreted to favor chemical sedimentation and coal deposits (Williams and Ferm, 1964); in the Dunkard basin, the thickest coal deposits per cycle are located inland of maximum marine transgression. Wanless and others (1970) suggested that river avulsions characterize sedimentation within the subsequent fluvial-deltaic systems, and also that sediment-supply shifts produced autocycles as smaller cycles within the minor allocycle. Coal splits and partings mainly occur during the transgressive phase of the minor allocycle, yet the associated stream avulsions caused both transgression and regression to occur simultaneously within a fluvial-deltaic apron.

The Allegheny model (Williams and Ferm, 1964; Ferm and Williams, 1965) and its later modifications (Ferm, 1970, 1975) claimed a dominant autocyclic rather than allocyclic control, but later workers disagreed (Busch and Rollins, 1984; Cecil, 1986). In the Eastern Interior basin, the Walshville River flowed through a vast swamp, with peat accumulating contemporaneously alongside the channel sands. The river maintained its position during marine regression and transgression for the cycle containing the Herrin No. 6 coal bed.

Appalachian examples of this model (Fig. 8) involve smaller rivers that generally persisted within relatively fixed smaller drainage basins, where abundant avulsions occurred throughout the allocycle. These events resulted in commercial coal beds and associated coal splits during the regional transgressive phase. Increasing with the subsidence rate in the ancient basin are the size and stacking of coal splits, growth faults, and proportion of delta subaqueous platform to delta-plain facies.

Paleoclimate is another important consideration during sedimentation and peat accumulation. Poorer quality and decreased thickness of coal beds occur in cycles containing paleosols that suggest drier climatic conditions (Donaldson and others, 1985; Cecil and others, 1985b; Phillips and Peppers, 1984). These findings suggest that the climates in the Appalachian and Eastern Interior basins were everwet in Early and Middle Pennsylvanian time, transitional in uppermost Middle Pennsylvanian time, driest during early Late Pennsylvanian time, and fluctuating wet/dry during late Late Pennsylvanian time (Cecil, 1990).

TABLE 3. CLASSIFICATION OF ALLOCYCLES FOR APPALACHIAN COAL MEASURES

Allocycle	Upper Pennsylvanian (10 m.y.)	Middle Pennsylvanian (14 m.y.)	Lower Pennsylvanian (10 m.y.)
Major Allocycle (Tectophase)			
Duration	7 m.y.	15 m.y.	12 m.y.
Thickness	300 m	550 m	460 m
Intermediate Allocycle			
Duration	3.5 m.y. (2 cycles)	2.5–3 m.y. (5 cycles)	3 m.y. (4 cycles)
Thickness	90 m	110 m	115 m
Minor Allocycle - major cycle of Heckel (1986), 5th order cycle of Busch and Rollins (1984)			
Duration	0.3–0.5 m.y.	0.5–0.6 m.y.	0.7 m.y.
Thickness	14–27 m av. 18 m	15–42 m av. 30 m	14–52 m av. 30 m
Shoreline Shift	190–800+ km	32–160 km	32–160 km

Note: From Donaldson and others, 1985. Data from the Pocahontas Basin for Lower and Middle Pennsylvanian and the Dunkard Basin for Upper Pennsylvanian rocks. Thickness values are from different parts of the basin (from the stratotype sections) and lack common denominators in differential subsidence, source-area proximity, and other variables.

Variants of the Allegheny model displayed in Figure 8 show the importance of tectonic setting, paleoclimate, and glacially controlled eustatic conditions during sedimentation.

During the regressive phase of allocycles, fluvial-dominant deltas involved perhaps only one-third of the total shoreline at a given time because of the small size of the rivers, their active deltas, and fluvial-deltaic aprons. Marine-dominant conditions mainly influenced the greater part of the total shoreline between active deltas. Marine dominance of the shoreline increased, particularly during transgression, because of reduced sediment discharge at the rivers' mouths. Cecil (1990) reasoned that climate also reduced sediment discharge in streams during hot everwet conditions in response to leached soils, vegetative cover, and/or sediment-denuded upland areas. Because most, if not all, minable coal beds resulted from peat that formed during these wet periods (Cecil, 1990), coal-forming conditions were optimum during regional transgression of the epeiric sea in times of increased precipitation.

River avulsions within the fluvial-deltaic apron occurred during regressive and transgressive allocycle phases. The resulting autocycle generally failed to preserve coal beds (certainly not minable ones) during the regressive phase because of base-level drop (or at least reduced rate of base-level rise, compared with the transgressive phase). The faster base-level rise and reduced sediment discharge during the transgressive phase of the minor

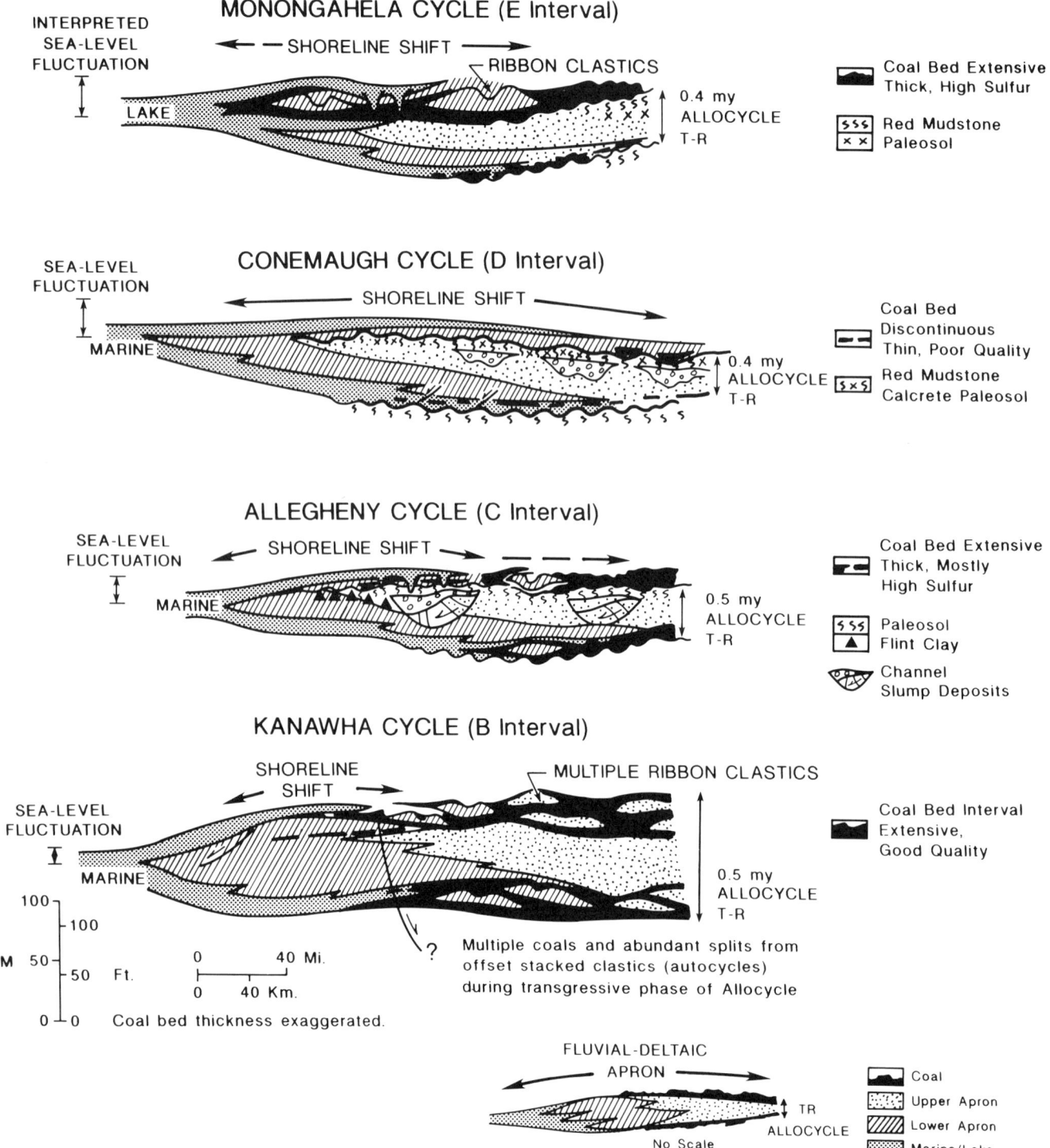

Figure 8. Representative allocycles for different stratigraphic units (Kanawha-Monongahela) in the central and northern Appalachians during Pennsylvanian time, with probable cycle duration based on the time scale of Harland and others (1982).

allocycle favored preservation of minable coal beds. Likewise, river avulsions and crevasse splays during this time account for the coal splits and partings commonly recognized. In fact, offset stacking of multiple coal-bed splits characterizes the more rapidly subsiding Pocahontas basin during accumulation of the Kanawha Formation, as tectonic control also is important. During Monongahela time in the more stable Dunkard basin, coal splits are rarer and not multiply stacked.

In the Black Warrior basin of Alabama, Hobday (1974) interpreted minable coal beds to have formed in the upper fluvial delta plain environment behind a marine wave-dominant shoreline. The Ouachita orogenic belt supplied sediment to the Black Warrior basin, causing northeastward progradation of the marine-dominant shoreline during earliest Pennsylvanian time. Later in the Early Pennsylvanian, fluvial deltas prograded westward from the southern Appalachian orogenic belt (Thomas, 1988). Epsman and others (1988) accept the barrier islands along the shoreline, but place the depositional environment for minable coals considerably inland in the upper delta plain (probably the upper part of the fluvial-deltaic apron).

In the Eastern Interior and Pocahontas basins, evidence for tidal influence is increasing. Cecil and others (1987, 1988) and Cecil and Englund (1989) suggested that minable coal beds of the Pocahontas and New River Formations formed from raised bogs (domed peats) that occurred on a tropical coastal plain in a tidally dominated setting; they selected equatorial parts of Indonesia as modern analogs. Tidal processes also were important in the Mansfield Formation of Lower Pennsylvanian rocks in Indiana (Kvale and others, 1989). R. Martino (personal communication, 1989) recognizes tidal influence from trace fossils and bimodal paleoflow directions in cross-bedded sandstones in upper parts of Middle Pennsylvanian Kanawha Formation cycles in southern West Virginia.

Although fluvial-dominant conditions probably characterized some shoreline depositional environments for Pennsylvanian rocks, marine-dominant conditions also affected at least part of the shoreline, mostly during the regional transgressions and less so during regional regressions.

FACTORS AFFECTING MINABLE COAL BEDS

The climate-change model for minable coal beds

Coastal-plain sedimentation in the nearshore fluvial-deltaic system is the interpreted environment of deposition for thick sequences of Carboniferous strata, some of which contain, but many of which lack, minable coal beds. Because similar tectonic and eustatic conditions were common to all these sequences, their minable coal beds cannot be ascribed only to these factors. Cecil (1990) points out that although depositional environments, tectonics, and eustatic conditions were important in forming and preserving minable coal beds, changes in climate may be most critical.

General paleoclimate for Pennsylvanian coal beds was interpreted for the Eastern Interior basin (paleobotany; White, 1913, 1925; Phillips and Peppers, 1984) and Appalachian basin (geochemistry; Cecil and others, 1985; stratigraphic associations; Donaldson and others, 1985). Paleomagnetic interpretations locate the Appalachian basin nearly straddling the Equator during much of Pennsylvanian time (Irving, 1964; Habicht, 1979). Phillips, Cecil, Donaldson, and their coworkers concluded that Early Pennsylvanian time and most of the Middle Pennsylvanian were everwet in the Appalachian basin. In contrast, Late Pennsylvanian, and earlier Late Mississippian times were seasonally dry. Cecil and others (1985) and Winston and Stanton (1989) also believe the upper Middle Pennsylvanian (Westphalian D) was transitional between the earlier everwet and later seasonally dry climates.

Evidence for Pennsylvanian everwet tropical conditions includes: (1) low-ash, low-sulfur coals associated with paleosols, suggesting acidic pH; (2) quartz-rich sandstones; (3) high-aluminum clays, indicating intense paleoweathering; (4) stream deposits showing uniform discharge; and (5) lithologies featuring mineralogically mature conditions. On the other hand, evidence for seasonally dry tropical conditions includes: (1) coal beds with high or variable sulfur and ash contents that are associated with "calcrete" paleosols, suggesting alkaline pH; (2) lithologies featuring less mineralogically mature conditions as well as calcareous sandstones; and (3) stream deposits with sedimentary structures indicating periodic high discharge. The interpretation that everwet tropical peats accumulated in raised bogs (domed), whereas seasonal peats accumulated in low-relief bogs (planar), makes paleoclimate a very important factor in understanding the geometry and composition of coal beds.

Similar criteria for Mississippian rocks in the Appalachian Basin suggest a wet paleoclimate for Early Mississippian time, changing to seasonal dry in Middle and Late Mississippian times. These long-term climatic changes for the eastern U.S. Carboniferous probably reflect migration of this lithospheric plate with respect to the Equator and/or an orographic effect from mountain building.

Cecil (1990) suggests that short-term climatic cycles have previously been ignored in evaluating Pennsylvanian coal-forming conditions, yet they may have significant effects on individual lithologies in variable sequences. In many ways the Pleistocene and Holocene are keys to the Pennsylvanian because major continental glaciation occurred during both times. During the Holocene, the climate has changed in tropical and subtropical western Africa from relatively wet to arid (Lezine and Casanova, 1989). Cecil compares this relatively rapid climate change with sedimentation rates and suggests that compound paleosols and alternating minable coal beds and freshwater limestones likewise could reflect these short-term climatic cycles. If correct, this concept will revolutionize our thinking about the Pennsylvanian coal measures.

Coal-bed geometry and distribution patterns in the Appalachian basin

On a regional scale, the best minable coals have characteristic positions, geometries, and trends within the depositional basins. Lower and Middle Pennsylvanian Pottsville Group coal beds in northern West Virginia may have accumulated on an alluvial plain as back-swamp peats in lenses elongated parallel to drainage channels (Presley, 1979).

Within the Dunkard and Pocahontas basins, most Pennsylvanian coal beds are associated with facies interpreted as lower coastal plain that are areally extensive yet discontinuous. The thickest minable coals are strike oriented, located in the upper part of the fluvial-deltaic apron, and commonly parallel the basinal structural trend. In detailed mapping (Haney and others, 1985; Donaldson, 1969; Donaldson and others, 1979), the thickest strike-trending coal beds contain relatively narrow belts of thinner coals that approximate the dip trend of the depositional basin, and also parallel the paleoslope and drainage. The more detailed the mapping, the more irregular the patterns become, probably reflecting poor drainage of swamp-covered coastal plains.

The thickness pattern of some coal beds suggests infilling by peat of abandoned stream or tidal channels. Examples of these patterns are the Pocahontas No. 6 (Padgett and Ehrlich, 1978), Fire Creek (Mullennex, personal communication, 1988), Beckley (Gwinn, 1950), and Pittsburgh (Mouyard, 1982). In each case, the coal bed is thickest in narrow, curvilinear dendroids that have tributary connections. The Pocahontas No. 6 coal bed remained confined to the interpreted abandoned drainage streams. However, in the other examples, coal beds extend beyond the original stream channels, which are identified by the thickened development of coal below mappable partings. Commonly, the coal beds are thickest alongside, rather than in, the interpreted drainage channels. In these examples, the thickest coal below mappable partings occurs laterally adjacent and parallel to the channel in the interpreted back-swamp setting between the levee and lake or bay. The Sewickley and Waynesburg coal beds particularly document this back-swamp origin (Donaldson and others, 1979).

Four different drainage patterns that developed during peat accumulation are recognized: (1) ribbon sandstones of the Walshville Channel (Nelson, 1983; McCabe, 1984; Trask and Palmer, 1986), Pittsburgh coal (Donaldson and others, 1979), and Douglas coal; (2) crevasse splays of the Stockton and Coalburg coal beds of the Kanawha Formation (Powell, 1979; Ferm and Staub, 1984) and the Redstone coal bed (Jake, 1981), causing coal-bed splits; (3) fault-controlled drainage as recognized by Weisenfluh and Ferm (1984) for the Pratt coal bed of Alabama; and (4) narrow, curvilinear trends of thin coal within otherwise thick coal beds such as the Pittsburgh that are overlain by siltstones and underlain by fluvial channel sandstones, indicating a persistence of drainage (Bean, 1982; Donaldson and Renton, 1984).

The coal splits and ribbon sandstones of the Appalachian basin are associated with channel avulsion during localized differential compaction and/or faulting, which changed the topography of the peat swamp with respect to stream and/or tidal channels. Drainage systems that postdate peat accumulation are abundant in the rock record. Their channel fills replaced the peat bed, causing coal-seam discontinuities (Donaldson and others, 1979) and soft-sediment deformation of the peat bed from loading pressures during early burial.

Cecil and others (1985a, b, 1987, 1988) have observed striking similarities between the geometries of the Pocahontas No. 2 and No. 3 coal beds of the Lower Pennsylvanian in southeastern West Virginia (Rehbein and others, 1981) to Recent peat deposits from Indonesia. They interpret the Pocahontas coal-bed geometries to be indicative of peat domes, whereas the absence of coal in linear areas is related to swamp drainage and estuarine channels of tidal rather than fluvial origin. We believe tidal influence is also important in abandoned parts of the fluvial-deltaic deltas during the transgressive phase of the minor allocycles of Middle and lower Upper Pennsylvanian coal beds in the Pocahontas and Dunkard basins.

Coal quality

Coal quality, especially sulfur and ash contents, influence the use and marketability of coal beds. Generally, bituminous coal must contain <25 percent ash and <3 percent sulfur prior to mining to be of commercial quality in the United States (Cecil and others, 1985b). The highest-quality bituminous coal beds contain <10 percent ash and <1 percent sulfur. In the Appalachians, such coals occur mainly in the Pocahontas and Black Warrior basins. Some anthracite beds are low in sulfur (<1 percent) but contain high ash (18 to 40 percent; Lyons, and 1984). Bituminous coals of the Dunkard and midcontinent basins typically are moderate to high in sulfur (> to 7 percent) and ash (>10 to 40 percent).

Ash in Pennsylvanian coals (Damberger and others, 1984) mainly consists of clay minerals (illite, smectite, kaolinite) with minor amounts of quartz, mica, carbonates (calcite, dolomite/ankerite, siderite), feldspars, and heavy minerals (iron disulfides and iron oxides). A certain amount of ash commonly is disseminated throughout the coal. Otherwise, the ash is concentrated within the coal bed in thin lenses and beds, laminae, or as fillings of fractures.

The origin of ash includes detrital influx into swamps during floods, authigenic inorganic material released from plants during degradation of peat, aeolian dust, and volcanic ash. Volcanic ash has been cited commonly in European Carboniferous coals as a major source of mineral matter. Recognition of volcanic ash in partings in U.S. Pennsylvanian coal beds are rare, but the existence of one has been thoroughly documented in the Middle Pennsylvanian Fire Clay coal of Kentucky and West Virginia (Chesnut, 1985).

High-sulfur Pennsylvanian coals contain mainly iron disulfides (particularly pyrite and marcasite), but low-sulfur coals contain nearly equal proportions of organic sulfur and iron sulfides. Coal beds having marine roof rocks in the Eastern Interior basin (Thiessen, 1920; Gluskoter and Hopkins, 1970), the Dunkard basin (Williams and Keith, 1963), and the Pocahontas basin (Horne and others, 1978; Cobb and Chesnut, 1989) have relatively high total sulfur content (generally more than 3 percent in the Eastern Interior basin). Also, coal beds underlying carbonate-rich strata commonly have high sulfur contents (Donaldson and Renton, 1984; Cecil and others, 1985b).

The geochemical coal model (Cecil and others, 1985b; Renton in Donaldson and others, 1985) reported that alkaline waters of overlying marine and nonmarine, carbonate-rich roof sediments drain downward and degrade underlying peat beds during deposition and early burial, promoting generation of syngenetic pyrite. Gluskoter and Hopkins (1970) recognized higher sulfur in the Herrin No. 6 coal bed (Eastern Interior basin) where its roof rock was marine compared with laterally adjacent nonmarine facies.

In the Appalachian basin, coal-quality maps are published for the following coal beds: Pocahontas No. 3 (Rehbein and others, 1981), Fire Clay (Haney and others, 1985), Lower Kittanning (Sentfle and Davis, 1982; Rimmer and Davis, 1984), Pittsburgh (Gomez and others, 1974; Donaldson and others, 1979; Donaldson and Renton, 1984), Redstone (Donaldson and Renton, 1984), and Waynesburg (Donaldson and others, 1979; Donaldson and Renton, 1984). In these studies, high-sulfur coal beds usually underlie roof rocks of marine or alkaline lake (freshwater carbonate) origin. However, they also are common where paleochannels with permeable sandstones occur in the overlying roof rocks (Cheek and Donaldson, 1969; Neavel, 1966; Donaldson and Renton, 1984; Burk and others, 1987).

Coal quality commonly is not homogeneous within a coal bed. In the Appalachians, Lower Pennsylvanian coal beds show the least within-bed variability in coal quality, and Upper Pennsylvanian coals exhibit the most (Cecil and others, 1985b). Upper Pennsylvanian coal beds (Pittsburgh, Redstone, Sewickley, Waynesburg) have high ash and sulfur stratified in layers, the majority of which are concentrated in the lowermost and uppermost 15 to 30 cm (Donaldson and Renton, 1984).

Although lateral coal-quality changes appear erratic and random, variograms indicate that Pittsburgh coal cores spaced at about 3 km allow for recognition of major thickness and sulfur-content trends that reflect curvilinear-belt patterns or interpreted ancient drainage (Donaldson and Renton, 1984). Whereas this spacing is sufficient for thickness/sulfur exploration in this coal bed, it is too wide for mine design. For optimum development, sampling sites must be <365 to 430 m apart to recognize variability in total sulfur. Detailed mapping within mines recognized very irregular patterns superimposed on the generalized regional trends. Lateral coal-bed sampling at intervals <365 m and Kriging statistics are necessary to determine small-scale sulfur-content trends within Dunkard basin coal beds (Donaldson and Renton, 1984).

SOME CONTINUING RESEARCH NEEDS

Invertebrate paleontology, paleobotany, and palynology have established the time-rock correlations currently used in Carboniferous coal-measure stratigraphy worldwide. The continued need to recognize shorter-duration time-rock units for correlation between coal basins remains a major challenge. Without this knowledge, the quantitative importance to the cyclic nature of the sequences is uncertain for the parameters of tectonism, eustacy, localized sediment-supply shifts, and climate. Basin analysis is no better than the criteria used, so continued new generations of improved criteria and evidence represent an ongoing quest.

Promising results are issuing from studies of modern raised and low-lying peat bogs as analogs of Pennsylvanian coal beds (Eble and others, 1989). Comparisons of geochemistry and petrology of different peats and coals already have suggested reasons for their sulfur and ash contents (Grady, 1989). Subsurface analyses of Holocene-Pleistocene peats and interbedded sediments from Indonesia are needed to show the depositional setting and processes most favorable for preservation of thick peat deposits in a foreland basin where both tropical wet and seasonal climates exist. Studies of other modern chemical deposits (carbonates, soils) for compound climatic effects within 20,000 to 40,000 years also are needed to establish criteria for recognizing similar short-term climatic conditions within Pennsylvanian coal basin rocks.

Within the foreland basin of the Appalachians, regional studies using closely spaced well data are needed to describe the precise parts of cycles displaying autogenic versus allogenic controls. Mapping of selected paleosol units and their facies equivalents in the Middle and Upper Pennsylvanian should augment other key beds in this analysis. Recent recognition of evidence for tidal processes in some sandstones should prompt reexamination of facies previously interpreted as deposits of a fluvial-dominant delta. Quartz-rich sandstones commonly have questionable source-area affinities and depositional environments, and need further research.

Mine-scale studies within the foreland basins need to document penecontemporaneous fault-block subsidence during peat accumulation. Likewise, patterns in coal-seam discontinuities and lateral changes in ash and sulfur within a coal bed need to be described and published. Coal partings and splits need to be mapped so the suspected widely varied conditions can be evaluated for different tectonic, depositional, and paleoclimatic settings throughout the Pennsylvanian of the central and eastern United States.

REFERENCES CITED

Arkle, T., Jr., 1974, Stratigraphy of the Pennsylvanian and Permian Systems in the central Appalachians, in Briggs, G., ed., Carboniferous of the southeastern United States: Geological Society of America Special Paper 148, p. 5–29.

Arkle, T., Jr., and others, 1979, The Mississippian and Pennsylvanian (Carboniferous) Systems in the United States: West Virginia and Maryland: U.S. Geological Survey Professional Paper 1110D, 35 p.

Arndt, H. H., Averitt, P., Dowd, S., Frendzel, D., and Gallo, P. A., 1968, Coal, in Mineral resources of the Appalachian Region: U.S. Geological Survey Professional Paper 580, p. 102–133.

Bean, H. P., 1982, Stratigraphic analysis of thickness, sulfur, and enclosing strata of the Pittsburgh and Redstone coals in northern West Virginia–southwestern Pennsylvania [M.S. thesis]: Morgantown, West Virginia University, 228 p.

Beaumont, C., Quinlan, G., and Hamilton, J., 1988, Orogeny and stratigraphy; Numerical models of the Paleozoic in the eastern interior of North America: Tectonics, v. 7, p. 389–416.

Beerbower, J. R., 1964, Cyclothems and cyclic depositional mechanisms in alluvial plain sedimentation: Kansas Geological Survey Bulletin 169, v. 1, p. 31–42.

Beuthin, J., 1989, Genetic identification and paleogemorphic interpretation of the sub-Pennsylvanian unconformity in northwestern West Virginia: Appalachian Basin Industrial Associates, v. 15, p. 107–131.

Boardman, D. R., II, and Heckel, P. H., 1989, Glacial-eustatic sea-level curve for early Late Pennsylvanian sequence in north-central Texas and biostratigraphic correlation with curve for midcontinent North America: Geology, v. 17, p. 802–805.

Boswell, R. M., and Donaldson, A. C., 1988, Depositional architecture of the Upper Devonian Catskill Delta complex; Central Appalachian Basin, United States, in McMillan, N. J., Embry, A. F., and Glass, D. J., eds., Devonian of the world, v. 2: Canadian Society of Petroleum Geologists, p. 65–84.

Bragonier, W. A., 1989, Stratigraphy of flint clays of the Allegheny and Pottsville Groups, western Pennsylvania, in Harper, J. A., ed., Geology in the Laurel Highlands, southwestern Pennsylvania: Guidebook of 54th Annual Field Conference of Pennsylvania Geologists, Department of Environmental Resources, Bureau of Topographic and Geologic Survey, Harrisburg, p. 69–88.

Branson, C. C., ed., 1962, Pennsylvanian System in the United States: American Association of Petroleum Geologist, A Symposium, 508 p.

Bristol, H. M., and Howard, R. H., 1971, Paleogeologic map of the sub-Pennsylvanian Chesterian (Upper Mississippian) surface in the Illinois Basin: Illinois State Geological Survey Circular 458, 14 p.

——, 1974, Sub-Pennsylvanian valleys in the Chesterian surface of the Illinois Basin and related Chesterian slump blocks, in Briggs, G., ed., Carboniferous of the southeastern United States: Geological Society of America Special Paper 148, p. 315–335.

Brown, L. F., Jr., Cleaves, A. W., II, and Erxleben, A. W., 1973, Pennsylvanian depositional systems in north-central Texas: Bureau of Economic Geology Texas Guidebook 14, 122 p.

Burk, M. K., Deshowitz, M. P., and Utgaard, J. E., 1987, Facies and depositional environments of the Energy Shale Member (Pennsylvanian) and their relationship to low-sulfur coal deposits in southern Illinois: Journal of Sedimentary Petrology, v. 57, p. 1060–1067.

Busch, R. M., and Rollins, H. B., 1984, Correlation of Carboniferous strata using hierarchy of transgressive-regressive units: Geology, v. 12, p. 471–474.

Cecil, C. B., 1986, Allocyclic and autocyclic conditions of coal formation, in Carter, L.M.H., ed., USGS research on energy resources-1986, Program and abstracts: U.S. Geological Survey Circular 974, p. 8–9.

——, 1990, Paleoclimate controls on stratigraphic repetition of chemical and siliciclastic rocks: Geology, v. 18, p. 533–536.

Cecil, C. B., and Englund, K. J., 1985, Geologic controls on sedimentation and peat formation in the Carboniferous of the Appalachian Basin, in Englund, K. J., and others, eds., Characteristics of the Mississippian–Pennsylvanian boundary and associated coal-bearing rocks in the southern Appalachians: U.S. Geological Survey Open-File Report, p. 27–33.

Cecil, C. B., and Englund, K. J., 1989, Origin of coal deposits and associated rocks in the Carboniferous of the Appalachian Basin, in Cecil, C. B., and Eble, C., eds., Carboniferous geology of the eastern United States; 28th International Geological Congress field trip guidebook T143: American Geophysical Union, p. 84–111.

Cecil, C. B., and 5 others, 1985, Paleoclimate controls on Late Paleozoic sedimentation and peat formation in the central Appalachian Basin (United States), in Phillips, T. L., and Cecil, C. B., eds., Paleoclimatic controls on coal resources of the Pennsylvanian Systems of North America: International Journal of Coal Geology, v. 5, p. 195–230.

Cecil, C. B., Supardi, and Neuzil, S. G., 1987, Domed peat swamps in Indonesia; A modern analog of coal formation: Geological Society of America Abstracts Programs, v. 19, p. 615.

Cecil, C. B., Dulong, F. T., Cobb, J. C., Supardi, Turnbull, P., 1988, Allogenic processes in the central Sumatra Basin; A modern analog for the origin of Lower Pennsylvanian coal-bearing strata in the eastern United States, in Carter, L.M.H., ed., U.S. Geological Survey Research on Energy Resources: U.S. Geological Survey Circular 1025, p. 8.

Cheek, R., and Donaldson, A. C., 1969, Sulfur facies of the Upper Freeport coal and northwestern Preston County, West Virginia, in Donaldson, A. C., ed., Some Appalachian coals and carbonates; Models of ancient shallow-water deposition: West Virginia Geological Survey, p. 279–307.

Chesnut, D. R., 1981, Marine zones of the upper Carboniferous of eastern Kentucky, in Cobb, J. C., Chesnut, D. R., Hester, N. C., and Hower, J. C., eds., Coal and coal-bearing rocks of eastern Kentucky; Guidebook and roadlog for Coal Division of Geological Society of America Field Trip 14: Kentucky Geological Survey, ser. 11, p. 57–66.

——, 1985, Source of the volcanic ash deposit (flint clay) in the Fire Clay coal of the Appalachian Basin; Dixeme Congress International de Stratigraphie et de Geologie du Carbonifere, Madrid, 1983: Compte Rendu, v. 1, p. 145–154.

Chesnut, D. R., Jr., 1989, Pennsylvanian rocks of the eastern Kentucky coal field, in Cecil, C. B., and Eble, C., eds., Carboniferous geology of the eastern United States; 28th International Geological Congress field trip guidebook T143: American Geophysical Union, p. 57–63.

Chesnut, D. R., Jr., and Cobb, J. C., 1989, Guide to the Carboniferous rocks of Kentucky, in Cecil, C. B., and Eble, C., eds., Carboniferous geology of the eastern United States; 28th International Geological Congress field trip guidebook T143: American Geophysical Union, p. 38–49.

Cobb, J. C., and Chesnut, D. R., Jr., 1989, Resource perspectives of coal in eastern Kentucky, in Cecil, C. B., and Eble, C., eds., Carboniferous geology of the eastern United States; 28th International Geological Congress field trip guidebook T143: American Geophysical Union, p. 64–83.

Collinson, C., Sargent, M. L., and Jennings, J. R., 1988, Illinois Basin region, in Sloss, L. L., ed., Sedimentary cover–North American Craton; U.S.: Boulder, Colorado, Geological Society of America, The Geology of North America, v. D-2, p. 383–426.

Craig, L. C., and others, 1979, Paleotectonic investigations of the Mississippian System in the United States: U.S. Geological Survey Professional Paper 1010, 569 p.

Damberger, H. H., 1989a, Overview of the Pennsylvanian in the Illinois Basin, in Cecil, C. B., and Eble, C., eds., Carboniferous geology of the eastern United States; 28th International Geological Congress field trip guidebook T143: American Geophysical Union, p. 20–26.

——, 1989b, The Pennsylvanian of the southern Illinois Basin; The nature of the sub-Pennsylvanian unconformity, in Cecil, C. B., and Eble, C., eds., Carboniferous geology of the eastern United States; 28th International Geological Congress field trip guidebook T143: American Geophysical Union, p. 17–19.

Damberger, H. H., Harvey, R. D., Ruch, R. R., and Thomas, J., Jr., 1984, Coal characterization, in Cooper, B. R., and Ellingson, W. A., eds., The Science and technology of coal and coal utilization: New York, Plenum Press, p. 7–45; also Illinois Geology Survey Division Reprint 1984, p. 7–45.

Davis, R. W., Plebuch, R. O., and Whitman, H. M., 1974, Hydrology and geology of deep sandstone aquifers of Pennsylvanian age in part of the western coal field region, Kentucky: Kentucky Geological Survey Report of Investigations 15, ser. x, 26 p.

Donaldson, A. C., 1969, Ancient deltaic sedimentation (Pennsylvanian) and its control on the distribution, thickness, and quality of coal, in Donaldson, A. C., ed., Some Appalachian coals and carbonates: Morgantown, West Virginia Geological and Economic Survey, p. 93–123.

——, 1974, Pennsylvanian sedimentation of central Appalachians, in Briggs, G., ed., Carboniferous of the southeastern United States: Geological Society of America Special Paper 148, 47–78.

Donaldson, A. C., and Eble, C., 1989, Morgantown area stops, in Cecil, C. B., and Eble, C., eds., Carboniferous geology of the eastern United States; 28th International Geological Congress field trip guidebook T143: American Geophysical Union, p. 104–111.

Donaldson, A. C., and Renton, J. J., 1984, A model for the evaluation of systematic variability in composition and thickness of high sulfur-high ash coals: U.S. Department of Interior/Office of Surface Mining 95084, 108 p.

Donaldson, A. C., and Shumaker, R. C., 1981, Late Paleozoic molasse of central Appalachians, in Miall, A. D., ed., Sedimentation and tectonics in alluvial basins: Geological Association of Canada Special Paper 23, p. 99–124.

Donaldson, A. C., Martin, R. H., and Kanes, W. H., 1970, Holocene Guadalupe Delta of Texas Gulf Coast, in Deltaic sedimentation: Society of Economic Paleontologists and Mineralogists Special Publication 15, p. 107–137.

Donaldson, A. C., Presley, M. W., and Renton, J. J., eds., 1979, Carboniferous coal guidebook: West Virginia Geological and Economic Survey, B-37-1, 301 p.; B-37-2, 1974 p.; B-37-3, 181 p.

Donaldson, A. C., Renton, J. J., and Presley, M. W., 1985, Pennsylvanian deposystems and paleoclimates of the Appalachians: International Journal of Coal Geology, v. 5, p. 167–193.

Duff, P.M.D., Hallam, A., and Walton, E. K., 1967, Cyclic sedimentation, developments in sedimentology: Amsterdam, Elsevier, 280 p.

Eble, C., Grady, W. C., and Gillespie, W. H., 1989, Palynology, petrography, and paleoecology of the Hernshaw–Fire Clay coal bed in the central Appalachian Basin, in Cecil, C. B., and Eble, C., eds., Carboniferous geology of the eastern United States; 28th International Geological Congress field trip guidebook T143: American Geophysical Union, p. 133–142.

Edmunds, W. E., 1988, The Pottsville Formation of the anthracite region: Pennsylvania Geological Survey Field Conference of Pennsylvania Geologist guidebook, p. 40–50.

Edmunds, W. E., and Eggleston, J., 1989, The Mississippian–Pennsylvanian boundary and associated strata in the northern basin; 28th International Geological Congress field trip T352, Part 1: American Geophysical Union, p. T352:5–37.

Edmunds, W. E., and 5 others, 1979, The Mississippian and Pennsylvanian Systems in the United States; Pennsylvania and New York: U.S. Geological Survey Professional Paper 1110-a-L, p. B1–B33.

Englund, K. J., 1974, Sandstone distribution patterns in the Pocahontas Formation of southwest Virginia and southern West Virginia, in Briggs, G., ed., Carboniferous of the southeastern United States: Geology Society of America Special Paper 148, p. 31–45.

——, 1979, Mississippian and Pennsylvanian (Carboniferous) Systems in the United States; Virginia: U.S. Geology Survey Professional Paper 1110-A-L, p. C1–C21.

Englund, K. J., Weber, J. C., Thomas, R. E., Windolph, J. F., Jr., and Dryden, J. W., 1983, Test drilling for coal in 1982–83 in the Jefferson National Forest, Virginia: U.S. Geological Survey Open-File Report 83–637, part 3, 249 p.

Englund, K. J., Windolph, J. F., Jr., and Thomas, R. E., 1986, Origin of thick low-sulfur coal in the Lower Pennsylvanian Pocahontas Formation, Virginia and West Virginia, in Lyons, P. C., and Rice, C. L., eds., Paleoenvironmental and tectonic controls in coal-forming basins in the United States: Geology Society of America Special Paper 210, p. 49–61.

Epsman, M. L., and 9 others, 1988, Geologic evaluation of critical production parameters for coalbed methane resources; Part 2, Black Warrior Basin: Gas Research Institute Contract 5087-214-1544, p. 53–100.

Ettensohn, F. R., 1985, The Catskill Delta complex and the Acadian orogeny; A model, in Woodrow, D. L., and Sevon, W. D., eds., Geological Society of America Special Paper 201, p. 39–50.

Ferm, J. C., 1970, Allegheny deltaic deposits, in Morgan, J. P., ed., Deltaic sedimentation, modern and ancient: Society of Economic Paleontologists and Mineralogists Special Publication 15, p. 246–255.

——, 1975, Pennsylvanian cyclothems of the Appalachian Plateau, a retrospective view, in McKee, E. D., and Crosby, E. N., eds., Paleotectonic investigations of the Pennsylvanian System in the United States: U.S. Geological Survey Professional Paper 853, part 2, p. 57–64.

Ferm, J. C., and Staub, J. R., 1984, Depositional controls of mineable coal bodies, in Rahmani, R. A., and Flores, R. M., eds., Sedimentology of coal and coal-bearing sequences: Association of Sedimentology Special Publication 7, p. 275–289.

Ferm, J. C., and Williams, E. G., 1965, Characteristics of a Carboniferous marine invasion in western Pennsylvania: Journal of Sedimentary Petrology, v. 35, p. 319–330.

Fuller, J. O., 1955, Source of Sharon Conglomerate of northeastern Ohio: Geological Society of America Bulletin, v. 66, p. 159–176.

Galloway, W. E., 1981, Depositional architecture of Cenozoic Gulf coastal plain fluvial systems, in Ethridge, F. C., and Flores, R. M., eds., Recent and ancient nonmarine depositional environments; Models for exploration: Society of Economic Paleontologists and Mineralogists Special Publication 31, p. 127–155.

Garner, H. F., 1974, The origin of landscapes: London, Oxford University Press, p. 646–667.

Giffin, J. W., 1978, Stratigraphy and petrography of the Livingston Limestone Member, Bond Formation (Missourian, Pennsylvanian), of east central Illinois and western Indiana [M.A. thesis]: Bloomington, Indiana University, 190 p.

Glick, E.E., 1975, Arkansas and northern Louisiana: U.S. Geological Survey Professional Paper 853-I, p. 157–175.

Gluskoter, H. G., and Hopkins, M. E., 1970, Distribution of sulfur in Illinois coals, in Smith, W. H., and others, eds., Depositional environments in parts of the Carbondale Formation; western and northern Illinois: Illinois State Geological Survey, Guidebook Series, v. 8, p. 89–95.

Gomez, M., Donaven, D. J., and Kent, B., 1974, Distribution of sulfur and ash in part of the Pittsburg seam, and probable mode of deposition: U.S. Bureau of Mines Report of Investigations 7827, 44 p.

Grady, W. C., 1989, A petrographic evaluation of environments of accumulation of the Pocahontas No. 3 coal bed in southwestern West Virginia, in Cecil, C. B., and Eble, C., eds., Carboniferous geology of the eastern United States; 28th International Geological Congress Field Trip Guidebook T143: American Geophysical Union, p. 127–133.

Greb, S. F., 1989, Structural controls on the formation of the sub-Absaroka unconformity in the U.S. Eastern Interior Basin: Geology, v. 17, p. 889–892.

Gwinn, J. E., 1950, Origin and practical implications of the structure of the Beckley coal Stansford Number 2 Mine, Stansford, West Virginia [M.S. thesis]: Morgantown, West Virginia University, 34 p.

Habicht, J.K.A., 1979, Paleoclimate, paleomagnetism, and continental drift: American Association of Petroleum Geologists Studies in Geology, no. 9, 31 p.

Haney, D. C., Cobb, J. C., Chesnut, D. R., and Currens, J. C., 1985, Structural controls on environments of deposition, core quality, and resources in the Appalachian Basin of Kentucky; Dixieme Congress International de Stratigraphie et de Geologie du Carbonifere, Madrid 1983: Compte Rendus, v. 1, p. 69–78.

Harland, W. B., and 5 others, 1982, A geologic time scale: Cambridge, England,

Cambridge University Press, 131 p.

Heckel, P. H., 1977, Origin of phosphatic black shale facies in Pennsylvanian cyclothems of Midcontinent North America: American Association of Petroleum Geologists Bulletin, v. 61, p. 1045–1068.

——, 1980, Paleogeography of eustatic model for deposition of Midcontinent Upper Pennsylvanian cyclothems, in Fouch, T. D., and Magathan, E. R., eds., Paleozoic paleogeography of west-central United States: Rocky Mountain Section, Society of Economic Paleontologists and Mineralogists West-Central United States Paleogeography Symposium 1, p. 197–215.

——, 1986, Sea-level curve for Pennsylvanian eustatic marine transgressive-regressive depositional cycles along Midcontinent outcrop belt, North America: Geology, v. 14, p. 330–334.

Hines, R. A., and Thomas, W. A., 1987, Foreland basin evolution of the Black Warrior Basin (Carboniferous) Alabama and Mississippi: Morgantown, Appalachian Basin Industrial Associates, West Virginia University Geology and Geography Department, v. B, p. 99–143.

Hobday, D. K., 1974, Beach-barrier-island facies in the Upper Carboniferous of northern Alabama, in Briggs, G., ed., Carboniferous of the southeastern United States: Geological Society of America Special Paper 148, p. 209–223.

Horne, J. C., 1968, Detailed correlation and environmental study of some Late Pennsylvanian units of the Illinois Basin [Ph.D. thesis]: Urbana, University of Illinois, 49 p.

Horne, J. C., Ferm, J. C., Caruccio, F. T., and Baganz, B. P., 1978, Depositional models in coak exploration and mine planning in Appalachian region: American Association of Petroleum Geologists Bulletin, v. 62, p. 2379–2411.

Horsey, C. A., 1981, Depositional environments of the Pennsylvanian Pottsville Formation in the Black Warrior Basin of Alabama: Journal of Sedimentary Petrology, v. 51, p. 799–806.

Houseknecht, D. W., 1980, Comparative anatomy of a Pottsville lithic arenite and quartz arenite of the Pocahontas Basin, southern West Virginia; Petrogenetic, depositional, and stratigraphic implications: Journal of Sedimentary Petrology, v. 50, p. 20–30.

Howard, R. H., 1979, The Mississippian-Pennsylvanian unconformity in the Illinois Basin; Old and new thinking, in Palmer, J. E., and Dutcher, R. R., eds., Depositional and structural history of the Pennsylvanian System of the Illinois Basin; Part 2, Invited papers; Field trip 9/Ninth International Congress of Carboniferous Stratigraphy and Geology: Illinois State Geological Survey Guidebook Series 15a, p. 34–43.

Irving, E., 1964, Paleomagnetism: New York, John Wiley and Sons, 399 p.

Jake, T. R., 1981, Deposystem analysis of the Upper Pennsylvanian Redstone Coal and adjacent units in Barbour County, West Virginia [M.S. thesis]: Morgantown, West Virginia University, 120 p.

Jennings, J. R., and Fraunfelter, H. G., 1986, Preliminary report on micropaleontology of strata above and below the upper boundary of the type Mississippian: Transactions of the Illinois Academy of Science, v. 79, p. 253–262.

Johnson, K. S., and 7 others, 1988, Southern midcontinent region, in Sloss, L. L., ed., Sedimentary cover–North American Craton; U.S.: Boulder, Colorado, Geological Society of America, The Geology of North America, v. D-2, p. 307–359.

Klein, G. deV., and Willard, D. A., 1989, Origin of the Pennsylvanian coal-bearing cyclothems of North America: Geology, v. 17, p. 152–155.

Kosanke, R. M., Simon, J. A., Wanless, H. R., and Willman, H. B., 1960, Classification of the Pennsylvanian strata of Illinois: Illinois State Geological Survey Report of Investigations 214, 84 p.

Krissek, L. A., Ketring, C. L., Jr., and Kulikonski, D. L., 1986, Lower Pennsylvanian sandstone of southeastern Ohio; Implications for sediment sources and depositional environments in the north-central Appalachian Basin: Morgantown, Appalachian Basins Industrial Associates, West Virginia University, Geology and Geography Department, v. 7, p. 109–141.

Krumbein, W. C., and Sloss, L. L., 1951, Stratigraphy and sedimentation: San Francisco, W. H. Freeman, 497 p.

Kvale, E. K., Archer, A. W., and Johnson, H. R., 1989, Daily, monthly, yearly tidal cycles within laminated siltstones of the Mansfield Formation (Pennsylvanian) of Indiana: Geology, v. 17, p. 365–368.

Lezine, A-M., and Casanova, J., 1989, Pollen and hydrological evidence for the interpretation of past climates in tropical west Africa during the Holocene: Quaternary Science Review, v. 8, p. 45–55.

Lyons, P. C., 1984, Carboniferous megafloral zonation of New England, in Sutherland, P. K., and Manager, W. L., eds., Neuvieme Congress International de Stratigraphie et de Geologie du Carbonifere, Compte Rendu, Biostratigraphy: Carbondale, Southern Illinois University Press, v. 2, p. 503–514.

Lyons, P. C., and Rice, C. L., eds., 1986, Paleoenvironmental and tectonic controls in coal-forming basins in the United States: Geological Society of America Special Paper 210, 200 p.

McCabe, P. J., 1984, Depositional environments of coal and coal-bearing strata, in Rahmani, R. A., and Flores, R. M., eds., Sedimentology of coal and coal-bearing sequences; International Association of Sedimentology Special Publication 7: Blackwell Scientific Publications, p. 13–42.

McCulloch, C. M., Diamond, W. P., Bench, B. M., and Duel, M., 1975, Selected geological factors affecting mining of the Pittsburgh coalbed: United States Bureau of Mines Report of Investigations 8093, 72 p.

McKee, E. D., and others, 1975, Paleotectonic investigation of the Pennsylvanian System in the United States: U.S. Geological Survey Professional Paper 853, Part 2, 191 p.

Meckel, L. D., 1967, Origin of Pottsville conglomerates (Pennsylvanian) in the central Appalachians: Geological Society of America Bulletin, v. 78, p. 223–258.

Merriam, D. F., ed., 1964, Symposium on cyclic sedimentation: Kansas Geological Survey Bulletin 169, 636 p.

Morris, R. C., 1974, Carboniferous rocks in the Ouachita Mountains, Arkansas; A study of facies patterns along the unstable slope and axis of a flysch trough, in Briggs, G., ed., Carboniferous of the southeastern United States: Geological Society of America Special Paper 148, p. 241–279.

Mouyard, D. P., 1982, Effects of structure and stresses on underground mining in the Pittsburgh coalbed of southwestern Pennsylvania [M.S. thesis]: Morgantown, West Virginia University, 76 p.

Neavel, R. C., 1966, Sulfur in coal; Its distribution in the seam and in mine products [Ph.D. thesis]: University Park, Pennsylvania State University, 351 p.

Nelson, W. J., 1983, Geologic disturbances in Illinois coal seams: Illinois State Geological Survey Division Circular 530, 47 p.

Nelson, W. J., and 5 others, 1991, Absaroka Sequence, in Leighton, M. W., Kolata, D. R., Oltz, D. F., and Eidel, J. J., eds., Interior cratonic basins: American Association of Petroleum Geologists, World Petroleum Basins, v. 1 (in press).

Padgett, G., and Ehrlich, R., 1978, An analysis of two tectonically controlled integrated drainage nets of mid-Carboniferous age in southern West Virginia, in Miall, A. D., ed., Fluvial sedimentology: Canadian Society of Petroleum Geologists, p. 789–799.

Palmer, J. E., and Dutcher, R. R., eds., 1979, Depositional and structural history of the Pennsylvanian System of the Illinois Basin; 9th International Congress of Carboniferous Stratigraphy and Geology guidebook for field trip 9, Part 1: Illinois State Geological Survey Guidebook 15, 116 p.

Phillips, T. L., and Peppers, R. A., 1984, Changing patterns of Pennsylvanian coal-swamp vegetation and implications of climate control on occurrence: International Journal of Coal Geology, v. 3, p. 205–255.

Potter, P. E., and Siever, R., 1956, Sources of basal Pennsylvanian sediments in the Eastern Interior Basin; Part 1, Cross-bedding: Journal of Geology, v. 64, p. 225–244.

Powell, L. R., 1979, Breathitt depositional systems in Martin County, Kentucky, in Donaldson, A. C., Presley, M. W., and Renton, J. J., eds., Carboniferous coal: West Virginia Geological and Economic Survey, p. 51–100.

Presley, M. W., 1979, Facies and depositional systems of Upper Mississippian and Pennsylvanian strata in the central Appalachians, in Donaldson, A. C., Presley, M. W., and Renton, J. J., eds., Carboniferous coal: West Virginia Geological and Economic Survey, p. 1–50.

Pryor, W. A., and Sable, E. G., 1974, Carboniferous of the Eastern Interior Basin, in Briggs, G., ed., Carboniferous of the southeastern United States: Geological Society of America Special Paper 148, p. 281–313.

Quinlan, G. M., and Beaumont, C., 1984, Appalachian thrusting, lithospheric flexure, and the Paleozoic stratigraphy of the Eastern Interior of North America: Canadian Journal of Earth Sciences, v. 21, p. 973–996.

Rahmani, R. A., and Flores, R. M., 1984, Sedimentology of coal and coal-bearing sequences of North America; A historical review, in Rahmani, R. A., and Flores, R. M., eds., Sedimentology of coal and coal-bearing sequences: Association of Sedimentology Special Publication 7, p. 3–10.

Rehbein, E. A., Henderson, C. D., and Mullenex, R., 1981, No. 3 Pocahontas Coal in southern West Virginia—Resources and depositional trends: West Virginia Geological and Economic Survey, Bulletin B-38, 41 p.

Rice, C. L., 1984, Sandstone units of the Lee Formation and related strata in eastern Kentucky: U.S. Geological Survey Professional Paper 1151-G, 53 p.

—— , 1985, Terrestrial vs marine depositional model; A new assignment of subsurface Lower Pennsylvanian rocks of southwestern Virginia: Geology, v. 13, p. 786–789.

Rice, C. L., and Schwietering, J. F., 1988, Fluvial deposition in central Appalachians during the Early Pennsylvanian: U.S. Geological Survey Bulletin 1839-A-D, p. B1–B10.

Rice, C. L., and others, 1979, Kentucky, in The Mississippian and Pennsylvanian (Carboniferous) Systems in the United States: U.S. Geological Survey Professional Paper 1110-F, p. F1–F32.

Rice, C. L., Currens, J. C., Henderson, J. A., Jr., and Nolde, J. E., 1987, The Betsie Shale Member; A datum for exploration and stratigraphic analysis of the lower part of the Pennsylvanian in the central Appalachian Basin: U.S. Geological 1834, 17 p.

Rightmire, C. T., Eddy, G. E., and Kirr, J., eds., 1984, Coalbed methane resources of the United States: American Association of Petroleum Geologists Studies in Geology Series 17, 378 p.

Rimmer, S. M. and Davis, A., 1984, A data base for the analysis of compositional characteristics of coal seams and macerals; lateral variability in mineralogy and petrology of the Lower Kittanning Seam, western Pennsylvania and eastern Ohio: University Park, The Pennsylvania State University Coal Research Section, 247 p.

Saunders, W. B., and Ramsbottom, W.H.C., 1986, The mid-Carboniferous eustatic event: Geology, v. 14, p. 208–212.

Schutter, S. R., and Heckel, P. H., 1985, Missourian (early Late Pennsylvanian) climate in midcontinent North America, in Phillips, T. L., and Cecil, C. B., eds., Paleoclimate controls on coal resources of the Pennsylvanian System of North America: International Journal of Coal Geology, v. 5, p. 111–138.

Senftle, J. T., and Davis, A., 1982, A data base for the analysis of compositional characteristics of coal seams and macerals; relationships between coal constitution, thermoplastic properties, and liquefaction behavior of coals and vitrinite concentrates from the lower Kittanning Seam: University Park, The Pennsylvania State University Coal Research Section, 223 p.

Shawe, F. R., and Gildersleeve, B., 1969, An anastomosing channel complex at the base of the Pennsylvanian System in western Kentucky: U.S. Geological Survey Professional Paper 650-D, p. 206–209.

Shumaker, R. C., 1986, Structural development of Paleozoic continental basins of eastern North America: Proceedings of the 6th International Conference on Basement Tectonics, p. 82–95.

Siever, R., 1951, The Mississippian–Pennsylvanian unconformity in southern Illinois: American Association of Petroleum Geologists Bulletin, v. 35, p. 542–581.

Simon, F. O., and Englund, K. J., 1983, Test drilling for coal in 1982–83 in the Jefferson National Forest; Part 4, Analyses of coal cores from the Valley coal fields: U.S. Geological Survey Open-File Report 83-626, 16 p.

Skehan, J. W., Rast, N., and Mosher, S., 1986, Paleoenvironmental and tectonic controls of sedimentation in coal-forming basins of southeastern New England, in Lyons, P. C., and Rice, C. L., eds., Paleoenvironmental and tectonic controls in coal-forming basins in the United States: Geological Society of America Special Paper 210, p. 9–30.

Slingerland, R., and Beaumont, C., 1989, Tectonics and sedimentation of the upper Paleozoic foreland basin in the central Appalachians, in Slingerland, R., and Furlong, K., leaders, Sedimentology and thermal-mechanical history of basins in the central Appalachian orogen; 28th International Geological Congress field trip guidebook T152: American Geophysical Union, p. T152, 4–24.

Sloss, L. L., ed., 1988, Sedimentary cover–North American Craton; U.S.: Boulder, Colorado, Geological Society of America, The Geology of North America, v. D-2, 506 p.

Sloss, L. L., Krumbein, W. C., and Dapples, E. C., 1949, Integrated facies analysis, in Longwell, C. R., ed., Sedimentary facies in geologic history: Geological Society of America Memoir 39, p. 91–123.

Sonnefield, R. D., 1981, Geology of northwestern Jackson County, Illinois, with special emphasis on the Caseyville Formation [M.S. thesis], Carbondale, Southern Illinois University, 84 p.

Swann, D. H., 1963, Classification of Genevievian and Chesterian (Late Mississippian) rocks of Illinois: Illinois State Geological Survey Report of Investigations 216, 91 p.

Thiessen, R., 1920, Occurrence and origin of finely disseminated sulfur compounds in coal: Transactions of the American Institute of Mining, Metallurgical, and Petroleum Engineers, v. 62, p. 913–926.

Thomas, W. A., 1977, Evolution of Appalachian–Ouachita salients and recesses from reentrants and promontories in the continental margins: American Journal of Science, v. 277, p. 1233–1278.

—— , 1988, The Black Warrior Basin, in Sloss, L. L., ed., Sedimentary cover–North American Craton; U.S.: Boulder, Colorado, Geological Society of America, The Geology of North America, v. D-2, p. 471–492.

Trask, C. B., and Palmer, J. E., 1986, Structural and depositional history of the Pennsylvanian System in Illinois, in Lyons, P. C., and Rice, C. L., eds., Paleoenvironmental and tectonic controls in coal-forming basins in the United States: Geological Society of America Special Paper 210, p. 63–77.

Treworgy, C. G., and Bargh, M. H., 1982, Deep-minable coal resources of Illinois: Illinois State Geological Survey Circular 527, 62 p.

Treworgy, J., and Norby, R., 1989, Overview in the Illinois Basin, in Cecil, C. B., and Eble, C., eds., Carboniferous geology of the eastern United States; 28th International Geological Congress Field Trip Guidebook T143: American Geophysical Union, p. 1–16.

Udden, J. A., 1912, Geology and mineral resources of the Peoria Quadrangle, Illinois: U.S. Geological Survey Bulletin 506, 103 p.

U.S. Bureau of Mines, 1974, The reserve base of bituminous coal and anthracite for underground mining in the eastern United States: Bureau Mines Information Circular 8655, 428 p.

Vail, P. R., Mitchum, R. M., Jr., and Thompson, S., III, 1977, Seismic stratigraphy and global changes of sea level, in Payton, C. E., ed., Seismic stratigraphy; Applications to hydrocarbon exploration: American Association of Petroleum Geologists Memoir 26, p. 83–97.

Veevers, J. J., and Powell, C. McA., 1987, Late Paleozoic glacial episodes in Gondwanaland reflected in transgressive-regressive depositional sequences in Euramerica: Geological Society of America Bulletin, v. 98, p. 475–487.

Wanless, H. H., 1947, Regional variations in Pennsylvanian lithology: Journal of Geology, v. 55, p. 237–253.

—— , 1975, Illinois Basin region, in McKee, E. D., and Crosby, E. J., coordinators, Paleotectonic investigations of the Pennsylvanian System in the United States; Part 1, Introduction and regional analyses of the Pennsylvanian System: U.S. Geological Survey Professional Paper 853-E, p. 71–95.

Wanless, H. R., and Shepard, F. P., 1936, Sea level and climatic changes related to late Paleozoic cycles: Geological Society of America Bulletin, v. 47, p. 1177–1206.

Wanless, H. R., and Weller, J. M., 1932, Correlation and extent of Pennsylvanian cyclothems: Geological Society of America Bulletin, v. 43, p. 1003–1016.

Wanless, H. R., and 9 others, 1970, Late Paleozoic deltas in the central and eastern United States, in Morgan, J. P., ed., Deltaic sedimentation, modern and ancient: Society of Economic Paleontologists and Mineralogists Special Publication 15, p. 215–245.

Weisenfluh, G. A., and Ferm, J. C., 1984, Geologic controls on deposition of the Pratt seam Black Warrior Basin, Alabama, U.S.A., *in* Rahmani, R. A., and Flores, R. M., eds., Sedimentology of coal and coal-bearing sequences: Association of Sedimentologists Special Publication 7, p. 317–330.

Weller, J. M., 1930, Cyclic sedimentation of the Pennsylvanian Period and its significance: Journal of Geology, v. 38, p. 97–135.

Weller, J., Herbert, L.G., and Dunbar, C. D., 1942, Stratigraphy of the fuseline-bearing beds of Illinois: Illinois Geological Survey Bulletin 67, 9–34.

White, D., 1913, Climates of coal-forming periods, *in* White, D., and Theissen, R., eds., The origin of coal: U.S. Bureau Mines Bulletin 38, p. 68–79.

—— , D., 1925, Environmental conditions of deposition of coal: American Institute of Mining and Metallurgical Engineers Transactions, v. 75, p. 3–34.

Williams, E. G., and Bragonier, W. A., 1974, Controls of Early Pennsylvanian sedimentation in western Pennsylvanian, *in* Briggs, G., ed., Carboniferous of the southeastern United States: Geological Society of America Special Paper 148, p. 135–152.

Williams, E. G., and Ferm, J. C., 1964, Sedimentary facies in the lower Allegheny rocks of western Pennsylvania: Journal of Sedimentary Petrology, v. 34, p. 610–614.

Williams, E. G., and Keith, M. L., 1963, Relationship between sulfur in coals and the occurrence of marine roof beds: Economic Geology, v. 58, p. 720–729.

Williams, E. G., Bergenback, R. E., Falla, W. S., and Udagawa, S., 1968, Origin of some Pennsylvanian underclays in western Pennsylvania: Journal of Sedimentary Petrology, v. 38, p. 1129–1193.

Winston, R. B., and Stanton, R. W., 1989, Plants, coal, and climate in the Pennsylvanian of the central Appalachians, *in* Cecil, C. B., and Eble, C., eds., Carboniferous geology of the eastern United States; 28th International Geological Congress field trip guidebook T143: American Geophysical Union, p. 118–126.

Wood, G. H., Jr., Kehn, T. M., and Eggleston, J. R., 1986, Depositional and structural history of the Pennsylvanian Anthracite region, *in* Lyons, P. C., and Rice, C. L., eds., Paleoenvironmental and tectonic controls in coal-forming basins in the United States: Geological Society of America Special Paper 210, p. 31–47.

Manuscript Accepted by the Society September 24, 1990

ACKNOWLEDGMENTS

We acknowledge past support of two federally funded projects, Department of Energy contracts No. EY-76-5-05-5065 and DOI/OSM-MRI-G1105090, which improved our understanding of variability in thickness and quality within a coal bed for the high-sulfur and high-ash coals in the Dunkard Basin. We thank John Renton, Bob Shumaker, Bill Grady, C. Blaine Cecil, Ray Boswell, Tom Jake, and Bill Gillespie for their valued opinions, the contributions of former graduate students; the suggestions of John Nelson, Heinz Damberger, Kenneth Englund, Phillip Heckel, Harold Gluskoter, and Pete Palmer as critical reviewers; and the drafting of Alison Hanham.

Printed in U.S.A.

The Geology of North America
Vol. P-2, Economic Geology, U.S.
The Geological Society of America, 1991

Chapter 34

Cretaceous and Tertiary coals of the Rocky Mountains and Great Plains regions

Romeo M. Flores
Branch of Coal Geology, U.S. Geological Survey, Box 25046, Denver Federal Center, Denver, Colorado 80225
Timothy A. Cross
Department of Geology and Geological Engineering, Colorado School of Mines, Golden, Colorado 80401

INTRODUCTION

The Rocky Mountains and Great Plains regions contain a large part of the total coal resource in the United States. These resources, which occur in Upper Cretaceous (Cenomanian to Maastrichtian) and lower Tertiary (Paleocene and Eocene) strata, range from lignite to anthracite ranks. Upper Cretaceous coals are principally bituminous to subbituminous, and lower Tertiary coals are generally subbituminous to lignite. Although coals in these regions have been mined over the past century, approximately 99 percent of the original resource remains (Averitt, 1975).

Upper Cretaceous and lower Tertiary coal-bearing strata occupy a belt about 885 km wide and 2,000 km long between the international borders with Canada and Mexico. Major coalfields generally occur within discrete structural basins that formed during Late Cretaceous to early Cenozoic Laramide deformation (Fig. 1). These basins vary from about 2,000 to 145,000 km^2 in area, and contain as much as 3,500 m of coal-bearing strata.

Following historical precedence, the major coal-bearing basins and coalfields are separated into three regional divisions in this chapter. The northern Rocky Mountains and Great Plains regions include the North Central coalfield in Montana and Wyoming and the Williston basin in North Central Dakota, respectively. Coals in this region are principally of early Tertiary age, but minor Late Cretaceous coals also are represented. The central Rocky Mountains region comprises the Wind River basin, Green River basin, Hams Fork coalfield, and Hanna basin in Wyoming; the Uinta-Piceance basin in Utah and Colorado; and the North Park and Denver basins in Colorado. Both Upper Cretaceous and lower Tertiary coals are represented in this region. The southern Rocky Mountains region includes the Kaiparowits and Henry Mountain coalfields in Utah, the Black Mesa coalfield in Arizona, the San Juan basin in New Mexico, and the Raton basin in Colorado and New Mexico. Most coals in this region are Upper Cretaceous, but the Raton basin contains predominantly lower Tertiary coals.

The characteristics of these coal basins are diverse with respect to degree of structural deformation, association with igneous activity, and depth of burial of coal-bearing strata. In most basins, coal-bearing strata continue from outcrop to deep structural positions. Structural attitudes of coal-bearing strata range from nearly horizontal in basin centers to nearly vertical along the flanks of some basins. Igneous intrusions are common in the central and southern Rocky Mountains regions, and contributed to increasing the rank of some coals in the Piceance and Raton basins. Deep burial may have contributed to increasing Paleocene coals in the Powder River basin to subbituminous rank compared with coeval lignites in the adjacent Williston basin.

In the past decade an increased sophistication in coal exploration concepts and methods has enhanced development of coals in the Rocky Mountains and Great Plains regions. Exploration methods have focused on and utilized empirical relations among occurrences of major coals, their associations with sedimentologic facies, and their positions within large-scale stratigraphic architecture. This paper reviews the occurrences and quality attributes of major coals in these regions, with emphasis on: (1) their stratigraphic and geographic distributions, (2) their associations with depositional systems and specific positions within the stratigraphic architecture, and (3) tectonic and paleoclimatic influences on peat accumulation and preservation. Because there are observed or inferred differences in the tectonic, paleoclimatic, and depositional settings of Upper Cretaceous and lower Tertiary coals, the attributes and origins of coals of the two periods are discussed separately.

GEOGRAPHIC AND STRATIGRAPHIC DISTRIBUTION OF CRETACEOUS COALS

Economic coals of Mesozoic age in the United States Rocky Mountains and Great Plains regions occur predominantly in Upper Cretaceous strata. Upper Cretaceous coal-bearing strata are thicker (as much as 4,900 m) in the southern and central

Flores, R. M., and Cross, T. A., 1991, Cretaceous and Tertiary coals of the Rocky Mountains and Great Plains regions, *in* Gluskoter, H. J., Rice, D. D., and Taylor, R. B., eds., Economic Geology, U.S.: Boulder, Colorado, Geological Society of America, The Geology of North America, v. P-2.

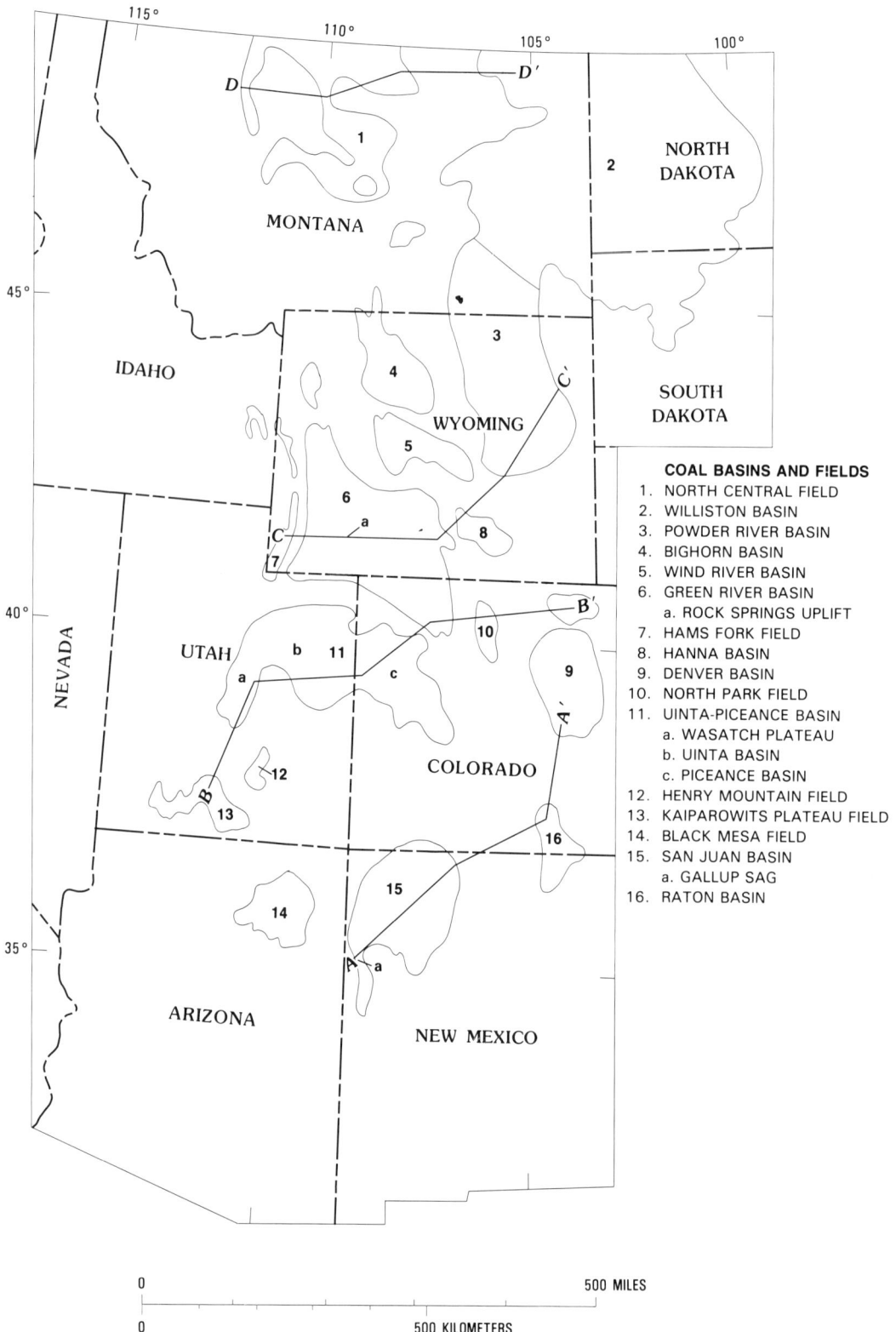

Figure 1. Map of the Rocky Mountains and Great Plains regions showing the major coal basins and coalfields. Lines A-A', B-B', C-C', and D-D' indicate locations of restored sections shown in Figures 2 to 5.

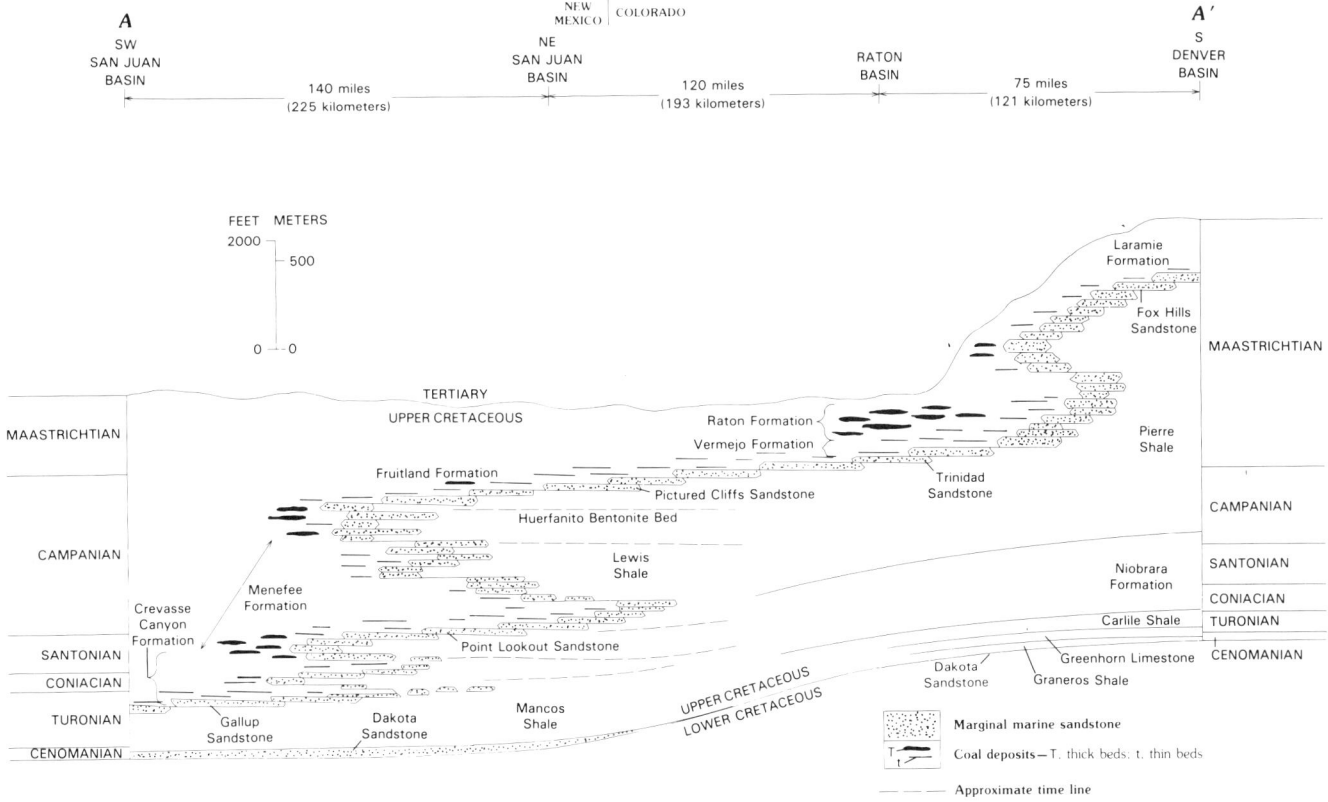

Figure 2. Diagrammatic restored section between the San Juan and Denver basins showing the relation of Upper Cretaceous coal-bearing and marine strata to major transgressive-regressive cycles. Modified from Weimer (1960). Location of restored section shown on Figure 1.

Rocky Mountains regions than in the northern Rocky Mountains region (as much as 3,050 m). Cumulative thicknesses of economic coals show a similar geographic distribution. the Mesaverde Group, as much as 1,800 m. thick, accounts for about 50 percent of the total coal-bearing interval, and contains coals as thick as 30 m. Coals are mainly of early to middle Campanian age in the southern Rocky Mountains region and of late Campanian through early Maastrichtian age in the central and northern Rocky Mountains regions. Stratigraphic and geographic distributions of Cretaceous coals, as compiled by Weimer (1960), are shown on Figures 2 through 5.

Important mineable coals in the southern Rocky Mountains region occur in the San Juan basin of New Mexico and Colorado, the Black Mesa coal field in Arizona, the Kaiparowits Plateau–Henry Mountains coalfield in Utah, and the Raton basin of Colorado and New Mexico (Fig. 2). In the San Juan basin, mineable coals occur within the Dakota Sandstone, Gallup Sandstone, Crevasse Canyon Formation, Menefee Formation, and Fruitland Formation. Coal seems are as thick as 12 m and 24 km in lateral extent in the Fruitland Formation. The principal coals in the Kaiparowits Plateau–Henry Mountains coalfield are found in the Straight Cliffs Formation and the Emery Sandstone Member of the Mancos Shale. In the Straight Cliffs Formation, coals occur as single and multiple seams averaging 3 m thick in a coal-bearing stratigraphic interval more than 15 m thick and 19 km in lateral extent. In the Emery Sandstone Member, the coal-bearing zone attains a maximum thickness of 30 m, with coal beds as thick as 4 m. In the Raton basin, mineable coals occur in the Vermejo and Raton Formations and are as thick as 3.5 m and about 10 km in lateral extent.

In the central Rocky Mountains region, significant mineable coals are found in the Uinta-Piceance basin of Utah and Colorado, in the Hams Fork coal region, Rock Springs uplift and Hanna basin of Wyoming, and in the Denver basin of Colorado (Figs. 3 and 4). The principal coals in the Uinta-Piceance basin are contained in the Blackhawk Formation exposed in the southwest Uinta basin, the Wasatch Plateau, and the Book Cliffs. Significant coals occur in the Adaville Formation in the Hams Fork coal region, in the Rock Springs and Almond Formations along the Rock Springs uplift, in the Williams Fork Formation along the Axial basin anticline, and in the Medicine Bow Formation in the Hanna basin. These coals vary from as thick as 4.5 m and about 88 km in lateral extent in the Rock Springs Formation, to as thick as 33 m in the Adaville Formation. The coal-bearing interval within the Adaville Formation can be traced for almost 160 km. Mineable coals are concentrated in the lower part of the Medicine Bow Formation and occur as discontinuous, multiple seams (as many as 15 beds) as thick as 3.3 m. In the Williams Fork Formation, total coal thickness is as much as 18 m from 2 to 15 coal seams in any one area. Only a few mineable coals occur in the Laramie Formation in the Denver basin. In the northern Rocky Mountains region, mineable coals occur in the Mesaverde and Meeteetse Formations in the Wind River and Bighorn basins of Wyoming, and in the Eagle Sandstone, Judith River Formation, and Two Medicine Formation in the North Central coal region of Montana (Fig. 5). These coals are as thick as 2 m and 27 km in lateral extent. In the Mesaverde Formation (or Group), mineable coals are as thick as 3.6 m and 5 km in lateral extent, but are more commonly less than 2 m thick. Some Lower Cretaceous coals in the Bear River Formation also are mined in Wyoming.

TECTONIC AND CLIMATIC SETTING OF UPPER CRETACEOUS COALS

During the late Mesozoic and early Cenozoic, oceanic plates of the Pacific ocean were converging with and subducting beneath the western margin of the North American continent (Atwater, 1970; Jurdy, 1984; Engebretson and others, 1985). Three major tectonic features were developed opposite the convergent margin. In the near-coastal position was an Andean-type volcanoplutonic arc. In the continental interior, varying from about 400 to 700 km from the convergent boundary, was a foreland fold and thrust belt, termed the Sevier orogenic belt in the United States (Armstrong, 1968). Immediately to the east of the foreland fold and thrust belt was a foreland basin in which Upper Cretaceous coal-bearing strata of the Rocky Mountains and Great Plains regions were deposited along its western margin. In response to loading of the lithosphere by the foreland fold and thrust belt and to a secular eustatic rise, the Western Interior seaway covered the foreland basin from the Albian through the Maastrichtian. This epicontinental seaway extended from the Arctic Ocean to the Gulf of Mexico and occupied an area more than twice that of the Mediterranean Sea (Fig. 6).

The foreland basin of the western United States displays two discrete phases of development, with corresponding differences in rates, geometries, and positions of subsidence. In the first phase, from before 92 to about 80 Ma, the foreland basin existed as a north-south trending, relatively narrow, asymmetric structural trough colinear with and immediately adjacent to the Sevier fold and thrust belt (Fig. 7A). During the basin's evolution, the structural and depositional axis migrated to the east in response to the eastward advance of thrust sheets. Jordan (1981) demonstrated that flexure of lithosphere in response to loading by thrust sheets and derived sediments is sufficient to explain the observed subsidence history and the asymmetry of this phase of foreland basin development. Coeval with this phase of foreland basin development was a subduction period of typical high- to moderate-angle

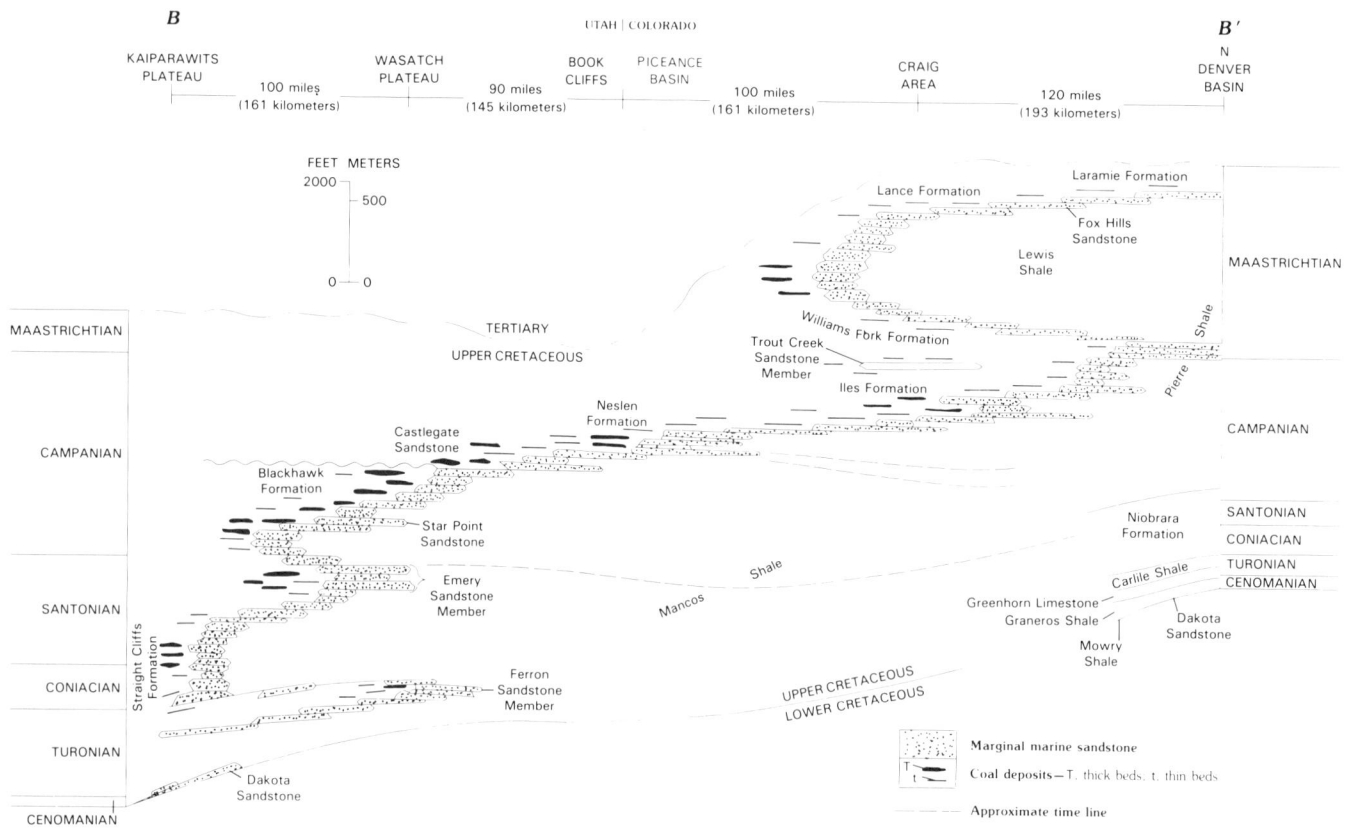

Figure 3. Diagrammatic restored section from the Kaiparowits coalfield to the Denver basin showing the relation of Upper Cretaceous coal-bearing and marine strata to major transgressive-regressive cycles. Modified from Weimer (1960). Location of restored section shown on Figure 1.

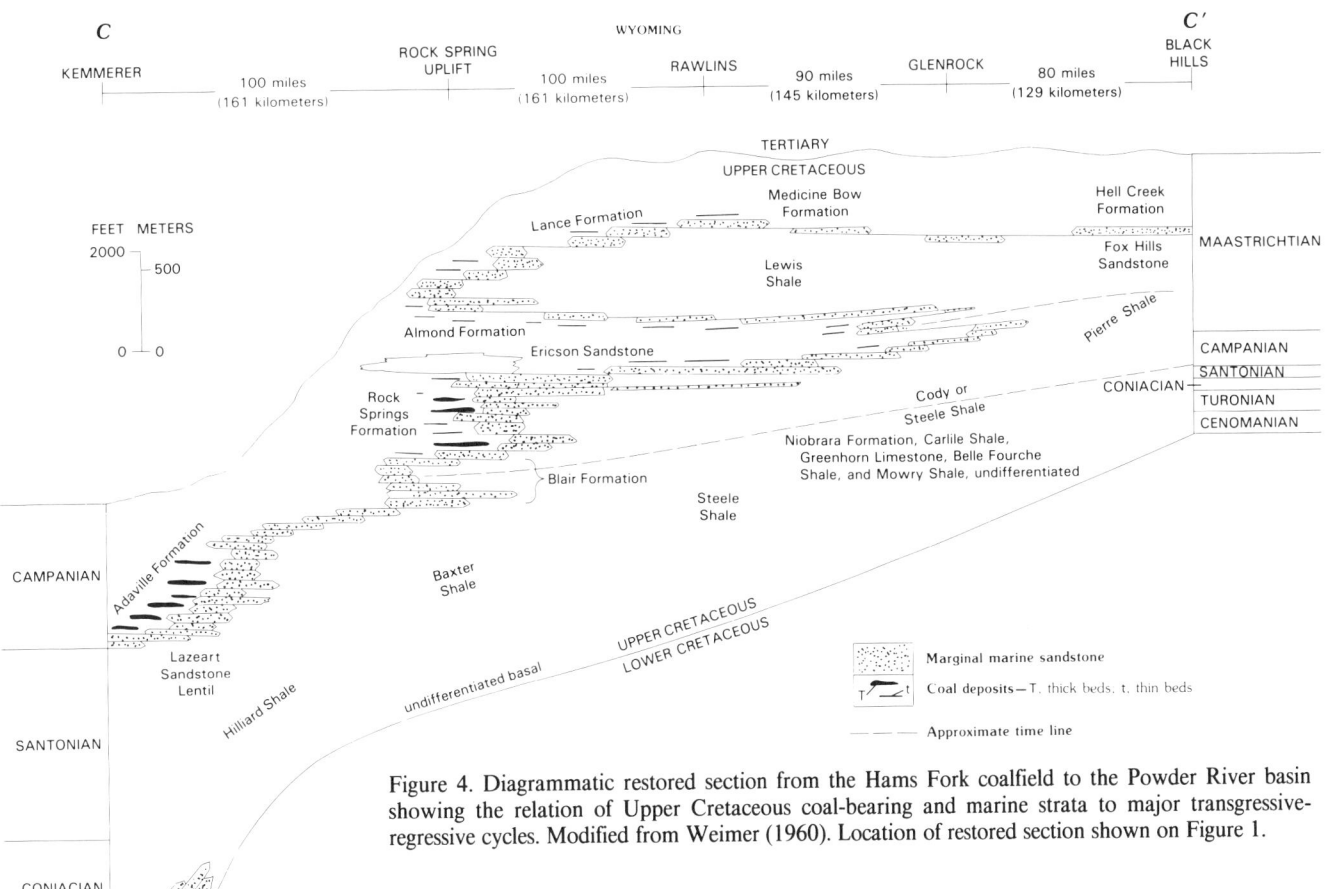

Figure 4. Diagrammatic restored section from the Hams Fork coalfield to the Powder River basin showing the relation of Upper Cretaceous coal-bearing and marine strata to major transgressive-regressive cycles. Modified from Weimer (1960). Location of restored section shown on Figure 1.

geometry along the coast of the western United States (Engebretson and others, 1985; Cross, 1986).

The second phase of foreland basin subsidence occurred during the Campanian and Maastrichtian (from about 76 to 65 Ma) and occupied a broad region centered about western Colorado and southern Wyoming (Cross and Pilger, 1978; Cross, 1986; Dickinson and others, 1988). To the north and presumably to the south of the Colorado-Wyoming locus of subsidence, the simple pattern of subsidence along a narrow structural trough colinear with an active fold and thrust belt was maintained into the Eocene. By contrast, fold and thrust deformation along the Sevier belt in Utah and southern Wyoming ceased by about 75 Ma (Cross, 1986). Isopachs of Campanian and Maastrichtian strata in the Colorado-Wyoming sector of the foreland basin are subcircular in map pattern and depicts centers of maximum thicknesses (3 km) displaced approximately 300 km to the east of the previous foreland basin structural axis (Fig. 7B). Jordan (1981) showed that supracrustal loading by thrust sheets and sediment is insufficient to cause this geometry and anomalous amount of subsidence. This phase of increased, regional subsidence in the Colorado-Wyoming sector was shown by Cross and Pilger (1978) to correspond to a period of low-angle subduction, which also caused the extinction of the corresponding portion of the coastal volcanoplutonic arc. They attributed the additional subsidence to sublithospheric loading and cooling caused by direct underplating of subducted oceanic lithosphere beneath the western United States. A more rigorous approach by Bird (1984) confirmed that the additional weight of the subducted plate would have depressed the overlying crust of this region by at least the observed amount of subsidence.

The previously noted occurrence of a thicker Upper Cretaceous stratigraphic section and thicker coals in the southern and central Rocky Mountains regions is attributed largely to the increased amount and rate of regional subsidence in the Colorado-Wyoming locus during the Campanian and Maastrichtian. Two other factors that contributed to the accumulation of thick and widespread peat deposits were a favorable climate and sea-level fluctuations that periodically or episodically provided additional space for sediment accumulation and preservation.

The Late Cretaceous was a time of warm, equable climate with a pole-to-equator temperature gradient much less than that of the present. Recent studies and reviews of late Cretaceous climate include those by Barron (1983, 1985), Hallam (1985), Ziegler and others (1985), and Parrish and Curtis (1982), and Parrish and others (1982, 1984). The Western Interior seaway spanned the middle to high latitudes throughout its history, and

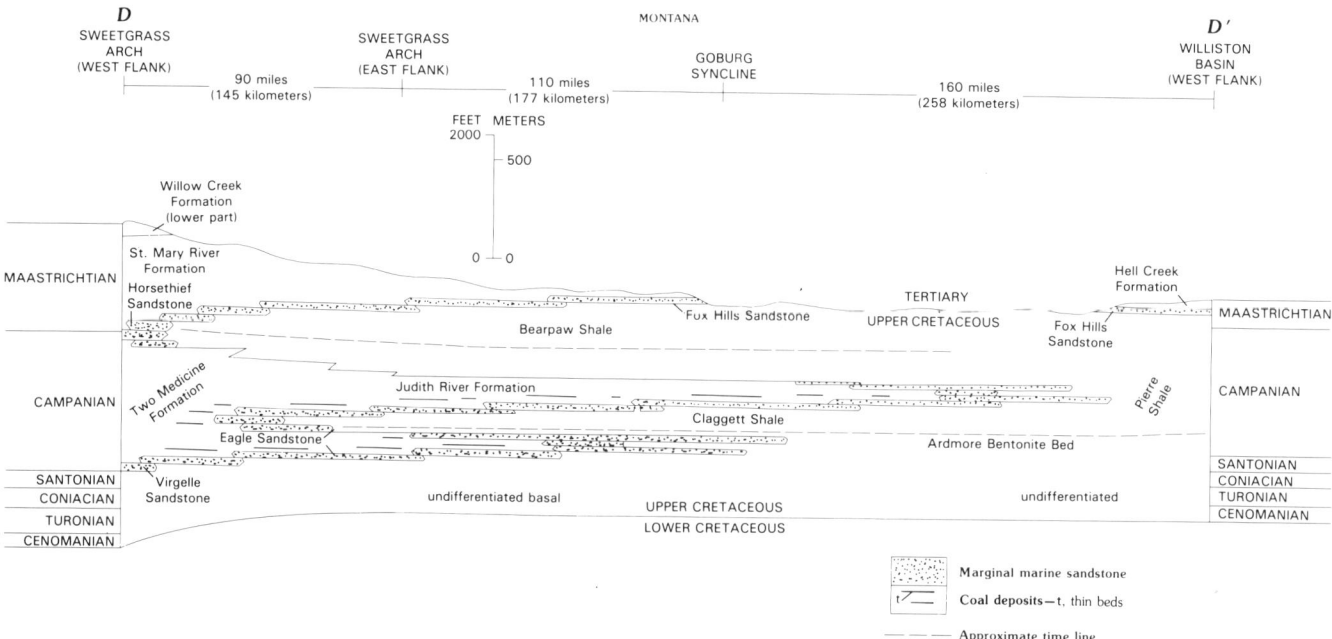

Figure 5. Diagrammatic restored section from the North Central coalfield to the Williston basin showing the relation of Upper Cretaceous coal-bearing and marine strata to major transgressive-regressive cycles. Modified from Weimer (1960). Location of restored section shown on Figure 1.

the U.S. portion lay between approximately 30° and 60° N during the Late Cretaceous (Ziegler and others, 1985). A probable consequence of its great length was that it occupied climatic zones extending from the low-pressure cells in the north to the subtropical high-pressure cells in the south.

Numerical simulations (Barron and Washington, 1982) and analog models (Parrish and others, 1982, 1984) have attempted reconstructions of atmospheric circulation over the Western Interior seaway. These models predict year-round westerlies or winter southwesterlies and summer northwesterlies over the U.S. portion of the seaway. Owing to the presumed existence of a rain shadow in the lee of the western United States Cordillera and to the absence of easterly winds blowing across the seaway, it is likely that precipitation was relatively low along the western side of the seaway. In addition, the models indicate a climatic seasonality, with most precipitation occurring during the winter in association with the mid-latitude winter storm track. If these reconstructions are reasonable facsimilies of the Late Cretaceous climate, low-moisture conditions in areas to the west of the coast, beyond the ameliorating influence of the seaway, would have prevented the development of temporally persistent swamps. The observed restriction of coals to coastal positions along the seaway may be partly a function of climatic control.

DISTRIBUTION OF COALS IN TRANSGRESSIVE-REGRESSIVE CYCLES

Upper Cretaceous coals in the Rocky Mountains and Great Plains regions accumulated in coastal mires along the western margin of the Western Interior seaway. Paleogeomorphic settings that have been ascribed to these coals include alluvial plain, active and abandoned delta plains, and mires of various types behind barred and nonbarred strandlines. Regardless of specific depositional settings that have been ascribed to Upper Cretaceous coals, there is a more fundamental association between the occurrence of volumetrically significant coals and their position in the stratigraphic architecture. Numerous studies of the past few decades (e.g., Sears and others, 1941; Weimer, 1960; Fassett and Hinds, 1971; Beaumont and others, 1971; Ryer, 1984) have demonstrated that most major coals occur at specific stratigraphic positions within the transgressive-regressive sequences that typify Upper Cretaceous strata of the Western Interior (Figs. 2 through 5). These studies have shown that virtually all coals occur at the top of, and landward of, shoreline facies within progradational events that constitute the transgressive-regressive sequences. However, the thickest, most extensive and/or greatest volume coals occur preferentially at stratigraphic positions where the shoreface facies of successive progradational events are stacked vertically.

Characteristics of transgressive-regressive cycles

McGookey (1972), Kauffman (1980), and Weimer (1984), among others, described five marine transgressive-regressive sequences, or T-R cycles, that range from about 3 to 7 m.y. in duration within latest Albian through Maastrichtian strata. Along the western margin of the Western Interior seaway, these cycles are represented by wedges of coastal-plain and shoreface strata that interfinger with offshore marine shales. Ryer (1984) noted the correspondence in temporal and spatial scales of these cycles with the third-order, unconformity-bounded depositional sequences of Vail and others (1977).

Ryer (1984) also summarized the results of numerous studies and observations, which demonstrate that these third-order cycles are composed internally of multiple asymmetrical, shallowing-upward cycles, each representing a discrete progradational event. Ryer equated these smaller progradational units with the fourth-order cycles of Vail and others (1977).

In Cretaceous strata of the Western Interior, these fourth-order progradational units often are arranged systematically in a regular geometric pattern shown diagrammatically in Figure 8. The direction and magnitude of facies offset across progradational event boundaries may be used to describe the internal stratigraphic architecture of third-order T-R cycles as a succession of seaward-stepping, vertically stacked, landward-stepping and vertically stacked progradational units. This systematic, hierarchical geometry is similar or identical to the genetic sequence of strata of Busch (1971, 1974), the depositional-episode of Frazier (1974), the PAC sequence of Anderson and Goodwin (1980) and Goodwin and Anderson (1985), the en echelon and stacked cycles of Ryer (1984), and the parasequence set of Van Wagoner (1985; Van Wagoner and others, 1988).

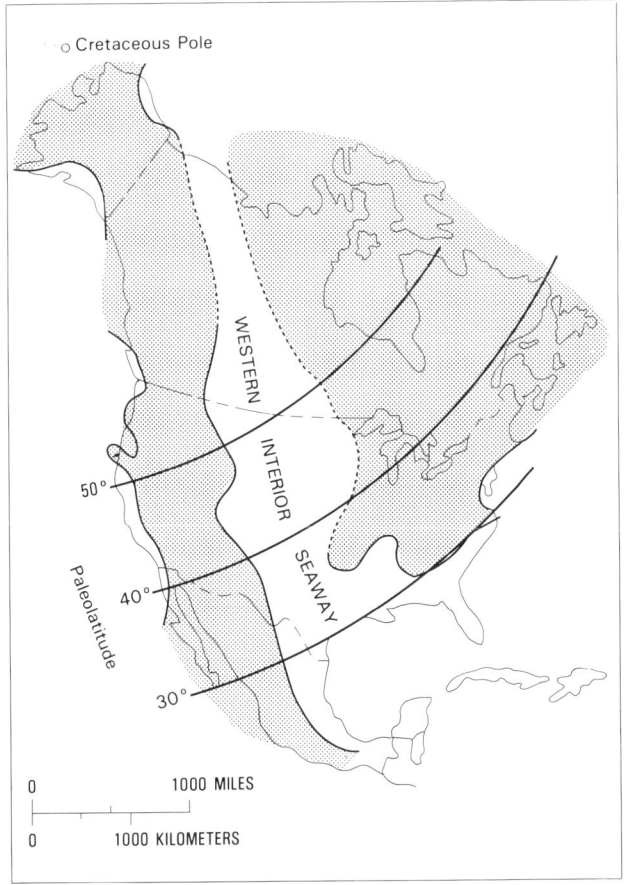

Figure 6. Generalized paleogeography of the Western Interior seaway during the Campanian (modified from Williams and Stelck, 1975). Paleolatitudes from Ziegler and others (1985).

Beginning with the most landward position attained by a progradational unit—equivalent to the "transgressive maximum" of many authors—the stratigraphic architecture of a typical Late Cretaceous third-order T-R cycle has the following characteristics. Progradation of the shoreface and adjacent environments causes deposition of progressively shallower water facies upon previously deposited deeper water facies. At a single geographic location, this produces a shallowing-upward succession of facies (Walther's Law). The termination of a progradational event usually is marked by a sharp lithostratigraphic break that represents a temporary landward shift in sites of sediment accumulation and, within the marine realm, a substantial decrease in rate of sediment accumulation and increase in water depth. In vertical profile at a single geographic location, the shallowest water facies developed at the termination of one progradational event is overlain directly by the deepest water facies of the next progradational event. Relative to the initial and final shoreline positions of the first progradational event, shorelines of the second event are both initiated and terminated in a more seaward position. Because the second event steps in a seaward direction, causing a seaward offset of identical facies across the boundary separating the two events, the second event is termed seaward-stepping. The next few progradational events may also step seaward; that is, each successive event begins and ends in a more seaward position than the preceding one. In vertical profile, a seaward-stepping progradational unit is recognized when the facies succession within it represents shallower water than the facies succession within the underlying unit. This seaward-stepping geometric pattern of progradational events is equivalent to the progradational parasequence set of Van Wagoner (1985; Van Wagoner and others, 1988) and may occur within a lowstand systems tract or the upper part of a highstand systems tract (Van Wagoner and others, 1988).

The next progradational events are stacked vertically in the most seaward position, or "regressive maximum," or a third-order T-R cycle. In vertically stacked progradational units, the geographic positions of the initial and final shorelines of each are approximately coincident, and the magnitude of facies offset across progradational event boundaries is slight. In vertical profile, a vertically stacked progradational unit is recognized when the facies succession within it is identical or similar to the facies succession within the underlying unit. This geometric pattern of vertically stacked progradational events is equivalent to the aggradational parasequence set and shelf-margin systems tract of Van Wagoner (1985; Van Wagoner and others, 1988).

The next progradational events are successively displaced landward relative to each other. That is, subsequent events are both initiated and terminated in more landward positions, and facies tracts are offset landward across progradational event boundaries. In vertical profile, a landward-stepping progradational unit is recognized when the facies succession within it represents deeper water than the facies succession within the underlying unit. This landward-stepping geometric pattern is equivalent to the retrogradational parasequence set and transgres-

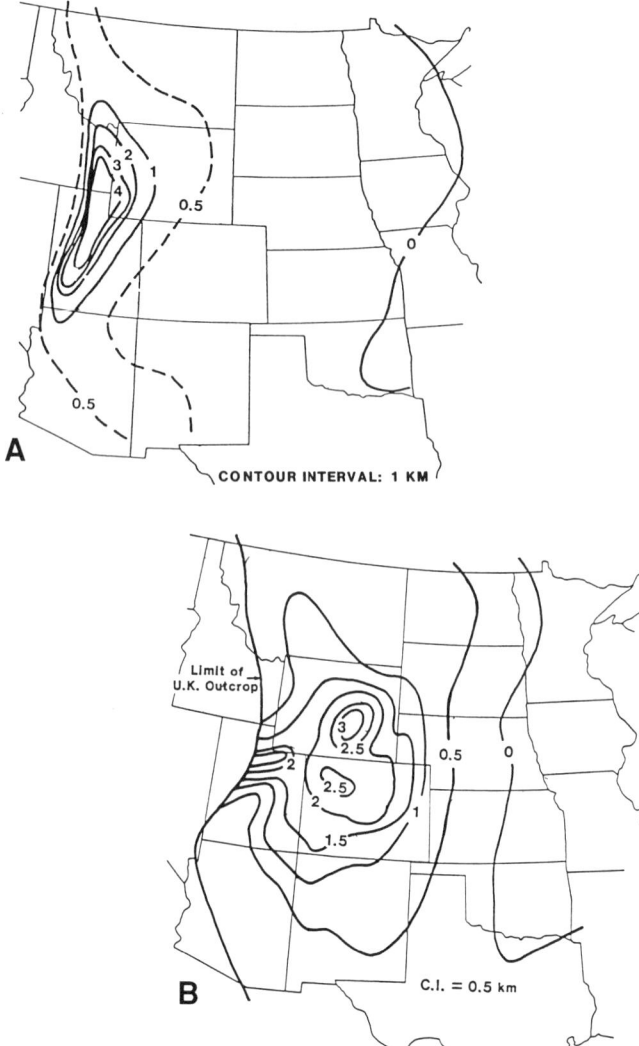

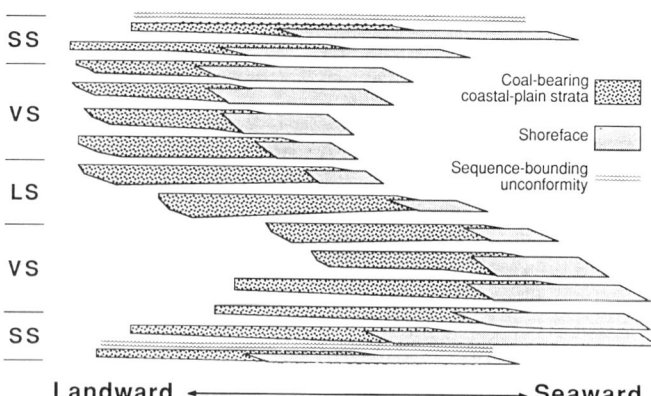

Figure 8. Stratigraphic architecture of a typical third-order transgressive-regressive (T-R) cycle in Upper Cretaceous strata of the Western Interior. Progradational events are arranged in seaward-stepping (SS), landward-stepping (LS), and vertically stacked (VS) geometries. Associated with this geometric stacking pattern are systematic changes in relative thickness and lateral extent of the coastal-plain and shoreface facies tracts. Only the lower, potentially coal-bearing portion of the coastal plain—including lower alluvial-plain and lower delta-plain environment—is depicted. Although coals commonly occur behind and above the landward limit of marine facies of any progradational event, they are best developed within the vertically stacked and/or slightly landward-stepping parts of a third-order T-R cycle.

Figure 7. Isopach maps of Cretaceous strata deposited in the foreland basin showing two spatially disjunct modes of subsidence (from Cross, 1986). A, Isopachs showing restored thickness of upper Albian through Santonian strata. B, Isopachs showing restored thickness of Campanian and Maastrichtian strata.

sive systems tract of Van Wagoner (1985; Van Wagoner and others, 1988).

Termination of the landward-stepping phase of a third-order T-R cycle is marked by a resumption of vertical stacking of progradational events at the transgressive maximum of a third-order T-R cycle. The landward and seaward limits of identical facies tracts within successive vertically stacked progradational events occupy essentially identical geographic positions, just as they did in the regressive maximum portion of the T-R cycle. This phase of vertical stacking corresponds to the aggradational parasequence set and the basal portion of the highstand systems tract of Van Wagoner (1985; Van Wagoner and others, 1988).

Variations in the balance between rates of sediment supply and subsidence produce variations in the internal stratigraphic architecture of third-order T-R cycles. Consequently, not all T-R cycles in Cretaceous strata of the Western Interior display the stacking patterns of progradational events shown in Figure 8 in an absolute sense. For example, if the ratio of sediment supply to subsidence increases—that is, the rate of sediment supply generally is increased or the rate of subsidence is decreased—all progradational units within a T-R cycle will generally step seaward where there is space available for sediments to accumulate. However, differences in the magnitude of seaward advance of successive progradational events—and concomitant differences in the magnitude of facies offset across progradational event boundaries—show relative changes that are analogous to the pattern shown in Figure 8. In this case, progradational units equivalent to the seaward-stepping units of Figure 8 would display the maximum facies offset; those equivalent to the landward-stepping units would have minimum facies offset or even be vertically stacked; and those equivalent to the vertically stacked units would have intermediate magnitudes of facies offset.

Controls on coal distribution in transgressive-regressive cycles

As depicted on the restored sections of Figures 2 through 5, coals occur at the top of, and landward of, shoreface platforms of most progradational events within third-order T-R cycles. However, most volumetrically significant coals are restricted to vertically stacked progradational units in both the transgressive and regressive maxima of third-order cycles.

Cross (1988) advanced an explanation for this observation through consideration of the interactions of eustatic fluctuations and tectonic subsidence that combine to produce geographically and temporally varying space in which sediments may accumulate. Numerical models of stratigraphic architecture by Pitman (1978), Morrow (1986), Cross (1988), and Jervey (1988), among others, have shown that both the asymmetric shallowing-upward cycles of fourth-order progradational events and their hierarchical arrangement within third-order cycles may be simulated with eustatic sine functions added to linear subsidence and linear or nonlinear sediment supply. These models showed further that the potential space in which sediment may accumulate—termed accommodation potential—is produced by the interaction of tectonic movement and eustatic change. When the rate of sea-level rise is maximum, the incremental addition of new space also is maximum. When the rate of sea-level fall is maximum, the incremental addition of new space is minimum. When the rate of sea-level change is zero, the incremental addition of new space is intermediate and equals the rate of tectonic movement. The accommodation potential at a point in space at a particular time is the cumulative sum of these incremental changes (additions and subtractions of space) through time, less the volume of sediment that actually accumulated.

The accommodation potential of a particular geographic location at a particular time limits the volume of sediment that may accumulate there. If the volume of sediment delivered is less than the accommodation potential, the basin may be (temporarily) underfilled. However, if the volume of sediment delivered is greater than the accommodation potential, the excess volume is bypassed to locations where accommodation potential remains. The models show that the hierarchical stacking geometry of progradational events within a basin is controlled by geographic variations in accommodation potential through time. In an otherwise balanced system, landward-stepping events occur at or near the maximum rate of increasing accommodation, when the sediment supply is insufficient to fill available space at a single geographic location; seaward-stepping events occur at or near the maximum rate of decreasing accommodation, when the sediment supply is greater than the sum of tectonic movement and eustatic change; and vertically stacked events occur at intermediate values of change in accommodation, when the sediment supplied by progradational events keeps pace with the accommodation potential.

Geographic changes in accommodation potential control not only the hierarchical stacking geometry of progradational events through time. They also control the volume of sediment that may accumulate in different facies tracts. Cross (1988) showed that in landward-stepping events, proportionally more sediment will accumulate in the shoreface and marine environments than in the coastal plain. Conversely, in seaward-stepping events, proportionally more sediment will accumulate in the coastal plain than in the shoreface and marine environments. These relations are illustrated schematically in Figures 8 and 9. Because sediment volumes are partitioned differentially into different facies tracts as

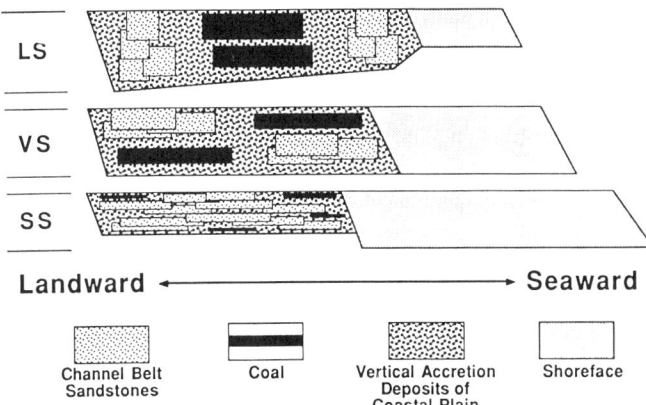

Figure 9. Stratigraphic architecture of a typical third-order transgressive-regressive (T-R) cycle in the Upper Cretaceous strata of the Western Interior. Progradational events are arranged in seaward-stepping (SS), landward-stepping (LS), and vertically stacked (VS) geometries. Associated with this geometric stacking pattern are systematic changes in channel-belt sandstones, coals, and vertical accretion deposits.

a function of the position of progradational events within the stacking hierarchy, major coals occur in parts of third-order cycles that represent a balance between the maximum potential for vertical aggradation and availability of sediment. This balance occurs most frequently in vertically stacked and slightly landward stepping progradational events.

Observed distributions of coal with transgressive-regressive cycles

Ryer (1984) summarized the association of major coals in Utah with respect to their position in the stratigraphic architecture. Major coals associated with vertically stacked progradational events of transgressive maxima include the Kolob-Alton (Dakota Sandstone) and Kaiparowits (Straight Cliffs Formation) coalfields. Major coals associated with vertically stacked progradational events of regressive maxima include the Henry Mountains (Ferron Sandstone Member), Emery (Ferron Sandstone Member), and Wasatch Plateau Deep (Emery Sandstone Member) coalfields. These coalfields represent approximately 60 percent of the known Cretaceous coal resources in Utah. The coal-bearing strata consist of as much as 60 m "carbonaceous interval," which includes from 3- to 6-m-thick coal seams.

The middle Turonian Ferron Sandstone Member of the Mancos Shale, studied by Ryer (1981), was deposited during a regressive maximum of third-order cycle, perhaps as a lowstand systems tract and/or a transgressive systems tract (Shanley and McCabe, 1989). The Ferron displays significant variations in coal development with respect to the position of progradational events within the hierarchical stacking pattern (Fig. 10). Although coals occur within all nine progradational events of the Ferron, they are thinnest, most discontinuous, and often poorest in quality within the basal seaward- and upper landward-stepping events. Coals of best quality and greatest volume occur within the six vertically stacked events, particularly within the upper two at the transition to the landward-stepping phase.

Occurrences of volumetrically significant coals in the southern Rocky Mountains region lend additional support for the ideas of volumetric partitioning of sediments into different facies tracts as a function of the position of progradational events within the stacking hierarchy. Beaumont and others (1971) described three major reversals in direction of strandline movements in the Mesaverde Group of the Gallup coalfield. The best coals occur within the vertically stacked progradational events, as shown on Figure 11, even though coals of lesser volume occur in most progradational events. These coals accumulated in swamps in the immediate vicinity of, and landward of, the maximum landward position of marginal marine deposits.

Fassett and Hinds (1971) described the occurrence of coals within terrestrial strata of the Fruitland Formation associated with shoreface strata of the Pictured Cliffs Sandstone in the San Juan basin, New Mexico. In an absolute sense, progradational events within these formations display a seaward-stepping geometric arrangement, although the magnitudes of facies offset between progradational events vary. These strata provide an example of the previous discussion concerning the importance of recognizing relative, in addition to absolute, geometric arrangements of progradataional events. Fassett and Hinds (1971) noted that although coals occur on top of, and landward of, shoreface platforms in all these progradataional units, the thickest and most widespread coals—as thick as 9 m and elongate parallel to the coast—occur where there is a stratigraphic rise; that is, where there is a subtle change from seaward-stepping to vertically stacked geometries within an overall seaward-stepping pattern. In basins where the balance between sediment supply and subsidence causes continual filling and seaward-stepping geometries of progradational events, the positions of maximum rate of increase in accommodation are represented by decreased magnitudes of seaward offset of progradational events, rather than by an absolute reversal in direction and landward-stepping geometries. Fassett (1986) subsequently compared coal occurrences in the Fruitland Formation and intertonguing Pictured Cliffs Sandstone with those in the slightly older Menefee Formation and intertonguing Point Lookout Sandstone in the San Juan basin. Although the stratigraphic architecture of the two units is similar—both display overall seaward-stepping geometries—coals in the older unit are not as well developed. This comparison provides a useful caution about the importance of other controls on peat accumulation and preservation; accommodation potential only provides the space for sediment accumulation, it does not independently control the lithologies of the sediments that accumulate.

Major coals of the central Rocky Mountains region also occur in positions of third-order T-R cycles where accommodation potential is maximum. Levey (1985) related coals to active progradation of deltas within the lower Campanian Rock Springs Formation in the Green River basin, Wyoming. The thickest and most extensive coals occur within slightly landward-stepping

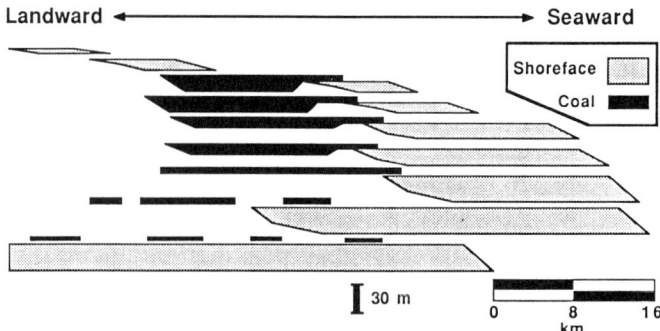

Figure 10. Diagrammatic cross section showing occurrences of coals in the Ferron Sandstone as they relate to the hierarchical stacking geometry of progradational events and positions of landward pinchouts of delta-front sandstones. Adapted from Ryer (1981).

progradational events in the lower three-quarters of the formation (Fig. 12). This geometry indicates long-term increase in accommodation potential within the coastal plain. Thinner progradational units and thinner coals occur at the top of the Rock Springs Formation, indicating a reduction in accommodation potential in the coastal plain and the initiation of seaward-stepping progradational events.

Lawrence (1982) described the occurrence of coals within the Adaville Formation (Santonian) in the Hams Fork coalfield, Wyoming. These coals occur within a succession of progradational events that begin as slightly seaward-stepping and end with pronounced seaward-stepping geometries (Fig. 13). There is a concomitant decrease in thickness of successive progradational units and associated coals within each event. This geometry indicates decreasing accommodation potential in the coastal plain of successive progradational events.

DEPOSITIONAL SYSTEMS OF UPPER CRETACEOUS COALS

Paleogeomorphic settings that have been ascribed to Upper Cretaceous coals include alluvial plain, active and abandoned delta plains, and coastal mires behind barred and nonbarred strandlines. Sediments supplied to these coastal, paralic, and marine environments were derived from the Sevier orogenic belt, which also was the site of the Cretaceous continental divide. Because the Sevier belt was relatively close to the coast of the seaway (variably from about 75 to 250 km), it is probable that rivers entering the seaway were numerous, short-headed, and emanated from relatively small drainage basins. An implication of this observation is that the importance of deltas, especially fluvial-dominated deltas, as major geomorphic elements of Late Cretaceous peat accumulation may be overestimated in the literature of the past few decades. This is not surprising when considered in historical perspective; depositional models of coal occurrence in vogue during that time emphasized deltas as one of the two major geomorphic environments in which peats may accumulate.

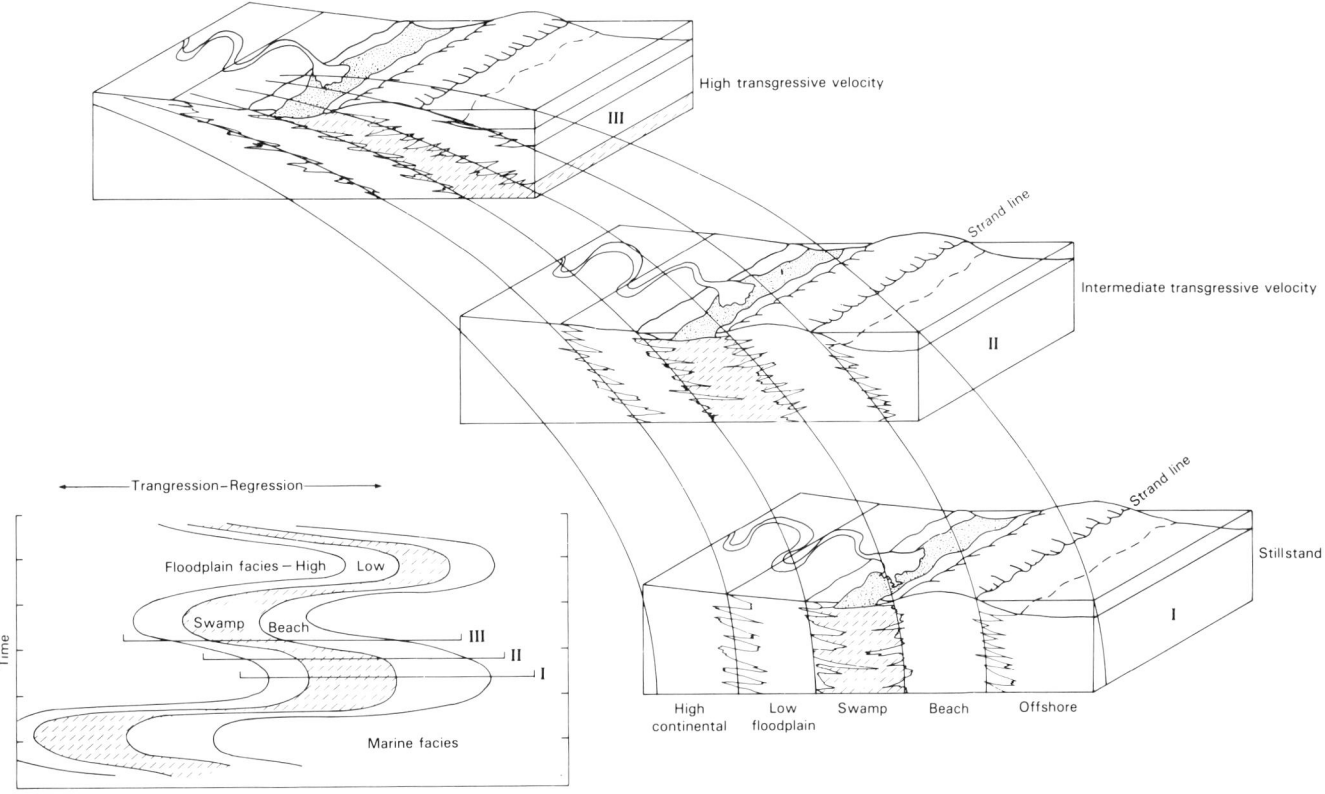

Figure 11. Effects of varying rates of shoreline movements on the accumulation of coals landward to marginal marine deposits. The block diagrams I, II, and III represent time slices shown in the small graph of the transgressive-regressive curves. Adapted from Beaumont and others (1971).

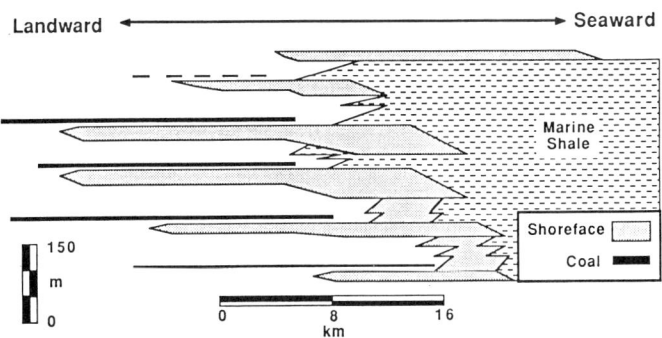

Figure 12. Diagrammatic cross section showing positions of coals above delta-front sandstones in the Rock Springs Formation. Adapted from Levey (1985).

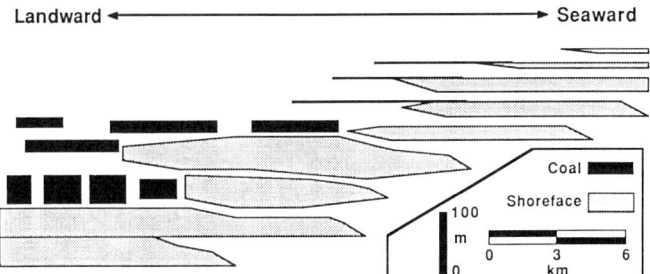

Figure 13. Diagrammatic cross section showing positions of coals above delta front sandstones in the Adaville Formation. Adapted from Lawrence (1982).

Coals associated with fluvial-deltaic systems

Both fluvial-dominated and wave-dominated delta systems have been ascribed to Upper Cretaceous coal-bearing strata in the Rocky Mountains and Great Plains regions. One of the best documented occurrences of coals in fluvial-dominated deltas is the Ferron Sandstone Member of the Mancos Shale described by Ryer (1981). Coals as thick as 10 m occur in lower delta-plain deposits in bands that parallel depositional strike and extend about 10 km landward from the pinchouts of delta-front sandstones. Ryer (1981, 1984) described as exploration model in which the landward pinchouts of delta-front sandstones are used to locate belts of coals landward of and parallel to the pinchouts.

Other examples of coals associated with fluvial-dominated deltas are described from vertically stacked progradational events in regressive maxima of third-order T-R cycles. Principal peat-forming environments have been identified as poorly drained swamps adjacent to channel-levee complexes and as flood-basin environments of the delta plain (Weimer, 1977; Siemers, 1978). Delta-plain strata overlie a coarsening-upward sequence of pro-delta and delta-front deposits. When not reworked by waves, the delta front is characterized by destructional deposits of bioturbated sediments and lenticular, distributary mouth bar-finger sandstones (Flores and Tur, 1982). The destructional phase of the delta front commonly is related to subdelta abandonment after which extensive peat swamps covered the abandoned lobe. In the Vermejo Formation and intertonguing Trinidad Sandstone of the Raton Basin, individual coal seams as thick as 4 m were formed in swamps overlying deposits of abandoned subdeltas, and multiple seams with a total thickness of 5 m accumulated in flood-basin swamps (Flores and Tur, 1982). These coals are elongate parallel to depositional dip.

Perhaps most of the major Upper Cretaceous coals in the Rocky Mountains region were associated with wave-dominated delta systems (e.g., Balsley, 1980; Lawrence, 1982; Levey, 1985). In comparison with the delta fronts of fluvial-dominated deltas, delta-front strata of wave-dominated deltas have fewer distributary channels and broader, sheet-like, delta-front sandstone platforms consisting of coalesced distributary mouth bar and intermouth bar sediments. Generally, the delta-front sandstones are widespread along depositional strike and, where continuous progradation occurred, they also are extensive along depositional dip. Thick, laterally continuous peats, only infrequently segmented by distributary channels, accumulated on the broad, delta-front platforms. Examples are coals in the Blackhawk Formation in Utah (Balsley, 1980) and the Adaville and Rock Springs Formations in Wyoming (Lawrence, 1982; Levey, 1985). In the Blackhawk Formation, coals as thick as 6 m (e.g., the Sunnyside coal) occur parallel to depositional strike in bands as much as 55 km wide.

The studies of Lawrence (1982) and Levey (1985) provide good contrasts in thickness and continuity of major coals associated with wave-dominated deltas. Levey (1985) recognized three types of coals in the Rock Springs Formation (Fig. 12). Those overlying delta-front sheet sandstones of the lower delta plain are as much as 7 m thick and greater than 800 km^2 in areal extent. One coal seam can be traced subparallel to depositional dip for more than 50 km. Coals formed in upper delta-plain swamps are as much as 5.3 m thick and cover up to 400 km^2. Coals formed in swamps of abandoned deltas are thinner, up to 2.5 m thick, and cover up to 80 km^2. By contrast, Lawrence (1982) reported thick but discontinuous coals from the Adaville Formation, which originated as back-beach swamps on the lower delta plain (Fig. 13). Individual coal seams are greater than 30 m thick, but only 1.5 to 4.5 km in lateral extent subparallel to depositional dip. Although there are pronounced differences in thickness and continuity of coals in these two formations, they were deposited in essentially identical environments. This suggests that environmental factors were not important controls on coal thickness and continuity. It is likely that the Adaville coals are products of higher accommodation potential and more pronounced vertical stacking.

Coals associated with barrier strand-plain systems

The other major Late Cretaceous coal-forming environments were coastal plains behind barred and nonbarred coasts. Descriptions of barrier systems indicate construction by waves in microtidal and macrotidal coastal settings. Like the delta-front sheet sandstones of the wave-dominated delta systems, the barrier strand-plain systems built extensive sandstone platforms upon which back-barrier swamps were established. Existing literature suggests that barred and nonbarred coasts were best developed during regressive maxima and least developed during transgressive maxima. However, it is likely that this relation will not be supported by future work.

Examples of major coals that accumulated in back-barrier swamps in a microtidal setting occur in the Almond Formation in Wyoming (Roehler, 1977; Flores, 1978); the Straight Cliffs (Johnson and Vaninetti, 1982) and Blackhawk Formations (Flores and others, 1984) in Utah; and the Mesaverde Group in New Mexico. Roehler (1977) interpreted coals as thick as 5 m in the Almond Formation as having accumulated in swamps behind coastal barriers during a transgressive maximum. A microtidal setting was indicated by the strong influence of longshore currents and dominance of flood over ebb deltas (Flores, 1978). Swamps were established on sediment-filled lagoons and protected from marine influence by disjunct, abandoned barrier ridges.

Barrier systems deposited during regressive maxima are represented in the Straight Cliffs and Blackhawk Formations, and the undivided Crevasse Canyon and Menefee Formations. Barrier ridges in these formations also were established in microtidal coasts and were associated with wave-dominated and fluvial-dominated deltas. Johnson and Vaninetti (1982) described coals as thick as 6 m from the Straight Cliffs Formation that were deposited in back-barrier swamps associated with wave-dominated deltas. In the undivided Crevasse Canyon and Menefee Formations of the Mesaverde Group, Cavaroc and Flores (1984)

TABLE 1. SUMMARIES OF COAL-QUALITY DATA FOR MAJOR UPPER CRETACEOUS COAL-BEARING ROCKS IN THE ROCKY MOUNTAIN REGION*

Geographic Area	Stratigraphic Unit	Major Coal Beds Represented	Number of Samples	Ash	Sulfur	Calorific Value (Btu/lb)	Apparent Rank
Northern Rocky Mountains							
Wind River Basin	Mesaverde	Signor; Welton	25	10.7 (4.2–18.9)	1.0 (0.3–3.0)	9,580 (7,260–11,470)	Subbituminous A (Sub. C-Sub. A)
Central Rocky Mountains							
Greater Green River basin	Mesaverde	No. 7; Wadge; Wolf Creek	88	8.2 (2.5–23.0)	0.7 (0.2–3.1)	10,520 (6,930–14,880)	HVC Bituminous (Lig. A-HVA bit.)
Overthrust belt, Wyoming	Mesaverde	Adaville	28	9.8 (3.3–24.4)	0.6 (0.3–1.8)	10,530 (8,560–12,330)	HVC Bituminous (Sub. C-HVB Bit.)
Uinta-Piceance Creek basins	Mesaverde	B; Blind Canyon; Cameo; Hiawatha; Castlegate	238	11.8 (2.9–45.8)	0.7 (0.1–3.6)	11,070 (5,780–15,090)	HVC Bituminous (Lig. A-Anth.)
Denver basin	Laramie	No. 3	9	5.4 (3.5–11.2)	0.4 (0.3–0.7)	9,220 (6,770–10,730)	Subbituminous B (Lig. A-Sub. A)
Southern Rocky Mountains							
Raton Mesa	Vermejo	Brookside; Little Johnny	16	11.3 (4.2–22.8)	0.7 (0.5–1.7)	11,670 (10,800–13,250)	HVB Bituminous (Sub. A-HVA Bit.)
San Juan basin	Mesaverde	Pueblo	30	14.0 (4.7–39.0)	0.8 (0.3–3.7)	10,250 (5,140–13,860)	HVC Bituminous (Sub. C-HVA Bit.)
	Fruitland	A; B; C	89	20.3 (9.6–38.9)	0.6 (0.3–1.4)	8,920 (4,850–13,870)	Subbituminous A (Lig. A-HVA Bit.)
Kaiparowits Plateau; Henry Mtns.	Dakota	Smirl	14	10.1 (4.8–20.9)	1.1 (0.3–1.8)	9,210 (7,430–9,980)	Subbituminous B (Sub. C-Sub. B)
	Straight Cliffs	Alvey; Rees	13	11.7 (4.4–27.3)	0.9 (0.4–2.6)	9,570 (7,640–10,920)	Subbituminous A (Sub. B-Sub. A)
Black Mesa	Wepo	Blue; Green; Red	10	7.7 (4.7–11.1)	0.5 (0.3–0.7)	11,020 (10,360–11,560)	Subbituminous A (Sub. A-HVC Bit.)

*Compiled by R. T. Hildebrand.

described coals as thick as 3 m that accumulated preferentially in back-barrier lagoonal swamps built upon abandoned fluvial-dominated deltas. Flores and others (1984) compared the geometry and distribution of major coals developed in barrier environments to those that accumulated in adjacent fluvial-dominated deltas in the Blackhawk Formation. The back-barrier coals are as thick as 5 m and are laterally continuous parallel to depositional strike. By contrast, deltaic coals are a maximum of 3 m thick and are elongate parallel to depositional dip.

A possible example of a coal-bearing barrier system deposited under macrotidal conditions occurs in the Vermejo Formation of New Mexico. Pillmore and Mayberry (1976) interpreted coals as thick as 1.5 m to have originated on a barrier strandplain, and Leighton (1980) recognized predominant ebb-tidal currents in these barrier deposits and suggested a macrotidal coast.

CHARACTER, RANK, QUALITY, AND UTILIZATION OF ECONOMIC UPPER CRETACEOUS COALS

Economic Upper Cretaceous coals in the Rocky Mountains and Great Plains regions range from lignite to anthracite (Table 1). (The definition of economic coal follows Averitt's [1975] categories of greater than 14 in [356 mm] for bituminous and

TABLE 2. MACERAL COMPOSITION OF SOME CRETACEOUS COALS IN THE ROCKY MOUNTAINS AND GREAT PLAINS REGIONS DETERMINED BY WHITE LIGHT ANALYSIS*

	Vitrinites	Liptinites	Inertinites	R_o†	N§
Hams Fork Coal Region					
Adaville coal	87	6	7	0.45–0.47	2
Southwest Uinta Basin					
Sunnyside coal	85	11	4	0.69–0.84	12
Hiawatha coal	78	4	18	0.46–0.63	31
Rock Canyon coal	81	5	14	0.65–0.78	7
Blind Canyon coal	80	9	11	0.48–0.65	9
Castlegate coals A-D	75	7	18	0.58–0.79	37
Rock Springs uplift					
Rock Springs coal	94	4	2	0.52	1
Denver basin					
Laramie coal	76	15	9	0.52	1
San Juan basin					
Dakota coal	54	42	4	0.74	1

*Maceral content in average percent. Data compiled from Pennsylvania State University data bank and from Smith (1986).
† R_o = vitrinite reflectance.
§ N = number of samples.

higher grade coals, and greater than 2.5 ft [0.76 m] for subbituminous coals and lignite.) Mineable coals in the southern Rocky Mountains region include lignite through high-volatile A bituminous. The central Rocky Mountains region contains mineable coals ranging from lignite to high-volatile A bituminous and, rarely, anthracite. Coals in the northern Rocky Mountains region are mainly subbituminous, although a few are bituminous in the northern Bighorn basin. The high rank of coals in the southern and central Rocky Mountains regions, including the presence of graphite in the eastern Piceance basin, may have been caused by heat from igneous intrusions (Amuedo and Bryson, 1977; Collins, 1977).

In the southern Rocky Mountains region, the mineable coals, based on arithmetic means of 172 samples, range form 0.3 to 3.7 percent sulfur, 4.2 to 39 percent ash, and 4,850 to 13,250 Btu. On the basis of arithmetic means of 363 samples, coals in the central Rocky Mountains region contain 0.1 to 3.6 percent sulfur, 2.5 to 45.8 percent ash, and 5,780 to 15,090 Btu. Only limited data are available for the sulfur, ash, and Btu contents of mineable coals in the northern Rocky Mountains region. Based on arithmetic means of 25 samples from the Wind River basin, these coals contain 0.3 to 3 percent sulfur, 4.2 to 18.9 percent ash, and 7,260 to 11,470 Btu.

Table 2 summarizes available information on the petrology of Upper Cretaceous coals. Petrologic data from 101 samples are derived principally form the Sunnyside, Castlegate, and Rock Canyon coals of the Blackhawk Formation in the southwestern Uinta basin (Wasatch Plateau–Book Cliffs coal region). The high vitrinite content of the Blackhawk, Adaville, and Rock Springs coals suggest that the peats originally contained a high wood content. Although based on a few samples, the Dakota and Laramie coals contain high liptinites that are characterized by high hydrogen-to-carbon ratios. The hydrogen-rich liptinites may prove these coals to be economically feasible for liquefaction.

The Sunnyside-Castlegate coal region in the southwest Uinta basin is the most important source of coking coal in the Rocky Mountains region. These coals are high volatile A and B bituminous in rank (Averitt, 1966). Doelling and Smith (1982) reported that coal mined in the Sunnyside area is semicoking and, when blended with suitable coals, can be used as metallurgical-quality coal.

Generally, coals of subbituminous and high-volatile C and B bituminous ranks are unsuitable for the manufacture of coke. However, Averitt (1966) identified a few localities in the Rocky Mountains region that contain coals of these ranks that will coke to some degree. The most important coking coal localities, with reserves as estimated by Goolsby and others (1979), are the Raton basin (20.5 billion short tons), the San Juan basin (1.78 billion short tons), and the Somerset-Crested Butte-Carbondale coal region of the eastern Piceance basin (0.45 billion short tons). The estimates of coking-coal reserves in the Raton basin include the undifferentiated Vermejo (Upper Cretaceous) and Raton (Upper Cretaceous and Paleocene) Formations, which contain high-volatile A and B bituminous blending coal for the production of coke. In the northern part of the San Juan basin, coking-coal reserve is from the high-volatile A and B bituminous coals from the Dakota, Menefee, and Fruitland Formations. Coking-coal reserve in the eastern Piceance basin is estimated from high-volatile A and B bituminous coals from the Iles and Williams Fork Formations.

GEOGRAPHIC AND STRATIGRAPHIC DISTRIBUTION OF CENOZOIC COALS

The more important Cenozoic coal-bearing strata in the Rocky Mountains and Great Plains regions are contained in Paleocene and Eocene strata. These coals are principally of lignite and subbituminous ranks and they constitute a large part of the total coal resource of the United States. The major coals are very thick and discontinuous, and occur in isolated structural basins. Averitt (1975) reported that Cenozoic coals generally are found less than 1,200 m below the surface in the northern Great Plains and the Powder River and Raton basins. In these areas, coal-bearing strata are nearly horizontal, and much of the reserve base is recoverable by strip mining.

In the southern Rocky Mountains region the Paleocene part of the Raton Formation that attains a maximum thickness of 600 m contains mineable coals in the Raton basin. In the central Rocky Mountains region the most important coal-bearing units are the Paleocene Fort Union and Eocene Wasatch Formations in the Uinta-Piceance, greater Green River, and North Park basins; the Upper Cretaceous and Paleocene Ferris and Paleocene to Eocene Hanna Formations in the Hanna basin; and, the Upper Cretaceous and Paleocene Denver Formation in the Denver basin (Fig. 1). The Fort Union and Wasatch Formations are as thick as 300 m and 1,000 m, respectively. The combined Ferris and Hanna Formations are as thick as 3,200 m, and the Denver Formation is as thick as 460 m. In the northern Rocky Mountains region the Paleocene Fort Union and Paleocene and Eocene Wasatch Formations contain economic coals in the Wind River, Bull Mountains, and Powder River basins. The Fort Union and Wasatch Formations in the Powder River basin have a maximum thickness of 2,000 m. The most prolific coal-bearing strata are the Tongue River Member of the Fort Union Formation, which is as thick as 600 m. In the Hams Fork coalfield and Bighorn basin, some economic coals occur in the Upper Cretaceous and Paleocene Evanston (as thick as 300 m) and Paleocene and lower Eocene Polecat Bench Formations (as thick as 2,450 m), respectively. In the Great Plains region of the Williston basin, the most important coal-bearing units are the Paleocene Fort Union and Sentinel Butte Formations. These formations have a combined thickness of as much as 570 m.

The Powder River basin in northeast Wyoming and southeast Montana contains the greatest number of anomalously thick coals (average of 20 m) and, as a result, has a concentration of coal resources larger than that of any basin of its size in the Rocky Mountains and Great Plains regions. These coals, which include the Wyodak-Anderson, Anderson-Dietz, and Big George coals of the Tongue River Member of the Paleocene Fort Union Formation, are subbituminous in rank. Throughout the central part of the basin, the Big George coal is as much as 62 m thick and contains as much as 113 billion tons of resource at a cutoff of 335 m of overburden (Pierce and others, 1982). Although the Big George coal contains the largest tonnage in a single continuous seam anywhere in the Rocky Mountains and Great Plains regions, it is not economically mineable at present. The Wyodak-Anderson coal, which is as thick as 38 m and contains at least 15 billion tons of reserve between its outcrop to the 60 m overburden line, is the most suitable coal to be recovered by strip mining (Averitt, 1975).

TECTONIC AND CLIMATIC SETTING OF CENOZOIC COALS

By the close of the Cretaceous, the Western Interior seaway had withdrawn from most of the western United States and southern Canada. In the United States, a remnant of the seaway occupied the central part of the Williston basin through most of the Paleocene. A major drop in global sea level near the end of the Maastrichtian was at least partly responsible for the withdrawal of the sea. It is likely that tectonism was an additional contributor. At about 69 Ma, Laramide deformation began and transformed the foreland basin into multiple, isolated basins and adjacent mountain ranges. Laramide structures are characterized by basement-cored uplifts and asymmetric anticlines, typically bounded by high-angle reverse and thrust faults. The intermontane basins also are generally asymmetric. In the southern Rocky Mountains, Laramide structures occupy a band about 200 km in breadth, parallel and adjacent to the eastern physiographic boundary of the Colorado Plateau. In the central Rocky Mountains, Laramide structures also form a band about 200 km in breadth that is displaced about 300 km north and northeast from the physiographic margin of the Colorado Plateau.

Fragmentation of the Rocky Mountains and Great Plains regions by Laramide deformation created topographic relief, localized depocenters and sediment sources, and developed fluvial drainage patterns that were quite different from the geomorphic character of the region during the Cretaceous. Intermontane sedimentary basins were occupied predominantly by either lacustrine or fluvial environments. Deposits of large lakes include lacustrine carbonates, kerogenous shale, claystone, and trona-halite in lake centers, and lacustrine carbonates, sandstones, and mudstones along the lake margins (Eugster and Hardie, 1975; Ryder and others, 1976). Deposits of adjacent fluvial systems are represented by conglomerate, sandstone, and mudstone. Intermontane basins filled predominantly by fluvial sediments contain fluvial channel sandstone, siltstone, and conglomerate; floodplain sandstone, siltstone, and claystone; and carbonaceous shale and coal of flood-plain mires. The large lakes occur in the west-central part of the Rocky Mountains region, whereas the coal-forming fluvial systems occur in the eastern Rocky Mountains and Great Plains regions.

Paleobotonical and paleopedological evidence suggests a succession of warming, cooling, and warming trends from early Paleocene to early Eocene, and cooling from late Eocene–Oligocene to Neogene in the Rocky Mountains and Great Plains regions (Brown, 1962; Leopold and MacGinitie, 1972; Wolfe, 1985). Paleocene floras from Montana, North Dakota, and Wyoming indicate a generally warm temperate (15°C mean annual temperature), moist climate (Hickey, 1977, 1980). However, Wolfe (1985) suggested that latest Paleocene–early Eocene mean annual temperature was about 22° to 23° C in the northern Rocky Mountains and Great Plains regions. On the basis of the floral assemblages from the Paleocene Tongue River Member of the Fort union Formation in the Powder River basin and upper Paleocene and lower Eocene Willwood Formation in the Bighorn basin, Wolfe (1985) and Wing (1981) interpreted a paleoclimatic change at present-day 50°N in the North American continental interior. Wolfe (1985) interpreted plant fossils from the middle to upper Paleocene Tongue River Member as representing mesothermal broad-leaved deciduous forests with temperatures of about 18° C, to megathermal/subtropical broad-leaved evergreen forest with temperatures of about 23° C. Wing (1981) interpreted the lower Eocene Willwood floral assemblages as representing mesothermal broad-leaved evergreen forest with temperatures of about 13° to 20° C. These paleoclimatic changes from late Paleocene to early Eocene at the same latitude probably reflect altitude differences that produced a cooler climate in the Bighorn basin. According to Wolfe (1985), mesothermal vegetation in the North American continental interior during the late Paleocene may have extended to at least 65 N, based on the occurrence of some broad-leaved evergreens.

Paleosols in the lower Eocene part of the Willwood Formation in the Bighorn basin exhibit changes that are consistent with increasing seasonal dryness and thickening of soil horizons through time (Wing and Brown, 1985). The warming and drying trend is supported by upward physiognomic and floristic changes observed by Wing (1981). Middle Eocene floras in the Green River Formation in southwest Wyoming were interpreted by MacGinitie (1969) and Leopold and MacGinitie (1972) to represent a seasonally dry, subtropical climate.

Wolfe (1985) suggested that at the end of the Eocene, a major cooling trend developed, and large areas of the Northern Hemisphere, including the Rocky Mountains and Great Plains regions, were occupied by microthermal broad-leaved deciduous forests representative of 0° to 13° C mean annual temperature. This cooling trend in the Rocky Mountains and Great Plains regions seems to be coincident with the absence of coal-forming fluvial environments in the Oligocene and Neogene. Apparently, only the generally warm temperate to subtropical climate during the Paleocene and Eocene supported widespread, luxuriant, forest swamps in fluvial environments, which in turn, produced thick peat deposits.

DEPOSITIONAL SYSTEMS OF CENOZOIC COALS

Most Cenozoic coals in the Rocky Mountains and Great Plains regions are associated with lacustrine and fluvial strata. In the Williston basin, Paleocene coals also accumulated along the margins of Cannonball sea during two periods of transgression. Forest swamps in which thick peats accumulated occurred mainly in marginal lacustrine and interfluvial areas. The structural basins and uplifts formed during Laramide deformation may have caused tectonic damming and internal drainage that maintained ground-water level near the surface (Stanley and Collinson, 1979). These structural basins were occupied by large lakes fed by internally draining rivers, and by through-flowing fluvial systems. Coals that formed in swamps associated with lakes are usually uneconomic, whereas coals that formed in swamps associated with fluvial systems are more likely to be economic.

Coals associated with lacustrine systems

During most of the Paleocene and Eocene, large lacustrine systems occupied structural basins in the west-central part of the Rocky Mountain region. The largest were prehistoric Lake Flagstaff in the Uinta basin in Utah, and prehistoric Lake Uinta in the Piceance basin in Colorado and the greater Green River basin in Wyoming (Ryder and others, 1976; Stanley and Collinson, 1979; Johnson, 1985). Peats accumulated in swamps along the margins of lakes and in flood plains of adjacent meandering rivers.

Stanley and Collinson (1979) suggested that the Paleocene Lake Flagstaff, which covered only the western part of the Uinta basin, was generally fresh, although it fluctuated between fresh and saline water. In this setting, decimeter-thick subbituminous coals formed in swamps associated with marginal lacustrine environments (Ryder and others, 1976; Fouch and Hanley, 1977). These algal-derived, kerogenous coals have organic carbon values of 50 to 70 percent and yield as much as 83 gallons of oil per ton (Fouch and Hanley, 1977).

During the Paleocene through early Eocene, the nearby Piceance basin was a separate hydrologic basin containing small freshwater lakes and fluvial systems that produced only thin coals in associated floodplain swamps. The Uinta and Piceance basins may have been connected briefly by a precursor of Lake Uinta (Bradley, 1931).

However, the transformation to Lake Uinta—a single saline lake that covered the Uinta and Piceance basins—was not complete until the middle Eocene (Johnson, 1985). During this time, kerogenous coals, similar to those of Lake Flagstaff, accumulated marginal lacustrine swamps along the southwest margin of the Piceance basin. In the Uinta basin, decimeter-thick kerogenous coals formed in swamps associated with nearshore open-lacustrine environments. These coals yield from 23 to 79 gallons of oil per ton and vary in organic carbon value from 54 to 70 percent (Fouch and Hanley, 1977).

A counterpart of Lake Uinta was the Eocene Lake Gosiute of the greater Green River basin. Freshwater stages of this lake, such as early Eocene Luman Lake, were accompanied by accumulation of both uneconomic and economic coals in marginal lacustrine swamps (Sklenar and Anderson, 1985; Roehler, 1986). Thin coal beds, up to a meter thick, of the lower Eocene Luman Tongue of the Green River Formation were interpreted by Sklenar and Anderson (1985) as originating in marginal-lacustrine swamps developed in interfluvial-deltaic systems in the southwest Washakie basin. Filling of the Luman Lake by fluvial-deltaic sediments transformed it into ephemeral shallow freshwater lake, swamp, and fluvial environments in which the lower Eocene Niland Tongue of the Wasatch Formation was deposited. In this setting, the economic Vermillion Creek coal, as thick as 3 m, accumulated in a marginal-lacustrine swamp in the Vermillion Creek basin (Roehler, 1986). Unlike the nearshore open and marginal lacustrine kerogenous coals of the Uinta and Piceance basins, the Vermillion Creek coal consists dominantly of woody and herbaceous plants (Nichols, 1986). After deposition of the Niland Tongue, freshwater stages of Lake Gosiute persisted through the deposition of the lower Eocene Tipton Shale Member of the Green River Formation. Subsequently, Lake Gosiute was transformed into a saline lake (Eugster and Surdam, 1973; Eugster and Hardie, 1975).

The paucity of economic coals associated with large alkaline-saline lakes such as the Lake Uinta and Lake Gosiute is a contrast to some major coals that formed in swamps associated with smaller, slightly saline lakes in the northern Bighorn basin, Wyoming. In this area, the Paleocene Fort Union Formation contains mineable coal deposits that accumulated in lacustrine-paludal environments bounded by fluvial systems (Hickey, 1980; Yuretich and others, 1984). Peterson (1984) interpreted a few of these coals to represent deposits in swamps associated with anastomosed fluvial systems.

Coals associated with fluvial systems

Laramide structural basins along the eastern side of the Rocky Mountains and in the Great Plains region are filled dominantly by fluvial strata. These coal-bearing basins include the Powder River basin in Wyoming and Montana, the Williston basin in Montana and North Dakota, the Hanna basin in Wyoming, the North Park and Sand Wash basins in Colorado, and the Raton basin in Colorado and New Mexico. Coals within these basins are highly variable in rank and thickness. For example, the Upper Cretaceous and Paleocene Raton Formation in the Raton basin contains moderately thick (up to 3.6 m) high-volatile A bituminous coal, whereas the Paleocene and Eocene Wasatch Formation in the Powder River basin contains an anomalously thick (75 m) lignite.

Within these basins, peats accumulated in mires associated with fluvial systems. Swamps were developed both on topographic highs (e.g., the raised platforms of abandoned meanderbelts) and on topographic lows (e.g., flood plains and lakes). Flores (1981, 1983) and Warwick (1985), among others, have suggested that some peats accumulated in raised swamps. Raised swamps may have developed either on elevated, relatively uncompactible meanderbelt sands, or as domed bodies independent of substrate topography. They also suggested that these swamps were gradually drowned as a result of autocompaction of the peat with concomitant rise of the ground-water table. After drowning and development of lakes, the topographically low areas were sites of fluvial invasion that choked further organic accumulation.

The facies association of economic coals in fluvial settings includes backswamp coals that are laterally and vertically juxtaposed to channel sandstones and conglomerates and floodplain–lacustrine deposits. Areally restricted, anomalously thick (25 to 40 m) coals such as the Anderson-Wyodak coal of the Tongue River Member of the Fort Union Formation in the Powder River basin, Wyoming, are discontinuous and lenticular bodies bounded by channel deposits (Figs. 14 and 15). The channel deposits (north-south– and east-west–oriented bodies) consist of fine-grained sandstones and siltstones (Warwick and Stanton, 1988). The channel sandstones probably represent deposits of a trunk-tributary fluvial system as proposed by Ethridge and others (1981). An anastomosed morphology of the trunk-tributary system is indicated by subparallel, north-south–oriented, minor channel sandstones (Fig. 16). Figure 14 shows these channel sandstones as narrow, stringer-like bodies (width:thickness = 1.2) encased by thick mudstone and coal, indicating vertical aggradation in confined channels. Similar interconnected channel sandstone stringers in the Tongue River Member were described by Flores and Hanley (1984) as deposits of suspended-load, anastomosed streams. The discontinuous swamps of these anastomosed fluvial systems are reflected by the pod-like distribution of the coals and are illustrated by the isopach map of the Wyodak-Anderson coal (Figs. 14, 15, and 16). The Wyodak-Anderson, which formed in these swamps, contains either carbonaceous shale or claystone partings in the upper 6 to 14 m of the coal bed (Warwick and Stanton, 1988). These detrital partings, which are limited to the coal-bed margins, represent sporadic flooding late in the swamp history from nearby channels. Despite the domed or raised condition of these swamps suggested by Flores (1981), fluvial, lacustrine, and crevasse-splay incursions may have contributed to their demise. It is suggested that fluvial incursions occurred via low-lying hollows or ponds that formed by rise of ground water influenced by differential compaction of peat. Coalescing of these ponds may have created larger bodies of water or lakes and rivulets that drowned the swamps. These processes and oxidation of peat contributed to the termination of the swamps.

The accumulation of peat above detrital partings in the upper part of the Anderson-Wyodak coal suggests that sediment carried into the raised swamp temporarily suspended vegetal growth and peat accumulation. However, the flooding also provided essential nutrients to the nutrient-starved swamp. This pattern of termination is repeated in the Anderson-Dietz coal bed (as thick as 25 m) of the Tongue River Member in the Powder River basin, Montana, and in the Felix coal bed (as thick as 10 m) of the Wasatch Formation in the Powder River basin. Moore and others (1986) suggested that flooding of the swamp that produced the Anderson-Dietz coal may have increased sulfur (as framboidal pyrite), ash, and degraded huminitic maceral content at the top. These changes in the characteristics of the coal reflect flooding by fresh water, which allowed increased microbial attack of the peat and sulfur fixation. Petrographic analyses of the Anderson-Dietz and Felix coals by Moore and others (1986) and Warwick (1985), respectively, indicate an overall composition of variable amounts of cellular and degraded huminitic macerals and a general increase of inertinite macerals toward the top. This variation in maceral content was interpreted by Warwick (1985) and Stanton and others (1986) as reflecting differential preservation due to changing pH conditions in a raised swamp. A high pH (>4.5) would be ideal for bacteria to degrade organic material. Thus, an increase of inertinites in the upper part of the coal suggests a highly acidic, nutrient-poor swamp that supported only a few stunted plant species. Satchel (1985) supported the raised swamp interpretation for the Anderson-Wyodak coal and palynologically identified it as a *Glyptostrobus* swamp. The geochemical and paleobotanical modes of peat accumulation in modern raised, or ombrotrophic, swamps in fluvial environments were documented by Anderson and Muller (1975).

Development of raised swamps may account for the unusually thick Lake de Smet coal of the lower Tertiary Wasatch Formation in the Powder River basin. This coal, which is as thick as 75 m and occupies an area 4 by 27 km, is bordered along its eastern margin by stacked channel sandstones and along its western margin by alluvial fan conglomeratic sandstones (Obernyer,

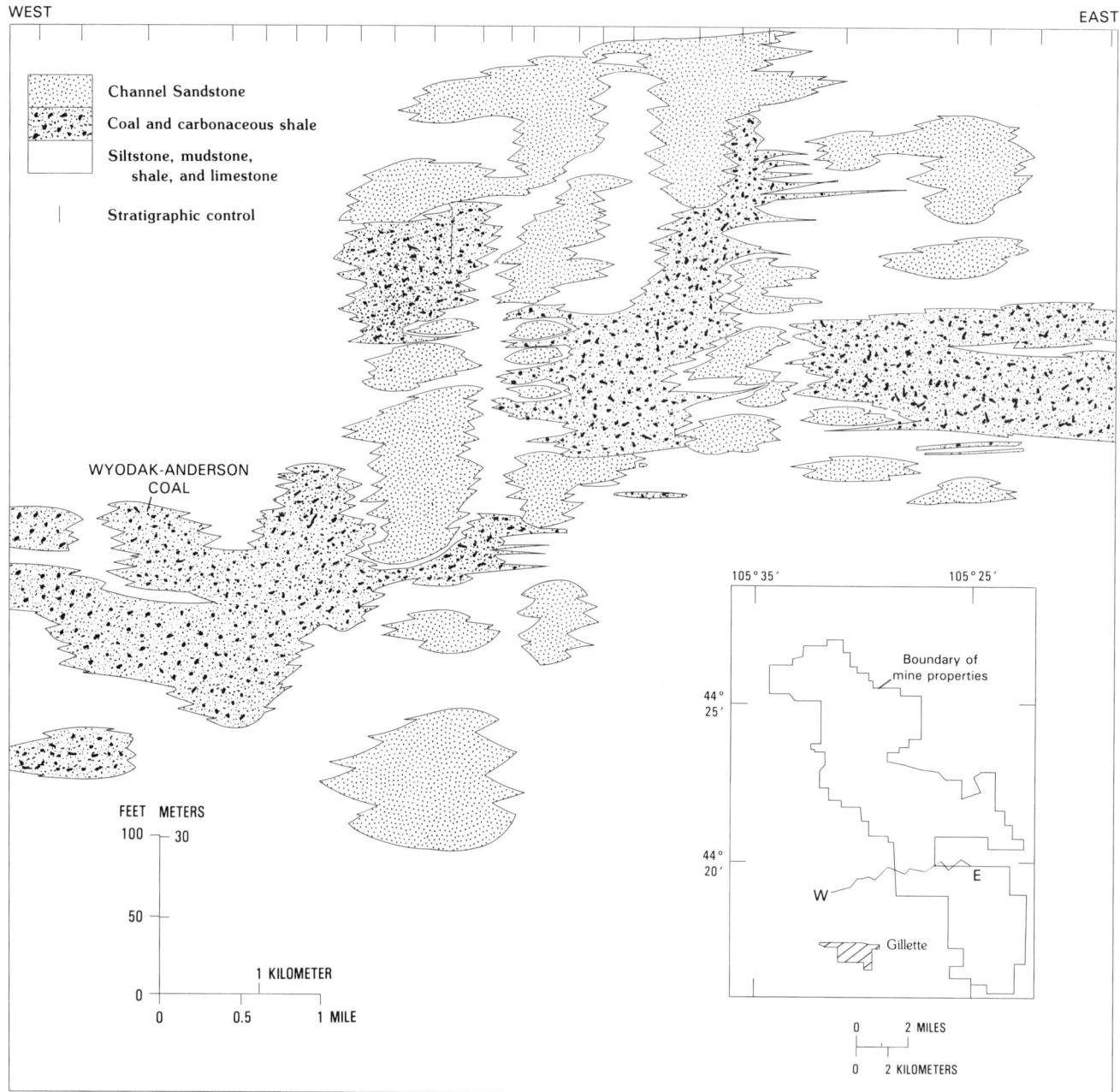

Figure 14. Diagrammatic cross section showing east-west stratigraphic and paleoenvironmental framework of the Wyodak-Anderson coal and associated sediments in the Tongue River Member. Modified from Warwick and Stanton (1986).

1978; Flores and Warwick, 1984). The presence of very coarse detritus along the Lake de Smet swamp margins may indicate that it was a raised swamp at the toe of an alluvial fan and supplied by nutrient-rich, ground-water discharge through the conglomeratic sands of the fan (Flores and Warwick, 1984).

Clastic sedimentation plays a significant role in controlling the thickness of coals that accumulated in low-lying swamps. The facies association of coals accumulated in this environment consists of crevasse and lake-splay sandstones and mudstones, and fossiliferous lacustrine limestones and carbonaceous shales. These facies are either laterally or vertically juxtaposed to the coal deposit. The low-lying swamps probably formed by subsidence caused by compaction of underlying fine-grained flood-plain–lacustrine deposits. The accumulation of peat in such a swamp is controlled by a delicate balance between subsidence and detrital influx. Accumulation of peat proceeds as long as the rates of subsidence and detrital influx are less than the rate of organic accumulation. A reversal of this condition results in drowning of the swamp, permitting rapid biodegradation of peat by oxygenated water and increasing ash content of the peat. The continuous

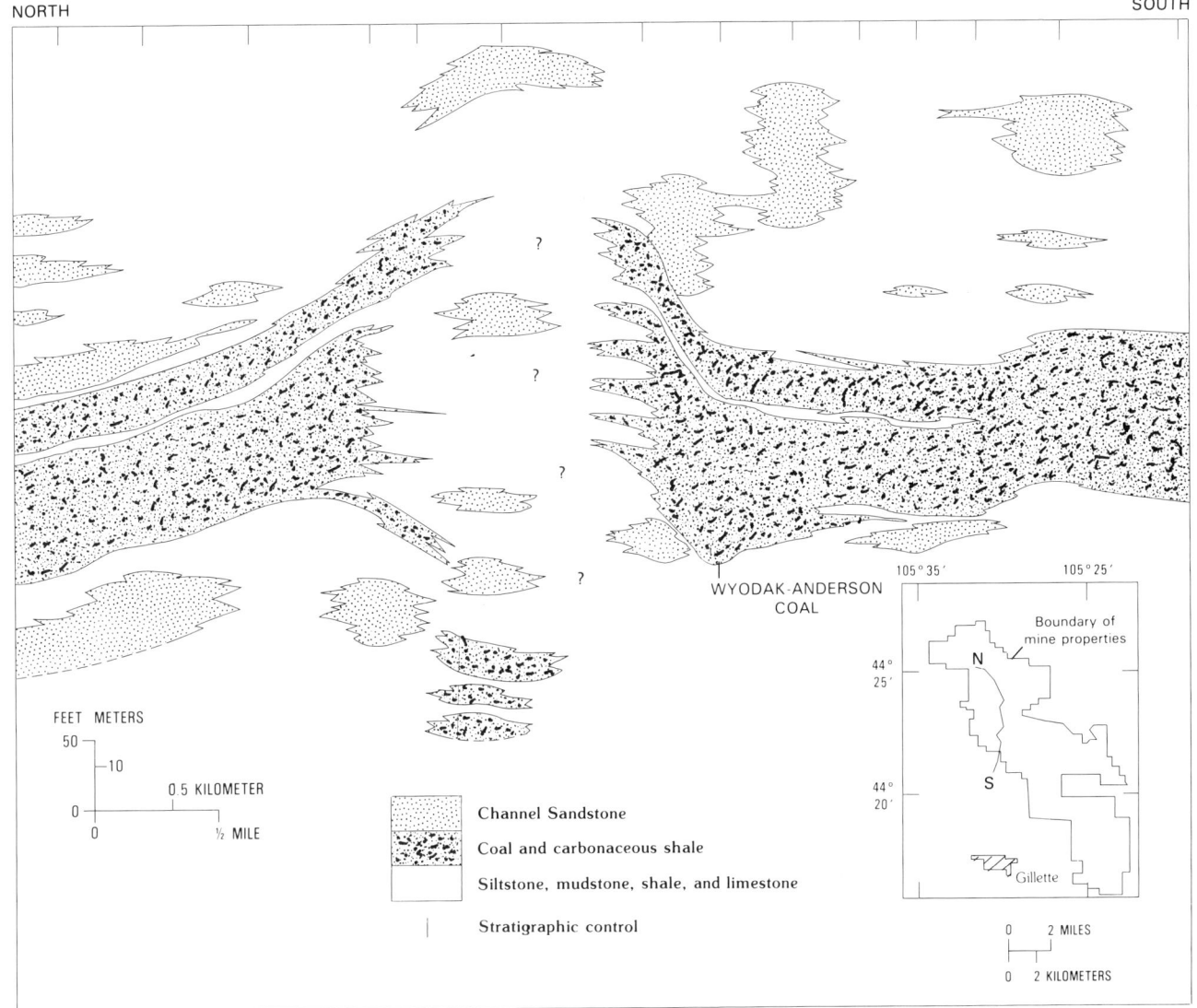

Figure 15. Diagrammatic cross section showing north-south stratigraphic and paleoenvironmental framework of the Wyodak-Anderson coal and associated sediments in the Tongue River Member. Modified from Warwick and Stanton (1986).

interplay of one of these factors over the other in the low-lying swamps accounts for the success or failure to accumulate economic coal deposits in this setting.

Low-lying swamps in the fluvial environments that produced economic coals include the York Canyon coal of the Paleocene part of the Raton Formation in the Raton basin, New Mexico (Flores, 1984), the Riach coal of the Paleocene to Eocene Coalmont Formation in the North Park basin, Colorado (Hendricks, 1978), the Paleocene Fort Union coals in the Sand Wash basin, Colorado (Beaumont, 1979), and the Beaulah-Zap coal of the Paleocene Sentinel Butte Formation in the Willison basin, North Dakota (Daly and others, 1985). These coals, which range in thickness from 3.6 m (York Canyon coal) to 20 m (Riach coal), were accumulated in low-lying swamps related to meandering fluvial systems. The most common overlying facies associated with these coals are crevasse-splay sediments. The absence of fossiliferous lacustrine sediments in this facies association supports the idea of termination by detrital floods instead of drowning by lakes that is common in raised swamps of the Powder River basin.

The raised and low-lying swamps that accumulated economic coals in the Rocky Mountains and Great Plains regions are mainly related to meandering and anastomosed fluvial systems. These fluvial systems can be characterized as laterally and vertically aggrading stream complexes. These modes of aggradation in the alluvial plain have curtailed areas of formation of poorly drained swamps. The lateral aggradation by the meandering fluvial system tends to erode and recycle peat deposits, and the vertical aggradation of the anastomosed fluvial system is prone to rapidly bury peat deposits. In both systems, however, aggradation

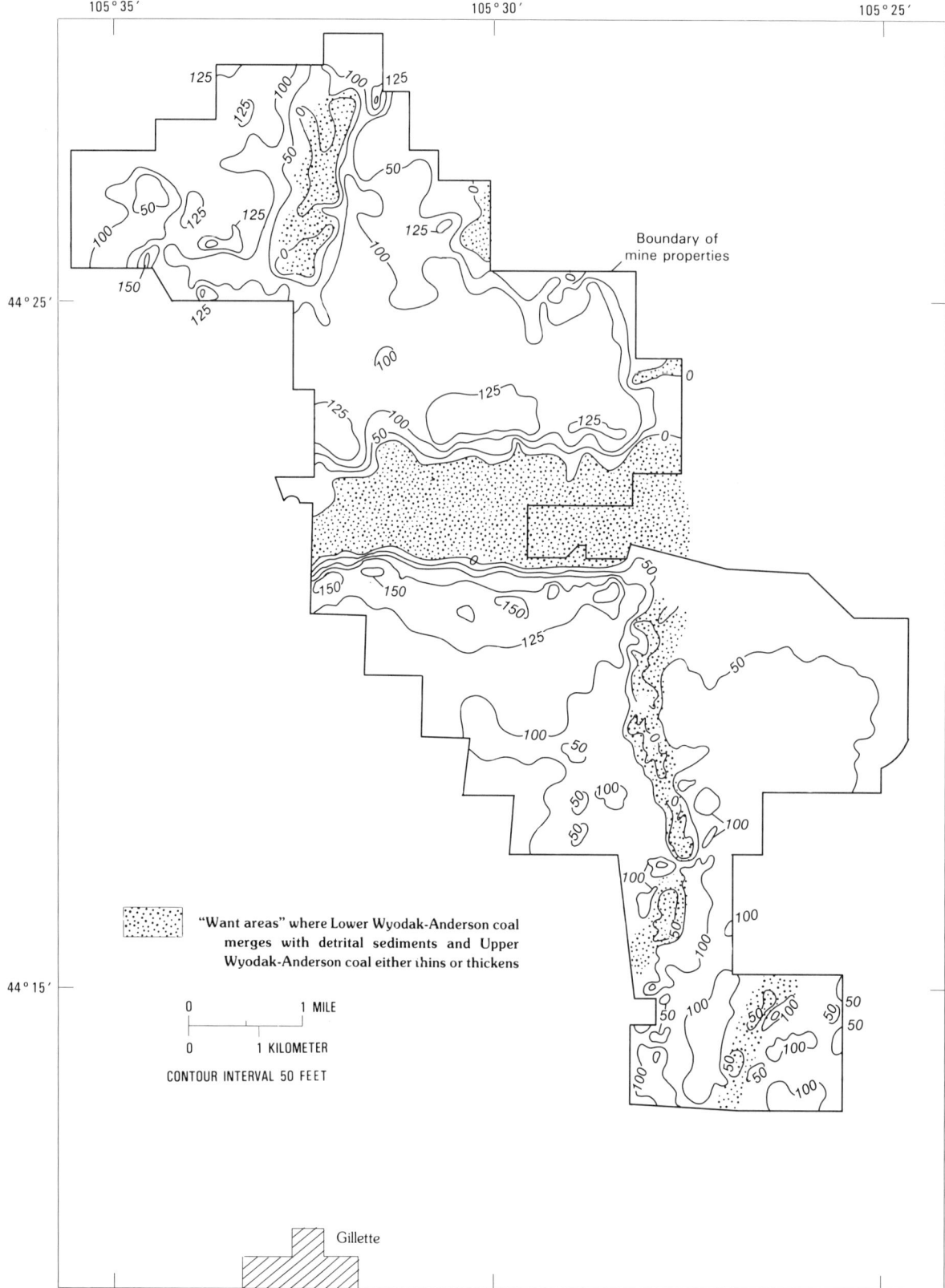

Figure 16. Isopach map showing discontinuity of the Wyodak-Anderson coal and "want" areas occupied by fluvial channel deposits. Modified from Warwick and Stanton (1986).

materializes by avulsion and abandonment of long courses and broad belts of streams, whose deposits, in turn, are transformed into detrital-free, reducing, poorly drained swamps. These swamps are isolated in the alluvial plain and may exist for a short or long period of time depending on the rates of sedimentation, avulsion, and subsidence. In these fluvial settings, avulsion by a crevasse splay probably is the most important process of river shifts into the swamps. That is, progradation of a crevasse splay into a swamp produces a complex of splays consisting of multiple diverging and converging channelized flows much like an anastomosed pattern. The concentration of flow along one of these anastomosed channels due to increased discharge, abandonment of accessory channels, and confinement by levee deposits, initiated the occupation of a major river channel into the swamp. Contemporaneous lateral erosion and deposition along this river channel, resulting in increased gradient, transformed the anastomosed streams to meandering streams. River anastomosis may be developed by damming the downstream reaches of fluvial systems by uplifts, as proposed in the Powder River basin by Flores and Hanley (1984).

CHARACTER, RANK, QUALITY AND UTILIZATION OF ECONOMIC CENOZOIC COALS

Economic Cenozoic coals in the Rocky Mountains and Great Plains regions range form lignite B to high-volatile A bituminous (Table 3). These coals range from 3 m thick for the

TABLE 3. SUMMARIES OF COAL-QUALITY DATA FOR MAJOR TERTIARY AND UPPERMOST CRETACEOUS COAL-BEARING ROCKS IN THE ROCKY MOUNTAIN REGION*

Geographic Area	Stratigraphic Unit	Age†	Major Coal Beds Represented	Number of Samples	Ash	Sulfur	Calorific Value (Btu/lb)	Apparent Rank
Northern Rocky Mountains								
Williston basin	Fort Union	P	Buelah-Zap; Dunn Center; Harmon	284	8.4 (3.5–25.9)	0.9 (0.1–8.4)	6,370 (2,810–8,290)	Lignite A (Lig. B-Sub. C)
Bull Mountains	Fort Union	P	Mammoth; McCleary	11	8.2 (5.6–10.4)	0.8 (0.5–1.2)	10,120 (8,930–10,990)	Subbituminous A (Sub. B-Sub. A)
Powder River basin	Fort Union	P	Anderson; Dietz; Canyon; Rosebud; Knoblock; Wyodak	636	7.0 (1.6–27.5)	0.6 (0.1–5.2)	8,080 (4,450–10,300)	Subbituminous C (Lig. B-Sub. A)
	Wasatch	PE	Felix	35	11.6 (3.5–27.9)	1.1 (0.3–6.4)	7,430 (5,680–8,810)	Subbituminous C (Lig. A-Sub. B)
Central Rocky Mountains								
Greater Green River basin	Fort Union	P	Bridger	18	12.8 (7.1–30.5)	0.6 (0.3–1.7)	7,170 (4,840–8,620)	Subbituminous C (Lig. B-Sub. C)
Hanna basin	Ferris	KT	No. 21, 24, 25	19	10.4 (3.9–20.8)	0.6 (0.3–1.3)	9,750 (6,520–11,150)	Subbituminous A (Lig. A-Sub. A)
	Hanna	P	No. 2, 79, 82	41	21.2 (5.4–41.1)	1.7 (0.4–3.6)	8,870 (6,330–11,350)	Subbituminous A (Sub. B-Sub. A)
North Park	Coalmont	PE	Riach; Sudduth	39	11.2 (2.1–37.0)	0.4 (0.2–1.4)	9,350 (5,500–11,280)	Subbituminous A (Lig. A-Sub. A)
Denver basin	Denver	KT	Comanche	12	18.2 (9.5–28.4)	0.4 (0.3–0.9)	5,790 (4,830–6,910)	Lignite A (Lig. A)
Southern Rocky Mountains								
Raton Mesa	Raton	KT	Allen; York Canyon	25	15.0 (5.8–27.6)	0.5 (0.4–0.9)	12,570 (10,170–14,230)	HVA Bituminous (HVA Bit.-LV Bit.)

*Compiled by R. T. Hildebrand.
†Age designations: KT = Cretaceous–Paleocene; P = Paleocene; PE = Paleocene–Eocene; E = Eocene.

high-volatile A bituminous in the Raton basin to 75 m thick for the lignite in the Powder River basin. Bituminous coals in the Raton basin were increased in rank by Tertiary igneous intrusion (Amuedo and Bryson, 1977). Low-grade lignites occur principally in the Powder River and Williston basins and the Great Plains of the northern Rocky Mountains region, and in the Denver basin of the central Rocky Mountains region. In both the central and northern Rocky Mountains regions, a large amount of the economic coals are subbituminous.

On the basis of arithmetic means of 25 samples, the coals in the southern Rocky Mountains region (Raton basin) range from 0.4 to 0.9 percent sulfur, 5.8 to 27.6 percent ash, and 10,170 to 14,230 Btu. In the central Rocky Mountains region, arithmetic means of subbituminous coals based on 129 coal samples range from 0.2 to 3.6 percent sulfur, 2.1 to 41.1 percent ash, and 4,830 to 11,350 But. The Wyodak-Anderson coals in the Powder River basin contain 0.5 percent average sulfur, 6 percent average ash, and 8,224 average Btu. In the northern Rocky Mountains region, arithmetic means of 628 coal samples range from 0.1 to 6.4 percent sulfur, 1.6 to 27.9 percent ash, and 4,450 to 10,990 Btu. Coals in the Great Plains region (Williston basin), based on 284 samples, range from 0.1 to 8.4 percent sulfur, 3.5 to 25.9 percent ash, and 2,810 to 8,290 Btu.

The petrology of Tertiary coals in the Rocky Mountains and Great Plains regions is summarized in Table 4. The microscopic organic constituents, or macerals, of Cenozoic coals contain three major suites: huminites-vitrinites, liptinites, and inertinites. Huminites are precursors of vitrinites in low-rank coals and are derived principally from woody tissues. Liptinites are derived from pollen, spores, algae, waxes, resins, latex, and other lipid-like plant particles. Inertinites are derived from plant material that has been strongly altered and degraded in the peat stage of coalification and include fusinites and semifusinites.

Petrographic characteristics of Cenozoic coals in the Rocky Mountains and Great Plains regions may be summarized on the basis of 104 complete analyses. The Hanna coals of the Paleocene to Eocene Hanna Formation in the Hanna basin, Wyoming, contain the most abundant vitrinites or huminites. The Zap coal of the Paleocene Sentinel Butte Formation in the Williston basin contains the least amount of huminites or vitrinites. Liptinites vary from 12 to 1 percent, inertinites range from 21 to 1 percent and vitrinite (huminite) reflectance varies from 0.31 to 0.86.

The maceral compositions of the Wyodak-Anderson and Dietz coals in the Powder River basin do not show any significant variation. Vitrinites (huminites) range from 82 to 93 percent, liptinites from 2 to 8 percent, and inertinites from 3 to 16 percent. Rich (1980), using blue light maceral analysis, measured as much as 38 percent liptinites and up to 40 percent inertinites from the Wyodak-Anderson coal. The presence of high liptinites in the Wyodak-Anderson coal is a positive indication of its potential for liquefaction. The high inertinites are caused by abundant fusinites (as much as 18 percent) and semifusinites (as much as 27 percent), perhaps indicating episodic fires in the Wyodak-Anderson swamp. Rich (1980) suggested that most of the wood-derived macerals may have been derived from cypress and sequoia as well as allied trees.

TABLE 4. MACERAL COMPOSITION OF SOME TERTIARY COALS IN THE ROCKY MOUNTAINS AND GREAT PLAINS REGIONS DETERMINED BY WHITE-LIGHT ANALYSIS, EXCEPT AS INDICATED BY DOUBLE ASTERISK*

	Vitrinites	Liptinites	Inertinites	R_o†	N§
Powder River basin					
Wyodak-Anderson coal	88	2	10	0.4–0.42	2
Wyodak-Anderson coal†	69	21	10	n.d.	22
Anderson-Dietz coal	90	3	7	0.35–0.44	12
Monarch (Canyon) coal	84	4	12	0.36–0.48	4
Wall coal	89	1	10	0.39–0.42	3
Cook coal	89	2	9	0.35–0.44	3
School coal	82	8	10	n.d.	1
Smith coal	89	3	8	n.d.	2
Arvada coal	93	4	3	n.d.	1
Felix coal	89	2	9	n.d.	9
Rosebud-McKay coal	80	4	16	0–0.55	4
Hanna basin					
Hanna #1 coal	90	7	3	0.42–0.54	29
Hanna #2 coal	96	2	2	0.47–0.48	2
Hanna #5 coal	95	3	2	0.44	1
Williston basin					
Beulah-Zap coal	83	1	16	n.d.	1
Zap coal	67	12	21	0.31	1
Vermillion Creek basin					
Vermillion Creek coal	91	8	1	0.45–0.48	3
Raton basin					
York Canyon coal	84	1	15	0.81–0.86	4
Boncarbo coal	85	3	12	0.76	2

*Maceral content in average percent. Data compiled from Pennsylvania State University data bank, from Dutcher and others (1982), and from data contributed by Fredrick J. Rich, Ronald W. Stanton, Timothy Moore, and Peter Warwick.
†R_o = vitrinite reflectance.
§N = number of samples.
**Maceral analysis in blue light.
n.d. = not determined.

REFERENCES CITED

Amuedo, C. L., and Bryson, R. S., 1977, Trinidad-Raton basins; A model coal resource evaluation program, *in* Murray, D. K., ed., Proceedings, 1976 Geology of Rocky Mountain Coal: Colorado Geological Survey Resources Series 1, p. 45–60.

Anderson, E. J., and Goodwin, P. W., 1980, Application of the PAC hypothesis to limestones of the Helderberg Group: Eastern Section, Society of Economic Paleontologists and Mineralogists Field Trip Guidebook, 32 p.

Anderson, J.A.R., and Muller, J., 1975, Palynological study of a Holocene peat and a Miocene coal deposit from NW Borner: Review of Paleobotany and Palynology, v. 19, p. 291–351.

Armstrong, R. L., 1968, Sevier orogenic belt in Nevada and Utah: Geological Society of America Bulletin, v. 79, p. 429–458.

Atwater, T., 1970, Implications of plate tectonics for the Cenozoic tectonic evolution of western North America: Geological Society of America Bulletin, v. 81, p. 3513–3536.

Averitt, P., 1966, Coking-coal deposits of the western United States: U.S. Geological Survey Bulletin 1222-G, 48 p.

—— , 1975, Coal resources of the United States, January 1, 1974: U.S. Geological Survey Bulletin 1412, 131 p.

Balsley, J. K., 1980, Cretaceous wave-dominated delta systems; Book Cliffs, east-central Utah: American Association of Petroleum Geologists Continuing Education Department Guidebook, 162 p.

Barron, E. J., 1983, A warm, equable Cretaceous; The nature of the problem: Earth Science Reviews, v. 19, p. 305–338.

—— , 1985, Numerical climate modeling; A frontier in petroleum source rock prediction; Results based on Cretaceous simulations: American Association of Petroleum Geologists Bulletin, v. 69, p. 448–459.

Barron, E. J., and Washington, W. M., 1982, Cretaceous climate; A comparison of atmospheric simulations with the geologic record: Palaeogeography, Palaeoclimatology, Palaeoecology, v. 40, p. 103–133.

Beaumont, E. A., 1979, Depositional environments of Fort Union sediments (Tertiary, northwest Colorado) and their relation to coal: American Association of Petroleum Geologists Bulletin, v. 63, p. 194–217.

Beaumont, E. C., Shomaker, J. W., and Kottlowski, F. E., 1971, Stratidynamics of coal deposition in southern Rocky Mountain region, U.S.A.: New Mexico Bureau of Mines and Mineral Resources Memoir 25, p. 175–185.

Bird, P., 1984, Laramide crustal thickening event in the Rocky Mountain foreland and Great Plains: Tectonics, v. 3, p. 741–758.

Bradley, W. H., 1931, Origin and macrofossils of the oil shale of the Green River Formation of Colorado and Utah: U.S. Geological Survey Professional Paper 168, 58 p.

Brown, R. W., 1962, Paleocene flora of the Rocky Mountains and Great Plains: U.S. Geological Survey Professional Paper 375, 119 p.

Busch, D.A., 1971, Genetic units in delta prospecting: American Association of Petroleum Geologists Bulletin, v. 55, p. 1137–1154.

—— , 1974, Stratigraphic traps in sandstones; Exploration techniques: American Association of Petroleum Geologists Memoir 21, 174 p.

Cavaroc, V. V., and Flores, R. M., 1984, Lithologic relationships of the Upper Cretaceous Gibson-Cleary stratigraphic interval; Gallup Coal Field, New Mexico, U.S.A., *in* Rahmani, R. A., and Flores, R. M., eds., Sedimentology of coal and coal-bearing sequences: International Association of Sedimentologists Special Publication 7, p. 197–215.

Collins, B. A., 1977, Coal deposits of eastern Piceance basin, *in* Murray, D. K., ed., Proceedings, 1976 Geology of Rocky Mountain Coal: Colorado Geological Survey Resources Series 1, p. 29–43.

Cross, T. A., 1988, Controls on coal distribution in transgressive-regressive cycles, Upper Cretaceous, Western Interior, U.S.A., *in* Wilgus, C. K., and others, eds., Sea-level changes; An integrated approach: Society of Economic Paleontologists and Mineralogists Special Publication 42, p. 371–380.

—— , 1986, Tectonic controls of foreland basin subsidence and Laramide-style deformation, western United States, *in* Homewood, P., and Allen, P., eds., Foreland basins: Oxford, Blackwell Scientific Publishers (in press).

Cross, T. A., and Pilger, R. H., Jr., 1978, Tectonic controls of Late Cretaceous sedimentation, Western Interior, U.S.A.: Nature, v. 274, p. 653–657.

Daly, D. J., Groenewald, G. H., and Schmit, C. R., 1985, Paleoenvironments of the Paleocene Sentinel Butte Formation, Knife River area, west-central North Dakota, *in* Flores, R. M., and Kaplan, S. S., eds., Cenozoic paleogeography of the west-central United States: Rocky Mountain Section, Society of Economic Paleontologists and Mineralogists Symposium 3, p. 171–185.

Dickinson, W. R., and 6 others, 1988, Paleogeographic and paleotectonic setting of Laramide sedimentary basins in the central Rocky Mountain region: Geological Society of America Bulletin, v. 100, p. 1023–1039.

Doelling, H. H., and Smith, M. R., 1982, Overview of Utah coal fields, *in* Gurgel, K. D., ed., Proceedings, 5th Symposium on the Geology of Rocky Mountain Coal, 1982: Utah Geological and Mineral Survey Bulletin 118, p. 1–26.

Dutcher, R. R., Crelling, J. C., and Cascia, M. C., 1982, Petrography and fluorescence properties of bituminous coals of the Spanish Peaks region, Colorado, *in* Gurgel, K. D., ed., Proceedings, 5th Symposium on the Geology of Rocky Mountain Coal, 1982: Utah Geological and Mineral Survey Bulletin 118, p. 179–186.

Engebretson, D. C., Cox, A., and Gordon, R. G., 1985, Relative motions between oceanic and continental plates in the Pacific basin: Geological Society of America Special Paper 206, 59 p.

Ethridge, F. G., Jackson, J. J., and Youngberg, A. D., 1981, Flood basin sequence of a fine-grained meanderbelt subsystem; The coal-bearing Wasatch and Upper Fort Union Formations, Powder River Basin, Wyoming, *in* Ethridge, F. G., and Flores, R. M., eds., Recent and Ancient nonmarine depositional environments; Models for exploration: Society of Economic Paleontologists and Mineralogists Special Publication 31, p. 191–289.

Eugster, H. P., and Hardie, L. A., 1975, Sedimentation in an ancient playa-lake complex: Geological Society of America Bulletin, v. 86, p. 319–334.

Eugster, H. P., and Surdam, R. C., 1973, Depositional environment of the Green River Formation of Wyoming; A preliminary report: Geological Society of America Bulletin, v. 84, p. 1115–1120.

Fassett, J. E., 1986, The non-transferability of a Cretaceous coal model in the San Juan Basin of New Mexico and Colorado, *in* Lyons, P. C., and Rice, C. L., eds., Paleoenvironmental and tectonic controls in coal-forming basins of the United States: Geological Society of America Special Paper 210, p. 165–171.

Fassett, J. E., and Hinds, J. S., 1971, Geology and fuel resources of the Fruitland Formation and Kirtland Shale of the San Juan basin, New Mexico and Colorado: U.S. Geological Survey Professional Paper 676, 76 p.

Flores, R. M., 1978, Barrier and back-barrier environments of deposition of the Upper Cretaceous Almond Formation, Rock Springs uplift, Wyoming: The Mountain Geologist, v. 15, p. 57–65.

—— , 1981, Coal deposition in fluvial paleoenvironments of the Paleocene Tongue River Member of the Fort Union Formation, Powder River basin, Wyoming and Montana, *in* Ethridge, F. G., and Flores, R. M., eds., Recent and ancient nonmarine depositional environments; Models for exploration: Society of Economic Paleontologists and Mineralogists Special Publication 31, p. 169–190.

—— , 1983, Basin facies analysis of coal-rich Tertiary alluvial deposits, northern Powder River basin, Montana and Wyoming, *in* Collinson, J. D., and Lewin, J., eds., Modern and ancient fluvial systems: International Association of Sedimentologists Special Publication 6, p. 501–515.

—— , 1984, Comparative analysis of coal accumulation in Cretaceous alluvial deposits, southern United States Rocky Mountain basins, *in* Scott, D. F., and Glass, D. G., eds., The Mesozoic of middle North America: Canadian Society of Petroleum Geologists Memoir 9, p. 373–385.

Flores, R. M., and Hanley, J. H., 1984, Anastomosed and associated coal-bearing fluvial deposits; Upper Tongue River Member, Paleocene Fort Union Formation, northern Powder River Basin, Wyoming, U.S.A., *in* Rahmani, R. A., and Flores, R. M., eds., Sedimentology of coal and coal-bearing sequences: International Association of Sedimentologists Special Publication 7, p. 85–103.

Flores, R. M., and Tur, S. M., 1982, Characteristics of deltaic deposits in the Cretaceous Pierre Shale, Trinidad Sandstone, and Vermejo Formation, Raton Basin, Colorado: The Mountain Geologist, v. 19, p. 25–40.

Flores, R. M., and Warwick, P. D., 1984, Dynamics of coal deposition in intermontane alluvial paleoenvironments, Eocene Wasatch Formation, Powder River basin, Wyoming, in Houghton, R. L., and Clausen, E. N., eds., Proceedings, 1984 Geology of Rocky Mountain Coal: North Dakota Geological Society Publication 84-1, p. 184–199.

Flores, R. M., Blanchard, L. F., Sanchez, J. D., Marley, W. E., and Muldoon, W. J., 1984, Paleogeographic controls of coal accumulation, Cretaceous Blackhawk Formation and Star Point Sandstone, Wasatch Plateau, Utah: Geological Society of America Bulletin, v. 95, p. 540–550.

Fouch, T. D., and Hanley, J. H., 1977, An interdisciplinary analysis of some potential petroleum source rocks in east-central Utah; Implications for hydrocarbon exploration in nonmarine rocks of the western United States: Rocky Mountain Section, American Association of Petroleum Geologists and Society of Economic Paleontologists and Mineralogists Abstracts with Programs, v. 36.

Frazier, D. E., 1974, Depositional-episodes; Their relationship to the Quaternary stratigraphic framework in the northwestern portion of the Gulf Basin: Austin, University of Texas Bureau of Economic Geology Geological Circular 74-1, 28 p.

Goodwin, P. W., and Anderson, E. J., 1985, Punctuated aggradational cycles; A general hypothesis of episodic stratigraphic accumulation: Journal of Geology, v. 93, p. 515–533.

Goolsby, S. M., Reade, N. S., and Murray, D. K., 1979, Evaluation of coking coals in Colorado: Colorado Geological Survey Resource Series 7, 72 p.

Hallam, A., 1985, A review of Mesozoic climates: Journal of the Geological Society of London, v. 142, p. 433–445.

Hendricks, M. L., 1978, Stratigraphy of the Coalmont Formation near Coalmont, Jackson County, Colorado, Colorado, in Hodgson, H. E., ed., Proceedings, 1977 Geology of Rocky Mountain Coal: Colorado Geological Survey Resource Series 4, p. 35–48.

Hickey, L. J., 1977, Stratigraphy and paleobotany of the Golden Valley Formation (Early Tertiary) of western North Dakota: Geological Society of America Memoir 150, 181 p.

——, 1980, Paleocene stratigraphy and flora of the Clark's Fork basin, in Gingerich, P. D., ed., Early Cenozoic paleontology and stratigraphy of the Bighorn basin, Wyoming: Ann Arbor, University of Michigan Papers on Paleontology, v. 24, p. 33–49.

Jervey, M. T., 1988, Quantitative geological modeling of siliciclastic rock sequences and their seismic expression, in Wilgus, C. K., and others, eds., Sea-level changes; An integrated approach: Society of Economic Paleontologists and Mineralogists Special Publication 42, p. 47–69.

Johnson, R. C., 1985, Early Cenozoic history of the Uinta and Piceance Creek basins, Utah and Colorado, with special reference to the development of Eocene Lake Uinta, in Flores, R. M., and Kaplan, S. S., eds., Cenozoic paleogeography of the west-central United States: Rocky Mountain Section, Society of Economic Paleontologists and Mineralogists Symposium 3, p. 247–276.

Johnson, S. R., and Vaninette, G. E., 1982, Multiple barrier island and deltaic progradational sequences in Upper Cretaceous coal-bearing strata, northern Kaiparowits Plateau, Utah, in Gurgel, K. D., ed., Proceedings, 5th Symposium on the Geology of Rocky Mountain Coal, 1982: Utah Geological and Mineral Survey Bulletin 118, p. 62–69.

Jordan, T. E., 1981, Thrust loads and foreland basin evolution, Cretaceous, western United States: American Association of Petroleum Geologists Bulletin, v. 65, p. 2506–2520.

Jurdy, D. M., 1984, The subduction of the Farallon plate beneath North America as derived from relative plate motions: Tectonics, v. 3, p. 107–113.

Kauffman, E. G., 1980, Major factors influencing the distribution of Cretaceous coal in the Western Interior United States, in Carter, L. M., ed., Proceedings, 1980 Geology of Rocky Mountain Coal: Colorado Geological Survey Resource Series 10, p. 1–3.

Lawrence, D. T., 1982, Influence of transgressive-regressive pulses on coal-bearing strata of the Upper Cretaceous Adaville Formation, southwestern Wyoming, in Gurgel, K. D., ed., Proceedings, 5th Symposium on the Geology of Rocky Mountain Coal, 1982: Utah Geological and Mineral Survey Bulletin 118, p. 32–49.

Leighton, V. L., 1980, Depositional environments and petrography of the Trinidad Sandstone and related formations, Raton area, New Mexico [M.S. thesis]: Fort Collins, Colorado State University, 105 p.

Leopold, E. B., and MacGinitie, H. D., 1972, Development and affinities of Tertiary floras in the Rocky Mountains, in Graham, A., ed., Floristic and paleofloristics of Asia and eastern North America: Amsterdam, Elsevier, p. 147–200.

Levey, R. A., 1985, Depositional model for understanding geometry of Cretaceous coal; Major coal seams, Rocky Springs Formation, Green River basin, Wyoming: American Association of Petroleum Geologists Bulletin, v. 69, p. 1359–1380.

MacGinitie, H. D., 1969, The Eocene Green River flora of northwestern Colorado and northeastern Utah: University of California Publications in the Geological Sciences, v. 83, 140 p.

McGookey, D. P., Compiler, 1972, Cretaceous System, in Mallory, W. W., ed., Geologic atlas of the Rocky Mountain region: Rocky Mountain Association of Geologists, p. 190–228.

Moore, T. A., Stanton, R. W., and Flores, R. M., 1986, Nature of petrographic variations in some Tertiary coal beds, Powder River basin, Montana and Wyoming [abs.]: American Association of Petroleum Geologists Bulletin (in press).

Morrow, D. W., 1986, The sea-level rise staircase on continental margins and the origin of upward-shoaling carbonate sequences: Bulletin of Canadian Petroleum Geology, v. 34, p. 284–285.

Nichols, D. J., 1986, Palynology of the Vermillion Creek coal bed and associated strata, in Roehler, H. W., and Martin, P. L., eds., Geological investigations of the Vermillion Creek coal bed in the Eocene Niland Tongue of the Wasatch Formation, Sweetwater County, Wyoming: U.S. Geological Survey Professional Paper 1314D, p. 47–43.

Obernyer, S., 1978, Basin-margin depositional environments of the Wasatch Formation in the Buffalo-Lake de Smet area, Johnson County, Wyoming, in Hodgson, H. E., ed., Proceedings, 1977 Geology of Rocky Mountain Coal: Colorado Geological Survey Resources Series 4, p. 49–65.

Parrish, J. T., and Curtis, R., 1982, Atmospheric circulation, upwelling, and organic-rich rocks in the Mesozoic and Cenozoic Eras: Palaeogeography, Palaeoclimatology, Palaeoecology, v. 40, p. 31–66.

Parrish, J. T., Ziegler, A. M., and Scotese, C. R., 1982, Rainfall patterns and the distribution of coals and evaporites in the Mesozoic and Cenozoic: Palaeogeography, Palaeoclimatology, Palaeoecology, v. 40, p. 67–101.

Parrish, J. T., Gaynor, G. C., and Swift, D.J.P., 1984, Circulation in the Cretaceous Western Interior seaway of North America; A review, in Stott, D. F., and Glass, D. J., eds., The Mesozoic of middle North America: Canadian Society of Petroleum Geologists Memoir 9, p. 221–231.

Peterson, J. W., 1984, Evolution of a coal-bearing fluvial sequence; Polecat Bench Formation [M.S. thesis]: Chicago, University of Illinois, 65 p.

Pierce, F. W., Kent, B. H., and Grundy, W. D., 1982, Geostatistical analysis of a 113-billion-ton coal deposit, central part of the Powder River basin, northeastern Wyoming, in Gurgel, K. D., ed., Proceedings, 5th Symposium on the Geology of Rocky Mountain Coal: Utah Geological and Mineral Survey Bulletin 118, p. 262–272.

Pillmore, C. L., and Maberry, J. O., 1976, The depositional environment and trace fossils of the Trinidad Sandstone, southern Raton basin, New Mexico: New Mexico Geological Society 27th Annual Field Conference Guidebook, p. 191–197.

Pitman, W. C., III, 1978, Relationship between eustacy and stratigraphic sequences of passive margins: Geological Society of America Bulletin, v. 89, p. 1389–1403.

Rich, F. J., 1980, Brief survey of chemical and petrographic characteristics of Powder River basin coals, *in* Glass, G. B., ed., Guidebook to the coal geology of the Powder River basin, Wyoming: Geological Survey of Wyoming Public Information Circular 14, p. 133–158.

Roehler, H. W., 1977, Lagoonal origin of coals in the Almond Formation in the Rock Springs uplift, Wyoming, *in* Murray, D. K., ed., Proceedings 1976 Geology of Rocky Mountain Coal: Colorado Geological Survey Resources Series 1, p. 85–89.

——, 1986, Paleoenvironments and sedimentation, *in* Roehler, H. W., and Martin, P. L., eds., Geological investigations of the Vermillion Creek coal bed in the Eocene Niland Tongue of the Wasatch Formation, Sweetwater County, Wyoming: U.S. Geological Survey Professional Paper 1314C, p. 27–45.

Ryder, R. T., Fouch, T. D., and Elison, J. H., 1976, Early Tertiary sedimentation in the western Uinta basin, Utah: Geological Society of America Bulletin, v. 87, p. 496–512.

Ryer, T. A., 1981, Deltaic coals from Ferron Sandstone Member of Mancos Shale; Predictive model for Cretaceous coal-bearing strata of Western Interior: American Association of Petroleum Geologists Bulletin, v. 65, p. 2323–2340.

——, 1984, Transgressive-regressive cycles and the occurrence of coal in some Upper Cretaceous strata of Utah, U.S.A., *in* Rahmani, R. A., and Flores, R. M., eds., Sedimentology of coal and coal-bearing sequences: International Association of Sedimentologists Special Publication 7, p. 217–227.

Satchel, L. S., 1985, Climate and depositional environment of *Glyptostrobus* forest swamps that formed thick low-ash coals in the Paleocene Powder River basin [abs.]: Society of Organic Petrology 2nd Annual Meeting, p. 11.

Sears, J. D., Hunt, C. B., and Hendricks, T. A., 1941, Transgressive and regressive Cretaceous deposits in southern San Juan basin, New Mexico: U.S. Geological Survey Professional Paper 193-F, p. 110–121.

Shanley, K. W., and McCabe, P. J., 1989, Sequence-stratigraphic relationships and facies architecture of Turonian-Campanian strata, Kaiparowits Plateau, south-central Utah: American Association of Petroleum Geologists Bulletin, v. 73, p. 410–411.

Siemers, C. T., 1978, Generation of a simplified working depositional model for repetitive coal-bearing sequences using field data; An example from the Upper Cretaceous Menefee Formation (Mesaverde Group), northwestern New Mexico, *in* Hodgson, H. E., ed., Proceedings, 1977 Geology of Rocky Mountain Coal: Colorado Geological Survey Resource Series 4, p. 1–22.

Sklenar, S. E., and Anderson, D. W., 1985, Origin and early evolution of an Eocene lake system within the Washakie basin of southwestern Wyoming, *in* Flores, R. M., and Kaplan, S. S., eds., Cenozoic paleogeography of the west-central United States: Rocky Mountain Section, Society of Economic Paleontologists and Mineralogists Symposium 3, p. 231–245.

Smith, A. D., 1986, Utah coal core methane desorption project, final report: Utah Geological and Mineral Survey Open-File Report 88, 59 p.

Stanley, K. O., and Collinson, J. W., 1979, Depositional history of Paleocene-lower Eocene Flagstaff Limestone and coeval rocks, central Utah: American Association of Petroleum Geologists Bulletin, v. 63, p. 311–323.

Stanton, R. W., Minkin, J. A., and Moore, T. A., 1986, Petrographic and physical properties of coal and rock samples, *in* Roehler, H. W., and Martin, P. L., eds., Geological investigations of the Vermillion Creek coal bed in the Eocene Niland Tongue of the Wasatch Formation, Sweetwater County, Wyoming: U.S. Geological Survey Professional Paper 1314F, p. 105–120.

Vail, P. R., Mitchum, R. M., and Thompson, S., 1977, Global cycles of relative changes of sea level, *in* Payton, C. E., ed., Seismic stratigraphy—Applications to hydrocarbon exploration: Tulsa, American Association of Petroleum Geologists, p. 83–98.

Van Wagoner, J. C., 1985, Reservoir facies distribution as controlled by sea-level change, *in* Mid-year Meeting Abstract and Poster Session, Golden Colorado: Society of Economic Paleontologists and Mineralogists, p. 91–92.

Van Wagoner, J. C., and 6 others, 1988, An overview of the fundamentals of sequence stratigraphy and key definitions, *in* Wilgus, C. K., and others, eds., Sea-level changes; An integrated approach: Society of Economic Paleontologists and Mineralogists Special Publication 42, p. 39–45.

Warwick, P. D., 1985, Depositional environments and petrology of the Felix coal interval (Eocene), Powder River basin, Wyoming [Ph.D. thesis]: Lexington, University of Kentucky, 333 p.

Warwick, P. D., and Stanton, R. W., 1988, Petrographic characteristics of the Wyodak-Anderson coal bed (Paleocene), Powder River Basin, Wyoming, U.S.A.: Organic Geochemistry, v. 12, p. 389–399.

Weimer, R. J., 1960, Upper Cretaceous stratigraphy, Rocky Mountain area: American Association of Petroleum Geologists Bulletin, v. 44, p. 1–20.

——, 1977, Stratigraphy and tectonics of western coals, *in* Murray, D. K., ed., Proceedings, 1976, Geology of Rocky Mountain Coal: Colorado Geological Survey Resources Series 1, p. 9–27.

——, 1984, Relation of unconformities, tectonics, and sea-level changes, Cretaceous of Western Interior, U.S.A., *in* Schlee, J. S., ed., Interregional unconformities and hydrocarbon accumulation: American Association of Petroleum Geologists Memoir 36, p. 7–35.

Williams, G. D., and Stelck, C. R., 1975, Speculations on the Cretaceous paleogeography of North America, *in* Caldwell, W.G.E., ed., The Cretaceous system of the Western Interior of North America: Geological Association of Canada Special Paper 13, p. 1–20.

Wing, S. L., 1981, A study of paleoecology and paleobotany in the Willwood Formation (early Eocene, Wyoming) [Ph.D. thesis]: New Haven, Connecticut, Yale University, 391 p.

Wing, S. L., and Bown, T. M., 1985, Fine scale reconstruction of late Paleocene-early Eocene paleogeography in the Bighorn Basin of northern Wyoming, *in* Flores, R. M., and Kaplan, S. S., eds., Cenozoic paleogeography of the west-central United States: Rocky Mountain Section, Society of Economic Paleontologists and Mineralogists Symposium 3, p. 93–105.

Wolfe, J. A., 1985, Distribution of major vegetational types during the Tertiary, *in* Carbon cycle and atmospheric CO^2; Natural variations Archean to Present: American Geophysical Union Geophysical Monograph 32, p. 357–375.

Yuretich, R. F., Hickey, L. J., Gregson, B. P., and Haia, Y. L., 1984, Lacustrine deposits in the Paleocene Fort Union Formation, northern Bighorn basin, Montana: Journal of Sedimentary Petrology, v. 54, p. 836–852.

Ziegler, A. M., and 5 others, 1985, Paleogeographic interpretation, with an example from the Mid-Cretaceous: Annual Review of Earth and Planetary Sciences, v. 13, p. 385–425.

Manuscript Accepted by the Society July 26, 1990

ACKNOWLEDGMENTS

We thank Ed Belt, Hal Gluskoter, and Tom Ryer for their extremely helpful critiques of this manuscript.

Printed in U.S.A.

Chapter 35

Tertiary coals of the Gulf Coast

John A. Breyer
Department of Geology, Texas Christian University, Fort Worth, Texas 76129

INTRODUCTION

The Paleocene and Eocene Series of the Gulf Coast contain lignite reserves equivalent to almost 50 billion barrels of crude oil. Preliminary estimates by Phillips Coal Company place lignite reserves at 22.5 billion short tons (Luppens, 1982). This figure includes lignite in seams at least 0.9 m thick and at depths of less than 61 m, the general limit of present surface mining practice in the Gulf Coast. Substantial lignite deposits also occur at greater depth. Texas alone has 34.8 billion short tons of lignite in seams at least 1.5 m thick at depths between 61 and 610 m (Kaiser and others, 1980). Rising energy costs in the 1970s made in situ gasification of this lignite seem viable, but declining energy costs in the 1980s have removed the impetus for developing deep lignite resources. Power plants tied to nearby surface mining facilities seem likely to remain the only way in which to utilize the energy in Gulf Coast lignite on a commercial scale. The world energy situation precludes deep lignite resources from becoming reserves at any time in the immediate future.

Texas has the largest reserves in the Gulf Coast, 11.5 billion short tons of lignite, followed by Mississippi, 5.0; Arkansas, 2.5; Alabama, 1.4; Louisiana, 1.1; and Tennessee, 1.0 (Luppens, 1982). Texas is also the leading producer, producing 39.8 million short tons of lignite in 1984 to rank sixth in the nation in coal production (Friedman and others, 1985). Almost all of this lignite fuels mine-mouth power plants, which now generate nearly 20 percent of the state's electricity. Louisiana is the only other active producer. The Dolet Hills mine on the Sabine uplift began operation in late 1985 and will supply lignite to a mine-mouth power plant (Friedman and others, 1986).

DISTRIBUTION OF LIGNITE DEPOSITS

Lignite occurs in the Paleocene and Eocene Series of the Gulf Coast in the Midwayan, Sabinian, Claibornian, and Jacksonian stages (Fig. 1). Most of the lignite in Texas and almost all of the lignite in Alabama, Mississippi, and Louisiana with potential for economic development occurs in the Wilcox Group. The principal lignite reserves in Arkansas and Tennessee also occur in the Wilcox Group. The outcrop of the Wilcox Group extends from the international border with Mexico across Texas, up the Mississippi Embayment, and down into Mississippi, Alabama, and Georgia (Fig. 2). The Wilcox thickens from its erosional edge near the updip limit of the Tertiary to over 3,000 m downdip. The Sabine uplift brings Wilcox strata to the surface on a large inlier in east Texas and northwestern Louisiana. The principal lignite reserves in Alabama and Louisiana are in Wilcox strata of Midwayan age. Most of the lignite reserves in Texas are in Wilcox strata of Sabinian age.

Midwayan Stage

The Midwayan Stage includes strata equivalent to the Clayton, Porters Creek, and Naheola Formations in the type area in Alabama (Murray, 1961). The Naheola Formation contains Alabama's principal lignite deposit, the Oak Hill lignite. The Naheola is of Wilcox lithology, consisting of lignitic sand and shale, and is less than 40 m thick. To the west, in the subsurface of Mississippi and Louisiana and in the vicinity of the Sabine uplift, the Midwayan includes a greater thickness of strata of Wilcox lithology (Murray, 1955, p. 684). The Hall Summit, Logansport, and Naborton Formations may be as much as 300 m thick (Fisher, 1961). The Chemard Lake lignite, Louisiana's principal lignite deposit, occurs at the top of the Naborton. Farther west, in Texas, the top of the Midwayan occurs near the transition from calcareous and argillaceous strata of typical Midway lithology to arenaceous and argillaceous strata of typical Wilcox lithology (Murray, 1961, p. 372). Most of the Wilcox section in Texas is of Sabinian age. The Dolet Hills mine in Louisiana, which produces from the Chemard Lake lignite on the Sabine uplift, is the only active mine in strata of the Midwayan Stage.

Sabinian Stage

The Sabinian Stage includes the main body of the Wilcox Group. In Alabama and eastern Mississippi the Wilcox consists of the Nanafalia, Tuscahoma, and Hatchetigbee Formations, all of Sabinian age (Toulmin, 1969). This section thickens into Mis-

Breyer, J. A., 1991, Tertiary coals of the Gulf Coast, *in* Gluskoter, H. J., Rice, D. D., and Taylor, R. B., eds., Economic Geology, U.S.: Boulder, Colorado, Geological Society of America, The Geology of North America, v. P-2.

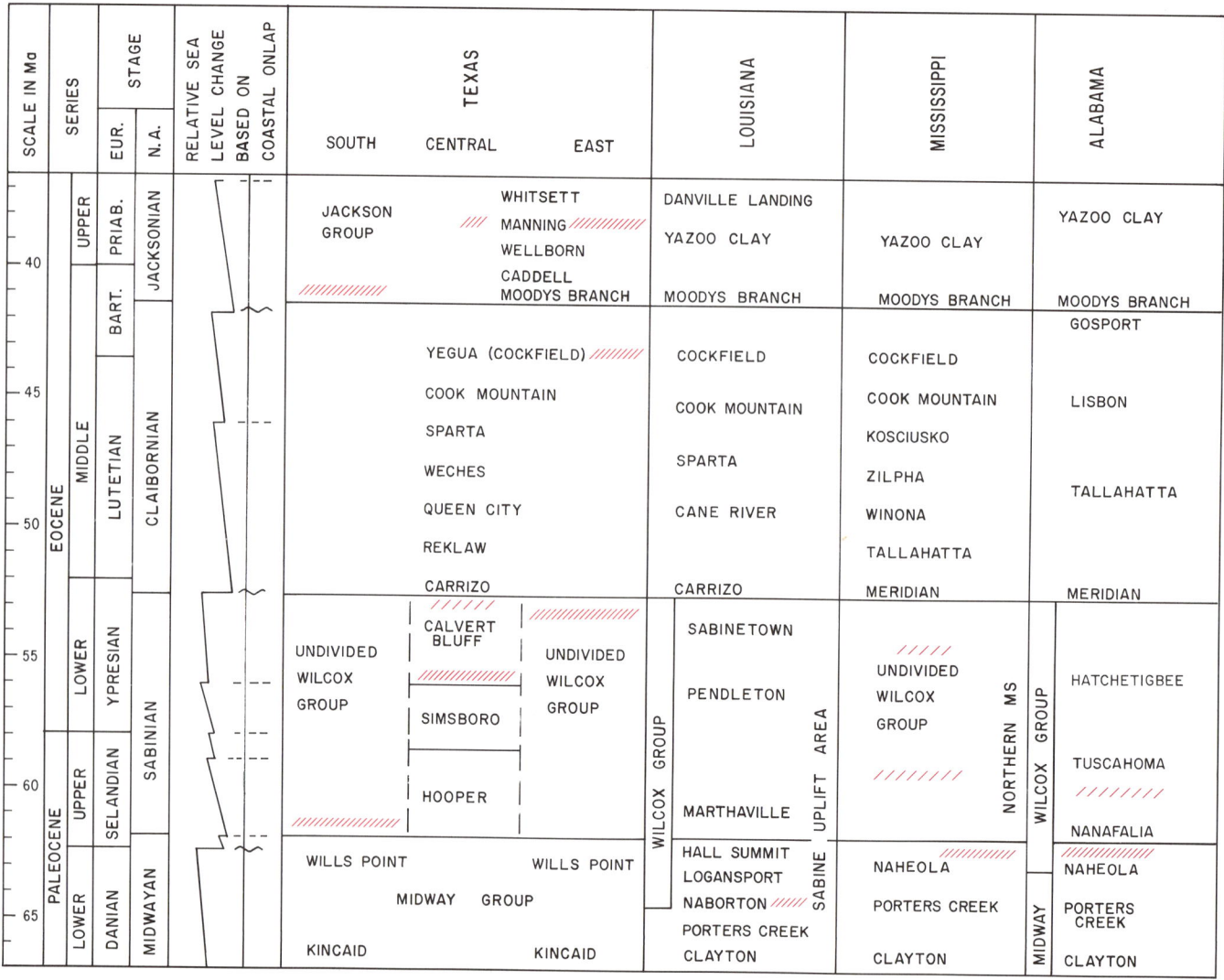

Figure 1. Stratigraphic distribution of lignite in the Paleocene and Eocene series of the Gulf Coast. Principal concentrations of lignite shown by closely-spaced diagonal lines (/////). Secondary accumulations shown by widely-spaced diagonal lines (/ / /). Calibration of series and stage boundaries with geochronologic scale follows Berggren and others (1985). Correlation of Gulf Coast stages in part from Siesser and others (1985). Sea-level curve adapted from Vail and Hardenbol (1979). Type 1 and Type 2 unconformities indicated by wavy and dashed lines, respectively. Stratigraphy mainly after Murray (1961). No attempt has been made to correlate formation-rank units with specific cycles of coastal onlap. The stage boundaries probably correlate with the cycles of coastal onlap as shown. Gulf Coast stratigraphers recognize three transgressive-regressive cycles within the Claiborne Stage. The global coastal onlap curve shows only two eustatic cycles in this interval and may need to be modified.

sissippi and Louisiana and is usually not divided into formations except in the vicinity of the Sabine uplift. In central Texas, between the Colorado and Trinity Rivers, the Wilcox Group consists of the Hooper, Simsboro, and Calvert Bluff Formations. Beyond the Trinity River, in east Texas, and below the Colorado River, in south Texas, the Wilcox Group is usually not divided into formations.

Lignite occurs throughout the Wilcox Group in Alabama and Mississippi, including seams as much as 12 m thick that occupy solutional or erosional depressions in places where the Naheola is missing and the Nanafalia overlies limestones in the Clayton Formation (Self and Williamson, 1977). Such seams, however, are of limited areal extent. Of the lignites in Alabama in strata of Sabinian age, those in the Tuscahoma have the best potential for economic development (Self, 1978). Potential economic lignite deposits also occur in undifferentiated Wilcox strata

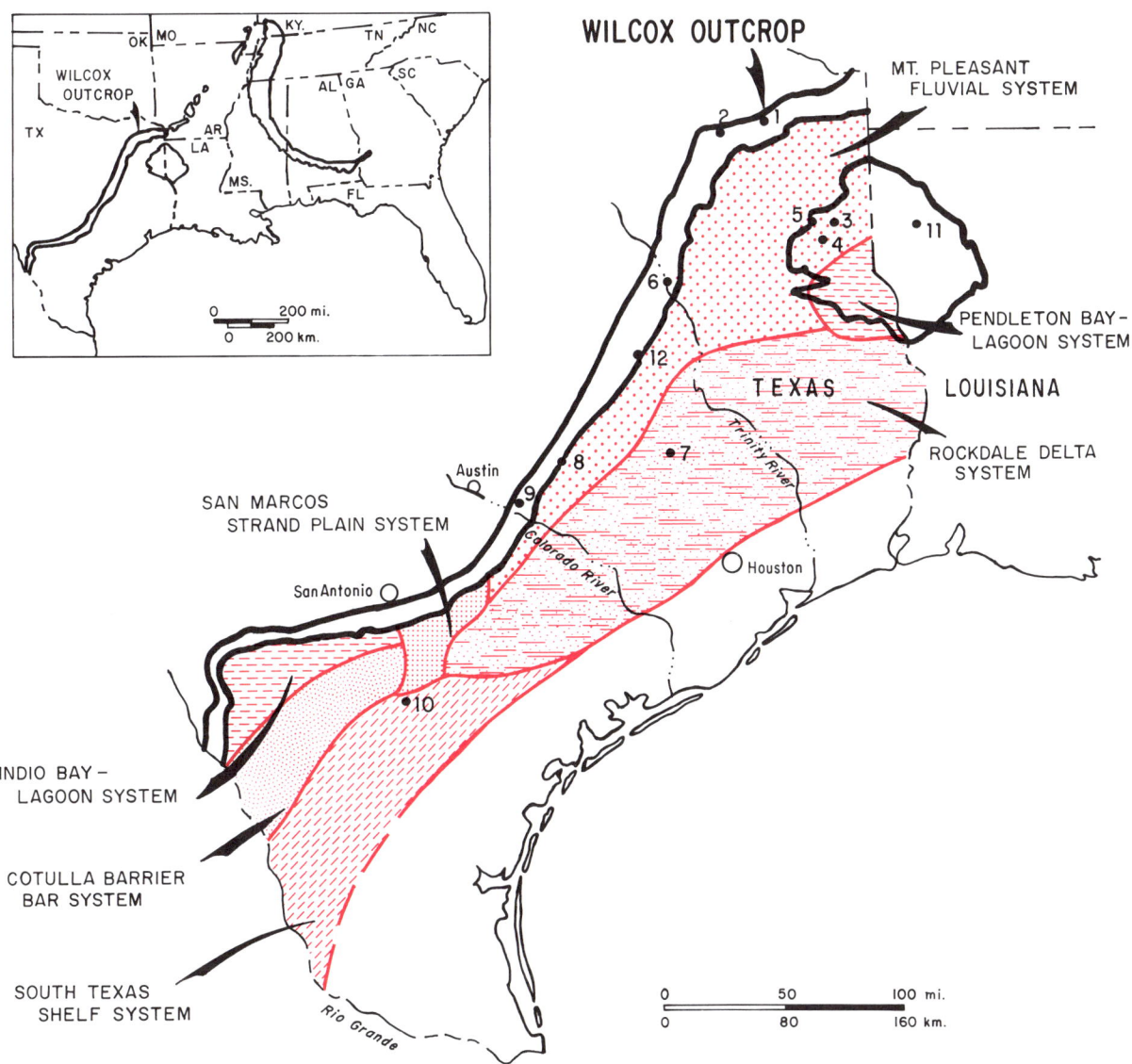

Figure 2. Map showing the position of the Wilcox outcrop and the location of active lignite mines: 1, Winfield mine; 2, Thermo mine; 3, Darco mine; 4, Martin Lake mine; 5, South Hallsville mine; 6, Big Brown mine; 7, Gibbons Creek mine; 8, Sandow mine; 9, Powell Bend mine; 10, San Miguel mine; 11, Dolet Hills mine; 12, Jewett mine. Depositional systems for the lower Wilcox after Fisher and McGowen (1967, their Fig. 1).

in Mississippi (Cleaves, 1980). Lignite is present in the Wilcox in the type area of the Sabinian Stage in Sabine Parish, Louisiana, but most of the state's shallow lignite is in Wilcox strata of Midwayan age on the Sabine uplift (Roland, 1978).

In Texas the Wilcox Group contains 16,570 million short tons of shallow lignite resources, 71 percent of the state's resources (Kaiser and others, 1980). Most of this lignite is of probable Sabinian age. In central Texas, commercial lignite deposits are most common in the lower part of the Calvert Bluff Formation but also occur near the top of the formation (Kaiser, 1978). This region contains 6,441 million short tons of lignite resources (Kaiser and others, 1980). East Texas has 3,912 million short tons of shallow lignite resources in the area of the Sabine uplift and 5,099 million short tons of resources along the Wilcox outcrop on the west side of the East Texas Embayment. This lignite is in the upper part of the Wilcox Group. Lignite resources in south Texas are smaller—1,078 million short tons—and occur in the lower part of the Wilcox Group (Kaiser and others, 1980).

Nine mines produce lignite from Wilcox strata of probable Sabinian age in Texas: the Winfield mine in Titus County, the Thermo mine in Hopkins County, the South Hallsville and Darco mines in Harrison County, the Martin Lake mine in Panola

County, the Big Brown mine in Freestone County, the Jewett mine in Leon County, the Sandow mine in Milam County, and the Powell Bend mine in Bastrop County (Fig. 2). All of these mines (except the Darco mine, which produces lignite for activated carbon, and the Powell Bend mine) supply lignite to mine-mouth power plants (Friedman and others, 1985). Lignite from the Powell Bend mine is shipped to a generating station in Fayette County for blending with western coal (Kaiser, 1985).

Claibornian Stage

The Claibornian Stage includes strata equivalent to exposures of the Claiborne Group in the type area in western Alabama. The Claiborne Group consists of the Tallahatta, Lisbon, and Gosport Formations. The Claibornian comprises the Meridian through Cockfield sequence in Mississippi and the Carrizo through Yegua sequence in Texas. Common practice places these sequences in the Claiborne Group also. Siesser (1984) questions this practice, emphasizing the distinction between lithostratigraphic and chronostratigraphic units.

Claibornian strata contain less lignite than do strata in the other Paleocene and Eocene stages of the Gulf Coast. The Lisbon and Gosport in Alabama contain a few thin, discontinuous lignite seams, but these have no potential for economic development (Self, 1978). Lignite occurs throughout the Claibornian in Mississippi and is common in the Cockfield (Williamson, 1978). However, the seams are thin and the ash content is high: the lignite probably has no potential for commercial utilization. The Yegua in east Texas contains 1,551 million short tons of lignite at depths of less than 61 m (Kaiser and others, 1980). This represents 7 percent of the state's shallow lignite resources and is the only significant occurrence of lignite in the Claibornian in Texas. Claibornian strata in Arkansas and Tennessee contain some lignite reserves (Luppens, 1982). At present there is no lignite production from strata of the Claibornian Stage.

Jacksonian Stage

The Jacksonian Stage includes strata deposited during a rise and fall of eustatic sea level in the late Eocene (Fig. 1). In the type area in Mississippi it consists of the Moodys Branch Formation and the Yazoo Clay. A major disconformity separates these shelf sediments from the nonmarine Cockfield Formation below. The Jacksonian sequence thickens to the west as the marine muds of the Yazoo are replaced in Texas by deltaic and paralic sediments (Murray, 1961; Fisher and others, 1970). In central Texas the Jacksonian consists of the Caddell, Wellborn, Manning, and Whitsett Formations. These formations and equivalent strata in south Texas are included in the Jackson Group, a practice that equates the group with the stage. A new group name would be appropriate here (see Siesser, 1984). Murray (1961) has suggested the name Fayette Group.

Only in Texas do Jacksonian strata contain significant lignite deposits; elsewhere the stage consists mainly of marine units.

In east Texas the Jacksonian Fayette delta system contains 4,499 million short tons of lignite at depths of less than 61 m, and in south Texas the Jacksonian contains 757 million short tons of lignite at depths of less than 61 m (Kaiser and others, 1980). These two concentrations of lignite hold 22 percent of the state's shallow lignite resources. Most of the Jacksonian lignite in east Texas occurs in delta plain sediments of the Manning Formation (Fisher and others, 1970). The Gibbons Creek mine in Grimes County produces lignite from these sediments. Jacksonian lignite in south Texas occurs in lagoonal and strandplain depositional systems in the lower part of the Jacksonian sequence. The Atascosa Mining Company produces 3.0 million short tons a year from the San Miguel lignite deposit in Atascosa and McMullen counties to fuel a mine-mouth power plant.

DEPOSITIONAL SYSTEMS AND LIGNITE DEPOSITS

Spectacular road cuts expose the Carboniferous of the Appalachian coalfields, and scenic panoramas lay open the coal-bearing Cretaceous strata of the Western Interior. Outcrops of Paleocene and Eocene strata in the Gulf Coast are more modest. The regional geology and the geology of the region's lignite deposits are known primarily from subsurface studies. Foremost among these is the study of the lower Wilcox in Texas by Fisher and McGowen (1967). It is the best-known example of the depositional systems approach to stratigraphic analysis. Fisher and McGowen (1967) relied mainly on data from electric logs to determine the depositional setting of the Wilcox Group. They used maps of net sand to identify seven depositional systems in the lower Wilcox in Texas (Fig. 2). Their work had immediate impact on the oil and gas industry, and similar studies came in quick succession: Galloway (1968), Fisher and others (1970), and Guevara and Garcia (1972). Kaiser (1974, 1978) led in using the depositional systems approach to locate and evaluate lignite resources.

Fisher and McGowen (1967) commented on the occurrence and distribution of lignite in the depositional systems composing the lower Wilcox in Texas. They ascribed regional variation in coal quality reported by Fisher (1963) to differences in depositional setting. McGowen (1968) and Fisher (1969a) characterized the lignite deposits of Wilcox deltaic, fluvial, and lagoonal depositional systems in more detail. Their work became the basis for models relating the thickness, extent, and quality of lignite deposits in the Paleocene and Eocene Series of the Gulf Coast to depositional setting.

Most workers consider deltaic depositional systems to be the major sites of accumulation of Gulf Coast lignite. Kaiser and others (1980), however, estimate that only 20 percent of the shallow lignite resources in Texas occur in deltaic depositional systems. They estimate that over 70 percent of the lignite resources in the state occur in fluvial depositional systems at the junction of the deltaic and alluvial plains. Here the distinction between "fluvial" and "deltaic" is difficult but may have impor-

TABLE 1. PROXIMATE ANALYSES OF TEXAS LIGNITE*
(as received basis)

Group: Area: Setting:	Wilcox East-central Fluvial	Wilcox Northeast Fluvial	Wilcox Sabine Uplift Fluvial	Jackson East Deltaic	Jackson South Lagoonal	Yegua East Fluvial
Moisture	32	33	33	41	23	37
Volatile matter	29	27	27	23	22	26
Fixed carbon	24	25	25	15	13	19
Ash	15	15	15	21	42	18
Sulfur	.96	.79	.94	1.29	—	.99
pyritic	.22	.29	.26	.40	—	.50
sulfate	.02	.01	.03	.02	—	.02
organic	.72	.49	.65	.87	—	.47
BTU/lb	6,593	6,499	6,441	4,729	3,972	5,761

*Data taken from Kaiser and others (1978, their Tables 8 and 9).

tant implications. Fielding (1985) lists differences in the size, shape, and thickness of coal seams in the two settings. The importance of deltas as sites for peat accumulation has probably been overstated (McCabe, 1984). Table 1 shows proximate analyses of lignites from some fluvial, deltaic, and strandplain/lagoonal depositional systems in Texas.

Lignite in Fluvial Depositional Systems

Lignites in fluvial depositional systems in the Paleocene and Eocene Series of the Gulf Coast represent backswamp peats that accumulated in hardwood swamps between alluvial ridges marking abandoned stream courses (Kaiser, 1978). The thickest and most continuous lignite seams are found at the junction of the upper delta plain and lower alluvial plain because swamps in interchannel basins there were larger and persisted longer than swamps higher on the alluvial plain or lower on the delta plain (Kaiser, 1978; Kaiser and others, 1978). The lignites alternate with coarsening-upward sequences produced by prograding crevasse splays or small deltas filling floodbasin lakes. The coarsening-upward sequences are usually about 15 m thick (Kaiser, 1978).

Most of the lignite in the Wilcox Group in Texas occurs in fluvial depositional systems (Kaiser and others, 1980). Lignite also occurs in fluvial depositional systems in the Yegua Formation in east Texas (Kaiser and others, 1980). Williamson (1978) associates sinuous, elongate lignite deposits in the Wilcox of Mississippi with fluvial depositional systems.

Lignite in Deltaic Depositional Systems

Fisher and McGowen (1967) identified two kinds of lignites in the lower Wilcox Rockdale delta system. They equated tabular lignites of limited areal extent with the interdistributary peats on the Mississippi delta described by Fisk (1960). They interpreted widespread tabular lignites as the landward equivalent of marine delta destructive units and thought these lignites formed from peats that accumulated on the upper part of the abandoned delta plain in marshes that were expanding as the sediments compacted. The Chemard Lake lignite, Louisiana's principal lignite seam, probably represents such a lignite (Johnston, 1982). Self (1978) suggests the widespread Oak Hill (Midwayan) lignite in Alabama formed as a blanket peat on a foundering delta lobe and interprets lenticular lignite seams in the Tuscahoma Formation (Wilcox) to have formed as interdistributary peats on an actively prograding delta.

Fisher (1968, 1969b) noted that the deposits of river-dominated deltas contained numerous peats or lignites and that the deposits of wave- or tide-dominated deltas did not contain significant peat or lignite. He cited the lower Wilcox Rockdale and Holly Springs delta systems, the Jackson Fayette delta system, and a delta system in the Yegua on the upper Gulf Coast of Texas as ancient examples of river-dominated deltas and gave the Mississippi delta as a modern example. He cited the upper Wilcox delta system in Texas east of the San Marcos Arch and the modern Apalachicola delta, Florida, as examples of wave-dominated deltas. Guevara and Garcia (1972) distinguished between the lignite-bearing deposits of river-dominated deltas in the Queen City Formation in east Texas and the deposits of wave-dominated deltas, which lack significant lignite, in the same formation in south Texas. Cleaves (1980) dismissed the deltaic facies of the Meridian Sandstone in south-central Mississippi as unlikely to contain significant lignite, because he interpreted them as the deposits of a wave-dominated delta.

The generalization that the deposits of wave- or tide-dominated deltas are not likely to contain significant coal deposits is not supported by work outside the Gulf Coast region. For example, much of the coal in the Cretaceous strata of the Western Interior is associated with wave- or tide-dominated deltaic shorelines (Flores, 1979; Ryer, 1981; among others). Kaiser and others (1980) interpret the delta system in the Yegua Formation in east Texas to be wave dominated, rather than river dominated as suggested by Fisher (1969b). The Yegua delta system contains significant lignite deposits but at depths of more than 610 m.

Kaiser (1978) and Kaiser and others (1978, 1980) identify the transition between the upper delta plain and the lower alluvial plain as the area most likely to contain commercial lignite deposits. They both cite the work of Frazier and Osanik (1969) and Frazier and others (1978) on the modern Mississippi delta to support their claim. They also give lignite in the lower part of the Calvert Bluff Formation along the Wilcox outcrop in central Texas and lignite in the upper part of the Wilcox Group on the Sabine uplift as ancient examples. Coates and others (1980) also cite Frazier and Osanik (1969), but following Horne and others (1978), they indicate that the best lignites occur at the transition between the upper delta plain and the lower delta plain and suggest prospecting at this position on the Holly Springs delta system in Louisiana and Mississippi.

Lignite in Lagoonal Depositional Systems

Lignite occurs in bay-lagoon depositional systems in the Wilcox Group in east Texas and south Texas (Fisher and McGowen, 1967) and in the Claiborne Group in Arkansas (Wielchowsky and others, 1977). Lignite in the Jackson Group in south Texas occurs in association with lagoonal or marsh sédiments capping barrier bar or strandplain sandstone (Kaiser and others, 1980). The San Miguel lignite is part of the Jacksonian south Texas lagoonal-coastal plain depositional system recognized by Fisher and others (1970). It formed from peat that accumulated in a coastal marsh adjacent to a lagoon behind a barrier island (Snedden and Kersey, 1981). The lignite is continuous along depositional strike. It is high in sulfur and ash with a low fixed carbon content and a low calorific value. Breyer and McCabe (1986) interpret lignite seams in the Wilcox near Uvalde in south Texas to represent peats that formed in swamps along the landward margin of a tidal channel complex some kilometers inland from the main coastline. They explain the coal-bearing sequence as the result of repeated regressions and transgressions with corresponding expansion and contraction of the peat swamps within an estuarine complex. Breyer (1987) interprets coarsening-upward sequences overlying lignite seams at the Big Brown mine, Freestone County, and Martin Lake mine, Panola County, in the Wilcox in east Texas to be estuarine deposits. The sequence at the Big Brown mine is interpreted as a floodplain deposit by Kaiser (1978) and Kaiser and others (1978, 1980), who assign the sediments at the Martin Lake mine a fluvial origin also.

LIGNITE EXPLORATION MODELS

Kaiser (1978) and Kaiser and others (1978) developed "exploration models" for locating areas in which to concentrate exploration drilling for shallow subsurface lignite (0–610 m). The models were not meant to be used to identify specific lignite deposits. The changing relationship between the area of maximum peat accumulation and the area of greatest sand deposition in different depositional settings forms the basis for the models (see below). The area of maximum peat accumulation is inferred from isopleths showing the number of lignite penetrations; lignite isopach maps are not made. The depositional setting is inferred from isolith patterns on maps of net sand or sand percent. The lignite isopleth and sand isolith maps are made using electric logs from oil and gas wells. Cleaves (1980) used these exploration models to identify areas in which to prospect for lignite in the Wilcox Group and Meridian Sandstone in northern Mississippi, and Coates and others (1980) used the same techniques to locate and evaluate lignite deposits in the Wilcox Group in west-central Louisiana.

The "exploration models" can be summarized as follows. High lignite counts overlap areas of high sand percent on the lower delta plain. The lignites represent blanket peats formed in marshes on foundering delta lobes. Jackson lignites on the Fayette delta system in southeast Texas provide an ancient example. High lignite counts coincide with areas of low sand percent at the junction of the alluvial and delta plains, where the thickest and most widespread lignites occur. Wilcox lignites in the Calvert Bluff Formation in central Texas and peats of the upper Des Allemands–Barataria and Atchafalaya interchannel basins provide ancient and modern examples, respectively. High lignite counts in the Calvert Bluff coincide with areas having less than 55 percent sand (Fig. 3). High lignite counts also overlap areas of low sand percent in fluvial depositional systems higher on paleoslope. Wilcox lignites in east Texas are an ancient example. Peats forming between stream courses in tropical Borneo are a possible modern example. Lignite is found in narrow concentrations updip of the axes of linear, strike-oriented sand bodies in strandplain and barrier bar depositional systems. Yegua–Jackson lignites in south Texas and thin marsh peats accumulating on the Nayarit strandplain, Mexico, provide ancient and modern examples, respectively.

DISCUSSION

A depositional system is the stratigraphic record of sedimentation in a particular depositional setting. The term "depositional systems approach" is used here to refer to a method of stratigraphic analysis in which isolith patterns on maps of net sand or sand percent for thick intervals of strata provide the principal basis for interpreting depositional setting. The depositional systems approach differs from subsurface facies analysis (see Cant, 1984) in resolution and focus. The former shows broad patterns of sedimentation and sand accumulation for thick intervals of strata.

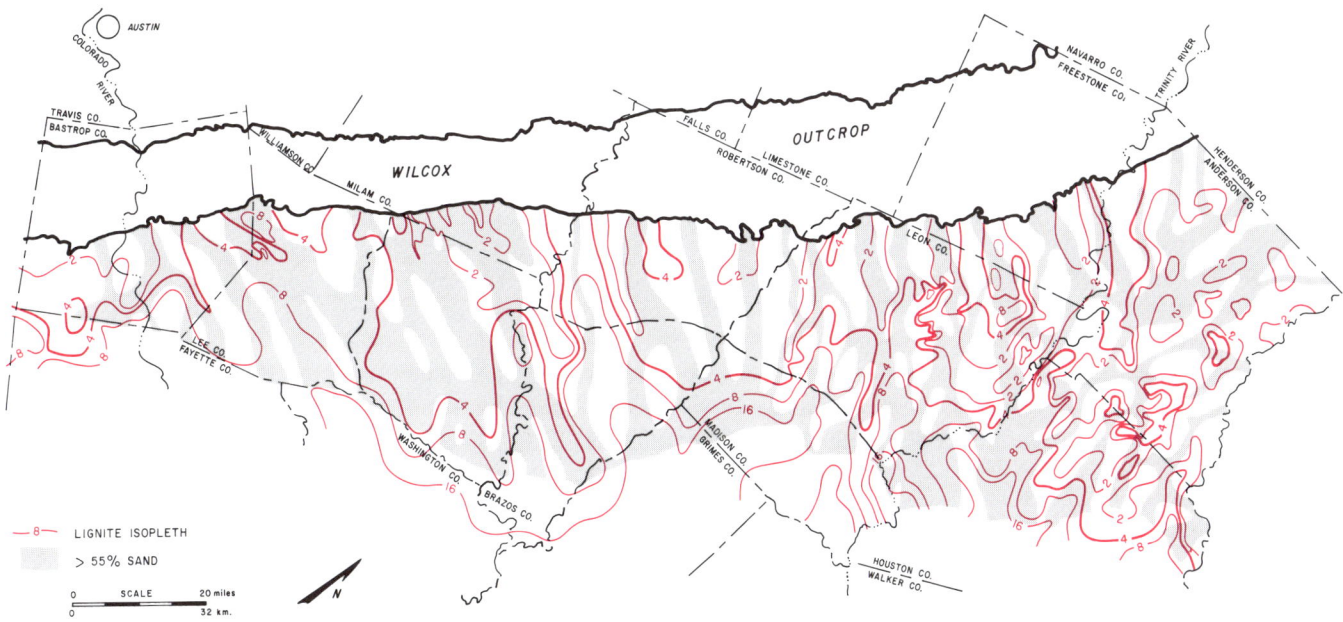

Figure 3. Areas of high sand percent and lignite isopleths for the Calvert Bluff Formation between the Colorado and Trinity Rivers, Texas. Areas of greater than 55 percent sand after Kaiser and others (1978, their Fig. 6); lignite isopleths from Kaiser and others (1978, their Fig. 7).

The latter has a higher resolution (providing information about the environment of deposition) but a narrower focus (dealing only with individual sands or thin intervals of strata). Isolith patterns on maps of net sand and sand percent for thick intervals of strata do not provide information that can be used to interpret environments of deposition except in special circumstances (see below).

Smith (1986) recognizes four depositional sequences in the lower and middle Wilcox in south Texas (Fig. 4). The composite section for the four sequences ranges in thickness from 400 m updip to 650 m downdip. Sand percent values for the composite section range from 20 to 60 percent. Isoliths on the sand percent map for the composite section have an ENE–WSW orientation and show only slight departure from this trend (Fig. 4). Sand percent maps of the individual sequences differ markedly from each other and from the map for the composite section (Smith, 1986). For example, the sand percent map for Sequence B shows a large, strike-oriented sand accumulation in western Karnes County and a smaller dip-oriented sand accumulation in southern Atascosa County and northern McMullen County (Fig. 4). Neither sand accumulation is evident on the sand percent map for the composite section. A large sand accumulation also occurs in western Karnes County in the overlying sequence, but the sand accumulation in Sequence A has a dip orientation (NW–SE) rather than a strike orientation (NE–SW) like the sand accumulation in Sequence B (Smith, 1986). The location and orientation of sand accumulations changes from one sequence to another. Most interpretations of depositional systems in the Paleocene and Eocene Series of the Gulf Coast involve intervals of strata closer in thickness to the composite section in this example than to the individual depositional sequences. The isolith patterns on maps for even these smaller intervals, however, show only broad patterns of sedimentation and sand accumulation, not sand body geometry (see below).

Figure 4 shows isopach maps for two sands in the middle part of the Wilcox Group in south Texas. Both sands are in Sequence B. The Wales Sand is the basal sand in the sequence. The isopach map for the Wales Sand shows distinct sand accumulations that thin and bifurcate toward the basin. The sand body geometry suggests a series of small deltas aligned along depositional strike. The Gasper Sand occurs in the middle part of Sequence B, some 75 m above the Wales Sand (Fig. 4). The isopach map for the Gasper Sand shows a linear sand body oriented along depositional strike. The sand passes into shale both updip and downdip. The sand probably formed in a barrier island setting, an interpretation supported by log curve shapes showing coarsening-upward sequences downdip and blocky sands updip. The small lobes of sand extending updip into shales behind the barrier might represent washover fans or tidal deltas. The environment of deposition of the sands in Sequence B cannot be inferred from the isolith patterns on the sand percent map for the sequence. Only if there had been no significant change in depositional setting within the interval would the isolith map for the sequence be useful in interpreting the environment of deposition of the sands.

Isolith maps showing the pattern of sand accumulation in a thick stratigraphic interval over a large area provide information

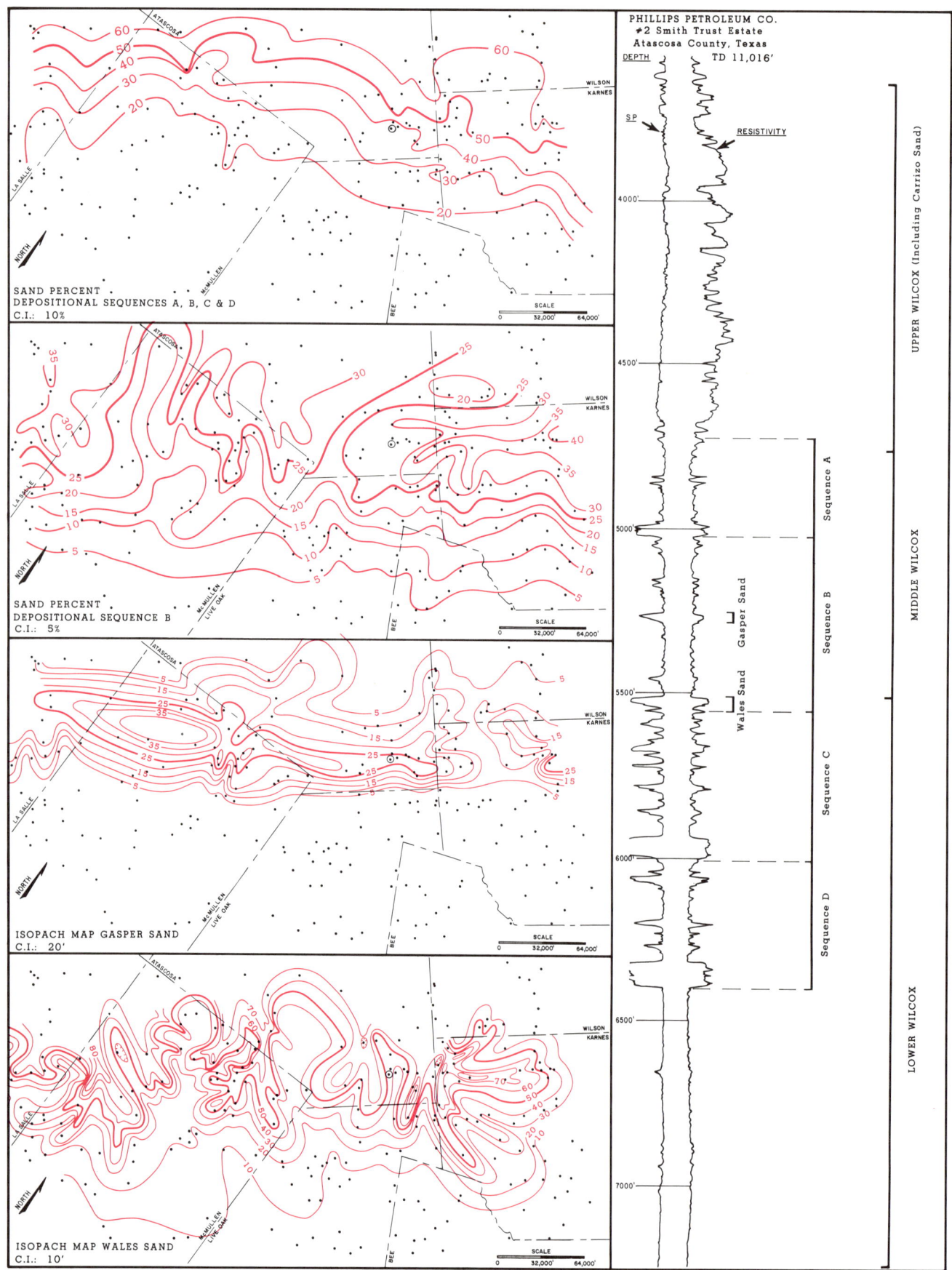

Figure 4. Sand percent and isopach maps for intervals of different thickness in the Wilcox Group in South Texas. Intervals included on each map are shown on the log for Phillips Petroleum Company, #2 Smith Trust Estate, Atascosa County. Black dots show well control for maps. Circled dot shows location of #2 Smith Trust Estate. Sand percent maps from Smith (1986). Isopach maps for Wales Sand and Gasper Sand courtesy of W. M. Smith.

about the regional geologic setting. The maps constrain interpretation but are not sufficient for interpretation at the level of resolution or with the degree of certainty attained using the facies analysis method. The maps show time-averaged patterns of sand accumulation; they do not show the geometry and orientation of individual sands. In most cases, the usefulness of the maps in facies studies will increase as the thickness of the interval for which they are being made decreases and approaches the thickness of individual sands (for example, see Tewalt and others, 1981).

McCabe (1984) cautions that a significant hiatus may separate the time of deposition of the overlying and underlying sediment. Only if the peat swamps had a geometry that excluded clastic sediment would coals and coal-bearing strata represent coexisting sedimentary environments. Determining the environment of deposition of coal-bearing sediments using facies analysis is not sufficient to determine the environment of deposition of the coal. Freshwater peats in the Okefenokee Swamp of Georgia are accumulating on Pleistocene beach ridges nearly 80 km inland from the present shore (McCabe, 1984). Similarly, determining the environment of deposition of a coal from a study of the palynomorphs or maceral types is not sufficient to determine the environment of deposition of the overlying and underlying clastic sediment.

Most coal petrographers ignore the rock in which the coal is found, and most sedimentologists dismiss the coal as "swamp" (McCabe, 1984). The coal industry needs high-resolution facies models linking clastic facies with coal facies. The Paleocene and Eocene Series of the Gulf Coast offer coal petrographers and sedimentologists the opportunity to formulate such models for lignites in a wide vareity of depositional settings.

REFERENCES

Berggren, W. A., Kent, D. V., Flynn, J. J., and Van Couvering, J. A., 1985, Cenozoic geochronology: Geological Society of America Bulletin, v. 96, p. 1407–1418.

Breyer, J. A., 1987, A tidal origin for coarsening-upward sequences above two Wilcox lignites in East Texas: Journal of the Geological Society of London, v. 144, p. 463–469.

Breyer, J. A., and McCabe, P. J., 1986, Coals associated with tidal sediments in the Wilcox Group (Paleogene), South Texas: Journal of Sedimentary Petrology, v. 56, p. 510–519.

Cant, D. J., 1984, Subsurface facies analysis, in Walker, R. G., ed., Facies models (second edition): Geoscience Canada Reprint Series 1, p. 297–310.

Cleaves, A. W., 1980, Depositional systems and lignite prospecting models; Wilcox Group and Meridian Sandstone of northern Mississippi: Gulf Coast Association of Geological Societies Transactions, v. 30, p. 283–307.

Coates, E. J., Groat, C. G., and Hart, G. F., 1980, Subsurface Wilcox lignite in west-central Louisiana: Gulf Coast Association of Geological Societies Transactions, v. 30, p. 309–332.

Fielding, C. F., 1985, Coal depositional models and the distinction between alluvial and delta plain environments: Sedimentary Geology, v. 42, p. 41–48.

Fisher, W. L., 1961, Stratigraphic names in the Midway and Wilcox Groups of the Gulf Coastal Plain: Gulf Coast Association of Geological Societies Transactions, v. 11, p. 263–295.

—— , 1963, Lignites of the Texas Gulf Coastal Plain: University of Texas at Austin, Bureau of Economic Geology Report of Investigations 50, 164 p.

—— , 1968, Basic delta systems in the Eocene of the Gulf Coast Basin: Gulf Coast Association of Geological Societies Transactions, v. 18, p. 48.

—— , 1969a, Variations in lignites of fluvial, deltaic, and lagoonal systems, Wilcox Group (Eocene), Texas: Geological Society of America Special Paper 121, p. 97.

—— , 1969b, Facies characterization of Gulf Coast Basin delta systems, with some Holocene analogues: Gulf Coast Association of Geological Societies Transactions, v. 19, p. 239–261.

Fisher, W. L., and McGowen, J. H., 1967, Depositional systems in the Wilcox Group of Texas and their relationship to occurrence of oil and gas: Gulf Coast Association of Geological Societies Transactions, v. 17, p. 105–125.

Fisher, W. L., Galloway, W. E., Proctor, C. V., and Nagle, J. S., 1970, Depositional systems in the Jackson Group of Texas; Their relationship to oil, gas, and uranium: Gulf Coast Association of Geological Societies Transactions, v. 20, p. 234–261.

Fish, H. N., 1960, Recent Mississippi River sedimentation and peat accumulation: Compte rendu, 4 Congres l'avancement des etudes de stratgraphie et de geologie du Carbonifere, Heerlen 1958, v. 1, p. 187–199.

Flores, R. M., 1979, Coal depositional models in some Tertiary and Cretaceous coal fields in the U.S. Western Interior: Organic Geochemistry, v. 1, p. 225–235.

Frazier, D. E., and Osanik, A., 1969, Recent peat deposits, Louisiana coastal plain, in Dapples, E. C., and Hopkins, M. E., eds., Environments of coal deposition: Geological Society of America Special Paper 114, p. 63–85.

Frazier, D. E., Osanik, A., and Elsik, W. C., 1978, Environments of peat accumulation, coastal Louisiana, in Kaiser, W. R., ed., Proceedings of the 1976 Gulf Coast Lignite Conference: University of Texas at Austin, Bureau of Economic Geology Report of Investigations 90, p. 5–20.

Friedman, S. A., Jones, R. W., and Jackson, M. L., 1985, Developments in coal in 1984: American Association of Petroleum Geologists Bulletin, v. 69, p. 1898–1902.

Friedman, S. A., Jones, R. W., Jackson, M.L.W., and Treworgy, C. G., 1986, Developments in coal in 1985: American Association of Petroleum Geologists Bulletin, v. 70, p. 1643–1649.

Galloway, W. E., 1968, Depositional systems of the lower Wilcox Group, north-central Gulf Coast basin: Gulf Coast Association of Geological Societies Transactions, v. 18, p. 275–289.

Guevara, E. H., and Garcia, R., 1972, Depositional systems and oil-gas reservoirs in the Queen City Formation (Eocene), Texas: Gulf Coast Association of Geological Societies Transactions, v. 22, p. 1–22.

Horne, J. C., Ferm, J. C., Caruccio, F. T., and Baganz, B. P., 1978, Depositional models in coal exploration and mine planning in Appalachian region: American Association of Petroleum Geologists Bulletin, v. 62, p. 2379–2411.

Johnston, J. E., 1982, Geologic aspects of Louisiana lignite, in Johnston, J. E., ed., Lignite of the Gulf Coastal Plain, Louisiana and Texas: Field Trip Guidebook published in conjunction with the 1982 meeting of the Geological Society of America, p. 42–64.

Kaiser, W. L., 1974, Texas lignite—near-surface and deep-basin resources: University of Texas at Austin, Bureau of Economic Geology Report of Investigations 79, 70 p.

——, 1978, Depositional systems in the Wilcox Group (Eocene) of east-central Texas and the occurrence of lignite, *in* Kaiser, W. R., ed., Proceedings of the 1976 Gulf Coast Lignite Conference: University of Texas at Austin, Bureau of Economic Geology Report of Investigations 90, p. 33–53.

Kaiser, W. R., 1985, Texas lignite; Status and outlook to 2000: University of Texas at Austin, Bureau of Economic Geology Mineral Resource Circular 76, 17 p.

Kaiser, W. R., Johnston, J. E., and Bach, W. N., 1978, Sand-body geometry and the occurrence of lignite in the Eocene of Texas: University of Texas at Austin, Bureau of Economic Geology Geological Circular 78-4, 19 p.

Kaiser, W. R., Ayers, W. B., and La Brie, L. W., 1980, Lignite resources in Texas: University of Texas at Austin, Bureau of Economic Geology Report of Investigations 104, 52 p.

Luppens, J. A., 1982, Exploration for Gulf Coast United States lignite deposits; Their distribution, quality, and reserves, *in* Johnston, J. E., ed., Lignite of the Gulf Coastal Plain, Louisiana and Texas: Fieldtrip guidebook published in conjunction with the 1982 meeting of the Geological Society of America in New Orleans, Louisiana, p. 3–21.

McCabe, P. J., 1984, Depositional environments of coal and coal-bearing strata, *in* Rahmani, R. A., and Flores, R. M., eds., Depositional models for coal and associated strata: International Association of Sedimentologists Special Publication no. 9, p. 1–30.

McGowen, J. H., 1968, Utilization of depositional models in exploration for nonmetallic minerals, *in* Brown, L. F., Jr., ed., Proceedings, Fourth Forum on Geology of Industrial Minerals: University of Texas at Austin, Bureau of Economic Geology Special Publication, p. 157–174.

Murray, G. E., 1955, Midway Stage, Sabine Stage, and Wilcox Group: American Association of Petroleum Geologists Bulletin, v. 39, p. 671–696.

——, 1961, Geology of the Atlantic and Gulf Coastal Province of North America: New York, Harper and Brothers, 692 p.

Roland, H. L., 1978, Louisiana lignite, *in* Kaiser, W. R., ed., Proceedings of the 1976 Gulf Coast Lignite Conference: University of Texas at Austin, Bureau of Economic Geology Report of Investigations 90, p. 66–78.

Ryer, T. A., 1981, Deltaic coals of Ferron Sandstone Member of Mancos Shale; Predictive model for Cretaceous coal-bearing strata of Western Interior: American Association of Petroleum Geologists Bulletin, v. 65, p. 2323–2340.

Self, D. M., 1978, Lignite resources of the Alabama-Tombigbee Rivers region, Alabama, *in* Kaiser, W. R., ed., Proceedings of the 1976 Gulf Coast Lignite Conference: University of Texas at Austin, Bureau of Economic Geology Report of Investigations 90, p. 59–65.

Self, D. M., and Williamson, D. R., 1977, Occurrence and characteristics of Midway and Wilcox lignites in Mississippi and Alabama, *in* Campbell, M. D., ed., Geology of alternate energy resources in the south-central United States: Houston Geological Society, p. 161–177.

Siesser, W. G., 1984, Gulf Coast Paleogene strata; Groups or stages?: Journal of Geology, v. 92, p. 439–446.

Siesser, W. G., Fitzgerald, B. G., and Kronman, D. J., 1985, Correlation of Gulf Coast provincial Paleogene stages and European standard stages: Geological Society of America Bulletin, v. 96, p. 827–831.

Smith, W. M., 1986, Exploration in a mature petroleum province; The updip Wilcox trend in South Texas [M.S. thesis]: Fort Worth, Texas Christian University, 49 p.

Snedden, J. W., and Kersey, D. G., 1981, Origin of the San Miguel lignite deposit and associated lithofacies, Jackson Group, South Texas: American Association of Petroleum Geologists Bulletin, v. 65, p. 1099–1109.

Tewalt, S. J., Bauer, M. A., and Mathew, D., 1981, Detailed evaluation of two Texas lignite deposits of deltaic and fluvial origins: Gulf Coast Association of Geological Societies Transactions, v. 31, p. 201–212.

Toulmin, L. D., 1969, Paleocene and Eocene guide fossils of the eastern Gulf Coast Region: Gulf Coast Association of Geological Societies Transactions, v. 19, p. 465–487.

Vail, P. R., and Hardenbol, J., 1979, Sea-level changes during the Tertiary: Oceanus, v. 22, p. 71–79.

Wielchowsky, C. C., Collins, G. F., Gerahain, L. K., and Calhoun, E. J., 1977, Frontier lignite exploration in the south-central United States, *in* Campbell, M. D., ed., Geology of alternative energy resources in the south-central United States: Houston Geological Society, p. 125–159.

Williamson, D. R., 1978, The Tertiary lignites of Mississippi, *in* Kaiser, W. R., ed., Proceedings of the 1976 Gulf Coast Lignite Conference: University of Texas at Austin, Bureau of Economic Geology Report of Investigations 90, p. 54–58.

MANUSCRIPT ACCEPTED BY THE SOCIETY MAY 18, 1987

Chapter 36

Coals of far-western United States

Aureal T. Cross
Department of Geological Sciences, Michigan State University, East Lansing, Michigan 48824

GENERAL DISTRIBUTION AND CHARACTERISTICS OF THE COALFIELDS AND COALBEDS

More than 100 coalfields or clusters of small coalfields, many in complexly faulted and folded structural basins, have been identified in the area west of the Rocky Mountain region in United States (Wood and Bour, 1988; Plate 9, and Fig. 1), an area of more than 1,500,000 km^2. Most of the fields are very small, but a few are more than 50 km^2. The distribution and generalized areas of most of these coalfields or grouped coalfields are shown in Figure 1, which has been slightly modified from the "Coal Map of North America" (Wood and Bour, 1988). One, the Centralia-Chehalis field in Washington (No. 384), occupies about 500 km^2. The Goose Creek basin (Nos. 135 and 230) in Idaho and Nevada, respectively, the Coos Bay coalfield (No. 287) in Oregon, and the Rogue River (Nos. 289 and 81) in Oregon and California, respectively, are examples of coalfields that occupy only portions of their respective structural basins, which are from 650 km^2 to more than 1,000 km^2.

Many of these coalfields have more than one minable bed. In one area in the Wilkinson-Carbonado coalfield (No. 381), on the west flank of the Cascade Mountains in Washington, in an extensively faulted area of synclines and anticlines, 22 stratigraphically sequential coal beds have been plotted over a distance of about 5 km (Hume and others, 1977). In the Green River district, 40 named coals are more than 1 m thick; five are more than 5 m.

About 20 of the coalfields in the western United States are estimated to contain more than 50 million short tons each of coal-in-place (Table 1; Wood and Bour, 1988). Several such fields are in Washington, three each are in western Montana and Oregon, two each in California, Idaho, and southwestern Utah, and one in southern Arizona. Nevada and the southwestern corner of New Mexico, as well as all seven states mentioned above, have several smaller fields each. The coal-in-place estimates for all the fields in the area west and south of the "Coherent Cratonic Terrane" of western United States (Fig. 1) total more than 50 billion short tons (Wood and Bour, 1988). These coals range in rank from lignite through subbituminous and bituminous to anthracite. Much of the resource is subbituminous to high-volatile B bituminous. Coking coals are found in several fields, especially in Washington (Beikman and others, 1984).

FACTORS CONTROLLING DEPOSITIONAL SITES OF WESTERN COALS

Tectonics, including plate movements, terrane accretion, and subduction, and associated volcanic activity, have exerted primary controls on the formation of basins in which coal-forming peat swamps and lakes developed in the western United States west of the margin of the coherent cratonic terrane (Fig. 1). There were no major seaways associated with the formation of most of the widely scattered coals west of the Rocky Mountains. This is in marked contrast to the formation of the more important Cretaceous coals on the coastal lowlands and deltas bordering the great interior seaway of North America during the Cretaceous Period.

Peat-forming swamp conditions in this western region developed in local wetlands and lakes in Tertiary alluvial basins that were rapidly filled with clastics derived from erosion of strike-slip translated borderlands, horst blocks, batholiths, or volcanic extrusives. Source headlands for some of these basins are very difficult to determine in some places because of uncertain time and magnitude of offsets along strike-slip and transcurrent faults bordering the basins or valleys, and difficulty in determining timing of lakes or alluvial plain ponds formed when drainage was blocked or choked by volcanic extrusives.

Coal-forming peats also accumulated in small areas of basin-and-range-type extensional tectonic basins both in the Great Basin and the structurally disturbed foreland west of the Rocky Mountains in Montana and Idaho. Some coals, especially in western Washington and Oregon, did form in coastal-margin deltas, lagoons, and lower alluvial-plain lowlands, but most of these were strongly controlled by tectonic movements.

The tectonic basins in which many of these coalbeds accumulated, as well as other types of coal-forming lowlands and the environments of peat formation in these basins will be discussed, with examples drawn from various areas in western United States.

TECTONIC SETTING OF THE COAL BASINS

The area of coal deposits discussed here includes the conterminous United States west of the western margin of the coher-

Cross, A. T., 1991, Coals of far-western United States, *in* Gluskoter, H. J., Rice, D. D., and Taylor, R. B., eds., Economic Geology, U.S.: Boulder, Colorado, Geological Society of America, The Geology of North America, v. P-2.

Figure 1. Map of far-western United States showing the location and general outline or size of most of the coalfields (from Wood and Bour, 1988). The western margin of the "Coherent Cratonic Terrane" (Irwin and Barnes, 1982) may also be defined as the "Eastern Limit of Imbricate Thrusts" (Hamilton, 1987, Fig. 5), as far north as northern Wyoming. Northward from southwestern Montana it continues as a nearly parallel line slightly east of the line marking the eastern limit of major Cenozoic normal faults (Hamilton, 1987, Fig. 5). For the more important or notable coal fields, the circles or outlines have been filled (blackened). Coal-field numbers are those on the "Coal Map of North America" (Wood and Bour, 1988; Plates 8 and 9) except numbers 445 (Shasta) and 446 (Saltdale) in California, and 447 (Waldo Hills) and 448 (Davis Creek) in Oregon. A few other locations have been corrected. Several allochthonous terranes (e.g., East Klamath terrane) are labelled, and approximate boundaries are shown by dotted lines. The Blue Mountains–Wallowa terrane of northeast Oregon and western Idaho (dotted line boundary) is not labelled. The Grindstone terrane (double-circle in central Oregon) is indicated.

ent cratonal terrane (Irwin and Barnes, 1982), depicted on Figure 1. The cratonic margin (indicated on Fig. 1 by a dotted-zone line) approximates the western passive margin of this part of North America during late Precambrian and early Paleozoic time. Some 500 to 700 km west of this, another line on Figure 1 (indicated by hachures) extends through central Nevada, southwestern Idaho, and near the eastern boundary of Washington. That line marks the approximate western position of the late Paleozoic cratonic margin (Ross and Ross, 1983). The zone between these two lines is an area of late Paleozoic to early Mesozoic shelf (Hamilton, 1987), a miogeocline that generally received clastic sediments from the east. The Great Basin occupies the southern portion of this miogeocline area (i.e., the southern part of the northern Basin and Range Province). The southern part of the Basin and Range Province occupies the eastward extension into southern Arizona and New Mexico, south of the boundary of the coherent cratonic margin (Fig. 1).

West of the late Paleozoic cratonic margin, the continent appears to be composed of a large number of suspect terranes of allochthonous origin but of uncertain original paleogeographic position in relation to North America; some are certainly exotic. Some, such as the Blue Mountains–Seven Devils terranes in northeastern Oregon and western Idaho, are a melange of Devonian to Triassic sedimentary rocks and Triassic to Jurassic volcanic sandstone, conglomerate, and argillaceous rocks containing fossil plants of wet lowlands or swamp origin. Another example is the Klamath Mountains of northern California and southwest Oregon, which are considered to include vestiges of clastic, volcanic, and carbonate rocks formed in late Paleozoic oceans. These are overlain by Mesozoic siliciclastics and some marine limestones.

Most of these terranes were probably formed nearby as oceanic ridges, island arcs, and oceanic crust, although some exotic terranes, transported great distances, have been identified (Newton, 1987). These blocks or slivers of accreted terranes do not appear to have introduced any exotic coal deposits into this area of the western United States. However, a discrete terrane of extensively folded late Paleozoic rocks containing some carbonaceous strata, the Grindstone-Twelvemile Creeks Paleozoic inlier, is present in southeast Crook County, Oregon (see double circle at about 44°N, 120°W, Fig. 1). This Late Paleozoic sequence is overlain by less folded Triassic-Jurassic rocks and, in some places, by erosional remnants of nearly unmodified Tertiary strata. The flora found in the Pennsylvanian(?)-age strata has been an enigma. The dominant plants and relative proportions of species present are atypical of all other North American and western European Carboniferous floral assemblages (Mamay and Read, 1956). This may be evidence of an exotic terrane from the western portion of the proto-Pacific realm.

These terranes and many others were joined to the North American craton at various times. Since their accretion, block faulting, along with parallel right-lateral strike-slip fault movements and rotation, have resulted in formation of several types of lacustrine basins in which coal-forming peat accumulated locally. Such tectonic activity resulted in closed topographic lows in some areas. Other topographic lows were associated with the later development of structurally controlled drainage patterns. Widespread, episodic volcanic activity also played a major role in the formation of Tertiary lake beds.

STRUCTURAL COAL BASINS OF COASTAL MARGINS: ALTERNATING MARINE AND TERRESTRIAL FILLING

The Centralia-Chehalis coal district of Washington (Loc. 384) represents an original structural basin that was alternately filled by several thousand meters of nearshore marine, coastal-margin brackish, and prograding fluvial-deltaic sediments through middle and late Eocene time and into the Oligocene. This north- and west-trending basin was divided into several segments by localized uplifts along the margins and by occasional extrusive flows from accompanying volcanism. Peat accumulated intermittently in these subbasins during deposition of the late Eocene clastic sediments of the McIntosh Formation and, more importantly, the Skookumchuck Formation. When sediment supply, mainly from the eastern margin, was low, or when downwarping of the basin was reactivated, the embayment enlarged and shallow seas (bays?) repeatedly spread fossiliferous marine sediments over the thick peat beds. Some transgressions lasted for considerable periods of time. This scenario is also identified in at least part of the Coos Bay, Oregon (Loc. 287), coal basin (Duncan, 1953).

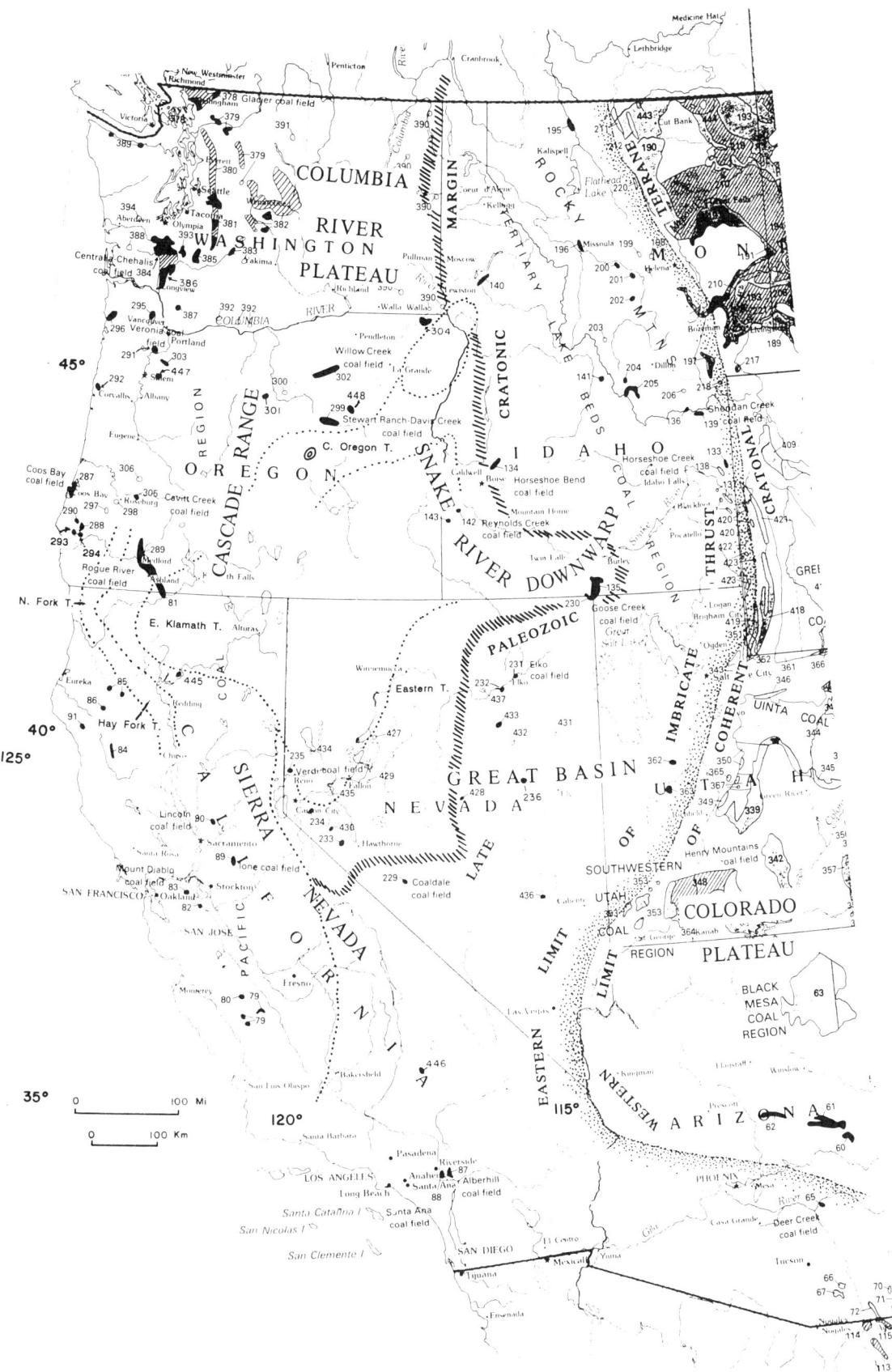

TABLE 1. COAL FIELDS WEST OF THE ROCKY MOUNTAIN REGION, U.S.A.*

State	Percent of Area	Area (km²)	No. of Coal Fields	Short tons (millions)	Important Fields (Coal Field No. From Fig. 1)
Arizona	36.4	107,400	8	155†	Deer Cr. (65), Pinedale (60-61), Fossil Cr. (62)
California	100	411,000	14 +3	285	Ione (89), Corrall Hollow (82, Mt. Diablo (83)
Idaho	100	216,400	11 +4	354	Horseshoe Cr. (133), Goose Cr. (135), Horseshoe Bend (134)
Montana	27.5	104,800	13	587†	Flathead (195), Missoula (196), unnamed (205)
Nevada	100	286,300	19	59	Coaldale (229), Elko (231), Carlin Canyon (232)
New Mexico	8.2	25,840	4	300†	Unnamed (263)
Oregon	100	251,200	22	572	Coos Bay (287), Rogue R. (289), Eden Ridge (288)
Utah	34.5	75,875	2 +2	140†	Unnamed (362, 363)
Washington	100	176,600	17 +20	50,680	Centralia-Chehalis (384), Roslyn-Manastash (382), Grand Ridge-Green R. (380), Wilkeson Carbonado (381), Glacier (378), Skagit (379).
Totals		1,657,800	110 +29	53,132	

*The area is shown on Figure 1 west of the boundary of the "Coherent Cratonal Terrane." Coal fields of the greatest comme potential at this time are identified for each state in the right-hand column. The numbers correspond with those on Figure 1 and with those given on the "Coal Map of North America" (Wood and Bour, 1988; Plates 8 and 9). Areas underlain by coal in California, Idaho, Nevada, Oregon, and Washington are given in column 3. Only parts of the coal-bearing areas of Arizona, Montana, New Mexico, and Utah are included in this chapter because the main coal-bearing areas lie east of the western limit of the "Coherent Cratonal Terrane" (Fig. 1). The number of coal fields in each state is given in column 4. In several states, several separate fields are included under one number. Column 5 shows the amount of "indicated" or "hypothetical" "coal-in-place" in millions of short tons of coal of all ranks, thicknesses, and depths.

†Tonnages given are from only those portions of each of the four states west of the "Coherent Crational Terrane" (Fig. 1). The tonnages are taken from the report of Wood and Bour (1988).

The age and environment of deposition of the late Eocene Skookumchuck Formation are identified by the fossil floras and faunas. Foraminiferal assemblages are closely comparable to those in the type Cowlitz Formation (upper Eocene) of southwestern Washington; to the Coaledo Formation in the Coos Bay area of Oregon (Mason, 1969); and to the Poway Conglomerate of California. These assemblages indicate a slightly younger age than the rich nearshore foraminiferal assemblage of the McIntosh Formation below the Northcraft Formation here (Snavely and others, 1958). The molluscan fauna (Snavely and others, 1958, p. 34–37) of nearshore marine habitats is particularly abundant in the calcareous sandstones and just above the coalbeds.

The Skookumchuck coal-bearing sedimentary rocks are as much as 1,000 m thick. Locally they lie discordantly upon the volcanic and nonmarine sedimentary rocks of the Northcraft Formation below. They are overlain conformably by basaltic sandstones and tuffaceous siltstones of the Oligocene Lincoln Formation. Coal beds grouped near the base and the top of the Skookumchuck Formation are much faulted, both laterally (mainly right-lateral, strike-slip separations of from 1 to 3 km) and vertically. The vertical faults, both normal and reverse, vary from a few centimeters displacement to a kilometer or more, but the offsets are usually less than 3 m in the coalbeds mined. Series of later-formed, closely spaced E-W– and NW-SE–plunging anticlines and synclinal basins have resulted in considerable change of dip along each seam, some to 90°. Major northwest-trending faults have offset or abruptly truncated many of these basins, making continuity of mining difficult. The most prominent of these fold and fault features were formed during a time of extensive deformation and volcanism in the early Miocene, although regional strike-slip faulting probably propagated up from earlier Late Cretaceous and pre-Eocene basement faults.

Coals of lignitic and subbituminous B rank in the Skookumchuck Formation here show considerable original plant structures, including logs and stumps. The lower group of coals generally includes only one minable coal. However, the upper group, lying above about 250 m of intervening sandstones and mudstones, and some dispersed carbonaceous and coaly zones, includes at least 13 beds that have been mined in one or more areas of this district. These beds range in thickness up to 15 m and are mostly divided by many partings of volcanic siltstone, claystone, high-ash coaly zones, or carbonaceous shales. Some benches of coal, as much as 2 m thick locally, contain small amounts of ash and sulfur. The various coal beds generally contain from 5 to 25 percent ash, 0.3 to 4.5 percent sulfur, 15 to 35 percent moisture, and 25 to 35 percent fixed carbon.

Extensive outcrops are exposed in the present topographically rugged terrain. A variety of surface and underground mining techniques have been used in extraction of the coal in the different subbasins due to diverse angles of dip and strike, drainage, thickness of coalbeds, nature, number, and magnitude of partings, etc. Only about 9 million short tons were mined commercially, mostly during the first half of the twentieth century. However, since the opening of the surface mining operations at Centralia, more than 4 million tons a year have been mined, on the average, totalling more than 75 million tons since 1971.

SYNDEPOSITIONAL TECTONISM IN DELTA PLAIN COALS

The Green River and Grand Ridge coalfields of Washington (Loc. 380, southern end) east of Tacoma, Washington, are located along the east side of the Puget Lowland. The Puget Lowland is bordered on the east here by the Straight Creek fault. That fault is interpreted as a strike-slip fault with right-lateral offset of 90 to 190 km (Johnson, 1985). The commercial coals here are interpreted as being derived from peats deposited in low-lying, upper delta-plain and lower alluvial-plain environments (Vine, 1969). The Naches Formation, which includes these coal-bearing strata, is interpreted to be contemporaneous with the Puget Formation farther north, on the basis of sandstone petrography (Frizzell, 1979). Johnson (1985) believes this Puget-Naches basin probably was continuous with the Chuckanut basin to the north, prior to the Eocene tectonism. The source area for the sediments now lies east of the Straight Creek fault. It is interpreted to be located 50 to 150 km farther south, at present, in the high-grade metamorphic terrain in eastern Washington. The sedimentary rocks of the lower delta plain and estuarine and lagoonal depositional sites have not been identified westward beyond the position of the Puget fault (?). Those sediments may have been displaced laterally northward, west of the Puget fault (?) zone.

The upper delta-plain deposits consist of channel sandstones, interdistributary mudstones, and coal. Paleocurrent data indicate that sediment transport during deposition of the Naches Formation rocks was to the west.

The basin was actively subsiding during part of the time represented, but there were strong variations in the rate of subsidence. Turner and others (1983) dated volcanic ash partings of the upper and lower boundaries of the Green River section isotopically in order to estimate the rate of sediment accumulation. The dates obtained indicate 50 cm per thousand years, although it might be twice that rate (Johnson, 1985). There are also several lines of evidence to indicate abrupt thinning and thickening of sedimentary units. These criteria, and the tight folding and faulting, even overturning of beds, along the Straight Creek fault on the eastern side of the Puget-Naches basin, indicate strong syndepositional tectonism during the deposition of the Green River section.

EXTENSIONAL BASINS WITH PREDOMINANT FILLING BY VOLCANIC EXTRUSIVES

Goose Creek basin (Locs. 135, 230) in southern Cassia County, Idaho, extends southwestward across Box Elder County, Utah, and into Elko County, Nevada, an area of about 600 km^2. The Goose Creek coalfield occupies only a portion of this north-trending, intermontane basin (Kiilsgaard, 1964), which formed by basin-and-range-type extensional tectonism in the Miocene (Hildebrand and Newman, 1985). North-trending, high-angle, normal faults border this basin on the east and west, with vertical displacements of several hundred meters. The boundary faults continued to propagate up through the accumulating Miocene sediment until about 8 Ma (late Miocene). Additional faults and minor folding developed in the basin fill due to late Miocene compressional forces as a result of downwarping of the crust northward toward the Snake River Plain. Stream flow, originally directed to the south during early valley-fill accumulation, was reversed toward the Snake River by this downwarping.

The Idavada volcanics that filled the valley during the Miocene, from about 15 to 8 Ma, are divided into lower, middle, and upper units. The peat-forming bogs and lake basins formed principally in the earlier two units. Volcanic sediments increased markedly upward through the section. These were mainly rhyolitic, vitric ashfalls and ash-flow tuffs in several beds, some as much as 35 m thick. The source was a developing volcanic field to the northwest along the Snake River Plain following the passing of a hot spot, along the Snake River downwarp (Fig. 1), that is now beneath the Yellowstone Plateau. Much of the later accumulation was from direct airfall, but earlier, the valley-fill was more evenly sourced by fluvial redistribution of ash falls and ash-flow tuffs and by coarse to fine clastics from the eroding mountain ranges on the east and west. These transported clastics were deposited as alluvial fans near the base of the mountains and as nearly flat valley-floor fills in the center of the basin.

Lakes and marshy areas and bogs developed on this valley floor. Peats developed intermittently in lake-margin environments. The peat surfaces were repeatedly inundated with fluvial sediments, especially during times of volcanic eruptions that may have been accompanied by increased local precipitation. In some instances, excessive ash falls probably choked the drainage channels, raising water levels in the basin, and extending the areas of the peat swamp development. The uneven, lenticular distribution of the lignite indicates such intermittent expansion and restriction of peat swamps on a flat flood plain. The climate has been interpreted by Axelrod (1964) as cool temperate and seasonally equable on the basis of diatom-rich lake beds and rich, lake-margin forests, marshes, and bogs around shallow, freshwater, acidic, open lakes.

Coals are lignitic, with high ash content, low sulfur, and high moisture. They are interspersed with many partings—as many as ten thin carbonaceous shales in 25 m of volcanic-rich sediments. In the Barrett lignite zone, traced from Trapper Creek to Hardister Creek, there are three organic-rich zones (lignites and carbonaceous shales) from 0.5 to 12 m thick, separated by 21 m to 75 m of fluvial clastics or volcanic extrusives (Mapel and Hail, 1959). Lignite mined near the north end of the basin is in thin, lenticular beds. Rates of sediment deposition here generally exceeded peat-bog growth rates, largely from widespread dispersal of subaerial volcanic ashfalls and ash-flow tuffs.

A similar deposit farther northwest, near the Idaho-Oregon boundary, is present in the Reynolds Creek basin, Owyhee County, Idaho. A 25-m lignite-bearing volcanic-ash sequence is the site of a late mid-Miocene pond or lake represented by lignite 1.35 m thick, lignitic shales, fluvial and airfall ash, and recycled siltstones from a granitic terrain. Near the base of the section, below the lignite sequence, a dense distribution of fossil water lily rhizomes (*Nelumbo*) and abundant leaves, flowers, and seed capsules, together with Chinese water chestnut (*Trapa*), and freshwater diatoms, indicate a cool-water, shallow lake with marshy environment surrounding the site. This developed successively upward into a peat-forming bog with typical plants, followed by cyclical repetition of peat development between volcanic sediment interbeds. Gradually the climate cooled, the lake plain was succeeded by a drier valley, and the dominant pollen vegetation preserved represents an upland coniferous forest on the slopes above the valley (Cross and Taggart, 1982).

CONTINENTAL BASINS: FAULT-BOUNDED GRABENS

Fault-bounded, nonmarine basins of several types contain coal beds in the western United States. The Roslyn, Manastash, and Taneum coalfields in Washington occupy such grabens in the Swauk, Manastash, and Chiwaukum basins. The 50-km-wide Swauk basin (Fig. 1, Loc. 382), on the eastern flanks of the Cascade Mountains, in Kittitas County, is a nonmarine basin lying within a regional network of right-lateral faults of Eocene age. Swauk Basin is an example of a fault-wedge basin developed between two NW-SE–trending, diverging, strike-slip faults: the Straight Creek fault zone on the west and the Leavenworth fault on the east (Johnson, 1985). This pull-apart wedge was filled rapidly during early to late Eocene with an estimated 7,500 to 10,000 m of fluvial arkosic sandstones and mudstones, and volcanic sediments (Tabor and others, 1984; Johnson, 1985). These were derived from horst blocks, mainly to the north and east, and from the Mt. Stuart batholith to the north. The transport system appears to have been a meandering river down the paleoslope.

The coal-bearing Roslyn Formation, 2,590 m thick, is of late-middle to early-late Eocene age. It lies unconformably on the Teanaway volcanics, which in turn, overlie unconformably the extensively folded Swauk Formation of about 5,000 m thickness. The Roslyn Formation contains at least eight minable coal beds. The Roslyn bed was, until 1970, the most important commercial coal bed in Washington. At least 57 mt was produced before 1960 (Averitt, 1975), much of it suitable for blending with coking coals. The rank varies from northwest to southeast in the field from high-volatile A bituminous to high-volatile B bituminous. The Roslyn coal yields about 12,000 Btu per pound, and contains less than 0.5 percent sulfur and 12 percent ash. These qualities generally characterize the other minable seams in the basin.

INTERMONTANE BASINS FORMED BY EXTENSIONAL TECTONICS

A few, thin Oligocene coals along the Flathead River of northwest Montana lie in the SE-NW–trending, sediment-filled, fault-bounded Kishenehn basin. This basin resulted from extensional tectonics of the basin-and-range type southwest of Glacier National Park. The Laramide orogenic activity and foreland basin thrusting had slowed and virtually ceased in the mid-Paleogene. Basin-and-range tectonics, perhaps resulting from back-arc spreading, created a large number of intermontane basins of various sizes throughout the area of western Montana and eastern Idaho. Many of these basins filled with large amounts of sediment from adjacent uplifts. Lakes intermittently occupied parts of some of these basins. The area marked "Tertiary Lake Beds Coal Region" on Figure 1 indicates this distribution.

The Kishenehn basin is an asymmetric graben lying between the Flathead listric normal fault on the east and the nearly vertical Nyack fault system near the southern end of the elongate basin. The basin has been filled with an estimated 3,350 m of alluvial fan, braided stream, flood basin, paludal, and lacustrine sediments to form the Kishenehn Formation during the Oligocene (Constenius, 1988). Facies relations indicate the source sediments to be from the footwall block of the Flathead fault system to the east. The transport direction was toward the southwest into the basin formed by the subsiding hanging-wall block. The North Fork of the Flathead River, where several of the coalbeds are located, now flows along the western side of the basin.

Several beds of lignitic coal are exposed at several localities along North Fork and Middle Fork of the Flathead River in the Kishenehn Formation. Representative of these are coalbeds located near the mouth of Coal Creek. Five lignitic coalbeds, 0.7 to 4.0 m thick, were mined here into the 1940s (Johns, 1970). Underground entries for a room-and-pillar operation were driven for several hundred feet in one 4-m-thick coalbed. The lignite in this area is in multiple beds from 0.3 to 4.5 m thick and aggregate to 10 m, exclusive of partings. Forty-eight km west of Columbia Falls Siding, eight seams, from 0.5 m to 3.5 m thick, dip at 45°; four of these eight seams aggregate 9 m. They dip 30 to 40°N and strike 55 to 60°W (Rowe, 1906).

COAL FORMED IN A PALEOVALLEY IN A CHANNEL COMPLEX

An interesting Pennsylvanian-Permian coal deposit, first reported in 1916, is located in a remote area of central Arizona (Loc. 62). The exposures of this coal are along Fossil Creek ("Fossil Canyon"), a small tributary of the Verde River, in a rugged area south of the Mogollon Rim near the Gila-Yavapai County line. McGoon (1962) reported three thin beds of weathered bituminous coal that yielded slightly more than 10,000 Btus per pound. This coal crops out in two areas along Fossil Creek and is interbedded with light gray shale and overlain by a coarse conglomerate. The coal has been tentatively correlated laterally with carbonaceous and coaly mudstones. Drill test holes about 32 km south of Holbrook and about 70 km east of Fossil Creek, also encountered carbonaceous and thin, coaly zones in a similar sequence. These coals appear to be ". . . flood-basin deposits near stream channels and in local swamps on the Supai delta which lay north of the main Pennsylvanian and Early Permian seas" (Beaumont and others, 1971, p. 177). A very recent summary of the complex stratigraphic relations indicates these strata to be either latest Pennsylvanian or earliest Permian on the basis of palynology and leaf floras. Blakey (1990) interprets these coal-bearing rocks to be accumulations in a northwest-trending stream channel, roughly 1 km wide and 10 to 25 m deep, filled with fine-grained quartz sandstone and limestone-pebble conglomerate. He has identified several fining-upward cycles, trough and epsilon cross-stratification, and associated plant debris, all indicative of fluvial origin.

COALS OF SPECIAL CHARACTER

Anthracite and anthracitic coals

Small amounts of anthracite coal have been produced at the extreme eastern end of the Chuckanut (Bellingham) basin near Glacier (Loc. 378), northwest of Mt. Baker. Several intensely folded, faulted, and crushed coalbeds have been identified; some have been mined. Estimates of total anthracite available range from 5 to 50 million tons (Averitt, 1975; Wood and Bour, 1988).

One coal in Skagit County, near Cokedale, Washington (Loc. 379), has a reported analysis of 86 percent fixed carbon, 3.8 percent volatile matter, 0.3 percent moisture, 0.62 percent sulfur, and 8.6 percent ash. Several other beds in this coalfield are of the same general character. These are separated from coals of the Puget Group in the Puget basin to the west by a severely deformed metamorphic belt. About 60 million tons of coal have been produced here from an area of 65 km^2, along the flanks of a southeast-plunging syncline on the east slope of the Cascade Range. It is estimated that three to four times that amount of coal is present in the deeper part of the syncline in the depth interval 300 to 1,000 m.

A small deposit of anthracite coal has also been reported from Pahranagat, Lincoln County, Nevada, but little information is available.

Coals rich in resins, waxes, and other liptinites

Some coals have concentrations of waxes, oil, and resins that were present in the peat-forming plants as a result of metabolic processes. Such liptinitic materials are very resistant to processes of decay and are enriched by selective fungal and bacterial destruction or oxidation of other plant tissues. Also, some plants have considerably more waxes as protective coverings than others, and some form considerable amounts of gums, tannins, oils, and resins, generally as waste products, that are stored in various tissues. These products are occasionally concentrated in peats, lignites, and coals.

One California lignite in the Ione Formation, near Ione, Amador County (Loc. 89), has long been noted for the concentrations of a high-grade wax. This middle Eocene lignite occurs in lens-shaped pods from 150 m to nearly 1,000 m diameter and from 3 to 7 m thick, in flat-lying host rocks, at less than 35 m depth (Jennings, 1957). This lignite is mined in open pits, and is processed by grinding. A solvent extraction removes waxes, wax esters, resins, and asphaltic substances. This preparation ultimately produces between 75 and 105 kg of hard, brown, montan wax per ton. The two fields near Ione occupy approximately 25 km^2. Another similar deposit, with three thick lignite beds within 35 m of the surface, is located several kilometers south near Buena Vista.

An unusual deposit of sapropelic coal of Pliocene age in southern Lyon County, west of Hawthorne, Nevada (Loc. 233), was developed earlier in the century as a possible petroleum resource. Sapropelic coal results from extensive degradation of higher plant matter, with the addition of a considerable amount of algal material, to form a virtually structureless slime greatly enriched in hydrogen. This Washington Mine Camp deposit is lusterless, dull, dark, and extremely hard when dry. The fresh coal is black in color and contains 30 percent volatile hydrocarbons, 40 percent fixed carbon, 16 percent ash, and 14 percent water (Horton, 1964).

ALTERNATIVE COAL PRODUCTION/EXTRACTION TECHNOLOGIES

Many of the western U.S. coals are at great depths, or in steeply dipping beds, or in fields of small size, and some are far from transportation facilities, population centers, or industrial complexes. Conventional transportation facilities may be unavailable or of prohibitive cost in rugged terrain or in remote areas. Water for coal washing/beneficiation, as well as for public use, is, in many regions, of severely limited availability. Environmental concerns in national forests, wilderness areas, or other public lands, or on private lands near population centers, may strongly limit mining developments, particularly surface mining that, in some coal fields, would be more efficient and economical in extracting the coal than previously developed subsurface mines.

An experimental hydraulic extraction test was conducted at a steeply pitching section of the Roslyn coalbed in western Washington, and some positive potential was demonstrated (Price and Badda, 1965).

An underground gasification project was conducted from 1980 to 1986 on multiple coalbeds in the Centralia-Chehalis coalfield near the Thurston-Lewis County line. These late Eocene subbituminous coals in the Skookumchuck Formation have been strip-mined extensively for steam-generated electric power plants. A large burn cavity was developed. This cavity was excavated and surveyed, and 14 sections analyzed by the Western Research Institute of Laramie, Wyoming.

The most promising new technology has been demonstrated in western Washington, which is the potential for coalbed methane extraction. Methane drainage of coal beds could be an important production technology for the recovery of the energy and chemicals in the complexly folded, faulted, and deep coals.

More than 25 coal-bed methane test wells, some to 1,300 m, in the Puget Group coal-bearing strata, were recently demonstrated to yield commercial quantities of pipeline quality methane

(99 percent methane; 950 Btu/ft^3). Pappajohn and Mitchell (1990, and personal oral communication, 1990) predict that the 6.0 billion tons of these coals (DOE estimate, 1981, by Choate and Johnson, 1980) should produce nearly 25 tcf of methane with average production of about 2.59 bcf/km^2. Coals of the Eocene Renton, Skookumchuck, and Carbonado Formations of the Puget Group, along the eastern side of the Puget Lowland Province, and coals in the Centralia-Chehalis district (at Localities 380, 381, 384, and 386, Fig. 1), have been tested in a preliminary way. The test holes in several areas penetrated multiple coalbeds with net thicknesses of 12 to 15 m or more of low-volatile to high-volatile A bituminous and subbituminous coals.

REFERENCES CITED

Averitt, P., 1975, Coal resources of the United States, January 1, 1974: U.S. Geological Survey Bulletin 1412, 131 p.

Axelrod, D. I., 1964, The Miocene Trapper Creek flora of southern Idaho: University of California Publications in Geological Sciences, 148 p.

Beaumont, E. C., Shomaker, J. W., and Kottlowski, F. E., 1971, Stratidynamics of coal deposition in southern Rocky Mountain region, U.S.A., in Shomaker, J. W., Beaumont, E. C., and Kottlowski, F. E., Strippable low-sulfur coal resources of the San Juan Basin: New Mexico Bureau of Mines and Mineral Resources Memoir 25, p. 175–185.

Beikman, H. M., Gower, H. D., and Dana, T.A.M., 1961 (reprinted, 1984, with addendum by Schasse, H. W., Walsh, T. J., and Phillips, W. M.), Coal reserves of Washington: Washington Department of Natural Resources, Division of Geology and Earth Resources, Bulletin 47, 115 p.

Blakey, R. C., 1990, Stratigraphy and geologic history of Pennsylvanian and Permian rocks, Mogollon Rim region, central Arizona and vicinity: Geological Society of America Bulletin, v. 102, p. 1189–1217.

Choate, R., and Johnson, C. A., 1980, Geologic overview; Coal and coalbed methane resources of the western Washington coal region: McLean, Virginia, TRW Inc., report prepared for the U.S. Department of Energy, Morgantown Energy Technology Center.

Constenius, K., 1988, Structural configuration of the Kishenehn basin delineated by geophysical methods, northwestern Montana and southeastern British Columbia: The Mountain Geologist, v. 25, no. 1, p. 13–28.

Cross, A. T., and Taggart, R. E., 1982, A Miocene *Nelumbo-Trapa* lake, southwestern Idaho [abs.]: Botanical Society of America Miscellaneous Series Publication 162, p. 57.

Duncan, D. C., 1953, Geology and coal deposits in part of the Coos Bay coalfield, Oregon: U.S. Geological Survey Bulletin 982-B, p. 53–73.

Frizzell, V. A., Jr., 1979, Petrology and stratigraphy of Paleogene nonmarine sandstones, Cascade Range, Washington [Ph.D. thesis]: Stanford, California, Stanford University, 151 p.

Hamilton, W. B., 1987, Plate-tectonic evolution of the western U.S.A.: Episodes, v. 10, p. 271–276.

Hildebrand, R. T., and Newman, K. R., 1985, Miocene sedimentation in the Goose Creek basin; South-central Idaho, northeastern Nevada, and northwestern Utah, in Flores, R. M., and Kaplan, S. S., eds., Cenozoic paleogeography of west-central United States: Rocky Mountain Section, Society of Economic Paleontologists and Mineralogists, p. 55–70.

Horton, R. C., 1964, Coal, in Mineral and water resources of Nevada: Committee on Insular Affairs, U.S. Senate Document 87, U.S. 88th Congress, 2nd session, p. 49–57.

Hume, D. B., and Associates, 1977, Carbon River coal: P.O. Box 602, Mercer Island, Washington 98040, D. B. Hume and Associates (unpublished proprietary section).

Irwin, W. P., and Barnes, I., 1982, Map showing relation of carbon dioxide-rich springs and gas wells to the tectonic framework of the conterminous United States: U.S. Geological Survey Miscellaneous Investigations Series Map I-1301, scale 1:5,000,000.

Jennings, C. W., 1957, Coal, in Jenkins, O. P., and Wright, L. A., Mineral commodities of California: California Department of Natural Resources, Division of Mines Bulletin 176, p. 153–164.

Johns, W. M., 1970, Geology and mineral deposits of Lincoln and Flathead Counties, Montana: Montana Bureau of Mines and Geology Bulletin 79, p. 155–156.

Johnson, S. Y., 1985, Eocene strike-slip faulting and nonmarine basin formation in Washington, in Biddle, K. T., and Christie-Blick, N., eds., Strike-slip deformation, basin formation, and sedimentation: Society of Economic Paleontologists and Mineralogists Special Publication 37, p. 283–302.

Kiilsgaard, T. H., 1964, Coal, in Mineral and Water resources of Idaho: U.S. 88th Congress, 2nd Session, Committee on Interior and Insular Affairs, p. 58–66.

Mamay, S. H., and Read, C. B., 1956, Additions to the flora of the Spotted Ridge Formation in central Oregon: U.S. Geological Survey Professional Paper 274-I, p. 211–226.

Mapel, W. J., and Hail, W. J., Jr., 1959, Tertiary geology of the Goose Creek District, Cassia County, Idaho, Box Elder County, Utah, and Elko County, Nevada, in Denson, N. M., and others, eds., Uranium in coal in the western U.S.: U.S. Geological Survey Bulletin 1055-H, p. 217–254.

Mason, R., 1969, Fossil fuel resources; Coal, in Mineral and Water Resources of Oregon, Section 1: U.S. 90th Congress, 2nd Session, Committee on Interior and Insular Affairs, p. 272–278.

McGoon, D. O., Jr., 1962, Occurrences of Paleozoic carbonaceous deposits in the Mogollon Rim region, east-central Arizona: New Mexico Geological Society 13th Annual Field Conference Guidebook, p. 89–91.

Newton, C. R., 1987, Biogeographic complexity in Triassic bivalves of the Wallowa terrane, northwestern United States; Oceanic Islands, not continents, provide the best analogues: Geology, v. 15, p. 1126–1129.

Pappajohn, S. P., and Mitchell, T. E., 1990, The potential commercial extraction of natural gas resources from economically unminable coal deposits in the Pacific Northwest: Geological Society of America Abstracts with Programs, v. 22, p. A–16.

Price, G. C., and Badda, F., 1965, Hydraulic coal mining research; Development mining in a steeply pitching coalbed, Roslyn, Washington: U.S. Bureau of Mines Report of Investigations 6685, 16 p.

Ross, C. A., and Ross, J. P., 1983, Late Paleozoic accreted terranes, in Stevens, C. H., ed., Pre-Jurassic rocks in western North American suspect terranes: Pacific Section, Society of Economic Paleontologists and Mineralogists, p. 7–22.

Rowe, J. P., 1906, Montana coal and lignite deposits: Missoula, Montana University Bulletin 37, Geological Series 2, 82 p.

Snavely, P. D., Jr., Brown, R. D., Jr., Roberts, A. E., and Rau, W. W., 1958, Geology and coal resources of the Centralia–Chehalis district, Washington: U.S. Geological Survey Bulletin 1053, p. 1–159.

Tabor, R. W., Frizzell, V. A., Jr., Vance, J. A., and Naeser, C. W., 1984, Ages and stratigraphy of lower and middle Tertiary sedimentary and volcanic rocks of the central Cascades, Washington; Application to the tectonic history of the Straight Creek fault: Geological Society of America Bulletin, v. 95, p. 26–44.

Turner, D. L., Frizzell, V. A., Jr., Triplehorn, D. M., and Naeser, C. W., 1983, Radiometric dating of ash partings in coal of the Eocene Puget Group, Washington; Implications for paleobotanical stages: Geology, v. 11, p. 527–531.

Wood, G. H., Jr., and Bour, W. V., III, 1988, Coal map of North America: U.S. Geological Survey Coal Map, scale 1:5,000,000.

Vine, J. D., 1969, Geology and coal resources of the Cumberland, Hobart, and Maple Valley Quadrangles, King County, Washington: U.S. Geological Survey Professional Paper 62, 67 p.

MANUSCRIPT RECEIVED BY THE SOCIETY NOVEMBER 29, 1990

The Geology of North America
Vol. P-2, Economic Geology, U.S.
The Geological Society of America, 1991

Chapter 37

Economic Alaskan coal deposits

Gary D. Stricker
U.S. Geological Survey, Box 25046, Denver Federal Center, Denver, Colorado 80225

INTRODUCTION

Alaskan coal is a large untapped national resource that probably exceeds 40 percent of the total coal resources in the United States. Estimates by various authors of the total tonnage of Alaska's coal resources have increased over the years along with our knowledge of Alaskan coal geology. Recent estimates of Alaska's coal resources are as large as 5,600 billion short tons of hypothetical coal resources (Wood and Bour, 1988; Merritt and Hawley, 1986; Ferm and Muthig, 1982; McGee and Emmel, 1979). Averitt (1975) estimated that the conterminous United States contains 3,700 billion short tons. In Alaska, little is presently exploited. In 1989, the only active coal mine in the state had an annual production of 1.5 million short tons (Green and Bundtzen, 1989). Future use of Alaskan coal, in all probability, will be hydrocarbon feedstock and export to Pacific rim countries and small-scale mining for local consumption.

Wood and Bour (1988) listed 50 coal fields and coal occurrences in Alaska (Plate 7, in pocket) ranging in age from Mississippian to early Tertiary. McConkey and others (1977, p. 91–97) ranked Alaska's coal fields according to the likelihood of development in the following order: Nenana (including Jarvis Creek), Beluga, Matanuska, Herendeen Bay (including Chignik), northern Alaska, and Bering River (Fig. 1). The state of Alaska and various private energy companies, in the last several years, have investigated most of the forementioned fields; those activities are summarized in a series of annual reports (U.S. Geological Survey, 1982 through 1988). Based on McConkey and others' (1977) ranking of the coal fields and on data from government and private industries studies, this chapter focuses on the above-mentioned fields. For each of these fields, I discuss the known geology, tectonic setting, environments of coal deposition, paleoclimate, and coal quality and quantity. For a more extensive review of the geology of Alaskan coal, see Wahrhaftig and others (1991).

NENANA, BELUGA, MATANUSKA COAL FIELDS

The Nenana, Beluga, and Matanuska coal fields are three of many isolated occurrences of coal-bearing rocks in southern and central Alaska. These fields may be depositionally related to a

Figure 1. Index map showing locations of coal fields mentioned in this report.

large river system that drained much of central Alaska and emptied into the Gulf of Alaska through the Cook Inlet. The large coal basins of the Cook Inlet, Susitna Lowlands, and Matanuska Valley (Fig. 1) appear to be located where this large river system flowed into Cook Inlet and joined the Pacific.

The Cook Inlet basin lies on the site of an arc-trench gap (Moore, 1974) between the Mesozoic magmatic arc represented by the Lower Jurassic Talkeetna Formation and Middle Jurassic Talkeetna batholith on the north and the ancient Pacific Ocean crust on the south where the Kenai and Chugach Mountains are located. A thick, mainly terrigenous sequence of Middle Jurassic to Late Cretaceous strata accumulated on this shelf and lies unconformably beneath the coal-bearing Tertiary sedimentary rocks (Kirschner and Lyon, 1973; Fisher and Magoon, 1978). Within the Cook Inlet region, only the Healy Creek, Beluga, and Matanuska coal fields are considered economically important (Plate 7 and Fig. 2).

Nenana coal field

More than half the coal mined in Alaska has been produced from the Nenana coal field (the rest is mostly from the Matanuska

---

Stricker, G. D., 1991, Economic Alaskan coal deposits, *in* Gluskoter, H. J., Rice, D. D., and Taylor, R. B., eds., Economic Geology, U.S.: Boulder, Colorado, Geological Society of America, The Geology of North America, v. P-2.

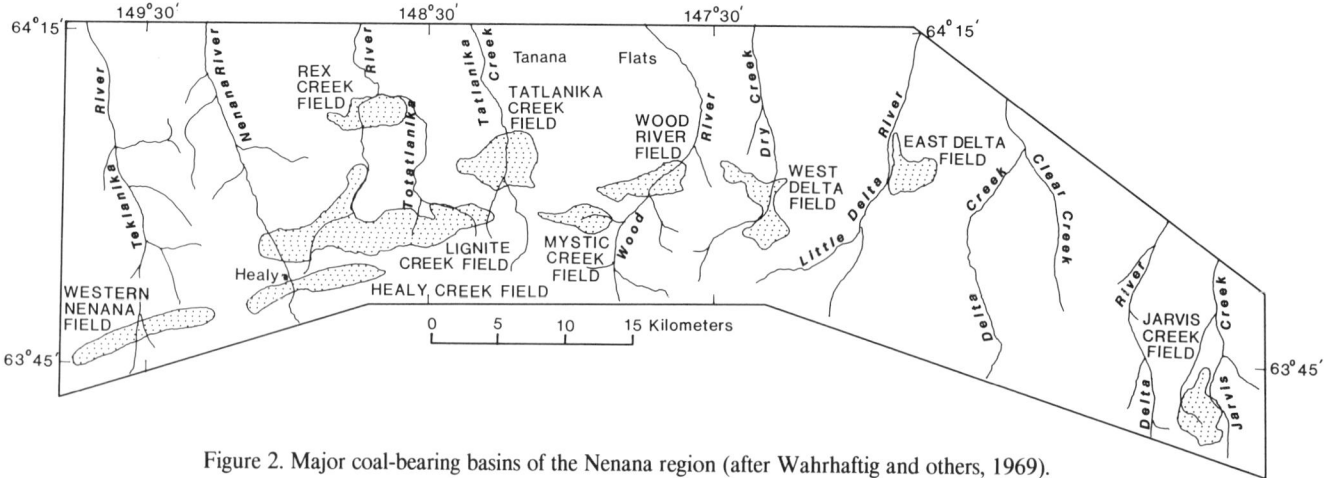

Figure 2. Major coal-bearing basins of the Nenana region (after Wahrhaftig and others, 1969).

field). The Nenana field consists of ten synclinal coal basins partially or completely separated by erosion. These basins extend about 200 km along the northern foothills of the Alaska Range and include, west to east, western Nenana, Healy Creek, Lignite Creek, Rex Creek, Tatlanika Creek, Mystic Creek, Wood River, West Delta, East Delta, and Jarvis Creek coal fields (Fig. 2). These coal fields are located north of the Alaska Range and appear to lie in areas where tributaries of the major river system draining central Alaska flowed across subsiding basins prior to the uplift of the Alaska Range (Wahrhaftig and others, 1991). The coal occurs in the Usibelli Group (Wahrhaftig, 1987), a sequence of five formations of poorly consolidated continental sedimentary rocks of late Eocene and early to late Miocene age (Fig. 3). The same coal-bearing units have been identified in all the synclinal basins in the Nenana coal field.

The oldest unit in the Usibelli Group is the Healy Creek Formation (Fig. 3), a fluvial sequence of lenticular beds of poorly sorted and basally scoured lenticular sandstone, conglomerate, siltstone, and claystone with coals that thicken, split, and pinch out abruptly. These sediments were deposited by migrating braided streams (Stanley and others, 1989). The Healy Creek Formation was deposited on a surface having a few hundred meters of relief; as a result, the thickness of the formation, as well as the number of coal beds, varies markedly. The coals are as thick as 20 m and have been mined extensively in underground and surface mines, both hydraulically and by truck and shovel.

Overlying the Healy Creek Formation is the Sanctuary Formation (Fig. 3), a thinly laminated shale (possibly varved) 40 m thick. This non-coal-bearing unit, which is assigned to the middle Miocene (Wolfe and Tanai, 1980), accumulated in a large, shallow lake.

The overlying Suntrana Formation, of middle Miocene age (Wolfe and Tanai, 1980), is as thick as 400 m and consists of 6 to 12 fining-upward sequences of conglomerates and cross-stratified sandstones overlain by mudstones, and finally, by coals as thick as 20 m. Stanley and others (1989) have interpreted the coarse clastics as having been deposited in stacked, high-energy fluvial channels and the mudstones as having been deposited in the quiet water of abandoned channels. The Suntrana Formation coals, when compared to the underlying Healy Creek Formation coals, are thicker and more laterally persistent. Alaska's only active coal mine is presently surface mining three seams in the Suntrana Formation.

Overlying the Suntrana Formation is the Lignite Creek Formation of middle Miocene age (Fig. 3). This unit, 150 to 240 m thick, is a multicycled, fining-upward sequence, similar to the underlying Suntrana Formation. The Lignite Creek Formation differs from the Suntrana; it has different sandstone lithology, fewer mud-filled abandoned channels, and thinner (typically less than 1.5 m), less laterally persistent coals.

The uppermost formation assigned to the Usibelli Group is the non-coal-bearing Grubstake Formation. This dark gray, laminated shale and claystone is interpreted to have accumulated in a lake formed by the damming of south-flowing streams by uplift of the Alaska Range (Wahrhaftig and others, 1969).

Wolfe and Tanai (1980), in studies of the flora in southern and south-central Alaska, report that the vegetation and trees in the Cook Inlet region and as far north as the Nenana coal field belong to a mixed northern hardward forest. They suggest that the mean annual temperature for this region was 6 to 7°C, with a temperature range of 26 to 27°C.

The total sulfur content of coals in the Usibelli Group ranges from 0.03 to 0.3 percent (Affolter and Stricker, 1987c), a content that ranks it among the lowest of any United States coal. Apparent rank ranges from lignite A to subbituminous B, with the mode being subbituminous C. Ash-content arithmetic mean is 12.6 percent (range of 6.5 to 37.5 percent), and heat-of-combustion (Btu/lb) mean is 8,030 (range of 6,130 to 9,210; Affolter and others, 1981).

Resources are: 1 billion short tons identified and 2 billion hypothetical for Healy Creek; 4.9 billion short tons identified and 7 billion short tons hypothetical for Lignite Creek; and 8 billion short tons identified and 14 billion short tons hypothetical for Nenana coal field (Merritt and Hawley, 1986).

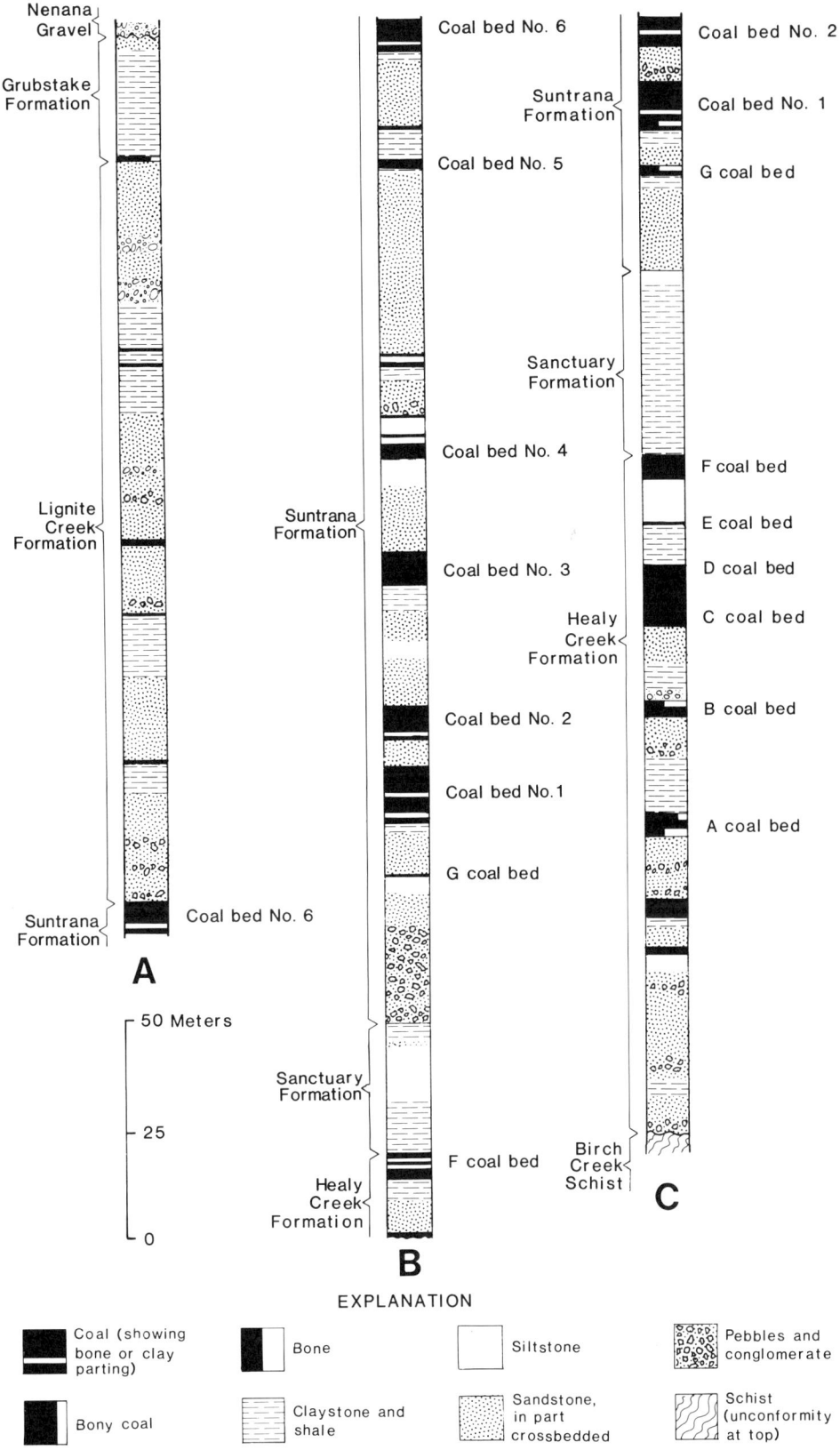

Figure 3. Type section of the Usibelli Group at Suntrana, Alaska (from Wahrhaftig and others, 1969; Wahrhaftig, 1987).

Beluga coal field

The main coal deposits of the Cook Inlet region are in the Kenai Group, which ranges in age from early Oligocene to Pliocene. The Hemlock Conglomerate at the base is overlain, in succession, by the Tyonek, Beluga, and Sterling Formations (Fig. 4; Calderwood and Fackler, 1972; Kirschner and Lyon, 1973; Magoon and others, 1976). The Kenai Group contains petroleum and natural gas reserves in the Cook Inlet petroleum province, and most of the information concerning distribution, lithology, and coal resources has resulted from drilling for oil and gas. Excellent basin-margin exposures of the coal-bearing rocks are found in the Beluga coal field and near Homer, Alaska. In the deepest part of the basin, total accumulation for the Kenai Group exceeded 7,800 m (Calderwood and Fackler, 1972; Fisher and Magoon, 1978) and consists of nearly continuous continental deposition from the Hemlock Conglomerate to the Sterling Formation. Within the Kenai Group, each formation has a different source terrane or depositional environment (Magoon and Egbert, 1986).

The Tyonek Formation (Fig. 4), which contains most of the coal resources in the Cook Inlet basin and the Beluga coal field, is as thick as 2,300 m (Calderwood and Fackler, 1972). Coal in the Tyonek Formation is concentrated along the northwestern margin of the Cook Inlet basin in beds as thick as 16 m. These coals are highly lenticular and only rarely can be correlated for distances more than a few kilometers. Hite (1976) suggested that the Tyonek Formation accumulated in poorly drained alluvial lowlands adjacent to tectonically active highlands that contained sporadic active volcanoes. The overlying Beluga Formation is as thick as 1,500 m (Fig. 4; Hartman and others, 1971) and contains numerous coal beds as thick as 2 m. Hayes and others (1976) have interpreted this formation as an alluvial fan complex with braided streams. Coals, because of the more active depositional system, are thinner than those in the underlying Tyonek Formation. The Sterling Formation, the uppermost formation in the Kenai Group, has a maximum thickness of 3,350 m (Calderwood and Fackler, 1972) and contains a few thin (<1 m) coal beds. Hayes and others (1976) interpreted the Sterling Formation as the product of a meandering stream complex in a rapidly subsiding basin. Because of the rapid deposition, time was insufficient to develop thick coal deposits.

The Tyonek Formation coals in the Cook Inlet basin and Beluga coal field range in apparent rank from subbituminous B to subbituminous C (Stricker and others, 1986; Merritt and others, 1987). In the Beluga coal field, the ash content ranges from 4.7 to 46.5 percent (mean of 14.9 percent), and the sulfur content range is one of the lowest reported for any United States coal (0.08 to 0.33 percent, mean of 0.16 percent). Based on coals thicker than 0.6 m to a depth of 3,000 m, McGee and O'Connor (1975) estimated the onshore and offshore hypothetical coal resources for the Cook Inlet basin to be 1.2 trillion metric tons. Of the 1.2 trillion metric tons, Affolter and Stricker (1987b) estimated 800

AGE	UNIT	FORMATION THICKNESS (IN METERS)	DESCRIPTION
CENOZOIC — QUAT.			Alluvium and glacial deposits
CENOZOIC — TERTIARY	Kenai Group	Sterling Formation 0–3,350	Massive sandstone and conglomerate beds with a few thin lignite beds
CENOZOIC — TERTIARY	Kenai Group	Beluga Formation 0–1,800	Claystone, siltstone, and thin subbituminous coal beds
CENOZOIC — TERTIARY	Kenai Group	Tyonek Formation 1,200–2,350	Sandstone, claystone, and siltstone interbeds and massive subbituminous coal beds
CENOZOIC — TERTIARY	Kenai Group	Hemlock Conglomerate 90–270	Sandstone and conglomerate
		OLDER TERTIARY ROCKS	

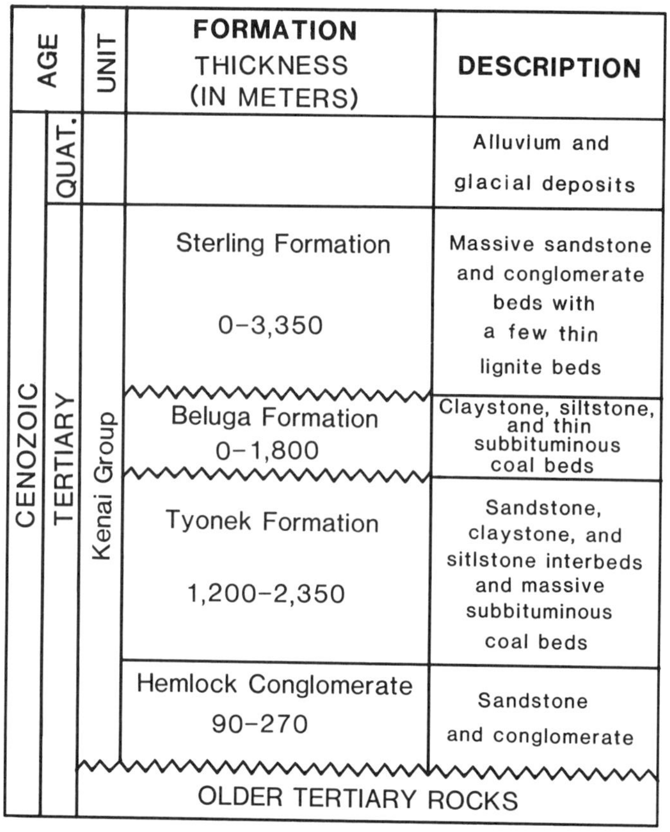

Figure 4. Stratigraphic nomenclature of the Tertiary Kenai Group, upper Cook Inlet region, Alaska (modified from Calderwood and Fackler, 1972).

billion metric tons to be offshore; therefore, 400 billion metric tons of hypothetical coal are estimated to be onshore.

Matanuska coal field

Paleocene coal of the Matanuska coal field occupies a graben in the Matanuska Valley (Fig. 5). The coal field is 9 to 12 km wide, 65 km long, and extends from Moose Creek on the west to Anthracite Ridge on the east. The Chickaloon Formation, a sequence 100 to 1,500 m thick, of claystone, siltstone, sandstone, minor conglomerate beds, and many beds of coal, rests unconformably on the marine clastic Matanuska Formation of Lower and Upper Cretaceous age (Barnes and Payne, 1956). Overlying the Chickaloon Formation is the Wishbone Hill Formation, a massive conglomerate 550 to 600 m thick (Fig. 6). Gabbro sills and dikes intrude the Chickaloon Formation, particularly in the eastern half of the field. The Matanuska Valley coal field is divided into five subfields (Fig. 5). Coals of the Chickaloon Formation range in rank from subbituminous in the western part of the field to anthracite in east, with a mean of high-volatile A bituminous coal. The Anthracite Ridge coals are associated with numerous dikes and sills and have been upgraded in rank as a result of localized heating. Total sulfur content ranges from 0.2 to 1.5 percent, with a mean of 0.45 percent (Barnes and Payne, 1956; Waring, 1934).

Historically, the Matanuska coal field has been one of two major fields (the other being Nenana) to have produced coal in Alaska. From 1915 to 1967, most of the production was from the Wishbone Hill district. Mining essentially ended when the Alaskan Railroad switched to diesel-electric engines and the Anchorage area military bases converted from coal to Cook Inlet gas-generated electric power. Resources for the Matanuska coal field have been estimated by various workers since 1906. Recently, Merritt and Belowich (1984) estimated a total of 380 million short tons of coal remaining in the Matanuska Valley.

HERENDEEN BAY AND CHIGNIK COAL FIELDS

The Upper Cretaceous Chignik Formation in the Chignik and Herendeen Bay coal fields (Fig. 7) on the Alaska Peninsula accumulated near an arc-trench gap (Burk, 1965). The coal fields are separated by about 160 km, and each is approximately 100 km^2 in size. Coals have been identified in the 500-m-thick Coal Valley Member (of Burk, 1965) of the Chignik Formation (Fig. 8). The Coal Valley Member is part of a cyclic, nearshore, marine-to-nonmarine, coal-bearing sequence (Deeterman, 1978). The coals accumulated in littoral and overbank-swamp environments as part of a fluvial to deltaic depositional system near an arc-trench gap (Dickinson, 1974). Mancini and Deeter (1977) suggested that during accumulation of the Coal Valley Member the Alaska Peninsula was characterized by a narrow shelf facing a major trench. The Chignik Formation is coal bearing only between Herendeen Bay and Hook Bay, indicating that sedimentation, subsidence, and climate were amenable for development and preservation of peat-accumulating swamps only in the Herendeen Bay-Chignik areas.

The late Campanian plant assemblage from the Alaska Peninsula near Herendeen Bay and Chignik was that of a microphyllous, broad-leaved evergreen forest (Wolfe and Upchurch, 1987). This plant assemblage indicates there was no precipitation deficit during the deposition of the Coal Valley Member. Wolfe and Upchurch (1987) estimate a mean annual temperature of 12°C with a mean annual range of 8°C for this region during the late Campanian.

Upper Cretaceous coal that accumulated in this arc-trench gap environment on the Alaska Peninsula has a low resource potential because the peat-accumulating swamps were not large and the ash content is high. The elevated tectonic activity in this depositional system has resulted in coals of relatively high rank. Conwell and Triplehorn (1978) and Gates (1944) reported that the coals in the Herendeen Bay and Chignik fields are high in ash (mean, 20 percent; range, 5 to 51 percent), low in sulfur (mean, 0.7 percent; range, 0.27 to 2.75 percent), and have a mean apparent rank of high-volatile B bituminous coal. Because of the narrow deposition shelf upon which peat was able to accumulate, hypothetical coal resources in the Herendeen Bay and Chignik fields are estimated by Merritt and McGee (1986) to be only 360 million short tons.

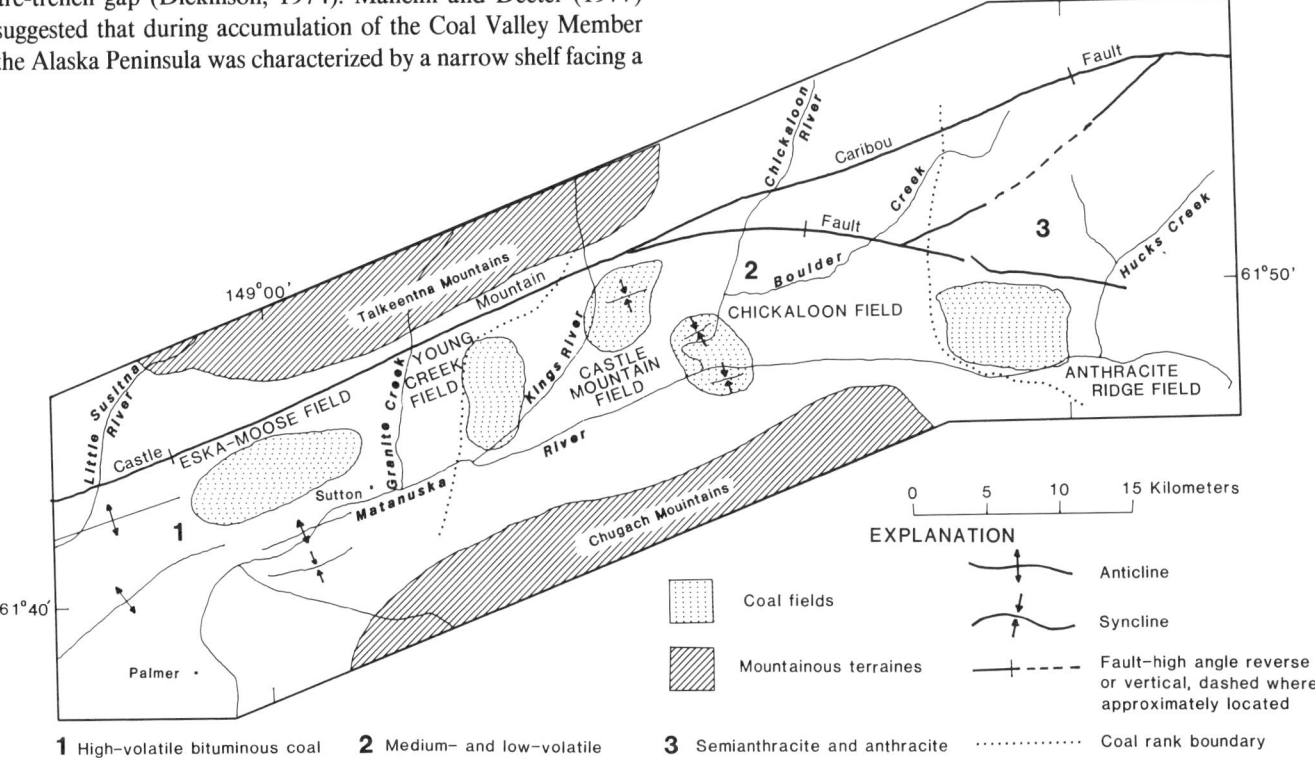

Figure 5. Major coal field subdivisions, rank, and geologic structure in the Matanuska Valley (modified from Merritt and Belowich, 1984).

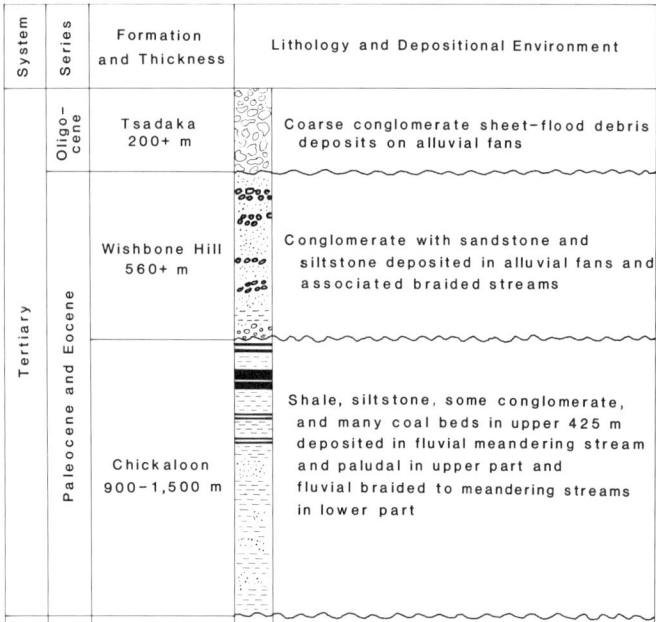

Figure 6. Stratigraphic nomenclature of the Tertiary rocks in the Matanuska Valley, Alaska (modified from Barnes and Payne, 1956; and Clardy, 1984).

NORTH SLOPE

Known coal deposits on the North Slope range from Mississippian to Tertiary in age. Most of this coal is related to a delta system that developed in the Early Cretaceous and continued into the early Tertiary, filling the Colville Basin, a deep assymetric basin or foredeep. During the Early Cretaceous, the beginning of the Brooks Range orogen recorded a shift in sediment dispersal from a predominantly northern source to a major southern and southwestern source: the Brooks Range orogenic belt. Coals accumulated in this delta system in the early Albian to Cenomanian Nanushuk Group, the Upper Cretaceous Colville Group, and the Upper Cretaceous to Pliocene Sagavanirktok Formation.

Nanushuk Group

The Nanushuk Group (Fig. 9) consists of an offlap, post-orogenic, molasse-type lithofacies, deposited on a passive continental margin. Sedimentary rocks of the Nanushuk Group are associated with the river-dominated Corwin delta in the western portion, and with the Umiat delta in the central portion of the North Slope (Ahlbrandt and others, 1979). Delta-front and shoreline strata of the Kukpowruk Formation in the west and the Tuktu Formation in the central portion of the North Slope grade upward and intertongue with the overlying nonmarine, coal-bearing Corwin and Chandler Formations (Fig. 9).

In the western North Slope, the Kukpowruk Formation, composed mainly of delta-front sandstones, ranges in thickness from 610 to 1,200 m in the outcrop belt in the northern foothills. The Corwin Formation consists of delta plain and alluvial plain shale, sandstone, conglomerate, and coal (Roehler and Stricker, 1979). This formation, while more than 3,450 m thick at Corwin Bluffs along the Arctic coast, thins eastward to zero in the subsurface near the Colville River.

In the central North Slope, the succession is more complex but, in general, nonmarine units overlie and intertongue with marine units. The marginal marine to marine Tuktu Formation intertongues with the delta-front and lower delta-plain Grandstand Formation (Fig. 9). The Grandstand Formation is overlain by and intertongued with the Killik Tongue of the Chandler Formation, a transitional or middle delta-plain sequence. Higher in the section, the Killik Tongue is overlain by a tongue of the Ninuluk Formation, which intertongues with the overlying Niakogon Tongue (Fig. 9). Molenaar (1985) indicated that the Seabee Formation, of the Colville Group, interfingers with both the Ninuluk Formation and Niakogon Tongue of the Chandler Formation.

Both deltas were river dominated, but the Umiat delta sediments reflect a higher degree of winnowing energy, as shown by a larger sand-to-mud ratio than the Colville delta (Ahlbrandt and others, 1979). The Umiat delta also apparently had lesser sediment volume and therefore a smaller source area than the Corwin delta. The area of demarcation between the two deltas is obscure; the Meade arch, extending southward from Point Barrow in Brookian time, probably did not play an active part in controlling the depocenters of the deltas. Molenaar (1981, 1985) suggested that the Corwin delta formed earlier than the Umiat delta and that the two merged during Albian time without specific demarcation. The Corwin delta continued to be the dominant depositional feature. Paleogeographic interpretations of Nanushuk deposition (Tetra Tech, 1982; Molenaar, 1981, 1985; Huffman

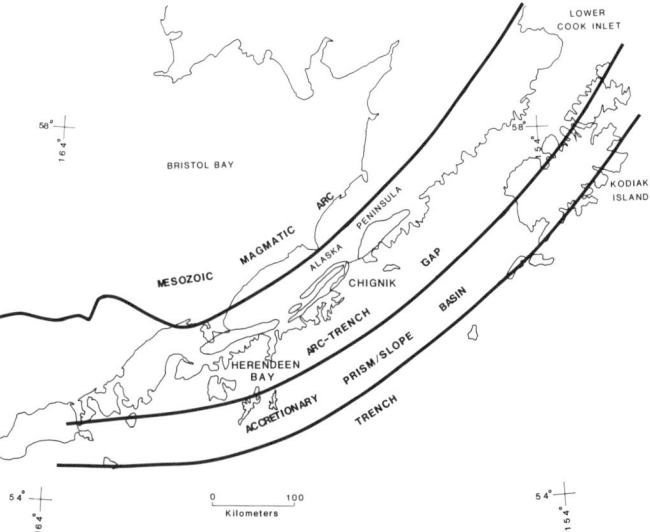

Figure 7. Tectonic setting and location of the Cretaceous coals of the Herendeen Bay and Chignik coal fields (modified from Mancini and others, 1978; and Wood and Bour, 1988).

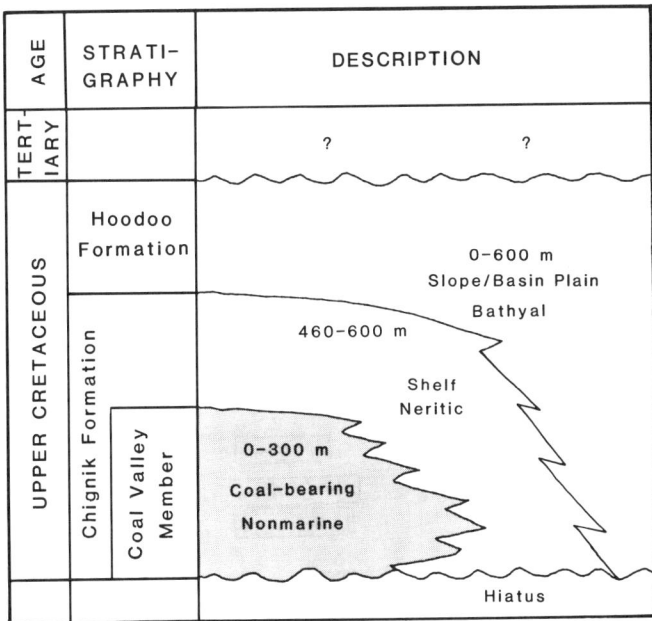

Figure 8. Generalized stratigraphic section of Cretaceous rocks in the Herendeen Bay and Chignik coal fields (modified from Mancini and others, 1978).

TABLE 1. ESTIMATES OF HYPOTHETICAL COAL RESOURCES FOR THE NANUSHUK GROUP COAL IN THE NORTH SLOPE

Rank	Attitude	Overburden (m)	Resource Estimate*
Subbituminous	Dips generally 15° or less	0–150	1,149
		150–300	20
		300–600	10
		>600	1
		Total	1,180
	Dips generally 15° or more	0–150	101
		150–300	5
		300–600	5
		>600	1
		Total	112
	Subbituminous Total (rounded)		1,290
Bituminous	Dips generally 15° or less	0–150	1,340
		150–300	0
		Total	1,340
	Dips generally 15° or more	0–150	571
		150–300	0
		Total	571
	Bituminous Total (rounded)		1,910
	North Slope Total (rounded)		3,200

*Reported in billions of short tons.

and others, 1984) stressed the dominant northeastward and eastward progradation of Nanushuk prodelta slope sediments. These studies also showed a strong northwestward concentration of sandstone in the upper part of the Nanushuk Group, from Umiat toward Point Barrow parallel to the paleoshoreline. This concentration may suggest that northwestward longshore currents moved sand from the Umiat delta along the active shelf of the Corwin delta front. This sand accumulation, interpreted to represent shoreline or offshore bar facies, may have been a controlling factor in restricting the development of the most prolific Nanushuk coal-generating delta environments to the western North Slope.

Spicer (1987) reported that the climate of the North Slope during Albian to Cenomanian time was cool with a mean annual temperature of 10 ± 3° C. Precipitation was sufficient to develop and accumulate thick peat deposits. Precipitation was also distributed throughout the year in a manner to preclude the oxidation and loss of organic material. Tree growth rings on the North Slope suggest a rapid change from summer to winter conditions during the Albian to Cenomanian (Spicer, 1987). This growth-ring pattern is consistent with the paleolatitude of approximately 80°N for the Colville Basin.

Sable and Stricker (1987), using all available data for the Nanushuk Group, estimated coal resources for the National Petroleum Reserve in Alaska portion of the North Slope. Using their methodology and all available data for the area of the known Nanushuk Group coal-bearing rocks, hypothetical coal resources for the Nanushuk Group on the North Slope are shown on Table 1. In summary, there are 1.3 trillion short tons of subbituminous and 1.9 trillion short tons of bituminous coal, for a total of 3.2 trillion short tons of hypothetical coal resources for the Nanushuk Group on the North Slope of Alaska.

The Nanushuk Group coals of the North Slope range in apparent rank from Lignite A to high-volatile A bituminous coal with a mean of high-volatile C bituminous coal (Fig. 10). Total sulfur content ranges from 0.1 to 2.0 percent and has a mean of 0.3 percent, and the ash content has a mean of 11.0 percent (Affolter and Stricker, 1987a).

Colville Group

The Upper Cretaceous Colville Group (Fig. 9), a Brookian sequence younger than the Nanushuk Group, is best exposed along the lower part of the Colville River and its tributaries. Most of its areal extent is in northeastern Alaska (Fig. 10). Coals of the Colville Group and the younger Cretaceous and Tertiary Sagavanirktok Formation have been studied less than those of the Nanushuk Group because they have shown less economic potential. The Colville Group contains coal, but outcrop descriptions indicate that most coal beds are thinner and of lower rank than those in the Nanushuk Group. Many of these coals are described as lignites and "bony coals."

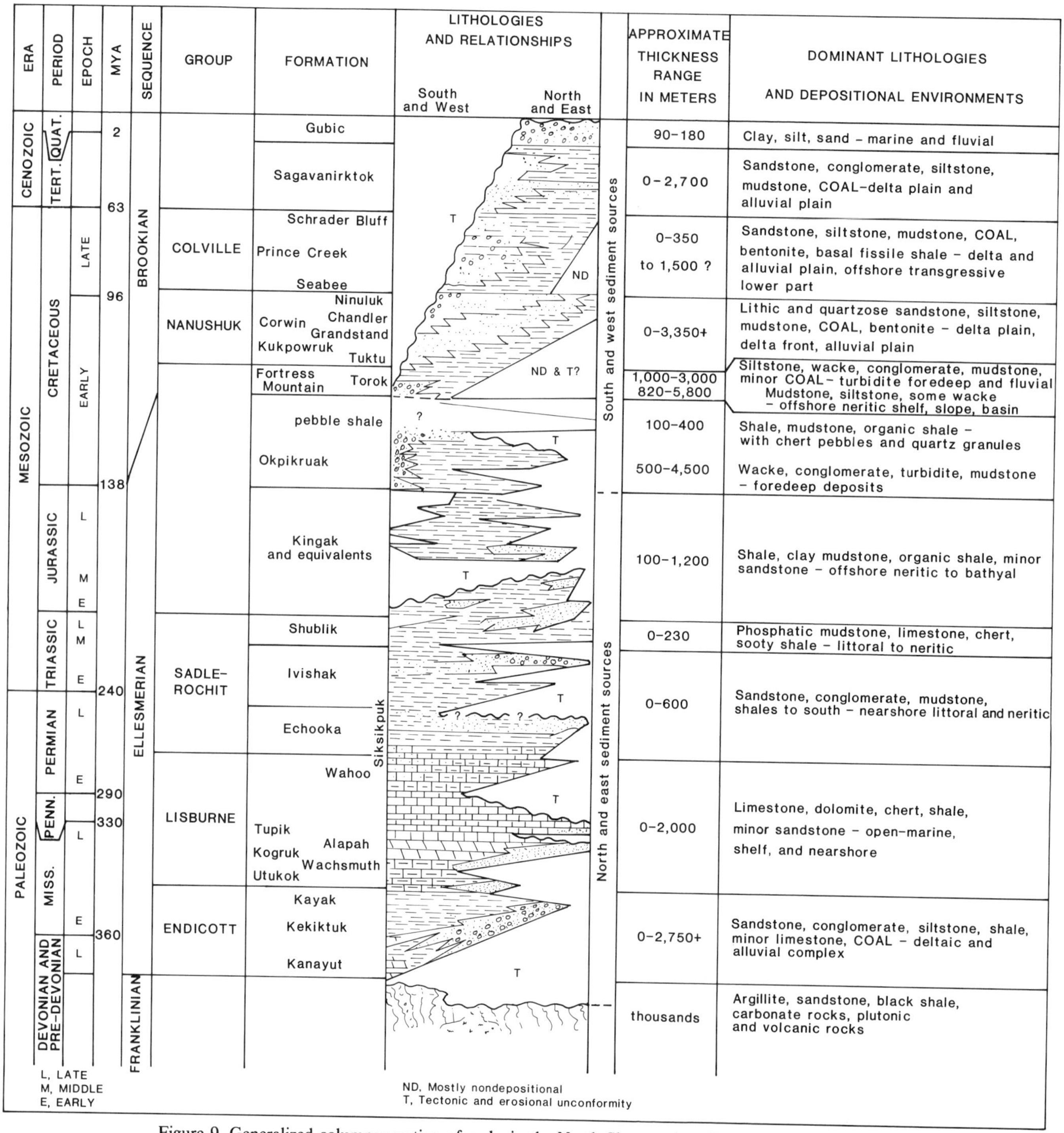

Figure 9. Generalized columnar section of rocks in the North Slope region (from Sable and Stricker, 1987).

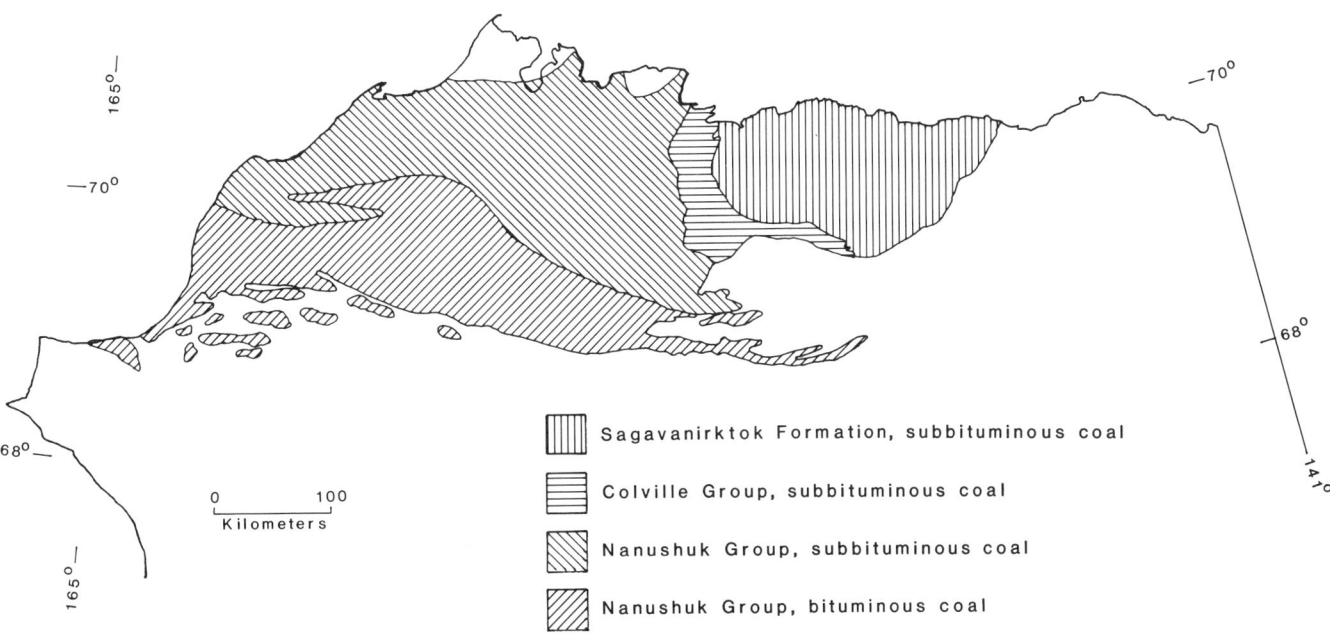

Figure 10. Distribution of coal rank in the North Slope region (modified from Sable and Stricker, 1987).

Sagavanirktok Formation

The Sagavanirktok Formation (Fig. 9), a thick sequence of sandstones, siltstones, mudstones, conglomerates, and coals, represents the final filling of the Colville Basin in the eastern North Slope during Late Cretaceous–early Tertiary time. Coal beds are distributed over an area of 15,000 km^2 (Fig. 10). West of the Sagavanirktok River the coal-bearing interval is as thick as 0.73 km. In the Prudhoe Bay area, the coal-bearing interval has been informally divided into an upper and lower coal zone (Roberts, 1991). The upper coal zone is as thick as 110 m and contains seven coal beds, and the lower coal zone is as thick as 260 m and contains 12 coal beds. The coals accumulated in alluvial and deltaic depositional environments. Apparent rank for these Sagavanirktok Formation coals range from lignite A to subbituminous B, with a mean of subbituminous C. Total sulfur content is low, with a mean of 0.37 percent (range of 0.08 to 2.02 percent) and a variable ash content of 1.16 to 46.72 percent (mean of 10.6 percent; Roberts and others, 1991).

Presently, no resource estimates are available for Tertiary coals in the eastern portion of the North Slope of Alaska. Affolter and Stricker (1987b) estimated the offshore hypothetical resources to be 300 billion short tons for the coal-bearing rocks in the Sagavanirktok Formation. My recent work indicates that the coal-bearing rocks of the Sagavanirktok Formation are thicker and of greater later extent onshore than offshore. Therefore, there should be at least as much hypothetical coal resource onshore as offshore in the eastern portion of the North Slope.

BERING RIVER COAL FIELD

The Bering River coal field contains low-volatile bituminous coal to meta-anthracite of unusually high rank (Fig. 11). The coals in the Kulthieth Formation (Fig. 12), of Eocene to Oligocene age (Plafker, 1967, 1987), crop out in a wedge-shaped area about 32 km east-west and from 3 to 8 km north-south. The Bering River coal field is located on the structurally complex Yakutat terrane (Plafker, 1983), near the subduction of the Pacific Plate beneath the North American Plate. The Yakutat terrane was several hundred kilometers south of its present position when these sediments were derived from upland areas of present-day British Columbia and southeastern Alaska.

The Kulthieth Formation is as thick as 2,800 m (Miller, 1957) and consists of cyclic fining- and coarsening-upward sequences. Turner and Whateley (1989) considered the lower part of the Kulthieth to consist of stacked, coarsening-upward, lower delta-plain deposits. The upper part of the Kulthieth is composed of fining-upward delta plain and lower alluvial plain sediments. Turner and Whateley (1989) noted that the thicker coals are found near the top of the fining-upward sequences.

Coals in the Bering River coal field are reported to be as thick as 9 m. However, the area is so extensively deformed that the term "pod" or "lens" is more applicable (Sanders, 1981). Sanders (1976) reported that the Bering River field may contain as much as 3.6 billion short tons of hypothetical coal resources.

The motion of the Yakutat terrane northward and docking and subduction of the terrane beneath the Chugach Mountains

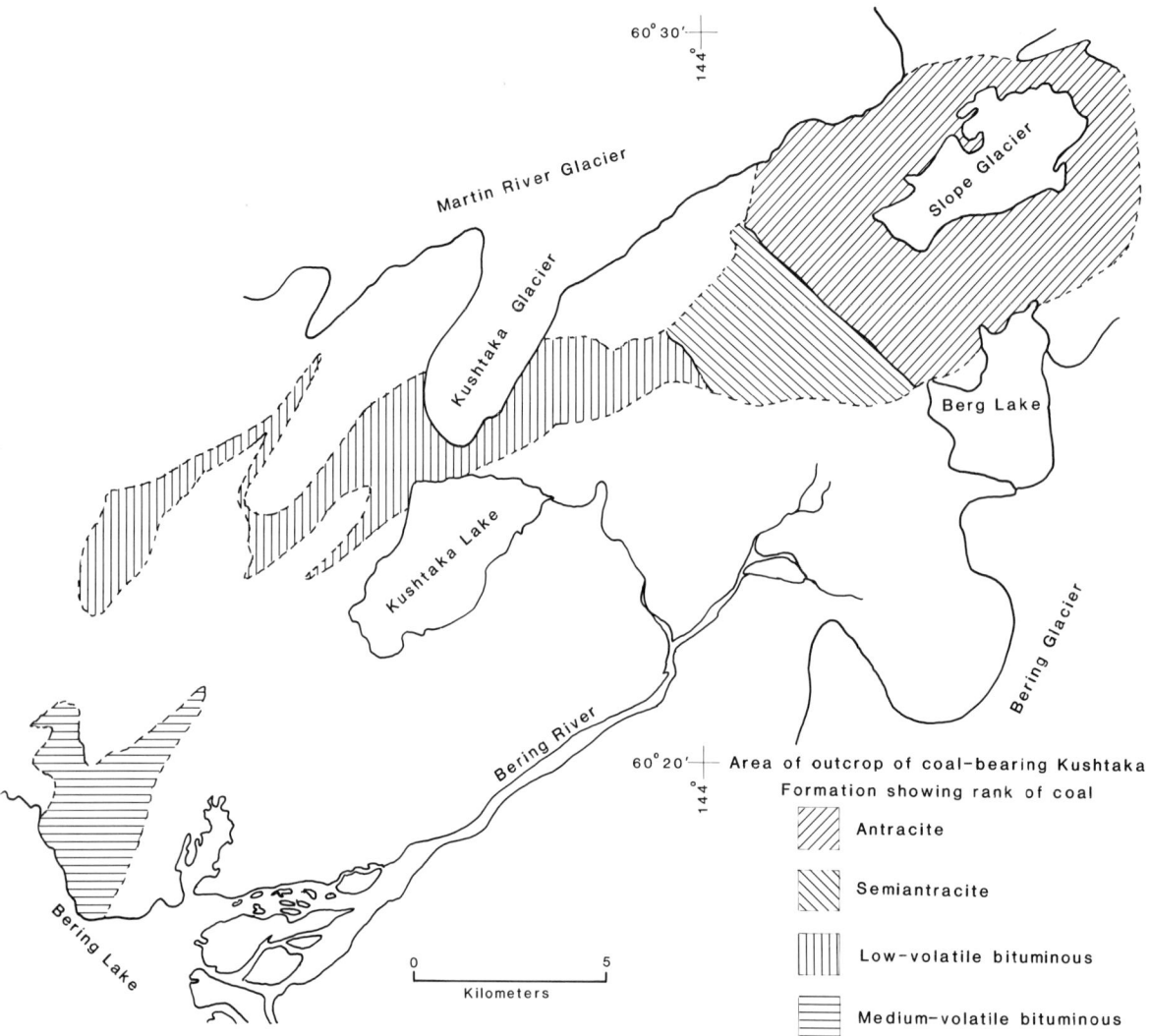

Figure 11. Distribution of coal rank in the Bering River coal field (modified from Barnes, 1951, 1967).

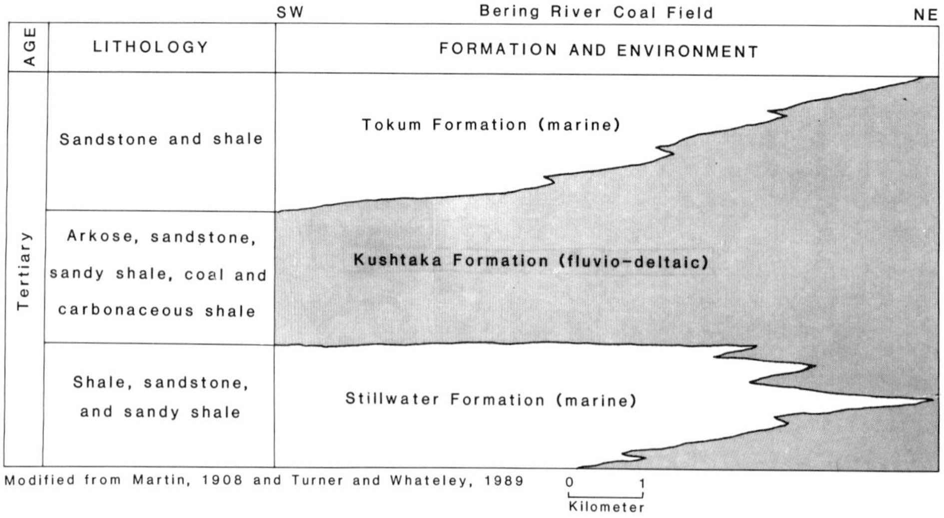

Figure 12. Generalized stratigraphic section of Tertiary rocks in the Bering River coal fields (modified from Martin, 1908; and Turner and Whateley, 1989).

intensely deformed the strata. Igneous intrusions at depth and numerous dikes and sills cut the coal-bearing sedimentary rocks and locally increased the rank of the coal. Bering River coals range in apparent rank from low-volatile bituminous coal at the west end of the field to meta-anthracite at the eastern end. Sulfur content ranges from 0.4 to 5.22 percent, with a mean of 1.2 percent.

SUMMARY

Coal in Alaska ranges in age from Mississippian to early Tertiary in age. Alaskan coals accumulated in many different environments of deposition. The coals with the greatest economic potential are found in the Tertiary Cook Inlet–Nenana regions, whereas most of Alaska's coal resources are in the Cretaceous strata on the North Slope. In general, the coals are bituminous to subbituminous in apparent rank and have the lowest sulfur contents of any United States coals. Merritt and Hawley (1986) have estimated that Alaska contains more than 170 billion short tons of identified and 5,600 billion short tons of hypothetical coal resources. With this large amount of coal, Alaska has the potential to play an important role in supplying the future energy needs of the United States.

REFERENCES CITED

Affolter, R. H., and Stricker, G. D., 1987a, Geochemistry of coal from the Cretaceous Corwin and Chandler Formations, National Petroleum Reserve in Alaska (NPRA), in Tailleur, I. L., and Weimer, R., eds., Alaskan North Slope geology: Society of Economic Paleontologists and Mineralogists and Alaska Geological Society Book 50, p. 217–224.

—— , 1987b, Offshore Alaska coal, in Scholl, D. W., Grantz, A., and Vedder, J. G., eds., Geology and resource potential of the continental margin of western North America and adjacent ocean basins: Houston, Texas, Circum-Pacific Council for Energy and Mineral Resources, Earth Science Series, v. 6, p. 639–647.

—— , 1987c, Variations in element distribution of coal from the Usibelli Mine, Healy, Alaska, in Rao, P. D., ed., Focus on Alaska's Coal, '86; Proceedings of the Conference: Mineral Industry Research Laboratory Report 72, p. 91–99.

Affolter, R. H., Simon, F. H., and Stricker, G. D., 1981, Chemical analyses of coal from the Healy, Kenai, Seldovia, and Utukok River Quadrangles, Alaska: U.S. Geological Survey Open-File Report 81-654, 88 p., scale 1:250,000.

Ahlbrandt, T. S., Huffman, A. C., Jr., Fox, J. E., and Pasternack, I., 1979, Depositional framework and reservoir quality studies of selected Nanushuk Group outcrops, North Slope, Alaska, in Ahlbrandt, T. S., ed., Preliminary geologic, petrologic, and paleontologic results of the study of Nanushuk Group rocks, North Slope, Alaska: U.S. Geological Survey Circular 794, p. 14–31.

Averitt, P., 1975, Coal resources of the United States, January 1, 1974: U.S. Geological Survey Bulletin 1412, p. 1–131.

Barnes, F. F., 1951, A review of the geology and coal resources of the Bering River coal field, Alaska: U.S. Geological Survey Circular 146, 11 p.

—— , 1967, Coal resources of Alaska: U.S. Geological Survey Bulletin 1242-B, 36 p.

Barnes, F. F., and Payne, T. G., 1956, The Wishbone Hill district, Matanuska coal field, Alaska: U.S. Geological Survey Bulletin 1016, p. 1–88.

Burk, C. A., 1965, Geology of the Alaska Peninsula; Island arc and continental margin: Geological Society of America Memoir 99, 250 p.

Calderwood, K. W., and Fackler, W. C., 1972, Proposed stratigraphic nomenclature for Kenai Group, Cook Inlet Basin, Alaska: American Association of Petroleum Geologists Bulletin, v. 56, p. 739–754.

Clardy, B. I., 1984, Bedrock geologic features of the Matanuska Valley, in Clardy, B. I., Hanley, P. T., Hawley, C. C., and Labelle, J. C., eds., Guide to the bedrock and glacial geology of the Glenn Highway, Anchorage to the Matanuska Glacier and Matanuska coal mining district: Anchorage, Alaska Geological Society Guidebook, p. 33–44.

Conwell, C. N., and Triplehorn, D. M., 1978, Herendeen Bay-Chignik coals, southern Alaska Peninsula: Alaska Division of Geological and Geophysical Surveys Report 8, 15 p.

Detterman, R. L., 1978, Interpretations of depositional environments in the Chignik Formation, Alaska Peninsula: U.S. Geological Survey Circular 772-B, p. 62–63.

Dickinson, W. R., 1974, Sedimentation within and beside ancient and modern magmatic arcs, in Dott, R. H., Jr., and Shaver, R. H., eds., Modern and ancient geosynclinal sedimentation: Society of Economic Paleontologists and Mineralogists Special Publication 19, p. 230–239.

Ferm, J. C., and Muthig, P. J., 1982, A study of United States coal resources: U.S. Department of Energy, Jet Propulsion Laboratory 82-14, 110 p.

Fisher, M. A., and Magoon, L. B., 1978, Geologic framework of lower Cook Inlet, Alaska: American Association of Petroleum Geologists Bulletin, v. 62, p. 373–402.

Gates, G. O., 1944, Part of the Herendeen Bay coal field, Alaska: U.S. Geological Survey Open-File Report, 6 p.

Green, C. B., and Bundtzen, T. K., 1989, Summary of Alaska's Mineral Industry in 1988: Alaska Division of Geological and Geophysical Surveys Public Data File 89-7, 6 p.

Hartman, D. C., Pressel, G. H., and McGee, D. L., 1971, Kenai Group of Cook Inlet, Alaska: Alaska Division of Geological and Geophysical Surveys Special Report 5, 4 p.

Hayes, J. B., Harms, J. C., and Wilson, T., Jr., 1976, Contrasts between braided and meandering stream deposits, Beluga and Sterling Formations (Tertiary), Cook Inlet, Alaska, in Miller, T. P., ed., Recent and Ancient sedimentary environments in Alaska: Alaska Geological Society Symposium Proceedings, p. J1–J27.

Hite, D. M., 1976, Some sedimentary aspects of the Kenai Group, Cook Inlet, Alaska, in Miller, T. P., ed., Recent and Ancient sedimentary environments in Alaska: Alaska Geological Society Symposium Proceedings, p. I1–I23.

Huffman, A. C., Jr., Ahlbrandt, T. S. Pasternack, I., Stricker, G. D., and Fox, J. E., 1984, Deposition and sedimentologic factors affecting the reservoir potential of the Cretaceous Nanushuk Group, central North Slope, Alaska, in Huffman, A. C., Jr., ed., Geology of the Nanushuk Group and related rocks, North Slope: U.S. Geological Survey Bulletin 1614, p. 61–74.

Kirschner, C. E., and Lyon, C. A., 1973, Stratigraphic and tectonic development of Cook Inlet petroleum province, in Pitcher, M. G., ed., Arctic geology: American Association of Petroleum Geologists Memoir 19, p. 396–407.

Magoon, L. B., and Egbert, R. M., 1986, Framework geology and sandstone composition, in Magoon, L. B., ed., Geologic studies of the Lower Cook Inlet COST No. 1 Well, Alaska Outer Continental Shelf: U.S. Geological Survey Bulletin 1596, p. 65–90.

Magoon, L. B., Adkison, W. L., and Egbert, R. M., 1976, Map showing geology, wildcat wells, Tertiary plant fossil localities, K-Ar age dates, and petroleum operations, Cook Inlet area, Alaska: U.S. Geological Survey Miscellaneous Investigations I-1019, 3 sheets, scale 1:250,000.

Mancini, E. A., and Deeter, T. M., 1977, Alaska Peninsula Late Cretaceous fore-arc deposition [abs]: American Association of Petroleum Geologists Bulletin, v. 61, p. 811.

Mancini, E. A., Deeter, T. M., and Wingate, F. H., 1978, Upper Cretaceous arc-trench gap sedimentation on the Alaska Peninsula, Alaska: Geology, v. 6, p. 437–439.

Martin, G. C., 1908, Geology and mineral resources of the Controller Bay region, Alaska: U.S. Geological Survey Bulletin 335, 141 p.

McConkey, W., Lane, D. Quinlan, C., Rahm, M., and Rutledge, G., 1977, Alaskan regional Energy Resources Planning Project-Phase 1; Vol. 1, Findings and analysis: Anchorage, Alaska Division of Energy and Power Development, 289 p.

McGee, D. L., and Emmel, K. S., 1979, Alaska coal resources: Alaska Division of Geological and Geophysical Surveys Open-File Report 79-1, 23 p. (unpublished report).

McGee, D. L., and O'Connor, K. M., 1975, Cook Inlet basin subsurface coal study: Alaska Division of Geological and Geophysical Surveys Open-File Report 74, 19 p.

Merritt, R. D., and Belowich, M. A., 1984, Coal geology and resources of the geology and resources of the Matanuska Valley: Alaska Division of Geological and Geophysical Surveys Open-File Report 84-24, 64 p.

Merritt, R. D., and Hawley, C. C., 1986, Map of Alaska's coal resources: Alaska Division of Geological and Geophysical Surveys Special Report 37, 1 sheet, scale 1:2,500,000.

Merritt, R. D., and McGee, D. L., 1986, Depositional environments and resource potential of Cretaceous coal-bearing strata at Chignik and Herendeen Bay, Alaska Peninsula: Sedimentary Geology, v. 49, p. 21–49.

Merritt, R. D., and 6 others, 1987, Southern Kenai Peninsula (Homer District) Coal-Resource Assessment and Mapping Project: Final report: Alaska Division of Geological and Geophysical Surveys Public Data File 87-15, 125 p.

Miller, D. J., 1957, Geology of the southeastern part of the Robinson Mountains, Yakataga District, Alaska: U.S. Geological Survey Oil and Gas Investigations Map M-187, 2 sheets, scale 1:63,360.

Molenaar, C. M., 1981, Depositional history and seismic stratigraphy of Lower Cretaceous rocks, National Petroleum Reserve in Alaska and adjacent areas: U.S. Geological Survey Open-File Report 81-1084, 42 p.

—— , 1985, Subsurface correlations and depositional history of the Nanushuk Group and related strata, North Slope, Alaska: U.S. Geological Survey Bulletin 1614, p. 37–60.

Moore, J. C., 1974, The ancient continental margin of Alaska, in Burk, C. A., and Drake, C. L., eds., The geology of continental margins: New York, Springer-Verlag, p. 811–816.

Plafker, G., 1967, Geologic map of the Gulf of Alaska Tertiary Province, Alaska: U.S. Geological Survey Miscellaneous Geologic Investigations Map I-484, scale 1:500,000.

—— , 1983, The Yakutat Block, an active tectonostratigraphic terrane in southern Alaska [abs.]: Geological Society of America Abstracts with Programs, v. 15, p. 406.

—— , 1987, Regional geology and petroleum potential of the northern Gulf of Alaska Continental Margin, in Scholl, D. W., Grantz, A., and Vedder, J. G., eds., Geology and resource potential of the continental margin of western North America and adjacent ocean basins: Houston, Texas, Circum-Pacific Council for Energy and Mineral Resources Earth Science Series, v. 6, p. 229–268.

Roberts, S. B., 1991, Cross section showing subsurface coal beds in the Sagavanirktok Formation, vicinity of Prudhoe Bay, North Slope, Alaska: U.S. Geological Survey Coal Investigations Map C-139A, 1 sheet (in press).

Roberts, S. B., Stricker, G. D., and Affolter, R. H., 1991, Stratigraphic sections and coal quality analyses from outcrops in the Late Cretaceous–Tertiary Sagavanirktok Formation, east-central North Slope, Alaska: U.S. Geological Survey Coal Investigations Map C-139B, 1 plate (in press).

Roehler, H. W., and Stricker, G. D., 1979, Stratigraphy and sedimentation of the Torok, Kukpowruk, and Corwin Formations in the Kokolik-Utukok River region, National Petroleum Reserve in Alaska: U.S. Geological Survey Open-File Report 79-995, 80 p.

Sable, E. G., and Stricker, G. D., 1987, Coal in the National Petroleum Reserve in Alaska (NPRA); Framework geology and resources, in Tailleur, I. L., and Weimer, R., eds., Alaskan North Slope geology: Society of Economic Paleontologists and Mineralogists and Alaska Geological Society Book 50, p. 195–216.

Sanders, R. B., 1976, Geology and coal resources of the Bering River field, in Rao, P. D., and Wolff, E. N., eds., Focus on Alaska's Coal '75; Proceedings of the Conference: Fairbanks, University of Alaska, School of Mineral Industry Report 37, p. 54–58.

—— , 1981, Coal resources of Alaska, in Rao, P D., and Wolff, E. N., eds., Focus on Alaska's coal '80; Proceedings of the Conference: Fairbanks, University of Alaska, School of Mineral Industry Report 50, p. 11–31.

Spicer, R A., 1987, Late Cretaceous floras and terrestrial environment of northern Alaska, in Tailleur, I. L., and Weimer, R., eds., Alaskan North Slope geology: Society of Economic Paleontologists and Mineralogists and Alaska Geological Society Book 50, p. 497–512.

Stanley, R. G., Flores, R. M., and Wiley, T. J., 1989, Contrasting depositional styles in Tertiary fluvial deposits of Nenana coal field, central Alaska [Abs.]: American Association of Petroleum Geologists Bulletin, v. 73, p. 415.

Stricker, G. D., Affolter, R. H., and Brownfield, M. E., 1986, Geochemical characterization of selected coals from the Beluga energy resource area, south-central Alaska; Site of a proposed coal mine, in Carter, L.M.H., ed., U.S. Geological Survey research on energy resources—1986: U.S. Geological Survey Circular 974, p. 65–66.

Tetra Tech, Inc., Energy Management Division, 1982, Petroleum exploration of NPRA, 1974–1981, final report: Houston, Texas, Tetra Tech Reports 8200 and 8202. (3 volumes, 4 boxes of geophysical maps and plates)

Turner, B. R., and Whateley, M.K.G., 1989, Tidally influenced coal-bearing sediments in the Tertiary Bering River coal field, south-central Alaska: Sedimentary Geology, v. 61, p. 11–123.

U.S. Geological Survey, 1982, 1982 Annual report on Alaska's mineral resources: U.S. Geological Survey Circular 884, p. 23–27.

U.S. Geological Survey, 1983, 1983 Annual report on Alaska's mineral resources: U.S. Geological Survey Circular 908, p. 19–22.

U.S. Geological Survey, 1984, 1984 Annual report on Alaska's mineral resources: U.S. Geological Survey Circular 940, p. 19–21.

U.S. Geological Survey, 1985, 1985 Annual report on Alaska's mineral resources: U.S. Geological Survey Circular 970, p. 21–23.

U.S. Geological Survey, 1986, 1986 Annual report on Alaska's mineral resources: U.S. Geological Survey Circular 983, p. 18–21.

U.S. Geological Survey, 1987, 1987 Annual report on Alaska's mineral resources: U.S. Geological Survey Circular 1012, p. 24–27.

U.S. Geological Survey, 1988, 1988 Annual report on Alaska's mineral resources: U.S. Geological Survey Circular 1023, p. 23–26.

Wahrhaftig, C., 1987, The Cenozoic section of Suntrana, Alaska, in Hill, M. L., ed., The Cordilleran Section of the Geological Society of America: Boulder, Colorado, Geological Society of America, Centennial Field Guide 1, p. 445–450.

Wahrhaftig, C., Wolfe, J. A., Leopold, E. B., and Lanphere, M. A., 1969, The coal-bearing group in the Nenana coal field, Alaska: U.S. Geological Survey Bulletin 1274-D, p. D1–D30.

Wahrhaftig, C., Bartsch-Winkler, S., and Stricker, G. D., 1991, Coal in Alaska, in Plafker, G., and Berg, H. C., eds., Geology of Alaska: Boulder, Colorado, Geological Society of America, The Geology of North America, v. G-1 (in press).

Waring, G. A., 1934, Core drilling for coal in the Moose Creek area, Alaska: U.S. Geological Survey Bulletin 857-E, p. 155–173.

Wolfe, J A., and Tanai, T., 1980, The Miocene Seldovia Point flora from the Kenai Group, Alaska: U.S. Geological Survey Professional Paper 1105, 52 p.

Wolfe, J. A., and Upchurch, G. R., 1987, North American nonmarine climates and vegetation during the Late Cretaceous: Palaeogeography, Palaeoclimatology, Palaeoecology, v. 61, p. 33–77.

Wood, G. H., Jr., and Bour, W. B., III, 1988, Coal map of North America: U.S. Geological Survey, 2 sheets, scale 1:5,000,000.

Manuscript Accepted by the Society September 25, 1990

Printed in U.S.A.

Index

[Italic page numbers indicate major references]

Abbot Formation, 529
Abo, 354
Abo reservoir, 351
Absaroka system, 393
Absaroka thrust, 393, 394, *397*
Acadian clastic wedge, 523
acanthite, 9
acrotelm, 469, 470, 471
Adaville Formation, 549, 556, 558, 560
Africa, 137
  central, 99
  southwest, 128
Aiken County, South Carolina, 194
Alabama, 192, 194, 196, 573
Alafia River, 154
alkali feldspar, 148
Alapah Limestone, 160
Alaska, 4, 6, 12, 18, 25, 26, *35*, *56*, 76, 95, *117*, 120, 131, 149, 153, *156*, *160*, 192, 198, *203*, 216, *447*, 483, *591*
  coals, *591*
  deposits, 6
  North Slope, *221*, *447*
  southeastern, 6, 95
Alaska Range, *592*
Alberhill area, California, 195
Alberta, Canada, 475, 483, 491, 515
Alberta Plains, 505, 518
Albion-Scipio field, 290, *294*
albite, 5, 10, 134, 200
Alcova Limestone, 377
Aleutian arc, 10
alginites, 509
Algoma mining district, *73*
Algoma type, 66, *67*, *73*, 80
  genesis, *73*
Alkali Gulch field, Colorado, 363
Allamore district, Texas, 200
Allard Lake, Canada, 136
Allegan Platform, 297
Alleghany clastic wedge, 523
Allegheny Valley district, Pennsylvania, 6, 7, 195
Alligator Rivers area, 121
allocycles, *534*
alluvial deposits, 4, 5
Almaden deposit, Spain, *145*
Almaden Mine, Spain, *145*
Almond Formation, 549, 558
Alston Block, England, 519
alteration
  hydrothermal, 29
  subsurface, *227*
alumina, 74, 147, 206
aluminum, *147*, 148, 149
alunite, 9, 10, 148
Amador County, California, 589
Amanoa, 493
Amarillo-Wichita uplift, 327, 331
Amberjack High, 431
American River, 4

Amherstburg Formation, 296
Amsden Formation, 377
Anaconda system, 79
Anadarko basin, *173*, *219*, *325*
  exploration history, *327*
  future potential, *336*
  stratigraphy, *327*
  thermal history, *331*
anatase, 136
andalusite, 205
Andarko basin, 179
  reserves, *336*
Anderson-Dietz coals, 561, 563
Anderson Mine, 118
Anderson-Wyodak coals, *563*
angiosperms, 495, 498
anhydrite, 166, 168, 175, 176, 296, 299, 358, 397, 399
Animas Formation, *359*
Animikie Basin, *70*
Animikie Group, *70*
anisotropy, 517
Ankareh Formation, 395
ankerite, 6, 9, 154
anomaly, geomorphic, 403
anorthite, 200
anorthosites, 148
  massif-type, 76
Anschutz Ranch field, 395, 398
Antelope shale, 422, 425
anthracite, 508, 517, 523, *589*, 594
Anthracite Depression, 517
anthracite region, *527*, *530*
Anthracite Ridge coals, 594
antimony production, 14
apatite, 77, 138, 139, 153, 474
  igneous, 153
Apex Mine, Nevada, *116*
aplite, 129
Appalachian basin, 169, 176, *218*, 222, *273*, *523*, 526, *527*, *529*, 534, 539, 541
  allocycles, *536*
  distribution patterns, *540*
  exploration, *279*
  future exploration, *285*
  history, *279*
  petroleum geology, *273*
  production, *283*
  resources, *283*
  stratigraphy, *273*
  thermal history, *276*
Appalachian belt, 98
Appalachian Carboniferous, 478
Appalachian deposits, *51*
Appalachian foreland trough, 526
Appalachian Mountains, 59, 172, 514, 517
  central, *523*
  northern, 121
Appalachian Piedmont, 200
Appalachian region, 195, 197, 204, 475

Canada, 18, 38
Applachian trough, 530, 532
aragonite, 190
Araucaria, 493
Araxa, Brazil, *133*, 139
Arbuckle Group, 328, 333
Arcadia Formation, 159
Arctic Coastal Plain, 216, *221*
Arctic National Wildlife Refuge, 221
Arctic platform, 447, *452*
Arctic Slope, 447
arenite, 397
Argentina, 144
argentite, 14
argillite, 14, 198
Arizona, 13, 24, *25*, *28*, *33*, 73, *117*, *118*, 131, 192, 196, 207, 549
  southern, 10
  western, 10
Ark-La-Tex area, 312
Arkansas, *119*, 148, 153, 179, 196, 198, 507, 573
  northern, 51
Arkansas Novaculite, 199
Arkoma basin, 507, 516, 526
Arkoma coalfield, *523*
Arkoma trough, 532
Aroostock County, Maine, 80
arsenopyrite, 6, 8, 12, 15
Arthrotaxites, 492
Artillery district, Arizona, 80
Artillery Peak, 118
asbestos, *201*, *202*
ash, volcanic, *472*, 475, *540*, 560, 578, 586, 587, 595
Aspen, Colorado, 55
Aspen Formation, 396, *397*, 400
Asphalto field, 425
Atacama Desert, 179
Atascosa County, 579
Athabasca Basin, Canada, 121
Atlantic City district, Wyoming, 73
Atlantic coast, 137
Atlantic Coastal Plain, *157*, *161*
Atlantic Continental Shelf, 140
Atoka Formation, 507
Atokan section, 347
Atokan shales, 350
Atolia, California, 127
Austin Chalk, 252, 313, 314
Austin District, Nevada, *116*
Austin strata, 305
Austinville, Virginia, 43
Austinville-Ivanhoe, Virginia, 51
Australia, 87, 97, 121, 128, 131, 134, 136, 137, 140, *142*, 144, 149
  brown coal, 505
Austria, 128
autocycles, *534*
Avery Island, Louisiana, 173
Axial basin anticline, 549
Aztec Wash field, 365
azurite, 16

Bacon Flat field, *403, 407,* 411, 414
Baden well, 330
Baker Mountains, 206
Bakersfield arch, 419, *425*
Balmat, New York, 46
Balmat deposit, *54*
Baltic area, 183
Banded series, *90*
bannerite, 139, 140
Baraga Group, 70, *71*
Baringer Hill, Texas, *104*
barite, 9, 11, 12, 14, 15, *56*, 58, *196, 197,* 198
Barker dome, 361
Barnwell Formation, 196
Barrett lignite zone, 587
Barrow arch, 447, 451, 452, *453,* 459
  extension, 460
Barrow field, 458
Barrow-Prudhoe group, *458*
  oil types, *458*
Barstow Formation, 209
barylite, 143
barytes, 197
Basal series, *88*
basalts, 8, 30
  flows, 405
  pillow, 34
base metals, 28
Basin and Range Province, 10, 12, 143, *221,* 403, *411,* 584
Basin Dakota field, 362, 363
Basin Fruitland field, 368
basins
  continental, *588*
  evaporite, 171
  extensional, *587*
  intermontane, *588*
bastnaesite, 138, 139, 140
bat caves, 153
Batesville, Arkansas, 82
Battle Mountain, Nevada, *12,* 17
Battle Spring Formation, *114*
bauxite, *147, 148, 149*
Baxter deposit, 204
Bayan Obo deposits, China, 139
Bayhorse deposit, Idaho, *204*
Bayhorse Dolomite, 204
beach ridges, peat, *472*
Bear Lodge Mountains, Wyoming, 141
Bear River Formation, 396, *397, 400,* 550
Bear Valley, Idaho, 132, 134, 141
Bearpaw Shale, 196
Beaufort County, North Carolina, 156
Beaufort-Mackenzie Delta, 301
Beaufort margin, 452, *453*
Beaufort Sea, 447
Beaufort shelf, 460
Beaulah-Zap coal, 565
Beaver Valley district, Pennsylvania, 195
Beckham County, Oklahoma, 330
Beckley coal, 527, 540
bedrock, 5, 97
Belden Shale, 55, 56
Belgian Anticline Field areaa, 420
Bell Canyon reservoir, 251
Bell Canyon Sandstone, 251, 353

Bell Creek field, 375, 380, 387
Belle Fourche arch, 377
Belle Fourche Shale, 378, 385
Belt basin, 18
Belt Supergroup, 14, 18
Beluga coalfield, 591, *594*
Beluga Formation, *594*
Bend conglomerate, 347
Bendelari monocline, 51
Benton Shale, 195
bentonite, *195,* 378, 381, 456
bentonite sodium, 196
Berea Sandstone, 298
Bering River coalfield, 591, *599*
Bernie Lake, Manitoba, 134
Bertha Rodgers well, 328
bertrandite, *142*
beryl, *142*
beryllium, 134, *142, 143, 144*
Betsie marine zone, 534
Betsie Shale, 529
Bevier coal bed, 532
Big Brown mine, Freestone, County, 576, 578
Big Bug district, Arizona, 17
Big George coals, 561
Big Lime 336
Big Mountain, 207
Big Muddy field, 375, 376
Big Sandy field, 281
Bighorn basin, 550, 561, 562
Bighorn Mountains, 376, 377, 383
Bingham, Utah, 17, 24, *28,* 32, 46
Bingham-Lark, Utah, 59
biotite, 15, 94
Birmingham district, Alabama, 74
Bisbee, Arizona, 59
bismuthinite, 9
Bisti field, 361, 362, 365
Biwabik Formation, 70, 71
Black coals, 560
Black Hawk Formation, 493
Black Hills, 196, 200, 206, 377, 383
  southern, 8
Black Hills uplift, 383
Black Lake field, Louisiana, 313
Black Mesa coalfield, 547, 549
Black Range, 206
Black River Group, 290, 292, 294
Black Warrior basin, 222, 273, 523, 527, 539, 540
black smokers, 59
Blackbird deposit, *99,* 101
Blackburn field, *403, 408,* 411, 413, *414*
Blackhawk Formation, 549, 558, 560
Blackwells Corner field, 419
Blaine Formation, 173
Blanco gas field, 220
Blanco Mesaverde field, 362, 363
Blanco Pictured Cliffs South, 363
bleaching, 15
Blue Mountains, 96
Blue Mountains–Seven Devils terranes, 584
Blue Ridge Mountains, 200, 206
blue-rock deposits, *160*
Bluejacket coal bed, 532
Bodega basin, 443

boehmite, 147
Bokan Mountain, Alaska, *117*
Bolivia, 127, 128, *130,* 131
Bolshe Tokmak, Soviet Union, 79
Bond Creek, Alaska, 28
Bone Springs reservoir, 351
Bone Valley Formation, 158, *159,* 207
Bonito field, 437
Bonne Terre Mine, 47, 48
Bonneterre, Missouri, 47
Bonneterre Formation, 46, 48, 49
Bonneville salt flats, 175
Books cliffs, 549
Boone Formation, 50, *51*
borates, hydrated, 178
bornite, 17, 18, 27, 111
Boron, California, 178
boron minerals, *178*
Bossier Shale, 303
boulangerite, 15
Boulder District, Colorado, 127
Boulder field, 365
bournonite, 15
Bowie, Arizona, 208
Bowlens Pyramid, North Carolina, 206
Bowling Green fault, 294
Bradford oil field, 280, 281, 285
Bramsche Massif, 518
Brandon Lignite, Vermont, 487
Brazil, 79, 131, 132, *133,* 136, 137, 140, 142, 144, 149, 183
Breathitt Formation, 529
Breccia Hill area, 55
breccias
  collapse, *51*
  deposits, *128*
  pipes, *117,* 120, 130
Brewster County, Texas, 205
brimstone, 176
brines, 16, 55, 128, *165, 172,* 179
Broad Top coalfield, Pennsylvania, 517
brockite, 140
Bromide Formation, 341
bromine, *179*
Brookian sequence, 447, 453, *456*
Brooks Range, 18, *56,* 156, 160, 186, 216, 447, *451,* 457, 596
Broom Creek Group, 377
Brown Niagaran, 294, 295
brown-rock deposits, *159*
Browns Canyon district, Colorado, 204
brucite, 180
Brunswick deposit, 59
Brushy Basin Member, 363
Buckner anydrite, 303
Buckner red beds, 303
Buena Vista Hills field, Nevada, 78, 425
Buffalo, Wyoming, 376
Buffalo Mine, South Africa, 139
Buffalo Wallow field, 333
Buick Mine, 49
Burlington Limestone, 199
Burma, 128
Burning Springs anticline, 280
Burnt Bluff Group, 296

Burro Canyon Formation, *358*, 369, 492
Burro Mountains district, New Mexico, 204
Bushveld Complex, South Africa, 76, 87
Bushy Basin Member, *112*
Butte, Montana, 4, 17, *32*, 65, 79
Butte Quartz Monzonite, 79
Buttonwillow depocenter, 419, 422, 425

Caballo deposits, New Mexico, 204
Cache Creek, California, 209
Caddell Formation, 576
cadmium, 45, 46
    production, 14
Calaveras Formation, 6, 81
calcite, 6, 9, 10, 11, 14, 15, 16, 115, 165, 190, 342, 397
calcium, 172, 178
calcium silicate, 207
California, 12, 26, 37, 59, 120, 131, 154, *156*, *160*, *174*, 180, 192, 194, 196, 198, 200, 205, 206, 207, *417*
    basins, *443*
    eastern, 10
    northern, 4, 6
    southeastern, 4, 15
    southern, 176
    western, *431*
California Coast Ranges, 10
calomel, 145
Calumet, Michigan, 31
Calvert Bluff Formation, 574, 575, 578
Calvert Bluff lignites, 493
Cambrian, *292*
Camrick field, 327
Canada, 87, 97, 128, 132, 133, 134, 135, 137, 144, 475
    coal deposits, *463*
Canada basin, 447, 457
Canadaway Creek, New York, 279
Canadian basin, western, 518
Canadian Rocky Mountains, 491, 507, 515
Canadian Shield, 532
Cane Creek anticline, 175
canfieldite, 129
Canning Formation, 456, 459
Canyon section, 349
Canyon Springs Sandstone Member, 378
Cape Fear Arch, 157
Cape Lisburne area, Alaska, 492
Cape Province, South Africa, 202
Capitan Reef, 353
Carajas district, Brazil, 66, 80
carbon, *396*
Carbonado Formation, 590
carbonates, 5, 6, 9, 13, 15, *35*, 55, 165, 301, 305, 330, 335, 339, 351, 397, 398, 404
    banks, 294
    beds, 273
    facies, 340
    reservoirs, 282, 313

Carboniferous, pre-, *455*
carbonization, 8
Careaga Formation, 440
Cargo Muchacho district, 15
Cargo Muchacho Mountains, 15
Carlile Shale, 378, 385, 388
Carlin deposit, *11*
Carlin mine, 12, 19
Carlsbad district, New Mexico, *175*
carnotite, 16, *104*
Carr Fork, Utah, 12, *31*
Carson Hill mine, 6
Carter County, Oklahoma, 327
Cartersville district, Georgia, 197, 199
Cascade Mountains, 583, 589
Caseyville Formation, 529
Casmalia field, 439
Casper Formation, 377
Casper Mountain, 377
Caspian Sea, 168
Cassa Group, 377
cassiterite, 129, *130*
Castile Formation, 168, 173, 175
Castle Hayne Formation, 156
Castle Rock area, Washington, 195
Castlegate coals, 560
Cat Canyon field, 435, 440
catagenesis, *227*
Catahoula Tuff, 116
Catlin plant, Nevada, 186
catotelm, 469
Catskill deltaic complex, 523
Cave Canyon, California, 204
Cave in Rock, Illinois, 204
Cave Peak, Texas, 37
cawk, 197
Cedar Hill Basal Coalfield, 368
Cedar Mountain Formation, 492
cellulose, 471
Celo Knob, North Carolina, 206
cement, *191*
Cement field, 336
Cenozoic, *422*, *560*, *562*, *567*
Central Appalachian basin, 526
Central Basin, Tennessee, 159
Central Basin Platform, 219, 342, *345*, 353
Central City District, Colorado, 32, *103*, *120*
Central district, Missouri, 198
Central district, New Mexico, 78
Central Kansas uplift, 330, 331
Central Stable Interior, 216
Centralia, 586
Centralia-Chehalis coal district, 583, *584*, 589, 590
cerargyrite, 16
cerium subgroup, 140
Cerro de Mercado, Mexico, 77
Cerro Prieto geothermal field, Baja California, 507
Chaco slope, 357, 366
Chacon field, 365
Chaffee Formation, 55
Chainman Sandstone, 403
Chainman Shale, *404*, 409, *411*, 414, 415
chalcocite, 18

chalcopyrite, 6, 8, 12, 15, 25, 27, 28, 34, 35, 59, 77, 99, 111
Chaloneria, 498
Chamberlain Creek, Arkansas, 196, 198
Chamberlain Creek Syncline, 199
Chamberlain deposit, South Dakota, 81
chamosite, 63, 69, 74
Chanac Formation, 425
Chandler Formation, 596
Chandler Valley, Alaska, 492
channel deposits, 563
Chapin Wash Formatin, 118
Chappel Formation, 342
Charles Formation, 174
Charleston, South Carolina, *154*, *157*
Charleston coals, 532
Chase Group, 331, 335
Chatham, Virginia, 119, 121
Chatam fault zone, 119
Chatham Sag, 296
Chatsworth district, Georgia, 200
Chattanooga Shale, 51, 159, 160, 325
Chemard Lake lignite, 573, 577
Cheney Ranch field, 422, 425
Cherokee County, South Carolina, 64
Cherokee Group, 331, 335
Cherokee sandstones, 335
Cherokee Shale, 51
Cherry Creek Group, 200
chert, 8, 66, 73, 198, 341, 404, 437, *438*, 440
Cheshire area, Connecticut, 196
Chester shales, 335
Chiba Peninsula, 179
Chickaloon Formation, *594*
Chignik Formation, *595*
Chile, 179
China, 79, *127*, 131, 133, 137, 140, 142, 144, 146, 186, 466
Chinle Formation, 111
Chino-Hanover, New Mexico, 59
Chinook-Malta-Glasgow, Montana, 196
Chipeta field, 365
Chiquito field, 365
Chiwaukum basin, 588
chlorite, 6, 10, 13, 14, 15, 105
Chocolate Mountains, 15
Chocolay Group, 70
Chowchilla gorge, 425
Christmas, Arizona, 31
Christmas Creek field, 395
Christmas Mountains, 205
chromite, 87, 88, *89*, *96*, 98
    production, 96
chromium, *87*, 95, 97, 101
    uses, 87
chrysoberyl, 143
chrysotile, 201, *202*
Chuckanut basin, 587, 589
Chugach Mountains, 591, 599
Chugwater Group, *377*
Chukchi platform, 447, 451, *452*
Chukchi Sea, continental shelf, 447
Chukchi shelf, 452, 460
Chuska Sandstone, *359*

Cimarron Salt, 173
Cincinnati arch, 273, 280
cinerites, 474
cinnabar, 11, 145, 146
circum-Pacific belt, 130
Cisco limestone, 348
Cisco section, 349
Cisco Shale, 350
Clackaman County, Oregon, 195
Claiborne Group, 75, *576*, 578
Clairbornian Stage, *576*
Clareton field, 375
Clareton Trend, 375, 380
Clark Mountain, 204
Clarksville district, Georgia, 206
Classopollis, 492
clastics, 35, 301, 339, 353, 404, 405, 422, 423, 447, *485*, 491
  sedimentation, 564
  wedges, 526, 532
Clavatipollenites, 492
Clay Spur Bentonite, 196, 378
clays, 11, 15, 129, 143, *192*, 195, 282, 398, 424, 427, 493, 539, 540
  ball, 193, *194*
  flint, 530
  high-alumina, 148
  minerals, 110, 195
  refractory, 193, *194*
  residual, 82, 138, 198
claystone, 367, 427, 561, 563, 594
Clayton Formation, 573, 574
Clayton Valley, Nevada, 179
Clearfield district, Pennsylvania, 195
Clearfork reservoir, 351, 354
Cliff House Sandstone, 368
Climax, Colorado, 29, 127, 128
clinoptilolite, 209
Clinton field, Ohio, 280
Clinton Formation, 74
Clinton member, 296
Clinton Sandstones, 282
Clinton Shale, 294
Coal Valley Member, *595*
Coalburg coal bed, 540
Coaledo Formation, 586
coalification, *503*
  by-products, *509*
  coalfields, *503*
  maps, 519
  patterns, *503*, *513*, *514*, *515*, 517
Coalinga, California, 202
Coalinga field, *419*, 426
Coalmont Formation, 565
coals, 114, 145, 183, 220, 378, *463*, 505, *523*, *547*, *550*, *573*, *583*, *588*, 596
  ages, *495*
  Alaska, *591*
  anthracite, 509, *589*
  ash content, *472*, 475
  Australian brown, 505
  backswamp, 563
  balls, *483*, *485*, *489*, *495*, *498*
  basins, *547*, *583*, *584*
  beds, 222, 285, 363, *368*, *369*, *523*, *527*, *530*, *539*, *540*

  bituminous, *505*, 508, 509, 516, 517, 523, *540*, 560, 563, 567, 568, 583, 597, 595, 597, 601
  boghead, 472, *498*
  botanic origins, *483*
  brown, 505
  Canada, *463*
  cannel, 472, *498*, 526
  central United States, *523*
  channel complex, *588*
  clastic sediments, *474*
  climate-change model, *539*
  coking, 560
  delta plain, *587*
  depositional environment, *469*
  distribution, *552*, *554*
  extraction technologies, *589*
  facies, *474*
  fixed carbon, *507*
  fluvial environments, *475*
  fluvial systems, *563*
  fluvial-deltaic systems, *558*
  fluvial-dominated shorelines, *476*
  future uses, 464, 466
  geology, *469*
  Great Plains region, *547*
  Gulf Coast, *573*
  heating value, *503*
  kerogenous, 562
  lacustrine systems, *562*
  lignitic, 586
  map of North America, *463*
  maturity, 505
  measures, *523*, *534*
  mineral origin, *472*
  moisture content, *503*
  natural history, *464*
  organic constituents, *508*, *509*
  paleobotany, *483*
  paleoecology, *483*, *487*
  palynology, *485*
  petrology, *560*, *568*
  prediction of quality, *518*
  production, 464, *589*, 595
  properties, *508*
  quality, *540*
  rank, 519
  reserves, 466, 560, 561
  resin-rich, *589*
  resources, *464*, 466, 561, *591*, *592*, 597, *601*
  Rocky Mountains, *547*
  seams, 473, 474, 475, 477, 479, 549
  strand-plain systems, *558*
  subbituminous, 505, 562, 583, 586, 597, 599, 601
  tidal sediments, *479*
  tissues, *483*
  United States, *463*, *523*, *583*
  use, 464
  volatile matter, 507
  wax-rich, *589*
Coast Ranges, 96, 160
Coast Ranges Province, 431
coastal margins, *584*
Coastal Plain province, 447
cobalt, 46, 47, 78, 87, 97, *99*
cobaltite, 99

Cobb coal group, 527
Cochabamba district, Bolivia, 202
Cody Shale, 381, 384, 385
Coeur d'Alene district, Idaho, 4, 13, *14*, 19, 44, 46, 59, 99
coffinite, 113, 118
Cologne, Germany, 505
Colorado, 12, *29*, 44, 80, *118*, 140, 153, 184, 192, 194, 200, 203, 206, *357*, 475, 549
  central, 55
  deposits, 55
  northwest, 475
Colorado Group, 493
Colorado mineral belt, 55
Colorado Plateau, 16, 18, *105*
Columbia River Basalt Group, 149
columbite, 132, 141
columbite-tantalite, 131
columbium, 131
Colville basin, 447, 451, 456, 457, 596
Colville Group, 459, 596, *597*
Colville Valley, Alaska, 492
Comstock fault, 10
Comstock Lode district, 3, 9, *10*
Comus Formation, 198
condensate, 455
conglomerates, 70, 252, 303, 330, 439, 456, 527, 529, 562, 594, 599
Connecticut, 200
Connoquenessing sandstones, 531
Converse County, Wyoming, 250
Cook Inlet coal basin, 591, 594
Cook Inlet petroleum province, 594
Cook Inlet region, 592, 594
Cooks Peak district, New Mexico, 204
Coos Bay, Oregon, 584
Coos Bay coalfield, 583
Coosawhatchie area, South Carolina, *158*
copper, 3, 4, 9, 12, 14, 16, *18*, *23*, *25*, *30*, 34, 35, 36, 38, 46, 58, 59, 78, 87, 94, 95, *99*, 101, 117, *591*
  Bingham (Utah), *28*
  Butte (Montana), *32*
  by-product, 32, 33, 37
  Carr Fork (Utah), *31*
  co-product, 18
  Copper Canyon (Nevada), 12, *31*
  districts, *26*
  Dos Pobres (Arizona), *27*
  Ducktown (Tennessee), *34*
  Duluth (Minnesota), *36*
  future resources, *39*
  Gap (Pennsylvania), *36*
  imports, *40*
  Keweenaw (Michigan), *30*
  locations, *25*
  native, *18*, *23*, *30*
  porphyry deposits, 13, *17*, 18, *17*, *23*, *25*, *27*
  production, 14, 17, *23*, 29, 31, 36, *39*
  prospect, 111
  redbed deposit, *30*
  replacement deposit, *33*
  reserves, *23*, *25*, 36

resources, *38*
Ruby Creek (Alaska), 35, *36*
San Manuel (Arizona), *25*
Sierrita (Arizona), *28*
skarn deposit, *31*
sources, *24*
Stillwater (Montana), *36*
substitutes, *39*
Superior (Arizona), *33*
Tintic (Utah), *33*
types, *25*
users, *25*
vein deposit, *32*
White Pine (Michigan), *35*
Copper Basin ore body, 17
Copper Canyon, Nevada, 12, *31*
Copper Canyon stock, 12
Copper Marl, 157, 158
copper molybdenum, 13, *17*
Copper Mountain, Wyoming, *119*, 121
Copper Ridge, Tennessee, 51
copper sulfides, 30, 88, 89
Copperopolis, California, 202
coral, 342
coral floras, 487
Corbin Sandstone, 529
corderoite, 145, 146
Cordilleran System, 216
Cordilleran thrust belt, *220, 391*
  exploration history, *393*
  geologic phases, *392*
  Wyoming-Idaho-Utah segment, 392
Cordova district, Alabama, 195
Cornwall, England, 128, 129, 130
Cornwall, Pennsylvania, 100
Cornwall mine, *78*
Cortez, Nevada, 55
Corwin area, Alaska, 492
Corwin Bluffs, 596
Corwin Delta, 492, *596*
Corwin Formation, 596
Costa Rica, 81
Cotton Valley Group, 303
Cotton Valley reservoir, 307, *309*
Cottonwood district, Arizona, 200
Cove Peninsula deposits, Australia, 148
covellite, 9, 111
Cowley facies, 330
Cowlitz County, Washington, 195
Cowlitz Formation, 586
Coyote Creek field, 387
Crandon ore body, 18
Crandon, Wisconsin, 34, 44, 46, *59*
Crawford thrust plate, 395
Creede deposit, 10
Crescent Formation, 81
Creta, Oklahoma, 36
Cretaceous, *312, 358, 363, 547*
  Early, *491*
  Late, *493*
  Lower, *378, 387*
  Middle, 492
  Upper, *378, 380, 388, 550, 556, 559*
Crevasse Canyon Formation, 370, 549, 558
Cripple Creek deposit, 10

cristobalite, 196
Crooks Gap, *114*
Crow Mountain Sandstone, 378
Croweburg coal bed, 532
crust, oceanic, 96
Crystal Mountain deposit, Montana, 204
Cuba, 96
Cuddapah district, India, 198
Culberson County, Texas, 176, 177
Cumberland saddle, 273, 281
cumulates, 93, 95, 96
Curtin Formation, 191
Cuyama basin, 440, 443
Cuyuna Range, Minnesota, 64, 70, 80
cyclothemic era, 534
cylinderite, 129
Cymric field, 419
Cymric-Mekittrick area, 426
Czechoslovakia, 146

Dago Peak stocks, 15
Daisy deposit, Nevada, 204
Dakota Group, 492
Dakota reservoir, 250
Dakota Sandstone, 112, 195, 250, 361, 363, *365, 379*, 549, 555
Dakota silt, 378
Dakota-Morrison producers, 363
Dalhart basin, 325
Dantzler Formation, 305
Danville basin, 119
Darby, Montana, 204
Darby-Prospect-Hogsback system, 383
Darco mine, Harrison County, 575
Darwin, California, 17, 34
Date Creek Basin, Arizona, *118*, 121
Datil Group, 80
Davidson Granodiorite, 10
Davis Formation, 46, 49
dawsonite, 148
Dead Horse Creek Field, 375
Dean updip, 354
Death Valley, California, 172
Death Valley National Monument, 178
Deep Basin, Alberta, 252
dehydration, 509
Del Rio district, Tennessee, 199
Delaware, 180
Delaware basin, 168, 175, 177, 249, 251, 340, 341, 345, *346*, 351
Delaware Mountain Group, 351, 353, 354
Delta depocenter, 422, 423
Delta Mine, *111*
deltaic systems, *577*
  fluvial, *558*
demethanation, 509
Denali, Alaska, 31
Densosporites, 498
Denver basin, 547, 549, 550, 561
Denver Formation, 561
Desloge, Missouri, 47, 48
detritus, 4, 276, 393, 564
Detroit River Group, 179, *296*
Devilwater/Gould shale, 422
Devonian, *296, 333, 340*
Devonian Formation, 186
diagenesis, *227, 247*

silica, 438, 438
Diamond Peak Sandstone, 403, *404*
Diaphorodendron, 487, 488, 490
diaspore, 9
Dickinson, Group, 73
dickite, 9
Dietz coals, 568
Difficulty Shale, 377
digenite, 18
Dillon, Montana, 185
Dillon-Ennis district, Montana, 200
Dinwoody Formation, 377
Dismal Swamps, 496
Disturbed Belt, 515
Dolet Hills mine, Louisiana, 573
dolomicrites, 352
dolomite, 11, 14, 33, 51, 70, 80, 165, 166, 175, 180, *190*, 216, 276, 292, 328, 330, 340, 341, 342, 358, 392, 398, 414, 437
dolostones, 55
dolowackestones, 352
Dominican Republic, 146
Donezella reefs, 347
Dortenhausen, West Germany, 186
Dos Pobres, Arizona, *27*
Dos Pobres ore body, 17
Douglas Group, 331
Downieville district, 6
Dragon Mine, 194
Drake oil well, Pennsylvania, 186, 279
Ducktown, Tennessee, 18
Duluth, Minnesota, *36*
Duluth Complex, Minnesota, 18, *92*, 101
Dundee Limestone, 289, 290, *296*
dunite, 96
Dunka Pit deposit, 94
Dunkard basin, 273, 490, 514, 526, 532, 533, 534, 540, 541
Dunkard basin coal beds, 541
Dunkard Group, 475
Duplin Marl, 158
DuQuoin monocline, 531
Dyer Member, 55
Dykersburg Shale, 476

Eagle Mesa field, 363
Eagle Mills red beds, 303
Eagle Mountains, California, 78, 205
Eagle Sandstone, 550
Eagle Springs, 413
Eagle Springs field, 403, *405*, 412, *414*
Eagleford sandstone, 305
Eagleford Shale, 252, 305, 312, 314
East Barrow field, Alaska, 449
East Coalinga field, 419, 426
East Gold Belt district, 6
East Grove gas field, Virginia, 282
East Liverpool district, 195
East Painter Reservoir, 395
East Ridge, Virginia, 206
East Salt Creek field, 375
East Texas Embayment, 575
East Texas field, Louisiana, 249, 301, 313
East Texas oil field, 186

East Texas salt basin, 251, 302
East Tinic, Utah, 208
East Umiat field, Alaska, 449
Eastern Interior basin, 523, 526, *529, 531*, 534, 539, 540, 541
Eastern Shelf, 351
Echooka Formation, 456
eclogite, 136
ecosystems
  peat accumulates, 469, 487
  peat-swamp, 498
Eden, Vermont, 202
Edison field, 419
Edwards, New York, 46
Edwards deposits, 314
Edwards district, *54*
Edwards Group, 305
Eel River basin, 443
El Laco, Chile, 77
El Paso Group, 340
El Portal, California, 197
El Vado fields, *366*
El Vado reservoir, *365*
El Vado Sandstone Member, 362, 363, 365, 369
Elatides, 492
electrum, 9, 12
Elk City field, 336
Elk Hills field, 422, 424, 425
Elk-Poca field, West Virginia, 281, 285
Elk Point Group, 174
Elko, Nevada, 185, 186
Elko County, Nevada, 198
Elko Formation, 185
Elko Shale, 412
Ellenburger fields, 345
Ellenburger gas, 345
Ellenburger Group, *340*
Ellenburger Hills, 340
Ellenburger oil, 345
Ellenburger suite, 325
Ellesmerian sequence, 447, 453, *455*, 459
Elliott Lake, Canada, 139, 140
Ellison Formation, 8
Elmworth Field, 249, *252*
Ely, Minnesota, 94
Ely, Nevada, 31
Ely Formation, 414
Ely Formation Carbonates, 407
Ely Greenstone, 73
Emery coalfield, 555
Emery Sandstone Member, 555
Eminence Formation, 46, 198
Emma-Travona vein, 79
Empire mine, 6
Endicott field, Alaska, 451
Endicott Group, 457, 459
Eneabba deposits, Western Australia, 139
Energy Shale, 476
Engelhardia, 493
English Colony field, 425
Entrada anomalies, *363*
Entrada Sandstone, 105, 112, 369
epidote, 6, 10
epithermal vein deposits, 3, 4, *8*, 11, 16, *18*

Erkelenz anomaly, 518
Eromanga Basin, Australia, 183
Ervey Member, 377
Esmeralda County, Nevada, 198
Essex County, Ontario, 294
Estonia, Russia, 183
Etchegoin Formation, 427
Eureka, Nevada, 17, 34, 55
Eureka, Utah, 194
Eureka County, Nevada, 11, 198
Eureka Quartzite, 404
europium, 139
Eutaw Formation, 196
euxenite, 134, 141
Evanston Formation, 561
evaporites, *165*, 301, 398
  continental, *169*
  economic deposits, 171
  marine, *165*
  mineral products, *172*
  nonmarine, *169*, 171
Everglade swamp, 471, 472, 474
Evergreen, Colorado, 206
exinites, 509
exploration techniques, *255, 256*

Fairbank Formation, 377
Fairbanks, 4
Fairweather Range, 95
Fall River Sandstone, 375, *379*, 380, 387
Falls Branch–Greenville district, Tennessee, 199
fanglomerates, 405
Farmington Sandstone Member, 361, 366, 368
faults
  detachment, 405
  growth, *303*
  high-angle, 143
  normal, 10, 419, 425, 452, 527, 586, 588
  Osburn, *14*
  reverse, 118, 417, 425, 434, 435, 452, 586
  thrust, 391, 393
Favret Formation, 405
Feather River, 4
Federal Republic of Germany, 146
Felder-Lamprecht deposit, 116
feldspar, *200*, 397, 398, 426
Felix coal bed, 563
ferberite, 126
Fernvale Limestone, 82
ferricrete, 97
Ferris Formation, 561
Ferron Sandstone Member, 477, 555, 558
Ferry Lake Formation, 305
Fiddler Creek field, 375
Fiddler Creek trend, 380
Findlay Arch, 294
Findlay segment, 280
Finland, 137
Fire Clay coal bed, Kentucky, 527, 530, 540, 541
fire-clay deposits, *195*
Fire Creek coal, 527, 540
Fire Creek Siltstone Member, 456

First Frontier, 380
First Wall Creek, 380
Fish Creek field, Alaska, 449
Fish Creek platform, 457
Flambeau deposit, 59
Flat River, Missouri, 47
Flathead River, 588
Fleming coal bed, 532
Fleming Formation, 196
Flin Flon deposit, 59
floras, peat-swamp, 485
Florida, 120, 137, 140, 141, 148, 153, *154*, *158*, 180, 196, 205
Florida–South Georgia district, 158
fluids
  hydrothermal, 91
  thermal, 172
Fluorine district, Nevada, 204
Fluorite Ridge district, New Mexico, 205
fluorite, 9, 11, 16, *18*, 29, 203
fluorspar, 59, *202, 203*
fluvial environments, *475*
fluvial systems, *563*
  deltaic, *558*
  depositional, *577*
flysch, 276
Foothills province, 451, 460
Forelle Limestone, 377
Forest City basin, 526, 529, 532
forests, peat-swamp, *485*
Fort Payne Chert, 160, 282
Fort Union coals, 565
Fort Union Formation, 378, *561*, 563
Fort Worth basin, 526, 532
Fossil Basin area, *391, 401*
  hydrocarbons, *391, 394, 397*
Fossil Creek, 588
Four Corners platform, 357, 361, *363*, 369
Foxen Formation, 435, 440
France, 142
Franciscan Complex, 81
Franciscan Lake fields, 366
franckeite, 129
Franklin, New Jersey, *54*
Franklin Furnace, New Jersey, 44, 46
Franklin Mountains, 340
franklinite, 54
Fraser River delta, British Columbia, 470, 473, 474, 475
Frederick City, Virginia, 64
Fredericksburg, 303, 314
Fredericktown, Missouri, *47*, 100
Fredonia, New York, 279
Freezeout Shale, 377
French Gulch district, 6
Fresno County, California, 197
Friday deposit, 95
Friedensville, Pennsylvania, 51
Frog Pond adit, 90
Front Range, Colorado, 10, 136
Frontier Formation, 375, 378, 380, 382, 384, 385, *388*, 396, *397*, 400, 401
Frontier reservoirs, *380*
Fruitland Formation, 549, 556
  coalbeds, *368, 369*
Fruitvale field, 419, 425

Frying Pan Shoals, 157
Fuller's earth, *195*
　deposits, *196*
　uses, 195
fumarole deposits, 130
Furnace Creek deposits, 178
Fushun, Manchuria, *183*, 186
fusain, 485
Fuson Shale, 385
Fusselman Formation, 340, 341
　345

gadolinite, 139
Galatia channel, *476*
galena, 6, 8, 12, 14, 15, 18, 43, 111
　by-product, 33
Gallup coalfield, 363, 556
Gallup Sandstone, 549
Gap, Pennsylvania, *36*
Gap deposit, 95
Garajas, Brazil, 79
Garfield Mine, *105*
garnet, 15
Garret Ranch Group, *407*, 413
Garret Ranch volcanics, 405
Garza Platform, *348*
Gas Hills, 114
gas, *213*, 221, *225*, 282, *283*, 327,
　335, *336*, 354, *361*, *366*, *369*,
　357, *450*
　accumulations, *437*, *450*, *453*, 460
　chemical nature, 225
　coal-bed, *222*
　deep, *222*
　dry, 423
　earth, 225
　exploration. See gas exploration
　fields, 116, 219, 392
　future potential, *221*
　generation from organic matter, *226*
　hydrates, 457
　methane-rich, 222
　migration. See gas migration
　origin, *225*, *226*
　potential resources, *369*
　production, 213, 220, 280, *238*,
　　290, *294*, 301, *307*, *309*, *312*,
　　314, *316*, *317*, *333*, *335*, *336*,
　　351, 357, *362*, *379*, *382*, *388*
　projects, development stages, *255*
　reserves, *309*, *313*, *317*, 327, *336*,
　　357, 419, 594
　reservoirs, 222, 252, 296, *340*, *347*
　resources, *214*, *222*, *283*, *299*
　seeps, 289
　shale, *222*
　temperature influence, *230*
　time influence, *230*
　transportation, *233*, 238
　wells, 221, 279, 280
　wet, 397
gas exploration, *170*, 515
　computer applications, *268*
　concepts, *255*
　geochemistry, *261*
　geophysical methods, *262*
　techniques, *255*, *256*, *259*
gas migration, *225*, *231*, 238, *307*,
　312, *315*

diffusion, *232*, 238
directions, *235*
distances, *235*
mechanisms, *232*
moving water, *233*, 238
gas province, *218*, 419
Gasconde Dolomite, 199
Gasper Sand, 579
Gatchell Sand, 426, 427
Gatesville Sandstone, 247
Gem stocks, *15*
General Mining Law (1872) 1
generation efficiency, *229*
geochemical techniques, *261*
Georgia, 137, 140, 141, 148, 149,
　156, *158*, 193, 196, 200, 205
Germany, 184
gersdorffite, 15
Gib Horn uplift, 383
Gibbons Creek mine, Grimes County,
　576
gibbsite, 147, 148
Gila district, New Mexico, 204
Gilman, Colorado, 55
Gilman district, Colorado, 17
Glacier National Park, 186, 393
Glacier Peak, Washington, 28
glauconite, 69
Glen Canyon, 105, 120
Glen Canyon Sandstone, 105
Glen Rose, 313
Glendo Shale, 377
Glenwood Shale, 292
Glorieta reservoir, 351, 354
Glyptostrobus, 487, 488, 495
　swamp, 563
Goat Mountain, 206
Goddard Shale, 330
goethite, 63, 71, 74, 79, 97
Gogebic Range, 70
Golconda, Nevada, 128
gold, 3, *5*, 7, 9, 12, *15*, 16, *23*, 27,
　28, 32, 34, 35, 59, 78
　alluvial, 4
　by-product, 3, *16*, 33
　disseminated epithermal, *11*
　genesis, 16
　iron formation, 7
　lode, 16
　mines, *6*
　native, 4, 8, 9, 12, 14
　occurrence, *4*
　placer, 4, 24
　production, 3, 6, *8*, 10, 13, 14, 17,
　　18, 19, 29, 32
　reserves, 8, 12, *19*
　resources, 17
　types, *3*
　uses, *3*
　western United States, *10*
Gold Hill deposit, Colorado, 100
Golden Trend field, 335
Golden Valley Formation, 493
Gomez field, 219
Goodnews Bay deposit, *98*
Goodsprings district, Nevada, 99
Goodsprings Dolomite, 204
Goose Creek basin, 583, 587
Goose Creek coalfield, 587

Goose Egg Formation, 377
Gordonia, 497
Gosport Formation, 576
gossans, *79*
Gouverneur district, New York, 199
Government-Tracy #1 well, 384
grabens, 403, 407
　fault-bounded, *588*
Grace mine, 78
grainstone, 398
Grand Canyon area, *117*
Grand Canyon breccia pipes, Arizona,
　117
Grand Hogback monocline, 105
Grandstand Formation, 596
granite, 118, 126, 128, 191, 328, 417
　peralkaline, 117
Grant Canyon, 414
Grant Canyon field, *403*, *407*, 411,
　414
Grant Range, 405
Grants mineral belt, New Mexico, *112*
Grass Valley district, *6*
gravels, *5*, *192*, 529
Graves Mountain, Georgia, 206
gravity, *267*
Grayburg reservoir, *352*, 354
graywacke, 5, 8, 71
Great Basin, 169, *221*, 584
　evaporite deposits, 171
　exploration history, *403*
　exploration model, *405*
　exploration techniques, *409*
　petroleum potential, *403*
　potential accumulations, *410*
Great Britain, 142
Great Divide basin–Crooks Gap area,
　114
Great Dyke, 87
Great Eastern deposit, 95
Great Falls arena, Montana, 493
Great Plains, 491, *547*, *562*
　coals, *547*
　northern, 222, 505
Great Plains province, 447
Great Salt Lake, Utah, 172, *174*, *175*,
　178, 180, 403
Great Salt Lake graben, 405
Great Valley, California, 417
Greeley field, 425
Green Cove Springs, Florida, 137,
　139
Green River Basin, Wyoming, 148,
　187, 222, 547, 556, 561, 562
Green River coalfield, 587
Green River district, 583
Green River Formation, 172, 177,
　184, 187, 209, 392, 562
greenalite, 69
Greenhorn Formation, 385
greenstones, 6, 34
　belts, 5, 8, 36, 59
Greenwood-Lansing limestone, 336
Gregg County, Louisiana, 251
Grenville limestone, 54
Greybull-Lovell, Montana, 196
Greybull-Lovell, Wyoming, 196
Grindstone-Twelvemile Creeks inlier,
　584

Groote Eylandt, Australia, 79, 81
Grubstake Formation, 593
Guadalupe field, 440
Guadalupian Series, 331
guano, 153
Guijarral Hills field, 426
Guilmette Formation, 304, 407, 411, 414, 415
Guinea, 149
Gulf basin, *301*
  exploration history, *301*
  future potential, *313*
  stratigraphy, *303*
  thermal history, *306*
Gulf Coast, 137, 223, 493, 505, *573*
  coals, *573*
  lignites, 493
Gulf Coast basin, 169, *173*, 303
Gulf Coastal Plain, 216
Gulf Coastal Region, 495
Gulf of California, 167
Gulf of Karabaghas, 168
Gunflint Formation, 70
Gunflint Range, 70, 71
Gunnison County, Colorado, 133
gypsum, 165, 166, 168, *176*, 299
  production history, *176*
  sources, *176*
Gypsum Spring Formation, 378
gyttjae, 472

Hackberry embayment, 305
halite, 165, 168, 174, 175
Hall Summit Formation, 573
halloysite, 193, *194*
Hamersley Range, 66, 67
Hamme, North Carolina, 127
Hammersley Range, 202
Hammersmith ironworks, 64
Hams Fork coalfield, 547, 549, 556, 561
Hanna basin, 547, 549, 561, 563
Hanna coals, 568
Hanna trough, *452*
Happy Jack Mine, *111*
Hardin Sandstone Member, 160
Harding pegmatite, New Mexico, 134
Harford County, Maryland, 200
Harrisburg Coal, Illinois, 476
Hart County, Georgia, 206
Hartnell No. 1 well, 435
Hartville Formation, 377
Hartville uplift, 376, 377, 384
Hartzog Draw field, 388
Harvard mine, 6
harzburgite, 96
Hat Creek coalfield, British Columbia, 505
Hatchetigbee Formation, 573
Havallah Formation, 81
Hawaii, 79, 137, 149, 153
Hawkins field, Texas, 313
Hawthorn Formation, 157, 158, 159, 196
Hawthorn Group, 158
Hayden Group, 377
Haynesville limestone, 303
Haynesville sandstone, 303
Hazard No. 4 coal, Kentucky, 474

Healy Creek coalfield, 591, 592
Healy Creek Formation, *592*
Heath Formation, 186
Hector, California, 208
Heidelberg field, Mississippi, 313
Hells Mesa Formation, 80
Helm field, 422
helvite, 143
hematite, 15, 16, 63, 64, 71, 74, *76*, 79, 115, 136
Hemlock Conglomerate, 594
Henderson, Colorado, 29
Henry Mountains coalfield, Wyoming, 547, 549, 555
Henry's Knob, 206
Henryetta coals, 531
Herendeen Bay coalfield, 591, *595*
Hermosa Formation, 175
Herrin Coal, 485
Herrin No. 6 coal bed, Illinois, 473, 476, 490, 531, 537, 541
hessite, 17
heulandite, *207*
Hiawatha coal, Utah, 478
Hickman County, 155
Hicks Dome, Illinois, 141
High Plains, 196
Hilight trap, 380
Hilliard Formation, 396, 401
Hocking Valley district, Ohio, 195
Hogback field, 365
Hogback Ridge field, 395
Holly Springs delta system, 577
Holston Limestone, 191
Homestake deposit, *8*
Homestake Formation, *8*
Homestake mine, *8*
Hooper Formation, 574
Hoover sandstone, 336
Horne deposit, 59
Horse Trap field, 395
Horseshoe Atoll, *348*, 351, 354
Horseshoe Canyon Formation, 479
Horseshoe field, 361, 365
Horseshoe Gallup field, 362
Hospan structure, 365
Hosston, 313
Hosston Formation, 303
Hot Springs, Arkansas, 119
Hot Springs County, Arkansas, 199
hot spot, 55
House Creek field, 388
Houston embayment, 302
Huasna basin, 443
Hudson Bay Basin, 289
Hue-Sagavanirktok system, 459
huebnerite, 126, 127
Hueco Mountains, 340
Hugoton Embayment, 325, 330, 331
Hugoton field, 335, *336*, 357
Hugoton-Panhandle gas field, 327
Hulett Sandstone Member, 378
Humboldt Range, 206
huminite, 509, 568
Huntersville Chert deposits, 533
Hunton field, 335
Hunton Group, 328, 333
Hunton reservoirs, 333, 335
Hutchinson Salt Member, 173

hydrocarbons, *213*, 289, 290, *314*, 332, 345, *357*, *401*, 403
  accumulations, *220*, 246, 248, 249, 250, *348*, 387, 391, 393, *394*, 397, 403, 407, *437*, 444
  aromatic, 512
  exploration techniques, 221, *255*, *256*, *259*
  gaseous, 397
  generation, 391, 399, *401*, 442
  liquid, 397
  migration, *307*, *312*, *315*, *346*, *350*, *354*
  occurrence, *391*, *431*
  parafinic, 512
  potential resources, *369*
  production, 219, *309*, *314*
  projects, *255*
  reserves, *309*, *313*, *354*
  reservoirs, 252, 296, *340*, *347*, *351*, 395
  resources, *214*
  seals, *246*, *309*, *313*, 333, *341*, *345*, *354*, 407, 408, 415, 425, 437
  trapping, *241*, *247*, *249*, 294
  traps, *309*, *313*, *317*, 333, *345*, *350*, *354*, 375, *379*, *398*, *405*, *414*, *425*
hydrodynamic forces, 366
hydrogen sulfides, 397
hydrostatic gas trap hypothesis, 366
hydrothermal deposits, *76*, *79*, *99*, 128
hydrothermal minerals, 141

Iberia Parish, Louisiana, 173
Idaho, 4, 18, 59, 131, 140, *155*, 160, 198, 203, 206, 207, *220*
  northern, 14
  southern, 10
  west-central, 4
  western, 6
Idaho batholith, 141
Idaho Springs, Colorado, 32
Idaho-Maryland mine, 6
Idavada volcanics, 587
Idria Mine, Yugoslavia, 145
Ignacio Blanco anticline, *368*
Ignacio Blanco field, 363, 368
ignimbrites, 413
Ikpikpuk basin, 452
Illinois, 194, 203, 204
Illinois Basin, 222, 289, 490, 491, 505
Illinois Basin Coal Field, 489, 490
Illinois-Kentucky district, 18, 204
illite, 9, 398, 443
illite-smectite, 105
ilmenite, 76, 135, *136*
Ima mine, Idaho, 127
imagery
  satellite, *260*
  thermal infrared, *260*
India, 79, 137, 140, 142, 144
Indian Creek Mine, 49
Indian Peak district, Utah, 205
Indiana, 185
Indiana-Ohio Platform, 289

indicators, organic maturity, *229*
Indonesia, 87, 130, 131
industrial minerals, 1, *189*
inertinites, 508, 568
influx, detrital, 540
Inigok-1 well, 452
Inner Foothills Belt, 507
Innsbruck, Austria, 186
Inspiration deposit, 28
Interstate field, 336
intrusions, synorogenic, *95*
Inyan Kara Group, 378, *379*, 382
iodine, *179*
    production, 179
    use, 179
Ione Formation, 589
Irati Formation, 183
Ireland, 146
iron, 12, 16, *63, 72*, 82
    abundance, *63*
    deposits, 64, *65, 75, 76, 78, 79*
    formation, 7, *63, 65, 66, 67, 69, 70, 73,* 80, 116
    imports, 64
    mines, 65
    mining districts, 63
    occurrence, *63*
    ore, 63, *64, 75*
    replacement deposits, *78*
    uses, *63*
iron chlorites, 69
Iron Hill, Colorado, 141
Iron King district, Arizona, 17
Iron Mountain, Missouri, *77*
Iron Post coal bed, 532
Iron River–Crystal Falls district, 70, 71
Iron Springs, Utah, *78*
ironstones, 63, *74*
    Clinton-Minette-type, *74*
    deposits, *74*
Ironwood Formation, 70
Israel, 184, 186
Italy, 136, 146
Itkilyariak Formation, 455
Ivishak Formation, 456

J-M Reef, *90*
Jackford Sandstone, 529
Jackpile sandstone, *112*
Jackson Fayette delta system, 577
Jackson Group, 116, 576, 578
Jackson-Plymouth district, 6
Jacksonburg Formation, *192*
Jacksonian Stage, *576*
Jago River, 458
Jago type, 458
Jamaica, 149
Jamestown district, Colorado, 204
Japan, 179
jarosite, 10
Jarvis Creek coalfield, 591, 592
jaspilite, 73
Jauchau Fu, China, 193
Jayfield, 309
Jefferson City Dolomite, 199
Jelm Formation, 377, 378
Jerome, Arizona, 17, 46, 59
Jersey Valley, Nevada, 208

Jessamine dome, 273
Jewett mine, Leon County, 576
Jiangxi Province, China, 139
John Day Formation, 209
Jordan, 184, 186
Juab County, Utah, 194, 205
Judith River Formation, 550
Julia Creek, Queensland, 183
Juneau deposit, 6
Jurassic, *307, 363, 378*
    Late, *491*

Kaiparowits coalfield, 547, 555
Kaiparowits Plateau, 549
Kaktovik basin, 453
Kalahari field, South Africa, 79
Kanawha Black Flint marine zone, 534
Kanawha Formation, 529, 539
Kangankunde, Malawi, 139
Kansas, 50, 336
kaolin, *193*
kaolinite, 9, 148, 426
Karen Creek Sandstone, 456
Karinopteris, 485
Karnes County, Texas, 116, 579
Karonge, Burundi, 139
Katahdin pluton, *95*
Kate Spring field, 403, *407*
Kavik field, 458
Kavik Member, 456
Kavik type, 458
Kayak Shale, 455
Kaycee-Midwest, Wyoming, 196
Kechika trench, 58
Kelvin Formation, 395
Kemik Sandstone, 456
Kenai Group, *594*
Kenai Mountains, 591
Kendrick marine zone, 534
Kendrick Shale, 529
Kennecott, Alaska, 17, 31
Kentucky, 185, 192, 194, 203, 204, 281, 283
Kern Bluff field, 419
Kern Front field, 419
Kern River field, 419, 426
Kern River Formation, 425
kerogen, *183*, 259, 400, 423, 440, 441, *442*
    oil/gas products, *227*
    types, *227*, 238
Kettleman Hills field, 425
Kettleman North Dome field, 426
Keweenaw, Michigan, *30*
Keyes dome, 325, 336
Keywest deposit, 95
Khibiny, Soviet Union, 139
Kidd Creek deposit, 59
Kikiktuk Conglomerate, 455
Killik Tongue, 596
King County, Washington, 195
King Lear Formation, 405, 412
Kingak Shale, 456, 458, *459*
Kingak type, 458
Kings Mountain, North Carolina, 201
Kings Mountain district, 206
Kingston Range, 200
Kirtland Shale, 359, 361, 366, 368
Kiruna, Sweden, 77

Kishenehn basin, *588*
Kishenehn Formation, 186
Kitty trap, 280
Klamath Mountains, 96, 584
Knox Group, 199
Knox suite, 325
Kobeh-Antelope area, 411
Kola Peninsula, Soviet Union, 139
Kolob-Alton coalfield, 555
komatiites, 96, 101
Kona, North Carolina, 201
Kootenai Formation, 491, 492
Korea, 128
Kramer deposit, California, 178
Kreyenhagen shale, *424*, 427
Krivoy Rog–Kursk anomaly, 66
Kuk River Valley, Alaska, 492
kukersite, 183
Kukpowruk Formation, *596*
Kultheith Formation, *599*
Kuparuk Formation, 456
Kuparuk reservoir, 457
Kuparuk River field, Alaska, 451, 457
kyanite, *205*

La Perouse Complex, 101
La Perouse intrusion, *95*
La Plata field, 365
LaBarge field, *393*
Labrador trough, Canada, 66
Lac Member, 378
Lac Tio deposit, 136
Ladson Formation, 157
Ladysmith, Wisconsin, 59
lagoons, 168, *578*
Laguna area, 113
lake beds, saline, 201
Lake de Smet coal, 563
Lake de Smet swamp, 564
Lake Flagstaff, 562
Lake George, Colorado, 143
Lake Gosiute, 562
Lake McDonald, 186
Lake of Larnaca, Cyprus, 167
Lake Superior region, *23*, 64, *70*
Lake Superior type, *66, 70*, 80
    genesis, *71*
Lake Tecopa, California, 209
Lake Uinta, 562
Lake Vermilion Formation, 73
lakes, closed-basin, 169
    See also specific lakes
Lakota Formation, 375, 378, *387*
Lamoille County, Vermont, 202
Lamotte Sandstone, 46
Lance Creek field, 375, 388
Lance Formation, 376, 378
Lander County, Nevada, 198
langbeinite, 175
Lansing Group, 331
Laramie Formation, 550
Laramie Range, Wyoming, 76
Laramie uplift, 376
Lasalle anticline, 526, 531
laterites, *79*, 87, 98
lavas, 405
laws, mining, *1*
leaching, 475

lead, 4, 12, 14, 15, *18*, 32, 34, 35, *43*, 45, 49, *55*, 58, *59*, 100
  belt, 18
  future reserves, *60*
  mantos, *55*
  metallic, 46
  production, 14, 29, 32, 43, *44*
  Southeast Missouri District, *46*
  substitutes, 43
  uses, *43*
Lead, South Dakota, 8
Leadville district, Colorado, 13, 17, 55
Leadville Limestone, 45, *55*
Leadwood, Missouri, 47, 48
Leavenworth fault, 588
Lebo Shale, 376
Ledge Sandstone Member, 456
Lee Member, 529
Lee Sandstone, 529
Leipers Limestone, 160
Lemhi Pass, 141
Leo sandstone, 379
Leonardian Series, 331
Lepidodendron, 485, 487, 488, 490, 496
Lepidophloios, 485, 487, 488, 490
leucoxene, 135, 137, 139
Lewis County, 155
Lewis fault, 393
Lewis reservoir, *381*
Lewis seals, *382*
Lewis Shale, 368, 376, 378, 380
Lewis traps, *382*
Lexington, Kentucky, 204
Lexington coal bed, 532
lignin, 471
Lignite Creek coalfield, 592
Lignite Creek Formation, *592*
lignites, 509, 513, 560, 563, 568, *573*, *574*, *576*, 583, *587*, 597
  deltaic depositional systems, *577*
  exploration models, *578*
  fluvial depositional systems, *577*
  lagoonal depositional systems, *578*
  reserves, *573*
  resources, *575*
Lik Creek, 18
Lima-Indiana field, 289, *294*
lime, *192*
limestone, 11, 14, 33, 50, 51, 52, 55, 74, 100, 145, 159, 166, 172, *190*, 198, 207, 216, 276, 301, 307, 325, 328, 330, 331, 340, 341, 342, 350, 351, 358, 377, 378, 392, 398, 414, 515, 532
  phosphatic, *159*
  reefs, 348
limonite, 15, 63, 64, 97
Lincoln County, Nevada, 589
Lincoln Formation, 586
Lincoln Hill, Nevada, 206
Lindrith West field, 365
Lion Rock unit, 437
liptinites, 508, 509, 560, 568, *589*
Lisbon Formation, 576
Lisbon Valley anticline, 111
Lisbon Valley District, Utah, *111*
Lisburne field, Alaska, 451, 456

Lisburne Group, 160, 455
Lisburne Hills, 451
lithium, 134, *178*, *179*
litter patterns, *496*
Little Maria Mountains, 207
Little Medicine Limestone Member, 377
Little River Series, 206
Live Oak County, 116
livingstonite, 145
Lodgepole field, 395
Logansport Formation, 573
logs, *256*
Lompoc field, 435
Long Valley, 141
loparite, 138, 139
Los Angeles basin, 220, 443
Lospe Formation, 434, 437, *439*
Lost Creek, Montana, 38, 114
Lost Hills field, 419, 425
Lost Hills reservoir, 425
Lost River, Alaska, 129, 131
  deposits, *204*
Louann Salt, 169, 173
Louden anticline, 531
Louisiana, 196, 573
Louisiana coastal plain, 472
Lower Glenwood, 292
Lower Hartshorne coalfield, 532
Lower Kittanning coal bed, 541
Lower Sandstone, 378
Lower Sunnyside coal, 497
Lucas Formation, 296
Luis Lopez district, New Mexico, 80
Luman Lake, 562
Luman Tongue, 562
lycopod, 485
Lycospora, 485, 490
Lykens Valley coal zone, 527
Lynn County, Texas, 353
Lyon County, Nevada, 589

Mackinaw mine, Washington, 100
Madison Group, 174
Madison Limestone, 377
magmatic deposits, *76*
Magmont Mines, 49
magnesium, 165, *179*
  uses, 180
magnesium chloride, 180
magnesite, 180
Magnet Cove, Arkansas, 119, 136
magnetics, *267*
magnetite, 15, 27, 63, 64, 69, 74, *76*, 136
magnetotellurics, *267*
Magnolia, 497
Magoffin Member, 529
Mahakam Delta, Indonesia, 301, 472
Mahoning Coal, 490
Main Orebody, 77
Main Stage veins, *32*, *79*
Main Tintic subdistrict, 14
Maine, 37, 95, 200
Major County, Oklahoma, 335
malachite, 16
Malaysia, 127, 130, 131, 137, 140
Maljasalmi district, Finland, 202
Malvern, Arkansas, 196

Mammoth, Arizona, 37
Manastash basin, 588
Manastash coalfield, 588
Mancos Shale, 111, 362, 363, 365, 368, 549, 555, 558
manganese, 16, 32, *63*, *79*, *82*, 128
  abundance, *63*
  co-product, 54
  genesis, *81*
  iron formation, *80*
  minerals, 80, 81
  occurrence, *63*
  ore grades, 63
  production, 32, *65*
  reserves, *81*
  sedimentary, *80*, *81*
  uses, *63*
  volcanogenic, *80*
manganese oxides, 9
Manistique Group, 296
Manitou Formation, 55
Manning Formation, 576
Manning Point seep, 458
Manning type, 458
Manono pegmatite, Zaire, 129
Mansfield Formation, 529, 539
mantos, *55*
Many Rocks field, 365
Maoming, Canton Province, 186
mapping, *257*, *261*
maps, *258*
Marathon Limestone, 340
Marathon-Ouachita thrust belt, 340, 350
marble, 46
marcasite, 12
marine deposits, *173*, *175*
marine salinas, *167*
Maritime Boundary Area, 318
Maritime Region, Canada, 475, *507*
marl, 378
Marmaton Group, 331, 336
Marquette Range, 64, 70
Marquette Range Supergroup, *70*
Marshall Sandstone, 299
Martin Lake mine, Panola County, 576, 578
Mary Lee coal group, 527
Mascot–Jefferson City, Tennessee, 51
Massachusetts, 153
Matador Uplift, 347
Matanuska coal basin, 591
Matanuska coalfield, 591, *594*
Matanuska Valley, 594
Mathers Ranch field, 333
Mauch Chunk, Pennsylvania, 121
Mauch Chunk red beds, 527
Maury Shale, 160
McAlester coals, 532
McDermitt, Nevada, 146
McDonald Shale, 422
McIntosh Formation, 584, 586
McKittrick field, 419, 425
McLaughlin deposit, 10
McLish shales, 345
McLure Shale, 427
McMullen County, 579
McMullin Ranch, 425
McNairy Sand, 137

Meade arch, 596
Meade-Crawford system, 393
Meade Peak Member, 160, 396
Meadow Creek field, 375
Meadows field, 365
Meddy Sandstone, 375, 382
Media field, 363
Medicine Bow Formation, 549
Medina field, New York, 280
Medullosa, 489, 490
Meek Group, 377
Meeteetse Formation, 550
Meigs-Attapulgus-Quincy district, 196
Melones fault zone, *6*
Menefee Formation, 361, 366, 369, 370, 478, 549, 556, 558
Menominee district, 70
Menominee Group, *70*
Menominee Range, 70
Meramec Lime, 330
Mercur district, Utah, *12*
mercury, *144, 145, 146*
Meridian Sandstone, 577, 578
Mesa field, 365
Mesa Range, 65, 70, 71
Mesabi ores, 65
Mesaverde fields, 366
Mesaverde Formation, 375, 378, 380
Mesaverde Group, 361, 362, 366, 549, 550, 558
Mesaverde reservoir, *381*
Mesaverde traps, *382*
Mescal Limestone, 202
mesothermal deposits, 55
Mesozoic, *420*
Mesquite mine, 15, 16
metacinnabar, 145
metagenesis, *227*
Metaline district, 18
Metaline Falls, Washington, 46
metamorphism, 15, 37
Metasequoia, 487, 488
methane, 512
    coal-bed, *369*
    drainage, 589
Mexico, 131, 146
Miami trough, 51
micas, 142, 426
Michigan, 31, 35, 70, 179, 192
    Upper Peninsula, 64
Michigan basin, 168, 169, *173*, 176, *219*, 222, *287*
    Cambrian age, *292*
    Devonian System, *296*
    history, *289*
    Mississippian System, *298*
    Ordovician System, *292*
    producing zones, *291*
    resources, *299*
    Silurian System, *294*
    thermal history, *290*
Michigan Formation, 298
Michigan River system, ancient, 526
Michipicoten area, Canada, 73
micrinite, 513
microcline, 200
microlite, 134
microporosity, 242

Mid-Continent area, 216, 217, 222, 336
Mid-Michigan Gravity High, 290
Midas thrust, 14
Middle Atlas Mountains, 184
Midland Basin, 219, 342, 345, *346*, 351
Midnite deposit, Washington, 120
Midway Sunset field, 419, 425, 426
Midwayan Stage, *573*
migration
    diffusion, *232*
    gas, *225, 231, 235,* 238, *307, 312, 315*
    hydrocarbon, *307, 312, 315, 346, 350, 354*
    mechanisms, *232, 235*
    oil, *232, 235,* 238, *307, 312, 315*
    primary, *231,* 234
    secondary, *231*
Mill City, Nevada, 127
Mill Ranch field, 333
Mille Lacs Group, *70*
Milne Point field, Alaska, 451
Minas Gerais, Brazil, 66, *132*, 136
Mineral coal bed, 532
Mineral County, Nevada, 198
Mineral Leasing Act (1920), 1
mineralization, sulfide, *94*
minerals
    alteration, 31
    authigenic, 397
    barium-rich, 197
    calc-silicate, 78
    detrital, 473
    evaporite, *165*
    gangue, 5, *9*, 12, 14, *26*, 33, 54, 59, 76, 77, 78, 112, 127, 130, 136, 157, 197
    high-alumina, 26
    hydrothermal, 141
    hypogene, 80
    industrial, *189*
    iron-rich, 67
    ore, 15, 27, 28, 33, 35, 36, 63, 110, 112, 127, 134
    sulfur-bearing, 397
    See also specific minerals
mines. See specific mines
Mineville, New York, 139
Mining Law (1872), 1
mining methods
    hydraulic, 5
    open pit, 39
    placer, 4
Minnamax deposit, 94
Minneapolis adit, 90, 91
Minnekanta Limestone, 377
Minnelusa, *379*
Minnelusa Formation, 375, *377, 379,* 382, 383, *384*
Minnesota, 18, 25, 36, 70
minnesotaite, 69
mires
    defined, 469
    floating, *471*
    low-lying, *471*
    raised, *470,* 475
    types, *469*

Misener sandstone, 335
Mission Canyon Formation, 397, *398*
    lithofacies, *398*
Mississippi, 196, 573
Mississippi embayment, 302, 516, 526
Mississippi Lime, 330
Mississippi River Delta, 472, 475, 477
Mississippi salt basin, 302
Mississippi Series, 331
Mississippi Valley deposits, 4
Mississippi Valley type (MVT), *45, 46, 53*
Mississippian, *298, 335, 523, 526*
Missouri, 44, 50, 100, 153, 192, 194, 195, 196
    southeastern, 18, 43
Moab, Utah, *175*
Moanda, Gabon, 79
Mocane-Laverne field, 327, 336
models
    climate-change, *539*
    exploration, *405*
    lignite exploration, *578*
    seismic, *266*
Mojave desert, California, 175
Molalla area, Oregon, 195
Molango, Mexico, 79, 81
molasse, 276
    sedimentation, *523*
molybdenite, 28, 29
molybdenum, 3, *23, 25,* 29, *36,* 38
    by-product, 37, 38
    Climax (Colorado), *29*, 207
    future resources, *39*
    Pine Creek (California), *137*
    porphyry deposits, 23, *29*
    production, *23,* 29
    Questa (New Mexico), *36*
    reserves, *23, 25,* 30, 37
    resources, *38*
    skarn deposit, *37*
    sources, *25*
    substitutes, *39*
    users, *25*
    vein deposits, *36*
monazite, 131, 138, 139, 140, 141, *142*
Monongahela coals, 533
Monongahela Group, 475
Monroe uplift, 302
Montana, 12, 18, 24, 25, 26, 29, 32, 36, 131, *155,* 160, 203, 206, 220, 550
    southeastern, 373
    southwestern, 88
    western, 4
Monte Amiata District, Italy, 145
Monterey Formation, 160, 220, 422, *423,* 426, 427, 431, 434, 437, *438, 441*
    fracture origin, *439*
    phosphatic facies, *440*
    reservoir rocks, *439*
montmorillonite, 9, 10, 195, 196, 443
Montoya Formation, *341*
Montoya/Sylvan Formation, 340
Montpelier, Idaho, 155

montroseite, 110
Monument No. 2 Mine, *111*
Monument Valley District, Arizona, *111*
Moodys, Branch Formation, 576
Moonlight Mine, *111*
Mooringsport deposits, 314
Mooringsport–Glen Rosa strata, 305
Morenci deposit, 28
Moreno Shale, *424*
Morris coals, 532
Morrison Formation, 109, 348, 363, 375, 378, 380, 491, 492
Morrow Formation, 347
Morrow shales, 350
Morton County, Kansas, 336
Mother Lode district, *6*
Mount Emmons, Colorado, 29
Mt. Pleasant, Tennessee, 155
Mount Pleasant field, 290
Mount Tolman, Washington, 30
Mountain Pass deposit, California, 199
Mountain Pass district, California, 138, *139*, 141
Mowry Shale, 196, 378, 385
Moxie pluton, *95*
Mrima Hill, Kenya, 139
Muddy No. 1 coal, 479
Muddy Sandstone, 378, 382, 387
Muddy Sandstone reservoir, *380*
muds, 276
　sapropelitic, 472
mudstones, 73, 160, 276, 307, 359, 367, 473, 563, 599
　facies, 348
Mulberry coal beds, 532
Mule Creek field, 375
Mulky coal beds, 532
Murphy marble belt, North Carolina, 200
Muscatatuck Group, 296
Muskegon field, 290
MVT. See Mississippi Valley type
Mystic coal beds, 532
Mystic Creek coalfield, 592

Naborton Formation, 573
Nacimiento, New Mexico, 36
Nacimiento Formation, *359*, 368
Naheola Formation, *573*
Nanafaha Formation, 573
Nanches Formation, 587
Nanushuk Group, 456, 459, 492
Narragansett basin, 527
Nashville dome, 159, 204, 273
National Petroleum Reserve, 597
natrolite, *207*
Naushuk Group, *596*
Navan deposit, 58
Navarro, 314
Navarro Group, 305
Nebraska
　northwestern, 121
　western, 175
Neganuee Formation, 70
Nei Monggol Autonomous Region, China, 139
Nelumbo, 588

Nemaha anticline, 529
Nemaha Ridge, 526
Nenahnezad field, 366
Nenana coalfield, *591*
nepheline, 148
Nevada, 10, 12, 26, 29, *31*, 59, *116*, 153, 160, 176, 196, *198*, 203, 205, 207
Nevada City district, 6
Nevada Dolomite, 409
New Almaden district, California, *145*
New Caledonia, 87, 96
New England, 200
　southeastern, *527*
New Idria body, California, 202
New Idria mine, California, 145
New Jersey, 23, 137, 153
New Market Limestone, 191
New Mexico, 13, 24, 26, 36, *112*, 131, 153, 203, *357*, 475, 549
　southeastern, 339
　southwestern, 10
New Rambler Mine, Wyoming, 99
New River Formation, 527, 529, 539
New South Wales, South Australia, 183
New York District, *54*
New York State, 44, 146, 153, 192, 207, 281
Newark Canyon Formation, 405, 412
Newark Supergroup, 169
Newcastle Sandstone, *375*, 387
Newcastle Sandstone reservoir, *380*
Newfoundland, 51
NGB. See Northern Gulf Basin
Niagara Group, 191, 290, *294*
Niakogon Tongue, 596
nickel, 18, 36, 46, 47, 48, 87, 94, 95, 97, *100*
　mine, 79
　uses, 87
nickel sulfides, 88, 89
Nicoya Complex, 81
Niger Delta, 301
Nikopol, Soviet Union, 79, 81
Niland Tongue, 562
niobium, *131*, 132, *133*
Niobrara Formation, 385
nitrite compounds, 179
nodules, 101
Nome, 4
nonmarine deposits, *174*, *175*
Norilsk deposits, 96, 101
Norilsk district, 87, 96
Norphlet Formation, 303
Norphlet reservoir, 307, *390*
North American craton, 51
North Belridge field, 419
North Carolina, 4, 37, 140, 141, 154, *156*, 193, *200*, 205
　wetlands, 487
North Central coal region, Montana, 550
North Central coalfield, 547
North Chukchi basin, 453
North Coles Levee field, 419
North Dakota, 477, 483
North Dome field, 425
North Kettleman Hills, 423
North Louisiana salt basin, 302

North Meadow Creek field, 375
North Park basin, 547, 561, 563, 565
North Slope, Alaska, *221*, *447*, *596*
　Barrow-Prudhoe group, *458*
　exploration history, *448*
　hydrocarbon production, *450*
　potential resources, *459*
　Simpson-Umiat group, *458*
　stratigraphic plays, *460*
　stratigraphy, *453*
　thermal history, *456*
North Slope petroleum province, *447*
North Star mine, 6
North Willow Creek field, 403
Northcraft Formation, 586
Northern Foothills, 447
Northern Gulf (of Mexico) Basin (NGB), *219*, *301*
Northern Territory, Australia, 183
Northern Trend, 294
Northgate district, Colorado, 204
Northview Shale, 51
Northwest Orebody, 77
Northwest Shelf, 350, 351
Norway, 128, 137
Nova Scotia, 51, 474, 476, 479
Nsuta, Ghana, 79
Nugget reservoir, *396*
Nugget Sandstone, 395, *397*
Nuttall Sandstone, 529
Nuwok basin, 453
Nymphaea, 497
Nyssa, 493, 495, 497
Nyssapollenites, 492

Oak Hill district, Ohio, 195
Oak Hill lignite, 573
Oakville Sandstone, 116
Occidental fault, 10
Occidental vein, 10
Ochoan Series, 175
offshore deposits, *156*
ocean deposits, *100*, 209
Ogallala Formation, 331
Ogden, Utah, 178
Ogilby deposit, California, 206
Ohio, 185, 192, 194, 281, 283
oil, *213*, *225*, 335, 348, *357*, *361*, *363*, *373*, *408*, *411*, *413*, *431*, *441*, *457*
　accumulations, 430, *437*, *450*, *453*, 455, 450
　chemical nature, *225*
　crude, 120, *225*
　exploration. See oil exploration
　future potential, *221*
　generation, *226*
　gravity, *443*
　heavy, *419*
　migration. See oil migration
　origin, *225*, *226*
　platforms, 437
　potential resources, *369*
　production, 213, 220, 278, 280, 282, *283*, 290, *294*, 301, *307*, *309*, *312*, *314*, *316*, *317*, 333, 335, 351, 357, 362, *379*, *382*, *388*, 393, 403, 407, *413*, 435, *460*

project development stages, *255*
reserves, *309, 313, 317, 336, 357,* 403, 419, 426, 437, 594
reservoirs, *340, 346*
resources, *214, 283, 299, 318*
seeps, 419, 435, 449
shale. *See* oil shale
temperature influence, *230*
time influence, *230*
transportation, *233,* 238
wells, 221
Oil Creek, Pennsylvania, 279
Oil Creek shale, 456
oil exploration, 515
    computer applications, *268*
    concepts, *255*
    geochemistry, *261*
    geophysical methods, *262*
    prospect analysis, *270*
    subsurface techniques, *256*
    surface techniques, *259*
    techniques, *255*
oil migration, *225, 231, 307, 312, 315*
    diffusion, *232,* 238
    directions, *235*
    distances, *235*
    mechanisms, *232,* 238
    moving water, *233,* 238
oil province, *218,* 419
oil shale, 148, *183*
    development history, *186*
    domestic deposits, *184*
    extraction, *187*
    foreign deposits, *183*
    occurrence, 183
    production potential, *187*
    reserves, 183
Ojo Alamo Sandstone, *359,* 368
Okefenokee Swamp, 469, 471, 475, 478, 497
Oklahoma, 50, 196, 200, 325
Old Lead belt, *47*
Olean Conglomerate, 529
Olive Hill district, Kentucky, 195
olivines, 92, 95, 180
Olympic Dam deposit, 121
Olympic Dam type, 140
Olympic Peninsula, 81
Onondaga, New York, 173
Onslow Bay, 157
Ontario, Canada, 73
oolites, 49
ooze, 431
Opeche Formation, 174, 377
ophiolites, 5, *96*
Ophir, Utah, 34
Oquirrh Mountains, *12*
Oracle Ridge, Arizona, 31
Orcutt field, 435, 439
Ordovician, *292, 340*
Oreana, Nevada, 206
Oregon, 6, 10, 59, 79, 96, 137, 206, 209
    northeastern, *4*
    southeastern, *4*
organic matter, *226, 227*
Oriskany gas, 281
Oriskany Sandstone, *281*

Orissa State, India, 79
Orleans County, Vermont, 202
Orocopia Mountains, 204
Orphan Lode Mine, 117
orpiment, 11
orthoclase, 200
Osburn fault, *14*
Osceola National Forest, 158
Osgood Mountain, Nevada, 38
Oswego, Oklahoma, 336
Otanmaki deposit, Finland, 136
Ouachita clastic wedge, 523
Ouachita fold belt, 219
Ouachita foreland trough, 529
Ouachita orogenic belt, 345, 539
Owens Lake, California, 128
Owl Creek Mountains, 119
Owyhee County, Idaho, 587
Ozark dome, 532
Ozark Mountains, 46
Ozona Uplift, 352
Ozone Arch, 347, 354

Paakila district, Finland, 202
Pacific Coast, 216
Pacific Ocean, 156
Paddy Run, Virginia, 64
Paduca Field, New Mexico, 249
Pahranagat, Nevada, 589
Paint River Group, 70, 71
Painter Reservoir field, 395, 398
Painter Reservoir/Clear Creek area, 398
Pala district, California, 200
paleobotany, *483*
paleoclimate, 537
paleoecology, *483, 487, 527*
Paleogene, *493*
paleosols, 532, 539, 562
paleovalleys, 526, 529, *588*
Paleozoic, *362*
Pall Mall mine, Tennessee, 199
palladium, 94
Palmetto Formation, 198
Paluxy, 313
Paluxy Formation, 205
palygorskite, 195
Panamint Range, 200, 207
Panasqueira deposit, 127
Panhandle gas field, 376
Panhandle-Hugoton field, 217, 219
Papers Wash field, 363
Paradox Basin, 109, *175*
Paradox Basin evaporites, 169
Paradox evaporite section, 111
Paradox Member, *175*
paragenesis, mineral, *165*
Paralycopodites, 485, 487, 488
Paris Basin, France, 184
Paris coals, 532
Paris-Willard system, 393
Park City, Utah, 17, 34, 55
Park City Formation, 377
Parkman Sandstone Member, 375, 378, 380, *381,* 529
partings, *473*
    detrital, 563
Partridge River Troctolite, 94
Paso Robles Formation, 440

Pea Ridge, Missouri, *77*
Peace River, 154
Peace River Formation, 158
Pearsall shale, 312, 313
peat marshes, *498*
peat swamps, *485,* 487, 490, *495, 497,* 498, 527, 558, 562, 563, 581
peats, 121, 365, 483, *503,* 533, 541, 560, 587
    accumulation, *469, 474,* 485, *487, 495,* 537, 540, 556, 562, 563, 564, 578, 584
    allochthonous, *472*
    back-swamp, 540
    beach ridges, *472*
    coal-ball, *483, 489,* 495, 498
    coal-forming, 583
    composition, 469
    freshwater, 473
    lake bottom, *472*
    marine, 473
    profile, 469
    seasonal, 539
    tree types, *488*
    tropical, 539
Peel Region, Netherlands, 507
pegmatites, *38, 104,* 128, 129, 132, 134, 143, *200*
Pelican River deposit, 59
Peltzer area, South Carolina, 206
Pennsylvanian, *335, 346, 377, 379, 386, 488, 523*
    Lower, *527*
    Middle, *529*
    Upper, *532*
Pennsylvania, 36, 37, 64, 78, 96, 192, 194, 280, 281, 283
    anthracite region, 532, 533
Peridotite zone, 89
permafrost, 457
permeability, *243,* 251, 253, 257, 280, 291, 294, 365, 397, *398, 409, 413, 414,* 424, 470
    oil-water relative, *245*
Permian, *336, 351, 377, 379, 386*
Permian Basin, 169, *173, 175,* 176, 216, *219,* 222, *339*
    hydrocarbon potential, *355*
    sedimentation, *339*
Peru, 127
Petaca area, New Mexico, 134, 200
petroleum
    accumulation, 456
    exploration, *256, 259, 261, 262, 268, 270*
    geology, *273*
    liquid, *225*
    potential, *403*
Petrolia seeps, 289
petrology, coals, *560, 568*
phenakite, 143
Philippines, 87, 96
Philipsburg, Montana, 33, 65, *79*
phillipsite, *207,* 209
phosophorite, 129, 160
phosphates, *120, 153,* 155, 156, *163*
    by-products, *163*
    development, *153*

discovery, *153*
offshore deposits, *156*
pellets, *160*
production, 153, *154*
reserves, *161*
resources, 156, *157*, *161*
sedimentary, *156*
substitutes, *163*
Phosphoria Formation, 109, *120*, 153, 157, *160*, 186, *396*, *399*
phosphorites, 153, *160*
phosphorous, 74
photography, aerial, *260*
Picacho mine, 15
Piceance basin, 547, *562*
  eastern, 560
Piceance Creek basin, 177, 184, 187, 222
Picher Field, 50, 51
Picket Pin zone, *92*
Pictou coalfield, Nova Scotia, 479
Pictured Cliffs field, 362, 366, 368
Pictured Cliffs Sandstone, 357, 361 362, 363, *366*, 556
Piedmont region, 4, 6, 206, 216
Pierre Shale, 82, 378
Pierrepont, New York, 46
Pierrepont deposits, *54*
Pilot Knob, Missouri, 77
Pine Butte Member, 378
Pine Creek, California, *37*, 127
Pine Creek Mine, 37
Pine Grove district, Utah, 29, 205
Pine Nut, Nevada, 30
Pine Valley, Nevada, 403, *408*, 411, 413, 414
Pineview, Utah, 394
Pineview field, 395, *396*, 398
Pioche, Nevada, 17, 55
pipe systems, breccia, *25*
Pismo basin, 443
Pitch Mine, Colorado, *118*, 120
pitchblende, *103*, 119
Pittsburgh coal bed, 533, 540, 541
placers, 4, 16, *98*, 128, 130, 132, 137, 140
Placerville, Colorado, 105
Placerville District, *105*
plagioclase, 90, 172
plagioclase feldspar, 143
plants, *488*
platinum, 87, 94
platinum-group elements, *87*
playas, 169
  stratigraphic, *379*, *386*, *388*
  structural, *382*, *388*
Pleasant Prairie field, 335
Pleasant Valley field, 426
Pleasant View, Nevada, 198
Pleasanton, Kansas City Group, 331
plutonism, 516
Pocahontas basin, 526, *527*, *529*, 529, 540, 541
Pocahontas Formation, 527, 539
  coal beds, 527, *540*, 541
podiform deposits, *96*
Podozamites, 492
Point Arena basin, 443
Point Arguello field, *437*

Point Barrow, 455, 457
Point Lookout Sandstone, 363, *366*, 556
Point Pedernals field, *437*
Point Sal Formation, 434, 437, *439*
Point Sal unit, 437
Point Thomson field, 455
Poison Canyon Tongue, 113
Poison Draw field, 382
Polecat Bench Formation, 561
polyhalite, 168
polymetallic vein deposits, 4, *13*, 16, 32, 33
Pony Express Limestance Member, 105
Poorman Formation, 8
Popo Agie Formation, 378
pore geometry, *242*, *245*
pore system, *241*
  classification, *242*
  displacement pressure, 246
  relative permeability, *244*
porosity, *241*, *243*, 251, *252*, 257, 280, 291, 294, 297, 313, 317, 335, 341, 352, 363, 365, 397, *398*, 403, 413, 414, *424*, 439, 455, 505
  dissolution, 242
  logs, *256*
porphyry, 3, *17*, *25*, *29*, 40
Port Huron field, 289
Porters Creek Clay, 196
Porters Creek Formation, 573
Portugal, 127, 128
Poso Creek field, 419
Post-Knox unconformity, 292
Postle field, 327
Potash Sulfur Springs complex, 119
potash
  production history, *174*
  uses, 174, *176*
potash feldspar, *201*
potassium, 165
potassium-mica, 11
Potosi Dolomite, 198
Pottsville Formation, 523, *527*, 529
Pottsville Group coal beds, 540
Poway Conglomerate, 586
Powder River Basin, *114*, 220, 250, *373*, *475*, 479, 497, 547, 561, 563, 568
  exploration history, *373*
  potential resources, *385*
  production history, *385*
  stratigraphy, 373, *377*
  thermal history, *382*
Powderhorn, Colorado, 133, 136, 141
Powell field, 280
Powell mine, Bastrop County, 576
Powell-Ross field, 288
Prairie du Chien Group, *292*
Prairie Formation, 174
Pratt anticline, 326
Pratt coal group, 527, 540
Precambrian, *54*
Precambrian Shield, 59
Preuss Formation, 399
Price deltaic complex, 523
Price Formation, 523

prospecting, surface geochemical, *262*
Prudhoe Bay area, 456, *459*, 599
Prudhoe Bay field, 221, 449, 450, 457, 458, 460
Psaronius, 488, 489, 490
Puckett field, 219
Puerto field, 365
Puget Formation, 587
Puget Group, 589, 590
Puget Lowland Province, 590
Puget-Nanches basin, 587
Puget Sound, 589
Pumpkin Buttes, 114
Pumpernickel Formation, 81
Pungo River Formation, 154, *157*
Purcell Supergroup, 58
Purgatoire Formation, 195
Purisima Point unit, 437
Putman field, 327, *336*
pyrite, 5, 6, 8, 9, 10, 11, 12, 14, 15, 34, 35, 54, 59, 63, 69, 77, 111, 115, 116
pyrochlore, 132, 140
pyrophyllite, 9
pyrrhotite, 8, 12, 14, 15, 35, 63

Quarternary, *314*
Quartz Hill, Alaska, 30
quartz, 5, 9, 11, 13, 14, 15, 29, 77, 127, 198, 397, 425, 474
quartz-alunite, 9
quartzite, 14, 16, 18, *35*, 117, 358, 392, 404, 408
Quebec district, 202
Queen City Formation, 577
Queen reservoir, 351, 353
Queensland, Australia, 186
Queenston delta, 276
Questa, New Mexico, 29, *36*

radar, *260*
radium, 103
Railroad Valley, Nevada, 403, 404, *405*, 411, 413
Rain Bow Mine, Utah, 205
Rainier National Park, Washington, 209
Ramona field, 365
Randsburg, California, 207
Rapitan Group, 66
Rapitan type, 66
rare earth elements, *137*
  production, 140
  reserves, 139, 140
  resources, *140*
  substitutes, 138
  uses, 137
Raton basin, 475, 547, 549, 560, 563, 565, 568
Raton Formation, 560, 563, 565
Rattlesnake field, 365
Raven Creek field, 386
Ray deposit, 28
realgar, 8, 11
Reclamation Group, 377
Recluse field, 375
Recluse trap, 380
red beds, *30*, *35*, 116, 117, 157, 273, *303*, 354, 382, 517, 532

Red Cave sandstones, 336
Red Dog Creek area, 18, 58
Red Dog deposit, *58*
Red Dog Mine, Alaska, 45
Red Fork sandstones, 335, 336
Red Mesa field, 363, 365
Red Mountain, 37, 96, 98
Red Mountain Formation, 74
Red Peak Formation, 377
Redstone coal beds, 533, 540, 541
Redwall Limestone, 117
Redwater Shale Member, 378
Reef Package, 90, 91
Reef Ridge Formation, 425
Reef Ridge Shale, 422, 427
reefs, stromatoporoid, 342
Reelfoot graben, 529
remote sensing, *259*
  exploration, *260*
Renton Formation, 590
replacement deposits, *13, 17, 18*
Republic anticline pool, 426
reservoirs, 315, *316, 413*
  carbonate, 282, 313
  deltaic, *381*
  diatomite, *424*
  gas, *252, 296, 340, 347*
  hydrocarbon, 395
  oil, *340, 346*
  permeability characteristics, *245*
  primary, 216, 217
  rock, *397*
  sandstone, 313, *379, 424*, 449
  shale, 285, *424*
residual, clay deposits, 82, 138
resinites, 509
resins, *589*
Retort Phosphatic Shale Member, 160, 396
Rex Creek coalfield, 592
Reynolds Creek basin, 587
Rhacophyton, 485
Rhamnites, 493
Rhine-Ruhr coal district, 518
rhodochrosite, 9, 14, 80
rhodonite, 80
rhyolites, 328
Riach coal, 565
Richfield Zone, *296*
Ricky Mountains province, 447
Riddle, Oregon, 79, 87, 98
Riddle deposit, *98*
Ridgeland high, 158
Rifle, Colorado, 105
Rifle District, *105*
Rifle Mine, *105*
Ringwood field, 335
Rio Grande embayment, 302, 303
Rio Grande Rift, 143
Rio Tinto deposit, 59
Rio Viejo field, 419, 424
Rising Fawn, Georgia, 194
Riverdale field, 422
Riverside County, California, 195
Riverton Iron Formation, 71
Road Hollow field, 395
Roberts Mountains Formation, 11
Roberts Mountains thrust, *411*
Rochester Canyon, Nevada, 206

Rock Canyon coals, 560
Rock Creek, Wyoming, 195
Rock Mountain coal basins, 505
Rock Springs coals, 560
Rock Springs Formation, 478, 549, 556, 558
Rock Springs uplift, 549
rocks
  carbonate, 46, *54, 190*, 219, 281, 285, 325, 369, 455
  clastic, 447, 455
  igneous, 455
  mafic, *36, 87*, 101
  metamorphic, 8, 13, *15*, 25, 76, 205
  reservoir, *397*
  sedimentary, 7, 8, 11, 18, 25, 28, 33, 36, 38, 65, 74, 76, 114, 116, 117, 120, 129, 158, 160, 165, 183, *190*, 195, 197, *208*, 216, 219, 243, 357, 376, 382, 447, 475, 527, 586, 587, 591, 596
  siliciclastic, 276, 285
  ultramafic, 87, 98, 101
  volcanic, 5, 25, 27, 28, 36, 71, 73, 76, 116, 405, 409
Rocky Mountains, 176, 220, 222, 252, 515, *547, 562*
  Canadian, 491, 507, 515
  central, *549, 556*, 560, 561
  coals, *547*
  northern, 373, 491, *549*, 560
  southern, *549, 556, 560*
Rocky Point field, 437
Rodessa, 313
Rodessa deposits, 314
Rodessa Formation, 305
Rogue River, 583
Rome, Oregon, 205
Rome Formation, 199
Rome trough, 219, 285, 533
root systems, *495*
Rose Hill district, Virginia, 282
Rosedale field, 425
Roseland, Virginia, 76, 136
Rosiclaire, Illinois, 43
Roslyn coalfield, 588
Roslyn Formation, 588
Ross Adams Mine, 117
Rotliegendes Formation, 241
Rough Creek graben area, 529
Roundtop Group, 377
Rowe coal bed, 532
Rozel Point field, 403, 405
Rozet field, 375
Ruby Creek, Alaska, 35, *36*
Ruby Range, 405
Ruhr coal basin, 517
Rundberg Mine, 116
Rusk County, Louisiana, 251
Russia, 186
Rustler Formation, 173, 175
Ruth Ledbetter well, 333
rutile, 135, *136*, 206
Rwanda, 144
Ryckman Creek field, 300, 395

Sabine Uplift, 251, 302, 573
Sabinian Stage, *573*
Sacramento basin, 220, *417*

gas province, 419
production history, *419*
Sadlerochit Group, 459
Sadlerochit reservoir, 451, 456, 457
Sag basin, interior, *287*
Sag River Sandstone, 456
Sagavanirktok Formation, 456, *459*, 596, 597, *599*
Sagavanirktok reservoir, 457
Sage Spring Creek field, 375
Saginaw Formation, 529
St. Anthony Mine, Arizona, 109
Saint Claire County, Michigan, 289
St. Francois Mountains area, 49
St. Lawrence County district, New York, 78
St. Peter Sandstone, 292
Salado Formation, 173, *175*
Salem anticline, 531
Salina basin, 526, 529
Salina-Forest City Basin, 289
Salina Group, 169, 173, 294
  evaporites, 168
  salts, 297
Salina Salt deposits, 533
Salinas basin, 443
saline deposits, 157
salinity, 168
Salmon, Idaho, 99
salt, 168, 169, 219, 296, 303, 399
  anticlines, 302, 309
  beds, 173
  bitter, 165, 168
  cakes, 177
  collapse, 297
  deposits, *172*
  diapirs, 302, 303
  domes, 173
  interior basins, 302
  mineral succession, *165*
  pans, *167*
  playa, 422
  production, *172*, 174
  rock, 173
  uses, *174*
Salt Chuck deposit, 95
Salt Creek anticline, 375
Salt Creek field, 382, 386, 388
Salt River region, Arizona, 202
Salt Wash Sandstone Member, *109*
Salton Sea brines, California, 172
Saltville, Virginia, 173
samarium, 139
San Andreas Fault Zone, 405, 417
San Andres reservoir, *352*, 354
San Francisco, Arizona, 32
San Joaquin basin, 220, *417*, 443
  central, *423*
  future exploration, *426*
  gas province, 419
  northern, *422*
  production history, *419*
  southern, *423*
  stratigraphy, *420*
  thermal history, *422*
  trap types, *425*
San Joaquin Formation, 427
San Joaquin Valley, California, 209
San Jose Formation, 359

San Juan Basin, *112*, *220*, 222, 475, 549, 547, 556, 560
   central basin, 357
   Chaco slope, 357, 366
   Four Corners platform, 357, 361, *363*, 369
   gas resource, *357*
   oil resource, *357*
   stratigraphy, *357*
   thermal maturity, *368*
San Juan Mountains, 80
San Juan 29-5 Unit 50 well, 357
San Juan volcanic field, 10
San Manuel, Arizona, *25*
San Marcos arch, 302, 303
San Miguel field, *437*
San Miguel lignite, 478, 578
San Rafael Swell uplift, 112
sanadine, 474
Sanctuary Formation, 592
Sand Wash basin, 563, 565
sand bars, *47*
sand spits, 168
Sandford Lake, New York, 76
sanding, 55
Sandow mine, Milam County, 576
sands, 116, *192*, 201, 241, 292, 327, 341, *345*, 351, 422, 425, 427, 455, 529
   accumulation, 579
   deposits, 351
Sandstone Member, 549
sandstones, *18*, *35*, *38*, *70*, *74*, 105, 111, 115, 118, 143, 191, 216, 219, 220, 249, 252, 273, 276, 280, 281, 299, 303, 325, 330, 347, 354, 367, 377, *381*, 387, 392, 414, 420, 437, 439, 455, 456, 473, 476, 515, 527, 529, 532, 539, 558, 561, 563, 599
   beach-barrier, 478
   beds, 16
   channel, 540, 563, 587
   deltaic, *388*
   marine shelf, *380*
   platforms, 558
   quartz-rich, *529*
   reservoirs, 313, *379*, *424*, 449
   sheet, 558
   shelf, *388*
sandstone silver, *16*
Sanford Lake District, New York, *136*
Santa Barbara Mine, Peru, 145
Santa Cruz basin, 443
Santa Lucia High, 431
Santa Margarita Formation, *160*, 420
Santa Maria basin, *220*, *431*, *434*, *435*, *437*, *441*, *443*
Santa Maria River–Little Pine fault system, 434
Santa Maria Valley field, 440
Santa Rita, New Mexico, 31, 46
Santa Ynez River fault, 434
Santee Limestone, 158
Santos Shale, 425, 426
saprolite, 94
Savannah River, *156*
Savannah River deposit, Georgia, *158*
Sawatch anticline, 55

Sawyer decision, 5
scheelite, 16, 126, 127, 128
Schmitz anticline, 368
Schuler Formation, 303
Schuyler, Virginia, 200
Schwartzwalder Mine, Colorado, *116*
Scotland, 186
Scurry field, 348
Searles Lake, California, 128, 172, *175*, *177*
seas, relict, 168
seawater, 180
   chemical analyses, 165
   evaporating, 165
   solar extraction, *174*
Second Frontier sandstone, 388
Secor coals, 532
Sedgwick basin, 325
sedimentation, *339*, 444, *526*, 539, 578
   clastic, 564
   molasse, 523
sediments, 80, *479*
   back-barrier, 478
   clastic, *474*
   iron-rich, *75*
   lacustrine, *479*
   marine chemical, 63
   tidal, *479*
seeps, 393
   oil, 419, 435, 449
seismic modeling, *266*
seismic theory, *262*
selenium, native, 115
Selma chalks, 313
Selwyn Basin, Canada, 58, 198
Sentinel Butte Formation, 561, 565, 568
sepiolite, 195
Sequoia, 487, 492, 493
sericite, 5, 6, 14
Serro do Navio, Brazil, 79
Seven Lakes field, 361
Seven Rivers reservoir, 351, 353
Sevier orogenic belt, *550*
Seward Peninsula, Alaska, 204
Sewell coal, 527
Sewickley coal beds, 533, 541
Shady Formation, 51
Shakan Bay, Alaska, 37
shale oil, 148, *183*
   development history, *186*
   domestic deposits, *184*
   extraction, *187*
   foreign deposits, *183*
   occurrence, 183
   production potential, *187*
   reserves, 183
shale, 5, *18*, *35*, 70, 71, 74, 114, 117, 118, 129, 143, 145, 148, 160, 172, 175, 186, 198, 219, 222, 246, 273, 276, 279, 281, 284, 303, 305, 314, 325, 328, 330, 335, 342, 347, 353, 358, 378, 384, 396, 399, 400, 411, 422, 423, 427, 437, *440*, 456, 459, 476, 529, 532, 561, 586, 592
   marine, 459
   pebble, 459

   reservoirs, 285, *424*
   siliceous, 441
Shandy Formation, 199
Shannon reservoir, *381*
Shannon Sandstone Member, 375, 378, 380, *381*, 388
Shannon seals, *381*
Shannon traps, *381*
Sharon-Brownsville paleovalley system, 526
Sharon Conglomerate, 529
Shawnee Group, 331
Shawnee town, Illinois, 172
Sheep Pass deposits, 405
Sheep Pass Formation, 405, 412, 414
Sheet Ground, ore, 51
Shinarump Conglomerate Member, *111*
Shinarump deposits, 111
Shirley Basin area, *114*
Shublik Formation, *160*, 456, 459
siderite, 9, 15, 63, 64, 69, *75*
Sierra Diablo, 340
Sierra Nevada, 4, 5, 81, 199
Sierra Nevada batholith, 5
Sierra Nevada belts, 96
Sierre Leone, 137
Sierrita, Arizona, *28*
Sierrita deposit, 28
Sigillaria, 487, 488, 489, 496
Sigsbee Deep, 303, 315
silica, *72*, 74
   biogenic, 438
silicates, 78, 135, 148, 197, *200*, 208
siliciclastics, 529, 532
sillimanite, 205, *206*
silts, 427
siltstones, 18, 246, 303, 325, 358, 382, 456, 476, 493, 540, 561, 563, 586, 599
Silurian, *294*, *333*, 340
silver, *3*, 12, 16, 28, 34, 35, 46, 55, 58, 59, 79
   by-product, 4, *16*, 18, 32, 33
   co-product, 4
   mantos, 55
   native, 11, 15, 18
   occurrence, *4*
   production, *3*, 9, 10, 13, 14, 17, 18, 19, 29, 32, 131
   reserves, 12, *19*
   resources, 17
   types, 4
   uses, *3*
Silver City, New Mexico, 17
Silver City fault, 10
Silver Peak, Nevada, 179
Silver Reef, Utah, 16
Silver Reef ore body, 16
silver sulfosalts, 9, 14, 15
Simonson Formation, 414
Simpson field, 449, 458
Simpson gas, 345
Simpson Group, 328, 340, 341
Simpson oil, 345
Simpson sandstones, 333
Simpson shales, 345
Simpson-Umiat group, *458*
Simsboro Formation, 574

Simsbury, Connecticut, 23
Sisquoc Formation, 431, 435, 437, *438*
Skagit County, Washington, 589
skarn deposits, *37*, *78*, 126
 gold-bearing, 12
 precious-metal, *12*
Skookum chuck Formation, 584, *586*, 590
Skull Creek Sandstone, 380
Skull Creek Shale, 378, 385
slag, titaniferous, 135, 137
slates, 6
Slaven Chert, 198
Slick Rock District, 110
Sligo, 303, 313
Smackover Formation, 179, 303
Smackover reservoir, 307, *309*
Snake Range, 405
Snake River Plain, *587*
Snuggedy Swamp, 475, 478, 479
soda ash, 177
sodium, 178
sodium bentonite, 196
sodium carbonate, *177*
sodium sulfate, *177*
 uses, *178*
solutes, 165, 169
Somerset district, Pennsylvania, 195
Somerset–Crested Butte–Carbondale coal region, 560
Sooner Trend, 335
Soudan Iron Formation Member, 73
Sour Zone, *296*
South Africa, 66, 79, 87, 120, 137, 144
South Barrow field, 449
South Belridge field, 419, 425, 426
South Carolina, 4, 140, 141, 148, 149, 153, *154*, *157*, 193, 194, 205, 475, 478, 479
South Cole Creek Field, Wyoming, 249, 375
South Dakota, 200
South Glenrock field, 375, 382
South Hallsville, Harrison County, 575
South Kawishiwi Intrusion, 94
South Korea, 128
South Meade well, 457
South Texas Coastal Plain, *116*
Southeast Missouri District, 18, *46*, *47*, *49*
Southland syncline, New Zealand, 209
Southwestern Interior basin, 523
Soviet Union, 66, 79, 87, 96, 128, 131, 137, 140, 144, 146, 148, 149
Spain, 146
Spar Lake, Montana, 36
Spearfish Formation, 174, 377
Spearhead Ranch field, 380, 388
Sphagnum moss, 471, 493
sphalerite, 6, 8, 12, 14, 15, 18, 35, *44*, 54, 59, 111
 by-products, 33
Spindletop Dome, Texas, 173, 301
Spirit River Formation, 252, 253
Spor Mountain, Utah, *118*, 143, 205

Spraberry updip, 354
Spraberry/Dean reservoir, 351
Springer Formation, *330*
Springer sandstones, 335
Springfield Coal, Illinois, 476
springs
 brine, *172*
 hot, 128
Spruce Pine district, North Carolina, 200, 201
Spruce Road deposit, 94
Sri Lanka, 137, 140
Stanley Shale, 199
stannite, 129
Star district, Utah, 205
Statenville Formation, 158
Steele Shale, 375, 378, 380, 381, 385
Steenkampskraal, South Africa, 139
Sterling Formation, *594*
Sterling Hill, New Jersey, 46, *54*
Stevens Formation, 426
Stevens turbidite sands, 425
stibnite, 11
Stigler coals, 532
Stillwater, Montana, 36
Stillwater Complex, 18
 Banded series, *90*
 Basal series, *88*
 J-M Reef, *90*
 Montana, 76, 87, *88*, 101
 Picket Pin zone, *92*
 Ultramafic series, *89*
stilpnomelane, 69
Stockade Beaver Shale Member, 378
Stockton Arch fault, 417, 419, 423
Stockton coal bed, 540
stockworks, *9*, 16, 25, 28, 130
stone, crushed, *191*
Stone Corral Formation, 173
Stonewall Formation, 174
Stoney Fork Member, 520
Stoney Point field, 299
Story, Wyoming, 376
Straight Cliffs Formation, 549, 555, 558
Straight Creek fault zone, 587, 588
Strand and Canfield Ranch, 425
strand-plain systems, *558*
stratabound deposits, *99*
stratigraphy, *273*, *303*, *327*, *357*, *377*, *420*, *453*, *488*
Strawberry, California, 38
Strawn section, 348, 349
Strawn shales, 350
Stray Sandstone, 290, 298, *299*
strueverite, 134
Stuart City, 313, 314
Sudbury Complex, Canada, 87
sulfates 168
sulfides, 6, 12, 14, *15*, 28, *91*, 94, *95*, 99, 111, 117, 130, 197, 198
 magmatic, 87
 massive deposits, *17*, *18*, *34*, 38, 41, 46, *59*
 PGE-bearing, 88, *90*
 volcanogenic massive, *58*
sulfosalts, 13, 14
sulfur, *473*, 477, 479, 527, 533, 540, 560, 578, 586, 592, 595, 599

production history, *176*
sources, *177*
uses, *177*
Sully Member, 82
Sulphide Queen orebody, 138, 141
Sunbury Shale, 299
Sundance Formation, 378
Sunniland limestone, 305
Sunnyside-Castlegate coal region, 560
Sunnyside coals, 560
Sunnyside Mine, 13
Suntrana Formation, *592*
Superior, Arizona, 17
Surinam, 149
Susitna Lowlands coal basin, 591
Sussex field, 375
Sussex reservoir, *381*
Sussex Sandstone Member, 375, 378, 380, *381*, 388
Sussex seals, *381*
Sussex traps, *381*
Suwannee River Mine, 159
swamps, 158
 back-barrier, 558
 clastic, 487, 490
 peat accumulating, 595
 peat, *485*, 487, 490, *495*, 527, 558, 562, 563, 581
 peat-forming, 583
 raised, *563*
Swannanoa-Burnsville belt, North Carolina, 206
Swanson deposit, Virginia, *119*, 121
Swauk basin, 588
Swauk Formation, 588
Sweetwater district, Tennessee, 197, *199*
Swiftcurrent Creek, 393
Sword discovery, 437
Sydney coalfield, Nova Scotia, 516
Sykesville district, Maryland, 99, 101
Sylvan Shale, 342, 345
Sylvania Sandstone, 179
sylvanite, 17
sylvite, 175
Syracuse, New York, 172

Table Mesa field, 365
talc, *199*
Talkeetna batholith, 591
Talkeetna Formation, 591
Tallahatta Formation, 576
Tallin-Leningrad area, Soviet Union, 186
Tanco Mine, Canada, 134
Taneum coalfield, 588
Tansill reservoir, 351, 353
tantalite, 134
tantalum, 132, *133*, 134, 135
tar accumulations, 440
tasmanite, 186
Tatlanika Creek coalfield, 592
Taxodiaceae, 495
Taxodium, 487, 488, 493, 495, 496, 497
Taylor, 314
Taylor Group, 305
Tchiatura, Soviet Union, 79
teallite, 129

Teapot Dome field, 388
Teapot Sandstone Member, 375, 378, 380, *382*
Tebo coal bed, 532
Teckla Sandstone Member, 375, 378, 380, *382*
tectonics, 444, *550, 561, 583*
  extensional, *588*
tectonism, *526*
  syndepositional, *587*
Tejon depocenter, 419, 422
Tejon-Grapevine field, 419
tellurides, 6, 9, 14
Temblor Formation, 420, 423, 424, 426
Tempiute, Nevada, 127
Temple Mountain District, Utah, *111*
Ten Section field, 425, 426
tennantite, 9
Tennessee, 24, *34*, 44, *155, 159, 161*, 185, 194, 196, 573
Tennessee salient, 536
Tensleep Formation, 377, 382
Tertiary, *114, 314, 358, 547, 573*
tetrahedrite, 6, 8, 9, 15
Texas, 153, 180, 192, 193, 194, 196, 203, 477, 478, 479, 573
  western, 339
Texas-Louisiana salt basin, 303
Texas Panhandle, 325, 327, 330, 336
Thailand, 127, 131, 140
thallium, 11
Thayness Formation, 395
The Dell, 118
Thermo mine, Hopkins County, 575
Thermopolis Shale, 378
Third Zone Sand, 439
Thirtyone Formation, 340, *342*, 346
Thomas Range, 118, 205
Thompson Creek, Idaho, 30
Thompsonipollis, 493
Thomson sand, 455
Thor Lake, Northwest Territories, 144
thorianite, 141
thorite, 140, *141*
thorium, *140, 141*, 142
Thornapple deposit, 59
thrust belts, 216
tiff, 197
Tigluckpuk Formation, 186
Timbered Hills Group, 328
tin, *128*, 129, *130, 131*, 134
Tinaquaic Sandstone Member, 437, *440*
Tincup, Colorado, 55
Tintic, Utah, 13, 46, 55, 59
Tintic district, Utah, *13*, 194
Tipton Shale, 562
titanium, 132, *135, 136, 137*
Titusville, Pennsylvania, 186, 279
Tobosa basin, 339, *340*, 346
Tocito field, 362, *365*
Tocito Sandstone, 369
Tocito Sandstone Lentil, 361, 362, 363, 368
Todilto Limestone Member, 112, *113*, 363
Tombstone, Arizona, 17, 34
Tomera Ranch field, 403

Tomstown Formation, 191
Tongue River Member, 376, 561, 563
Tonkawa sand, 336
Tonopah, Nevada, 4
tonsteins, *474*
Toolebue Formation, 183
topaz, 206, 207
Topaz Mountain tuff, 118
Torok Formation, 456, 459
Torok-Nanushu system, 459
Torrivio Sandstone Member, 365
Tortugas deposit, New Mexico, 205
tourmaline, 16
Tradewater Formation, 529
Trail Ridge deposit, Florida, 137
Transcontinental arch, 532
transportation, gas and oil, *233*
Transvaal-Griquatown, South Africa, 66
Transvaal Province, South Africa, 201
Trap Spring field, 403, 405, *407*, 410, 411, 413, *414*
Trapa, 588
trapping
  fracture, 426
  hydrocarbon, *241, 247*
traprock, 191
traps, *309, 313, 317*, 333, *345, 350*
  anticlinal, 425
  classification, 247
  diagenesis, *247, 426*
  fault, 425
  fault-block, 403
  Fusselman Formation, 345
  hydrocarbon, *241, 309, 313, 317*, 375, *379, 398, 405, 414*
  hydrodynamic, 426
  mechanisms, *249*, 294
  stratigraphic, 425
  submarine channel, 425
Traverse Group, *296*
Traverse Limestone, *297*
Trenton Group, 289, 294
Trenton Limestone, 280, 289, 290
Tri-state District, 44, *50*
Triassic, *111, 377*
Trinity, 303, 305
tripolite, *342*
troctolite, 94
Trommald Formation, 70
Troodos ophiolite, Cyprus, 81
Troy Canyon intrusive, 414
Troy deposit, 18
Troy mine, *18*
Trucial Coast, Persian Gulf, 167
Tubb reservoir, 351, 354
tuffs
  ash-flow, 413
  beds, 201
Tuktu Formation, 596
Tulare Lake area, 427
Tulare pool, 426
Tule Elk field, 426
Tulip Creek Formation, 341
Tulip Creek shales, 345
Tullock Member, 376
tungsten, *126, 127, 128*
turbidites, 71, 276, 318, 425, 426
Turkey, 87, 96, 146

Turner Sandy Member, 378, 381, 388
Tuscahoma Formation, 573, 577
Tuscaloosa reservoirs, Louisiana, 313
Tuscaloosa sandstone, 305, 314
Tuscaloosa shale, 305
Tuscarora intrusion, 94
Tuscarowas Valley district, Ohio, 195
Twiggs Clay Member, 196
Twiggs County, Georgia, 194
Twin Buttes, Arizona, 31
Twin Buttes deposit, 28
Twin Creek Formation, 395, 399
Twin Creek reservoir, 396
Two Medicine Formation, 550
Tynagh deposit, 58
Tyonik Formation, *594*
Tyrol area, Austria, 186
Tyrone, New Mexico, 28

Uganda, 144
Uinta basin, 560, *562*
Uinta-Piceanca basin, 547, 549
ultramafic series, *89*
Umiat basin, 452
Umiat Delta, *492, 596*
Umiat field, 458
underclays, 532
Union Island field, 425
United States
  central, *523*
  coal deposits, *463*
  eastern, *523*
  evaporite basins (marine), *173*
  far-western, *583*
  northeastern, *54*
  northwestern, *160*
  western foreland basin, *550*
  western, *161*
  See also specific states
Upper Hartshorne coals, 532
Upper Mississippi district, 18
Upper Sundance, 378
uraninite, 112, 113, 115, 117, 118, 139
uranium, 16, 38, *103*, 139, 158
  genesis, 114
  production, *103*, 111, *112*
  reserves, 112, 118
  resources, *120*
  uses, 103
Uravan mineral belt, Colorado, *104, 109*
Uravan mineral belt, Utah, *104, 109*
Urkut district, Hungary, 79
Usibelli Group, *592*
USSR. *See* Soviet Union
Utah, 12, 24, 26, *28, 31, 33*, 111, *117, 155*, 160, 203, 207, *220*, 391, 477, 549
  north-central, 12
  western, 10
Ute dome, 361
Ute Indian Reservation, southern, 368

Vado Sandstone, Member, 368
Val Verde Basin, 340, 345
Valentine Member, 191
Vallecitos Trough, 417
Valley and Ridge Province, 282, 517

Valley Coal Fields, 523
Valmy Formation, 198
Van field, Texas, 313
Van Horn Mountains, 205
vanadium, 16, *103*, *104*
    production, *103*, 105, 111
    resources, *120*
    uses, 103
Vananda, Montana, 196
veins, 139, *197*, 204
    epithermal, 3, 4, *8*, *9*, 11, 16
    gold-quartz, *5*
    polymetallic, *13*, *17*
    tungsten, *127*
Venezuelan oil, 120
Ventura–Santa Barbara basin, 443
Verde field, 365
Verde River, 588
Vermejo Formation, 559, 560
Vermilion district, Minnesota, 73
Vermilion Range, 73
Vermillion Creek basin, 562
Vermillion Creek coal, 562
Vernon Parish, Louisiana, 196
Viburnum trend, 47, *49*
Victoria, Nevada, 31
Vinini Formation, 411
Viola Group, 328, 333
Virgilian Series, *331*
Virginia, 4, *119*, 205, 523
Virginia Formation, 94
Virginia Range, 10
vitrinite, 509, 516
    reflectance, *505*
Vlotho Massif, 518
volcanic extrusives, *587*
Volcanic Slate Belt, 206
volcanism, 9, 26, 35
volcanogenic deposits, *80*
Vulcan Formation, 70

Wabaunsee Group, 331
Wah Wah Mountains, 205
Walakpa field, Alaska, 449
Wales Sand, 579
Walker Lane fault, 405
Walshville channel, *476*
Walshville River, 537
Wanakah Formation, 112, 363
Wasatch Formation, 114, 376, 561, 562, *563*
Wasatch Mountain uplift, 392
Wasatch Plateau, 549, 555
Washakie basin, 562
Washington, 10, 26, 28, 59, 79, *117*, 206
    northeastern, 18
Washington County district, Missouri, *197*, *198*
Washington County, Georgia, 194
Washington Mine Camp, 589
Washita, 303, 314
Washita County, Oklahoma, 328
Washita Creek field, 333
Washita Group, 252, 305
Washville Channel, 531, 540
Water Men intrusion, *94*
water
    dilute inflow, 170
    juvenile, 172
    meteoric, 366
Watonga-Chickasha Trend, 335
Watonga field, 327
waxes, *589*
Waynesburg Coal, West Virginia, 476
Waynesburg coal bed, 533, 541
weathering, 72, 82, 87, *97*, 119, 130, 148
    chemical, 72, 98
    reactions, *169*
Webster, North Carolina, 98
Weches Formation, 75
wedge
    clastic, 380
    sedimentary, 391, 392
Weir-Pittsburg coal bed, 532
Weisner Formation, 199
Well Draw field, 382, 388
Wellborn Formation, 576
Wellington Formation, 173
Wells Cargo mine, Nevada, 204
wells, heavy oil, 419
    See also specific wells
Wendover, Utah, 175
Wendover Group, 377
West Delta coalfield, 592
West Edmond field, 335
West Fork adit, 90, 91
West Gold Belt district, 6
West Kitty field, 375
West Mayfield field, 333
West Moore field, 335
West Rozel field, 403, 405
West Shasta district, California, 17, 34
West Springs Creek Formation, 328
West Sussex field, 375
West Virginia, 194, 283, 476, 483
    northern, 540
Western Interior, 553
Western Interior basin, 523, 526, *529*, *532*, 534, 536
Western Interior Coal Region, 489, 490, 491, 495
Western Middle Anthracite field, 517
Westwater Canyon Member, *112*
Wet Mountains, Colorado, 141, 197
Wheat Field, Texas, *249*
Wheeler coal bed, 532
Wheeler County, Texas, 333
White Canyon area, Utah, 111
White Mountain deposit, California, 205, 206
White Pine, Michigan, *18*, *35*
White River Formation, 376
White River Valley, 405
White Rock, 49
white-rock deposits, *160*
Whitebreast coal bed, 532
Whitney Canyon, 393, 395
Whitney Canyon/Carter Creek field, 395, 397
Whitney Canyon field, 398
Whitsett Formation, 116, 576
Wichita-Albany reservoir, 351, 354
Wichita Mountains, 328
Wichita reservoir, 351
Wiggins arch, 302

Wilcox Group, 479, *573*, *576*, *578*
Wilcox Rockdale delta system, 577
Wilkinson-Carbonado coalfield, 583
Wilkinson County, Georgia, 194
William field, 299
Williams Fork Formation, 549
Willis Mountain, Virginia, 206
Williston Basin, *174*, 288, 547, *561*, *563*, 565, 568
    coalfield, 493
Willmax pool, 426
Willsboro, New York, 207
Willwood Formation, 561, 562
Wilrich Member, 252
Wilson Springs, Arkansas, *119*
Wilson Springs deposit, *119*
Wind River Basin, 114, 547, 550
Wind River Range, 73
Winfield mine, Titus County, 575
Winterboro, Alabama, 200
Wisconsin, northern, 18
Wisconsin dome, 532
Wisconsin-Illinois District, 43
Wishbone Hill district, 595
Wishbone Hill Formation, 594
witherite, 197
wodginite, 134
Wolfcamp reservoir, 351
Wolfcamp shale, 350
wolframite, 126, 127, 130
wollastonite, *207*
Wood mine, 96
Wood River coalfield, 592
Wood River District, 58
Woodbine aquifer, 251
Woodbine sand, 252
Woodbine sandstone, 305, 314
Woodbine shale, 305
Woodbridge, New Jersey, 194
Woodford Shale, 328, 331, 333, 340, 342, *345*
Woodrat Mountain, 206
Woodrow Mine, *113*
Woodruff Formation, 186, 411
Wristen Formation, 346
Wygal Sandstone Member, 420
Wyodak-Anderson coals, 561, *563*, 568
Wyoming, 26, 73, 119, 153, *155*, 160, 184, 200, 220, 391, 478, 550
    deposits, 115
    northeastern, 373
Wyoming basin, *114*

xenotime, 138, 140

Yakutat terrane, *599*
Yates field, 352
Yates reservoir, 351, 353
Yavapai Series, 73
Yazoo Clay, 576
Yegua Formation, 576, 578
Yellow Chief Mine, Utah, 118, 121
Yellow Pine District, Idaho, 128
Yellowjacket Formation, 99
Yellowstone National Park, Wyoming, 208
Yerrington, Nevada, 26

York Canyon coal, 565
Yorktown Formation, *157*
Yowlumne field, 424
yttrium, 137, 140
Yuba River, 4, *5*
Yugoslavia, 146, 184, 186
Yukon River region, 4

Zaire, 87, 129
Zambia, 87

Zap coal, 568
zebra rock, 55
zeolites, 10, 165, *207*, *209*
 natural, 208
 production, 208
Zilch Formation, 425
Zimbabwe, 97
zinc, 4, 12, *18*, 32, 34, 35, *43*, *44*, *45*, 46, 49, *54*, *55*, 58, *59*, 79, 100
 by-product, 44

co-product, 44
future reserves, *60*
mantos, *55*
production, 14, 29, 32, *44*, 54
uses, *44*
zircon, 139, 474
Zone of Unconformity, 292
Zuni Mountains deposit, New Mexico, 204
zunyite, 9

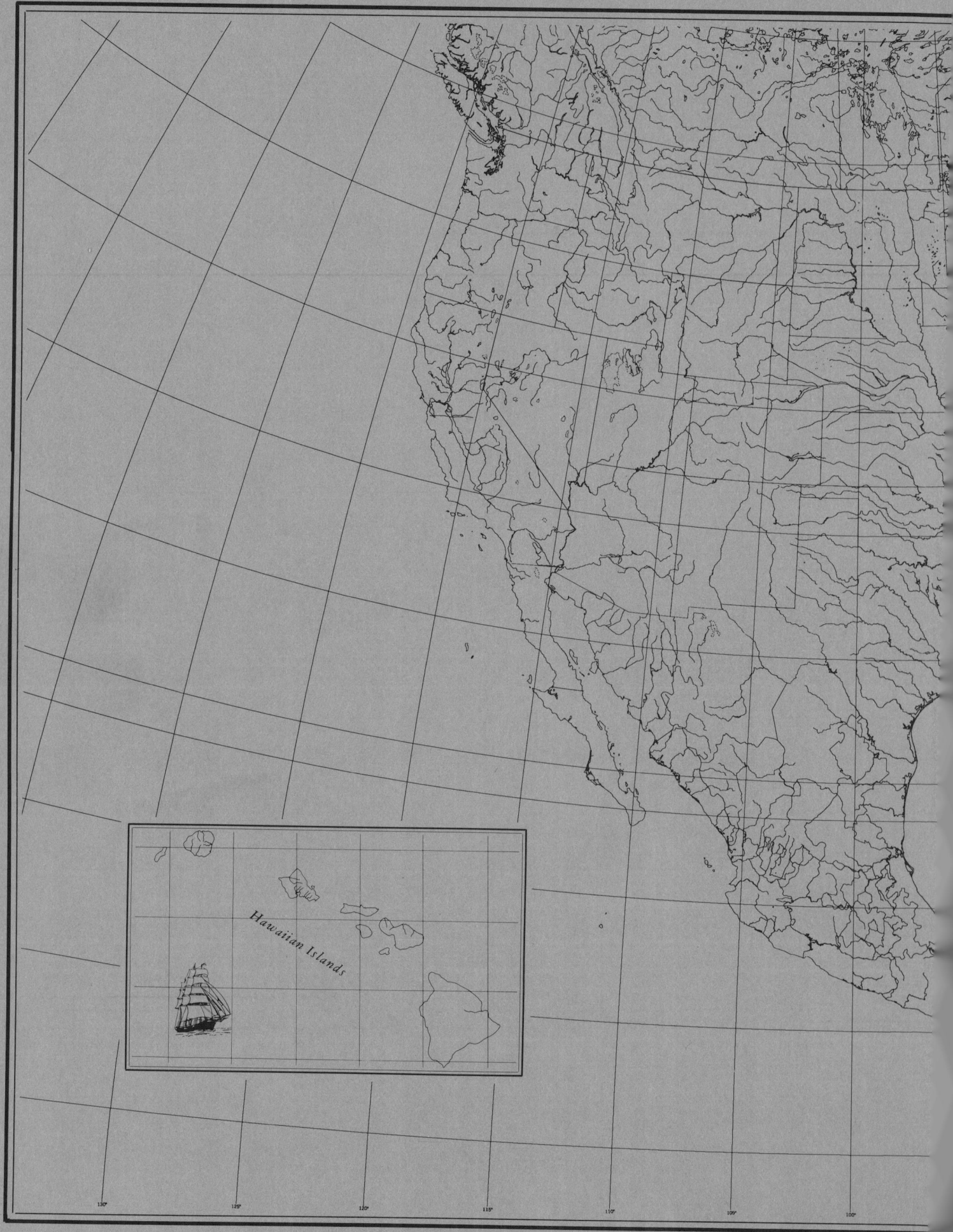